Botany FOR DEGREE

BOTANY
for Degree Students

FIFTH EDITION

A. C. DUTTA M.SC.
*Formerly Head of the Departments
of Botany and Biology
Cotton College, Gauhati*

CALCUTTA
OXFORD UNIVERSITY PRESS
DELHI BOMBAY MADRAS

Oxford University Press, Walton Street, Oxford OX2 6DP

NEW YORK TORONTO
DELHI BOMBAY CALCUTTA MADRAS KARACHI
KUALALUMPUR SINGAPORE HONG KONG TOKYO
NAIROBI DAR ES SALAAM
MELBOURNE AUCKLAND

and associates in
BERLIN IBADAN

© Oxford University Press 1979

First published	1964
Second edition	1968
Third edition	1970
Fourth edition	1974
Fifth edition	1979
Third impression with a glossary of Arabic & Nepali names of plants.	1982
Fifth impression with amendments	1983
Fourteenth impression	1993

Printed in India by offset
at Swapna Printing Works Private Limited
52 Raja Rammohan Roy Sarani, Calcutta 700 009
and published by Neil O'Brien, Oxford University Press,
5 Lala Lajpat Rai Sarani, Calcutta 700 020

PREFACE TO THE FIFTH EDITION

A large fourth edition of *Botany for Degree Students* was published in 1974 but to meet the increasing demand in India and abroad this was followed by a second impression in 1976 and a third impression in 1977. The current (fifth) edition now appears in a much more improved and enlarged form as a result of thorough revision of the whole text. The revision involves rewriting and often enlarging several topics on the basis of recent researches, and also addition of some new topics as prescribed by several Universities. Obviously many additions and alterations have to be made throughout the whole text. Major changes have, however, been effected in Histology, Physiology, Ecology, Cryptogams and Economic Botany. In this edition special emphasis has been laid on nucleic acids (DNA and RNA), cell-wall, protein synthesis (specially the role played by DNA and RNA), photosynthesis, respiration, ATP and several other topics. Further, anatomical structures of *Casuarina* stem showing primary growth, of *Vitis* stem showing secondary growth, of *Leptadenia* stem and *Bignonia* stems showing anomalous secondary growth have been newly introduced with illustration and description in each case. A detailed description of *Calamites* has also been added. Thus the book, as it stands now with its manifold improvements in different directions, will, it is expected, adequately meet the the needs of a wider circle of students preparing for their Degree Examinations (pass standard).

For many valuable suggestions received for all-round improvement of this edition the author expresses his sincere and grateful thanks to Prof C. S. Bhutada of G. S. College of Science, Khamgaon, Prof A. K. Ghosh of Calcutta University, Calcutta, Prof C. S. Belekar of M. M. College of Science, Nagpur, Prof P. N. Purekar and Prof (Mrs) H. M. Kamdar of Shri Shivaji Education Society's Science College, Nagpur, Dr Rajkumar Beliram of Hislop College, Nagpur, Dr B. G. Deshpande of St Francis de Sales College, Nagpur, Prof T. R. Bandre of Sindhu Mahavidyalaya, Nagpur, Prof S. B. Dalal of Vidarbha Mahavidyalaya, Amravati, Prof D. G. Lakhe of Shri Shivaji Mahavidyalaya, Udgir, Prof K. K. Choudhary of Shri Shivaji Science College, Chikhli, Prof S. R. Kalantri of Adarsha Mahavidyalaya, Dhamangaon, Dr K. Sankaran Unni of Government Post Graduate College, Chhindwara, Prof V. K. Sharma of Punjab Agricultural College, Ludhiana, Prof T. Seshagiri Rao of S. K. B. Ramraj College, Amalapuram, Prof A. M. Halder of G. M. College, Sambalpur, and also several others.

The *Glossary of Names of Plants* deals with names in *eleven* Indian languages in addition to English and Botanical names.

In this connection the author offers his sincere thanks to Dr (Mrs) S. Chitaley of Institute of Science, Nagpur for Gujarati names, Prof Y. B. Raje and Miss Tara W. Kelkar of Sir Parashurambhau College, Pune for Marathi names, Dr B. Samantarai, formerly of Ravenshaw College, Cuttack and Prof S. Naik of G. M. College, Sambalpur for Oriya names, Prof Hukamchand, formerly of Government College, Dharmsala for Punjabi names, Prof M. S. S. Rao of Basaveshvar Science College, Bagalkot for Kannada names, and Prof B. K. Nyer, formerly of Gauhati University, Gauhati for Malayalam and Tamil names; names of plants in other languages have been compiled by the author from different sources.

The illustrations including the additions to the current edition are mostly the author's own drawings and photographs; while FIGS. 20, 21 and 22 of Part II, FIGS. 14, 15, 54, 59, 154, 169-70 and 188 of Part V, FIGS. 16, 17, 18, 24, 25 and 26 of Part VI have been redrawn from other publications (as detailed in the preface of the first edition) with the kind permission of the publishers, for which due acknowledgements are again made in their respective places in the text.

Suggestions are again invited.

Satribari Road A.C.D.
Gauhati 781008, Assam

EXTRACT FROM THE
PREFACE TO THE FIRST EDITION

BOTANY for Degree Students is an enlarged version of the author's *A Class-book of Botany*, eleventh edition, with necessary modifications and considerable additions. The book, as the name implies, is meant to meet the need of students preparing for the three-year or two-year degree examination (pass standard) of Indian Universities. With this end in view an attempt has been made to cover as far as possible the syllabuses for the above courses. The author has tried to present the subject-matter in an easily understandable way without sacrificing the essential details and general principles and yet avoiding redundant matter and unnecessary complications. The book, as it stands, is expected to meet adequately the need of the students for whom it is meant.

The illustrations, as incorporated in the book, are mostly the author's own drawings and photographs, either from *A Class-book of Botany*, eleventh edition, or drawn afresh; while others, as listed below, have been adapted from the following publications with the permission of the publishers to whom sincere thanks of the author are due. Thus, with the permission of McGraw-Hill Book Company, Inc., FIG. II/11 has been redrawn from *Introduction to Plant Anatomy* by A. J. Eames and L. H. MacDaniels, edition and copyright 1925; FIGS. II/19, II/20 and II/21 redrawn from *Fundamentals of Cytology* by L. W. Sharp, edition and copyright 1943; FIGS. V/13 and V/14 redrawn from *Cryptogamic Botany* by G. M. Smith, edition and copyright 1938; and FIGS. V/44, V/49, V/126, V/154, V/160-1 and VI/16 from *Plant Morphology* by A. W. Haupt, edition and copyright 1953. With the permission of The University of Chicago Press FIGS. VI/14, VI/15, VI/20D, VI/22 and VI/23A-F have been redrawn from *Gymnosperms: Structure and Evolution* by C. J. Chamberlain, edition and copyright 1935.

Satribari Road
Gauhati, Assam
May, 1963

A.C.D.

CONTENTS

Introduction x

PART I *Morphology*

Chapter
1. The Root 1
2. The Stem 10
3. The Leaf 33
4. Defensive Mechanisms in Plants 60
5. The Inflorescence 64
6. The Flower 71
7. Pollination 99
8. Fertilization 108
9. The Seed 111
10. The Fruit 126
11. Dispersal of Seeds and Fruits 136

PART II *Histology*

Chapter
1. The Cell 143
2. The Tissue 198
3. The Tissue System 216
4. Anatomy of Stems 228
5. Anatomy of Roots 241
6. Anatomy of Leaves 248
7. Secondary Growth in Thickness 252
8. Anomalous Secondary Growth in Thickness 263
9. Healing of Wounds and Fall of Leaves 266H

PART III *Physiology*

Chapter
1. General Considerations 268
2. Soils 279
3. Chemical Composition of the Plant 286
4. Absorption of Water and Mineral Salts 298
5. Conduction of Water and Mineral Salts 304
6. Manufacture of Food 319
7. Special Modes of Nutrition 343
8. Translocation and Storage of Food 348
9. Enzymes 356
10. Digestion and Assimilation of Food 360
11. Respiration and Fermentation 361
12. Metabolism 378
13. Growth 379
14. Movements 389
15. Reproduction 400

CONTENTS

PART IV *Ecology*
Chapter 1 Preliminary Considerations 411
2 Ecological Groups 418
3 Types of Vegetation in India 428
4 Phytogeographical Regions of India 430

PART V *Cryptogams*
Chapter 1 Divisions and General Description 439
2 Algae 441
3 Bacteria 515
4 Fungi 524
5 Lichens 587
6 Bryophyta 591
7 Pteridophyta 621

PART VI *Gymnosperms*
Chapter 1 General Description 659
2 Cycadales 664
3 Coniferales 669
4 Gnetales 681

PART VII *Angiosperms*
Chapter 1 Origin and Life-history 688
2 Principles and Systems of Classification 691
3 Selected families of Dicotyledons 706
4 Selected Families of Monocotyledons 767

PART VIII *Evolution and Genetics*
Chapter 1 Organic Evolution 783
2 Genetics 794

PART IX *Economic Botany*
Chapter 1 General Description & Economic Plants 810

PART X *Palaeobotany*
Chapter 1 General Description 844
2 Fossil Plants 846

Glossary of Names of Plants 863

Index 881

INTRODUCTION

1. Botany. The science that deals with the study of living objects goes by the general name of **biology** (*bios*, life; *logos*, discourse or science). Since both animals and plants are living, biology includes a study of both. Biology is, therefore, divided into two branches: **botany** (*botane*, herb) which treats of plants, and **zoology** (*zoon*, animal) which treats of animals.

2. Scope of Botany. The subject of botany deals with the study of plants from many points of view. This science investigates the internal and external structures of plants, their functions in regard to nutrition, growth, movements and reproduction, their adaptations to the varying conditions of the environment, their distribution in space and time, their life-history, relationship and classification, the laws involved in their evolution from lower and simpler forms to higher and more complex ones, the laws of heredity, the uses that plants may be put to and, lastly, the different methods that can be adopted to improve plants for better uses by mankind.

3. Origin and Continuity of Life. Life itself is mysterious, and its origin still remains shrouded in mystery. It is assumed, however, that many millions of years ago life first came into existence in water as a speck of protoplasm (*protos*, first; *plasma*, form) from inorganic or non-living materials as a result of certain chemical and physical changes in them under adventitious circumstances. Protoplasm is, therefore, the first-formed living substance. It is interesting to note the following facts: protoplasm is not formed afresh and, therefore, no new life comes into being or can be created; protoplasm is, however, continuous from generation to generation through reproduction; and but for evolution life would have still remained in the first-formed, one-celled stage. [see also pp. 783-4].

4. Importance of Green Plants. Green plants are essential for the existence of all kinds of life, including even human life. Their importance in this respect lies, *first*, in the fact that they are the only natural agents which are able to purify the atmosphere by absorbing carbon dioxide gas from it and releasing from their body (by the breaking down of water) an almost equal volume of pure oxygen to it; and *second*, the green plants prepare food such as starch, the chief constituent of rice, wheat, millets, etc., from the raw materials—carbon dioxide obtained from the air, and water and inorganic salts obtained from the soil. Both these functions, viz., purification of the atmosphere and manufacture of food, are the monopoly of green plants, and are performed

INTRODUCTION

by the green corpuscles or chloroplasts of the leaf during the daytime, sunlight being the source of energy. Animals being devoid of chloroplasts have no such power. It is evident, therefore, that animals, including human beings, are deeply indebted to plants for these basic needs, viz., oxygen for respiration and food for nutrition and energy. In these respects chloroplasts may be said to hold a strategic position so far as the living world is concerned. As a matter of fact, the existence of man would be impossible and unthinkable without plants.

5. **Uses of Plants.** The primary necessities of man are threefold: food, clothing and shelter. All these are extensively supplied by the plant kingdom. The most essential need of man is of course food. This food primarily comes from plants in the form of cereals (rice, wheat, maize, oat, rye and barley), millets (smaller grains), pulses, vegetables, fruits, some of the vegetable oils, etc. For clothing again plants are indispensable sources of fibres, coarse and fine, for the manufacture of garments. Then again shelter from the inclemencies of the weather and protection against natural enemies have been sought after from time immemorial. In this respect the value of wood, bamboo, cane, reed, thatch grass, etc., is inestimable. With advance in knowledge man has tried to tap plants as sources for his comforts and varied uses. Thus a host of other useful products have been obtained from the plant world through his knowledge and the proper application of it. (See part IX Economic Botany).

6. **Characteristics of Living Objects.** We do not know what life really is. It is something mysterious and we are not in a position to define it. All living objects have, however, certain characteristics by which they can be distinguished from the non-living. These characteristics are as follows:

(1) **Life-cycle.** All living objects follow a definite life-cycle of birth, growth, reproduction, old age (senescence) and death.

(2) **Cellular Structure.** All living organisms are composed of characteristic types of structural units, called cells. A cell is an organized mass of living substance, called protoplasm with a nucleus in it, surrounded by a membrane or wall. This cellular structure is an exclusive feature of all living organisms.

(3) **Protoplasm.** Life cannot exist without protoplasm. It is the actual living substance in both plants and animals, and it is, as Huxley defined it, the physical basis of life. It performs all the vital functions. Protoplasm is a highly complex mixture of proteins and a variety of other chemical compounds occurring in particular proportions and in particular patterns and interacting in a harmonious and consistent manner. The property we call life depends on the co-ordinated action of all these substances.

(4) **Respiration.** All living beings—plants and animals—respire continuously day and night, and for the process of respiration they take in *oxygen gas* from the atmosphere and give out an almost equal volume of *carbon dioxide gas*. Respiration is an *energy-releasing* process, i.e. the energy that is stored up in the food and other materials is released by this process and made use of by the protoplasm for its manifold activities.

(5) **Reproduction.** Living beings—plants and animals—possess the power of reproduction, i.e. of giving rise to new young ones like themselves. Non-living objects have no such power. They may mechanically break down into a number of irregular parts; but living objects follow certain definite modes of reproduction periodically and give rise to the offspring of the same kind.

(6) **Metabolism.** Metabolism is a phenomenon of life. It includes constructive (or anabolic) and destructive (or catabolic) changes that the protoplasm is constantly undergoing.

(7) **Nutrition.** A living organism requires to be supplied with food. The chemical constituents of food are much the same in plants as in animals. These are ultimately digested and assimilated by the protoplasm for its own nutrition and growth. A regular supply of food is thus an indispensable factor to the living organisms.

(8) **Growth.** All living objects—plants and animals—grow. Certain non-living bodies may also grow, as does a crystal. But they differ in many respects. The growth of non-living objects is external, while that of living objects is internal.

(9) **Movements.** Movements are commonly regarded as a sign of life. Movements in plants, however, are restricted, as most of them are fixed to the ground; while most animals move freely. Moving plants and fixed animals are not, however, uncommon among the lower organisms. Movements in plants and animals may be *spontaneous* or *induced*.

(*a*) **Spontaneous movement** is the movement of an organism or of an organ of a plant or an animal *of its own accord*, i.e. without any external influence. This kind of movement is regarded as a characteristic sign of life. Spontaneous movement is evident in animals, and in plants it is exhibited by many algae, e.g. *Euglena, Chlamydomonas, Volvox*, many desmids and diatoms, *Oscillatoria*, etc. Among the 'flowering' plants the best example of spontaneous movement is exhibited by the Indian telegraph plant (*Desmodium gyrans*; see FIG. III/47). Besides, the streaming movements of protoplasm in the cells of many higher plants are distinctly visible under the microscope.

(*b*) **Induced movement or irritability**, on the other hand, is the movement of living organisms or of their organs in response to external stimuli. Protoplasm is sensitive to a variety of

INTRODUCTION

external stimuli, and when a particular stimulus is applied the reaction is usually in the form of a movement, as seen in sensitive plant (*Mimosa*), sensitive wood-sorrel (*Biophytum*), sundew (*Drosera*), Venus' fly-trap (*Dionaea*), etc. Leaves of many plants again close up in the evening when the light fails and open again in the morning. This is spoken of as 'sleep' movement. Irritability, however, is more pronounced in animals than in plants.

7. **Differences between the Living and the Non-living.** It is very difficult to trace the absolute differences between the living and the non-living. Certain points, however, may be cited by way of general differences between the two. Protoplasm is the physical basis of life; so all objects containing protoplasm are regarded as living. Non-living objects are conspicuous by its absence. Thus the presence or absence of protoplasm makes a fundamental difference between the animate and the inanimate, and the various life-processes carried on by the protoplasm, such as respiration, metabolism, nutrition, growth, movements and reproduction, are characteristics of the living objects only. Non-living objects, however, may show movements and growth in a certain sense. Thus non-living objects like machines are seen to move when induced by external forces. Very minute particles embedded in a liquid are also seen to vibrate with great rapidity; this vibration is called **Brownian movement** as it was first observed by Robert Brown in 1828 while examining pollen grains under the microscope. Non-living objects like crystals and corals may also grow, but there is a difference in the mode of growth between the living and the non-living as already discussed. All nerves and tissues undergo fatigue on repeated stimulation from which they recover after a period of rest. Non-living objects like metals may also undergo similar fatigue when worked for a prolonged period, and they can be poisoned or stimulated by drugs, as the late Sir J. C. Bose has experimentally proved. Thus no hard and fast line of distinction can be drawn between the living and the non-living.

8. **Distinctions between Plants and Animals.** Higher plants and higher animals are readily distinguished from one another by their possession of definite organs or members, particularly the organs of locomotion in the latter case, for the discharge of definite functions; but difficulty is experienced where the lower *unicellular* plants and animals are concerned. In fact, no hard and fast line of distinction can be drawn between plants and animals. The distinguishing features in general, however, are as follows:

(1) **Growth.** The regions of growth are localized in the case of plants, lying primarily at the extremities—root-apex and stem-apex—and also in the interior, i.e. growth is both apical and intercalary; while in the case of animals growth is not localized to any definite regions, i.e. all parts grow simultaneously. Moreover, in plants growth proceeds until death; while in animals growth ceases long before death.

(2) **Chlorophyll.** Chlorophyll is highly characteristic of plants with the exception of fungi and total parasites. Chlorophyll and plastids are conspicuous by their absence in animal cells.

(3) **Cell-wall.** Both plants and animals are cellular in composition. Each plant cell, however, is surrounded by a distinct but dead wall, called the *cell-wall*. The cell-wall is almost universally present in all plants and is the most conspicuous part of the cell. The cell-wall, however, is always absent from an animal cell. In it, and also in plant cell, an extremely thin membrane called the *plasma membrane* made of fat and protein occurs covering each cell.

(4) **Cellulose.** The cell-wall of the plant cell is made up of a chemical substance, called *cellulose*; pure cellulose, however, is not found in fungi. But it is altogether absent from the animal body.

(5) **Food.** Green plants absorb raw food materials from outside —water and inorganic salts from the soil and carbon dioxide from the air—and prepare organic food substances out of them, primarily in the leaf, with the help of chlorophyll in the presence of sunlight. Animals being devoid of chlorophyll have no power of manufacturing their own food. They have solely to depend on plants for this primary need. It is also to be noted that plants take in food in solution only, whereas animals can ingest solid food.

(6) **Utilization of Carbon dioxide.** Plants possess the power of utilizing the carbon dioxide of the atmosphere. It is only the green cells that have got this power. Thus during the daytime the green cells of the leaf absorb a volume of carbon dioxide from the surrounding air, manufacture sugar, starch, etc., out of this carbon dioxide and water, and give out an almost equal volume of oxygen by the break-down of the water, H_2O (and not carbon dioxide, CO_2). Animals do not possess this power of utilizing the carbon dioxide or of manufacturing food.

(7) **Movements.** Plants grow fixed to the ground or attached to some support, and as such they cannot bodily move from one place to another, except some lower types of plants; while animals move freely in search of food and shelter, and manoeuvre when attacked; some animals, of course, grow attached to some object.

INTRODUCTION

(8) **Organs.** Various organs such as the organs of locomotion, respiratory organs, excretory organs, etc., have reached a stage of perfection in animals for efficient functioning; while in plants the corresponding organs are simple in construction or even altogether absent.

9. **Divisions of the Plant Kingdom.** Various schemes for classifying the vast plant kingdom comprising all categories of plants into several groups, bigger and smaller, have been formulated from time to time, each having its own merits and demerits. The earlier workers had divided the plant kingdom into two main groups, viz. 'flowerless' or 'seedless' plants called **Cryptogamia** (*kryptos*, concealed; *gamos*, marriage) and 'flowering' or 'seed-bearing' plants called **Phanerogamia** (*phaneros*, visible) or **Spermatophyta** (*sperma*, seed). Cryptogams have been further divided into three groups, viz. **Thallophyta, Bryophyta** and **Pteridophyta**, and phanerogams into two groups, viz. **Gymnospermae** (naked-seeded plants), e.g. cycads and conifers, and **Angiospermae** (closed-seeded plants)—the latter into **Dicotyledonae** (embryo with two cotyledons) and **Monocotyledonae** (embryo with one cotyledon). A tentative scheme based on modern natural classifications, satisfying our needs, may be as follows. It may be noted that the use of the terms *sub-kingdom*, *division* and *class* to represent the bigger groups of plants is somewhat arbitrary—in fact, these are matters of opinion and convenience.

Sub-kingdom A. **Thallophyta** (plants not forming embryos)

Division I. **Phycophyta** or **Algae**: class (1) Cyanophyta or blue-green algae; class (2) Euglenophyta or euglenoids; class (3) Bacillariophyta or diatoms; class (4) Chlorophyta or green algae; class (5) Phaeophyta or brown algae; and class (6) Rhodophyta or red algae.

Division II. **Mycophyta** or **Fungi**: class (1) Schizomycophyta or bacteria; class (2) Myxomycophyta or slime fungi; class (3) Eumycophyta or true fungi; (*a*) Phycomycetes or alga-like fungi, (*b*) Ascomycetes or sac fungi, and (*c*) Basidiomycetes or club fungi.

Sub-kingdom B. **Embryophyta** (plants forming embryos)

Division I. **Bryophyta** (plants without vascular tissues): class (1) Hepaticae or liverworts; class (2) Anthocerotae or horned liverworts; and class (3) Musci or mosses.

Division II. **Tracheophyta** (plants with vascular tissues):

Class (1) Psilotopsida, e.g. *Psilotum*
Class (2) Lycopsida, e.g. *Lycopodium*
Class (3) Sphenopsida, e.g. *Equisetum*   } Pteridophyta
Class (4) Pteropsida   : Filicinae

: Gymnospermae   } Spermatophyta
: Angiospermae

Dicotyledons and Monocotyledons. Angiosperms have been divided into two big classes: dicotyledons (*di*, two) and monocotyledons (*monos*, single), primarily on the basis of the number of cotyledons (first introduced by John Ray of Cambridge in 1686 and later followed by others). Other morphological distinctions are: in dicotyledons the primary root persists and gives rise to the tap root, while in monocotyledons the primary root soon perishes and is replaced by a cluster of fibrous roots; as a rule the venation of the leaf is reticulate (net-like) in dicotyledons, while it is parallel in monocotyledons (with but few exceptions in both); and dicotyledonous flower has commonly a pentamerous symmetry, while monocotyledonous flower a trimerous symmetry (see also pp. 233, 241 and 703).

10. Number of Species on Record

1	Algae	20,000
2	Fungi	90,000
3	Bacteria	2,000
4	Lichens	15,000
5	Bryophyta	23,725
	(*a*) Liverworts (8,750)	
	(*b*) Mosses (14,975)	
6	Pteridophyta (ferns & allies)	9,000
7	Gymnosperms	700
8	Angiosperms	199,000
	(*a*) Dicotyledons (159,000)	
	(*b*) Monocotyledons (40,000)	
	TOTAL	**359,425 species**

11. Branches of Botany. Botany, like every other science, may be studied from two aspects—the *pure* and the *applied* or *economic*. Pure botany deals with the study of plants as they form a part of nature, and applied botany as it is applied to the well-being of mankind. The subject as a whole may be divided into the following branches:

(1) **Morphology** (*morphe*, form; *logos*, discourse or study). This deals with the study of forms and features of different plant organs such as roots, stems, leaves, flowers, seeds and fruits. The study of external structures of such organs is otherwise known as *external morphology*, and that of internal structures as *internal morphology*. The latter may be histology or anatomy.

(2) **Histology** (*histos*, tissue). The study of the detailed structure of tissues making up a particular organ is called **histology**. The study of gross internal structure of a plant organ, as seen in a section, is called **anatomy** (*ana*, asunder; *lemnein*, to cut). Cytology (*kytos*, cell) dealing with the cell-structure with special

INTRODUCTION

reference to the behaviour of the nucleus is a newly-established branch of histology.

(3) **Physiology** (*physis*, nature of life). This deals with the various functions that the plants perform. Functions may be *vital* or *mechanical*; vital functions are performed by the living matter, i.e. the protoplasm, and the mechanical functions by certain dead tissues without the intervention of the protoplasm; as, for example, bark and cork protect the plant body, and certain hard tissues strengthen it. It is to be noted that structure and function are correlated, i.e. a particular structure develops in response to a particular function.

(4) **Ecology** (*oikos*, home). This deals with the interrelationships between the plants and the environment surrounding them.

(5) **Plant Geography.** This deals with the distribution of plants over the surface of the earth and the factors responsible for such distribution.

(6) **Taxonomy or Systematic Botany.** This deals with the description and identification of plants, and their classification into various natural groups according to the resemblances and differences in their morphological characteristics.

(7) **Organic Evolution.** This deals with the sequence of descent of more complex, more recent and more advanced types of plants and animals from the simpler, earlier and more primitive types through successive stages in different periods of the earth.

(8) **Genetics.** This deals with the facts and laws of inheritance (variation and heredity) of parental characters by the offspring.

(9) **Palaeobotany** (*palaios*, ancient). This deals with the ancient forms of plants preserved in the form of fossils in the earth's strata in past geological ages.

(10) **Applied or Economic Botany.** This deals with the utilization of plants and plant products for the well-being of mankind, and the various scientific methods employed for their improvement. (See part IX). It has also several branches: (*a*) **Agronomy** dealing with the cultivation of field crops for food and industry; (*b*) **horticulture** dealing with the cultivation of garden plants for flowers and fruits; (*c*) **plant pathology** dealing with the diagnosis, cure and prevention of plant diseases, mainly in field crops and other useful plants, commonly caused by fungi and bacteria, and also deficiency diseases; (*d*) **pharmacognosy** dealing with the study of medicinal plants with special reference to preparation and preservation of drugs; (*e*) **forestry** dealing with the study and utilization of forest plants for timber and other forest products; and (*f*) **plant breeding** dealing with the cross-breeding of plants evolving newer and more improved types with desired characteristics, e.g. higher yield and better quality.

12. Parts of an Angiospermic Plant (FIG. 1). Parts of the plant body mainly concerned with the nutrition and growth are called *vegetative parts,* and they comprise the root system and the shoot system (partly). The root system with its differentiated parts performs two primary functions: fixation and absorption. The shoot system on the other hand may be vegetative or reproductive. The vegetative shoot consisting of main stem, branches and leaves has mainly three functions: support, conduction, and food manufacture (primarily by leaves). The reproductive shoot is the flower with its differentiated organs, and is essentially concerned with the **reproduction of the plant.**

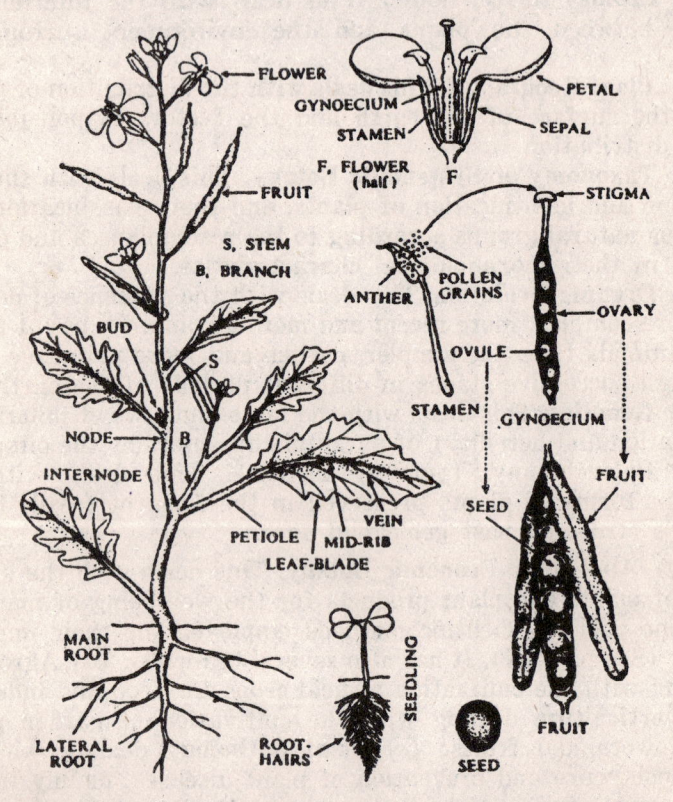

FIG. 1. Parts of an angiospermic plant (mustard plant).

13. A Short History of Botany. Botany as a subject originally developed out of practice and study of agriculture and medicine in prehistoric times. It was from the 16th century that several attempts were made to systematize the knowledge of the plant kingdom in the shape of classification on scientific lines. From the 18th century classification began to take a definite shape. The 19th century was specially remarkable for the development of the science of Botany in its many branches.

INTRODUCTION

Ancient History. From the records available, meagre though they are, it is known that early civilized men, especially the Assyrians and the Egyptians, as long ago as 5,000-3,400 B.C., possibly earlier, were acquainted with the uses of certain plants, particularly those with agricultural and medicinal values. They used to cultivate wheat, barley, etc., and grow roses, grapes, pomegranates, dates, figs, etc., and also certain medicinal plants. They also knew the art of pollinating date-palms. **Herodotus**, a Greek historian, who travelled to Egypt about 465 B.C., gave a good account of Egyptian agriculture. It is also known that the Chinese as early as 2,500 B.C. or even earlier used to grow rice, tea, oranges and some medicinal plants. In Greece about 340 B.C., a considerable advance was made in the study of plants, particularly by **Theophrastus** (372-287 B.C.), a disciple of Aristotle (384-322 B.C.). In his book *History of Plants*, Theophrastus described about 500 plants of food and medicinal value and other economic uses, with their morphological characteristics, particularly the habit—herbs, shrubs and trees, and annuals, biennials, and perennials. Mention may also be made of **Dioscorides**, a Cicilian Greek, and **Pliny**, a meritorious Roman writer, both of 1st century A.D. The former, a learned physician, described about 600 plants in his *Materia Medica*, while the latter had to his credit *Natural History* complete in 37 volumes. Pliny died at the age of 56 while watching the eruption of Vesuvius in 79 A.D. A period of great upheavals—religious, political and social—followed and there was a definite setback in the pursuit of scientific knowledge. So far as ancient India is concerned, the system of Ayurvedic medicine may be traced back to the Vedic period (6,000 B.C.). There is mention of SOMA and of vegetable drugs for the cure of pthisis and leprosy in the *Rig Veda* which is the earliest book in the library of man (3,000 B.C.). The Vedic Aryans were acquainted with about 100 medicinal plants (2,000 B.C.). Records show again that the Indo-Aryans were acquainted with several medicinal plants, and that surgery was highly developed then. The rise of Buddhism (5th century B.C.) gave further impetus to the study of medicine in ancient India, but because of AHIMSA preached by Lord Buddha, surgery suffered a setback. The edict of Asoka (2nd century B.C.) provided for the establishment of hospitals in all the principal towns and cities of India. The works of **Sushruta** (5th century B.C.) and **Charaka** (1st century A.D.), however, deserve special mention. The writings of these two ancient celebrities are considered to be standard works on Hindu medicine, anatomy and surgery. **Sushruta** was a specialist in surgery, but his works also included medicine (he mentioned 700 medicinal plants), pathology, midwifery, opthalmology, etc. **Charaka** was, however, a specialist in drugs and their uses. The Ayurvedic system further flourished under the care of eminent scholars such as Nagarjuna, Bagbhatta, Madhaba, Chakrapani Dutta and others, and attained its full glory between the 5th and the 11th centuries A.D. The agricultural system was also fairly advanced in India during 3,500-2,000 B.C.

16th Century. The Renaissance began from this century; medical botany still dominated. **Brunfels** (1464-1534), a German botanist, wrote his book on medicinal plants in three volumes in 1530. His work was a link between ancient and modern botany. **Fuchs** (1501-66), a German medical botanist, wrote his *Historia of Plants* in 1542. **Turner** (1515-68), an English physician and botanist, wrote his *A New Herbal* in three volumes in 1551, 1562 and 1568. He was regarded as the father of English botany at that time. **Caesalpino** (1519-1603), an Italian botanist and medical man, wrote his *De Plantis* in 1583 describing about 1,520 plants. In 1596, **Kaspar Bauhin** (1560-1624), an Italian botanist, described about 2,700 plants.

17th Century. In 1620 and 1623, **Bauhin** published his life-long work in the form of two books containing 6,000 species. He introduced binomial names for several species and formulated a system of classification largely based on natural affinities of plants, mainly depending on texture and form. He was the first to get rid of medical superstition in classifying plants. He could not recognize, however, the importance of flowers as a basis of classification. **John Ray** (1627-1705) of Cambridge showed outstanding merit in classifying plants in 1686 and later into (*a*) Imperfectae (algae, fungi, mosses and ferns) and (*b*) Perfectae (dicotyledons and monocotyledons). Ray also published *Flora of British Isles* in 1690. **Tournefort** (1656-1708), a French botanist and contemporary of Ray, introduced for the first time certain characters of flowers and fruits in classifying plants in 1695 without, however, ignoring the habit, and revived the concept of genera and species. About 8,000 species came to be known then.

18th Century. Progress continued. A very prominent figure of this century was a Swedish botanist **Carolus Linnaeus** (1707-78), professor of botany at Uppsala. His monumental work on taxonomy and binomial nomenclature (1735 and later) proved to be invaluable to future generations. He is regarded as the father of botany. His *Sexual System of Classification* published in 1735, artificial though it is, is a brilliant piece of work on taxonomic botany. His great achievement, however, lay in the foundation of a precise binary system of nomenclature for each and every species of plants. **De Jussieu** (1748-1836), a French botanist, introduced several changes in Linnaean system. He classified plants in 1789 into 100 orders (families) and placed them under (*a*) Acotyledons (cryptogams of today including, however, some aquatic 'flowering' plants), (*b*) Monocotyledons and (*c*) Dicotyledons, and introduced the terms hypogyny, perigyny and epigyny. His system formed a basis for subsequent progress in the elaboration of natural systems.

19th Century. In the earlier part of this century the systems of classification proposed were those of **De Candolle** in 1819, **Robert Brown** in 1827, **Endlicher** in 1836, **Lindley** in 1845. Then two great English botanists, **Bentham** (1800-84) and **Hooker** (1817-1911), came together and laid the foundation of a natural system of classification known as 'Bentham and Hooker's System' (1862-83) which is widely followed. They have given an account of 202 families of angiosperms. *The Flora of British India* published by them over a long period (1872-97) is a masterpiece on Indian flora. Some years after the publication of Darwin's *Origin of Species by Natural Selection* in 1859 attempts were made to classify plants on the basis of evolution and to establish a phylogenetic tree. Thus many new systems of classification were formulated, particularly in Germany; for example, **Prantl's** in 1883, **Eichler's** in the same year, **Engler's** in 1886. Engler and Prantl published *Die Naturlichen Pflanzenfamilien* (1887-1909) and Engler and Gilg's very useful publication *Syllabus der Pflanzenfamilien* (1892) gives a comprehensive account of 297 families of 'flowering' plants in addition to systematic classification of cryptogams.

20th Century. In the present century **Hutchinson's** (an English botanist, born in 1884) phylogenetic system appeared in 1926 (Dicotyledons) and 1934 (Monocotyledons) with several modifications of Engler's system. In his revised second edition of *Families of Flowering Plants* published in 1959, Hutchinson described 411 families (342 of dicotyledons and 69 of monocotyledons). Other important systems that have appeared during the current century are those of **Wettstein**, an Austrian, in 1901 and later a posthumous improved edition in 1930-35; of **Bessey**, an American, in 1915; of **Rendle**, an Englishman, in 1904 (1st volume) and 1925 (2nd volume); and of **Tippo**, an American, in 1942.

Part 1   MORPHOLOGY

Chapter 1   THE ROOT

Tap Root System. The primary root and its branches form the tap root system of the plant. The primary or tap root normally grows vertically downwards to a shorter or longer depth, while the branched roots (secondary, tertiary, etc.) grow obliquely downwards, or in many cases spread horizontally outwards. The primary root may be sparingly or profusely branched according to the need of the plant. The tap root system is normally meant to absorb water and mineral salts from the soil, to conduct them upwards to the stem and to give proper anchorage to the plant, but in order to perform some specialized functions it becomes modified into distinct shapes.

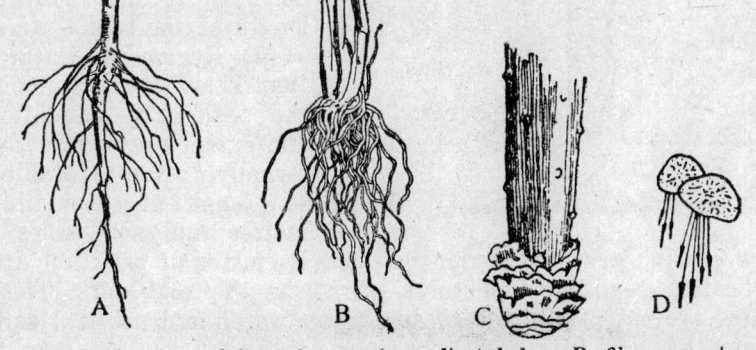

FIG. 2. *A*, tap and lateral roots in a dicotyledon; *B*, fibrous roots in a monocotyledon; *C*, multiple root-cap in screwpine (*Pandanus*); *D*, root-pocket in duckweed (*Lemna*).

Regions of the Root (FIG. 3). The following regions may be distinguished in a root from the apex upwards. There is of course no line of demarcation between one region and another, and each tends to merge into the next.

(1) **Root-cap.** Each root is covered over at the apex by a sort of cap or thimble known as the root-cap which protects the tender apex of the root as it makes its way through the soil. Due to the impact of the hard soil particles the outer part of the root-cap wears away and newer cells formed by the underlying growing tissue are added to it. The root-cap is, however, usually absent in the aquatic plant.

(2) **Region of Cell Division.** This is the growing apex of the root lying within and a little beyond the root-cap and extends to a length of one to a few millimetres. The cells of this region are

very small and thin-walled, and contain a dense mass of protoplasm. The characteristic feature of this region is that the cells undergo repeated divisions, and hence this region is otherwise called the **meristematic region** (*meristos,* divided). Some of the newly formed cells contribute to the formation of the root-cap and others to the next upper region.

(3) **Region of Elongation.** This lies above the meristematic region and extends to a length of a few millimetres (1 to 5 mm. or a little more). The cells of this region undergo rapid elongation and enlargement, and are responsible for growth in length of the root.

(4) **Region of Maturation.** This region lies above the region of elongation and extends upwards. Externally, often extending to a length of a few millimetres and sometimes a few centimetres, this region produces a cluster of very fine and delicate thread-like structures known as the **root-hairs.** These hairs are essentially meant to absorb water and mineral salts from the soil. Internally, the cells of this region are seen to undergo maturation and differentiation into various kinds of primary tissues. Higher up it gradually merges into the region of secondary tissues. Above the root-hair region lateral roots are produced in *acropetal* succession.

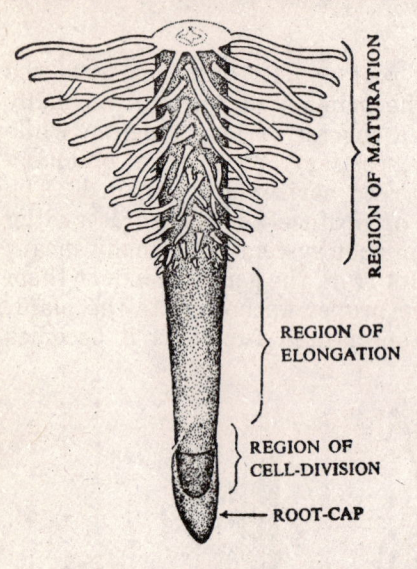

FIG. 3. Regions of the root.

Characteristics of the Root. There are certain distinctive characteristics of the root by which it can be distinguished from the stem. These are as follows:

(1) The root is the descending portion of the axis of the plant, and is not normally green in colour.

(2) The root does not commonly bear **buds** except in sweet potato (*Batatas*), wood-apple (*Aegle*), *Trichosanthes* (B. PATAL; H. PARWAL), Indian redwood (*Dalbergia*), lemon and ipecac. Such plants are sometimes propagated by root-cuttings, e.g. ipecac.

(3) The root ends in and is protected by a cap- or thimble-like structure known as the **root-cap** (FIG. 3); while the stem ends in

THE ROOT

a bud. A distinct multiple root-cap is seen in the aerial root of screwpine (*Pandanus;* FIG. 2C).

In water plants like duckweed (*Lemna*), water lettuce (*Pistia*), water hyacinth (*Eichhornia*), etc., a loose sheath which comes off easily is distinctly seen at the apex of each root. This is an anomalous root-cap, called the **root-pocket** (FIG. 2D).

(4) The root bears **unicellular hairs** (FIG. 4B), while the stem or the shoot bears mostly **multicellular hairs** (FIG. 4C). Root-

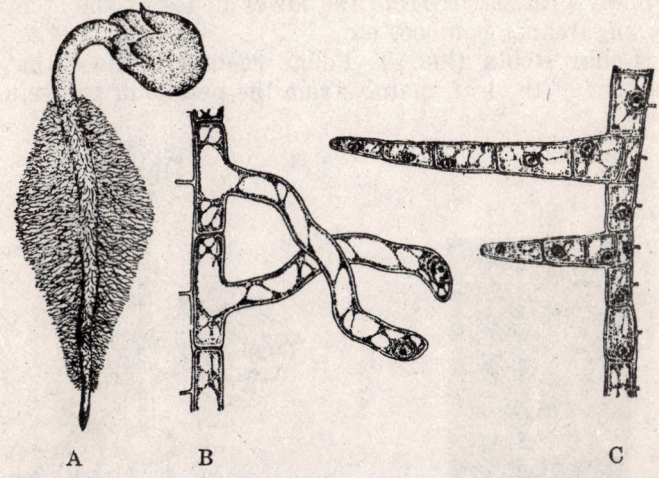

FIG. 4. *A*, root-hairs in mustard seedling; *B*, two root-hairs (magnified)—unicellular; *C*, two shoot-hairs (magnified)—multicellular.

hairs occur in a cluster in the tender part of the root a little behind the apex. Shoot-hairs, on the other hand, are of various kinds and they remain scattered all over the surface of the shoot. Root-hairs absorb water and mineral salts from the soil, while shoot-hairs prevent evaporation of water from the surface of the plant body and afford protection.

(5) Lateral roots always develop from an inner layer (pericycle; see FIG. II/66); so they are said to be **endogenous** (*endo*, inner; *gen*, producing). Branches, on the other hand, develop from a few outer layers; so they are said to be **exogenous** (*exo*, outer).

(6) **Nodes** and **internodes** are always present in the stem, although they may not often be quite distinct; but in the root these are absent.

Adventitious Root System. Roots that grow from any part of the plant body other than the radicle are called **adventitious roots**. They may develop from the base of the stem replacing the primary root or in addition to it, or from any node or internode of the stem or the branch, or even from the leaf under special circumstances. Adventitious roots are of various kinds and have diverse functions—normal and specialized. Those with normal functions may be of the following types:

(1) **Fibrous Roots** (FIG. 2B). Fibrous roots of monocotyledons are all adventitious roots. They may be given off in clusters from the base of the stem, as in onion, tuberose, etc., or from the nodes and sometimes internodes of branches creeping along the ground, as in many grasses, or from the lower nodes of the stem, as in maize, sugarcane, bamboo, etc.

(2) **Foliar Roots** (FIG. 5). Foliar roots are those that come directly out of the leaf, mainly from the petiole or the vein. Such

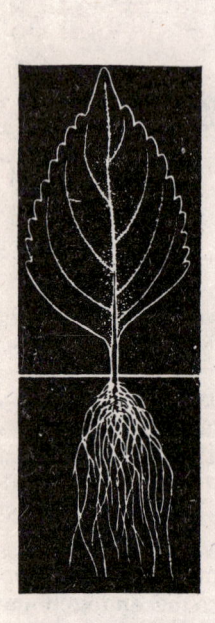

FIG. 5. Foliar (adventitious) roots in *Pogostemon*.

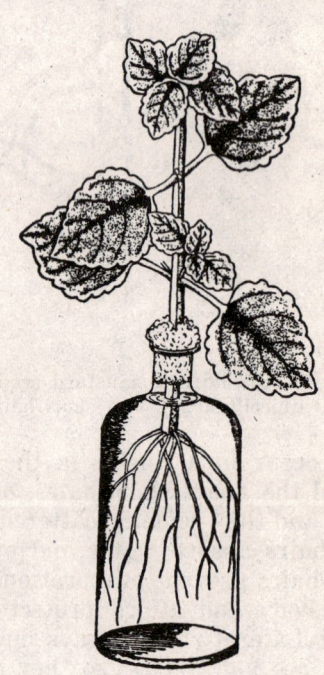

FIG. 6. Adventitious roots in *Coleus*.

roots may sometimes arise spontaneously, or more commonly as a result of injury (e.g. when incised), or they may be induced to grow by the application of certain chemicals, called *hormones*, which are growth-promoting substances. Thus *Pogostemon* leaf, when treated with a synthetic hormone, e.g., indole-butyric acid, is seen to produce a cluster of roots from the petiole. Such roots are not common in nature but can be induced to grow by the above treatment.

THE ROOT

(3) **Adventitious roots** are also given off by many plants from their nodes and sometimes from the internodes as they creep on the ground, as in Indian pennywort (see FIG. III/55), woodsorrel (see FIG. 32), etc. Such roots are also produced in many cases from branch-cuttings when these are put into the soil, as in rose, sugarcane, China rose, marigold, tapioca, etc., or kept partially immersed in water in a bottle, as in garden croton (*Codiaeum*), *Coleus* (FIG. 6), etc. Adventitious roots also grow from foliar buds developing on leaves, as in sprout leaf plant (*Bryophyllum*; see FIG. 16A) and elephant ear plant (*Begonia*; see FIG. 16B).

Modified Roots

A. *TAP ROOT MODIFIED* (*for storage of food*)
(1) **Fusiform Root** (FIG. 7A). When the root (hypocotyl) is swollen in the middle and gradually tapering towards the apex and the base, being more or less spindle-shaped in appearance, it is said to be fusiform, e.g. radish. In radish it is really the hypo-

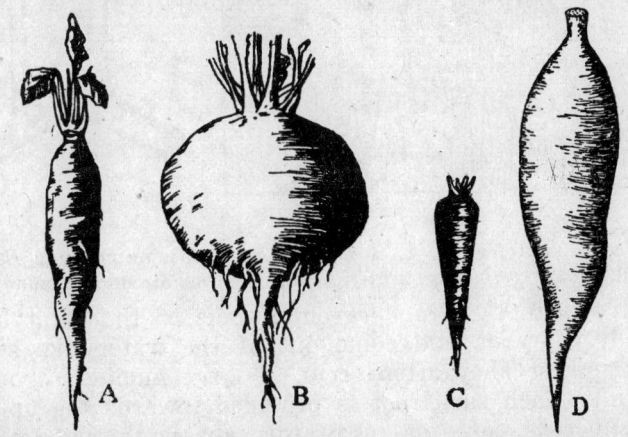

FIG. 7. Modified Roots. *A,* fusiform root of radish; *B,* napiform root of turnip; *C,* conical root of carrot; *D,* tuberous root of *Mirabilis.*

cotyl and the base of the stem that swell; only the tapering end is the root proper. (2) **Napiform Root** (FIG. 7B). When the root is considerably swollen at the upper part (usually the hypocotyl), becoming almost spherical, and sharply tapering at the lower part, it is said to be napiform, e.g. turnip and beet. In turnip it is the hypocotyl that swells and becomes spherical; while in beet the hypocotyl and the root together become swollen.
(3) **Conical Root** (FIG. 7C). When the root is broad at the base and gradually tapers towards the apex like a cone, it is said to be conical, e.g. carrot. In carrot it is the root proper that swells.

(4) Tuberous or **Tubercular Root.** When the root is thick and fleshy but does not take a definite shape, it is said to be tuberous or tubercular, as in four o'clock plant (*Mirabilis;* FIG. 7D).

B. *BRANCHED ROOT MODIFIED* (*for respiration*)

(5) Pneumatophores. Many plants growing in marshy places and salt lakes, occasionally inundated by tides, as in the Sundarbans, develop special kinds of roots, called respiratory roots or **pneumatophores** (FIG. 8), for the purpose of respiration. Such

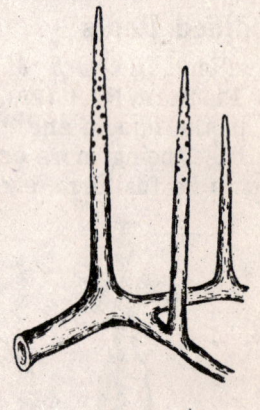

FIG. 8. Pneumatophores. *A*, two plants with pneumatophores; *B*, pneumatophores growing vertically upwards from an underground root.

roots grow from the underground roots of the plant but rise vertically upwards and come out of the water like so many conical spikes. They often occur in large numbers around the tree-trunk. Each such root is provided towards the upper end with numerous pores or respiratory spaces through which air is taken in for respiration. Examples are seen in *Rhizophora* (B. KHAMO), *Heritiera* (B. SUNDRI), etc.

C. *ADVENTITIOUS ROOTS MODIFIED*

(*a*) *For Storage of Food.* **(1) Tuberous** or **Tubercular Root** (FIG. 9A). This is a swollen root without any definite shape, as in sweet potato (*Ipomoea batatas*). Tuberous roots, whether tap or adventitious, are produced singly and not in clusters. **(2) Fasciculated Roots** (FIG. 9B). When several tubercular roots occur in a cluster or fascicle at the base of the stem, they are said to be fasciculated, as in *Dahlia, Ruellia* and *Asparagus.* **(3) Nodulose Root** (FIG. 9C).

THE ROOT

When the slender root becomes suddenly swollen at the apex, it is said to be nodulose, as in mango ginger (*Curcuma amada*), turmeric (*C. domestica*), *Asparagus sprengeri*, arrowroot (*Maranta*) and some species of *Calathea*. (4) **Moniliform or Beaded Root** (FIG. 10A). When there are some swellings in the root at frequent intervals, it is said to be moniliform or beaded, as in *Portulaca*, Indian spinach (*Basella*), *Momordica*, wild vine (*Vitis trifolia*) and some grasses. (5) **Annulated Root** (FIG. 10B). When the root has a series of ring-like swellings on its body, it is said to be annulated, as in ipecac (*Psychotria*)—a medicinal plant.

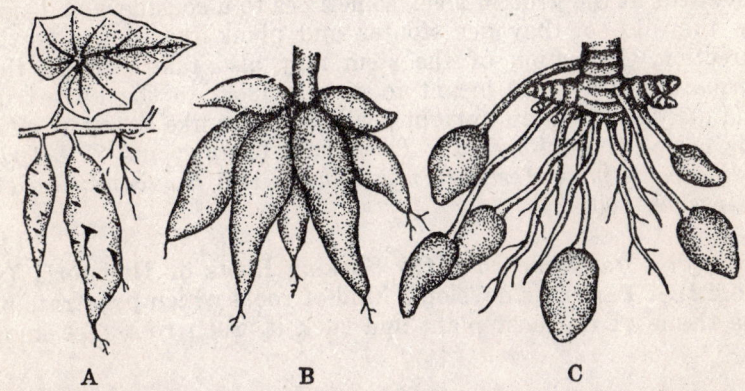

FIG. 9. Adventitious Roots. *A*, tuberous roots of sweet potato; *B*, fasciculated roots of *Dahlia*; *C*, nodulose roots of mango ginger.

(b) *For Mechanical Support.* (6) **Prop** or **Stilt Roots** (FIG. 11). In plants like banyan, india-rubber plant, screwpine, *Rhizophora*, etc., a number of roots are produced from the main stem and often from the branches. These roots grow vertically or obliquely downwards and penetrate into the soil. Gradually they get stouter and act as pillars supporting the main stem and the branches or the plant as a whole. Such roots are known as **prop** or **stilt roots**. The big banyan tree in the Indian Botanic Garden near Calcutta has produced over 900 such roots from its branches.

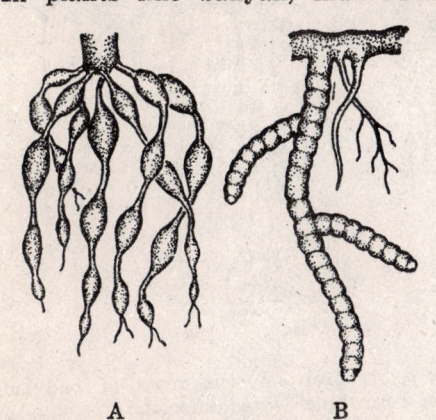

FIG. 10. Adventitious Roots (*contd.*). *A*, moniliform roots of *Momordica*; *B*, annulated roots of ipecac.

Its age is about 200 years, and the circumference of the crown is well over 360 metres. (7) **Climbing Roots** (FIG. 12A). Plants like betel (*Piper betle*), long pepper (*P. longum*), black pepper (*P. nigrum*), *Pothos*, Indian ivy (see FIG. 17), etc., produce climbing roots from their nodes and often from the internodes to ensure a foothold on neighbouring objects. (8) **Buttress Roots.** In certain large forest trees some of the stout roots around the base of the main trunk show prolific abnormal growth, particularly on their upper side. They at first grow obliquely downwards from the base of the trunk and then spread horizontally outwards at the ground level, sometimes to a considerable length. As they do so, they get stouter and plank-like in the vertical direction. A portion of the stem may also take part in their formation. They are meant to give support to the huge trunk and maintain it in an upright position. Trees like kapok (*Ceiba*), silk cotton tree (*Bombax*), *Terminalia catappa* and *T. belerica*, *Adina cordifolia*, *Heritiera macrophylla* and *H. acuminata*, etc., bear such roots.

(c) *For Vital Functions*. (9) **Sucking Roots** or **Haustoria** (see FIG. 22B). Parasites develop a kind of roots which penetrate into the tissue of the host plant and suck it. Such roots are known

FIG. 11. Adventitious Roots (*contd.*). *A*, prop or stilt roots of banyan (*Ficus bengalensis*); *B*, the same of screwpine (*Pandanus odoratissimus*).

as sucking roots or haustoria (sing. haustorium). Parasites, particularly non-green ones, have to live by sucking the host plant, i.e. by absorbing food from it with the help of their

THE ROOT

sucking roots. Common examples are dodder (*Cuscuta*; see FIG. 22A), *Cassytha*, broomrape (*Orobanche*; see FIG. 23A),

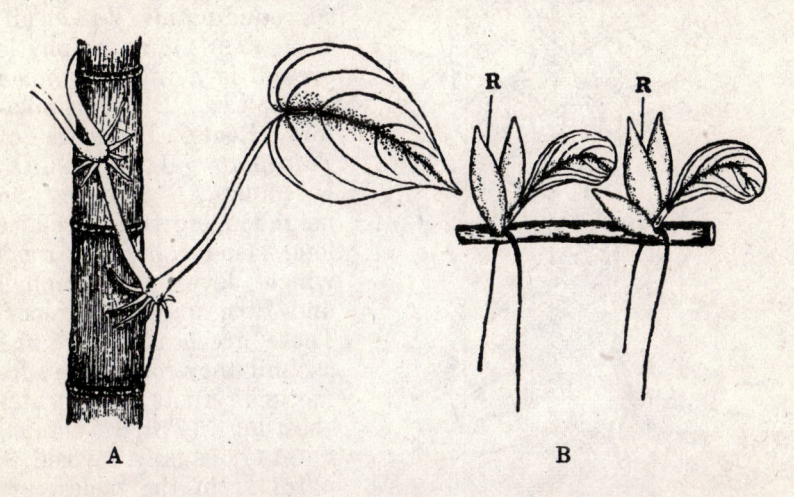

FIG. 12. Adventitious Roots (*contd.*). A, climbing roots of betel (*Piper betle*); B, respiratory roots (R) of *Jussiaea repens*.

mistletoe (*Viscum*; see FIG. 24) and *Loranthus*. (10) **Respiratory Roots** (FIG. 12B). In *Jussiaea* (B. KESSRA), an aquatic plant, the floating branches develop adventitious roots which are soft, light, spongy and colourless. They usually develop above the level of water and serve to store up air. Thus they facilitate respiration. (11) **Epiphytic Roots** (FIG. 13). There are certain plants, commonly the orchids, which grow perched on branches of trees. Such plants are known as **epiphytes** (*epi*, upon; *phyta*, plants). They never suck the supporting plant as do the parasites. So instead of sucking roots they develop a special kind of aerial

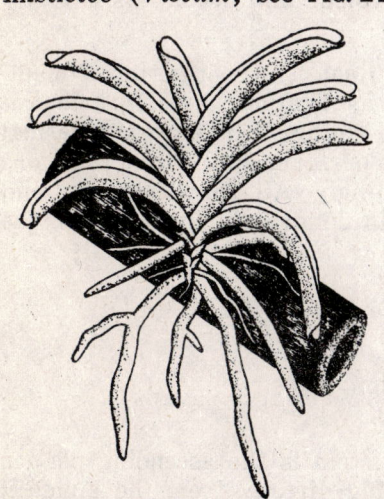

FIG. 13. Epiphytic roots of *Vanda* (an orchid).

roots which hang freely in the air. Each hanging root is surrounded by a spongy tissue, called **velamen** (see FIG. 26).

With the help of this velamen the hanging root absorbs moisture from the surrounding air. *Vanda* (B. & H. RASNA), an epiphytic orchid, is a fairly common example. (12) **Assimilatory Roots.** Branches of *Tinospora* (B. GULANCHA; H. GURCHA) climbing on neighbouring trees produce long, slender, hanging roots which develop chlorophyll and turn green in colour. These green roots are the assimilatory roots. The submerged roots of water chestnut (*Trapa natans*; FIG. 14), usually formed in pairs from the nodes, are green in colour and perform carbon-assimilation. The hanging roots of epiphytic orchids also often turn green in colour and act as assimilatory roots.

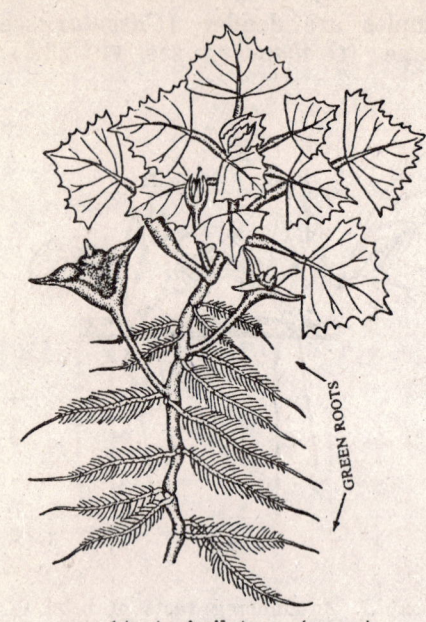

FIG. 14. Assimilatory (green) roots of *Trapa natans*.

Functions and Adaptations of the Root. The root performs manifold functions—*mechanical* such as **fixation**, and *physiological* such as **absorption, conduction** and **storage**. These are the normal functions of the root. Roots also perform specialized functions and they adapt themselves accordingly. All these functions and adaptations have been discussed in detail in connexion with the modified roots.

Chapter 2   THE STEM

Characteristics of the Stem. The stem is the ascending portion of the axis of the plant, developing directly from the plumule, and bears leaves, branches and flowers. When young, it is normally green in colour. The growing apex is covered over and protected by a number of tiny leaves which arch over it (FIG. 15). The stem often bears multicellular hairs of different kinds; it branches exogenously; and it is provided with nodes and internodes which may not be distinct in all cases. Leaves and branches normally develop from the nodes. When the stem or the branch ends in a vegetative bud it continues to grow upwards or sideways: If, however, it ends in a floral bud the growth ceases.

Forms of Stems. There is a variety of stem structures adapted to perform diverse functions. They may be aerial or underground. Aerial stems may be erect, rigid and strong, holding themselves in an upright position; while there are some too weak to support themselves in such a position. They either trail on the ground or climb neighbouring plants and other objects. Some stems remain permanently underground and from there periodically give off aerial shoots under favourable conditions; such stems are meant for food storage and perennation (see pp. 21-4).

1. **Erect or Strong Stems.** The unbranched, erect, cylindrical and stout stem, marked with scars of fallen leaves, is called **caudex**, as in palms. The jointed stem with solid nodes and hollow internodes is called **culm**, as in bamboo. Some herbaceous plants, particularly monocotyledons, have no aerial stem. The underground stem in them produces an erect unbranched aerial shoot bearing either a single flower or a cluster of flowers; such a flowering shoot is called **scape**, as in tuberose, onion, aroids banana, etc.

2. **Weak Stems.** Weak-stemmed plants are commonly described as (1) trailers, (2) creepers, and (3) climbers. **Trailers** are those plants whose thin and long or short branches trail on the ground, with or without rooting at the nodes. When such plants lie prostrate on the ground they are said to be (*a*) **prostrate** or **procumbent**, e.g. *Oxalis* and *Evolvulus*. When the branches of such plants after trailing for some distance tend to rise at their apex they are said to be (*b*) **decumbent**, e.g. *Tridax* (see FIG. VII/43). When the plants are much branched and the branches spread out on the ground in all directions, they are said to be (*c*) **diffuse**, e.g. *Boerhaavia*. Weak-stemmed plants with their long or short branches creeping along the ground and rooting at the nodes are said to be **creepers**; a creeping stem may be a runner, stolon, offset or sucker according to its varied nature (see FIGS. 32-5). **Climbers** are those plants that attach themselves to any neighbouring object, often by means of some special devices, and climb it to a long or short distance, e.g. pea, passion-flower, gourd, vine, etc. (see pp. 14-17).

Nodes and Internodes. The place on the stem or branch where one or more leaves arise is known as the **node**, and the space between two successive nodes is called the **internode**. Sometimes nodes and internodes are very conspicuous, as in bamboos and grasses; in others they are not always distinct.

Buds

A bud is a young undeveloped shoot consisting of a short stem and a number of tender leaves arching over the growing apex. In the bud the internodes have not yet developed and thus the leaves remain crowded together forming a compact structure. The lower leaves of the bud are older and larger than those higher up. A cabbage cut longitudinally gives a good idea of a bud—the development of the leaves in *acropetal* succession and the condensed shoot with the growing apex. The normal position of a bud is at the apex of the stem or the branch and in the axil of a leaf. The bud, in the former case, is known as the **terminal** or **apical bud**, and in the latter as the **axillary bud**. Sometimes some extra buds develop by the side of the axillary bud; these are known as the **accessory buds**. Sometimes buds appear at various other parts of a plant such as the root (radical buds), as in sweet potato, or the leaf (foliar buds), as in sprout leaf plant (*Bryophyllum*; FIG. 16A), *Kalanchoe* (see FIGS. III/56-7), *Crassula,* elephant ear plant (*Begonia;* FIG. 16B), *Scilla,* walking fern (*Adiantum*; see FIG. III/54), and sometimes water lily (*Nymphaea*), or at different positions of the stem and the branches (cauline buds). Such buds are known as **adventitious** because of their abnormal position.

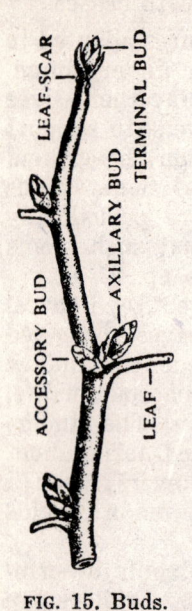

FIG. 15. Buds.

Protection of the Bud. Since buds have to give rise to flowers, leaves and branches it is imperative that these should be protected against external injuries—sun, rain, fungi, insects, etc., and this protection is afforded in different ways. (1) The young leaves of the bud normally overlap each other, and remain variously rolled or folded to protect themselves and the growing apex against sun and rain. (2) They may be covered by hairs, or in some cases they remain bathed in resinous or gummy secretions. (3) They may be enclosed by some dry and scaly outer leaves, called **bud-scales**, as in banyan, jack, *Magnolia,* iron-wood tree (*Mesua;* B. NAGESWAR; H. NAGKESAR), etc. (4) There may be a coating of wax or cutin on the leaf-surface to check evaporation of water and to prevent the leaves and the growing apex from getting wet.

Modification of the Bud. Vegetative buds may be modified into tendrils (see FIG. 20A), as in passion-flower and vine, or into

THE STEM

thorns (see FIG. 38), as in *Duranta, Carissa,* wood-apple (*Aegle*), etc. Sometimes these may become modified into special reproductive bodies known as **bulbils** (see p. 30). Floral buds may likewise be modified into tendrils (see FIGS. 36B & 37), as in Sandwich

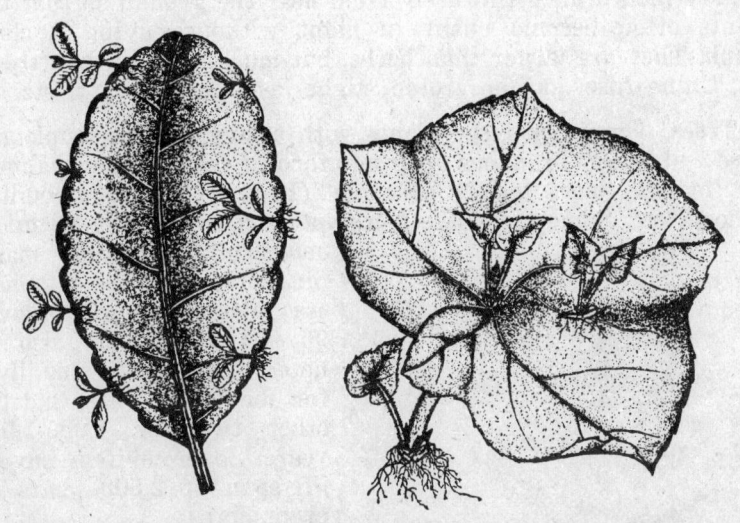

FIG. 16. *A*, foliar buds and adventitious roots of *Bryophyllum pinnatum;* *B*, the same of *Begonia*.

Island climber (*Corculum=Antigonon*) and balloon vine (*Cardiospermum*), or into bulbils (see p. 30) for the purpose of reproduction.

Habit of the Plant. The nature of the stem, the height the plants attain, and the duration and mode of their life determine their habit.

1. **Herbs.** These are small plants with soft stems. According to the duration of their life they may be classified as (1) **annuals**, (2) **biennials**, and (3) **perennials**. (1) **Annuals** are those plants that attain their full growth in one season, living for a few months or at the most for one year only, producing flowers, fruits and seeds within this period, e.g. sunflower, mustard, rice, pea, bean, etc. (2) **Biennials** are those plants that live for two years. They attain their full vegetative growth in the first year and produce flowers and seeds in the second year after which they die off. Common examples are cabbage, radish, beet, carrot, turnip, etc. (in tropical climates they behave like annuals); and (3) **perennials** are those plants that persist for a number of years; the aerial parts of such plants may die down every year at the end of the flowering season but next year again new shoots

develop from the underground stem after a few showers of rain, e.g. *Canna*, ginger (*Zingiber*), arrowroot (*Maranta*), etc.

2. Shrubs. These are medium-sized plants with hard and woody stems which branch profusely from near the ground so that the plants often become bushy in habit without having a clear trunk. They are larger than herbs, but much smaller than trees, e.g. China rose, garden croton, night jasmine, *Duranta*, etc.

3. Trees. These are large plants with a single stout trunk and hard and woody branches profusely formed (except most palms), e.g. mango, jack, teak, *Casuarina* (B. & H. JHAU), country almond, etc. Some trees like *Eucalyptus*, redwood tree (*Sequoia sempervirens*) and mammoth tree (*Sequoia gigantea*) attain a height of over 90 metres. It may also be noted that *Eucalyptus* lives for about 300 years, and the other two for 1,000-1,500 years. Some conifers have a life-span of 2,500 years or even more.

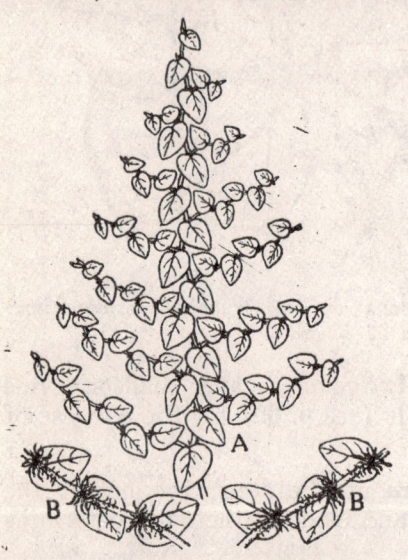

FIG. 17. Indian ivy (*Ficus pumila*)— a rootlet climber. *A*, upper side; *B*, lower side.

4. Climbers. These have thin and long stems with diffuse branches; they climb by means of some special organs of attachment or by their twining stem.

(1) **Rootlet Climbers.** Such plants climb by means of small adventitious roots which often form small adhesive discs or claws to act as holdfasts or secrete a sticky juice, as in betel (*Piper betle*; see FIG. 12A), long pepper (*Piper longum*), *Piper chaba*, ivy (*Hedera helix*), Indian ivy (*Ficus pumila*; FIG. 17), wax plant (*Hoya*), *Pothos*, etc.

(2) **Hook Climbers.** The flower-stalk of Artabotrys (B. & H. KANTALI-CHAMPA) produces a curved hook (FIG. 18C), which facilitates to some extent the climbing of the branches. Often **prickles** and **thorns** are curved and hooked in certain plants. Thus in cane (*Calamus*; FIG. 19A) a long slender axis beset with numerous sharp and curved hooks is produced from the leaf-sheath. Climbing rose (FIG. 19B) and *Pisonia* are provided with numerous curved prickles for the purpose of climbing (and also

THE STEM

for self-defence). Glory of the garden (*Bougainvillea*; FIG. 18A), and *Uncaria* (FIG. 18B) climb by curved hooks (thorns). In cat's nail (*Bignonia unguis-cati*; see FIG. 65) the terminal leaflets become modified into very sharp and curved hooks.

(3) **Tendril Climbers.** These are plants which produce slender, leafless, spirally-coiled structures known as **tendrils**, and climb

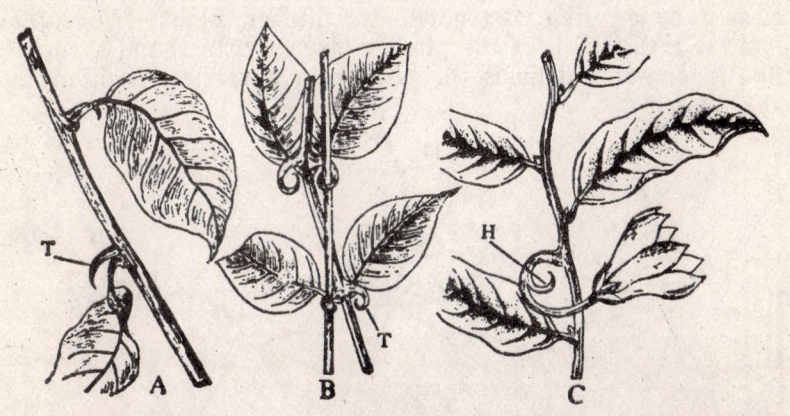

FIG. 18. Hook and Thorn Climbers. *A*, Glory of the garden (*Bougainvillea*); *T*, thorn; *B*, *Uncaria*; *T*. hooked thorn: *C*. *Artabotrys*; *H*, hook.

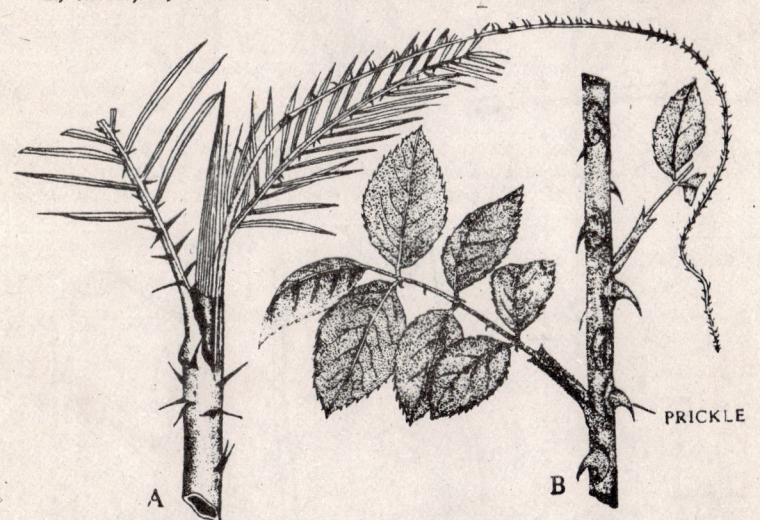

FIG. 19. Prickle Climbers. *A*, cane; *B*, rose.

objects with the help of them; tendrils twine themselves round some support, and help the plants concerned to support their weight and climb easily. Tendrils may be modifications of the stem, as in passion-flower (FIG. 20A), vine, Sandwich Island climber (*Corculum;* see FIG. 36B), balloon vine (*Cardiospermum;* see FIG. 37), etc., or of leaves, as in pea (FIG. 20B), wild pea (*Lathyrus;* FIG. 20C), *Naravelia* (see FIG. 64), etc.

(4) **Leaf Climbers.** The petiole (i.e. the leaf-stalk) of *Clematis* (FIG. 21A) and that of garden nasturtium are sensitive to contact and coil round any neighbouring object helping the plant to climb. In glory lily (*Gloriosa;* FIG. 21B) the leaf-apex becomes closely coiled like a tendril. In pitcher plant (*Nepenthes;* FIG. 21C) the stalk of the pitcher often twists round a support like a tendril and holds the pitcher in a vertical position (see also FIG. 69).

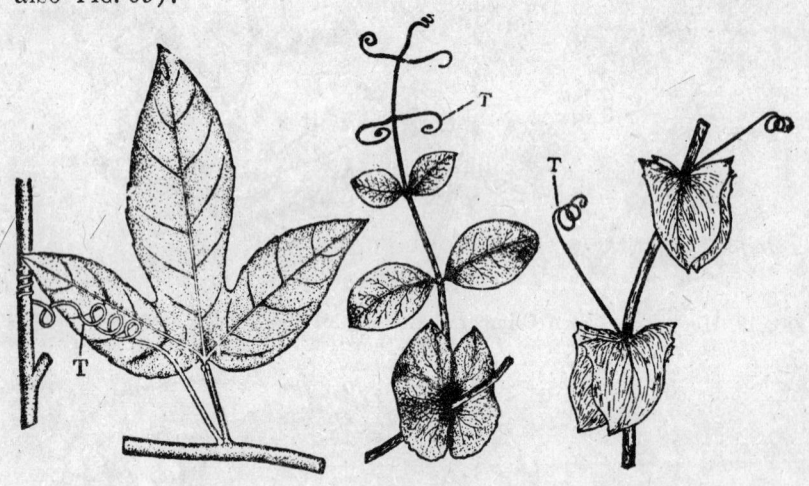

FIG. 20. Tendril Climbers. *A*, passion-flower; *B*, pea; *C*, wild pea (*Lathyrus*); *T*, tendril.

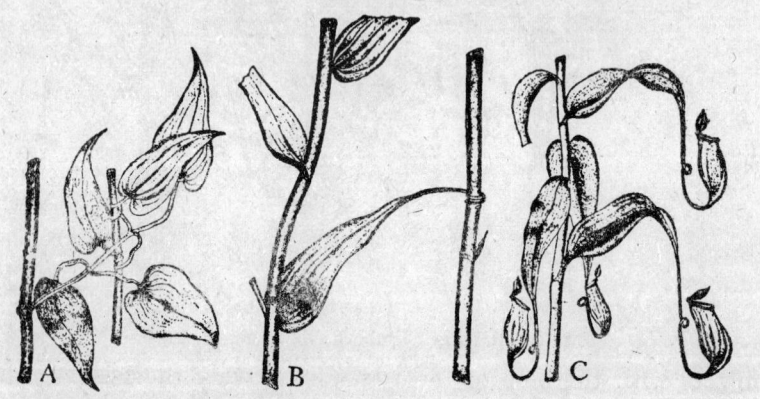

FIG. 21. Leaf Climbers. *A*, *Clematis;* *B*, glory lily (*Gloriosa*); *C*, pitcher plant (*Nepenthes;* see also FIG. 69).

(5) **Stem Climbers** or **Twiners.** These are plants with long

THE STEM

and slender stems and branches; they climb by twining bodily round trees, shrubs and hedges, e.g. country bean (*Dolichos*), railway creeper (*Ipomoea*), *Clitoria*, Rangoon creeper (*Quisqualis*), etc. They have no special organs of attachment like the climbers proper. Some of the climbers twine clockwise (dextrorse), e.g. white yam (*Dioscorea alata*) or anticlockwise (sinistrorse), e.g. wild yam (*D. bulbifera*), while others are indifferent in the direction of their movement.

(6) **Lianes.** These are very thick and woody perennial climbers, commonly met with in forests. They twine themselves round tall trees in search of sunlight, and ultimately reach their tops. There they get plenty of sunlight and produce a canopy of foliage. Common examples are woodrose (*Ipomoea tuberosa*), *Hiptage madablota* (B. MADHABILATA; H. MADHULATA), camel's foot climber (*Bauhinia vahlii*; B. LATA-KANCHAN; H. CHAMBULI; see FIG. 183), nicker bean (*Entada gigas*; B. & H. GILA), etc.

Special Types of Plants. Green plants normally prepare their own carbohydrate food and nourish themselves; such plants are said to be **autophytes** or **autotrophic plants** (*autos*, self; *phyta*, plants; *trophe*, food) or self-nourishing. There are, however, many plants which draw their organic food from different sources; such plants are said to be **heterophytes** or **heterotrophic plants** (*heteros*, different). These are of various kinds.

1. **Parasites.** These are plants that grow upon other living plants or on animals, and absorb the organic food from the hosts by their sucking roots called **haustoria**. Common examples of different types of phanerogamic parasites are as follows:

(1) Total stem-parasites, e.g. dodder (*Cuscuta*; FIG. 22A).

(2) Partial stem-parasites, e.g. mistletoe (*Viscum;* FIG. 24), *Loranthus, Cassytha,* and *Arceuthobium*.

(3) Total root-parasites, e.g. broomrape (*Orobanche indica*; B. BANIA-BAU; H. SARSON-BANDA; FIG. 23A)—parasitic on roots of potato, tomato, brinjal, mustard, tobacco, etc., often doing considerable damage to these crops—common in Bihar and Uttar Pradesh, *Aeginetia indica*—parasitic on roots of grasses and other plants in Khasi Hills, *Balanophora* (FIG. 23B)—parasitic on roots of forest trees (*B. dioica* is common in Khasi Hills), *Sapria* (1 sp.)—parasitic on roots of various plants in Arunachal and Nagaland, and *Rafflesia* (6 sp.)—parasitic on *Vitis* roots in Java and Sumatra.

(4) Partial root-parasites, e.g. sandalwood tree (*Santalum*) and *Striga*—both found abundantly in Karnataka, the latter also in Maharashtra.

Arceuthobium minutissimum is a minute, green, leafless parasite. Among the dicotyledons this is known to be the minutest plant. It has been found to grow on *Pinus excelsa*, almost within its bark, in the Kumaon Himalayas at an altitude of 3,260 metres.

Cassytha filiformis, as the name implies, is a thread-like, twining, leafless parasite, outwardly looking like *Cuscuta*, but it is very slender, much branched and matted. It has also some minute scales but unlike *Cuscuta*, which is golden-yellow in colour, *Cassytha* is pale green to dark green in colour. The plant grows extensively on hedges in Madras coasts and also in Sundarbans coasts.

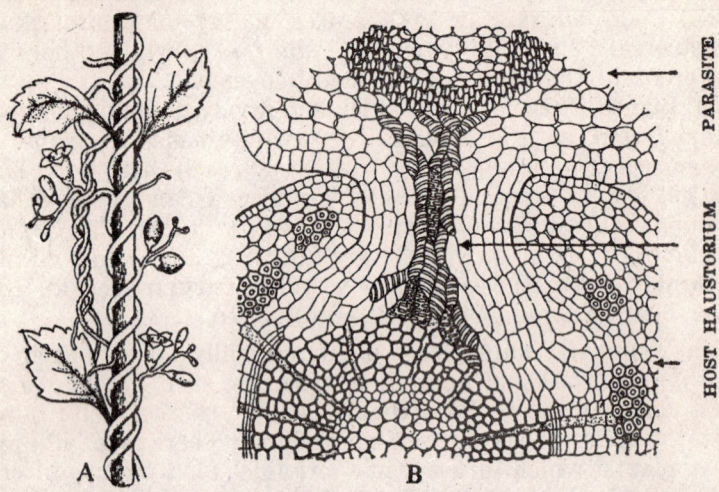

FIG. 22. *A*, dodder (*Cuscuta*)—a total stem-parasite; *B*, a section through dodder (and the host plant) showing the sucking root (haustorium).

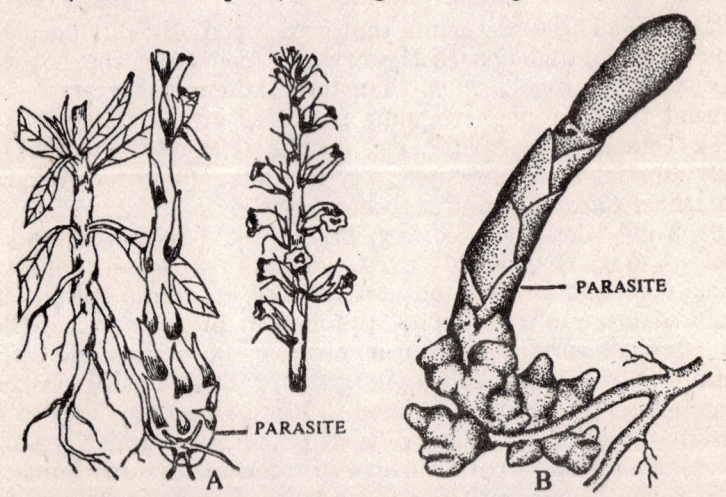

FIG. 23. *A*, broomrape (*Orobanche*)—a total root-parasite; *B*, *Balanophora*—a total root-parasite.

Striga lutea is a small, green plant, 25-30 cm. in height. It is parasitic on the roots of some of the field crops such as *Sorghum*, sugarcane, maize, etc., in South India. It produces abundant minute seeds, and thus spreads rapidly. As it is weeded out every year while ploughing the fields, it seldom gets a

THE STEM

chance to do damage to any crop. There are five species of *Striga* common in South India and in Maharashtra, e.g. *S. orobanchoides, S. densiflora,* etc.

Rafflesia arnoldi (FIG. 25) is a very interesting plant, inasmuch as it bears the most gigantic flower in the world, measuring 50 cm. in diameter and weighing 8 kg. The plant was first discovered in 1818 by Sir Stamford Raffles, while making a tour in the interior of Sumatra, and was named after him. Altogether 12 species have been discovered in Sumatra, Java and the neighbouring islands. The flower is of a livid, fleshy colour, and the smell is like that of putrid meat. Another point of interest is that its stem and root are reduced to a network of slender threads which penetrate into the root of the host plant, ramify through it and draw food from it. Here and there the thread-like stem bears flower-buds within the host, which burst out and open into full-fledged flowers of this size. The flowers are unisexual.

Sapria himalayana found only in the Aka and Dafla hills of Arunachal and in the Naga hills of Nagaland is identical with *Rafflesia arnoldi* in all respects, and also belongs to the same family, i.e. *Rafflesiaceae,* but the flowers are smaller in size, measuring 15-30 cm. in diameter. They are, however, the biggest in India.

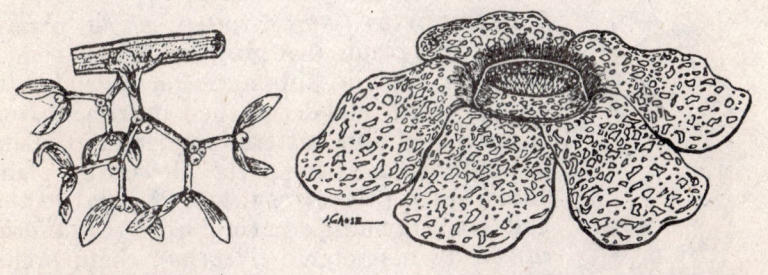

FIG. 24. Mistletoe— a partial stem-parasite.

FIG. 25. *Rafflesia*— a total root-parasite.

2. **Epiphytes** (*epi,* upon; *phyta,* plants). These are plants that grow upon other plants (see FIG. 13), but do not suck them, i.e. do not absorb food from them, as do parasites. They usually develop three kinds of roots, viz. *clinging roots, absorbing roots*

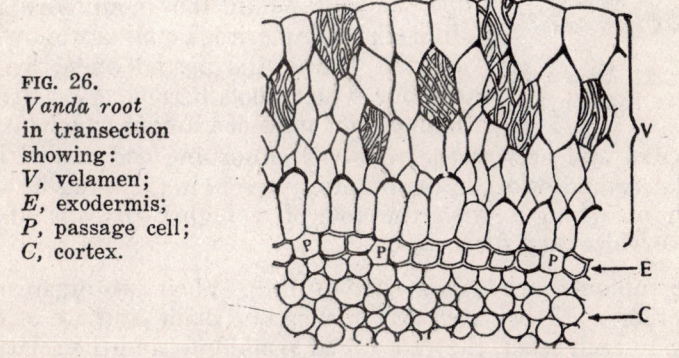

FIG. 26. *Vanda root* in transection showing:
V, velamen;
E, exodermis;
P, passage cell;
C, cortex.

and *hanging roots.* The clinging roots grow into cracks and crevices in the bark of the supporting plant and fix the epiphyte in proper position on the branch; besides, they act as reservoirs of humus which accumulates in the network formed by such roots. The absorbing roots developing from the clinging roots project into the humus and draw food from it. The hanging roots are provided with an outer covering of a special absorptive tissue, called **velamen** (FIG. 26), which usually consists of 4 or 5 layers of oblong-polygonal cells. The cells are dead containing air or water only, and their walls develop fibrous thickenings. There are also minute pits in the walls. The velamen acts as a sort of sponge and absorbs moisture from the surrounding air and also water trickling down the root. Examples are found in many orchids, e.g. *Vanda* (B. & H. RASNA–see FIG. 13) and some ferns. Banyan, peepul, etc., are, in their earlier stages, often epiphytic on date-palm and other trees.

3. **Saprophytes** (*sapros,* rotten; *phyta,* plants). These are plants that grow in places rich in decaying organic substances of vegetable or animal origin, and derive their nutriment from them. Fungi and bacteria are either parasites or saprophytes. Among the 'flowering' plants Indian pipe (*Monotropa*; FIG. 27), *Burmannia* and some orchids, e.g. coral root (*Corallorhiza*), bird's nest orchid (*Neottia*), chain orchid (*Pholidota*), etc., are good examples of saprophytes, *Monotropa uniflora,* and a few species of *Burmannia,* e.g. *B. disticha, B. candida,* etc., grow in the Khasi Hills—*Monotropa* at an altitude of 1,800 metres and *Burmannia* of 1,400 metres. Another interesting saprophyte is snowball (*Sarcodes*); it is allied to *Monotropa,* and grows in the mountains around California in America. Total saprophytes are colourless, while the partial ones are green in colour. Their roots become associated with a filamentous mass of a fungus which takes the place of and acts as the root-hairs, absorbing food material from the decomposed organic substances present in the soil. The association of a fungus with the root of a higher plant is known as **mycorrhiza** (see next page).

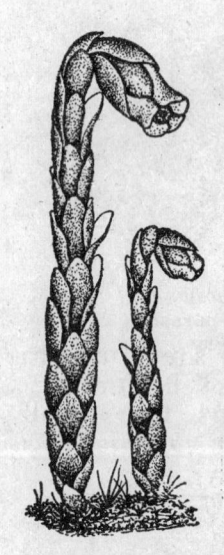

FIG. 27. *Monotropa*—a saprophyte.

4. **Symbionts** (*syn,* together; *bios,* life). When two organisms live together, as if they are parts of the same plant, and are of mutual help to each other, they are called **symbionts,** and the relationship between the two is expressed as **symbiosis.** Lichens are typical examples. These are associations of algae and fungi, and commonly occur as thin round greenish patches on tree-trunks and

old walls. The alga in a lichen being green prepares food and shares it with the fungus, while the latter absorbs water and mineral salts from the surrounding medium, and also affords protection to the alga. Some of the mycorrhizas are also good examples of symbiosis.

Mycorrhiza. Mycorrhiza (fungus-root) is the association of a fungus with the root of a higher plant. This association was first discovered by Frank in 1885, who found it to be a regular feature in many species of plants, particularly forest trees (beech, oak, etc.), many conifers (pine, etc.), saprophytic phanerogams, orchid seedlings, etc. The fungi concerned may belong to various classes. The infected root does not elongate but frequently branches profusely, and no root-hairs are formed. Two types of mycorrhiza are seen: (*a*) **endotrophic,** in which the fungus is internal, usually living within the cortical cells of the root, as in many orchids, and (*b*) **ectotrophic,** in which the fungus is external, growing attached to the surface of the root, as in conifers and most other plants. There are also certain types of mycorrhiza in which the fungus growing outside gradually penetrates inside. The biological relationship between the fungus and the root is not very clear in all cases and divergent opinions have been given. The relationship may range from true parasitism to genuine symbiosis. In the latter case the fungus absorbs water, mineral salts and nitrogenous organic substances from the soil, and in some cases it even fixes free nitrogen of the air. The fungus also helps respiration of the root, as in pine. In return it receives food from the root. The mycorrhizal fungus particularly benefits certain plants. Thus it is seen that orchid seeds often do not germinate if they are not infected by a particular fungus, and the pine seedlings and orchid seedlings are slow in growth and become weak in the absence of a similar infection.

5. **Carnivorous Plants** (see part III, chapter 7). Carnivorous plants are those that capture insects and small animals, and feed upon them, absorbing only the nitrogenous compounds from their bodies. Such plants are green in colour and prepare their own carbonaceous food, while they partially depend on insects and other animals for nitrogenous food. Some of the examples are sundew, butterwort, Venus' fly-trap, *Aldrovanda,* pitcher plant, bladderwort, etc.

Modifications of Stems

The main functions that the modified stems perform are: (*a*) perennation, i.e. surviving from year to year through bad seasons by certain underground stems; (*b*) vegetative propagation by certain horizontal sub-aerial branches spreading out in different directions; and (*c*) specialized functions by certain metamorphosed aerial organs.

1. **Underground Modifications of Stems.** For the purpose of perennation stems develop underground and lodge there permanently, lying in a dormant, defoliated condition for some time and then giving off aerial shoots annually under favourable conditions. They are always thick and fleshy, having a heavy deposit of reserve food material in them. Developing underground they simulate roots in their general appearance, in being non-green in

colour and in lying buried in the soil, but are readily distinguished from the latter by the presence of (*a*) nodes and internodes, (*b*) scale-leaves, and (*c*) buds (axillary and terminal). The main function of this group of modified stems is, as already stated, (*a*) perennation; but they are also meant (*b*) to store up food material and (*c*) to propagate, i.e. to multiply plants vegetatively. The various types met with in this group are as follows:

(1) **Rhizome** (FIG. 28). Rhizome is a prostrate, thickened stem, creeping horizontally under the surface of the soil. It is

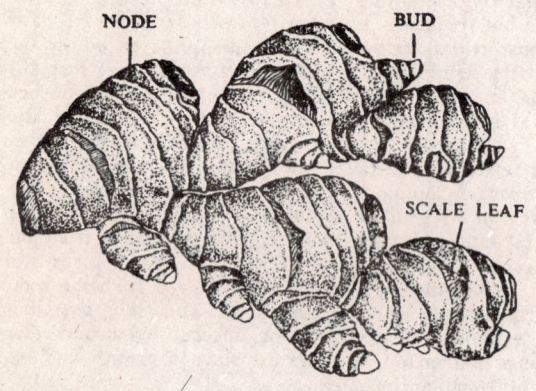

FIG. 28. Rhizome of ginger.

provided with distinct nodes and long or short internodes; it bears some scaly leaves at the nodes; it possesses a bud in the

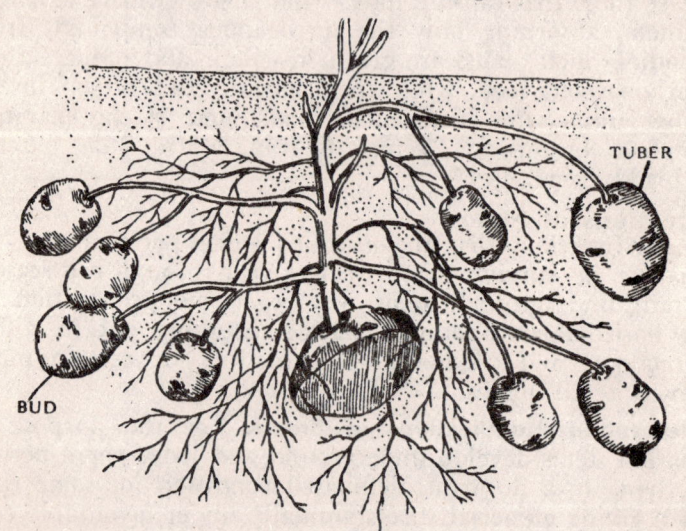

FIG. 29. Tubers of potato.

axil of the scaly leaf; and it ends in a terminal bud. Some slender, adventitious roots are given off from its lower side. The rhizome

THE STEM

may be unbranched or sometimes the axillary buds grow out into short, stout branches. It remains dormant underground and then with the approach of the vegetative season the terminal bud and sometimes also some of the axillary buds grow into the aerial shoots. Its direction is normally horizontal, but sometimes it grows in the vertical direction (**rootstock**), as in *Alocasia* (B. MANKACHU; H. MANKANDA). Examples of rhizome are seen in *Canna*, ginger, turmeric, arrowroot, water lily, lotus, ferns and many aroids.

(2) **Tuber** (FIG. 29) This is the swollen end of a special underground branch (tuber means a swelling). The underground branch arises from the axil of a lower leaf, grows horizontally outwards and ultimately swells up at the apex. It has on its surface a number of 'eyes' or buds which grow up into new plants. Adventitious roots which are abundantly formed in other underground stems are usually absent from a tuber. A tuber is often very much swollen owing to a heavy deposit of food material, becoming almost spherical, e.g. potato. Jerusalem artichoke (*Helianthus tuberosus*; B. & H. HATICHOKE) with edible tubers is another example.

(3) **Bulb** (FIG. 30). This is an underground modified shoot (rather a single, often large, terminal bud) consisting of a

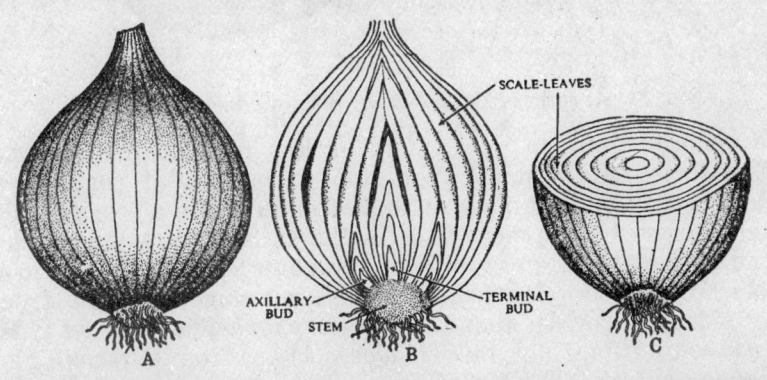

FIG. 30. Bulb of onion. *A*, an entire onion showing the lower part of the bulb with adventitious roots, and outer dry scale-leaves with distinct veins; *B*, an onion cut longitudinally; and *C*, an onion cut transversely.

shortened convex or slightly conical stem, a terminal bud and numerous scale-leaves (which are the swollen bases of foliage leaves). The scale-leaves, often simply called scales, grow from the upper surface of the stem or around it, while a cluster of adventitious (fibrous) roots are given off from its base. The inner scales are commonly fleshy, the outer ones dry. The scales may occur surrounding the stem in concentric rings, as in onion, leek, garlic, tuberose, most lilies, etc., or they may be narrow, partially overlapping each other by their margins only, as in tulip and certain lilies. The former type of bulb is most common and is said to be *tunicated* or *coated*, while the latter is rather rare and is said to be *scaly*. The fleshy scales store food (sugar in onion and mostly starch in others), while the dry scales give protection. The bulb is vertical in direction and its terminal bud gives rise to the aerial shoot. Some axillary buds may also be produced in the axils of fleshy scales. These may develop into aerial shoots and finally form daughter bulbs or they may remain dormant. The daughter bulbs grow in the following season.

(4) **Corm** (FIG. 31). This is a condensed form of rhizome consisting of a stout, solid, fleshy, underground stem growing in

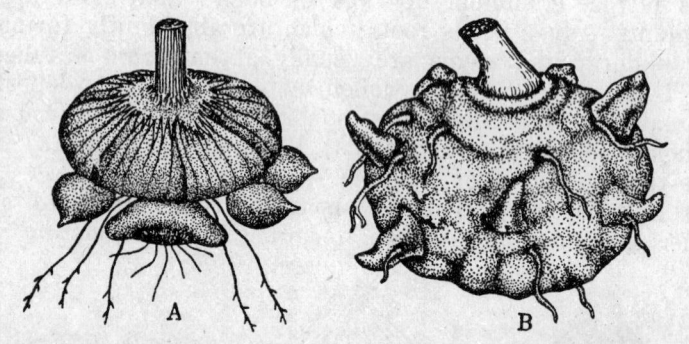

FIG. 31. *A*, corm of *Gladiolus*; *B*, corm of *Amorphophallus*.

the vertical direction. It is more or less rounded in shape or often somewhat flattened from top to bottom. It contains a heavy deposit of food material and often grows to a considerable size. It bears one or more buds in the axils of scale-leaves, and some of these buds grow up into daughter corms. Adventitious roots normally develop from the base but sometimes also from the sides. Corm is found in *Amorphophallus* (B. OL; H. ZAMIKAND), *Gladiolus*, taro (*Colocasia*), saffron (*Crocus*) and meadow saffron (*Colchicum*).

2. **Sub-aerial Modifications of Stems.** These are meant for vegetative propagation and are of the following kinds:

(1) **Runner** (FIG. 32). This is a slender, *prostrate* branch with long or short internodes, creeping on the ground and rooting at the nodes. The runner arises as an axillary bud and creeps some distance away from the mother plant, then strikes roots and grows into a new plant. Many such runners are often produced by the mother plant and they spread out on the ground on all sides. They may break off from the mother plant and grow up as independent daughter plants. Examples are seen in wood-

THE STEM

sorrel (*Oxalis*), *Marsilea*, strawberry (*Fragaria*), Indian pennywort (*Centella=Hydrocotyle*), etc.

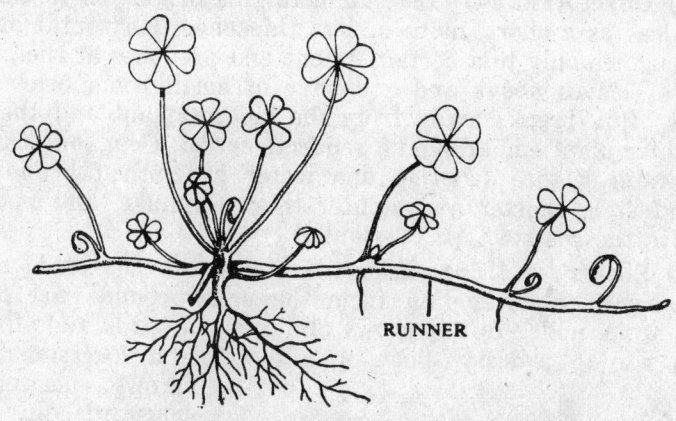

FIG. 32. Runner of wood-sorrel (*Oxalis*).

(2) **Stolon** (FIG. 33). Like the runner this is also a slender lateral branch originating from the base of the stem. But at

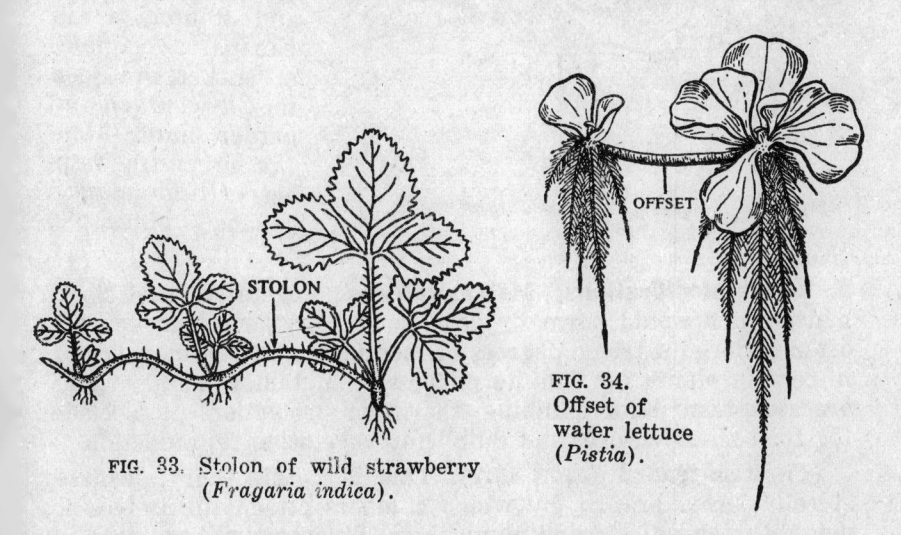

FIG. 33. Stolon of wild strawberry (*Fragaria indica*).

FIG. 34. Offset of water lettuce (*Pistia*).

first it grows *obliquely* upwards to some extent and then it bends down to the ground, striking roots at the tip and producing a bud. The latter soon grows into a daughter plant. A stolon may grow further in the same way, producing roots and buds at successive stages. Many such stolons, each provided with long or short internodes, may grow out of the mother plant and spread out in different directions. Common examples are peppermint (*Mentha piperita*), wild strawberry (*Fragaria indica*), *Oenanthe stolonifera*, etc.

(3) **Offset** (FIG. 34). Like runners this originates in the axil of a leaf as a short, more or less thickened, horizontal branch. It elongates only to a certain extent and produces at the apex a tuft of leaves above and a cluster of small roots below. The offset often breaks away from the mother plant and then the daughter plant embarks on a separate career. Common examples are water lettuce (*Pistia*) and water hyacinth (*Eichhornia*). An offset is shorter and stouter than a runner, and is found only in the rosette type of plants.

(4) **Sucker** (FIG. 35). Like the stolon the sucker is also a lateral branch developing from the underground part of the stem at its node. But it grows obliquely upwards and directly gives rise to a leafy shoot or a new plant. Occasionally it grows horizontally outwards only to a certain extent, but soon it turns up, as in *Chrysanthemum*, or it may be shorter and stouter, as in banana. Examples of sucker are seen in *Chrysanthemum*, garden mint (*Mentha arvensis*), raspberry (*Rubus idaeus*) banana, pineapple, bamboo, etc.

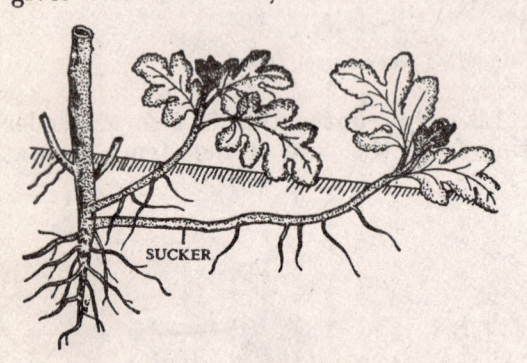

FIG. 35. Suckers of *Chrysanthemum*.

3. **Aerial Modifications: Metamorphoses.** Vegetative and floral buds, which would normally develop into branches and flowers, often undergo extreme degrees of modification (metamorphosis) in certain plants for definite purposes. Metamorphosed organs are stem-tendril for climbing, thorn for protection, phylloclade for food manufacture, and bulbil for vegetative reproduction.

(1) **Stem-tendril** (FIGS. 36-7). This is a thin, wiry, leafless, spirally-curled branch, by which climbers attach themselves to neighbouring objects and climb them. Stem-tendrils are seen in vine (*Vitis*), passion-flower, etc. In passion-flower (FIG. 36A) the axillary bud is modified into the tendril, and in *Vitis* it is the terminal bud that becomes so modified. Sometimes, as in Sandwich Island climber (*Corculum=Antigonon*; FIG. 36B) and balloon vine (*Cardiospermum*; FIG. 37), floral buds are modified into tendrils. In *Gouania* and *Serjania*, extensive woody climbers

THE STEM

(lianes), some of the branches end in strong watch-spring-like tendrils for the support of such plants.

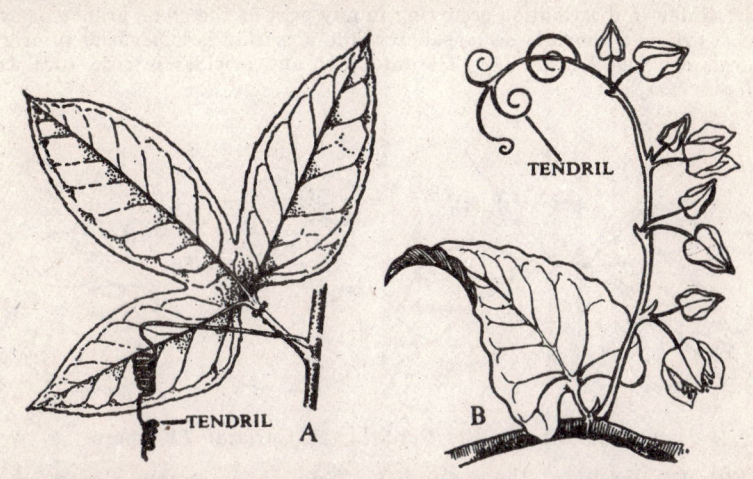

FIG. 36. Stem-tendrils. *A*, tendril of passion-flower (*Passiflora*); *B*, tendrils of Sandwich Island climber (*Corculum=Antigonon*).

(2) **Thorn** (FIG. 38). The thorn is a hard, often straight and pointed structure. It is regarded as a modified branch because it arises in the axil of a leaf or sometimes at the apex of a branch, which is the normal position of a bud. Thus in lemon, *Duranta*, pomegranate, *Vangueria* (B. & H. MOYNA), etc., the axillary bud is modified into a thorn, and in *Carissa* (B. KARANJA; H. KARONDA) the terminal bud is modified into a pair of thorns. The thorn sometimes bears leaves, flowers and fruits, as in *Duranta* and prune (*Prunus*), and not infrequently it becomes branched, as in *Flacourtia* (B. &. H. PANIALA).

FIG. 37. Tendrils of balloon vine (*Cardiospermum*). *T*, a tendril.

Differences between Thorns and Prickles. Both the thorns and the prickles are primarily defensive organs being sharp and pointed; they also sometimes act as climbing organs. Their morphological differences are: a thorn is a modification of an axillary bud or sometimes of a terminal bud, as in *Carissa* and may bear leaves, flowers and fruits, and may also be branched; whereas a prickle is a mere outgrowth, never bears leaves, flowers and fruits, and is unbranched. A thorn is axillary or terminal in position; whereas a prickle is irregular in distribution occurring in any part of the stem, branch or even leaf. Further a thorn is deep-seated; while a prickle is superficial in origin. Thorns are found in *Carissa, Duranta,* etc., and prickles in rose, coral tree (*Erythrina*), etc.

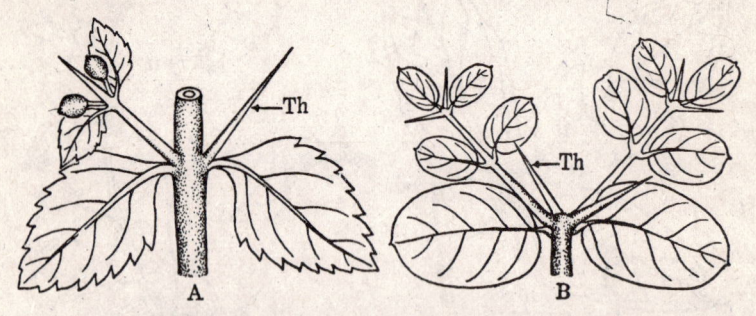

FIG. 38. Thorns. *A, Duranta; B, Carissa; Th,* thorn.

(3) **Phylloclade** (FIGS. 39-40). This is a *green,* flattened or cylindrical stem or branch *of unlimited growth,* consisting of a

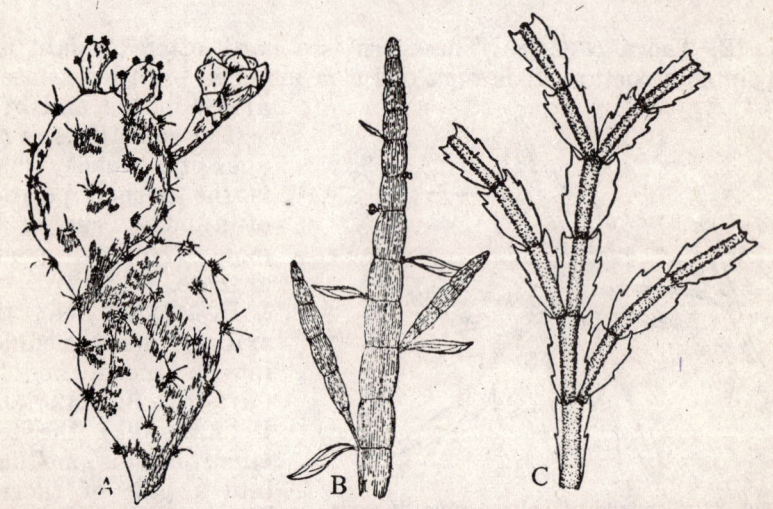

FIG. 39. Phylloclades. *A,* prickly pear (*Opuntia dillenii*); *B,* cocoloba (*Muehlenbeckia platyclados*); *C,* Christmas cactus (*Epiphyllum truncatum*).

succession of nodes and internodes at long or short intervals. The phylloclade characteristically develops in many xerophytic plants where the leaves often grow out feebly, or fall off early, or are modified into spines, evidently reducing evaporating surfaces. The phylloclade then takes over all the functions of the leaves, particularly photosynthesis. It also often functions as a

THE STEM

storage tissue, retaining plenty of water and mucilage. Further, because of strong development of cuticle it can reduce transpiration to a considerable extent. Common examples are cacti, such as prickly pear (*Opuntia dillenii;* FIG. 39A), night-blooming cacti (*Cereus* and *Phyllocactus;* FIG. 40B), *Epiphyllum truncatum* (FIG. 39C), etc., cocoloba (*Muehlenbeckia;* FIG. 39B), *Casuarina* (B. & H. JHAU), several species of *Euphorbia*, e.g. *E. tirucalli* (phylloclades cylindrical; FIG. 40A), *E. antiquorum* (phylloclades flattened), etc. The phylloclade is otherwise called *cladophyll*.

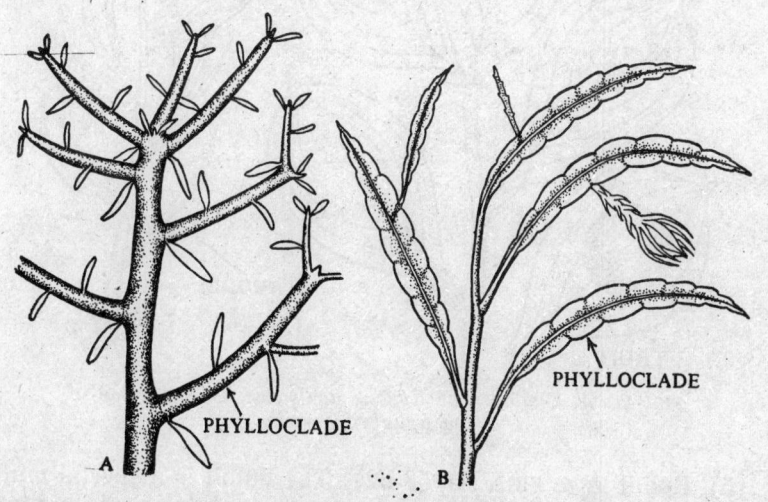

FIG. 40. *A*, Phylloclades of *Euphorbia tirucalli; B*, the same of *Phyllocactus latifrons*.

(4) **Cladode** (FIG. 41). In some plants one or more short, green, cylindrical or sometimes flattened branches *of limited growth* develop from the node of the stem or branch in the axil of a scale-leaf. Such a branch is known as the cladode. *Asparagus* is a typical example (FIG. 41B). Here the cladode is cylindrical, and consists of one internode only. A very interesting example is butcher's broom (*Ruscus aculeatus;* FIG. 41A), a small shrub. In this plant the green, flat, leaf-like organs (branches), each arising in the axil of a scale-leaf, are the cladodes. The cladodes bear male or female flowers (*Ruscus* is dioecious) from a point (representing a node) half-way up on their surface in the axil of another scale-leaf. The female flower subsequently produces a large red berry. In *Phyllocladus* (a conifer) the green, flat, leaf-like 'short shoots', each developing in the axil of a scale-leaf on the 'long shoot', are the cladodes. The flat, green, floating

blade (stem) of duckweed (*Lemna*; see FIG. 2D) is also regarded by many as a cladode. Similarly the frond of *Wolffia*, a minute, rootless, floating plant, is regarded as a cladode. *Wolffia*, it may be noted, is the smallest 'flowering' plant, with its frond as small as a sand grain.

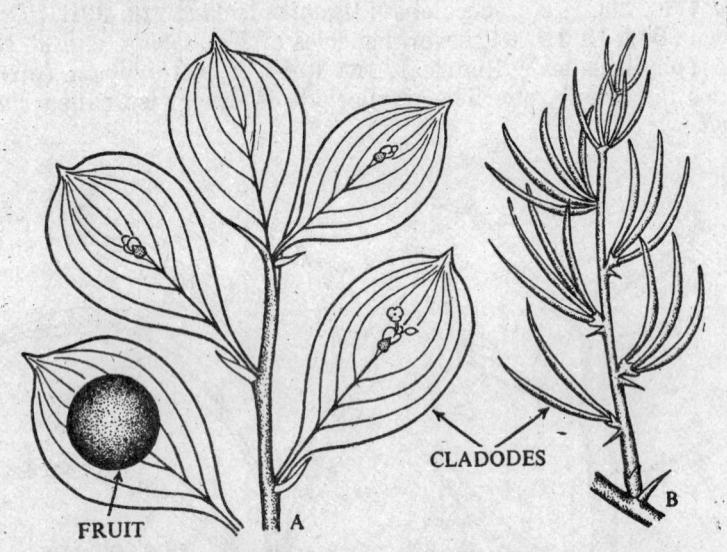

FIG. 41. *A*, Cladodes of *Ruscus aculeatus*; *B*, the same of *Asparagus racemosus*.

(5) **Bulbil** (see FIGS. III/58-62). The bulbil is a special multicellular body essentially meant for the reproduction of the plant. It may be the modification of a vegetative bud or of a floral bud. In any case it detaches itself from the mother plant and grows up into a new independent one. Bulbils are seen in *Dioscorea bulbifera, Oxalis repens, Globba bulbifera, Agave americana*, onion (*Allium cepa*), *Lilium bulbiferum*, etc.

Branching
The mode of arrangement of the branches on the stem is known as **branching**. There are two principal types of branching, *viz*. **lateral** and **dichotomous**.

A. LATERAL BRANCHING
When the branches are produced laterally, that is, from the sides of the main stem, the branching is called **lateral**. The lateral branching may be **racemose** or indefinite or monopodial (*monos*, one; *pod-*, foot or axis) and **cymose** or definite.

THE STEM

1. **Racemose Type.** Here the growth of the main stem is indefinite, that is, it continues to grow indefinitely by the terminal bud and give off branches laterally in *acropetal* succession, i.e. the lower branches are older and longer than the upper ones (FIG. 42A). Branching of this type is also called **monopodial** because there is a single continuous axis, as in *Casuarina*, mast tree (*Polyalthia*), pine (*Pinus*), etc. As a result of this branching the plant takes on a conical or pyramidal shape.

2. **Cymose Type.** Here the growth of the main stem is definite, that is, the terminal bud does not continue to grow, but lower down, the main stem produces one or more lateral branches which grow more vigorously than the terminal one. The process may be repeated over and over again. As a result of cymose branching the plant spreads out above, and becomes more or less dome-shaped. Cymose branching may be of the following kinds:

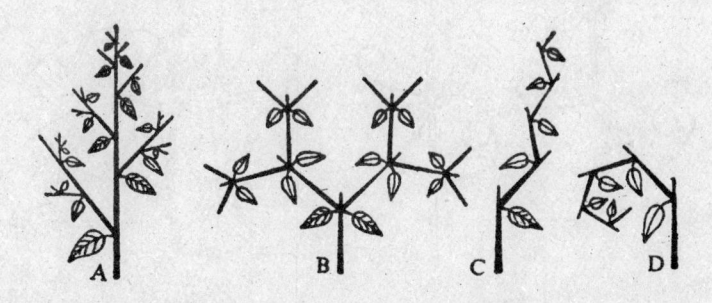

FIG. 42. Types of Branching. *A*, racemose type; *B*, true (biparous) cyme; *C*, scorpioid cyme; *D*, helicoid cyme.

(1) **Uniparous Cyme.** If, in the cymose type, only one lateral branch is produced at a time, the branching is said to be uniparous or monochasial. The uniparous type of branching is otherwise called **sympodial** (*syn*, together or united; *pod-*, foot) because there is a succession of daughter axes (false axes) fused together in course of development of the plant (FIGS. 43-4). It has two distinct forms: (*a*) **helicoid** or one-sided cyme (FIG. 42D), when successive lateral branches develop on the same side, forming a sort of helix, as in *Saraca* (B. ASOK; H. SEETA ASHOK), and (*b*) **scorpioid** or alternate-sided cyme (FIG. 42C), when successive lateral branches develop on alternate sides, forming a zigzag, as in vine (*Vitis vinifera*), wild vine (*V. trifolia*), *Cissus quadrangularis* (B. & H. HARHJORA; FIG. 44), etc. In them the apparent or false axis (**sympodium**) is a succession of lateral axes, and the tendrils are modified terminal vegetative buds (FIGS. 43-4).

(2) **Biparous Cyme.** If, in the cymose branching, two lateral axes develop at a time, it is called biparous or dichasial (FIG. 42B). Examples are seen in mistletoe (*Viscum*; see FIG. 24), four o'clock plant (*Mirabilis*), temple or pagoda tree (*Plumeria*), *Ervatamia* (=*Tabernaemontana*), *Datura*, *Carissa* (see FIG. 38B), etc. Sometimes it so happens that the terminal bud remains undeveloped or soon dies off, the branching then looks like a dichotomy, often called *false dichotomy*.

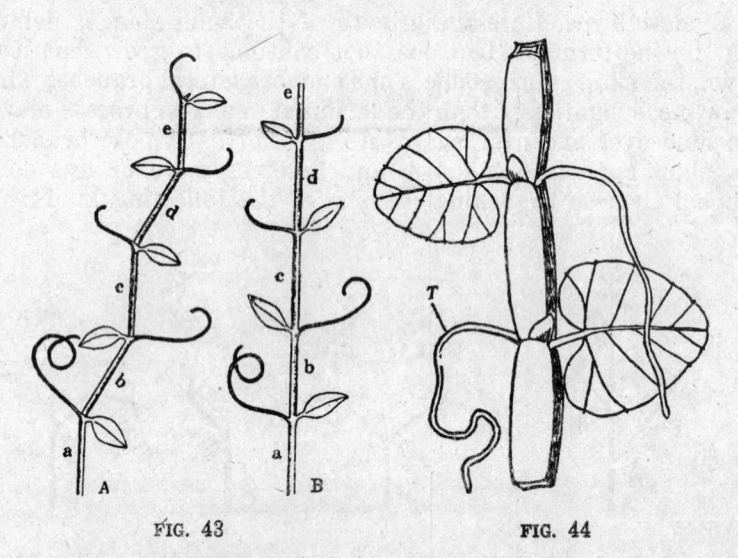

FIG. 43.

FIG. 44

FIG. 43. Sympodial Branching. *A*, scorpioid type showing terminal tendrils and lateral axes; *B*, the same straightened out after growth; *a-e* are respective axes of sympodium. FIG. 44. Sympodial branching of *Cissus quadrangularis* (B. & H. HARHJORA); *T*, a tendril.

(3) **Multiparous Cyme.** If more than two branches develop at a time, the branching is said to be multiparous or polychasial, as in *Croton sparsiflorus* and some species of *Euphorbia*.

B. DICHOTOMOUS BRANCHING

When the terminal bud bifurcates, that is, divides into two, producing two branches in a forked manner, the branching is termed **dichotomous**. Dichotomous branching is common among the 'flowerless' plants, as in *Riccia* (FIG. 45), *Marchantia* (see FIG. V/118), *Lycopodium phlegmaria* (see FIG. V/151A) and *L. squarrosum* (see FIG. V/150). Among the 'flowering' plants examples are afforded by *Hyphaene* (a kind of palm), screwpine (*Pandanus*), *Canscora* (a weed), etc.

THE LEAF

Functions of the Stem. The main purpose of the stem is to bear leaves and flowers and spread them out on all sides for proper functioning—the leaves to get the adequate amount of sunlight for manufacture of food material, and the flowers to attract insects from a distance for the purpose of pollination and reproduction. Other functions are support of the branches which push forward the leaves and the flowers, conduction of water, mineral salts and prepared food through the plant body, storage of water and food in many cases, and manufacture of food by the green shoot.

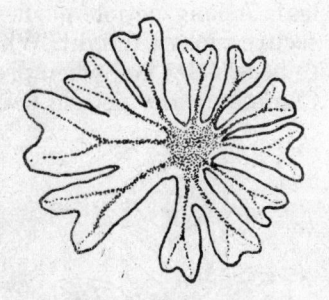

FIG. 45. Dichotomous branching in *Riccia*.

Chapter 3 THE LEAF

The leaf may be regarded as the flattened, lateral outgrowth of the stem or the branch, developing *exogenously* (i.e. from superficial tissues) from a node and having a bud in its axil. In this respect the leaf is a partial stem or branch having a limited growth. It is normally green in colour and is regarded as the most important vegetative organ of the plant since food material is prepared in it. Leaves always follow an *acropetal* order of development and are *exogenous* in origin.

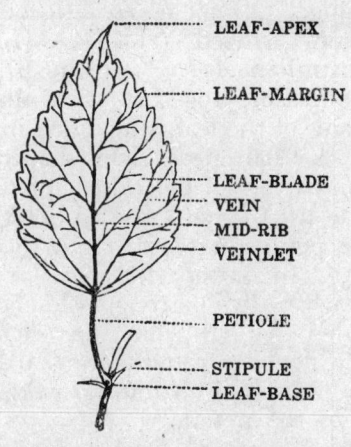

FIG. 46. Parts of a leaf.

Parts of a leaf (FIG. 46). A typical leaf consists of the following parts, each with its own function.

1. **Leaf-base** is the part attached to the stem. In monocotyledons the leaf-base commonly expands into a **sheath** which partially or wholly clasps the stem, while in many dicotyledons the leaf-base bears two lateral outgrowths known as the **stipules**. In *Leguminosae* and also in many other plants the leaf-base is swollen, and then it is known as the **pulvinus** (FIG. 47A).

2. **Petiole** is the stalk of the

leaf. A long petiole pushes out the leaf-blade and thus helps it to secure more sunlight. When the petiole is absent the leaf is said to be **sessile**; and when present it is said to be **petiolate** or stalked. Commonly the petiole is cylindrical being terete or grooved, but

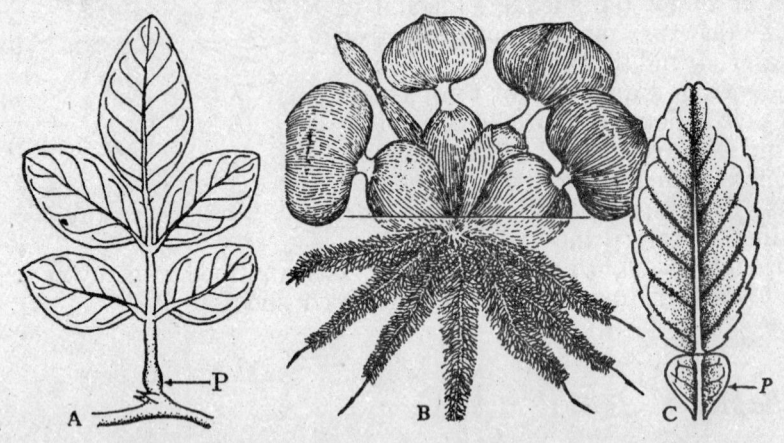

FIG. 47. *A, Clitoria* leaf showing pulvinus (P); *B*, water hyacinth (*Eichhornia*) leaf showing bulbous petiole; *C*, pummelo (*Citrus*) leaf showing winged petiole (*P*).

in many cases the petiole shows certain peculiarities. Thus in water hyacinth or lilac devil (FIG. 47B) it swells into a spongy bulb, often called pseudo-bulb, containing innumerable air chambers for facility of floating; while in *Citrus* (e.g. orange, pummelo, etc.), it becomes winged (FIG. 47C). In Australian *Acacia* (see FIG. 67) the petiole together with the rachis of the leaf is modified into a flattened sickle-shaped lamina or blade, called phyllode. In *Clematis* (see FIG. 21A) the petiole is tendrillar in nature. In sarsaparilla (*Smilax*; FIG. 48) two strong, closely-coiled tendrils, one on each side, develop from the leaf-stalk.

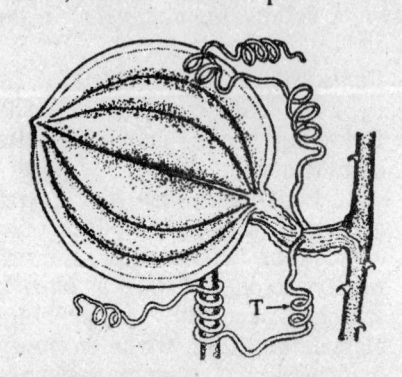

FIG. 48. Tendrils (*T*) of *Smilax*.

They are formed, as now known on the basis of anatomical work, by chorisis (splitting or branching) of the petiole.

3. **Leaf-blade** or **lamina** is the green, expanded portion. A strong vein, known as the **mid-rib**, runs centrally through the leaf-blade from its base to the apex; this produces thinner lateral

THE LEAF

veins which in their turn give rise to still thinner veins or **veinlets**. The lamina is the most important part of the leaf since this is the seat of food-manufacture for the entire plant. Its external and internal organization is well adapted for this purpose as well as for other functions it has to perform.

When the lobes at the base of the leaf partially enclose the stem, the leaf is said to be **auriculate** (*auricle*, lobe), as in madar (*Calotropis*), *Sonchus*, etc.; when completely, it is called **amplexicaul** (*amplexus*, embrace; *caulis*, stem), as in grass, wheat and cauline leaves of *Emilia*; when incompletely, it is called **semi-amplexicaul**, as in buttercup, palms, etc.; when the lobes meet across the stem and fuse together so that the latter seems to pass through the leaf-blade, the leaf is said to be **perfoliate** (*per*, through; *folium*, a leaf), as in *Canscora perfoliata*, *Aloe perfoliata*, etc. When two sessile opposite leaves meet each other across the stem and fuse together, they are said to be **connate**, as in *Canscora diffusa* (B. DANKUNI) and wild honeysuckle (*Lonicera flava*). In some cases, as in *Laggera pterodonta*, *Laggera alata*, *Canscora decurrens*, *Crotalaria alata* and some of the thistles, the petiole and the leaf-base become winged, and this wing extends down the stem so that the latter also seems to be winged; a leaf of this nature is said to be **decurrent**.

Types of Leaves. 1. **Foliage leaves** are ordinary green, flat, lateral appendages of the stem or the branch, borne at the node. 2. **Cotyledons** or **seed leaves** are attached to the axis of the embryo of the seed. As the seed germinates they usually turn green and become leaf-like. 3. **Cataphylls** or **scale-leaves** are reduced forms of leaves, stalkless and often brownish; they are the bud-scales, scales on the rhizome, and also on other parts of the plant body. 4. **Hypsophylls** or **bract leaves** are special leaves developing at the base of the inflorescence or of individual flowers or florets; they are commonly

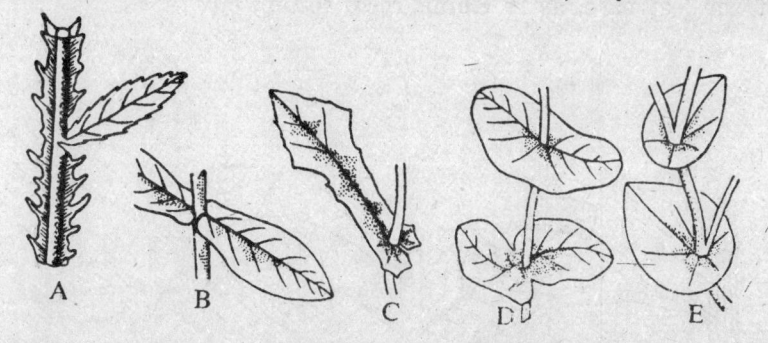

FIG. 49. Sessile Leaves. *A*, decurrent leaf of *Laggera pterodonta*; *B*, auriculate leaves of madar (*Calotropis*); *C*, amplexicaul leaf of *Emilia sonchifolia*; *D*, connate leaves of *Lonicera flava*; *E*, perfoliate leaves.

the different kinds of bracts and also **bracteoles** (otherwise called **prophylls**). 5. **Stipules** are the lateral appendages of the leaf borne at its base. Like the bracts they are also of various kinds. 6. **Ligules** are minute scaly outgrowths borne at the upper end of the leaf-sheath, as in *Gramineae*. They are very rare in dicotyledons. 7. **Floral leaves** are members of a flower, forming into two accessory whorls (calyx and corolla), and two essential whorls (androecium and gynoecium). 8. **Sporophylls** are the spore-bearing leaves concerned in asexual reproduction of plants. They may be ordinary vegetative leaves bearing spores, as in the common ferns, or metamorphosed leaves separating into a distinct reproductive region (called cone, spike or strobilus), as in *Lycopodium, Selaginella* and *Equisetum,* and also gymnosperms, or into distinct flowers, as in angiosperms. In gymnosperms and angiosperms the stamens and carpels are the sporophylls.

Stipules

Stipules are the lateral appendages of the leaf borne at its base. Their function is to protect the young leaves in the bud, and when green they manufacture food material in the same way as leaves. When stipules are present the leaf is said to be **stipulate,** and when absent **exstipulate.** Stipules are present in many families of dicotyledons, but they are absent or very rare in monocotyledons. A few cases of the latter class showing free lateral stipules are *Dioscorea, Hydrocharis* and *Potamogeton*. They appear as outgrowths of the leaf-sheath, and are interpreted as stipules. Sometimes, as in butterfly pea (*Clitoria*; B. APARAJITA; H. APARAJIT), a small stipule is present at the base of each leaflet. Such a small stipule is otherwise known as a **stipel.**

Kinds of Stipules (FIG. 50). According to their shape, position, colour and size, the stipules are of the following kinds:

(1) **Free Lateral Stipules** (see FIG. 46). These are two free stipules, usually small and green in colour, borne on the two sides of the leaf-base, as in China rose, cotton, etc.

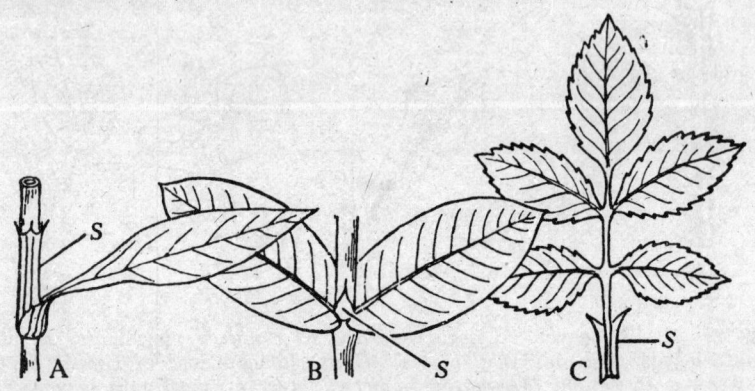

FIG. 50. Kinds of Stipules. *A,* ochreate stipule (*S*) of *Polygonum; B,* interpetiolar stipule (*S*) of *Ixora; C,* adnate stipule (*S*) of rose.

(2) **Scaly Stipules.** These are small dry scales, usually two in number, borne on the two sides of the leaf-base, as in *Desmodium,* e.g. Indian telegraph plant (*Desmodium gyrans*).

(3) **Adnate Stipules** (*C*). These are the two lateral stipules that grow along the petiole up to a certain height, adhering to

it and making it somewhat winged in appearance, as in rose, groundnut or peanut, strawberry and lupin.

(4) **Interpetiolar Stipules** (B). These are the two stipules that lie between the petioles of opposite or whorled leaves, thus alternating with the latter. They are seen in *Ixora* (B. RANGAN; H. GOTAGANDHAL), *Anthòcephalus* (B. & H. KADAM), *Vangueria* (B. & H. MOYNA), etc. Sometimes, as in cape jasmine (*Gardenia*; B. & H. GANDHARAJ), *Randia, Pavetta*, etc., the two stipules are axillary in position, each lying between the petiole and the stem; they are then known as the **intrapetiolar stipules**.

(5) **Ochreate Stipules** (A). They form a hollow tube encircling the stem from the node up to a certain height of the internode in front of the petiole, as in *Polygonum*, sorrel (*Rumex*) and buckwheat (*Fagopyrum*).

(6) **Foliaceous Stipules** (see FIG. 63A-B). These are two large, green, leafy structures, as in pea (*Pisum*), wild pea (*Lathyrus*), some species of passion-flower (*Passiflora*) and *Cassia auriculata*

(7) **Bud-scales**. These are scaly stipules which enclose and protect the vegetative buds, and fall off as soon as the leaves unfold. They are seen in banyan, jack, *Magnolia*, iron-wood tree (*Mesua*; B. NAGESWAR; H. NAGKESAR), etc.

(8) **Spinous Stipules** (FIG. 51). In some plants, as in gum tree (*Acacia*), Indian plum (*Zizyphus*), sensitive plant (*Mimosa*), caper (*Capparis*), wood-apple (*Aegle*), etc., the stipules become modified into two sharp pointed structures known as spines, one on each side of the leaf-base. Such spinous stipules give protection to the leaf against the attack of herbivorous animals. [Note that the last three kinds of stipules are regarded as modified ones.]

FIG. 51. Spinous stipules (S) of Indian plum (*Zizyphus*).

Leaf-Blade

Apex of the leaf (FIG. 52). The apex of the leaf is said to be (1) **obtuse**, when it is rounded, as in banyan (*Ficus bengalensis*); (2) **acute**, when it is pointed in the form of an acute angle, but not stiff, as in China rose; (3) **acuminate** or **caudate**, when it is drawn out into a long slender tail, as in peepul (*Ficus religiosa*) and lady's umbrella or Chinese hat (*Holmskioldia*); (4) **cuspidate**, when it ends in a long rigid sharp (spiny) point, as in date-palm, screwpine and pineapple; (5) **truncate**, when it ends abruptly as if cut off in a straight line, as in Indian sago palm (*Caryota urens*) and *Bauhinia anguina* (a large climber); (6) **retuse**, when the obtuse or truncate apex is furnished with a shallow notch, as in water lettuce (*Pistia*); (7) **emarginate**, when the apex is provided with a deep notch, as in *Bauhinia* (B. KANCHAN; H. KACHNAR) and wood-sorrel (*Oxalis*);

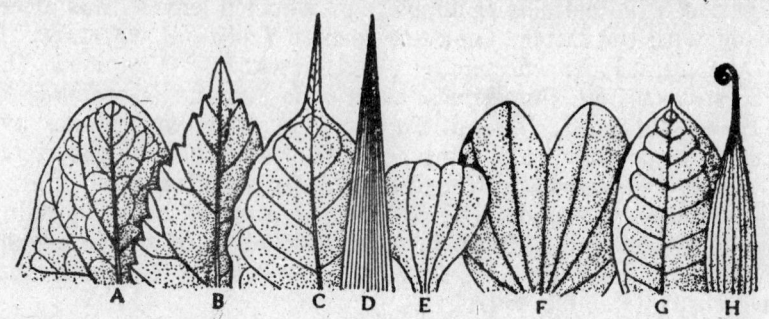

FIG. 52. Apex of the leaf. *A*, obtuse; *B*, acute; *C*, acuminate; *D*, cuspidate; *E*, retuse; *F*, emarginate; *G*, mucronate; and *H*, cirrhose.

(8) **mucronate**, when the rounded apex abruptly ends in a short point, as in *Ixora* (B. RANGAN; H. GOTAGANDHAL); and (9) **cirrhose** (*cirrus*, a tendril or a curl), when it ends in a tendril, as in glory lily, or in a slender, curled, thread-like appendage, as in banana.

Margin of the Leaf. The margin of the leaf may be (1) **entire**, i.e. even and smooth, as in mango, jack, banyan, etc.; (2) **repand**, i.e. shallowly wavy or undulating, as in mango; (3) **sinuate**, i.e. deeply undulating, as in mast tree (*Polyalthia;* B. DEBDARU; H. ASHOK) and some garden crotons; (4) **serrate** i.e. cut like the teeth of a saw and the teeth directed upwards, as in China rose, rose, and margosa (*Azadirachta;* B. & H. NIM or NIMBA); (5) **biserrate**, i.e. doubly serrate (each tooth serrated again); (6) **serrulate**, i.e. minutely serrate; (7) **dentate**, i.e. the teeth directed outwards at right angles to the margin of the leaf, as in melon and water lily; (8) **runcinate**, i.e. serrated with the teeth pointed backwards; (9) **crenate**, i.e. the teeth round, as in sprout leaf plant (*Bryophyllum*) and Indian pennywort (*Centella*); (10) **fimbriate**, i.e. fringed with fine segments; (11) **ciliate**, i.e. fringed with hairs; and (12) **spinous**, i.e. provided with spines.

Surface of the Leaf. The leaf is said to be (1) **glabrous**, when its surface is smooth and free from hairs or outgrowths of any kind; (2) **rough**, when the surface is somewhat harsh to touch; (3) **glutinous**, when the surface is covered with a sticky exudation, as in tobacco; (4) **glaucous**, when the surface is green and shining; (5) **spiny**, when it is provided with spines; and (6) **hairy**, when it is covered, densely or sparsely, with hairs. A hairy surface may be (*a*) **pubescent**, when it is covered with short, soft, straight hairs; (*b*) **pilose**, i.e. thinly covered with long, soft hairs; (*c*) **villous**, i.e. thickly covered with long, soft hairs; (*d*) **tomentose**, i.e. densely covered with short, soft, more or less tangled hairs like cotton; (*e*) **floccose**, i.e. cottony with locks of hairs easily detachable; (*f*) **hispid**, i.e. beset with rigid or bristly hairs; (*g*) **hirsute**, i.e. covered with long, coarse, stiff hairs.

Shape of the Leaf (FIG. 53). *A*, Acicular, when the leaf is long, narrow and cylindrical, i.e. needle-shaped, as in pine, onion, etc. *B*. Linear, when the leaf is long, narrow and flat, as in many

THE LEAF

grasses, tuberose, *Vallisneria,* etc. *C.* **Lanceolate,** when the shape is like that of a lance, as in bamboo (*Bambusa*), oleander (*Nerium*), mast tree (*Polyalthia*), etc. *D.* **Elliptical** or **oval,** when the leaf has more or less the shape of an ellipse, as in

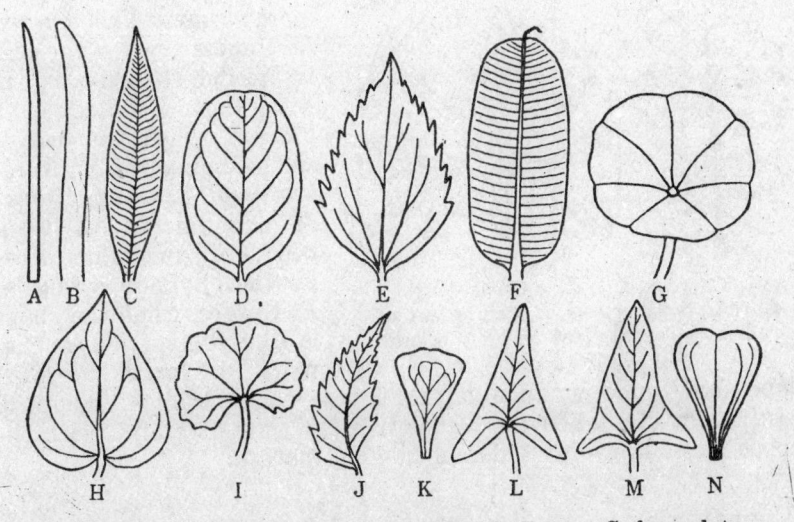

FIG. 53. Shape of the Leaf. *A,* acicular; *B,* linear; *C,* lanceolate; *D,* elliptical or oval; *E,* ovate; *F,* oblong; *G,* rotund or orbicular; *H,* cordate; *I,* reniform; *J,* oblique; *K,* spathulate; *L,* sagittate; *M,* hastate; and *N,* cuneate.

Carissa (B. KARANJA; H. KARONDA), periwinkle (*Vinca*), guava (*Psidium*), rose-apple (*Syzygium*), etc. *E.* **Ovate,** when the blade is egg-shaped, i.e. broader at the base than at the apex, as in China rose, banyan etc. When the leaf is inversely egg-shaped it is said to be **obovate,** as in country almond (*Terminalia*) and jack (*Artocarpus*). *F.* **Oblong,** when the blade is wide and long, with the two margins running straight up, as in banana. *G.* **Rotund** or **orbicular,** when the blade is circular in outline, as in lotus, garden nasturtium, etc. *H.* **Cordate,** when the blade is heart-shaped, as in betel (*Piper betle*), *Peperomia,* etc. When the leaf is inversely heart-shaped it is said to be **obcordate,** as in wood-sorrel (*Oxalis*). *I.* **Reniform,** when the leaf is kidney-shaped, as in Indian pennywort. *J.* **Oblique,** when the two halves of a leaf are unequal, as in *Begonia.* In margosa (*Azadirachta;* B. & H. NIM) and Indian cork tree (*Millingtonia;* B. & H. AKASNIM), Persian lilac (*Melia;* B. GHORA-NIM), etc., the leaflets are oblique. *K.* **Spathulate,** when the shape is like that of a spatula, i.e. broad and somewhat rounded at the top and narrower towards the base, as in sundew (*Drosera*) and *Calendula.*

L. **Sagittate**, when the blade is shaped like an arrow, as in arrowhead (*Sagittaria*) and some aroids. *M.* **Hastate**, when the two lobes of a sagittate leaf are directed outwards, as in water bindweed (*Ipomoea*) and *Typhonium*. *N.* **Cuneate**, when the leaf is wedge-shaped, as in water lettuce (*Pistia*). *O.* **Falcate**, when the leaf is sickle-shaped, as in *Eucalyptus globulus* and *Arundinaria falcata* (a bamboo). In Australian *Acacia* the phyllode is falcate. *P.* **Lyrate**, when the shape is like that of a lyre, i.e. with a large terminal lobe and some smaller lateral lobes, as in radish, mustard, etc. *Q.* **Pedate**, when the leaf is like the claw of a bird, with the lobes spreading outwards, as in *Vitis pedata*.

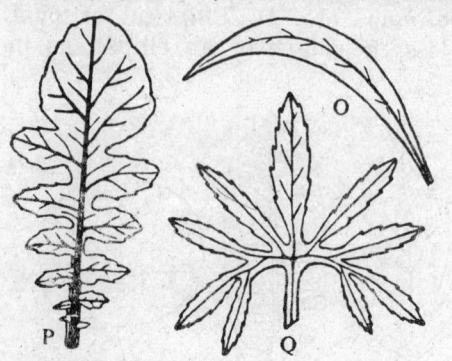

FIG. 53 (*contd.*). *O*, falcate leaf of *Eucalyptus globulus*; *P*, lyrate leaf of radish; *Q*, pedate leaf of *Vitis pedata*.

Venation

Veins are rigid, linear structures which arise from the petiole and the mid-rib and traverse the leaf-lamina in different directions; they are really vascular ramifications made of conducting and mechanical tissues—the former serving to distribute the

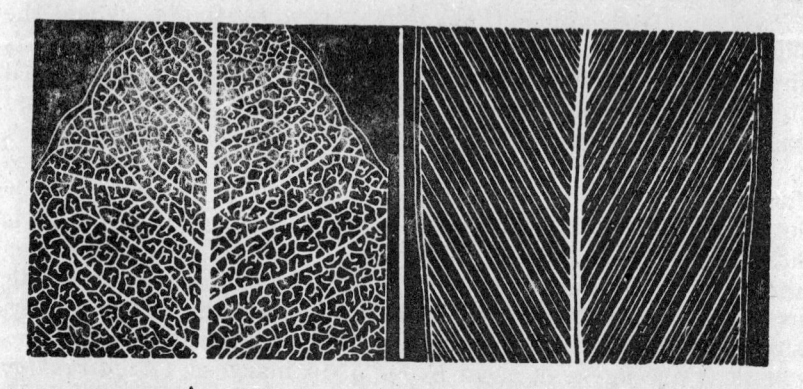

FIG. 54. Systems of Veins. *A*, reticulate venation in a dicotyledonous leaf; *B*, parallel venation in a monocotyledonous leaf.

water and dissolved mineral salts through the lamina and to carry away the prepared food from it, and the latter to give the

THE LEAF

necessary amount of strength and rigidity to the thin, flat leaf-lamina.

The arrangement of the veins and the veinlets in the leaf-blade is known as **venation**. There are two principal types of venation, viz. **reticulate**, when the veinlets are irregularly distributed, form-

<pre>
         A   B   C
</pre>

FIG. 55. *A*, leaf of *Dioscorea* (a monocotyledon) showing reticulate venation; *B*, leaf of *Smilax* (a monocotyledon) showing reticulate venation; *C*, leaf of *Calophyllum* (a dicotyledon) showing parallel venation.

ing a network; and **parallel**, when they run parallel to each other. The former is characteristic of dicotyledons and the latter of monocotyledons.

Exceptions. Among monocotyledons, yams (*Dioscorea*), *Smilax*, aroids, etc., show reticulate venation (FIG. 55A-B), and among dicotyledons, *Calophyllum* (B. & H. SULTANA-CHAMPA; FIG. 55C) and a few others show parallel venation.

I. RETICULATE VENATION

1. **Pinnate** or **Unicostate Type** (*unus*, one; *costa*, a rib). In this type of venation there is a strong mid-rib or costa; this gives off lateral veins which proceed towards the margin or apex of the leaf, like plumes in a feather (FIG. 56A). These are then connected by smaller veins which pass in all directions, forming a network, as in peepul (*Ficus*), mango (*Mangifera*), guava (*Psidium*), etc.

2. **Palmate** or **Multicostate Type** (*multi*, many). In this type there are a number of more or less equally strong ribs which arise from the tip of the petiole and proceed outwards or upwards. There are two types: (*a*) **divergent** (FIG. 56B), when the main veins diverge towards the margin of the leaf, as in papaw,

gourd, castor, China rose, etc.; and (*b*) **convergent** (FIG. 56C), when the veins converge to the apex of the leaf, as in Indian plum (*Zizyphus*), bay leaf (*Cinnamomum*), etc.

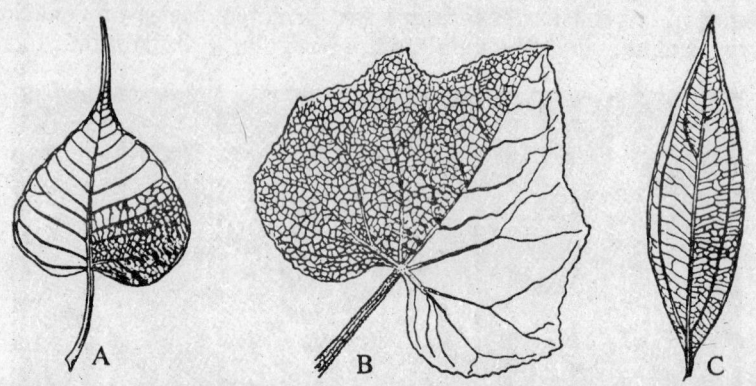

FIG. 56. Types of Reticulate Venation. *A*, pinnate type in peepul (*Ficus*) leaf; *B*, palmate (divergent) type in cucumber (*Cucumis*) leaf; *C*, palmate (convergent) type in bay leaf (*Cinnamomum*).

II. PARALLEL VENATION

1. Pinnate or **Unicostate Type** (FIG. 57A). In this type of venation the leaf has a prominent mid-rib, and this gives off lateral veins which proceed parallel to each other towards the margin or apex of the leaf-blade, as in banana, ginger, *Canna*, turmeric, etc.

2. Palmate Type. Two forms are also met with here: (*a*) the veins arise from the tip of the petiole and proceed (diverge) to-

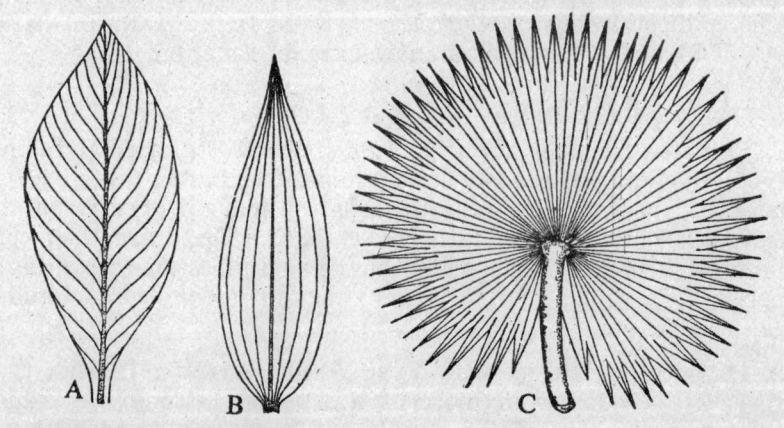

FIG. 57. Types of Parallel Venation. *A*, pinnate type in *Canna* leaf; *B*, palmate (convergent) type in bamboo leaf; *C*, palmate (divergent) type in pamyra-palm leaf.

THE LEAF

wards the margin of the leaf-blade in a more or less parallel manner (**divergent type**; FIG. 57C), as in fan palms such as palmyra-palm; and (*b*) a number of more or less equally strong veins proceed from the base of the leaf-blade to its apex in a somewhat parallel direction (**convergent type**; FIG. 57B), as in water hyacinth, grasses, rice, bamboo, etc.

Functions of the Veins. Veins are rigid structures and their mechanical functions are to give necessary strength to the leaf-blade so that it may not get torn or crumpled when a strong wind blows, and at the same time to help the leaf-blade to keep flat so that its surface may be evenly illuminated by sunlight. A very important physiological function of the veins is to carry water and inorganic salts into the leaf-blade and finally the prepared food material from the leaf into the main body of the plant, particularly the storage organs.

Incision of the Leaf-blade. In the pinnately-veined leaf the incision or cutting of the leaf-blade proceeds from the margin towards the mid-rib (**pinnate type**), and in the palmately-veined leaf it passes towards the base of the leaf-blade (**palmate type**).

First Series: Pinnate Type. (1) **Pinnatifid**, when the incision of the margin is half-way or nearly half-way down towards the mid-rib, as in poppy. (2) **Pinnatipartite**, when the incision is more than half-way down towards the mid-rib, as in radish, mustard, etc. (3) **Pinnatisect**, when the incision is carried down to near the mid-rib, as in some ferns, *Quamoclit* (B. KUNJA-LATA or TORULATA; H. KAMALATA), *Cosmos*, etc. (4) **Pinnate compound**, when the incision of the margin reaches the mid-rib, thus dividing the leaf-blade into a number of segments or leaflets, as in pea, gram, gold mohur, *Cassia*, etc.

Second Series: Palmate Type. (1) **Palmatifid**, as in passion-flower, cotton, etc. (2) **Palmatipartite**, as in castor, papaw, etc. (3) **Palmatisect**, as in tapioca (*Manihot*), hemp (*Cannabis;* B. &. H. GANJA) and some aroids, e.g., snake plant (*Arisaema;* see FIG. 83). (4) **Palmate compound**, when the incision is carried down to the base of the leaf-blade, as in silk cotton tree (*Bombax*) and *Gynandropsis* (see FIG. 62A-B).

Compound Leaves: Pinnate and Palmate

Simple Leaf and Compound Leaf. A leaf is said to be **simple** when it consists of a single blade which may be entire or incised (and, therefore, lobed) to any depth, but not down to the mid-rib or the petiole; and a leaf is said to be **compound** when the incision of the leaf-blade goes down to the mid-rib (rachis) or to the petiole so that the leaf is broken up into a number of segments, called leaflets, these being free from one another, that is, not connected by any lamina, and more or less distinctly jointed (articulated) at their base. A bud (axillary bud) is present in the axil of a simple or a compound leaf, but it is never present in the axil of the leaflet of a compound leaf. There are two types of compound leaves, viz. **pinnate** and **palmate**.

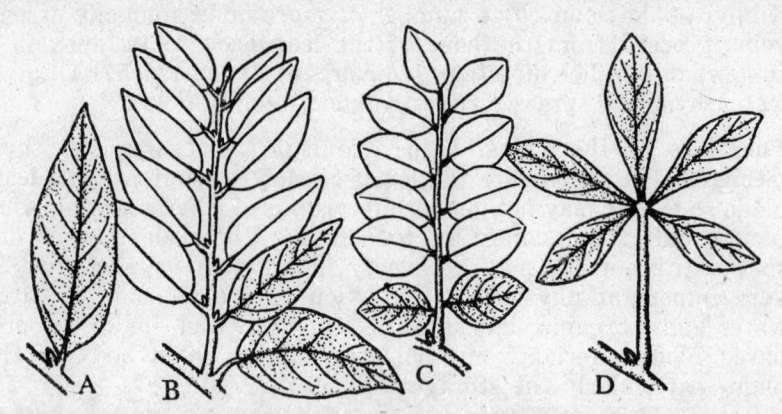

FIG. 58. *A*, a simple leaf; *B*, a branch; *C*, a pinnately compound leaf with the leaflets articulated to the mid-rib; *D*, a palmately compound leaf with the leaflets articulated to the petiole. Note the position of the bud in each case.

Compound Leaf and Branch. A compound leaf may be distinguished from a branch by the following facts: (1) a compound leaf never bears a terminal bud; whereas a branch always does so; (2) a compound leaf, like a simple one, always bears a bud (axillary bud) in its axil, but itself does not arise in the axil of another leaf; whereas a branch does not bear an axillary bud, but itself occupies the axillary position of a leaf—simple or compound—developing directly from the said bud; (3) the leaflets of a compound leaf have no axillary buds; whereas the leaves (simple) borne on a branch have a bud in their axil; and (4) a branch is always provided with nodes and internodes; while the rachis of a compound leaf is free from them.

1. **Pinnately Compound Leaf.** A pinnately compound leaf is defined as the one in which the mid-rib, known as the **rachis**, bears *laterally* a number of leaflets, arranged alternately or in an opposite manner, as in tamarind, gram, gold mohur, rain tree, sensitive plant, gum tree (*Acacia*), *Cassia*, etc. It may be of the following types:

(1) **Unipinnate.** When the mid-rib of the pinnately compound leaf bears the leaflets directly, it is said to be unipinnate, as in rose, margosa (B. & H. NIM or NIMBA), etc. When the leaflets are even in number the leaf is said to be **paripinnate** (FIG. 60A), as in tamarind, *Abrus, Sesbania, Saraca, Cassia*, etc., and when the leaflets are odd in number the leaf is said to be **imparipinnate**

THE LEAF

(FIG. 60B), as in rose, margosa (*Azadirachta*), Chinese box (*Murraya*), etc.

The pinnate leaf is said to be unifoliate, when it consists of only one leaflet, as in *Desmodium gangeticum, Aphania danura* and *Bauhinia*; bifoliate or unijugate (one pair), when of two leaflets, as in *Balanites* (B. HINGON; H. HINGOL; FIG. 59) and sometimes in rose; trifoliate or ternate, when of three leaflets, as in bean, *Erythrina* and *Vitis trifolia*. It may similarly be quadrifoliate, pentafoliate or multifoliate, according as the leaflets are four, five or more in number. Leaflets also may vary in number on the same plant.

FIG. 59. Bifoliate leaf of *Balanites*.

(2) **Bipinnate** (FIG. 60C). When the compound leaf is twice pin-

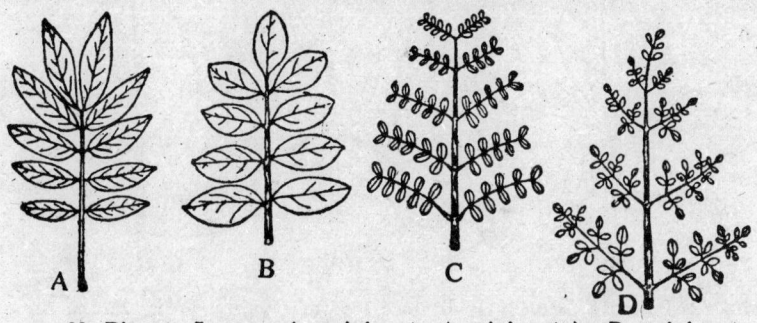

FIG. 60. Pinnate Leaves. *A*, unipinnate (paripinnate); *B*, unipinnate (imparipinnate); *C*, bipinnate; *D*, tripinnate.

nate, i.e. the mid-rib produces secondary axes which bear the leaflets, it is said to be bipinnate, as in dwarf gold mohur (*Caesalpinia*), gum tree (*Acacia*), sensitive plant (*Mimosa*), etc.

(3) **Tripinnate** (FIG. 60D). When the leaf is thrice pinnate, i.e. the secondary axes produce the tertiary axes which bear the leaflets, it is said to be tripinnate, as in drumstick (*Moringa*; B. SAJINA; H. SAINJNA) and *Oroxylum* (B. SONA; H. ARLU).

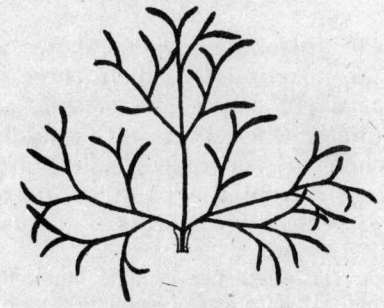

FIG. 61. Decompound leaf of coriander.

(4) **Decompound** (FIG. 61). When the leaf is more than thrice pinnate, it is said to be decompound, as in anise (*Foeniculum*), carrot (*Daucus*), coriander (*Coriandrum*), *Cosmos*, etc.

2. Palmately Compound Leaf. A palmately compound leaf is defined as the one in which the petiole bears *terminally*, articulated to it, a number of leaflets which seem to be radiating from a common point like fingers from the palm, as in silk cotton tree (*Bombax*), lupin (*Lupinus*), *Gynandropsis* and *Polanisia* (=*Cleome*).

According to the number of leaflets it may be **unifoliate**, when a single leaflet is articulated to the petiole (rare); in *Citrus* (shaddock or pummelo, orange, lemon, etc.) the unifoliate leaf with a joint at the junction of the blade and the petiole is now regarded as a simple leaf—possibly derived from a compound

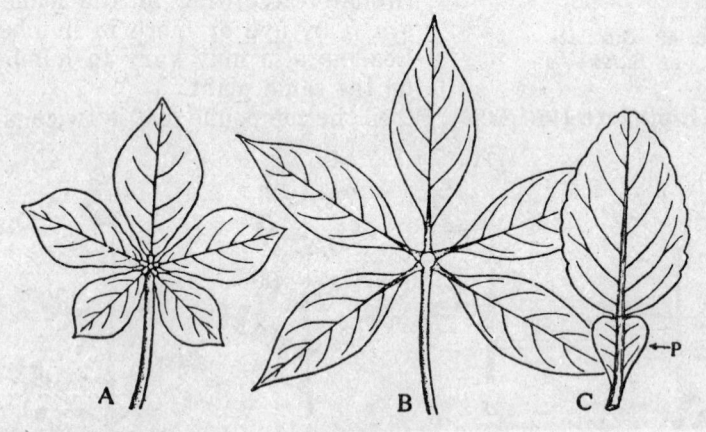

FIG. 62. Palmate Leaves. *A*, digitate leaf of *Gynandropsis; B*, the same of silk cotton tree (*Bombax*): *C*, unifoliate leaf of pummelo (*Citrus*). *P*, winged petiole.

leaf; **bifoliate**, when there are two leaflets so articulated (rare); **trifoliate**, when three as in wood-apple (*Aegle*), *Crataeva* (B. BARUN; H. BARNA), and wood-sorrel (*Oxalis*); **quadrifoliate**, when four (*rare*); **multifoliate** or **digitate** (FIG. 62A-B), when five or more leaflets are so jointed and spreading like fingers from the palm, as in silk cotton tree (*Bombax*), lupin (*Lupinus*), *Gynandropsis*, *Polanisia* (=*Cleome*), etc.

Note. **Trifoliate Leaves.** Trifoliate leaf of pinnate type can be distinguished from trifoliate leaf of palmate type by the following fact: in the former, as seen in coral tree (*Erythrina*) and country bean (*Dolichos lablab*) the petiole prolongs into a mid-rib (or rachis) and the terminal leaflet is articulated to its apex; while in the latter all the three leaflets are directly borne by (or articulated to) the apex of the petiole, as in wood-apple (*Aegle*).

Duration of the Leaf. The leaf varies in its duration. It may fall off soon after it appears, and then it is said to be (1) **caducous**; if it lasts one season, usually falling off in winter, it is (2) **deciduous** or **annual**; and if it persists for more than one season, usually lasting a number of years, it is (3) **persistent** or **evergreen**.

Some Descriptive Terms. (1) **Peltate Leaf.** The leaf-blade and the petiole usually stand on one and the same plane. In some cases, however, as in lotus, water lily, garden nasturtium, etc., the petiole is attached to the centre of the blade at a right angle to it; such a leaf is said to be peltate. (2) **Dorsiventral Leaf.** When the leaf is flat, with the blade placed horizontally, showing a distinct upper surface and a lower surface, as in most dicotyledons, it is said to be dorsiventral (*dorsum*, back; *venter*, belly or front). A dorsiventral leaf is more strongly illuminated on the upper surface than on the lower and, therefore, this surface is deeper green in colour than the lower. In internal structure also there is a good deal of difference between the two sides (see FIG. II/67). (3) **Isobilateral Leaf.** When the leaf is directed vertically upwards, as in many monocotyledons, it is said to be isobilateral (*isos*, equal; *bi*, two; *later-*, side). An isobilateral leaf is equally illuminated on both the surfaces and, therefore, the leaf is uniformly green and its internal structure is also uniform from one side to the other (see FIGS. II/68-9). (4) **Centric Leaf.** When the leaf is more or less cylindrical and directed upwards or downwards, as in pine, onion, etc., the leaf is said to be centric. A centric leaf is equally illuminated and, therefore, evenly green on all sides. (5) **Cauline Leaf.** Commonly leaves are directly borne by the aerial parts of the stem and the branches; such leaves are said to be cauline (*caulis*, a stem). (6) **Radical Leaf.** In some cases, as in pineapple (*Ananas*), Indian aloe (*Aloe*; B. GHRITAKUMARI; H. GHIKAVAR), American aloe (*Agave*), many lilies (*Lilium*), tuberose (*Polianthes*), etc., a cluster of leaves arises from the short underground stem, as if from the root; such leaves are said to be radical (*radix*, a root). Radical leaves are rather common among monocotyledons. Both radical and cauline leaves may be borne by the same plant, as seen in mustard, radish, etc.

Modifications of Leaves

Leaves of many plants which have to perform specialized functions become modified or metamorphosed into distinct forms. These are as follows:

1. **Leaf-tendrils** (FIGS. 63-4). In some plants leaves are modified

into slender, wiry, often closely coiled structures, known as tendrils. Tendrils are always climbing organs and are sensitive to contact with a foreign body. Therefore, whenever they come in contact with a neighbouring object they coil round it and help the plant to climb. The leaf may be partially or wholly

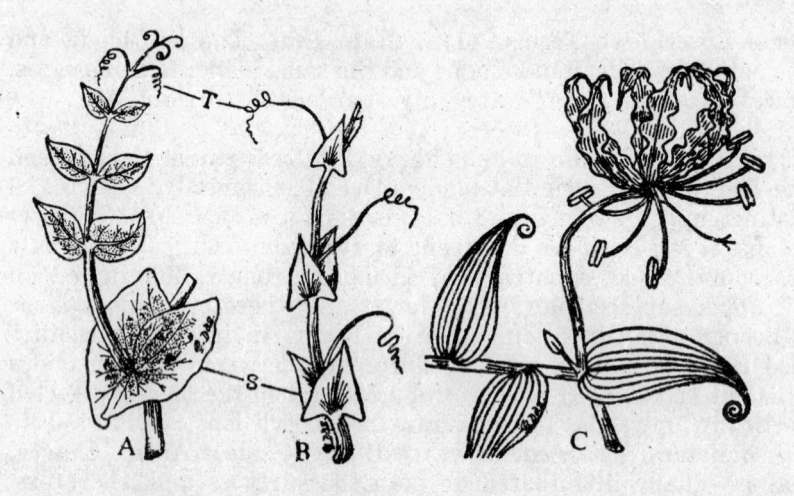

FIG. 63. Modified Leaves: Leaf-tendrils. *A*, leaf of pea (*Pisum*) with upper leaflets modified into tendrils; *B*, portion of wild pea (*Lathyrus*) stem; *T*, tendrils; *S*, stipules; *C*, portion of glory lily (*Gloriosa*) stem with the leaf-apex modified into a tendril.

modified. Thus in pea (*Pisum*; FIG. 63A), *Lathyrus sativus* (B. & H. KHESARI), golden shower or Venus' flower (*Bignonia venusta*), etc., only the upper leaflets turn into tendrils. *Bignonia venusta*, it may be noted, is an ornamental climber bearing orange-coloured flowers in huge trusses. In traveller's joy (*Naravelia*; FIG. 64) it is only the terminal leaflet that is converted into a long coiling tendril. In lentil (*Lens culinaris*; B. & H. MASUR) the rachis of the leaf ends in a tendril. In glory lily (*Gloriosa*; FIG. 63C) the leaf-apex ends in a closely coiled tendril. In pitcher plant (*Nepenthes*; see FIG. 69) the petiole often acts as a tendril holding the pitcher in an upright position. In sarsaparilla (*Smilax*) the leaf-stalk is modified into two tendrils (see FIG. 48). The above are cases of partial modification. In wild pea (*Lathyrus aphaca*; FIG. 63B), however, the whole leaf is transformed into a single tendril; while the two foliaceous stipules take over the functions of the leaf.

Hooks. In cat's nail (*Bignonia unguis-cati*; FIG. 65), an elegant climber, the terminal leaflets become modified into three, very sharp, stiff and curved hooks, very much like the nails of a cat. These hooks cling to the bark of

THE LEAF

a tree and act as organs of support for climbing. The plant thus easily climbs to the top of a lofty tree.

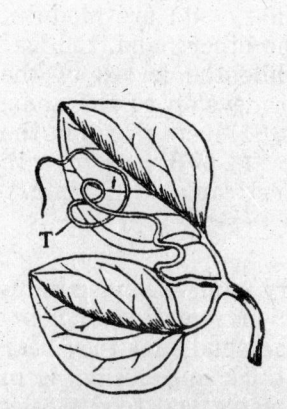

FIG. 64. Leaf of *Naravelia* with the terminal leaflet modified into a tendril.

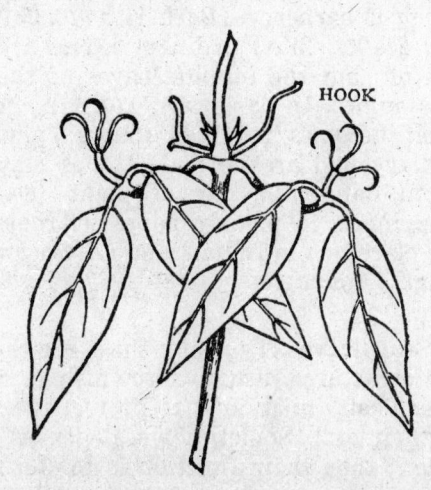

FIG. 65. Cat's nail (*Bignonia unguis-cati*) with hooks.

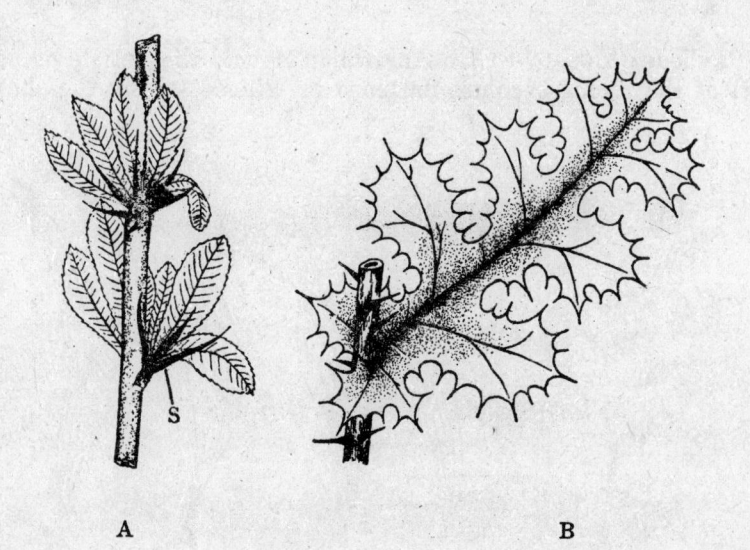

FIG. 66. *A*, barberry (*Berberis*): primary leaves modified into spines (*S*); *B*, leaf of prickly poppy (*Argemone*) showing spines.

2. Leaf-Spines (FIG. 66).

Leaves of certain plants become modified for defensive purposes into sharp, pointed structures known as **spines**. That spines are modifications of leaves is evident from the fact that they occupy the same position as the leaves and that they often bear a bud in their axil, as seen in the flowering shoot of barberry (*Berberis;* FIG. 66A). In prickly pear (*Opuntia;* see FIG. 39A) ordinary leaves are feebly developed and soon fall off, but the minute leaves of the axillary bud are modified into spines. In barberry (FIG. 66A), on the other hand, the leaf itself becomes modified into a spine; while the leaves of the axillary bud are normal. Spines may also develop at the apex, as in date-palm, dagger plant (see FIG. 81), etc., or on the margin, as in prickly poppy (*Argemone;* FIG. 66B), or in both the places, as in Indian aloe (*Aloe*) and American aloe or century plant (*Agave*).

3. **Scale-leaves.** Typically these are thin, dry, stalkless, membranous structures, usually brownish in colour or sometimes colourless. Their function is to protect the axillary bud that they bear in their axil. Sometimes scale-leaves are thick and fleshy, as in onion; then their function is to store up water and food. Scale-leaves are common on underground stems, saprophytes, parasites, *Ficus, Casuarina* (B. & H. JHAU), *Tamarix* (B. & H. BANJHAU), etc.

4. **Phyllode** (FIGS. 67-8). In Australian *Acacia* the petiole or any part of the rachis becomes flattened or winged taking the shape

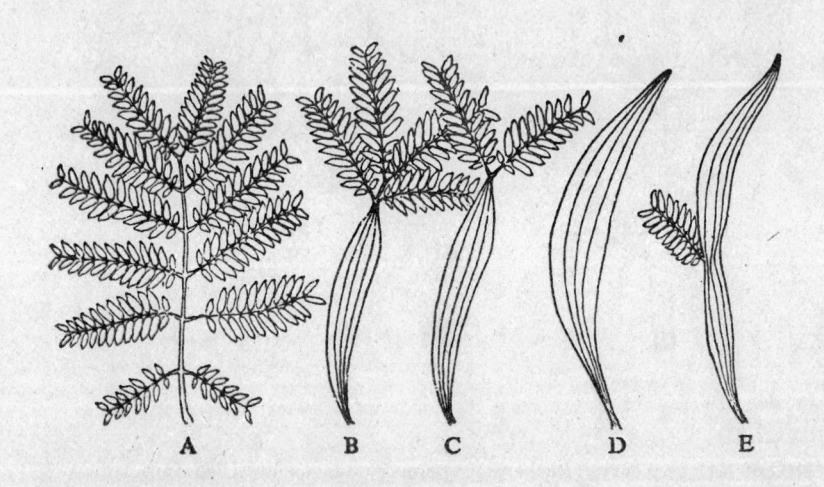

FIG. 67. Development of phyllode in Australian *Acacia*. *A*, pinnately compound leaf; *B-C*, petiole developing into phyllode; *D*, phyllode; and *E*, petiole and rachis developing into phyllode.

THE LEAF

of the leaf and turning green in colour. This flattened or winged petiole or rachis is known as **phyllode**. In some species, as in *A. moniliformis*, the normal bipinnate leaf develops in the seedling stage only, but soon falls off; later only the phyllodes develop throughout the life of the plant. In other species, as in *A. melanoxylon*, the adult plant bears both bipinnate leaves and phyllodes in all stages of development. In all cases the leaflets fall off soon, and the phyllodes take over their functions, particularly photosynthesis. Further, the edge of the phyllode is normally turned upwards, thus avoiding direct sunlight. This mechanism reduces evaporation of water. There are about 300 species of Australian *Acacia*, all showing the phyllode. In Jerusalem thorn (*Parkinsonia aculeata*), a small prickly tree, the primary rachis of the bipinnate leaf ends in a short spine, while each secondary rachis is a phyllode being green and often much flattened. The phyllode performs the functions of the leaflets which are very small and often fall off early. Some species of *Oxalis* (e.g. *O. bilimbi*) also develop phyllodes, particularly in their younger stage.

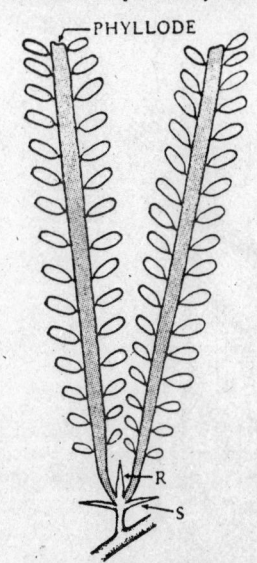

FIG. 68. Phyllodes of *Parkinsonia aculeata*. R, spiny rachis; S, a pair of spiny stipules.

5. **Pitcher** (FIG. 69). In the pitcher plant (*Nepenthes*) the leaf becomes modified into a **pitcher**. There is a sort of slender stalk which often coils like a tendril holding the pitcher vertical, and the basal portion is flattened like a leaf. The pitcher is provided with a lid which covers its mouth when the pitcher is young. The function of the pitcher is to capture and digest insects. The morphology of the leaf of pitcher plant is that the pitcher itself is the modification of the leaf-blade, the inner side of the pitcher corresponding to the upper surface of the leaf; the lid arises as an outgrowth of the leaf-apex. The slender stalk, which coils like a tendril, stiffening as it does so, is the petiole. The laminated structure, which looks like and behaves as the leaf-blade, develops from the leaf-base.

Another peculiar modification of the leaf into a kind of **pitcher** is found in a climbing epiphyte, called *Dischidia rafflesiana* (FIG. 70), common in Assam. The pitcher usually varies in size from 5 to 8 cm. in length and 2 to 2·5 cm. in width. It is, however, not carnivorous by nature. As in *Nepenthes*, it has an opening—a basal one, but no lid, and instead of this there is a sort of tongue projecting inwards. A root enters the cavity of the pitcher and becomes much branched. After a shower of rain, water flows down into

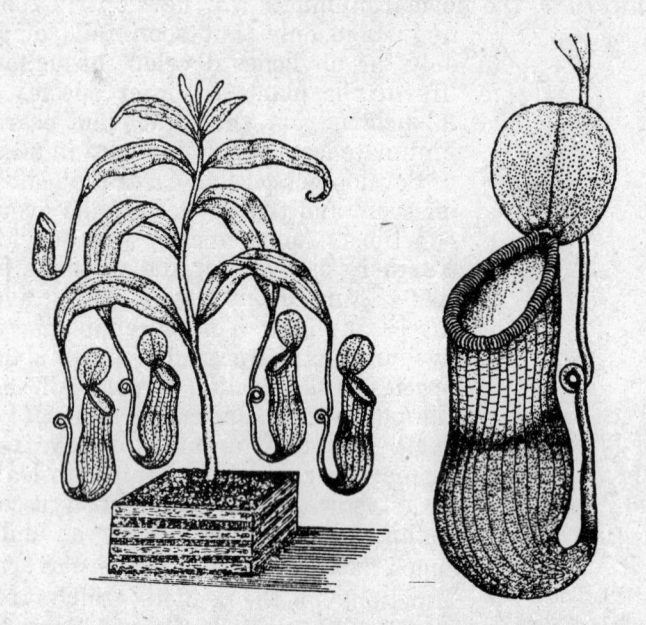

FIG. 69. *A*, pitcher plant (*Nepenthes*); *B*, a pitcher

the pitcher. Prior to this, debris is collected there by ants. All this is then absorbed by the root.

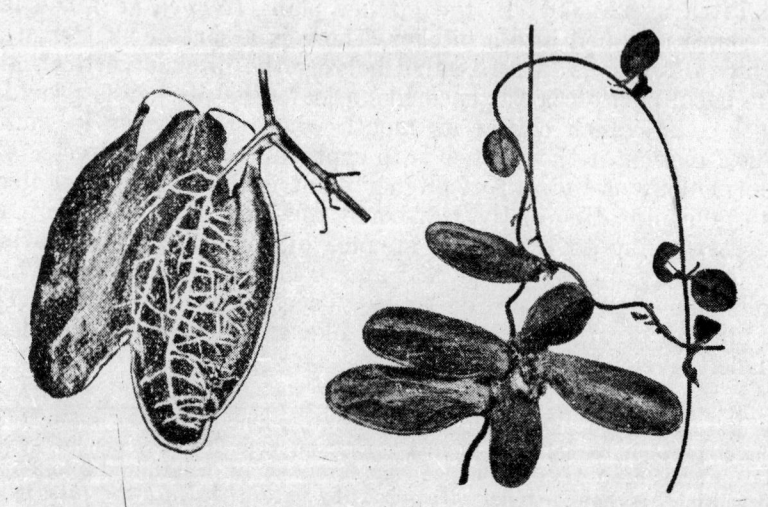

FIG. 70. *Dischidia rafflesiana;* left, a pitcher opened out.

THE LEAF

6. Bladder (FIG. 71). In bladderwort (*Utricularia*), a rootless free-floating or slightly submerged weed common in many tanks, the

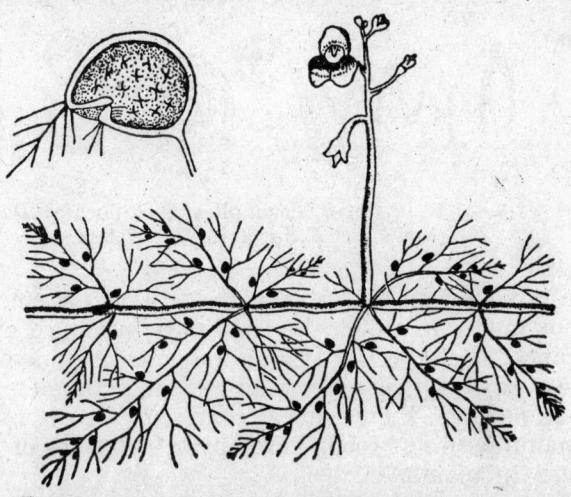

FIG. 71. Bladderwort (*Utricularia*) with many small bladders; *top*, a bladder in section (magnified).

leaf is very much segmented. Some of these segments are modified to form bladder-like structures, with a trap-door entrance which allows aquatic animalcules to pass in, but never to come out.

Prefoliation. The way in which leaves are arranged in the bud is known as **prefoliation**. This is considered from two standpoints, viz. *first*, the way in which each individual leaf is rolled or folded in the bud (**ptyxis**); and *second*, the way in which the leaves are arranged in the bud with respect to each other (**vernation**). The arrangement of foliage leaves in the vegetative bud and that of the floral leaves in the floral bud are nearly the same, and as such the same terms are used to explain the identical types in both cases.

Ptyxis. Rolling or folding of individual leaves may be as follows:

(A) **Reclinate**, when the upper half of the leaf-blade is bent upon the lower half, as in loquat (*Eriobotrya joponica*).

(B) **Conduplicate**, when the leaf is folded lengthwise along the mid-rib, as in guava, sweet potato and camel's foot tree (*Bauhinia*; B. KANCHAN; H. KACHNAR).

(C) **Plicate** or **Plaited**, when the leaf is repeatedly folded longitudinally along ribs in a *zigzag* manner, as in fan- or palmyra-palm.

(D) **Circinate**, when the leaf is rolled from the apex towards the base like the tail of a dog, as in ferns.

(E) **Convolute**, when the leaf is rolled from one margin to the other, as in banana, aroids and Indian pennywort (*Centella*=*Hydrocotyle*).

(F) **Involute**, when the two margins are rolled on the upper surface of the leaf towards the mid-rib or the centre of the leaf, as in water lily, lotus, Sandwich Island climber (*Corculum*=*Antigonon*) and *Plumbago* (B. CHITA; H. CHITRAK).

(G) **Revolute,** when the leaf is similarly rolled on its lower surface, as in oleander and country almond.

(H) **Crumpled,** when the leaf is irregularly folded, as in cabbage.

FIG. 72. Ptyxis. *A,* reclinate; *B,* conduplicate; *C,* plicate; *D,* circinate; *E,* convolute; *F,* involute; *G,* revolute.

Phyllotaxy

The term **phyllotaxy** (*phylla,* leaves; *taxis,* arrangement) means the various modes in which the leaves are arranged on the stem or the branch. The object of this arrangement is to avoid shading one another so that the leaves may get the maximum amount of sunlight to perform their normal functions (see p. 57-8), particularly manufacture of food. Three principal types of phyllotaxy are noticed in plants.

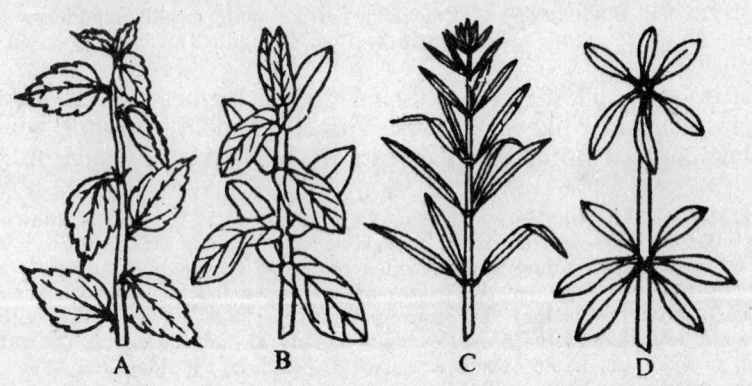

FIG. 73. Types of Phyllotaxy. *A,* alternate phyllotaxy of China rose; *B,* opposite phyllotaxy of madar (*Calotropis*); *C,* whorled phyllotaxy of oleander (*Nerium*); *D,* ditto of devil tree (*Alstonia*).

(1) **Alternate** or **Spiral** (FIG. 73A), when a single leaf arises at each node, as in tobacco, China rose, mustard, sunflower, etc.

(2) **Opposite** (FIG. 73B), when two leaves arise at each node standing opposite each other. In opposite phyllotaxy one pair of leaves is most commonly seen to stand at a right angle to the next upper or lower pair. Such an arrangement of leaves is said to be **decussate.** This is seen in *Ixora,* sacred basil (*Ocimum*), madar (*Calotropis*), guava (*Psidium*), etc. Sometimes, however, a pair of leaves is seen to stand directly over the lower pair in the same plane. Such an arrangement of leaves is said to be **superposed,** as in Rangoon creeper (*Quisqualis*).

THE LEAF

(3) **Whorled** (FIG. 73C-D), when there are more than two leaves at each node and these are arranged in a circle or whorl, as in devil tree (*Alstonia*), oleander (*Nerium*), *Allamanda, Vangueria* (B. & H. MOYNA), etc. Sometimes both opposite and whorled phyllotaxes may be seen in the same plant.

Alternate Phyllotaxy. The leaves in this case are seen to be spirally arranged round the stem. Now, if an imaginary spiral line be drawn from the base of one particular leaf, and this line be passed round the stem through the bases of the successive leaves, it is seen that the spiral line finally reaches a leaf which stands vertically over the starting leaf. The imaginary spiral line, thus drawn, is known as the **genetic spiral**, and the vertical line, i.e. the vertical row of leaves, known as the **orthostichy** (orthos, straight; *stichos*, line).

(1) **Phyllotaxy $\frac{1}{2}$ or 2-ranked** or **distichous** (FIG. 75). In grasses, elephant grass (*Typha*), traveller's tree (*Ravenala;* FIG. 74), ginger, *Vanda* (see FIG. 13), *Belamcanda, Iris,* etc., the *third* leaf stands over the *first*, and the genetic spiral makes *one* complete revolution to come to that leaf, and it involves two leaves (leaving out of consideration the first or the third leaf). The fourth leaf stands over the second, the fifth over the first and the third, and so on. Thus there are only two orthostichies, i.e. leaves are arranged in two rows or ranks. Phyllotaxy is, therefore, 2-ranked or distichous (*di*, two; *stichos*, line). If now the position of the leaves is marked out on a circle or helix, these are seen to be placed at half the distance of the circle, the leaves being equidistant from each other. The phyllotaxy is said to be half and represented by the fraction $\frac{1}{2}$, the numerator indicating the number of turns of the genetic spiral, and the denominator the number of intervening leaves. The genetic spiral makes one complete turn in this case, subtending an angle of 360° in the centre of the circle, and it involves two leaves; so the **angular**

FIG. 74. Traveller's tree (*Ravenala*) showing distichous phyllotaxy.

divergence, that is, the angular distance between any two consecutive leaves, is ½ of 360°, i.e. 180°.

(2) **Phyllotaxy ⅓ or 3-ranked or tristichous** (FIG. 76). In sedges (B. & H. MUTHA) the fourth leaf stands vertically over the first

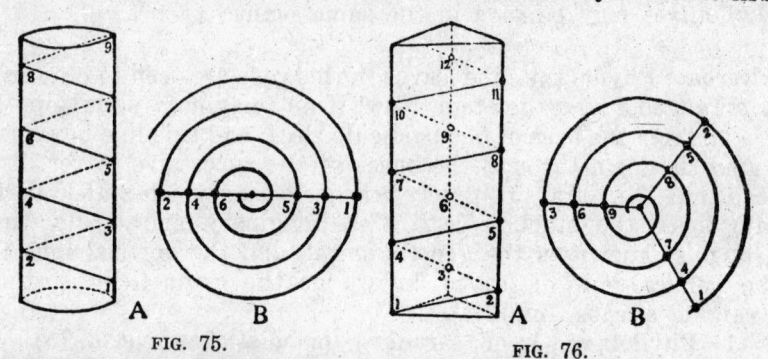

FIG. 75.
FIG. 76.

Phyllotaxy and Angular Divergence. FIG. 75. A, phyllotaxy ½; B, angular divergence 180°. FIG. 76. A, phyllotaxy ⅓; B, angular divergence 120°.

one, and the genetic spiral makes one complete turn to reach that leaf, and it involves three leaves. The fifth leaf stands over the second, the sixth over the third, and the seventh over the fourth and the first. Thus there are three orthostichies, i.e. leaves are arranged in three rows or ranks. If now their position be marked out on a circle or helix, these are seen to be placed at one-third the distance of the circle; so the phyllotaxy is ⅓ or 3-ranked or tristichous. The **angular divergence** is ⅓ of 360°, i.e. 120°.

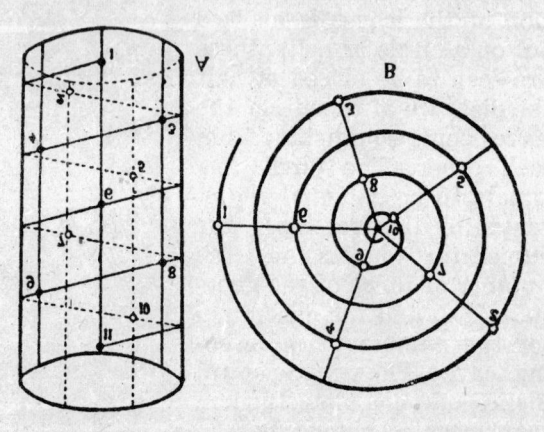

FIG. 77.
A, phyllotaxy ⅖;
B, angular divergence 144°

(3) **Phyllotaxy ⅖ or 5-ranked or pentastichous** (FIG. 77). In China rose the sixth leaf stands over the first, and the genetic spiral completes *two* circles to come to that particular leaf. The

seventh leaf stands over the second, the eighth over the third, the ninth over the fourth, the tenth over the fifth, and the eleventh over the sixth and the first. Thus there are five orthostichies, i.e. leaves are arranged in five rows, and because two turns of the genetic spiral involve five leaves, the latter are seen to be placed at two-fifths the distance of the circle. Phyllotaxy is, therefore, $\frac{2}{5}$ or 5-ranked or pentastichous. This is the commonest type of alternate phyllotaxy. The **angular divergence** in this case is $\frac{2}{5}$ of 360°, i.e. 144°.

(The same fraction can also be arrived at by adding separately the numerators and the denominators of the two previous cases, e.g. $\frac{1+1}{2+3}=\frac{2}{5}$. The next case will, therefore, be $\frac{1+2}{3+5}=\frac{3}{8}$ and so on. Fractions higher than $\frac{3}{8}$ are not commonly met with.)

Leaf Mosaic. In the floors, walls and ceilings of many temples and decorated buildings we find setting of stones and glass pieces of variegated colours and sizes into a particular pattern. This pattern is known as mosaic. Similarly, in plants we find the setting or distribution of leaves in some definite patterns. Each such pattern of leaf-distribution is known as **leaf mosaic**. Leaves are in special need of sunlight for manufacture of food material, and this being so, they tend to fit in with one another and adjust themselves in such a way that they may secure the maximum amount of sunlight with the minimum amount of overlapping. Thus in climbers bearing a dense mass of leaves, as in ivy (*Hedera helix*), Indian ivy (*Ficus pumila;* see FIG. 17), railway creeper (*Ipomoea palmata*), etc., leaves are disposed in the pattern of a tile-roof. In plants with a rosette of radical leaves or with whorls of leaves it is seen that the upper leaves alternate with the lower ones: In plants with crowded leaves, as in *Acalypha, Begonia,* garden nasturtium, etc. these are distributed like the glass pieces fitting into a mosaic, with the smaller leaves fitting into the interspaces of the broader ones. Crowded leaves of prostrate plants like wood-sorrel, Indian pennywort, etc., also form a more or less perfect mosaic.

FIG. 78. Leaf mosaic of *Acalypha*.

Functions of the leaf. Normal functions of the green leaf are threefold: (1) **manufacture of food** by the chloroplasts in the presence of sunlight out of carbon dioxide and water obtained from the air and the soil respectively; (2) **interchange of gases** —carbon dioxide and oxygen—between the atmosphere and the plant body, the former for manufacture of food by green cells only and the latter for respiration by all the living cells; (3) **evaporation of water**, mainly through the lower surface of the leaf. Besides, certain leaves have some subsidiary functions; for example, storage of water and food by fleshy leaves of Indian aloe, *Portulaca*, etc., fleshy scales of onion, lilies, *Amaryllis*, etc.; vegetative propagation by *Bryophyllum* (see FIG. 16A), *Begonia* (see FIG. 16B), *Kalanchoe* (see FIGS. III/56-7), walking ferns (see FIG. III/54), etc.

Heterophylly. Many plants bear different kinds of leaves on the same individual plant. This condition is known as heterophylly

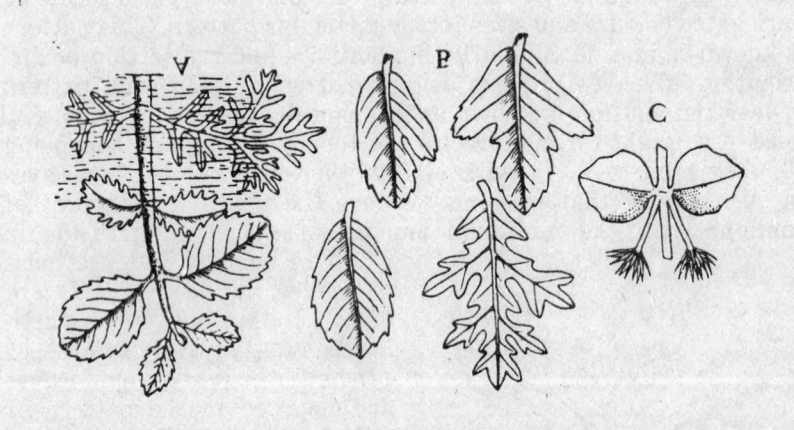

FIG. 79. Heterophylly. *A, Cardanthera triflora; B, Artocarpus chaplasha; C, Hemiphragma heterophyllum* with needle-like and broad leaves.

(*heteros*, different, *phylla*, leaves). Heterophylly is found in many aquatic plants, particularly in those growing in running water. Here the floating or aerial leaves and the submerged leaves are of different kinds; the former are generally broad, often fully expanded, and undivided or merely lobed; while the latter are narrow, ribbon-shaped, linear or much dissected. Heterophylly in water plants is thus an adaptation to two different conditions of the environment. Among them water crowfoot (*Ranunculus aquatilis*), *Cardanthera triflora* (FIG. 79A), etc., show heterophylly, with the submerged leaves much segmented and the floating (or aerial) leaves undivided or merely lobed. In water plantain (*Alisma plantago*) and arrowhead (*Sagittaria*; FIG. 80) submerged leaves are sometimes narrow and ribbon-shaped, while the upper ones are entire or broadly lobed. All transitions of leaf-forms are found in *Limnophila heterophylla*. Some land plants also exhibit this phenomenon without any apparent reason. Among them *Sterculia villosa*, chaplash (*Artocarpus chaplasha;* FIG. 79B), jack (*A. heterophyllus*) in early stage, *Ficus heterophylla* (B. BHUI-DUMUR), etc., show leaves varying from entire to variously lobed structures, and *Hemiphragma heterophyllum* (FIG. 79C), a prostrate herb common in Darjeeling and Shillong, bears two kinds of leaves—those on the main stem are ovate and entire, and those on short axillary branches are needle-shaped.

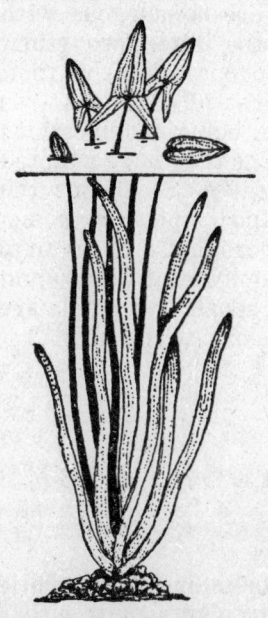

FIG. 80. Arrowhead (*Sagittaria*) showing heterophylly.

Homology and Analogy. Homology is the morphological study of modified organs from the standpoint of their origin, and analogy is the study of organs from the standpoint of their identical structure and function; in other words, organs which resemble one another in their origin, and are, therefore, morphologically the same, whatever be their structure and function, are said to be *homologous* with one another, and organs which resemble one another in their structure and are adapted to the performance of identical functions, although their origin is different, are said to be *analogous* with one another. Thus, all **tendrils**, whatever be their origin, are analogous with one another, being structurally the same and performing the same function; but tendrils of passion-flower (see FIG. 36A) are homologous with axillary buds, i.e. modifications of the latter, and tendrils of pea (see FIG. 63A) are homologous with leaflets. Likewise **thorns** and **spines** are analogous structures being defensive in function but thorns are modifications of axillary or terminal buds and are, therefore, homologous with them, while spines are homologous with leaves being modifications of them wholly or partly. It will further be noted that the thorn of *Duranta* and the tendril of passion-flower are homologous structures. In addition, **modified stems** (e.g. rhizome and tuber) and **modified roots** (e.g. fusiform root and napiform root) are analogous structures, being adapted to the performance of an identical function, i.e. storage of food; but it must be noted that the former two (rhizome and tuber) are homologous with the stem, being modifications of it, while the latter two (fusiform root and napiform root) are homologous with the root, being modifications of it. Again we find that **phylloclades** (see FIGS. 39-40) are homologous with the stem, being modifications of it, but they are analogous with the leaves as they have adapted themselves to perform the functions of leaves. So far as the **flower** is concerned, we find that stamens and carpels are homologous with microsporophylls and megasporophylls of gymnosperms and *Selaginella* and with the sporophylls of ferns, and ultimately with vegetative leaves. Similarly sepals and petals are modified vegetative leaves and are homologous with them.

Chapter 4 DEFENSIVE MECHANISMS IN PLANTS

The animal kingdom as a whole is directly or indirectly parasitic upon the plant kingdom, and this being so, plants must either fall a victim to various classes of animals, particularly the herbivorous ones, which live exclusively on a vegetable diet, or they must be provided with special organs or arms of defence, or have other special devices to repulse or avoid the attack of their enemies. Being fixed to the ground they cannot, of course, manoeuvre when attacked by animals.

I. ARMATURE

1. Thorns, Spines, Prickles and Bristles. These are all sharp, pointed, hard structures, specially developed to ward off herbivorous animals. Small spinous plants, commonly called thistles, —glove thistle (*Echinops*), for example,—are beset with numerous spines and prickles all over their body so that no animal ever dares attack them.

(1) **Thorns** (see p. 27) are modifications of branches, and originate from deeply-seated tissues of the plant body. They are straight and hard, and can pierce the body of thick-skinned animals. Plants like *Vangueria* (B. & H. MOYNA), lemon (*Citrus*), pomegranate (*Punica*), *Duranta, Carissa* and many others are well provided with thorns for self-defence.

(2) **Spines** (see p. 49-50) are modifications of leaves or parts of leaves, and serve the purpose of defence. These are seen in pineapple, date-palm, prickly poppy, American aloe (*Agave*), dagger

DEFENSIVE MECHANISMS IN PLANTS

plant or Adam's needle (*Yucca*), etc. In dagger plant (*Yucca*; FIG. 81) each leaf ends in a very sharp-pointed spine, and is directed obliquely outwards. It acts as a sort of dagger or pointed spike, protecting the plant against grazing animals.

(3) **Prickles** (see pp. 27-8) are also hard and pointed like thorns, but are usually curved and have a superficial origin; they are further irregularly distributed on the stem, branch or leaf. They are found in rose, coral tree (*Erythrina*), silk cotton tree (*Bombax*), *Prosopis* (B. & H. SHOMI), etc. Cane (*Calamus*) and *Pisonia* (B. BAGH-ANCHRHA), which are large climbing shrubs, are elaborately armed with numerous sharp prickles and spines for self-defence (as well as for climbing.)

(4) **Bristles** are short, stiff and needle-like hairs, usually growing in clusters, and not infrequently barbed. Their walls are often thickened with deposition of silica or calcium carbonate. Bristles are commonly met with in prickly pear (*Opuntia*; see FIG. 39A) and in many other cacti.

FIG. 81. Dagger plant or Adam's needle (*Yucca*).

2. **Stinging Hairs.** Nettles (B. BICHUTI; H. BARHANTA) develop stinging hairs on their leaves or fruits or all over their body. Each hair (FIG. 82) has a sharp siliceous apex which breaks off even when touched lightly. The sharp point penetrates into the body, and inflicts a wound. The acid poison of the hair contained in its bulbous base is instantly injected into the skin of the victim's body. This evidently causes a sharp burning pain which is often attended with inflammation. The secretion of the poison is caused by the sudden pressure exerted on the swollen base of the hair. There are various kinds of nettles, e.g. *Laportea*

(=*Fleurya*) *interrupta,* devil or fever nettle (*Laportea crenulata*), *Tragia involucrata, Girardinia zeylanica, Urtica dioica,* cowage (*Mucuna prurita*; B. ALKUSHI; H. KAWANCH), etc.

3. **Glandular Hairs.** Many plants produce glandular hairs on their leaves, branches and fruits. These glandular hairs secrete a sticky substance which is a kind of gum. If any animal feeds upon such a plant the glands stick to its mouth, and the animal finds it difficult to brush them off. Plants bearing glandular hairs are thus never attacked by grazing animals. Glands of this nature are borne by plants like tobacco (*Nicotiana*), *Boerhaavia, Jatropha* and *Plumbago.*

4. **Hairs.** A dense coating of hairs or presence of stiff hairs on the body of the plant is always repulsive to animals as these hairs stick on to their throat and cause a choking sensation, e.g. cud-weed (*Gnaphalium*), *Aerua* and many gourd plants (*Cucurbita*).

II. OTHER DEVICES OF DEFENCE

5. **Poisons.** Many plants secrete poisonous and irritating substances; such plants are carefully avoided by animals which can by intuition distinguish between poisonous and non-poisonous ones.

FIG. 82.
A stinging hair.

(1) **Latex** is the milky juice secreted by certain plants. It always contains some waste products, and often irritating and poisonous substances so that it causes inflammation and even blisters when it comes in contact with the skin. Plants like madar (*Calotropis*), spurges (*Euphorbia*), oleander (*Nerium*), yellow oleander (*Thevetia*), periwinkle (*Vinca*), *Ficus* (e.g. banyan, fig, peepul), etc., contain latex in latex cells; while plants like papaw (*Carica papaya*), poppies, e.g. opium poppy (*Papaver somniferum*), garden poppy (*P. orientale*), prickly or Mexican poppy (*Argemone mexicana*), and some plants of *Compositae,* e.g. *Sonchus,* contain latex in latex vessels.

(2) **Alkaloids** are in many cases extremely poisonous, and a very minute quantity is sufficient to kill a strong animal. There are various kinds of them found in plants, e.g. strychnine in nux-vomica (*Strychnos*), morphine in opium poppy (*Papaver*), nicotine in tobacco (*Nicotiana*), daturine in *Datura,* quinine in *Cinchona,* etc.

(3) **Irritating Substance.** Plants like many aroids, e.g. taro (*Colocasia*), *Amorphophallus,* etc., possess needle-like or otherwise sharp and pointed crystals of calcium oxalate, i.e. raphides. These crystals, when such plants are fed upon, prick the tongue

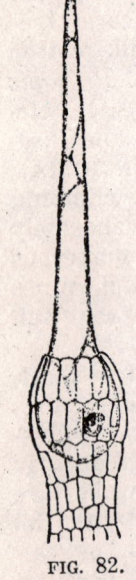

and the throat and cause irritation. Therefore, such plants never fall victims to the attack of grazing animals.

6. Bitter Taste and Repulsive Smell. These are also effective mechanisms to ward off animals. *Paederia foetida* emits a bad smell so that no animal likes to go near it. Plants like sacred basil (*Ocimum*), mint (*Mentha*), *Blumea lacera, Gynandropsis,* etc., also emit a strong disagreeable odour. The fetid smell of the inflorescence of *Amorphophallus* (see FIG. 137) is very offensive and nauseating. Margosa (*Azadirachta*), bitter gourd (*Momordica*), *Andrographis*, etc., have a bitter taste and are avoided by animals.

7. Waste Products. Apart from latex, alkaloids, etc., as mentioned before, the presence of many other waste products such as tannin, resin, essential oils, and silica also keeps plants free from the attack of animals.

8. Mimicry. Certain plants also protect themselves against grazing animals by imitating the general appearance, colour, shape or particular feature of another plant or animal which has developed a special weapon of defence; for instance, there are certain aroids (e.g. varieties of *Caladium*) which resemble multi-coloured and variously spotted snakes. Leaves are also variously spotted and striped in many species of bowstring hemp (*Sansevieria*; B. MURGA; H. MARUL) and other allied plants. Herbivorous animals, possibly mistaking them for snakes or some other deadly creatures, carefully avoid them. The inflorescences of devil's spittoon (*Amorphophallus bulbifer;* B. BAN-OL) coming out of the ground look like hoods of snakes, at least from a distance. In another aroid, called snake or cobra plant (*Arisaema;* FIG. 83), common in Shillong and Darjeeling during the rains, the spathe is greenish-purple in colour and it expands over the spadix like the hood of the cobra. This act of imitating the appearance, colour or any particular feature of another plant or animal is called **mimicry** (*mimikos,* imitative).

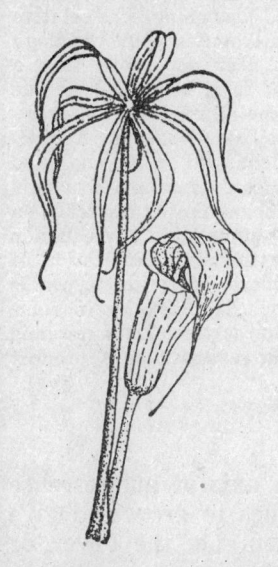

FIG. 83. Snake or cobra plant (*Arisaema*).

Plants have also to protect themselves against the attack of many parasitic fungi and gnawing insects, and also against the scorching rays of the sun; this they do by developing a thick cuticle, cork and bark.

Chapter 5 THE INFLORESCENCE

The reproductive shoot bearing commonly a number of flowers, or sometimes only a single flower, is called the **inflorescence**. It may be terminal or axillary, and may be branched in various ways. Thus depending on the mode of branching different kinds of inflorescence have come into existence, and these may primarily be classified into two distinct groups, viz. **racemose** or **indefinite** and **cymose** or **definite**.

Origin. It is not possible to trace the origin and phylogeny of inflorescence from the primitive to the recent type. This is particularly so because our knowledge regarding the phylogeny of angiosperms is still insufficient. The speculation in this regard is based on the following considerations, and three theories have been advanced to explain the origin of inflorescence. *1st Theory.* Nageli (1883) and Pilger (1922) were of the opinion that the primitive type of inflorescence was a panicle and other forms were derived from it. This view was also supported by Goebel in 1931. *2nd Theory.* Parkin (1914) believed that the primitive type was indicated by a solitary flower and other types with many flowers derived from it as a result of lateral growth of additional branches. It is, however, generally accepted that solitary condition represents reduction and suppression. *3rd Theory.* Ricket (1944) considered the dichasium as the primitive type. According to him a 3-flowered dichasium was the most primitive and a many-flowered dchasium was formed from it as a result of repeated branching. The relative primitiveness of dichasium or panicle is, however, a disputed point. Analogy of angiosperms with pteridosperms suggests but does not prove that a paniculate type is the more primitive. In any case it is believed that one form, dichasium or panicle, has been derived from the other. Ricket suggested the following evolutionary stages from the primitive dichasium. According to him the lines of development are: (*a*) the dichasium on the one hand gives rise to such types as verticillaster, cyme, helicoid cyme and scorpioid cyme by structural modifications; and (*b*) the dichasium (or panicle) on the other hand gives rise to such types as raceme, spike, etc., by reduction of individual dichasia. According to Ricket's view the old classification of inflorescences into racemose (indefinite) and cymose (definite) types is no longer valid. He cited the case of umbel which is *indefinite* in some families, as in *Umbelliferae,* and *definite,* as in certain *Liliaceae* and *Amaryllidaceae.* In his scheme a solitary flower is regarded as a product of reduction.

Kinds of Inflorescences

1. Racemose Inflorescences. Here the main axis of inflorescence does not terminate in a flower, but continues to grow and give off flowers laterally in *acropetal* succession, i.e. the lower or outer flowers are older than the upper or inner ones, or, in other words, the order of opening of flowers is *centripetal*. The various forms of racemose inflorescence may be described under three heads: *first,* those in which the main axis is elongated; *second,* those in which the main axis is shortened; and *third,* those in which the main axis becomes flattened, concave or convex.

THE INFLORESCENCE

A. WITH THE MAIN AXIS ELONGATED

(1) **Raceme** (FIG. 84A). The main axis in this case is elongated and it bears laterally a number of flowers which are all stalked, the lower or older flowers having longer stalks than the upper or younger ones, as in radish (*Raphanus*), mustard (*Brassica*), dwarf gold mohur (*Caesalpinia*), Indian laburnum (*Cassia*), etc. When the main axis of the raceme is branched and the lateral branches bear the flowers, the inflorescence is said to be a compound raceme or **panicle** (see FIG. 89), as in gold mohur (*Delonix*).

The main axis of the inflorescence together with the lateral axes, if present, is known as the **peduncle**. The stalk of the individual flower of the inflorescence is called the **pedicel** (see FIG. 94). In the case of the solitary flower its stalk is regarded and termed as the peduncle. In some flowers such as China rose, gold mohur, etc., the peduncle and the pedicel may, however, be clearly marked out due to the presence of an articulation on the floral axis. When

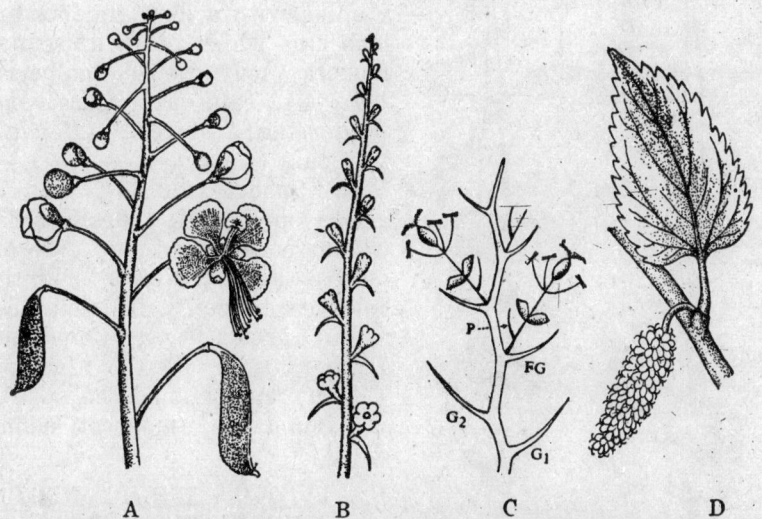

FIG. 84. Racemose Inflorescences. *A*, raceme of dwarf gold mohur; *B*, spike (diagrammatic); *C*, spikelet of a grass (diagrammatic); G_1, first empty glume; G_2, second empty glume; *FG*, flowering glume or lemma; and *P*, palea; *D*, female catkin of mulberry.

the peduncle of an inflorescence is short and dilated forming a sort of convex platform, as in sunflower, or becoming hollow and pear-shaped, as in fig (*Ficus*), it is often called a **receptacle**. The unbranched, often leafless, peduncle arising out of the underground stem in the midst of radical leaves and ending in a single flower, as in lotus, or in an inflorescence, as in onion, tuberose etc., is known as the **scape** (see also p. 11).

(2) **Spike** (FIG. 84B). Here also the main axis is elongated and the lower flowers are older, opening earlier than the upper ones, as in raceme, but the flowers are sessile, that is, without any

stalk. Examples are seen in tuberose (*Polianthes*), *Adhatoda* (B. BASAK; H. ADALSA), amaranth (*Amaranthus*), chaff-flower (*Achyranches*), etc.

(3) **Spikelets** (FIG. 84C). These are very small spikes with one or a few flowers (florets). Spikelets are arranged in a spike, raceme or panicle, and may be sessile or stalked on the main inflorescence. Each spikelet bears at its base two minute scales or bracts called *empty glumes*; slightly higher up it bears a third bract called *flowering glume* or *lemma*; and opposite to the lemma it bears a small 2-nerved bracteole called *palea*. Each flower of the spikelet remains enclosed by the lemma and the palea. Flowers and glumes are arranged on the spikelet in two opposite rows. Spikelets are characteristic of *Gramineae*, e.g. grasses, paddy, wheat, sugarcane, bamboo, etc.

(4) **Catkin** (FIG. 84D). This is a spike with a long and pendulous axis which bears unisexual flowers only, e.g. mulberry (*Morus*), cat's tail (*Acalypha sanderiana*), birch (*Betula*) and oak (*Quercus*).

(5) **Spadix** (FIG. 85). This is also a spike with a fleshy axis, which is enclosed by one or more large, often brightly coloured bracts, called spathes, as in aroids (e.g. *Colocasia, Typhonium,* etc.), banana (*Musa*) and palms. The spadix is found in monocotyledons only.

B. *WITH THE MAIN AXIS SHORTENED*

(6) **Corymb** (FIG. 86A). Here the main axis is comparatively short, and the lower flowers have much longer stalks or pedicels than the upper ones so that all the flowers are brought more or less to the same level, as in candytuft (*Iberis*) and wallflower (*Cheiranthus*).

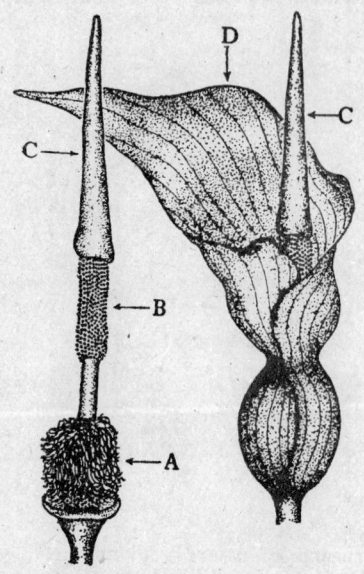

FIG. 85. Spadix of an aroid (*Typhonium*); *A*, female flowers; *B*, male flowers; *C*, appendix; and *D*, spathe.

(7) **Umbel** (FIG. 86B-C). Here the primary axis is shortened, and it bears at its tip a group of flowers which have pedicels of more or less equal lengths so that the flowers are seen to spread out from a common point. In the umbel there is always a whorl

THE INFLORESCENCE

of bracts forming an involucre, and each flower develops from the axil of a bract. Commonly the umbel is branched (**compound umbel**) and the branches bear the flowers, as in anise or fennel, coriander, cumin, carrot, etc. Sometimes, however, it is simple or unbranched (**simple umbel**), the main axis directly bearing the flowers, as in Indian pennywort (*Centella*) and wild coriander (*Eryngium*). Umbel is characteristic of coriander family or *Umbelliferae*. Umbel is a near approach to capitulum.

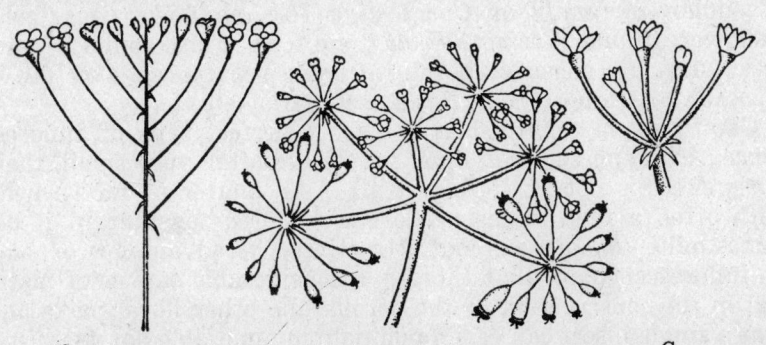

FIG. 86. *A*, corymb (diagrammatic); *B*, a compound umbel; *C*, a simple umbel.

C. WITH THE MAIN AXIS FLATTENED

(8) **Head** or **Capitulum** (FIG. 87). Here the main axis or receptacle is suppressed, becoming almost flat, and the flowers

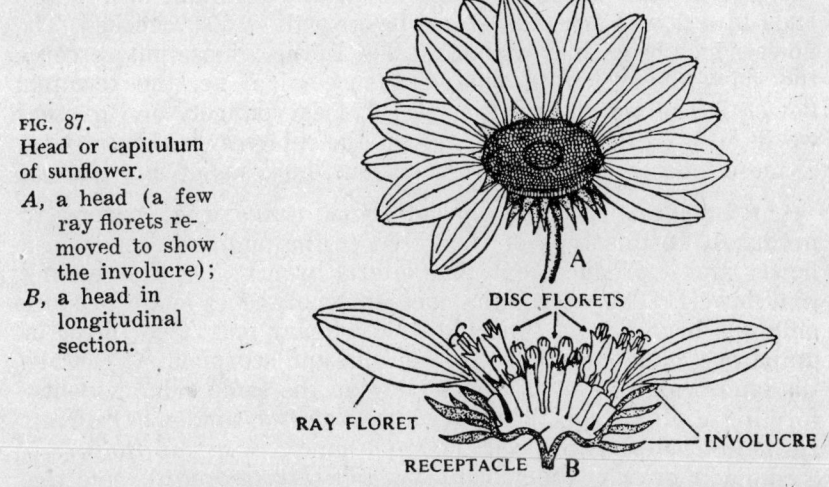

FIG. 87. Head or capitulum of sunflower.
A, a head (a few ray florets removed to show the involucre);
B, a head in longitudinal section.

(florets) are also without any stalk so that they become crowded together on the flat surface of the receptacle. In it the outer flowers are older and open earlier than the inner ones. Although the whole inflorescence looks like a single flower, it really consists of a clustered mass of small sessile flowers (florets) usually of two kinds—**ray florets** (marginal strap-shaped ones) and **disc florets** (central tubular ones). The head may also consist of only one kind of florets. The inflorescence is surrounded at the base by one or more whorls of often green bracts forming an *involucre* (see p. 75). A head or capitulum is characteristic of sunflower family or *Compositae* (e.g. sunflower, marigold, safflower, *Zinnia, Cosmos, Tridax*, etc.). It is also found in gum tree (*Acacia*), sensitive plant (*Mimosa*), *Anthocephalus* (B. & H. KADAM), *Adina* (B. & H. KELI-KADAM), etc.

Capitulum is regarded as the most perfect type of inflorescence. Although the individual florets are often very small, their mass effect is not negligible at all. As a matter of fact a head with often a few dozens of florets clustered together in it becomes quite conspicuous and attractive. The advantages of such an inflorescence are that there is a considerable saving of material in the construction of the corolla and other floral parts and that a single insect can easily pollinate innumerable florets within a very short time without having to fly from one flower to another. The ultimate advantage is that this mass pollination helps the setting of seeds in most heads for reproduction, multiplication in number and continuity of species.

2. **Cymose Inflorescences.** Here the growth of the main axis is soon checked by the development of a flower at its apex, and the lateral axis which develops below the terminal flower also ends in a flower and, therefore, its growth is also checked. The flowers may be with or without stalks. In the cymose inflorescence the flowers develop in *basipetal* succession, i.e. the terminal flower is the oldest and the lateral ones younger, or, in other words, the order of opening of the flowers is *centrifugal*. Cymose inflorescence may be **uniparous, biparous** or **multiparous.**

(1) **Uniparous** or **Monochasial Cyme** (*unus*, one; *parere*, to produce). In this type of inflorescence the main axis ends in a flower and it produces only one lateral branch at a time ending in a flower. The lateral and succeeding branches again produce only one branch at a time like the primary one. Two forms of uniparous cyme may be seen—helicoid and scorpioid. (*a*) When the lateral axes develop successively on the same side, evidently forming a sort of helix, as in *Begonia, Hamelia,* sundew (*Drosera*), rush (*Juncus*), several species of *Solanum*, e.g. *S. sisymbrifolium*—a common prickly undershrub, day lily (*Hemerocallis*), etc., the cymose inflorescence is said to be a **helicoid** (or one-sided) **cyme** (FIG. 88C). (*b*) On the other hand when the lateral branches develop on alternate sides, evidently forming a zigzag, as in heliotrope

THE INFLORESCENCE

(*Heliotropium*), forget-me-not (*Myosotis*), hound's tongue (*Cynoglossum*), *Geranium*, henbane (*Hyoscyamus*), *Crassula*, *Freesia*, bird of paradise (*Strelitzia*), etc., the cymose inflorescence is said to be a **scorpioid** (or alternate-sided) **cyme** (FIG. 88B). The scorpioid cyme is otherwise called **cincinnus.**

In *monochasial* cyme successive axes may be at first zigzag or curved, but subsequently become straightened out due to their rapid growth thus forming the so-called central axis, otherwise known as *pseudo-axis*. This type of inflorescence is called a **sympodial cyme.** This may be distinguished from the racemose type by examining the position of a bract to a flower; in a sympodial cyme a bract appears opposite to a flower, while in a racemose type a bract appears at the base of a flower.

(2) **Biparous** or **Dichasial Cyme** (*bi*, two; *parere*, to produce). In this type of inflorescence the main axis ends in a flower and at the same time it produces two lateral younger flowers or two lateral branches. The lateral and succeeding branches in their turn behave in the same manner (FIG. 88A). This is **true cyme.** Examples are seen in pink, jasmine, teak, night jasmine, *Ixora*, *Bougainvillea*, etc.

(3) **Multiparous** or **Polychasial Cyme.** In this kind of cymose inflorescence the main axis, as usual, ends in a flower, and at the same time it again produces a number of lateral flowers around. There being a number of lateral flowers developing more or less simultaneously, the whole inflorescence looks like an umbel, but is readily distinguished from the latter by the opening of the middle flower first. This is seen in madar (*Calotropis*), blood-flower (*Asclepias*), etc.

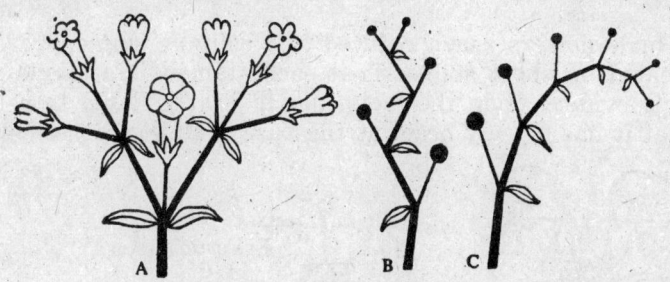

FIG. 88. Cymose Inflorescences. *A*, biparous cyme; *B*, scorpioid cyme; *C*, helicoid cyme.

Compound and Mixed Forms. When the main axis of the inflorescence is branched and the branches bear the flowers, the inflorescence is said to be compound; for example, when raceme is branched it is called a compound raceme or **panicle** (FIG. 89), as in gold mohur (*Delonix*), margosa (*Azadirachta*), dagger plant (see FIG. 81), etc. Similarly, other compound forms are also present such as **compound spike**, as in wheat; **compound spadix**, as in palms; **compound corymb**, as in candytuft; **compound umbel**, as in coriander; and **compound head**, as in globe thistle (*Echinops*). Frequently mixed inflorescences can be found.

3. **Special Types.** The following types may be noted.

(1) **Cyathium** (FIG. 90). This is a special kind of inflorescence found in *Euphorbia*, e.g. poinsettia (B. & H. LALPATA) and spurges (B. & H. SIJ), and also in jew's slipper (*Pedilanthus*; B. RANGCHITA; H. NAGDAMAN). In cyathium there is a cup-shaped involucre, often provided with nectar-secreting glands. The involucre encloses a single female flower (reduced to a pistil) in the centre, seated on a comparatively long stalk, and a num-

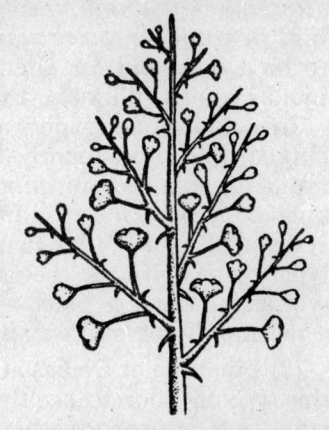

FIG. 89. A panicle.

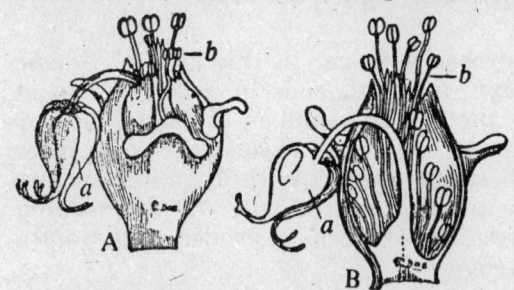

FIG. 90.
Cyathium
of poinsettia.
A, cyathium;
B, the same in longitudinal section;
(a) female flower;
(b) male flower.
Note the involucre.

ber of male flowers (each reduced to a solitary stamen) around this, seated on short stalks. That each stamen is a single male flower is evident from the facts that it is articulated to a stalk and that it has a scaly bract at the base. The flowers follow the

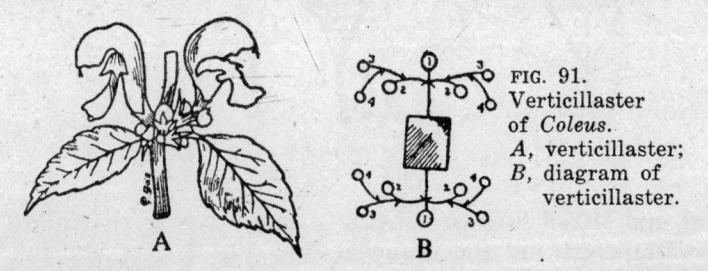

FIG. 91.
Verticillaster of *Coleus*.
A, verticillaster;
B, diagram of verticillaster.

centrifugal (cymose) order of development. The female flower in the centre matures first, and then the stamens (male flowers) just surrounding it, and ultimately the marginal ones.

THE FLOWER

(2) **Verticillaster** (FIG. 91). This is a condensed form of cymose inflorescence with a cluster of sessile or almost sessile flowers in the axil of a leaf, forming a false whorl at the node. The first axis gives rise to two lateral branches and these branches and the succeeding ones bear only one branch each on alternate sides. This kind of inflorescence is characteristic of basil family or *Labiatae*, e.g. *Coleus, Leonurus*, etc. In sacred basil (*Ocimum*; B. & H. TULSI), sage (*Salvia*) and a few others the verticillaster is, however, reduced to a dichasial cyme, succeeding branches remaining undeveloped.

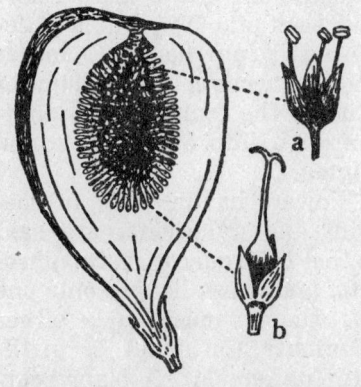

(3) **Hypanthodium** (FIG. 92). When the fleshy receptacle forms a hollow cavity, more or less pear-shaped, with a narrow apical opening guarded by scales, and the flowers are borne on the inner wall of the cavity, the inflorescence is a hypanthodium, as in *Ficus* (e.g. banyan, fig, peepul, etc.). Here the female flowers develop at the base of the cavity and the male flowers higher up towards its mouth.

FIG. 92. Hypanthodium of fig (*Ficus*).
a, male flower; *b*, female flower.

Chapter 6 THE FLOWER

The **flower** is a specialized shoot of limited growth, bearing reproductive organs—**microsporophylls** (or **stamens**) and **mega-**

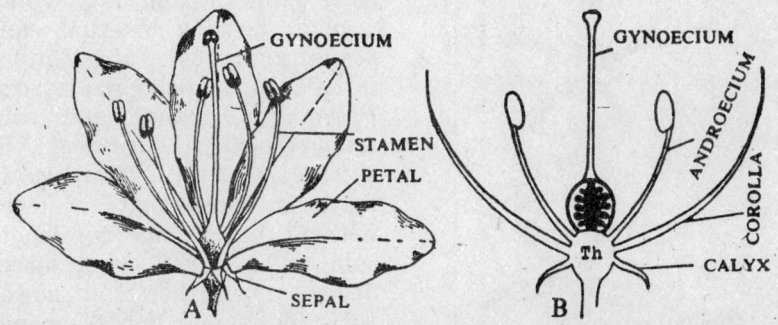

FIG. 93. *A*, parts of a flower; *B*, a flower in longitudinal section showing the position of the whorls on the thalamus (*Th*).

sporophylls (or **carpels**) or only one, often with two accessory whorls—**calyx** and **corolla**, sometimes only one or even none at all. The flower serves as a means of sexual reproduction. A typical or complete flower consists of *four* whorls—two lower *accessory* whorls—**calyx** and **corolla**, and two upper *essential* or *reproductive* whorls—**androecium** and **gynoecium** or **pistil**. The individual units of calyx are **sepals**, of corolla **petals**, of androecium **stamens** or microsporophylls, and of gynoecium **carpels** or megasporophylls. The term **perianth** is collectively used for undifferentiated calyx and corolla and its members are called **tepals**. The four whorls develop in an ascending order from the swollen suppressed end (**thalamus**) of the floral axis or stalk (**pedicel**).

Androecium is the male whorl and each stamen of it is differentiated into **filament**, **anther** and **connective**. Gynoecium or pistil is the female whorl differentiated into **ovary**, **style** and **stigma**.

Flowers having both androecium and gynoecium are said to be **bisexual** or **hermaphrodite**, and those having only one of them **unisexual**, either **staminate** (or male) or **pistillate** (or female). A plant bearing both male and female flowers is said to be **monoecious**, e.g. gourd, and a plant bearing either male flowers or female flowers is said to be **dioecious**, e.g. mulberry, papaw, palmyra-palm, etc.; while a plant bearing bisexual, unisexual and even neuter flowers is said to be **polygamous**, e.g. *Polygonum*, mango and wild mangosteen (*Diospyros*; B. GAB; H. KENDU). Further it may be noted that a flower without calyx and corolla is said to be naked or **achlamydeous**, as in betel; a flower with only one whorl **monochlamydeous**, as in *Polygonum*; and a flower with both the whorls **dichlamydeous**.

The flower is said to be

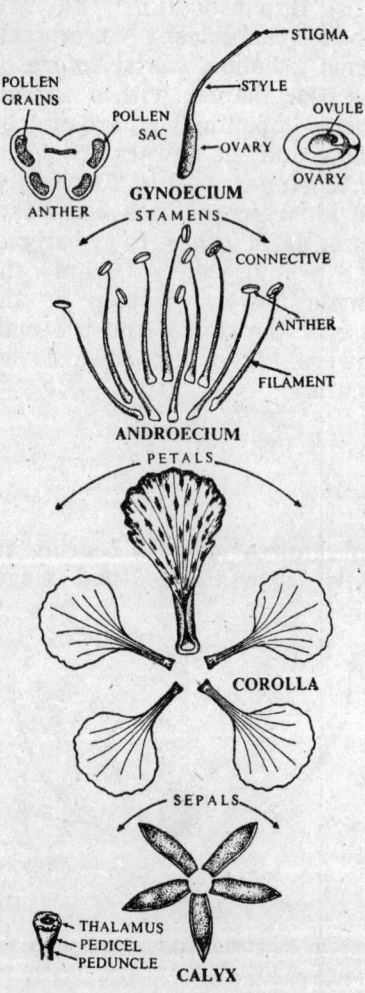

FIG. 94. Flower of gold mohur (*Delonix regia*) dissected out.

THE FLOWER

cyclic when sepals, petals, stamens and carpels are arranged in circles or whorls round the thalamus, as in most flowers, and **acyclic** when these are spirally arranged, as in water lily, *Magnolia* (B. & H. DULEE-CHAMPA), *Michelia* (B. CHAMPA; H. CHAMPAK), etc. The flower may be **hemicyclic** also when some parts are cyclic and others acylic, as in rose.

Thalamus

The thalamus (FIG. 93B), also called torus or receptacle, is the suppressed swollen end of the flower-axis on which are inserted the floral leaves, viz. the sepals, petals, stamens and carpels. In most flowers this thalamus is very short; but in a few cases it becomes elongated, and then it shows distinct nodes and internodes. Thus the internode between the calyx and the corolla, when elongated, is known as the **anthophore** (*anthos*, a flower; *phore*, a stalk), as in *Silene*. In *Gynandropsis* (FIG 95A) and passion-flower (FIG. 95B) the internode between the corolla and the androecium is considerably elongated, and is known as the **androphore** (*andros*, male) or gonophore. In *Capparis* (FIG. 96A). *Gynandropsis* (FIG. 95A), *Pterospermum* (B. MOOCH-KANDA; H. KANAKCHAMPA; FIG. 95C), drumstick (*Moringa*;

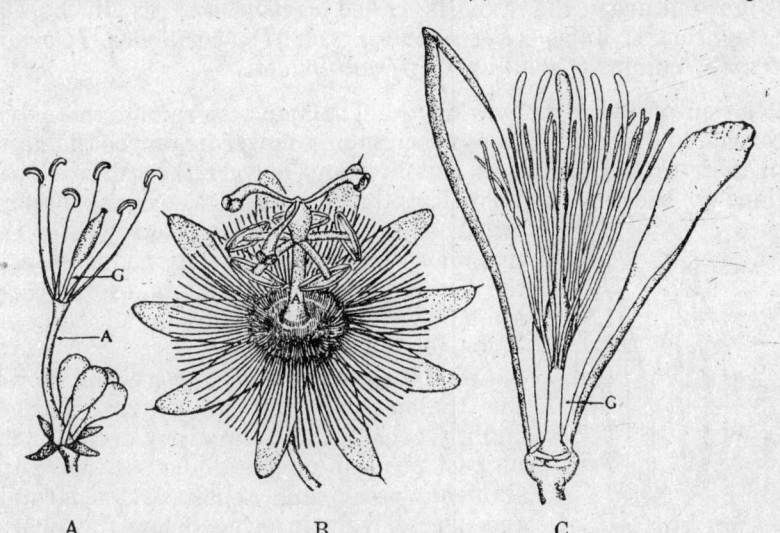

FIG. 95. Thalamus. *A,* flower of *Gynandropsis;* A, androphore; G, gynophore; *B,* passion-flower; A, androphore; *C,* flower of *Pterospermum;* G, gynophore (with the staminal tube adnate to it).

B. SAJINA; H. SAINJNA), etc., the axis between the androecium and the gynoecium is elongated, and is known as the **gynophore** (*gyne*, female). When both androphore and gynophore develop they are together known as the **androgynophore**, as in *Gynan-*

dropsis. In *Magnolia* and *Michelia* the thalamus is fleshy and elongated, and bears the floral leaves spirally round it. In rose (FIG. 96B) it is concave and pear-shaped. The thalamus of lotus

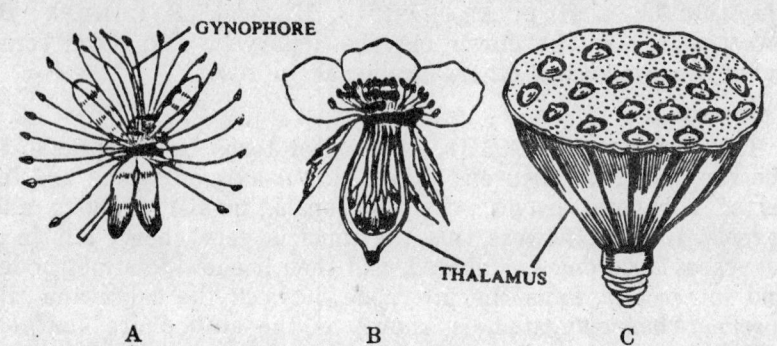

FIG. 96. Thalamus (*contd.*). *A*, flower of *Capparis;* *B*, rose (in section); *C*, lotus.

(*Nelumbo*; FIG. 96C) is spongy and top-shaped. When the thalamus becomes prolonged upwards into a slender axis with the carpels remaining attached to it at first and separating from it on maturity the axis is called **carpophore**, as in balsam (*Impatiens*), anise (*Foeniculum;* FIG. 97), coriander (*Coriandrum*), cumin (*Cuminum*), *Geranium*, etc.

Position of Floral Leaves on the Thalamus. Normally the calyx, corolla, androecium and gynoecium of a flower lie on the thalamus in their proper sequence. But in many flowers the relative positions of the first three whorls in respect of the ovary become disturbed due to the unusual growth of the thalamus. The relative positions, as seen in the flowers of different plants, are of three kinds, viz. hypogyny, perigyny and epigyny (FIGS. 98-100).

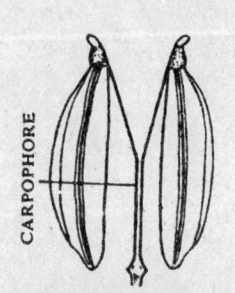

FIG. 97. Fruit of anise (*Foeniculum*).

(1) **Hypogyny.** In a hypogynous flower the thalamus is conical, convex, flat or slightly concave, and the ovary occupies the highest position on the thalamus; while the stamens, petals and sepals are separately and successively inserted below the ovary. The ovary is said to be *superior* and the rest of the floral members *inferior*. Examples are seen in mustard, brinjal, China rose, *Magnolia*, etc.

(2) **Perigyny.** In a perigynous flower the margin of the thalamus grows upward to form a cup-shaped structure, called the calyx-tube, enclosing the ovary but remaining free from it, and carrying with it the sepals, petals and stamens. The ovary is

THE FLOWER

said to be *half-inferior*. In some perigynous flowers the ovary may be partially sunken in the thalamus. Examples are seen in rose, primrose, *Prunus*, e.g. plum, peach, prune, etc., crepe flower

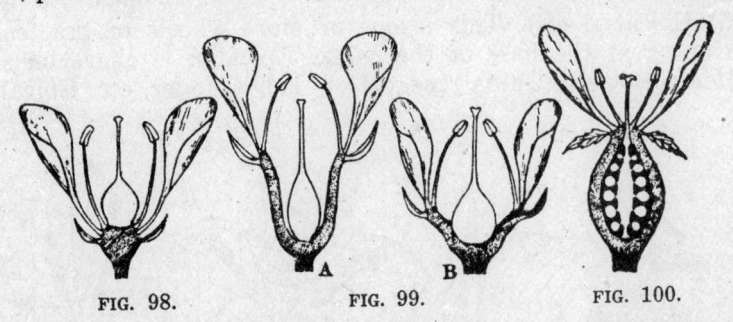

FIG. 98.   FIG. 99.   FIG. 100.

Position of Floral Leaves on the Thalamus. FIG. 98. Hypogyny. FIG. 99. Perigyny (two types—A & B). FIG. 100. Epigyny.

(*Lagerstroemia*), and sometimes in *Leguminosae* (e.g. pea, bean, gold mohur, etc.).

(3) **Epigyny.** In an epigynous flower the margin of the thalamus grows further upward, completely enclosing the ovary and getting fused with it, and bears the sepals, petals and stamens above the ovary. The ovary in this case is said to be *inferior*, and the rest of the floral members *superior*. Examples are seen in sunflower, guava, gourd, cucumber, apple, pear, etc.

Bracts

Bracts (FIG. 101) are special leaves from the axil of which one or more flowers arise. When a small leafy or scaly structure is present on the flower-stalk (pedicel) in any part of it, it goes by the name of **bracteole**. Bracts vary in shape, size, colour and duration. They may be of the following kinds.

(1) **Foliaceous (or Leafy) Bracts.** These are green, flat and leaf-like in appearance, as in *Adhatoda, Acalypha, Gynandropsis*, etc.

(2) **Spathe** (A-B). This is a large, sometimes very large, and commonly boat-shaped bract enclosing a cluster of flowers or even an inflorescence (spadix), as in banana, aroids, palms, maize cob, etc.

(3) **Petaloid Bracts** (C). These are brightly coloured bracts looking somewhat like petals, as in glory of the garden (*Bougainvillea*). In poinsettia (*Poinsettia pulcherrima*) the petaloid bracts, red in colour, take the shape of leaves.

(4) **Involucre** (D). This is one or more whorls of bracts, normally green in colour, present around a cluster of flowers.

Involucre is characteristic of *Compositae*, e.g. sunflower, marigold, *Cosmos*, etc. It is also present in *Umbelliferae*, e.g. coriander, anise, carrot, etc.

(5) **Epicalyx** (E). This is one or more whorls of bracteoles developing at the base of the calyx. Epicalyx is characteristic of *Malvaceae*, e.g. China rose, cotton, lady's finger, etc. Epicalyx

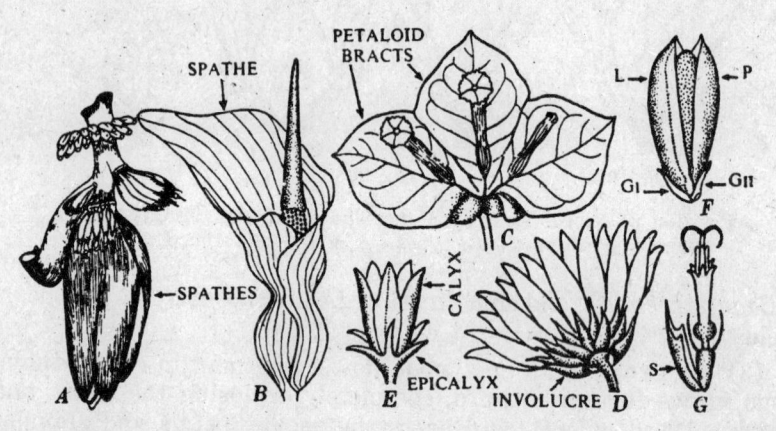

FIG. 101. Bracts and Bracteoles. *A*, spathes of banana; *B*, spathe of an aroid (*Typhonium*); *C*, petaloid bracts of *Bougainvillea*; *D*, involucre of sunflower; *E*, epicalyx (bracteole) of China rose; *F*, glumes of paddy grain (GI, GII, empty glumes; L, lemma or flowering glume; P, palea—a bracteole); *G*, scaly bracteole (S) of a disc floret of sunflower.

is also present in many plants of *Rosaceae*, e.g. strawberry (*Fragaria*).

(6) **Scaly Bracteole** (G). At the base of the individual florets of head or capitulum of *Compositae* there is often a thin, membranous, awl-shaped scaly bracteole.

(7) **Glumes** (F). These are special bracts—small, dry and scaly—found in the spikelet of *Gramineae*. The bracts take the form of two minute scales called *empty glumes* at the base, a flowering glume called *lemma*, and a bracteole called *palea* (see also p. 66).

Flower is a Modified Shoot.

The following facts may be cited to prove that the *thalamus* is a modified branch; *sepals*, *petals*, *stamens* and *carpels* are modified vegetative leaves; and the *flower* as a whole a modified vegetative bud.

(1) The thalamus represents the axis of the floral whorls with internodes between them normally remaining undeveloped or

THE FLOWER

suppressed; but in some flowers the thalamus becomes elongated showing distinct nodes and internodes (see FIGS. 95-7), as in *Gynandropsis*, passion-flower, *Capparis, Pterospermum*, etc. The thalamus may, therefore, be regarded as a modified branch.

(2) The thalamus sometimes shows monstrous development, i.e. after bearing the floral members it prolongs upwards and bears ordinary foliage leaves. The thalamus thus behaves as a branch. Examples are sometimes seen in rose (*Rosa;* FIG. 102), larkspur (*Delphinium*), *Calendula*, pear (*Pyrus*), etc.

(3) The foliar nature of sepals is evident from their similarity to leaves as regards structure, form and venation; in fact, in *Mussaenda* (FIG. 103) one of the sepals becomes modified into a distinct leafy structure, often coloured or sometimes white. The origin of petals is controversial. Some are of opinion that the petals are related to the sepals, while others consider the petals to have been derived from stamens. In green rose the petals are leaflike in structure and green in colour. But stamens and carpels are unlike leaves in all respects.[1] The homology of stamens and

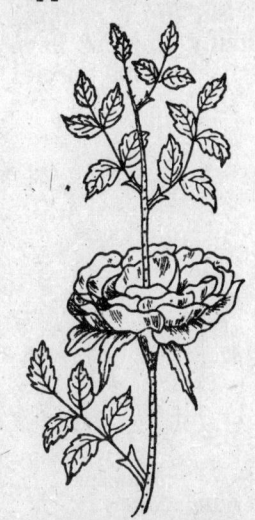

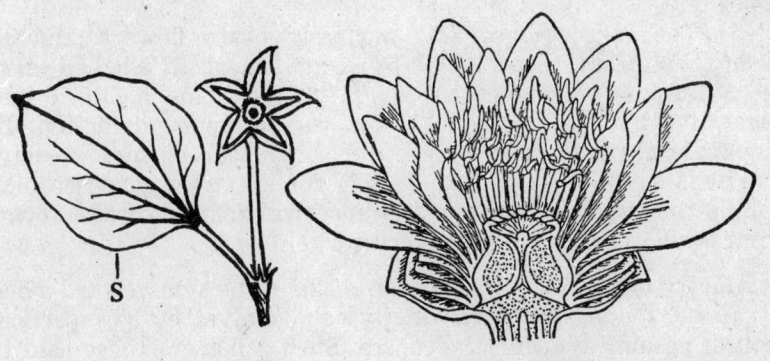

FIG. 102. Rose showing monstrous development of the thalamus.

FIG. 103.   FIG. 104.

FIG. 103. *Mussaenda* flower showing a modified (petaloid and leaf-like) sepal (S). FIG. 104. Water lily flower showing transition of floral parts.

carpels with leaves can be made out from certain flowers. Thus water lily (FIGS. 104-5) shows a gradual transition from sepals

---

[1] The foliar nature of the stamen is explained by the way that it is formed by the inrolling of the margins of the lamina towards the mid-rib, having

to petals and from petals to stamens. The cultivated rose shows many petals; while in wild roses (e.g. dog rose—*Rosa canina*) there are only five. The explanation is that many stamens have gradually become modified into petals. Similarly, in some species

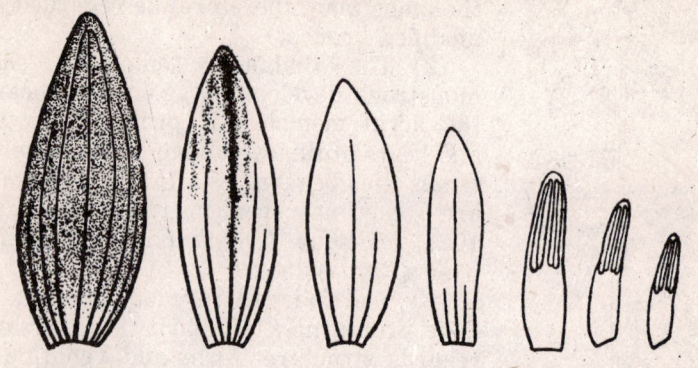

FIG. 105. Transition of floral parts in water lily flower.

or varieties of *Hibiscus*, e.g. *Hibiscus mutabilis* (B. STHAL-PADMA; H. GULAJAIB) some or many of the stamens have passed into petals.

(4) A floral bud like a vegetative bud is either terminal or axillary in position. The arrangement of sepals, petals, etc., on the thalamus is much the same as that of the leaves on the stem or the branch, being either whorled, alternate (spiral) or opposite.

(5) The inflorescence axis normally bears flowers. But in *Globba bulbifera* (see FIG. III/58) and American aloe (*Agave*; see FIG. III/60) some of the floral buds become modified into vegetative buds, called bulbils, for vegetative reproduction. In pineapple also the inflorescence axis bears one or more vegetative buds or bulbils (see FIG. III/62) for the same purpose. Such bulbils thus show a reversion to ancestral forms, i.e. the forms from which they have been derived.

Symmetry of the Flower. A flower is said to be symmetrical when it can be divided into two exactly equal halves by *any* vertical section passing through the centre. Such a flower is also said to be **regular** or **actinomorphic** or radially symmetrical. Examples are found in mustard, *Datura*, brinjal, chilli, etc. When a flower

---

two parts (filament representing the mid-rib, and the anther the lamina) according to some; three parts including the connective according to others. The foliar nature of the carpel may be made out from certain flowers; for example, in pea (see FIG. 127) the carpel folds to form the ovary, but when opened by the ventral suture it looks like a leaf. This view was first expressed by De Candolle, a French systematist, in 1827.

can be divided into two similar halves by *one* such vertical section only, it is said to be **zygomorphic** or **monosymmetrical** or **bilaterally symmetrical**, as in pea, bean, rattlewort (*Crotalaria*; B. ATASHI; H. JHUNJHUNIA), gold mohur (*Delonix*), *Cassia*, etc., and when it cannot be divided into two similar halves by any vertical plane whatsoever, it is said to be **asymmetrical** or **irregular**.

A flower is also said to be symmetrical when its whorls have an equal number of parts or when the number in one whorl is a multiple of that of another. Such a symmetrical flower is said to be **isomerous** (*isos* equal; *meros*, a part). An isomerous flower may be *bimerous, trimerous, tetramerous* or *pentamerous*, according as the number of parts in each whorl is 2, 3, 4 or 5 or any multiple of it. Carpels, however, often do not fit into this symmetry and may, therefore, be ignored. **Trimerous** flowers are characteristic of monocotyledons, and **pentamerous** flowers, and also **tetramerous** flowers, of dicotyledons. When the number in all the whorls is neither the same nor any multiple, the flower is said to be **heteromerous** (*heteros*, different).

Calyx

Calyx is usually green (*sepaloid*), sometimes coloured otherwise (*petaloid*). In its symmetry the calyx may be regular, zygomorphic or irregular. It may again be **polysepalous** (sepals free) or **gamosepalous** (sepals united). Calyx may be modified into **pappus**, as in *Compositae*. In *Mussaenda* (see FIG. 103) one of the sepals becomes distinctly leafy, large, often yellow or orange, sometimes scarlet or white.

Duration. If the calyx falls off as soon as the floral bud opens it is said to be **caducous**, as in poppy. The calyx is said to be **deciduous** if it falls off when the flower withers. But sometimes it remains adherent to the fruit; then it is known as **persistent**. A persistent calyx may assume a withered appearance, as in cotton, or it may continue to grow and form a sort of cup at the base of the fruit, as in brinjal, or it may be inflated enclosing the fruit, as in balloon vine (see FIG. 37), gooseberry and wild gooseberry, or it may be quite fleshy, as in *Dillenia* (B. & H. CHALTA), and rozelle (*Hibiscus sabdariffa*; B. MESTA; H. PATWA), or it may be fleshy and coloured forming the outer envelope of the fruit, as in *Duranta*.

Corolla

Corolla may also be **regular** or radially symmetrical, **zygomorphic** or bilaterally symmetrical, or **irregular** (see above). Like the calyx again, the corolla may be **gamopetalous** or **polypetalous**, according as the petals are united or free. In the former case the petals may be united partially or wholly.

Forms of Corollas

I. *REGULAR AND POLYPETALOUS*

(1) **Cruciform** (FIG. 106A). The cruciform corolla consists of four free petals, each differentiated into a claw and a limb, and these are arranged in the form of a cross, as in *Cruciferae*, e.g. mustard, radish, cabbage, cauliflower, candytuft, etc.

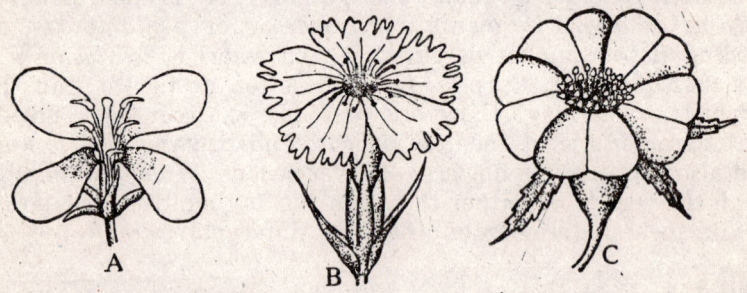

FIG. 106. Forms of Corollas. *A*, cruciform; *B*, caryophyllaceous; *C*, rosaceous.

(2) **Caryophyllaceous** (FIG. 106B). This form of corolla consists of five petals with comparatively long claws, and the limbs of the petals are placed at right angles to the claws, as in pink (*Dianthus*).

(3) **Rosaceous** (FIG. 106C). This form consists of five petals, as in the previous case, but these have very short claws or none at all, and the limbs spread regularly outwards, as in rose, tea, prune, etc.

II. *REGULAR AND GAMOPETALOUS*

(1) **Campanulate** or **Bell-shaped** (FIG. 107A). When the shape of the corolla resembles that of a bell, as in gooseberry (*Physalis*), bell flower (*Campanula*), wild mangosteen (*Diospyros*; B. GAB; H. KENDU), etc., it is said to be campanulate.

(2) **Tubular** (FIG. 107B). When the corolla is cylindrical or tube-like, that is, more or less equally expanded from base to apex, as in the central florets of sunflower, it is said to be tubular.

(3) **Infundibuliform** or **Funnel-shaped** (FIG. 107C). When the corolla is shaped like a funnel, that is, gradually spreading outwards from a narrow base, as in thorn-apple (*Datura*), *Ipomoea*, e.g., water bindweed (B. & H. KALMI-SAK), railway creeper, morning glory, etc., it is said to be infundibuliform.

THE FLOWER

(4) **Rotate** or **Wheel-shaped** (FIG. 107D). When the tube of the corolla is narrow and short and the limb of it is at a right

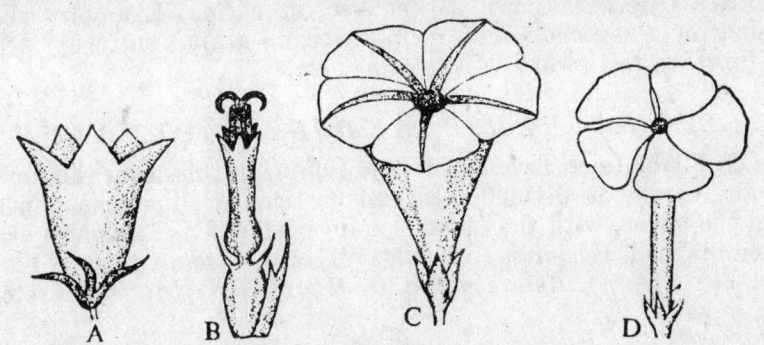

FIG. 107. Forms of Corollas (*contd.*). *A*, campanulate; *B*, tubular; *C*, funnel-shaped; *D*, rotate.

angle to the tube, the corolla having more or less the appearance of a wheel, as in jasmine (*Jasminum*), night jasmine (*Nyctanthes*), brinjal, etc., it is said to be rotate.

(5) **Hypocrateriform** or **Salver-shaped**. Sometimes in a rotate type the corolla-tube is seen to be comparatively long, and the corolla as a whole more or less salver-shaped, as in periwinkle (*Vinca*), *Ixora*, *Ipomoea quamoclit* (B. KUNJALATA; H. KAMLATA), etc. Such a corolla is said to be hypocrateriform or salver-shaped.

III. ZYGOMORPHIC AND POLYPETALOUS

Papilionaceous or **Butterfly-like** (FIG. 108A). The general appearance is like that of a butterfly. It is composed of the five petals, of which the outermost one is the largest and known as

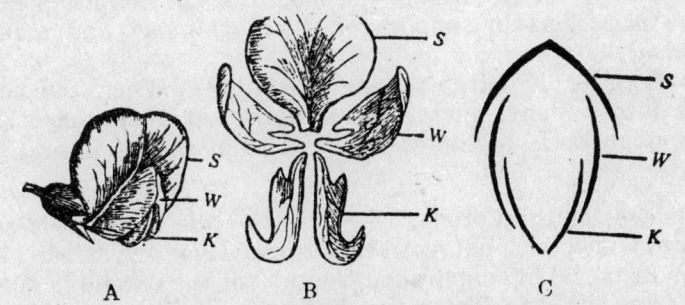

FIG. 108. *A*, papilionaceous flower of pea; *B*, petals of the same opened out; *C*, vexillary aestivation of papilionaceous corolla; *S*, standard or vexillum; *W*, wing; *K*, keel.

the **standard** or **vexillum**; the two lateral ones, partially covered by the former, are somewhat like the two wings of a butterfly and known as the **wings** or **alae**, and the two innermost ones, apparently united to form a boat-shaped cavity, are the smallest and are together known as the **keel** or **carina**. Examples are found in *Papilionaceae*, e.g. pea, bean, gram, butterfly pea (*Clitoria*), rattlewort (*Crotalaria*), etc.

IV. ZYGOMORPHIC AND GAMOPETALOUS

(1) **Bilabiate** or **Two-lipped** (FIG. 109A). In this form the limb of the corolla is divided into two portions or lips—the upper and the lower, with the mouth gaping wide open. Examples are seen in basil (*Ocimum;* see FIG. VII/59), *Leonurus* (see FIG. VII/60), *Leucas, Asteracantha* (=*Hygrophila*), *Adhatoda*, etc.

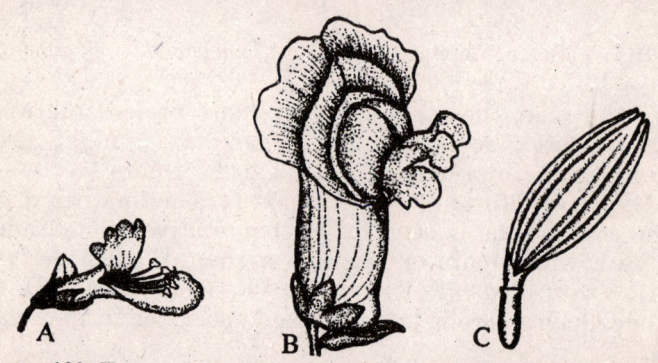

FIG. 109. Forms of Corollas (*contd.*). *A*, bilabiate; *B*, personate; *C*, ligulate.

(2) **Personate** or **Masked** (FIG. 109B). This is also two-lipped like the previous one, but in this case the lips are placed so near each other as to close the mouth of the corolla. The projection of the lower lip closing the mouth of the corolla is known as the *palate*, as in snapdragon (*Antirrhinum*) and toad-flax (*Linaria*).

(3) **Ligulate** or **Strap-shaped** (FIG. 109C). When the corolla forms into a short, narrow tube below, but is flattened above like a strap, as in the outer florets of sunflower, it is said to be ligulate.

Appendages of the Corolla. The corolla or the perianth is sometimes provided with outgrowths or appendages of various kinds; as, for instance, in snapdragon the tube of the corolla is slightly dilated on one side like a pouch or sac; it is then said to be **saccate** or **gibbous** (FIG. 110A). In some cases, as in balsam (*Impatiens*), garden nasturtium (*Tropaeolum*), larkspur (*Delphinium*), orchids, etc., the perianth is prolonged into a tube, known as the **spur** (FIG. 110B-D), and it (the perianth) is then

THE FLOWER

said to be spurred. The spur contains nectar. In many flowers of *Ranunculaceae, Compositae, Labiatae, Rubiaceae*, etc., a special gland known as the **nectary** develops containing nectar (see also p. 101).

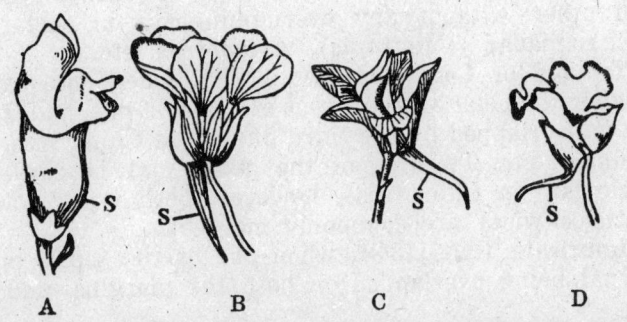

FIG. 110. Appendages of Perianth. *A*, saccate corolla (*S*) of snapdragon; *B*, flower of garden nasturtium; *C*, flower of larkspur; *D*, flower of balsam; *S*, spur.

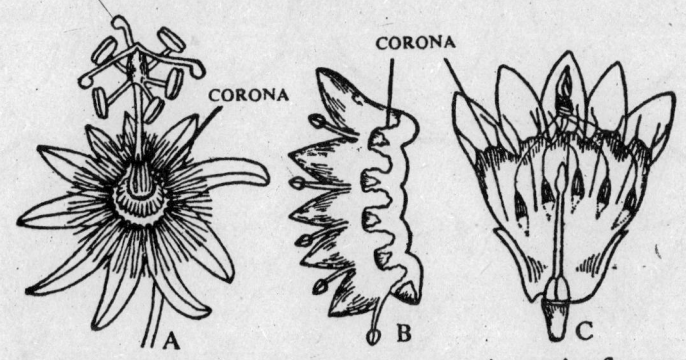

FIG. 111. Appendages of Corolla: Corona. *A*, passion-flower; *B*, flower of dodder; *C*, flower of oleander.

Sometimes, by a transverse splitting of the corolla, an additional whorl may be formed at its throat. This additional whorl may be made up of lobes, scales or hairs, free or united, and is known as the **corona** (*crown*). The corona may be well seen in passion flower (*Passiflora*; FIG. 111A), dodder (*Cuscuta*; FIG. 111B), and oleander (*Nerium*; FIG. 111C). A beautiful, cup-shaped corona is seen in daffodil (*Narcissus*). The corona adds to the beauty of the flower and is thus an adaptation to attract insects for pollination.

Aestivation

The mode of arrangement of the sepals or of the petals, more particularly the latter, in a floral bud with respect to the members of the same whorl (calyx or corolla) is known as **aestivation**. Aestivation is an important character from the viewpoint of classification of plants, and may be of the following types:

(1) **Valvate** (FIG. 112A), when the members of a whorl are in contact with each other by their margins, or when they lie close to each other without any overlapping, as in custard-apple (*Annona*), madar (*Calotropis*), *Artabotrys*, etc.

(2) **Twisted** or **Contorted** (FIG. 112B), when one margin of the sepal or the petal overlaps that of the next one, and the other margin is overlapped by the third one, as in China rose, cotton, lady's finger, etc. Twisting of the petals may be clockwise or anticlockwise. In China rose, however, both types (clockwise and anticlockwise) are commonly met with.

(3) **Imbricate** (FIG. 112C), when one of the sepals or petals is internal being overlapped on both the margins, and one of

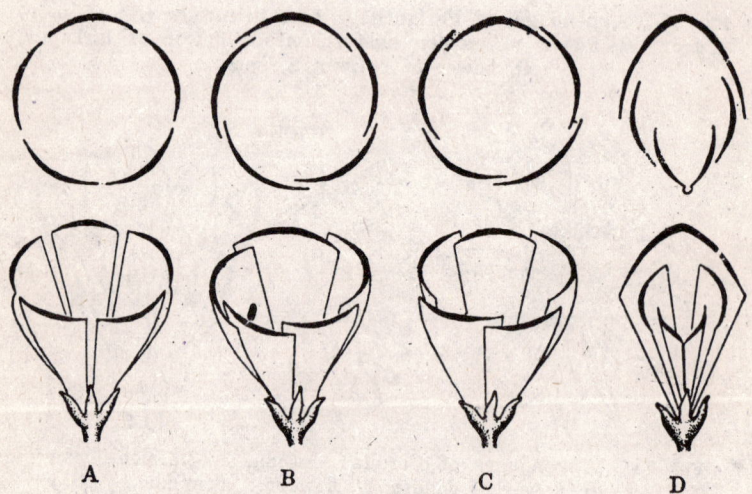

FIG. 112. Aestivation of Corolla. *A*, valvate; *B*, twisted; *C*, imbricate; *D*, vexillary.

them is external and each of the remaining ones is overlapped on one margin and it overlaps the next one on the other margin, e.g. *Cassia*, gold mohur (*Delonix*), dwarf gold mohur (*Caesalpinia*), etc.

(4) **Vexillary** (FIG. 112D), when there are five petals, of which the posterior one is the largest and it almost covers the two lateral petals, and the latter in their turn nearly overlap the two anterior or smallest petals. Vexillary aestivation is universally found in all papilionaceous corollas (see also FIG. 108), as in pea family or *Papilionaceae*, e.g. pea, bean, butterfly pea (*Clitoria*), rattlewort (*Crotalaria*), etc.

(3) Androecium

Androecium (*andros,* male) is composed of a number of **stamens** or microsporophylls. Each stamen consists of **filament, anther** and **connective** (FIG. 113). Each of the two anther-lobes has two chambers or loculi, called the **pollen-sacs** or *microsporangia*; thus there are altogether four loculi in each anther (FIG. 114). But in many cases there are only two, or even one. Each chamber of the anther is filled with pollen grains or *microspores*. The filament corresponds to the petiole of the leaf, the anther to the leaf-blade, and the connective to the mid-rib. The connective is attached to the *back* of the anther; evidently the other side of the anther is its *face*. The face often has a longitudinal groove running along the whole length of the anther. When the face is turned inwards, the anther is said to be **introrse,** and when outwards, the anther is **extrorse.** A sterile stamen (without pollen grains) is known as a **staminode,** as in pink (*Dianthus*), noon flower (*Pentapetes*), *Pterospermum* (see FIG. 95C), elengi (*Mimusops*; B. BAKUL; H. MULSARI), etc.

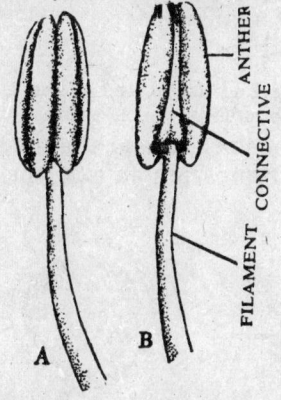

FIG. 113. Two stamens. *A*, face of the anther showing four pollen-sacs; *B*, back of the anther showing connective.

The Pollen. Pollen grains are very minute in size, usually varying from 10 to 200 μ (microns) and are like particles of dust. Each pollen grain consists of a single microscopic cell, and possesses two coats: the **exine** and the **intine.** The exine is a tough, cutinized layer, which is often provided with spinous outgrowths or reticulations of different patterns, sometimes smooth. The intine, however, is a thin, delicate, cellulose layer lying internal to the exine. In pine the pollen grain is provided with two distinct wings. When the pollen grain has to germinate the intine grows out into a tube, called the **pollen-tube** (FIG. 116), through some definite thin and weak slits or pores, called **germ pores,** present in the exine (FIG. 115). Sometimes the pore is covered by a distinct lid which is pushed open by the growth of the intine. At first each pollen grain contains only one nucleus; this divides to form two nuclei, of which the larger one is known as the vegetative nucleus or **tube-nucleus** and the smaller one the **generative nucleus.** As the pollen-tube grows it carries with it at its apex the tube-nucleus and the generative nucleus, the former going ahead of the latter. The generative nucleus soon divides and two male reproductive units are formed, which are known as the **male gametes**. The tube-nucleus then gets disorganized. (It may be noted that in

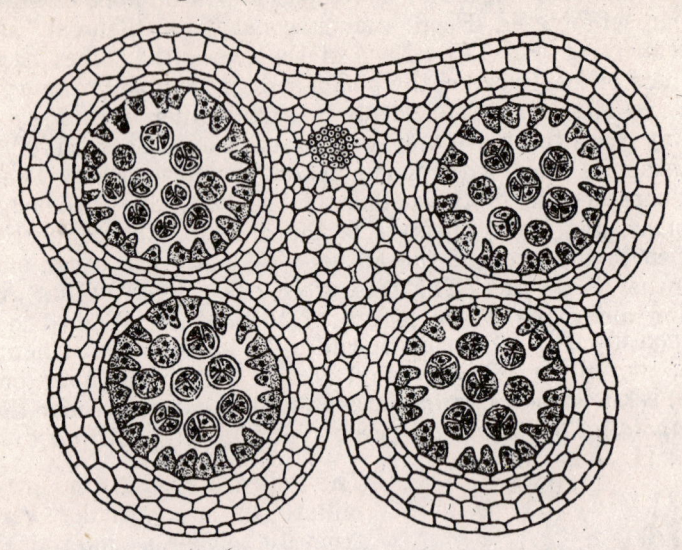

FIG. 114. An anther in transection showing four loculi and pollen grains in tetrads. Each tetrad separates into four pollen grains.

angiosperms the pollen-tube with the tube-nucleus and the generative nucleus, or the two male gametes, represents an extremely reduced male gametophyte.)

FIG. 115. Pollen grains. *A*, an entire grain; *B*, a grain in section showing tube-nucleus (bigger one) and generative nucleus (smaller one).

FIG. 116. Growth of the pollen-tube.

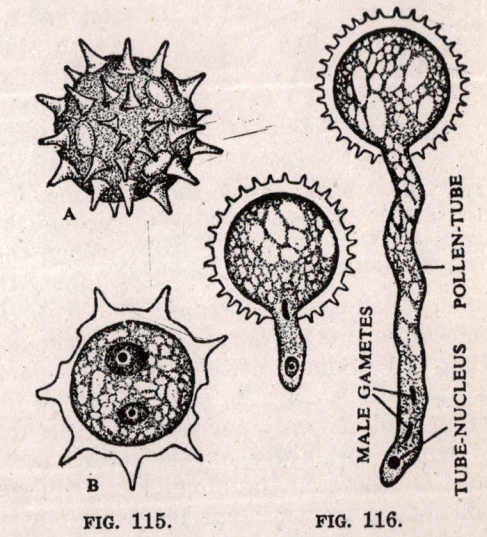

FIG. 115.   FIG. 116.

Development of Pollen Grains (FIG. 117). At an early stage of the anther four vertical rows of cells, one at each lobe, with dense

THE FLOWER

protoplasmic contents become apparent (A). Each cell of the row divides into an inner larger cell—the **sporogenous cell** or **archesporium**, and an outer smaller cell—the **parietal cell** (B). The latter divides tangentially into 2 or 3 layers (C). These (parietal) cells again divide repeatedly by radial walls and extend around the sporogenous cell (D) which also begins to divide forming a central group of cells (E). These (sporogenous) cells grow, separate from one another and become the **pollen (spore) mother**

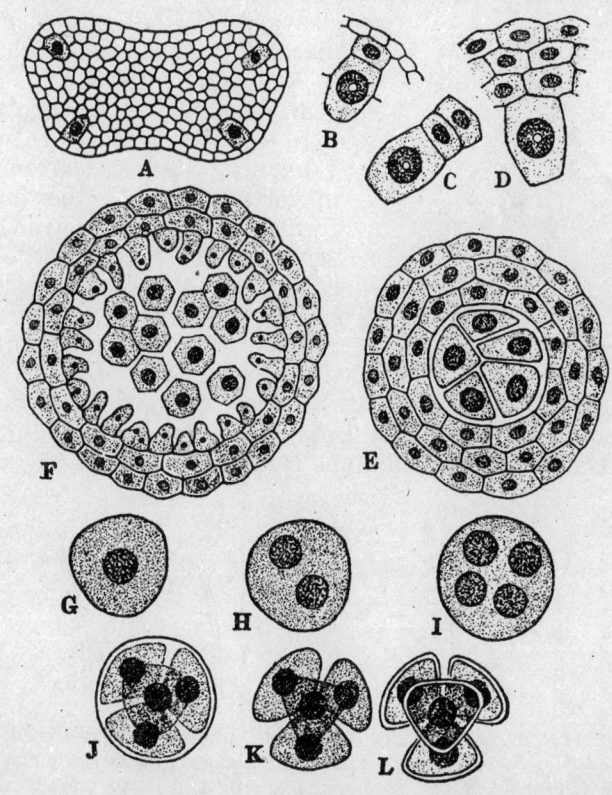

FIG. 117. *A-F*, development of pollen-sac; *G-L*, development of pollen grains. (Explanation in the text.)

cells. The innermost layer of parietal cells, directly abutting upon the sporogenous cells and later the pollen mother cells, is called the **tapetum** (F). The cells of the tapetum are more or less wedge-shaped and contain one or more nuclei. It is a nutritive tissue supplying food to the pollen grains as they develop. Ultimately the tapetum becomes disorganized.

The nucleus of each pollen mother cell (G) divides twice so that four nuclei are formed in it. Of the two successive divisions the first one (H) is meiosis and the second one (I) is mitosis so that each nucleus has half (*n*) the usual number (2*n*) of chromosomes. The four nuclei so formed are arranged in a tetrahedral manner, and cleavage of the cytoplasm occurs separating the nuclei into four distinct segments—the pollen cells (J). The wall of the mother cell disappears (K) and each pollen cell secretes a thick outer wall—the **exine**, and a thin inner wall—the **intine** (L). The four mature cells separate from one another and form four pollen grains. In monocotyledons, however, the cleavage of the cytoplasm takes place by two planes at right angles to each other, and not in a tetrahedral manner, as in dicotyledons.

In elephant grass (*Typha*), rush (*Juncus*), sundew (*Drosera*) and in certain orchids, the four cells formed in a group (tetrad) do not separate, but remain more or less coherent. In *Asclepiadaceae*, e.g. madar (*Calotropis*), milkweed (*Asclepias*), *Hoya*, etc., and in *Orchidaceae*, e.g. *Vanda* (see FIG. 13),

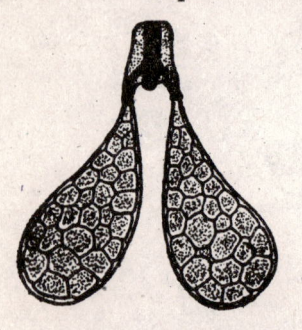

FIG. 118. Pollinia of madar (*Calotropis*).

Dendrobium, *Orchis*, etc., the pollen cells of each pollen-sac instead of separating into loose pollen grains are united into a mass known as the **pollinium** (FIG. 118). Pollen cells may also

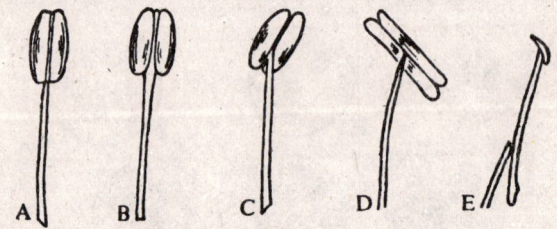

FIG. 119. *A*, basifixed or innate; *B*, adnate; *C*, dorsifixed; *D*, versatile; *E*, elongated connective of sage (*Salvia*) separating the two anther-lobes (upper one fertile and lower one sterile).

be in compound forms, i.e. united in small masses—pollen masses, each consisting of 8-32 or more cells, as in *Mimosa* and *Acacia*.

Attachment of the Filament to the Anther (FIG. 119). There are four principal ways in which the filament is attached to the anther. The anther is said to be (1) **basifixed** or **innate**, when the filament is attached to the base of the anther, as in mustard, radish, sedge, water lily, etc.; (2) **adnate**, when the filament runs up the whole length of the anther from the base to the apex, as in *Michelia*, *Magnolia*, etc.; (3) **dorsifixed**, when it is attached

THE FLOWER

to the back of the anther, as in passion-flower; and (4) **versatile**, when it is attached to the back of the anther at one point only so that the latter can swing freely in the air, as in grasses, palms, spider lily (*Pancratium*), etc. In sage (*Salvia*; FIG. 119E) the filament is attached to the elongated connective separating the two anther-lobes, of which the upper one is fertile and the lower one sterile.

Cohesion and Adhesion. The terms 'adhesion', 'adnate', and 'adherent' are used to designate the union of members of different whorls, e.g. petals with stamens, or stamens with carpels; and 'cohesion', 'connate', and 'coherent' to designate the union of members of the same whorl, e.g. stamens with each other, and carpels with each other.

Cohesion of Stamens (FIG. 120). Stamens may either remain free or they may be united (coherent). There may be different degrees of cohesion of stamens, and these may be called (*a*) the

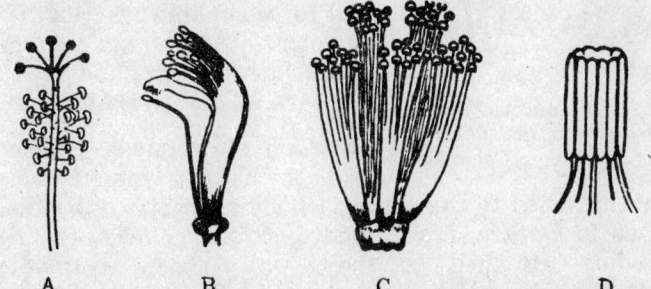

FIG. 120. Cohesion of Stamens. *A*, monadelphous; *B*, diadelphous; *C*, polyadelphous; *D*, syngenesious.

adelphous condition when the stamens are united by their filaments only, the anthers remaining free; (*b*) the syngenesious condition when the stamens are united by their anthers only, the filaments remaining free; or (*c*) the synandrous condition when the stamens are united by both the filaments and the anthers. Accordingly the following types are to be noted:

(1) **Monadelphous Stamens** (*monos*, single; *adelphos*, brother). When all the filaments are united together into a single bundle but the anthers are free, the stamens are said to be monadelphous (A), as in *Malvaceae*, e.g. China rose (*Hibiscus*), lady's finger (*Abelmoschus*), cotton (*Gossypium*), etc. In them the filaments are united into a tubular structure, called the staminal tube, ending in free anthers.

(2) **Diadelphous Stamens** (*di*, two). When the filaments are united into two bundles, the anthers remaining free, the stamens are said to be diadelphous (B), as in *Papilionaceae*, e.g. pea,

bean, gram, butterfly pea (*Clitoria*), rattlewort (*Crotalaria*), etc. In them there are altogether ten stamens of which nine are united into one bundle and the tenth one is free.

(3) **Polyadelphous Stamens** (*polys*, many). When the filaments are united into a number of bundles—more than two—but the anthers are free, the stamens are said to be polyadelphous (C), as in silk cotton tree (*Bombax*), castor (*Ricinus*), lemon (*Citrus*), etc.

(4) **Syngenesious Stamens** (*syn*, together or united; *gen*, producing). When the anthers are united together into a bundle or tube, but the filaments are free, the stamens are said to be syngenesious (D), as in *Compositae*, e.g. sunflower, marigold, etc.

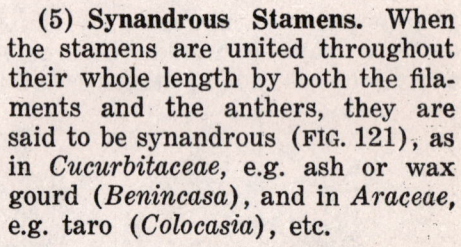

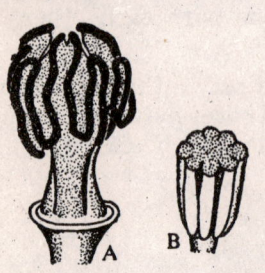

(5) **Synandrous Stamens**. When the stamens are united throughout their whole length by both the filaments and the anthers, they are said to be synandrous (FIG. 121), as in *Cucurbitaceae*, e.g. ash or wax gourd (*Benincasa*), and in *Araceae*, e.g. taro (*Colocasia*), etc.

FIG. 121. Synandrous stamens. *A*, ash or wax gourd (*Benincasa*); *B*, taro (*Colocasia*).

Adhesion of Stamens. Stamens are said to be (1) **epipetalous**, when they are attached to the corolla wholly or partially by their filaments, as in *Datura, Ixora*, potato, sunflower, etc., (2) **epiphyllous**, when attached to the perianth, as in *Liliaceae*; and (3) **gynandrous**, when united with the carpels, either wholly or by their anthers only, as in *Calotropis, Asclepias*, orchids, etc.

Length of Stamens (FIG. 122). In *Labiatae*, e.g. *Ocimum, Leonurus, Leucas*, etc., there are four stamens, of which two are long and two short; such stamens are said to be (1) **didynamous** (*di*, two; *dynamis*, strength). In *Cruciferae*, e.g. mustard, radish, turnip, rape, etc., there are six stamens, of which the inner four are long and the outer two short; such stamens are said to be (2) **tetradynamous** (*tetra*, four). Sometimes different kinds of flowers, some with longer stamens and others with shorter stamens, are borne by the same plant (*dimorphic* stamens).

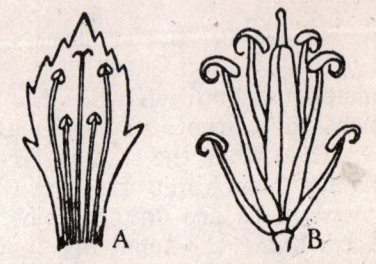

FIG. 122. Length of Stamens. *A*, didynamous; *B*, tetradynamous.

THE FLOWER

Dehiscence of the Anther. Dehiscence of the anther may be (1) **longitudinal**, as in China rose, cotton, *Datura*, etc.; (2) **transverse**, as in basil; (3) **porous** (by pores), as in *Solanum*, e.g. potato, brinjal, etc.; and (4) **valvular** (by valves), as in *Cinnamomum*, e.g. cinnamon, camphor, bay leaf, etc.

Appendages of Stamens. Both filaments and anthers may have outgrowths or appendages in the form of hairs, scales, etc. Thus in *Osbeckia* anthers are beaked, while in oleander (*Nerium*) these are provided with long hairy appendages twisted together into a cone over the stigma (see FIG. 111C). Similarly, filaments may have appendages. When these are in the form of a regular whorl they form the **staminal corona**. Thus in madar (*Calotropis*) the filaments form a whorl of horn-like corona. In spiderwort (*Tradescantia*) there is a circle of hairs. In spider lily (*Pancratium*), eucharis lily (*Eucharis*) the corona is formed as a membranous cup adherent to the filaments at their base.

(4) Gynoecium or Pistil

Gynoecium (*gyne*, female) or **pistil** (*pistillum*, a pestle) is composed of one or more **carpels** or megasporophylls which bear female spores or megaspores (embryo-sacs). The pistil may be

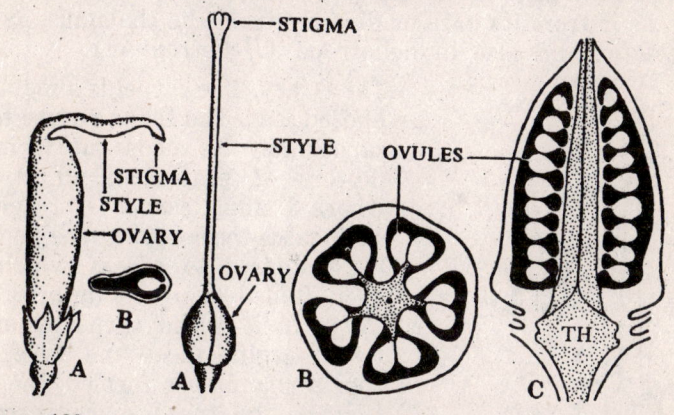

FIG. 123. FIG. 124.

FIG. 123. Pistil. *A*, simple pistil of pea; *B*, one-chambered ovary of the same. FIG. 124. *A*, syncarpous pistil; *B*, five-chambered ovary of the same (in transection); *C*, ovary of the same in longi-section. *TH*, thalamus.

simple (made of one carpel) or **compound** (made of two or more carpels). In a compound pistil the carpels may be free, as in lotus (*Nelumbo*), *Michelia*, rose, stonecrop (*Sedum*—a pot herb), *Magnolia, Artabotrys, Unona* etc., when the pistil is said to be **apocarpous** (*apo*, off or free; *karpos*, fruit or ovary; FIG. 125), or all the carpels may be united together, as is more commonly found, when the pistil is said to be **syncarpous** (*syn*, together or united; FIG. 124). Each pistil consists of three parts—**stigma**, style and ovary (FIG. 124A). The ovary contains one or more little, roundish or oval, egg-like bodies which are the rudiments of seeds and are known as the **ovules** (FIG. 124B-C). Each ovule encloses a large oval cell known as the **embryo-sac** (see FIG.

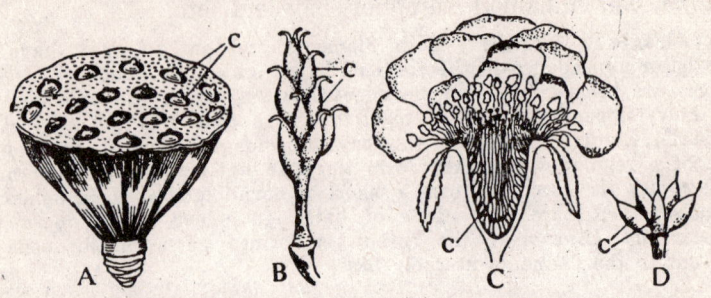

FIG. 125. Apocarpous Pistil. *A*, lotus; *B*, *Michelia;* *C*, rose; *D*, stonecrop (*Sedum*). *c*, carpels.

133). A sterile pistil is known as the **pistillode**. In position the style may be **terminal, lateral** or **gynobasic**. The gynobasic style (see FIG. 126) rises from the depressed centre of the four-lobed ovary, as if from its base or directly from the thalamus, as seen in *Labiatae* and also in heliotrope (*Heliotropium*).

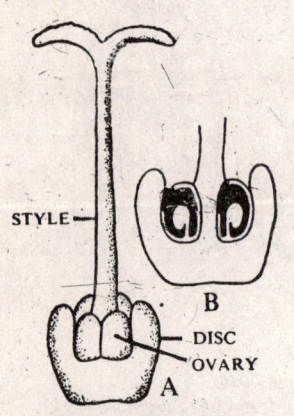

FIG. 126. *A*, gynobasic style of basil (*Ocimum*); *B*, the same in section.

The Ovary. The carpel is a metamorphosed leaf. The foliar nature of the carpel may be made out from the flowers of pea, bean, gram, etc., where a single carpel is present. In such cases the carpel or the pod may be compared to a leaf which has been folded along its midrib (FIG. 127). In a folded carpel when the two margins meet and fuse together a chamber is formed, the junction of the fused margins of the carpel being known as the *ventral suture,* and the midrib along which the carpel is folded being known as the *dorsal suture*. Along the ventral suture a ridge of tissue, called the **placenta**, develops and bears the ovules. The closed chamber formed by the folding of the carpel, enclosing the ovules, is the **ovary**. In apocarpous pistil, as in buttercup (*Ranunculus*), the ovary is also formed in the above way (FIG. 128). In syncarpous pistil, however, the carpels may be united by their margins only, forming an one-chambered ovary (FIG. 129), as in orchids, papaw, etc., or the carpels may be folded inwards, their

THE FLOWER

margins meeting in the centre, thus resulting in a few- to many-chambered ovary (usually as many as the carpels) with a central axis (FIG. 130), as in lily (*Lilium*), China rose (*Hibiscus*), etc. In gymnosperms, however, the carpels are not closed up to form the ovary and, therefore, there is no stigma, style or ovary. In them the ovules are borne, freely exposed, along the margins of open carpels.

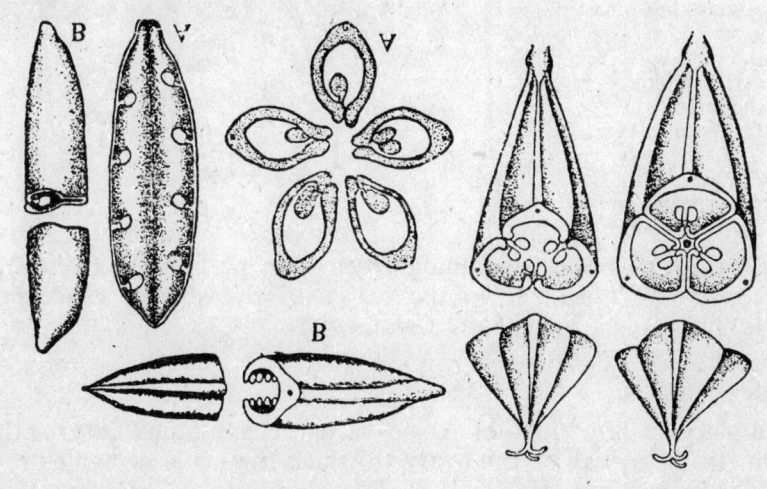

FIG. 127   FIG. 128   FIG. 129   FIG. 130

Development of the Ovary. FIG. 127. *A*, a single carpel opened out with the ovules on the margins; *B*, one-chambered ovary formed by the folding of the carpel with ovules at the ventral suture. FIG. 128. *A*, one-chambered ovaries formed by five free carpels of an apocarpous pistil; *B*, one of the five ovaries (carpels). FIG. 129. One-chambered ovary formed by the union of margins of three carpels of a syncarpous pistil. FIG. 130. Three-chambered ovary formed by the infolding of three carpels and their margins meeting at the centre.

Carpels in Syncarpous Pistil. In a syncarpous pistil it is often difficult to determine the number of carpels. To obviate this difficulty the following points should be noted: (1) the number of stigmas or of stigmatic lobes; (2) the number of styles; (3) the number of lobes of the ovary; (4) the number of chambers (loculi) of the ovary; (5) the number of placentae in the ovary; and (6) the number of groups of ovules in the ovary. It is seen that in most cases the number of parts, as mentioned above, corresponds to the number of carpels making up the syncarpous pistil.

Cohesion of Carpels (Syncarpy). The carpels may be united either throughout their whole length, as in most syncarpous pistils; or, they may be united in the region of the ovary alone, styles and stigmas remaining free, as in pink (*Dianthus*), linseed (*Linum*) and *Plumbago* (B. CHITA; H. CHITRAK); or in the region of the ovary and the style, stigmas remaining free, as in cotton and China rose; or in the region of the style and the

FIG. 131. Cohesion of carpels.
A, pistil with free stigmas in China rose;
B, the same with free styles in pink;
C, the same with free ovaries in oleander;
D, the same with free ovaries and styles in madar.

stigma, ovaries only remaining free, as in periwinkle (*Vinca*), oleander (*Nerium*); or in the region of the stigma (and the style partly), as in madar (*Calotropis*).

Placentation

The **placenta** is a ridge of tissue—a parenchymatous outgrowth—in the inner wall of the ovary to which the ovule or ovules remain attached.

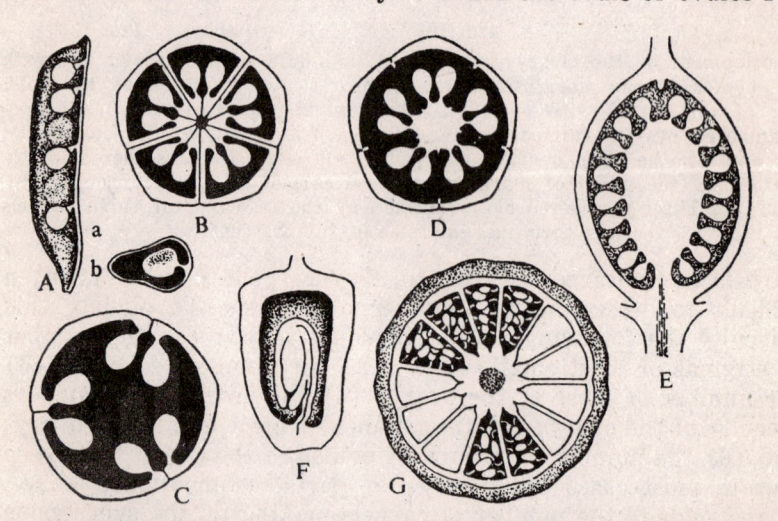

FIG. 132. Types of Placentation. A, marginal—a, in longi-section; b, in transection; B, axile; C, parietal; D, central; E, free-central; F, basal; and G, superficial.

main attached. The placentae most frequently develop on the margins of carpels, either along their whole line of union, called

THE FLOWER

the **suture**, or at their base or apex. The manner in which the placentae are distributed in the cavity of the ovary is known as **placentation**. As a rule the origin of an ovule or a group of ovules determines the position of the placenta.

Types of Placentation (FIG. 132). In the simple ovary (of one carpel) there is one common type of placentation, known as **marginal**, and in the compound ovary (of two or more carpels united together) placentation may be **axile, parietal, central, free-central, basal,** or **superficial.**

(1) **Marginal.** In marginal placentation (A) the ovary is one-chambered and the placenta develops along the junction of the two margins of the carpel, called the *ventral suture*, as in Leguminosae (e.g. pea, gram, bean, gold mohur, *Cassia, Mimosa,* etc.). The line, or suture, corresponding to the mid-rib of the carpel, is known as the dorsal suture. No placenta develops here.

(2) **Axile.** In axile placentation (B) the ovary is two- to many-chambered—usually as many as the number of carpels—and the placentae bearing the ovules develop from the central axis corresponding to the confluent margins of carpels, and hence the name axile (lying in the axis), as in lemon, orange, China rose, lady's finger, tomato, potato, etc.

(3) **Parietal** (*parietis*, wall). In parietal placentation (C) the ovary is one-chambered, and the placentae bearing the ovules develop on the inner wall of the ovary. Their position corresponds to the confluent margins of carpels, and their number corresponds to the number of carpels, as in papaw (*Carica*), poppy (*Papaver*), prickly poppy (*Argemone*), orchids, etc. In *Cruciferae*, e.g. mustard, radish, etc., the placentation is also parietal although the ovary is two-chambered. In them the ovary is at first unilocular but soon a false partition wall develops across the ovary dividing it into two chambers, while the seeds remain attached to a wiry framework called the *replum*.

(4) **Central.** In central placentation (D) the septa or partition walls in the young ovary soon break down so that the ovary becomes one-chambered and the placentae bearing the ovules develop from the central axis, as in *Caryophyllaceae*, e.g. pink (*Dianthus*), *Polycarpon* (B. GIMA-SAK), soapwort (*Saponaria*), etc. Remnants of partition walls may often be seen in the mature ovary.

(5) **Free-central.** In free-central placentation (E) the placenta arises from the base of the ovary, projects far into its cavity as a swollen central axis and bears the ovules all over its surface. Since the placenta lies free in the single chamber of the ovary, the placentation is said to be free-central. This is seen in primrose (*Primula*).

(6) Basal. In basal placentation (F) the ovary is unilocular and the placenta develops directly on the thalamus, and bears a single ovule at the base of the ovary, as in *Compositae*, e.g. sunflower, marigold, *Cosmos*, etc.

(7) Superficial. In superficial placentation (G) the ovary is multilocular, carpels being numerous, as in axile placentation, but the placentae in this case develop all round the inner surface of the partition walls, as in water lily (*Nymphaea*).

The Ovule

Structure of the Ovule. Each ovule (FIG. 133) is attached to the placenta by a slender stalk known as (1) the **funicle**. The point of attachment of the body of the ovule to its stalk or funicle is known as (2) the **hilum**. In the inverted ovule, as shown in FIG. 133, the funicle continues beyond the hilum alongside the body of the ovule forming a sort of ridge; this ridge is called (3) the **raphe**. The upper end of the raphe which is the junction of the integuments and the nucellus is called (4) the **chalaza**. The main body of the ovule is called (5) the **nucellus**, and it is surrounded by *two* coats termed (6) the **integuments**. But in many families with gamopetalous corolla there is only *one*. In

FIG. 133.
An anatropous ovule in longitudinal section.

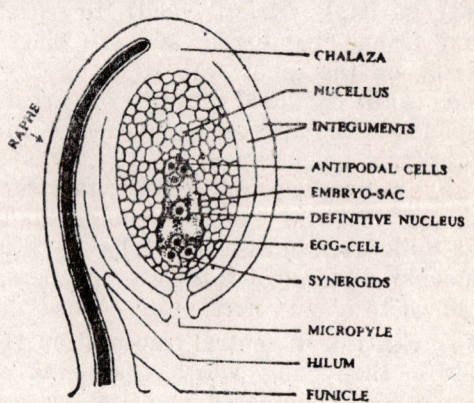

parasites like sandalwood (*Santalum*) and *Loranthus* there is no integument. A small opening is left at the apex of the integuments; this is called (7) the **micropyle**. Lastly, there is a large, oval cell lying embedded in the nucellus towards the micropylar end; this is (8) the **embryo-sac**, that is, the sac that bears the embryo, and is the most important part of the ovule.

Development and Structure of the Embryo-sac. The embryo-sac, as shown in FIG. 133, develops in the following way (FIG. 134).

The ovule at first arises as a tiny protuberance from the placenta in the cavity of the ovary (A). In it even at a very early stage a cell, i.e. the mother cell of the embryo-sac, becomes evident in the nucellus (B). This mother cell increases in size and *divides twice*

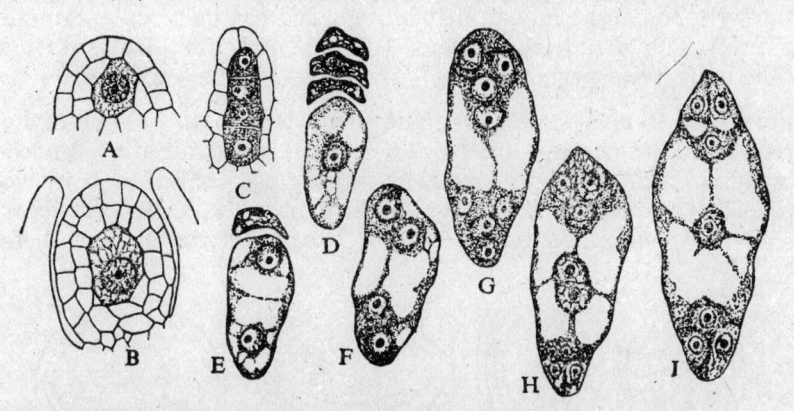

FIG. 134. Development of the embryo-sac. *A-H*, are stages in its development; *I*, fully developed embryo-sac.

to form a row of four megaspores, known as the linear tetrad (C). It is to be noted that of the two successive divisions of the mother cell the first one only is the reduction division (see FIG. II/22). Of the four cells, so formed, each with half (n) the usual number ($2n$) of chromosomes, the three upper ones degenerate and appear as dark caps, while the lowest one functions (D). The nucleus of this cell divides and the two daughter nuclei move to the two poles (E). These again divide so that the number is increased to four (F). Each of these nuclei divides again so that altogether eight nuclei are formed in the embryo-sac, four at each end (G). The embryo-sac increases in size. Then one nucleus from each end or pole passes inwards, and the two polar nuclei fuse together somewhere in the middle, forming the **definitive nucleus** (H). The remaining three nuclei at the micropylar end, each surrounded by a very thin wall, form the **egg-apparatus**, and the other three at the opposite or chalazal end, lying in a group or sometimes in a row, often surrounded by very thin walls, form the **antipodal cells** (I).

Of the three cells constituting (1) the **egg-apparatus**, one is the female gamete known as (*a*) the **egg-cell** (or **ovum** or **oosphere**), and the other two are known as (*b*) the **synergids**. Synergids are more or less pear-shaped, and the egg-cell which is enlarged lies below them. The egg-cell on fertilization gives rise to the embryo; while the synergids are ephemeral structures, getting disorganized soon after fertilization or sometimes even before or during the process. The **antipodal cells** appear to have no definite function; so sooner or later they also get disorganized. They may, however, be nutritive in function. They possibly represent vegetative cells of the extremely reduced female gametophyte. (In angiosperms the fully formed embryo-sac with the eight nuclei in it is regarded as the female gametophyte.) The **definitive nucleus** (FIG. 133) on fertilization (now called the **endosperm nucleus**) gives rise to the endosperm.

Forms of Ovules (FIG. 135). The ovule is said to be (1) **orthotropous** (*ortho,* straight; *tropos,* a turn) or **straight**, when the ovule is erect or straight so that the funicle, chalaza and micropyle lie on one and the same vertical line, as in *Polygonaceae,* e.g. *Polygonum,* sorrel (*Rumex*), etc., *Piperaceae,* e.g. betel

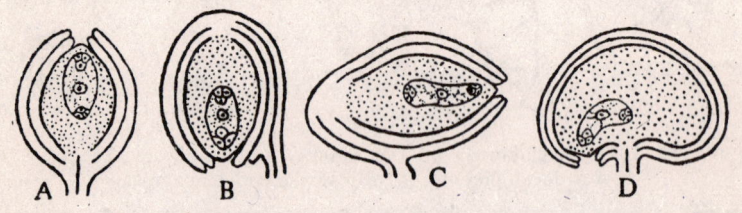

FIG. 135. Forms of Ovules. *A,* anatropous; *B,* orthotropous; *C,* amphitropous; *D,* campylotropous.

(*Piper betle*), long pepper (*Piper longum*), black pepper (*Piper nigrum*), and *Casuarinaceae,* e.g. beef-wood tree (*Casuarina*); (2) **anatropous** (*ana,* backwards or up), or **inverted**, when the ovule bends along the funicle so that the micropyle lies close to the hilum; the micropyle and the chalaza, but not the funicle, lie on the same straight line; this is the commonest form of ovule; (3) **amphitropous** (*amphi,* on both sides) or **transverse**, when the ovule is placed transversely at a right angle to its stalk or funicle, as in duckweed (*Lemna*); and (4) **campylotropous** (*kampylos,* curved) or **curved**, when the transverse ovule is bent round like a horse-shoe so that the micropyle and the chalaza do not lie on the same straight line, as in several members of the following families, viz. *Capparidaceae,* e.g. caper (*Capparis*), *Cruciferae,* e.g. mustard (*Brassica*), *Caryophyllaceae,* e.g. *Polycarpon, Chenopodiaceae,* e.g., *Beta, Chenopodium,* and also in some members of *Gramineae.*

Position of the Ovule within the Ovary. An ovule may be (1) **ascending**, i.e. directed upwards, as in sunflower family or *Compositae* and basil family or *Labiatae*; (2) **pendulous**, i.e. turned downwards from the apex, as in *Euphorbia, Anemone,* coriander, anise, rose, etc.; (3) **suspended**, i.e. turned obliquely downwards from the side; and (4) **horizontal**, i.e. turned horizontally inwards from the side.

Chapter 7 POLLINATION

In 1694 Camerarius, as a result of his experimental work on mulberry, maize, castor, etc., carried out in Tubingen, established for the first time the fact that pollination is essential for the production of the seed, and thus proved the existence of sexuality in plants. He distinguished the stamen with the pollen grains as the male organ and the pistil with the style and the ovary as the female organ. But his work was not widely known, and no further advance was made for several years. Koelreuter from 1764 to 1806 in Berlin, Leipzig and other places pollinated as many as 310 flowers of many different species. He is regarded as the first scientific hybridizer. He realized the importance of insects and wind as pollinating agencies. He made also some plant hybrids. The actual process (fertilization) leading to seed-production, however, remained unravelled. But he made a very interesting observation. By microscopic examination of the pollen grains on the stigma he noticed that something—an oily substance—escaped from the grains and this, mixed with the oil secreted by stigma, worked down into the style and entered into the ovary and there produced an embryo. Konrad Sprengel published in 1793 an account of his observations on many common wild flowers and made it clear for the first time that various adaptations of flowers are meant to achieve cross-pollination by means of insects. He further concluded from his observations on dichogamy and dicliny that nature does not intend that flowers should be self-pollinated. Further elaborate experimental work on pollination should also be noted: Charles Darwin's *The Effect of Cross and Self Pollination in the Vegetable Kingdom*; Hermann Muller's *The Fertilisation of Flowers*; Paul Knuth's *Handbook of Flower Pollination*; and Anton Kerner's *The Natural History of Plants*.

Pollination is the transference of pollen grains from the anther of a flower to the stigma of the same flower or of another flower of the same or sometimes allied species, often through various agencies such as wind, insects, etc. Pollination is of two kinds, viz. (1) **self-pollination** or **autogamy** (*autos,* self; *gamos,* marriage) and (2) **cross-pollination** or **allogamy** (*allos,* different). Pollination taking place within a single flower (evidently bisexual) or between two flowers borne by the same parent plant is self-pollination; the latter method is otherwise called **geitonogamy** (*geiton,* neighbour), i.e. pollination between neighbouring flowers of the same plant. It will be noted that in self-pollination only one parent plant is concerned to produce the offspring. On the other hand pollination taking place between two flowers (bisexual or unisexual) borne by two separate plants of the same or allied species is cross-pollination; this is otherwise called **xenogamy** (*xenos,* stranger). In cross-pollination two parent plants are concerned and, therefore, a mingling of two sets of parental characters takes place, and this results in healthier offspring (see p. 105). Both the methods are, however, widespread in nature.

A. SELF-POLLINATION OR AUTOGAMY

The following adaptations are commonly met with in flowers to achieve self-pollination.

1. **Homogamy** (*homos*, the same). This is the condition in which the anthers and the stigmas of a *bisexual* flower mature at the same time. Under this condition some of the pollen grains may reach the stigma of the same flower through the agency of insects or wind or by the sudden bursting of the anther, thus effecting self-pollination.

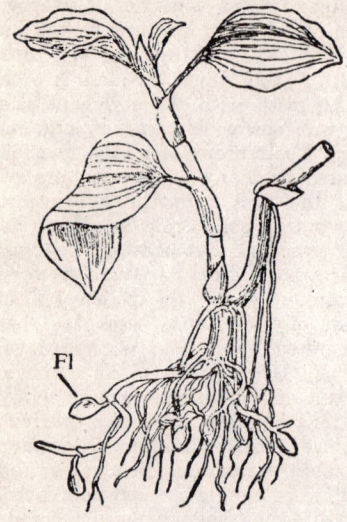

FIG. 136. *Commelina bengalensis*. *Fl*, underground flower.

2. **Cleistogamy** (*kleistos*, closed). There are many bisexual flowers which never open. They are called cleistogamous or closed flowers, and self-pollination is the rule in them. Cleistogamy is seen in the underground flowers of *Commelina bengalensis* (FIG. 136), and also in some species of pansy (*Viola*), balsam (*Impatiens*), sundew (*Drosera*), wood-sorrel (*Oxalis*), sage (*Salvia*), *Juncus*, *Cardamine*, *Stellaria*, *Parochetus* (common in Darjeeling), etc.

B. CROSS-POLLINATION OR ALLOGAMY

Cross-pollination is brought about by external agents such as insects (bees, flies, moths, etc.), animals (birds, snails, etc.), wind and water. Cross-pollination is the rule in unisexual flowers, while in bisexual flowers it is of general occurrence. Nature favours cross-pollination and, therefore, adaptations in flowers to achieve it through external agents are many and varied.

1. **Entomophily** (*entomon*, an insect; *philein*, to love). Pollination by insects is of very general occurrence among flowers. Entomophilous or insect-loving flowers have various adaptations by which they attract insects and use them as conveyors of pollen grains for the purpose of pollination. Principal adaptations are **colour**, **nectar** and **scent**.

    Colour. One of the most important adaptations is the **colour** of the petals. In this respect the brighter the colour and the more irregular the shape of the flower the greater is the attraction. Sometimes, when the flowers themselves are not conspicuous, other parts may become coloured and showy to attract insects. Thus in *Mussaenda* (see FIG. 103) one of the sepals is modified into a large white or coloured leafy structure which serves as an 'advertisement' flag to attract insects. In some cases bracts become highly coloured and attractive, as in glory of the garden (*Bougainvillea*), poinsettia (*Poinsettia pulcherrima*), etc. Spathes are often brightly coloured, as in bananas and aroids.

Nectar. Another important adaptation is the **nectar**. Nearly all flowers with gamopetalous corolla secrete nectar which is a positive attraction to the cleverer insects like bees. Nectar is contained in a special gland, called nectary, and sometimes in a special structure, called the spur (see p. 82). The nectary occurs at the base of one of the floral whorls, and as the bees which are very active pollinating agents collect the nectar from the nectary or the spur, they incidentally bring about pollination.

Scent. The third adaptation is the **scent**. Most of the nocturnal flowers are insect-loving and they emit at night a sweet scent which attracts insects from a distance. At night when the colour fails, the scent is particularly useful in directing the insects to the flowers. Thus nocturnal flowers are mostly sweet-smelling. Common examples are night jasmine (*Nyctanthes*), queen of the night (*Cestrum*), jasmines, Rangoon creeper (*Quisqualis*), etc. Sometimes the smell that is offensive and nauseating to human beings is immensely liked by certain small insects. Thus the appendix of the mature inflorescence of *Amorphophallus* (B. OL; H. ZAMIKAND; FIG. 137) and *Arum* emits a stinking smell, more offensive than that of putrid meat; this always attracts a swarm of carrion-flies, and pollination is achieved through them.

The pollen grains of entomophilous flowers are either sticky or provided with spinous outgrowths. The stigma is also sticky. Pollen grains and nectar sometimes afford excellent food for certain insects.

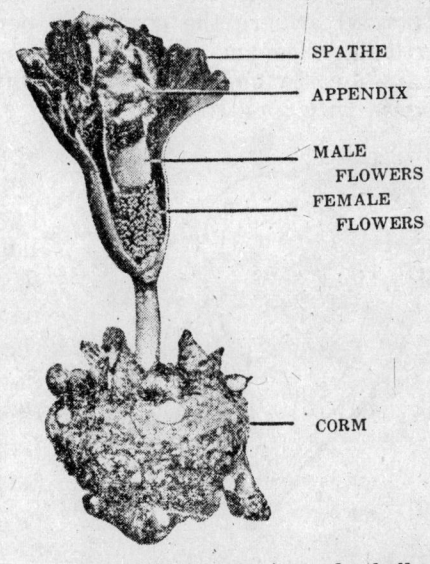

FIG. 137. Spadix of *Amorphophallus companulatus* (B. OL; H. ZAMIKAND).

Sometimes insects visit the flowers in search of shelter from sun and rain, and incidentally bring about pollination.

Pollination in Dagger Plant (*Yucca*). In *Yucca* (see FIG. 81) the adaptation of the flower and its dependence on a special moth *Pronuba* to achieve cross-pollination through it may be noted. The white egg-shaped flowers borne on a large panicle open fully at night and become scented. The stigma protrudes far beyond the anthers and thus self-pollination is prevented. The female moth becomes active at night. It visits a flower and collects a load of pollen grains, pressing them into a sort of ball. It then visits a second flower, carrying the load on its body. With its long ovipositor it bores a number of holes in the wall of the ovary, each hole usually close to an ovule and deposits eggs there, often singly. At the end of oviposition the moth goes to the top of the stigma. Here it deposits the ball of pollen grains and presses it into stigmatic depression with its tongue. Thus cross-pollination is effected. After fertilizaton the larvae and the seeds begin to develop simultaneously, some of the ovules acting as food for the larvae. As the larvae grow into insects they bore their way out through the wall of the fruit. A few seeds drop through these holes but the majority of the seeds are liberated only after the dehiscence of the fruit.

Special Adaptations. In sunflower, marigold, *Cosmos, Anthocephalus* (B. & H. KADAM), gum tree (*Acacia*), etc., where the individual flowers are small and inconspicuous, they are massed together into a dense inflorescence which evidently becomes much more showy and attractive. Dense inflorescence has another advantage; flowers being close together have every chance of being pollinated (see p. 68).

In snapdragon (*Antirrhinum*; see FIG. 109B) and toad-flax (*Linaria*), where the corolla is personate, only the insects of particular size and weight can open the mouth of the corolla. Again long-tongued insects are only useful in the case of the flowers with long corolla-tube.

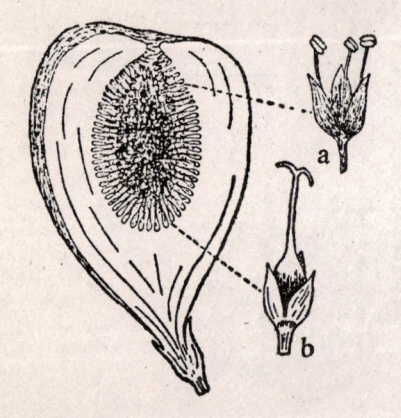

FIG. 138. Fig (*Ficus*) cut lengthwise. Note the apical pore guarded by scales. *a*, male flower; *b*, female flower.

In *Ficus* (e.g. banyan, peepul, fig, etc.) the insects enter the hollowed out chamber of the fleshy receptacle through its narrow apical opening and as they crawl over the unisexual flowers inside the chamber they bring about pollination (FIG. 138). Female flowers lie at the base of the cavity and open earlier, while male flowers lie near the apical opening and open later so that pollen grains have to be brought over from another inflorescence.

POLLINATION

Pollination in *Compositae* Although there is a vast assemblage of forms in *Compositae* the mechanism of cross-pollination is more or less uniform in most cases. Honey (nectar) is secreted around the base of the style within a ring-shaped nectary, adequately protected by the corolla-tube. The latter varies in lengths within wide limits, allowing only the right types of insects to get to the nectary either bodily or by its proboscis. When the flower is still young, the style with the two closed stigmas just projects beyond the anther-tube. Later, as the anthers mature they discharge their pollen grains into the anther-tube. Now as the hairy style elongates it sweeps the pollen grains and pushes them upwards to the upper end of the anther-tube and deposits them there. At this stage the insect visitors get dusted with pollen grains all over their body. They fly or crawl over to other flowers with these pollen grains. The style further elongates and the two, often large, stigmas open and spread out over the flower, with their upper receptive surfaces exposed. From this vantage position the stigmas receive the pollen grains from the visiting insects. Thus, the flowers of a head become easily and quickly cross-pollinated. In default, the stigmas roll back and receive the pollen grains from the same or adjoining flowers. This ensures pollination, self- or cross-, and setting of seeds. It will further be noted that in *Centaurea*, e.g. sweet sultan (*C. moschata*) the stamens are irritable. Coming in contact with the pollinating insects (or any hard object) they contract and twist, forcing out the pollen grains to the top of the anther-tube.

Pollination in Sage (*Salvia*). A very interesting case of cross-pollination by insects is seen in this plant (FIG. 139). There are two stamens in the flower, with the two anther lobes of each widely separated by the elongated curved connective which plays freely on the filament. The upper lobe is fertile and the lower one sterile. In the natural position the connective is upright. When the insect enters the tube of the corolla it pushes the lower sterile anther lobe of each stamen; the connective swings round with the result that the upper fertile lobe comes down and strikes the back of the insect and dusts it with pollen grains. The flower is protandrous, and later when the stigma matures it bends down and touches the back of another insect-visitor coming with pollen grains from another flower. Thus pollination is brought about. This is a special mechanism for cross-pollination.

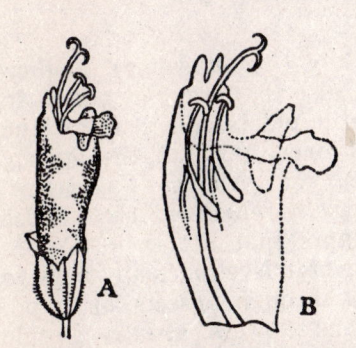

FIG. 139. Sage (*Salvia*). *A*, entire flower; *B*, showing elongated connective.

2. Anemophily (*anemos*, wind).
In some cases pollination is brought about by wind. Anemophilous or wind-loving flowers are small and inconspicuous. They are never coloured or showy. They do not emit any smell nor do they secrete any nectar. The anthers produce an immense quantity of pollen grains, wastage during transit from one flower to another being considerable. They are also light and dry; sometimes, as in pine, they are pro-

vided with wings for facility of distribution by wind. Stigmas are comparatively large and protruding, sometimes branched and often feathery. Examples are afforded by grasses, bamboos, cereals, millets, sugarcanes, sedges, pines, and several palms. Anemophily is well illustrated by maize or Indian corn plant (*Zea mays*).

Pollination in Maize (FIG. 140). The plant bears a large number of male flowers (spikelets) in a terminal panicle, and lower down it bears 1 or 2 female spadices, with a tuft of fine, long and silky threads—the styles—hanging from them. When the anthers burst a cloud of dust-like pollen grains is seen floating in the air close round the plant. Some of them are caught by the protruding stigmas, and thus pollination is effected. By far the greater majority of pollen grains are, however, blown away and wasted.

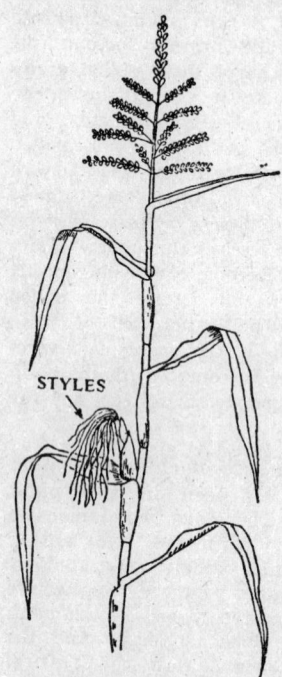

FIG. 140. Anemophily in maize plant. Male flowers in a panicle (on the top) and female flowers in a spadix (at the bottom).

3. **Hydrophily** (*hydor*, water). Pollination may also be brought about in some aquatic plants, particularly the submerged ones, through the agency of water, e.g. *Naias, Vallisneria, Hydrilla, Ceratophyllum*, etc. Those aquatic plants that lift their flowers above the water level normally achieve pollination through insects or wind. Hydrophilous flowers are as a rule small and inconspicuous.

Pollination in *Vallisneria* (FIG. 141). The mode of pollination in it is as follows. *Vallisneria* is dioecious and submerged. The male plant bears a large number of very minute male flowers in a small spadix surrounded by a spathe and borne on a short stalk, while the female plant bears solitary female flowers, each on a long slender stalk. The stalk quickly elongates and lifts the female flower to the surface of the water. The spathe bursts and the male flowers are released from the spadix, while still closed, and float on the surface of the water; the perianth expands giving buoyancy to them. Some of the floating male flowers come in contact with the female flowers. The anthers dehisce, and some of the sticky pollen grains adhere to the margins and surfaces of the trifid stigmas which then close up. After pollination the stalk of the female flower becomes spirally coiled and thus pulls the female flower down into the water. The fruit develops and matures under water a little above the bottom.

POLLINATION

4. Zoophily (*zoon,* animal). Birds, squirrels, bats, snails, etc., also act as useful agents of pollination; for example, birds and sometimes squirrels bring about pollination in coral tree (*Erythrina*), silk cotton tree (*Bombax*), rose-apple (*Syzygium*), *Bignonia*,

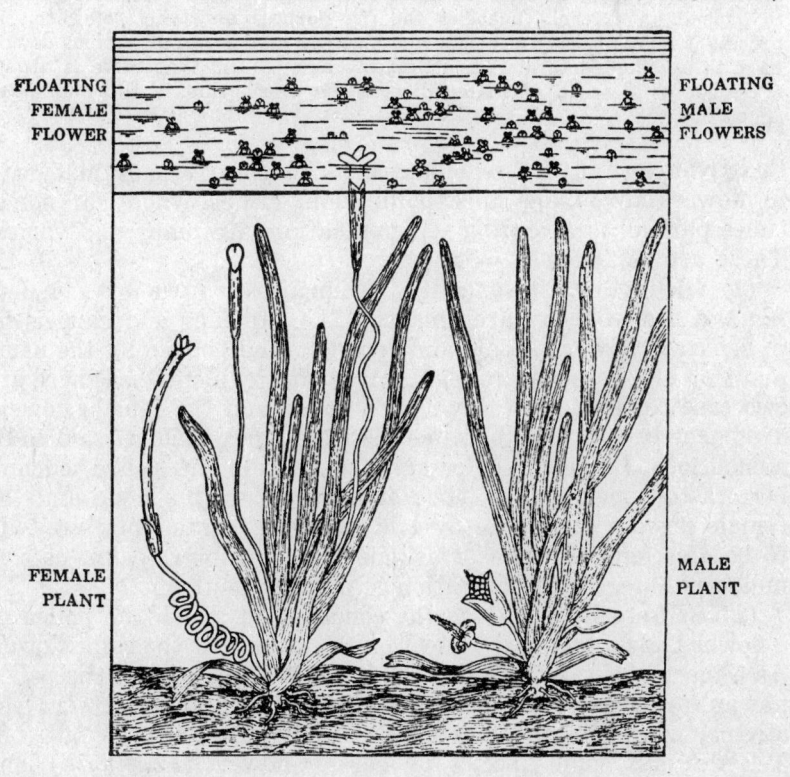

FIG. 141. Hydrophily in *Vallisneria*. Female plant with a floating flower, a submerged flower (-bud) and a fruit (15 cm. long) maturing under water. Male plant with three spadices—young (covered by spathe), mature (with the spathe bursting) and old (after the escape of the male flowers). Male flowers are now seen floating on water.

etc.; bats in *Anthocephalus* (B. & H. KADAM); and snails in certain large varieties of aroids and in snake plant (*Arisaema* —see FIG. 83). (Insects are also instrumental in bringing about pollination in them).

Advantages and Disadvantages of Self- and Cross-Pollinations. Self-pollination has this advantage that it is almost certain in a bisexual flower provided that both its stamens and carpels have matured at the same time. Continued self-pollination generation after generation has, however, this disadvantage that it results in weaker progeny. The advantages of cross-pollination are many, as first shown by Charles Darwin in 1876: (a) it always results in much healthier offspring which are better adapted to the struggle for existence; (b) more abundant and viable seeds are produced by this method; (c) germinating capacity is much better; (d) new varieties may also be produced by the method of cross-pollination; and (e) the adaptability of the plants to their environment is better by this method. The disadvantages of cross-pollination are that the plants have to depend on external agencies for the purpose and, this being so, the process is more or less precarious and also less economical as various devices have to be adopted to attract pollinating agents, and that there is always a considerable waste of material (pollen) when wind is the pollinating agent.

Contrivances for Cross-Pollination. The contrivances met with in flowers favouring cross-pollination, either wholly or sometimes partially preventing self-pollination, are many and varied. These are as follows:

(1) **Dicliny or Unisexuality.** Diclinous (*di*, two or asunder; *kline*, a bed) flowers are unisexual, i.e. stamens and carpels lie in separate flowers—male and female, either borne by the same plant or by two separate plants. It is thus evident that there are two cases of dicliny: (*a*) when the male and the female flowers are borne by one and the same plant, it (the plant) is said to be **monoecious** (*monos*, single; *oikos*, a house), e.g. gourd, cucumber, castor, maize, jack, etc.; and (*b*) when the male and the female flowers are borne by two separate plants, they are said to be **dioecious** (*di*, two or asunder), e.g. palmyra-palm, papaw, mulberry, etc. Cross-pollination is evident in them.

(2) **Self-sterility.** This is the condition in which the pollen of a flower has no fertilizing effect on the stigma of the same flower. As a matter of fact it is seen, as in some orchids, that the pollen has an injurious effect on the stigma of the same flower; the stigma, when the pollen is applied to it, dries up and falls off. Tea flowers, some species of passion-flower (*Passiflora*) and mallow (*Malva*) are also self-sterile. Pollen applied from another plant of the same or allied species is only effective in such cases. Cross-pollination is thus the only method in them for the setting of seeds.

(3) **Dichogamy** (*dicha*, in two). In many bisexual flowers the anther and the stigma often mature at different times. This condition is known as **dichogamy.** As the anther and the stigma come to maturity at different times, dichogamy often stands as a barrier to self-pollination. There are two conditions of dichogamy: (*a*) **protogyny** (*protos*, first; *gyne*, female) when the gynoecium matures earlier than the anthers of the same flower; here the stigma receives the pollen grains brought from another flower. Common examples are *Ficus* (fig, banyan, peepul, etc.), four o'clock plant (*Mirabilis*), *Magnolia*, custard-apple (*Annona*), some palms, etc.; and (*b*) **protandry** (*protos*, first; *andros*, male) when the anthers mature (burst and discharge

their pollen) earlier than the stigma of the same flower; here the pollen grains are carried over to the stigma of another flower. Common examples are China rose, cotton, lady's finger, sunflower, marigold, coriander, wood-sorrel (*Oxalis*), rose, etc. Protandry is more commonly found than protogyny.

(4) **Heterostyly** (*heteros*, different). There are some plants which bear flowers of two different forms (FIG. 142). One form

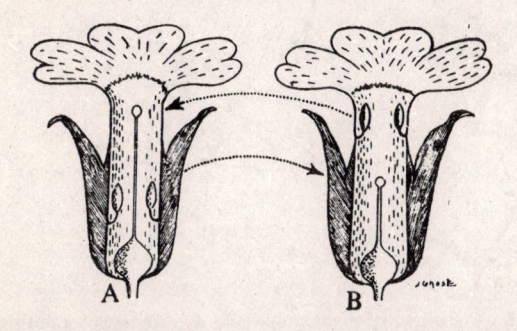

FIG. 142.
Dimorphic heterostyly in primrose.
A, a flower with long style;
B, a flower with short style.

bears long stamens and a short style, and the other form bears short stamens and a long style. This is known as *dimorphic* (*di*, two; *morphe*, form) *heterostyly*. Similarly there may be cases of *trimorphic heterostyly*, that is, stamens and styles of three different lengths borne by three different forms of flowers. In all such cases cross-pollination is effective only when it takes place between stamens and styles of the same length borne by different flowers (*legitimate pollination*). Dimorphic heterostyly is seen in primrose, buckwheat (*Fagopyrum*), wood-sorrel (*Oxalis*), linseed (*Linum*) and *Woodfordia* (B. DHAINPHUL). Trimorphic heterostyly is found in some species of *Oxalis* and *Linum*.

(5) **Herkogamy** (*herkos*, a fence or barrier). In many flowers there are often certain adaptations of the floral parts which act as a fence or barrier to self-pollination and thus favour cross-pollination by insects. The two organs may lie at some distance from each other; the anthers may be inserted within the corolla-tube and the style far exserted, or the anthers far exserted and the style inserted, or the anthers may be facing outwards, or they may be sheltered or hooded by the petals or by the petaloid style, as in *Iris*.

The stamens and the style may also move away from each other so as to keep the anthers and the stigma at a preventable distance. Thus in bleeding heart (*Clerodendrum thomsonae*; FIG. 143A-B)—a garden climber bearing flowers with white calyx and red corolla—the stamens stand erect and the style bends down on one side. After the anthers burst and the pollen grains are removed by insects the filaments roll up in close coils on one side of the flower. The style now stands erect and takes the

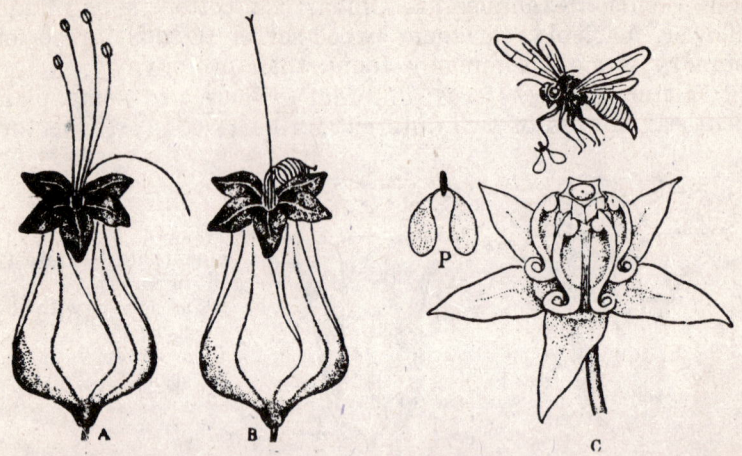

FIG. 143. Pollination in *Clerodendrum thomsonae*. *A*, stamens standing erect; *B*, style standing erect later; *C*, pollination in madar (*Calotropis*); an insect carrying a pair of pollinia from a flower; *P*, a pair of pollinia.

original position of the stamens. In this position the bifid stigma directly receives the pollen grains from the body of the incoming insect.

The relative position of anthers and stigma may be such as to prevent self-pollination. Thus we find that the pollinia of orchids and madar (*Calotropis*) develop in a position whence they are not able to reach the stigma of the same flower by themselves. They also remain fixed in their position by adhesive discs, and can only be carried away by insects, evidently to another flower. The peculiar arrangement of stamens and pistil in sage (*Salvia*) to achieve cross-pollination has been already discussed (see p. 103). Heterostyly (FIG. 142) is also an effective mechanism to achieve the same end. In pansy (*Viola tricolor*) the stigma is guarded by a flap or lid.

Chapter 8 FERTILIZATION

Fertilization is the fusion of two dissimilar sexual reproductive units, called *gametes*. In the 'flowering' plants the process of fertilization (FIG. 144), first discovered by Strasburger in 1884, is as follows. After pollination, that is, after the pollen grains

FERTILIZATION

reach the stigma, the intine of each grows out into a tube, called the **pollen-tube** (see FIG. 116), through some thin or weak spot or *germ pore* in the exine (see FIG. 115). Normally one pollen-tube is formed from each pollen grain but there are cases showing multiple tubes, as in *Malvaceae* and *Cucurbitaceae*. Of the many tubes in them the one with the generative nucleus and the tube-nucleus is only functional. Although the growth of the pollen-tube as far as the micropyle was first observed by Amici in *Yucca* in 1830 the details were worked out much later by Strasburger. The tube, as it grows, penetrates the stigma and pushes its way through the style and the wall of the ovary or alongside it, carrying with it the tube-nucleus and the generative nucleus. The generative nucleus soon divides forming two male gametes, while the tube-nucleus gets disorganized sooner or later. Sometimes, however, the generative nucleus divides even before pollination. As the tube elongates the two male gametes move to the tip of the pollen-tube and lie there in a mass of cytoplasm. The tube then turns towards the micropyle, passes inwards through it, and at length reaches the embryo-sac close to the egg-cell. This phenomenon is called **porogamy** and is the normal method. Sometimes, however, as in *Casuarina* (B. & H. JHAU), walnut (*Juglans regia*) and birch (*Betula*) the pollen-tube enters the embryo-sac through the base (chalaza) of the ovule

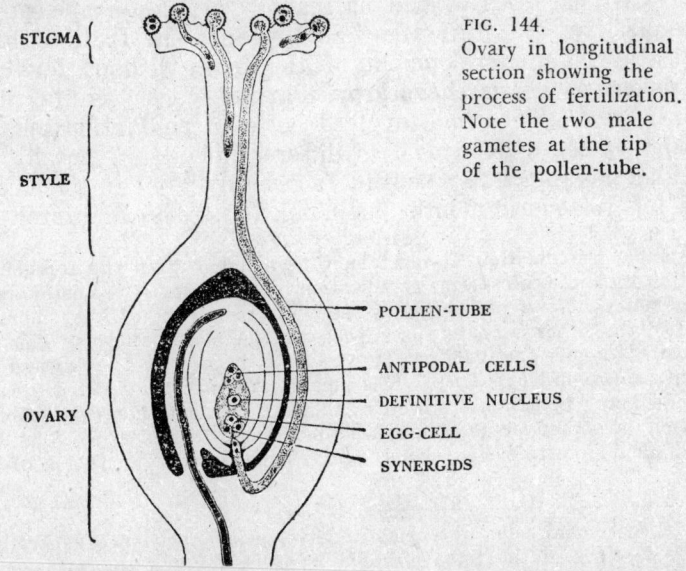

FIG. 144.
Ovary in longitudinal section showing the process of fertilization. Note the two male gametes at the tip of the pollen-tube.

or even pierces the integuments. This phenomenon is called **chalazogamy** and was first discovered by Treub in 1891. The growth of the pollen-tube is no doubt stimulated by proteins and sugars secreted by the stigma and the style but the factor controlling the direction of its movement towards the embryo-sac is yet unknown; it may be due to some sort of chemotactic attraction by the egg-cell. After the pollen-tube penetrates into the embryo-sac the tip of the tube dissolves and the male gametes are set free. Of the two gametes one fuses with the **egg-cell** (first observed by Strasburger in 1884), while the other pushes farther into the embryo-sac and fuses with the two polar nuclei or their fusion product, i.e. the **definitive nucleus** (first observed by Nawaschin in 1898). Thus fertilization (actually double fertilization which is the rule in angiosperms) is completed. The fusion of a male gamete with the two polar nuclei is often termed *triple fusion*. Synergids are ephemeral structures, and do not seem to be essential for the process of fertilization. They get disorganized soon after fertilization, or sometimes even before or during the process. Antipodal cells have no positive function; so they may disappear even before fertilization.[1] After fertilization the egg-cell clothes itself with a cell-wall and becomes known as the **oospore**. The oospore gives rise to the embryo, the ovule to the seed, and the ovary as a whole to the fruit, and the definitive nucleus, now called the **endosperm nucleus**, to the endosperm. If fertilization fails for some reason or other, the ovary simply withers and falls off. In certain cultivated varieties of banana, papaw, orange, grape, apple, pineapple, etc., the ovary may develop into the fruit without fertilization. The development of the fruit without fertilization is spoken of as **parthenocarpy**. Parthenocarpic fruits rarely contain seeds. The time involved between pollination and fertilization varies a good deal in different plants. Generally the time taken is from a few hours to a few days, but in some cases from a few to several months, as in oak (*Quercus*) and birch (*Betula*).

Double Fertilization. It must have been noted from the foregoing description that in angiosperms fertilization occurs twice: (*a*) one of the two male gametes of the pollen-tube fuses with the ovum of the embryo-sac, and (*b*) the other gamete fuses with the definitive nucleus. This process is called double fertilization. It was first discovered by Nawaschin in 1898 in *Lilium* and *Fritillaria*. This amazing discovery attracted great attention and many investigators soon established the fact that this double fertilization is of universal occurrence among the angiosperms. The significance of double fertilization is not clearly understood. The fusion of one of the

---

[1] Antipodal cells, synergids and two polar nuclei are generally regarded as vestiges of the female prothallus cells, the egg-apparatus or the group of four polar nuclei as the vestige of an archegonium (synergids representing its neck cells), and the embryo-sac as a much reduced female prothallus (Strasburger, 1879 and Porsch, 1907). The above explanation is, however, still controversial.

male gametes with the ovum results in the formation of the embryo; while the fusion of the other male gamete with the definitive nucleus (product of fusion of two polar nuclei) results in the formation of the endosperm, and not in the formation of a sister embryo, as expected. It has been suggested that the presence of a second polar nucleus might be a disturbing element, and as such the definitive nucleus (endosperm nucleus) develops, as a result of *triple fusion*, into the endosperm instead of growing into an embryo.

Reduction Division. Pollen grains and embryo-sac are formed by the process of reduction division from their respective mother cells (see FIG. 117 and FIG. 134) and, therefore, the male gamete of the pollen-tube and the female gamete or egg-cell of the embryo-sac contain n chromosomes, i.e. half as many chromosomes as those of the mother cells. Then, when these two gametes fuse together the number of chromosomes becomes doubled in the oospore ($n+n=2n$). This is how a constant number of chromosomes is maintained from generation to generation.

Chapter 9   THE SEED

Development of the Seed. After fertilization a series of changes takes place in the ovule, and as a result the seed is formed. The fertilized egg-cell or ovum grows and gives rise to the embryo, and the definitive nucleus to the endosperm; other changes also take place in the ovule.

1. Development of the Embryo. (*a*) *Dicotyledonous Embryo* (FIG. 145). After fertilization the **egg-cell** or **ovum** secretes a cellulose wall round itself and becomes the **oospore**. The latter divides into two cells—an *upper* (away from the micropyle) and a *lower* (towards the micropyle). The lower one lying towards the micropyle further divides in one direction into a row of cells, called the **suspensor**. The suspensor, as it elongates, pushes the developing embryo deep into the embryo-sac, and it also acts as a feeding organ to the embryo during the formation of the latter. For this purpose the basal cell of the suspensor often enlarges and acts as an absorbing organ. The suspensor, however, becomes disorganized as the radicle is formed. The terminal cell of the suspensor lying next to the embryonal mass is called the **hypophysis cell** (*hypo*, below; *physis*, growth). It divides and gives rise to the apex of the radicle. The upper cell lying away from the micropyle is called the **embryonal cell**. It enlarges and divides by three walls at right angles into eight cells (octants or compartments), four cells lying towards the suspensor forming the posterior octants and the other four cells lying away forming the anterior octants. Each octant then divides by a wall parallel to its curved surface. Thus a surface (superficial) layer of cells and a central mass of cells, the latter being known as the **embryonal mass**, are formed. The surface layer divides in one direction by radial walls only and remains single-layered; finally it gives

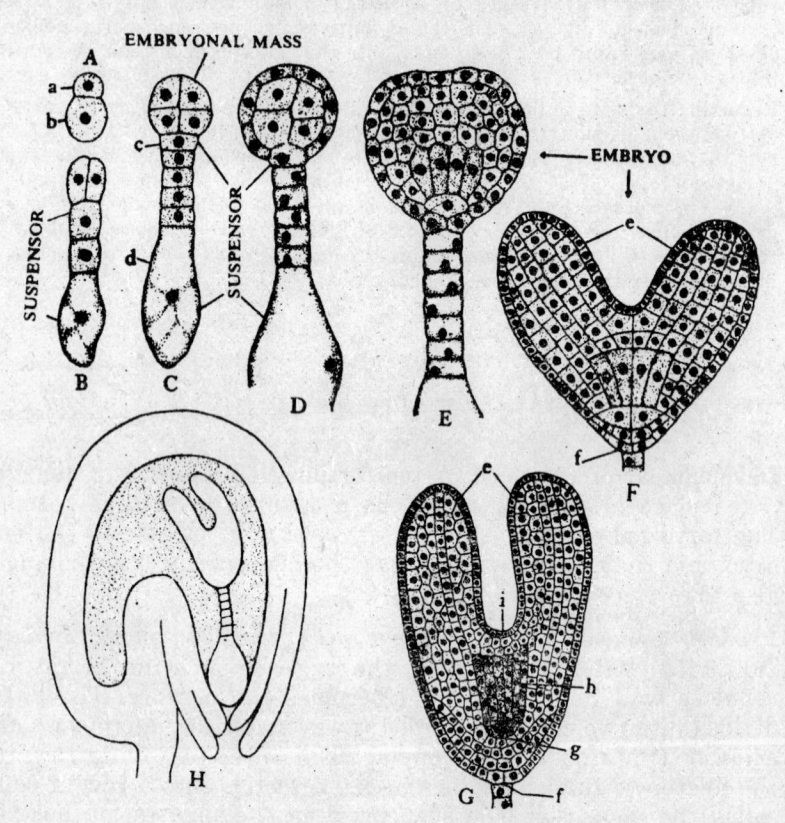

FIG. 145. A-H, development of dicotyledonous embryo. *a*, embryonal cell; *b*, suspensor cell; *c*, hypophysis cell; *d*, basal cell of the suspensor; *e*, cotyledons; *f*, root-cap; *g*, root-tip; *h*, hypocotyl; and *i*, stem-apex; H, embryo within the seed.

rise to the dermatogen, i.e. the outer layer of the stem-apex and the root-apex. The embryonal mass gives rise to the whole of the embryo except the root-tip. The cells of the embryonal mass divide repeatedly and the various parts of the embryo become differentiated. Thus it is seen that the **plumule** and the **two cotyledons** are derived from the anterior octants, and the main part of the **radicle** and the **hypocotyl** from the posterior octants. The apex of the radicle, as already stated, is derived from the hypophysis cell.

(*b*) *Monocotyledonous Embryo* (FIG. 146). Although the monocotyledonous embryo follows the same general trend of development as the dicotyledonous one the former has only *one* cotyledon against the latter's *two*. Besides, in some monocotyledons the cotyledon is terminal and the stem-apex is lateral, although in others the reverse is the case as in dicotyledons. There is also a considerable variation in the details of development of monocotyledonous embryos. A typical case may be as follows. The oospore divides into two cells. The upper one (lying towards the micropyle) enlarges considerably and forms a massive suspensor. The lower cell is the embryonal cell. It soon divides into two. The terminal cell of these two by repeated divisions in different planes gives rise to the single cotyledon. The other cell also divides in the same way and gives rise to the stem-apex, the hypocotyl and the root-tip.

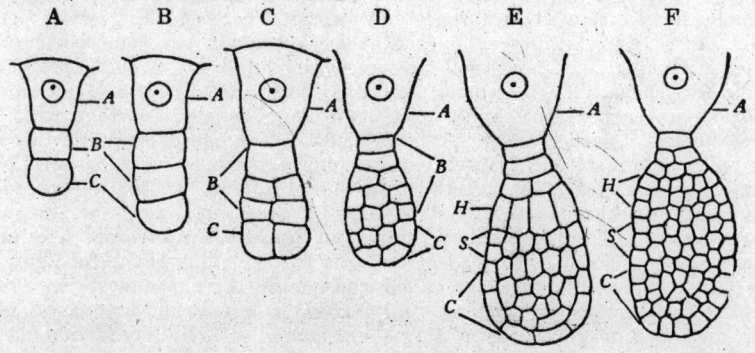

FIG. 146. A-F, development of monocotyledonous embryo. A, 3-celled pro-embryo; *A*, basal cell; *B*, middle cell; *C*, terminal cell; at successive stages (B-F) *A* gives rise to the suspensor; *B* to the stem-apex (*S*), hypocotyl (*H*) and root-tip with some accessory cells above (E-F); and *C* to cotyledon.

2. **Development of the Endosperm.** In all families of 'flowering' plants with the exception of *Orchidaceae* and *Podostemaceae* as a result of triple fusion (see p. 110) the definitive nucleus, now called the endosperm nucleus, tends to grow into a food storage tissue called the **endosperm**, usually triploid ($3n$), sometimes tetraploid ($4n$). In due course the endosperm may be a permanent feature of the seed (as in all endospermic seeds), or it may represent only a temporary phase, i.e. its development does not proceed far, and whatever be the extent of development it becomes completely absorbed by the growing embryo (as in all non-endospermic seeds). Usually the endosperm nucleus begins its division and even a large portion of the endosperm may be formed before the oospore begins to divide. As a general rule two polar nuclei fuse first to form the definitive nucleus; one of the male gametes of the pollen tube then fuses with the definitive nucleus to form the endosperm nucleus. But certain variations have been observed in some cases. Thus (*a*) in *Fritil*-

laria, Tulipa, Nicotiana, Zea, etc., the three nuclei (two polar nuclei and one male gamete) may unite simultaneously, or, (b) the micropylar polar nucleus may unite with the male gamete first, followed by the fusion of the antipodal polar nucleus, as in *Monotropa,* or, (c) the reverse may be the case, as in *Lilium, Adonis,* etc., i.e. the antipodal polar nucleus unites with the male gamete first, or (d) in extreme cases, as in *Limnocharis, Oenothera, Adoxa,* etc., where the antipodal polar nucleus is absent, the other polar nucleus directly fuses with the male gamete.

There are two main types of **endosperm development,** viz. the **nuclear type** and the **cellular type.** There is a third type also, called the **helobial type,** noticed in some cases.

(a) **Nuclear Type.** In this type the endosperm nucleus immediately after double fertilization divides repeatedly without corresponding wall formation. As a result a number of free nuclei appear in the embryo-sac (free nuclear phase). In a large embryo-sac several hundreds of such free nuclei may be formed, while in a small sac divisions are few. Soon a large central vacuole appears in the embryo-sac, and the cytoplasm together with the free nuclei is pushed to the periphery of the sac. Here the nuclei become more or less evenly distributed with an aggregate mass at each pole. Now cell-wall formation starts usually from the periphery, gradually proceeding inwards, i.e. centripetally (see FIG. II/24). Sometimes, however, the wall formation may be synchronous throughout the sac. In any case a cellular mass of tissue laden with food material is formed. This is the food storage tissue of the seed, otherwise called endosperm. As the endosperm grows it fills up the nucellus. During wall formation a few nuclei may often be included within a cell, or a single nucleus of a cell may divide, i.e. a multinucleate stage may appear at an early stage. Later, however, the enclosed nuclei fuse so that finally each cell is left with a single nucleus.

(b) **Cellular Type.** In this type as the endosperm nucleus divides there is corresponding formation of cell-wall round each nucleus. The free nuclear phase, which is characteristic of the previous type, is absent here. Within the type, however, there is some difference in the direction of formation of the first (primary) wall with respect to the embryo-sac. Thus the first wall may be longitudinal, as in *Adoxa,* or it may be transverse, as in many species of *Annonaceae,* or it may be oblique, as in some species of *Boraginaceae.*

(c) **Helobial Type.** In this type after the first division of the endosperm nucleus a transverse wall is formed between the two newly formed nuclei, separating the embryo-sac into two unequal portions. The smaller portion lies towards the antipodal end, while the bigger portion occupies the major area of the sac. In the latter portion, by free nuclear divisions as in the nuclear type, a number of free nuclei appear. Formation of cell-wall then proceeds inwards, finally giving rise to the endosperm. In the smaller portion the nucleus may not divide at all or it divides a few times only. If wall formation takes place here, a small mass of cells may be seen adhering to this end of the endosperm.

Endosperm Haustoria. In several plants special sucking organs or haustoria have been found to develop from the micropylar or chalazal end of the endosperm, or sometimes even from both the ends. *Micropylar haustoria* often appear as unicellular outgrowths from the upper end of the endosperm. They enlarge

THE SEED

and grow through the micropyle, and finally branch and spread over the ovule. The endospermal nuclei migrating into them undergo repeated free nuclear divisions. This type is seen in certain members of *Labiatae*. A massive micropylar haustorium is seen in *Impatiens*. *Chalazal haustoria* develop from the lower end of the endosperm. This end elongates and assumes a tubular form. It penetrates into the nutritive tissue at the base (chalazal end) of the ovule, and acts as a sucking organ or haustorium. This type is found in *Proteaceae*, e.g. *Grevillea* and *Macadamia*. In silver oak (*Grevillea*) the haustorium takes the form of a tubular, twisted and worm-like organ (called vermiform appendage). It invades the nutritive tissue at the chalazal end and acts as a very efficient sucking organ. In nut tree (*Macadamia*) the tubular haustorium is provided with several lobes for the same purpose. Endosperm haustoria formed from both the ends of the endosperm are found in *Scrophulariaceae* and *Acanthaceae*.

3. **Other Changes in the Ovule.** The two integuments develop into two **seed-coats,** of which the outer one is called the *testa* and the inner one the *tegmen*. In some seeds, as in water lily (*Nymphaea*), nutmeg (*Myristica*), wild mangosteen (*Diospyros*; B. GAB; H. KENDU), etc., there is usually found an outgrowth of the funicle, which grows up around the ovule and more or less completely envelops the seed; an outgrowth of this nature is called an **aril**; the bright-red, deeply lobed and aromatic mace (B. JAITRI) of nutmeg (*Myristica fragrans*) is the aril, and so also is the flesh of litchi and *Baccaurea* (B. LATKAN; H. LUTKO). In *Pithecolobium* (B. & H. DEKANI-BABUL) also the aril is fleshy and edible. A small outgrowth formed at the micropyle may also be seen in some seeds, as in castor, balloon vine (*Cardiospermum*), etc.; this is known as the **caruncle** (see FIG. 150). In several cases, as in *Scitamineae*, e.g. banana (*Musa*), ginger (*Zingiber*), *Canna*, etc., in *Nyctaginaceae*, e.g. four o'clock plant (*Mirabilis*), etc., the nucellus of the ovule is not completely used up but it persists in the seed as a thick or thin nutritive tissue like the endosperm, with the embryo lying embedded in it. This nutritive tissue which is in fact a remnant of the nucellus is called the **perisperm**. In some cases, as in castor (see FIG. 150), in *Piperaceae*, e.g. pepper (*Piper nigrum*), in *Nymphaeaceae*, e.g. water lily (*Nymphaea*), etc., both the perisperm and the endosperm are present in the seed.

A. DICOTYLEDONOUS SEEDS

Parts of Exalbuminous Seeds (gram, pea, country bean, gourd, tamarind, mango, sunflower, etc.). **Seed-coats** consist of two layers or integuments, united or free, the outer being called **testa** and the inner **tegmen,** and are provided with **hilum** (representing the point of attachment with the stalk), **micropyle** (a minute opening above the hilum) and **raphe** (a ridge formed by the funicle or stalk in many seeds). **Embryo** lying within consists of an **axis** and two fleshy **cotyledons** laden with food material. The pointed end of the axis is the **radicle** and the feathery or leafy end the **plumule**. As the seed germinates the radicle gives rise to the root and the plumule to the shoot.

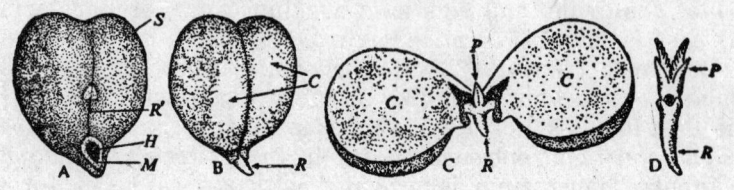

FIG. 147. Gram seed (*Cicer arietenum*). A, entire seed; B, embryo (after removal of the seed-coat); C, embryo with the cotyledons unfolded; and D, axis of embryo. S, seed-coat; R', raphe; H, hilum; M, micropyle; C, cotyledons; R, radicle; and P, plumule.

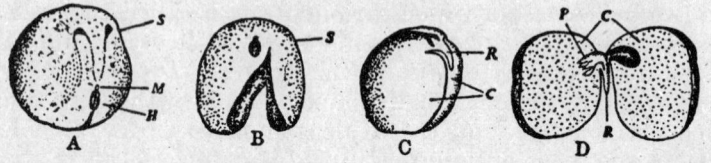

FIG. 148. Pea seed (*Pisum sativum*). A, entire seed; B, seed-coat with hilum and micropyle; C, embryo (after removal of the seed-coat); and D, embryo with the cotyledons unfolded. S, seed-coat—testa (it encloses a thin membranous tegmen); M, micropyle; H, hilum; R, radicle; C, cotyledons; P, plumule.

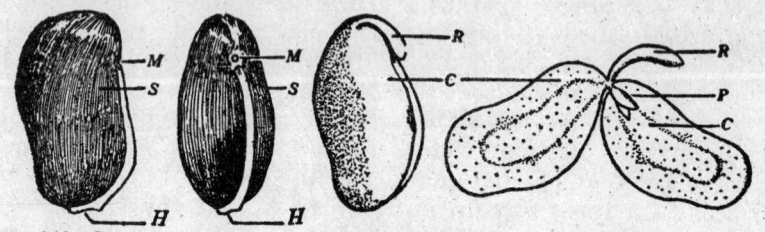

FIG. 149. Country bean seed (*Dolichos lablab*). M, micropyle; S, seed-coat; H, hilum; R, radicle; C, cotyledons; P, plumule.

Parts of Albuminous Seeds (castor, papaw, custard-apple, four o'clock plant, etc.). **Castor Seed. Seed-coat** or **testa** is the outer hard, blackish and mottled shell. The outgrowth formed at the micropyle is the **caruncle**; it absorbs moisture and helps germination. **Hilum** is almost hidden by the caruncle. The **raphe** is prominent. **Perisperm** is the thin white papery membrane surrounding the endosperm. This is a remnant of the nucellus, and is a nutritive tissue (see also p. 115). **Endosperm** is the fleshy food storage tissue, rich in oil, lying immediately within perisperm.

THE SEED

Embryo lies embedded in the endosperm and consists of an **axis** with a distinct **radicle** pointing outwards and an undifferentiated **plumule**, and two thin, flat **cotyledons** with distinct veins.

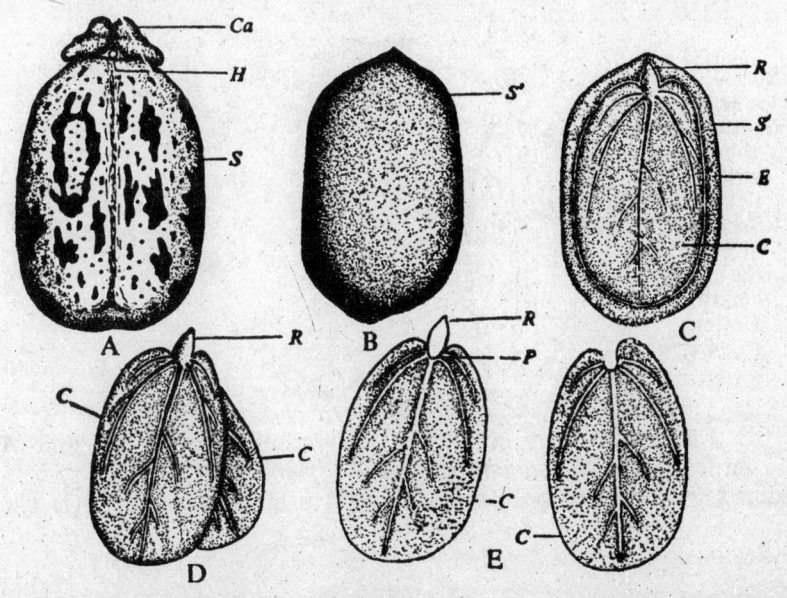

FIG. 150. Castor seed (*Ricinus communis*). *A*, an entire seed; *B*, the seed enclosed by the perisperm; *C*, the same split edgewise showing the embryo lying embedded in the endosperm; *D*, embryo; and *E*, the same with the two cotyledons separated. *Ca*, caruncle; *H*, hilum; *S*, testa; *S'*, perisperm; *R*, radicle; *E*, endosperm; *C*, cotyledon; and *P*, plumule.

B. MONOCOTYLEDONOUS SEEDS

Parts of Albuminous seeds (rice, maize, wheat, onion, palms, etc.).

Rice Grain and Maize Grain. The grain in each case is a small one-seeded fruit called caryopsis. Each grain remains enclosed by a brownish husk which consists of four parts, called *glumes*, arranged in two rows; the two minute ones at the base are *empty glumes*, while the two bigger ones enclose a flower; of these two the outer and slightly bigger one is called the *flowering glume* or *lemma*, while the inner and slightly smaller one, partially enveloped by the former, is called the *palea*. Each grain consists of the following parts:

(1) **Seed-coat** is the brownish membranous layer adherent to the grain. This layer is made up of the seed-coat and the wall of the fruit fused together.

(2) **Endosperm** forms the main bulk of the grain and is the food storage tissue of it, being laden with reserve food material.

particularly starch. In a longitudinal section of the grain it is seen to be distinctly separated from the embryo by a definite layer known as the *epithelium*.

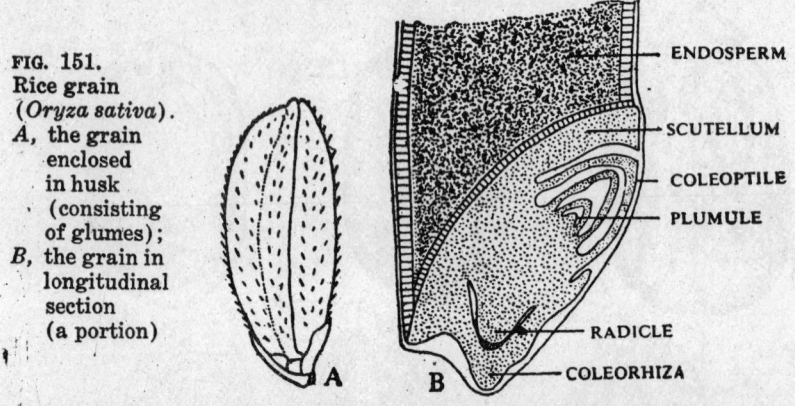

FIG. 151. Rice grain (*Oryza sativa*). A, the grain enclosed in husk (consisting of glumes); B, the grain in longitudinal section (a portion)

(3) **Embryo** is very small and lies in a groove at one end of the endosperm. It consists of only (*a*) *one* shield-shaped cotyledon known as the **scutellum**, and (*b*) a short axis with (i) the

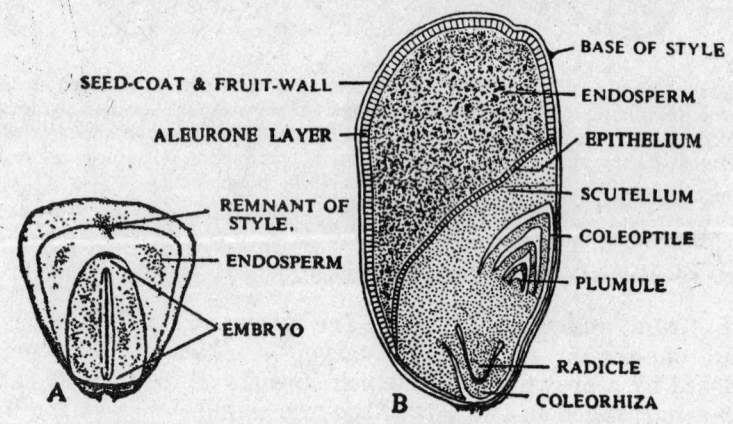

FIG. 152. Maize grain (*Zea mays*). A, the entire grain; B, the grain in longitudinal section.

plumule, and (ii) the **radicle** protected by the **root-cap**. The plumule as a whole (growing point and foliage leaves) is surrounded by a protective sheath called the plumule-sheath or **coleoptile**; similarly the radicle (including the root-cap) is surrounded and protected by a root-sheath called the **coleorhiza**. The surface layer of the scutellum lying in contact with the endosperm is the **epithelium**; its function is to digest and absorb food material stored in the endosperm.

THE SEED

In cereals (e.g. rice, wheat, maize, barley and oat), millets and other plants of the grass family the cotyledon is known as the **scutellum**. It supplies the growing embryo with food material absorbed from the endosperm with the help of the **epithelium**.

Monocotyledonous seeds are mostly albuminous; a few exalbuminous ones are orchids, water plantain (*Alisma*), arrowhead (*Sagittaria*), *Naias*, etc.

Germination

The embryo lies dormant in the seed, but when the latter is supplied with moisture the embryo becomes active and tends to grow and develop into a small seedling. *The process by which the dormant embryo wakes up, grows out of the seed-coat and establishes itself as a seedling is called* **germination**. The embryo grows by absorbing food material stored up in the cotyledons, or in the endosperm when it is present. Two kinds of germination will be noticed: epigeal and hypogeal.

1. **Epigeal Germination** (FIGS. 153-55). In some seeds such as tamarind, cucumber, cotton, gourd, castor, papaw, etc., the coty-

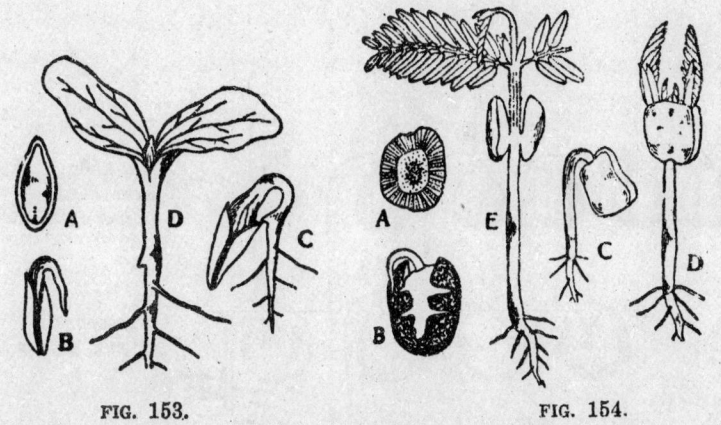

FIG. 153.   FIG. 154.

Epigeal Germination. FIG. 153. Gourd seed (*Cucurbita pepo*). FIG. 154. Tamarind seed (*Tamarindus indica*).

ledons are seen to be pushed upwards by the rapid elongation of the **hypocotyl** (*hypo*, below), i.e. the portion of the axis lying immediately below the cotyledons. Germination of this kind is called **epigeal** or **epigeous** (*epi*, upon; *ge*, earth). In most such cases, as the cotyledons come up above the ground they become flat, green and leaf-like in appearance; while in other such cases, particularly when the cotyledons are very thick, as in tamarind, sword bean, etc., they do not turn leafy, but gradually shrivel up and fall off.

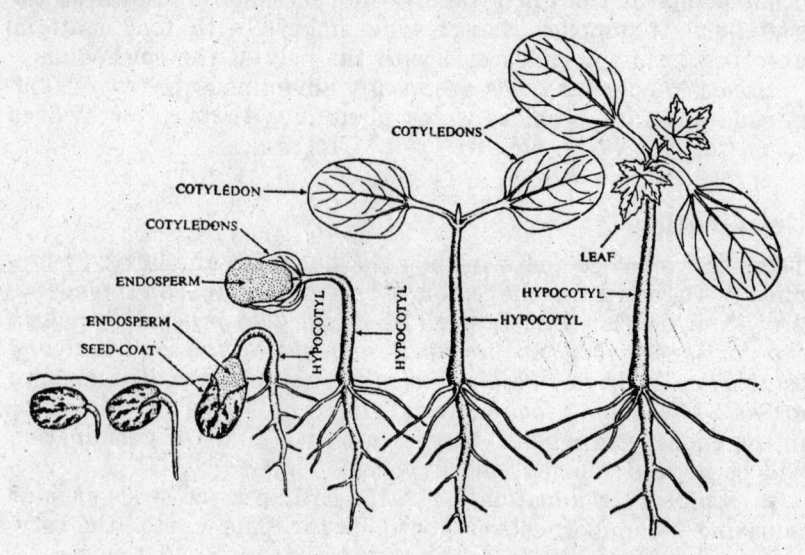

Epigeal Germination. FIG. 155. Castor seed (*Ricinus communis*).

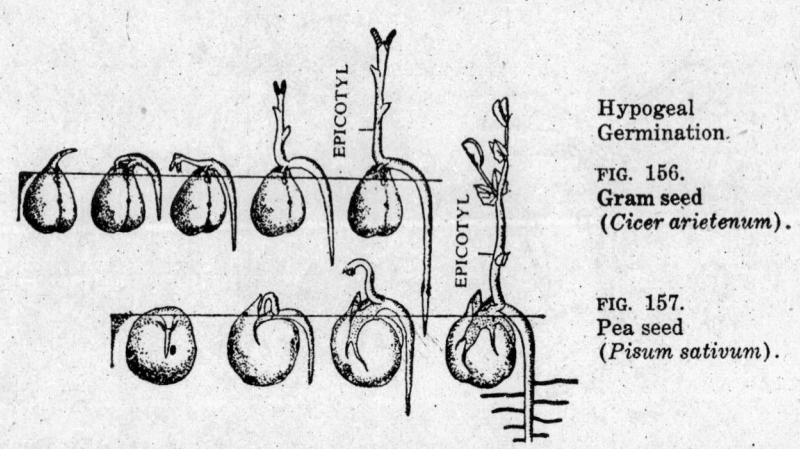

Hypogeal Germination.

FIG. 156. **Gram seed** (*Cicer arietenum*).

FIG. 157. Pea seed (*Pisum sativum*).

2. **Hypogeal Germination** (FIGS. 156-57). In other seeds such as gram, pea, mango, litchi, jack, broad bean (*Vicia*), groundnut, etc., the cotyledons are seen to remain in the soil or just on its surface. In such cases the **epicotyl**, i.e. the portion of the axis lying immediately above the cotyledons, elongates and pushes the plumule upwards. The cotyledons do not turn green, but gradually dry up and fall off. Germination of this kind is called **hypogeal** or **hypogeous** (*hypo*, below; *ge*, earth).

THE SEED

Hypogeal Germination of Monocotyledonous Seeds (FIGS. 158-59).

Monocotyledonous seeds are mostly albuminous and in their ger-

Hypogeal Germination.

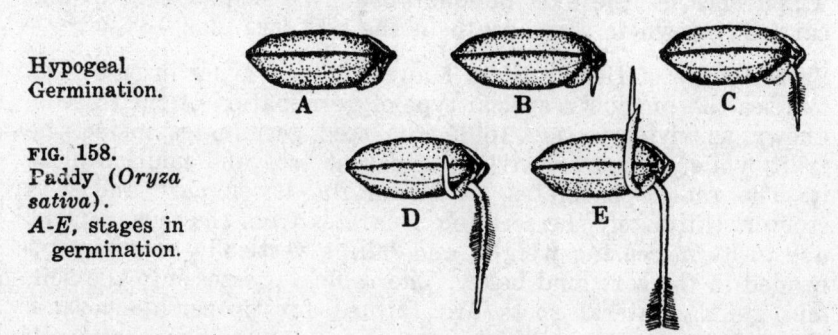

FIG. 158.
Paddy (*Oryza sativa*).
A-E, stages in germination.

mination the cotyledon and endosperm remain buried in the soil; germination is, therefore, hypogeal (except in the case of onion; see FIG. 162). In the germination of monocotyledonous seeds like paddy (unhusked rice) and maize, the cotyledon (or scutellum) absorbs the food material stored up in the endosperm. On germi-

Hypogeal Germination.
FIG. 159.
Maize-grain (*Zea mays*).
Pl, plumule;
Cl, coleoptile;
Cr, coleorhiza.

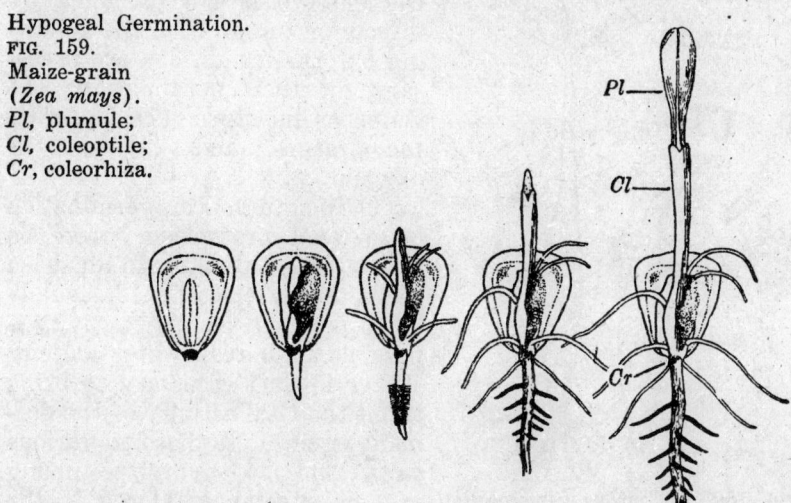

nation the radicle makes its way through the lower short, collar-like end of the sheath called the root-sheath or **coleorhiza**; while the plumule breaks through the upper distinct cylindrical portion of the sheath, called the plumule-sheath or **coleoptile** (FIG. 159). The radicle grows downwards and at first it develops into the primary root. In most cases the primary root soon perishes and a cluster of fibrous roots appears from the base. The plumule grows upwards. The first leaf soon emerges out of the plumule-sheath and others follow in succession. In the germination of many palms, e.g. date-palm and palmyra-palm (but not coconut-palm) a part of the cotyledon extends into a sheath, long or short, which encloses the axis of embryo a little behind the tip and carries it down to some depth in the soil (see FIGS. 163-64).

Special Type of Germination. Many plants growing in salt-lakes and sea-coasts show a special type of germination of their seeds, known as **vivipary** (FIG. 160). The seed germinates inside the fruit, while still attached to the parent tree and nourished by it. The radicle elongates, swells in the lower part and gets stouter. Ultimately the seedling separates from the parent plant due to its increasing weight, and falling vertically becomes embedded in the soft mud below. The radicle presses into the soil, and quickly lateral roots are formed for proper anchorage. Examples are seen in *Rhizophora* (B. KHAMO), *Sonneratia* (B. KEORA), *Heritiera* (B. SUNDRI), etc.

Conditions Necessary for Germination. Dry seeds retain their viability for many months and even years depending on their nature, provided, of course, that the embryo is not damaged by insects or fungi. Then the following external conditions are necessary for their germination: (1) **water** or **moisture**, (2) **moderate temperature**, and (3) **air** or **oxygen**.

(1) **Moisture.** For germination of a seed protoplasm must be saturated with water. In air-dried seeds the water content is usually 10-15 per cent. No vital activity is possible at this low water content. Water is thus necessary to bring about the vital activity of the dormant embryo; to dissolve various salts and to hydrolyse many organic substances stored in the cotyledons or in the endosperm; to facilitate necessary chemical changes; and to help the embryo to come out easily by softening the seed-coat.

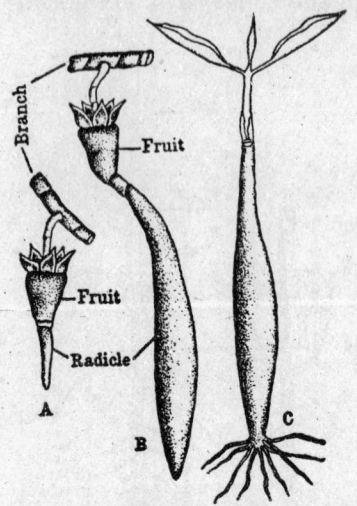

FIG. 160. Viviparous germination. *A-B*, stages in germination; *C*, seedling.

(2) **Temperature.** A suitable temperature is necessary for the germination of a seed. Protoplasm functions normally within a certain range of temperature. Within limits which vary according to the nature of the seed, the higher the temperature the more rapid is the germination.

THE SEED

(3) **Air.** Oxygen is necessary for respiration of a germinating seed. By this process a considerable amount of energy stored in the food material is liberated and made use of by the protoplasm. Respiration in the germinating seed is very vigorous as the active protoplasm requires a constant supply of oxygen, and hence the seed sown deeply in the soil shows very little or no sign of germination.

It may be noted in this connexion that **light** is not an essential condition of germination. In fact seeds germinate more quickly in the dark. Some seeds—tomato, for example—will not germinate unless they are kept in the dark. For subsequent growth, however, light is indispensable. Seedlings that are grown continually in the dark elongate rapidly but become frail, develop no chlorophyll and bear only pale, undeveloped leaves. Such seedlings are said to be *etiolated* (see FIG. 111/43).

Three Bean Experiment (FIG. 161). That all the conditions mentioned above are essential for germination can be shown by a simple experiment, known as the **three bean experiment**. Three air-dried seeds are attached to a piece of wood, one at each end and one in the middle. This is then placed in a beaker, and water is poured into it until the middle seed is half immersed in it. The beaker is then left in a warm place for a few days. From time to time water is added to maintain the original level. It is seen that the middle bean germinates normally because it has sufficient moisture, oxygen and heat. The bottom bean has sufficient moisture and heat, but no oxygen. It may be seen to put out only the radicle, but further development is checked for want of oxygen. The top bean having sufficient oxygen and heat only but no moisture does not show any sign of germination.

FIG. 161. Three bean experiment.

This experiment evidently shows that moisture and oxygen are indispensable for germination; the effect of temperature is only indirectly proved. It can, however, be directly proved in the following way. Other conditions remaining the same, if the temperature be considerably lowered or increased by placing the beaker with seeds in a freezing mixture or in a bath with high constant temperature it will be seen that none of the beans will germinate. Thus suitable temperature is also an essential condition of germination.

ADDITIONAL MONOCOTYLEDONOUS SEEDS

1. **Onion Seed** (FIG. 162). **Structure.** The seed is small, black, roughly semi-circular in shape, flattened on one side and grooved at the narrow end. Its outer black covering is (*a*) the **seed-coat** or **testa**. Cut the seed lengthwise and observe (*b*) the **endosperm** which is the whitish mass within the seed-coat, and (*c*) the **embryo** which is the slender, elongated, colourless, curved body lying embedded in the endosperm. The embryo consists of (*i*) a **single cotyledon** and (*ii*) a **radicle**. The bigger portion of the curved

body looped at the end is the cotyledon, and the narrow end of it directed towards the pointed end of the seed is the radicle. The plumule which is very minute and undifferentiated lies hidden laterally in the region of the

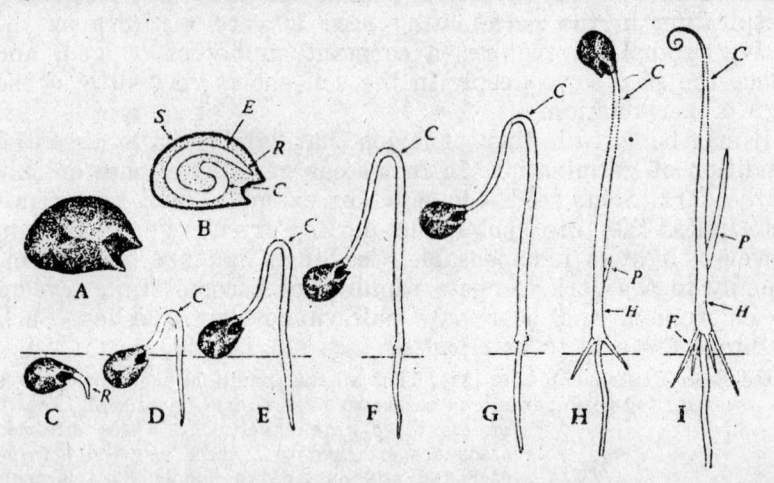

FIG. 162. Onion seed (*Allium cepa*). *A*, an entire seed; *B*, the seed in longitudinal section; *C-I*, stages in germination. *S*, seed-coat; *E*, endosperm; *R*, radicle; *C*, cotyledon; *P*, plumule; *H*, hypocotyl; and *F*, fibrous roots.

very short hypocotyl, and is distinguishable clearly only on the germination of the seed.

Germination. As germination takes place the radicle comes out through the pointed end of the seed and grows downward (C). The cotyledon elongates, emerges out of the seed except for its looped end and forms a distinct arch or loop (D). It turns green in colour and further elongates as a leaf-sheath lifting the seed from the soil (E). The germination is **epigeal**. The end of the cotyledon still coils within the seed and functions as an absorbing organ drawing food from the endosperm and supplying the same to the growing parts. The root elongates further and the cotyledon turns deep green in colour functioning as the first leaf (F-G). The cotyledon grows further and almost straightens out bearing the seed on the top (H). A slight swelling appears at the base of the hypocotyl and a few fibrous roots push out from this region. A little higher up the plumule soon pierces the leaf-sheath and grows upward as a slender body. It turns green and forms the second leaf of the seedling. The coiled end of the cotyledon withers and the seed-coat drops now or a little later (I). By this time the endosperm has already become exhausted.

2-3 Date-palm Seed and Palmyra-palm Seed (FIGS. 163-64). The stony covering of the date-palm seed represents the seed-coat and the endocarp of the fruit; in the palmyra-palm seed the shell is the endocarp, while the inner brownish layer is the seed-coat. In both the seed-coat is adherent to the endosperm. Other parts in both are: a large **endosperm** filling up the cavity of the seed, and a small undifferentiated **embryo** lying embedded in the endosperm on one side.

When germination begins the single **cotyledon** enlarges, and a portion of it breaks through the seed-coat in the form of a **sheath** enclosing the axis **of the embryo** inside it. This **sheath** of the cotyledon elongates and carries

THE SEED

down with it the axis of the embryo. After it has gone into the soil the **radicle** of the axis comes out piercing the root-sheath or **coleorhiza**, grows

FIG. 163.
Date-palm seed
(*Phoenix sylvestris*).
A, seed in section;
B, germinating seed in section;
C, seedling;
S, seed-coat and inner wall of the fruit;
E, endosperm;
Em, embryo (undifferentiated);
C, cotyledon;
Sh, sheath of cotyledon;
Cl, coleoptile;
Cr, coleorhiza.

FIG. 164.
Palmyra-palm seedling
(*Borassus flabellifer*).
M, mesocarp (fibrous);
E, endocarp (stony);
S, endosperm with seed-coat;
C, cotyledon (spongy);
Sh, sheath;
Cl, coleoptile;
Cr, coleorhiza.

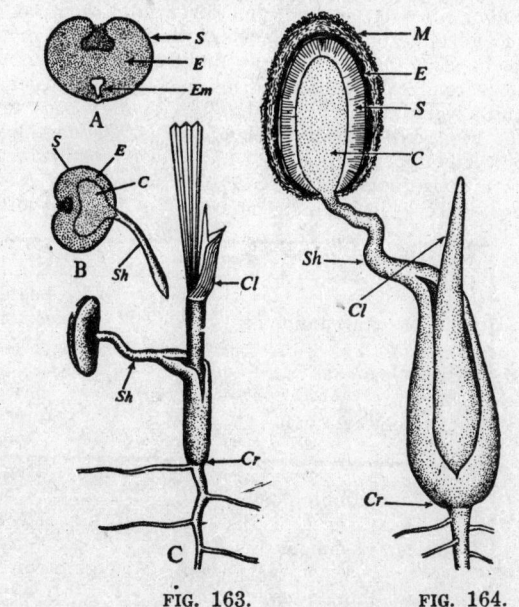

FIG. 163.   FIG. 164.

downwards and produces the root. Eventually the plumule bursts the plumule-sheath or **coleoptile** and grows upwards into the air, forming the shoot.

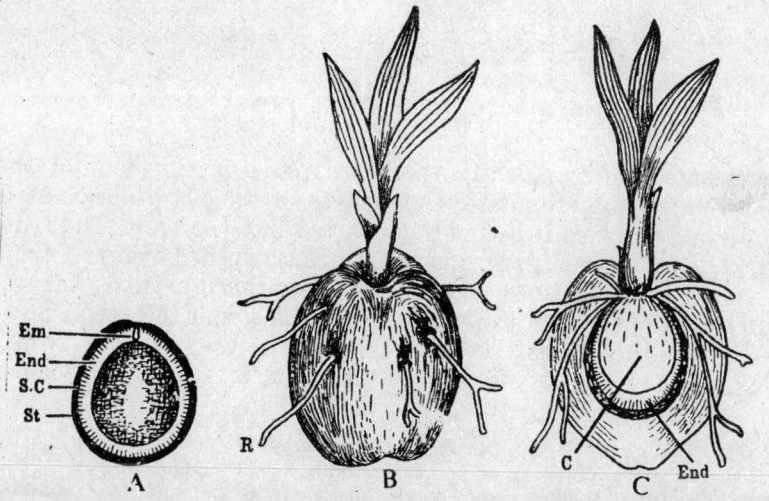

FIG. 165. Coconut-palm seed (*Cocos nucifera*). A, the seed cut lengthwise; B, germinating fruit; C, the same cut lengthwise. Em, embryo; End, endosperm; S.C, seed-coat; St, stone; R, root; C, cotyledon.

4. **Coconut-palm Seed** (FIG. 165). On removing the fibrous coat and breaking open the shell a black covering—a thin layer—adherent to the endosperm is seen; this is the **seed-coat**. The white, thick mass is the **endosperm**. The embryo lies as a small, undifferentiated body at one of the three 'eyes'. In the germination of the coconut-palm seed a regular sheath is not formed, as in other palms, and the undifferentiated embryo germinates *in situ*. Its lower end extends and forms a single **cotyledon** which gradually enlarges and swells into a spherical, white, spongy body and fills the whole cavity of the seed. With the growth of the cotyledon the endosperm is seen to thin out. The upper end of the embryo develops into a small **shoot** with a number of fibrous **roots** produced at its base. These roots pierce the thick fibrous coat of the fruit, and come out in different directions.

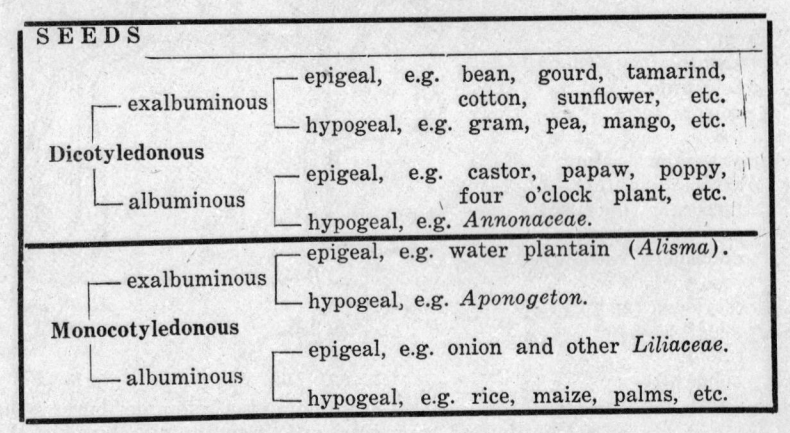

```
SEEDS
                    ┌─ exalbuminous ┌─ epigeal, e.g. bean, gourd, tamarind,
                    │               │            cotton, sunflower, etc.
                    │               └─ hypogeal, e.g. gram, pea, mango, etc.
Dicotyledonous
                    │               ┌─ epigeal, e.g. castor, papaw, poppy,
                    └─ albuminous   │            four o'clock plant, etc.
                                    └─ hypogeal, e.g. Annonaceae.

                    ┌─ exalbuminous ┌─ epigeal, e.g. water plantain (Alisma).
                    │               └─ hypogeal, e.g. Aponogeton.
Monocotyledonous
                    │               ┌─ epigeal, e.g. onion and other Liliaceae.
                    └─ albuminous
                                    └─ hypogeal, e.g. rice, maize, palms, etc.
```

Chapter 10 THE FRUIT

Development of the Fruit. Apart from the development of the seed fertilization stimulates the growth of the ovary also.* As it grows and matures it becomes converted into the fruit. The fruit may, therefore, be regarded as a mature or ripened ovary. A fruit consists of two portions, viz. the **pericarp** (*peri*, around; *karpos*, fruit) developed from the wall of the ovary, and the **seeds** developed from the ovules. The pericarp may be thick or thin; when thick, it may consist of two or three parts: the outer, called

---

*** Auxin Theory of Fruit Development.** It may be noted that auxin (a growth hormone) plays an important role in the development of the fruit from the ovary. It has been found that there is an increase in auxin content in the ovary immediately after pollination, growth of the pollen-tube and fertilization. Lund (1956) actually found that an enzyme secreted by the pollen-tube can convert the amino-acid *tryptophan* into auxin.

THE FRUIT

epicarp, forms the skin of the fruit; the middle, called *mesocarp*, is pulpy in fruits like mango, peach, plum, etc., and the inner, called *endocarp*, is often very thin and membranous, as in orange, or it may be hard and stony, as in many palms, mango, etc. In many cases, however, the pericarp is not differentiated into these three regions.

When only the ovary of the flower grows into the fruit, it is commonly known as the **true fruit**, but often it is found that other floral parts such as the thalamus, receptacle, or calyx may

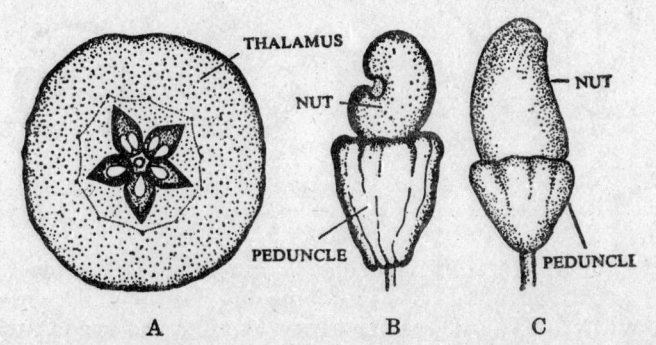

FIG. 166. A, apple (*Malus*) cut transversely; B, cashew-nut (*Anacardium*); C, marking nut (*Semecarpus*).

also grow and form a part of the fruit; such a fruit is known as the **false** or **spurious fruit** or pseudocarp. Thus in *Dillenia* (B. & H. CHALTA) the calyx is persistent and fleshy forming the prominent and the only edible part of the fruit. In apple (FIG. 166A) and pear the thalamus grows round the inferior ovary and becomes fleshy in the fruit. In cashew-nut (*Anacardium*; FIG. 166B) the peduncle and the thalamus grow and become swollen and fleshy forming an edible fruit-like body which is a false fruit or pseudocarp, while the actual fruit which is an edible reniform nut, developing from the ovary, is seated on the swollen peduncle. Similarly in marking nut (*Semecarpus*; FIG. 166C) the peduncle becomes fleshy with the actual nut seated on its top; the nut is not edible but it is used by washermen to mark cotton clothes.

Dehiscence of Fruits (FIG. 167). There are many ways of dehiscence of fruits to liberate their seeds. It may be of the following types. (1) **Transverse** (C), as in cock's comb (*Celosia*), purslane (*Portulaca*), etc. (2) **Porous** (B), i.e. by pores, as in poppy (*Papaver*), bath sponge (*Luffa*), etc. (3) **Valvular**, i.e. bursting partially or completely into pieces called valves. It may be of the following kinds: (*a*) **sutural** (A), i.e. opening by one or both the sutures, as in periwinkle, pea, bean, etc.; (*b*) **loculicidal** (D), i.e.

splitting through the back of the loculus (chamber), as in cotton, lady's finger, *Ruellia, Andrographis,* etc.; (c) **septicidal** (E), i.e. dehiscing through the septa (partition walls), as in linseed

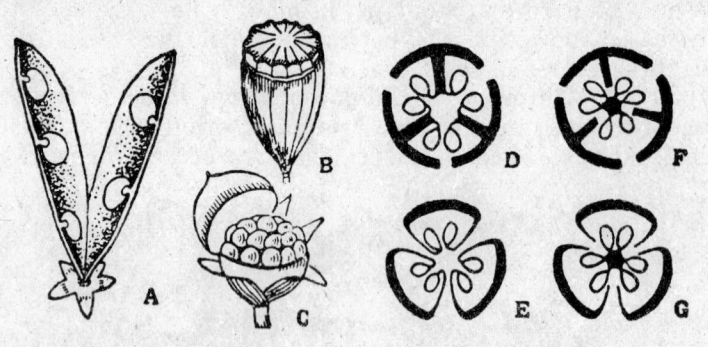

FIG. 167. Dehiscence of fruits. *A*, sutural (pea); *B*, porous (poppy); *C*, transverse (cock's comb); *D*, loculicidal; *E*, septicidal; *F-G*, septifragal.

(*Linum*), devil's cotton (*Abroma*), mustard (*Brassica*), etc., and (d) **septifragral** (F-G), i.e. dehiscing loculicidally, or septicidally, with the valves falling away leaving the seeds attached to the central axis, as in thorn-apple (*Datura*), toon (*Cedrela*), *Pterospermum,* etc.

Classification of Fruits

Fruits, whether true or spurious, may be broadly classified into three groups, viz. **simple, aggregate** and **multiple** or **composite.**

A. Simple Fruits. When only one fruit develops from the single ovary (either of simple pistil or of syncarpous pistil) of a flower with or without accessory parts (true or spurious fruits as explained above), it is said to be a **simple fruit.** A simple fruit may be dry or fleshy. Dry fruits may again be dehiscent or indehiscent or schizocarpic (in which the carpel or carpels split into one-seeded parts).

I. *DEHISCENT OR CAPSULAR FRUITS*

(1) **Legume** or **Pod** (FIG. 168A). This is a dry *monocarpellary* fruit developing from a superior, one-chambered ovary and dehiscing by both the sutures, as in *Papilionaceae,* e.g. pea, bean, pulses, rattlewort (*Crotalaria*; B. ATASHI; H. JHUNJHUNIA), etc.

(2) **Follicle** (FIG. 168B). This also is a dry, *monocarpellary,* superior, one-chambered fruit like the previous one, but it dehisces by one suture only. Simple follicle is rare; it is sometimes

THE FRUIT

seen in madar (*Calotropis*). Most commonly, however, follicles develop in an aggregate of two, three or many fruits (see pp. 133-34).

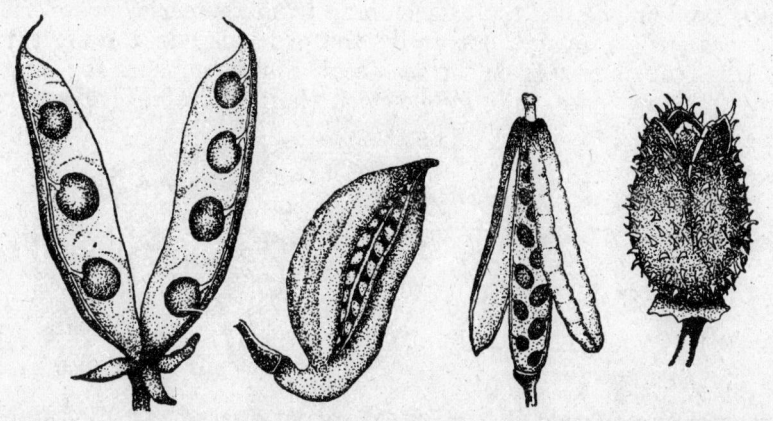

FIG. 168. Fruits. *A*, legume or pod of pea; *B*, follicle of madar (*Calotropis*); *C*, siliqua of mustard; *D*, capsule of thorn-apple (*Datura*).

(3) **Siliqua** (FIG. 168C). This is a long, narrow, many-seeded fruit developing from a superior, *bicarpellary* ovary with two parietal placentae. It dehisces from below upwards along the two ventral sutures into two valves, leaving a two-ribbed wiry framework called the *replum* with seeds attached to it, and a *false septum* across the replum. It may be noted that the ovary is at first one-chambered but after the ovules appear a thin papery longitudinal membrane grows inwards from the placenta at each suture. The two membranes, thus formed, meet and give rise to the false septum, dividing the ovary into two chambers, while the two suture ribs form the framework, i.e. the replum. Siliqua is commonly found in *Cruciferae*. A short, broad and flat siliqua is otherwise called a **silicula**. It is found in some members of Cruciferae, as in *Alyssum*, candytuft (*Iberis*), shepherd's purse (*Capsella*), *Senebiera*, etc.

(4) **Capsule** (FIGS. 168D and 169A). This is a many-seeded, uni- or multilocular fruit developing from a superior (or sometimes inferior), *bi-* or *polycarpellary* ovary, and dehiscing in various ways. All dehiscent fruits developing from a syncarpous ovary are commonly known as capsules, e.g. poppy (*Papaver*), cotton (*Gossypium*), lady's finger (*Abelmoschus*), *Datura*, etc.

II. INDEHISCENT OR ACHENIAL FRUITS

(1) **Caryopsis** (FIG. 169C). This is a small, dry, one-seeded fruit developing from a superior, *monocarpellary* ovary, with

the pericarp fused with the seed-coat. Examples are found in *Gramineae*, e.g. rice, wheat, maize, etc.

(2) **Achene** (FIG. 169B). An achene is a small, dry, one-chambered and one-seeded fruit developing from a superior or inferior *monocarpellary* ovary; but unlike the previous one, the pericarp of this fruit is free from the seed-coat. Simple achenes are found only in a few cases, as in four o'clock plant (*Mirabilis*), hogweed

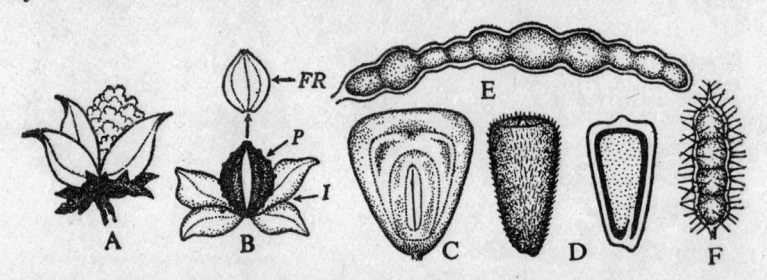

FIG. 169. Fruits (*contd.*). *A*, capsule of cotton; *B*, achene of *Mirabilis* (*I*, involucre; *P*, dry persistent perianth; *FR*, fruit—achene); *C*, caryopsis of maize, *D*, cypsela of sunflower; *E*, lomentum of *Acacia*; *F*, the same of *Mimosa*.

(*Boerhaavia*) and buckwheat (*Fagopyrum*). Most commonly, however, achenes develop in an aggregate (see p. 134).

(3) **Cypsela** (FIG. 169D). This is a dry, one-chambered and one-seeded fruit developing from an inferior, *bicarpellary* ovary with the pericarp and the seed-coat free, as in sunflower family or *Compositae*, e.g. sunflower, marigold, *Cosmos*, etc.

(4) **Samara** (FIG. 170B). This is a dry, indehiscent, one- or two-seeded fruit developing from a superior, *bi-* or *tricarpellary* ovary, with one or more flattened wing-like outgrowths, e.g. *Hiptage* (B. MADHABILATA; H. MADHULATA; FIG. 170B), yam (*Dioscorea*; see FIG. 175B), ash (*Fraxinus*; FIG. 176A), etc. In samara the wings always develop from the pericarp, and the fruit splits into its component parts, each enclosing a seed. In sal tree (*Shorea*; B. & H. SAL; FIG. 170E), *Hopea* (FIG. 170D), wood-oil tree (*Dipterocarpus*; B. & H. GARJAN; see FIG. 177A), etc., the fruit is also a winged one; but here the wings are the dry, persistent sepals. Winged fruit of this nature is known as a **samaroid**.

(5) **Nut.** This is a dry, one-chambered and one-seeded fruit developing from a superior, *bi-* or *polycarpellary* ovary, *with the pericarp hard and woody*, e.g. cashew-nut (*Anacardium*; see FIG. 166B), water chestnut (*Trapa*; see FIG. 14), chestnut (*Castanea*), oak (*Quercus*), beech (*Fagus*), etc.

Coconuts and palmyra-palms are drupes because in them it is the endocarp that becomes hard and woody (and not the whole of the pericarp), and

THE FRUIT

areca- or betel-nuts and date-palms are (one-seeded) berries because in them the pericarp is soft (fibrous in areca-nuts and pulpy in date-palms); it is the seed that is stony (and not the pericarp).

III. SPLITTING OR SCHIZOCARPIC FRUITS

(1) **Lomentum** (FIG. 169E-F). This is a type of dry, indehiscent legume constricted or partitioned between the seeds into a number of one-seeded compartments. The fruit splits transversely along the constrictions or partitions into one-seeded pieces, as in gum tree (*Acacia*), sensitive plant (*Mimosa*), Indian laburnum (*Cassia fistula*), nicker bean (*Entada gigas*; B. & H. GILA), *Desmodium*, e.g. Indian telegraph plant (*Desmodium gyrans*), etc.

(2) **Cremocarp** (FIG. 170A). This is a dry, indehiscent, two-chambered fruit developing from an inferior, bicarpellary ovary. When ripe, the fruit splits apart into indehiscent, one-seeded pieces, called *mericarps*. The mericarps remain attached to the prolonged end (*carpophore*) of the axis. Cremocarp is the characteristic fruit of *Umbelliferae*, e.g. coriander (*Coriandrum*), cumin (*Cuminum*), anise or fennel (*Foeniculum*), carrot (*Daucus*), etc.

(3) **Double Samara** (FIG. 176D). In maple (*Acer*), the fruit develops from a superior, bicarpellary ovary, and when mature it splits into two samaras, each with a wing and a seed. Such a fruit is called a double samara.

(4) **Regma** (FIG. 170F). This is a dry, indehiscent fruit developing from a syncarpous pistil. It splits away from the central

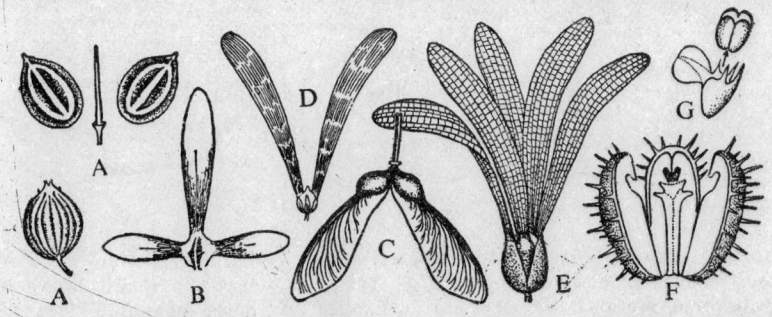

FIG. 170. Fruits (contd.). A, cremocarp of coriander; A¹, the same splitting away from the axis (carpophore) into two mericarps; B, samara of *Hiptage*; C, double samara of maple (*Acer*); D, samaroid of *Hopea*; E, the same of *Shorea*; F, regma of castor; G, carcerule of *Ocimum*.

axis into as many parts, called *cocci*, as there are carpels, each part containing one or two seeds. The seeds are liberated later on the decay of the dry pericarp. Common examples are castor (*Ricinus*), *Euphorbia, Jatropha, Geranium*, etc.

(5) **Carcerule** (FIG. 170G). This is small, dry, indehiscent, four-chambered fruit developing from a superior, bicarpellary pistil. This is the characteristic fruit of *Labiatae*. The fruit remains enclosed by the persistent calyx, and later splits into four chambers, each enclosing a seed or nutlet.

IV. FLESHY FRUITS

(1) **Drupe** (FIG. 171A). This is a fleshy, one- or more-chambered and one- or more-seeded fruit developing from a *monocarpellary* or *syncarpous* pistil, with the pericarp differentiated into the epicarp which forms the skin of the fruit, the mesocarp which is often fleshy, and the *endocarp which is hard and stony*, and hence this fruit is also known as **stone-fruit**, e.g. mango (*Mangifera*), plum and peach (*Prunus*), coconut-palm (*Cocos*), palmyra-palm (*Borassus*), country almond (*Terminalia*), etc.

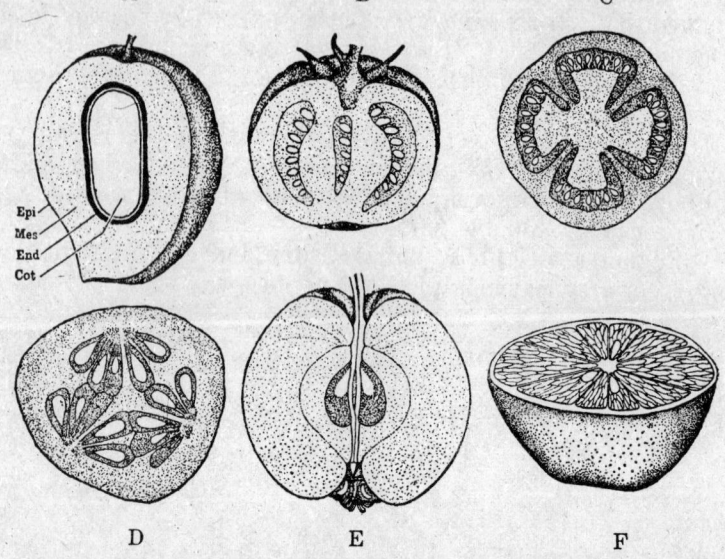

FIG. 171. Fruits (*contd.*). *A*, drupe of mango; *Epi*, epicarp; *Mes*, mesocarp; *End*, endocarp; *Cot*, cotyledon; *B-C*, berry of tomato in longitudinal and transverse sections; *D*, pepo *of* cucumber; *E*, pome of apple (see also FIG. 166A); E, hesperidium of orange.

(2) **Bacca** or **Berry** (FIG. 171B-C). This is a superior (sometimes inferior), indehiscent, usually many-seeded, fleshy or pulpy fruit developing from a *single carpel* or more commonly from a *syncarpous* pistil, with axile or parietal placentation, e.g. tomato, gooseberry, grapes, brinjal, banana, guava, papaw, etc.

In berry the seeds at first remain attached to the placentae, but afterwards they separate from them (placentae) and lie free in the pulp. Sometimes a one-seeded berry may be found, e.g. date-palm. In *Artabotrys* (B. & H. KANTALI-CHAMPA) one-seeded berries develop in an aggregate (see FIG. 172E); epicarp, mesocarp and endocarp may be also distinguished in such a berry, but it differs from the drupe in having *no stony endocarp*.

(3) **Pepo** (FIG. 171D). This is also a fleshy or pulpy, many-seeded fruit like the berry but it develops from an interior, one-celled or spuriously three-celled, *syncarpous* pistil with parietal placentation. This is the characteristic fruit of *Cucurbitaceae*, e.g. gourd, cucumber, melon, water melon, squash, etc. In pepo the seeds, lying embedded in the pulp, remain attached to the placentae.

(4) **Pome** (FIG. 171E). This is an inferior, two- or more-celled, fleshy, *syncarpous* fruit surrounded by the thalamus. The fleshy edible part is composed of the thalamus, while the actual fruit lies within. Examples are found in apple and pear.

(5) **Hesperidium** (FIG. 171F). This is a superior, many-celled, fleshy fruit developing from a *syncarpous* pistil with axile placentation. Here the endocarp projects inwards forming distinct chambers, and the epicarp and the mesocarp, fused together, form the loose or tight skin (rind) of the fruit as in *Citrus*, e.g. orange, lemon, etc.

(6) **Balausta** (FIG. 171G). This is a special type of inferior, many-chambered and many-seeded fruit developing from a *syncarpous* pistil with usually two whorls of basal carpels lying within the receptacle but during the development of the ovary the outer carpels become tilted up and superposed. The result is that two layers of chambers are formed, with the outer or upper ones occupying a parietal position. The pericarp of the fruit is tough and leathery, and the chambers made of thin walls of carpels. The testa of the seed is filled with an acid juice while the tegmen is horny. Balausta is the characteristic fruit of pomegranate.

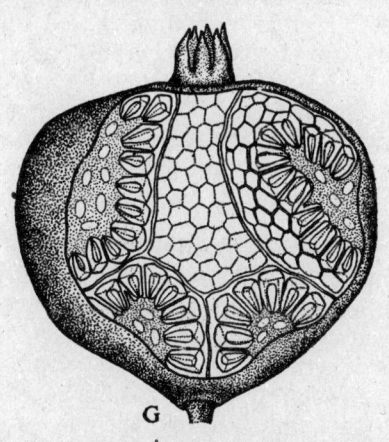

FIG. 171G. Balausta of pomegranate (*Punica granatum*).

B. Aggregate Fruits (FIG. 172). An aggregate fruit is a collection of simple fruits (or fruitlets) developing from the apocarpous pistil (free carpels) of a flower. An aggregate of simple fruits borne by a single flower is otherwise known as an 'etaerio', and the common forms of etaerios are: (1) an **etaerio of follicles**, as in madar (*Calotropis*), *Daemia extensa*, and periwinkle (*Vinca*) with two follicles, in larkspur (*Delphinium*) and

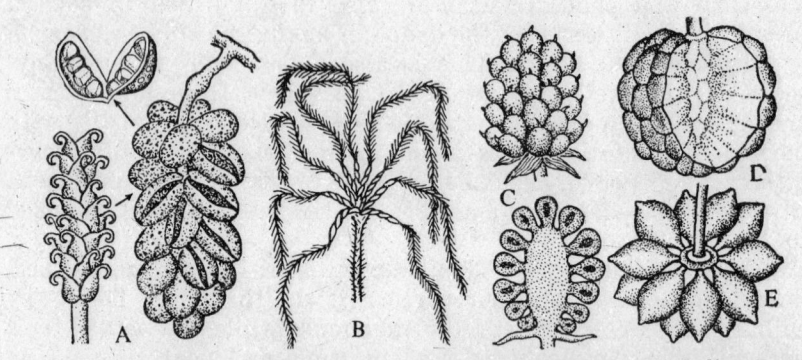

FIG. 172. Aggregate fruits. *A*, an etaerio of follicles in *Michelia*; *B*, an etaerio of achenes in *Naravelia*; *C*, an etaerio of drupes in *Rubus*; *D*, an etaerio of berries in custard-apple (*Annona*); *E*, the same in *Artabotrys*.

aconite (*Aconitum*) with three follicles, and in *Michelia* with numerous follicles; (2) an **etaerio of achenes**, as in *Clematis*, *Naravelia*, *Ranunculus*, strawberry, rose and lotus; (3) an **etaerio of drupes**, as in raspberry (*Rubus*), with small drupes or drupels (also called drupelets) remaining aggregated together on a fleshy thalamus; (4) an **etaerio of berries**, as in custard-apple (*Annona squamosa*) and bullock's heart (*A. reticulata*), where the berries lie embedded in the fleshy thalamus; while in *Artabotrys* (B. & H. KANTALI-CHAMPA) and *Polyalthia* (B. DEBDARU; H. DEVADARU or ASHOKA) the berries are distinct and separate.

C. Multiple or **Composite Fruits** (FIG. 173). A multiple or composite fruit is that which develops from a number of flowers juxtaposed together, or in other words, from an inflorescence. Such a fruit is otherwise known as an *infructescence*.

(1) **Sorosis** (FIG. 173A-B). This is a multiple fruit developing from a spike or spadix. The flowers fuse together by their succulent sepals and at the same time the axis bearing them grows and becomes fleshy or woody, and as a result the whole inflorescence forms into a compact mass, e.g. pineapple, screwpine and jack-fruit. Mulberry is also a sorosis, with the fleshy part made of loosely attached sepals.

(2) **Syconus** (FIG. 173C). The syconus develops from a hollow, pear-shaped, fleshy receptacle which encloses a number of minute male and female flowers. The receptacle grows, becomes fleshy

THE FRUIT

and forms the so-called fruit. It really encloses a number of true fruits or achenes which develop from the female flowers lying within the receptacle at its base, as in *Ficus*, e.g. fig, banyan, peepul, etc.

FIG. 173. Multiple fruits. *A*, sorosis of pineapple (*Ananas*); *B*, the same of mulberry (*Morus*); *C*, syconus of fig (*Ficus*).

Some Common Fruits and their Edible Parts. Apple (pome)—fleshy thalamus. **Banana** (berry)—mesocarp and endocarp. **Cashew-nut** (nut)—peduncle and cotyledons. **Coconut-palm** (fibrous drupe)—endosperm. **Cucumber** (pepo)—mesocarp, endocarp and placentae. **Custard-apple** (etaerio of berries)—fleshy pericarp of individual berries. **Date-palm** (1-seeded berry) —pericarp. *Dillenia* (special)—accrescent calyx. **Fig** (syconus)—fleshy receptacle. **Jack** (sorosis)—bracts, perianth and seeds. **Grape** (berry)— pericarp and placentae. **Guava** (berry—thalamus and pericarp). **Indian plum** (drupe)—mesocarp including epicarp. **Litchi** (1-seeded nut)—fleshy aril. **Maize, oat, rice and wheat** (caryopsis)—starchy endosperm. **Mango** (drupe) —mesocarp. **Melon** (pepo)—mesocarp. **Orange** (hesperidium)—juicy placental hairs. **Palmyra-palm** (fibrous drupe)—mesocarp. **Papaw** (berry)— mesocarp. **Pea** (legume)—cotyledons. **Pear** (pome)—fleshy thalamus. **Pineapple** (sorosis)—outer portion of receptacle, bracts and perianth. **Pomegranate** (special)—juicy outer coat of the seed. **Pummelo or shaddock** (hesperidium)—juicy placental hairs. **Strawberry** (etaerio of achenes)— succulent thalamus. **Tomato** (berry)—pericarp and placentae. **Wood-apple** (a special type called amphisarca)—inner endocarp and placentae.

Chapter 11 DISPERSAL OF SEEDS AND FRUITS

If seeds and fruits fall directly underneath the mother plant and the seedlings grow up close together they soon exhaust the soil of its essential food constituents. Besides, the available space, light and air (oxygen) under such a condition fall far short of the demand. A struggle for existence thus ensues for want of an adequate quantity of their requirements, the consequence of which may be fatal to all of them. Further when they grow together they may easily be a prey to the attack of herbivorous animals. To guard against these contingencies seeds and fruits have developed many special devices for their wide distribution with the result that some of them at least are likely to meet with favourable conditions for germination and growth. It is thus evident that the risk of a species of plant becoming extinct is reduced to a minimum.

1. **Seeds and Fruits dispersed by Wind.** Seeds and fruits have various adaptations which help them to be carried away by the wind to a shorter or longer distance from the parent plant.

(1) **Wings.** Seeds and fruits of many plants develop one or more appendages in the form of thin flat membranous wings, and

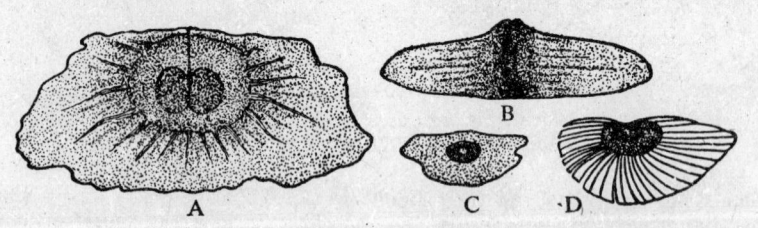

FIG. 174. Winged Seeds. *A, Oroxylum; B, Cinchona; C, Stereospermum; D, Lagerstroemia.*

the former also are light and dry; these devices help them to float in the air and facilitate their dispersion by wind. Thus we find that seeds of *Oroxylum* (FIG. 174A), *Cinchona* (FIG. 174B), *Stereospermum* (FIG. 174C), *Lagerstroemia* (B. & H. JARUL; FIG. 174D), drumstick (*Moringa*; B. SAJINA; H. SAINJNA; FIG. 175A), and *Tecoma* are provided with wings. Similarly many fruits are also provided with one or more wings for the same purpose, e.g. yam (*Dioscorea*; FIG. 175B), ash (*Fraxinus*; FIG. 176A), *Terminalia myriocarpa* (B. HOLOK; FIG.

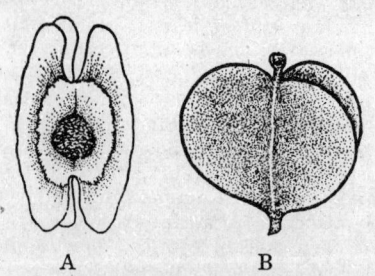

FIG. 175. *A*, winged seed of drumstick (*Moringa*): *B*, winged fruit of yam (*Dioscorea*).

176B), *Hopea* (FIG. 176C), maple (*Acer*; FIG. 176D), wood-oil tree (*Dipterocarpus*; B. & H. GARJAN; FIG. 177A), *Hiptage* (B. MADHABILATA; H. MADHULATA; FIG. 177B), *Dodonaea* (a hedge plant), and sal tree (*Shorea*; B. & H. SAL; FIG. 177C).

(2) **Parachute Mechanism.** In many plants of *Compositae* the calyx is modified into hair-like structures known as **pappus**

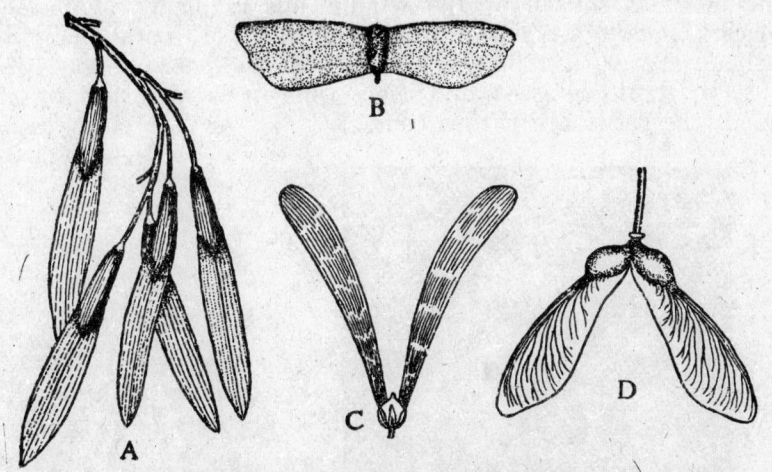

FIG. 176. Winged fruits. *A*, ash (*Fraxinus*): *B*, *Terminalia myriocarpa*; *C*, *Hopea*; *D*, maple (*Acer*).

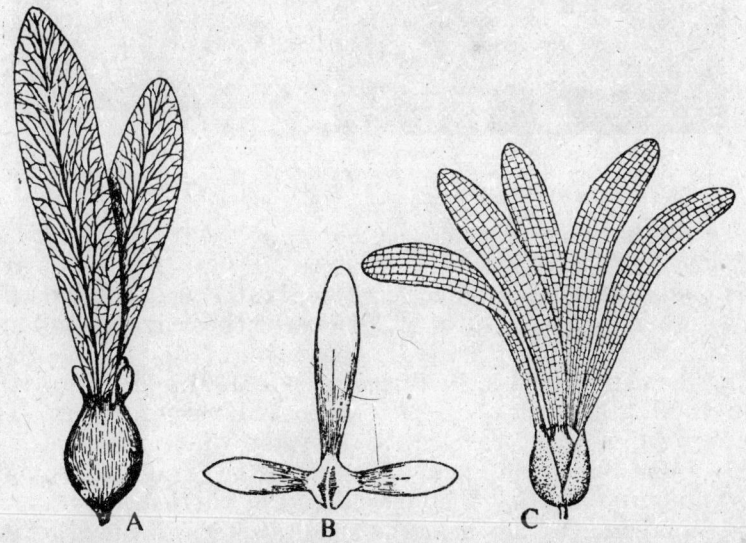

FIG. 177. Winged Fruits (*contd.*). *A*, *Dipterocarpus*; *B*, *Hiptage*; *C*, *Shorea*.

(FIG. 179A). This pappus is persistent in the fruit, and opens out in an umbrella-like fashion. As the fruit gets detached from the parent plant the pappus acts like a parachute and helps it to float in the air. The fruit is often seen being carried by air current to a great distance.

(3) **Censer Mechanism.** Seeds of certain plants can only be scattered by the wind after the dehiscence of the fruit; in such cases the seeds are often not discharged from the fruit unless the latter is shaken by the wind. Thus in poppy (*Papaver*), prickly poppy (*Argemone*), bath sponge or loofah (*Luffa*), pelican flower (*Aristolochia gigas;* B. HANSALATA; FIG. 178), etc., the fruit dehisces, and then, when it is disturbed by the wind, the seeds are thrown out.

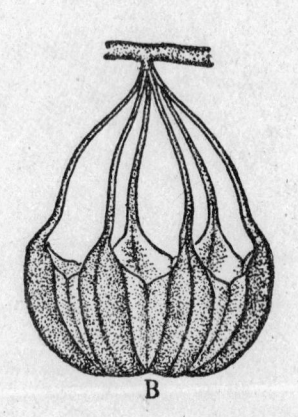

FIG. 178. *A*, pelican flower (*Aristolochia gigas*) with duck-shaped flowers; *B*, a fruit of the same like a hanging basket.

(4) **Hairs.** Seeds of madar (*Calotropis;* FIG. 179B), milkweed (*Asclepias*), *Holarrhena* (B. KURCHI; H. KARCHI), devil tree (*Alstonia;* FIG. 179C) and cotton (FIG. 179D) are provided with hairs either in 1 or 2 tufts or all over their body. These hairs aid the distribution of seeds by the wind.

(5) **Persistent Styles.** In *Clematis* (FIG. 180A) and *Naravelia* (FIG. 180B) the styles are persistent and very feathery. The fruits are thus easily carried away by the wind.

(6) **Light Seeds and Fruits.** Some seeds and fruits are so light and minute in size that they may easily be carried away by the gentlest breeze. Orchids bear the smallest seeds in the vegetable kingdom. In them millions of dust-like seeds are produced in a capsule. Seeds (fruits) of some grasses are also very small and light. Seeds of *Cinchona* (the quinine-yielding plant) are also

DISPERSAL OF SEEDS AND FRUITS

very small, thin and extremely light, and further provided with a membranous wing (see FIG. 174B). There are about 2,500 seeds to a gramme. In *Bucklandia*, a handsome tree of the Khasi Hills, about 250 seeds, mostly winged, weigh a gramme. In *Terminalia myriocarpa* (HOLOK), a timber tree of Assam, about 180 seeds weigh a gramme.

FIG. 179. Pappus and Hairy Seeds. *A*, pappus of a *Compositae* fruit; *B*, hairy seed of madar (*Calotropis*); *C*, hairy seed of devil tree (*Alstonia*); *D*, hairy seed of cotton.

2. **Seeds and Fruits dispersed by Water.** Seeds and fruits to be dispersed by water usually develop floating devices in the form of spongy or fibrous outer coats. The fibrous fruit of coconut is capable of floating long distances in the sea without suffering any injury. Hence coconut forms a characteristic vegetation of sea-coasts and marine islands. The same is the case with double coconut or *coco de mer* (*Lodoicea maldivica*; FIG. 181), a native of the Seychelles Islands. The plants are dioecious and

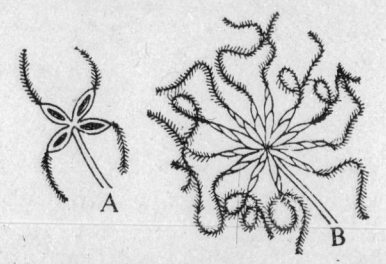

FIG. 180. Persistent styles. *A*, fruits of *Clematis*; *B*, fruits of *Naravelia*.

begin to flower after 30 years of growth. The female plant bears the largest, two-lobed seed and fruit weighing 18 kg. or sometimes even more and measuring often about a metre in length. The fruit takes 6-10 years to ripen. Fruits were seen floating in the Indian Ocean long before the tree was discovered. In lotus (see FIG 96C) the spongy thalamus, bearing the fruits on its hemispheric top, floats bodily about in water, and drifts according to the currents of the water or wind. Sometimes seeds are small and light, and can float on water, e.g. seeds of water lily; these are also provided with an *aril* which encloses air. Seeds and fruits of river-side plants are regularly carried downstream by currents.

FIG. 181. Double coconut seed (*Lodoicea*).

3. **Seeds dispersed by Explosive Fruits.** Many fruits burst with a sudden jerk, with the result that seeds are scattered a few yards away from the parent plant. Common examples of explosive fruits are afforded by balsam (*Impatiens*), wood-sorrel (*Oxalis*), night jasmine (*Nyctanthes*), castor (*Ricinus*), etc. Ripe fruits of

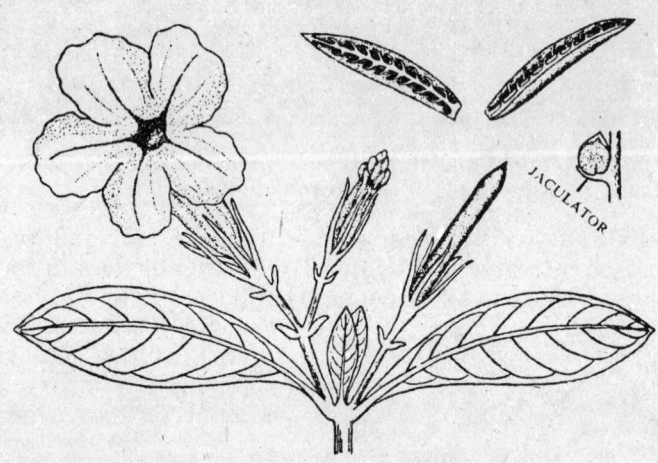

FIG. 182. *Ruellia tuberosa*; note the explosive fruit.

balsam, when touched, burst suddenly. The valves roll up inwards, and the seeds are ejected with great force and scattered in all directions. Many plants of *Acanthaceae* bear explosive fruits which under wet or dry conditions dehisce suddenly from the apex to the base and throw out the seeds with

DISPERSAL OF SEEDS AND FRUITS

some force. In many such cases the seeds are further provided with jaculators (curved hooks) which straighten out suddenly and help in their ejectment. Thus the dry fruits of *Ruellia tuberosa* (FIG. 182) coming in contact with water, usually after a shower of rain, burst suddenly with a noise into two valves, and the seeds are scattered on all sides. Similarly the mature fruits of *Andrographis, Barleria, Acanthus,* etc., burst suddenly when the air is dry, and the seeds are ejected. The cracking sound of the bursting fruits of *Phlox* and *Barleria* is distinctly audible on a bright sunny day.

A very interesting example of bursting fruits is found in camel's foot climber (*Bauhinia vahlii*; B. CHEHUR or LATA-KANCHAN; H. CHAMBULI; FIG. 183). Its long pods, sometimes more than 30 cm. in length, explode violently with a loud noise, scattering the seeds in all directions.

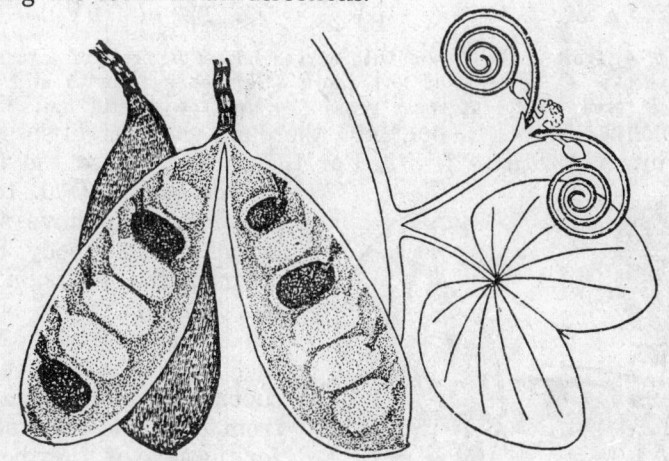

FIG. 183. *Bauhinia vahlii*; note the explosive fruit.

4. Seeds and Fruits dispersed by Animals. Many seeds and fruits are dispersed through the agency of animals including human beings, and have thus certain adaptations for the purpose. (*a*) *Hooked fruits.* Many fruits are provided with hooks, barbs, spines, bristles, stiff hairs, etc., on their body, by means of which they adhere to the body of woolly animals as well as to the clothing of mankind, and are often unwarily carried by them to distant places. Thus fruits of *Xanthium* and *Urena* (FIG. 184A-B) are covered with numerous curved hooks. Spikelets of love thorn (*Chrysopogon;* FIG. 184C) and seeds (fruits) of spear grass (*Aristida;* FIG. 184D) are provided with a cluster of minute, stiff hairs pointing upwards. Flower clusters of *Pupalia* (FIG. 185B) bear stellate hooked bristles spreading outwards. Tiger's nail (*Martynia annua=M. diandra;* B. BAGH-

NAKHI; H. SHERNUI; FIG. 185C) is a very interesting case. Its seed is provided with two, very sharp-pointed, stiff and bent hooks for effective dispersion by woolly animals. In *Tribulus*

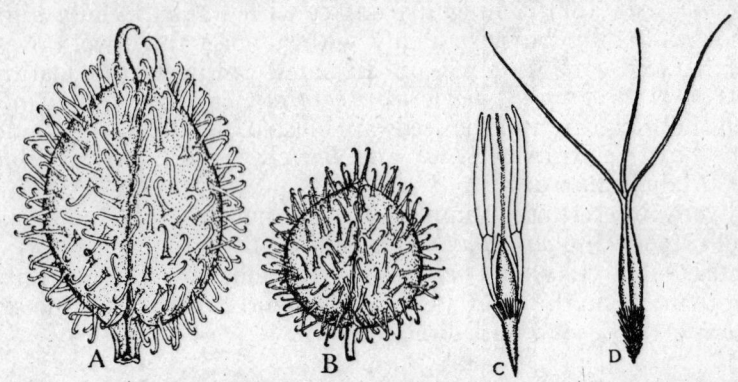

FIG. 184. *A*, fruit of *Xanthium* with curved hooks; *B*, fruit of *Urena* with curved hooks; *C*, spikelet of love thorn (*Chrysopogon*) with stiff hairs; *D*, seed (fruit) of spear grass (*Aristida*) with stiff hairs.

(B. GOKHRI-KANTA; H. GOKHRU) there are sharp rigid spines on the fruit. (*b*) *Sticky fruits*. For the same purpose the fruits of *Boerhaavia* (FIG. 185A) and *Plumbago* have sticky glands on their body. Seeds of mistletoe (*Viscum;* see FIG. 24) are very sticky. (*c*) *Fleshy fruits*. Many such fruits having conspicuous colour are carried from one place to another for the sake of their beauty. (*d*) *Edible fruits*. Human beings and birds are active and useful agents in distributing such fruits. Commonly birds feed upon the pulpy portion of fruits like guava, grape, fig, etc., and pass out the undigested seeds with the faeces; seeds then germinate and grow up into new plants. Seeds of many pulpy fruits are also distributed by birds over wide areas. Jackals feed upon dates, plums, etc., and the seeds germinate after passing through their alimentary canal. Bats and squirrels are also useful agents for the dispersal of seeds and fruits. There is always wide distribution of seeds and fruits, even from one country to another, through the agency of mankind.

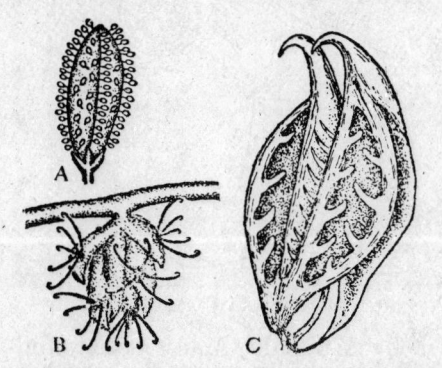

FIG. 185. *A*, fruit of *Boerhaavia* with sticky glands (see also FIG. II/41B); *B*, flowers of *Pupalia* with hooked bristles; *C*, seed of tiger's nail (*Martynia*) with a pair of sharp, curved hooks.

Part II   HISTOLOGY

Chapter 1   THE CELL

A Short History. The study of histology dates from the year 1665 when plant cells were discovered for the first time. It was **Robert Hooke** (1635-1703), an Englishman, who first studied the internal structure of a thin slice of bottle cork with the help of a microscope improved by himself. He discovered for the first time a porous structure in it very much like a honey-comb, and to each individual cavity of such a structure he applied the term **cell**. It was then only the cell-wall that was noticed, this being the prominent part of the cell. Jansen, a spectacle maker of Middleburg in Holland, first invented compound microscope in 1590. **Leeuwenhoek** (1632-1723), a cloth merchant of Delft in Holland, at the age of 21 in the year 1653 developed a mania for grinding lenses. He pursued this work with zeal and assiduity and within 20 years (1653-73) achieved marvellous fineness, accuracy and perfection in his lenses. He made about 400 compound microscopes and gave a demonstration of them before the Royal Society in 1667. He was the first to discover bacteria, protozoa and other minute forms of life—'the wretched beasties' as he called them—under his own microscope.

The prominent workers of that time, who studied plant tissues under the microscope, were **Grew** and **Malpighi**. Grew (1641-1712), an English physician and botanist, published his first paper on plant tissues in 1671 and the second one in 1682. Malpighi (1628-94), an Italian physician, studied the various tissues of vascular plants, and published his first paper in 1675. Grew and Malpighi discovered parenchymata, fibres and vessels. Malpighi further discovered stomata.

In 1838-39 **Schleiden** (1804-81), a German botanist, and **Schwann** (1810-82), a German zoologist, proved definitely that both plants and animals are cellular in character, and founded the **cell theory**. **Von Mohl** (1805-72) and **Nageli** (1817-91), working independently, distinguished in 1844 the two main parts of a cell: the cell-wall and the contents. To the granular semi-fluid contents filling up the cavity of the plant cell the name **protoplasm** was proposed by Von Mohl in 1846. Cohn in 1850 established the fact that plant protoplasm and animal protoplasm are identical. De Bary (1831-88), investigating plant cells, and Max Schultze, investigating animal cells, established the **protoplasm theory** about the year 1861, namely, that the cells or units of plants and animals are tiny masses of protoplasm, each mass containing a **nucleus**. The cell-wall of plant cell, though prominent under the microscope, is of secondary importance, being formed later by the protoplasm for its own advantage.

In 1831 the nucleus of the cell was first discovered by **Robert Brown** (1773-1858). In 1846 **Nageli** (1817-91) first showed that new cells arise by division of the pre-existing cells. In 1880 the first satisfactory account of the structure of the nucleus and its mode of division was, however, given by **Strasburger** (1844-1912) although in 1849 the division was roughly observed by **Hofmeister**. The early discovery by Strasburger of the constant number of chromosomes in a species was finally confirmed by Boveri in 1900. In 1884 it was recognised by Strasburger, Weismann (1834-1914) and others that the nucleus is concerned in the problem of inheritance of characters. Between 1880 and 1892 **Flemming** gave a detailed account of the longitudinal splitting of the chromosomes and their equal distribution among the daughter cells. In 1888 Strasburger discovered reduction division in angiosperms and in 1891 he observed the same phenomenon in mosses and ferns. In 1905 the term meiosis was applied to reduction division by **Farmer** and **Moore**. Polyploidy and aneuploidy were discovered between 1912 and 1927—triploidy ($3n$) by **Miss Lutz** in 1912, tetraploidy ($4n$) by **Digby** in the same year, pentaploidy ($5n$) by **Nawaschin** in 1925 and 1927, aneuploidy ($2n-1$, $2n+1$, $4n+1$) by **Tackholm** in 1922, and by **Blakeslee** in 1924.

Further important early work on plant anatomy may only be just mentioned: **Sanio's** work on the activity of the cambium and the formation of the secondary tissues (1863), **De Bary's** work on comparative anatomy (1877) and on tissue systems (1884), **Jeffrey's** theory of stelar structure (1868), **Nageli's** theory of single-celled apical meristem (1858), **Hanstein's** histogen theory of apical meristem (1868), **Schmidt's** tunica-corpus theory (1924), **Haberlandt's** work on physiological plant anatomy (1884), etc.

Electron microscope, the latest in the series of microscopes, magnifying objects more than 2,00,000 X, was invented by two German scientists, **Knoll** and **Ruska**, in 1932. Exciting pictures of cells began to be revealed by this instrument, much improved in the meantime, since early 1950. Many of the solid grains and rods of the cytoplasm, as seen under a compound microscope, have now proved to be complex structures under an electron microscope. In view of such new revelations the biologists are now forming a new concept of cell organization.

Cell Structure. Cells are the fundamental structural and functional units that the plant body or the animal body is composed of. A plant **cell** may be defined as a unit or independent, tiny or microscopic mass of protoplasm enclosing in it a denser spherical or oval body, called the **nucleus**, and bounded by a distinct wall, called the **cell-wall**. Protoplasm and nucleus are living, while the cell-wall is non-living, the latter having been formed by the protoplasm, primarily for its own protection. The living parts of a cell (protoplasm, nucleus and other living bodies) together constitute the **protoplast** of the cell—a collective and convenient term introduced by Hanstein in 1880. A plant cell thus consists of a protoplast (representing the living parts) and a cell-wall (forming a non-living framework round the protoplast to maintain its shape and firmness and to afford necessary protection). The cell-wall has also to regulate the flow of materials into and out of the cell.

Cells vary widely in shapes and sizes. In shape they may commonly be spherical, oval, polygonal, rectangular or considerably elongated. When young, they are often spherical or of like nature. Usually they are very minute in size invisible to the naked eye. The average size of fully developed rounded or polygonal cells varies between 1/10th and 1/100th of a millimetre. Sometimes, as in fleshy fruits and pith, they may be as big as 1 mm. or even bigger, or as small as 1/200th mm. or even smaller. Among the known cells bacteria are the smallest, usually

THE CELL

ranging between 1/100th and 1/1,000th mm. Still smaller in size are the viruses which defy microscopic observation. Fibrous cells are considerably elongated; they chiefly vary in length from 1 to 3 mm., but in woody stems they may often be as big as 6 mm. or even 8 mm. In some fibre-yielding plants such as flax, hemp and rhea the fibrous cells may grow to a length of 20 to 550 mm. Still larger cells are the latex cells.

The Protoplast

The protoplast is the living unit—actually the physiological unit —of a cell, while the **protoplasm** is the essential living material that comprises the different parts of it (the protoplast). Protoplasm is the only substance that is endowed with life and, therefore, plants and animals containing this substance in their body are regarded as living. As the protoplasm dies the cell ceases to perform any function for the plant or the animal, which then as a whole becomes inert and dead. Protoplasm is thus fittingly described as the *physical basis of life*. As the protoplast as a whole has to perform manifold functions of a cell, such as manufacture of food, nutrition, growth, respiration, reproduction, etc., it is differentiated into distinct living (protoplasmic) bodies: (1) cytoplasm, (2) nucleus, and in special cells (3) plastids. These are only masses of protoplasm differentiated into distinct bodies (FIG. 1C) to perform specialized functions. It must, however, be noted that these living bodies are never newly formed in the cells but always develop from pre-existing ones and that one kind of living body cannot give rise to another kind.

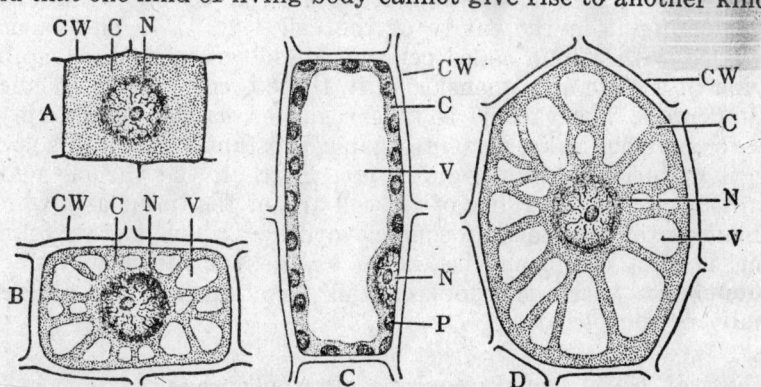

FIG. 1. Plant cells. A, a very young cell; B, a growing cell with many small vacuoles; C, a mature cell with a large vacuole; D, a mature cell with many vacuoles. *CW*, cell-wall; *C*, cytoplasm; *N*, nucleus; *V*, vacuole; *P*, plastid (chloroplast).

1. Cytoplasm. The protoplasmic mass of a cell leaving out the nucleus and the plastids is otherwise called cell-protoplasm or

cytoplasm. When the cell is young the cytoplasm fills in the space between the cell-wall and the nucleus. The surface of the cytoplasm forms into an extremely thin and delicate membrane known as the **plasma membrane** or **ectoplasm.** This plasma membrane is also hyaline but non-granular and somewhat firmer in consistency than the rest of the cytoplasm, and it lies adpressed against the cell-wall. This is a very important layer controlling the entrance and exit of substances into the cell and out of it. The inner granular mass of the cytoplasm is often called **endoplasm.** Besides, the cytoplasm encloses numerous minute granules whose nature is obscure; these are known as the *microsomes*. They may represent pieces of the endoplasmic reticulum (see FIG. 6B). The fluid portion or matrix of the cytoplasm is known as **hyaloplasm.** When the cell is very young (FIG. 1A) it remains completely filled with the cytoplasm, but as the cell grows a large number of small non-protoplasmic but fluid-filled cavities of varying sizes, apparently like little bubbles, called **vacuoles** (*vacuus*, empty) appear in the cytoplasm. As the cell enlarges (FIG. 1B) all these small vacuoles begin to fuse together, and finally in the mature cell they form one large central vacuole which occupies the major part of the cell-cavity. The cytoplasm under this condition is pushed outwards as a thin layer alongside the cell-wall, with the nucleus and the plastids lying embedded in this layer (FIG. 1C). Or, sometimes, instead of one, a number of comparatively small vacuoles persist in a mature cell, and then the cytoplasm occurs as delicate strands around those vacuoles and also as a very thin layer lining up with the cell-wall; these strands are seen to radiate from around the nucleus, often suspending it in the cavity of the cell (FIG. 1D). The vacuole is filled with a fluid called **cell-sap.** Dissolved in the cell-sap, or lying in a state of suspension in it, there occur various chemical compounds. The vacuole is thus regarded as the storehouse of water, various salts, certain organic substances (mainly soluble food materials), anthocyanins, etc. Further, the vacuole maintains necessary turgidity of the cell and of the tissue as a whole. The layer of cytoplasm in contact with the vacuole and surrounding it as a membrane is known as the **vacuole membrane** or **tonoplasm.** Like the ectoplasm this membrane is also differentially permeable.

Physical Nature of Protoplasm. Protoplasm is a transparent, foamy or granular, slimy, semi-fluid substance, somewhat like the white of an egg. It is never homogeneous but contains granules of varying shapes and sizes, suspended in solution and, therefore, it looks finely granular under the microscope. There are of course many dissolved materials in it. Although often semi-fluid it may be fluid or viscous. It completely fills up the cavity of the young cell, but in a mature cell it may occur as a thin layer against the cell-wall (FIG. IC) or it may occur as delicate strands around the vacuoles (FIG. ID), as already described. In its *active* state the protoplasm remains saturated with water which makes up 75-90 per cent of the content. With decreasing water content its vital activity diminishes and gradually comes to a standstill, as in dry seeds. Protoplasm coagulates on heating, and when killed it loses its transparency. Protoplasm responds to the action of external stimuli such as a needle- or pin-prick, electric shock, application of particular chemicals, sudden variation of temperature or of light, etc. On stimulation the protoplasm contracts but expands again when the stimulating agent is removed. This 'contractility' involving both contraction and expansion, first demonstrated by Kuhne in 1864 in the staminal hair of spiderwort (*Tradescantia*), is an inherent power of protoplasm.

Protoplasm is semi-permeable in nature, i.e. it allows only certain substances and not all to enter its body. This property is, however, lost when the protoplasm is killed.

Under normal conditions the protoplasm of a living cell is in a state of slow but constant motion. In many cases, however, it shows distinct movements of different kinds (see pp. 148-50).

Chemical Nature of Protoplasm. The principal component of protoplasm is water. Otherwise it is a highly complex mixture of a variety of organic compounds, of which proteins are predominant, and they are of various kinds. A complex type of protein called **nucleoprotein** is constant in the cytoplasm and the nucleus. Nucleoprotein (see p. 151) is composed of nucleic acids and certain proteins. There are two important types of nucleic acids—DNA (deoxyribonucleic acid) and RNA (ribonucleic acid). Of them RNA is constant in the cytoplasm and nucleolus, and DNA in the chromosomes.

Structure. Protoplasm, as it works in the cell, undergoes dynamic physical changes and, therefore, its ultimate structure cannot be known with any amount of certainty. So from time to time its structure has been variously expressed as (*a*) **fibrillar** consisting of interlacing fine fibres or fibrils (Flemming, 1882), (*b*) **alveolar** or **foamy** consisting of a froth of minute bubbles (Butschli, 1878), and (*c*) **granular** consisting of fine grains or granules more or less uniformly dispersed throughout the protoplasm (Hanstein, about 1886).

The exact chemical composition of the living protoplasm, however, cannot be determined because any attempt to analyse it kills it outright with some unknown changes in it. Besides, it undergoes continual changes and its composition is not, therefore, constant. Further, protoplasm always encloses many foreign substances in its body in varying quantities and, therefore, it is not possible to get it in a pure state. Analysis of the dead protoplasm reveals a long list of elements present in it. Of these oxygen (O)—about 65%, carbon (C)—about 18.3%, hydrogen (H)—about 11%, and nitrogen (N)—about 2.5%, are most conspicuous. Other elements making up the remaining 3% evidently occur in very small quantities; these are chlorine (Cl), sulphur (S), phosphorus (P), silicon (Si), calcium (Ca), magnesium (Mg), potassium (K), iron (Fe) and sodium (Na). In traces are also present the following: zinc (Zn), manganese (Mn), aluminium (Al), copper (Cu), boron (B) and molybdenum (Mo) and often a few others. Active protoplasm contains a high percentage of water—usually varying from 75-90%, and remains saturated with it. Leaving out this water the solid matter of the protoplasm contains the following: proteins—40-60%; fatty substances (true fats and lipids, particularly lecithin)—12-14%; carbohydrates—12-14%; and inorganic salts—5-7%.

Colloidal State of Protoplasm. The constituents of protoplasm, mainly proteins, form a **colloidal system** (see Part III, Chapter 1). In a colloidal state a matter is divided into very fine particles—aggregations of molecules, and not individual molecules as in a solution, which are dispersed through a continuous medium, mostly a liquid or a semi-solid. In the case of protoplasm the dispersion medium is water with dissolved salts and other substances in it, and the colloidal particles of proteins are dispersed through this medium. A colloid exists as a mixture and not as a compound but the colloidal particles are mostly not filterable through a fine parchment membrane. In many colloids the particles are stable, while in others they are unstable, settling down under slight chemical or physical change. A colloidal mixture may be fluid or viscous. A fluid colloid is called a **sol**, and a viscous colloid a **gel**. Protoplasm mostly exists in the state of a sol but may change to a gel under certain conditions, the changes being reversible (sol$\rightleftharpoons$gel). But when it dies it cannot change from one state to the other. The colloidal system of protoplasm is believed to be responsible for its various life-processes.

Tests. (*a*) Iodine solution stains protoplasm **brownish yellow**. (*b*) Dilute caustic potash dissolves it. (*c*) Millon's reagent (nitrate of mercury) stains it **brick-red**; the reaction is hastened by heating.

Movements of Protoplasm. Protoplasm shows movements of different kinds. Naked masses of protoplasm, not enclosed by the cell-wall, show two kinds of movement—**ciliary** and **amoeboid**. The protoplasm, enclosed by the cell-wall, shows a streaming movement within it, which is spoken of as *cyclosis*. Cyclosis is of two kinds—**rotation** and **circulation**.

(1) **Ciliary Movement** (FIG. 2A) is the *swimming* movement of free, minute, protoplasmic bodies such as the zoospores of many algae and fungi, bacteria, antherozoids of mosses and ferns, etc., provided with one or more special organs of motion

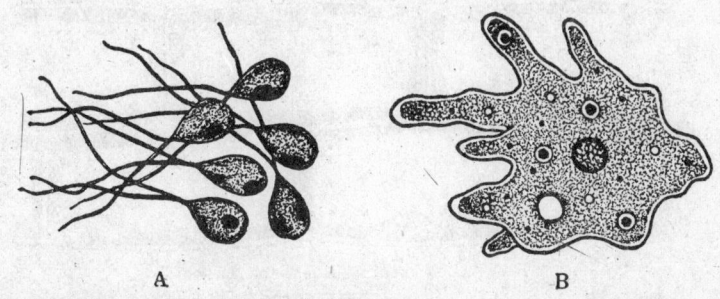

FIG. 2. Movements of Protoplasm. *A*, ciliary movement; *B*, amoeboid movement.

in the form of whip-like structures, called **cilia** or **flagella**. By the vibration of these cilia such ciliary bodies move or swim freely and rapidly in water.

(2) **Amoeboid Movement** (FIG. 2B) is the creeping movement of a naked mass of protoplasm. This protoplasmic mass moves or creeps by the protrusion of one or more parts of its body in the form of false feet or pseudopodia (*pseudos*, false; *podos*, foot) and their withdrawal at the next moment, very much like the animalcule *Amoeba*. Cell-wall being absent, the protoplasmic mass has no definite shape, and is capable of engulfing solid particles of food. Amoeboid movement is exhibited by many slime fungi and large solitary zoospores and gametes of certain algae and fungi.

(3) **Rotation** (FIG. 3A). When the protoplasm moves or streams within a cell alongside the cell-wall, clockwise or anti-clockwise, round a large central vacuole, the movement is expressed as rotation. The direction of movement is constant so far as a particular cell is concerned. As the protoplasm rotates, it carries in its current the nucleus and the plastids. Rotation is distinctly seen in *Vallisneria, Hydrilla, Chara* and *Nitella*, and also in many other aquatic plants.

(4) **Circulation** (FIG. 3B). When the protoplasm moves or *streams* in different directions within a cell round a number of small vacuoles, the movement is called circulation. In this process the protoplasmic mass around the nucleus radiates in different directions in the form of delicate strands; each strand then moves round one or more vacuoles and finally comes back

to the nucleus. Circulation is very distinctly seen in the purplish staminal hairs of *Commelina obliqua* (B. JATA-KANSHIRA; H. KANJURA). It is also seen in the staminal hairs of spiderwort (*Tradescantia*) and *Rhoeo discolor,* and in the young shoot-hairs of gourd, elephant ear plant (*Begonia*) and in many other land plants.

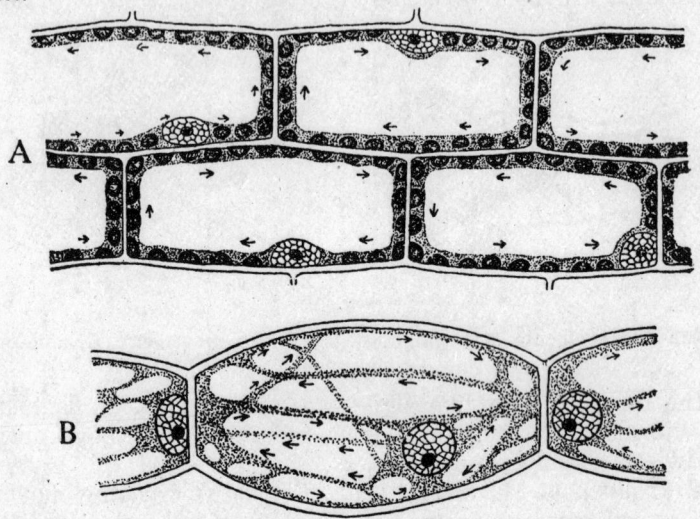

FIG. 3. Movements of Protoplasm (*contd.*). *A,* rotation in the leaf of *Vallisneria*; *B,* circulation in the staminal hair of *Commelina obliqua.*

2. **Nucleus.** Embedded in the cytoplasm there is a specialized protoplasmic body, usually spherical or oval in shape and much denser than the cytoplasm itself; this is the **nucleus.** Its shape depends to some extent on the nature of the cell in which it occurs. In the young cell it occupies a median position and is almost always spherical or oval; but in the long cell it may become correspondingly elongated. In the mature cell with the formation of the vacuole it lies in the lining layer of the cytoplasm and may become flattened against the cell-wall. Nuclei are universally present in all living cells. In the higher plants only a single nucleus is present in each cell (FIG. 4A); in the latex tissue (see FIG. 40), and in many algae and fungi numerous nuclei are often seen in a single cell. In the lower organisms like bacteria and blue-green algae, however, true nuclei are absent, but there is corresponding nuclear material. Nuclei may vary widely in size from $1\,\mu$ to $500\,\mu$ (microns). Their usual size, however, is $5\text{-}25\,\mu$ (microns). A nucleus can never be newly formed, but multiplies in number by the division of the pre existing one.

Structure. Each nucleus (FIG. 4B) is surrounded by a thin, transparent membrane known as (1) the **nuclear membrane** which

FIG. 4A. Cellular structure and nuclei in onion scale.

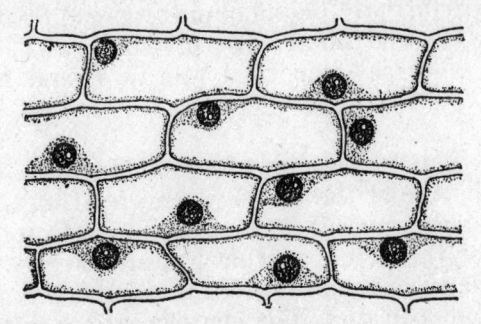

separates the nucleus from the surrounding cytoplasm. The electron microscope reveals that the membrane is of a double nature, and that it is provided with numerous pores through which transport of materials takes place between the nucleus and the cytoplasm (see FIG. 6B). The shape of the nucleus depends partly on its membrane and partly on the shape of the cell in which it occurs. Within the membrane, completely filling up the space, there is a dense but clear mass of protoplasm known as (2) the **nuclear sap** or **nucleoplasm** or **karyolymph**. Suspended in the nucleoplasm there are numerous, fine, crooked threads, loosely connected here and there, forming a sort of network, called (3) the **nuclear reticulum** or **chromatin network**. The modern idea is that the chromatin threads, otherwise called **chromosomes**, occur as separate and independent structures in the nucleus and that they are made of nucleoprotein (see below). Besides, one or more highly refractive, relatively large and usually spherical bodies without any membrane, but much denser than the nucleoplasm, also occur in the nucleus; they are known as (4) the **nucleoli**. They remain attached to certain chromosomes in their particular regions, and are regarded as parts of those chromosomes.

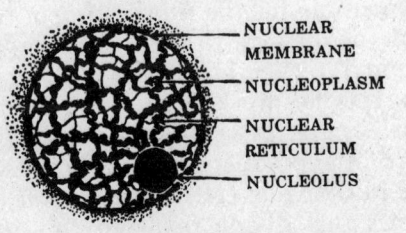

FIG. 4B. Nuclear structure.
NUCLEAR MEMBRANE
NUCLEOPLASM
NUCLEAR RETICULUM
NUCLEOLUS

Chemical Composition. Chemically the nucleus is more or less similar to that of the cytoplasm. The nucleus, however, is predominantly composed of nucleoproteins which are phosphorus-containing nucleic acids (see next page) and certain proteins.

Nucleic acids occur in the nucleus to the extent of 15-30% of its dry weight. Besides, the nucleus also contains some amount of lipids, particularly phospholipid. Inorganic salts such as those of Ca, Mg, Fe and Zn are also present in the nucleus in small quantities, and play important role in several biochemical processes.

Functions. The nucleus and the protoplasm are together responsible for the life of the cell. Experiments and observations have proved that the nucleus is the controlling centre of all the vital activities of the cell, particularly assimilation and respiration. If the nucleus be removed from a cell the protoplasm ceases to function and soon dies. The electron microscope has further revealed that it is the DNA of the chromosomes that is primarily responsible for the functioning of the nucleus. The DNA sends a code or message to the cytoplasm, particularly to the ribosomes, through nucleotides and RNA and induces the cytoplasm to take up specific work. The principal functions performed by the nucleus are, however, as follows: (1) The nucleus takes a direct part in reproduction, asexual or sexual. (2) The nucleus takes the initiative in cell division, i.e. it is the nucleus and the chromosomes that divide first and this is followed by the division of the cell. (3) The nucleus, more particularly the chromosomes, are the *bearers* of hereditary material, i.e. DNA. It is in fact the DNA of the chromosomes that is the sole hereditary material, and the characteristics of the parent plants are transmitted to the offspring through the medium of this DNA. (See also Chemistry of Chromosomes, p. 185).

Nucleic Acids[1]. Nucleic acids are wonderful discoveries of the modern times. They are universally present in the nucleus and in the cytoplasm of all living cells, and are now definitely known to form the chemical basis of life. They are very complex organic compounds made of phosphate, pentose sugar (ribose, as in RNA or deoxyribose, as in DNA) and nitrogen bases (purine and pyrimidine; see p. 155). Nucleic acid molecules are very large, even larger than protein molecules, and consist of infinite numbers of *repeating* nucleotide units linked in any sequence into a long chain. They are thus high polymers of nucleotides, and have very high molecular weights. Depending on the sequence of nucleotide units in the chain the nucleic acids may be of infinite structural

---

[1] Much has been known about the importance of nucleic acids through the brilliant researches carried out by several investigators extending over a period of 30 years. In 1868 Miescher, a German chemist, first isolated from human pus cells a peculiar phosphorus-containing compound which he called nuclein (later renamed as nucleic acid). Its importance was not, however, recognized for long many years. In 1944 Avery, MacLeod and McCarthy working on bacteria and viruses at the Rockefeller Institute, New York showed for the first time that

varieties. A **nucleotide** is a molecular unit (monomer) of a nucleic acid molecule (macro-), and consists of three subunits: a phosphate, a pentose sugar (ribose or deoxyribose) and a nitrogen base (purine or pyrimidine). Phosphate and sugar alternate as links (or subchains) in the chain while nitrogen base projects inward from sugar link. There are a few types of nucleotides, each with a specific nitrogen base. They may also occur free in the cytoplasm as ATP, DPN, TPN, CoA (coenzyme A), etc. A nucleotide is formed when a phosphate group is added to a nucleoside. A **nucleoside** is a compound consisting of two subunits: a pentose sugar and a nitrogen base. It is the precursor of a nucleotide. There are two kinds of nucleic acids, viz. DNA (deoxyribonucleic acid) and RNA (ribonucleic acid); the latter occurs in three forms: messenger RNA (mRNA), transfer RNA or soluble RNA (tRNA or sRNA) and ribosomal RNA (rRNA), as detailed on p. 341. A summary of nucleic acid formation may be given thus: pentose sugar+nitrogen base → nucleoside; nucleoside+phosphate group → nucleotide; nucleotide+nucleotide +......... → nucleic acid.

DNA and RNA. *Occurrence*. DNA occurs almost exclusively in the chromosome, and to a small extent only, as is now known, in the chloroplasts and in the mitochondria. RNA occurs mostly in the cytoplasm (about 90% of cell's RNA occurs here), nucleolus and ribosomes, and to some extent in the chromosomes, of course, in three different forms, as already mentioned. A big portion of the RNA formed in the nucleolus, possibly under the control of DNA, moves to the surrounding cytoplasm. DNA or RNA, with certain protein in each case, is the predominant constituent of most virus particles (see FIG. V/64). This is also true of many bacterial cells. *Chemistry*. DNA and RNA are close chemical relatives. The principal difference between the two lies

---

DNA is the sole genetic material (confirmed by others within the next few years). In 1953 Wilkins, a bio-physicist of King's College, London worked out the molecular structure of nucleic acids. In the same year (i.e. 1953) Watson and Crick (co-winners of 1962 Nobel Prize along with Wilkins) of the Cambridge University worked out the famous *double helix* model of DNA, and the role played by DNA and RNA in the synthesis of proteins. Kornberg in 1956 isolated from *Escherichia coli* (an intestinal bacterium) an enzyme *DNA-polymerase* which could duplicate DNA *in vitro*. For this wonderful discovery Kornberg was awarded Nobel Prize in 1956. This enzyme is now found to be present in the living cells. Recently Hargobind Khorana, Marshall Nirenberg and Robert Holley (Nobel Prize winners in 1968) investigating independently on nucleic acids at the Massachusetts Institute of Technology, U.S.A., worked out suitable methods for synthesis of nucleic acids in the laboratory, leading to the understanding of genetic codes. Their work may possibly lead to the manufacture of synthetic proteins at no distant future. Very recently, after nine years' research, Khorana has actually succeeded in artificially creating an exact copy of a bacterial gene and transplanting the same with success for normal behaviour. There may thus be a possibility of curing some of the hereditary diseases carried through defective gene. Fritz, another geneticist of the same Institute, however, asserts that creation of a mammalian gene, much more complicated as it is, is a far cry.

in the kind of pentose sugar present in their molecules. RNA contains a 5-carbon atom (pentose) sugar 'ribose', whereas DNA contains 'deoxyribose', also a 5-carbon (pentose) sugar but it has one less oxygen atom in its molecule than in RNA. *Structure.* Both occur as macromolecules but RNA molecules are single-stranded, while DNA molecules (FIG. 5) are double-stranded (with but few exceptions in each case). DNA and RNA bases are almost the same excepting that RNA has uracil instead of thymine of DNA. *Functions.* DNA is the sole genetic material (analogous to genes) migrating intact from generation to generation through the reproductive units or gametes, and is responsible for the development of specific characters in the successive generations. DNA is the controlling centre of all the vital activities of the living cell, and is responsible for all biosynthetic processes including protein synthesis. Biologists now believe that all secrets of life are confined to and controlled by DNA of the living cell. RNA, under the instructions of DNA, is directly connected with the synthesis of proteins (see DNA and Protein Synthesis, part III, chapter 6).

DNA Molecule (FIG. 5). The DNA molecule of a single chromosome is very long and complex (macromolecule), forming the backbone of each chromosome. In 1953 Watson and Crick (see footnote on p. 153) investigating nucleic acids proposed a *double helix* model of DNA molecule (Watson-Crick model), universally accepted since then. According to them DNA occurs as a double-stranded molecule, with the two strands profusely coiled and entwined about each other throughout their whole length. The structure is like a ladder twisted in a helical fashion. Each spiral strand is made of groups (micromolecules) of deoxyribose sugar (a 5-carbon or pentose sugar) alternating with groups of phosphate, and an infinite number of cross-links connecting the two strands (like the rungs of

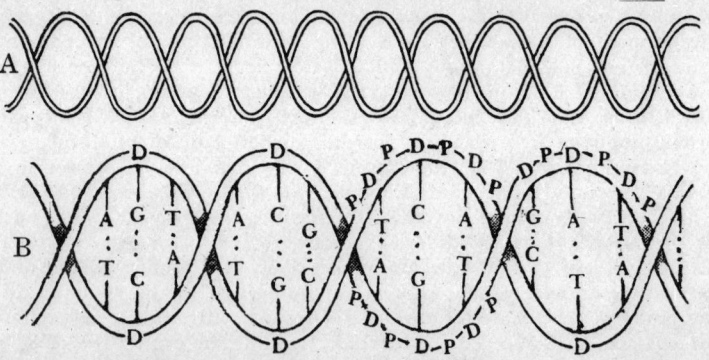

FIG. 5. *A*, Watson-Crick model of DNA molecule; the two strands are twisted about each other; *B*, a portion of the same magnified; note the distribution of deoxyribose sugar (*D*), phosphate (*P*) and cross-links— thymine-adenine (*T·A*) and guanine-cytosine (*G·C*).

THE CELL

a ladder). Each pair of cross-links is made of two distinct types of nitrogenous bases—**purines** and **pyrimidines**, each attached to a sugar. Each pair of bases is loosely linked by hydrogen bonds. Altogether there are two purines (adenine and guanine) and two pyrimidines (thymine and cytosine). It is the rule that a specific purine always pairs with a specific pyrimidine as alleles (i.e. complementary pairs), e.g. adenine with thymine (A-T) and guanine with cytosine (G-C). It may be noted that each base is a part of a nucleoside (see p. 153). It is important to note that the pairs of nitrogenous bases occur in infinite sequences in a DNA molecule, enabling the latter to coin an infinite number of chemical codes (messages or informations) and transmitting the appropriate codes through its working partner, RNA, to the surrounding cytoplasm for its manifold activities. The above may be summarized thus: a DNA strand (FIG. 6B) is made of four types of nucleotides—PDT, PDA, PDG and PDC, evidently including four types of nucleosides—DT, DA, DG and DC, and also four kinds of nitrogenous bases—T, A, G and C. Although such bases combine in only four specific pairs— T-A, A-T, G-C and C-G they may occur in infinite sequences in a DNA molecule.

3. **Plastids**. Besides the nucleus, the cytoplasm of a cell encloses many small specialized protoplasmic bodies, usually discoidal or spherical in shape; these are called **plastids** (see FIG. 3A). Their average size is 4μ to 6μ. Each plastid is bounded by a double membrane. The ground substance or matrix of the plastid, as first investigated by Pringsheim in 1874, is called the **stroma** which is a proteinaceous material. Lying embedded in the stroma there is a large number of granules called **grana**; each granum consists of a varying number of discs (see FIG. 6C). The stroma is colourless, whereas the granules contain the pigment or colouring matter. Plastids are living, and multiply in number by division of the pre-existing ones (Sanio, 1864). Although plastids were known from the time of Von Mohl about the middle of the nineteenth century, their origin could not be traced for many years. According to Guilliermond and his students (1920-24) mitochondria (see p. 158) present in the embryo cells of phanerogams become partly differentiated into plastids and partly remain as they are. This view is now discarded. According to recent studies the plastids arise from pre-existing bodies called **proplastids** present in the embryonic cells. As the cells grow, the plastids also grow, multiply in number by divisions and assume their characteristic forms. The plastids occur in cells which have to perform specialized functions, and are always absent from blue-green algae, fungi and bacteria. Plastids are of three types, viz. **leucoplasts, chloroplasts** and **chromoplasts**. One form of plastids can change into another; as for example, leucoplasts change into chloroplasts when the former are exposed to light for a prolonged period; similarly, chloroplasts change into leucoplasts in the continued absence of light; similar changes may take place in chromoplasts. In the young tomato fruit the leucoplasts gradually change into chloroplasts and the latter into chromoplasts as the fruit ripens.

(1) **Leucoplasts** (*leucos*, white). These are colourless plastids. Leucoplasts occur most commonly in the storage cells of roots

and underground stems; they are also found in other parts not exposed to light. They vary in shape, being frequently spherical, discoidal or rod-like. Leucoplasts convert sugar into starch in the form of minute grains for the purpose of storage. Larger leucoplasts, specially acting as starch-storing bodies, are known as **amyloplasts**. There is another type of colourless plastids called **elaioplasts**. They are concerned with the formation and storage of fats in their body, as does cytoplasm of the cell. Elaioplasts are also capable of forming starch. They are common in liverworts and also found in many monocotyledons and certain dicotyledons.

(2) **Chloroplasts** (*chloros*, green). These are green plastids, their colour being due to the presence of a *pigment* to which the name **chlorophyll** was given by Caventou in the year 1818; sometimes the green colour may be masked by other colours, Chloroplasts are only found in parts exposed to light; they occur abundantly in green leaves, and also to some extent in green parts of the shoot. They are mostly spherical or discoidal in shape, but in some bryophytes and algae they may assume peculiar forms. *Functions*. They work *only in the presence of sunlight* and perform some very important functions with the help of their chlorophyll. They absorb carbon dioxide from the air; manufacture sugar and starch from this carbon dioxide and the water absorbed from the soil; and liberate oxygen (by splitting the water) which escapes to the surrounding air.

Chlorophyll, as worked out by Willstatter and Stoll (1906), is a mixture of four different pigments, viz. chlorophyll *a* (blueblack), chlorophyll *b* (green-black), carotene (orange-red) and xanthophyll (yellow). Chlorophyll *a* and chlorophyll *b* are associated with each other in the chloroplast, but carotene and xanthophyll may also occur without chloroplast in any part of the plant. Chlorophyll as a whole can be easily extracted with alcohol, benzene, acetone, ether, or chloroform and the leaves then become colourless. The chlorophyll solution appears deep green in transmitted light, but blood-red in reflected light. This is the physical property of chlorophyll called *fluorescence*. A mixture of the two pigments—carotene and xanthophyll—which are always associated with chlorophyll, can be easily separated from the chlorophyll solution by shaking it with a small quantity of benzene and allowing the solution to settle for a few minutes; the benzene floats on the top (green solution) carrying chlorophyll, while alcohol settles at the bottom (yellow solution) retaining carotene and xanthophyll. Instead of benzene, ether or olive-oil may be used. Chlorophyll is not soluble in water, even under prolonged boiling. Chlorophyll forms about 8% of the dry weight of the chloroplast, while carotene and xanthophyll about

2%. *Functions.* It is definitely known that chlorophyll absorbs light energy from sunlight and initiates the process of photosynthesis. It does not, however, undergo any chemical change in the process but acts as a catalyst. *Origin.* Chlorophyll has its origin in a colourless substance called *leucophyll*; the latter is first converted into proto-chlorophyll (or chlorophyllogen) which is finally transformed into chlorophyll in the presence of light.

Chemical Composition of Chlorophyll

Chlorophyll *a* —$C_{55}H_{72}O_5N_4Mg$ Carotene —$C_{40}H_{56}$
Chlorophyll *b* —$C_{55}H_{70}O_6N_4Mg$ Xanthophyll —$C_{40}H_{56}O_2$

(3) **Chromoplasts** (*chroma,* colour). These are variously coloured plastids—yellow, orange and red. They are mostly present in the petals of flowers and in fruits, and the colouring matters (pigments) associated with them are **xanthophyll** (yellow) and **carotene** (orange-red). Various other colours are formed as a result of combinations of red, yellow and green. The function of pigments occurring in flowers is to attract insects for cross-pollination (see pp. 100-01).

Carotenoids. A number of pigments—yellow, orange and sometimes red—are found in plants; these are collectively called carotenoids. They may be divided into two main groups—**carotenes** and **xanthophylls**. They are always associated with chlorophyll, but may also occur independently in any part of the plant body. They are not connected with photosynthesis. All carotenoids are insoluble in water but dissolve readily in ethyl ether. They are not easily destroyed by heat or light. A **carotene** is a hydrocarbon, i.e. it consists of carbon and hydrogen, its formula being $C_{40}H_{56}$. It is readily soluble in petroleum ether. A **xanthophyll** on the other hand is an oxidation product of carotene, i.e. it has oxygen in addition, its formula being $C_{40}H_{56}O_2$. It is readily soluble in ethyl alcohol but not in petroleum ether.

Anthocyanins. Colours of most violet or purple and blue flowers and also of many red and brown ones are, however, due to pigments—**anthocyanins**, *dissolved in the cell-sap.* Anthocyanins occur in flowers, in coloured roots, e.g. beet-root, and in coloured stems, e.g. balsam stem. They also occur in the variegated leaves of garden crotons and amaranth, and in the young red leaves of many plants, e.g. mango, country almond, etc., and frequently mask the chlorophyll. They possibly serve as a screen to the chloroplasts, protoplasm, etc., and protect them against strong sunlight. Occurring in flowers they, of course, serve to attract insects for pollination. They are soluble in water and alcohol. When a coloured leaf is boiled with water, anthocyanins are extracted and the leaf then appears greenish due to the presence of chlorophyll. Anthocyanins are capable of changing their colour. The colour depends upon the reaction of the sap of the cells in which they occur; the colour is red when the sap is acid, and it is blue when the sap is alkaline in reaction. There are various other pigments also in the plant body.

Other Cytoplasmic Bodies (FIG. 6). (1) **Centrosome** is a minute body occurring close to the nucleus of animal cells and in those of lower plants (many algae and fungi). It is not found in seed plants. The centrosome has usually two central granules called **centrioles** which are deeply stainable. During nuclear division the centrioles pass on to the two opposite ends of the cell and organize the nuclear spindle. They are also associated with the formation of the cilia in motile ciliate gametes.

(2) **Mitochondria** are very small bodies, usually $0.2\text{-}2\mu \times 3\text{-}5\mu$, present often in very large numbers in the cytoplasm of both plant and animal cells. They occur in the form of rods or filaments, or as somewhat spherical or sausage-shaped bodies. They are mostly distributed throughout the cytoplasm, but in certain cells they cluster round the nuclear membrane or beneath the plasma membrane. They are, however, absent in bacteria and blue-green

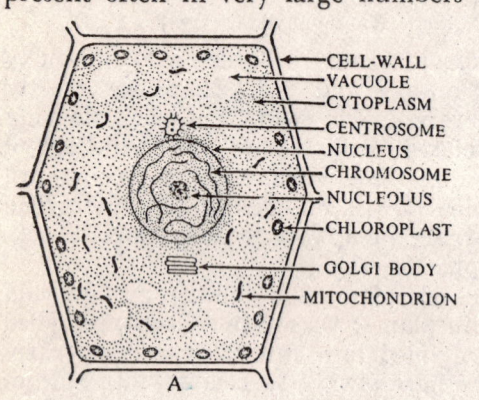

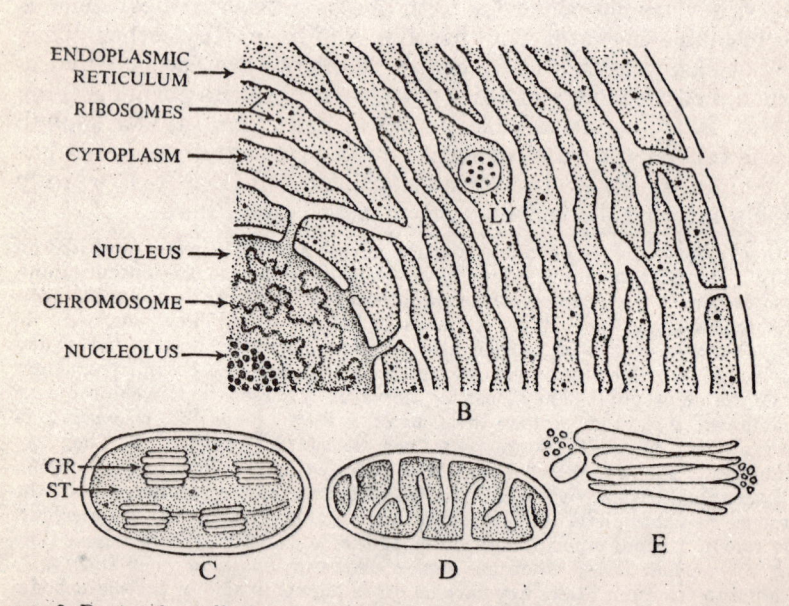

FIG. 6. Parts of a cell. *A*, as seen under a compound microscope; *B-E*, as seen under an electron microscope; *B*, a portion of the cell (*LY*, lysosome); *C*, a chloroplast (*GR*, granum; *ST*, stroma); *D*, mitochondrion; *E*, Golgi body.

THE CELL

algae. The electron microscope reveals the following organization of a mitochondrion: (*a*) an outer membrane, (*b*) an inner membrane which is thrown into folds (called *cristae*) into the cavity of the mitochondrion, and (*c*) a granular matrix filling up the cavity. The mitochondria are largely composed of protein and phospholipid (a phosphorus-containing fatty substance). They have certain characteristic staining properties and can be demonstrated only after special treatment. They are easily destroyed by acid-containing fixatives. They multiply by division, and are carried down to the next generation through the reproductive cells. Mitochondria are now regarded as very important constituents of plant and animal cells inasmuch as they provide *energy* for vital activities of the living cells, being closely associated with respiration and photosynthesis. Most of the oxidative enzymes of the respiratory cycle, particularly of the Krebs cycle, are located in them. Several oxidative enzymes concerned with the breakdown of glucose and other food materials during respiration, releasing energy, are also located in them. ATP (adenosine triphosphate), an energy-rich phosphate compound, commonly used to phosphorylate glucose to 'active' glucose-phosphate, is formed in the mitochondria. Since mitochondria are abundant in the young metabolic tissue they may also be connected with other physiological activities. They are in fact the powerhouses of the cell, from which emanates the necessary energy for vital work. According to some investigators the mitochondria are regarded as precursors of different types of plastids. According to recent studies this view is not, however, held valid.

(3) **Golgi Bodies** occur as a packet of tiny, elongated, flattened sacs in the cytoplasm of certain types of cells as seen after a special method of preparation. They appear as net-like bodies under the compound microscope. Under the electron microscope the Golgi body is seen to consist of (*a*) a stack of parallel, flattened sacs, each with two membranes, (*b*) some conspicuous vacuoles between the two membranes and (*c*) clusters of very small vesicles which are pinched off ends of sacs. This whole structure is otherwise known as *Golgi complex*. The Golgi body was first noted in animal cells (nerve cells of cat and owl) by Golgi, an Italian physician, in 1898. Although Golgi bodies are common in animals many plant cells are now known to contain them. In gland cells of animals they are associated with secretions (certain enzymes, hormones, etc.), and also storage of proteins, but in plant cells their significance is still obscure. Golgi bodies are composed of protein and phospholipid in almost equal quantities.

(4) **Endoplasmic Reticulum** is an intricate network of tube-like structures distributed extensively throughout the cytoplasm, as revealed by the electron microscope only, and first elaborately described by Porter (an American) in 1960-61. It is noted that some of the tubes of the reticulum connect the nuclear membrane and after extending through the cytoplasm open on the cell-membrane. The tubes in the first instance increase the surface area of the cytoplasm for metabolic activities of the cell. Further they appear to be associated with enzyme formation, protein synthesis, storage and transport of metabolic products. In nuclear division they contribute to the formation of the cellplate and the new nuclear membrane around each daughter nucleus.

(5) **Ribosomes.** Associated with the membrane of the endoplasmic reticulum and also occurring free in the hyaloplasm are seen extremely minute particles in abundance; they are called ribosomes. They are composed of nucleoprotein, particularly RNA (ribonucleic acid) and protein. Ribosomes are the main seats of protein synthesis.

(6) **Lysosomes.** The electron microscope also reveals other tiny, usually spherical cytoplasmic particles with an outer membrane and dense contents; they are called lysosomes. They have been found in many animal cells and in the meristematic cells of a few plants. They are associated with intracellular digestion. Several enzymes, particularly acid hydrolases (but not oxidative enzymes), are seen to remain confined to them.

The Cell-wall

Formation and Structure of the Cell-wall. The cell-wall is a constant feature of all plant cells. Each cell is bounded by a non-living wall, thick or thin, according to the nature of the cell, forming an elastic or semi-rigid framework around it. Such a framework around the cell is called the **cell-wall.** It maintains the form of the cell and affords necessary protection to the protoplast. Besides, the cell-walls form the skeleton of the plant body and are responsible for its strength, rigidity and flexibility. The cell-wall is a laminated structure, i.e. it consists of layers laid down by the protoplasm, one against another. In a young cell immediately following nuclear division in it a thin cell-plate (see FIG. 20 I) is formed across its centre (equator). It extends to the lateral walls in the form of a complete layer or septum, thus dividing the cell into two daughter cells. This original thin layer which later stands as the middle layer of the cell-wall is called the **middle lamella.** It is composed of calcium pectate (calcium salt of pectic acid), and acts as a cementing material firmly holding together the contiguous cells. As the cells enlarge, a thin wall is deposited on either surface of the middle lamella by the protoplasm of each cell. This wall is called the **primary wall,** and it is composed of cellulose, hemicellulose and pectose in varying proportions. The primary wall is very thin and elastic in nature so that it can keep pace with the growth of the cell. This may be the only wall in several types of cells even when they mature, as in most parenchyma, collenchyma, cambium, mesophyll, etc. But in certain other cells, as they mature, the cell-wall may thicken by the addition of new layers laid down by the protoplasm of each cell on each surface of the primary wall. This later-formed, thickened wall is called the **secondary wall,** and it is composed exclusively of pure cellulose. The secondary wall is tough and has high tensile strength. It is apparent that the cell-wall as a whole connecting two adjoining cells now comprises a middle lamella, two primary walls and two secondary walls in their sequence of development. In special cases, as in woody tissues, e.g. tracheids, vessels, wood fibres, bast fibres, stone cells, etc., the secondary wall becomes further thickened on each side due to deposition of lignin, and this thickening may take special patterns (see FIG. 8). With the development of the lignified secondary walls the cells often lose their protoplasmic contents and become dead. The whole wall including the primary wall and the middle lamella may often be strongly lignified. It is further seen that the cytoplasm of one cell is connected with that of the adjoining cell by fine cytoplasmic strands which extend through extremely minute pits that are left in the cell-wall during its formation. These cytoplasmic strands are called **plasmodesmata** (sing. plasmodesma; FIG. 7). They were first discovered by Tangl in 1879 and later more elaborately studied by Strasburger in 1901. According to Pfeffer (1896) and Sharp (1926) the strands are connected with the transmission of stimuli from cell to cell, and also translocation of nutritive materials, particularly in storage tissues.

The electron microscope reveals a complex structure of the cell-wall. The wall is seen to be made of an interwoven network of extremely fine strands. These strands are the chain molecules of cellulose and occur in the cell-wall in bundles of 100-170 (up to 200) to form the smallest structural unit called *micelle*. Approximately 20 micelles may be associated together to form a larger structural unit called the *microfibril*. Micelles and microfibrils are detectable only in electron micrograph. A *fibril* considered to be an aggregation of 25 microfibrils is the smallest unit visible under a light microscope. The wall develops in successive layers, and each layer consists of a fine network of cellulose strands. The primary wall consists of one layer in which the micelles extend transversely or somewhat obliquely with reference to the long axis of the cell, forming a loose network with its meshes filled with pectic compounds. This arrangement helps elongation of the cell-wall with the growth of the cell. The secondary wall, when formed, commonly consists of three layers, sometimes more, on each side of the primary wall, in which the microfibrils are laid down in different directions in the successive layers, forming a compact network in each case, evidently adding to the greater strength of the cell-wall. At this stage nine layers may be counted in the cell-wall as a whole. Cellulose, no doubt, is the chief constituent of the cell-wall but as the wall grows in thickness various new substances (see pp.166-67) are freshly deposited in the meshes of the network.

Plasma Membrane. This is an extremely thin hyaline living membrane covering the cytoplasm as a surface layer of it. It is mainly made of fat and protein. In plant cells it lies adpressed against the cell-wall and is hardly distinguishable from it except under special treatment, but in animal cells (cell-wall being absent) this membrane forms the boundary of each cell. The plasma membrane plays a very important role in the physiology of the cell. It has a selective power unlike the dead cell-wall, allowing only certain materials to pass through it into the cell and out of it. Large molecules of proteins, fats and carbohydrates cannot pass through it. Such a membrane having a selective transmitting power is said to be semipermeable or differentially permeable. The electron microscope reveals some pores in the membrane. These pores may help the diffusion process of materials that have to enter the cell or leave it. This outer plasma membrane is otherwise called ectoplasm; while a similar plasma membrane surrounding the vacuole is called tonoplasm (see p. 146). This membrane is also semipermeable in nature like the outer one.

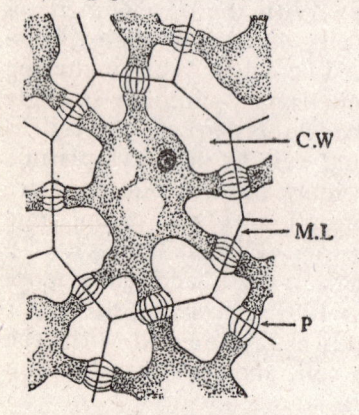

FIG. 7. Cells from the endosperm of date seed. *C.W*, cell-wall (hemicellulose, see p. 169); *M.L*, middle lamella; *P*, plasmodesma.

Secondary Thickening of the Cell-wall (FIG. 8). The secondary thickening of the cell-wall may be more or less uniform all round the cell, almost always showing a stratified appearance. But in those cells which have ultimately to grow up into vessels (see FIG. 37) and tracheids (see FIGS. 34-6) it may be localized in particular portions of the wall in special patterns, and is due to the deposit of a chemically complex and hard substance, called *lignin*, in the meshes of the cellulose network. Such thickening takes place only after the aforesaid elements have grown and attained their full dimensions. The thickening being localized, a portion of the wall may remain unthickened. The patterns of thickening may be as follows:

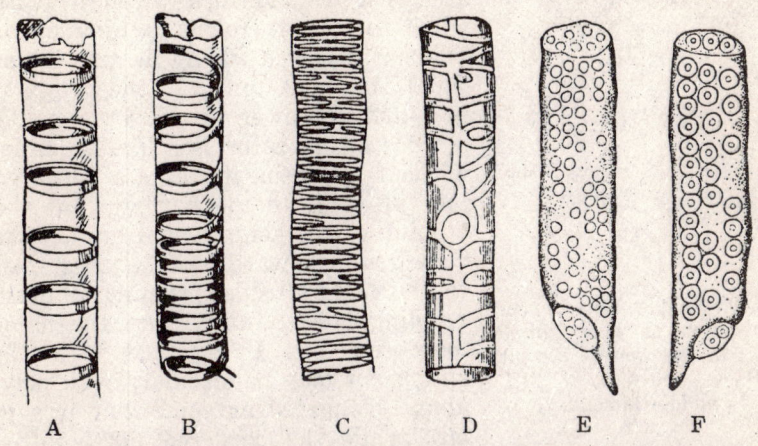

FIG. 8. Thickening of the Cell-wall. *A*, annular; *B*, spiral; *C*, scalariform; *D*, reticulate; *E*, pitted (with simple pits); *F*, pitted (with bordered pits).

(1) **Annular** or **ring-like** (A), when the deposit of lignin is in the form of rings which are placed, one a little above the other, in the interior of the original cell-wall, the remaining portion of the wall being unthickened.

(2) **Spiral** (B), when the thickening takes the form of a spiral band.

(3) **Scalariform** or **ladder-like** (C), when the thickening matter or lignin is deposited transversely in the form of rods or rungs of a ladder, and hence the name scalariform or ladder-like. Unthickened portions of the wall appear as transverse pits, while the thickened spaces between them give a ladder-like appearance to the wall.

(4) **Reticulate** or **netted** (D), when the thickening takes the form of a network, evidently leaving a number of irregular unthickened spaces in the wall.

(5) **Pitted** (E-F), when the whole inner surface of the cell-wall is more or less uniformly thickened, leaving here and there some small unthickened areas or cavities. These unthickened areas are called **pits**, and are of two kinds, viz. (*a*) **simple pits** and (*b*) **bordered pits**. Pits are formed in pairs lying against each other on the opposite sides of the wall. The portion of the original wall separating the two opposing pits is called the *closing mem-*

brane. The closing membrane in the bordered pits shows a slight swelling or thickening in the middle, called **torus** (FIG. 11 B-C). When the area of a pit is uniform throughout its whole depth, it forms a simple pit (FIGS. 9-10); and when this area is unequal, broader towards the original wall and narrower towards the cavity of the cell, more or less like a funnel without the stem, it forms a bordered pit (FIG. 11). In the bordered pit the adjoining thickening matter of the wall grows inwards and arches over the pit from all sides forming an overhanging border, and hence the name 'bordered' pit. In surface view the simple pit may be circular, oval, polygonal, elongated or somewhat irregular; while the bordered pit is often circular or oval. Pits are areas through which diffusion of liquids takes place more easily. In bordered pits this diffusion is regulated to a great extent by the torus which, when pushed

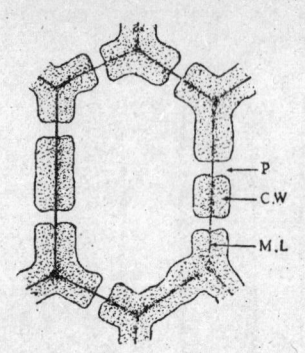

FIG. 9. Simple Pits. A cell in section showing simple pits in its wall; *P*, pit; *C.W*, cell-wall; *M.I.* middle lamella.

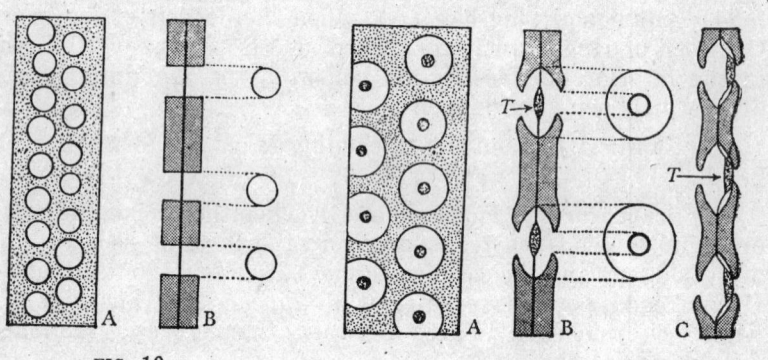

FIG. 10.

FIG. 11.

FIG. 10. Simple Pits. *A*, cell-wall with simple pits (surface view); *B*, the same (sectional view). FIG. 11. Bordered Pits. *A*, cell-wall with bordered pits (surface view); *B*, the same (sectional view); *C*, the same showing the torus (*T*) pushed against the pit blocking it.

from one side, blocks the pit (FIG. 11C). Through simple pits, contained in the living cells, diffusion of protoplasm also takes place. Bordered pits are abundantly found in the tracheids of conifers (e.g. pine; FIGS. 12, 34 & 36) and in the vessels of angiosperms; simple pits are also found in them, but they are more frequent in some of the living cells, and occur largely in the wood parenchyma, medullary rays, phloem parenchyma, companion cells, etc. Fibres are often provided with simple oblique pits and sometimes also with bordered pits, and stone cells with simple, branched pits.

Chemical Nature of the Cell-wall. The chemical substances which enter into the composition of the cell-wall are mainly pectin, cellulose, lignin, cutin, suberin and mucilage. Many mineral matters may also be introduced into the cell-wall.

FIG. 12. Tracheids with bordered pits of pine stem (diagrammatic).

Cellulose. Cellulose, an insoluble carbohydrate, is universally present in the cell-walls of all plant cells with the exception of fungi. In fact it is the chief constituent of the cell-wall. Associated with cellulose are other compounds deposited at different stages of wall formation (see p. 160). In the primary wall, as already mentioned, cellulose is associated with hemicellulose and pectose, making the wall soft and elastic. In the secondary wall formed later almost pure cellulose occurs, making the wall stiff but flexible. Later still cellulose may in special cases be associated with lignin, cutin, suberin, etc. Cellulose is a soft, elastic and transparent substance, and is readily permeable to water but insoluble in it. Seed fibres such as cotton and kapok are made of pure cellulose, while bast fibres and woody tissues are predominantly made of cellulose impregnated with lignin (lignocellulose). Chemically cellulose is a polysaccharide represented by the formula $(C_6H_{10}O_5)n$. Cellulose molecules are mostly long straight chains of D-glucose units. *Origin.* Sponsler is strongly of opinion from his X'ray studies in 1929 that cellulose is directly transformed from glucose in the presence of protoplasm. *Uses.* Cellulose is a very important substance in many respects. Many micro-organisms utilize cellulose as food. They secrete an enzyme *cellulase* to hydrolyse it. It forms a major part of the food for the herbivorous animals. They, however, have to depend for its digestion on the enzyme secreted by such bacteria which dwell in their digestive canal. Cellulose, however, cannot be digested by human beings. Economically articles like paper, cellophane, gun-cotton, celluloid, rayon (artificial silk), lacquer, etc., which have worldwide uses, are prepared from cellulose.

Hemicellulose is a mixture of different organic compounds (and not chemically allied to cellulose). It occurs in the cell-walls which may sometimes become heavily thickened. There is a thick deposit of it in the cell-walls of the endosperm of date seed (see FIG. 7) and certain other palm seeds. It is stored there as a reserve food for the use of the embryo, and is often called **reserve cellulose**. When the seed germinates it becomes converted into glucose and other compounds by the action of the enzyme *cytase*. In the vegetable ivory palm (*Phytelephas*) the reserve cellulose is very hard and is called vegetable ivory. Billiard balls, buttons, handles of sticks and umbrellas, etc., are made from it. It is also abundant in many stony seeds and fruits, and in woody tissues of certain trees, e.g. apple tree.
Pectic compounds occur in plants in three forms: insoluble pectose (protopectin), soluble pectin and insoluble pectic acid. They occur in the cell-walls, and one form may change into the other in the plant body. Pectin swells in water into mucilage, as the gelatinous sheath of many algae. It is present in many fruits and vegetables, and is responsible for the setting of jellies made from certain fruits. Protopectin present in the primary cell-walls acts as a binding or cementing material holding together the cells of the plant body. Protopectin and hemicellulose associated with cellulose, as deposited in the corners of collenchymatous cells, make them elastic.

Lignin. Lignin is deposited in the meshes of the network formed by cellulose microfibrils in the cell-wall (see p. 161), and is responsible for considerable thickening and strengthening of the secondary cell-wall, as in xylem elements. It may however, be present in the middle-lamella, primary wall and secondary wall, as in woody tissues, i.e. the whole wall may be lignified. Lignin is a hard and chemically complex substance. Its exact chemical composition is not known; possibly it is a mixture of several organic compounds and occurs in different forms in the hard and woody tissues. Lignified cells are usually thick-walled and always dead. Although hard, lignin is permeable to water. Sclerenchyma, sclereids, bast fibres, tracheids, wood vessels and wood fibres are common lignified structures. The function of lignified tissues is mechanical, i.e. they contribute to the rigidity of the plant body.
Cutin. Cutin is a mixture of some waxy substances. Associated with cellulose and often some pectic compounds it forms a definite layer, sometimes of considerable thickness, called the **cuticle**, on the skin (outer surface of the epidermal layer) of the stem, leaf and fruit. Cutin makes the cell-wall impermeable or very slightly permeable to water. Its function, therefore, is to prevent or check evaporation of water from the exposed surfaces of the plant body.

Suberin. Cell-walls of certain tissues may be charged with another waxy substance, called suberin. Like cutin it is also a mixture of some waxy substances, and is, therefore, allied to cutin. The constituent fatty acids are, however, different in the two cases. Suberin occurs in the walls of cork cells, and also in the endodermis and exodermis of roots of several plants. Being waxy in nature it makes the cell-wall almost impervious to water and, therefore, like cutin it also prevents or checks evaporation of water. The bark of cork oak (*Quercus suber*) on the Mediterranean coasts is the source of bottle cork used for this purpose.

Mucilage. Mucilage is a slimy substance widely distributed in plants in their different parts. Chemically it is a complex carbohydrate. Its physical property is that it absorbs water greedily, retains it tenaciously and forms a viscous mass; but when dry, it is very hard and horny. It is insoluble in alcohol. Mucilage is copious in the fleshy leaves of Indian aloe (*Aloe*; B. GHRITAKUMARI; H. GHIKAVAR). It is also abundant in the flowers of China rose, in the fruits of lady's finger, in the branches and leaves of Indian spinach (*Basella*; B. PUIN; H. POI), and in the seeds of linseed (*Linum*), flea seed (*Plantago*; B. ISOBGUL; H. ISOBGOL), Lallimantia (B. & H. TOPMARI), etc. Such seeds, when wetted, swell up and become mucilaginous. Mucilage also occurs in the fleshy leaves of desert plants.

In the majority of fungi and also in certain algae (but not in higher plants) the cell-wall is made of *chitin*—a substance somewhat allied to cellulose. Chitin, however, is peculiar to animals.

Besides, various mineral crystals such as silica, calcium oxalate and calcium carbonate may be introduced into the cell-walls, but not as integral parts of them. In several plants certain organic compounds like tannin, resin, gum, lipids or fatty substances, organic acids, etc., may also enter the cell-walls.

Micro-chemical Tests of the Cell-wall

Reagents	Cellulose	Lignin	Cutin and Suberin	Mucilage
1 Iodine solution	pale yellow	deep yellow	deep yellow	—
2 Chlor-zinc-iodine	blue or violet	yellow	yellowish brown	—
3 Iodine solution + sulphuric acid or zinc chloride	blue	brownish	deep brown	violet
4 Aniline sulphate (acid)	—	bright yellow	—	—
5 Phloroglucin (acid)	—	violet red	—	—
6 Caustic potash solution (concentrated)	—	—	yellow and brown	—
7 Potash + chlor-zinc-iodine	—	—	violet	—
8 Chlorophyll solution	—	—	green	—
9 Sudan IV	—	—	red	—
10 Methylene blue	—	—	—	deep blue

Cell Inclusions (non-living)

Various chemical substances appear in the plant body as products of metabolism or as by-products. These are called **ergastic substances** and include a number of compounds of varied nature. They may occur in the vacuole (particularly the soluble ones), or in the cytoplasm (particularly insoluble ones) or even in the cell-wall (particularly some of the waste products). The various such substances are carbohydrates, proteins, and fats and oils, and these constitute the *food* of plants and animals, supplemented by vitamins and essential minerals. A number of other compounds, whose utility is only imperfectly known are also formed in several plants as by-products, commonly called *waste products*. They are tannins, essential oils, resins, gums, etc. (see pp. 176-80).

1. Carbohydrates. These are substances containing carbon, hydrogen and oxygen. Of these, hydrogen and oxygen occur in the same proportion as they do in water. The general chemical formula is $C_x(H_2O)_y$. When these substances are heated they become charred, forming a black mass. This black mass is carbon. The water escapes and the carbon is left behind. From the economic standpoint carbohydrates are very important. Many of them are extensively consumed as food, many are largely employed in various industries as in the manufacture of fabrics and paper, and many are used for production of alcohol.

Classification of Carbohydrates. (*a*) **Monosaccharides** (sugars) are the simplest carbohydrates, i.e. they cannot be hydrolysed further into simpler ones. They may be (i) *pentoses* with five carbon atoms—($C_5H_{10}O_5$), of which the common ones found in plants are **arabinose** (occurring as a constituent of glycosides) and **ribose** (occurring as a constituent of nucleic acids), and (ii) *hexoses* with six carbon atoms—$C_6H_{12}O_6$, of which the common ones widespread in plants are **glucose** and **fructose.**

(*b*) **Disaccharides** (sugars) are represented by the formula $C_{12}H_{22}O_{11}$; these are most abundant in higher green plants, e.g. **sucrose** and **maltose.**

(*c*) **Polysaccharides** (non-sugars) are represented by the formula $C_6H_{10}O_5$; these are condensation products of simple sugars, e.g. **inulin, starch, dextrin** and **glycogen** (cellulose and hemicellulose also belong to this group).

(*d*) **Compound carbohydrates** (non-sugars) are complex carbohydrate-molecules, e.g. **gums, mucilages, tannins** and **glycosides.**

Hydrolysis of Carbohydrates. The more complex forms of carbohydrates may be readily hydrolysed by acids or by hydrolytic enzymes into simpler forms, soluble and easily diffusible. Thus starch on boiling with sulphuric acid is converted into glucose through certain intermediate stages, as follows: starch (blue with iodine) + $H_2O \rightarrow$ less diffusible dextrin (red with iodine) + $H_2O \rightarrow$ more diffusible dextrin (no colour with iodine) + $H_2O \rightarrow$ maltose + $H_2O \rightarrow$ glucose. In the germinating seeds and elsewhere starch undergoes similar hydrolysis under the action of certain hydrolytic enzymes such as *diastase* (*amylase*) and *maltase*, as follows:

$$2n(C_6H_{10}O_5) + nH_2O + diastase \rightarrow nC_{12}H_{22}O_{11} \text{ (maltose)};$$
$$C_{12}H_{22}O_{11} + H_2O + maltase \rightarrow 2C_6H_{12}O_6 \text{ (glucose)}$$

(1) **Sugars.** Sugars are sweet, crystalline, white and soluble substances and are of various kinds such as grape-sugar or **glucose** (a reducing sugar) chiefly found in grapes; fruit-sugar or **fructose** (a reducing sugar) found in many fruits associated with glucose; it may be readily formed from glucose or by hydrolysis of sucrose; cane-sugar or **sucrose** (a non-reducing sugar) chiefly found in sugarcanes and beet-roots; and malt-sugar or **maltose** (a reducing sugar) formed by the action of *diastase* (an enzyme) on starch and, therefore, commonly present in germinating seeds, particularly cereals. Grape-sugar is the simplest of all carbohydrates and is formed in the leaves by chloroplasts *in the presence of sunlight*. Other forms of carbohydrates are derived from it under the action of specific enzymes. Glucose travels in the plant body as such until it reaches the storage tissues, where it is mostly converted into starch, an insoluble carbohydrate, and deposited for a shorter or a longer period. This starch may again be converted into sugar. The chemical formula of grape-sugar is $C_6H_{12}O_6$, and that of cane-sugar $C_{12}H_{22}O_{11}$. Glucose contents of grapes are 12-15% or more, of apples 7-10%, and of plums 3-5%; sucrose contents of sugarcanes 10-15% and of beet-roots 10-20%.

Test for Glucose. Add **Fehling's solution** or an alkaline solution of **copper sulphate** to it, and boil, a yellowish red precipitate of cuprous oxide is formed. Test for Sucrose. Boil sucrose solution with 1 or 2 drops of sulphuric acid, and then apply the test for glucose.

(2) **Inulin** (FIG. 13). Inulin is a soluble carbohydrate, and occurs in solution in the cell-cap. Like starch it is easily converted into a form of sugar. Inulin is present in the tuberous roots of *Dahlia* and some other plants of *Compositae*. When pieces of *Dahlia* roots are steeped in alcohol or glycerine for 6 or 7 days, preferably more, inulin becomes precipitated in the form of spherical crystalline masses. A section is then prepared

from one of the pieces and examined under the microscope. Inulin may also be precipitated by cutting rather thick sections from fresh material and keeping them in strong alcohol for about an hour. Under the microscope fully-formed inulin crystals are seen to be star- or wheel-shaped, and half-formed ones more or less fan-shaped. These crystals are deposited mostly across the cell-walls, and occasionally only in the cell-cavity. Sometimes these crystals are so large that they extend through many cells. Inulin has the same chemical composition as starch, viz. $(C_6H_{10}O_5)n$. When precipitated, inulin is easily recognized by its peculiar form.

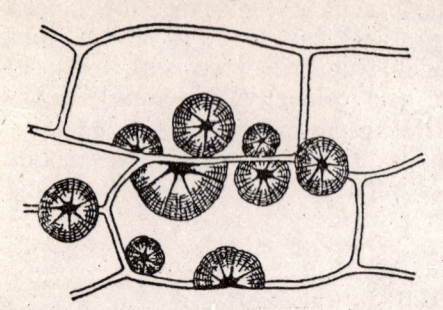

FIG. 13. Inulin crystals in the tuberous root of *Dahlia*.

(3) **Starch** (FIGS. 14-15). This is an insoluble carbohydrate occurring as a reserve food in the form of minute grains. Starch grains are of universal occurrence in plants with the exception

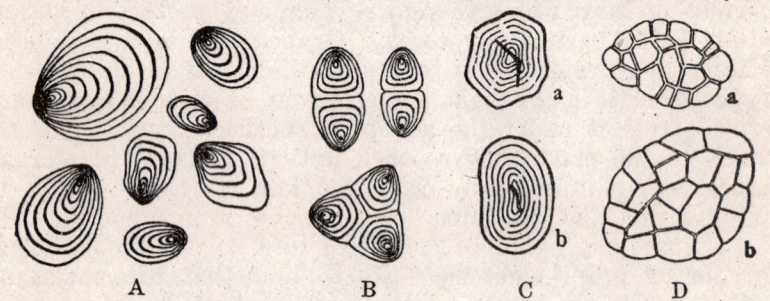

FIG. 14. Starch Grains. *A*, simple eccentric grains in potato; *B*, compound grains in the same; *C*, *a*, simple concentric grain in maize; *b*, ditto in pea; *D*, *a*, compound grain in rice; *b*, ditto in oat.

of fungi. They occur in almost all parts of a plant, but in storage tissues they are specially abundant. Cereals and millets, which constitute the staple food of mankind, are specially rich in starch. When required for nutrition it is converted into glucose. Starch grains are of various forms; they may be *rounded and flat*, as in wheat; *polygonal*, as in maize; nearly *spherical*, as in pea and bean; or more usually *oval*, as in potato; or rarely *rod-* or *dumb-bell-shaped*, as in the latex cells (see FIG. 40A). They also vary very much in size, the largest known being about 100μ or (1/10 mm.) in length, as in the rhizome of *Canna*, and the smallest about 5μ (or 1/200 mm.) in length, as in rice. In potato they are of varying sizes. Starch may be synthesized in the green cells by the chloroplasts, called *assimilation starch*, or in the non-green cells by the leucoplasts (amyloplasts), called *storage starch*. In any case it is always formed from glucose. Large and well-formed starch grains are always found in the storage organ. Starch is made of two components: amylose and amylopectin. Amylose is more soluble in water than amylopectin. The former turns deep blue with iodine solution, while the latter turns light blue with iodine solution. Waxy starch consists almost wholly of amylopectin which becomes viscous in solution. Mealy starch contains a high percentage of amylose. Both amylose and amylopectin are derived from glucose.

In the starch grain a definite roundish or elongated scar, called the **hilum**, may be observed; it represents the centre of origin of the grain. Around the hilum a variable number of layers or striations of different densities are alternately deposited. Each starch grain has thus a *stratified* appearance. When the layers are laid

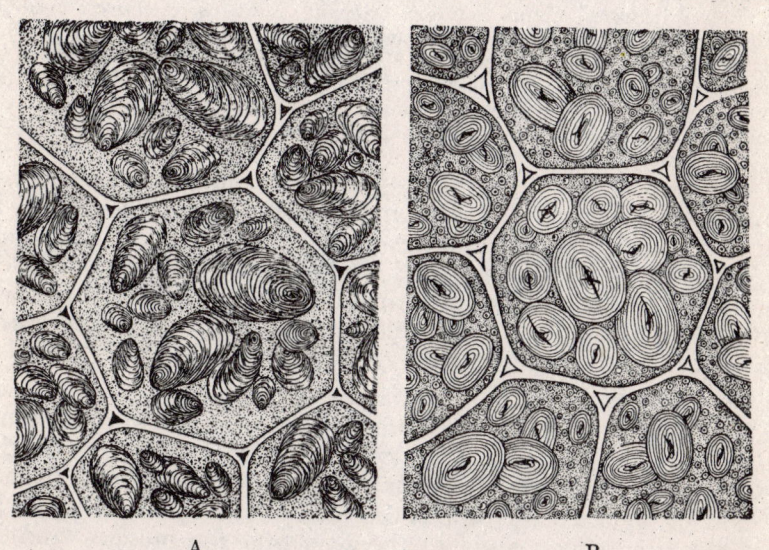

A   B

FIG. 15. Starch Grains (*contd.*). *A*, section through a potato tuber showing a few cells with eccentric grains; *B*, section through a cotyledon of pea showing a few cells with concentric grains (and small granules of protein).

down on one side of the hilum, as in potato, the grain is said to be **eccentric**, and when these are deposited concentrically round the hilum, as in wheat, maize, pea, bean and many pulses, the grain is said to be **concentric**. The former is more commonly met with than the latter. Starch grains may occur singly with one hilum, when they are said to be **simple**; sometimes, however, two or more grains occur together in a solid group with as many hila as there are grains in it: this group then is said to form a **compound** grain. It is seen that leucoplast simultaneously begins to secrete two or more grains very close to one another and subsequently as fresh layers are added to the grains these become pressed together into a mass; this mass is the compound grain; often the whole mass remains enveloped by a few common outer layers secreted by the leucoplast. Compound grains are found in the endosperm of rice and oat (FIG. 14D); a few compound grains are also often found in potato (FIG. 14B) and sweet potato. Starch has the same chemical composition as cellulose and inulin, viz. $(C_6H_{10}O_5)n$. It is insoluble in water and alcohol. Rice contains 70-80% of starch; wheat about 70%; maize about 68%; barley 66-65%; arrowroot 20-30%; and potato 20%.

Test for Starch. Starch turns **blue** to **black** when treated with **iodine solution**, the density of the colour depending on the strength of the reagent.

Uses of Starch. Apart from its use as food for both plants and animals including human beings, starch has a variety of industrial uses. When boiled in water, it forms a thin solution or paste which is extensively used in laundry, textile industry, paper industry, China clay industry, etc., as a sizing and cementing material. Starch is also widely used in the preparation of toilet powders, commercial glucose (by hydrolysis) and industrial alcohol (by fermentation) on a large scale. Sources of commercial starch mainly are potato, maize, tapioca, rice, wheat, sago-palm and arrowroot.

(4) **Dextrin.** This is formed as an intermediate product in the hydrolysis of starch and also in the synthesis of the latter. It does not, however, accumulate in the plant body as such, but occurs only as a transitory product. When starch is gently boiled or hydrolysed with a dilute mineral acid, it becomes converted into a whitish or yellowish, amorphous powder (dextrin) which dissolves in water and forms an adhesive paste (called British gum). It is used as a substitute for natural gums.

(5) **Glycogen.** This is a common form of carbohydrate occurring in fungi. In yeast, a unicellular fungus, it occurs to the extent of about 30% of the dry weight of the plant. It is not found in higher plants but is the major carbohydrate reserve in animals occurring mainly in the liver and the muscles, and is, therefore, sometimes called 'animal starch'. It occurs in the form of granules in the cytoplasm of the cell. Glycogen is a white amorphous powder and dissolves in hot water. It is coloured **reddish brown** with **iodine solution.** The colour disappears on heating and reappears on cooling. Its chemical formula is $(C_6H_{10}O_5)n$.

2. **Nitrogenous Materials.** The nitrogenous reserve materials occurring in plants are the various kinds of **proteins** and **amino-compounds** (amines and amino-acids).

(1) **Proteins.** Proteins[1] are very complex, nitrogenous, organic compounds essentially containing carbon (C), hydrogen (H), oxygen (O), and nitrogen (N). All plant proteins also contain sulphur (S), and many complex ones contain phosphorus (P) too. Of all the organic compounds, with the exception of protoplasm and nucleus, proteins are the most complex in their chemical composition, and various kinds of them are found in the plant body. In plants some proteins occur in solution and some as solids (either crystalline or amorphous), while most of them occur in a colloidal state. Some proteins are soluble in dilute salt solution, while all are soluble in weak acid or alkaline solution. Proteins are found in plenty in the storage tissue, less so in active (growing) tissue, and they are practically absent from mature, inactive tissue. A common form of insoluble or sparingly soluble protein abundantly found in the endosperm of the castor seed is the **aleurone grain** (FIG. 16). Each aleurone grain is a solid, ovate or rounded body, and encloses in it a crystal-like body known as the **crystalloid,** and a rounded mineral body called the **globoid.** The crystalloid occupies the wider part of the grain and is protein in nature, while the globoid occupies the narrower part and is a double phosphate of calcium and magnesium. The occurrence of crystalloid and globoid is not always constant in the aleurone grain. There may be one or more of them, or sometimes none at all. Aleurone grains vary in size. When they occur with starch they are very small, as in pea; but in oily seeds they are very much larger, as in castor.

Fatty seeds usually contain a higher percentage of proteins than starchy seeds, e.g. rice contains only 7% of proteins, wheat 12%, while sunflower seeds contain proteins as high as 30%. Starchy seeds of leguminous plants, however,

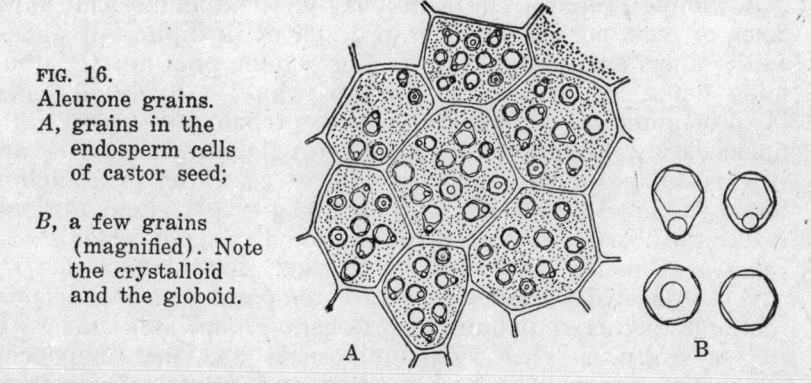

FIG. 16.
Aleurone grains.

A, grains in the endosperm cells of castor seed;

B, a few grains (magnified). Note the crystalloid and the globoid.

---

[1] Average percentage composition may be given thus: carbon—50-54%; hydrogen—about 7%; oxygen—20-25%; nitrogen—16-18%; sulphur—0.4%; and phosphorus—0.4%.

contain as high a percentage of proteins as fatty seeds, e.g. in the pulses there is an average of about 25% of proteins; in soya-bean (*Glycine max = Glycine soja*) protein contents are 35% or more; and, in groundnuts 31% of easily assimilable proteins.

Uses of Proteins. Proteins materially contribute to the building up of the plant body and the animal body. Half the dry weight of the protoplasm is made of proteins. The nucleus is predominantly made of nucleoprotein. All enzymes are proteins. Proteins are also a source of energy. In fact proteins are indispensable for sustaining life. It has been estimated that of the total proteins consumed by man plants account for 65% and animals 35%. An intake of about 75 gms of proteins per day may be considered adequate for human beings. For human diet, however, the amino-acids of animal proteins (milk, meat, fish and egg) are in better balance than those of plant proteins.

Tests for Proteins. (1) Proteins are coloured yellowish brown with strong **iodine solution** (see No. 5). (2) Some of them, e.g. albumins and globulins, coagulate on heating. (3) **Xanthoproteic reaction**—by adding some strong nitric acid a white precipitate is formed; on boiling it turns yellow. After cooling add a little strong ammonia and the yellow colour changes to orange. (4) **Millon's reaction**—add Millon's reagent (nitrate of mercury) and a white precipitate is formed; on boiling it turns brick-red. (5) **Biuret reaction**—an excess of caustic soda followed by a few drops of copper sulphate gives a violet colour which deepens on heating. (6) Treat a thin section of castor endosperm with 90% alcohol for 3-4 minutes and then with strong iodine solution. Mount it in thick glycerine and note under the microscope that the aleurone grains and the crystalloids turn deep brown, while the globoids remain colourless. Add 1% or 2% caustic soda solution to a fresh section and note that the aleurone grains get dissolved, while the globoids remain unaffected. Treat another section with dilute acetic acid and see that only globoids get dissolved.

Classification of Proteins.

A precise classification of proteins could not be devised yet because of our imperfect knowledge of them, their infinite numbers, similarity in general composition, and identical chemical reactions. However, a general classification based on their solubility and hydrolytic products is as follows:

A. **Simple Proteins.** On hydrolysis they yield amino-acids only. Some of them are soluble in pure water or in dilute salt water, while others are insoluble in it. The soluble ones are (1) albumins found in barley and in egg white; (2) histones and (3) protamines, both found in nuclei, probably associated with nucleic acids. The insoluble ones are (4) globulins mainly occurring as storage proteins in seeds and in egg white; (5) glutelins found in cereal grains, as in wheat and rice (stickiness of flour is due to them); and (6) prolamines found in many plant seeds, e.g. zein of maize, hordein of barley, and gliadin of wheat.

B. **Conjugated Proteins.** They are composed of a simple protein and another compound (a prosthetic group) associated with it. On hydrolysis they yield amino-acids and the component prosthetic group. They are (1) nucleoproteins (proteins + nucleic acids) occurring in DNA, RNA and viruses; (2) glycoproteins (proteins + carbohydrates) occurring in cell membranes; (3) lipoproteins (proteins + lipids, e.g. lecithin) found

THE CELL

in cell membrane, nuclear membrane and vacuolar membrane, and in egg yolk; (4) chromoproteins (proteins+a coloured pigment); (5) metalloproteins (proteins+a metal) as in many enzymes; and (6) phosphoproteins (proteins+phosphoric acid), e.g. casein of milk.

Hydrolysis of Proteins. When proteins are subjected to hydrolysis, i.e. when treated with a mineral acid or acted on by enzymes such as trypsin, they are finally resolved into amino-acids out of which they have been built up, after passing through intermediate stages, as follows: proteins→metaproteins→proteoses→peptones→polypeptides→amino-acids.

(2) **Amino-compounds.** Amino-acids and amines are the simplest forms of all nitrogenous food materials, and occur in solution in the cell-sap. They are abundantly found in the growing regions of plants, less frequently in storage tissues. When translocation is necessary, proteins become converted into amines and amino-acids. The amines and amino-acids travel to the growing regions where the protoplasm is very active, and they are directly assimilated by it. They are also the initial stages in the formation of proteins. They contain carbon, hydrogen, oxygen, and nitrogen, and in the amino-acids, cystine, cysteine and methionine sulphur is also present. Some of the other amino-acids are glycine, alanine, glutamic acid, aspartic acid, leucine, lysine, tyrosine, etc.

3. **Fats and Oils.** Fats and oils occur to a greater or less extent in all plants. They occur in the form of minute globules in the protoplasm of the living cells where they are formed, and cannot travel from cell to cell. In the 'flowering' plants often special deposits of them are found in seeds and fruits. But in starchy seeds and fruits there is very little fat. Fats and oils are composed of carbon, hydrogen and oxygen, but the last two do not occur in the same proportion as they do in water—the proportion of oxygen being always much less than in the carbohydrates. They contain no nitrogen. They are insoluble in water, but very readily soluble in ether, petroleum and chloroform. Comparatively few of them are soluble in alcohol, e.g. castor oil. Fats are synthesized in living bodies from fatty acids (mainly oleic, palmitic and stearic acids) and glycerine under the action of the enzyme *lipase*. Both these products, viz. fatty acids and glycerine, are derived from carbohydrates (sugar and starch) during respiration. They form an important reserve food with a considerable amount of *energy* stored in them. Their energy value is more than double that of the carbohydrates. When fats are decomposed the energy stored in them is liberated and made use of by the protoplasm for its manifold activities. Digestion

of fats into fatty acids and glycerine is also brought about by the enzyme *lipase*. Fats that are liquid at ordinary temperature are known as oils. In plants fats are usually present in the form of oils. Oils are of two kinds, viz. *fixed* or *non-volatile*, as described above, and *essential* or *volatile* (see p. 177)

A large number of them are used for food, for manufacture of soap, toilet products and oil-paints, for illumination, lubrication, etc., and are, therefore, of considerable economic importance, e.g. coconut oil, olive oil, sesame or gingelly oil, castor oil, groundnut oil, linseed oil, mustard oil, cotton seed oil, etc.

Tests for Fats and Oils. (1) **Osmic acid** (1% aqueous solution) stains them black. (2) Alcoholic solution of **Sudan III, Sudan IV** or **Sudan Red** stains them red. (3) Alcoholic solution of **alkanet** (or alkannin) stains them red, but the stain develops only after an hour or so. (4) Pressed against a paper they leave a permanent greasy mark on it.

Other Cell Products. It must also be noted that some of the cell products like vitamins, hormones and enzymes play important role in the physiology of the cell; these are described in detail later. Besides, a number of by-products, commonly known as waste products, also appear in different plants or in different parts of the same plant, either in isolated cells or in groups of cells; what role they play in the physiology of the cell is not, however, clear. Some of the common ones are as follows:

1. **Tannins.** These are a heterogeneous group of complex compounds widely distributed in plants. They commonly occur dissolved in the cell-sap, either in single isolated cells or in small groups of cells in almost all parts of the plant body. They are also found in the cell-walls, often abundantly in certain dead tissues, as in the bark and the heart-wood. In the leaves, young and old, and in many unripe fruits tannins are abundant. As the fruits ripen tannins disappear; they become converted into glucose and other substances. They are abundant in the fruits of myrobalans, e.g. emblic myrobalan (*Emblica officinalis*; B. AMLAKI; H. AMLA), chebulic myrobalan (*Terminalia chebula*; B. HARITAKI; H. HARARA), and beleric myrobalan (*Terminalia belerica*; B. BAHERA; H. BHAIRAH). Tea leaves contain about 18% of tannin. Catechu, a kind of tannin, is obtained from the heart-wood of *Acacia catechu*, and is also present in betel- or areca-nut. Tannin is a bitter substance, and that is why 'very strong' tea and fruits of myrobalans taste bitter. It is aseptic, i.e. free from the attack of parasitic fungi and insects. The presence of tannin makes the wood hard and durable. Tannins

have a variety of uses. Mixed with iron salts they are used in the manufacture of ink. They are extensively used in tanning, i.e. converting hide into leather. They are also used for various medicinal purposes. They turn **blue-black** with an **iron-salt** such as ferric chloride.

2. **Essential Oils.** These are volatile oils, and occur mostly in glands, known as *oil-glands* (see FIG. 41A). The transparent spots in the leaves of sacred basil, pummelo or shaddock, lemon, lemon grass (*Cymbopogon*), *Eucalyptus*, etc., and those in the rind of fruits like orange, lemon, shaddock, etc., are all oil-glands. They are also present in the petals of flowers of many plants, as in rose, jasmines, etc. The fragrant odour of such flowers is due to the presence of essential oils formed in them. They differ from fatty oils in their chemical composition as well as in being volatile. They are sufficiently soluble in water to impart to it their taste and odour. Being volatile, essential oils are obtained from plants by distillation, whereas fixed oils may be obtained by mere pressure. Like fatty oils, these are readily soluble in ether, petroleum, etc., and give the same tests (see p. 176). They are soluble in alcohol, but fixed oils are not. They commonly occur in mere traces; in some cases, however, they may occur to the extent of 1-2%. There are some 200 essential oils of commercial value. Some of the common ones are lemongrass oil, eucalyptus oil, clove oil, lavender oil, sandalwood oil, thyme oil, etc.

3. **Resins.** These are mostly found in the stems of conifers, and occur in abundance in special canals or ducts, known as *resin-ducts* (see FIG. 27). They are yellowish solids, insoluble in water but soluble in alcohol, turpentine and spirit. When present in the wood, resins add to its strength and durability. They occur associated with a small quantity of turpentine which is removed by distillation, and the residue is pure resin.

4. **Gums.** Gums are formed in various kinds of plants as a result of breakdown of cell-walls and cell-contents (gummosis). They are insoluble in alcohol but soluble in water, readily swell up in it, and form a viscous mass. They are found in many phanerogamic plants, and are of various kinds. *Acacia senegal* yields the best gum-arabic of commerce. Gums also occur in mixtures with resins.

5. **Mineral Crystals.** The common forms of crystals consist of silica, calcium carbonate and calcium oxalate. They occur

either in the cell-cavity or in the cell-wall. Of them crystals of calcium oxalate are most common, and are very widely distributed among various plants.

(1) **Silica** occurs as an incrustation on the cell-wall or it lies embedded in it. It is abundantly found in the leaves and stems of horsetail (*Equisetum*) several grasses, e.g. lemon grass, rice straw, wheat straw, etc., and in several algae with silicified structures, e.g. diatoms. Wheat straw contains about 72% of silica, rye straw about 50% and *Equisetum* about 71%.

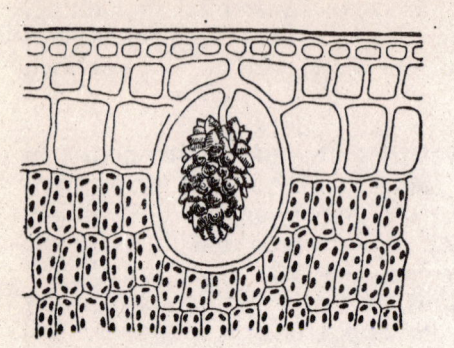

FIG. 17. Mineral Crystals. Cystolith in the leaf of india-rubber plant.

(2) **Calcium carbonate** occurs in the form of a crystalline mass, often pear-shaped in appearance, in the leaf of *Ficus* (e.g. india-rubber plant, banyan, etc.). This crystalline mass is called **cystolith** (FIG. 17), and it remains enclosed in a large cell. In its formation the inner cellulose wall of the upper (multiple) epidermis extends inwards into the cell in the form of a stalk, and calcium carbonate is deposited on it as small crystals; finally the crystalline mass or cystolith looks like a bunch of grapes suspended from a stalk. When cystolith is dissolved, the cellulose matrix on which crystals are deposited shows stratifications and

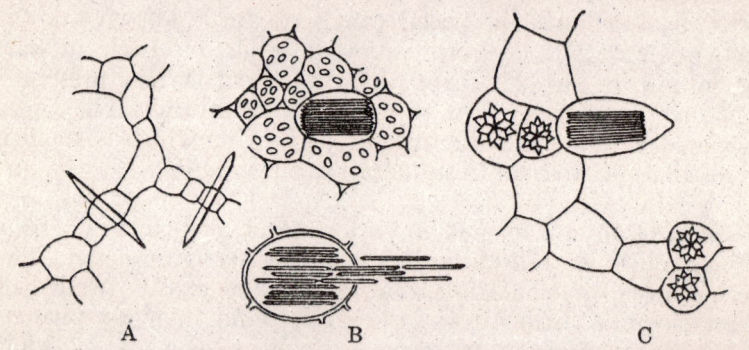

FIG. 18. Mineral Crystals (*contd.*). *A*, solitary raphides (two) in the petiole of water hyacinth; *B*, a bundle of raphides in the same; *bottom*, needles (raphides) shooting out; *C*, sphaero-crystals (four) and a bundle of raphides in taro (*Colocasia*).

radial striations. Cystolith is found in many plants of *Acanthaceae, Moraceae, Urticaceae, Ulmaceae, Cannabinaceae*, etc.

THE CELL

(3) **Calcium oxalate** occurs as crystals of various forms. (a) raphides, (b) conglomerate- or sphaero-crystals, and (c) octahedral and other forms. (a) **Raphides** (FIG. 18) are needle-like crystals occurring singly or in bundles. These are found in many plants in smaller or larger quantities, but are specially common in water hyacinth (*Eichhornia*), balsam (*Impatiens*), water lettuce (*Pistia*) and aroids such as *Colocasia, Alocasia, Amorphophallus,* etc. They are frequently shut off by a cell-wall from coming in contact with the protoplasm. (b) **Conglomerate crystals** or **sphaero-crystals** (FIG. 18C) are clusters of crystals which radiate from a common centre, and hence have a more or less star-shaped appearance. They are found in water lettuce (*Pistia*), taro (*Colocasia*), etc. (c) **Octahedral, cubical, prismatic** and **rod-like crystals** (FIG. 19) of calcium oxalate are also common in plants; they can be readily seen in the dry scales of onion.

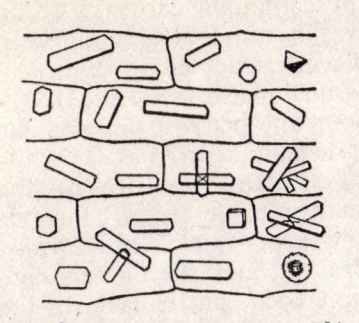

FIG. 19. Mineral Crystals (*contd.*). Various forms of calcium oxalate crystals in the dry onion scale.

Tests. (a) 50% nitric acid solution (or any other mineral acid) dissolves both calcium carbonate and oxalate crystals, but bubbles of CO_2 gas are evolved only in the case of carbonate crystals. (b) 30% acetic acid solution readily dissolves calcium carbonate crystals only, but not the oxalate crystals.

6. **Latex.** This is a milky juice found in latex cells and latex vessels (see FIG. 40). Latex occurs as an emulsion consisting of a variety of chemical substances. Among the nutritive materials sugars, starch grains (rod- or dumb-bell-shaped), proteins and oils are often found, and among the waste products gum, resin, tannin, alkaloids, rubber, etc., are common. Latex also contains some salts, enzymes, and often some poisonous substances, as in yellow oleander (*Thevetia*). The function of latex is not clear; perhaps in some way it is associated with nutrition, healing of wounds and protection against parasites and animals. Latex of *Hevea* is the source of rubber, and that of papaw (*Carica papaya*) contains a digestive enzyme called *papain*. Latex is commonly milky, as in *Ficus, Euphorbia*, etc., sometimes coloured (yellow, orange or red), as in opium poppy (*Papaver*), prickly poppy (*Argemone*), etc., or even watery, as in banana (*Musa*).

7. **Organic Acids.** Living cells give an acid reaction. The sour taste of many fruits, particularly of unripe ones, is due to the

presence of some kinds of such acids in them. Several organic acids are formed in plants and they contain a carboxyl group (-COOH). They are formed as a result of respiration in both anaerobic and aerobic phases. Oxalacetic acid, citric acid, aconitic acid, etc., are formed during aerobic phase (see FIG. III/34) but they do not accumulate in the plant body in appreciable quantities (with the exception of citric acid, as in *Citrus*); while oxalic acid (as in *Oxalis* and *Rumex*), malic acid (as in the leaves of *Bryophyllum*, *Cicer*, and many unripe fruits), tartaric acid (as in tamarind, pineapple and grape), etc., are formed as end-products of anaerobic respiration, and they accumulate in considerable quantities in the plant body, particularly in the leaves and the fruits. Many succulent plants also contain some such organic acids in abundance. Fatty acids and amino-acids are very important organic acids directly concerned in metabolism; for example, fatty acids combine with glycerol to form fats and oils, and different amino-acids combine to form proteins. All these organic acids are mostly the result of glucose breakdown during respiration, and the reactions are in many cases reversible.

7. **Alkaloids.** These are complex nitrogenous substances, and occur combined with some organic acids, mostly in seeds and roots of some plants. They may be by-products of nitrogen metabolism in plants. They have an intensely bitter taste and many of them are extremely poisonous. A few of them are liquids; the majority of them are, however, crystalline solids which are insoluble or sparingly soluble in water, but readily so in alcohol. Alkaloids are generally, but not universally, formed in the roots, and from there translocated to certain organs of a plant. There are over 200 known alkaloids found in plants, of which a few may be mentioned here. These are quinine and cinchonine in the root and stem of *Cinchona*, strychnine in the ripe seeds of nux-vomica (*Strychnos*), morphine in unripe fruits of opium poppy (*Papaver*), atropine in the leaves and roots of deadly nightshade (*Atropa belladonna*), nicotine in the leaves of tobacco (*Nicotiana*), daturine in the leaves and seeds of *Datura*, caffeine in the seeds of coffee and in the leaves of tea, solanine in the stem and ripe fruits of bitter-sweet (*Solanum dulcamara*), etc. The role played by the alkaloids in the physiology of plant cells is not known.

Contents of the Vacuole. The vacuolar sap may contain as much as 98% of water, and dissolved in it some of the reserve materials, secretory products and waste products. Although the cell-sap is specially rich in soluble substances, it often contains some solid bodies in amorphous or crystalline condition. (1) **Inorganic salts** such as chlorides, sulphates, nitrates and phosphates are always present in the cell-sap. (2) **Organic acids** such as oxalic, malic, citric, tartaric, etc., and their various salts are fairly common in the cell-sap. (3) **Soluble carbohydrates** are specially common in the cell-sap, e.g. grape-sugar in grapes, cane-sugar in sugarcane, inulin in *Dahlia*, etc. (4) Of the nitrogenous materials **soluble proteins, amines** and **amino-acids** occur dissolved in the cell-sap, particularly in the cells of the growing regions and to some extent only in the storage cell. (5) **Anthocyanins** frequently occur in the petals of flowers and also in the young leaves of some plants. (6) Some of the **enzymes** are also found in the cell-sap. (7) In some plants **mucilage** may be found in the cell-sap. (8) Some of the waste products are also common in particular cells. These are **tannins, latex, alkaloids** and **glucosides.** Glucosides are substances which on decomposition give rise to a kind of sugar together with other products—which are chiefly aromatic. Common glucosides are saponin of soap-nut, indican of the indigo-dye, amygdalin of bitter almond, etc. It is not known what role they play in the metabolism of cells. In many cases, however, glucosides are distinctly poisonous.

Formation of New Cells

Plants, however big and complicated their body may ultimately be, begin their existence as a single cell—the embryonic or egg-cell. But they have to grow from the initial stage to their normal size and form. Growth in them is initiated by the formation of new cells and their enlargement, and two processes are closely associated in this direction: first, division of the nucleus (mitosis) and second, division of the cell (cytokinesis). The division of the egg-cell begins often immediately after fertilization, sometimes after a resting period, and continues throughout the life of the plant, but later with the formation of tissues and organs cell-formation becomes mostly restricted to the meristematic regions such as the root-tip and the stem-tip. There are different methods of cell-formation, as found in different parts of the plant body or in different plants. The methods are as follows:

1. **Somatic Cell Division.** Cell division leading to the development of the vegetative body (soma) of the plant is known as somatic cell division. It includes the division of the nucleus, called **mitosis** (*mitos,* thread) or **karyokinesis** (*karyon,* nut or nucleus) or **indirect nuclear division,** and the division of the cytoplasm, called **cytokinesis.**

Mitosis (FIG. 20). In mitosis (first worked out by Strasburger, a German botanist, in 1875, and later more elaborately by Fleming, a German biologist, during 1879-82) the behaviour of chromosomes is the most important feature. In this process the metabolic nucleus (A) passes through a complicated system of changes, most essentially the longitudinal doubling of chromosomes and the even distribution of longitudinal halves among the two daughter nuclei, which can be studied in properly fixed and stained preparations of the root-tip or the stem-tip. The changes comprise four stages: prophase, metaphase, anaphase and telophase.

Prophase. This is the longest mitotic stage. The metabolic nucleus contains in its karyolymph numerous crooked, often coiled, delicate threads called **chromonemata**, not recognizable as distinct entities. The first sign of the prophase is the appearance of a certain number of distinct, slender threads called **chromosomes** (B). The term chromosome was first introduced by Waldeyer in 1888. The chromosomes, particularly the longer ones, are more or less spirally coiled. Close scrutiny shows that the individual chromosomes are always longitudinally double, with the two threads remaining closely coiled about each other throughout their length, and each longitudinal half of the chromosome is called a **chromatid**. Chromosomes are composed of nucleoproteins, and are the vehicles of genes or heredi-

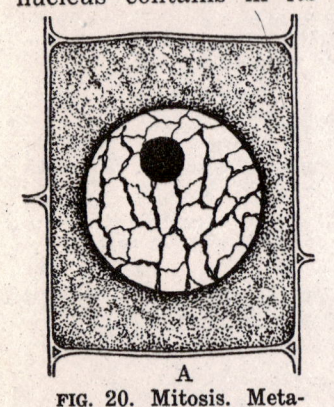

FIG. 20. Mitosis. Metabolic nucleus.

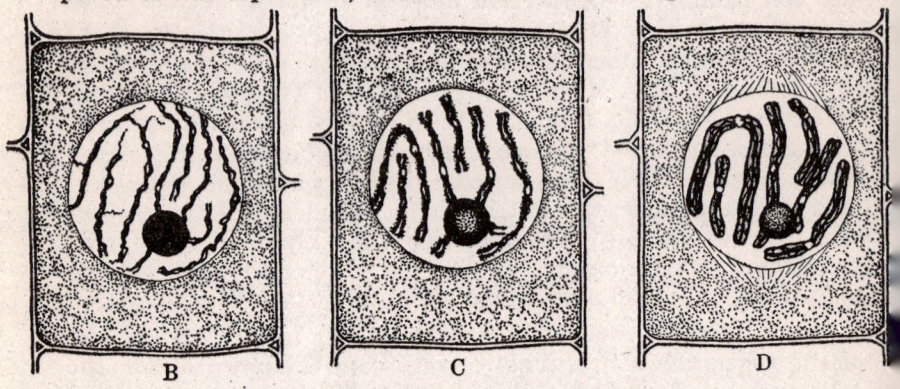

FIG. 20 *B-D*. Mitosis (*contd.*). Prophase.

tary factors. As prophase proceeds the chromosomes relax their coils and thicken somewhat (C). Their double nature becomes more apparent. The outlines of the chromotids present a slightly irregular, hairy appearance. Soon, however, they lose their hairiness and become thicker and smoother. It is also seen that each chromatid divides longitudinally into two. Thus a chromo-

Figs. 20 A-J redrawn after Fig. 40 in Fundamentals of Cytology *by L. W. Sharp by permission of McGraw-Hill Book Company. Copyright 1943.*

THE CELL

some at this stage consists of four threads (chromonemata), two belonging to a chromatid. Further, a chromosomal substance accumulates in a sheath or **matrix** round each chromatid and the two threads become closely coiled in it (D). In well-fixed chromosomes some unstained gaps or constrictions are seen; these are the attachment regions, called **centromeres**. The matrix soon becomes more apparent, and the nucleoli lose their staining power, decrease in size, and finally disappear at the end of the prophase. The nucleus then rapidly passes into the next stage—the metaphase.

Metaphase. The nuclear membrane disappears and a spindle-like body known as the **nuclear spindle** (usually bipolar, in some cases multipolar or even monopolar) consisting of very delicate fibres or fibrils makes its appearance (E). The mode of origin of the spindle varies considerably. It may be formed out of the nuclear sap or karyolymph (nuclear origin) or more frequently out of cytoplasm (cytoplasmic origin). Commonly in root-tips it appears as two opposite polar caps outside the nuclear membrane (as in D). The membrane then disappears and the spindle extends into the nuclear area. The chromosomes move to the equatorial plane of the spindle and stand there clearly apart from one another. A feature of mitotic metaphase is that the centromeres of the chromosomes are lined up along the equator, while the arms of the chromosomes extend outwards into the cytoplasm (*cf.* meiotic metaphase, p. 185). In the mitotic cycle the chromosomes are most sharply defined and best observed at this stage. The chromatids are now seen to come even more

FIG. 20. *E-F*.
Mitosis (*contd.*).
E, metaphase;
F, anaphase.

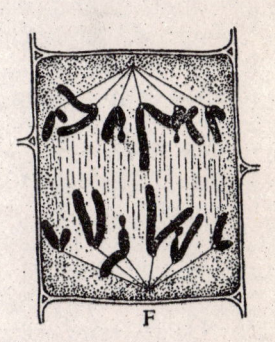

closely together. The centromere of each chromosome divides so that each chromatid has its own centromere. The centromeres of each pair of chromatids become attached to the spindle fibres, called *tractile fibres*, passing to the opposite poles (E). The number of chromosomes is normally constant for a particular species of plants and this number is also normally even, expressed as $2n$ (or $2x$) or **diploid**. Chromosome numbers cover a wide range but 24 seems to be a common figure. It may also be noted that in plants the chromosomes range in lengths from 0.1μ to 30μ. In a few cases they are slightly bigger.

Anaphase. At the end of the metaphase the centromeres of each pair of chromatids appear to repel each other. They diverge and move ahead towards the two opposite poles along the course of tractile fibres (F). The movement of the chromatids is auto-

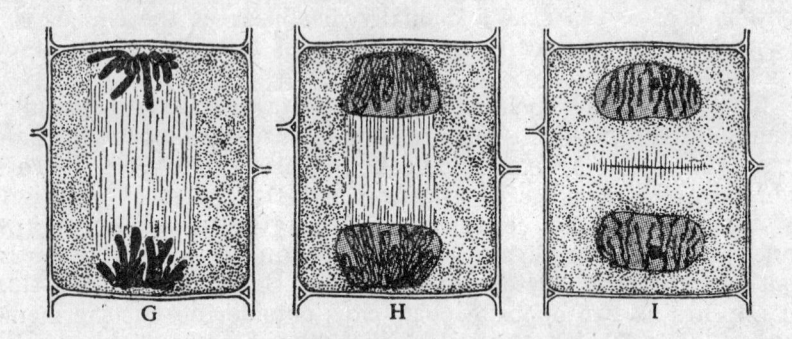

FIG. 20 *G-I*. Mitosis (*contd.*). Telophase.

nomous. The causes of this movement are, however, not clearly understood. The chromatids soon become separated from each other. The spindle may also undergo elongation and thus help the complete separation of the two sets of chromatids. It is apparent from the above that the two longitudinal halves (chromatids) of a chromosome go to the two opposite poles of the spindle. Anaphase covers the shortest period in mitosis.

Telophase. At each pole the chromatids (now regarded as chromosomes) form a close group (G). The polar caps of the spindle disappear and a nuclear membrane is formed round each group of chromosomes (H). Nucleoli reappear at definite points on certain chromosomes. The spindle body disappears and so does the matrix. The chromosomes reorganize as two nuclei. The nuclear sap reappears and each nucleus increases in size (1). It passes into the metabolic stage or prepares for the next division. The duration of mitosis varies considerably in different plants and also in the same plant under different conditions. In this respect temperature is an important factor. Generally the time taken for complete division is from $\frac{1}{2}$ to 2 or 3 hours. Regarding temperature effect Sharp gives the following figures in connexion with mitosis in the staminal hair of *Tradescantia* about 30 minutes at 45°C., 75 minutes at 25°C., and 135 minutes at 10°C.

Cytokinesis or the *division of the cytoplasm and the formation of the cell-wall.* Cytokinesis has recently been the

THE CELL

subject of considerable investigation. The division of cytoplasm appears to take place in one of two ways: by the formation of a new cell-wall in the equatorial region or by furrowing (i.e. by cleavage of the cytoplasm). The former process, known as the cell-plate method, is the usual one in vegetative cell. It usually begins in the telophase when new cellulose particles are gradually deposited in the equatorial zone, and soon these particles fuse together to form a delicate membrane, dividing the cytoplasm into two new cells (J). In the latter process, as in the formation of pollen grains in the anther (see FIG. I/117J), constrictions or furrows appear in the ectoplast and these gradually proceed within, dividing the cytoplasm into two parts.

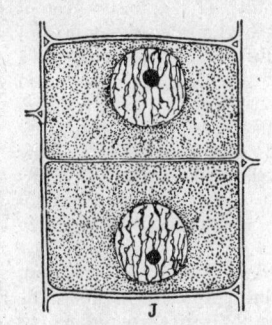

FIG. 20 J. Mitosis (contd.). Cytokinesis.

Importance. The importance of mitosis lies in the fact that by this complicated process of nuclear division, the DNA, which is the chief component of the chromosomes, is distributed equally among the two daughter nuclei formed. Thus the daughter nuclei become qualitatively and quantitatively similar to the mother nucleus. The DNA is the sole genetic (hereditary) material of the chromosomes and because of even distribution of this material among the two daughter nuclei they possess all the characteristics and qualities of the mother nucleus.

Chemistry of Chromosomes. The main chemical components of the chromosomes are the nucleic acids and certain special types of proteins. The nucleic acids occur in the chromosomes in the form of DNA and to some extent as RNA (see p. 153). The nature and quantity of the DNA present in the diploid cells are constant for a particular species of plants or animals, and in the haploid cells just half this quantity occurs. The DNA of the chromosomes is the sole genetic material (see p. 153). Although DNA is chemically the same in all living organisms, each species of plants or animals is characterized by its own specific type of DNA. This is why a species cannot normally go out of bounds, or in other words, a particular species always gives rise to the same species. The diversity in species may, however, be due to an infinite number of combinations of nitrogenous bases (purines and pyrimidines) in the DNA molecules (see Genes, Part VIII, Chapter 2).

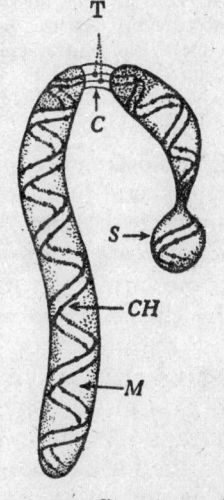

Structure of the Chromosome (FIG. 21). The structure of the chromosome may best be studied in its metaphase or anaphase stage, preferably in the latter. It must be noted that the chromosomes always appear in their characteristic shapes and sizes,

FIG. 21. Structure of a chromosome at anaphase stage. T, tractile fibres; C, centromere; S, satellite; CH, chromonema; M, matrix.

Fig. 21 redrawn after Fig. 57 in Fundamentals of Cytology *by L. W. Sharp by permission of McGraw-Hill Book Company. Copyright 1943.*

apart from their constant number, in the succeeding divisions of the nucleus so far as a particular species of plants is concerned. Most commonly the anaphasic chromosomes of somatic cells lie within limits of 1-20μ in length. Long chromosomes of course give a better idea of their constitution. As previously stated, a chromosome consists of two parts: usually two spiral threads, called **chromonemata**, twisted about each other, and a chromosomal **matrix**, often very clear at certain stages of mitosis. The two chromonemata are sometimes so closely associated that their double nature cannot be clearly made out. This has led to a controversy as to their true nature— single or double, or more according to some. The evidence is, however, in favour of two chromonemata. Along the whole length of each chromonema there is a series of granules, called **chromomeres**, which look like beads in a chain. These are, however, more clear in meiosis rather than in mitosis. The attachment region or **centromere** (also called kinetochore or primary constriction) is a very important part of the chromosome. Tractile fibre extends from it to the pole. The position of centromere is always constant in a given chromosome, as seen in successive divisions. It is a clear achromatic zone. The chromonemata are continuous through this zone, and each of them may contain a minute granule or spherule (also called kinocome). The portions of the chromosome lying on the sides of centromere are called **arms**; the arms may be equal or unequal depending on the position of the centromere. Commonly one of the chromosomes of a nucleus has in one of its arms, sometimes in both, a secondary constriction which is connected with the organization of the nucleolus. The small distal segment of the chromosome, thus formed, is called the **satellite.** Its position is also fixed in the chromosome so that it (the chromosome) may be recognized in the succeeding divisions of the nucleus. Abundant matrices, particularly after staining, often obscure this feature. Besides, it is seen that some parts of the chromosome, often close to the centromere, are much denser and more strongly stainable (chromatic) than other parts. The significance of this difference is not clearly understood. It is, however, suggested that the denser parts are connected with the formation of nucleic acid.

2. **Meiosis** or **Reduction division** (FIG. 22). Meiosis (*meiosis*, reduction) was first worked out by Strasburger in 1888, and later more elaborately by Farmer and Moore in 1905. Meiosis is a complicated process of nuclear division whereby the chromosome number is reduced to half (n) in the four nuclei so formed by this method. Supposing then that the mother nucleus bears 12 chromosomes ($2n$), each of the daughter nuclei will have only 6. Reduction division takes place in all sexually reproducing organisms at some time in their life-cycle. Sexual reproduction means the fusion of a male gamete with a female one resulting in the formation of a zygote from which the offspring develops in due course. Had the gametes contained the same number of chromosomes as their parents the offspring would have an ever-increasing number of chromosomes from generation to generation apart from the peculiar composition of the latter. In consequence the offspring would have developed into new, peculiar and distinct types since chromosomes are the bearers of hereditary characters and meiosis is the mechanism for the transmission of these characters. Thus in all sexually reproducing plants and animals the gametes are **haploid** (n) to

compensate for the chromosome doubling $(n + n = 2n)$ in the zygote as a result of fertilization. The DNA (see p. 185) of the chromosomes, which is the sole hereditary material, is also distributed in equal halves among the gametes. In higher plants showing an alternation of generations meiosis occurs as soon as a plant enters into the gametophytic phase in its life-cycle and, therefore, during the formation of spores from the spore mother cell. In lower plants, on the other hand, meiosis occurs immediately after fertilization or on the germination of the zygote.

Process. Meiosis comprises two successive divisions of the mother nucleus (meiocyte), of which *division I* is reduction division whereby the chromosome number $(2n)$ is reduced to half or (n), and *division II* is mitotic in nature. This being so, the four nuclei that are formed by this process have the same reduced number (n) of chromosomes.

Division I (FIG. 22A). In this division also the nucleus passes through the same phases, as in mitosis, but there are certain

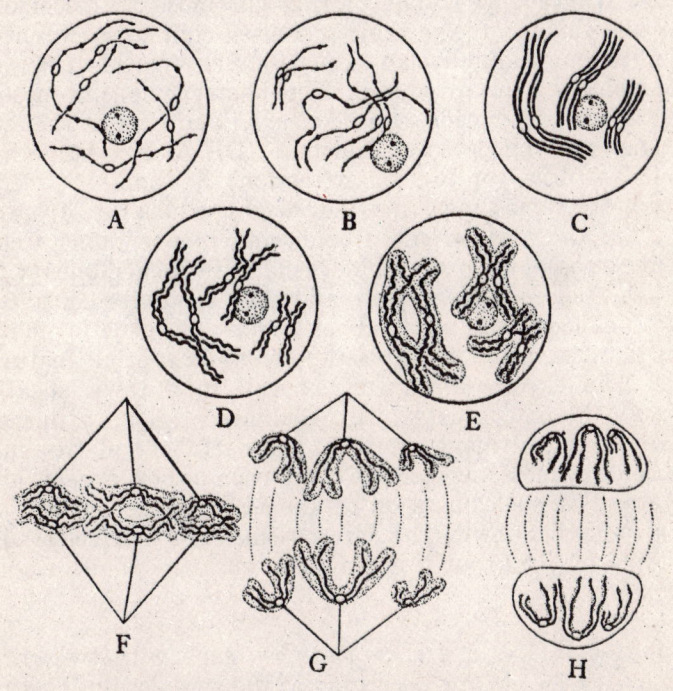

FIG. 22A. Meiosis. Division I. Prophase—A, leptotene; B, zygotene; C, pachytene; D, diplotene; E, diakinesis; F, metaphase; G, anaphase; H, telophase.

Figs. 22 A-B redrawn after Fig. 76 in Fundamentals of Cytology *by L. W. Sharp by permission of McGraw-Hill Book Company. Copyright 1943.*

special features of meiosis distinct from mitosis. **Prophase I.** This phase of meiosis is a prolonged and complicated one, and can be subdivided into the following five stages. **Leptotene (A).** At the early prophase the nucleus increases in volume, and the chromosomes appear as long, slender and single threads (not double, as in mitosis) in diploid number. It will, however, be noted that the threads are present in exactly identical pairs, one being paternal and the other maternal. They now begin to coil. In each thread there appear a number of small beadlike granules known as the **chromomeres** which represent the tightest coils of the chromosome. **Zygotene (B).** It is seen that the identical chromosomes develop a strong attraction for each other, and they soon come together in pairs part by part throughout their whole length except in very long chromosomes. This pairing, called **synapsis**, is only in the nature of close association (bivalent condition) but not their actual fusion. The pairing chromosomes are *homologous*, being derived from the same parent chromosome. They now begin to shorten and thicken. **Pachytene (C).** The pairing chromosomes become still shorter and thicker due to their increased coiling. Each chromosome splits longitudinally, and thus four chromatids (two from each homologue) are produced. The homologous chromosomes, each with two chromatids, can be clearly identified at this stage. Pachytene is a long process. **Diplotene (D).** At this stage a sort of repulsive force (or loss of attraction) develops between the homologous chromosomes, and they begin to separate from each other. They, however, remain connected in one point (usually in shorter ones) or more points (usually in longer ones); these points are known as **chiasmata**, and they are quite conspicuous. At each chiasma a physical exchange of genetic material ('genes') takes place by *crossing over*—a special feature of meiosis. The chromosomes further coil in a thick sheath or matrix and become shorter still. **Diakinesis (E).** At this stage the chromosome bivalents move to the periphery of the nucleus and become almost separated from each other except at the chiasmata. The nucleolus soon disappears and the nuclear membrane also breaks down. The chromosomes are thus released into the cytoplasm. At about this time the nuclear spindle begins to be formed.

Metaphase I (F). The chromosomes (each with two chromatids) move to the equatorial region of the spindle. But this phase is distinct from the mitotic metaphase. Here the centromeres of the chromosome pairs become attached to spindle fibres near the equator, clearly apart from each other and facing opposite poles, while their arms lie towards the equator, i.e. the chromo-

THE CELL

somes are not lined up along the equator, as in mitosis (see p. 183).

Anaphase I (G). The two centromeres of the homologous chromosomes repel each other and move towards the opposite poles of the spindle along the course of the tractile fibres, each carrying a chromosome (paternal or maternal but not both) with it to one pole. This finally results in the reduction of the chromosome number from diploid to haploid. This is the essential feature of meiosis.

Telophase I (H). As in mitosis, the chromosomes (each with a pair of chromatids) form a compact group at each pole. The nucleolus reappears and a nuclear membrane is formed around each polar group of chromosomes. The two daughter nuclei, thus formed, evidently contain haploid or (n) chromosomes, each with a pair of chromatids. A cell-membrane may or may not be formed at this stage, separating the two nuclei. Almost immediately or a little later each nucleus passes on to division II.

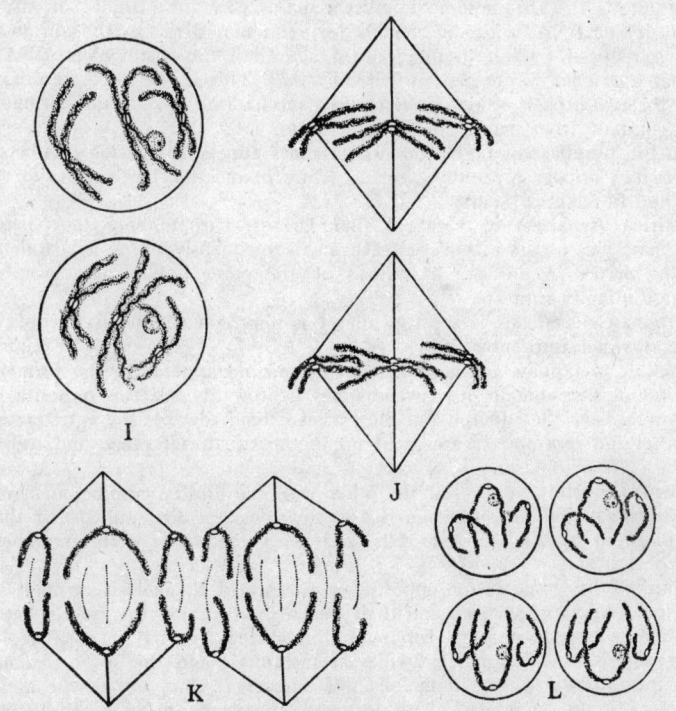

FIG. 22B. Meiosis (*contd.*). Division II. *I*, prophase; *J*, metaphase; *K*, anaphase; and *L*, telophase.

Division II (FIG. 22B). The phases are almost the same as found in mitosis. In **prophase II** (I) of meiosis the two chromatids of a chromosome remain distinctly separate and loose except at the centromeres. The chromosomes, however, are long and coiled. Soon they become shorter and thicker by further coiling. This phase ends with the disappearance of the nucleolus and the nuclear membrane, and the appearance of the spindle fibres. In **metaphase II** (J) the chromosomes take up an equatorial position in the newly-formed spindle. This phase is short. Further, the chromosome pair separates, each having its own centromere. In **anaphase II** (K) the two sister chromosomes of each pair begin to move towards the opposite poles, being drawn by their centromeres. In **telophase II** (L) a nuclear membrane is formed around each polar group of chromosomes, and the nucleolus reappears. Finally by cytokinesis *four* cells are formed, each nucleus having haploid or (n) chromosomes. It will be noted that the first division of meiosis is *reductional*, while the second division is *equational*.

Differences between Mitosis and Meiosis. 1. Mitosis occurs in somatic (meristematic) cells, while meiosis occurs in reproductive cells, resulting in the formation of spores or gametes.

2. In mitosis the chromosome number remains constant (diploid or $2n$) with full quantity of DNA, while in meiosis (or reduction division) the chromosome number is reduced to half (haploid or n) with half the quantity of DNA, i.e. mitosis is *equational*, while meiosis is *reductional*. This is the essential difference between the two processes. Meiosis is really a mechanism to keep the chromosome number constant from generation to generation.

3. In the prophase of both, the chromosomes appear in specific numbers but in mitosis they appear in double threads, while in meiosis they appear in single threads but in identical pairs.

4. Pairing (synapsis) of identical (homologous) chromosomes (one paternal and one maternal) occurs in meiosis (each pair subsequently acting as a unit), but no pairing occurs in mitosis. In mitosis on the other hand each chromosome splits longitudinally into two.

5. Prophase of mitosis is short, while it is a prolonged one in meiosis and, therefore, divided into sub-stages.

6. Meiotic metaphase is different from mitotic metaphase. In the former the centromeres of the homologous chromosomes lie towards the two opposite poles of the spindle near the equator and their arms extend towards the equator, while in the latter the centromeres are lined up in the equatorial plane and the arms extend into the cytoplasm.

7. Mitosis continues one after the other with the diploid number of chromosomes, whereas meiosis is completed in two divisions (the first only being the reduction division) resulting in four cells (each with haploid or n chromosomes) in a group (tetrad).

8. Haploid (n) gametes (of opposite sexes) formed by meiosis normally fuse in pairs in sexual reproduction, and their fusion product, i.e. the zygote, becomes diploid ($2n$). Normally no such fusion of diploid cells (nuclei) takes place.

9. In mitosis the centromere divides at metaphase and the sister chromatids move to the opposite poles; in meiosis the centromere does not divide and the homologous chromosomes, each with its own centromere, move to the opposite poles. In both the processes the centromere moves first and draws the chromosome.

10. Chromomeres are more prominent in the meiotic chromosome than in the mitotic one.

11. Chiasma and crossing over (exchange of genes) are normal and exclusive features of meiosis.

12. In mitosis the chromosomes are equally apportioned to the daughter nuclei, i.e. the latter are qualitatively and quantitatively the same as the mother nucleus, while in meiosis the four threads of a chromosome go to the four cells (and their assortment is also a matter of chance, i.e. it is not known which thread will go to which cell). Thus meiosis results in four new types of cells.

Polyploidy. Normally plants and animals have in their somatic cells two sets of chromosomes (paternal and maternal), expressed as diploid or $2n$, while they have in their reproductive cells one set of chromosomes, expressed as haploid or n. Increase in the number of chromosomes in certain tissues or entire organisms in multiples of the basic or haploid number is expressed as polyploidy. This is of wide occurrence in nature, particularly in plants (wild or cultivated), rarely in animals. There are, however, organisms with more than two sets or less than two sets of chromosomes in their somatic cells. The term *heteroploidy* is sometimes used to include all types of variations found in chromosome numbers. All such variations may be discussed under the following heads: euploidy and aneuploidy.

Euploidy (*eu*, true or even; *ploid*, unit). This represents cases where the somatic chromosome complements[1] are exact multiples of the haploid number, characteristic of a particular species. Among euploids, a *monoploid* organism has but a single genome[2] per nucleus instead of a double set. This, however, is not common, more so in animals. Among plants this has been found in *Sorghum, Triticum, Hordeum, Datura*, etc. Such individuals are small, weak and generally sterile. They have very irregular meiosis because of the absence of homologues. More commonly, however, the individuals are *triploid* having 3 sets of chromosomes, *tetraploid* with 4 sets, *hexaploid* with 6 sets, and *octoploid* with 8 sets. Other multiples are also found. Organisms having 3 or more sets of chromosomes are, however, commonly known as polyploids.

Polyploidy. As stated above, individuals, cells or tissues having 3 or more sets of chromosomes are called polyploids. Among them *triploidy* ($2n+n=3n$) was first found by Miss Lutz (1912) in *Oenothera*; later this has been found in many cultivated plants, e.g. hyacinths, chilli, tomato, rice, maize, *Datura*, etc. Such plants show certain variations and are distinct from the normal types. *Tetraploidy* ($2n+2n=4n$) was first found by Digby (1912) in *Primula*; later this has been found to be widespread, and is perhaps the most important of all polyploid types. Gigantism and great vigour (heterosis) are frequently associated

---

[1] A chromosome **complement** is a group of chromosomes making up a nucleus, irrespective of their number and kind. [2] A complement is said to be a **genome** when it consists of a set of chromosomes inherited as a unit from one parent, whatever be their number, form and function.

with this condition, e.g. giant *Chrysanthemum*, marigold, rose, etc. Similarly, *hexaploidy* (6n), *octoploidy* (8n) and other multiples have been found. It may be noted that different varieties of cultivated wheat show tetraploidy (4n), pentaploidy (5n) and hexaploidy (6n). Nawaschin (1925 & 1926) found 3n, 4n and 5n in *Crepis*. Polyploidy is often differentiated into two major types on the basis of the source of chromosomes: autopolyploidy and allopolyploidy. **Autopolyploids** are organisms in which the multiple genomes are identical. The increase in chromosome number is due to doubling of chromosomes of a single homozygous diploid. On the other hand **allopolyploids** are organisms in which the multiple genomes are different, coming together from different sources through crossing or hybridization. They are, therefore, found in hybrids only. The increase in chromosome number may be due to crossing between two diploid gametes, or multiplication of chromosomes in the hybrids. There are several cases on record. A classical example is *Raphanobrassica* (a tetraploid with 36 chromosomes) which is a hybrid produced by Karpechenko (1927) in Russia as a result of a cross between radish (*Raphanus*) and cabbage (*Brassica*), each with 9 chromosomes. Other examples of allopolyploidy are the several hybrids of *Rosa*, *Primula*, *Papaver*, *Iris*, etc.

Aneupolyploidy (*aneu*, uneven). This represents cases where the somatic chromosome complements are not exact multiple of the basic number. There is addition or loss of usually 1 or 2 chromosomes. Thus when the number is less it is expressed as *hypoploidy*, and when more it is *hyperploidy*. These conditions are due to unequal distribution of chromosomes during anaphase. One daughter cell may get an additional chromosome, while the other has to lose one. An organism (individual) lacking one chromosome of a diploid number is called *monosomic* (2n−1). A trisomic individual has two sets of chromosomes plus one extra chromosome (2n+1). A *tetrasomic* has two sets plus two extra chromosomes (2n+2), and a *double tetrasomic* has (2n+2+2).

Effect of Polyploidy. Polyploid plants are usually healthier, stronger, and larger than their diploid counterparts. This may be due to increase in size and chromatin contents of the polyploid cells. Smith (1943) working on *Nicotiana* found in a series consisting of haploids, diploids, triploids and tetraploids that with increase in each additional set of chromosomes there was a corresponding increase in the width of the corolla tube, size of leaves, thickness of parts of plants including pollen grains, guard cells, and the cells of the root-tip. In octoploids, however, he observed a reduction in the size of certain organs. Blakeslee and Warmke (1938) found that many tetraploid plants possess larger seeds, larger pollen grains, wider and thicker leaves, larger inflorescences, larger floral parts, and stouter fruits. In polyploidy some differences are also detected in certain physiological properties; for example, vitamin contents of fruits (e.g. vitamin C in tomato, and vitamin A in maize) and vegetables are on the increase. In sugar beet (*Beta vulgaris*) sugar contents increase; and in tobacco the percentage of nicotine is seen to be higher (see also p. 194).

Origin of Polyploidy. All biologists agree that diploid condition of an organism is primitive and that polyploids are derived from diploids and are more advanced. The question, however, arises: 'How does polyploidy originate'? The possible ways for initiation of polyploidy are as follows. (1) *Doubling of chromosome number in the somatic tissue*. This may be due to some irregularities in certain somatic cells affecting normal mitosis. For example, (a) cell-wall may not be formed after the division of the chromosomes during mitosis; the result is a tetraploid nucleus ($2n+2n=4n$). The gamete mother cell formed this way may give rise to diploid ($2n$) gametes. (b) Spindle may not develop properly, and a single nuclear membrane may surround the two diploid nuclei; the restitution nucleus may thus have four sets of chromosomes (tetraploid or $4n$). (c) Two diploid gametes may also fuse, resulting in a tetraploid somatic cell. (2) *Doubling of chromosome number in the formation of reproductive cells* (*gametes*). The irregularities leading to increase in chromosome number in such cells may take the following patterns: (a) Failure of meiotic division (one or both) may lead to the formation of diploid gametes. (b) Cell-wall may not be formed after division of the chromosomes during meiosis; thus no reduction in chromosome number takes place. (c) Production of gamete mother cell with two nuclei; their homotypic spindles may fuse together, producing diploid gametes. (d) The tetraploid nucleus of the gamete mother cell may give rise to diploid gametes by meiosis. The above conditions may arise spontaneously in nature, and can also be induced artificially. If a tetraploid condition arises in a bud primordium a $4n$ branch will result. A triploid ($3n$) somatic cell may be produced by the union of a diploid ($2n$) gamete with a normal haploid (n) gamete. A triploid plant is usually sterile and have lower vitality. Other multiples may arise by repeated duplication of chromosomes or by intercrossing among polyploids.

Induced Polyploidy. Polyploidy can be artificially induced by drastic changes in the environment by the use of different agents. (1) Sudden change in temperature; extreme heat or cold brings about polyploidy. (2) Wounding the plant leading to the formation of callus; it is seen that in many cells of the callus the walls are not formed after mitosis, and as a result doubling of chromosomes takes place; buds formed from them develop into polyploid shoots (Winkler, 1916). (3) Application of the alkaloid *colchicine* (Blakeslee, 1937 and Nebel, 1938); other chemicals like veratrine sulphate (Whitkus and Berger, 1944), benzene, acenaphthane, sulfanilamide, chloral hydrate, heteroauxin, etc. (4) Infection by bacteria, insects and other infective agents. (5) Changes in osmotic pressure. (6) Irradiation by X-ray.

Importance of Polyploidy. Polyploidy is now considered to be very important from the standpoint of evolution, plant breeding and geographical distribution. Their economic and ornamental values in many cases are no less important. Among cultivated plants polyploidy has led to extensive varieties or species, e.g. rice, wheat, maize, tobacco, cotton, mustard, sugarcane, etc. Polyploid plants have many advantages over diploid ones. This is particularly true of tetraploidy. With additional chromosomes and increased number of genes and certain new physiological properties (see p. 193) polyploid forms tend to be far better than diploid forms in many respects. Such plants are more liable to variation and mutation, and may ultimately lead to the evolution of new species. It is seen that in many polyploids the cell size increases, while the growth rate decreases. Polyploidy may lead annuals to become biennials or perennials (Muntzing, 1936). In 1934 Sharp reported that a certain polyploid variety of maize is perennial, while the diploid variety is annual. Polyploidy thus may be one of the methods by which new species are evolved. *Raphanobrassica* (see p. 192) with $4n$ chromosomes formed by allopolyploidy breeds true and is regarded as a new species. Similarly Muntzing crossed two species of *Galeopsis* and produced a hybrid with $4n$ chromosomes, which is as good as a new species. It is believed that in nature several new species have arisen by allopolyploidy. Besides, many of them become large and show hybrid vigour (heterosis), and have certain other favourable properties (see pp. 192-93). Their commercial value is being gradually exploited. Another advantage is that polyploidy can be artificially induced, and many polyploid plants are seen to be very fertile. In many cases they are also disease-resistant. As such, plant breeders are specially interested in this line of research work. A special feature of polyploidy is giganticism (gigantism). This means the polyploid plants normally show more vigorous growth, often resulting in much larger size; such plants also develop larger vegetative and reproductive parts (see pp. 192-93). Thus the tetraploid variety of jute (*Corchorus*) with $4n$ chromosomes may be twice as tall as the diploid variety. Sometimes giganticism is associated with increase in the size of the chromosomes. A gigas polyploid form reproducing vegetatively, e.g. sugarcane, is a decided advantage. Polyploidy is also common in many garden plants, e.g. *Hibiscus*, *Chrysanthemum*, marigold, rose, snapdragon, etc. Besides, the sturdier nature of many polyploids helps them to extend over wide areas against adverse conditions of life. *Saccharum spontaneum* with gradual increase in chromosome number is known to have invaded new extensive areas.

3. **Amitosis** or **Direct Nuclear Division** (FIG. 23). In this case the nucleus elongates to some extent and then it undergoes constriction, i.e. it becomes narrower and narrower in the middle or at one end, and finally it splits into two. The nuclei so formed are often of unequal sizes. Amitosis may or may not be followed by the division of the cell. Amitosis is of frequent occurrence in the lower organisms like algae and fungi. In the higher plants it is seen to occur in old cells and in those showing distinct signs of degeneration.

4. **Free Cell Formation** (FIG. 24). This is a modification of indirect nuclear division. It differs from the latter in that the cell-wall is not formed immediately after the division of the nucleus. In this process by repeated mitotic divisions a large number of nuclei are formed within the mother cell. When the

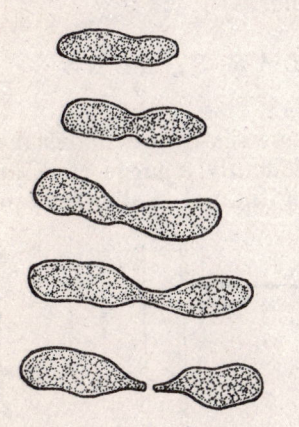

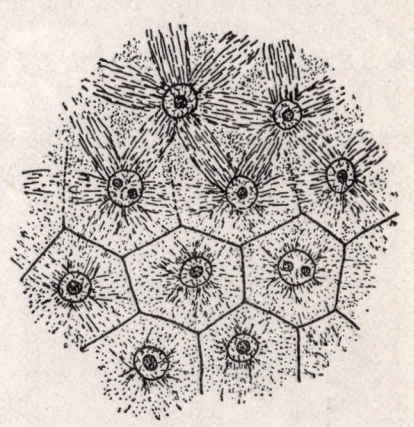

FIG. 23. Amitosis.

FIG. 24. Free cell formation in the development of endosperm.

divisions of the nuclei cease, the cytoplasm aggregates round them, and a cell-wall is formed round each nucleus. The formation of the cell-wall gradually proceeds from one side to the other, resulting in a regular tissue (combination of cells). The endosperm, i.e. the food storage tissue of the seed, is formed by this method.

In some cases mitosis may be immediately followed by the formation of cell-wall resulting in free cells, e.g. ascospores, or the nuclei may remain free without wall-formation, e.g. zoospores, or repeated nuclear divisions may result in a large number of free nuclei within the mother cell which quickly elongates and often becomes branched, e.g. latex cell; such a structure is called a **coenocyte**. *Vaucheria* is another good example of a coenocyte.

5. **Budding** (FIG. 25). This is seen in yeast—a unicellular fungus. In this plant the cell forms one or more tiny outgrowths on its body. The nucleus undergoes direct division (amitosis) and splits up into two. One of them passes on to one outgrowth. The outgrowth increases in size and is ultimately cut off from the mother yeast as a new independent cell (a new yeast plant). This process of cell-formation is known as *budding*. Often budding continues one after the other so that chains and even sub-chains of cells are formed. Ultimately all the cells separate from one another.

FIG. 25. Budding in yeast.

VESSELS, INTERCELLULAR SPACES AND CAVITIES

Vessels. Sometimes it is only the cross-walls of a cylindrical row of cells that get dissolved, and consequently a large continuous opening is formed, the side or longitudinal walls remaining

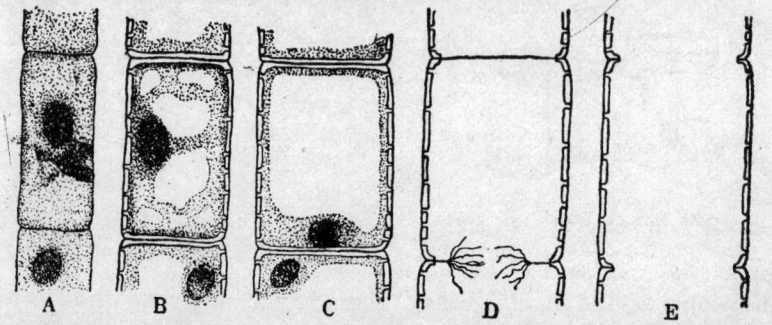

FIG. 26. Development of a vessel. *A-E* are stages in its development.

intact. The constituent cells thus together give rise to a tube- or pipe-like structure. This pipe-like structure without partition walls is known as the **vessel** (FIG. 26). The ends of the constituent cells, or vessel segments, are mostly narrow with the perforation oblique (see FIG. 37 E-F). Vessels are of different kinds. They are thick-walled, lignified and dead. Lignin is deposited as a thickening material on the inner surface of the side-walls in various patterns, and according to the mode of thickening (see p. 159) vessels may be of the following types—**annular, spiral, scalariform, reticulate,** and **pitted** (see FIG. 37). Vessels are carriers of water and raw food materials, i.e. they conduct these substances from the root to the leaf. Being thick-walled and lignified they also serve to give strength to the plant body.

Intercellular Spaces. When the cells are young they remain closely packed without any empty space or cavity between them; but as they grow, their walls split at certain points, giving rise to small cavities or empty spaces; these are intercellular spaces. They remain filled with air or water.

Schizogenous Cavities. Bigger cavities are also often formed by the splitting up of common walls and the separation of masses of cells from one another; these are schizogenous (*schizein*, to split) cavities. Intercellular spaces and these cavities form an intercommunicating system so that gases and liquids can easily diffuse from one part of the plant body to the other. Most resin-ducts in plants are schizogenous cavities (FIG. 27).

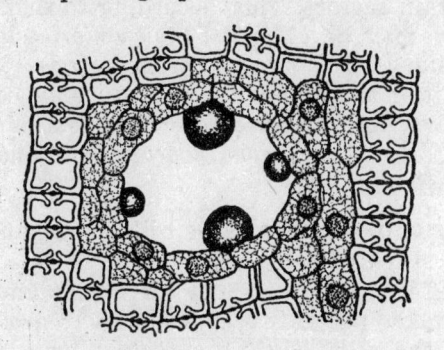

FIG. 27. Resin-duct of pine stem with resin.

Lysigenous Cavities. Sometimes, during the development of a mass of cells, their walls break down and dissolve, and as a consequence large irregular cavities appear; these are known as **lysigenous** (*lysis*, loosening) cavities. These cavities are meant for storing up water, gases, essential oils, etc., and thus act as glands (see FIG. 41A).

Chapter 2 * THE TISSUE

Cells grow and assume distinct shapes to perform definite functions. Cells of the same shape grow together and combine into a group for the discharge of a common function. Each group of such cells gives rise to a tissue. *A tissue is thus a group of cells of the same type or of the mixed type, having a common origin and performing an identical function.* Tissues may primarily be classified into two groups: **meristematic** and **permanent**.

Meristematic Tissues (*meristos*, divided). These are composed of cells that are in a state of division or retain the power of dividing. These cells are essentially alike and isodiametric being spherical, oval or polygonal; their walls thin and homogeneous; the protoplasm in them abundant and active with large nuclei; and the vacuoles small or absent.

Classification of Meristems. Meristems are classified in different ways on the basis of certain factors. Thus according to their origin and development they may be classified as *promeristem* or *primordial meristem, primary meristem,* and *secondary meristem.* According to their position in the plant body they may be classified as *apical, intercalary,* and *lateral.* According to their functions they may be classified as *protoderm, procambium,* and *ground* or *fundamental meristem.*

Promeristem consists of a group of meristematic cells representing the earliest or youngest stage of a growing organ. It is in fact the stage from which differentiation of later meristems and finally of permanent tissues takes place. It occupies a small area at the tip of the stem and the root. The promeristem by cell divisions gives rise to the primary meristem, and is, therefore, the earliest stage or originator of the latter.

Primary meristem is derived from the promeristem, and still fully retains its meristematic activity. As a matter of fact, its cells divide rapidly and become differentiated into distinct tissues —the *primary permanent tissues* which make up the fundamental structure of the plant body. It is primarily the growing apical region of the root and the stem. It is further to be noted that the cambium of the stem is also a primary meristem although it gives rise to the *secondary permanent tissues*. Another fact to be noted in this connexion is that the cambial cells divide mainly in one plane (tangential), while those of the primary meristem divide in 3 or more planes.

Secondary meristem, on the other hand, appears later at a certain stage of development of an organ of a plant. It is always lateral lying along the side of the stem and the root. It is seen that some of the primary permanent tissues become meristematic, i.e. they acquire the power of division, and constitute the secondary meristem, e.g. the cambium of the root, the interfascicular cambium of the stem, and the cork-cambium of both. All lateral meristems (primary and secondary) give rise to the *secondary permanent tissues*, and are responsible for growth in thickness of the plant body.

As stated before, on the basis of position the meristems may be apical, intercalary, and lateral. (*a*) **Apical meristem** lies at the apex of the stem and the root, representing their *growing regions*, and is of varying lengths, usually ranging from a few millimetres to a few centimetres. It includes the promeristem and the primary meristem, gives rise to the *primary permanent tissues*, and is responsible for growth in length of the plant body. It should be noted that the promeristem consists of a group of meristematic cells in phanerogams (mostly), while in pteridophytes (mostly) it is represented by a single cell. (*b*) **Intercalary meristem**, when present, lies between masses of permanent tissues, either at the base of the leaf, as in pine, or at the base of the internode, as in some grasses and horsetail (*Equisetum*), or sometimes below the node, as in mint (*Mentha*). It is a detached portion of the apical meristem, separated from the latter due to growth of the organ. Like the apical meristem the intercalary meristem, when present, gives rise to the primary permanent tissues. It is generally short-lived, either disappearing soon or becoming transformed into permanent tissues. (*c*) **Lateral meristem**, e.g. cambium of the stem, lies laterally in strips of elongated cells, extending from the apical meristem, as in the stems of dicotyledons and gymnosperms. It divides mainly in the tangential direction, giving rise to the *secondary permanent tissues* to the inside and outside of it, and is responsible for growth in thickness of the plant body.

Theories regarding Apical Meristem. (*a*) **Apical cell theory.** Nageli (1858) first coined the term 'meristem' and said that the apical meristem consists of a single apical cell in all plants, and that the sequence of cell divisions is responsible for the formation of different members of the plant body. Nageli's 'single cell' theory is no doubt true of the thallophytes and vascular cryptogams but his assumption that this is applicable to all cryptogams and phanerogams has proved to be wrong. His 'single cell' theory was supported by Hofmiester but he expressed doubt regarding its applicability to all cases, particularly phanerogams. (*b*) **Histogen theory.** In 1870 Hanstein formulated his 'histogen' theory. According to this theory the apical meristem of angiosperms is divisible into three zones, each consisting of a variable number of layers, and these zones he called dermatogen (outer), periblem (middle) and plerome (inner) respectively, giving rise to epidermis, cortex and stele. Further, he formed the idea that there was a clear differentiation into outer layer (dermatogen) and an inner core even in the embryo of the phanerogams. Sachs accepted this view but on a physiological basis he proposed three systems derived from the above, viz. epidermal, fascicular (vascular) and fundamental. Sachs, however, recognized an apical single cell in most cryptogams and zonal layers in phanerogams. (c) **Tunica-corpus theory.** In 1924 Schmidt proposed the 'tunica-corpus' theory. According to this theory there are two

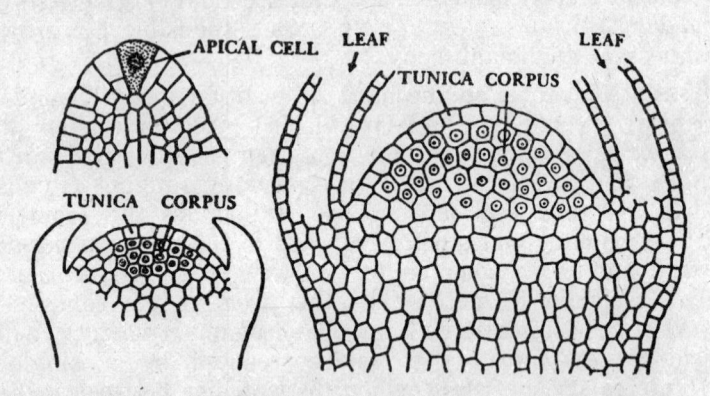

FIG. 28. Apical meristem. *Top*, apical cell in a pteridophyte; *bottom*, tunica and corpus in an angiosperm.

zones in the apical meristem. Tunica is the outer zone consisting of one or more peripheral layers of small uniform cells normally showing anticlinal divisions, and corpus is the undifferentiated mass of larger cells enclosed by the tunica; its cells vary in number from a few to many, divide in many planes and, therefore, they are more or less irregular in shape and arrangement. The epidermis arises from the outer layer of tunica, while the remaining tissues arise from the corpus (or partly from the tunica). This view is now generally accepted.

Haberlandt (1914) introduced another system on a physiological basis to explain the differentiation of the apical meristem. According to him the promeristem differentiates into **protoderm** which gives rise to the epidermal tissue system, **procambium** which gives rise to the vascular tissue system, and **ground or fundamental meristem** which gives rise to the ground tissue system.

Apical Meristems of Stem and Root

1. **Stem Apex** (FIG. 29A). A median longitudinal section through the apex of a stem, when examined under the microscope, shows that the apical meristem or growing region is composed of a small mass of usually rounded or polygonal cells which are all essentially alike and are in a state of division; these meristematic cells constitute the **promeristem** (or primordial meristem). The cells of the promeristem soon differentiate into three regions, viz. dermatogen, periblem and plerome. The cells of these three

THE TISSUE

regions grow and give rise to primary permanent tissues in the mature portion of the stem. The section further shows on either side a number of outgrowths which arch over the growing apex;

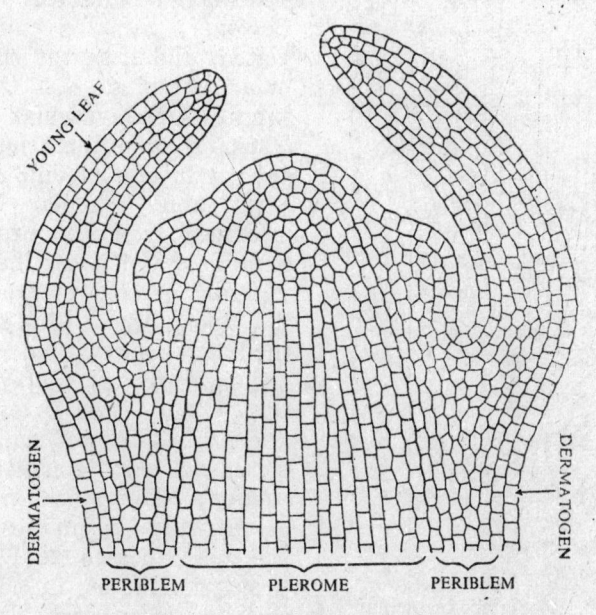

FIG. 29A. Stem apex in median longitudinal section.

these are the young leaves of the bud, which cover and protect the tender growing apex of the stem.

(1) **Dermatogen** (*derma*, skin; *gen*, producing). This is the single outermost layer of cells. It passes right over the apex and continues downwards as a single layer. The cells divide by *radial* walls only, i.e. at right angles to the surface of the stem and increase in circumference, thus keeping pace with the increasing growth in volume of the underlying tissues. The dermatogen gives rise to the skin layer or epidermis of the stem.

(2) **Periblem** (*peri*, around; *blema*, clothing or covering). This lies internal to the dermatogen, and is the middle region of the apical meristem. At the apex it is single-layered but lower down it becomes multi-layered. It forms the cortex of the stem, which is often, particularly in dicotyledons, differentiated into hypodermis, general cortex and endodermis.

(3) **Plerome** (*pleres*, full). This lies internal to the periblem, and is the central region of the stem apex. At a little distance behind the apex certain groups or strands of cells show a tendency to elongate. These groups or strands of elongated cells are said to

form the **procambium**. The procambial strands ultimately become differentiated into bundles of vessels and sieve-tubes, i.e. into vascular bundles. A portion, however, remains undifferentiated, and it forms the fascicular cambium, i.e. the cambium of the vascular bundle. Plerome is differentiated into the pericycle, medullary rays, pith and vascular bundles (derived from the procambial strands), and forms the central cylinder or *stele* of the stem.

2. **Root Apex** (FIG. 29B). A median longitudinal section through the apex of the root shows that it is covered over and protected by a many-layered tissue which constitutes the **root-cap**. The apical meristem or growing region lies behind the root cap (see FIG. I/3). The promeristem, as in the stem, soon differentiates into three regions, viz. dermatogen, periblem, and plerome. In many roots, however, these three regions are not clearly marked.

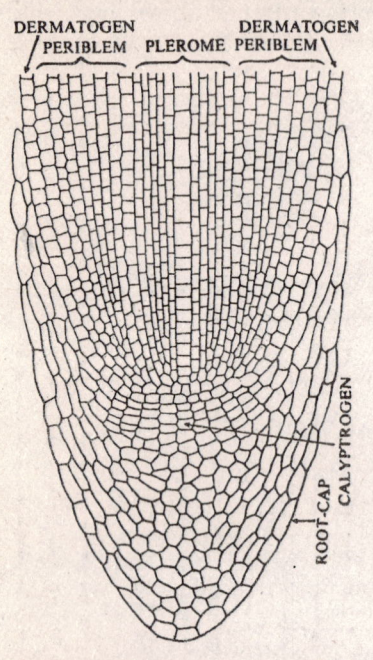

FIG. 29B. Root apex in median longitudinal section.

(1) **Dermatogen.** As in the stem, this is also single-layered, but at the apex it merges into the periblem; just outside this the dermatogen cuts off many new cells, thus forming a small-celled tissue known as the **calyptrogen** (*calyptra*, cap; *gen*, producing). The calyptrogen is also meristematic, and by repeated divisions of its cells gives rise to the **root-cap**. As the root passes through the hard soil, the root-cap often wears away but is then renewed by the underlying calyptrogen. In some plants the dermatogen directly gives rise to the root-cap without the intervention of the calyptrogen. The walls of outer cells of the root-cap may be modified into mucilage which helps the root to push forward in the soil more easily. The root-cap is absent from aquatic plants, although an analogous structure, called *root-pocket*, is conspicuous in many of them (see p. 3). Sometimes, as in dicotyledons generally, the dermatogen continues upwards as a single outermost layer (epiblema) of the root; but in monocotyledons generally the dermatogen is exhausted in the formation of the root-cap so that the outermost layer of the root is

derived from the outermost layer of the periblem. At a little distance from the root-tip the outermost layer bears a large number of *unicellular root-hairs*. Root-hairs are mostly absent from aquatic plants.

(2) **Periblem.** As in the stem, this is also single-layered at the apex and many-layered higher up. In monocotyledons generally the outermost layer of the periblem forms the outermost layer of the root. Periblem forms the middle region or cortex of the root.

(3) **Plerome.** The plerome's structure and function are practically the same as those of the stem. But here some procambial strands give rise to bundles of vessels (xylem) and others to bundles of sieve-tubes (phloem) in an alternating manner (see FIG. 49).

Permanent Tissues. These are composed of cells that have lost the power of dividing, having attained their definite form and size. They may be living or dead and thin-walled or thick-walled. Permanent tissues are formed by differentiation of the cells of the meristems and may be **primary** or **secondary**. The primary permanent tissues are derived from the apical meristems of the stem and the root, and the secondary permanent tissues from the lateral meristems, i.e. cambial layers. In dicotyledons and gymnosperms the cambium is present, and by its activity the secondary growth takes place in these cases; while *in monocotyledons there is no cambium* and hence no secondary growth.

Classification of Permanent Tissues. In their earlier stages the cells are more or less similar in structure, but as the division of labour increases they gradually assume various forms and give rise to **permanent tissues**. These may be classified as *simple* and *complex*. A simple tissue is made up of one type of cells forming a homogeneous or uniform mass, and a complex tissue is made up of more than one type of cells working together as a unit. To these may be added another kind of tissue—the secretory tissue.

I. SIMPLE TISSUES

1. **Parenchyma** (FIG. 30A). Parenchyma consists of a collection of cells which are more or less isodiametric, that is, equally expanded on all sides. Typical parenchymatous cells are oval, spherical or polygonal in shape. Their walls are thin and made of cellulose; they are usually living. Parenchymatous tissue is of universal occurrence in all the soft parts of plants. Its function is mainly storage of food material. When parenchymatous tissue contains chloroplasts it is called *chlorenchyma*; its function is to manufacture food material. A special type of parenchyma develops in many aquatic plants and in the petiole of *Canna* and banana. The wall of each such cell grows out in several places

like rays radiating from a star and is, therefore, stellate or star-like in general appearance. These cells leave a lot of air

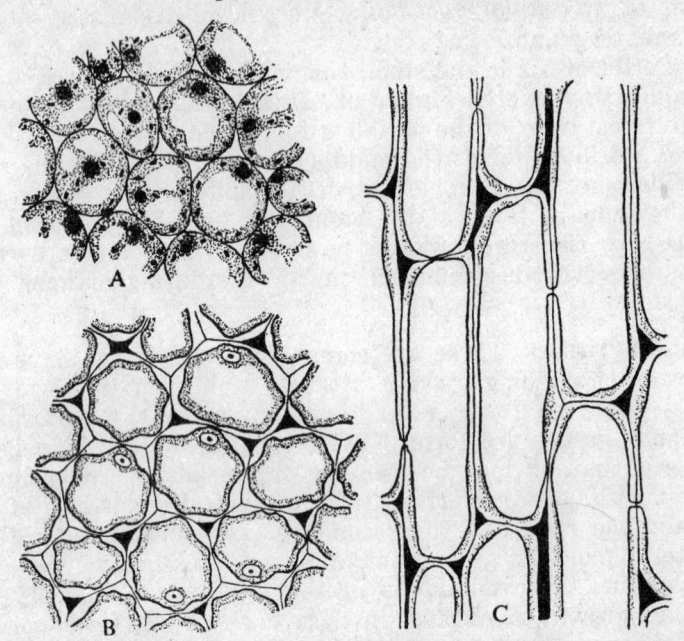

FIG. 30. *A*, parenchyma; *B*, collenchyma in transection; *C*, collenchyma in longitudinal section.

cavities between them, where air is stored up. Such a tissue is often called *aerenchyma* (FIG. 31).

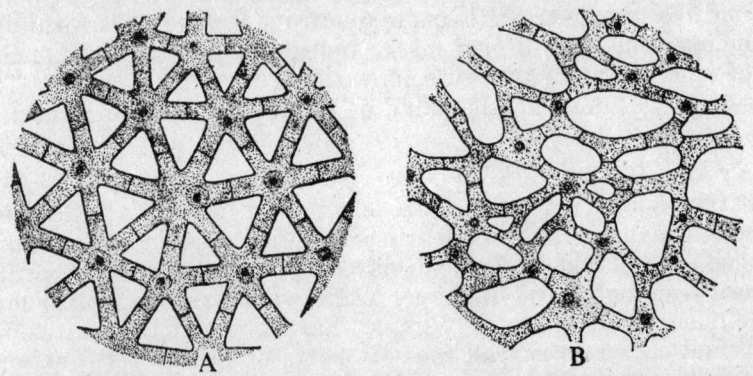

FIG. 31. *A*, aerenchyma in the petiole of banana; *B*, the same in the petiole of *Canna*.

2. **Collenchyma** (FIG. 30 B-C). This tissue consists of somewhat elongated parenchymatous cells with oblique, slightly rounded or tapering ends. The cells are much thickened at the corners

against the intercellular spaces. They look circular, oval or polygonal in a transverse section of the stem. The thickening is due to a deposit of cellulose, hemicellulose and protopectin; although thickened, the cells are never lignified. They are provided with simple pits here and there in their walls. Their thickened walls have a high refractive index and, therefore, this tissue in section is very conspicuous under the microscope. Collenchyma is found under the skin (epidermis) of herbaceous dicotyledons, e.g. sunflower, gourd, etc., occurring there in a few layers with special development at the ridges, as in gourd stem. It is absent from the root and the monocotyledon except in special cases. The cells are living and often contain a few chloroplasts. Being flexible in nature collenchyma gives tensile strength to the growing organs, and being extensible it readily adapts itself to rapid elongation of the stem. Containing chloroplasts it also manufactures sugar and starch. Its functions are, therefore, both mechanical and vital.

3. **Sclerenchyma** (FIG. 32). Sclerenchyma (*scleros*, hard) consists of very long, narrow, thick and lignified cells, usually pointed at both ends. They are fibre-like in appearance, and hence they are also called sclerenchymatous fibres, or simply **fibres**. Their walls often become so much thickened that the cell-cavity is nearly obliterated. They have simple, often oblique, pits in their walls. The middle lamella is conspicuous in sclerenchyma. Sclerenchymatous cells are abundantly found in plants, and occur in patches or definite layers which are seen to dovetail into each other in the longitudinal direction. Sometimes also they occur singly among other cells. They are dead cells, and serve a purely mechanical function, that is, they give necessary strength, rigidity, flexibility and elasticity to the plant body and thus enable it to withstand various strains. Their average length is 1 to 3 mm. in angiosperms, and 2 to 8 mm. in gymnosperms. In special cases, as in hemp (*Cannabis*; B. & H. GANJA), rhea (*Boehmeria nivea*), flax (*Linum*), etc., the fibres are of excessive lengths ranging from 20 mm. to 550 mm. Such long, thick-walled cells make excellent textile

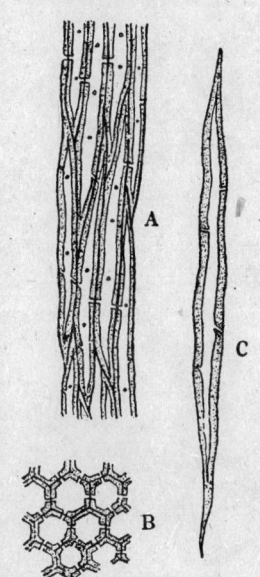

FIG. 32. Sclerenchyma. *A*, fibres as seen in longitudinal section; *B*, the same as seen in transection; and *C*, a single fibre.

fibres of commercial importance. Other common plants yielding long fibres are jute, coconut, Indian or sunn hemp (*Crotalaria juncea*; B. SHONE; H. SAN), Madras or Deccan hemp (*Hibiscus cannabinus*), sisal hemp (*Agave sisalana*), bowstring hemp (*Sansevieria*; B. MURGA; H. MARUL), rozelle (*Hibiscus sabdariffa*; B. MESTA; H. PATWA), etc.

Sclereids (FIG. 33). Sometimes here and there in the plant body special types of sclerenchyma develop to meet local mechanical needs. They are known as **sclereids** or **sclerotic cells**. They may occur in the cortex, pith, phloem, hard seeds, nuts, stony fruits, and in the leaf and stem of many dicotyledons and also gymnosperms. The flesh of pear, sometimes of guava, is gritty because of the presence of such cells (also called *grit cells*) in them. The cells though very thick-walled, hard and strongly lignified (sometimes cutinized or suberized) are not long and pointed like sclerenchyma, but are mostly isodiametric, polyhedral, short-cylindrical, slightly elongated or irregular in shape; commonly

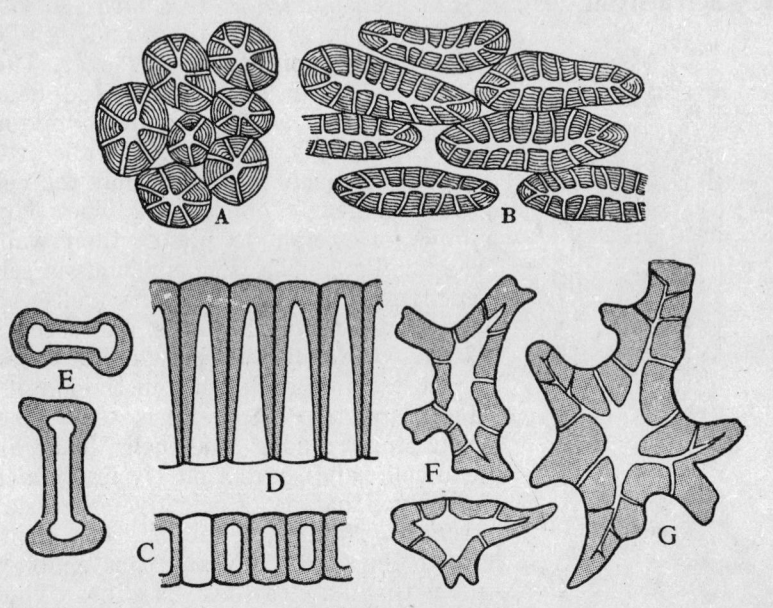

FIG. 33. Sclereids. *A*, brachysclereids (grit cells) in the flesh of pear; *B*, the same in coconut shell (endocarp); *C*, macrosclereids in the epiderms of onion scale; *D*, the same in the seed coat of *Phaseolus*; *E*, osteosclereids (two) in the seed coat of pea; *F*, astrosclereids (two) in tea leaf; *G*, the same in *Tsuga* (a conifer).

they have no definite shape. They are dead cells (seldom living), and have a very narrow cell-cavity which may be almost obliterated owing to excessive thickness of the cell-wall. Their walls are provided with many simple pits which may be branched or unbranched. Further, their thickened walls often show distinct lamellation. They may be somewhat loosely arranged or closely packed; they may also occur singly. They contribute to the firmness and hardness of the part concerned.

Types of Sclereids (FIG. 33). The following four common types may be noted. (1) **Brachysclereids** (*brachys*, short; A-B) or stone cells are more or less isodiametric in nature, and commonly found among masses of parenchyma in different parts of the plant body, as in the pith, cortex, bark, phloem, fleshy pericarp of certain fruits (e.g. pear), coconut shell, etc. (2) **Macrosclereids** (*makros*, long; C-D) are rod-like or columnar cells, with truncated ends, forming a solid palisade-like epidermal layer. They are common in the bark and in the seed-coats of many leguminous plants such as pea (*Pisum*), black gram (*Phaseolus*) and other pulses. They are also found in the protective scales of onion, garlic, etc. (3) **Osteosclereids** (*ostoon*, bone; E) are also columnar cells but they are dilated or lobed at one or both ends, somewhat like bones. They are commonly found in the seed coats and fruit walls, and also in the leaves of some dicotyledonous plants. (4) **Astrosclereids** (*astron*, star; F-G) are irregular stone cells, i.e. they are branched in an irregular way, with radiating arms of varying lengths, showing a stellate or star-like appearance. Evidently they assume various forms, depending on the nature of branching. They are found in the leaves of certain dicotyledons, as in tea leaf, and also in *Tsuga* (a conifer) and *Gnetum*; they are also found in the bark of certain conifers, as in *Abies* and *Larix*.

II. COMPLEX TISSUES

1. Xylem. Xylem or wood is a conducting tissue and is composed of elements of different kinds, viz. (*a*) tracheids, (*b*) vessels or tracheae (sing. trachea), (*c*) wood fibres and (*d*) wood parenchyma. Xylem as a whole is meant to conduct water and mineral salts upwards from the root to the leaf, and to give mechanical strength to the plant body.

(*a*) **Tracheids** (FIGS. 34-6). These are elongated, tube-like cells with hard, thick and lignified walls and large cell-cavity. Their ends are tapering, either rounded or chisel-like, less frequently pointed. They are dead, empty cells with their walls provided with one or more rows of bordered pits. Tracheids may also be annular, spiral, scalariform or pitted (with simple pits). In transverse section they are mostly angular, either polygonal or rectangular. Tracheids (and not vessels) occur alone in the wood of ferns and gymnosperms, whereas in the wood of angiosperms they occur associated with the vessels. Their walls being lignified and hard, tracheids give strength to the plant body but their main function is conduction of water from the root to the leaf.

(*b*) **Vessels** or **Tracheae** (FIG. 37). Vessels are cylindrical, tube-like structures; they are formed of a row of cells, placed end to end, from which the transverse partition walls break down (see pp. 193-94). A vessel or trachea is thus a tube-like series of

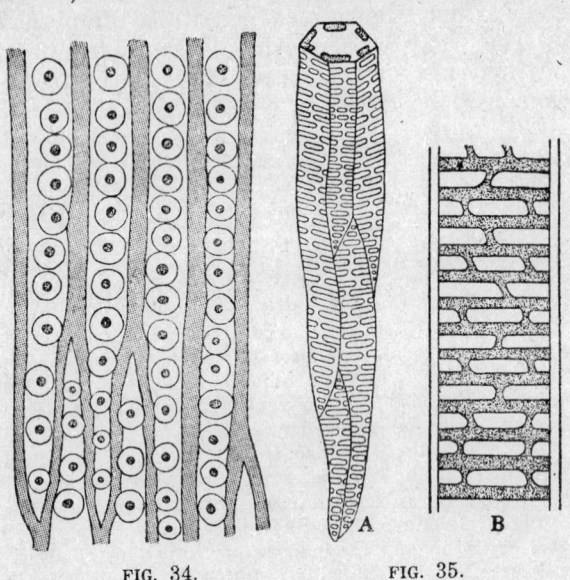

FIG. 34. Tracheids of pine stem (in radial section) with bordered pits. FIG. 35. *A*, a scalariform tracheid of fern; *B*, a portion of the wall of the same magnified.

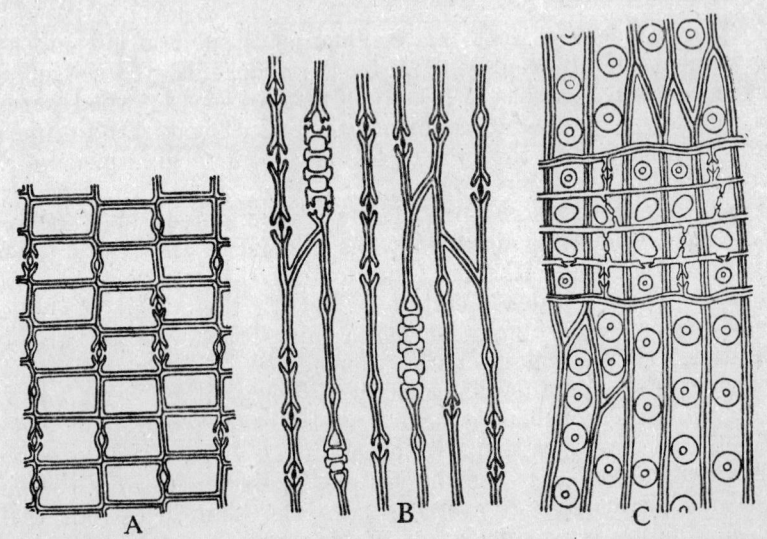

FIG. 36. Tracheids with bordered pits. *A*, pine stem in transection; *B*, the same in tangential (longitudinal) section; *C*, the same in radial (longitudinal) section.

THE TISSUE

cells, very much like a series of water pipes forming a pipe line. Their walls are thickened in various ways, and according to the mode of thickening vessels have received their names such as **annular, spiral, scalariform, reticulate,** and **pitted**. Associated with the vessels are often found some tracheids. Vessels and tracheids form the main elements of the wood or xylem of the vascular bundle (see FIG. 48). They serve for conduction of

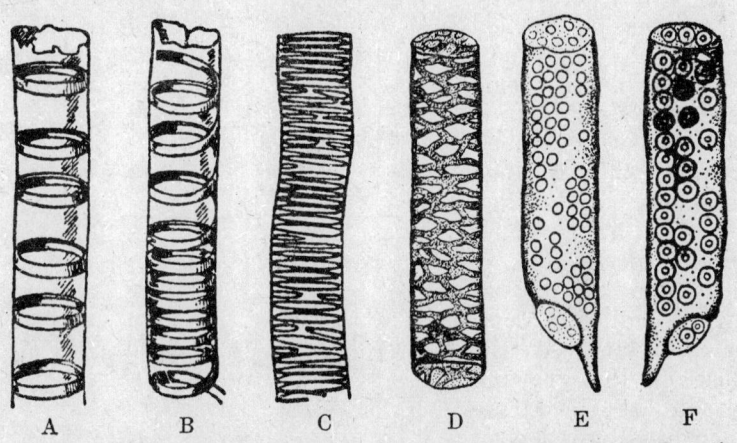

FIG. 37. Kinds of Vessels. *A*, annular; *B*, spiral; *C*, scalariform; *D*, reticulate; *E*, a vessel with simple pits; *F*, a vessel with bordered pits.

water and mineral salts from the roots to the leaves. They are dead, thick-walled and lignified, and as such they also serve the mechanical function of strengthening the plant body.

(*c*) **Wood Fibres.** Sclerenchymatous cells associated with wood or xylem are known as wood fibres. They occur abundantly in woody dicotyledons and add to the mechanical strength of xylem and of the plant body as a whole.

(*d*) **Wood Parenchyma.** Parenchymatous cells are of frequent occurrence in xylem, and are known as wood parenchyma. The cells are alive and generally thin-walled. The wood parenchyma assists, directly or indirectly, in the conduction of water upwards through the vessels and the tracheids; it also serves for food storage.

2. **Phloem.** Phloem or bast is another conducting tissue, and is composed of the following elements: (*a*) **sieve-tubes,** (*b*) **companion cells,** (*c*) **phloem parenchyma** and (*d*) **bast fibres** (rarely). Phloem as a whole is meant to conduct prepared food materials from the leaf to the storage organs and the growing regions.

(*a*) **Sieve-tubes** (FIG. 38-9). Sieve-tubes are slender, tube-like structures, and are composed of elongated cells, placed end on end. Their walls are thin and made of cellulose; the transverse partition walls are, however, perforated by a number of pores. The transverse wall then looks very much like a sieve, and is called the **sieve-plate**. The sieve-plate may sometimes be formed in the side (longitudinal) wall. In some cases the sieve-plate is not transverse (horizontal), but inclined obliquely, and then different areas of it become perforated. A sieve-plate of this nature is called *a compound plate*. At the close of the growing season the sieve-plate is covered by a deposit of a colourless, shining substance in the form of a pad, called **callus** or **callus pad**. This consists of a carbohydrate, called *callose*. In winter the callus completely clogs the pores; but in spring, when the active season begins, it gets dissolved. In old sieve-tubes callus forms a permanent deposit. The sieve-tube contains no nucleus, but has a lining layer of cytoplasm which is continuous through the pores. Sieve-tubes are used for the longitudinal transmission of prepared food materials —proteins and carbohydrates —from the leaves to the storage organs in the downward direction, and later from the storage organs to the growing regions in the upward direction. A heavy deposit of food material is found on either side of the sieve-plate with a narrow median portion.

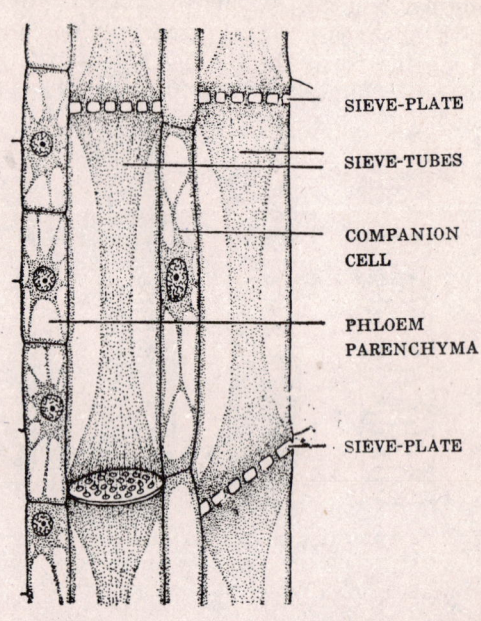

FIG. 38. Sieve-tissue in longitudinal section.

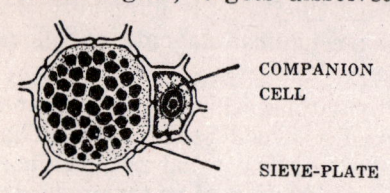

FIG. 39. Sieve-tube in transection.

(*b*) **Companion Cells.** Associated with each sieve-tube and connected with it by pores is a thin-walled, elongated cell known as the **companion cell**. It is living, containing protoplasm and an elongated nucleus. The companion cell is present only in angiosperms (both dicotyledons and monocotyledons). It assists the sieve-tube in the conduction of food.

(c) **Phloem Parenchyma.** There are always some parenchymatous cells forming a part of the phloem in all dicotyledons, gymnosperms and ferns. The cells are living, and in shape often cylindrical. They store up food material and help in the conduction of the same. Phloem parenchyma is, however, absent in most monocotyledons.

(d) **Bast Fibres.** Sclerenchymatous cells occurring in the phloem or bast are known as bast fibres. These are generally absent in the primary phloem but are of frequent occurrence in the secondary phloem.

Position of Phloem. Normally in angiospermic stems the phloem lies external to the xylem. But in several families of dicotyledons such as *Acanthaceae, Apocynaceae, Asclepiadaceae* (e.g. *Leptadenia*; see FIG. 85), *Combretaceae, Convolvulaceae, Cucurbitaceae, Myrtaceae, Solanaceae, Compositae*, etc., a part of the *primary phloem*, called **intraxylary phloem** or **internal phloem** is seen to occur, often in small groups or strands, internal to the primary xylem around the pith, either in association with this xylem or detached from it. The internal phloem is similar to the normal external phloem in origin, structure, and composition but its elements are fewer in number and do not increase as a result of cambial activity. Sometimes small groups of *secondary phloem* formed by the cambium are seen embedded in the secondary xylem. This is called **interxylary phloem** or **included phloem.** Among dicotyledons this is found in *Acanthaceae, Asclepiadaceae* (e.g. *Leptadenia*; see FIG. 85),*Salvadoraceae* (e.g. *Salvadora*; see FIG. 84), *Combretaceae, Cucurbitaceae, Nyctaginaceae,* (e.g. *Mirabilis*; see FIG. 82), *Loganiaceae* (e.g. *Strychnos*), etc., and among monocotyledons in some arborescent types showing secondary growth, e.g. *Dracaena* (see FIG. 89), *Yucca* (see FIG. 1/81), *Agave, Aloe,* etc. It will be noted that some dicotyledonous families have both the types of phloem.

III. SECRETORY TISSUES

1. **Laticiferous Tissue.** This consists of thin-walled, greatly elongated and much branched ducts (FIG. 40) containing a milky juice known as *latex* (see p. 179). Laticiferous ducts are of two kinds: latex vessels and latex cells. They contain numerous nuclei which lie embedded in the thin layer of protoplasm lining the cell-wall which is usually thin and made of cellulose. They occur irregularly distributed in the mass of parenchymatous cells, and their function also is not clearly understood. They may act as food storage organs or as reservoirs of waste products. They may also act as translocatory tissues.

Latex vessels (FIG. 40B) are the result of fusion of many cells. They are formed from rows of elongated meristermatic cells from which partition walls soon get dissolved, as in wood vessels. They grow more or less as parallel ducts, and in the mature portion of the plant they anastomose with one another *by the fusion of their branches, forming a network*. Latex vessels are found in poppy (*Papaver*), e.g. opium poppy, garden poppy and prickly poppy, and also in some species of sunflower family or *Compositae*, e.g. *Sonchus*.

Latex cells (FIG. 40A), on the other hand, although much branched like the latex vessels, are really single or independent

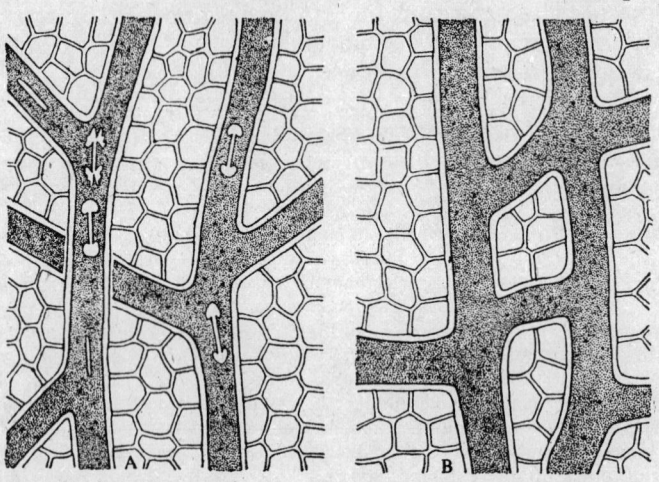

FIG. 40. Laticiferous tissue. *A*, latex cells; *B*, latex vessels.

units. They originate as minute structures and then with the growth of the plant elongate and branch, ramifying in all directions through the tissues of the plant, but *without fusing together to form a network*. They are coenocytic in nature (see p. 196). Latex cells are found in madar (*Calotropis*), spurges (*Euphorbia*), oleander (*Nerium*), yellow oleander (*Thevetia*), periwinkle (*Vinca*), *Ficus* (e.g. banyan, fig, peepul), etc.

Latex cells may be isolated and studied in the following way. Slice off a big portion of the pith from a piece of *Euphorbia neriifolia* stem. Break transversely the cortical portion of the stem with a light incision on its surface and then pull it apart. Some slender threads will be seen to come out. Mount them carefully in the usual way and examine under the microscope. A number of elongated, tubular, branched cells will be clearly seen. These are the latex cells. By irrigating with iodine solution rod- and dumb-shaped starch grains may be distinctly seen in the latex cells. A longitudinal section through the stem will, however, show that the latex cells lie embedded in parenchyma, as shown in FIG. 40A.

2. Glandular Tissue. This tissue is made of **glands** which are special structures containing some secretory or excretory products. Glands may consist of single isolated cells or small groups of cells with or without a central cavity. They are of various kinds and may occur as *external glands* on the epidermis or as *internal glands* lying embedded in other tissues in the interior of the plant body. They are parenchymatous in nature, and

THE TISSUE

contain abundant protoplasm with a large nucleus. They contain different substances and have manifold functions.

Internal glands are (1) oil-glands (FIG. 41A) secreting essential oils, as in the fruits and leaves of orange, lemon, pummelo, etc.; (2) mucilage-secreting glands, as in the betel leaf; (3) glands secreting gum, resin, tannin, etc (see FIG. 27); (4) digestive glands secreting enzymes; and (5) water-secreting glands known as **hydathodes**.

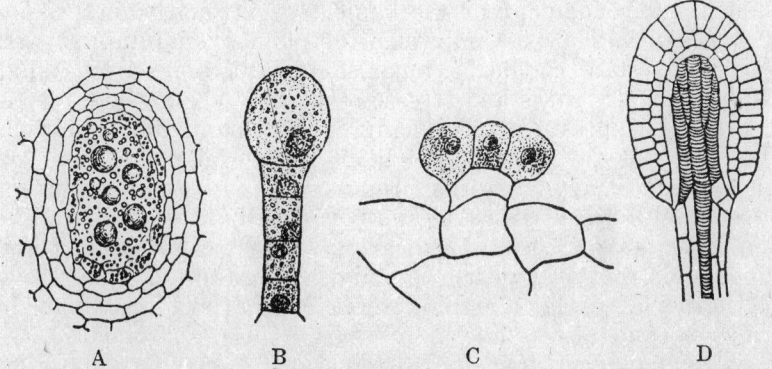

FIG. 41. Glands. *A*, an oil-gland of orange rind; *B*, a glandular hair of *Boerhaavia* fruit; *C*, a digestive gland of butterwort (*Pinguicula*); *D*, digestive gland of sundew (*Drosera*).

Hydathodes (FIG. 42) are special structures through which exudation of water takes place in liquid form. They are mainly found in aquatic plants and in some herbaceous plants growing in moist places. They occur at the tip of the leaf or on its margin at the apices of veins and are made of a group of living cells having numerous intercellular spaces filled with water, but few or no chloroplasts. They represent modified bundle-ends. These cells, called *epithem cells*, open out into one or more sub-epidermal chambers; these in turn communicate with the exterior through an open **water stoma** or **water pore**. The water stoma structurally resembles an ordinary stoma (see FIG. 44A), but is usually larger and has lost the power of movement. Hydathodes are commonly seen in water lettuce (*Pistia*), water hyacinth (*Eichhornia*), garden nasturtium (*Tropaeolum*), rose (*Rosa*), balsam (*Impatiens*), aroids, many grasses, etc.

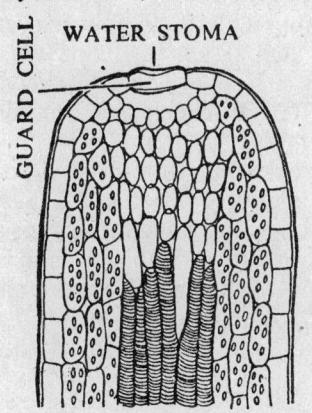

FIG. 42. Hydathode of water lettuce (*Pistia*).

External glands occur as outgrowths and are in the nature of short hairs tipped by glands, known as glandular hairs. External glands are: (1) water-secreting hairs or glands, also called hydathodes; (2) **glandular hairs** (FIG. 41B) secreting gummy substances, as in tobacco (*Nicotiana*), *Plumbago* (B. CHITA; H. CHITRAK), and *Boerhaavia* (B. PUNARNAVA; H. THIKRI); (3) glandular hairs secreting irritating, poisonous substances, as in nettles (see FIG. I/82); (4) nectar-secreting glands or nectaries, as in many flowers; and (5) enzyme-secreting glands (FIG. 41 C-D), as in carnivorous plants.

The Mechanical System. For the existence and stability of a plant the development of mechanical or strengthening tissues is a necessity. They not only help the plant to maintain a particular shape as it should be, but also to withstand various kinds of mechanical strains and stresses that they are often subjected to under the prevailing or changing environmental conditions. The stem has to bear the weight of the aerial parts, often extensive and heavy, and the branches to withstand transverse stresses with a load of leaves and periodically flowers and fruits. The leaves have to bear the shearing stresses often against high winds. The root system has to stand a longitudinal compression by the aerial parts, a pulling force by the swaying stem and branches, and also a lateral pressure by the surrounding soil. It is thus evident that the distribution of mechanical tissues should be such as to help the plant body to stand various strains to the maximum degree. In default the whole plant body is bound to collapse.

Kinds of Mechanical Tissues. (1) **Sclerenchyma** is the most important and efficient mechanical tissue widely distributed in plants for the specific purpose of mechanical strength. It includes (*a*) **bast fibres**, sometimes of excessive lengths, and (*b*) **wood fibres** associated with wood or xylem. (2) **Collenchyma** mostly associated with young growing parts, finally forming a permanent structure of those organs. Collenchyma though not as efficient as sclerenchyma has many advantages of its own: it has power of growth; it offers no resistance to the growing organs; and it is flexible. Further it does not interfere with secondary growth but sometimes helps this growth. Both sclerenchyma and collenchyma are remarkable for their tensile strength. (3) **Vessels** and **tracheids** also give mechanical strength to the plant body. (4) **Sclereids** often meet local strain (see pp. 206-7).

Principles Governing the Distribution of Mechanical Tissues. The principle governing the construction of mechanical tissues and their distribution in the plant body is 'maximum of strength with minimum expenditure of material', as in the case of a building or a bridge. Therefore the mechanical strength of the tissues is correlated with their economical distribution according to certain principles to give maximum efficiency. The main principles involved are: (1) inflexibility, (2) inextensibility, and (3) incompressibility.

1. *Inflexibility.* To achieve this the ideal arrangement is to have the mechanical tissues in the form of girders as in the case of engineer's architecture. A simple girder is the one in which there is no flange and in cross-section it is like I. A compound girder is the one in which there may be one flange like T or sometimes two flanges. Where much strength is not needed simple girders may do but in some cases many of them may be required to maintain necessary strength. More efficient are of course the compound girders. This principle of distribution of mechanical tissues applies to stems which evidently have to bear the enormous weight of the branches, leaves, flowers and fruits. Stems are also swayed back and forth (bending stress) by the wind, sometimes very strong. They are thus subjected to alternate stretching and compressing on all their sides. In them, therefore, the best position for the strengthening tissues is close to the periphery in the form of separate girders. In dicotyledons it is further of utmost importance that the mechanical tissues do not obstruct secondary growth in thickness.

Stems, as said above, are subjected to bending stress in any direction, and to meet the need collenchyma and sclerenchyma appear in suitable places. The distribution of collenchyma has been already mentioned (see p. 202). The distribution of sclerenchyma is, however, more diversified. A few cases may be mentioned. (1) Long strands of vascular bundles are associated with and strengthened by fibrous tissue, as in most cases. (2) Patches of sclerenchyma lying associated with phloem and forming the *hard bast*, as in sunflower stem. (3) Pericyclic sclerenchyma forming a hollow cylinder, as in *Cucurbitaceae*. In monocotyledons there being no secondary growth sclerenchyma occurs in various forms. (4) Sclerenchymatous sheath, complete or incomplete, encircling a vascular bundle, is a common feature. (5) Patches of sclerenchyma of varying sizes associated with the vascular bundles, as in *Colocasia*. (6) Sub-epidermal patches, as in *Cyperus*. (7) Isolated patches in the cortical region, as in bamboo. (8) Patches tangentially connected, as in many palms. (9) Simple hollow cylinder, broken or unbroken, in the cortical region, as mostly in *Liliaceae*, e.g. *Asparagus* (see FIG. 59).

2. *Incompressibility.* Roots are subjected to longitudinal compression by the load they carry overhead, and to the lateral pressure exerted by the surrounding soil. These forces are met by the roots by the development of a solid wood cylinder in or around the centre. At an early stage the lignified wood vessels make the root as a whole sufficiently incompressible. Later with the progress of secondary growth the new thick-walled elements—new wood vessels and wood fibres—form a solid central column which proves efficient for the mechanical strength of the root. In some dicotyledons, as in broad bean (*Vicia faba*), hard bast is present. Roots are wanting in collenchyma. The scattered arrangement of mechanical tissues, as in the case of the stem, may prove to be fatal to the root since individual strands are liable to break under very heavy pressure. In monocotyledonous roots the vascular bundles provide for necessary strength. There are cases, however, as in *Pandanus*, where each bundle is surrounded by a strong sclerenchymatous sheath, and in addition there are strands of sclerenchyma in it. In aroids the pith is often lignified and thickened, and in orchids the conjunctive tissue is often so (see FIG. 64).

3. *Inextensibility.* Roots are also subjected to the pulling force exerted by the swaying stem. As in the previous case, the centralization of vascular bundles and the associated mechanical tissues forming a solid central column seems to be very effective to withstand longitudinal tension due to the

bending of the stem. Sometimes, however, as in the maize root, an additional fibrous cylinder develops in the peripheral region.

It may also be noted that the leaves are subjected to shearing stresses, and to guard against them sclerenchyma is distributed in various ways according to their need, commonly as patches flanking the upper epidermis and the lower, and also in the mesophyll. Vascular bundles may also be strengthened by sclerenchymatous sheath, complete or incomplete.

Chapter 3 THE TISSUE SYSTEM

On the principle of division of labour, tissues are arranged in three systems, each taking a definite share in the common life-work of the plant. Each system may consist of only one tissue or a combination of tissues which may structurally be of similar or different nature, but perform a common function and have the same origin. The three systems are: (I) the **epidermal tissue system**, (II) the **ground** or **fundamental tissue system**, and (III) the **vascular tissue system**.

I. THE EPIDERMAL TISSUE SYSTEM

Epidermis. The epidermal tissue system is derived from the dermatogen of the apical meristem and forms the **epidermis** (*epi*, upon; *derma*, skin) or the outermost skin layer which extends over the entire surface of the plant body, and is continuous except for certain openings (stomata and lenticels). At surface view the cells of the epidermis are somewhat irregular in outline (see FIG. 43), varying in shape and size, but closely fitted together without intercellular spaces. They, however, appear more or less rectangular in cross-section. Epidermis is mostly single-layered, but sometimes, as in the leaves of india-rubber plant, banyan, oleander, etc., it becomes few-layered, called *multiple epidermis*. Epidermal cells are parenchymatous in nature with comparatively small amount of cytoplasm lining the cell-wall and a large vacuole filled with colourless cell-sap. In some plants the cells may contain anthocyanins or chromoplasts, but not chloroplasts except in guard cells, ferns, submerged plants and a few others. In the leaves and young green shoots the epidermis, however, possesses numerous **stomata**

through which an interchange of gases takes place between the plant and the atmosphere. Epidermal cells soon die off and become filled with various substances such as tannin, silica particles, gum, mucilage, crystals, etc. Inner and radial walls of epidermal cells are thin, while the outer walls are thick and usually impregnated with cutin or suberin. Cutinization or suberization sometimes extends to the radial walls also. The cutinized layer called the **cuticle** acts as a hard varnish-like coating and protects the inner cells against loss of water, mechanical injury and potential parasites (see below). Excretion of wax in the form of rods, scales, grains, etc., prevents further loss of water. In many plants epidermal cells often bear outgrowths, known as **hairs** or **trichomes**. These may be unicellular or multicellular, simple or branched, soft or sharp and stiff. Besides, the epidermis may also bear stinging hairs (see FIG. I/82), as in nettles, glandular hairs (see FIG. 41B), as in *Boerhaavia* (B. PUNARNAVA; H. THIKRI), tobacco, *Plumbago* (B. CHITA; H. CHITRAK), etc. A dense coating of hairs, as in *Gnaphalium* and *Aerua*, is another feature of the epidermis.

The outermost layer of the root is called the **epiblema** or **piliferous layer**. It is mainly concerned with the absorption of water and mineral salts from the soil. Thus to increase the absorbing surface, which has been estimated to be 5 to 20 times greater, the outer walls of most of its cells extend outwards and form tubular unicellular prolongations called the **root-hairs**. The epiblema is neither cutinized nor is it provided with stomata.

Functions. (1) The primary function of the epidermis is the protection of the internal tissues against mechanical injury, excessive heat or cold, fluctuations of temperature, attack of parasitic fungi and bacteria, and against leaching effect of rain; this is due to the presence of cuticle, hairs, tannin, gum, etc. (2) Prevention of excessive evaporation of water from the internal tissues by the development of thick cuticle, wax and other deposition, cutinized hairs, scales, multiple epidermis, etc., is another important function of the epidermis. (3) Protection of the plant against intense illumination (i.e. strong sunlight) and against excessive radiation of heat is mainly afforded by strong cuticle and cutinized hairs, particularly a dense coating of hairs. (4) The epidermis has also to protect the plant against the attack of herbivorous animals; this the epidermis does with the help of sharp and stiff hairs as in some cucurbits, a dense coating of hairs as in *Gnaphalium*, stinging hairs as in nettles (see FIG. I/82), glandular hairs as in *Boerhaavia* (see FIG. 41B), silica particles as in many grasses (lemon grass, for example), *Equisetum*, etc., and raphides (see FIG. 18) as in many aroids. (5) The epidermis also acts as a store-house of water, as in desert plants. (6) The epidermis sometimes has some minor functions such as photosynthesis, secretion, etc.

Stomata. *Structure and Behaviour.* Stomata (*stoma*, a mouth) are very minute openings (FIG. 43) formed in the epidermal layer in green aerial parts of the plant, particularly the leaves. Roots and non-green parts of the stem are free from them. Each stoma is surrounded by two semi-lunar cells, known as the *guard cells*. The term 'stoma' is often applied to the stomatal

FIG. 43. Stomata (surface view) in epidermal layer.

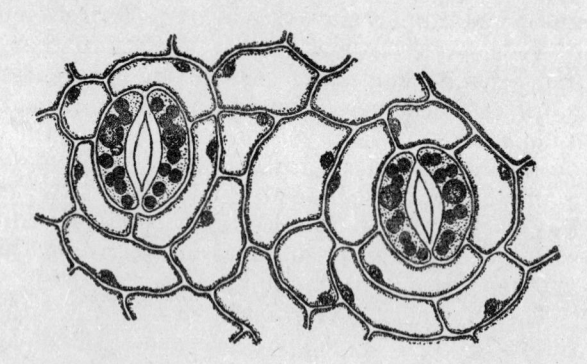

opening plus the guard cells. The guard cells are living and always contain chloroplasts, and their inner walls are thicker and outer walls thinner. They guard the stoma or the passage, i.e. they regulate the opening and closing of it like lips. Sometimes the guard cells are surrounded by two or more cells which are distinct from the epidermal cells; such cells are called *accessory cells*. In dicotyledonous leaves the stomata remain scattered, while in monocotyledonous leaves they occur in parallel rows. Under normal conditions the stomata remain closed at night, i.e. in the absence of light, and they remain open during the daytime, i.e. in the presence of light. In most plants the stomata open fully only in bright light, but in certain plants the stomata do so in diffuse light. Commonly they open fully in the morning and close towards the evening. They may close up at daytime when very active transpiration (evaporation of water) takes place from the surface of the leaf under certain conditions such as dryness of the air, blowing of dry wind and deficient supply of water in the soil. The intensity of light markedly affects the degree of stomatal opening. The opening and closing of the stomata are due to the movement of the guard cells. When the guard cells become turgid, i.e. full of water, they expand and bulge out in the outward direction, and the stoma is open; when the guard cells become flaccid by losing water the stoma is closed.

The turgidity or flaccidity of the guard cells is due to the presence of sugar or of starch in them. In light the sugar manufactured by the chloroplasts of the guard cells accumulates in them and, being soluble, increases the concentration of the cell-sap. Under this condition the guard cells absorb water from the neighbouring cells and become turgid, and the stoma opens. In darkness on the other hand the sugar present in the guard cells becomes converted into starch—an insoluble compound. The concentration of the cell-sap is, therefore, lower than that of the neighbouring cells. Under this condition the guard cells lose water and shrink and the stoma closes. The transformation of sugar into starch at night and vice versa at daytime is due to acidity and alkalinity of the cell-sap of the guard cells. At night photosynthesis being in abeyance carbon dioxide accumulates in the guard cells and the cell contents become weakly acid. Under this condition sugar becomes converted into starch. During the daytime carbon dioxide is utilized in photosynthesis and thus the cell contents become slightly alkaline. Under this condition the starch becomes converted into sugar.

There is another hypothesis, called **colloidal hypothesis**, put forward to explain the movement of the guard cells. According to this, the cell contents become alkaline as a result of the effect of sunlight on guard cells and this causes the colloids present in them to swell, apart from the fact that this causes the transformation of starch into sugar. The swelling of the colloids according to this theory causes the guard cells to bulge out and the stoma to open. At night the acidity of the guard cells increases and causes the colloids to shrink again, and thus to close the stoma, apart from the fact that this increased acidity brings about the conversion of sugar into starch.

Functions and Distribution. Stomata are used for interchange of gases between the plant and the atmosphere—oxygen for respiration and carbon dioxide for manufacture of carbohydrates. For the facility of diffusion of these gases each stoma

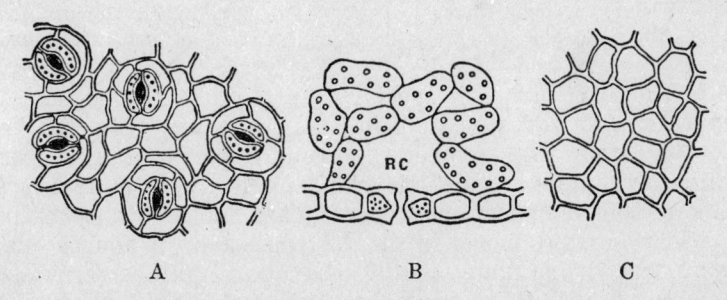

FIG. 44. Stomata in betel leaf. *A*, lower epidermis with numerous stomata (surface view); *B*, section of leaf (a portion of the lower side): *RC*, respiratory cavity internal to a stoma; *C*, upper epidermis with no stoma (surface view).

opens internally into a small cavity known as the *respiratory cavity* (FIG. 44B) which in its turn communicates with the system of intercellular spaces and air-cavities. Stomata are also the organs through which evaporation of water normally takes place, and the plant thus gets rid of the excess quantity. Stomata are most abundant in the lower epidermis (FIG. 44A) of the dorsiventral leaf (see p. 47); none (or sometimes comparatively few) are present in the upper (FIG. 44C), e.g. in sunflower leaf the average numbers in lower and upper surfaces are 325 and 175, in pea leaf 216 and 101, in gourd leaf 269 and 28, etc. In the isobilateral and centric leaves (see p. 47) stomata are more or less evenly distributed on all sides (see FIGS. 68-9). In the floating leaves, as in those of the water lily, stomata remain confined to the upper epidermis alone; in the submerged leaves no stoma is present. In desert plants and in those showing xerophytic adaptations, e.g. American aloe or century plant (*Agave*; FIG. 45), oleander (*Nerium*; FIG. 46), pine (*Pinus*; see FIG. VI/10), etc., one or more stomata are situated in grooves or

FIG. 45. Sunken stoma in the leaf of American aloe (*Agave*).

FIG. 46. Sunken stomata in the leaf of oleander (*Nerium*).

pits in the leaf. This is a special adaptation to reduce excessive evaporation, as the stomata sunken in pits are protected from gusts of wind. The number of stomata per unit area varies within wide limits. In ordinary land plants there is an average of about 100 to 300 stomata per square millimetre, sometimes much less or many more. In the floating leaves of aquatic plants stomata may be as many as 400 per square millimetre, while in submerged leaves there are none. In desert plants on the other hand there may be only 10 to 15 stomata per square millimetre; while there are cases with about 1,300 stomata in the same space.

II. THE GROUND OR FUNDAMENTAL TISSUE SYSTEM

This system forms the main bulk of the body of the plant, and extends from below the epidermis to the centre (excluding the vascular bundles). It is partly derived from the periblem and partly from the plerome. Its primary functions are manufacture and storage of food material; it also has a mechanical function. This system consists of various kinds of tissues, of which parenchyma is most abundant; other tissues are sclerenchyma and collenchyma, and sometimes also laticiferous tissue and glandular tissue. It is differentiated into the following zones and sub-zones.

1. **Cortex.** The cortex is the zone that lies between the epidermis and the pericycle, varying in thickness from a few to many layers. In dicotyledonous stems (see FIG. 53) it is usually differentiated into the following sub-zones: (a) **hypodermis**—a few layers of collenchyma or sometimes sclerenchyma; (b) **general cortex** or cortical parenchyma—a few layers of thin-walled parenchymatous cells with or without chloroplasts, but often with intercellular spaces; and (c) **endodermis**—a single wavy layer, not often very conspicuous; it is also called *starch sheath* as it often has numerous starch grains. In monocotyledonous stems (see FIG. 57), owing to the scattered arrangement of vascular bundles, the cortex is not marked out into the cortex proper and the endodermis; sclerenchyma is, however, often present as hypodermis. In roots (see FIG. 61) the cortex consists of (a) many layers of thin-walled parenchymatous cells (general cortex) often with conspicuous starch grains in them, and leaving a lot of intercellular spaces between them; (b) a distinct circular layer of endodermis; and (c) sometimes hypodermis —a few external layers of fairly big, often radially elongated, parenchymatous cells.

Functions. In stems the cortex primarily functions as a protective tissue; its secondary functions are storage, photosynthesis, etc. In roots the cortex is essentially a storage tissue. It is also the pumping station of the root where the individual cells by their alternate expansion and contraction act as microscopic pumps forcing the water absorbed by the root-hairs into the xylem vessels.

Endodermis is the inner limiting layer of the cortex and is formed of vertically elongated cells; in cross section the endodermis appears as a single layer of barrel-shaped cells without intercellular spaces. The layer is wavy in stems and often not readily distinguishable or even altogether wanting; while in roots it is circular and well defined. The cells are living containing abundant protoplasm, large nuclei and often starch grains— hence this layer is also called **starch sheath**. Some cells of the endodermis may contain mucilage, tannin, gum, etc. The outer walls of endodermal cells are thin, while radial and inner walls are often thickened, being suberized or cutinized and also sometimes lignified, particularly in roots. The thickened walls are provided with numerous simple pits. Sometimes the thickening takes the form of a band or strip surrounding each cell; this band, first recognized by Caspary in 1865, is called the **Casparian strip**. It may be made of lignin or suberin. Among the thick-walled cells of the endodermis, as in many roots, often occur, opposite to protoxylem vessels, some small thin-walled cells; these are called **passage cells** (see FIG. 62). Through them the sap absorbed by root-hairs enters the xylem vessels.

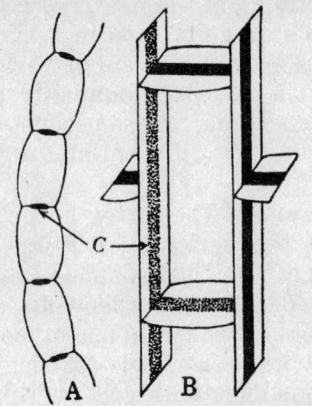

FIG. 47. Casparian strip (C) in endodermis; A, endodermis in tran- section; B, diagram of an endodermal cell (longitudinal view).

Endodermis is well developed in the roots of all plants, in the stems of pteridophytes and herbaceous dicotyledons, and in the leaves of gymnosperms; it is, however, absent or indistinct in the stems of woody plants and in the leaves of angiosperms.

Functions. Functions of the endodermis are somewhat obscure. Some regard it as a water-tight jacket between the xylem and the surrounding tissues. It may act as an air-dam preventing diffusion of air into the vessels and thus clogging them. It may be a 'diffusion layer' preventing loss of water, mineral salts and food from the vascular bundles. It may be a storage tissue containing starch grains, as in dicotyledons. It may be connected with the osmotic pressure that develops in the root-cortex. It may serve as a passage for water from the cortex of the root to the protoxylem.

2. Pericycle. This forms a multi-layered zone between the endodermis and the vascular bundles and occurs as a cylinder encircling the vascular bundles and the pith, as in dicotyledonous stems. In may consist wholly of sclerenchyma forming a continuous zone, as in gourd (*Cucurbita*) stem but more commonly it is made of both parenchyma and sclerenchyma, the latter forming isolated strands in it (the pericycle). Each such strand associated with the phloem or bást of the vascular bundle in the form of a cap is known as the **hard bast**, as in sunflower stem. In the roots and stems of some aquatic plants the pericycle is absent, and it is not distinguishable in the monocotyledonous stems. In the stems of pteridophytes the pericycle is single-layered, while in those of gymnosperms it is multi-layered. In the roots of angiosperms and pteridophytes the pericycle is a single layer of very small, thin-walled and somewhat barrel-shaped cells. In some monocotyledonous roots, however, the pericycle may be a few layers thick and even lignified. In gymnosperms it is multiseriate and thin-walled.

Functions. In all roots the pericycle is the seat of origin of lateral roots (see FIG. 66). In dicotyledonous roots it further gives rise to secondary meristems—a portion of the cambium (see FIG. 75) and later the whole of the cork-cambium (see (FIG. 77). In all stems the pericycle is the seat of origin of adventitious roots. Otherwise its functions are mechanical, secretion, storage, etc.

3. **Pith** and **Pith Rays.** The **pith** or **medulla** forms the central core of the stem and the root and is usually made of large-celled parenchyma with abundant intercellular spaces. In the dicotyledonous stem the pith is often large and well developed, while in the monocotyledonous stem, owing to scattered distribution of vascular bundles, it is not distinguishable; in the dicotyledonous root the pith is either small or absent, bigger vessels having met in the centre; while in the monocotyledonous root a distinct pith is present. It is often parenchymatous, but sometimes sclerenchymatous. In the dicotyledonous stem the pith extends outwards to the pericycle between the vascular bundles. Each extension which is a strip of parenchyma is called the **pith ray** or **medullary ray**. It is not present as such in the root. The cells of the pith and the pith ray are usually larger than those of the cortex and enclose numerous intercellular spaces.

Functions. They serve to store food material. The function of the sclerenchymatous pith is, of course, mechanical. The medullary ray further transmits water and food material outwards to the peripheral tissues, and is the seat of origin of a strip of cambium (i.e. the interfascicular cambium; see FIG. 72B) prior to secondary growth.

III. *THE VASCULAR TISSUE SYSTEM*

This system consists of a number of vascular bundles which are distributed in the **stele**. The stele is the central cylinder of the stem and the root (and the pine leaf) surrounded by the endodermis and consists of vascular bundles, pericycle, pith and medullary rays. Each bundle is made up of **xylem** and **phloem** with a cambium, as in dicotyledonous stems, or without a cambium, as in monocotyledonous stems, or of only one kind of tissue —xylem or phloem, as in roots. The function of this system is to conduct water and raw food materials from the roots to the leaves, and prepared food materials from the leaves to the storage organs and the growing regions. The elements of a vascular bundle are derived from the *procambial strands* of the plerome, which show a tendency to elongate even at an early stage. The vascular bundles may be regularly arranged in a ring, as in the stems of dicotyledons, gymnosperms and in all roots, or they may be scattered in the ground tissue, as in the stems of monocotyledons.

Elements of a Vascular Bundle (FIG. 48). A vascular bundle of a dicotyledonous stem, when fully formed, consists of three well-

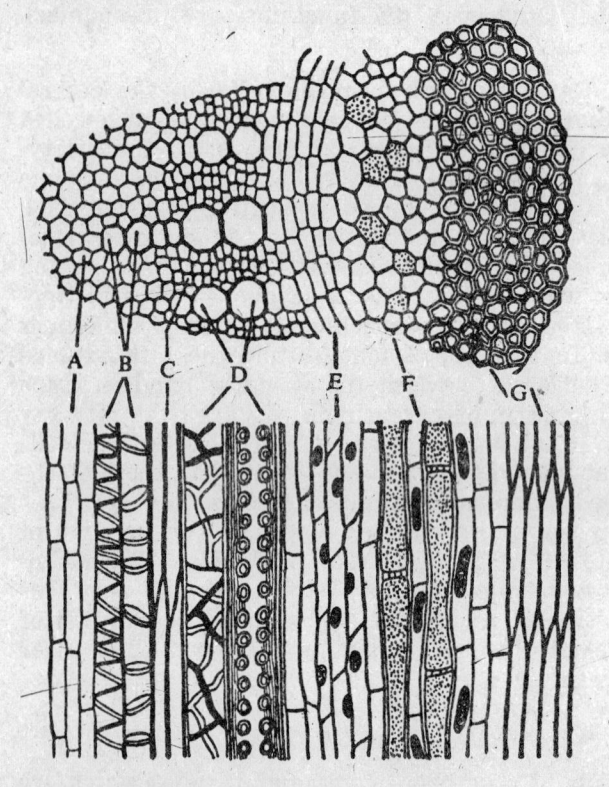

FIG. 48. Vascular bundles of sunflower stem in transverse and longitudinal sections.

A, wood parenchyma;

B, protoxylem (annular and spiral vessels);

C, tracheids and wood fibres;

D, metaxylem (reticulate and pitted vessels);

E, cambium;

F, phloem (sieve-tubes, companion cells and phloem parenchyma);

G, sclerenchyma (hard bast).

defined tissues: (1) xylem or wood, (2) phloem or bast, and (3) cambium. They have different kinds of tissue-elements.

(1) **Xylem** or **Wood**. This lies towards the centre, and is composed of the following elements: (1) tracheae or vessels, (2) some tracheids, (3) a number of wood fibres, and (4) a small patch of wood parenchyma. Vessels are of various kinds (see FIG. 37) such as *spiral, annular, scalariform, reticulate* and

pitted (with simple or bordered pits). Some tracheids also lie associated with the vessels. Wood fibres and wood parenchyma are ordinary sclerenchymatous and parenchymatous cells lying associated with the wood or xylem. They are provided with simple pits in their walls. Sometimes in the secondary xylem the wood parenchyma becomes thick-walled and lignified. Xylem vessels and tracheids are used for the conduction of water and mineral salts from the roots to the leaves and other parts of the plant; xylem parenchyma assists them in their task and also serves for food storage, and wood fibres give proper rigidity to the xylem. Except for the wood parenchyma all the other elements of xylem are dead and lignified, and hence their secondary function is to give mechanical strength to the plant.

The first-formed xylem or **protoxylem** consists of *annular, spiral* and *scalariform* vessels; it lies towards the centre of the stem and its vessels have smaller cavities. The later-formed xylem or **metaxylem** consists of *reticulate* and *pitted* vessels and some *tracheids*; it lies away from the centre and its vessels have much bigger cavities. Of course all transitional stages are noticed between protoxylem and metaxylem. The development of xylem is *centrifugal* in the stem, or, in other words, it is said to be *endarch* (meaning inner origin).

(2) **Phloem** or **Bast.** This lies towards the circumference, and consists of (1) sieve-tubes, (2) companion cells, and (3) phloem parenchyma. Companion cells and phloem parenchyma are provided with simple pits, particularly in the walls lying against the sieve-tubes. Phloem as a whole is used for translocation of prepared food materials (soluble proteins, amines and aminoacids, and soluble carbohydrates) from the leaves to the storage organs in the downward direction, and later from there to the different growing regions in the upward direction. Sieve-tubes, constant in all higher plants, are the main channels through which this translocation takes place. Companion cells and phloem parenchyma, when present, assist the sieve-tubes in this task, and they also transmit many of the soluble food materials sideways to the surrounding tissues. All the elements of phloem are made of cellulose, and are living. Primary phloem hardly ever contains bast fibres but it may be capped by a patch of sclerenchyma, called the **hard bast,** as seen in the sunflower stem (see FIG. 53).

The outer portion of phloem consisting of narrow sieve-tubes constitutes the **protophloem,** and the inner portion consisting of bigger sieve-tubes constitutes the **metaphloem.**

(3) **Cambium.** This is a thin strip of primary meristem lying in between xylem and phloem. It consists of one or a few layers of thin-walled and roughly rectangular cells. Although cambial

cells look rectangular in transverse section, they are much elongated with often oblique ends. They become flattened tangentially, i.e. at right angles to the radius of the stem.

Note. The wood of gymnosperms and ferns consists exclusively of tracheids; no vessels are present there. The primary wood in them consists of annular and spiral tracheids, and the secondary wood (in gymnosperms only) consists of pitted tracheids (with bordered pits). Also no companion cells are formed in these two divisions of plants. In monocotyledons there is seldom any phloem parenchyma to be found; cambium is also absent from them. In all roots xylem forms one bundle, and phloem another.

Types of Vascular Bundles. According to the arrangement of xylem and phloem, the vascular bundles are of the following types:

(1) **Radial**, when xylem and phloem form separate bundles and these lie on different radii alternating with each other, as in roots (FIG. 49). The radial bundle is the most primitive type of all vascular bundles.

(2) **Conjoint**, when xylem and phloem combine into one bundle. There are different types of conjoint bundles.

(a) **Collateral**, when xylem and phloem lie together on the same radius, xylem being internal and phloem external. When in a collateral bundle the cambium is present, as in all dicoty-

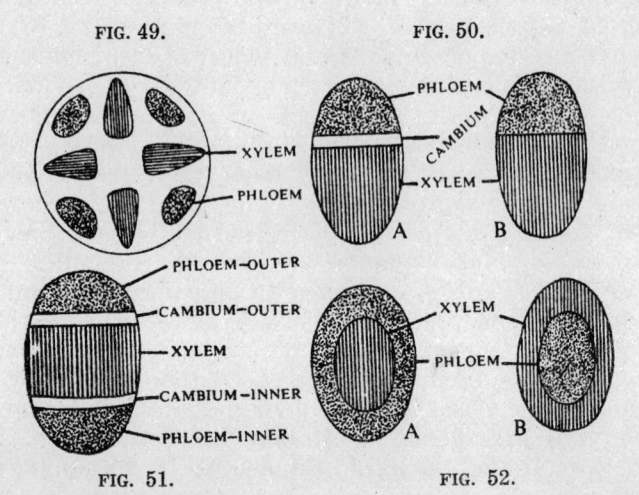

FIG. 49.   FIG. 50.

FIG. 51.   FIG. 52.

Types of Vascular Bundles. FIG. 49. Radial. FIG. 50. Collateral—*A*, open; *B*, closed. FIG. 51. Bicollateral. FIG. 52. Concentric—*A*, xylem central (amphicribral); *B*, phloem central (amphivasal).

ledonous stems, the bundle is said to be *open* (FIG. 50A), and when the cambium is absent it is said to be *closed* (FIG. 50B), as in monocotyledonous stems.

(b) **Bicollateral** (FIG. 51), when in a collateral bundle both phloem and cambium occur twice—once on the outer side of xylem and then again on the inner side of it. The sequence is outer phloem, outer cambium, xylem, inner cambium and inner phloem. Bicollateral bundle is characteristic of *Cucurbitaceae*; it is also often found in *Solanaceae, Apocynaceae, Convolvulaceae, Myrtaceae*, etc. A bicollateral bundle is always open.

(c) **Concentric**, when one kind of vascular tissue (xylem or phloem) is surrounded by the other. Evidently there are two types according to whether one is central or the other one is so. When phloem lies in the centre and is surrounded by xylem (FIG. 52B), as in some monocotyledons, e.g. dragon plant (*Dracaena*; see FIG. 84), dagger plant (*Yucca*; see FIG. I/81, *Cordyline*, sweet flag (*Acorus*; B. & H. BOCH), etc., the concentric bundle is said to be **amphivasal**.[1] When on the other hand xylem lies in the centre and is surrounded by phloem (FIG. 52A), as in many ferns, *Lycopodium, Selaginella* and some aquatic angiosperms, the concentric bundle is said to be **amphicribral**. A concentric bundle is always closed.

Primary Meristems and Tissue Systems

PROMERISTEM   PROMERISTEM

```
→dermatogen ──→ epidermis ──────────→ epidermal ←─ protoderm ←─
                                       tissue
                                       system

                           ┌─ hypodermis ─
→periblem ──→ cortex ──→  │  general cortex
                           └─ endodermis
                                                  → ground ← ground ←
                           ┌─ pericycle             tissue    meristem
                           │  pith ray              system
                           │  pith        ─
→plerome ──→ stele ──→    │  vascular ── → vascular ← procambium ←
                           │  bundles        tissue
                           │  (phloem,       system
                           │  cambium and
                           └─ xylem)
```

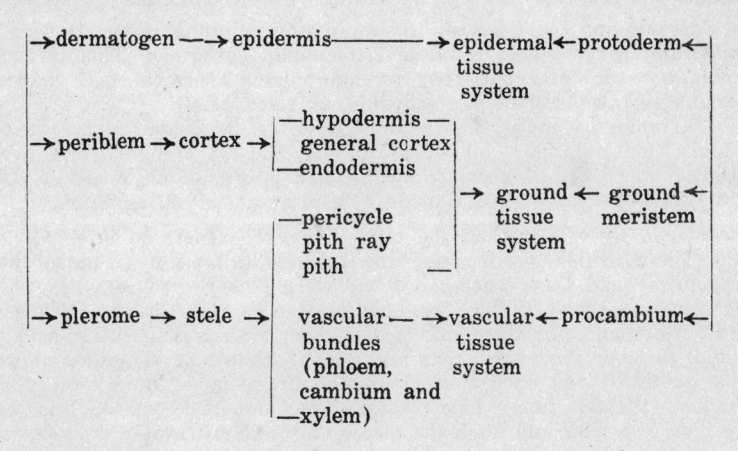

[1] Haberlandt used the term *hadrome* for xylem, and the term *leptome* for phloem. Accordingly an amphivasal bundle is said to be **leptocentric**, and an amphicribral bundle **hadrocentric**.

Chapter 4 ANATOMY OF STEMS

Double Staining and Permanent Preparation of Microscopic Sections. For free-hand sections of ordinary material such as stem, root and leaf the following procedure may be adopted by a beginner. Of the various combinations of stains the following are recommended to start with. It must be noted that only thin and uniform sections can be properly stained.

A. Safranin and Haematoxylin. Aqueous solution of safranin (0.5-1%) or alcoholic solution of it (1% in 50% alcohol) and Delafield's haematoxylin (prepared according to standard formula) may be used. Avoid overstaining with the latter; if necessary, dilute it with water in a watch glass before use. Safranin stains lignified elements deep red, while haematoxylin stains cellulose elements purplish. The schedule given below may be followed.

1. Safranin—5-10 minutes. 2. Washing in water (or 50% alcohol followed by water)—5 minutes (to remove most of the stain from cellulose elements). 3. Haematoxylin (better diluted)—1-2 minutes, the stain deepens in water at the next stage. 4. Washing in water—a few minutes. 5. Dehydration (by passing through grades of alcohol)—30%, 50%, 75%, 85% or 90% and 100% (absolute alcohol)—½-1 minute in each, with another change in 100% for ½-1 minute for complete dehydration. Note that incomplete dehydration will result in fogginess at the next stage. 6. Clove oil—2-5 minutes or more. Clove oil is a clearing reagent. 7. Canada balsam (dissolved in xylol)—mount a section in it on a clean slide (after removing excess clove oil from the section with a piece of blotting paper). 8. Cover the section with a cover-glass by gently sliding it down with the help of forceps or needle. 9. Label and keep the slide flat for drying up away from dust.

B. Safranin and Light Green. Safranin stains lignified elements deep red, while light green stains cellulose elements bright green. This is a good combination of stains and easy to manipulate. There is least chance of overstaining also. Follow the schedule as given below.

1. Safranin, as above. 2. Washing in water. 3. Dehydration, as above. 4. Light Green (in clove oil)—2-5 minutes (0.2 gm. of light green powder dissolved in 50 c.c. of absolute alcohol and 50 c.c. of clove oil). 5. Clove oil or xylol—1-2 minutes. 6. Canada balsam and the rest, as above.

Maceration. By this method the tissues may be separated into individual cells. The principle lies in dissolving the middle lamella (made of pectic compounds) and then teasing out the cells. The procedure is as follows (Schultze's method). Cut the material into small pieces to the thickness of match-sticks and put them into a test-tube. Pour strong nitric acid just enough to cover the pieces; then add a few crystals of potassium chlorate. Heat gently (in an open space to avoid disagreeable fumes) for 4 or 5 minutes until the fumes have ceased. Pour out the contents into water in a dish or beaker and wash the pieces thoroughly in water. For examination a piece may be placed on a slide, crushed under cover-glass and then teased with needles. Stain, and mount in glycerine. Cover and examine the isolated elements. The washed pieces may be preserved in 3-4% formalin for future use.

Dicotyledonous Stems

I. YOUNG SUNFLOWER STEM (FIG. 53)

At first note the distribution of different zones and sub-zones, the arrangement of vascular bundles more or less in a ring close

ANATOMY OF STEMS

to the periphery, and a very large pith occupying the major part of the stem. Then study the tissues in detail.

1. Epidermis. This forms the outermost layer, and consists of a single row of cells, flattened tangentially and fitting closely along their radial walls, with a well-defined cuticle extending over it. Here and there it bears some multicellular hairs and a few stomata, but no chloroplasts; the guard cells of the stomata, however, contain chloroplasts.

2. Cortex. This lies below the epidermis and consists of external collenchyma, central parenchyma and internal endodermis or starch sheath. (*a*) **Hypodermis** (collenchyma)—this lies immediately below the epidermis, and consists of some 4 or 5 layers of collenchymatous cells. These cells are specially thickened at

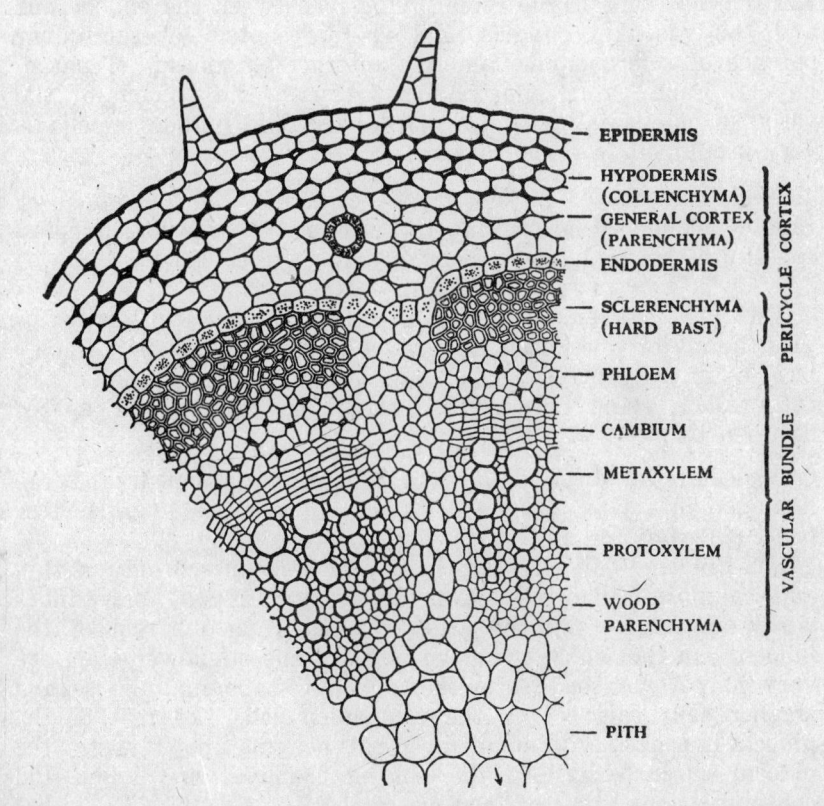

FIG. 53. Young sunflower stem (a sector) in transection. Note the resin-duct in the cortex, and the medullary ray in between the two vascular bundles.

the corners against the intercellular spaces. The thickening is due to a deposit of cellulose impregnated with pectin. The cells are living and contain a number of chloroplasts. (b) **General cortex** or **cortical parenchyma**—this lies internal to the hypodermis and consists of a few layers of thin-walled, large, rounded or oval, parenchymatous cells. It may be reduced to 1 or 2 layers outside the vascular bundle. There are conspicuous intercellular spaces in it. Some isolated resin-ducts, each surrounded by a layer of small thin-walled living cells, are also seen here and there in it. (c) **Endodermis**—this is the innermost layer of the cortex and demarcates it from the stele. The cells are more or less barrel-shaped and fit closely without intercellular spaces. Endodermis is conspicuous outside the patch of sclerenchyma, but loses its identity on either side. It almost invariably contains numerous starch grains and is also known as the *starch sheath*.

3. **Pericycle.** This is the region lying in between the endodermis and the vascular bundles, and is represented by semi-lunar patches of sclerenchyma and the intervening masses of parenchyma. Each patch lying associated with the phloem of the vascular bundle is called the **hard bast**. The middle lamella is very prominent in this tissue.

4. **Medullary Rays.** A few layers of fairly big polygonal or radially elongated cells, lying in between two vascular bundles, constitute the medullary rays.

5. **Pith.** This is very elaborate in the sunflower stem, and occupies the major portion of it. It extends from below the vascular bundles to the centre, and is composed of rounded or polygonal, thin-walled living cells with conspicuous intercellular spaces between them.

6. **Vascular Bundles.** These are collateral and open, and are arranged in a ring. Each bundle is composed of (1) **phloem** or **bast**, (2) **cambium** and (3) **xylem** or **wood**.

(1) **Phloem.** This lies externally and is composed only of thin and cellulose-walled elements. It consists of (a) **sieve-tubes** which appear as slightly larger cavities than the rest of the phloem. On the whole the sieve-tubes of the sunflower stem are very narrow. Associated with each sieve-tube may be seen a smaller cell; this is (b) the **companion cell**. The rest of the phloem is packed with small-celled parenchyma known as (c) the **phloem parenchyma**. All the phloem elements are living, and contain various kinds of food material.

(2) **Cambium.** Passing inwards, a band of thin-walled tissue is seen, whose cells are regularly arranged in radial rows and are roughly rectangular in shape, very small in size and thin-walled. (If the section be cut through a comparatively old

portion of the stem the cambium is seen to be continuous from one vascular bundle to another, and the division of its cells noted both inside and outside. This indicates the beginning of secondary growth.)

(3) **Xylem** or **Wood**. This lies internally and consists of the following elements: (*a*) **Wood Vessels**. Some large thick-walled elements, arranged in a few radial rows, can be easily recognized in the wood; these are the wood vessels. The development of xylem is *centrifugal*, or in other words, it is said to be *endarch* (meaning inner origin). The smaller vessels constituting the *protoxylem* lie towards the centre, and the bigger ones constituting the *metaxylem* lie away from the centre. Protoxylem consists of annular, spiral and scalariform vessels, and metaxylem of reticulate and pitted vessels. Their walls are always thick and lignified. (*b*) **Tracheids**. Surrounding the metaxylem vessels and lying in between them some small, thick-walled cells can be seen; these are the tracheids. In transverse section of the stem they are hardly distinguishable from the wood fibres which lie mixed up with them. (*c*) **Wood Fibres**. These appear somewhat irregular and polygonal in section. They are thick-walled and lignified, and stained like the wood vessels. Excepting the vessels nearly the whole of the wood is packed with these elements. (*d*) **Wood Parenchyma**. A patch of thin-walled cells seen on the inner side of the bundle surrounding the protoxylem is the wood parenchyma. The cells of the wood parenchyma retain their protoplasm.

II. *YOUNG RANUNCULUS STEM* (FIG. 54)

Note the zones, sub-zones and nature of tissues as labelled against the sketch. Note also that a number of air-cavities develop in the cortex (the plant being sub-aquatic in habit) and that the cambium is absent; some feebly-developed cambium-like cells represent the remnant of the procambium. Study the tissues in detail with reference to FIG. 54.

III. *YOUNG CUCURBITA STEM* (FIG. 55-6)

Note that it is hollow, and has usually five ridges and five furrows, and that the vascular bundles are usually ten in number and arranged in two rows, those of the outer row corresponding to the ridges and those of the inner to the furrows (FIG. 55). Then study the tissues in detail (FIG. 56).

1. **Epidermis**. This is the single outermost layer passing over the ridges and furrows; it often bears many long and narrow multicellular hairs.

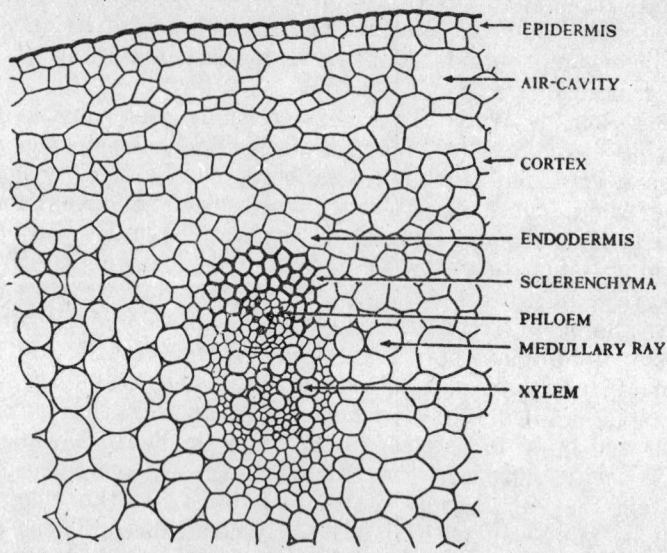

FIG. 54. Young *Ranunculus* (*R. sceleratus*) stem (a sector) in transection. Note that the cambium is absent or very feebly developed

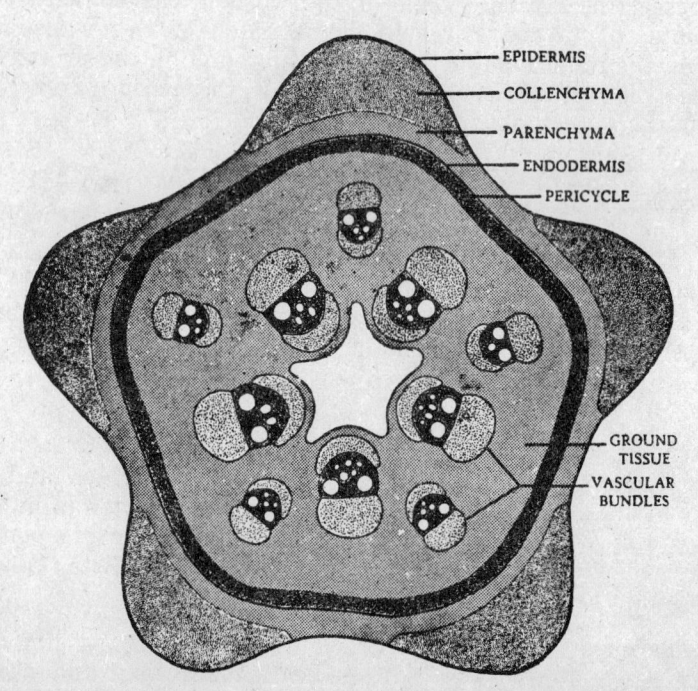

FIG. 55. Young *Cucurbita* stem in transection, as seen under a pocket lens.

2. **Cortex.** This consists of hypodermis externally, general cortex or cortical parenchyma in the middle, and endodermis internally. (*a*) **Hypodermis (collenchyma)** lies immediately below the epidermis, and consists of six or seven (sometimes more) layers of collenchymatous cells in the ridges, and in the furrows only two or three layers, sometimes none. Collenchyma contains some chloroplasts. (*b*) **General cortex** or **cortical parenchyma** forms a narrow zone in the middle, two or three layers thick; in the furrows it passes outwards right up to the epidermis through the collenchyma. Chloroplasts are abundant in it. (*c*) **Endodermis** is the innermost layer of the cortex, lying immediately outside the pericyclic sclerenchyma. This layer is wavy in outline and contains starch grains.

3. **Pericycle.** Below the endodermis there is a zone of sclerenchyma which represents the pericycle. This zone consists of four

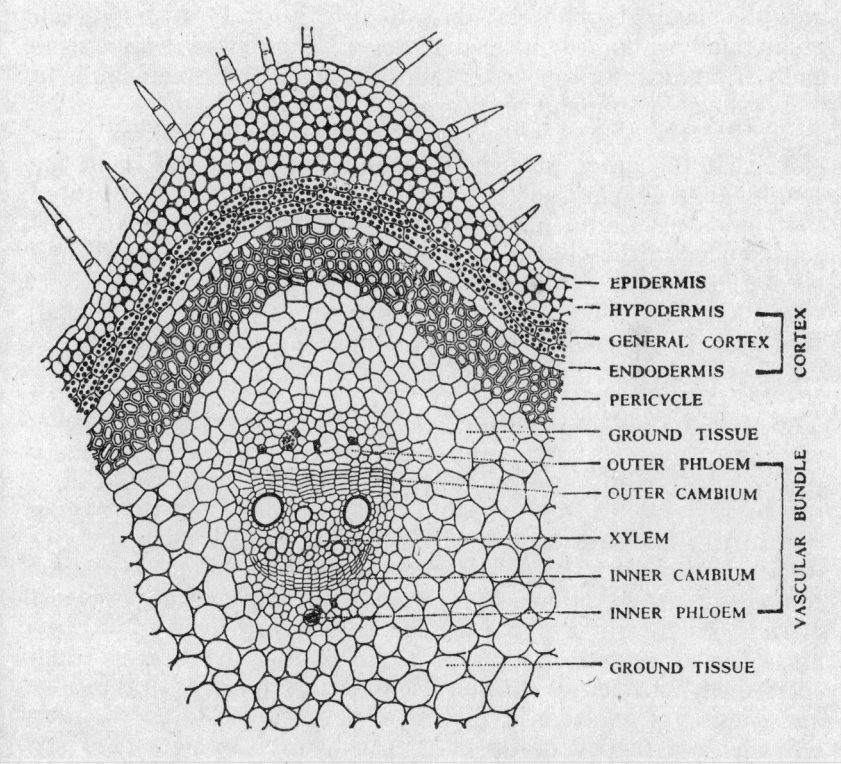

FIG. 56. Young *Cucurbita* stem (a sector) in transection.

or five layers of thick-walled, lignified cells which are polygonal in shape.

4. **Ground Tissue.** This is the continuous mass of thin-walled parenchymatous cells extending from below the sclerenchyma to the pith cavity; in this tissue lie embedded the vascular bundles.

5. **Vascular Bundles.** These are *bicollateral*, usually ten in number, and are arranged in two rows. Each bundle consists of (1) **xylem,** (2) **two strips of cambium,** and (3) **two patches of phloem.**

(1) **Xylem** occupies the centre of the bundle, and consists of *protoxylem* (smaller vessels on the inner side) and *metaxylem* (slightly bigger vessels a little higher up); the larger vessels still higher up represent the beginning of secondary xylem. There may be some tracheids and wood fibres but wood parenchyma is abundant. Vessels are not arranged in radial rows, as in the sunflower stem.

(2) **Cambium.** This tissue occurs in two strips—the outer and the inner, one on each side of xylem, forming a narrow strip between phloem and xylem to the inside, and between xylem and phloem to the outside; its cells are thin-walled and rectangular, and arranged in radial rows. The outer cambium is many-layered and is more or less flat, while the inner cambium is few-layered and curved (crescent-shaped).

(3) **Phloem** occurs in two patches—the outer and the inner. Note that the outer phloem is plano-convex and the inner one semilunar in shape. Each patch of phloem consists of sieve-tubes, companion cells, and phloem parenchyma. Sieve-tubes are very conspicuous in the phloem of the *Cucurbita* stem. Here and there sieve-plates with perforations in them may be distinctly seen (see FIGS. 38-9). The rest of the phloem is made up of small, thin-walled cells, which constitute the phloem parenchyma.

IV. YOUNG CASUARINA STEM (FIG. 57)

The stem is wavy in outline with distinct ridges and furrows, varying in number, commonly 7-12. Assimilatory tissue develops in the ridges, while stomata in large numbers remain confined to the lateral walls in the furrows. Hairs grow out from the base of the furrow and are of two kinds—simple and branched, with usually two short basal cells and a long terminal cell. The different tissues are as follows. **Epidermis** is the single outermost layer of cells; on the ridge it has a thick cuticle with warty protuberances but is devoid of stomata, while in the furrow it is interspersed with numerous stomata. **Hypodermis** develops in the ridge and consists of one to a few layers of parenchyma followed internally by a T-shaped group of sclerenchyma; the stem of T extends far deep towards a leaf trace bundle, thus dividing the assimilatory tissue into two groups, more or less symmetrical. **Assimilatory tissue** (chlorenchyma) occurs on each side of T and consists of a few rows of radially elongated palisade cells. **Leaf**

ANATOMY OF STEMS

trace bundles occur in a ring, each opposite to a ridge, lying embedded in a mass of thin-walled parenchyma (ground tissue). Each bundle is feebly developed with xylem on the inside and phloem on the outside, capped by a small patch of sclerenchyma. A wavy layer of **endodermis** (outer) passes over the bundle. **Transfusion tracheids** occur on either side of the bundle and are conspicuous. **Cauline bundles** occur in a ring in the central cylinder, alternating with the leaf trace bundles, and while still young they are separated by distinct, rather broad, medullary rays. The

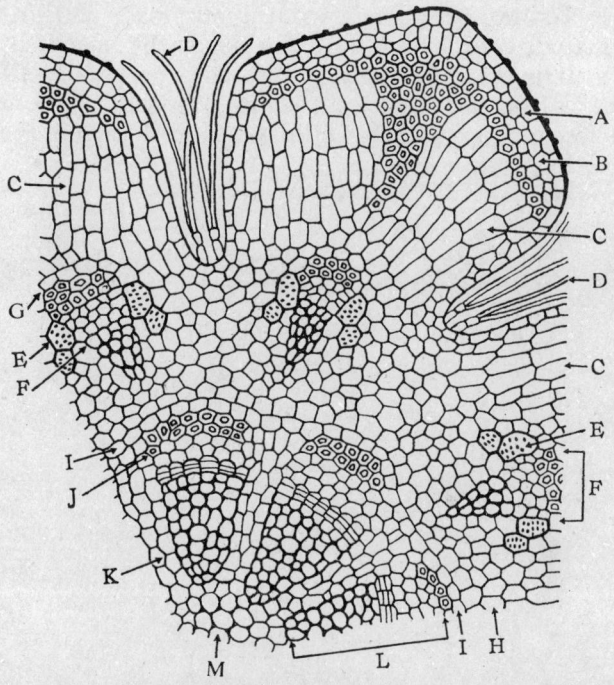

FIG. 57. Young *Casuarina* stem (a sector) in transection. *A*, epidermis; *B*, hypodermis with T-shaped sclerenchyma; *C*, palisade tissue (assimilatory tissue); stomata remain confined to the epidermal layer of the furrow; *D*, hair; *E*, transfusion tracheid; *F*, leaf trace bundle (note xylem and phloem, capped by sclerenchyma); *G*, endodermis (outer); *H*, ground tissue; *I*, endodermis (inner); *J*, hard bast; *K*, medullary ray; *L*, cauline bundle (note xylem, cambium and phloem, capped by hard bast); and *M*, pith.

bundles, however, soon become compact with the progress of secondary growth. They are collateral and open, consisting of xylem, cambium and phloem, capped by a sheath of sclerenchyma (hard bast). A wavy layer of **endodermis** (inner) passes over the bundles. **Pith** consists of a mass of parenchyma in the centre.

Monocotyledonous Stems
I MAIZE OR INDIAN CORN STEM (FIG. 58)

1. **Epidermis.** This is a single outermost layer with a thick cuticle on the outer surface. Here and there in the epidermis a few stomata may be seen.

2. **Hypodermis (Sclerenchyma).** This forms a narrow zone of sclerenchymatous cells, usually two or three layers thick, lying below the epidermis.

3. **Ground Tissue.** This is the continuous mass of thin-walled parenchymatous cells, extending from below the sclerenchyma to the centre. It is not differentiated into cortex, endodermis, pericycle, etc., as in a dicotyledonous stem. The cells of the ground tissue enclose numerous intercellular spaces between them.

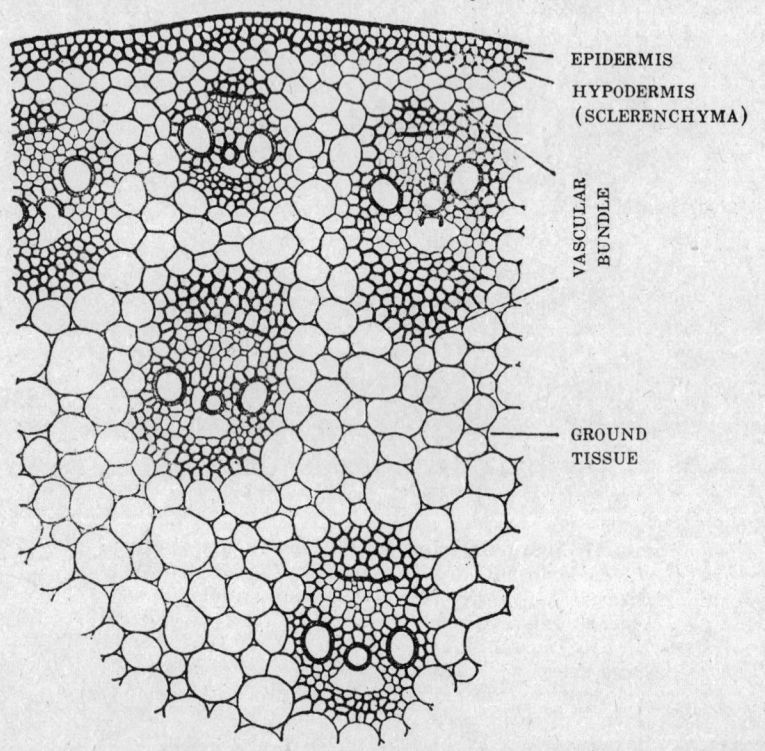

FIG. 58. Maize or Indian corn stem (a sector) in transection.

4. **Vascular Bundles** (FIG. 59). These are collateral and closed, and lie scattered in the ground tissue; they are more numerous, and lie closer together nearer the periphery than the centre. The peripheral ones are also seen to be smaller in size than the

ANATOMY OF STEMS

central ones. Each vascular bundle is somewhat oval in general outline and is more or less completely surrounded by a **sheath** of sclerenchyma which is specially developed on the two sides—

[*Continued on page* 238*]*

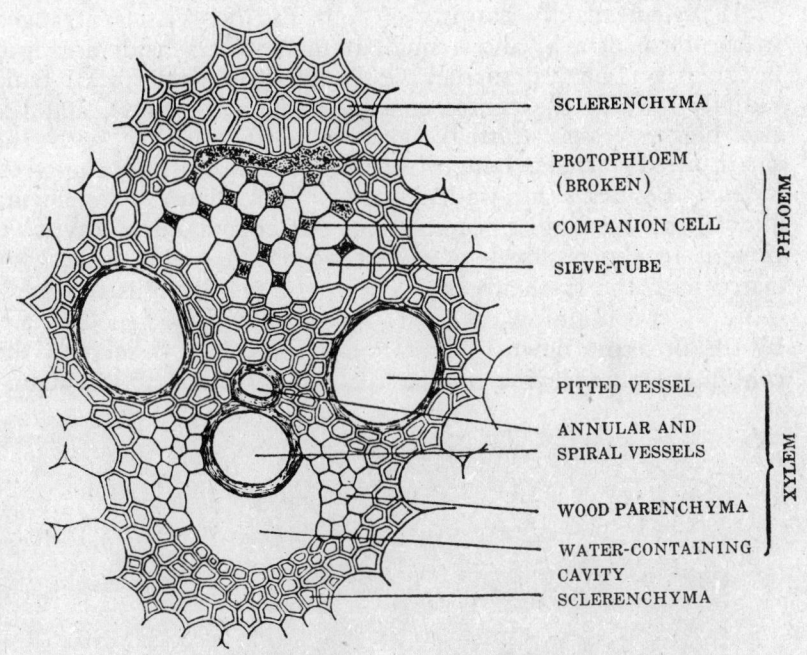

FIG. 59. A vascular bundle of maize stem (magnified).

Differences between Dicotyledonous and Monocotyledonous Stems

	Dicotyledonous stem (e.g. sunflower)	Monocotyledonous stem (e.g. maize)
1. Hypodermis	collenchymatous	sclerenchymatous
2. General Cortex	a few layers of parenchyma	a continuous mass of parenchyma up to the centre (ground tissue) without differentiation into distinct tissues
3. Endodermis	a wavy layer	
4. Pericycle	a zone of parenchyma and sclerenchyma	
5. Medullary Ray	a strip of parenchyma in between vascular bundles	not marked out
6. Pith	the central cylinder	not marked out
7. Vascular Bundles	(a) collateral and open	collateral and closed
	(b) arranged in a ring	scattered
	(c) of uniform size	larger towards the centre
	(d) phloem parenchyma present	it is absent
	(e) usually wedge-shaped	usually oval
	(f) bundle sheath absent	strongly developed

Continued from page 237]

upper and lower. The bundle consists of two parts, viz. xylem and phloem.

(1) **Xylem** mainly consists of four distinct vessels arranged in the form of a Y, and a small number of tracheids arranged irregularly. The two smaller vessels (annular and spiral) lying radially towards the centre constitute the *protoxylem,* and the two bigger vessels (pitted) lying laterally together with the small pitted tracheids lying in between them constitute the *metaxylem.* Besides, thin-walled wood (or xylem) parenchyma almost surrounding a conspicuous water-containing cavity is present in the protoxylem; a few wood fibres also occur associated with the tracheids in between the two big pitted vessels. This water-containing cavity has been formed lysigenously, i.e. by the breaking down of the inner protoxylem vessel and the contiguous parenchyma during the rapid growth of the stem.

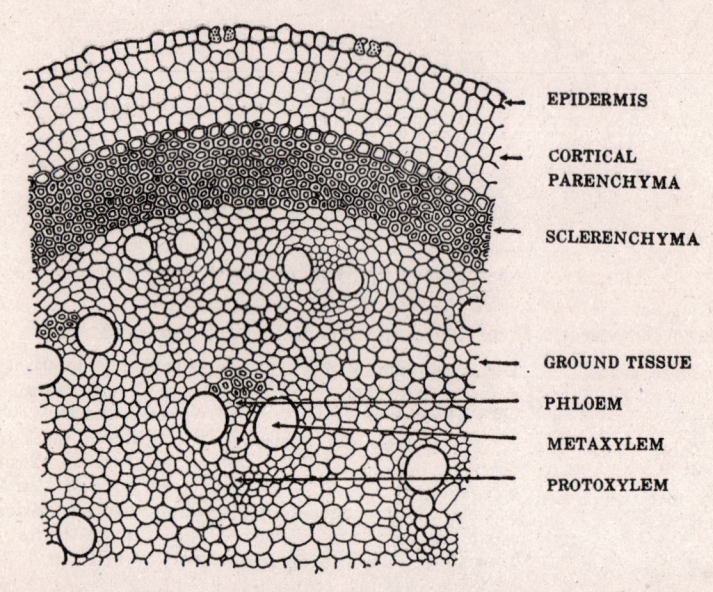

FIG. 60. *Asparagus* stem (a sector) in transection.

(2) **Phloem** consists exclusively of sieve-tubes and companion cells; no phloem parenchyma is present in the monocotyledonous stem. The outermost portion of the phloem, which is a broken mass, is the *protophloem* and the inner portion is the *metaphloem.* The former soon gets disorganized, and the latter shows distinct sieve-tubes and companion cells.

ANATOMY OF STEMS 239

II. ASPARAGUS STEM (FIG. 60)

In a transverse section of the stem note the tissues as labelled against the sketch, and study them in detail.

III. FLOWERING STEM (SCAPE) OF CANNA (FIG. 61)

1. Epidermis. This is the outermost layer consisting of a single row of very small, polygonal cells flattened tangentially. The outer walls of the epidermis are cutinized.

2. Ground Tissue System. From below the epidermis to the centre the whole mass of tissues, leaving out the vascular bundles, constitutes the ground tissue system. It is differentiated into (a) **cortex** consisting of two layers of fairly large polygonal cells; (b) **chlorophyllous tissue** consisting of one or two layers of

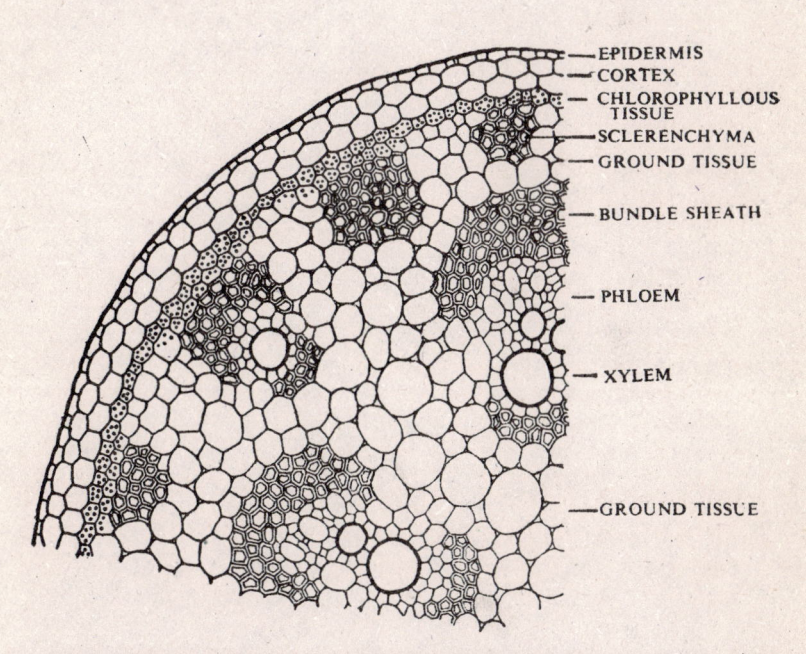

FIG. 61. Flowering stem (scape) of *Canna* (a sector) in transection.

chloroplast-bearing cells, intruding inwards here and there; (c) several patches of **sclerenchyma** of different sizes lying against he chlorophyllous tissue; and (d) **ground tissue** consisting of a continuous mass of large, thin-walled, parenchymatous cells, containing starch grains and enclosing numerous intercellular spaces between them.

3. **Vascular bundles.** These are numerous and are of different sizes, lying scattered in the ground tissue. Each bundle is closed and collateral. It is incompletely surrounded by a sheath of sclerenchyma (**bundle sheath**), with a distinct patch of it on the outer side in the form of a cap, and a thin strip on the inner side; seldom a regular and complete sheath is formed encircling the vascular bundle. Each bundle consists of (*a*) xylem on the inner side, and (*b*) phloem on the outer. **Xylem** consists of a large prominent spiral vessel, with often one or two smaller ones, also spiral in nature, lying usually on its outer side, and some parenchyma. **Phloem** consists of sieve-tubes and companion cells.

Chapter 5 ANATOMY OF ROOTS

Dicotyledonous Roots

I. YOUNG GRAM ROOT (FIG. 62)

1. **Epiblema or Piliferous Layer.** This is a single outermost layer of thin-walled cells; outer walls of most of these cells extend outwards and form unicellular root-hairs. This layer is used for absorption of water and solutes from the soil and, therefore, it has no cuticle. Root-hairs increase the absorbing surface of the root.

2. **Cortex.** This consists of many layers of thin-walled rounded cells, with numerous intercellular spaces between them. The cells of the cortex contain leucoplasts and store starch grains. The epiblema is, in some cases, only short-lived; as it dies off, a few

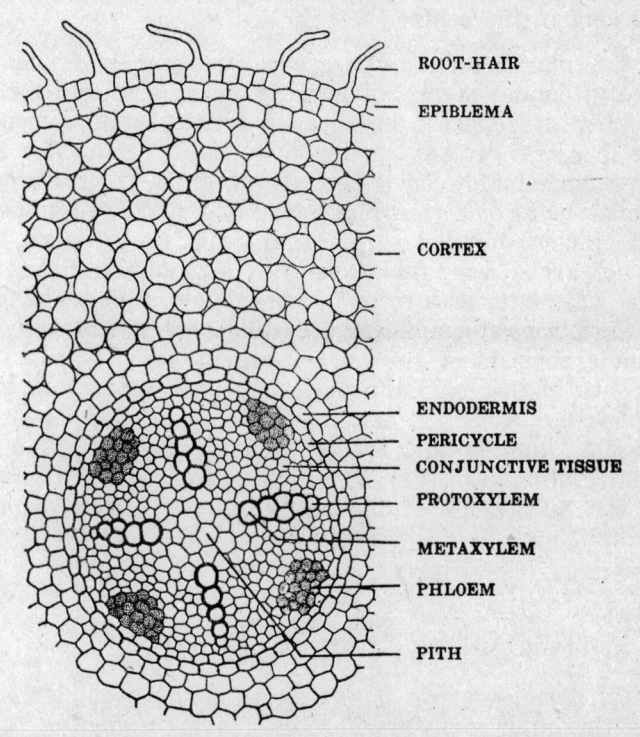

FIG. 62. Young dicotyledonous root (gram seedling) in transection.

outer layers of the cortex become cutinized and form the *exodermis* of the root.

3. **Endodermis.** This is a single ring-like layer of barrel-shaped cells which are closely packed without intercellular spaces. The radial walls of this layer are often thickened, and sometimes this thickening extends to the inner walls also, and not infrequently the walls abutting upon the protoxylem are provided with simple pits. Endodermis is the innermost layer of the cortex and occurs as a ring (or cylinder) around the stele. Here and there, particularly lying against the protoxylem, small thin-walled cells are often found in the endodermis; these are the *passage cells*.

4. **Pericycle.** This lies internal to the endodermis, and is a single circular layer like the latter; its cells are very small and thin-walled, but contain abundant protoplasm.

5. **Conjunctive Tissue.** The parenchyma lying in between xylem and phloem bundles constitutes the *conjunctive tissue*.

6. **Pith.** This occupies only a small area in the centre of the root. But soon the pith becomes obliterated owing to the wood vessels meeting in the centre.

7. **Vascular Bundles.** These are arranged in a ring, as in the dicotyledonous stem, but here xylem and phloem form an equal number of separate bundles, and their arrangement is *radial* see p. 226). Protoxylem lies away from the centre so that the development of wood is *centripetal,* or in other words, xylem is said to be *exarch* (meaning outer origin). The number of xylem (or phloem) bundles varies from two to six (di-, tri-, tetr-, pent-, or hex-arch), very seldom more. The cambium makes its appearance only later as a secondary meristem. **Phloem bundle** consists of sieve-tubes, companion cells and phloem parenchyma. **Xylem bundle** consists of protoxylem which lies towards the circumference abutting on the pericycle, and metaxylem towards the centre, i.e. xylem is exarch. Protoxylem is composed of small vessels (annular and spiral) and metaxylem of bigger vessels (reticulate and pitted). The metaxylem groups often meet in the centre, and then the pith is obliterated (it gets disorganized).

II. *YOUNG RANUNCULUS ROOT* (FIG. 63)

1. **Epiblema**—the single outermost layer.

2. **Exodermis**—a few layers internal to the epiblema, representing the outer zone of the cortex (corresponding to the hypodermis of the stem).

3. **Cortex**—Several layers of rounded or oval cells, leaving a lot of intercellular spaces between them.

ANATOMY OF ROOTS

4. **Endodermis**—the innermost layer of the cortex; the layer is distinct and cylindrical, with thick-walled cells except the *passage cells* lying against the protoxylem.

5. **Pericycle**—a single ring-like layer internal to the endodermis; the cells are small and thin-walled.

6. **Conjunctive tissue**—the parenchyma in between xylem and phloem.

7. **Vascular bundles**—radial, with 4 or 5 xylem bundles and as many phloem bundles; xylem is exarch and the metaxylem vessels meet in the centre. The pith is absent.

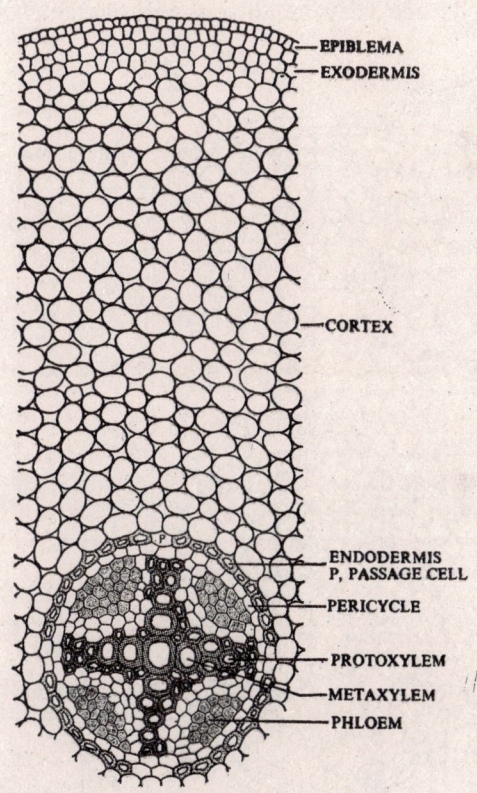

FIG. 63. Young dicotyledonous root (*Ranunculus sceleratus*) in transection.

Monocotyledonous Roots

I. *AMARYLLIS ROOT* (FIG. 64)

1. **Epiblema** or **Piliferous Layer.** This is the single outermost layer with a number of unicellular root-hairs.

2. **Cortex.** This is a many-layered zone of rounded or oval cells with intercellular spaces between them. As the epiblema dies off a few outer layers of the cortex become cutinized and form the *exodermis*.

3. **Endodermis.** This is the innermost layer of the cortex and forms a definite ring around the stele. Radial walls and often the inner walls of the endodermis are considerably thickened. Cells of the endodermis are barrel-shaped. *Passage cells* are often present in this layer, lying against the protoxylem.

4. **Pericycle.** This is the ring-like layer lying internal to the endodermis. Its cells are very small and thin-walled.

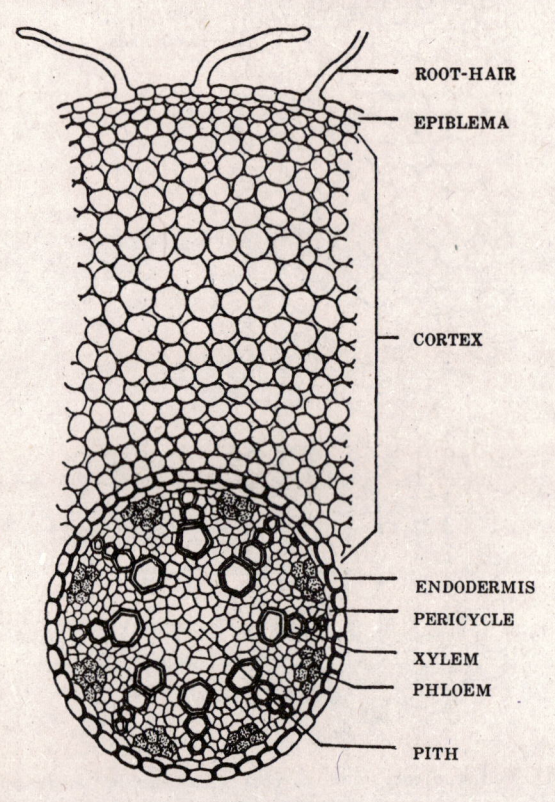

FIG. 64. Monocotyledonous root (*Amaryllis*) in transection.

5. **Conjunctive Tissue.** The parenchyma in between xylem and phloem bundles is known as the *conjunctive tissue*.

6. **Pith.** The mass of parenchymatous cells in and around the centre is the pith. It is well developed in most monocotyledo-

nous roots. In some cases the pith becomes thick-walled and lignified.

7. **Vascular Bundles.** Xylem and phloem form an equal number of separate bundles, and they are arranged in a ring. The arrangement is *radial* (see p. 226). Bundles are numerous (polyarch); rarely, as in onion, they are limited in number. The development of wood is *centripetal*, or, in other words, xylem is said to be *exarch*. **Phloem bundle** consists of sieve-tubes, companion cells and phloem parenchyma. **Xylem bundle** consists of protoxylem which lies abutting on the pericycle, and metaxylem towards the centre, i.e. xylem is *exarch*. Protoxylem consists of annular and spiral vessels, and metaxylem of reticulate and pitted vessels. A few isolated, big vessels may often be seen in the pith.

II. *VANDA (ORCHID) ROOT* (FIG. 65)

1. **Velamen**—a few layers of outer absorbtive tissue, spongy in

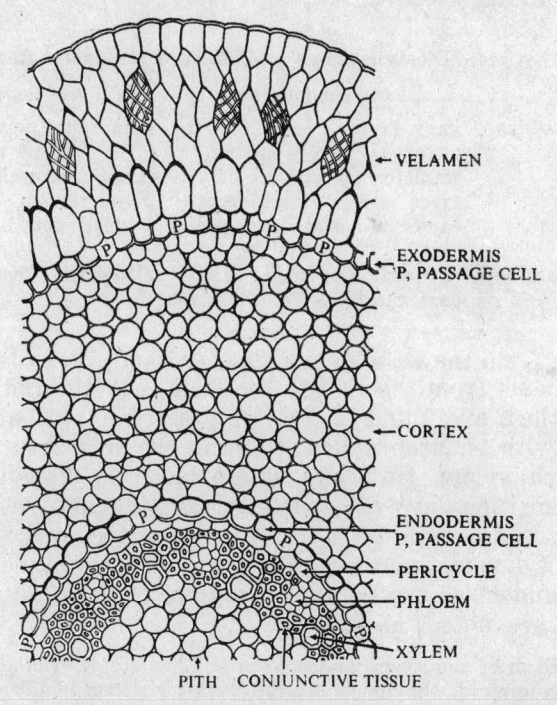

FIG. 65. Monocotyledonous root (*Vanda*) in transection.

nature (see p. 20); it is derived from the dermatogen of the root, and its outermost layer is sometimes called *limiting layer*.

2. **Exodermis**—a single layer of thick-walled cells, with unthickened *passage cells* here and there.

3. **Cortex**—several layers of rounded or oval parenchymatous cells; a few outer layers contain chloroplasts.

4. **Endodermis**—a single layer of somewhat barrel-shaped cells; inner walls, and to some extent also radial walls, are specially thickened and suberized; unthickened *passage cells* are present, lying against the protoxylem.

5. **Pericycle**—a single layer of cells, lying internal to the endodermis.

6. **Conjunctive tissue** lying in between xylem and phloem is thick-walled and lignified.

7. **Vascular bundles** are numerous and radial, with exarch xylem and small patches of phloem alternating with xylem.

8. **Pith** is well-developed and parenchymatous in nature, sometimes becoming sclerified.

Differences between Dicotyledonous and Monocotyledonous Roots

	Dicotyledonous root	Monocotyledonous root
1. Xylem bundles	vary from 2 to 6 (di- to hexarch), rarely more	numerous (polyarch), rarely a limited number.
2. Pith	small or absent	large and well-developed.
3. Pericycle	gives rise to lateral roots, cambium and cork-cambium	gives rise to lateral roots only.
4. Cambium	appears later as a secondary meristem	altogether absent.

Transition from the Root to the Stem (FIG. 66). Vascular bundles are continuous from the root to the stem, but different arrangements of them are found in the two cases. In the stem they are collateral with endarch xylem, while in the root they are radial with exarch xylem. How and where has this transition taken place? Transition involves splitting, twisting and reorientation of xylem or phloem or both, and it takes place in the region of the hypocotyl, sometimes a little lower or a little higher up. This region is known as the **transition region.** Usually four types of transition are noted, as follows.

Type. I. Each of the xylem and phloem bundles divides radially into two. As they pass upwards they move laterally. As the xylem bundles move, they twist round by 180°, and each comes to lie on the inner side of the adjacent phloem bundle, evidently in an inverted position, i.e. the centripetal (exarch) xylem in the root now becomes centrifugal (endarch) in the stem, or in other words, the radial bundles become collateral. The number of

ANATOMY OF ROOTS

collateral bundles in the stem is just double that of the phloem bundles in the root. This type is rather common and is characteristically found in *Phaseolus, Cucurbita, Tropaeolum*, etc.

Type II. In this type each of the xylem bundles divides radially into two, while each phloem bundle remains more or less in its own position without any division. The two strands of xylem bundle then move laterally by 180° and come to lie on the inner side of the two adjacent phloem bundles. It is thus evident that each phloem bundle receives two xylem strands which soon fuse. The number of collateral bundles, as formed in the stem, corresponds with that of the phloem bundles, as seen in the root. This type is found in *Mirabilis, Fumaria, Dipsacus*, etc.

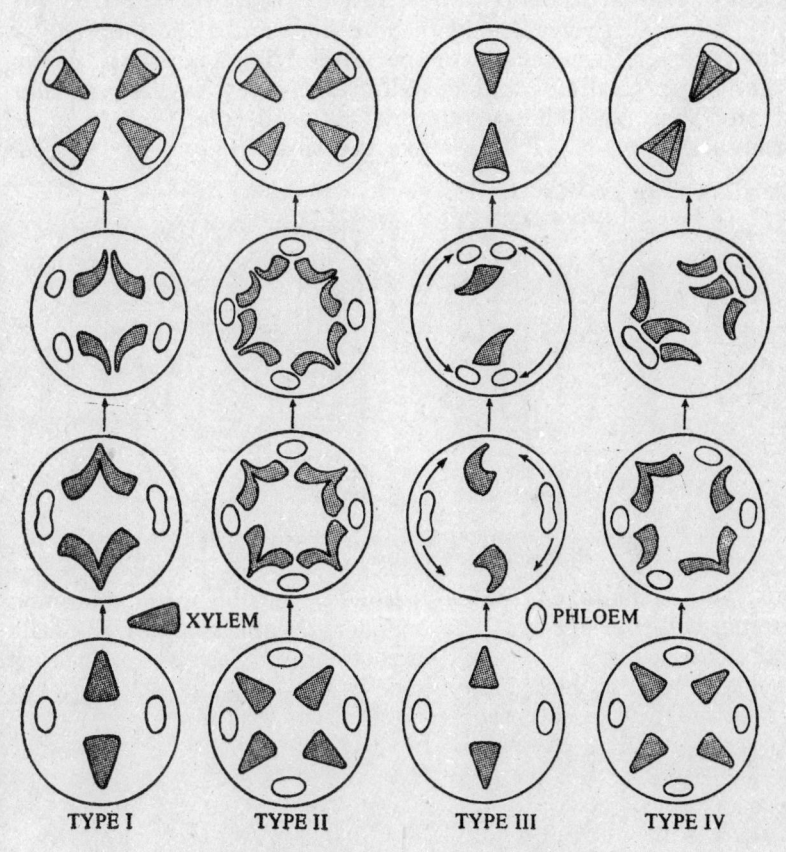

FIG. 66. Transitional stages from the root (bottom) to the stem (top)—four types (diagrammatic).

Type III. In this type the phloem bundles divide radially and change their position laterally, while the xylem bundles do not divide nor do they shift. They, however, twist round by 180°, as in the previous cases, so that the exarch xylem now becomes endarch. The halves of the phloem bundles move laterally in the opposite direction and come to lie on the outer side of the xylem bundle and fuse. The number of collateral bundles in the stem thus remains the same as that of the phloem bundles in the root. This type is found in *Lathyrus, Medicago, Phoenix*, etc.

Type IV. In this type half the number of xylem bundles divide, while the other half remain undivided. The divided strands move laterally, twisting round as they do so by 180°, and join the undivided strand which in the meantime has already become inverted. The phloem bundles do not divide but unite in pairs on the outer side of the three xylem strands. The collateral bundle is now made of three xylem strands and two phloem strands. The number of bundles in the stem is half that of the phloem bundles in the root. This is a rare type found only in certain monocotyledons.

Origin of Lateral Roots (FIG. 67). Lateral roots originate from an inner layer; so they are said to be *endogenous*. The inner layer is the pericycle. The cells of the pericycle lying against the protoxylem begin to divide tangentially, and a few layers are thus cut off. They push the endodermis outwards and tend to grow through the cortex. At this stage the three regions of the root

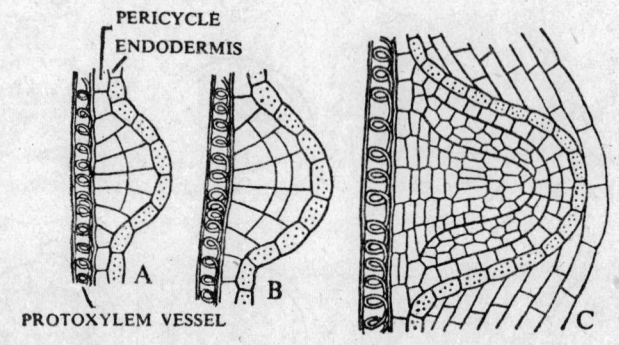

FIG. 67. Origin of a lateral root. *A, B* and *C* are stages in its formation from the pericycle.

apex, viz. dermatogen (or calyptrogen), periblem and plerome, become well marked out. The endodermis and some of the cells of the cortex form a part of the root-cap, but as the root passes through the soil this portion soon wears off, and the root-cap is renewed by the calyptrogen.

Chapter 6 ANATOMY OF LEAVES

I. DORSIVENTRAL LEAVES (FIG. 68)

A section cut through the blade of such a leaf (see p. 47) at a right angle to one of the veins reveals the following internal structure.

1. Upper Epidermis. A single layer of cells with a thick cuticle which checks excessive evaporation of water from the surface.

It also protects the internal tissues from mechanical injury. It usually contains no chloroplasts except in some aquatic plants. Stomata are also usually absent or few except in special cases.

2. Lower Epidermis. Similarly a single layer but with a thin cuticle. It is, however, interspersed with numerous stomata, the two guard cells of which contain some chloroplasts; none are present in the epidermal cells. Internal to each stoma a large cavity, known as the *respiratory cavity*, may be seen. The lower epidermis of the leaf is meant for the exchange of gases (oxygen and carbon dioxide) between the atmosphere and the plant body. Excess of water contained in the plant body also evaporates mainly through the lower epidermis.

3. Mesophyll. The ground tissue lying between the two epidermal layers is known as the mesophyll. It is differentiated into (1) **palisade parenchyma** and (2) **spongy parenchyma**.

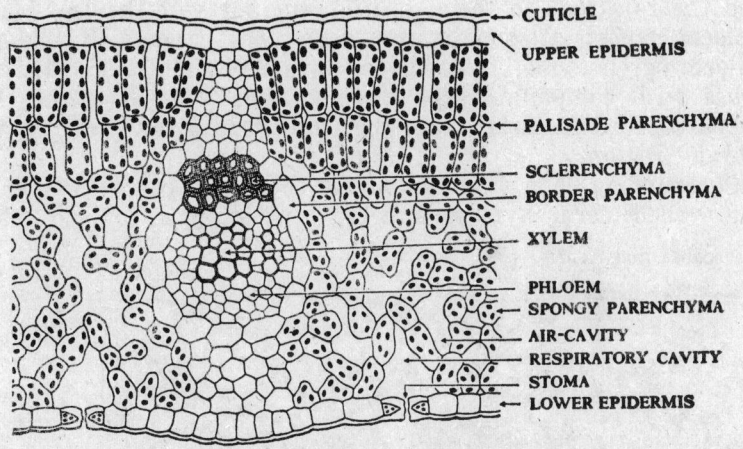

FIG. 68. A dorsiventral leaf in section.

(1) **Palisade parenchyma** consists of usually one to two or three layers of elongated, more or less cylindrical cells, closely packed with their long axes at right angles to the epidermis, leaving only narrow intercellular spaces here and there. They contain numerous chloroplasts which are arranged alongside the cell-walls. The function of palisade parenchyma as a whole is to manufacture sugar and starch in the presence of sunlight, i.e. during the day.

(2) **Spongy parenchyma** consists of oval, rounded, or more commonly irregular cells, loosely arranged towards the lower epidermis, enclosing numerous large intercellular spaces and air-cavities. They, however, fit closely around the vein or the

vascular bundle. The cells contain a few chloroplasts, and manufacture sugar and starch to some extent only. Spongy cells help diffusion of gases through the empty spaces left between them.

4. **Vascular Bundles.** Vascular bundles (or veins) ramify through the leaf-blade for facility of distribution of water and mineral salts among the green cells, and collection of prepared food material from these cells. As they pass from the base of the leaf-blade towards its apex or margin they get reduced in size as well as in the number of their elements. Each vascular bundle (vein) consists of xylem always lying towards the upper **epidermis and phloem always towards the lower. Xylem** consists of various kinds of vessels (particularly annular and spiral), tracheids, wood fibres and wood parenchyma. Towards the apex of the vein xylem is represented by only a few narrow annular and spiral tracheids, or even by a single spiral tracheid; other elements disappear. Xylem conducts and distributes the water and the raw food material to different parts of the leaf-blade. **Phloem** consists of some narrow sieve-tubes, companion cells and phloem parenchyma. Towards the apex a few undeveloped sieve-tubes with companion cells may be seen. Phloem carries the prepared food material from the leaf-blade to the growing and storage regions.

Surrounding each vascular bundle there is a compact layer of thin-walled parenchymatous cells, containing a few to many

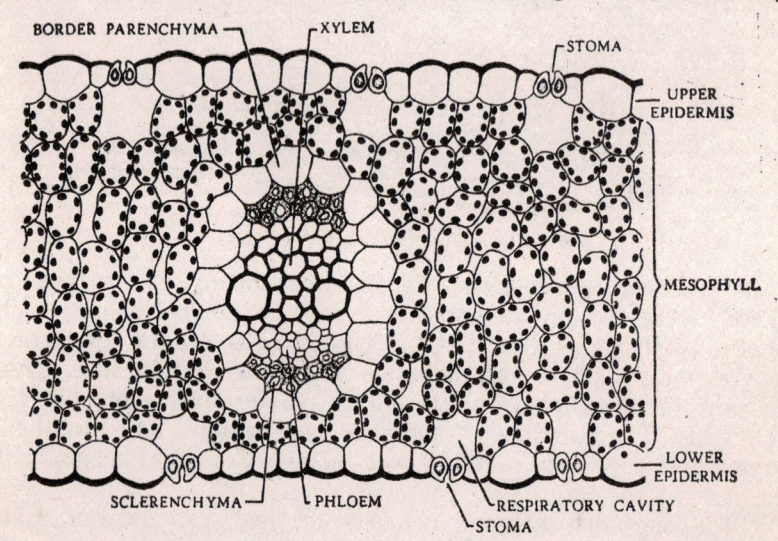

FIG. 69. An isobilateral leaf (a lily leaf) in section.

chloroplasts or none at all; this layer is known as the **border parenchyma** or **bundle sheath.** The cells of this layer are elongated running parallel with the course of the vascular bundle and extending right up to the end of it. The bundle sheath may also extend radially towards the upper or the lower epidermis or towards both as *bundle sheath extensions*. The border parenchyma takes part in conduction between the vein and the mesophyll; it often photosynthesizes actively; and it may act as a starch sheath.

The distribution of **sclerenchyma** is rather irregular in leaves; sometimes it forms patches here and there in the mesophyll; at other times it forms a continuous zone connecting two or more vascular bundles, or it occurs as a patch flanking a vascular bundle, or it extends from the epidermis, upper or lower, to one or more bundles. Frequently, however, it occurs as one or two patches lying associated with xylem or phloem or both (FIGS. 69-71). Sometimes sclerenchyma occurs as a complete or incomplete sheath (**sclerenchymatous sheath**) surrounding a vascular bundle (vein). In any case sclerenchyma gradually disappears towards the end of the vein.

II. *ISOBILATERAL LEAVES* (FIGS. 69-70)

A section cut at a right angle to one or more veins of any of such leaves (see p. 47) reveals the following internal structure.

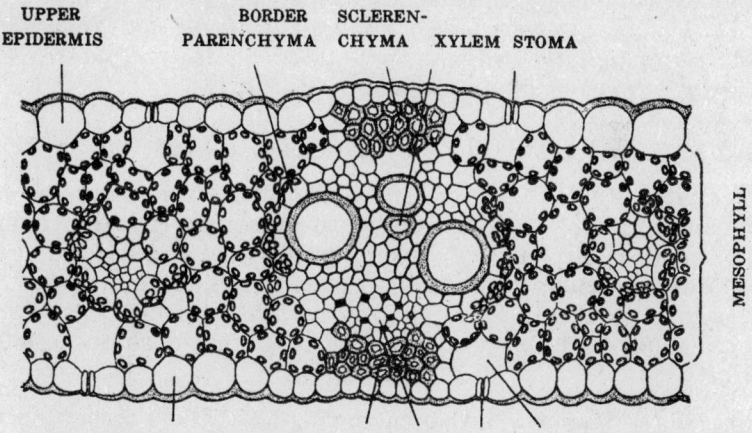

FIG. 70. An isobilateral leaf (maize leaf) in section.

The structure is more or less uniform from one surface to the other. The epidermis on either side contains more or less an equal number of stomata, and is also somewhat uniformly thickened and cutinized. The mesophyll is not normally differentiated into palisade and spongy parenchyma, but mostly consists of spongy cells only, in which the chloroplasts are evenly distributed. Instead of spongy cells the mesophyll may consist of palisade cells only, as in many shade plants. In some cases the mesophyll is seen to be differentiated into spongy parenchyma in the centre and palisade parenchyma on either side. The vascular bundle, border parenchyma and sclerenchymatous sheath are much the same as in a dorsiventral leaf.

Chapter 7 SECONDARY GROWTH IN THICKNESS

(I) Dicotyledonous Stem

In perennial dicotyledons (shrubs and trees) after the primary tissues are fully formed, the **cambium** becomes active and begins to cut off new (secondary) tissues in the stelar region. Sooner or later another strip of meristem, the **cork-cambium**, makes its appearance in the peripheral region and begins to form other secondary tissues, viz. cork, etc., in that region. All these secondary tissues are added on to the primary ones, and as a result the stem increases in thickness. *This increase in thickness, due to the addition of secondary tissues cut off by the cambium and the cork-cambium in the stelar and extra-stelar regions, respectively, is spoken of as* **secondary growth.**

A. *ACTIVITY OF THE CAMBIUM*

Cambium Ring. It is seen that some of the medullary ray cells, mostly in a line with the **fascicular cambium** (i.e. the cambium of the vascular bundle), become meristematic and form a strip of **interfascicular cambium** (i.e. the cambium in between two vascular bundles). This joins on to the fascicular cambium on either side and forms a complete ring known as the **cambium ring.**

Secondary Tissues. The cambium ring as a whole becomes actively meristematic and gives off new cells both externally and internally. Those cut off on the outer side are gradually modified into the elements of phloem; these constitute the **secondary phloem.** The secondary phloem consists of sieve-tubes, companion cells and phloem parenchyma and often also some bands or patches of bast fibres. Many of the textile fibres of commerce such as jute, hemp, flax, rhea (or ramie), etc., are the bast fibres of secondary phloem.

The new cells cut off by the cambium on its inner side are gradually modified into the various elements of xylem; these constitute the **secondary xylem**. The secondary xylem consists of scalariform and pitted vessels, tracheids, numerous wood fibres arranged mostly in radial rows, and some wood parenchyma. The cambium is always more active on the inner side than on the outer. Consequently xylem increases more rapidly in bulk than phloem, and soon forms a compact mass. As a matter of fact, the secondary xylem forms the main bulk of the plant body after secondary growth. In consequence of continued formation of secondary xylem and the pressure exerted by it the cambium, phloem and surrounding tissues are gradually pushed outwards, and for the same reason some of the primary tissues get crushed. The primary xylem, however, remains more or less intact in or around the centre.

Here and there the cambium forms some narrow bands of parenchyma, radially elongated and passing through the secondary xylem and the secondary phloem; these are the **secondary medullary rays**. They are one, two or a few layers in thickness, and one to many layers in height.

Annual Rings (FIGS. 71-2). In regions where climatic variations are of pronounced nature the activity of the cambium is not

FIG. 71.
Cut surface of stem showing annual rings.

FIG. 72.
An annual ring in section (magnified).

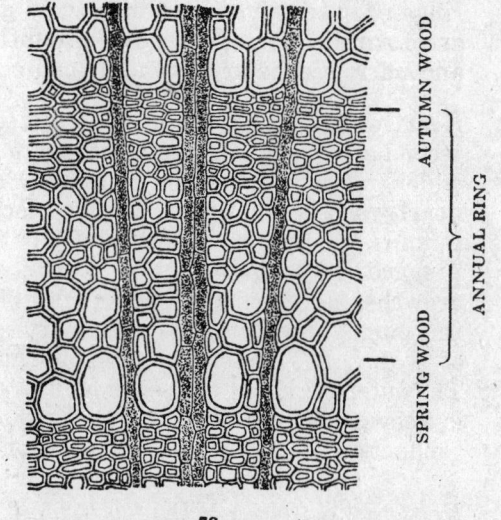

uniform throughout the year. In spring or during the active vegetative season with greater production and activity of foliage leaves, the need to transport sap is acute, and hence at this time of the year the cambium is more active and forms a greater number of vessels with wider cavities (large pitted vessels). In winter or during the inactive period, however, when there is less demand for transporting sap, the cambium is less active and forms elements of narrower dimensions (narrow pitted vessels, tracheids and wood fibres). The wood thus formed in the spring is called the **spring wood** or early wood, and that formed in winter is called the **autumn wood** or late wood. These two kinds of wood appear together, in a transverse section of the stem, as a concentric ring known as the **annual ring** or **growth ring** (FIG. 72), and successive annual rings are formed year after year by the activity of the cambium. There is a sharp contrast between the late autumn wood and the early spring wood, and this makes the successive rings distinct even to the naked eye. Annual rings are readily seen with the naked eye in the trunk of a tree which has been cut down transversely (FIG. 71). Each annual ring corresponds to one year's growth, and by counting the total number of annual rings the age of the plant can be approximately determined, as in pine and many timber trees. The number of annual rings may, however, vary in many plants. In some trees large spring vessels are arranged more or less in a ring; the wood then is said to be **ring-porous**. In others the vessels have equal diameters and are uniformly distributed throughout the whole wood; the wood then is said to be **diffuse-porous**. Annual rings of successive years may vary greatly in width. Wide rings are formed under favourable conditions of growth of the tree, and narrow ones are formed when conditions are unfavourable.

Heart-wood and Sap-wood. In old trees the central region of the secondary wood is filled up with tannin and other substances which make it hard and durable. This region is known as the **heart-wood** or **duramen**. It looks black, owing to the presence of tannins, oils, gums, resins, etc., in it. The vessels often become plugged with *tyloses* (see FIG. 90), which are balloon-like ingrowths, developing from the adjoining parenchyma, through the pits. The function of the heart-wood is no longer conduction of water, but simply to give mechanical support to the stem. The outer region of the secondary wood which is of lighter colour is known as the **sap-wood** or **alburnum**, and this alone is used for conduction of water and salt solutions from the root to the leaf.

B. *ORIGIN AND ACTIVITY OF THE CORK-CAMBIUM*

The formation of new tissues by the cambium exerts a considerable pressure on the peripheral tissues of the stem. The epidermis becomes considerably stretched and gets ruptured

SECONDARY GROWTH IN THICKNESS

here and there, often breaking down altogether. Sclerenchyma and collenchyma become much flattened tangentially. Cortex is also similarly affected but it persists for a long time because of the elastic nature of the cell-walls and the power of accommodation of its cells. To replace or to reinforce the peripheral protective tissues, particularly the epidermis, a strip of secondary meristem, called the **cork-cambium** or **phellogen** (*phellos*, cork: *gen*, producing), arises in that region to give rise to new (secondary) tissues for better protection of the stem at the secondary stage. The cork-cambium commonly originates in the outer layer of collenchyma. It may also arise in the epidermis

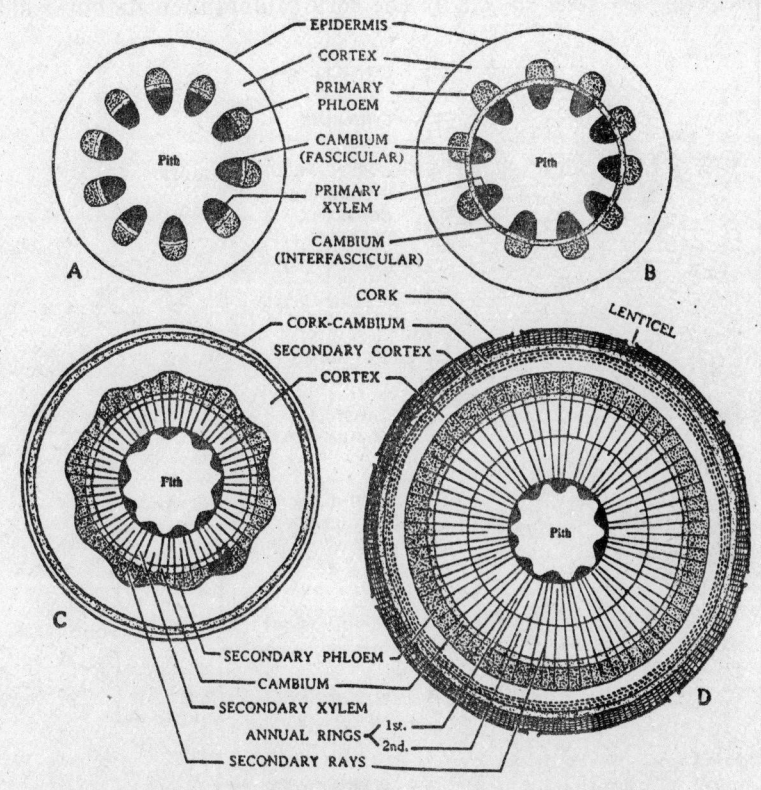

FIG. 73. Diagrams showing stages in the secondary growth of a dicotyledonous stem up to two years.

itself, or in the deeper layers of the cortex. In the formation of the cork-cambium the outer layer of collenchyma becomes meristematic; it divides and forms a thin strip of cork-cambium consisting of a few rows of narrow, thin-walled and roughly rectangular cells; these cells are living and active. The cork-cambium takes on meristematic activity and begins to divide and give off new cells on both sides, forming the **secondary cortex** on the inner side and the **cork** on the outer side.

Secondary Cortex. The cells that are cut off on the inner side are parenchymatous in nature. These constitute the **secondary cortex** or **phelloderm**. The cells of the secondary cortex generally contain chloroplasts and carry on photosynthesis. Sometimes they are thick-walled, but made up of cellulose and provided with pits. The cells of the secondary cortex are arranged in a few rows, and are added on to the primary cortex.

Cork. The new cells cut off by the cork-cambium on its outer side

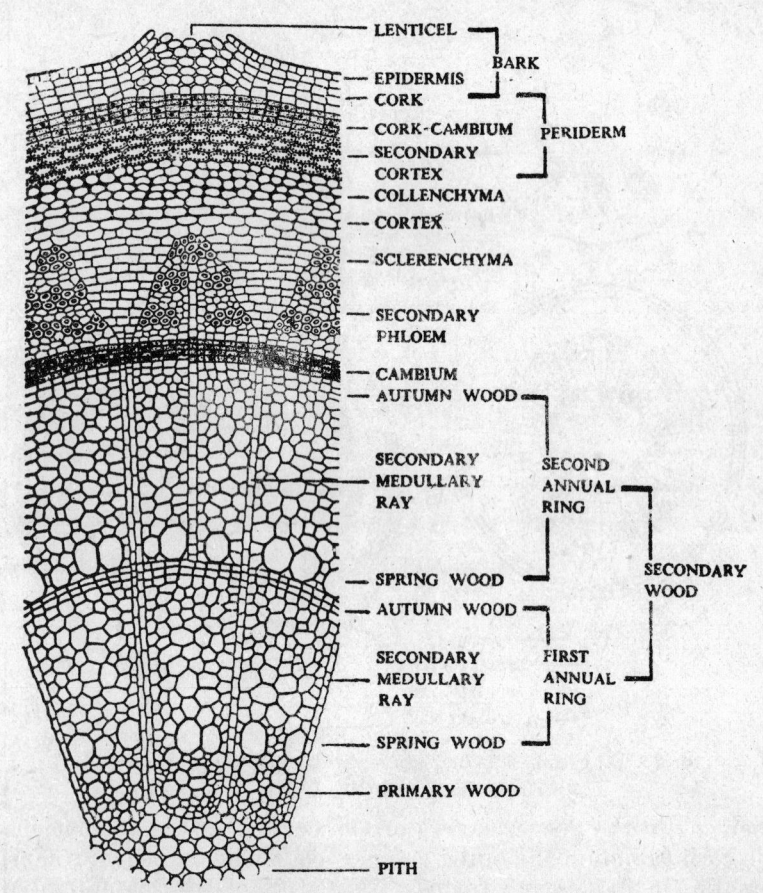

FIG. 74. A two-year-old dicotyledonous stem (a sector) in transection showing secondary growth in thickness.

are roughly rectangular in shape and soon become suberized. They form the **cork** or **phellem** of the plant. The cork tissue of cork oak (*Quercus suber*), a Mediterranean plant, is of considerable thickness, and is the source of bottle cork. When this is removed from the tree a fresh strip of cork is produced by the underlying cork-cambium. Cork cells are dead, suberized and thick-walled. They are arranged in a few radial rows, without leaving intercellular spaces between them, and are usually brownish in colour. Being suberized the cork is impervious to water, and it thus cuts off the outer tissues from the supply of water and food material. Consequently, they soon die off and act as the bark of the plant. Both cork and bark are protective tissues (see pp. 262-63).

All the new tissues formed at the peripheral region, viz. the cork or phellem, the cork-cambium or phellogen and the secondary cortex or phelloderm, are together known as the **periderm**.

Bark. All the dead tissues lying outside the active cork-cambium constitute the bark of the plant. It, therefore, includes the epidermis, lenticels and cork, and sometimes also hypodermis and a portion of the cortex, depending on the position of the cork-cambium, that is, the deeper the origin of the cork-cambium the thicker is the bark.

When the cork-cambium appears in the form of a complete ring, the bark that is formed comes away in a sheet; such a bark is known as the **ring-bark**, as in *Betula* (B. BHURJJA-PATRA); and when it appears in strips the resulting bark comes away in the form of scales; such a bark is, therefore, known as the **scale-bark**, as in guava. The function of the bark is protection (see pp. 262-63).

Lenticels (FIG. 74). These are aerating pores formed in the bark, through which exchange of gases takes place. Externally they appear as scars or small protrusions on the surface of the stem. A section through one of the scars shows that the lenticel consists of a loose mass of small, thin-walled cells (**complementary cells**). At each lenticel the cork-cambium, instead of producing compact rows of cork cells, usually forms oval, spherical or irregular cells which are very loosely arranged, leaving a lot of intercellular spaces between them. The lenticel commonly develops below a stoma, and as its cells increase in number and size the epidermis gets ruptured.

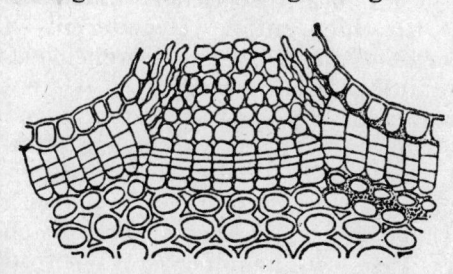

FIG. 75. A lenticel, as seen in transection.

Communication is thus established between the atmosphere and the internal tissues of the plant. The gases can then easily diffuse in and out through the lenticel. To facilitate the diffusion of gases empty spaces are left between the different rows, upper and lower, of the cork and the cork-cambium. The lenticel may be closed in winter by the formation of cork; this, however, gets ruptured as the new active season begins.

Tissues in Primary Structure	Meristematic Tissues	Tissues in Secondary Structure
Epidermis	.. .. ..	1 Bark (a) Epidermis (b) Lenticels (c) Cork
Collenchyma →	Cork-cambium →	2 Cork-cambium 3 Secondary cortex
└──→	.. .. ..	4 Collenchyma
Primary cortex	.. .. ..	5 Primary cortex
Endodermis	.. .. ..	6 Endodermis
Pericycle	.. .. ..	7 Pericycle (sclerenchyma)
Medullary rays →	Interfascicular cambium	
Primary phloem	.. .. ..	(gets crushed)
	↓	8 Secondary phloem 9 Cambium ring
Cambium ────→	Cambium ring ───→	10 Sec. wood (annual rings) (a) Spring wood (b) Autumn wood 11 Sec. medullary rays
Primary wood	.. .. ..	12 Primary wood
Pith	.. .. ..	13 Pith

SECONDARY GROWTH IN VITIS STEM (FIG. 76)

Grape vine (*Vitis vinifera*) is a liane type of plant, and it shows the internal structure of a typical dicotyledon. **Epidermis**—a single outermost layer with thick cuticle. **Hypodermis**—a few layers of collenchyma. **Cortex**—several layers of parenchyma. **Pericycle**—many-layered, containing isolated strands of sclerenchyma. **Vascular bundles**—numerous but distinct and arranged in a ring; each bundle is collateral and open with xylem, cambium and phloem from inside to outside; vessels are characterized by wide diameter; tyloses often present; phloem with distinct sieve-tubes and associated cells, and capped by a patch of sclerenchyma (hard bast). **Primary medullary rays**—broad (multiseriate), clearly separating the vascular bundles; crystals occur in many cells as transparent dots. **Pith**—large and parenchymatous.

Secondary growth soon begins with the activity of the vascular cambium. It produces secondary xylem on the inside and secondary phloem on the outside, as usual. Xylem consists of series of wide vessels and some amount of parenchyma. Secondary medullary rays are distinct and extend up to the secondary phloem. Secondary phloem consists of bands of sieve-tubes with companion cells and parenchyma, alternating with tangential bars of sclerenchyma, a few layers thick. Soon cork-cambium arises in the primary phloem and begins to cut off layers of cork on the outside. As a consequence all the tissues outside the cork-cambium die out and collapse, forming the periderm of the stem.

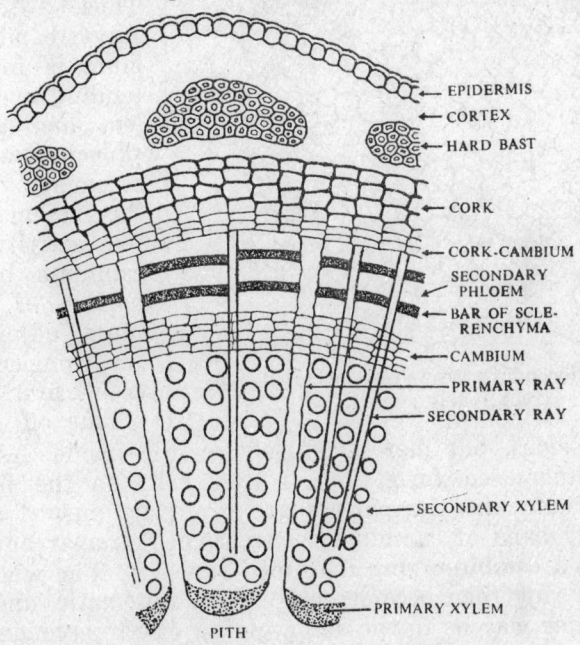

FIG. 76. Secondary growth in *Vitis vinifera* stem.

(II) Dicotyledonous Root

As in the stem, the secondary growth in thickness of the root is due to the addition of new tissues cut off by the cambium and the cork-cambium in the interior as well as in the peripheral region. In the root the secondary growth commences a few centimetres behind the apex.

A. ORIGIN AND ACTIVITY OF THE CAMBIUM

The conjunctive tissue just flanking the phloem on its inner side becomes meristematic and by dividing gives rise to a strip of cambium. It is evident that there are as many strips of cambium as there are phloem bundles. The cells of the conjunctive tissue lying in between xylem and phloem bundles also become meristematic so that the strips of cambium are seen to extend outwards between phloem and xylem. Then the portion of the pericycle abutting on the protoxylem becomes meristematic; it divides and forms a strip of cambium there, joining with the earlier-formed cambium strips on either side of xylem. Thus a continuous wavy band of **cambium** is formed, extending over the xylem and down the phloem (FIG. 77). The secondary growth then commences with the activity of this cambium band. The portion of the cambium adjoining the inner phloem becomes active first. It begins to cut off new cells

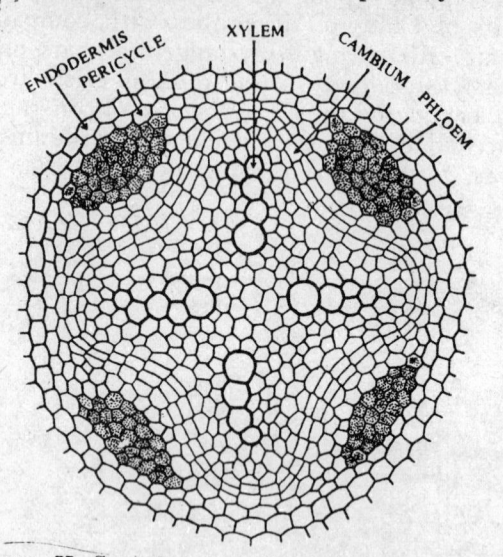

FIG. 77. Secondary growth of dicotyledonous root (early stage).

on both sides, but more profusely on the inside. As a result of this increased formation of new cells on the inner side the cambium and the phloem are gradually pushed outwards. The wavy band of cambium soon becomes circular or ring-like, and thus a **cambium ring** is formed (FIG. 78). The whole of the cambium ring then becomes actively meristematic, and behaves in the same way as in the stem, giving rise to secondary xylem on the inside and secondary phloem on the outside.

Secondary Xylem. The new cells cut off by the cambium on the inner side gradually become differentiated into the elements of xylem and all these new elements together constitute the secondary xylem. The cambium is always more active on the inner side than on the outer, and consequently secondary wood increases more rapidly in bulk than secondary phloem; in fact, the secondary wood forms the main bulk of the plant body after secondary growth. It is made of numerous large vessels with comparatively thin walls, abundance of wood parenchyma, but few wood fibres. As more wood is added, cambium and phloem are gradually pushed farther out. As the root lies underground it is not subjected to variations of aërial conditions; consequently annual rings, which are so characteristic of woody stems, are rarely formed in the root. Even when the root has increased considerably in thickness the primary xylem bundles still remain intact and can be recognized under the microscope in several cases. Against the protoxylem the cambium forms distinct and widening radial bands of parenchyma, which constitute the **primary medullary rays**. These extend up to the secondary

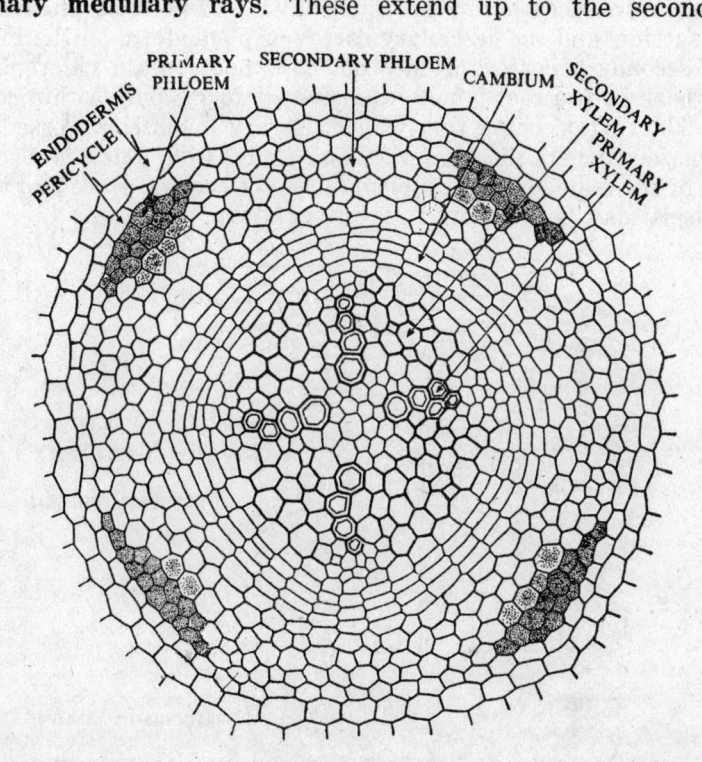

FIG. 78. Secondary growth of dicotyledonous root (later stage).

phloem. Other smaller and thinner medullary rays are also formed later by the cambium. Medullary rays are larger and more prominent in the root than in the stem.

Secondary Phloem. The new elements cut off by the cambium on the outer side becomes gradually modified into the elements of phloem, and all these together constitute the secondary phloem. It consists of sieve-tubes with companion cells and abundant parenchyma, but less bast fibres (except in special cases). The secondary phloem is much thinner than the secondary xylem. The primary phloem soon gets crushed.

B. ORIGIN AND ACTIVITY OF THE CORK-CAMBIUM

When the secondary growth has advanced to some extent, the single-layered pericycle as a whole becomes meristematic and divides into a few rows of thin-walled, roughly rectangular cells; these constitute the **cork-cambium** or **phellogen**. The cork-cambium may also arise, in some cases, in the phloem. As in the stem, it produces a few brownish layers of **cork** or **phellem** on the outside, and the **secondary cortex** or **phelloderm** on the inside. The secondary cortex of the root does not contain chloroplasts. The **bark** of the root is not extensive; it forms only a thin covering. The cortex, being thin-walled, is very much compressed and ultimately it gets disorganized, and sloughs off. Such is also the fate of the endodermis. Epiblema dies out earlier. Here and there **lenticels** may be developed, as in the stem.

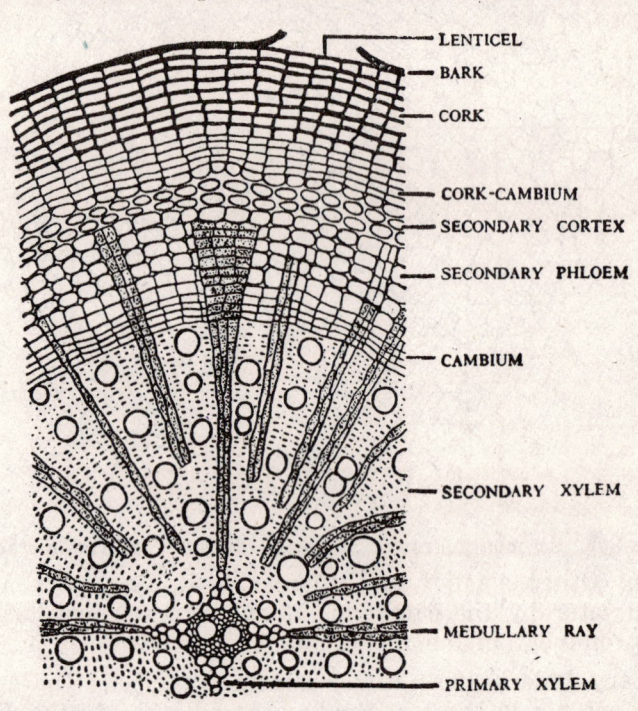

FIG. 79. A dicotyledonous root (a sector) in transection showing secondary growth in thickness.

Functions of Cork and Bark. Cork and bark are the protective tissues of plants. They are meant to check evaporation of water, to guard the plant body against variation of external temperature, and to protect it against the attack of parasitic fungi and insects.

(1) **Cork.** Sooner or later, in shrubs and trees, the epidermis is reinforced or sometimes replaced by the cork which then takes on the functions of the former, being essentially a protective

tissue. The cork is always much thicker than the epidermis, and, as such, it can afford greater protection than the epidermis. The renewal of the cork by the underlying cork-cambium is a decided advantage in this respect. All the cork cells are suberized, and thus the cork acts as a waterproof covering to the stem. Loss of water by evaporation is, therefore, prevented or greatly minimized. The cork tissue also protects the plant against the attack of parasitic fungi and insects. Cork cells, being dead and empty containing air only, are bad conductors of heat. This being so, sudden variation of outside temperature does not affect the internal tissues of the plant. Cork is also made use of by the plant for the healing of wounds.

(2) **Bark.** Since bark is a mass of dead tissues lying in the peripheral region of the plant body as a hard dry covering, its function is protection. It protects the inner tissues against the attack of fungi and insects, against loss of water by evaporation, and against the variation of external temperature. In many plants the bark sloughs off, and then all these functions are performed by the cork part only.

Protective Tissues. It is to be noted that there are three tissues in plants, namely (1) the **epidermis**, (2) the **cork**, and (3) the **bark**, which develop for the specific purpose of protection. At an early stage of the plant the epidermis alone affords the necessary protection (see pp. 217-18), but in shrubs and trees at a later stage the epidermis becomes reinforced or even replaced by the cork and the bark for the same purpose.

Chapter 8 ANOMALOUS SECONDARY GROWTH IN THICKNESS

(I) Dicotyledonous Stems

A large number of dicotyledonous plants of varying habits including many lianes show anomaly in their growth in thickness, sometimes resulting in peculiar structures, particularly in respect of xylem and phloem. Since cambium is responsible for growth in thickness the anomalous structure may be directly correlated with its irregular behaviour and also often its abnormal position. The anomalous growth varies considerably in different plants. (1) In some plants, as in *Bignonia, Bauhinia, Thunbergia, Aristolochia*, etc., the cambium is normal in position but its behaviour is irregular, giving rise to secondary xylem and secondary phloem in incompatible proportions and arrangements. (2) In other plants, as in *Amaranthus, Boerhaavia, Mirabilis, Tinospora*, etc., (and also in *Gnetum*) the fascicular cambium of the primary bundles does not function or is least active; in them one or more accessory cambia arise in the peripheral region outside the primary bundles and give rise to secondary xylem and secondary phloem in an irregular manner. (3) In still others, as in *Piperaceae*, e.g. *Piper betle, P. nigrum*, etc., the anomalous structure is attributable to the development of distinct medullary bundles around the pith and cortical bundles towards the periphery.

1. **Anomalous Growth in *Amaranthus* Stem** (FIG. 80). In the young stem the primary (medullary) vascular bundles remain scattered in the ground tissue. They are numerous and collateral

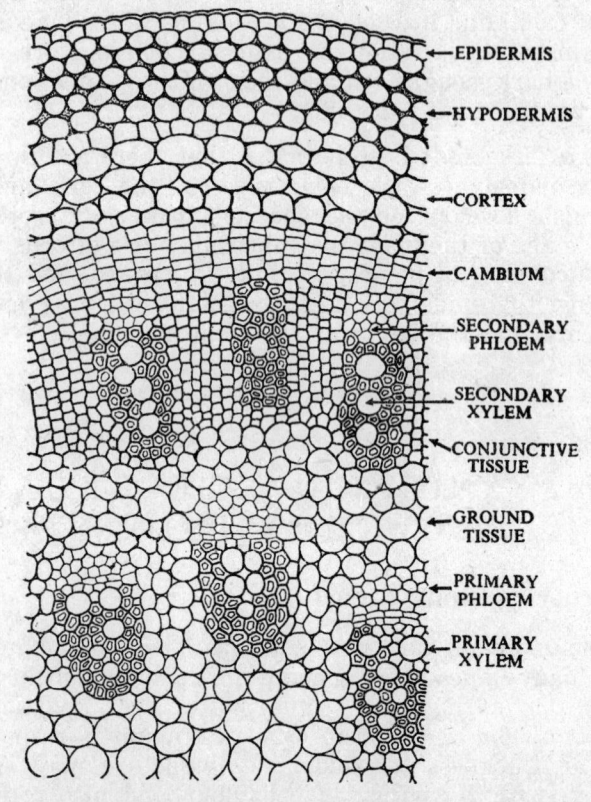

FIG. 80. Anomalous growth in *Amaranthus* stem.

with the cambium in them either feebly developed and functionless, or absent. The region of the pericycle just outside the outer primary bundles soon becomes meristematic and forms into a few-layered cambium. The cambium soon becomes active and begins to cut off secondary xylem and secondary phloem in the form of collateral bundles towards the inside only. In addition the cambium also cuts off several layers of parenchyma (conjunctive tissue) towards the inside, which soon becomes thick-walled and lignified. All the bundles of the secondary origin lie embedded in this tissue. On the outer side the cambium produces a little parenchyma, sometimes none at all. The endodermis soon gets distorted. As the secondary bundles increase in number, they together with the conjunctive tissue form a compact secondary structure. The compact structure so formed by the secondary tissues is occasionally interrupted by thin strips of parenchyma. There is very little or no formation of periderm in *Amaranthus*.

2. **Anomalous Growth in *Boerhaavia* Stem** (FIG. 81). The primary structure consists of (a) epidermis with thick cuticle and some stomata; (b) hypodermis (collenchyma) below the epidermis, interrupted by the underlying cortex, usually below a stoma; (c) cortex (chlorenchyma) in several layers with

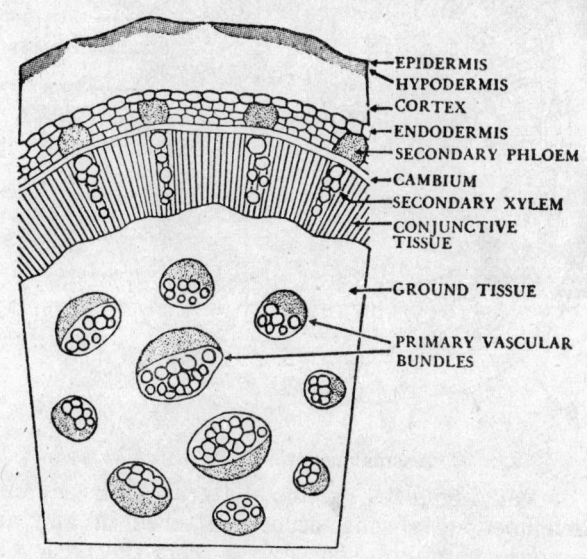

FIG. 81. Anomalous growth in *Boerhaavia* stem.

abundant chloroplasts; (d) endodermis clearly defined; (e) pericycle sometimes with strands of sclerenchyma; (f) vascular bundles; and (g) pith. Vascular bundles—two large bundles on the two sides of the pith, and a number of small bundles (6-14) just outside, arranged in a second or middle ring. The bundles, particularly the bigger ones, show only a limited amount of growth in thickness by their fascicular cambium. Soon secondary growth begins. Cambium arises secondarily from the pericycle or from certain layers outside the primary bundles, and becomes active. It cuts off a peripheral ring (third or outer ring) of several collateral bundles (secondary), each consisting of xylem on the inner side, and phloem on the outer, with the fascicular cambium lying in between. Soon the interfascicular cambium becomes active and begins to produce rows of cells internally, which soon become thick-walled and lignified, called the *conjunctive tissue*. The former also produces some amount of parenchyma externally. A little later cork and lenticel develop outside the hypodermis.

3. Anomalous Growth in *Mirabilis* Stem (FIG. 82). In the primary structure a large number of primary (medullary) vascular

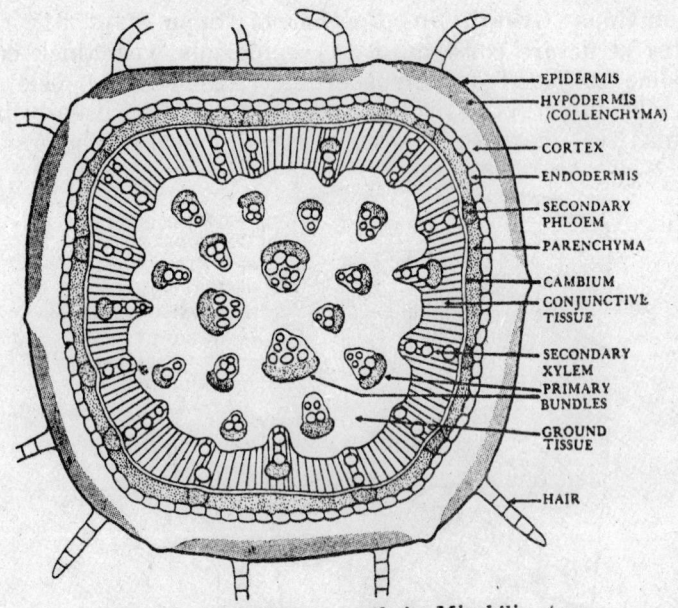

FIG. 82. Anomalous growth in *Mirabilis* stem.

bundles of varying sizes, each collateral with cambium either feebly developed or absent, occur scattered in and around the pith. Secondary cambium soon arises from the pericycle or from layers outside the primary bundles. It becomes active and begins to produce secondary bundles—xylem on the inner side and a little phloem on the outer. Depending, however, on the irregular activity of the cambium here and there small strands of secondary phloem may be formed centripetally, evidently lying embedded in the secondary xylem; this embedded phloem is called **interxylary phloem** or **included phloem**. Besides, the cambium also forms rows of cells on the inside, which soon become thick-walled and lignified, called the *conjunctive tissue*. It appears as a distinct band connecting the groups of secondary xylem.

4. Anomalous Growth in *Piper* Stem (FIG. 83). The anomalous structure is due to the presence of distinct medullary (pith) bundles and cortical (peripheral) bundles. The former occur

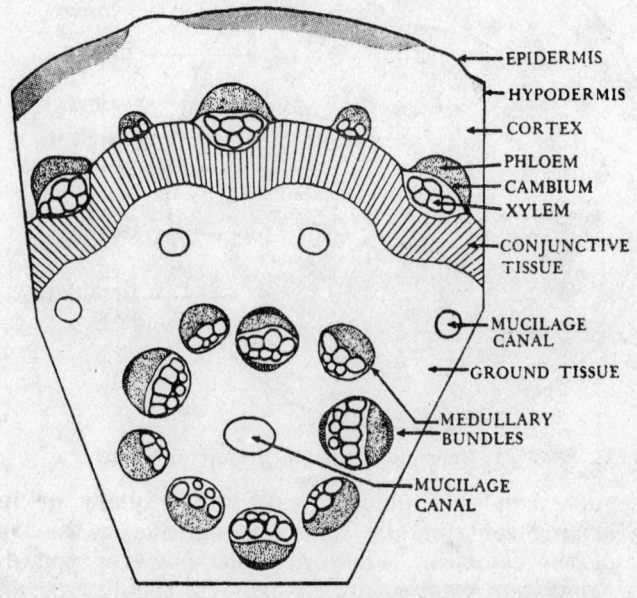

FIG. 83. Anomalous growth in *Piper* stem.

around the pith in a somewhat irregular ring (as in *Piper betle*) or a regular ring (as in *Piper nigrum*). These bundles are collateral with little or no cambium in them. Xylem or phloem or both may have a thin strip of sclerenchyma on the outer or inner margin or on both. There are some distinct mucilage canals (up to 12) in the ground tissue, and also a large one in the pith. Cortical bundles of different sizes occur in large numbers (many of them are leaf-trace bundles) towards the periphery in a somewhat irregular ring. Each bundle is collateral, consisting of xylem and phloem, with little or no cambium. Phloem may have a thin cap of sclerenchyma on the outer margin. The cortical bundles are bounded internally by a broad wavy band of thick-walled and lignified cells, called the *conjunctive tissue*. Towards the periphery occur cortex made of parenchyma, hypodermis made of collenchyma interrupted by the underlying parenchyma, and epidermis externally.

5. Anomalous Growth in *Salvadora* Stem (FIG. 84).

Vascular bundles occur in a circle separated by broad medullary rays which become lignified in the region of xylem. Vessels are of different sizes occurring in irregular clusters mixed with parenchyma and tracheids. The most interesting feature is the

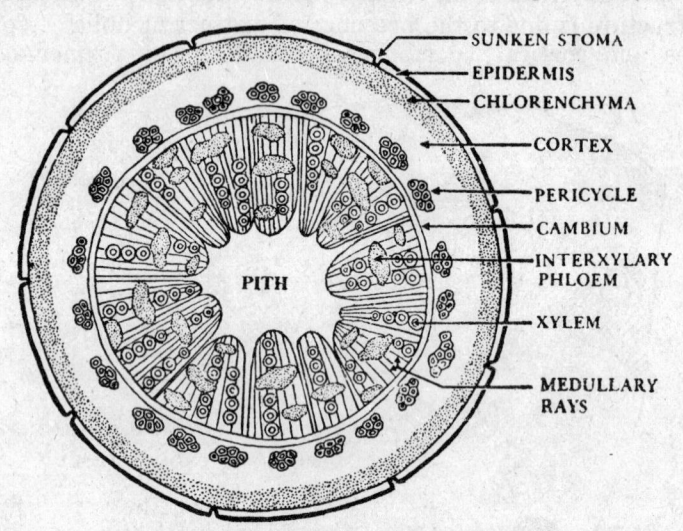

FIG. 84. Anomalous growth in *Salvadora* stem.

presence of strands of phloem, called **interxylary or included phloem**, arising centripetally in the xylem due to the irregular activity of the cambium. These strands, however, get disorganized as the stem begins to mature. Externally, pericycle is multiseriate and includes several strands of fibrous cells, particularly opposite the bundles; cortex forms a narrow zone of parenchyma and contains numerous crystals; chlorenchyma with abundant chloroplasts; and epidermis with thick cuticle. Soon cork arises in the periphery. Centrally there is a broad parenchymatous pith.

6. Anomalous Growth in *Leptadenia* Stem (FIG. 85).

Leptadenia, a much-branched, often leafless, shrub of *Asclepiadaceae*, shows anomalous secondary growth. The primary structure consists of (*a*) **epidermis** with a thick cuticle; (*b*) **hypodermis** (collenchyma) in 1 or 2 layers below the epidermis; (*c*) **cortex** (chlorenchyma) in a few layers of thin-walled cells containing chloroplasts; (*d*) **endodermis** in a single conspicuous layer of somewhat barrel-shaped cells, representing the innermost layer of the cortex; (*e*) **pericycle** composed of a few layers of thin-walled parenchyma with patches of sclerenchyma in it; (*f*) **primary vascular bundles** occur in a ring and are bicollateral in nature; and (*g*) **pith** in the centre. Secondary growth begins with the formation of a complete ring of cambium and its activity, cutting off a profuse amount of secondary xylem towards the inside, and a thin band of secondary phloem towards the outside. The cambium, however, is irregular in its behaviour, giving rise to an anomalous secondary structure. It so happens that here and there certain cells of the cambium ring instead of forming secondary xylem elements begin to produce for a while elements of secondary phloem on the inner side, contrary to their normal behaviour. Soon, however, these cambial cells revert to their normal activity, producing secondary xylem again inwards. The result is the appearance of a number of secondary phloem islands in the compact mass of secondary xylem. Each such phloem patch lying embedded in the secondary xylem is called **interxylary** or **included phloem** (see p. 211). The primary phloem soon gets crushed, while the primary xylem is pushed inwards; the inner primary phloem of the bicollateral bundle still remains associated with the primary xylem on the inner side, and is known as the **intraxylary** or **internal phloem** (see p.   ). The secondary xylem consists of rows of vessels and tracheids and some amount of parenchyma, with narrow but distinct medullary rays; while the secondary phloem consists of sieve-tubes with companion cells and phloem parenchyma (but no phloem fibres). There is a small pith in the centre, consisting of parenchymatous cells.

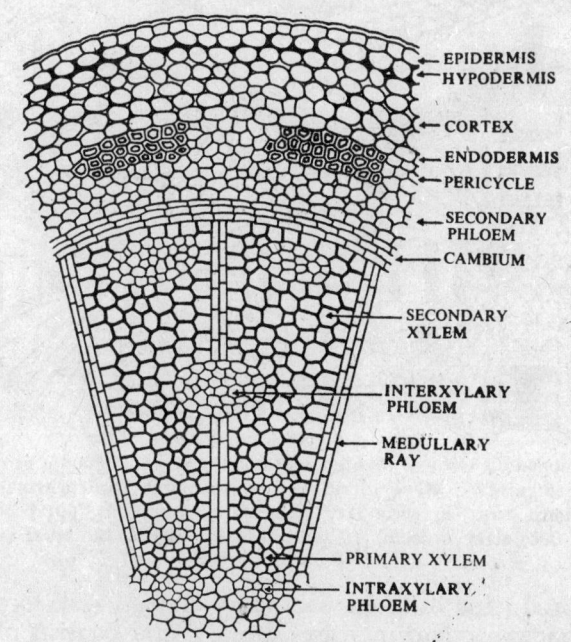

FIG. 85. Anomalous growth in *Leptadenia* stem.

7. Anomalous Growth in *Bignonia* Stems (FIGS. 86-7). Several species of *Bignonia* are liane types of plants, and they show different kinds of anomalous secondary growth. Two types are described below. In the young stem the primary structure is normal and similar to other typical dicotyledons. In most cases the stem has a wavy outline. **Epidermis**—a single-layer with a thick cuticle. **Hypodermis**—2 or 3 layers of collenchyma. **Cortex** (chlorenchyma)—a few layers of parenchyma with chloroplasts, often with isolated strands of sclerenchyma. **Pericycle** containing patches of sclerenchyma. **Primary vascular bundles** arranged in a ring around the central pith; they, however, get obliterated with the progress of secondary growth. Cambium is also normal in position but its behaviour is abnormal, soon giving rise to anomalous secondary structure, mainly furrowed xylem and wedged-shaped phloem. With the activity of the cambium the secondary growth starts at an early stage of the stem. Initially the secondary growth is normal for a while, forming secondary xylem towards the inside and secondary phloem towards the outside. Soon, however, the cambium begins to show its abnormal

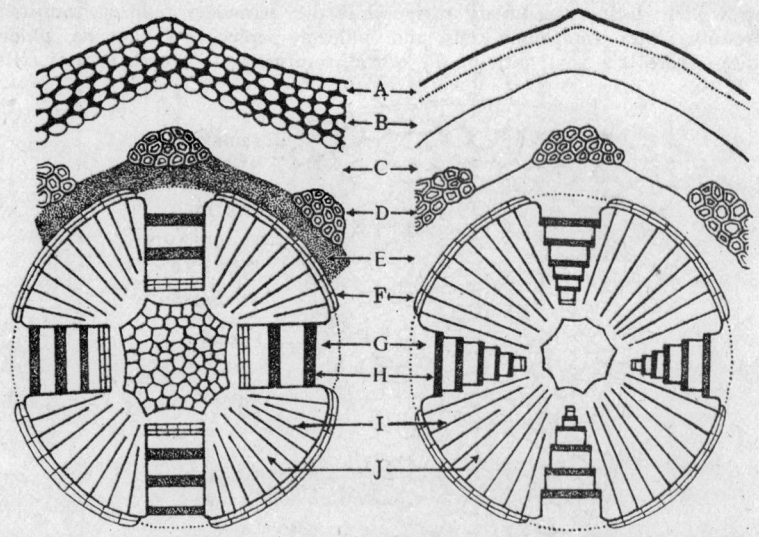

FIGS. 86-7. Anomalous Growth in *Bignonia* Stems. FIG. 86. *Bignonia venusta*. FIG. 87. *B. unguis-cati*. *A*, epidermis; *B*, hypodermis (collenchyma); *C*, cortex; *D*, hard bast; *E*, secondary phloem; *F*, cambium (outer strip); *G*, wedge of secondary phloem; *H*, sclerenchyma bar; *I*, cambium (inner strip); *J*, secondary xylem; *K*, medullary ray; and *L*, pith.

behaviour, producing uneven secondary xylem and secondary phloem. Secondary xylem increases rapidly, and consists of wide vessels and distinct medullary rays. As secondary growth proceeds, it is seen in several species that four longitudinal furrows appear in a crosswise manner in the secondary xylem and extend to a considerable depth. Soon secondary phloem is wedged into the furrows of xylem. In some species, as in *Bignonia venusta* (FIG. 86), the wedges of phloem may be of uniform width throughout their whole length, with a strip of cambium (inner) on the inside of each furrow; while in some other species, as in *Bignonia unguis-cati* (FIG. 87), the wedges may widen from inside to outside in a step-like manner. In the same way additional wedges of smaller depths may be formed later. Each wedge is strengthened by bars of sclerenchyma. It will thus be noted that the cambium ring splits into four strips—four bigger ones in their normal position outside the secondary xylem, and four smaller strips, each at the base of a furrow. Each strip of cambium in the furrow produces a wedge of secondary phloem, while each outer strip of cambium produces a thin ring of secondary phloem.

(II) Dicotyledonous Root

Anomalous Growth in Beet Root (*Beta vulgaris*; FIG. 88). Morphologically the beet root consists of three parts: root, hypocotyl and swollen portion of the stem. A section through the upper part of the root reveals the following anomalous structure. The

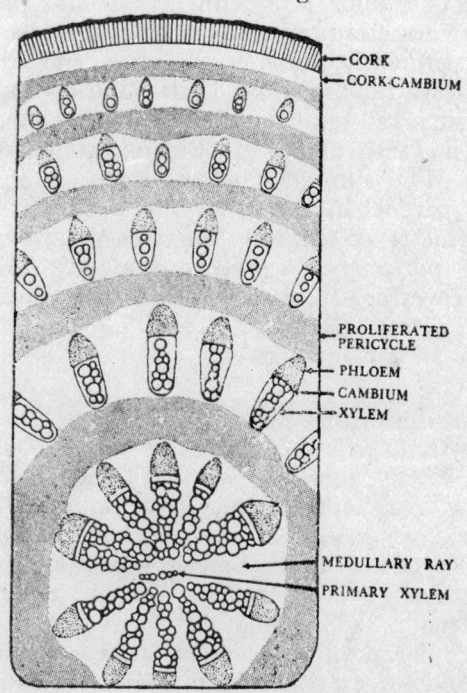

FIG. 88. Anomalous growth in beet root (*Beta vulgaris*).

main feature of the root is the succession of growth rings of vascular bundles formed by separate cambia arising successively by proliferation (i.e. repeated multiplication) of the pericycle. The cambia, apparently appearing as circular bands, form separate vascular bundles and the intervening parenchyma (*conjunctive parenchyma*) in between the bundles. The first cambium forms the first growth ring of compact vascular bundles around the diarch primary xylem lying in the centre. Against each protoxylem there is also a wide band of medullary ray. A second cambium arising similarly by proliferation of the pericycle forms almost immediately a second ring of separate vascular bundles in the same way. Successive cambial layers appear and function likewise. In the meantime the pericycle also proliferates and produces, each time, a wide band of parenchyma which acts as a storage tissue together with a major part of the phloem parenchyma. It is thus seen that rings of vascular bundles alternate with the bands of storage parenchyma (proliferated pericycle). Several such growth rings are formed one after the other in quick succession. It will be noted that all the cambia are normal in behaviour producing xylem and phloem in the usual way, and are simultaneously active but less so gradually outwards. Consequently, smaller bundles and thinner rings appear in the outward direction. Each bundle consists of abundant parenchyma, a few lignified xylem elements, and a patch of phloem (with plenty of phloem parenchyma in it) on the outer side. The simultaneous proliferation of the cambia, pericycle and parenchyma of the bundle leads to the rapid increase in diameter of the root. Soon cork-cambium and cork appear in the periphery. A broken mass of cells forms the external dark covering. It may be noted that spinach (*Spinacia oleracea*) and carrot (*Daucus carota*) also show an almost similar type of anomalous growth.

(III) Monocotyledonous Stem

Secondary growth in monocotyledons is rather rare. It is commonly seen in woody monocotyledons such as *Dracaena, Yucca, Aloe, Agave,* etc. Exceptionally a large amount of secondary growth in thickness is seen in most species of *Dracaena*. One plant of *Dracaena draco* in the Canary Isles measured 14 metres in girth at the base and was 6,000 years old when it was destroyed by a storm in 1868. It may be noted that the stout stems of palms are not the result of secondary growth but are the result of protracted primary growth by a primary thickening meristem occurring beneath the apical meristem. Although often very stout no cambium is formed in them and, therefore, there is no secondary growth in such plants.

Secondary Growth in *Dracaena* Stem (FIG. 89). The primary structure is a typically monocotyledonous one with numerous closed and collateral or concentric (amphivasal type; see p. 227) vascular bundles lying scattered in the ground tissue. Secondary growth in it begins with the formation of a secondary meristematic tissue—the cambium—in the parenchyma outside the primary bundles. This parenchyma divides tangentially and forms a band of cambium, a few layers in thickness. The cambium thus formed is more active on the inner side. It begins

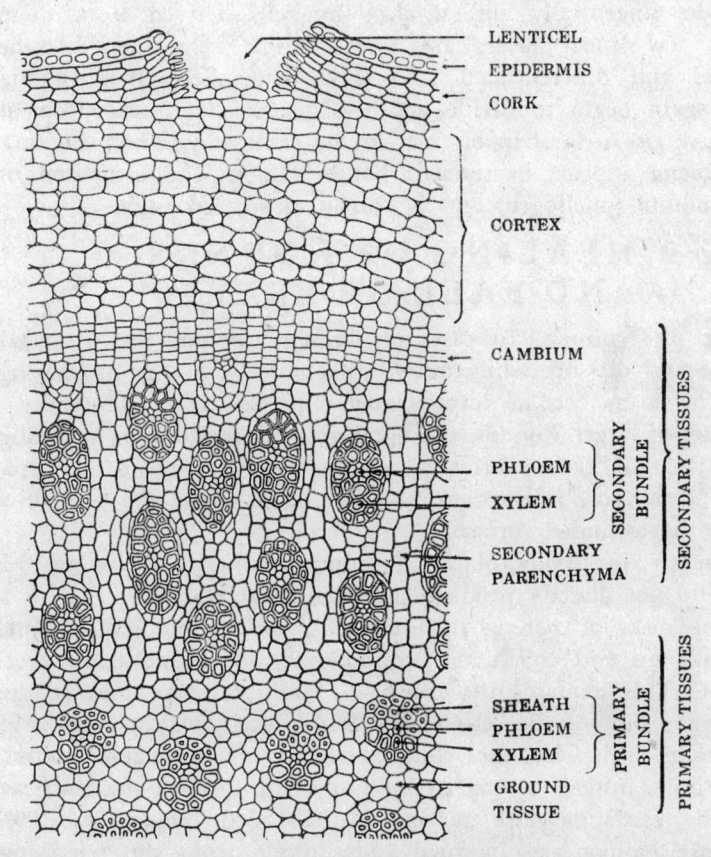

FIG. 89. Secondary growth in monocotyledonous stem (*Dracaena*).

to cut off new cells towards the inside, which soon become differentiated into distinct vascular bundles (secondary) and thick-walled, often lignified parenchyma (secondary). On the outer side the cambium only produces some amount of thin-walled parenchyma which may contain some crystals. While the primary bundles remain scattered the secondary ones are somewhat radially seriated and so also the surrounding secondary parenchyma. The vascular bundles are oval in transection, and concentric with phloem in the centre surrounded by xylem (**amphivasal**). In some species of *Dracaena* the vascular bundles are, however, collateral. Phloem consists of short sieve-tubes, companion cells and phloem parenchyma; while xylem consists of long tracheids with a small amount of thick-walled (lignified) wood parenchyma. After the secondary growth has proceeded to some extent the peripheral parenchyma becomes meristematic and begins to divide tangentially and so also the cells derived from them until a few linear layers are formed. The cells then become suberized and differentiated into cork. Some deeperlying parenchyma again begin to divide and the new layers formed from it again give rise to a strip of cork in the same way. Thus the cork in *Dracaena* appears in seriated bands without the formation of cork-cambium (phellogen) and is known as **storied cork.**

Chapter 9 HEALING OF WOUNDS AND FALL OF LEAVES

Healing of Wounds. In cases of simpler wounds, the wounded cells die and dry up, while outer walls or cells of the underlying uninjured layers become impregnated with protective substances.

In cases of larger wounds, the outermost uninjured layer of living parenchymatous tissue forms a meristem (phellogen) which produces one or more layers of cork—the **wound cork;** the cork then protects the wounded surface.

Frequently in woody plants the uninjured cells adjoining the wound do not directly produce the cork tissue, but give rise to a succulent mass of parenchymatous cells, called the **callus.** This callus fills up and covers the wound, and not infrequently overgrows it. This explains the origin of knots in some trees. If the cambium is injured the cells of the callus often form a fresh strip of cambium which becomes connected with the original cambium.

Sometimes, instead of any fresh layer being formed, the tracheae or wood vessels develop tracheal plugs, called **tyloses** (FIG. 90), which are balloon-like ingrowths developing from the adjoining parenchyma through pits. Tyloses plug the lumen of the vessels, while other elements simply dry up. Latex, if present, coagulates. In this way loss of water is prevented from the exposed surface.

Fall of Leaves. In deciduous trees and shrubs the leaf falls in the dry season, when the absorption of water by roots is minimized and the evaporation of water from the surfaces of leaves is enhanced. Both the conditions prevail in winter or in a prolonged dry summer and, therefore, the leaf is seen to fall at that time. The immediate structural cause of he leaf-fall is he formation of a layer of cork across the base of the petiole and the development of a well-defined *separation layer*, called the **abscission layer**, just external to the cork. The living parenchymatous cells lying across the base of the petiole, and also those of the vascular bundles, become meristematic and form a layer of cork, or these living cells become suberized directly, forming the cork layer without any division. In either case this cork is later reinforced by a fresh strip of cork formed by the underlying cork-cambium. In some cases the cork-cambium directly produces a few layers of cork at the base of the petiole. The cork being suberized and the vessels getting constricted owing to the lateral pressure of the cork, the leaf is cut off from the supply of water; it dries up and dies. The separation layer, or the abscission layer, lying just external to the cork, turns yellowish and gets disorganized. Cellulose of this layer becomes converted into pectin which dissolves, and thus the cells become separated from one another. The leaf then remains supported by the vessels only, and it breaks off mechanically at the abscission layer, either under its own weight or when disturbed by the wind. The vessels are clogged with gum and tyloses, and the exposed surface is covered with cork. Thus exudation of sap is prevented. When the leaf falls, a scar is left on the stem; this is called the **leaf-scar.**

FIG. 90. Vessels with tyloses.

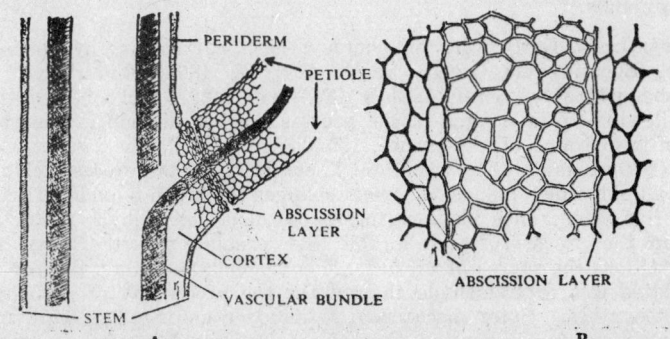

FIG. 91. *A*, formation of abscission layer at the base of the leaf; *B*, a portion of the abscission layer in surface view some time before leaf-fall.

Part III PHYSIOLOGY

Chapter 1 GENERAL CONSIDERATIONS

A Short History. Experimental physiology actually began from the time of **Stephens Hales (1677-1761)**, curate of Teddington in Middlesex. He was the first to devise experimental methods with his knowledge of physics and statistics to find out the movement of sap through the plant body, rate of transpiration current, suction due to transpiration, loss of weight due to transpiration, root pressure and capillarity as factors in the ascent of sap. Hales was the first to show that leaves make use of air to form a part of body substance, and he also showed that apical regions are the most active regions of growth. These and several other experiments were the first contributions to our knowledge of plant physiology. His *Vegetable Staticks* published in 1727 is a famous work.

For half a century there was no further incentive for work in this new branch of study. **Priestley (1733-1804)**, a chemist at Warrington (England) investigating the composition of different kinds of air, found in 1771 that green plants grown in an atmosphere rich in carbon dioxide (bad air) produced in course of several days a large quantity of oxygen (pure air), as tested by a burning candle. His 'mouse and twig' experiment was very interesting in this respect. A mouse kept in a jar containing 'bad' air (by burning a candle in it, evidently releasing carbon dioxide) soon died. But a mouse kept under the same condition with a twig of mint enclosed in the jar survived, and the twig also remained healthy. Priestley, however, could not recognize the conditions that led to the gas exchanges between the twig and the mouse. It was very unfortunate that Priestley could not pursue his work. He met with strong opposition, his house was mobbed, and he, barely escaped with his life and fled to America. **Ingenhousz (1730-99)**, a physician educated in Holland, who later migrated to London in 1765, was interested from the medical point of view in the composition of the air. Priestley's work attracted his notice. He reported in 1779 that *green plants exposed to light* for a few hours absorb carbon dioxide (bad air) unfit for respiration and exhale oxygen and thus purify the air. He further proved that at night plants vitiate the air by their respiration, as do animals.

Photosynthesis. Early work on photosynthesis carried out by **De Saussure (1804), Boussingault (1864), Sachs (1862 & 1864), Timiriazeff (1875), Engelmann (1881), Pfeffer (1881 & 1896), Kny (1897)** and others established the following initial stages in the process: carbon dioxide is absorbed and oxygen is liberated (De Saussure, 1804; Sachs, 1882); the two volumes are equal (Boussingault, 1864); sunlight is essential for the process; chlorophyll absorbs light, which induces chemical changes in carbon dioxide (Timiriazeff, 1872); plastids cannot photosynthesize without chlorophyll; water is also fixed in the process; starch is the first visible product (Sachs, 1862 & 1864); CO_2 is the source of all organic compounds in plants. **Baeyer** in 1870 formulated the 'formaldehyde theory' for the production of carbohydrates in photosynthesis (later discarded). **Willstatter** and **Stoll** in 1918 modified Baeyer's view and suggested that sugar is formed in several stages (at least six). **Warburg** in 1919 working on *Chlorella* agreed with Willstatter and Stoll but introduced 'light reaction' and 'dark reaction' (which he called Blackman reaction) in the intermediate processes of photosynthesis. Since then much spadework has been done and only recently have the main facts regarding photosynthesis been elucidated (see text).

Nitrogen Assimilation. In 1840 Liebig, an agricultural chemist, formed the idea that plants obtain their nitrogen from the air and that ammonia is the source. Boussingault (1802-87) carried out a long series of experiments from 1837 to 1852 (and later) and came to the conclusion that plants do not utilize free nitrogen of the air and that nitrate of the soil is the principal source. He reported, however, in 1838 that certain plants like lupin, *Trifolium*, etc., grown in calcined soil to which only distilled water was added, showed a gain in weight in nitrogen; while wheat grown under the same condition showed no such gain. Boussingault's views were corroborated by Lawes and Gilbert in 1861 and later by Hellriegel, Russell and others. The importance of nitrogen in plant growth was determined by Sachs and Knop in 1860 and 1865 by their water culture experiments. Hellriegel and Wilfarth in 1887 first discovered the fixation of nitrogen by symbiotic bacteria in the root-nodules of leguminous plants. The formation of ammonia from plant proteins in the soil by the action of certain soil organisms (especially *Bacillus mycoides* and also several fungi) was made known by the work of Marchal in 1893 and later by others. Winogradsky's (1856-1934) elucidation of nitrification in 1890-91 was most valuable. He established the fact that one type of oval bacteria which he named *Nitrosomonas* is responsible for oxidation of ammonia to nitrite, and a second type of rod-shaped bacteria which he named *Nitrobacter*, for further oxidation of nitrite to nitrate. Jensen in 1898 showed that denitrification is due to the action of a group of putrifying bacteria called *Pseudomonas*.

The Colloidal System. Protoplasm exists in a colloidal condition (see p. 148) and various physiological processes are attributable to the colloidal nature of the cell contents. Most soils also contain materials in colloidal state. As a matter of fact, the colloids play an important part in the physiology of plants and of animals.

In course of his investigations on *diffusion* in solution, Thomas Graham (1861) found that soluble substances (inorganic or organic) could be divided into two classes—crystalloids and colloids according to the rates at which their solutions passed through a parchment membrane (dialyser). Substances such as salts, sugar, urea, etc., which diffuse readily, were termed *crystalloids* because of the fact that they generally exist in crystalline form; on the other hand, substances such as gelatine, albumen, gum, silicic acid, starch, etc., which diffuse at a very slow rate, were termed *colloids* (meaning glue-like). This distinction, as was later realized by Graham and others, is not rigid since many crystalline substances can be obtained in colloidal solution, e.g. sodium chloride in benzene. Further X-ray studies have shown that particles in colloidal systems are often truly crystalline in character. Consequently instead of the term colloid or colloidal substance it is the practice to refer to the colloidal state or to a colloidal system.

General Properties of the Colloidal System. In a colloidal solution the particles are either very large molecules or aggregates of a large number (even thousands) of molecules, still not visible under the microscope. If, however, the colloidal particles grow further in size, they become visible under the microscope. A colloidal solution is essentially a two-phase system: a disperse phase or discontinuous phase consisting of the discrete particles, and a dispersion medium or continuous phase consisting of the medium (solid, liquid or gaseous) in which the particles are distributed. When the dispersion medium is water, the colloidal solution is commonly called *hydrosol.* In a true solution, however, the particles are of molecular size and there is no true surface of separation between the disperse phase and the dispersion medium. Colloidal solutions consisting of large insoluble particles lying in a state of suspension in the dispersion medium are called *suspensoids.* Suspensoids do not play any significant role in plant physiology. Suspensoids are, however, common in soils. Two immiscible liquids—water and oil, for example—may form an emulsion. Particles in both the suspensoid and the emulsoid slowly separate out of the dispersion medium under the influence of gravity. Emulsions can be stabilized by adding a third substance called *emulsifier,* e.g. casein in milk stabilizing fat globules. When a colloidal solution resembles a solid or a jelly-like substance, it is called a **gel**; gelatine, agar agar, pectin, silicic acid, etc., easily form gels; common fruit jelly is a familiar example of gel. When the colloidal solution looks like a liquid, it is called a **sol**. A gel and a sol may be reversible, and protoplasm is a reversible colloid (see p. 148). Colloidal solutions in water are termed *hydrosols.* Similarly there may be alcosols, benzosols, etc.

Colloidal solutions with a liquid as dispersion medium fall into two classes: *lyophobic* (liquid-hating) and *lyophilic* (liquid-loving). When water is the dispersion medium the corresponding terms used are *hydrophobic* and *hydrophilic.* Gelatine, agar agar, gum, silicic acid, various albumens, starch, soap and many dyes, etc., which directly pass into colloidal solutions when brought in contact with water, are examples of hydrophilic colloids; they are also called reversible colloids. Insoluble substances like metals, metal sulphides, metal hydroxides and other substances on the other hand, which do not readily yield colloidal solutions when brought in contact with water, are examples of hydrophobic colloids; they are also called irreversible colloids.

The characteristic property of the disperse system is attributable to the enormous surface area of the disperse phase. A solid block reduced to colloidal particles enormously increases the exposed surface area. One of the most important results of the large surface area is the adsorption of ions and other materials by the particles. This adsorption may lead to the formation of electric charges on the particles which prevent them from collecting into larger aggregates. The surface of the colloidal particles is the seat of chemical energy and various chemical reactions take place here. The adsorption is somewhat selective, and is an important factor in plant physiology, particularly with reference to the cytoplasmic membrane (ectoplasm).

Optical Properties. The presence of colloidal particles although not detectable under a microscope can be made evident by optical means. Thus, if an intense beam of light be passed through a colloidal solution, the particles in it scatter the light and the beam is rendered visible indicating the presence of particles which are larger than molecules but too small to be separated by filtration. The phenomenon of scattering the light by the particles is known as the Tyndall effect after the name of its discoverer. The Tyndall effect has been better demonstrated by the ultramicroscope invented by Siedentopf and Zsigmondy. The presence of individual particles is made evident by this instrument as flashes of scattered light. It does not, however, reveal the shape, colour, or relative size of the particles.

Brownian Movement. Careful ultramicroscopic examination of a colloidal solution reveals that the particles of the disperse phase are in constant, rapid, zigzag motion called the Brownian movement after the name of its discoverer, Sir Robert Brown. Particles within the range of microscopic visibility also show this phenomenon. Brownian movement counteracts the force of gravity acting on the colloidal particles and is thus responsible to a certain extent for the stability of the colloidal solutions. Brown first noted this movement while examining pollen grains suspended in water. Brownian movement is caused by molecular impacts, i.e. bombardment by the molecules of the dispersion medium on the colloidal particles on any one side of them at any given moment.

Electric Properties. An important property of colloidal solution is that their particles carry an electric charge, either positive or negative, and therefore move towards one or the other electrode when a solution is placed in an electric field. This migration of colloidal particles under the influence of an electric field is called *cataphoresis*.

Flocculation. Flocculation of many colloidal systems, e.g. white of egg, is the change into an irreversible gel condition brought about by various means such as increased frequency of collisions of particles resulting in the formation of larger particles or masses, or by the application of heat or cold, or by the addition of a dehydrating agent such as alcohol. Flocculation is most commonly initiated by the introduction of electrolytes. An important property of many colloidal systems is their sensitivity to small quantities of electrolytes. The presence of a small quantity of ionizable substances causes the particles of many colloidal systems to coagulate so that a visible precipitate is readily formed. The term flocculation, or coagulation or precipitation is employed to designate such a condition. It may be noted that very small quantities of an electrolyte may cause flocculation of a large volume of the solution.

Diffusion. This is the movement of molecules or ions of a solute or a solvent, be it a liquid or a gas, from the region of its higher concentration to that of its lower concentration. Diffusion continues until an equilibrium is reached. The molecules or ions are in continuous motion following straight pathways at different speeds (according to their specific nature and the surrounding conditions) deflected, however, only by collision. Molecules or ions enter plant cells and move from one to the other by following the simple law of diffusion from the region of higher concentration to that of lower concentration. So far as the diffusion of soil solution into the root-hairs is concerned, the process is not considered to be of primary importance. Diffusion is, however, the basic phenomenon of osmosis and imbibition.

Imbibition. Imbibition is the phenomenon of soaking up water by certain materials, particularly in dry or semi-dry conditions. Fibres, pieces of wood, some proteins, sponges, etc., are some such materials. The cell-wall and the protoplasm are also able to absorb water by imbibition, and it plays an important role in the physiology of plant life. Imbibition can only occur when there is an affinity between the two. Thus cotton fibres imbibe water, while rubber does not. In this process the constituent particles of a particular substance take up water by *surface attraction* and increase in volume. For this reason seeds soaked in water are seen to swell up. The amount of attraction of dry cell-walls and of protoplasm for water is often very great and a considerable imbibition force may be developed within the plant body. As a result of imbibitional pressure the seed-coat of a germinating seed bursts. Germinating seeds kept in a closed vessel often burst it with tremendous pressure. Imbibition is believed to be an important force concerned in the ascent of sap. Imbibition also plays an important part (together with osmosis) in the intake of soil-water by the root-hairs. In an imbibing system it is the rule that the water always moves with some force from a saturated region to a drier region.

Osmosis. It has been observed that there are certain membranes which when used to separate a solvent (water, for example, which is the only important solvent in plants) and a solute (salt or sugar in water) allow the solvent on the one side to pass through them freely but at the same time resist the solute on the other side so that only a minute quantity of the latter can cross through. On account of this property of selective transmission such membranes are said to be semi-permeable or differentially permeable. Parchment paper, fish- or animal-bladder and egg-membranes are some such membranes. So far as the plant cells are concerned, the ectoplasm (and not the cell-wall) acts as the differentially permeable membrane. When weak and strong solutions are separated by such a membrane, there is a net transfer of the solvent from the weaker solution to the stronger one. *This process of selective transmission of a liquid in preference to another or a solvent in preference to the solute through a semi-permeable membrane is termed* **osmosis**. By this process of osmosis one liquid passes on to the side of the other liquid or the solvent passes from the side of the weaker to that of the stronger solution and goes on accumulating there. This process continues until the *hydrostatic* pressure (turgor pressure; see p. 276) due to the accumulated flow of the liquid or solvent has attained a value sufficient to stop further flow. This excess pressure, which is just sufficient to stop the flow through the membrane, is called the **osmotic pressure** of the stronger solution. It has been found that this pressure is proportional to the concentration (or density) of the solution, or, in other words, the greater the concentration (or density) of a solution the greater would be the osmotic pressure of it (the solution), as first shown by Pfeffer in 1877. The magnitude of osmotic pressure varies considerably in different cells, and even in the same cell under different conditions. Commonly it is 10-20 atmospheres, seldom lower than 3.5 atmospheres. In cells containing sugar the osmotic pressure may be as high as 40 atmospheres. In cells containing a high percentage of sodium chloride, as in halophytic plants, an osmotic pressure amounting to over 100 atmospheres may be generated. It may be noted that the living protoplasm has the ability to adjust within certain limits the osmotic condition of the cell in response to the changing factors of the environment; the protoplasm achieves this by changing according to its need the salt and sugar contents of the cell-sap. A familiar example of osmosis is this: raisins immersed in water are seen to swell up as a result of endosmosis, and at the same time a small quantity of the high percentage of sugar contained in them is found in the water outside as a result of exosmosis. Similarly, grapes immersed in strong solution of sugar or salt (say, 25% or 30%) are seen to shrink.

Suction Pressure. Many substances take up water vigorously or slowly. Evidently there is some force which exerts pressure (suction) on the water to drive it into the absorbing material. Such a pressure is called suction pressure. It varies in degrees according to the nature of the absorbing material and also on other conditions. Suction pressure is a measure of water-absorbing power of a cell. This is specially important in the case of root hairs (and also other contiguous cells of the plant body) where the suction pressure of the cell-sap often exceeds that of the soil water, and as a consequence exerts a pull and thus causes entry of water from the soil particles. Suction pressure also brings about movement of water from cell to cell, as from the root hairs to the cortical cells and finally to the vessels. The causes of suction pressure are to be sought in osmotic pressure and turgor pressure. The magnitude of suction pressure (also called diffusion-pressure deficit) is the difference between the osmotic pressure and the turgor pressure (Stiles, 1922): O.P.−T.P.− = S.P. The turgor pressure may be equal to the osmotic pressure, and then no further movement of water takes place, i.e. the suction pressure becomes nil. As the turgor pressure of a cell decreases, its suction pressure increases.

Experiment 1. Physical process of osmosis (FIG. 1). Take a wide thistle-funnel with a narrow long stem and close its mouth with parchment paper or fish-bladder. Fill it with strong salt solution a little above its neck and introduce it, stem upwards, into a beaker containing water. Mark the level of the solution in the

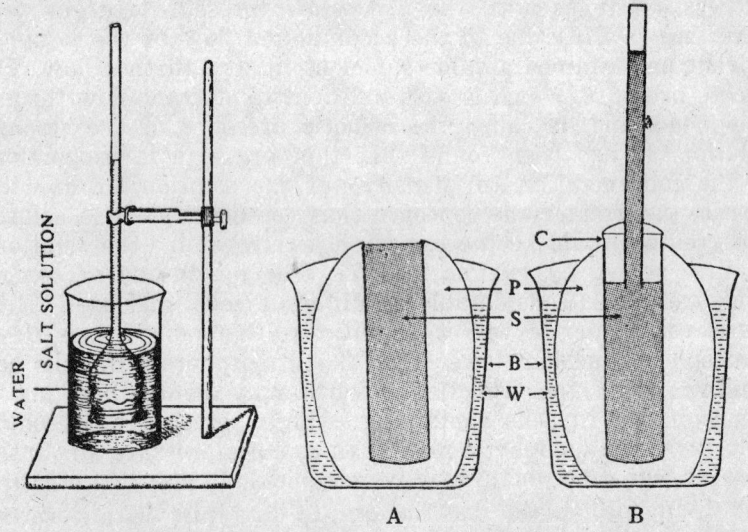

FIG. 1.   FIG. 2.
Experiments on Osmosis. FIG. 1. Physical process of osmosis.
FIG. 2. Physiological process of osmosis. A, experiment with potato tuber; B, the same with potato osmometer. C, cork; P, potato tuber; S, sugar solution; B, beaker; W, water.

stem of the thistle-funnel. After a few hours note that the level of the solution in the stem has gone up. This rise is due to the accumulation of water in the funnel as a result of more rapid flow of the water into it by osmosis (endosmosis) through the membrane. This rise of water is seen to continue until the level has gone sufficiently high up to exert a hydrostatic pressure on the membrane which then stops further net transfer of water by osmosis. This value of the hydrostatic pressure is equal to the osmotic pressure of the solution. At the same time a small quantity of salt also passes out through the membrane.

Experiment 2. Physiological process of osmosis (FIG. 2A). (*a*) Take a large potato tuber and size it into a bowl in the following way. Remove the skin from it and slice off the bottom to make it flat. Scoop out the central flesh with a knife or scalpel or borer into a deep hollow cavity with the wall comparatively thin. Place the potato bowl in a small beaker and pour some strong salt or sugar solution into it so as to cover more or less three-fourths of the cavity. Then pour water into the beaker almost to its brim. The water may be coloured with a few drops of eosin. Within a short time the solution in the cavity is seen to increase in volume turning reddish and to overflow it soon after, as a result of endosmosis. The presence of a small quantity of salt or sugar may also be detected in the water of the beaker as a result of exosmosis.

(*b*) The experiment may be carried out in a modified form with a **potato osmometer** (FIG. 2B). The potato tuber, after skinning the sides and slicing off the bottom, may be scooped out into a deep hollow cylindrical cavity, and salt or sugar solution poured into it. A soft cork of appropriate size with a glass tube fitted in it may then be gently but tightly pressed in. Place the potato tuber in a beaker and fill it (the beaker) with water. Within a short time the salt or sugar solution is seen to rise in the glass tube, soon overflowing it. The actual rise is one to a few metres, depending on a number of factors.

Importance of Osmosis in Plant Life.
(1) Root-hairs absorb water from the soil by the process of osmosis; at least this entry is controlled by osmosis. (2) From the root-hairs cell to cell osmosis takes place until the cortical cells of the root become saturated with water. Similar cell to cell osmosis takes place throughout the body of the plant. It is, however, now asserted that it is the suction pressure (and not the osmotic pressure) that is fundamentally responsible for the movement of water from cell to cell. (3) The osmotic pressure generated in the root-cortex is responsible for forcing the water into the xylem vessels, and possibly upwards through them at least to some height. (4) The living cells surrounding xylem draw water from it by this process, and so do the mesophyll cells of the leaf at the upper end of xylem, prior to transpiration. (5) Osmosis makes the cells turgid. This turgid condition gives a certain amount of rigidity to the young soft parts of the plant body, and is also an essential condition of growth. Enlargement of meristematic cells at the root-apex and stem-apex is due initially to osmosis. (6) Various movements, turgor movements particularly, such as those exhibited by the leaflets of Indian telegraph plant (*Desmodium gyrans*; see FIG. 47), sensitive plant (*Mimosa pudica*; see FIG. 53), sensitive wood-sorrel (*Biophytum sensitivum*; see

FIG. 52), sleep movement by most species of *Leguminosae*, opening and closing of stomata, bursting of many fruits and sporangia, etc., are largely due to osmotic phenomena. (7) By plasmolysis, which is an osmotic phenomenon, it is possible to determine the osmotic pressure of a cell (see experiment 3).

Turgidity. As a cell absorbs more and more water resulting in its accumulation in the vacuole, a certain pressure is exerted on the surrounding protoplasm and the cell-wall. As a consequence the protoplasm is forced outward against the cell-wall and the latter also becomes much stretched. The stretched cellulose wall being elastic tends to return to its original shape and thus in its turn exerts a pressure upon the fluid contents of the cell. *A cell thus charged with water with its wall in a state of tension is said to be turgid, and the condition is designated as* **turgidity** *or* **turgor**. It will be noted that in a fully turgid cell two pressures are involved: outward and inward. The outward pressure exerted on the cell-wall by the fluid contents of the cell is called the **turgor pressure**, and the inward pressure exerted on the cell-contents by the stretched cell-wall is called the **wall pressure**. Normally these two pressures counter-balance each other and a state of equilibrium is maintained between them. Three factors influence the turgidity of a living cell, viz. (1) formation of osmotically active substances inside the cell, (2) an adequate supply of water, and (3) a semi-permeable membrane.

Importance. Turgid condition is necessary for the transit of nutrient solutions from cell to cell; this is so because of the difference in the concentration of the cell-sap between one cell and the other. Turgidity is also necessary for growth; in fact, it is always the initial stage of growth. Rapid growth of certain organs of plants is principally due to turgidity, i.e. full expansion of the cells of those organs, and not to their rapid multiplication. Turgidity is also responsible for various movements of different organs of the plant. Thus movements of the guard cells of the stomata are due to changes in the turgidity of these cells, and similarly, the rising and the falling of the leaf and the leaflets of sensitive plant (see FIG. 53), Indian telegraph plant (see FIG. 47), etc., are brought about by alterations in the turgidity of the cells of the pulvinus. Turgidity of the cells of the root-cortex is responsible for forcing the water into the xylem vessels. Turgidity also gives a certain amount of rigidity to the plant, particularly to the growing regions and the soft leaves which easily wilt in strong sunlight, and also to other soft parts composed of only thin-walled parenchyma without any mechanical tissue.

Plasmolysis (FIG. 3). If a section of a plant organ or a *Hydrilla* leaf or a coloured petal or a *Spirogyra* filament be immersed in a *hypertonic* solution[1] (say, 5-10% sucrose solution) and after a few minutes observed under the microscope, it will be seen that the cell as a whole contracts and more obviously the protoplasm together with the nucleus and the plastids gradually shrinks away from the cell-wall and forms a rounded or irregular mass in the centre; while the space between the cell-wall and the protoplasmic mass becomes filled with the sugar solution. It will be noted that while the cell-wall is freely permeable to the solution the protoplasmic membrane is selectively or differentially permeable to it. The reason for such shrinkage of the protoplasm is that the sugar solution being of greater osmotic value than the cell-sap, the cell loses water by outward osmosis. As the water moves out of the cell, the protoplasm and the cell-wall are no longer in a state of tension. Further loss of water evidently results in the shrinkage of the protoplasm. *This shrinkage of the protoplasm from the cell-wall under the action of some strong solution—stronger than that of the cell-sap—is known as* **plasmolysis**. Protoplasm in one type of cells commonly follows the same pattern of shrinkage on plasmolysis. If the sugar-solution be replaced by pure water, soon after plasmolysis, the protoplasm is seen to return to its normal position and the vacuole reappears (deplasmolysis). Potassium nitrate solution (10%) is a very good reagent to bring about plasmolysis readily and is, therefore, useful for general class-work.

Plasmolysis is a vital phenomenon. It explains on the one hand the phenomenon of osmosis, and on the other it shows the permeability of the cell-wall and semi-permeability of the outer layer of the protoplasm—the ectoplasm—to the entrance of certain substances. Plasmolysis also shows that the protoplasm can retain the osmotically active substances of the sap. This is evident from the fact that the plasmolysed protoplasm turns back to its original position when the sugar solution is replaced by pure water. The phenomenon of plasmolysis also indicates whether the cells are living or dead. When a tissue is killed by boiling in water or by dipping into absolute alcohol or a strong formalin solution for a few seconds, the cells show no plasmolysis. From plasmolysis it is also possible to determine the osmotic pressure of cells (see experiment 3).

---

[1] A solution is said to be *hypertonic*, *isotonic* or *hypotonic* when its osmotic pressure is greater than, equal to, or less than that of the cell being examined.

Experiment 3. Determination of osmotic pressure of plant cells by plasmolytic method (De Vries' method, 1888). Sucrose solutions are most commonly used for this purpose because the osmotic pressures of different molar concentrations[1] of sucrose have been worked out with considerable amount of accuracy. First, a series of sucrose solutions of different molar concentrations are prepared. Then sections of plant tissues or entire material in particular cases (e.g. *Hydrilla* leaf or *Spirogyra* filament) are immersed in the graded solutions—0.1 M to 1.0 M—for 10 to 15 minutes and observed under a microscope. It will be found that in one particular solution about 50% of the cells have just plasmolysed. This *incipient* plasmolysis indicates that the solution has the same osmotic pressure as that of the cell-sap, and is said to be *isotonic* with it. The isotonic solution having been found, it is now possible to determine the osmotic pressure of the cells. It is known that one molar solution of sucrose (according to Morse, 1914) is equivalent to 26.64 atmospheres of pressure at 20°C. If the isotonic solution in the above experiment be 0.5 M at 20°C, then the osmotic pressure of the cells immersed in that solution would be equal to 13.32 atm. It must be noted that the osmotic pressures of different groups of cells or those of the same group under different conditions vary considerably. Most of the values are, however, commonly within the range of 10-20 atm.

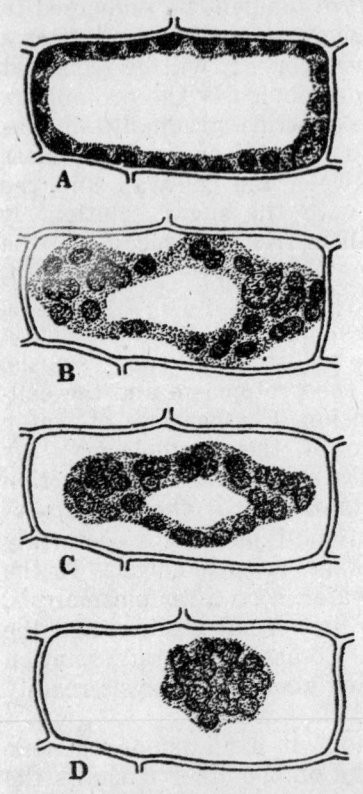

FIG. 3. Plasmolysis in a cell of *Vallisneria* leaf under the action of 10% potassium nitrate solution; *A*, a normal cell; *B-D*, stages in plasmolysis.

---

[1] A molar solution is prepared by dissolving one mol, i.e. one gram molecular weight of a substance in 1 litre (1,000 c.c.) of distilled water at 20°C. The molecular weight of sucrose ($C_{12}H_{22}O_{11}$) is $12 \times 12 + 22 \times 1 + 11 \times 16 = 342$. Thus 342 gms. of sucrose dissolved in 1 litre of water will give a normal molar solution or 1 M solution. This may be diluted to solutions of 0.1 M, 0.2 M, 0.3 M, etc. Molar solutions are commonly used in experiments on osmosis.

Chapter 2   SOILS

Since water and mineral salts are almost exclusively obtained from the soil for their utilization later in the plant body, a knowledge of soil science in different aspects is an essential prerequisite to the study of plant physiology.

Soil Formation. Soils are formed by the disintegration and decomposition of rocks due to weathering (action of rain-water, running streams, glaciers, wind, alternate high and low temperatures, atmospheric gases, etc.) and the action of soil organisms such as many bacteria, fungi, protozoa, earthworms, etc., and also interactions of various chemical substances present in the soil. Although soils are normally formed from underlying rocks in a particular region, these may be transported to long distances by agencies like rivers, glaciers, strong winds, etc.

Physical Nature. Physically the soil is a mixture of mineral particles of varying sizes—coarse and fine—of different degrees, some angular and others rounded, with certain amount of decaying organic matter in it. The soil has been graded into the following types according to the size of the particles :

Coarse particles	..	2 -.2 mm. form coarse sand
Smaller particles	..	.2 -.02 mm. form sand
Finer particles	..	.02-.002 mm. form silt
Very fine particles	..	less than .002 mm. form clay

Types of Soils and their Physical Properties. The property of the soil (both physical and chemical) largely depends on the size of the particles that a particular type of soil is composed of, and is mainly determined by the proportion of clay present in it. On this basis soils may be classified into the following types : (1) **sandy soil** containing more or less 10% each of clay and silt with a large proportion of sand ; (2) **clay soil** containing 40% or more of clay ; and (3) **loam** containing 30-50% of silt, a small amount of clay (5-25%), the rest being sand. The physical properties of soils are : porosity (30-50% of the pore space of the soil volume required for water and air is suitable for most crops), capillary action, hygroscopicity, weight, colour, temperature, quantity of humus present (see p. 284), etc. The important physical properties of the above types of soils are as follows: (1) **Sandy soil** is well aerated, being porous ; but as it allows easy percolation of water through large pore spaces left between its particles it quickly dries up and often remains dry Capillarity decreases in this soil ; such a soil can retain only 25-31 parts of water by weight. The soil is always light. (2) Clay soil on the other hand is badly aerated and it easily becomes waterlogged. It has, however, a great capacity for retaining water (40 parts by weight), the particles being very fine. Drainage in this soil is difficult, while capillarity increases considerably. It is heavy and easily becomes compact, and cracks when dried up. Clay particles are mainly made of oxides of aluminium and are bound up with certain important minerals such as K, Ca and Mg. A considerable amount of plant food is, however, available in this soil. (3) **Loam** is the best soil for vigorous plant growth and is most suitable for agricultural crops because all the important physical conditions are satisfied—porosity for better aeration and for percolation (downward movement) of excess water, and capillarity for upward movement of sub-soil water. It can retain 50 parts of water by weight. At the same time it is rich in organic food. The proportions of the above constituents of the soil can be approximately determined by stirring a small lump of soil in a beaker to which an excess of water has been added, and then pouring the contents into a measuring cylinder. When allowed to settle, it is seen that sand particles collect at the bottom, silt higher up, and clay on the top—these in distinct layers. The fine portion of the clay, however, remains suspended in water. Their proportions are then determined and percentages calculated. Humus (see p. 284) mostly floats on water.

There are other kinds of soils also. Some of the common ones are: *calcareous soil* containing over 20% of calcium carbonate which is useful in neutralizing organic acids formed from humus: it is whitish in colour; the presence of calcium carbonate may be detected by adding strong hydrochloric acid to a small sample of soil when effervescence is noticed in it either with naked eyes or under a pocket lens; *laterite soil* containing a high percentage of iron and aluminium oxides; it is reddish, brownish or yellowish in colour; *peat soil* containing a high percentage (even up to 80% or 90%) of humus; it is dark in colour, porous and light; the floating garden of Kashmir is made of peat soil; it can absorb water to the extent of several times its own weight.

Physical Properties of Soils. The physical properties of soils largely depend on the size of the particles that such soils are composed of. Regarding their sizes the extremes are gravel and coarse sand at one end and clay at the other. Loam (see above) satisfies most of the physical conditions favourable for plant growth. Besides, it contains a good amount of organic matter. The physical properties are :

Porosity. Porosity is a very important factor. Irregularity in the size of the soil particles and their arrangement always leave some space, called *pore space*, between them, however compact the soil may be. Porosity of the soil is essential because it makes room for water and air. It helps percolation of water normally through bigger pores, and capillary retention of water normally through smaller pores. Loam with more or less even distribution of finer and coarser particles is regarded as the best in respect of the above conditions. A good soil contains 30-50% of pore space of the soil volume.

Soil Water. Ordinarily two-thirds of the pore space occupied by water and one-third occupied by air are found to be suitable for normal growth of most crop plants. An excess of water in the soil chokes its pore space and is, therefore, harmful to plants. Conversely a very low percentage of water in the soil results in wilting of plants. The excess water is commonly removed by gravitational pull, drainage and evaporation. The water loosely held by the small soil particles by capillary force, with mineral salts dissolved in it, is the water absorbed by the root-hairs (see pp. 299-300). So the *water-holding capacity* of the soil is of primary importance, and it mainly depends on the fineness of the soil particles. Ordinary agricultural soil takes up about 50 parts of water and this is good enough for normal plant growth.

Soil Air. Free space must be available for diffusion of gases—carbon dioxide and oxygen—through soils, the former to escape into the atmosphere above and the latter to come in close contact with all parts of roots, protozoa, earthworms, etc., in order that they may respire and remain alive and active. Soil air is usually richer in CO_2 and poorer in O_2 than the atmospheric air. But in poorly aerated soil the concentration of CO_2 may be as high as 10% and that of O_2 as low as 10%, as against 0.03% and 20% of CO_2 and O_2 respectively in the atmospheric air. Growth of most plants is retarded under this condition. Proper aeration of the soil is, therefore, a necessity for normal growth of plants.

Capillarity. The capillary power of soils to draw water from below upwards depends on their texture. The maximum rise is exhibited by medium-sized grains such as silt, and not by finer-grained or coarser-grained soils. It has been estimated that over a period of 18 days the capillary rise is 63 cm. in the case of sand, 84 cm. in the case of clay and 252 cm. in the case of silt (loam). The soils were air-dried in all cases. Initially, however, for a period of one hour sand shows more rapid capillary movement.

Experiment 4. Water content of the soil. To find out the water content of the soil the following procedure may be adopted. Collect from a depth of 0.3 to 1 metre a small sample of soil by digging the earth and keep it in a stoppered jar. Take out a small lump from it and weigh it. Heat it at 110°C. for a while, stirring the mass occasionally. All the water will be driven out by then. After cooling take the weight of the soil again. To make sure that all the water has been driven out, heat the soil again. A constant weight will indicate the loss of all the water from the soil. The difference in weight will indicate the quantity of water originally present in the soil. Then calculate the water content on a percentage basis.

Experiment 5. Water holding capacity of the soil by capillarity. Crush a lump of air-dry soil to break up the clay aggregates, but do not grind it. Take a circular brass box (5½ cm. in diameter by 1½ cm. in height) which is perforated at the bottom. Place a filter paper at the bottom of the box, and transfer the soil in small quantities at a time, gently pressing it after each addition until the box is nearly full. Place it in a petri dish and add water to the dish to a depth of 1 cm. After a time add water again, if necessary, to restore the above depth and maintain it. After a period of 12-24 hours weigh the box after wiping the outside of it and deducting the weight of the filter paper. Then heat the box at 110°C. to drive off the water. Cool in a desiccator and weigh it again, deducting the weight of the filter paper. The weight of the box may be determined before or after the experiment. The weight of the soil after saturation with water *minus* that taken after heating will indicate the moisture content of the soil, i.e. the amount of water held by the soil particles. Then calculate the moisture content on a percentage basis.

Chemical Nature. Chemically the soil water contains, dissolved in it, a variety of **inorganic salts** such as nitrates, sulphates, phosphates, chlorides, carbonates, etc., of potassium (K), calcium (Ca), magnesium (Mg), sodium (Na) and iron (Fe), and of the 'trace' elements like boron (B), manganese (Mn), copper (Cu), zinc (Zn), aluminium (Al), molybdenum (Mo), etc. These salts, when analysed, are generally calculated in terms of oxides, and these often occur in the soil in very low percentages—less than 1; 'trace' elements mostly occur in .002-.0001%. Further, many of these remain in the soil in a variety of complex chemical forms and are not available to plants. In nature, however, the nutrient salts are very widely distributed. In the absence of any of the required compounds the plant suffers. A certain amount **of organic compounds,** chiefly proteins and their decomposition products, derived from the waste products of animals and dead bodies of plants and animals as a result of oxidation by soil bacteria and fungi, is present in the soil. **Humus** contains a certain amount of organic food (see p. 284). **Acidity and alkalinity** of the soil are also of considerable importance for normal plant growth (see below).

Acidity and Alkalinity of the soil are no less important for growth and distribution of plants than the availability of plant food in the soil and the physical condition of it. Soils containing a certain amount of lime (calcium carbonate) are alkaline, while soils containing a certain amount of humus, as in marshes and forests, are acid. These conditions may, however, be altered by the addition of one or the other, as the case may be. Some plants grow well in a neutral soil; but others require more or less acid or alkaline condition of the soil. There are some species, *Acacia nilotica*, for example, which are indifferent to this condition of the soil. Most of the field crops, such as maize, barley, tomato, potato, etc., prefer a slightly acid soil; while *Musa, Rhododendron, Erica, Rumex*, etc., require distinctly acid soils for their normal growth. Leguminous crops, however, always prefer a slightly alkaline soil. Some plants such as beet, lucerne, *Asparagus*, etc., grow well in neutral soil. Saline soil is the requirement of certain plants, e.g. seablite (*Suaeda maritima* and *S. fruticosa*), *Salicornia brachiata*, saltwort (*Salsola foetida*), *Acanthus ilicifolius*, etc.

The acidity or alkalinity of the soil may be expressed in terms of pH. The pH value of the soil may be determined by a simple and easy method—the chemical indicator method. This depends upon the colour reaction of the soil given by certain chemical indicators when the pH is within certain limits. For plant growth and for field experiments the most useful range of the pH value of the soil is between 4 and 9. The colour reactions vary from red (pH 4) to deep blue (pH 9) through yellow and green according to acidity or alkalinity. The neutral lies at 7. Paddy grows on silty soils between pH 5.5 and 7.5; potato grows on more sandy soils and can tolerate a more acid condition of pH 4.8; while *Luffa acutangula* can tolerate even a stronger acid soil of pH 4.5. Thus from an agricultural standpoint the pH is of special importance. This gives the agriculturist an indication as to whether he should add lime to the soil or add more organic acidic garbage, acidic phosphate, etc.

Soil Organisms. Various kinds of bacteria and fungi are present in the soil, the former sometimes occurring to the extent of a few million individuals per gram of soil, particularly in the region of organic matter, and many of them are useful agents of soil fertility. Thus nitrifying bacteria convert proteins of dead plants and animals into nitrates, and it is a fact that but for the activity of such bacteria the proteins would have ever remained locked up in the soil as such without being used. Then there are nitrogen-fixing bacteria, ammonifying bacteria, sulphur bacteria and a host of other types in the soil. Fungi are also abundant in the soil, particularly in the acid soil often replacing bacteria. Like the bacteria they are also useful agents in decomposing proteins. Many higher plants, particularly in forests, utilize mycorrhizal fungi (see p. 21) to absorb water and mineral salts from the soil rich in humus. Many algae are also present in the soil. It is now definitely known that many of the blue-green algae fix atmospheric nitrogen in the soil. Among animals the soil-dwellers like many protozoa, earthworms, rats, etc., are useful agents in altering the soil. The burrowing animals make the soil loose for better aeration and percolation of water.

Humus. Humus is a dark-coloured substance present in many soils. It consists of organic (vegetable) matter, mainly cellulose and lignin combined with proteins, in various stages of decomposition in the soil, derived from dead roots, trunks, branches and leaves, under the action of various types of soil bacteria and fungi. Humus usually forms a surface layer, sometimes of some depth, as in forests and swamps. It is of considerable importance to plants both chemically and physically. The nitrogenous organic compounds of the humus are acted on by various bacteria and fungi and finally converted into nitrates which are absorbed by plants. Humus, therefore, is a source of plant food. Physically, however, it is more important since it gives the soil a loose texture ensuring better aeration, and being colloidal in nature like clay particles it has also a great capacity for imbibing and retaining water to the extent of 190 parts of its own weight. Thus, added to sandy soil it increases its water-holding capacity and added to clay soil it loosens its compactness and increases porosity for better aeration. Soil containing 5-15% of humus is suitable for agricultural crops. It is the seat of most of the bacterial processes in the soil.

Experiment 6. Humus content of the soil. To find out the humus content of the soil proceed as follows (ignition method). After heating a lump of soil at 110°C. to drive off the water cool it in a desiccator and then take its weight. Next in a platinum crucible burn the dehydrated soil at a high temperature for about an hour, occasionally stirring the mass. During ignition fumes are seen to escape. (Organic matters become converted into ammonia, oxides of nitrogen or free nitrogen, sulphur dioxide and carbon dioxide and escape as such). After complete combustion cool it in a desiccator and then weigh it again. The loss in weight almost approximately represents the quantity of humus originally present in the soil sample. Then calculate the humus content of the soil on a percentage basis. The residue left after combustion is the incombustible or inorganic matter present in the soil.

Fertility of Soil. A soil may be regarded as fertile when all the conditions—physical, chemical and biotic—are satisfied. Absence of any one of them acts as a limiting factor and thus affects normal growth of the plant, and the crop as a whole suffers. Composition of the soil lying within the following limits (given in terms of volumes) may be considered good: mineral particles —50-70%, pore space (containing water and air)—30-50%, and organic matter (humus)—5-15%, with the following essential elements occurring in the following proportions: N—0.1-0.5%, P—0.08-0.5%, K—1.5-3.0%, Ca—0.1-2.0%, Mg—0.3-1.0%,

S—0.01-0.14% and Fe—a trace. Oxygen and carbon (the latter as CO_2) are of course obtained from the air.

Fertilizers. Ordinarily the soil contains the necessary salts required by plants. Deficiency, however, sometimes occurs in one or more of them, particularly in nitrogen, phosphorus, potassium and calcium, mainly due to gravitational pull of the soil water, heavy drainage, and intake by roots, and to make good this deficiency the use of fertilizers or manures becomes a necessity. Fertilizers are certain chemical substances which when properly added to the soil make it fertile, i.e. enable it to produce more abundantly. Production may be doubled or even trebled by proper use of chemical fertilizers. Manuring of the field for better crop production may be done by any of the following three methods. (1) Artificial manuring is done by introducing into the soil particular compounds or their mixtures in suitable proportions according to deficiencies. Commonly ammonium sulphate, urea, superphosphate, leaf-compost, bonemeal, oil-cakes, etc., are used as chemical fertilizers. (2) Farmyard manuring is done by adding decomposed cowdung and organic refuses to the soil. (3) Green (natural) manuring is done by growing one or more types of leafy vegetables, preferably mixed with certain nodule-bearing leguminous plants (see FIG. 5A) and finally ploughing the whole lot into the field.

Sulphate of ammonia is now extensively used as a chemical fertilizer for many of the field crops like rice, barley, potato, sugarcane, tea, orange, cabbage, cauliflower, turnip, mustard, etc. This chemical becomes quickly nitrified in the soil in course of a few days and changed to calcium nitrate. It however, makes the soil acid and, therefore, unsuitable for many crops. The remedy, however, lies in adding lime to the soil. Ammonium sulphate is not washed out of the soil even by torrential rains. Nitrate of soda is also another source of nitrogen for various field crops.

Urea, $CO(NH_2)_2$, is a good source of nitrogen for normal, even vigorous, growth of many higher plants as well as a large number of soil bacteria and fungi. Urea

---

The Fertilizer Corporation of India has set up a number of Fertilizer Units in the country to produce large quantities of chemical fertilizers such as urea, ammonium sulphate, ammonium nitrate and phosphates in order to secure maximum agricultural production to feed her growing population. The production of fertilizers has rapidly increased to 26.70 lakh tonnes (1977-78). This meets 70-80% of her domestic needs. The six main operating units are (1) the Sindri Unit (1951) in Bihar, (2) the Nangal Unit (1961) in Punjab, (3) the Trombay Unit (1965) in Maharashtra, (4) the Gorakhpur Unit (1969) in Uttar Pradesh, (5) the Namrup Unit (1969) in Assam, and (6) the Durgapur Unit (1974) in West Bengal. The new projects are the Korba Unit in Madhya Pradesh, the Panki Unit in Uttar Pradesh, the Talcher Unit in Orissa, the Ramagundam Unit in Andhra, the Barauni Unit in Bihar, and the Haldia Unit in West Bengal. At present India uses only about 16 kg. of fertilizers per hectare per year, as against 314 kg. or even much more in most of the advanced countries.

is present in small quantities in some seed plants but it is fairly abundant in some fungi. Urea may originate in the tissues from the amino-acid *arginine* (formed from ornithine). In the soil urea is rapidly hydrolysed to ammonia and carbon dioxide by the action of the enzyme *urease*. Some plants may, however, directly absorb urea in small quantities, and utilize it to form some amino-acids and finally protein in their tissues. Radioactive C^{14} or N^{15} used in urea has been traced in certain amino-acids and proteins. Urea is, therefore, a valuable fertilizer, containing 46% of nitrogen Foliar application of urea, as shown by Webster (1955), is also effective in many cases.

Chapter 3 CHEMICAL COMPOSITION OF THE PLANT

The various elements that have entered into the composition of the plant body may be determined by **chemical analyses**, and those essentially required by the plant determined by **water culture experiments.**

1. **Chemical Analyses.** By chemical analyses of a plant we can find out the various elements that have entered into its composition. For this purpose a representative sample of the plant (i.e. a sample representing all parts of the plant body) is taken and it is dried at 110°C. or so. All the water that the plant contains is thus driven off. Then, by careful weighing, the proportion of water to the total weight of the plant is determined. Plants in general are found to contain a high percentage of **water**—in woody parts about 50 per cent, in soft parts about 75 per cent, in succulent parts from about 85 to 95 per cent, and in water plants 95 to 98 per cent. When the plant is *charred* we get charcoal. The main bulk of this charcoal is **carbon**; in fact, almost half the dry weight of the plant is carbon. The dried plant is then carefully burnt over a flame at a temperature of about 600°C. On thus burning, the **organic compounds** such as the proteins, carbohydrates, fats and oils, etc., which constitute often over 90% of the dry matter of plants, being combustible, are converted into carbon dioxide, water vapour, sulphur dioxide and ammonia or free nitrogen, and escape as such. These gases may be collected by proper methods and their composition studied. *Proteins* when analysed are seen to contain **carbon (C), hydrogen (H), oxygen (O), nitrogen (N),** and often **sulphur (S),** and **phosphorus (P)**; *carbohydrates* and *fats and oils* contain only the first three elements. The residue left after the above treatment consists only of **inorganic compounds** which are incombustible, and is known as **ash**. The percentage of ash varies

greatly in different plants and also in different parts of the same plant, usually lying within the range of 1%-15%. Analyses of the ash show that, of the 92 well-known chemical elements occurring in nature, about 40, possibly more, are present in it. Most of these elements occur in very minute quantities and their presence too is not very constant. The following, however, are constantly found in the ash of the plant, occurring of course in varying proportions in different plants: potassium (K), calcium (Ca), magnesium (Mg), iron (Fe) and sodium (Na) among the metals, and sulphur (S), phosphorus (P), chlorine (Cl) and silicon (Si) among the non-metals. In addition certain other elements found in the ash in traces only are: boron (B), manganese (Mn), zinc (Zn), copper (Cu) molybdenum (Mo), aluminium (Al), etc.; these are known as 'trace' elements (see pp. 297-98).

Chemical analyses of the plant body (including the combustible material and the ash) show that, of the various elements present in it in easily detectable and measurable quantities, the following 13 elements are constant in all plants: potassium, calcium, magnesium, iron and sodium among the metals, and carbon, hydrogen, oxygen, nitrogen, sulphur, phosphorus, chlorine and silicon among the non-metals. Besides, some of the 'trace' elements which are now known to be constant in green plants are: boron, manganese, zinc, copper and molybdenum. Aluminium, though not constant, is also very widespread in plants. The average chemical composition of the plant body may be given thus: carbon—45.0%, oxygen—42.0%, hydrogen—6.5%, nitrogen—1.5% and ash—5.0% (after Maximov).

2. **Water Culture Experiments.** Water culture experiments are carried out to ascertain which elements are essentially required by plants for their normal growth, and which only incidentally absorbed. These experiments further help us to understand the forms (chemical compounds) in which they are best taken up, the particular concentration of the solute, and the source of supply (soil or air) of these elements. Water culture experiments consist in growing some seedlings in water containing some known salts in particular proportions, known as **normal culture solution**, and studying the effect produced on them (seedlings) regarding growth and development. Normal culture solutions of various compositions are used (Sachs 1860, Knop 1865, Pfeffer 1887). The following composition has been worked out by Knop as forming the normal water culture solution, that is, the solution required by the seedlings for their normal growth.

Knop's normal culture solution

Potassium nitrate, KNO_3	1 gm.
Acid potassium phosphate, KH_2PO_4	1 gm.
Magnesium sulphate, $MgSO_4$	1 gm.
Calcium nitrate, $Ca(NO_3)_2$	4 gms.
Ferric chloride solution, $FeCl_3$	a few drops
Water	1,000 c.c.

This is a stock solution of 0.7% strength. To make a 0.1% solution, which is suitable for water culture experiments, add 6,000 c.c. of water to the stock solution.

Experiment 7. Water culture experiments. A series of bottles or jars of the same size and shape are fitted each with a split cork. A number of seedlings of the same kind and more or less of the same size are taken. The bottles marked A, B, C, D, etc., are filled with culture solutions of known composition. Through the split cork a seedling is introduced into each bottle. The bottles are wrapped with black paper and exposed to light. Arrangements should be made for proper aeration of the roots. It is desirable that the culture solution should be renewed fortnightly. The following table shows the nature of solutions used and the effect produced on the seedlings.

Solutions used	Observations
A with normal culture solution	Growth of the seedling is normal.
B the same minus potassium salts	Growth becomes checked; leaves lose their colour; the seedling withers; and carbohydrate formation is slow.
C the same minus calcium salts	Root system does not develop properly; leaves become yellowish, spotted and deformed; and the seedling becomes short and weak, and is liable to be easily diseased.
D the same minus magnesium salts	Chlorophyll is not formed; the seedling becomes stunted in growth; and carbohydrate formation is slow.
E the same minus iron salts	Seedling becomes chlorotic.
F the same minus phosphorus compounds	Growth is slow and the seedling begins to weaken.
G the same minus sulphur compounds	Leaves yellowish and stem slender.
H the same minus nitrogen compounds	Seedling is weak and straggling, and leaves yellowish.

Inference. Water culture experiments prove conclusively that a plant can grow satisfactorily only when it is supplied with K, Ca, Mg, Fe and H, O, N, S, P. These experiments thus help us to understand that these elements (together with C) are essential, while others are non-essential being only incidentally absorbed; they show further that these elements are absorbed in soluble compounds, in suitable proportions and in very dilute solutions, occurring in the soil; that free oxygen and carbon dioxide are obtained from the air (and not from the soil)—oxygen for the respiration of the living cells and carbon dioxide for the manufacture of food by the green cells; that free nitrogen of the air is of no use to the plant; and that the plant must be exposed to light. Chemical analyses give us no clue to any of the above-mentioned facts.

Sand or Charcoal Culture Experiments. To obviate many difficulties in water culture experiments, it has become the growing practice with physiologists to take to sand or charcoal culture. Charcoal is thoroughly washed and powdered. In the case of sand, it is washed, dried and then ignited to remove organic impurities. Normal culture solution is added to any of the two media and growth of the seedling studied. The effect produced on the seedling under the exclusion of a particular element is studied in the same way as in water culture experiments.

FIG. 4. Water culture experiments; *left*, in normal solution; *right*, in the same minus one of the essential elements.

Essential and Non-essential Elements. Chemical analyses of the plant reveal the presence of a long list of elements in it, while water culture experiments prove that *ten* elements (including carbon which is obtained from the air) are essential for normal growth of all plants. Of the trace elements boron, manganese, zinc, copper and molybdenum are also now regarded as essential (see pp. 297-98). Thus the total number of elements now considered essential are 15 (see list below). Other elements present in the plant body are non-essential. It should, however, be noted that certain plants require for their normal growth one or more elements other than the established 15 essential elements.

Classification of Elements
Essential: metals—K, Ca, Mg and Fe.
non-metals—C, H, O, N, S and P.
Non-essential: metal—Na.
non-metals—Cl and Si.
Trace (essential): metals—Mn, Zn, Cu and Mo.
non-metal—B.

Role played by the Elements in the Plant Body. (1) Potassium is abundantly present in the growing regions. It is essentially a constituent of the protoplasm and is closely connected with its vital activity. It is, however, absent from the nucleus and the plastids. Potassium is known to act as a catalytic agent in the synthesis of carbohydrates and proteins; starch grains are not formed in the absence of potassium. Potassium helps the growth of the plant and enables it to produce healthy flowers, seeds and fruits. In the absence of potassium the growth of the plant is checked; the stem becomes slender and the leaves lose their colour and gradually wither. Potassium nitrate and potassium chloride are the usual forms in which potassium is absorbed by plants. Water lettuce (*Pistia*), when burnt, yields about 12% of potash, and is used as a valuable manure.

(2) **Magnesium** helps in the synthesis of phosphorus-containing lipid substances (phospholipids) which are important constituents of protoplasm. It is present in the chlorophyll to the extent of about 5.6% by weight and, therefore, in its absence chlorophyll is not formed, and the plant becomes stunted in growth. It is present to a considerable extent in the seeds of cereals and leguminous plants.

(3) **Calcium** is always present in green plants. It occurs in the cell-wall, particularly in the middle lamella, as calcium pectate. It is useful in neutralizing acids which would otherwise have a toxic effect on plants. It helps to maintain the semi-permeability of the protoplasm. It is also anti-toxic to various poisonous substances. It also promotes the growth of roots. Plants like lemon, orange, shaddock, etc., grow well in a soil rich in calcium (lime). Fruits in general, and stone-fruits in particular, require plenty of calcium (lime) for their normal development. The stone of the stone-fruit very often does not form in the absence of lime in the soil. In general plants become stunted in growth in the absence of calcium. Many plants, however, cannot stand a high amount of calcium in the soil, and they become chlorotic in consequence.

(4) **Iron** is essential for the formation of chlorophyll although it is not present as a constituent. It may be associated with the plastids. Iron is always present in the protoplasm and in the chromatin of the nucleus.

(5-6) **Sulphur and Phosphorus.** Sulphur is a constituent of an amino-acid, cystine, which is one of the compounds forming plant protein. It is an important constituent of mustard oil. It is contained in the living substance, viz. the protoplasm. In its absence leaves become chlorotic and the stem slender. Phosphorus is always present in the nucleoprotein, a constituent of nucleus, and in lecithin, a constituent of protoplasm; it promotes nuclear and cell divisions, and is concerned in carbohydrate breakdown in respiration. Phosphorus aids nutrition and hastens maturity and ripening of fruits, particularly of grains. It promotes the development of the root system. Underground

CHEMICAL COMPOSITION OF THE PLANT

organs like radish, beet and potato require phosphorus for their normal development. Sulphur is absorbed as sulphates of some metals and phosphorus as phosphate of calcium or potassium.

(7) **Carbon** forms the main bulk—45% or even more—of the dry weight of the plant. It is the predominant constituent of all organic compounds which are, in fact, known as compounds of carbon. Carbon is absorbed from the atmosphere as carbon dioxide. Although carbon dioxide occurs in the air to the extent of only 0.03%, still air is the only source of all the carbon for the plant, as proved by water culture experiments. It is to be noted that there is a regular circulation of carbon dioxide and oxygen between the plant and the atmosphere, and two processes are connected with it: one is photosynthesis and the other is respiration. In photosynthesis *green* plants take in carbon dioxide from the atmosphere *during the daytime* and give off oxygen. (The oxygen that is given off in the process is, however, released from water). There is thus a tendency of the atmosphere to become poorer in carbon dioxide and richer in oxygen. In the reverse process, i.e. in respiration, *all* plants and animals take in oxygen from the atmosphere *at all times* and give off carbon dioxide. In the combustion of coal and wood also carbon dioxide is given out to the atmosphere. Thus the atmosphere has a tendency to become richer in carbon dioxide and poorer in oxygen. It is evident, therefore, that by these two processes the total volumes of these gases are kept constant in the air. The circulation of carbon by the above two processes through the green plants (and animals and non-green plants, and also by the chemical combustion of non-living material, e.g. coal) and the atmosphere is spoken of as the **carbon-cycle**.

(8) **Nitrogen.** Although nitrogen occurs to the extent of about 78 parts in every 100 parts of air by volume, it is not as a rule utilized by plants in its free state. It may enter the plant body through the stomata with other gases, but it comes back unused. Although nitrogen is so abundant in the air, it occurs in the dry substance of the plant to the extent of 1-3% only. Nevertheless, it is indispensable to the life of the plant, as it is an essential constituent of proteins, chlorophyll and protoplasm. Nitrogen is essential for growth, more particularly of the leaves. Leafy herbs like lettuce suffer considerably in the absence of nitrogen in the soil. In the absence of this element leaves become yellowish. An excess of nitrogen causes vigorous growth of vegetative parts, specially the leaves, but delays reproductive activity. Plants become readily susceptible to the attacks of fungi and insects in an excess of nitrogen.

Nitrogen of the Soil. The amount of nitrogen in the soil varies from 0.096 to 0.21% (average Indian soil contains about 0.05% of nitrogen); still the soil is the main source of nitrogen for the plant. Here it exists as both inorganic and organic compounds. The chief forms of *inorganic compounds* are the nitrates and nitrites of potassium and calcium, and also ammonia and ammonium salts, e.g. ammonium carbonate—$(NH_4)_2CO_3$; while the *organic compounds* are mainly the decomposition products of proteins and also urea. A portion of the ammonia gas formed in the soil as a result of putrefaction of the above compounds may diffuse into the air, but most of it combines with other substances and forms ammonium salts, e.g. ammonium carbonate. Ammonia or ammonium salt is normally made available for the use of green plants after conversion into nitrate by the action of certain micro-organisms—the nitrifying bacteria—which live in the soil. This process of conversion is called **nitrification**, and it represents a very important phase in the nitrogen cycle (FIG. 6). In this process, as shown by Winogradsky in 1889, the ammonia or ammonium salt in the soil is oxidized into nitrate in two stages: (*a*) ammonia or any of its salts is first acted on by the nitrite-bacteria (*Nitrosomonas*) and oxidized into nitrite (NO_2), e.g. KNO_2 or $Ca(NO_2)_2$; and (*b*) the nitrite thus formed is then acted on by the nitrate-bacteria (*Nitrobacter*) and further oxidized into nitrate ($-NO_3$), e.g. KNO_3 or $Ca(NO_3)_2$. The nitrate thus produced is readily absorbed by the green plants. In acid soils, however, ammonia is the chief form in which nitrogen is readily absorbed by many higher plants. Most bacteria and some fungi and algae readily assimilate ammonia. It is also to be noted that a portion of the nitrates (and also nitrites) may be acted on by certain anaerobic bacteria—the denitrifying bacteria (*Bacterium denitrificans*, for example)—and reduced to free nitrogen (N_2) which then escapes into the air. The process is called **denitrification**. The denitrifying bacteria being anaerobic, their activity is not much in evidence in well-aerated soils, as in well-ploughed fields.

The chief forms of organic compounds of nitrogen are amino-acids, amines and also urea (see p. 285). Dead bodies of animals and plants containing various proteins are decomposed by several groups of putrefying (ammonifying) bacteria and certain fungi present in the soil. In the first stage, *in the absence of oxygen*, the proteins are reduced to amino-acids and then to ammonia (**ammonification**) by the putrefying bacteria and fungi; and in the second stage, *in the presence of oxygen*, the ammonia undergoes nitrification, as stated before. The nitrate, thus produced, is readily absorbed by green plants.

Test for Nitrates. The presence of nitrates in the plant tissue or in the soil

CHEMICAL COMPOSITION OF THE PLANT

is easily detected with diphenylamine A 0.5% solution of it in strong sulphuric acid turns nitrates blue.

Ammonia of the Air. It has been suggested that ammonia of the air may be an important source of nitrogen for the soil. It is absorbed by some of the constituents of the soil, nitrified there and made available to plants in the form of nitrate. It is a known fact that acid soils always absorb ammonia from the air. Sea water releases ammonia during evaporation, which soon condenses on the surface of the soil, particularly cultivated soil. Ammonia of the air may, therefore, be regarded as one of the sources of nitrogen for the soil, particularly in the neighbourhood of the sea.

Fixation of Atmospheric Nitrogen. Under certain circumstances the gaseous nitrogen of the air may combine with other elements and is ultimately made available to the plants as compounds of nitrogen in the soil. The methods by which nitrogen may be fixed are as follows: *physico-chemical*—(1) discharge of electricity in the atmosphere (Boussingault 1837); *bio-chemical*—(2) activity of certain saprophytic bacteria (Winogradsky 1893, Beijerinck 1901); (3) activity of symbiotic bacteria (Hellreigel and Wilfarth 1887); and (4) activity of blue-green algae (P. K. De 1944).

1. Nitrogen Fixation by Electric Discharge. The free nitrogen of the air to some extent becomes available to the green plants by the discharge of electricity (lightning) during a thunderstorm. Under the influence of electricity nitrogen of the air combines with oxygen to form nitric oxide—$N_2+O_2=2NO$ (nitric oxide). This nitric oxide at once unites with oxygen of the air and forms nitrogen peroxide—$2NO+O_2=2NO_2$ (nitrogen peroxide). The nitrogen peroxide, thus produced, is then dissolved by falling rain forming nitrous acid (HNO_2) and nitric acid (HNO_3)—$2NO_2+H_2O=HNO_2+HNO_3$, and washed down into the soil. Here they combine with some metal like potassium or calcium, and form respectively nitrite and nitrate of potassium or calcium. Nitrate is directly absorbed by plants; while nitrite is oxidized into nitrate by nitrate-bacteria. On an average rain-water brings down to the soil about 4 kilograms of nitrogen per year per hectare.

2. Nitrogen Fixation by Saprophytic Bacteria of the Soil. Various types of nitrogen-fixing bacteria present in the soil have the power of fixing free nitrogen of the soil-air in their own bodies in the form of amino-acids and finally building up proteins from them. After the death of these bacteria the proteins are released to the soil. In due course these are acted on by the nitrifying bacteria and finally transformed into nitrates which are then made use of by the green plants. But it must be noted that the amount of free nitrogen fixed by the saprophytic bacteria is much less than that fixed by the symbiotic bacteria. There are two distinct groups of saprophytic bacteria —aerobic and anaerobic. Several species of *Clostridium* (anaerobic) first discovered and named by Winogradsky (1893) and *Azotobacter* (aerobic) first discovered and named by Beijerinck (1901) are typical of these two groups. These bacteria are widely distributed in soils. The efficiency of nitrogen-fixation by these bacteria depends on the oxidation of carbohydrates (particularly sugars) in the soil as a source of energy. The chemistry of nitrogen fixation representing different intermediate stages is not, however, definitely known. But it is certain that molecular nitrogen is reduced to ammonia (NH_3). Ammonia is toxic to plants and, therefore, it occurs in a very dilute solution and is rapidly synthesized into some form of amino-acid (e.g. glutamic acid).

3. Nitrogen Fixation by Symbiotic Bacteria: Nodule Bacteria of *Leguminosae*. Agriculturists have noted for a long time that leguminous plants such as pulses grown in a soil make it fertile and lead to an increase in the yield of cereals. On an experimental basis Boussingault first proved in 1851-52 that leguminous plants

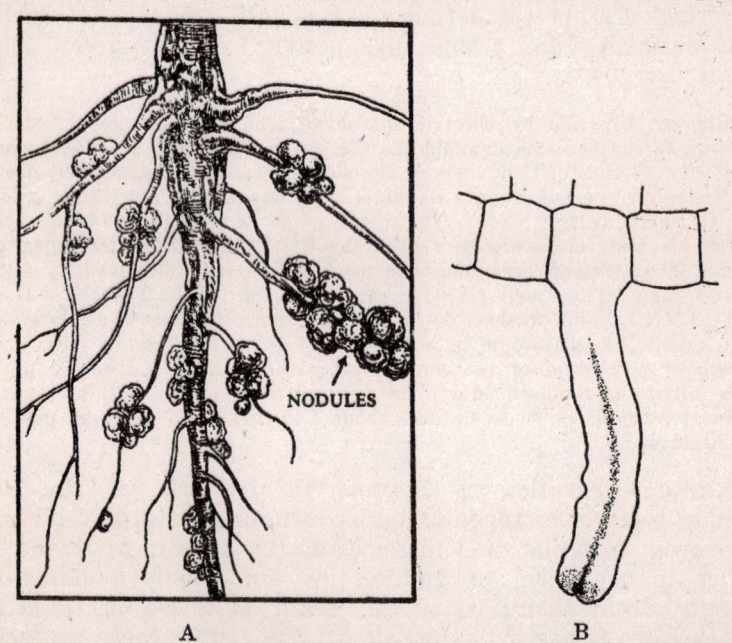

FIG. 5. *A*, nodules of a leguminous plant; *B*, a root-hair infected with bacteria. Note the bacterial thread.

use free nitrogen (N_2) of the air for their normal growth. It was later discovered by Hellriegel and Wilfarth in 1887 that the roots of these plants possess some swellings, called **nodules** or **tubercles**, which are infected with some types of nitrogen-fixing bacteria, particularly the different species of *Rhizobium*, and these bacteria have the power of fixing the free nitrogen of the soil air in the said nodules. It is to be particularly noted that neither the leguminous plants nor the bacteria can fix nitrogen by themselves. It is now known that such bacteria are present in the nodules of most plants (but not all) of *Leguminosae*, particularly of *Papilionaceae*, and in the roots of a few other plants. The mode of infection of the root by these bacteria and of nodule formation is as follows. Bacteria enter through the tip of the root-hair. After penetrating into it they form into a sort of thread consisting of innumerable bacterial cells held together by mucilage. This thread passes down the hair and reaches the cortex of the root perforating the cell-walls. Bacteria then multiply in number and colonize the cortex. Cortical cells are stimulated to grow, perhaps due to the secretion of some stimulant by these bacteria, and thus give rise to small swellings or nodules of varying sizes and fix in them the nitrogen of the air in the form of some amino-compounds. The molecular nitrogen is first reduced to ammonia which then rapidly changes to certain amino-acids (e.g. glutamic acid). A portion of the amino-compounds thus formed is absorbed into the plant body, another portion is excreted out of the nodules and the remaining portion remains locked up in the nodules. Thus the soil becomes richer in nitrogen, more particularly so, if the nodule-bearing leguminous plants are ploughed into the soil. The leguminous plants supply the bacteria with carbohydrates, and the bacteria supply the former with nitrogenous food; so this is a case of **symbiosis** (see p. 20). It is, however, to be noted that the intermediate chemical changes leading to the formation of amino-compounds in the nodules are not clearly understood.

4. Nitrogen Fixation by Blue-green Algae. It is now definitely known that certain members of Myxophyceae, particularly several species of *Nostoc* and *Anabaena*, which are common in many soils, apart from their aquatic habitat, have the power of fixing free nitrogen of the air. A part of nitrogenous compounds fixed in their body is excreted into the surrounding soil, while the remaining part is released to the soil after their death. In tropical agricultural soils, as in water-logged rice-fields, these algae are particularly common and they contribute to the fertility of such soils.

Nitrogen Cycle (FIG. 6). Although plants are continually absorbing compounds of nitrogen from the soil it should not be supposed that the nitrogen contents of the soil would sooner or later become exhausted. Under natural conditions the soil soon becomes replenished of this element. This is so because of the fact that there is a regular circulation of nitrogen through the air, soil, plants and animals. Nitrogen in the soil is, therefore, inexhaustible. We have already seen how the free nitrogen of the air is brought down into the soil as ultimate products of nitrite and nitrate of some metals.

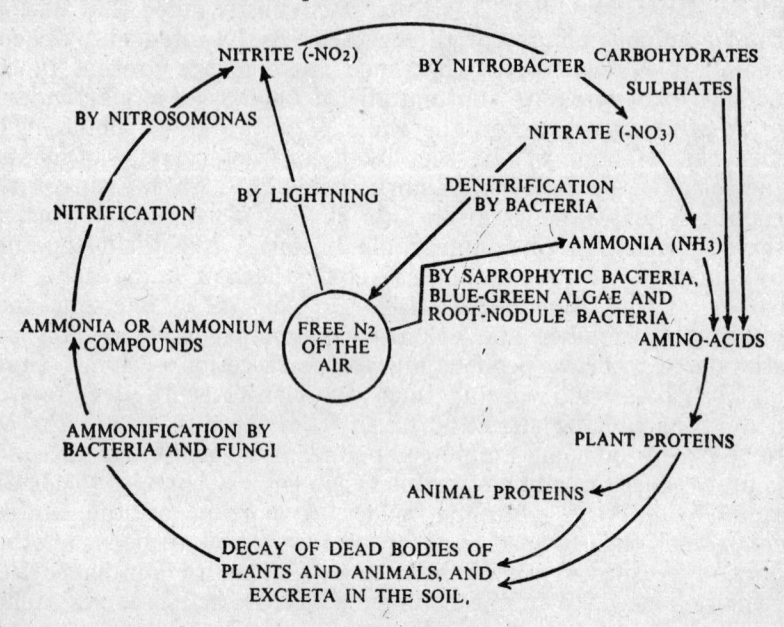

FIG. 6. Nitrogen cycle.

Nitrates are absorbed by plants and reduced to ammonia and then to amino-acids. Finally proteins are made out of them. Plant proteins are taken up by animals. After the death and decay of animals and plants the proteins and also other nitrogenous organic compounds contained in their body are again converted into nitrates in several stages (ammonification and nitrification) and absorbed as such by plants again. At the same time a portion of the nitrates and also nitrites present in the soil is disintegrated by denitrifying bacteria into free nitrogen or oxides of nitrogen which then escape into the surrounding air. This free nitrogen is again brought down into the soil from the air. It will be noted that ammonia holds a key position in protein metabolism, i.e. in the synthesis of proteins as well as in their breakdown. Ammonia is, however, toxic to plants and, therefore, it becomes quickly metabolized.

Rotation of Crops. The fixation of atmospheric nitrogen in the soil is of very great agricultural importance. Most crops absorb the nitrogenous compounds from the soil and impoverish it. Leguminous plants, on the other hand, enrich it in nitrogen when their nodule-bearing roots are left in the soil. Thus leguminous crops such as pulses, *Sesbania cannabina* (B. DHAINCHA), cow pea (*Vigna sinensis*) etc., are grown in the field in rotation with the non-leguminous crops such as cereals (rice, wheat, maize, barley, oats, etc.) and millets. For the same reason certain leguminous plants—*Tephrosia* and *Derris*, for example—are grown in tea gardens as natural fertilizers (and also for shade). Root crops such as turnip, radish, beet, etc., take plenty of potash, calcium and nitrogen from the soil.

Trace- or Micro-elements. It is now definitely known that at least five 'trace' elements such as boron, manganese, zinc, copper and molybdenum are essential for normal growth of plants. The absence of any of them in the culture solution or in the soil leads to abnormal growth and to certain plant diseases. It may also be mentioned that aluminium though not recognized as essential is very widely distributed among plants. The normal culture solution, however, does not include any of the 'trace' elements, and still the plant grows. What is the explanation? It is likely that some of the so-called pure chemicals used by the early workers contained traces of these elements. Distilled water commonly used in water culture experiments might have contained traces of them. Traces of some of these elements might have dissolved out of the glass bottles and other vessels containing the chemicals and the solutions. Besides, seeds themselves are likely to contain these elements in their cotyledons or endosperm. Thus the sources of contamination of the water culture solution being many, such elements crept into the solution undetected. In recent times by using extra-pure chemicals, re-distilled water and special glass bottles it has been proved beyond doubt that the five elements, mentioned above, are indispensable and essential for all green plants.

Boron. This is possibly required by all plants. The beneficial effect of boron has been proved in a number of cases, e.g. maize, tomato, potato, tobacco, lemon, beet, turnip, mustard, cotton, etc. Cauliflower is in particular need of boron, while cereals have a very low requirement for it. Boron helps the formation of root-nodules in leguminous plants, and it improves the yield of sugar in beet. In its absence beet suffers from 'heart rot', tobacco from 'top rot', potato from 'leaf roll'. Fruits like apple and pear also suffer in its absence. In general, in its absence the growth of the plant is retarded and the leaves become spotted. Its deficiency or absence particularly affects the storage organs of plants (not so much the green tissues) and also the apical meristems, i.e. the root-tip and the stem-tip, which become brittle and die off.

Manganese. Absence of this element or its deficiency results in drying up of leaves, weak growth of the plant, poor bloom, chlorosis and certain diseased conditions of leaves. There is always an appreciable amount of manganese in orange, lemon and tomato. There is a relationship between manganese and oxidation enzymes. Manganese is also connected with the synthesis of chlorophyll. Cabbages require manganese, while, as stated above, cauliflower is in need of boron. In pines and allied plants there is comparatively a high percentage of manganese. This element particularly benefits the leguminous plants, cereals and potatoes.

Zinc. Absence of zinc results in stunted growth of the leaf and the shoot, mottling of leaves, drying back of the growing tips and also various physiological diseases. Cells of leaves also do not utilize carbohydrates in respiration in the absence of zinc. The beneficial effect of zinc has been already proved in a number of cases, e.g. cereals, lettuce, pea, bean, lupin, beet, potato, kohl-rabi, tomato, and many fruit trees. *Citrus* fruits are particularly benefited by it. Zinc occurs more abundantly in green tissues than in other parts and helps the formation of chloroplasts, as proved by Reed in 1935. Zinc also helps synthesis of auxin (indole-acetic acid) and several enzymes.

Copper. Plants deficient in copper lack in chlorophyll-formation. Barley, wheat and oat grains do not form in the absence of this element in the root medium. The beneficial effect of copper has also been already proved in flax, carrot, pea, bean, tomato, etc. Deposit of copper after spraying helps the formation of starch in the underlying tissues. Copper is a constituent of certain enzymes. Except very dilute solutions copper salts are highly toxic.

Molybdenum. Molybdenum is known to be essential for plant growth. The first sign of the deficiency of this element is the formation of chlorotic or necrotic areas in the leaf. It has been claimed that this element is required in the cells for the reduction of nitrate to ammonia for protein synthesis. On this basis it is considered to enter into the composition of enzymes. There is also evidence that this element is required for nitrogen-fixation by *Azotobacter* and *Rhizobium*.

Aluminium. Aluminium has been found in the ash of many plants in a very small percentage, specially in wheat, maize, rye, bean, lentil, carrot, cabbage, turnip, lettuce, sunflower, etc. It is, however, found in a large quantity in the ash of *Lycopodium*. Aluminium is found in almost all parts of the plant body, more so in the root and the leaf. It occurs mainly in the protoplasm and the nucleus. Aluminium in very low concentration stimulates growth, while in higher concentration it is toxic. It influences the colour of the flower.

Chapter 4 ABSORPTION OF WATER AND MINERAL SALTS

Roots and leaves are the main absorbing organs of plants. Roots absorb water and dissolved mineral salts from the soil, and leaves take in oxygen and carbon dioxide from the atmosphere.

Water and Inorganic Salts. Green plants absorb water and inorganic salts from the soil by the unicellular root-hairs which pass irregularly through the interstices of the soil particles and come in close contact with them. Absorption is also actively carried on by the tender-growing region of the root. Maximum absorption of soil water takes place through the root-hairs and also the zone of cell enlargement, while maximum absorption of inorganic salts takes place through the zone of cell division (see FIG. I/3). Water is absorbed in large quantities, always in excess of the requirements of the plant. Small quantities of various soluble inorganic salts such as nitrates, chlorides, sulphates, phosphates, etc., dissolved in the soil water are absorbed in a state of *very dilute solution*. It must, however, be noted that the absorption of water and that of salts are independent of each other. While a large volume of water is absorbed by most plants, the intake of salts may be comparatively small. The absorption of water is not correlated to the accumulation of salts in the cells. The absorbing surfaces of cells must be such as to allow ready passage of water and dissolved salts through them. In this connexion the membranes of an absorbing cell may be recalled to mind: the membranes are the cell-wall, the plasma membrane (ectoplasm) and the tonoplasm. The cell-wall is easily permeable to water and is also minutely perforated, while the other two membranes although very thin and delicate and possibly made of phospholipids possess the property of selective permeability.

The substances that are absorbed from the soil may be classified into two groups: the first group consists of water (and also sugars so far as other absorbing cells of the plant body are concerned) which undergo no or little ionization and they may enter the cells by following the simple laws of diffusion and other physical processes, while the second group consists of mineral salts which undergo extensive ionization. The ionized particles of such salts are taken up by the cells where they accumulate, sometimes in heavy concentration; the ions may travel as such or they combine into suitable compounds.

Availability of Soil Water. A portion of the soil water moves downward in response to the force of gravity, rapidly through sandy soil and slowly through loam or clay soil, and carries down or sometimes even washes out a considerable quantity of essential food elements. This moving water percolates through the interspaces of large soil particles and is not of any use to the plant as root-hairs cannot absorb it. Then again each soil particle holds some water on its surface as an extremely thin film by the force of imbibition (see p. 272); this water is known as *hygroscopic water*. It is held so tenaciously by the soil particle that the root-hair cannot dissociate it from the particle. This hygroscopic water also is not of any use to the plant. Surrounding each soil particle there is a thin or sometimes thick film of water, loosely held to it by capillary force; this is known as *capillary water*. It also occurs in the spaces between the soil particles. This capillary water together with the various nutrient salts dissolved in it can be absorbed by the root-hairs and is, therefore, regarded as the principal source of supply of water to the plant. Capillary water may move from particle to particle in any direction. If this capillary water diminishes in quantity the plant suffers and even death may occur due to wilting. It must,

---

Gases. Of the various gases present in the air it is only the oxygen and the carbon dioxide that diffuse into the plant body and are finally utilized by the plant. Other gases may similarly diffuse into the plant body, but they are returned unused. Oxygen is utilized by all the living cells of the plant for respiration at all times; but carbon dioxide is utilised by only the green cells for the manufacture of carbohydrates during the daytime only.

[1]Composition of the Air. Of 100 parts of air by volume nitrogen occupies 78%, oxygen 21%, carbon dioxide 0.03%, and other gases such as hydrogen, ammonia, ozone, aqueous vapour, etc., occur in traces only.

however, be noted that capillarity cannot raise water for more than 1 or 2 metres from the deeper water level of the soil, except near tanks, lakes and rivers; capillarity still remains useful in the cases of deep-rooted plants.

Soil and Root. The repeated branching of the root, its ramifications in all directions and penetration downward, coupled with the production of root-hairs in enormous numbers, help the plant to absorb a huge quantity of water, etc., from the soil. It has been estimated that the total length of the root system of a plant may extend up to several kilometres, and in some cases a few hundred kilometres. Many millions of root-hairs, each coming in intimate contact with many soil particles, may be formed by a single plant, thus enormously increasing the absorbing area of the root. The plant has thus developed an elaborate system for the intake of water and mineral salts from the soil. It is the root-hairs and the tender-growing regions of the roots that are utilized by plants for the purpose of absorption; the older parts of the roots, being impervious to water, are of no use in this respect. Root-hairs vary in length from a few millimetres to a few centimetres, and are composed of cellulose and pectic compounds. Soil particles strongly adhere to the root-hairs because of the presence of the pectic compounds in their walls.

Absorption of Water. Water adheres to the soil particles with some force (see p. 299), particularly so when there is scarcity of it in the soil. Clay and humus retain water very tenaciously. There must then be some stronger forces for dissociation of this water from the soil particles and its uptake into the root-hairs. The forces concerned are diffusion (see p. 272), imbibition (see p. 272), suction pressure (see pp. 273-74) and osmosis (see p. 273). Of them suction pressure and osmosis are particularly important. **Parts played by root-hairs (FIG. 7).** In the case of root-hairs which contain some sugars and salts in solution, the cell-sap is stronger than the surrounding soil water. The two fluids (cell-sap and water) are separated by the cell-membrane (cellulose cell-wall + plasma membrane). As

FIG. 7. A root-hair with soil particles adhering to it. Each particle is surrounded by a film of capillary water.

a consequence osmosis is set up. There is a flow of water from the soil into the root-hairs through the intervening cell-membrane (endosmosis). Osmosis, however, is not in this case a purely physical process. Although the cell-wall is permeable to both the water and the solutes, the plasma membrane is but differentially and selectively permeable, allowing the water to flow in, while stopping the sugars and salts of the cell-sap from flowing out. This selective permeability is characteristic of the plasma membrane. The same membrane, however, varies in its permeability under different conditions.

Absorption of Mineral Salts (Ionic Theory or Electrolytic Dissociation Theory). Several workers carrying on experiments on the physiological process of absorption of salts over a prolonged period (1917-44) have shown that inorganic salts are absorbed in the form of **ions**[1] although certain compounds, as experimentally proved by Osterhout and others, may as such enter the plant cell through the plasma membrane by the physical process of diffusion from the region of higher concentration of the soil solution to that of lower concentration of the cell-sap. It has, however, been seen in many cases that the concentration of the cell-sap is higher than that of the soil solution. This being so, the process of absorption cannot be explained on the basis of simple diffusion (Stiles, 1924). As a matter of fact many mineral salts which may undergo extensive ionization do not follow simple laws of diffusion. Hoagland in 1936 has definitely proved that it is the ions (and not the undissociated molecules of salts in the soil solution) that make their entrance into the cell, independent of the rate of absorption of water, from the region of lower concentration (soil solution) to that of higher concentration (cell-sap). The physico-chemical nature of the plant cell is very complex, changing continually in response to its environment, and at the same time the soil itself is a heterogeneous medium. So the forces concerned must be of varied nature. As already proved by many workers on the basis of experiments conducted by them, absorption of salts takes place in the form of ions (+ and −) produced by electrolytic dissociation (or ionization) of molecules of different salts. Further, component ions of the salts are taken up individually

---

[1] Ions are atoms or groups of atoms which carry either a positive charge of electricity or a negative charge. When an ionizable material in water is subjected to electrolysis molecules of it break up into two or more ions of different kinds—those charged with positive electricity are said to be electropositive ions such as K^+, Na^+, Ca^{++}, Mg^{++} and also H^+, and those charged with negative electricity are said to be electro-negative ions such as Cl^-, Br^-, NO_3^-, $H_2PO_4^-$, OH^-, and SO_4^{--}. The process is reversible as the following examples will show: $NaCl \rightleftharpoons Na^+ + Cl^-$; $HCL \rightleftharpoons H^+ + Cl^-$. The breaking up of molecules may not always be complete.

and independently of one another. The special feature of the living cells is that they can accumulate individual ions (and not salts) in them to a concentration far exceeding that of the surrounding medium. Actually several workers (notably Hoagland, 1944) have proved it by experimental work on *Nitella* and other plants.

Passive and Active Absorption. In modern research a passive absorption of salts and an active absorption are distinguished according to their dependence on non-metabolic energy and metabolic energy. One speaks of passive or non-metabolic absorption when the forces driving the salts through the membrane originate in the environment of the cell, i.e. these forces are physical and non-metabolic, and one speaks of active or metabolic absorption when it is dependent on metabolic energy which originates in the cell as a result of metabolic activity (particularly respiration) within it. Active uptake, as explained below, is known to be the principal method of salt absorption although some salts are absorbed, sometimes rapidly for a time, by the passive method. The interaction between the cell and its environment is essential to maintain a certain concentration within the cell in order to sustain life. By passive transport an exchange of ions takes place between the external solution (soil colloids readily yield ions on electrolysis) and the cell. It is known that the cell membrane, possibly in all cases, maintains differences in electric potential between the inner side and the outer side, evidently acting as a driving force. This influences passive uptake of ions through the membrane into the peripheral or outer plasm (see below). The ions in this phase may move freely and even out of the cell. It may be noted that ions move by diffusion through the cell-wall and the cytoplasm in their water phase, and that the plasma membrane has the ability to select and permit the entrance of certain ions and greatly restrict others. By passive uptake soon, however, an equilibrium is reached between the outer plasm and the external medium. Ions may move upwards through the transpiration current along with the mass flow of water. This being so, transpiration may help in the absorption of ions. By active transport which is slow but steady ions are brought into the inner or central plasm, i.e. from the region of lower concentration to the region of higher concentration. There is supposed to be a dividing line or membrane in the cytoplasm, though not clearly demarcated, between the outer plasm and the inner plasm. This membrane is regarded as impermeable to the exchange of free ions between the two sides. This leads to the conception of a 'specific carrier' which can pick up ions from the outer plasm and release them to the inner plasm through the so-called membrane. Since this 'carrier' moves in one direction only from the outer to the inner, ions once released into the inner plasm cannot leach out of the cell and thus cannot be exchanged for those in the external solution. Evidently the ions may accumulate there for any length of time. The 'carrier' concept has received support from many investigators. It has been suggested that *lecithin,* a phospholipid, may act as such a 'carrier'. Active transport is closely connected with metabolic energy in the form of ATP (an energy-rich phosphate compound) formed in the living cell. The chemical energy required for active transport of ions is believed to be supplied by ATP. ATP in its turn receives this energy from glucose as a result of oxidation of the latter in root respiration. It is known that young roots respire vigorously. Synthesis of lecithin also depends on the availability of ATP. Thus absence of ATP in the cell interferes with active transport. Specific enzymes may also help the passage of certain ions through the cell membrane. The concentration of ions in the cells is not even in all cases—the maximum accumulation being K^+ ions and also some other cations (see footnote, p. 301). Ions of both the electric charges must be taken up by the cell in order to maintain an electric balance both inside and outside of it; for example, a negative ion released by the ectoplasm establishes a difference of potential between the two media. Thus to equalize the charge the soil solution yields a positive ion to the ectoplasm. In fact, an interchange of ions takes place between the cell and the surrounding solution.

Conditions. Absorption of salts depends on a number of conditions, viz. aerobic root respiration, amount of light, rate of transpiration, permeability of the plasma membrane, metabolic activity of the cell, influence of temperature, hydrogen-ion concentration, etc.

Experiment 8. Absorption of water. (*a*) An interesting experiment may be carried out in the following way. Dip the end of a white-flowered lupin branch into a coloured solution (preferably water coloured with eosin) and watch. Within a very short time it will be seen that the white flowers turn pinkish—the colour of eosin—as a result of absorption of coloured water by the roots or by the cut end of the branch. *Peperomia* plant may also be similarly used, and streaks of red noticed through the stem.

(*b*) To demonstrate the **rate of absorption** proceed in the following way. Arrange the experiment, as shown in FIG. 15, and mark the level of water in the graduated tube. Note every few hours the gradual fall of the water level. At the end calculate the rate of absorption per unit of time. The experiment may be repeated under different conditions of light and temperature, and the rates of absorption compared.

Chapter 5 CONDUCTION OF WATER AND MINERAL SALTS

Root Pressure

The water that is absorbed from the soil by the root-hairs, whether by the process of osmosis or imbibition, gradually accumulates in the tissue of the cortex. As a result of this accumulation of water the cells of the cortex become fully *turgid*. Under this condition their walls which are composed of cellulose exert pressure on the fluid contents and force out a quantity of them towards the xylem vessels, and the cortical cells become *flaccid*. They again absorb water and become turgid, and this process of alternate expansion and contraction continues. Thus

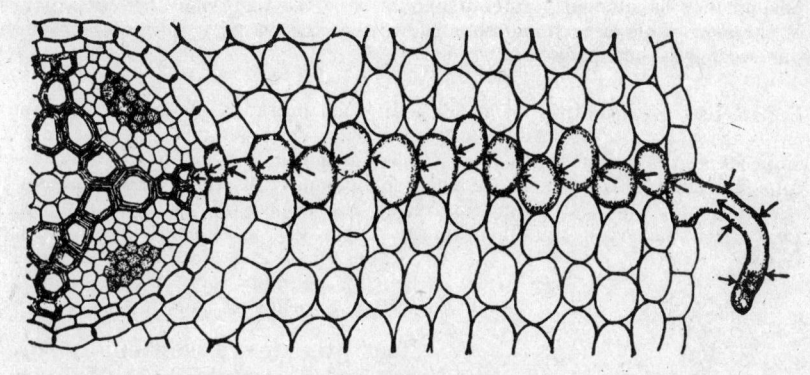

FIG. 8. A root in transection showing the course of water from the root-hairs to the xylem.

an intermittent pumping action goes on in cortex of the root, and this pumping action naturally gives rise to a considerable pressure. As a result of this pressure the water is forced into the xylem vessels through the passage cells and the unthickened areas and pits that the endodermis and the vessels are provided with. Besides, the lignified walls of the vessels are also permeable by water. **Root pressure** *is thus explained as the pressure exerted by the cortical cells of the root upon their liquid contents under a fully turgid condition, forcing a quantity of them into the xylem vessels and through them upwards into the stem.*

Experiment 9. Root pressure (FIG. 9). Cut across the stem of a healthy plant (preferably a pot plant) a few cm. above the ground in the morning, and fix to it, by means of a rubber tubing, a T-tube. Pour some water into the tube and freely water the soil. Fill a **manometer** (i.e. the U-tube with a long arm and a bulb) partially with mercury, as shown in the figure. Connect the manometer to the T-tube through a rubber cork. Insert a cork fitted with a narrow glass tube to the upper end of the T-tube. Make all the connexions air-tight by applying melted paraffin-wax. Seal the bore of the narrow tube and note the level of mercury in the long arm of the manometer. *Observation.* After a few hours note the rise of mercury-level in the long arm; also note the rise of water-level in the T-tube. *Inference.* The rise of mercury is certainly due to accumulation of water in the T-tube and the pressure exerted by it. This

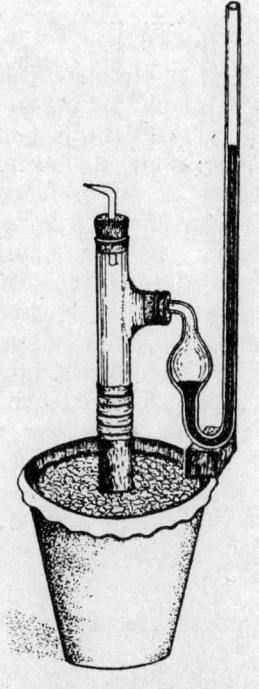

FIG. 9. Experiment on root pressure (qualitative).

phenomenon is evidently due to exudation of water from the cut surface of the stem. This experiment thus shows that the water is *forced up* through the stem by root pressure.

Experiment 10. Quantity of exudate in root pressure. Arrange the apparatus, as shown in FIG. 10. The T-tube fitted with a bent side-tube and a stopcock at the upper end is fixed to the cut end of a stem through a rubber tube and tied or otherwise properly bandaged. The tubes are filled with water. Water the soil freely and keep the apparatus in a shady place. Leave it undisturbed till the next day or the day after. Note the quantity of water that has accumulated in the measuring cylinder, say, within 24 or 48 hours as a result of root pressure.

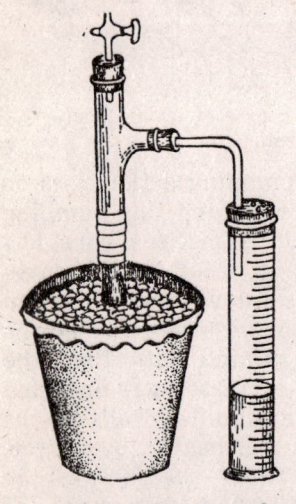

FIG. 10. Quantity of exudate in root pressure.

Root pressure is continually forcing up the water through the xylem (or wood), but it is difficult to determine the process when active transpiration is in progress. The water accumulates in the vessels only when transpiration is in abeyance. Sometimes it so happens that certain plants when cut, pruned, tapped or otherwise wounded, show a flow of sap from the cut ends or surfaces, quite often with considerable force. This phenomenon is commonly known as *bleeding*, and is often seen in many land plants in the spring, particularly grape vine, some palms, sugar maple, etc. Although the flow of sap is ordinarily slow, a considerable quantity of it exudes within a period of 24 hours in certain plants. Thus in some palms, when tapped, there may be a flow of sap to the extent of 10-15 litres per day. The sap in such plants contains sugar in addition to organic and inorganic salts.

Conditions Affecting Root Pressure. (1) **Temperature**. Temperature of the air as well as of the soil affects root pressure. The warmer the air and the soil, the greater is the activity of the root. (2) **Oxygen**. There must be an adequate supply of oxygen to the roots in the soil for respiration; otherwise their activity diminishes and may soon come to a standstill. (3) **Moisture in the soil**. A certain amount of moisture must be present in the soil. Within certain limits, the more the better. (4) **Salt in the soil**. Preponderance of salts, making the soil saline, greatly interferes with the absorption of water.

Transpiration

Plants absorb a large quantity of water from the soil by the root-hairs. Only a very small part (1-2%) of this water is retained in the plant body for the building-up processes, while the most part (98-99%) of it is lost in the form of water vapour. Transpiration *is the giving off of water vapour from the internal tissues of living plants through the aerial parts such as the leaves, green shoot, etc., under the influence of sunlight.* It is not a simple process of evaporation since it is influenced by the vital activity of the protoplasm and some structural peculiarities of the transpiring organs (see pp. 312-13). A detached leaf is seen to lose water much more rapidly than the one still attached to the plant, and this loss has been found to be 5 or 6 times greater. The total quantity of water that evaporates from a single plant is considerable. It has been estimated that the loss of water from a single sunflower plant during a period of 144 days is 27,000 c.c. This means there is a daily average loss of 187.5 c.c.

Mechanism of Transpiration. Water evaporates at all temperatures, and since the parenchymatous cells are charged with water, it continues to evaporate from these cells and collect in the intercellular spaces so long as these are not saturated with water vapour. From there the water vapour escapes into the atmosphere either through the stomata or through the thin cuticle. The former is called **stomatal transpiration**, and the latter **cuticular transpiration**. Stomatal transpiration is the rule amounting to 80-90% and is many times (approximately about ten times) in excess of cuticular transpiration under ordinary conditions of light, temperature and humidity. At night the stomata remain closed and transpiration is checked. Since water vapour is given off in transpiration, the process markedly affects the humidity of the air around. The air under big leafy trees is moist and cool for the same reason. In dorsiventral leaves the lower surface has always a much larger number of stomata; often none or sometimes comparatively few are present in the upper. Consequently the lower surface transpires water more vigorously than the upper. In isobilateral leaves, however, stomata are more or less evenly distributed on both the surfaces. The guard cells no doubt regulate transpiration to a considerable extent by partially or fully opening the stoma or by closing it altogether according to circumstances. But it cannot be said that transpiration always takes place at the maximum rate through fully-open stomata. As a matter of fact in some plants half-open stomata are as efficient as fully-open ones. Recent investigations have shown that the degree of stomatal opening cannot always be directly correlated with the rate of transpiration. Even when stomata are fully open, transpiration is greatly influenced by the water vapour in the respiratory cavities and the intercellular spaces. In woody plants and in many fruits transpiration takes place through the lenticels (see FIG. II/74). These organs help transpiration as they always remain open, and the water vapour escapes through the loose mass of cells (i.e. the complementary cells) of each lenticel (**lenticular transpiration**).

Experiment 11. Transpiration: bell-jar experiment. Transpiration can be easily demonstrated in the following way. A pot plant with its soil-surface covered properly with a sheet of oil-paper is enclosed in a bell-jar and maintained at room temperature for some time. It is then seen that the inner wall of the bell-jar becomes bedewed with moisture.

Experiment 12. Unequal transpiration from the two surfaces of a dorsiventral leaf (FIG. 11). Soak small pieces of filter paper or thin blotting paper in 5% solution of cobalt chloride (or cobalt nitrate) and dry them over a flame. The property of cobalt papers is that they are deep blue when dried, but in contact with moisture they turn pink. Place two dried cobalt papers, one on the upper and the other on the lower surface of a thick, healthy leaf, and cover them completely with glass slides (or with a leaf-clasp, as shown in the figure), and clamp them properly to the leaf. Then quickly seal the sides with vaseline. It will be seen that the cobalt paper on the lower surface of the leaf turns pink sooner than the one of the upper surface. This change in coloration takes place within a few minutes. This evidently shows that the leaf transpires water more vigorously from the lower surface than from the upper. This is due to the occurrence of a large number of stomata on the lower surface, none or few being present on the upper.

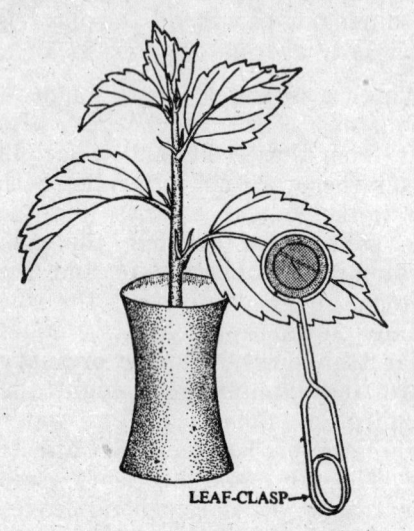

FIG. 11. Unequal transpiration from the two surfaces of a leaf.

Experiment 13. Quantitative estimation of unequal transpiration from the two surfaces of a dorsiventral leaf. (*a*) Fit up the apparatus (Garreau's potometer), as shown in FIG. 12. The two small test-tubes containing dehydrated calcium chloride are weighed before introducing them into the small bell-jars. The leaf is placed in between the two jars and the sides smeared with vaseline; other connexions are also made air-tight. The two bent tubes at the two ends are partially filled with oil. The experiment is carried out in bright light. After exposure for a few hours the test-tubes are re-weighed. The difference between the initial and final weights in each case will indicate the quantity of water absorbed by calcium chloride, evidently lost by transpiration from a unit area of the leaf within a specified time. It will be noted that the loss of water from the lower surface is much greater than that from the upper surface.

(b) Two long-stalked leaves of *Begonia* or garden nasturtium are taken and weighed separately. The upper surface of one leaf is coated with vaseline, while the lower surface of the other leaf is similarly coated. The two leaves are then exposed to sunlight for an hour or so. They are then separately re-weighed. The difference in weight will indicate in each case the quantity of water lost by transpiration. Thus a comparative idea may be had of cuticular transpiration from the upper surface and stomatal transpiration from the lower surface of the leaves.

Experiment 14. Rate of transpiration under varying external conditions. Cut three healthy leafy twigs and immediately put each in a light-weight bottle or conical flask half-filled with water, and pour a few drops of olive oil to prevent evaporation of water. Take the weight of each and expose one to direct sunlight, another to subdued light and keep the third one in a dark room, each for a specified period (say, two hours). Then take the weight of each again. The one exposed to sunlight will show the maximum rate, the second one much less, and the third one very little or no transpiration.

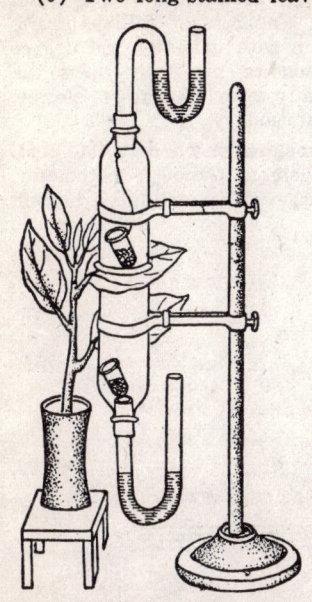

FIG. 12. Quantitative estimation of unequal transpiration with Garreau's potometer.

Experiment 15. Transpiration in relation to stomatal aperture. Proceed as in experiment 14 and then take a leaf from each. Peel off or slice off epidermis and put it immediately into absolute alcohol to fix the stomatal aperture. Then after washing and mounting in the usual way take micrometric measurements of at least ten stomata in each case, and find out the average width. Correlate the width with the rate of transpiration.

Experiment 16. To find out the degree of stomatal opening with Darwin's porometer (FIG. 13). The porometer is an ingenious and interesting device which gives a comparative idea about the degree of stomatal opening under different external conditions. The small jar of the apparatus at the end of the rubber-tube is glued to the leaf-surface (a thick smooth isobilateral leaf of *Crinum* or *Amaryllis* will serve the purpose well), and the long arm of the T-tube dipped into mercury. With a suction pump fixed

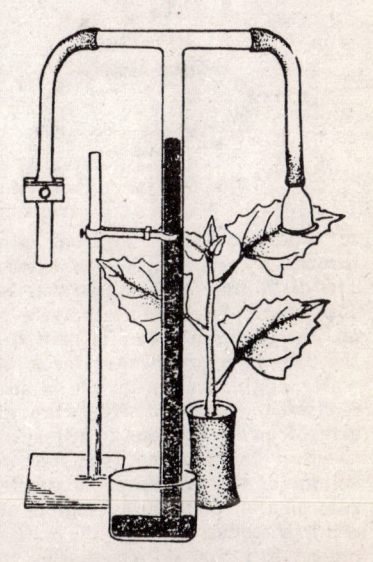

FIG. 13. Darwin's porometer (see experiment 16).

to the distal end of the other rubber-tube the mercury is lifted to a desired height in the long arm of the T-tube, and the rubber-tube is clipped. The experiment may be carried out in the morning, noon and evening (and also on a bright sunny day or a cloudy day). The rate of falling of the mercury-column within a specified time in each case indicates the degree of stomatal opening. Evidently when the stomata are fully open the mercury-column falls quickly, when only partially open the rate is slower, and when closed the column remains almost stationary.

Experiment 17. Measurement of the rate of transpiration current (FIG. 14). This experiment is best carried out with the help of **Ganong's potometer**, as depicted in the figure. The apparatus is filled with water, and a branch

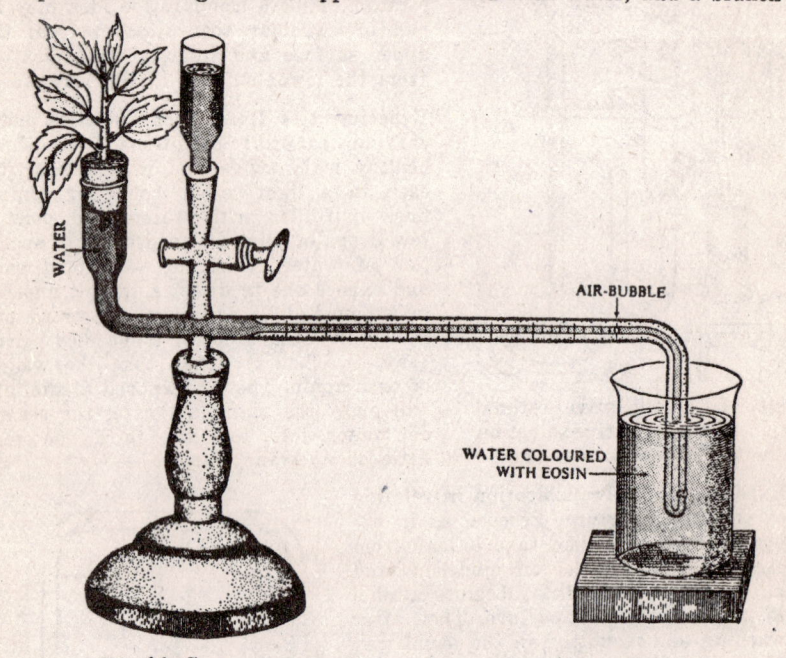

FIG. 14. Ganong's potometer to demonstrate the rate of transpiration current.

cut under water is inserted into the upper wide end of the apparatus through a cork, and the connexions made air-tight by applying paraffin-wax. The distal end of the apparatus is dipped into water contained in a beaker. The water of the beaker may be coloured with eosin. As transpiration goes on, the coloured water is seen to enter the tube. Then remove the end of the tube from the beaker for a while and allow air to enter it. Dip it into water again. An air-bubble is seen to form at the distal end of the tube; it rises and slowly travels through the horizontal arm of the potometer as a result of suction due to transpiration. Note the time that the bubble takes to cover the journey from one end of the graduation to the other. The volume of the graduated tube being known (or separately worked out), the rate of transpiration current is easily determined. By opening the stopcock which is connected with the water-reservoir on the top the bubble may be pushed back and the experiment re-started.

Experiment 18. Relation between transpiration and absorption (FIG. 15)

CONDUCTION OF WATER AND MINERAL SALTS 311

A wide-mouthed bottle with a graduated side-tube and a split india-rubber cork are required for this experiment. A small rooted plant is introduced through the split cork into the bottle which is filled with water. The level of water is noted in the side-tube, and 1 or 2 drops of oil poured into it to prevent evaporation of water from the exposed surface. The connexions are, of course, made air-tight. The whole apparatus is weighed on a compression (or pan) balance (FIG. 16) and the weight noted. It is seen after a time that the water-level in the side-tube has fallen, indicating the volume of water that has already been absorbed by the plant. The apparatus is then re-weighed. The difference in weight evidently shows the amount of water that has transpired from the leaf-surfaces. If the experiment be continued for a period of 24 hours it will be seen that the volume of water (in c.c.) absorbed is almost equal to the amount of water (in grams) lost by transpiration (1 c.c. of water=1 gm.). In this way the relation between transpiration and absorption can be worked out for the various hours of the day and under diverse external conditions. It will be noted that the experiment also shows separately the 'rate of absorption' and the 'rate of transpiration'.

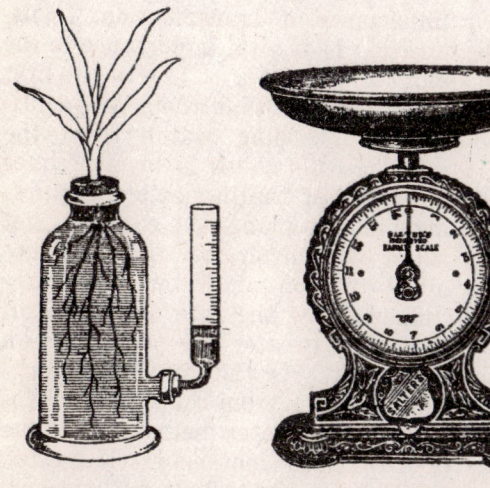

FIG. 15. Relation between transpiration and absorption.

FIG. 16. Compression (or pan) balance.

FIG. 15.   FIG. 16.

Experiment 19. Suction due to transpiration (FIG. 17). Take a Darwin's potometer (i.e. the tube with a side arm, as shown in the figure) and fix to its lower end a long narrow glass tube. Completely fill the tubes with water and insert a leafy shoot, with the cut end kept under water, into one of the arms of the potometer through a rubber cork. Close the other end with a cork. Make all the connexions air-tight by applying melted paraffin-wax. Dip the lower end of the tube into mercury in a beaker. As transpiration goes on water is absorbed, and within a few hours mercury is seen to rise in the tube to some height. This rise of mercury indicates the suction exerted by transpiration. The physiological processes connected with this phenomenon are: (*a*) osmosis in the mesophyll cells by which water is withdrawn from the tracheids in the veinlets; (*b*) evaporation of water from the mesophyll cells through the leaf-surface, bringing about concentration of their cell-sap; and (*c*) transpiration pull exerted on the column of water in the long tube, resulting in its absorption into the branch and the rise of mercury in consequence.

Transpiration Ratio. The term *transpiration ratio* (sometimes called *transpiration coefficient*) is widely used to express the ratio between the total amount of water that transpires from the plant body during the growing season and the total amount of dry matter that accumulates in it during this period. Since the difference between the amount of water transpiring from a plant and that absorbed by it is not great, the term *water requirement* is also often used to express the ratio between the total amount of water absorbed by a plant and the total amount of dry matter formed in it at the end of the growing season. Thus the transpiration ratio or the water requirement represents the number of grams (or kilograms) of water that transpires (or is absorbed) to produce one gram (or kilogram) of dry matter. The ratio varies from plant to plant and also in the same plant under different conditions. The ratio has already been worked out for a number of crop plants and weeds. Values obtained for some of the common field crops are: 368 for maize, 513 for wheat, 636 for potato, 646 for cotton, 683 for sunflower, 831 for lucerne (or alfalfa), etc., meaning that to produce one gram of dry matter the above quantity of water (in gm. or c.c.) has transpired (or has been absorbed) in each case.

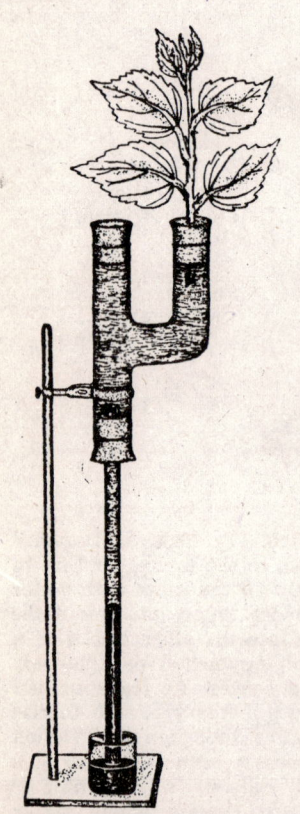

FIG. 17. Suction due to transpiration with Darwin's potometer.

Importance of Transpiration. Transpiration is of vital importance to the plant in many ways. (1) In the first place we find that roots are continually absorbing water from the soil, and this water is several times in excess of the immediate requirement of the plant; the excess is got rid of by transpiration. (2) The rate of absorption of water is greatly influenced by the rate of transpiration. The greater the transpiration the greater is the rate of absorption of water from the soil. (3) Absorption of water helps the intake of raw food materials (inorganic salts) from the soil. It is, however, not a fact that the greater the transpiration the greater is the rate of absorption of inorganic salts from the soil. As a matter of fact, the intake of salts is independent of the quantity of water absorbed. (4) Transpiration secures concentration of the cell-sap and thereby helps osmosis. (5) As a result of transpiration from the leaf-surface a suction force (see experiment 19) is generated which helps the ascent of water to the top of lofty trees. (6) Transpiration also helps the distribution of water throughout the plant body. (7) As a result of transpiration, plants become cool as a considerable amount of latent heat is lost in converting water from liquid to gaseous state. (8) Lastly, transpiration has some ecological significance in some plants. As the water evaporates hygroscopic salts are left on the surface of the leaf. These salts absorb moisture from the atmosphere, and do not allow the leaf or the plant as a whole to dry up. In the face of all these advantages the fact remains that excessive transpiration is a real danger to plant life. Many plants are often seen to dry up and die when excessive transpiration takes place for a prolonged period without adequate supply of water to the root medium.

FACTORS WHICH AFFECT TRANSPIRATION

(1) **Light.** Light is the most important factor. Transpiration normally takes place in light and, therefore, during the daytime. This is due to the fact that in the presence of light stomata remain fully open and evaporation of water takes place normally through them. At night stomata remain closed and consequently transpiration (except for a little cuticular transpiration) is markedly checked. Variations in the intensity of light, as on a bright day and a cloudy day, bring about different degrees of stomatal opening and so have a marked effect on transpiration. During the daytime again heat-rays of the sun directly falling upon the leaves enhance to a great extent the rate of transpiration.

(2) **Humidity of the Air.** There is an increase or decrease in the rate of transpiration according as the air is dry or moist. When the atmosphere is very dry it receives moisture very readily, but when it becomes very moist or saturated it can receive no more water vapour. Loss of water by transpiration is then very slight. Even though the stomata remain open at daytime, transpiration is greatly influenced by water vapour in the air.

(3) **Temperature of the Air.** The higher the temperature, the greater is the transpiration; at high temperatures the water evaporates more freely than at low temperatures. When the two factors, viz. dryness of the air and high temperature combine, transpiration is markedly enhanced.

(4) **Wind.** During high wind transpiration becomes very active because the water vapour is instantly removed and the area around the transpiring surface is not allowed to become saturated.

Adaptations to Reduce Excessive Transpiration. *Anatomical.* A thick cuticle and sometimes a multiple epidermis develop to check excessive evaporation of water. The loss of water from an apple with the cuticle removed is far greater than that from the one with the cuticle intact. The presence of cutinized hairs, scales, rods, etc., minimizes transpiration to a greater extent. A dense coating of hairs or of wax or 'bloom' on the surface is very efficient in this respect. Latex also checks transpiration. Stomata may be closed temporarily even in daytime when excessive transpiration is taking place from the leaf surface. In desert plants stomata remain sunken in pits (see FIGS. II/45-6), and these stomata are also very much reduced in number. After a time in shrubs and trees cork is formed to act as a waterproof covering and to afford protection of other kinds (see pp. 256-57). Cork cells being suberized are impervious to water and, therefore, loss of water by transpiration is prevented. A peeled potato transpires more quickly than an unpeeled one. Later still in the life of these plants bark is formed as a hard, dry covering to carry on identical functions.

Morphological. The leaf-area is often very much reduced; in extreme cases leaves are modified into spines. The size of the plant is also often reduced. Leaves may be rolled up or variously folded exposing minimum surface for transpiration. They may also assume a drooping or vertical position to avoid strong sunlight. It is further seen that deciduous trees shed their leaves in winter as a protection against excessive transpiration, while evergreen trees have their leaves well covered with cuticle.

Exudation of Water. The excess of water is also got rid of in many herbaceous plants by a process, commonly called *exudation* or *guttation*. Thus in balsam, rose, water lettuce, grape vine, many aroids (e.g. taro), garden nasturtium, sunflower, *Chrysanthemum*, *Canna*, many grasses, etc., it is seen that drops of water accumulate at the apex or margin of the leaf in the early morning. The water has escaped through the water stomata (or water pores) and hydathodes that have developed in that region (see p. 210). That the water is not dew-drops is evident from the fact that the drops are regularly arranged at the ends of veins, and that chemical analysis shows the presence of organic and inorganic salts. Exudation normally takes place during a warm and damp night. In some plants a considerable quantity of water exudes every night. The cause of exudation is to be sought in the wall pressure (see p. 276) that develops in the fully turgid parenchymatous cells that lie in the adjoining parts. Conditions necessary for the process are: abundant supply of water, a suitable temperature, activity of the living cells of the root, and also other conditions which check transpiration. At very low temperatures practically no exudation takes place.

CONDUCTION OF WATER AND MINERAL SALTS

Experiment 20. Escape of water in liquid form from the leaf: guttation. Fix a branch of a plant (e.g. balsam) to the short arm of a J-tube and apply paraffin-wax to the connexion to make it air-tight. Partially fill the J-tube with water leaving no air-gap between the water-column and the connexion. Then pour mercury into the long arm of the J-tube almost filling it. The excess water will flow out. The mercury-column will compress the water-column in the short arm and as a result it will be seen within a short time that drops of water have accumulated on the leaf-margin at the ends of veins.

Transpiration and Exudation. (1) In transpiration water escapes in the form of vapour; while in exudation water escapes in liquid form.

(2) The water that escapes in transpiration is pure; while the water that escapes in exudation contains minerals in solution.

(3) In transpiration water escapes through the stomata and to some extent through the cuticle; while in exudation water escapes through the hydathodes (see FIG. II/42) and water stomata (or water pores). Ordinarily stomata are distributed all over the surface (commonly lower) of the leaf; while hydathodes and water stomata develop at the margin or apex of the leaf at the end of a vein.

(4) Transpiration is regulated by the movement of the guard cells, partially or wholly opening or closing the stomatal aperture; while exudation cannot be so regulated, the guard cells of the water stomata having lost the power of movement.

(5) Transpiration normally takes place in the presence of sunlight and, therefore, during the daytime; while exudation takes place in the absence of transpiration and, therefore, at night.

(6) Transpiration secures concentration of sap, and also keeps the plant cool by dissipating the excess heat absorbed from sunlight; while exudation has no such effect.

Ascent of Sap

The water absorbed from the soil by the root-hairs is conducted upwards to the leaves and the growing regions of the stem and the branches. A cut branch of lupin bearing white flowers dipped into eosin solution shows a gradual change in the coloration of flowers from white to pink within a few minutes. In herbaceous plants the height this water has to reach is small, but in some trees such as *Eucalyptus*, some conifers, etc., which may attain a height of 90m. or even more, the distance to be traversed by this column of water is considerable, and the water has to resist a considerable pressure to reach that height. The rate at which the transpiration current flows upwards through the vessels varies a good deal in different plants, and at different times in the same plant. Generally speaking, the rate is about 1 to 2 metres per hour in healthy trees. Two questions naturally arise in this connexion; what is the path of movement of sap and what are the factors responsible for the ascent of sap?

Path of Movement of Sap. The path of movement of sap may be determined in one of the following two ways: (*a*) a small herbaceous plant (e.g. *Peperomia*) or a small branch of a plant (e.g. lupin) may be stood in eosin solution. After a short time sections, cross and longitudinal, are prepared from it at different heights and examined under the microscope. Sections will show the presence of coloured solution only in the vessels and tracheids. Therefore, these are the elements through which movement of sap, or **transpiration current** as it is called, takes place. (*b*) All the peripheral tissues right up to the phloem and cambium may be removed in the form of a ring (girdling) from a branch, leaving the xylem intact. In a cut branch treated similarly the pith may also be crushed. It is seen that no wilting of leaves takes place. As it is only the xylem that remains intact we may conclude that the ascent of sap takes place through it. This is known as the 'ringing' experiment.

Factors Responsible for the Ascent of Sap. Various theories have been advanced from time to time to explain the ascent of sap, but none has proved satisfactory yet. It is believed that root pressure forces up the water to a certain height, and that transpiration exerts a suction force on this column of water from above. In short, it may be said that root pressure gives a 'push' from below and transpiration a 'pull' from above. In this respect transpiration is a more powerful factor. Probable theories regarding the ascent of sap are as follows:

A. *PHYSICAL THEORIES*

(1) **Root Pressure.** Root pressure is regarded as one of the forces responsible for the ascent of sap. Many plants are seen to eject water with great force (bleeding) when the stem is cut above the ground. This phenomenon has been explained as due to the osmotic pressure which operates in the root-cortex to produce the root pressure. Root pressure may be adequate to force up water in herbs, shrubs and low trees, and that too in the absence of transpiration. The process can hardly generate 2 atmospheres of pressure and the maximum height to which a column of water may be raised by this pressure is only about 19m.; whereas a pressure amounting to 10-20 atmospheres is required to send the sap to the top of lofty trees, sometimes 90m. or even more in height. Root pressure is thus inefficient

in this respect. The process is also slow and cannot keep pace with the water lost by transpiration; further, root pressure is lowest when transpiration is highest; in fact, during active transpiration the water in the vessels is under a negative pressure. In many plants root pressure is absent or feeble at certain times of the year. Besides, water still rises through the stem if the roots are decapitated and the cut end of the stem dipped into water.

The role of root pressure in the ascent of sap has, however, been emphasized by White (1938). He has experimentally shown that excised tomato roots exude water with a pressure amounting to 6 atmospheres or even more. This pressure is sufficient to raise a column of water to a height of 54m. or even more.

(2) **Transpiration Pull and Force of Cohesion.** The understanding of the subject of ascent of sap was immensely advanced by the **cohesion theory** of Dixon and Joly, Irish plant physiologists, in 1895, concerning the tensile strength of the water column in the vessels due to a strong force of attraction (cohesion) between the water molecules, osmosis in the mesophyll cells of the leaf, and transpiration pull due to evaporation of water from the leaf-surface. According to the 'cohesion' theory the water molecules cohere together and form into a long continuous column in the vessels extending from the root to the leaf without air-bubbles anywhere in it. The water molecules cohere so strongly to one another that the column does not break or form bubbles anywhere in its entire length even under a state of very high tension due to transpiration pull, as further proved by Bode in 1923 by his microscopic observations of *Cucurbita, Impatiens, Tradescantia,* etc. This was again supported by Preston in 1958. Even if the water column breaks, its continuity is maintained by the vapour phase. The cohesive power of water, as has been experimentally proved by them, may maintain a very long column of water under tension greater than 100 atmospheres of pressure. Apparently the water column behaves as a solid column. It is to be noted that only a tension of 10 atmospheres, possibly 20 atmospheres considering the frictional resistance of the walls, is required for the ascent of sap to a height of 104m. The next operative factor is osmosis. It has been estimated that osmotically active mesophyll cells can draw up water from the ends of vessels against a pressure of 10-20 atmospheres. Finally transpiration plays its part as a powerful factor. A strong suction force, as already proved by Askenasy in 1880, is generated as a result of transpiration from the leaves. Evidently a pull is exerted on the end of the water-column, and the whole column is bodily pulled up like an iron rod which can be lifted by one hand. Cohesion theory has been strongly

supported by Dixon in 1914 and 1924, Fisher in 1948 and Briggs in 1950.

(3) **Capillarity.** The level of water inside a capillary tube is always higher than the level outside; the smaller the bore of the tube, the higher will be the rise of water in it. Xylem vessels may be regarded as so many capillary tubes extending from the root to the leaf, but from the known diameter of the vessels it is obvious that the rise of water can hardly exceed a metre or so. Further, the conducting elements in gymnosperms are so many tracheids with numerous transverse septa (and not vessels). Capillarity again implies free surface, but this is not found in plants as the vessels end in parenchymatous cells, and the water in the vessels is not in direct communication with the soil-water.

(4) **Imbibition Theory.** Sachs (1874) suggested that water moves along the walls of xylem vessels (and not through their cavities) due to the imbibition force (see p. 272), and this is responsible for the ascent of sap in plants. But when the cavities of the vessels are artificially blocked with oil, air or gelatin the branches are seen to wilt showing thereby that the amount of water absorbed by this process cannot at all keep pace with the amount of water lost by transpiration. The force of imbibition is no doubt great but the movement of water by this process is slow.

B. *VITAL THEORIES*

(1) **Vital Force.** Activity of living cells, e.g. wood parenchyma and medullary ray cells surrounding xylem, has been held by Godlewski (1884) to be responsible for the rise of sap through the plant body. The role played by the living cells is like that of relay pumps. The living cells take water from the vessels at a particular level and then force it again into the vessels at a higher level, and the sap thus rises. Strasburger (1891), however, refuted the idea of vital force by killing the living cells by the application of heat as well as by poisonous chemicals. He was definite that the forces concerned are physical rather than physiological. The vessels of the root no doubt withdraw water from the adjoining living cells.

(2) **Pulsation Theory.** According to the late Sir J. C. Bose (1923) the ascent of sap is due to active *pulsation* of the internal layer of the cortex abutting upon the endodermis. This he proved with the help of a fine electric probe which was thrust into the stem layer by layer; the probe was connected with a galvanometer. When it reached that particular layer the pulsating activity was suddenly exhibited; on either side of this layer the activity suddenly disappeared. His conclusion was that due to

the pulsating activity of the living cells of this layer a sort of pumping action is set up, and this is responsible for the *physiological* propulsion of the sap upwards through the stem. Conduction of water takes place through this layer even in the absence of root pressure and transpiration. Xylem vessels being dead and inactive no pulsation is exhibited by them, and these were regarded by him as only reservoirs of water; mechanical transport of water according to him is only possible through them to some extent. The cortex injects water into them and withdraws it from them according to circumstances. All the living cells exhibit pulsation to a greater or less extent, but the activity of the internal cortex is exceptionally great. Anatomical and experimental evidence, however, does not support this view.

Chapter 6 MANUFACTURE OF FOOD

I. CARBOHYDRATES

Photosynthesis (the name first proposed by Barnes in 1898). *Photosynthesis (photo, light; synthesis, building up) consists in the building up of simple carbohydrates such as sugars in the green leaf by the chloroplasts in the presence of sunlight (as a source of energy) from carbon dioxide and water absorbed from the air and the soil respectively.* The consensus of opinion is that glucose is formed first and all other carbohydrates are derived from it. The process is accompanied by a liberation of oxygen (see experiment 21). The volume of oxygen liberated has been found to be equal to the volume of carbon dioxide absorbed. But it is to be noted that all the oxygen liberated in the process is released exclusively from water (H_2O) and not from carbon dioxide (CO_2). Oxygen escapes from the plant body through the stomata. This formation of carbohydrates, commonly called **carbon-assimilation**, is the monopoly of green plants only, chlorophyll being indispensable for the process. By this process not only are simple carbohydrates formed but also a considerable amount of *energy* (initially obtained from sunlight as radiant energy) is transformed by green cells into chemical energy and stored up as such in organic substances formed. It must be noted that photosynthesis takes place only in the green cells and, therefore, mainly in the leaf and to some extent also in the green shoot. Under favourable conditions of light intensity and temperature the rate of photosynthesis increases enormously and a tremendous amount of CO_2 is absorbed from the air for this process, so much so that on a windless day the CO_2 content of the air over a field crop may drop to 0.01% from the normal 0.03%.

Mechanism of Photosynthesis.[1] Photosynthesis is the biological process by which some energy-rich carbon-containing compounds are produced from carbon dioxide and water by the *illuminated* green cells, liberating oxygen as a by-product. It is essentially an oxidation-reduction[2] process by which hydrogen is transferred from water to carbon dioxide through a 'carrier' substance, the nature of which is still imperfectly understood. In the process the volume of CO_2 absorbed is almost equal to that of O_2 liberated in most green plants (Boussingault, 1864). Since glucose or fructose appears to be the first carbohydrate formed in photosynthesis the overall reaction may be represented thus- $-6CO_2+12H_2O \rightarrow C_6H_{12}O_6+6H_2O+6O_2$. This overall reaction does not, however, explain the sequence of reactions involved in the process. It is not a single and simple reaction between CO_2 and H_2O to produce the end-product but a very highly complex and self-multiplying process in which a number of reactions, both photochemical and chemical (enzymic) occur. The complexity is further increased by the fact that in this process the most important energy transformation takes place, i.e. the radiant (light) energy absorbed by chlorophyll is transformed into *stable,* high potential, chemical energy which is available for all vital activities of the living cells. Although minute details of photosynthetic mechanism are not yet known, some important facts have, however, been elucidated, e.g. absorption of light energy by chlorophyll, splitting of water molecules with the liberation of oxygen and production of a reducing agent for the reduction and fixation of carbon dioxide, conversion of light energy to chemical energy, formation of a detectable quantity of phosphoglyceric acid as a stable intermediate compound, and finally its conversion to carbohydrates.

Nature of Reactions. Photosynthesis as a whole fundamentally consists of an oxidation-reduction series, in which one compound (water in this case) is oxidized (hydrogen transferred or donated) and another compound (CO_2 in this case) reduced (hydrogen added or accepted). Plant physiologists soon came to realize that the production of carbohydrates from H_2O and CO_2 consists of two types of reactions—one for which light is essential and another which can proceed in the dark.

---

[2] Oxidation-Reduction Reactions. Oxidation means (*a*) addition of oxygen, making the compound richer in it, or (*b*) removal of hydrogen, or (*c*) removal of electrons. Respiration is mainly of oxidation type. Conversely reduction means (*a*) addition of hydrogen, or (*b*) removal of oxygen, or (*c*) addition of electrons. Photosynthesis is mainly of reduction type.

Light and Dark Reactions (Blackman Reaction). In 1905 Blackman, a British plant physiologist, while working on photosynthesis under different intensities of light and different concentrations of sodium bicarbonate ($NaHCO_3$) as a source of CO_2, first suggested that photosynthesis involves two distinct phases —the first phase involves certain chemical reactions for which light energy is indispensable (light reactions), and the second phase involves a series of chemical reactions leading ultimately to carbohydrates and other organic compounds for which chemical energy (and not light) is required (dark reactions). The main, possibly the only, light reaction is the splitting of water at the initial stage of photosynthesis with the formation of a reducing agent. The dark reactions are a series of reactions involving the reduction of carbon dioxide to organic compounds by one or more reducing agents; as a matter of fact all the reactions from CO_2-reduction to final product (glucose) formed in a stepwise manner are dark reactions (see fixation of carbon dioxide, (p. 324). For continuity of the process, however, some product of light reaction is essential. By the use of light in flashes (intermittent light) Warburg working on *Chlorella,* a unicellular green alga, has shown in 1919 that the rate of photosynthesis is greater than the rate of the process in continuous light, of course, of the same intensity, and that at least three dark reactions participate in the process. The existence of light and dark reactions (Blackman reaction, as Warburg first termed it) in photosynthesis has been further proved by later workers (Emerson and Arnold in 1932, Briggs in 1935, and others in 1951-2).

Role Played by Light and Chlorophyll. Light, as is well-known, is an essential condition for initiating the process of photosynthesis. The part played by it in any photo-chemical reaction is not, however, completely understood, and consequently a clear and complete explanation of the action of light in photosynthesis cannot be given. Suffice to say that light energy absorbed by chlorophyll is effective in the splitting of water molecules at the initial stage of photosynthesis (see photolysis of water p. 322) and in this process it becomes transformed into chemical energy for further action, i.e. reduction and fixation of CO_2 in dark reactions. It has been estimated that an average green leaf absorbs about 80-85% of light; in photosynthesis, however, only about 1% of light is utilized. A portion of light may be reflected from the leaf-surface and from the chloroplasts, a portion is lost in heat, and another portion is used in transpiration. Experiments have proved that only certain rays (not all) of light are used for photosynthesis (see p. 335).

Chlorophyll is indispensable for photosynthesis. It is, however, not known what exact role it plays in the process excepting that (a) it absorbs light energy from sunlight and becomes 'activated' or 'excited', i.e. its energy level increases above normal to initiate the process of photosynthesis; (b) the extra energy stored in chlorophyll now goes to break up water molecules at the initial stage of photosynthesis; (c) a part of this energy goes to form a reducing agent, TPN in all probability, which accepts hydrogen from water for reduction of CO_2; (d) another part of this energy goes to synthesize ATP; and (e) chlorophyll acts as a catalytic agent; the amount of chlorophyll or the constituent pigments remain unaltered even by prolonged period of photosynthesis.

Photolysis of Water. The initial and essential part of photosynthesis lies in photolysis, i.e. splitting of water molecules into its two components. This is essentially a *light reaction* taking place in the body of the chloroplast. In fact all photosynthetic reactions, as shown by Arnon and his colleagues in 1954, occur in the grana of the chloroplast. In this respect the chloroplast may be regarded as a complete unit. Under the influence of light energy and the catalytic action of chlorophyll water, a substance of low energy value, is split up into oxygen (O_2) and hydrogen ($2H.$). Oxygen escapes to the atmosphere, while hydrogen atoms (or their electrons) combine with a reducing agent, most likely TPN, formed in the chloroplast. TPN becomes reduced to $TPNH_2$, a substance of high energy value. This further results in the synthesis of ATP from ADP. It is possibly during these reactions that light energy is converted to chemical energy to drive further reactions that immediately follow (see fixation of carbon dioxide, p. 324). It should be specially noted that all the oxygen evolved in photosynthesis is released from water (see Hill Reaction, below). Further, the volume of oxygen released in photosynthesis has been found to be equal to the volume of carbon dioxide absorbed. This may be represented by the following equation:

$$6CO_2 + 12H_2O \xrightarrow[\text{chlorophyll}]{\text{light energy}} C_6H_{12}O_6 + 6H_2O + 6O_2$$

Hill Reaction. Robin Hill, an English biochemist, carried on experiments with isolated chloroplasts suspended in water. He showed in 1937 and 1939 that such chloroplasts or even their fragments evolve oxygen when illuminated, even in the absence of CO_2, provided that a suitable hydrogen-acceptor is available. Hill used ferric oxalate for this purpose. Other compounds like ferricyanides, chromates and quinones (e.g. benzoquinone) have been used with similar results. With isolated chloroplasts it was not, however, found possible to make further advance in the process of photosynthesis. It was further evident from Hill reaction that the source of O_2 released in photolysis is H_2O (and not CO_2 as previously believed). Hill reaction is essentially the photolytic splitting of water with the release of O_2 and possibly also the formation of a reducing agent to act as hydrogen-acceptor. That all the oxygen evolved in photosynthesis is derived from H_2O (and not CO_2) has been definitely proved by Ruben and Kamen, American chemists, in 1941 and 1943 by using radioactive oxygen, O^{18}, in water, H_2O^{18}. When CO_2 molecule was tagged with radioactive O^{18} no trace of it was found in O_2 evolved.

Transfer of Hydrogen. It has been already mentioned that in photosynthesis water is the specific hydrogen-donor and TPN (as evidence goes now) is the hydrogen-acceptor. Hydrogen (or electrons) now available from H_2O is transferred to CO_2 through some intermediate 'carriers' for reduction of the latter (CO_2) in dark reactions. Experiments carried out so far have shown that a number of complex enzymes of the oxidation-reduction group take part in the whole series of reactions. It is now known that some of them (e.g. hydrogenase, dehydrogenase, cytochrome, DPN, thioctic acid, vitamin K, etc.) act as intermediate 'carriers' of hydrogen. They have been found in green cells although not in the chloroplasts in all cases, and in some cases enzyme reactions have been proved *in vitro*.

ATP—adenosine triphosphate. ADP—adenosine diphosphate. TPN—triphosphopyridine nucleotide=NADP—nicotinamide adenine dinucleotide phosphate. NAD—nicotinamide adenine dinucleotide=DPN—diphosphopyridine nucleotide.

ATP. ATP is an active energy-rich phosphate compound consisting of adenine, ribose sugar and three phosphate bonds, two of them being high-energy bonds. It is now definitely known that ATP provides most of the chemical energy for the activities of the living cell. For many biochemical reactions, particularly in respiration, photosynthesis, synthesis of proteins, fats, nucleic acids, etc., the source of chemical energy is ATP. These reactions are activated by the transfer of an energy-rich phosphate bond of ATP (as first stressed by Lipmann, a Nobel Prize winner, in 1941) when ATP becomes ADP. Again with the addition of a phosphate bond + energy ADP becomes ATP. This interconversion goes on continually in the living cell. The chemical energy needed for the conversion of ADP to ATP comes from oxidation of glucose (a substance much richer in chemical energy than ATP) during respiration and stored in the mitochondria. It may be noted that 1 molecule of glucose yields on complete oxidation 38 molecules of ATP. Light energy absorbed by chlorophyll also brings about synthesis of ATP in the chloroplast. In this process light energy becomes converted to chemical energy.

Fixation of Carbon Dioxide. This is the second phase of photosynthesis. Smith in 1943 showed that nearly the whole of carbon (about 97%) taken up in photosynthesis could be traced in sugar and its derivatives such as sucrose and starch. In recent years it has been possible by the use of radioactive isotopes, particularly radioactive carbon, C^{14}, in carbon dioxide, i.e. $C^{14}O_2$, to trace to some extent at least the sequence of reactions through which the 'marked' C^{14} passes on its way to the final products formed during the process of photosynthesis. This is called the 'tracer' method, and the 'marked' element the 'tracer' element. The sequence has definitely proved that the reactions are not so simple as earlier workers thought. Besides, the reactions are so rapid that the detection of the unstable intermediates is extremely difficult. As stated before, CO_2 fixation takes place in dark reactions but the energy required for the process is the potential chemical energy already stored in the green cells, and that water is the specific hydrogen-donor for reduction of CO_2. By using carbon dioxide which contains radioactive carbon of atomic weight 14, i.e. $C^{14}O_2$ it has been possible to trace the 'marked' carbon through some of the successive reactions. Thus in 1949 Calvin and Benson in California and later Gaffron and Fager in Chicago in 1951, by exposing *Chlorella*, a unicellular green alga, to light in the presence of radioactive $C^{14}O_2$, succeeded in tracing it through some of the intermediate stages of photosynthesis. By the use of this technique they have found that when the period of exposure to light is shortened to a few seconds and the material immediately immersed in boiling alcohol, the 'marked' C^{14} appears in phosphoglyceric acid in a detectable quantity. It may be noted that most of the isotopic carbon is detected in the carboxyl group (COOH) of this compound. Phosphoglyceric acid (PGA) is, therefore, the first stable intermediate product formed during the process of photosynthesis. Further, as the period of exposure to light is increased to 30-60 seconds or a little more the 'marked' C^{14} is seen to appear in a number of compounds, particularly phosphoric acid derivatives of sugars (hexose phosphates), and later in glucose or fructose (possibly the former) in a free state, following the main pathway of carbon dioxide fixation in photosynthesis. As said before, all the reduction steps from carbon dioxide to sugar are dark reactions. It has been suggested (Calvin, Bassham and others, 1954) that 5-carbon ribulose phosphate is activated by ATP forming ribulose diphosphate (also a 5-carbon compound). The latter combines with CO_2 and is carboxylated to an unknown 6-carbon compound. This unknown compound splits into two parts—one part of it being the stable and detectable 3-carbon phosphoglyceric acid (PGA). A portion of it reforms ribulose phosphate which is again fed into the chain of reactions to maintain the continuity of the process. The other part of the aforesaid 6-carbon compound may enter the mitochondria and follow the pathway of cellular respiration. The unstable intermediate products are, however, not known with any amount of certainty. The next step is the conversion of phosphoglyceric acid (a 3-carbon compound) to a hexose sugar—glucose or fructose (a 6-carbon compound), but how? It may be that the union of two 3-carbon compounds (derivatives of phosphoglyceric acid) results in the formation of a 6-carbon compound, i.e. glucose or fructose. It is more likely, however, that the reduction of phosphoglyceric acid finally to glucose or fructose takes place through a series of reactions under the action of specific enzymes. It is now known that the reactions are essentially a reversal of what we find in the glycolytic phase of cellular respiration (see Pathway of Glycolysis, Chapter 11) via 3-phosphoglyceraldehyde, as follows: 3 phosphoglyceric acid ⟶ 3-phosphoglyceraldehyde ⟶ fructose-1, 6 diphosphate → glucose-6-phosphate → glucose → starch or sucrose. The reduction of PGA evidently requires the use of a reducing agent (NADH, also called DPNH) in the presence of ATP formed in the light reaction. Thus PGA becomes reduced to 3-phosphoglyceraldehyde (PGAL) and later to a 6-carbon hexose (fructose or glucose) phosphate. The hexose is then dephosphorylated under the influence of specific enzymes, and finally fructose or glucose, possibly the latter, appears in a free state. Glucose and fructose are reversible. Later still, the 'marked' C^{14} appears in starch or sucrose. It may be noted that the conversion of glucose to starch, sucrose and other forms of carbohydrates is based on the use of glucose-1-phosphate, rather than free glucose, the phosphate **group** being donated by ATP under the action of the enzyme *phosphorylase*. Several intermediate reactions are involved in the whole process, each reaction being controlled by a specific enzyme. It is important to note that phosphoglyceric acid holds a pivotal position from which carbohydrates are formed in one direction following the main pathway of CO_2 fixation in photosynthesis, while in the other direction phosphopyruvic acid, pyruvic acid, many stable organic acids (fatty acids and some amino-acids, e.g. alanine, aspartic acid, etc.) are formed, and finally fats and proteins are elaborated. The above reactions are reversible.

Isotopes. Different kinds of atoms of a particular element having different atomic weights but identical chemical properties are called isotopes. For example, there are different kinds of carbon atoms with atomic weights of 10, 11, 12, 13 and 14 respectively. All isotopes are not radioactive. Some of the heavy elements like uranium and radium are naturally radioactive, while the stable isotopes of many elements can be artificially made radioactive. Thus the atomic weight of stable carbon is 12, while that of the artificially made radioactive carbon may be C^{10}, C^{12}, C^{13} or C^{14}. All radioactive elements disintegrate by the loss of charged particles and by radiation. The life of such an element is, therefore, unstable **varying** in duration from a second or even less to thousands of years. From biological standpoint radioactive elements have proved to be of immense value in elucidating certain intricate problems of plant-life and animal-life because it has been possible to trace such 'marked' elements through successive stages in the plant body or the animals body. Thus radioactive carbon can be traced in the intermediate and final products formed in the process of photosynthesis. This evidently helps us to ascertain the nature of products that appear during the process. Atomic weights of some of the radioactive elements used in plant research are C^{11}, C^{14}, O^{18}, S^{35}, P^{32}, K^{42} and Ca^{45}, their corresponding stable forms being C_{12}, O_{16}, S_{32}, P_{31}, K_{39} and Ca_{40}.

Production of Oxygen and Starch in Photosynthesis. Oxygen and starch are two important criteria for testing photosynthesis. Thus they are commonly made use of for experimental purposes. As already stated, oxygen is derived from water, H_2O, as a by-product in the initial stage of photosynthesis (Hill, 1937 & 1939), while starch is derived from sugar (glucose) as the final

MANUFACTURE OF FOOD

product in the process (Sachs, 1862 & 1864). Oxygen escapes from the leaf (see experiment 21) but starch accumulates in its mesophyll cells (see experiment 23). Starch may be detected in the following way. In the evening collect one or more leaves and bleach them with methylated spirit. Then dip them into iodine solution. They are seen to turn bluish-black in colour indicating the presence of starch grains. It is better to treat the leaves further with benzol for a few minutes. The brownish colour of the leaves due to the action of iodine solution on protoplasm and cellulose disappears, and the bluish-black colour of starch grains stands out clearly. Starch is insoluble in water. At night it is converted to sugar by the action of the enzyme *diastase* and translocated to storage organs (Sachs, 1864). In the storage tissues sugar is reconverted into starch by the leucoplasts. Starch may again be converted into sugar when necessity arises. This interconversion is independent of light and chlorophyll and may, therefore, take place anywhere in the plant body.

Experiment 21. Photosynthesis: to show that oxygen is given off during photosynthesis (FIG. 18). Place some green submerged water plants (e.g. *Hydrilla*) in a large beaker filled with water. Add a small quantity of soda water or a pinch of soda bicarbonate as a source of carbon dioxide. Cover the plants under water with a glass funnel, and invert over the funnel under water a test-tube filled with water. It is better to cut the stems and tie up the shoots into a bundle. The cut ends should be projected upwards into the funnel. *Observation.* When exposed to bright light a stream of small gas bubbles is seen to rise upwards through the cut ends of stems and collect at the upper end of the test-tube, displacing the water. *Inference.* That the gas is oxygen can be proved in the following way. Close the test-tube with the thumb under water, and invert it over a dish containing a quantity of pyrogallate of potash (5% pyrogallic acid to which an excess of caustic potash has been added). Then with the help of a bent tube introduce into the test-tube a quantity of this solution. The pyrogallate solution coming in contact with the gas will absorb it and will, therefore, rise and completely fill up the test-tube. The pyrogallate solution absorbs oxygen. The gas in the tube is, therefore, oxygen.

FIG. 18. Evolution of oxygen bubbles in photosynthesis of submerged water plants (*Hydrilla*).

Experiment 22. Analysis of gases with Ganong's photosynthometer. To analyse the gases—CO_2 and O_2—and to find out their quotient this instrument is very suitable. First fill up the graduated tube with water and then introduce into it a known volume of CO_2 from a generator by displacing water. Enclose a green leaf in the bulb of the photosynthometer, arrange the whole apparatus as shown in the figure, and expose it to bright sunlight for a few hours. Remove the graduated tube to a cylinder containing KOH solution. Note the rise of the solution in the tube indicating the volume of CO_2 still not used up by the leaf. The total volume of CO_2 being known, the actual volume of CO_2 absorbed by the leaf within a specified time can be easily calculated. Then transfer the graduated tube to pyrogallate of potash and note further rise of the solution in it. This will indicate the volume of O_2 given out by the leaf in the same period. It will be noted that the two volumes—$CO_2:O_2$—almost maintain a quotient of unity.

Experiment 23. Photosynthesis: to demonstrate that starch is formed in photosynthesis (FIGS. 20-2). Select a healthy green leaf of a plant *in situ* and cover a portion of it on both sides with two uniform pieces of black paper, fixed in position with two paper clips or soft wooden clips, either in the morning before the sun rises or the previous evening, so that the experiment is performed with a starch-free leaf. Or, keep a healthy, green pot plant in a dark room for 1 or 2 days so that its leaves become starch-free, and then cover a portion of a leaf of this plant as described above. To make sure that there is no starch, collect a few neighbouring leaves in the morning, decolorize them with alcohol and dip them into iodine solution. It will be seen that they do not turn black. Evidently all the leaves are starch-free. Now expose the plant to bright light for some time, preferably till the evening. Then collect the leaf, decolorize it with alcohol, and test it with iodine solution for starch grains. It will be seen that only the exposed portions turn blue or black.

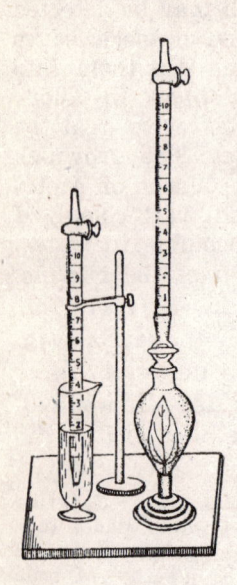

FIG. 19. Ganong's photosynthometer (see experiment 22).

A very interesting experiment known as the **starch print** (FIG. 21) may be carried out in the following way. A stencil (which may be a blackened thin tin plate or a black paper) with the letters S T A R C H punched or cut in it is used for this purpose, the procedure being the same as described under experiment 23. Later, when the leaf is decolorized and treated with iodine solution, the print of S T A R C H will stand out boldly

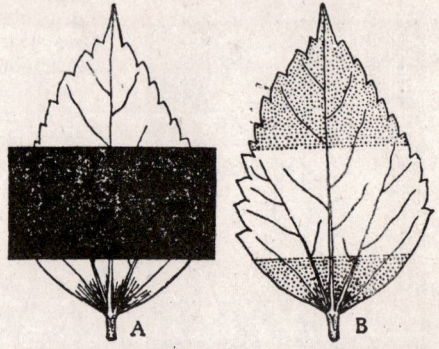

FIG. 20. Formation of starch grains in photosynthesis of land plants. *A*, leaf partially covered with black paper; *B*, covered portion without starch grains, while uncovered portions with plenty of them.

MANUFACTURE OF FOOD

in black on the bleached leaf owing to the formation of starch grains with the access of light and their turning black in contact with iodine solution.

FIG. 21. Starch print in photosynthesis.

Instead of loose black paper or stencil a **light-screen** (FIG. 22) may be used to cover a portion of the leaf. The advantage of the light-screen is that it allows free ventilation and at the same time cuts off all light.

FIG. 22. A light-screen.

Experiment 24. To find out **the quantity of photosynthate in the given area of a leaf.** (Sachs' half-leaf method.) (*a*) Cover a selected number of symmetrical leaves with black paper or black cloth the previous day to make them starch-free. Cut out one-half of each leaf very close to the mid-rib and place the cut-out halves on a graph paper to find out their area as correctly as possible. Then expose the plant with the remaining halves to bright sunlight for some hours. Cut out those halves again very close to the mid-rib. Kill and dry both sets of half-leaves to get rid of all water, and weigh them separately. The difference in weight will indicate the quantity of photosynthetic products formed in the given area of leaves within a specified period.

(*b*) **Ganong's leaf-area cutter** (FIG. 23). With this instrument cut out an equal number of circular discs (say, 20, each equivalent to 1 sq. cm.)—one set of discs from a plant kept in darkness and another set from the same plant exposed to sunlight for a few hours. Kill and

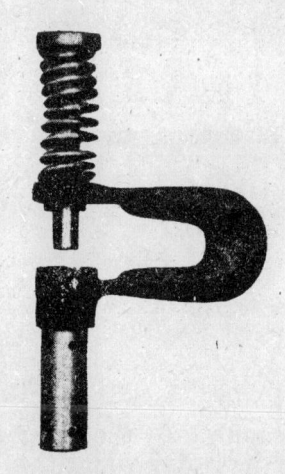

FIG. 23 Ganong's leaf-area cutter.

dry them separately and find out the weight in each case. The increased weight in the latter set will indicate the quantity of photosynthate formed in a given area within a specified period.

Experiment 25. To show that plants cannot photosynthesize unless carbon dioxide is available: Moll's experiment (FIG. 24).

(a) Arrange, as shown in FIG. 24, with some KOH solution in the bottle. The leaf should be starch-free (see experiment 23). After exposure to sunlight for some hours the leaf is decolorized and tested with iodine solution for starch grains. The portion of the leaf outside the bottle with access of atmospheric CO_2 is the only part that turns black. A branch or a pot plant, starch-freed, may be used instead of a single leaf.

FIG. 24. Moll's experiment on photosynthesis. (The bottle contains some KOH solution).

(b) Cover a healthy green pot plant, starch-freed, with a bell-jar stood on a large flat dish containing mercury (or distilled water), as shown in FIG. 25. Place a small pot with a little caustic potash solution inside the bell-jar to absorb all the enclosed carbon dioxide and fill the U-tube with soda-lime. Expose the apparatus to direct sunlight for some hours. Afterwards test the leaves for starch grains in the usual way. In the absence of CO_2 these are not formed.

(c) Replace the pot plant by *water plants* in a beaker, as in FIG. 18. No evolution of gas (oxygen) bubbles is noticed.

Experiment 26. To show that chlorophyll is essential for photosynthesis. Select a variegated garden croton, tapioca, or *Coleus*. Cut out a small branch from it and dip the cut-end into water in a bottle. Keep it in a dark room for 1 or 2 days to free the leaves of starch grains. Then mark the green portions in 1 or 2 leaves, expose the branch to bright sunlight for a whole day. In the evening collect the marked leaves, decolorize them with methylated spirit and dip them into iodine solution for a few minutes. Note that only the green portions of the leaf turn black indicating the presence of starch grains; while the non-green portions turn yellowish. It is, therefore, evident that without chlorophyll photosynthesis cannot take place.

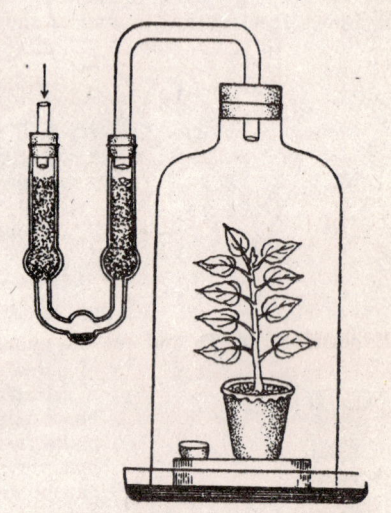

FIG. 25. Green plants cannot photosynthesize under the exclusion of carbon dioxide.

Factors Affecting Photosynthesis.
Light intensity, temperature, carbon dioxide concentration, and water supply are the chief *external conditions* for photosynthesis and its rate; while the chief *internal factors* are chlorophyll content, accumulation of photosynthetic products and protoplasmic factor.

MANUFACTURE OF FOOD

(1) **Light.** This is the most important factor for photosynthesis. Formation of carbohydrates cannot take place unless light is admitted to the chloroplasts. Naturally the process is in abeyance during the night. If a leaf or any part of it be covered with black paper starch grains are not formed there (see experiment 23). The rate of photosynthesis also varies according to the intensity of light. In very weak light no starch grains are formed except in shade-loving plants. Too intense light also has the same effect. It is, however, a fact that a very small part of the solar energy is used by the green plant for photosynthesis.

(2) **Carbon dioxide.** Carbon dioxide of the air is the source of all the carbon for the various organic products formed in the plant body, such as sugar, starch, etc., and, therefore, the process is in abeyance if carbon dioxide be not available to the plant (see experiment 25). Under favourable conditions of light and temperature if carbon dioxide concentration rises from 0.03% in the air to 0.1% or even more (say 2.5 to 5%, with corresponding increase in light intensity) carbohydrate formation greatly increases. Higher concentration of CO_2 is harmful to plants.

(3) **Temperature.** Photosynthesis takes place within a wide range of temperature. It goes on even when the temperature is below the freezing point of water, but the maximum temperature lies somewhere at 45°C. The optimum temperature, i.e. the most favourable temperature for photosynthesis, may be stated to be 35°C. Both maximum and optimum temperatures, however, vary in different species of plants and in those growing under different climatic conditions. Many cacti can, however, function at a temperature of 50°C.

(4) **Water.** Water is indispensable for photosynthesis because the process starts with the splitting of water into oxygen and hydrogen under the action of light and chlorophyll. Dryness of the soil and of the leaf thus retards photosynthesis. Water makes the photosynthetic cells turgid and active. It is, however, a fact that less than 1% of the water absorbed by the roots is utilized in photosynthesis.

(5) **Chlorophyll.** This is essential for photosynthesis; the plastids are powerless in this respect without the presence of chlorophyll. For the same reason non-green parts of plants cannot photosynthesize (see experiment 26). Fungi, and saprophytic and parasitic phanerogams have altogether lost this power, being devoid of chlorophyll.

(6) **Accumulation of Photosynthetic Products.** Any heavy accumulation of these products, whether sugar or starch, in the assimilating cells causes the process of photosynthesis to slow down and even to come to a standstill.

(7) Protoplasmic Factor. Internal factor other than chlorophyll plays a part in photosynthesis. It is called the protoplasmic factor (Briggs, 1920). But how it acts is not clearly understood. It is likely this factor is enzymic in nature.

Two other factors may be considered in this connexion. One is **oxygen** and the other is **potassium**. In the complete absence of oxygen photosynthesis cannot take place. But variation in oxygen percentage seems to have no effect on the rate of photosynthesis. Oxygen contents reduced to 2% or increased to 50% does not appreciably alter the rate. But in its complete absence the process comes to a standstill. Potassium helps synthesis of carbohydrates, and, therefore, in its absence starch grains are not formed. Potassium does not enter into the composition of carbohydrates but acts as a catalyst helping in their synthesis.

Limiting Factors in Photosynthesis. The principle of limiting factors was enunciated by Blackman, a British plant physiologist, in 1905. He showed that **light intensity, carbon dioxide concentration** and **temperature** are the limiting factors in photosynthesis, i.e. the ultimate rate of photosynthesis is determined by each such factor. Just as the strength of a chain is determined by its weakest link, so the rate of photosynthesis is controlled and limited by the least favourable factor, i.e. the weakest factor becomes the most important factor. Earlier workers tried to find out the maximum rate of photosynthesis under the action of any one factor at a time. But Blackman for the first time considered all the factors simultaneously while examining the effect of any one factor on the rate of photosynthesis. Under the normal condition of 0.03% concentration of CO_2 in the air and within a certain range of temperature photosynthesis may start with low intensity of light. Supposing then that the concentration of carbon dioxide remains constant, it is seen that with increasing intensity of light the rate of photosynthesis steadily increases until a particular maximum is reached. Without further increase in CO_2 concentration the rate of photosynthesis shows no further progress even if the light intensity is increased. It is evident then that at this point carbon dioxide acts as a limiting factor. Supposing again that light intensity remains constant, it is seen that with increasing concentration of CO_2, say, from 0.03% to 0.5-1% or even more, the rate of photosynthesis will rise again up to a certain point. Further increase in CO_2 concentration under the same condition of light intensity will have no influence on the rate of photosynthesis. At this point light acts as a limiting factor. Different concentrations of CO_2 are correlated with different intensities of light, i.e. more intense light is required to utilize more concentrated carbon dioxide. These two factors balance each other, and photosynthesis progresses rapidly. But then there is a limit beyond which each factor inhibits the rate and may even check the process altogether. Within a certain range the effect of temperature on the rate of photosynthesis shows very little variation. But it has been noted that with low intensity of light, even if the concentration of CO_2 be high, the rate of photosynthesis does not appreciably increase with increase in temperature. With high light intensity, however, even if the CO_2 concentration be low, the rate of photosynthesis increases with increasing temperature up to a maximum of 35°C. for most plants. With high light intensity and high CO_2 concentration the rise of temperature may even double the rate of photosynthesis. Blackman theory of limiting factors has been supported by Wilmott (1921) working on *Elodea* by his own bubble-counting technique, by Warburg (1919) working on *Chlorella*, by Harder (1921 and 1923) working on a number of aquatic plants, and also by many others.

The effect of the above factors may be easily determined by simple experiments, as first shown by Harder in 1921. A cut branch of *Hydrilla* with its cut end projected upwards is taken in a test tube in water. It is then exposed to different intensities of light, and the number of bubbles counted for each illumination per unit of time. With the addition of soda bicarbonate in different concentrations the experiments are repeated, and the bubbles counted again for each concentration. The experiments may be repeated by gradually increasing the temperature. The data obtained give a good idea about the rate of photosynthesis under the action of individual factors and combinations of factors.

Conditions Necessary for the Formation of Chlorophyll.

A number of factors, both internal and external, are responsible for the formation of chlorophyll. In the absence of any of them chlorophyll synthesis is in abeyance.

(1) **Light.** Normally chlorophyll does not develop in most plants in the absence of light, and continued absence of light also decomposes it into protochlorophyll (see p. 155). But in many algae, mosses, ferns and in the seedlings of many conifers and certain angiosperms chlorophyll may develop in the dark and the plants become pale green in colour. With the access of light, however, more chlorophyll develops in them and the plants become deeper green in colour. In most cases a low intensity of light is quite effective in inducing chlorophyll formation. It may also be noted that very strong light decomposes chlorophyll, particularly in shade-loving plants.

(2) **Temperature.** Chlorophyll develops within a wide range of temperature, the maximum rate usually varying from 26-30°C.; very high temperature, 45-48°C., decomposes chlorophyll.

(3) **Iron and Magnesium.** In the absence of the salts of these

metals chlorophyll is not formed, and seedlings assume a sickly yellow appearance. In this condition they are said to be *chlorotic*. Although both iron and magnesium are required for the formation of chlorophyll it is only magnesium that enters into its composition.

(4) **Manganese.** It is also believed that manganese is necessary, even essential, for the formation of chlorophyll.

(5) **Nitrogen.** Nitrogen enters into the composition of chlorophyll and, therefore, in its absence chlorophyll fails to develop.

(6) **Water.** Leaves, when drying up in the absence of water, are seen to lose their green colour. Desiccation thus brings about decomposition of chlorophyll. In prolonged droughts leaves of many plants, particularly of grasses, turn brownish in colour.

(7) **Oxygen.** It is also necessary for the formation of chlorophyll. Seedlings fail to develop chlorophyll in the absence of oxygen, even when these are exposed to sunlight. Chlorophyll formation is, therefore, an oxidation process.

(8) **Carbohydrates.** Cane-sugar, grape-sugar, etc., are also necessary for the formation of chlorophyll. Etiolated leaves without soluble carbohydrates in them develop chlorophyll and turn green in colour when floated on sugar solution.

(9) **Heredity.** This is a powerful factor and determines the formation of chlorophyll in the offspring. Familiar examples are garden crotons, aloes, aroids (e.g. *Caladium*), amaranth, etc.

Chemistry of Chlorophyll. Chlorophyll, as it exists in the chloroplasts, is a mixture of four different pigments, as given below.

1. Chlorophyll a, $C_{55}H_{72}O_5N_4Mg$—a blue-black micro-crystalline solid.
2. Chlorophyll b, $C_{55}H_{70}O_6N_4Mg$—a green-black micro-crystalline solid.
3. Carotene, $C_{40}H_{56}$—an orange-red crystalline solid.
4. Xanthophyll, $C_{40}H_{56}O_2$—a yellow crystalline solid.

Extraction of Chlorophyll and Separation of Pigments. Pieces of green leaves (preferably grass leaves) are bruised in a mortar with sand particles and transferred to a flask with about 100 c.c. of 80% acetone. The flask is shaken for a few minutes and then allowed to stand. Acetone takes up all the colour (deep green) which is then filtered off through a filter-funnel. The solution shows fluorescence (see p. 154). It is poured into a separating funnel, and petroleum ether and methyl alcohol added to it. The solution is stirred with a glass rod and allowed to settle. Within 2 or 3 minutes the solution separates out into two distinct layers—the upper (call it A) being deep green (containing chlorophyll a and carotene—both soluble in petroleum ether), and the lower (call it B) light greenish yellow (containing chlorophyll b and xanthophyll—both soluble in methyl alcohol). From the separating funnel the liquid B is collected in a test-tube, leaving the liquid A intact in the funnel. KOH solution and methyl alcohol are added separately to both the liquids (A and B) which are then shaken for a few minutes. The two liquids are then allowed to settle. Almost immediately each liquid separates out into two layers. In A the upper layer is deep green containing chlorophyll a, and the lower layer is yellowish brown containing carotene. In B the upper layer is greenish containing chlorophyll b, and the lower is light yellow containing xanthophyll. Finally the four pigments are separated through the separating funnel and collected in four test-tubes. The nature of each pigment may be verified by spectrum analysis.

Effect of Rays of Light on Photosynthesis. It is a known fact that white light is composed of seven colours arranged in the following order; red, orange, yellow, green, blue, indigo and violet. Although photosynthesis normally takes place in white light, it is only a few of the above rays that are required for this function. Experiments carried out by Sachs (1864), Timiriazeff (1875), Engelmann (1881) and others have shown that chlorophyll normally utilizes a large portion of the red ray and also the blue-violet rays, and to some extent only the orange ray. Green and other rays are not utilized to any extent (see experiments 27 a-c).

Experiment 27. To find out the rays of light utilized in photosynthesis.

(a) By covering a green starch-free pot plant with a double-walled bell-jar which is filled at a time with a red solution (aniline red), orange-yellow solution (potassium dichromate), blue solution (copper sulphate and ammonia; this transmits blue and violet rays), or green solution (ammoniacal copper sulphate and potassium dichromate), and exposing the same to bright sunlight for some hours the leaves may be tested for starch grains. The intensity of black colour which is the effect of iodine treatment gives a comparative idea about the quantity of starch formed in different rays. In the case of water plants the similar effect of rays of light may be studied from the rate of evolution of oxygen bubbles.

(b) Canong's light-screen (large form). The proper use of this instrument gives a good idea about the rays of light utilized in photosynthesis. Of the five vials in it four are filled each with red, orange, green and blue solution, and the fifth vial with water. A starch-free broad green leaf of a plant is inserted through the hole of the light-screen to lie flat underneath the five vials. Evidently different rays of light will fall on the leaf in strips. After exposure to sunlight for some hours the leaf is bleached and treated with iodine solution. The intensity of coloration will indicate which rays have been utilized for photosynthesis.

(c) **Spectrum analysis of chlorophyll solution.** A direct-vision spectroscope may be used for this purpose. The spectrum analysis will show a broad dark band in the region of red, another in blue-violet, but only a small one in orange. Therefore, these are the rays mainly utilized by chlorophyll for photosynthesis.

Photosynthesis by Autotrophic Bacteria. Autotrophic bacteria may be photosynthetic or chemosynthetic. Photosynthetic bacteria develop a pigment closely related to chlorophyll. They are purple sulphur bacteria and green sulphur bacteria. Both are anaerobic. **Purple sulphur bacteria** develop a purple pigment (*bacteriochlorophyll*) which may be a substitute for chlorophyll. According to Van Niel (1941), an American microbiologist, they utilize light as a source of energy to decompose hydrogen sulphide (H_2S)—a poisonous gas (and not water as in normal photosynthesis)—into hydrogen and sulphur. The hydrogen thus released is used to reduce CO_2 to carbohydrate in a series of dark reactions, as in green plants. Sulphur (and not oxygen) is liberated in the process. In this case H_2S oxidation and CO_2 reduction are closely interrelated according to the following overall equation—$6CO_2 + 12H_2S \rightarrow C_6H_{12}O_6 + 6H_2O + 12S$. Similarly, **green sulphur bacteria** containing the pigment *bacterioviridin* use light-energy for the oxidation of H_2S with the liberation of hydrogen which is then used for reduction of CO_2 to carbohydrate in a stepwise manner. This may indicate a primitive process of photosynthesis from which higher forms of plants have developed the use of water (H_2O) as the 'donor' of hydrogen in the normal photosynthetic process.

Chemosynthesis. Colourless autotrophic bacteria are able to synthesize carbohydrate without chlorophyll and, therefore, without light as a source of energy. These bacteria are aerobic, and the energy required for the metabolic processes is derived from the oxidation of certain inorganic compounds present in their environment. The energy released by this oxidative process is used to convert CO_2 through several intermediate reactions to carbohydrate and other organic compounds. Chemosynthesis thus involves transformation of one kind of chemical energy (and not light-energy) to another. Chemosynthetic bacteria do not use water (H_2O) either, as the 'donor' of hydrogen to reduce CO_2, as in normal photosynthesis and, therefore, no liberation of oxygen takes place in this process. This may be an evidence that oxygen released in normal photosynthesis comes from water (H_2O) and not from carbon dioxide (CO_2). Common chemosynthetic bacteria are sulphur bacteria, iron bacteria and nitrifying bacteria.
(*a*) **Sulphur bacteria** grow in sulphur springs and in stagnant water containing hydrogen sulphide. They are filamentous cells. They obtain their energy for the synthetic processes, as first shown by Winogradsky (1887), by the oxidation of sulphur compounds present in water. Thus they oxidize hydrogen sulphide (H_2S) and use the energy released to reduce CO_2 to carbohydrate and other organic compounds. Sulphur becomes deposited in the bacterial cells, and this deposit is used again as necessity arises. The reaction may be expressed by the following equation—
$2H_2S + O_2 \rightarrow 2S + 2H_2O + 32.5$ cal. During the process some of the sulphur deposit is further oxidized into sulphuric acid (H_2SO_4).
(*b*) **Iron bacteria** are filamentous types and grow in lakes and marshes, and in water containing ferrous iron. They obtain their energy for synthesis of organic compounds by the oxidation of

MANUFACTURE OF FOOD

ferrous hydroxide to ferric hydroxide. Such water is often reddish in colour. Bog iron ore and deposits of iron oxides in lakes and marshes are due to their activity. (c) **Nitrifying bacteria** live in the soil and are of two types. As already discussed (see p. 292), they transform ammonia (NH_3), formed during the process of protein decay in the soil, first into nitrites by mobile, spherical nitrite bacteria (*Nitrosomonas*)—$2NH_3 + 3O_2 \rightarrow 2HNO_2 + 2H_2O + 79$ cal., and then into nitrates by non-mobile, rod-shaped nitrate bacteria (*Nitrobacter*)—$2HNO_2 + O_2 \rightarrow 2HNO_3 + 21.6$ cal. The energy thus liberated by the oxidation of ammonia is used to make their own metabolic products, together with CO_2 absorbed from the air.

II. PROTEINS

Nature of Proteins. These are very complex organic nitrogenous compounds found in plants and animals (see also p. 173). Analyses of plant proteins show that carbon, hydrogen oxygen, nitrogen, sulphur and sometimes phosphorus enter into their composition, but we know little about their molecular structure. Protein molecules are often very large and extremely complex consisting of thousands of atoms and are composed of several chains of amino-acid molecules arranged in a definite order. Proteins are of high molecular weights ranging from several thousands to several millions (as a comparison may be cited examples of glucose and sucrose, having 180 and 342 respectively). Proteins are also very complex in their chemical composition. As a matter of fact, the structural formulae of not more than a few proteins have been determined so far. The first formula of a protein was established by Dr Frederick Sanger of Cambridge University in 1954 after ten years' hard work. The protein is *insulin,* and its molecular formula has been determined to be $C_{254}H_{377}N_{65}O_{75}S_6$. Its molecule is, however, small compared to most proteins but it is made of a sequence of 51 amino-acids, evidently with repetition of many of them. Some of the other formulae known are: of *zein* of maize—$C_{736}H_{1161}N_{184}O_{208}S_3$ and of *gliadin* of wheat—$C_{685}H_{1068}N_{196}O_{211}S_5$. Besides the elements mentioned above, small quantities of sodium, potassium, magnesium and iron are also present. Various kinds of proteins are found in plants (see pp. 174-5). Amino-acids are the initial stages in the formation of proteins, and they are also the degradation products of the latter. Proteins are the main constituents of the living materials, and occur to the extent of nearly half the total dry weight of them. They are linked with the metabolism of the cell and its vital activities; with high rate of protein synthesis there is correspondingly a high rate of respiration; and proteins are essential for growth and repair of the body (see also p. 174).

Synthesis of Proteins. Proteins are normally formed from nitrates absorbed from the soil. But the chemical reactions leading to the formation of these complex compounds are only imperfectly known. It has been already mentioned that phosphoglyceric acid holds a central position from which chains of reactions start— one leading to the formation of carbohydrates, while others to various organic acids, fats and proteins through phosphopyruvic acid, as revealed by radioactive carbon. It is likely that synthesis of the various products mentioned above proceeds simultaneously in different directions. Protein synthesis mostly takes place in the meristematic and storage tissues; some proteins are also formed in all active cells of the plant body. The whole process of protein synthesis takes place in three different stages.

(A) **Reduction of Nitrates.** Nitrogen is an essential constituent of all proteins. We know that the nitrate of the soil is the main source of supply of nitrogen to the higher plants. After the nitrate is absorbed into the plant body it is first reduced to nitrite by the enzyme *nitrate reductase*, and then nitrite reduced to ammonia (NH_3) by *nitrite reductase* through some unknown intermediate compounds, as follows: $-NO_3 \rightarrow -NO_2 \rightarrow NH_3$. The detailed chemical stages in the reduction series are, however, yet unknown. Meyer and Schultze in 1894 first suggested that the reductive stages might be as follows: nitrate→nitrite→hyponitrite→hydroxylamine→ammonia. Hydroxylamine has been actually detected in plant cells in small quantities although not hyponitrite. The reduction of nitrate to ammonia usually takes place in the root and in the leaf. At the next step ammonia is directly incorporated into α-ketoglutaric acid (see below). This is now known to be the main channel of entry of inorganic nitrogen into the system of nitrogen assimilation. Ammonia thus holds a key position in the pathway of protein synthesis. This is supported by the fact that a stable isotope N^{15} used in an inorganic nutrient solution can be traced in ammonia ($N^{15}H_3$) and later in α-ketoglutaric acid.

(B) **Synthesis of Amino-acids.** As stated before, amino-acids are the initial stages leading finally to the synthesis of proteins. About 20 amino-acids are known to be constituents of plant proteins. An amino-acid contains an acidic or carboxyl group (-COOH) and a basic or amino group ($-NH_2$) attached to a central C with a side chain or R group. The general structure of an amino-acid is shown on p. 339. The group R, it may be noted, is different in different amino-acids. As already mentioned, ammonia enters into α-ketoglutaric acid and reacts with it. The keto acid, chemically almost similar to an amino-acid, is an important intermediate product formed during the aerobic phase of respiration (see Krebs Cycle, FIG. 34). Ammonia combines with the keto acid, and under the action of the enzyme *glutamic hydrogenase*,

MANUFACTURE OF FOOD

which is widespread in plants, the first amino-acid makes its appearance in the form of *glutamic acid,* as demonstrated by Vickery in 1940 by the use of radioactive ammonia, i.e. $N^{15}H_3$. The glutamic acid thus formed holds a central position from which several other amino-acids are formed by a process called transamination (see below). Wilson in 1954 detected as many as 17 amino-acids formed by this process. Some amino-acids may also be formed otherwise; for example, NH_3 may directly combine with oxalacetic acid (and also fumaric acid) to produce *aspartic acid,* and with pyruvic acid to produce *alanine.* Amino-acids are mainly synthesized in the green leaf and in the root. There is experimental proof that the reduction of nitrate to ammonia and synthesis of amino-acids usually take place in the green leaf in the presence of light. Evidently, the synthetic process is correlated with photosynthesis which supplies necessary carbon, hydrogen and oxygen (80-85% of amino-acids are non-nitrogenous). It is, however, not known how exactly this takes place. Synthesis of amino-acids in the root appears to be connected with root respiration which supplies necessary energy.

Transamination. As stated before, glutamic acid holds a central position from which several other amino-acids may be formed by this process. Transamination involves transfer of an amino group ($-NH_2$) of the glutamic acid to the carboxyl group ($-COOH$) of any of the keto acids under the action of specific enzymes called *transaminases* with the formation of corresponding new amino-acids, as first demonstrated by Engel in 1934 and later by several others. Some amino-acids may also be formed by transfer of amino group of the glutamic acid or of the keto acid to oxalacetic acid to produce *aspartic acid,* and also to pyruvic acid to produce *alanine.* It will be noted that different transaminases work to produce different amino-acids. There are twenty different amino-acids that take part in the synthesis of proteins. The simplest amino-acid is *glycine* represented by the formula NH_2-CH_2-COOH; other important but complex ones are *alanine, leucine, aspartic acid, glutamic acid, cystine, tyrosine,* etc. The amino-acid *cystine* which is formed in all plants also contains sulphur. It may be noted that animals do not normally utilize ammonia to produce an amino-acid.

$$R-CH-COOH \qquad \text{e.g.} \quad H-CH-COOH \qquad CH_3-CH-COOH$$
$$\quad\;\; | \qquad\qquad\qquad\qquad\;\; | \qquad\qquad\qquad\qquad\;\; |$$
$$\;\; NH_2 \qquad\qquad\qquad\quad\; NH_2 \qquad\qquad\qquad\quad\; NH_2$$
$$\text{(general formula)} \qquad\quad \text{(glycine)} \qquad\qquad\quad \text{(alanine)}$$

(*C*) **Synthesis of Proteins.** Amino-acids are the precursors of all proteins and, therefore, they must be available to the living cells for final elaboration into proteins. Protein molecules are very large and complex, and hundreds of them may occur in a single cell. A protein molecule may finally be formed by linkage of hundreds or thousands of amino-acid molecules which may practically be arranged in an infinite variety of chains. Linkage of different amino-acids in several long chains under the action of specific enzymes results in the synthesis of an infinite variety of proteins (see p. 174.), each species having its own characteristic types. The arrangement of particular amino-acids in specific sequences in the chain, with often repetition of one or more of them in it, is a pre-determining factor responsible for making a particular kind of protein. It follows, therefore, that the omission of a single required amino-acid in the chain directly affects protein synthesis. Apart from enzymes, the reactions in protein synthesis are activated by ATP. It may be noted that a compound formed as a result of combination of two or few amino-acids in short chains is called a *peptide*. It is rather rare in plant or animal cells. When many amino-acids combine the resulting compound is called a *polypeptide*. Thus by linkage of several amino-acid molecules proteins are formed in the following order: amino-acids → polypeptides → peptones → proteoses → proteins. Emil Fischer first suggested (1899-1906) that proteins are formed by condensation of numerous amino-acids. He mentioned 18 such amino-acids. It may be noted that protein synthesis and protein breakdown, the latter under the action of proteolytic enzymes or by a reversal of the synthetic process, go on simultaneously in plant cells. The amino-acids, thus formed by breakdown, may again combine into proteins. All plant proteins contain sulphur, and more complex ones like nucleoprotein also contain phosphorus. Complex proteins contain sulphur, and more complex ones like nucleoprotein contain phosphorus also. Sulphur and phosphorus are obtained from the soil as sulphates and phosphates. As stated before, amino-acids are mainly formed in the root and in the leaf. They travel from there to distant tissues, and protein synthesis mostly occurs in the meristematic (root-tip and stem-tip) and storage tissues, and to some extent in most living cells. Proteins, however, do not travel as such. It may further be noted that protein synthesis is mostly localized in the ribosomes occurring in plenty in the cytoplasm (see p. 160). Some proteins may also be formed in the nucleus.

DNA and Protein Synthesis. Recent brilliant researches on nucleic acids (DNA and RNA) by a number of investigators (see p. 152) have established the fact that DNA of the chromosome is the sole genetic material controlling all the biosynthetic processes of the cell including protein synthesis. To understand how DNA works in this respect a preliminary acquaintance with the structure of the DNA molecule with its purine and pyrimidine bases (see FIG. II/5) is essential. The DNA of the chromosome prepares a master plan for the specific kinds of proteins and their quantities to be built up in particular cells of a plant. It must be noted that the synthesis of proteins in a cell is due to the co-ordinated action of DNA, RNA and ribosomes—the last two occurring abundantly in the cytoplasm of a cell (see FIG. II/6). RNA occurs in three forms: 'messenger' RNA or mRNA, 'transfer' or 'soluble' RNA or tRNA or sRNA, and 'ribosomal' RNA or rRNA (as follows). As stated before, protein molecules are made of long chains of amino-acid molecules, and one kind of protein is distinguished from another by the number and kinds of amino-acids and their sequence of arrangements in each protein molecule. It is important to note that the purine and pyrimidine bases of the DNA and RNA molecules occur in infinite sequences. The specific order or sequence of these bases in the double-stranded DNA molecule determines the specific order of amino-acid molecules which will finally be linked to synthesize a particular kind of protein (Hoagland, 1955). The necessary information to guide the synthesis of a protein is known to reside *in coded form* in the particular sequence of bases. DNA, however, does not take a direct part in protein synthesis but RNA acts as its working partner in the process. DNA remains permanently in the chromosome within the safe environs of the nucleus and organizes, as said before, a master plan for the whole process of protein synthesis, and works through *codes* (or chemical messages). DNA coins a particular code and transmits it to 'messenger' RNA (mRNA) which is now used as a messenger carrying the code from DNA to the cytoplasm and finally to a ribosome in it or to a small group of ribosomes called *polysome*. mRNA, it may be noted, is formed by *transcription* of one of the two strands of a DNA molecule, which serves as a template for the synthesis of single-stranded mRNA molecule. The second type of RNA—'transfer' or 'soluble' RNA (tRNA or sRNA)—occurs as comparatively small molecules abundantly in the cytoplasm in a free state. Possibly it originates in the nucleolus under the control of DNA and then moves out to the surrounding cytoplasm. sRNA receives necessary information from mRNA (through pairing of their bases), and accordingly it selects a particular amino-acid, possibly by chemical attraction, out of a heterogenous mixture of compounds present in the cell, with of course the aid of a highly specific enzyme, and carries it (the amino-acid) to the ribosome. It is further known that the amino-acid molecule is attached to a specific location on the sRNA, and that there is a separate such RNA for each amino-acid (Hoagland and his colleagues, 1959). Obviously there are 20 kinds of sRNA for the known 20 amino-acids. The third type of RNA—'ribosomal' RNA or rRNA—occurs in the ribosome. Ribosomes are distributed in the cytoplasm as very tiny particles in large numbers (see p. 158); they are the seats of protein synthesis. Ribosome consists of nucleoprotein, i.e. a nucleic acid (RNA in this case) and a protein in almost equal quantities. The process of protein synthesis is now as follows. DNA coins a particular code (or message) for the purpose of synthesizing a specific protein in a cell, and transmits this code to mRNA which then moves out of the nucleus to a ribosome in the cytoplasm. Several amino-acids, each carried by a sRNA, are brought to the ribosome. It is now for the rRNA to do the rest of the work. It brings about the association of mRNA and sRNA, and receives the code and the amino-acids from them respectively. rRNA now deciphers the code, i.e. translates it into the language of physiology, and accordingly brings about linkage of specific amino-acid molecules in definite sequences into long chains. The result is the synthesis of a specific protein molecule with the amino-acid molecules arranged in a definite order in it. Of course a specific enzyme is indispensable to catalyze each reaction, and the energy required for the whole process is furnished by ATP, an energy-rich phosphate compound. Protein molecules become very large and complex, depending on the number and frequency of amino-acid molecules that enter into their composition, forming long chains. Further, on the basis of the infinite sequences of the two bases (purine and pyrimidine; see FIG. II/5) in the long and complex DNA molecule it may be safely assumed that DNA may coin an infinite number of codes through them (the bases) and thus produce an endless variety of proteins. Although DNA is the same in all plants it produces only specific kinds of proteins for each species of plants. It may further be noted that mRNA after release from a ribosome may move to several other ribosomes (particularly in higher plants). It is also suggested that mRNA may be associated with several ribosomes by its long strand. In any case it soon becomes disorganized. rRNA moves back to the cytoplasm and repeats the process.

Investigations carried out during the last 30 years have established the fact as stated above, for which several investigators were awarded the Nobel Prize in different years. This is, however, a part of the whole story The mechanism of protein degradation still remains to be solved; not much work has been done in this direction either. The methods controlling increased output and less breakdown of proteins in plants, particularly crop plants, so vital from the standpoint of man's protein need, still remain to be achieved, probably at no distant future. Intensive research is progressing in this direction.

III. FATS AND OILS

Fats and oils are lipid (fatty) substances formed in the living cells of both plants and animals. There is only physical difference between the two—at ordinary temperature fats are solids, as mostly in animals, and oils are liquids, as mostly in plants. Fatty oils often occur in abundance in seeds and fruits. All fats and oils are composed of carbon, hydrogen and oxygen, the last one occurring in a low percentage. When fats and oils are hydrolysed by acids or alkalis or by the action of lipase extract, glycerol (glycerine) and one or more fatty acids are the products formed, indicating the compounds out of which the former are built up. Although the different stages in the synthesis of fats and oils are only imperfectly known, they are no doubt the final condensation products of glycerol and fatty acids under the reverse action of lipase. Fats and oils are insoluble in water, and cannot, therefore, diffuse out of the cells in which they are formed. It is known that both glycerol and fatty acids appear in the living cells as a result of carbohydrate breakdown (particularly glucose and fructose) under the action of some enzymes (aldolase, etc.) during the anaerobic phase of respiration, passing through several complicated reactions. The reactions are reversible—from carbohydrate to fats and vice versa. Several fatty acids are formed in plants, e.g. palmitic, stearic, oleic, lauric, linoleic, linolenic, etc. Glycerol and fatty acids do not accumulate in cells but are immediately utilized in fat synthesis. Finally, it may be stated that molecules of glycerol and one or more fatty acids (degraded products of carbohydrates) condense into fat molecules under the action of lipase. With the increase of fatty oils in seeds, it is noted that carbohydrates diminish in quantity. This supports the view that fats and oils are formed from carbohydrates, and three stages are known to be involved in the whole process, viz. (*a*) synthesis of fatty acids, (*b*) synthesis of glycerol, and finally (*c*) synthesis of fats and oils by condensation of the first two.

Chapter 7 SPECIAL MODES OF NUTRITION

Green plants are **autotrophic** (*autos*, self; *trophe*, food) or self-nourishing, that is, they are able to manufacture carbohydrates from raw or inorganic materials and thus nourish themselves. Non-green plants on the other hand are **heterotrophic** (*heteros*, different). Such plants cannot prepare carbohydrates and nourish themselves. They get their supply of carbohydrate food from different sources. They can, however, prepare other kinds of food. Heterotrophic plants are **parasites** when they depend on other living plants or animals, and they are **saprophytes** when they depend on the organic material present in the soil or in the dead bodies of plants and animals.

1. **Parasites** (see pp. 17-9). Total parasites such as dodder (*Cuscuta*), broomrape (*Orobanche*), etc., are never green, and consequently they have no power to prepare their own food. They draw in all their nourishment from the host plant on which they are parasitic. Partial parasites like mistletoe (*Viscum*), *Loranthus*, *Cassytha*, etc., on the other hand, are green in colour and are, therefore, not entirely dependent on the host plant. **Parasitic phanerogams** develop haustoria or sucking roots which go into the vascular bundles of the host plant and absorb from them the prepared food materials and water. Fungi and bacteria are either total parasites or saprophytes. Parasitic fungi send their mycelia into the tissue of the host plant. The mycelia ramify in all directions, and absorb the necessary food materials. Parasitic bacteria infect living plants and animals and absorb food from their bodies.

2. **Saprophytes** (see p. 20). Saprophytic phanerogams such as Indian pipe (*Monotropa*; see FIG. I/27), some orchids, saprophytic fungi and saprophytic bacteria grow on decaying animal or vegetable matter, and absorb the organic food from it.

3. **Symbionts** (see p. 20). Two organisms living in close association with each other, being of mutual benefit, are called symbionts, and the condition is known as symbiosis. Lichen, mycorrhiza, etc., are illustrative examples.

4. **Carnivorous Plants.** These plants are known to capture lower animals of various kinds, particularly insects. They digest the prey and absorb the nitrogenous products (proteins) from their body. Being green in colour they can manufacture their own carbohydrate food. Altogether over 450 species of carnivorous plants have till now been discovered representing 15 genera belonging to 5 or 6 families; of them over 30 species occur in India.

Classified according to their systematic position

(a) *Droseraceae* (105 sp.), e.g. *Drosera* (100 sp.)—cosmopolitan, *Dionaea* (1 sp.)

—the United States, *Drosophyllum* (1 sp.)—Morocco, Spain and Portugal, and *Aldrovanda* (1 sp.)—Europe, Asia and Australia.

(b) *Sarraceniaceae* (17 sp.), e.g. *Sarracenia* (10 sp.)—Atlantic North America, *Darlingtonia* (1 sp.)—California, and *Heliamphora* (4 sp.)—Guiana.

(c) *Nepenthaceae* (68 sp.), e.g. *Nepenthes* (67 sp.)—Malaysia, Australia, India and Madagascar.

(d) *Cephalotaceae* (1 sp.), e.g. *Cephalotus* (1 sp.)—Australia.

(e) *Lentibulariaceae* (170 sp.), e.g. *Utricularia* (120 sp.)—cosmopolitan, and *Pinguicula* (35 sp.)—north temperate regions and temperate Himalayas.

Classified according to their mode of catching the prey

(a) Plants with sensitive glandular hairs secreting a sweet viscid glistening substance, e.g. sundew (*Drosera*), butterwort (*Pinguicula*) and *Drosophyllum*.

(b) Plants with special sensitive hairs—trigger hairs—on the leaf-surface, e.g. Venus' fly-trap (*Dionaea*) and *Aldrovanda*. *Aldrovanda* is an aquatic plant.

(c) **Plants** with leaves modified into pitchers, e.g. pitcher plant (*Nepenthes*)—leaf partly modified into pitcher, other pitcher plants (*Sarracenia*, *Darlingtonia*, *Heliamphora* and *Cephalotus*)—entire leaf modified into pitcher.

(d) Plants with leaf-segments modified into bladders, e.g. bladderwort (*Utricularia*). Bladderworts are mostly aquatic; a few tropical ones are terrestrial.

(1) **Sundew** (*Drosera*; FIG. 26)—100 sp. Only 3 species, viz. *D. peltata*, *D. burmanni* and *D. indica*, have been found in India. They are small herbs. Each leaf is covered on the upper surface with numerous glandular hairs known as the **tentacles**. Each gland, which is reddish in colour, secretes a kind of viscous fluid which glitters in the sun like dewdrops and hence the name 'sundew'. The gland is sensitive and reacts only to chemical stimulus; this means that the movement of the tentacles is initiated by the presence of nitrogenous substances. No movement is exhibited by contact with any foreign object. When any insect mistaking the glistening substance for honey alights on the leaf, it gets entangled in the sticky fluid, and the tentacles stimulated by the digestible compounds present in the body of the insect bend down on it from all sides and cover it. When it is suffocated to death the process of digestion begins. The tentacles remain bent over the insect until all the nitrogenous compounds contained in its body have been absorbed. Digestion is extracellular in all carnivorous plants. The glands secrete an enzyme, called *pepsin hydrochloric*

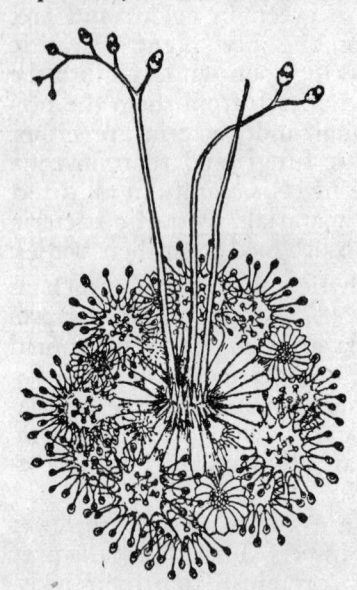

FIG. 26. Sundew (*Drosera*).

acid, which acts on the insect and changes the proteins of its body into soluble and simple forms. The products of digestion are then absorbed by the leaf. The carbonaceous materials are rejected in the form of waste products.

If the tentacles be poked with any hard object, they show no movement nor does any secretion of the enzyme take place. On the other hand, if a bit of raw meat be placed on the leaf, the tentacles bend over it and the glands begin to secrete the enzyme.

(2) **Butterwort** *(Pinguicula;* FIG. 27) —35 sp. Only 1 specie has been recorded from India, and that is *Pinguicula alpina.* It grows in mossy beds in the alpine Himalayas at an altitude of 3,000 to 4,000 m. The species of butterwort are small herbs with a scanty development of roots. The surface of the leaf is covered with numerous glands—some sessile and some stalked, the former being water-secreting and the latter mucilage-secreting in nature. When any small insect alights on the leaf, it gets caught by the sticky glands. Stimulated by the presence of proteins in the insect body the margins of the leaf roll inwards enclosing it. The sessile glands then begin to secrete *pepsin hydrochloric acid* which carries on the digestion of proteins. Digested products are then absorbed by the plant. After digestion and assimilation the leaf unrolls again. Carnivory in butterwort was studied by Darwin who found that meat, egg-white, cartilage, small seeds, pollen grains and other substances containing nitrogen, when placed on the leaf, caused secretion of the enzyme, while those containing no nitrogen excited no secretion.

FIG. 27. Butterwort *(Pinguicula).*

(3) **Venus' Fly-trap** *(Dionaea muscipula;* FIG. 28)—1 sp. The plant is a native of the U.S.A. It is herbaceous in nature and grows in damp mossy places. Each half of the leaf-blade is provided with three long pointed hairs—trigger hairs—placed triangularly on the leaf-surface. The hairs are extremely sensitive from base to apex. The slightest touch to any of these hairs is sufficient to bring about a sudden closure of the leaf-blade, the mid-rib acting as the hinge. The upper surface of the leaf is thickly covered with reddish digestive glands. When the insect is caught, or any nitrogenous material such as meat, fish, etc., placed on the leaf, it closes suddenly and the glands begin to secrete the enzyme *pepsin hydrochloric acid.* The enzyme then

brings about the digestion of proteins contained therein. The acid is stronger in this case than in most other carnivorous plants.

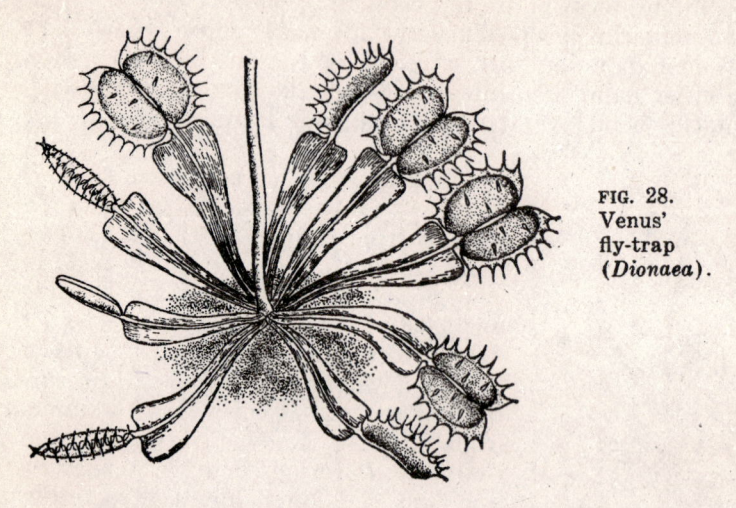

FIG. 28. Venus' fly-trap (*Dionaea*).

(4) **Water fly-trap** (*Aldrovanda vesiculosa*; FIG. 29-30)—1 sp. This plant is very widely distributed over the earth. It has been found in abundance in the salt-lakes of the Sundarbans, salt-marshes south of Calcutta, the freshwater 'jheels' of Bangladesh and in several tanks in Manipur. *Aldrovanda* may be regarded

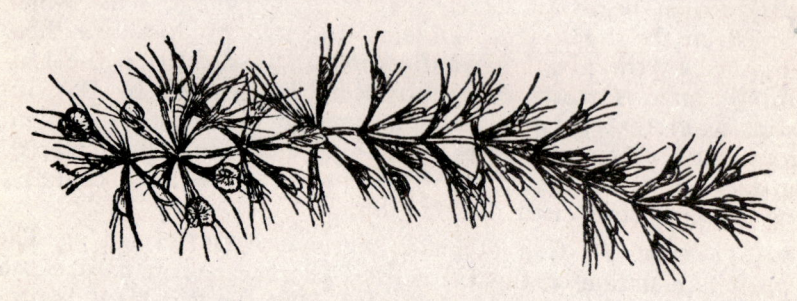

FIG. 29. Water fly-trap (*Aldrovanda*).

as a miniature *Dionaea* in some respects. It is a rootless, free-floating plant with whorls of leaves. The mechanism for catching the prey is practically the same as that of *Dionaea*, but instead of only six sensitive hairs there is a number of them here on either side of the mid-rib, and the leaf is protected by some bristles. There are, of course, numerous digestive glands on the upper surface of the leaf, and the margins beset with minute teeth pointing inwards.

SPECIAL MODES OF NUTRITION

(5) **Pitcher Plant** (*Nepenthes*; FIG. 31 and I/69)—67 sp. Only one species (*Nepenthes khasiana*) has been found in North-East India (in the Garo Hills and Khasi-Jaintia Hills of Meghalaya). Pitcher plants are climbing herbs or undershrubs which often climb by means of the tendrillar stalk of the pitcher. The pitcher itself is the modification of the leaf-blade, the tendrillar stalk supporting the pitcher is the modification of the petiole, and the laminated structure that of the leaf-base. Each pitcher varies from 10 to 20 cm. or even more in height. When it is young the mouth of the pitcher remains closed by a lid which afterwards opens and stands more or less erect. Below the mouth the inside of the pitcher is covered with numerous smooth and sharp hairs, all pointing downwards. Lower down, the inner surface is studded with numerous large digestive glands, each with a hood hanging down over it. Animals, as they enter, slip down the smooth surface, having lost their footing, and get drowned in the fluid that partially fills the cavity of the pitcher. After their death the process of digestion commences. The digestive power of the pitcher of *Nepenthes* was first discovered by Hooker in 1874. The digestive agent, secreted by the glands, is in the nature of a *trypsin*, as was first shown by Vines in 1877. It not only digests the proteins into peptones, but also changes the latter into amines. Amines are readily absorbed by the pitcher. Bits of egg-white, meat, etc., dropped into it, as was first found by Hooker, are seen to be dissolved and ultimately absorbed in the form of amines. Carbohydrates and other materials remain undigested in the pitcher as waste products.

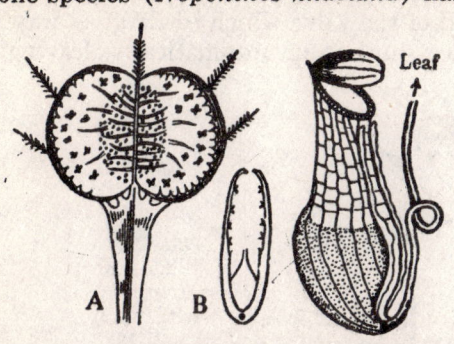

FIG. 30.

FIG. 31.

FIG. 30. *Aldrovanda*; *A*, an entire leaf open; *B*, section of a closed leaf. FIG. 31. A pitcher of pitcher plant. See also FIG. I/69.

(6) **Bladderwort** (*Utricularia*; FIG. 32)—120 sp. Over 20 species have been found to occur in India—*U. flexuosa* is a very common one. They are mostly floating or slightly submerged, rootless, aquatic herbs; there are a few terrestrial species also, e.g. *U. wallichiana* growing in moist hill slopes in Shillong. The leaves are very much segmented, and these simulate roots excepting that they are green in colour. Some of these segments become transformed into *bladders*. Each bladder is about 3-5 mm. in diameter and is provided with a trap-door entrance. The trap-door acts

as a sort of valve which can be pushed open inwards from outside, but never from inside to outside. Very small aquatic animals enter by bending the free end of the valve which easily gives way. After their entrance the valve shuts itself automatically, leaving

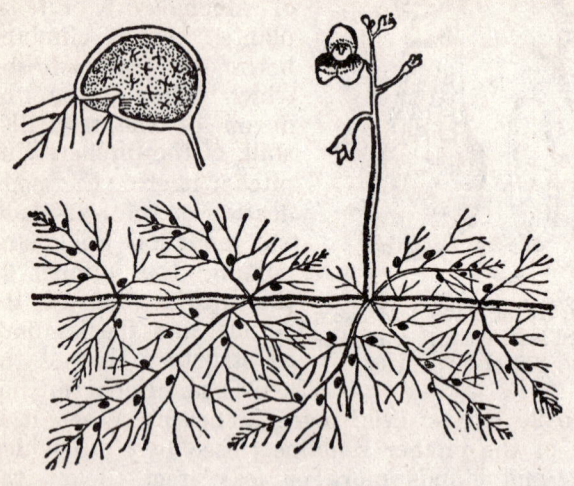

FIG. 32. Bladderwort (*Utricularia*) with many small bladders; *top*, a bladder in section (magnified).

no chance for them to escape. The inner surface of the bladder is dotted all over with numerous digestive glands which vary somewhat in shape. Their function is to secrete the digestive enzyme and absorb the digested products. A bit of raw meat, pushed inside the bladder, is found to disappear within a few days.

Chapter 8 TRANSLOCATION AND STORAGE OF FOOD

Translocation

Food materials are mostly prepared in the leaves. From there they are translocated to the storage organs and the growing regions which often lie at a considerable distance. There are definite and distinct channels extending through the whole length of the plant body for the translocation of raw or inorganic salts and prepared or organic food materials. As already discussed (see p. 314), an upward movement of inorganic salts takes place through the xylem. For downward transmission of food materials prepared in the leaves these are first rendered soluble and diffusible by enzymic actions—carbohydrates converted into sugars, and proteins into amino-acids and amines. After reaching the storage organs from the leaves sugars are mostly reconverted into insoluble starch, and amino-acids and amines into different kinds of insoluble proteins. Two questions naturally arise in this connexion: what is the pathway of translocation and what is the mechanism involved in the process? Although an intensive work has been carried out by Dixon (1923), Mason and his associates (1929), Curtis (1935) and many others for about three decades after a prolonged gap since the work of Hartig (1858), Hanstein (1860) and Sachs (1874), the position is far from clear yet in many respects. As a matter of fact, a mass of conflicting views has been expressed by the recent workers which evidently indicate that much remains to be done yet.

Pathway of Translocation. That phloem is the conduction channel was definitely proved by the early workers on the basis of 'ringing' experiments, first introduced by Malpighi in the late 17th century, and later by Hartig (1858), Hanstein (1860) and Sachs (1874). 'Ringing' experiments, it may be noted, also proved to be a useful tool in the hands of the recent workers. With improved technique, chemical analyses and use of radio-active elements they have made many new revelations in this aspect. (a) Chemical analyses of the contents of the phloem have revealed the presence of many food substances in it, such as sugars (predominantly cane-sugar in many cases) among the carbohydrates, and mostly amino-acids and amines and also some soluble proteins among the nitrogenous materials. This fact itself is no proof of conduction but it leads to this view. It has also been noted that certain insects and aphids sucking the phloem of the tender shoot and the leaf-stalk show the presence of sugar in their body. (b) By 'ringing' or 'girdling' experiments it has been shown by several workers that the food materials mainly move through the phloem (sieve-tubes and companion cells). If a ring of bark completely encircling a branch be removed down to the wood (evidently including phloem), swellings and later buds appear just above the ring or girdle due to the accumulation of food which cannot move farther down, the phloem having been removed. If such a girdle be made at the base of a flowering shoot, large flowers and fruits may be produced for the same reason. Similarly, if a ring of bark be removed from the lower end of a cut leafy branch (say, of garden croton) and the cut end dipped into water, roots are seen to develop only above the girdle where food has accumulated, the phloem portion being removed with the ring. The 'ringing' experiments prove that it is the phloem that is primarily responsible for transporting the food downward.

From the above and other experiments the following facts

become apparent. Phloem is definitely the principal conduction channel for food. Carbohydrates and proteins travel in the form of sugars (pre-eminently cane-sugar) and amines and amino-acids respectively. Translocation is through the sieve-tubes and to some extent through the companion cells and phloem parenchyma. It will be noted that all the elements of phloem are living. Food materials can easily pass through the perforated sieve-plates. They mostly move downwards from the leaf, and later upwards from the root (see below). The protoplasmic threads may facilitate the movement through the pores of the sieve-plates. Phloem elements are also permeable, and many minute pits develop in their walls. Food also moves through them laterally to other tissues (xylem parenchyma, medullary rays, etc.). Food substances formed in the mesophyll of the leaf and rendered soluble there move towards the vascular bundles. They pass through the border parenchyma and enter into the phloem. Phloem extends in long strands from one end of the plant body to the other so that any soluble compound can be transported from the leaves to other organs, particularly the storage ones and the growing regions. In the storage organs the food accumulates in the form of insoluble complex proteins and starch. Later during the period of active growth, as in the spring, when new buds, flowers and fruits are in the offing, the various forms of stored food are rendered soluble, and, therefore, suitable for travelling. Now an upward movement of the soluble food materials takes place through the phloem at linear rates up to 100 cm. per hour, and finally they are brought to the growing organs (Curtis, 1935). At this time of active growth a part of the food also moves upwards through the xylem (Fischer, 1915; Atkins, 1916; Dixon, 1923). The movement of the solutes in the phloem is bidirectional, i.e. downward and upward, either at the same time or at different times. That this is so has been proved by Biddulph and his colleague in 1944 by growing a plant in an atmosphere containing radioactive carbon dioxide, $C^{13}O_2$. It has been seen that the carbohydrate formed in the leaf, descending down the petiole, moves simultaneously in both the directions—upward and downward—through the stem. This view is also shared by the later workers. There is also a diurnal variation in the rate of movements of the solutes in different plants.

A Short Historical Account. Since the work of Hanstein (1860) and Sachs (1874) there was a time lag. Sachs held the view that there is a mass movement of the solutes (dissolved organic substances) and the solvent (water) through the phloem from the higher concentration region (leaf) to the lower concentration region (root) aided by turgor and tension of the sieve-tubes and the associated cells. In 1923 Dixon while working on the ascent of sap held the view that the anatomical structure of the phloem does not warrant the movement of organic solutes through it and that the flow of such substances must take place through the transpiration current in the xylem under certain tensile forces. This view has not been held valid by the well-known workers like Curtis (1935), Mason and Maskell (1928-29), Czapek (1905 onwards) and others who have strongly defended the view that the food is conducted through the phloem, as was proved by the early workers. Mason and Maskell clearly stated in 1929 that the inorganic salts absorbed by the roots move upward in the transpiration current through the xylem, and that the food elaborated in the leaf and converted into sugars (pre-eminently cane-sugar), amino-acids, etc., is transported through the phloem. Convincing proof of this fact was offered by Hoagland and Stout in 1939 by using radioactive K, Na, etc., in inorganic salts. Curtis in 1935, however, experimentally proved that certain inorganic salts like nitrates and phosphates also move upwards through the phloem. This view was supported by Gustafson in 1937.

Mechanism of Translocation. The transport of solutes through the phloem seems to be a vital process, rather than a purely physical process. With a segment of the twig chilled, or with the supply of oxygen cut off, the rate of translocatory process is very much affected. Further, if segments of phloem tissue be killed with hot wax or steam above and below a node translocation is seen to come to a standstill. The possible mechanism of solute movement in the phloem may be as follows:

1. **Cytoplasmic Streaming.** Curtis in 1923 maintained that food materials in solution are transported through the living cells of the phloem. As the cytoplasmic strands in the sieve-tubes move, although hardly noticeable in most cases, they carry the organic solutes with them, upward or downward, at least to some distance. In his view the transport of food greatly depends on the activity of the living cells. The streaming of the protoplasm accelerates the diffusion of the solute molecules. Further, the physiological conditions that affect the streaming, as he found, also affect the rate of translocation of the solutes. His theory seems to explain the mechanism of translocation. But this streaming, if at all it takes place, is very slow and cannot account for the rather rapid translocation of food through the phloem. Mason and Phyllis have rejected this view.

2. **Diffusion Theory.** Mason and Maskell reported in 1929 that nitrates move up in the transpiration current, and the elaborated food materials—sugars and amino-acids—move down the phloem. On the basis of the 'ringing' experiments carried out by them they found that the total inorganic nitrogen still increased in the leaf during the daytime and decreased at night and that sugars and amino-acids accumulated above the girdle. From the nature and rate of diffusion of sugars and nitrogenous compounds they held the view that the movement of the solutes in the phloem is not due to any mass flow in it. They inferred that the solute movement is due to a kind of diffusion process, not

clearly understood, at an accelerated or decelerated rate, depending on the energy released by respiration as a result of metabolism and the activity of the protoplasm, also inexplicable. Even then the rate of diffusion is far too slow to adequately explain the rapid transport of the solutes through the sieve-tubes.

3. Gradient Pressure. Munch in 1929 was of the view that the movement of the solvent (water) and the solutes (organic food materials) takes place simultaneously at the same rate. According to him there is an area of high turgor pressure in the leaf and an area of low turgor pressure in the root. The former is due to the formation of osmotically active substances like sugar in the leaf by photosynthesis, and the latter is due to the presence of osmotically inactive substances like starch (sugar either converted into starch or utilized in metabolism). Phloem, as is known, is continuous from the leaf to the root. Because of the difference in the turgor pressure (turgor gradient) there is a mass flow of the solutes and the solvent from the region of higher concentration (leaf) to the region of lower concentration (root) through the phloem. The cytoplasmic strands extending through the seive-plates are supposed to facilitate the movement through them. Later in the season, as in spring, higher turgor pressure develops in the root cells as a result of conversion of starch into sugar, and the flow starts in the reverse direction, i.e. upwards. To maintain the turgor pressure in the leaf cells there is a continuous flow of water into them through the xylem. The movement according to this view is unidirectional at a time. This view has, however, been strongly criticised by Mason and his colleagues in 1936.

Although Munch's theory gives a plausible explanation for the mechanism of solute movement in the phloem, it is open to some serious doubts. Thus Crafts in 1931 and 1938 while believing in the simultaneous transport of all organic solutes and water in the phloem through the lumen of the sieve-tubes as well as through the walls raised strong objections to Munch's hypothesis on the following grounds. (1) The resistance offered by the cytoplasmic strands extending through the sieve-plate, almost plugging the pores, is not conducive to the mass flow of the solutes. (2) Innumerable sieve-plates occurring at frequent intervals may in fact act as so many road blocks, interfering with the speed of the mass flow. (3) Turgor pressure cannot be so high a driving force as to account for the movement of the solutes over such a long distance—leaf to root, particularly at the rate known. (4) The movement as far as known is bidirectional at a time. (5) Sometimes translocation has been noted to take place towards the region of higher turgor pressure. (6) Activity of the protoplasm is not accounted for in this theory although it is known that killing the phloem markedly affects or even stops translocation. Munch's hypothesis no doubt fits in well as a physical theory. But it is presumed that the protoplasm plays a more prominent role than hitherto known. Finally it may be said that translocation of food is a complex problem which remains to be solved yet.

Storage

Food is prepared in excess of the immediate need of plants. This surplus food exists in plants in two conditions—either *suitable for travelling* or *suitable for storage*. The travelling form is characterized by *solubility*, and the storage form by *insolubility* in the cell-sap.

Storage Tissues. Tissues meant for storage of food have thin cellulose walls. The cells are mostly parenchymatous in nature. If the walls are thick they are provided with many simple pits in them. Storage cells are also living so that the protoplasm can secrete the necessary enzymes and render the food materials soluble or insoluble, according as translocation or storage is required. All parts made of large-celled parenchyma always contain a certain amount of stored food. Cortex of the root is particularly rich in it, and so also the large pith of the monocotyledonous root. There is also a quantity of food stored up in the endodermis, pith, medullary rays and xylem parenchyma of the stem. Border parenchyma of the leaf has also a store of food in it.

Storage Organs. Food materials are stored up in the endosperm or in the thick cotyledons of the seed for the development and growth of the embryo. In the fleshy pericarp of the fruit there is a considerable amount of food stored up. Food is specially stored up in the fleshy roots such as the fusiform, napiform, conical and other roots, and in the underground modified stems such as the rhizome, tuber, corm, etc. All fleshy stems and branches, as in many cacti and spurges (*Euphorbia*), succulent leaves, as in Indian aloe (*Aloe vera*), American aloe (*Agave*), purslane (*Portulaca*), etc., and fleshy scales of onion always contain a store of food in them. The swollen stem-base of kohl-rabi (*Brassica caulorapa*) and the gouty stem of *Jatropha podogarica* also contain stored food. Stores of food may also be detected in the growing regions and in the floral organs.

Forms of Stored Food. The various forms in which the food materials are stored in these different organs and tissues may now be considered. The food materials may be carbohydrates, proteins, or fats and oils (see also pp. 165-73).

I. CARBOHYDRATES

Starch (see FIGS. II/14-5). This is of universal occurrence in plants with the exception of fungi. As soon as sugar is formed in the leaf it is converted into starch. This starch is in the form of minute bodies without any definite structure. At night, when carbon-assimilation is in abeyance, starch is converted into sugar which travels down to the storage organs and is reconverted there into starch by the agency of leucoplasts. The starch grains deposited by the leucoplasts are much bigger in size and have a distinct, stratified appearance. Starch is insoluble in water, and is stored up as such for a longer or shorter period. When growth of the plant is taking place starch is again converted into sugar which then travels from the storage organs to the growing regions, and is eventually made use of by the protoplasm for its nutrition and growth.

Glycogen. In fungi the carbohydrate is stored in the form of glycogen. It is allied to starch and is readily converted into sugar. It is a white amorphous powder, and dissolves in hot water. It is coloured reddish brown by iodine solution [see also p. 169].

Inulin (see FIG. II/13). This is a soluble carbohydrate and has the same chemical composition as starch. It is found in the underground parts of some *Compositae*, *Liliaceae* and *Amaryllidaceae*. It may be converted into some form of sugar.

Sugars. Grape-sugar is the first carbohydrate formed in the green leaves of plants. But as soon as it is formed it is usually converted into starch. It is only in a few cases such as grape, onion, etc., that grape-sugar is stored up as such. Cane-sugar occurs as reserve food in sugarcane, banana, pineapple, beet, etc. Grape contains 12-15% of glucose and sugarcane contains 10-15% of sucrose.

Hemicellulose (see FIG. II/7). This occurs as a thickening matter in the cell-wall of the endosperm of the date seed. It may be converted into some form of sugar by the action of the enzyme *cytase*.

II. *NITROGENOUS MATERIALS*

Proteins. These are the most complex and at the same time the most important substances. Proteins are the chief constituents of protoplasm and nucleus, and consist of carbon, hydrogen, oxygen, nitrogen and sometimes sulphur and phosphorus; but their exact chemical composition is not known. There are various kinds of proteins found in plants as food, and they occur in three distinct conditions, viz. as definite oval or rounded granules known as the aleurone grains; as amorphous proteins (i.e. with no definite shape); and as soluble proteins which occur in solution in the cell-sap. Aleurone grains (see FIG. II/16) occur abundantly in seeds associated with starch, as in pea, bean, etc., or with oil, as in castor. When they occur with oil they are fairly big in size. The other two kinds of proteins occur in bulbs, tubers and in other storage organs. **Amino-acids** and **Amines**. These are much simpler nitrogenous substances than proteins, and occur in solution mostly in the growing regions. Less frequently they occur in the storage tissue. (See also p. 172).

III. FATS AND OILS

Fats and oils are found in all groups of plants and in almost all living cells of the plant body. In angiosperms they are specially common in seeds and fruits. When oil is in great preponderance usually very little carbohydrate is present. Similarly, when starch occurs in abundance little oil is found. Big aleurone grains often accompany fats and oils, as in castor seed. At ordinary temperature fats are solid and oils are liquid. They are not soluble in water or in alcohol (except castor oil); all of them are, however, soluble in ether, petroleum, chloroform, etc. They occur in the form of globules in the protoplasm often saturating it. They are formed from fatty acids and glycerine. (see also pp. 172-73).

It is known that these substances enter into the composition of protoplasm; but they undergo many chemical changes before they are utilized by the protoplasm. They are decomposed by the enzyme *lipase* into fatty acids and glycerine. By this decomposition a large amount of heat is liberated. When oily seeds germinate a considerable quantity of fatty acids may be detected in them, while glycerine disappears almost immediately. But these substances are finally converted into sugar. Fatty acids and glycerine may also be translocated to the growing regions where they are utilized by the protoplasm. Both of them may readily pass through cell-walls.

Food Stored in the Seed. There is always a considerable amount of food stored up in the cotyledons and in the endosperm of the seed for the use of the embryo as it grows. Food materials occur there in insoluble forms and these are first digested, i.e. rendered soluble and chemically simpler under the action of specific enzymes (see next chapter), and then utilized by the growing parts of the embryo for various purposes such as nutrition and growth of the protoplasm, cell-formation, development of the embryonic parts and also for vigorous respiration. Common forms of such food materials are the following. (1) **Starch** is a very common form of carbohydrate stored up in the seed. Cereals such as rice, wheat, maize, oat, barley, etc., are particularly rich in starch. (2) **Hemicellulose** is deposited as thickened cell-walls of the endosperm of many palm seeds, e.g. date palm, betel-nut-palm, nipa-palm, vegetable ivory-palm, etc., and also in some other seeds, e.g. coffee, mangosteen, etc. (3) **Oils** are deposited in most seeds to a greater or less extent. There is a special deposit of them in seeds like groundnut, gingelly, coconut, castor, safflower, etc. (4) **Proteins** also occur in all seeds in varying quantities. In seeds like pulses they occur in high percentage. Soya-bean contains 35% or more of proteins. Oily seeds also contain a high percentage of proteins, e.g. castor seed. During germination of the seed storage proteins break down under the action of proteolytic enzymes into amino-acids which then migrate to the embryo, and synthesis of new proteins takes place in it.

Chapter 9 • ENZYMES

Enzymes are organic (biological) catalysts, each being a certain kind of protein, secreted by the living cells to bring about thousands of biochemical reactions in various metabolic processes such as photosynthesis, respiration, digestion of food, etc., in both plants and animals. Their significance is that they initiate a particular biochemical reaction in a substrate and actually accelerate and regulate the rate of this reaction by their surface energy, without themselves undergoing any chemical change in the process. So enzymes are regarded as biological catalysts which catalyze thousands of chemical compounds, often within the cells. Each enzyme is a special protein, usually bringing about one particular reaction. It is apparent that but for the activity of the enzymes life would have been very slow or even still. Their importance in plant life and animal life thus cannot be overestimated. It is now definitely known that the mitochondria and the chloroplasts are the main seats of enzyme synthesis. All the enzymes connected with the oxidation of pyruvic acid in the Krebs cycle of respiration are synthesized in the mitochondria; while those required for fixation of CO_2 in photosynthesis (dark reactions) are synthesized in the chloroplasts. Further, it may be noted that the synthesis of each enzyme is controlled by a gene. This is the 'single gene—single enzyme' hypothesis of Beadle and Tatum, 1959. A large number of enzymes are concerned with hydrolysis. This means that the elements of H_2O enter into such a reaction. They are mostly the digestive enzymes. The substance on which an enzyme acts is called the *substrate*. All living cells secrete enzymes, always in small amounts, to act on a number of products simultaneously or one after the other in a regulated manner effecting reaction after reaction. The term 'enzyme' was first applied by Kuhne in 1876 to the plant juice which, as he found, could bring about digestion of certain chemical compounds. In 1897 Buchner first discovered that extract from crushed yeast cells could bring about fermentation in sugar solution. In 1903 he isolated the first enzyme. (Buchner was a winner of Noble Prize). Since then many other enzymes have been discovered, and it is now definitely known that thousands of chemical reactions in the plant body and the animal body are due to them. The enzyme 'urease' was first isolated by Sumner in 1926 in crystalline form. Since then several enzymes have been obtained in pure crystalline form. Enzymes are very complex chemical substances made of large and heavy molecules. They occur in the form of colloids in the protoplasm which may secrete one or more enzymes at a time. All enzymes are destroyed by high temperatures. Apart from mitochondria several other structures of the plant cell such as the nucleoli, plastids, ribosomes and lysosomes are also concerned in the manufacture of specific types of enzymes. They are soluble in water, alcohol, dilute glycerine, and in dilute acid or alkali. Some enzymes consist entirely of proteins, e.g. certain proteolytic enzymes and amylases which act directly on the substrate. There are, however, many enzymes which have two components acting together for any enzymic activity—a protein part and a non-protein part (a prosthetic group or a coenzyme). When the non-protein part is firmly attached to the protein part and is not separable from it, it (the non-protein part) is called a **prosthetic group,** and when it can be readily separated from the protein part by dialysis, it is called a **coenzyme**. Neither is active without the presence of the protein. All the oxidizing-reducing enzymes have a prosthetic group. The latter may be a metal or an organic compound with a metal. Thus the enzymes tyrosinase and ascorbic acid oxidase have a protein portion and a copper atom. Cytochrome oxidase, catalase and peroxidase contain iron (as iron porphyrin). Likewise zinc, manganese, cobalt, nickel, magnesium, calcium, potassium, etc., may form prosthetic groups of certain enzymes. In such cases the prosthetic groups act as specific *activators*. Some oxidizing enzymes (respiratory and fermenting) contain organic compounds as prosthetic groups. The coenzyme components of many enzymes include a variety of compounds. The protein portion and the coenzyme may occur separately in the same cell but they must be linked together or closely associated for the whole enzyme to be effective. The same prosthetic group or coenzyme may combine with different kinds of proteins and form different kinds of enzymes. Several vitamins (e.g. vitamin B complex) are now considered as forming coenzyme components of many enzymes. Similarly NAD (formerly called DPN), NADP, ATP, FAD, coenzyme A, etc. act as coenzymes. **Zymogen**. Cells sometimes do not directly produce an enzyme but at first a chemical precursor of an enzyme, known as zymogen or pro-enzyme, is secreted by the protoplasm. The zymogen then becomes converted into an active enzyme. Many kinds of zymogens have been identified so far, and they are now known to combine with different protein components as prosthetic groups or coenzymes to form different kinds of enzymes.

Properties of Enzymes. (1) *Specificity* Each enzyme is a specific catalyst, i.e. for a particular substance there is a particular enzyme; for instance, the enzyme that acts upon starch will **not** act on protein or any other substance. This is expressed as 'lock and key' action. Some enzymes (e.g. emulsin) can, however, act on a number of substances having a certain molecular pattern. Conversely, different enzymes may act on the same substrate with the formation of different end-products. Thus pyruvic acid may be converted into a number of compounds under the influence of different enzymes. (2) *Catalytic Property*. The enzyme acts as a catalytic agent: this means that the presence of the enzyme induces some chemical reaction in a particular substance without itself undergoing any chemical change. Thus the enzyme may be regarded as an organic catalyst. After the chemical change is over, it is released unchanged and used over again. (3) *Inexhaustibility*. The enzyme is never exhausted while it works, i.e. a small quantity of it can act on an almost unlimited supply of the substance provided that the products of its action are removed from the seat of its activity. Finally of course the enzyme breaks down and disappears. (4) *Colloidal Nature*. Enzymes occur in the protoplasm in the form of colloids. Enzyme molecules are very large and fall within the colloidal system. They have high molecular weights, and being very large in size diffuse slowly and can in many cases be more or less easily separated by dialysis. (5) *Reversible Action*. Some enzymes can bring about reactions in both directions. For example, the enzyme lipase can bring about synthesis of fats from glycerine and fatty acid, and can under certain other conditions decompose fats into glycerine and fatty acid. This is also true of some proteolytic enzymes, dehydrogenases and transferring enzymes. (6) *Sensitivity to Heat*. The enzymes in liquid medium are destroyed at a temperature of 60-70°C. In dry seeds and spores the enzymes can often stand a temperature of 100-120°C., at least for some time. The enzymes extracted from plant tissues can also stand this high temperature. (7) *Activators and Inhibitors*. There are certain compounds (particularly the prosthetic group) which accelerate the activity of certain enzymes. There are others which inhibit their action or even destroy them.

Classification of Enzymes. Enzymes have been classified in different ways by different authors. They are too numerous, and our knowledge about the chemical structure of most of them is still imperfect. This being so, a perfect classification has not been possible yet. A simple classification based on the kinds of reactions that they catalyze may be, as follows.

I. HYDROLYTIC ENZYMES OR HYDROLASES

A. Carbohydrases. (1) *Diastase* (or *amylase*) hydrolyses starch to dextrin and maltose. (2) *Maltase* hydrolyses maltose to glucose. (3) *Invertase* (or *sucrase*) hydrolyses sucrose to fructose and glucose. (4) *Inulase* hydrolyses inulin to fructose. (5) *Cellulase* hydrolyses cellulose to cellobiose to glucose. (6) *Cytase* (or *hemicellulase*) hydrolyses hemicellulose to glucose. (7) *Emulsin* hydrolyses glucosides to glucose and a non-sugar.

B. Proteolytic Enzymes. (*a*) *Proteoses*: (1) *Pepsin* hydrolyses proteins to peptones. (2) *Trypsin* hydrolyses proteins to polypeptides and amino-acids. (3) *Papain* hydrolyses a wide variety of proteins to polypeptides and amino-acids. (*b*) *Peptidases*: (4) *Erepsin* hydrolyses polypeptides to amino-acids. (*c*) *Amidases*: (5) *Asparaginase* hydrolyses asparagine to aspartic acid and ammonia. (6) *Glutaminase* hydrolyses glutamine to glutamic acid and ammonia.

C. Esterases. (1) *Lipase* breaks down fats into fatty acids and glycerine. Lipase also brings about reversible reaction, i.e. it can synthesize fats from fatty acids and glycerine. (2) *Pectase* converts pectin to pectic acid. (3) *Tannase* converts tannin to glucose and digallic acid.

II. OXIDIZING-REDUCING ENZYMES

D. Oxidases are a group of respiratory enzymes which bring about oxidation of certain chemical compounds in the presence of oxygen. *Cytochrome* (cell-colour), a pigmented member of this group closely related to chlorophyll, acts as a hydrogen-carrier. During certain oxidation-reduction processes certain compounds become oxidized with the transfer of hydrogen to cytochrome (hydrogenated or reduced). Under the influence of *cytochrome oxidase* the reduced cytochrome is again oxidized (dehydrogenated) and it goes back to its original condition. Other common oxidases are *tyrosinase* and *ascorbic acid oxidase*.

E. Catalases decompose hydrogen peroxide, H_2O_2, to water and molecular oxygen. They are very widely distributed in plants. They do not allow H_2O_2, a toxic compound, to accumulate in cells.

F. Peroxidases decompose H_2O_2 to water and active oxygen causing oxidation of many compounds. They are very widely distributed in plants.

G. Dehydrogenases act on many organic compounds, particularly organic acids; for example, they convert ethyl alcohol to acetaldehyde; malic acid to oxalacetic acid; lactic acid to pyruvic acid, etc. These enzymes are probably present in all plants and they act by transferring hydrogen from one compound to another, accompanied by electron transfer. The compound that releases hydrogen (i.e. dehydrogenated or oxidized) is said to be a *hydrogen-donor* and the one that receives hydrogen (i.e. hydrogenated or reduced) is said to be a *hydrogen-acceptor*. Their action is reversible.

H. Carboxylases convert pyruvic acid to acetaldehyde and carbon dioxide; oxalacetic acid to pyruvic acid and carbon dioxide; amino-acids to amines and carbon dioxide.

I. Zymase, an enzyme complex, consists of a complex mixture of several enzymes, coenzymes and also inorganic ions, and is concerned in oxidation-reduction processes; a familiar example is the conversion of glucose to ethyl alcohol and carbon dioxide. Harden (1923, 1932) discovered over a dozen enzymes and a number of coenzymes in the zymase complex. Zymase works in the presence of inorganic phosphate.

III. TRANSFERRING ENZYMES OR TRANSFERASES

J. Phosphorylases (in the presence of inorganic phosphate—H_3PO_4) catalyze starch or glycogen to glucose phosphate; and sucrose to fructose phosphate + glucose phosphate.

K. Transphosphorylases (e.g. *hexokinase, phosphohexokinase, phosphopyruvate,* etc.) transfer the phosphate group from one compound to another; for example, glucose or fructose+phosphate to glucose- or fructose-phosphate; glucose-phosphate to fructose-phosphate; phosphoglyceric acid to phosphopyruvic acid, phosphopyruvic acid to pyruvic acid+phosphate, etc.

Chapter 10 DIGESTION AND ASSIMILATION OF FOOD

Digestion

The reserve materials are generally insoluble in water or cell-sap and also indiffusible but when translocation is necessary they are rendered soluble and diffusible by the action of specific enzymes. It is only in the soluble forms that food materials are absorbed by the protoplasm. *This rendering of insoluble and complex food substances into soluble and simpler forms suitable for translocation through the plant body and assimilation by the protoplasm is collectively known as* **digestion**.

The process of digestion is chiefly intracellular, that is, it takes place inside the cell. Extracellular digestion occurs in a few cases, as in carnivorous plants, parasites, fungi and bacteria. In such cases the digestive agent or enzyme is secreted by the protoplasm outside the body, where it digests or splits up the complex food materials; the products of digestion are then absorbed by the cells. Digestion, like all other physiological functions, is performed by the protoplasm. For this purpose it secretes different kinds of digestive agents or **enzymes** to act on different kinds of food substances. Enzymes concerned in intracellular digestion are called *endoenzymes,* and those concerned in extracellular digestion *exoenzymes.*

Assimilation

Assimilation *is the absorption of the simplest products of digestion of foodstuff by the protoplasm into its own body and conversion of these products into the similar complex constituents of the protoplasm (the term 'assimilate' means to make similar).* As a result of assimilation the protoplasm increases in bulk, and the cell-walls are built up. Assimilation is a constructive process by which the protoplasm is continually reconstructing itself out of the nutritive substances such as sugar and simple products of proteins supplied to it. The various kinds of carbohydrates are converted into sugar, particularly glucose, and the various complex proteins are converted on rapid hydrolysis into peptones, polypeptides and amino-acids. These simplest products of digestion travel to the growing regions where the protoplasm is very active. Glucose is mostly broken down by the living cells during their respiration, releasing energy; a part of glucose supplies new material, particularly cellulose, for growth of cells; and another part of it is precipitated as starch for future use. The digested products of proteins, viz. peptones, etc., are directly assimilated by the protoplasm into its own body, and new complex protoplasmic proteins and nucleoproteins are built out of them. We know that the protoplasm itself is a living substance composed of very complex proteins. The food is, therefore. changed into complex protoplasmic proteins. The protoplasm being living, it is natural to suppose that food is changed into 'live' proteins, or, in other words, food passes from non-life into life, that is, protoplasm. This is the goal of nourishment. How this change takes place we do not know. We know only that the protoplasm has the power of bringing it about.

Chapter 11 RESPIRATION AND FERMENTATION

Respiration

Respiration *is essentially a process of oxidation and decomposition of organic compounds, particularly simple carbohydrates such as glucose, in the living cells with the release of energy.* The most important feature of respiration is that by this oxidative process the *potential* energy stored in the organic compounds in living cells is released in a stepwise manner in the form of active or *kinetic* energy under the influence of a series of enzymes and is made available, partly at least, to the protoplasm for its manifold vital activities such as manufacture of food, growth, movements, reproduction, etc. Often a considerable amount of energy escapes from the plant body in the form of heat, as seen in germinating seeds. The reserve food materials that undergo oxidation are mostly simple carbohydrates, principally glucose, and sometimes also, particularly in the absence of glucose, other substances such as complex carbohydrates, proteins and fats; these are of course first hydrolysed and then oxidized. The main facts associated with respiration are: (1) consumption of atmospheric oxygen, (2) oxidation and decomposition of a portion of the stored food resulting in a loss of dry weight as seen in the seeds germinating in the dark, (3) liberation of carbon dioxide and a small quantity of water (the volume of CO_2 liberated being equal to the volume of O_2 consumed), and above all (4) release of energy by the breakdown of organic food. The overall chemical reaction may be stated thus: $C_6H_{12}O_6 + 6O_2 = 6CO_2 + 6H_2O +$ Energy (sugar+oxygen=carbon dioxide+water+energy). This shows that for oxidation of one molecule of sugar six molecules of oxygen are used and that six molecules each of CO_2 and H_2O are formed. By burning sugar at a high temperature CO_2 and H_2O are also formed, but in the living cells this process is carried on by a series of enzymes at a comparatively low temperature. Oxidation may be complete, as shown in the formula, with the formation of carbon dioxide and water as end-products, the former escaping from the plant body and the latter getting mixed up with the general mass of water in the cells; or it may be incomplete with the formation some organic acid or ethyl alcohol and carbon dioxide, as shown by the equation: $C_6H_{12}O_6 = 2C_2H_5OH + 2CO_2$ (sugar = ethyl alcohol + carbon dioxide). In respiration of plants the oxygen gas, after entering through the stomata and lenticels diffuses through the intercellular spaces into the living cells and slowly oxidizes not only glucose and other carbohydrates but also, though less frequently, other organic materials—like fats, proteins, organic acids and even protoplasm under extreme conditions. The carbon dioxide that is formed in respiration diffuses through the intercellular spaces and finally escapes through the stomata and the lenticels into the surrounding air; a portion of it may be retained in the cells and used for photosynthesis. In submerged aquatic plants the surrounding water containing dissolved air supplies the necessary gases for respiration and carbon-assimilation.

All the living cells of the plant, however deeply seated they may be, must respire day and night in order to live. If the supply of air is cut off by growing the plant in an atmosphere devoid of oxygen, it soon dies. Growing organs such as the floral and vegetative buds, the germinating seeds, and the stem- and root-tips, respire actively; while adult organs do so comparatively slowly. The gases normally enter the plant through the stomata. But these are closed at night. So, to facilitate the interchange of gases concerned, special structures are developed on the branches. These are the lenticels. Unlike the stomata they remain open. For easy diffusion of gases in the interior of the plant body, there is developed a network of air-cavities and intercellular spaces which are connected throughout and are continuous with the stomata and the lenticels (see experiment 28).

In respiration the cells are continually exhaling carbon dioxide, and in photosynthesis the green cells are continually making use of this gas during the daytime and giving off oxygen. But the latter process goes on more vigorously than the former, and practically masks it. Thus in green parts of plants the composition of the intercellular air varies, becoming much richer in oxygen during the daytime, but richer in carbon dioxide during the night. The amount of nitrogen varies very little, as this gas is not made use of by the protoplasm. Since in respiration plants are continually exhaling carbon dioxide, the atmosphere has a tendency to become richer in this gas, especially at night. Carbon dioxide is a suffocating gas, and is likely to vitiate the atmosphere. But then during the day the green plants absorb it for photosynthesis and give out oxygen which goes back to the atmosphere. The atmosphere is thus purified, and the composition of the air remains constant.

Experiment 28. Aeriferous system of the plant (FIG. 33). Take a wide-mouthed bottle and a cork of appropriate size with two holes bored in it. Partially fill the bottle with water and insert the long petiole of a selected broad leaf (e.g. *Begonia*) into the water through a hole in the cork. Through the other hole introduce a bent tube with its inner end clear above the surface of water. All the connexions are made air-tight by applying paraffin-wax, and the air drawn out from the bottle through the bent tube, preferably with a vacuum pump. As this is done a series of minute bubbles is seen to escape through the cut end of the petiole and rise upwards. Repeat the experiment after applying a coat of vaseline on the surfaces of the leaf and see that no air-bubbles appear. This escape of air-bubbles indicates that there is an intercommunicating system of intercellular spaces connected with the stomata, forming as a whole the aeriferous system of the plant.

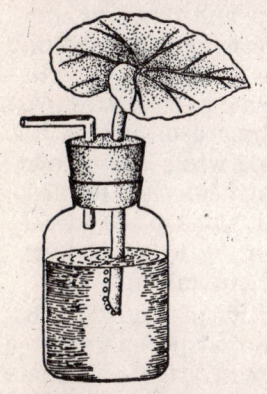

FIG. 33. Experiment on aeriferous system.

Aerobic and Anaerobic Respiration (*aer*, air; *an*, not; *bios*, life). Normally free oxygen is used in respiration resulting in complete oxidation of stored food and formation of carbon dioxide and water as end-products; this is known as **aerobic respiration**. A considerable amount of energy is released by this process, as represented by the equation—$C_6H_{12}O_6 + 6O_2 = 6CO_2 + 6H_2O + 674$ kg.-cal. (sugar+oxygen=carbon dioxide+water+674 kg.-cal. of energy). Under certain conditions, as in the absence of free oxygen, many tissues of higher plants, seeds in storage, fleshy fruits and succulent plants like cacti temporarily take to a kind of respiration, called **anaerobic respiration**, which results in incomplete oxidation of stored food and formation of carbon dioxide and ethyl alcohol, and sometimes also various organic acids such as malic, citric, oxalic, tartaric, etc. Very little energy is released by this process to maintain the activity of the protoplasm. This may be represented by the equation—$C_6H_{12}O_6 = 2C_2H_5OH + 2CO_2 + 28$ kg.-cal. (sugar=ethyl alcohol+carbon dioxide+28 kg.-cal. of energy). It is otherwise known as *intramolecular respiration* because in this process intramolecular oxidation of sugar and other compounds takes place without the use of free oxygen. Anaerobic respiration may continue only for a limited period of time, at most a few days, after which death ensues, evidently due to low production of energy and accumulation of toxic substances in the cells. In certain micro-organisms (certain bacteria, yeast and some other fungi), however, the fundamental process of energy-release is anaerobic respiration. Anaerobic respiration resulting in the production of alcohol is otherwise called alcoholic fermentation.

Mechanism of Respiration. Chemical changes in respiration (from the breakdown of glucose to the release of CO_2 and H_2O) are more or less definitely known. It is known that the whole process is controlled by a group of complex enzymes, the respiratory enzymes, of different kinds, working step by step, and that it is complete in two distinct phases: an anaerobic phase and an aerobic phase. The first phase which is **incomplete oxidation of glucose to pyruvic acid** through a chain of intermediate reactions is called **glycolysis**, and it takes place in the absence of oxygen (to be more precise it does not require oxygen); while the second phase which is complete oxidation of pyruvic acid, thus formed, to CO_2 and H_2O is called **Krebs cycle**, and it takes place in the presence of oxygen. Several reactions occur in the whole process, each reaction being controlled by a specific enzyme.

Anaerobic Phase. At this phase of respiration a simple carbohydrate like glucose or its isomer fructose is first phosphorylated (a phosphate group added by ATP, an energy-rich phosphate compound, now known to be formed in the mitochondria); ATP in this process is converted into ADP. This is a very important step initiating the process of respiration and leading to the formation of phosphoglyceric acid and finally pyruvic acid. Other common reserve materials used in respiration are starch and sucrose. Each, however, is first hydrolysed to glucose, as follows:

starch—$2n(C_6H_{10}O_5) + nH_2O \xrightarrow{\text{amylase}} nC_{12}H_{22}O_{11}$ (maltose), $C_{12}H_{22}O_{11} + H_2O \xrightarrow{\text{maltase}} 2C_6H_{12}O_6$ (glucose); and sucrose—$C_{12}H_{22}O_{11} + H_2O \xrightarrow{\text{sucrase}} C_6H_{12}O_6$ (glucose) $+ C_6H_{12}O_6$ (fructose). Glucose with the supply of ATP becomes activated in the presence of the enzyme *hexokinase*, and glucose-6-phosphate is formed. Glucose-6-phosphate is immediately converted to fructose-6-phosphate under the action of the enzyme *hexose phosphate isomerase*. Phosphorylation further continues with the supply of ATP, and thus in the presence of the enzyme *phosphofructokinase* fructose-6-phosphate is again phosphorylated to fructose-1, 6-diphosphate by the addition of a molecule of ATP. Then under the action of the enzyme *aldolase* the above compound splits into two 3-carbon fragments, one of which is 3-phosphoglyceraldehyde. The other fragment does not take part in respiration but it may be converted to 3-phosphoglyceraldehyde which follows the pathway of anaerobic respiration. The next step in the series is the conversion of 3-phosphoglyceraldehyde to 1, 3-diphosphoglyceric acid under the action of the enzyme *phosphoglyceraldehyde dehydrogenase;* this reaction involves addition of inorganic phosphate to form ATP from ADP for the next reaction and also reduction of NAD to $NADH_2$. Then under the action of the enzyme *phosphoglyceric kinase* plus ADP 1, 3-diphosphoglyceric acid is converted to 3-phosphoglyceric acid, and simultaneously ADP converted to ATP. Then by a shift (migration) in the position of the phosphate group in the presence of the enzyme *phosphoglyceromutase* it is further converted to 2-phosphoglyceric acid. It will be noted that the transfer of energy from one compound to another takes place through the transfer of a phosphate molecule from one to the other. The next step in the series of reactions results in the formation of phospho-enol-pyruvic acid from 2-phosphoglyceric acid; this is in fact a dehydration reaction (removal of H_2O) in the presence of the enzyme *enolase*. Finally with the removal of all phosphate from this compound by the action of the enzyme *pyruvic kinase* and ADP the pyruvic acid stands as the end-product in glycolysis, and ADP is converted to ATP for further action. The following points may be noted: (1) 1 mol. of glucose yields 2 mols. of pyruvic acid; (2) phosphate group as a source of energy is mainly donated by ATP; (3) there is synthesis of 8 mols. of ATP for each mol. of glucose converted to pyruvic acid; (4) each reaction is catalysed by a specific enzyme; (5) all the reactions are reversible; and (6) pyruvic acid holds an intermediate key position between the two phases of respiration.

Pathway of Glycolysis

glucose ($C_6H_{12}O_6$)
$\Downarrow$ hexokinase+ATP
glucose-6-phosphate+ADP
$\Downarrow$ hexose phosphate isomerase+ADP
fructose-6-phosphate
$\Downarrow$ phosphofructokinase+ATP
fructose-1, 6-diphosphate+ADP
$\Downarrow$ aldolase
3-phosphoglyceraldehyde (a 3-carbon compound)
$\Downarrow$ phosphoglyceraldehyde dehydrogenase+DPN
1, 3-diphosphoglyceric acid+$DPNH_2$
$\Downarrow$ phosphoglyceric kinase+ADP
3-phosphoglyceric acid+ATP
$\Downarrow$ phosphoglyceromutase
2-phosphoglyceric acid
$\Downarrow$ enolase
phosphoenolpyruvic acid
$\Downarrow$ pyruvic kinase+ADP
pyruvic acid ($C_3H_4O_3$)+ATP

Anaerobic Oxidation of Pyruvic Acid. Pyruvic acid, as it is formed, holds a key position from which under different conditions reactions proceed in different directions in different tissues and organisms. In the absence of O_2 pyruvic acid is converted to acetaldehyde and CO_2 under the action of pyruvic carboxylase. Acetaldehyde is now the starting point from which reactions lead to the production of a number of compounds. It is known that several bacteria and some fungi thrive normally under anaerobic conditions. In them it is seen that pyruvic acid in the absence of O_2 becomes converted to certain compounds under the action of different enzymes with the release of a certain amount of energy, small though, for the activity of such micro-organisms. Thus in certain bacteria pyruvic acid is reduced to lactic acid as the end-product; while it is butyric acid in certain other bacteria. In the fermentation of glucose by yeast cells pyruvic acid is reduced to ethyl alcohol and CO_2. The energy released in all such cases is of low order. In higher plants the anaerobic respiration may continue only for a short period because the end-products formed in this process are mostly toxic. In them the pyruvic acid may be converted to acetaldehyde, oxalacetic acid, some amino-acids, fatty acids and several other organic acids, e.g. oxalic acid, malic acid, tartaric acid, lactic acid, etc., under the action of different enzymes.

Aerobic Phase. This immediately follows the first phase with the supply of oxygen, and most of the pyruvic acid is completely oxidized to CO_2 and H_2O with the liberation of maximum amount of energy. The ATP formed in this process is about four times greater than that formed in glycolysis. Several organic acids and some amino-acids (derived from α-ketoglutaric acid) are formed during the process of aerobic respiration. A series of reactions takes place in a stepwise manner in the whole process under the action of distinct complex enzymes. In fact, for each reaction there is a specific enzyme. Some such enzymes have been extracted from plants and these have been found to bring about reactions *in vitro* similar to those *in vivo*. The sequence of reactions in this phase may be divided into two groups. The first is the oxidative decarboxylation (removal of a molecule of CO_2) of pyruvic acid to form acetyl-coenzyme A. The second is the citric acid cycle or Krebs cycle.

Formation of Acetyl-Coenzyme A. With the access of sufficient oxygen the pyruvic acid ($C_3H_4O_3$) undergoes oxidative decarboxylation and is converted to acetyl-coenzyme A. The chemical process leading to this conversion is, however, very complex, involving a series of reactions. Several complex enzymes and 4 or 5 coenzymes are involved in the whole process. Briefly speaking, under the action of *pyruvate decarboxylase* the pyruvic acid reacts with TPP (thiamine pyrophosphate) with the result that 'activated' TPP-acetaldehyde complex is formed, and a molecule of CO_2 released. The 'activated' acetaldehyde portion of this complex reacts with lipoic acid under the action of *lipoyl reductase*, and acetyl-lipoic acid complex is formed. Next the acetyl group is released from this complex and transferred to coenzyme A (CoA) under the action of *transacetylose* forming acetyl-coenzyme A (acetyl-CoA, also called 'active' acetate—C_2H_3O·CoA). It serves as a connecting link between the anaerobic phase (glycolysis) and the aerobic phase (Krebs cycle) of respiration.

Krebs Cycle or Citric Acid Cycle. Krebs cycle consists of a series of chemical reactions under aerobic conditions, proceeding step by step in a cyclic order. This cycle of reactions, each regulated by a specific enzyme, is collectively called **citric acid cycle** or **Krebs cycle,** as first worked out H. A. Krebs, an English biochemist, in 1943 (FIG. 34). This is otherwise called **tricarboxylic acid cycle** since citric acid and some other compounds of this cycle have three carboxyl groups (—COOH) in them. It may be noted that the cycle as a whole turns only in one direction. The first step in the cycle is the combination of acetyl portion of acetyl-coenzyme A with **oxalacetic acid** ($C_4H_4O_5$) under the action of the enzyme *citrogenase* to form **citric acid** ($C_6H_8O_7$), and coenzyme A is released. With the formation of citric acid reactions follow in quick succession. By dehydration process (removal of H_2O) citric acid is converted to **cis-aconotic acid** ($C_6H_6O_6$) and immediately by hydration process (addition of H_2O) the latter converted to **isocitric acid** ($C_6H_8O_7$). Both the reactions are catalyzed by one and the same enzyme *aconitase*. (Cis-aconitic acid may, however, be omitted in the process). Dehydrogenation (oxidation) of isocitric acid in the presence of *isocitric acid dehydrogenase* results in the formation of **oxalosuccinic acid** ($C_6H_6O_7$). TPN (triphosphopyridine nucleotide) as a coenzyme is essential for this process. In the Krebs cycle, it may be noted, this is the *first oxidation step*. It may further be noted that each of the above four acids (citric to oxalosuccinic) is a 6-carbon compound and, therefore, no release of CO_2 takes place in any of the above reactions. Oxalosuccinic acid is a keto acid, and is readily decarboxylated (a CO_2 molecule released) to α-**ketoglutaric acid** ($C_5H_6O_5$) under the action of a *carboxylase*. α ketoglutaric acid plays a very useful role in the metabolism of plants in many ways,

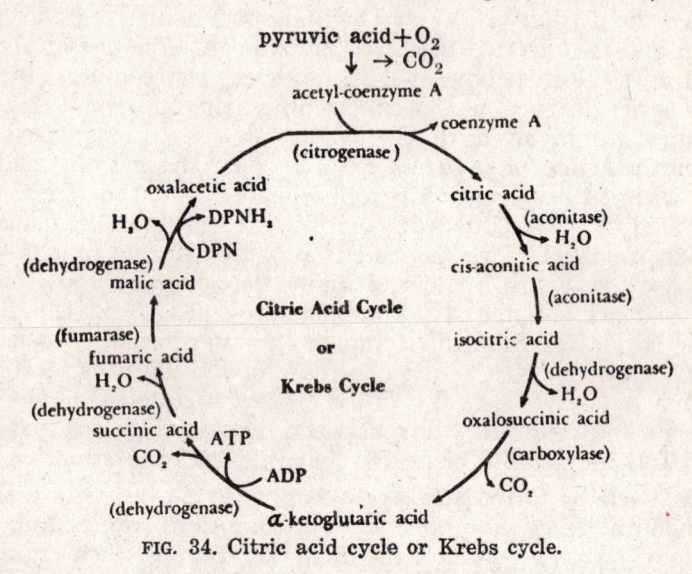

FIG. 34. Citric acid cycle or Krebs cycle.

particularly in the synthesis of some amino-acids (see pp. 338-9). By further oxidation and decarboxylation of α-ketoglutaric acid in the presence of *α-ketoglutaric dehydrogenase* **succinic acid** ($C_4H_6O_4$) is formed rapidly through succinyl-CoA. This is the *second oxidation step*. Oxidation of succinic acid ('2H' removed) leads to the formation of **fumaric acid** ($C_4H_4O_4$). The reaction is catalyzed by *succinic dehydrogenase* with FAD (flavin adenine dinucleotide) firmly attached to it as a prosthetic group. This is the *third oxidation step*. By hydration process fumaric acid is converted to **malic acid** ($C_4H_6O_5$) in the presence of *fumarase*.

Then oxidation of malic acid by NAD (nicotinamide adenine dinucleotide) in the presence of *malic dehydrogenase* leads to the formation of **oxalacetic acid** ($C_4H_4O_5$). In the process '2H' is removed and NAD reduced to $NADH_2$. This is the *fourth oxidation step*. Oxalacetic acid is the end-product of the Krebs cycle. After it is regenerated it again reacts with acetyl-coenzyme A, and the cycle kept going without interruption. Oxalacetic acid is present in plant cells for the purpose, and it may also be formed from pyruvic acid, aspartic acid, glutamic acid, etc. Necessary enzymes and ATP required for the whole process are formed in the mitochondria. Aerobic respiration thus appears to be restricted to the mitochondria. It may be noted that complete oxidation of 1 mol. of glucose yields 2 mols. of pyruvic acid, 6 mols. of CO_2 and 38 mols of ATP, as follows.

glycolysis	—	8 ATP
pyruvic acid to acetyl-CoA	2 CO_2	6 ATP
Krebs cycle:		
(a) isocitric acid to oxalosuccinic acid	—	6 ATP
(b) oxalosuccinic acid to α-ketoglutaric acid	2 CO_2	—
(c) α-ketoglutaric acid to succinyl CoA	2 CO_2	6 ATP
(d) release from succinyl CoA	—	2 ATP
(e) succinic acid to fumaric acid	—	4 ATP
(f) malic acid to oxalacetic acid	—	6 ATP
	6 CO_2	38 ATP

Electron Transport System. As stated before, oxidation of a compound means removal of electrons from it, usually accompanied by removal of hydrogen, and reduction means addition of electrons to a compound, usually accompanied by addition of hydrogen. The metabolic pathway through which the electrons pass from one compound to another is called the electron transport system or cytochrome system. In oxidation-reduction processes specific enzymes are always connected with the transport of electrons from one compound to another. In respiration the electron transport system consists commonly of NAD, FAD, coenzyme Q (a quinone compound) and cytochromes (pigmented bodies containing iron). NAD and FAD are first reduced (hydrogen or electrons received): NAD → $NADH_2$; FAD → $FADH_2$; later the energy released in the oxidation of $NADH_2$ and $FADH_2$ is used in the conversion of ADP to ATP (ADP + inorganic phosphate + energy = ATP). Hydrogen ions are released to the cytoplasm, while electrons pass down through coenzyme Q to the cytochromes (a, b and c); in the oxidation of cytochromes more ATP is synthesized in the cells. It may be noted that the synthesis of ATP from ADP in the mitochondria at the expense of chemical energy through oxidation-reduction reactions of the electron transport system is otherwise called *oxidative phosphorylation* (as opposed to *photophosphorylation* which is synthesis of ATP in the chloroplasts at the expense of light energy). For each pair of electrons passed from one compound to another 3 ATP are usually formed (only 2 ATP in certain reactions). In respiration, as stated before, most of the energy is released through the Krebs cycle. With the reduction of cytochromes the electrons pass down to oxygen and the 'activated' oxygen freely combines with hydrogen (already released to the cytoplasm) to form H_2O.

Experiment 29. Aerobic respiration (FIG. 35A). Respiration in plant can be experimentally proved in a very simple but efficient way by the following method. Appliances required for this experiment are: a flask with a bent bulb, called

respiroscope (an ordinary long-necked flask will also do), a beaker, a suitable stand with a clamp, a quantity of mercury (according to the size of the beaker),

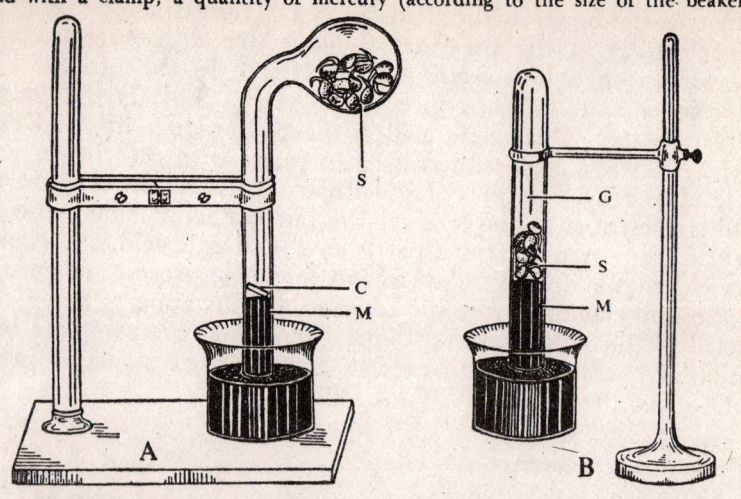

Experiments on Respiration. FIG. 35. *A*, aerobic respiration; *B*, anaerobic or intramolecular respiration. *S*, seeds; *C*, caustic potash stick; *M*, mercury; *G*, gas.

caustic potash stick and some germinating seeds or opening flower-buds. Introduce some germinating seeds into the respiroscope. Pour a quantity of mercury into the beaker and invert the respiroscope over it. The respiroscope is fixed in this vertical position with a stand and a clamp. The air enclosed in the flask is thus cut off from the surrounding atmosphere. With the help of the forceps introduce into the respiroscope a small piece of caustic potash stick. This will float on mercury inside the respiroscope. Leave the apparatus in this position for some hours, preferably till the next day. *Observation.* It will be seen on the following day that the level of mercury in the respiroscope has risen to the extent of nearly one-fifth of its total volume. *Inference.* Since caustic potash absorbs carbon dioxide we may conclude that the gas absorbed is carbon dioxide.

The atmospheric connexion being cut off, it may be safely assumed that this CO_2 gas must have been exhaled by the germinating seeds. The gas (4/5th in volume) still remaining in the respiroscope is nitrogen, while the oxygen originally occupying 1/5th of the total volume must have been absorbed by the seeds in the process of respiration since there is no chance for it to escape. It may also be noted that the rise of mercury will be immediate and quick if caustic potash stick be used at the end of the experiment.

Note. The experiment may be carried out in the following way. Instead of mercury, water may be poured into the beaker, and the respiroscope with the germinating seeds inverted over it. A short test tube containing a small piece of caustic potash stick may be floated on water inside the respiroscope. Subsequently the water level is noticed rising in the respiroscope.

Experiment 30. Anaerobic respiration (FIG. 35B). Completely fill a short narrow test-tube with mercury (*M*), close it with the thumb and invert it over mercury contained in a beaker. Keep the tube in a vertical position with a suitable stand. Take some germinating seeds, and remove the seedcoats from them to get rid of the enclosed air (oxygen). With the help of the forceps hold the skinned seeds under the test-tube, and release them one after another. As soon as released the seeds rise to the closed end of the tube. Introduce in this way five or six seeds.

They are now free from oxygen. Prior to their introduction it is better to soak the seeds in distilled water. This keeps the seeds moist. Note on the following day that the mercury column has been pushed down, owing to the exhalation of a gas (G) by the seeds. Within one or two days nearly the whole of the mercury is seen to be pushed out of the tube. Introduce a small piece of caustic potash stick into the test-tube with the help of the forceps. It floats on mercury, and coming in contact with the gas absorbs it quickly. The mercury rises again and fills up the test-tube. The gas evidently is carbon dioxide.

Experiment 31. To prove that carbon dioxide is released in respiration of a green plant. Set up the apparatus, as shown in FIG. 36. *A* contains soda-lime; *B* contains KOH solution; *C* encloses a pot plant—the bell-jar stands on a

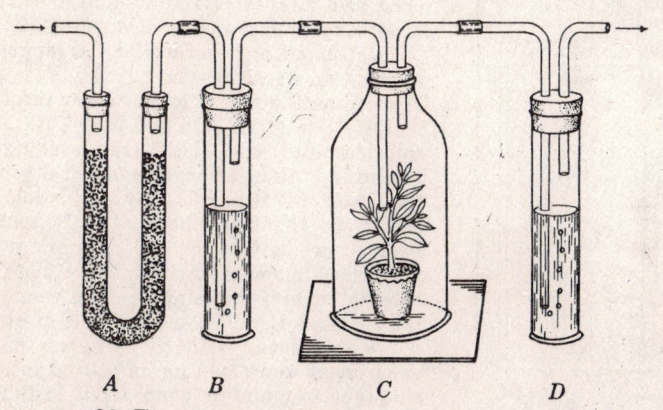

FIG. 36. Experiment on respiration of a green plant.

smooth flat plate stuck with vaseline to prevent leakage of air; *D* contains baryta (or lime) water. Other connexions are also made airtight. The experiment is carried out in a semi-dark room, with the bell-jar covered with a black cloth. The bent tube on the extreme right is connected to an aspirator bottle (not shown in the figure). The stopcock of the aspirator bottle is slightly opened. As a result a partial vacuum is produced and a slow current of air, freed of CO_2, is drawn in through the other end of the series. After some time, as the air-current slowly passes into *D*, it will be seen that the baryta water in it turns milky (barium carbonate is formed). Evidently this CO_2 has been released by the plant.

Note that carbon dioxide is absorbed by caustic potash, baryta water and lime water, but only the latter two turn milky in contact with carbon dioxide.

Experiment 32. The volume of CO_2 evolved in respiration is approximately equal to the volume of O_2 absorbed. (*a*) Set up the apparatus, as shown in FIG. 37. The small test-tube in *A* contains KOH solution; *B* is without it. The beaker at the bottom contains water in each case. Both the respiroscope-tubes contain some germinating seeds. The stopcock is closed. After some time it is seen that the water rises in *A*, the CO_2 released by the seeds being absorbed by KOH. In *B* there is no rise of water as no vacuum has been produced in it. *B*, therefore, shows that the total volume of air in it (whatever be the gaseous interchanges) is constant. As we know, in respiration O_2 and CO_2 only are involved, it may be taken for granted that the volumes of the two gases are equal. Then replace *B* by a new respiroscope.

tube (or use the same one after washing) without seeds and dip it into pyrogallate of potash (instead of water). Soon the solution, as it absorbs O_2, rises to the same extent as in *A*. We may, therefore, conclude that in respiration the volume of CO_2 given out is equal to the volume of O_2 absorbed.

Note that if starchy seeds like pea, gram, etc., are used the respiratory quotient is unity; if, however, oily seeds like castor be used the R.Q. is less than unity, i.e. they absorb more O_2 and give out less CO_2.

(*b*) **Ganong's respirometer.** Arrange the apparatus, as shown in FIG. 38. Put 2 cc. of germinating seeds (or any other respiring material) into the bulb and pour 10% KOH solution into the side- (or levelling- or reservoir-) tube after opening the hole at the neck of the bulb. By adjusting the height of the side-tube the KOH solution may be brought to the 100 cc. mark in the graduated manometer tube. Now close the hole by turning the stopper. The volume of the enclosed air is 100 cc. and it is at atmospheric pressure. With the progress of the experiment more solution is poured into the side-tube to maintain equal levels in the two tubes. As respiration takes place, CO_2 liberated by the seeds is absorbed by the KOH solution, and it rises and stands at the 80 cc. mark. Evidently 1/5th of the air (which is oxygen) has been absorbed by the seeds during respiration, and this is the volume of CO_2 exhaled by the seeds during the process, as indicated by the rise of KOH solution. If, instead of KOH solution, saturated common salt solution be used its level is seen to remain constant in the tubes, evidently indicating that the volume of oxygen absorbed is the same as that of carbon dioxide exhaled.

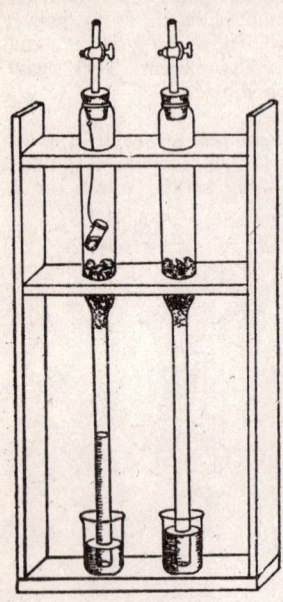

FIG. 37. Respiratory quotient—$CO_2:O_2$ (see experiment 32*a*)

Respiration is a destructive process consisting in the decomposition of some of the food materials, more particularly glucose, brought about by the action of specific enzymes secreted by the protoplasm; nevertheless, it is highly beneficial to the life of the plant for the reason that respiration sets free *energy* by which work is performed. This energy is absolutely necessary for the various synthetic processes, growth, movements, etc. If we think of the enormous development of a large tree we can at once realize what a vast amount of energy has been utilized in constructing that body. A considerable amount of energy, of course, escapes from the plant body in the form of heat. During vigorous respiration heat is generated. A thermometer thrust into a mass of germinating seeds will show a marked rise of temperature (see experiment 33). This production of heat is an easily observed form of energy. **Respiration results in a loss of dry**

weight of the plant. This is believed to be due to escape of carbon dioxide.

Experiment 33. Heat generated in respiration (FIG. 39). Take two thermoflasks and fill one (A) of them with germinating seeds and the other (B) with the same killed by boiling for a few minutes and then soaked in 5% formalin to prevent any fermentation in the flask generating heat. Insert a sensitive thermometer in each, as shown in the figure, and pack

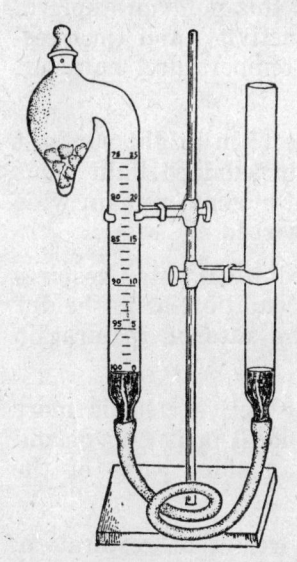

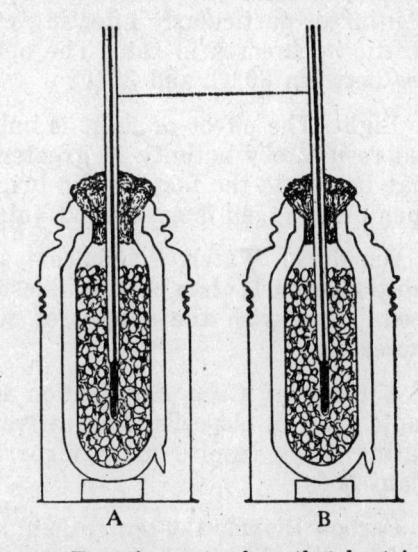

FIG. 38. Ganong's respirometer (see experiment 32b)

FIG. 39. Experiment to show that heat is generated in respiration (see text).

the mouth of the flask with cotton. It is better to place, half immersed in the lump of seeds, a small test tube containing a small piece of caustic potash stick. Wait for some time and note a remarkable rise of temperature in the case of flask A containing germinating seeds; while the flask B containing killed seeds shows no rise of temperature (the dotted line indicating the original temperature). This evidently proves that heat is evolved in respiration.

CONDITIONS AFFECTING RESPIRATION

1. Oxygen. Presence of oxygen is essential for respiration. But it must be noted that it is required for the reactions of the Krebs cycle only (and not for the anaerobic phase—glycolysis). Therefore, there are marked changes in the rate of respiration under varying conditions of oxygen concentration, affecting R.Q. values (see p. 373). If the concentration goes below 5% the process rapidly falls off. Under this condition more CO_2 is evolved than O_2 absorbed. If oxygen supply be cut off, only CO_2 is produced as a result of anaerobic respiration. With gradual increase in oxygen concentration there is corresponding steady increase in the rate, and the R.Q. value approaches unity, i.e. $CO_2:O_2=1$. But the rate does not go far beyond the normal rate as at atmospheric concentration of O_2.

2. **Temperature.** This affects markedly the rate of respiration. The minimum rate is reached at 0°C or even at 10°C. With the rise of temperature the rate increases and the maximum is reached at 45°C. or even at 40°C. Beyond this point protoplasm is injured, particularly affecting enzyme activity, and thus respiration decreases in rate. The optimum temperature, however, lies between 30°C. and 35°C.

3. **Light.** The effect of light is only indirect; in bright sunlight the respiratory activity is greater than in subdued light. This may be due to the fact that in bright light stomata remain wide open and oxygen is easily and quickly absorbed.

4. **Supply of Water.** Protoplasm saturated with water respires more vigorously than that in a desiccated condition, as in the dry seed. Thus with the supply of water the rate of respiration increases.

5. **Vitality of Cells.** Respiration in young active cells is more rapid than in old cells. Vegetative buds, floral buds and germinating seeds respire more vigorously than older parts of the plant body.

6. **Carbon Dioxide Concentration.** If, as a result of respiration, carbon dioxide be allowed to accumulate inside the plant as a result of stomatal closure, or around the plant, respiration slows down and comes to a standstill. If this carbon dioxide is removed, respiration again goes on normally. CO_2 concentration has a depressing effect on respiration.

7. **Nutritive Materials.** Food materials, more particularly soluble carbohydrates, affect respiration to a considerable extent. With the supply of oxygen these materials become quickly broken down.

Respiratory Quotient (R.Q.). There is a relation between the volume of carbon dioxide evolved and the volume of oxygen consumed in the process of respiration, and the ratio of these two volumes is known as respiratory quotient. Thus when sugars are consumed in respiration, as mostly in cereals, the respiratory quotient (R.Q.), i.e. the ratio of $CO_2 : O_2$ is equal to unity. This means that for one molecule of CO_2 given out one molecule of O_2 is used. Sugars are by no means the only compounds consumed in respiration. Fats, proteins, organic acids and other materials are also consumed, and the R.Q. in such cases may vary greatly from unity. When fats are consumed in respiration, as in the oily seeds, the R.Q., i.e. the ratio of $CO_2 : O_2$ is less than unity. Fats are poorer in oxygen than are sugars and, therefore, correspondingly more oxygen is required for combustion of fats, or, in other words, less CO_2 is evolved for a particular volume of O_2 taken up. Likewise, when proteins are used in respiration, the R.Q. is less than unity since the proportion of oxygen to carbon in them is less than in carbohydrates. When organic acids are oxidized the R.Q. is found to be more than unity; such compounds are relatively rich in oxygen in comparison with carbohydrates. At night succulent plants such as cacti absorb oxygen without giving out carbon dioxide; in them instead of carbon dioxide organic acids (malic, citric, oxalic, etc.) are formed as a result of incomplete oxidation of sugar. During the daytime, however, they give out CO_2. The concept of respiratory quotient is important because it serves as a clue to the nature of compounds used in respiration, and also to the nature of the process itself.

RESPIRATION AND PHOTOSYNTHESIS

1. In respiration plants utilize oxygen and give out **carbon dioxide**; while in photosynthesis plants utilize **carbon dioxide and give out oxygen**; that is, one process is just the reverse of the other.

2. Respiration is a destructive (catabolic) process, but photosynthesis is a constructive (anabolic) process. In the former process sugar is broken down into CO_2 and H_2O with the liberation of energy; while in the latter process CO_2 and H_2O are utilized to build up sugar with the storage of energy. Respiration is thus a *breaking-down process*, and photosynthesis a *building-up process*.

3. The intermediate chemical reactions in the breakdown of sugar in respiration (anaerobic phase) and those in the synthesis of sugar in photosynthesis are much the same. In both the processes phospho-glyceric acid holds a pivotal position from which reactions proceed in different directions under different conditions.

4. Respiration is performed by all the living cells of the plant at all times, i.e. it is independent of light and chlorophyll; while photosynthesis is performed only by the green cells, and that, too, only in the presence of sunlight. Although photosynthesis persists only for a limited period, this process is much more vigorous than respiration.

5. The seat of respiration is the mitochondria present in the living cells of both plants and animals; while the seat of photosynthesis is the chloroplasts, partly at least, present in the green cells of plants only.

6. Respiration results in a loss of dry weight of the plant due to breaking-down of food materials and the production of carbon dioxide which escapes from the plant body; but photosynthesis results in a gain in dry weight due to the formation of sugar, starch, etc., which accumulate in the plant body.

Fermentation

Fermentation is the incomplete oxidation of sugar (particularly glucose) into alcohol and carbon dioxide brought about by several species of bacteria and yeast (and certain other fungi) in the absence of oxygen. Although fermentation by yeast cells leading to the production of commercial alcohol was known for a long time, it was Louis Pasteur who first elaborately studied the process, and showed clearly about the year 1857 that it is a process of life, and not of death. Pasteur actually applied many of his discoveries to the making of different kinds of liquors by using different species and strains of yeast. Other fermentative processes brought about by different types of bacteria were also elaborately studied by him. It later came to be known through the work of Buchner in 1897 that the change of sugar into alcohol and carbon dioxide is due to the action of *zymase* (an enzyme complex) secreted by the micro-organisms, and not due to their direct action on sugar. Under the action of specific enzymes (see pp. 364-65) sugar becomes converted to pyruvic acid. The latter is then reduced to ethyl alcohol (through acetaldehyde) and carbon dioxide by the enzyme *carboxylase*. Fermentation processes are mostly anaerobic, and although much less energy is liberated by such processes the organisms living under anaerobic conditions have to depend on this energy for their vital activities. Fermentation is most readily seen in date-palm juice, where sugar is broken up by the yeast cells in the absence of oxygen into alcohol and carbon dioxide, the frothing on the surface of the liquid being due to the formation of this gas. Fermentation may be defined as an enzyme action on sugar in the absence of free oxygen, splitting it (sugar) into carbon dioxide and alcohol and sometimes organic acids. The process is analogous to anaerobic respiration and may be represented by identical formula—$C_6H_{12}O_6 + $ zymase $= 2C_2H_5OH$ (ethyl alcohol) $+ 2CO_2 + $ zymase $+ $ energy (28 kg.-cal.). The equation shows that 1 molecule of glucose produces 2 molecules ethyl alcohol and 2 molecules of carbon dioxide. The amount of energy released in fermentation is, no doubt, small but it suffices for the activity of the anaerobic micro-organisms. Other common examples of fermentation are: acetic acid fermentation which is the souring of alcoholic liquors such as wine, beer, etc., due to the conversion of alcohol to acetic acid (vinegar) by acetic acid bacteria, lactic acid fermentation which is the souring of milk due to the conversion of milk sugar into lactic acid by lactic acid bacteria, butyric acid fermentation which is the rancidity of butter due to the conversion of sugar into butyric acid by butyric acid bacteria, etc. Some of the fermentation processes are of great commercial importance, e.g. production of alcohol by different species and strains of yeast, manufacture of vinegar (acetic acid) from alcohol by species of *Acetobacter*, manufacture of flavoured butter and cheese, retting of jute and flax, tanning (conversion of hide into leather), etc. It may be noted that the organic products formed by the fermenting micro-organisms act as toxic to them. This checks their unlimited growth and multiplication in nature.

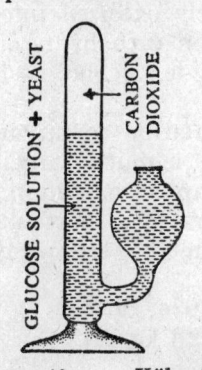

Experiment 34. Fermentation. Experiment on fermentation may be easily carried out with the help of a Kühne's fermentation vessel (FIG. 40) and the procedure is as follows. Prepare a 5% glucose solution in water and add to it a small quantity of fresh brewers' yeast or collect a quantity of date-palm juice early in the morning. Completely fill the closed upright tube of the fermentation vessel with this solution or the juice and partially fill the side-tube up to or below the base of the bulb. Warm the solution to about 25°C., and leave the apparatus in a warm place for a few hours. It will be seen that a gas collects at the upper end of the closed tube displacing the solution which now accumulates in the bulb of the side-tube and overflows it. The evolution of gas continues so long as the sugar in the solution is not exhausted. Then introduce a small piece of caustic potash stick into the fermentation vessel and gently shake it to dissolve the piece, taking care that the gas collected in the closed tube does not escape. Place the fermentation vessel again in the vertical position. Within a few minutes it will be seen that the gas is absorbed by the solution which rises and again fills up the tube. The gas evidently is carbon dioxide evolved during fermentation.

FIG. 40. Kühne's fermentation vessel (see experiment 34).

Respiration and Fermentation. Both respiration and fermentation are oxidation processes by which energy stored up in carbohydrates and other compounds is set free, and carbon dioxide given out. The two processes depend upon whether free oxygen is available or not, and to what extent oxidation (complete or incomplete) proceeds. Recent researches show that respiration in higher plants takes place in two stages: an anaerobic phase resulting in incomplete oxidation of sugar, and an aerobic phase resulting in complete oxidation of the intermediate products formed in anaerobic respiration, into carbon dioxide and water. Fermentation is that type of anaerobic respiration in which alcohol is produced (Palladin). Generally speaking, respiration includes only aerobic respiration; while fermentation is regarded as a synonym for anaerobic respiration. On this basis the distinctions between respiration and fermentation are as follows:

(1) Respiration takes place in the presence of free oxygen; while fermentation takes place in the absence of free oxygen which is derived from rearrangement of molecules of some carbohydrates, and thus fermentation is also known as intramole-

cular respiration. The initial or anaerobic phase is the same in both the processes.

(2) Respiration takes place in all the living cells of the plant body and at all times in the presence of carbohydrates; while fermentation takes place only in the presence of some easily available carbohydrates such as different sugars under the action of certain micro-organisms such as yeast, bacteria, fungi, etc.

(3) In respiration carbohydrates are completely oxidized into carbon dioxide and water; while in fermentation carbohydrates are incompletely oxidized forming carbon dioxide, alcohol and various other products.

(4) Respiration is more efficient than fermentation so far as release of energy is concerned; in respiration a much larger amount of energy is liberated as a result of complete oxidation; while in fermentation much less energy is liberated as a result of incomplete oxidation as may be seen from the following equations:

Respiration—$C_6H_{12}O_6 + 6O_2 = 6CO_2 + 6H_2O + 674$ kg.-cal.
Fermentation—$C_6H_{12}O_6 = 2C_2H_5OH + 2CO_2 + 28$ kg.-cal.

Chapter 12 METABOLISM

Two series of chemical changes or processes are simultaneously going on in a plant cell, one leading finally to the construction or building-up of the protoplasm, and the other to its decomposition or breaking-down. These two series of processes, which are constructive on the one hand and destructive on the other, are together known as **metabolism**. Metabolism takes place only in the living cells, and is one of the characteristic signs of life. The processes that lead to the construction of various food materials and other organic compounds and finally of protoplasm are together known as **anabolism**, and those processes leading to the destruction or breaking-down as **catabolism**.

Anabolism. The main anabolic or constructive changes are: formation of sugars and other carbohydrates, formation of proteins and formation of fats and oils. These changes are regarded as anabolic because the protoplasm continually reconstructs itself with these nutritive substances. By anabolism a considerable amount of potential energy is stored in the substances for future use by the protoplasm.

Catabolism. Side by side with anabolism, catabolic or destructive changes or processes are also going on in the living cells of the plant body. The main catabolic processes are: digestion, respiration and fermentation. By these processes complex food substances are gradually broken down into simpler products, e.g. various carbohydrates into glucose, various proteins into amines and amino-acids, and fats and oils into fatty acids and glycerine. The potential energy already stored up in them is released by catabolism into kinetic energy for manifold activities of the protoplasm. Carbon dioxide and water are formed as a result of complete oxidation of glucose in aerobic respiration, and alcohols and organic acids as a result of incomplete oxidation of glucose in anaerobic respiration or fermentation. Amino-acids sometimes result from the decomposition of protoplasm. Besides, the secretory products such as enzymes, vitamins, hormones, cellulose, nectar, etc., are also the results of catabolic processes. Catabolism also results in the formation of many *by-products* in plants. The various waste products such as tannins, essential oils, gums, resins, etc., belong to this category. These being useless to the protoplasm or even harmful are removed from the sphere of protoplasmic activity, and mostly stored up in special cells, bark, old leaves, heart-wood and glands. In this sense the various kinds of waste products may also be regarded as excretory products. Besides, many other substances such as some of the vegetable (organic) acids, aromatic compounds, anthocyanins, lignin, cutin, etc., are formed in the plant body as a result of catabolic processes.

Chapter 13 GROWTH

The growth of a plant is a complex phenomenon associated with numerous physiological processes—both constructive and destructive. The former lead to the formation of various nutritive substances and the protoplasm, and the latter to their breakdown. The protoplasm assimilates the protein food and increases in bulk (see p. 360-61); while the carbohydrates are mainly utilized in respiration and in the formation of the cell-wall substance, viz. cellulose. The cells divide and numerous new cells are formed; these increase in size and become fully turgid, and the plant grows as a whole. Growth is thus a complex vital phenomenon brought about by the protoplasm. *It may be defined as a permanent and irreversible increase in size and form attended by an increase in weight;* sometimes at an early stage of growth a loss in weight is noticed; as, for example, when a potato tuber sprouts it shows a loss of weight in the beginning due to transpiration and respiration. But that is soon made good as new materials begin to be formed by the sprouting shoot. Growth is usually very slow in plants, and is difficult to detect and measure accurately within a short space of time without the help of a suitable instrument. There are however, certain plants which show very rapid growth; for instance, some climbers like morning glory and wood-rose grow at the rate of about 20 cm. per day; young shoots of giant bamboo show a growth of over 40 cm. per day; while tendrils of some *Cucurbita* show an extraordinary growth of 6 cm. per hour. The growth of a plant, however slow it may be, can be accurately measured with the help of an instrument, called the **auxanometer**.

Experiment 35. Growth in Length of the Shoot. The **auxanometer** is an instrument by means of which a small increase in length can be magnified many times. From this total known magnification recorded by the auxanometer the actual length attained by a plant within a certain specified time can be easily calculated. Two types of auxanometer are in common use: the first and the simplest type is the **lever auxanometer** or **arc indicator** (FIG. 41), and the second is the **pulley auxanometer** or simply **auxanometer** (FIG. 42). In both the principle is the same.

(a) In the **lever type** there is a movable lever or indicator fixed to a wheel round which passes a cord. One end of the cord is tied round or gummed

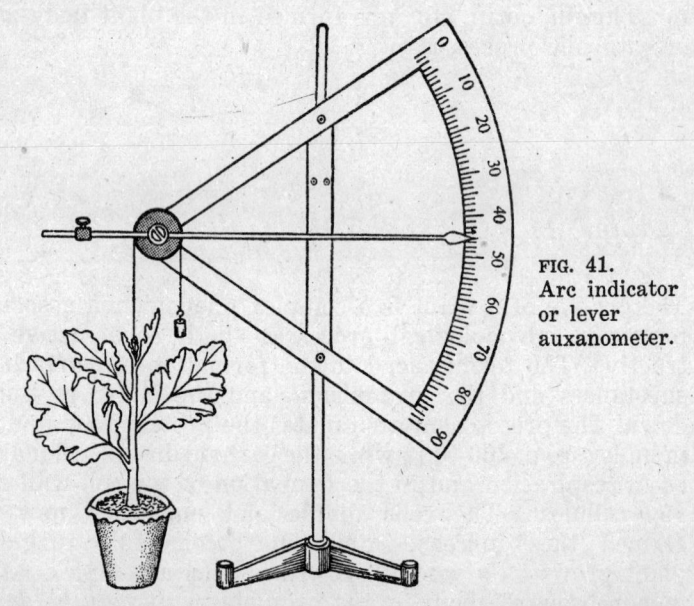

FIG. 41. Arc indicator or lever auxanometer.

to the apex of the stem, and from the other end a small weight is suspended to keep the cord taut. As the stem increases in length the wheel slowly rotates under the weight suspended and the indicator moves down the graduated arc. The growth in length of the plant is thus recorded by the instrument on a magnified scale. From the record thus obtained the actual increase in length of the stem is calculated; for instance, if the lever has transversed a distance of 45 cm. in 24 hours, and the magnification is 90 times, the actual growth in the same period is $45/90$ cm., i.e. 0.5 cm. or 5 mm. and, therefore, in 1 hour the actual growth of the plant is $5/24$ mm., i.e. 0.2 mm.

(b) With the **pulley auxanometer** a permanent record of growth within a specified time is obtained on a smoked paper which is wrapped round a drum or cylinder. The drum is rotated by means of a clockwork mechanism. A cord with one end attached to the plant is passed round a small wheel, and a small weight suspended from the other end. Round the bigger wheel, which is fixed to the smaller one, passes another cord with two small weights at the two ends. There is a horizontal pointer attached to the cord with the tip in contact with the smoked paper. As growth takes place and the drum rotates, the pointer, being attached to the cord of the bigger wheel, leaves a mark on the paper on a magnified scale. After a period of growth the paper is removed and dipped into varnish, and the smoke or soot becomes fixed on the paper. If the growth proceeds continuously a diagonal curve is traced on the paper (FIG. 42). If growth ceases a horizontal curve is the result. Magnification being known it is easy to calculate the actual growth within a certain specified time.

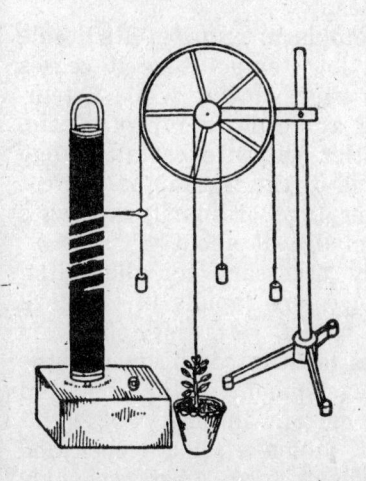

FIG. 42. A pulley auxanometer.

Crescograph. The late Sir Jagadish C. Bose constructed a very delicate apparatus known as crescograph which is an electric device. With the help of this apparatus the growth of a plant can be magnified one thousand to ten thousand times and accurately measured. At this high magnification it has been found possible to measure the progress of growth even in seconds.

Conditions Necessary for Growth.
Since growth is brought about by the protoplasm, the conditions necessary for growth are the same as those that maintain the activity of the protoplasm, as follows:

(1) **Supply of Nutritive Materials.** Growth can only take place when the protoplasm of the growing region is supplied with nutritive materials. The protoplasm assimilates these materials and builds up the body of the plant. Food materials are also a source of energy.

(2) **Supply of Water.** An adequate supply of water is absolutely necessary to maintain the turgidity of the growing cells. Turgidity is the first step towards growth. The protoplasm can only work when it is saturated with water. An abundant supply of water makes good the loss due to transpiration. It is a fact,

however, that only a small quantity is required for actual growth.

(3) **Supply of Oxygen.** Supply of free oxygen is indispensable for the respiration of all living cells. Respiration is an oxidation process by which the *potential* energy stored in the food is released in the form of *kinetic* energy and made use of by the protoplasm for its manifold activities.

(4) **Suitable Temperature.** The protoplasm requires a suitable temperature for its activities; at a low temperature it ceases to perform its functions or does so very slowly; while a temperature of 45° to 50°C coagulates and kills the protoplasm. The protoplasm maintains its activities within a certain range of temperature. This is said to be due to the **thermotonic** effect of temperature. The optimum temperature ordinarily averages from 28° to 30°C, and the minimum lies at about 4°C.

(5) **Light.** Light is not absolutely necessary for the initial stages of growth. In fact, plants grow more rapidly in the dark than in the light. Although light has a retarding effect on growth, the protoplasm maintains its healthy condition and the plant becomes sturdy with normal development of the stem and the green leaves when there is a certain intensity of light. Moreover, as we have already learnt, stomata remain open and chloroplasts function normally preparing food substances only

FIG. 43. Effect of light and darkness on growth of seedlings. *Left*, gourd seedlings; *right*, gram seedlings; *A*, grown in light; *B*, grown in darkness.

in the presence of light. All this is said to be due to the **phototonic** or stimulating effect of light. Continued absence of light is very harmful to plants. Plants grown in the dark or in very weak light have delicate, soft and slender **stems and branches** with elongated internodes, are pale-green or pale-yellow in colour and sickly in appearance, and seldom produce **flowers and fruits**;

leaves of such plants are small, pale-yellow in colour and often remain undeveloped; and their roots are seen to be poorly developed. Plants showing these characteristics are said to be **etiolated** (FIG. 43B). The relative length of day and night has a profound influence on the production of flowers and fruits (see photoperiodism, p. 388B). The effect of particular rays of light on the growth of the plant can be easily seen when a pot plant is grown within a double-walled bell-jar which has been filled up with red, yellow, green or any other solution (see also p. 335). The effect of unilateral light on growth and movements is discussed on p. 394.

(6) **Force of Gravity.** This factor determines the direction of growth of particular organs of the plant body (see p. 394). The root grows towards the force of gravity, and the stem away from it.

Phases of Growth (FIG. 45). Growth does not take place throughout the whole length of the plant body, but is localized in special regions called *meristems* which may be apical, lateral, or intercalary. The growth in length is due to gradual enlargement and elongation of the cells of the apical meristems (root-apex and stem-apex) and in dicotyledons and gymnosperms the growth in thickness is due to the activity of the lateral meristems, i.e. interfascicular cambium, fascicular cambium and cork-cambium. If the history of growth of any organ of a plant be followed three phases can be recognized in it.

(1) **The Formative Phase.** This is restricted to the apical meristem of the root and the stem. The cells of this region are constantly dividing and multiplying in number. They are characterized by abundant protoplasm, a large nucleus and a thin cellulose wall.

(2) **The Phase of Elongation.** This lies immediately behind the formative phase. The cells no longer divide in this phase, but increase in size; they begin to enlarge and elongate until they reach their maximum dimension. In the root this phase occupies a length of a few millimetres, and in the stem a few centimetres. In some of the climbers it may occupy a much longer space than this.

(3) **The Phase of Maturation.** This phase lies further back. Here the cells have already reached their permanent size; the thickening of the cell-wall takes place in this phase.

Grand Period of Growth. Every organ of the plant body, in fact every cell that the organ is composed of, shows a variation in the rate of its growth. The growth is at first slow, then it accelerates until a maximum is attained, then it falls off rather quickly, and gradually slows down until it comes to a standstill. This growth of an organ or a cell or the plant as a whole extending over the whole period is called the **grand period of growth**. Within the grand period variations in growth occur due to external and other causes. There is thus the *diurnal variation of growth*. Light in-

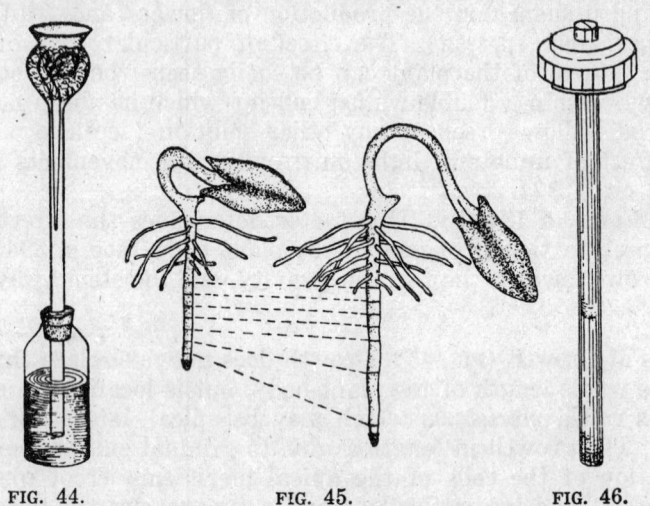

FIG. 44. FIG. 45. FIG. 46.
FIG. 44. Measurement of growth of root. FIG. 45. Phases of growth of root. FIG. 46. A space marker wheel.

hibits growth, and too intense light even checks it altogether. Thus plants grow quicker during the night than during the day. There is also *seasonal variation of growth*; during winter the growth of many plants is checked or becomes very slow, but during spring growth proceeds rapidly.

Experiment 36. Distribution and Rate of Growth. (*a*) Root. With the help of a space marker wheel (FIG. 46) mark the root of a germinating seed with transverse lines equidistant from each other. Waterproof Indian ink should be used for the purpose. The germinating seed is then packed in wet cotton and placed in the bulb of a thistle funnel which is allowed to stand in a bottle containing water. The stem of the funnel is covered with black paper to prevent curvature of the root owing to light. After a few days it is seen that the lines at a little distance behind the tip become widely separated from each other, while those higher up and at the tip remain more or less intact. This evidently shows that the growth is fastest behind apex. The *rate of growth* may easily be measured from day to day with the help of a scale. (*b*) Stem. The distribution of growth in the stem is also found out in a similar way. For this the thistle funnel need not be used. The stem shows the same phases of growth as the root. The rate of growth of the stem may be accurately measured with the help of an auxanometer. (*c*) Leaf. The distribution of growth of a flat organ like the leaf may be found out with the help of an instrument called the **space marker disc**.

Hormones. It is now definitely known that certain chemical substances formed in very minute quantities as a result of metabolism inside the plant body have a profound influence on the *growth* of the plant organs and on the various kinds of *tropic movements* exhibited by such organs; they also have a marked effect on certain physiological processes. They are known as the **hormones** or **phytohormones** or **growth hormones**. Of the various kinds of plant hormones discovered to date **auxins** (auxin-A and auxin-B) and heteroauxin (indole-acetic acid—IAA, first obtained from human urine) are known best so far as their occurrence, composition and actions are concerned. They are formed in one part of the plant body, chiefly in the apical meristem of the stem and the root and also in the tip of the leaf, and transported from there to another part to produce a particular physiological effect there. Their movement is strictly longitudinal, normally from the apex downwards (Went, 1928 and Thimann, 1934). Upward translocation is also possible through transpiration current. The presence of hormones was first demonstrated by experimental methods.[1] It has now been possible to extract them from plants by appropriate chemical methods. At low concentration they stimulate growth, while at high concentration they retard growth. Auxins cause the formation of roots in stem-cuttings and in grafting. They are responsible for fruit development, seed

---

[1] The term hormone was first used by Starling in 1906 in connexion with certain secretions in the animal body. In the case of plants the action of hormones (auxins) was first demonstrated on oat coleoptile by Boyes-Jensen (1910), Stark (1920) and more elaborately by Went (1926) and Thimann (1934). The tip of the oat coleoptile was cut off some millimetres behind the apex; evidently the growth of the coleoptile was checked. But when the decapitated tip was affixed on the cut surface it was found that the growth was resumed at almost the normal rate. Evidently the tip contained something which when transmitted to the main coleoptile induced growth. Further experiments carried out in 1928 and later clarified the point to a greater extent. The cut-off tip of an oat coleoptile was placed on agar (3%) for about an hour. The agar was then cut into small blocks, and one such block placed on the cut surface of the coleoptile. It was then seen that the growth of the coleoptile was normal, as if nothing had happened. Evidently something was transmitted from the cut-off tip to the agar medium and thence to the main coleoptile. That something is now known to be an auxin. Further experiments were performed. Another agar block treated in the same way was next placed on the cut surface on one side and it was seen that the growth was more rapid on this side, resulting in a curvature of the coleoptile. Evidently the auxin accelerated growth of this side. By 1934 at least three such growth substances or hormones were isolated: auxin a, auxin b and heteroauxin. All of them are widely distributed in plants. Hormones originally occur in plant tissues in certain forms as their precursors. These become catalyzed into active hormones by specific enzyme action; for example, the amino-acid 'tryptophan' is the precursor of heteroauxin (indole-acetic acid—IAA). About 1942 and later it was found that certain synthetic compounds (not formed in plants) induced similar reactions in plants as the natural hormones; these are indole-butyric acid, naphthalene-acetic acid, etc.

germination, seedling growth and growth of plant organs; they also stimulate cell divisions and cell elongation, and influence certain physiological processes; besides, the role of hormones in tropic responses has now been well established. Thus *phototropism* and *geotropism* are now explained on a hormonal basis. In the case of phototropism hormones in course of their downward movement from the apex accumulate in the shaded side which then grows more rapidly than the illuminated side, and thus a phototropic curvature takes place, i.e. the shoot bends towards the light. In the case of geotropism hormones accumulate on the lower side of the horizontally placed shoot. Thus the growth of this side is stimulated, and the stem bends upwards. In the horizontally placed root hormones similarly accumulate on the lower side but the root reacts in a different way. It is probable that the growth of the root is stimulated by lower concentration of the hormones. Therefore, the particular concentration that has stimulated the growth of the stem checks the growh of the root on its lower side. As a consequence the root bends downwards. Hormones responsible for the development of the root, stem, leaf, flower, fruit, etc., have also been discovered; for example, *rhizocaline* made in the leaf is necessary for the root formation, *caulocaline* formed in the root is necessary for the stem elongation, *phyllocaline* formed in the cotyledon is necessary for leaf growth, *florigen* formed in the leaf induces flower formation, *wound hormone* (traumatic acid) is necessary for the healing of wounds, etc. Traumatic acid induces cell division in the undamaged cells bordering a wound, resulting in callus formation which covers the wound and heals it. **Kinins** (kinetin and analogous compounds) first isolated from yeast DNA as *kinetin* (possibly a degradation product of DNA) promote cell division (as was first noted in tobacco pith culture), cell enlargement, seed germination, seedling growth, development of roots and buds, breaking of dormancy, etc. In some cases kinetin works better in association with IAA. Kinins have not been isolated from plants as such although they appear to be present in them in some form or the other.

Gibberellins. Gibberellins were first isolated from a fungus *Fusarium moniliforme* (=*Gibberella fujikuroi*) by Japanese scientists (Sewada in 1912, Kurosawa in 1926, Yabuta in 1938, and several others later), and their effects on many higher plants studied. Some 78 years back Japanese farmers first noted in their rice fields that many rice plants had grown abnormally tall and then died out without producing grains (bakanae disease of rice or foolish seedling disease), evidently resulting in heavy losses to the farmers. Japanese scientists soon started investigations on the cause of this abnormal growth. They succeeded in tracing the above-mentioned fungus as the cause of the disease and finally isolating from it at least three gibberellins (gibberellic acid and others as natural plant hormones). Since this discovery was made, several compounds (hormones) similar to gibberellins have been found to occur in higher plants. Experiments with them on a number of plants have shown much quicker growth of stems and leaves. Further investigations have revealed many other physiological effects of gibberellins, covering far wider grounds than auxins, viz. seed germination, seedling growth, flowering, fruit formation, parthenocarpy, cambial activity, cell elongation, breaking dormancy or dwarfism, better responses to light and temperature, etc.

Vitamins. Vitamins (*vita*, life) are a heterogenous group of organic (biological) products of plants, which have proved to be invaluable for normal growth and development of the body, maintenance of health, vigour, nervous stability, and proper functioning of the digestive system. They are also essential in preventing or curing certain deficiency diseases such as scurvy, beriberi, rickets, malnutrition, loss of appetite, poor physical growth, eye infection, nervous breakdown, etc., caused by the absence of the vitamins in the food or their faulty absorption due to intestinal troubles. For some centuries scurvy (livid spots on the skin and general debility) was a dreaded disease among sailors. Vasco da Gama lost 100 sailors on this account during his voyage round the Cape of Good Hope in 1498. Lind's discovery in 1757 of the value of *Citrus* fruits (oranges and lemons) and green vegetables in the treatment of scurvy, as emphasized in his book *A Treatise on the Scurvy*, came very opportunely. Captain Cook who included fruits and vegetables in the rations for sailors lost none during his voyage to Australia in 1772. About the year 1793 it was definitely found that the use of orange or lemon juice completely dispelled scurvy from the navy. Evidently it contained something (now known to be vitamin C) which was responsible for the prevention and cure of the disease. It was only from the year 1906 that investigations on vitamins were made from the biological standpoint. In 1910 Hopkin's work in England on rats fed on pure foodstuff and that with added fresh milk or fresh fruit juice definitely proved the importance of vitamins for healthy growth. Up till now several vitamins have been discovered and their value established. It is known that they do not take any direct part in body-building nor are they a source of energy. But they play a very important role in the metabolism of proteins, fats and carbohydrates, and their proper assimilation into the body. They also help proper utilization of certain minerals in the body. Some of the vitamins act as prosthetic groups or coenzymes and help synthesis of certain enzymes, or are even components of them, controlling many important biochemical reactions in both plants and animals. Vitamins are, however, always required in minute quantities for a particular effect, and are used up in the metabolic processes. Vitamins are mostly synthesized by plants and stored up in their different organs. Some of the vitamins are also formed by several bacteria and certain fungi. Animals, including human beings, cannot synthesize them in their body (a partial exception being vitamin D). Plants are, therefore, the sources of vitamins for them. It has now been possible to synthesize some of the vitamins, particularly vitamins A, C and D, on a commercial scale. Some common and important vitamins are as follows:

Vitamin A is a growth-promoting and anti-infective vitamin; soluble in fats and oils (sparingly soluble in water); fairly resistant to heat. It is especially good for children; essential for healthy growth, physical fitness and good vision; helps development of bones and teeth, and healthy skin; builds resistance to bacterial infections of the lungs and the intestines; helps normal functioning of the different organs of the body; its deficiency results in eye-diseases, particularly night-blindness, loss of weight, skin diseases, nervous weakness and respiratory diseases. Carotene of plants is the source of this vitamin, and animals can synthesize vitamin A in their body by taking food containing carotene. Excess of vitamin A is stored in the liver and utilized when deficiency occurs. Main sourcess of this vitamin are carrot, green leafy vegetables, cereals (particularly in their pericarp), sprouting pulses, many fruits (particularly yellow ones, e.g. tomato, mango, orange, banana, apple, papaw, etc.), milk, butter, meat, liver of mammals, egg-yolk, fish, fish-liver oils (e.g. cod-liver oil and halibut-liver oil), etc.

Vitamin B consists of a group of closely allied vitamins, commonly called *vitamin B complex*. As a whole it is indispensable for proper maintenance of health throughout life. Its deficiency results in digestive troubles, loss of appetite, diarrhoea, constipation, beriberi and neuritis. Its absence also interferes with proper metabolism of carbohydrates. Some of the vitamins of this group, e.g. thiamine, riboflavin, pyridoxine, niacin, etc., form important constituents (prosthetic groups or coenzymes) of certain enzymes. The important members of this group are the following. **Vitamin B_1** (or thiamine—an anti-beriberi and anti-neurotic vitamin) is soluble in water and, therefore, easily removed with rice-broth; not easily destroyed by heat. It prevents and cures beriberi, heart disease, neuritis and some forms of anaemia. Beriberi was for a long time a dreaded disease in the rice-eating countries, viz. India, Malaya, China and Japan. Polished rice (evidently something, now known to be this vitamin removed from its pericarp) was found to be the cause of this disease which resulted in immense suffering and often innumerable deaths. **Vitamin B_2** (or riboflavin) is slightly soluble in water; relatively stable to heat but destroyed by light. It is formed by all parts of the plant. It promotes growth and removes digestive troubles; prevents and cures nervous and general debility, low vitality, soreness of tongue and lips, and some form of eye-disease. It works in combination with other B vitamins. **Vitamin B_6** (or pyridoxine) is soluble in water; resistant to heat. It is distributed in all parts of the plant. It has clinical value in treating certain nervous and muscular disorders and certain types of anaemia. It helps metabolism of proteins and fats. **Vitamin B_{12}** is sparingly soluble in water; resistant to very high temperature. It helps to form red blood and thus cures symptoms of pernicious anaemia, and also diseases of the nervous system. It is linked up with protein metabolism. Several species of bacteria have the ability to synthesize this vitamin. **Nicotinic Acid** (niacin) is soluble in water; resistant to very high temperature. It is widely distributed in plants, and is common in milk, meat and yeast. It prevents pellagra (a kind of skin disease), frequent or persistent diarrhoea, loss of appetite, and mental disease like insanity. **Folic Acid** is sparingly soluble in water. It helps to form red blood corpuscles. Its deficiency may cause macrocytic anaemia and retardation of growth. **Biotin** (vitamin H) is soluble in water; resistant to heat. It cures scaling-off of the skin, muscular pain, distress of the heart and loss of appetite. Vitamin B complex is very widely distributed in plants in almost all their parts, and occurs in nearly all natural foods; rich sources, however, are dried yeast, whole grains (unpolished), pulses, most vegetables (particularly green ones like spinach, lettuce and cabbage), many fruits (e.g. tomato, orange, banana, apple, etc.), nuts, milk, cheese, egg-yolk, meat, liver, fish, etc.

Vitamin C (or ascorbic acid—an anti-scurvy vitamin) is soluble in water; sensitive to heat and, therefore, lost by cooking; destroyed when exposed to sunlight. It prevents scurvy, mental depression, swelling and bleeding of gums, degeneration of teeth, cold, and sore mouth; it raises power of resistance to infection; it helps healing of wounds and formation of blood, and it removes intestinal trouble. Vitamin C is a general activator of metabolic processes. It cannot be stored in the body and, therefore, a daily supply of this vitamin from outside is a necessity. It is found in high concentrations in most fresh fruits (particularly orange, lemon. pummelo, tomato, pineapple, emblic myrobalan, banana, guava, papaw, etc.), green vegetables (e.g. lettuce, spinach, cabbage, etc.), sprouting pulses and cereals, and in milk.

Vitamin D (or calciferol—an anti-ricket vitamin) is soluble in fats and oils; cannot stand strong light; otherwise sufficiently stable. Its deficiency causes rickets and dental caries in children, and osteomalacia (softening of bones) in adults, Absence of this vitamin inhibits proper absorption of calcium and phosphate. It is essential for normal development of bones and teeth, and for general growth. It is commonly found in dried yeast, milk (irradiated), butter, egg-yolk, fish and fish-liver oils. Vitamin D can also be produced in the human body by the action of ultra-violet ray (in sunlight or electricity) on the skin. Ergosterol, commonly prepared from yeast but known to be widely distributed in plants and animals, on irradiation, i.e. on exposure to sunlight, becomes transformed into vitamin D.

Vitamin E (or anti-sterility vitamin) is soluble in fats and oils; resistant to heat and light but destroyed by ultra-violet ray. Its deficiency causes sterility in animals (not yet definitely proved in the case of human beings), degeneration of muscles, and falling-out of hairs. It is found in green vegetables, germinating grains, wheat embryo, milk, egg-yolk, meat, etc.

Vitamin K is fat-soluble. Its deficiency causes a lowering of pro-thrombin value of the blood, i.e. does not help proper bloodclotting in wounds and cuts. It is found in green vegetables (concentrating mostly in the chloroplasts) and also in fresh fruits, and in yeast.

Vernalization and Photoperiodism. The influence of temperature and of day-length on the sexual reproduction of plants, particularly in the case of annuals and biennials, has been investigated for a period of over 40 years by several investigators. The outcome of these investigations is the discovery of very important phenomena known as vernalization and photoperiodism. It is interesting to note that some plants, mostly of the temperate regions, require a period of low temperature before flowering takes place, while there are plants which require treatment at a much higher temperature. But for this condition the plants would not produce flowers. The same effect and even more, as noted below, have been found if the soaked seeds are treated with a particular temperature for a certain period during their early stage of germination. The procedure is as follows. Seeds are soaked in water and allowed to germinate till only the radicle has emerged. Further growth is artificially arrested by the temperature method for a period varying from a few to several days for different kinds of seeds. For certain types of seeds the temperature requirements usually vary from 1° (or even less) to 10°C. This applies to certain varieties of wheat. For other types, e.g. rice, millets, cottons and soya-bean the temperature requirement is much higher—20° to 25°C, and the treatment period extends over several days. The seeds are then dried and sown afterwards in the usual way. The effect of such a treatment is the arrest of growth during germination, but accelerated growth of the seedlings, and early flowering. This method of inducing earlier flowering by pre-treatment of seeds with a certain temperature differing in different groups of plants was developed in Russia by Lysenko (1932), and is known as **vernalization**. (The Russian term is Jarovizacija meaning pre-sowing treatment). Practical benefits derived from this method are to induce earlier flowering and earlier maturing of the crop, to escape frost, drought and flood, and to extend cultivation to regions with very low temperature. Thus this method has helped the Russian farmers to grow crops in Siberia where for the ten months in the year the soil remains ice-bound and unfit for any kind of cultivation; during the remaining two months only early maturing crops can be grown.

The influence of day-length on reproduction was studied in America by Garner and Allard (1920). According to them some plants require a day-length longer than 12 hours for flowering, while others require less than 12 hours for this purpose. The former are known as long-day plants and the latter as short-day plants. There are also some plants which are day-neutral as they flower at any day-length. The relation of the time of flowering to the daily length of the period of illumination is known as **photoperiodism**. Photoperiodism has helped in the control of flowering of a large number of agricultural and horticultural plants. Artificial shortening of day-length by shading, or lengthening of the day-length by electric illumination has induced plants to flower earlier than the normal ones. By reducing the day-length by shading, Dr S. M. Sircar of Calcutta University (now Director of Bose Research Institute, Calcutta) has been able to induce flowering of a winter variety (AMAN) of rice in 50 days against the normal flowering in 140 days. The phenomenon of photoperiodism is of considerable practical importance in growing plants at seasons where the required day-length is not available under natural conditions. Flowering of various annual and biennial plants at different seasons of the year is mainly due to the seasonal day-length. In agricultural research this is of particular benefit as by artificial control of day-length (daily illumination) two crop varieties which normally flower at different seasons can be made to flower simultaneously so that cross-pollination for the purpose of crop-improvement can be effected.

Chapter 14   MOVEMENTS

Living beings are distinguished from non-living ones by their power of movement. Protoplasm is sensitive to various external agencies such as heat, light, electricity, gravity, certain chemicals, etc., which act as stimuli, and plants or plant organs often respond to such stimuli by movement of their body in a particular direction taking up thereby a convenient position. The capacity of plants or their particular organs to receive stimuli from without and to respond to them is spoken of as **irritability**. Irritability expresses itself in some kind of movement and is a decided advantage to the plant since by this movement it can adjust itself according to the conditions of the environment.

Conditions Necessary for Movements. (1) **Water.** An adequate supply of water to the organs concerned is essential for certain kinds of movements; a turgid condition of the cells is indispensable for the purpose. (2) **Temperature.** Movement can only occur within a certain range of temperature. (3) **Oxygen.** A steady supply of air (oxygen) is necessary for respiration of the living cells of the organs concerned. Respiration releases energy for work. (4) **Hormones.** Hormones (see pp. 383-85) are now known to have a profound influence on growth and certain kinds of movements. (5) **Non-fatigue.** Continued stimulation brings about fatigue. No response can be evoked from a fatigued organ or tissue.

Kinds of Movements. Plants show different kinds of movements, and these may be broadly classified as (*I*) movements of locomotion and (*II*) movements of curvature.

I. MOVEMENTS OF LOCOMOTION

Movements of protoplasm within the cell, free movements of naked masses of protoplasm and those of unicellular or multicellular organs and entire organisms are expressed as movements of locomotion. These movements may again be (1) spontaneous (or autonomic) and (2) induced (or paratonic).

1. **Spontaneous Movements of Locomotion** are the movements of the protoplasm or of minute free organs or entire organisms *of their own accord*, that is, without the influence of external factors; they may be due to some internal causes, not clearly understood. Common instances are: ciliary movement of free ciliate protoplasmic bodies such as the ciliate gametes and zoospores; amoeboid movement of free non-ciliate protoplasmic masses; rotation or circulation of protoplasm within the cell; oscillating movement of *Oscillatoria* (see FIG. V/2); and brisk movements of many unicellular algae like desmids and diatoms.

2. **Induced Movements of Locomotion** are the movements of minute free organs or entire organisms *induced by external factors* which may be in the nature of certain chemical substances, light and heat. These factors act as stimuli. Movements thus influenced by external stimuli are otherwise called **taxes** or **taxisms**, and depending on the nature of the stimulus taxic (or tactic) movements may be (1) chemotaxis when influenced by chemical substances, (2) phototaxis when influenced by light, and (3) thermotaxis when influenced by temperature.

(1) **Chemotaxis.** Chemotaxis is the movement of free organs or organisms brought about by the presence of certain chemical substances. There are certain bacteria which are strongly attracted by the presence of free oxygen; they move towards the source of supply. Chemotaxis is best exhibited by the male gametes (antherozoids) of many 'flowerless' plants. Thus in mosses cane-sugar is secreted by the archegonium for the purpose of attracting the antherozoids towards it. In ferns malic acid is secreted for the same purpose.

(2) **Phototaxis.** Phototaxis is similarly the movement of free organs or organisms in response to the stimulation of light. Algae afford very good examples of phototaxis. They move towards the source of weak light being attracted by it. Very strong light, however, repels them; they turn away from it. Another striking example of phototaxis is afforded by the chloroplasts of the leaf. Too intense light decomposes chlorophyll and, therefore, under this condition the chloroplasts arrange themselves one over the other alongside the lateral walls of the palisade cells of the leaf. This arrangement or movement of the chloroplasts is called **apostrophe**. In subdued light, however, they arrange themselves alongside the outer and inner walls. This arrangement or movement of the chloroplasts is called **epistrophe**.

(3) **Thermotaxis.** Thermotaxis is similarly the movement of free organs or organisms in response to the stimulation of heat. If there is a difference in temperature they are seen to move towards the warmer side. Protoplasm shows more rapid rotation or circulation if the tissue be gently warmed. Thus if a section from a *Vallisneria* leaf be slightly warmed over a burning match-stick and then examined under the microscope the protoplasm is seen to rotate more rapidly.

II. MOVEMENTS OF CURVATURE

Higher plants being fixed to the ground are incapable of any locomotion. Some of their organs, however, show different kinds of movement. Thus these organs may move and change their position or direction by means of curvature. As the organs move they take up an advantageous position in order to perform their functions more effectively. Movements in curvature may be mechanical or vital. Vital movements are broadly of two kinds; spontaneous (or autonomic) and induced (or paratonic).

1. **Mechanical Movements.** Mechanical movements are exhibited by certain non-living organs of plants, e.g. bursting of explosive fruits (see p. 140); bursting of sporangia of ferns and some other structures. Some fruits burst suddenly when they

dry up, e.g. *Phlox, Barleria, Bauhinia vahlii* (see FIG. I/183), *Andrographis*, etc., and others do the same when they absorb water, e.g. *Ruellia* (see FIG. I/182). The dry long awn of certain grasses—wild oat (*Avena sterilis*), for example—coming in contact with water begins to twist and roll. The elaters of *Equisetum* spore are very hygroscopic; when the air is moist they roll up spirally round the spore, and when the air is dry they uncoil and stand out stiffly from the spore. Mechanical movements of this nature having a definite relation with moisture (by imbibition or by loss of it) are otherwise known as **hygroscopic** movements.

2. **Spontaneous Movements.** Spontaneous movements are movements of certain living organs of plants *of their own accord*, that is, without the influence of external factors. Such movements may be of two kinds; (1) movement of variation and (2) movement of growth.

(1) **Movement of Variation.** The movement of variation is the movement of *mature* organs due to *variation in the turgidity* of the cells making up those organs. It is fairly rapid. The spontaneous movement of variation is rather rare; in the majority of plants the movement of cellular organs is induced by external factors. The spontaneous movement is, however, very remarkably exhibited by the *pulsation*, i.e. rising and falling of the two lateral leaflets of Indian telegraph plant (*Desmodium gyrans*; FIG. 47), the terminal leaflet, however, remaining fixed in its position. Normally these two leaflets move up and down from morning till evening, i.e. so long as sunlight is available; but sometimes they continue to move till late hours at night depending on the energy that they have conserved from the sunlight during the daytime.

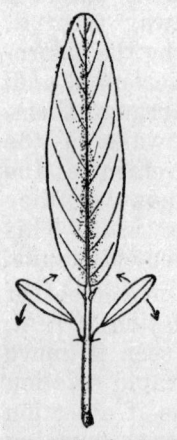

FIG. 47. A leaf of Indian telegraph plant.

Turgor Movements. These are movements exhibited by certain plant organs due to changes in the turgidity of certain cells of these organs, commonly the pulvinus in the case of the leaf. Turgor movements are often rapid and may occur again and again. They may be spontaneous (movement of variation), as exhibited by Indian telegraph plant, or may be induced by contact, light, heat, etc., as exhibited by sensitive plant, sensitive woodsorrel, carambola (*Averrhoa*; B. KAMRANGA; H. KAMRAKH), most species of *Leguminosae*, some species of wood-sorrel (*Oxalis*), and by the opening and closing of the stomata.

(2) **Movement of Growth.** The movement of growth is the movement of *growing* organs *due to unequal growth* on different

sides of those organs. It is very slow. This kind of movement is seen in some trailers and creepers. In them at one time the growth is comparatively rapid on one side of the stem and then it passes on to the opposite side. The stem tip then moves from one side to the other. In such a case the stem, as it elongates, moves in a zigzag course. Movement of this kind is known as (1) **nutation**. If the growth passes regularly around the stem, it then moves in such a way as to form a spiral, as in tendrils and twiners. Movement of this kind is said to be (2) **circumnutation**. Another kind of growth movement is exhibited by most of the young leaves. In them at the initial stage the growth is more rapid on the undersurface and, therefore, they remain rolled or folded on the upper surface. This kind of growth movement is called (3) **hyponasty**. Later due to more rapid growth on the upper surface the leaves open and become flat and straight. This is called (4) **epinasty**. A striking example is afforded by fern leaves which are at first closely coiled due to hyponastic growth, and later they uncoil and become straight due to epinastic growth.

3. Induced Movements. Movements of certain living organs of plants may be induced by external factors which act as stimuli. Induced movements are broadly of two kinds: (*a*) tropic and (*b*) nastic. The stimuli may be in the nature of (1) contact, (2) light, (3) gravity, (4) temperature, (5) certain chemical substances, and (6) moisture.

(*a*) **Tropic Movements** or **Tropisms.** Tropic movements of plant organs are, like taxisms, always directive, i.e. the direction of movement is determined by the direction from which the stimulus is applied, and the organs move either towards the source of the stimulus or away from it. The nature of the stimuli such as (1) contact, (2) light, etc., as stated above, and the corresponding tropic movements are as follows:

(1) **Contact with a Foreign Body.** The movement of an organ stimulated by contact with a solid object is called **haptotropism** or **thigmotropism**. Twining stems and tendrils are good examples of haptotropism. These organs are sensitive to contact with a foreign body but the reaction is rather slow and, therefore, the contact must be of long duration to bring about the movement. Some tendrils, however, respond very quickly, often within a few minutes. When such organs come in contact with any support or any hard object the growth of the contact side is checked, while the other side continues to grow. The result is that the organs slowly coil round that object. This is a mechanism for climbing. Some climbers move clockwise, e.g. white yam (*Dios-*

corea alata); while others anti-clockwise, e.g. wild yam (*D. bulbifera*). If the direction be artificially altered, growth becomes arrested. Haptotropism is also exhibited by the tendrillar leaf-apex of glory lily (*Gloriosa*) and the petiole of pitcher plant (*Nepenthes*), garden nasturtium (*Tropaeolum*) and virgin's bower (*Clematis*).

(2) **Light.** The movement of plant organs as determined by the direction of incidence of rays is called **heliotropism** or **phototropism**. Some organs grow towards the light and are said to be *positively heliotropic*, as the shoot; while others grow away from it and are said to be *negatively heliotropic*, as the root. Dorsiventral organs such as leaves, runners, etc., grow at right angles to the direction of incidence of rays so that their upper surface is exposed to light; such organs are said to be *diaheliotropic*. Positive heliotropism is seen markedly in potted plants, particularly the seedlings, when these are grown in a closed room or box (**heliotropic chamber**; FIG. 48) with one open window on one side. They all tend to grow towards the window, i.e. towards the source of light, and in the case of the box they ultimately come out through it. The cause of phototropic curvature is now explained on the basis of *hormones* (see pp. 383-85). The effect of this unilateral light may be eliminated by placing a pot plant on a **clinostat** (FIG. 50) in the vertical direction and rotating it. The plant is seen to grow vertically upwards and not to bend towards the window. The flower-stalk of groundnut (*Arachis hypogaea*; FIG. 49) grows towards light, but after pollination it becomes negatively heliotropic and positively geotropic like the root. The stalk bends down and quickly elongates pushing the fertilized ovary into the ground where gradually the ovary ripens into a pod (fruit). In *Eucalyptus* it is seen that the edge of the leaf is turned towards intense light and when the light is diffuse the surface is exposed to it. Some species of *Trifolium* also exhibit the same phenomenon.

FIG. 48. Heliotropic chamber.

(3) **Force of Gravity.** The movement of plant organs in response to the force of gravity is called **geotropism**. Geotropism has a marked effect on the direction of growth of plant organs. The primary root is seen to grow towards the centre of gravity and the primary shoot away from it. The former is, therefore, said to be *positively geotropic*, and the latter *negatively geotropic*. The lateral roots and the branches usually grow at right angles

MOVEMENTS

to the force of gravity and are said to be *diageotropic*. That the direction of growth is determined by the stimulating action of the force of gravity is clearly seen in a seedling which has been

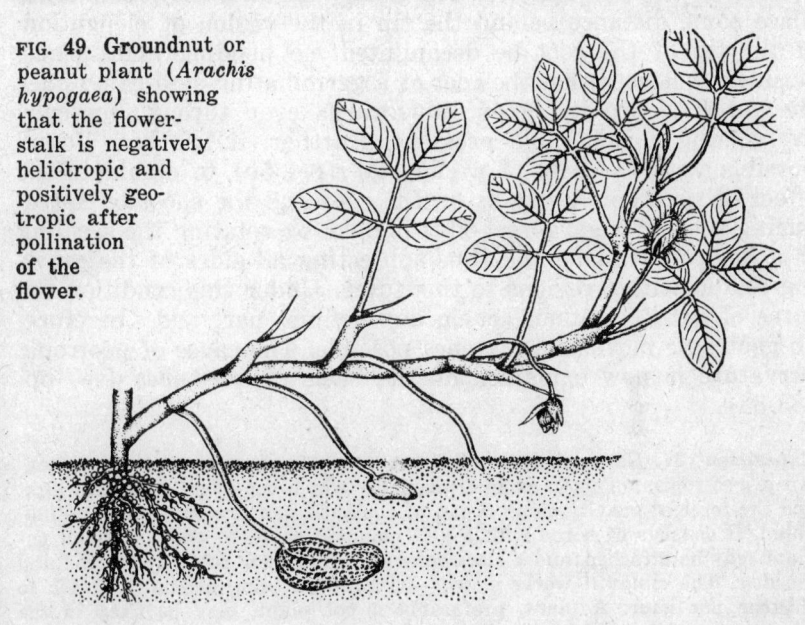

FIG. 49. **Groundnut** or peanut plant (*Arachis hypogaea*) showing that the flower-stalk is negatively heliotropic and positively geotropic after pollination of the flower.

placed in a horizontal position away from light. Both the stem and the root undergo curvature in their growing region behind

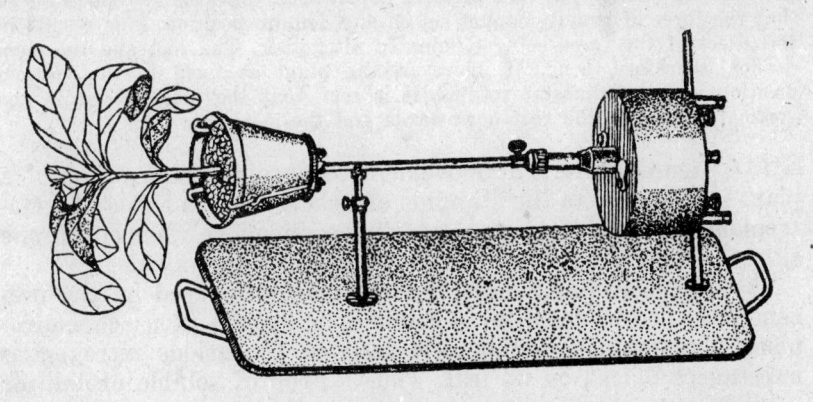

FIG. 50. Clinostat in the horizontal position to eliminate the effect of force of gravity.

the apex, passing through an angle of 90°; the root curves and grows vertically downwards, and so does the stem upwards. It is the very tip of the root, for a distance of 1 to a few mm. in length, that is sensitive to this stimulus; this is the region of cell division (see FIG. I/3). The actual bending, however, takes place some distance behind the tip in the region of elongation. If the tip of the root be decapitated, no bending takes place. Besides, it is seen that the root of a germinating seed can, under the force of gravity, grow downwards even through mercury overcoming considerable pressure. Further, it has been found possible with the help of a **clinostat** (FIG. 50) to eliminate the effect of geotropic stimulus on the root and the shoot by introducing a centrifugal force. This is done by rotating the seedling in a horizontal plane and thus subjecting all sides of the growing and sensitive regions to this force. Under this condition the force of gravity cannot act on any definite part and, therefore, no geotropic movement becomes possible. The cause of geotropic curvature is now explained on the basis of *hormones* (see pp. 383-85).

Experiment 37. Geotropism. A **clinostat** (FIG. 50) may be used to demonstrate geotropism. This is an instrument by which the *effect* of lateral light and the force of gravity on an organ of a plant—root or stem—can be eliminated. It consists of a rod with a disc mounted on it, to which a small pot plant may be attached, and a clockwork mechanism for rotating the rod and the disc. The clinostat works slowly—its rotation being ordinarily 1/4th to 4 turns per hour. A plant, preferably a pot plant, may be fixed in the clinostat in any position—vertical, horizontal or at an angle—and made to rotate by clockwork mechanism in the clinostat. When the plant is horizontal, the root and the stem grow horizontally, instead of the root curving downwards and the stem upwards. This is due to the fact that all sides of the growing axes are in turn directed downwards, upwards and sideways so that the force of gravity cannot act on any definite position. This results in the effect of the force being eliminated altogether. The root and the stem cannot, therefore, bend. If, however, the plant be fixed in the vertical position and the clinostat rotated, it is seen that the plant grows in the vertical direction—the root downwards and the stem upwards.

(4) **Temperature.** The movement or curving of an organ of a plant in response to the stimulus of heat or cold is called **thermotropism**. If a closed box containing seedlings be warmed on one side, it is seen that they curve towards the warm side.

(5) **Chemical Substances.** The movement induced by the presence of certain chemical substances is spoken of as **chemotropism**. The tentacles of sundew respond to various nitrogenous substances placed on its leaf. Thus a drop of soluble protein or a bit of raw meat induces movement only after a portion of it has been absorbed. On absorption the protoplasm gets stimulated and it sends a motor impulse to the surrounding tentacles which bend down on the protein or meat from all sides. The pollen-tube grows towards the ovule being stimulated by the sugary sub-

stance secreted by the stigma, the style and the ovary. Sucking roots of parasites and hyphae of parasitic fungi penetrate into the tissue of host plant in response to the stimulus of certain chemical substances contained in it. Similarly, respiratory roots of many plants growing in estuaries curve upwards from the soil into the air, that is, towards the source of oxygen which stimulates them (see FIGS. IV/2-3).

(6) **Moisture.** The movement of an organ in response to the stimulus of moisture is known as **hydrotropism**. Roots are sensitive to variations in the amount of moisture. They show a tendency to grow towards the source of moisture, and are said to be *positively hydrotropic*. It is seen that roots of seedlings growing in a hanging basket made of wire-netting and filled with moist sawdust at first project downwards coming out of the basket under the influence of the force of gravity, but they soon turn back in response to the stimulus of moisture (moist sawdust of the basket) and pass again into the basket having formed loops.

Experiment 38. Hydrotropism. A porous clay funnel, covered around with a filter paper, is placed on a wide-mouthed glass bottle (or hyacinth glass) filled with water, as shown in FIG. 51. The filter paper is thus kept moist. The porous funnel is filled with dry saw-dust and the soaked seeds arranged in a circle, each near a pore. It is necessary to add a few drops of water now and then to the seeds to help their germination. As they germinate, it will be seen that the roots, instead of going vertically downwards in response to the force of gravity, pass out through the pores towards the moist filter paper, and grow downwards alongside the paper into the bottle. Roots thus show movements towards moisture, or, in other words, they are positively hydrotropic.

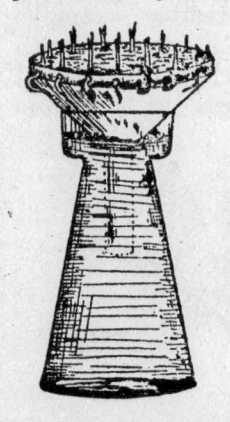

FIG. 51. Experiment on hydrotropism.

(*b*) **Nastic Movements** or **Nasties**. Like tropisms nastic movements of plant organs are induced by stimuli such as contact, light and heat, but these movements are not directive, i.e. the direction of movement in such cases is not determined by the direction from which the stimulus is applied, or, in other words, from whichever direction the stimulus acts it equally affects all parts of the organs, and they always move in the same direction. Their direction is largely determined by the structure or anatomy of the organs concerned. Nastic movements are mostly exhibited by flat dorsiventral organs like leaves and petals. The following kinds of nasties are common:

(1) **Seismonasty.** The movement brought about by mechanical stimuli such as contact with a foreign body, poking with any

hard object, drops of rain, a gust of wind, etc., is called **seismonasty**. Movements of the leaves (leaflets) of sensitive plant (*Mimosa pudica*; FIG. 53), sensitive wood-sorrel (*Biophytum sensitivum*; FIG. 52), *Neptunia*, carambola (*Averrhoa*), etc., are familiar examples. Leaflets of such plants close up when touched. It is also to be noted that the degree of movement varies according to the intensity of the stimulus applied; for example, when the leaf-apex of the sensitive plant is lightly touched, only a few pairs of leaflets close up; when rather roughly touched or pinched all the leaflets react in the same way from the apex downwards; and when roughly handled, all the leaflets close up simultaneously and the leaf as a whole droops. Venus' fly-trap (*Dionaea*; see FIG. 28) is another very interesting example. The two lobes of the leaf of this plant are each provided with three hairs which are extremely sensitive to touch, and in contact with a foreign body, particularly a flying insect, the two lobes

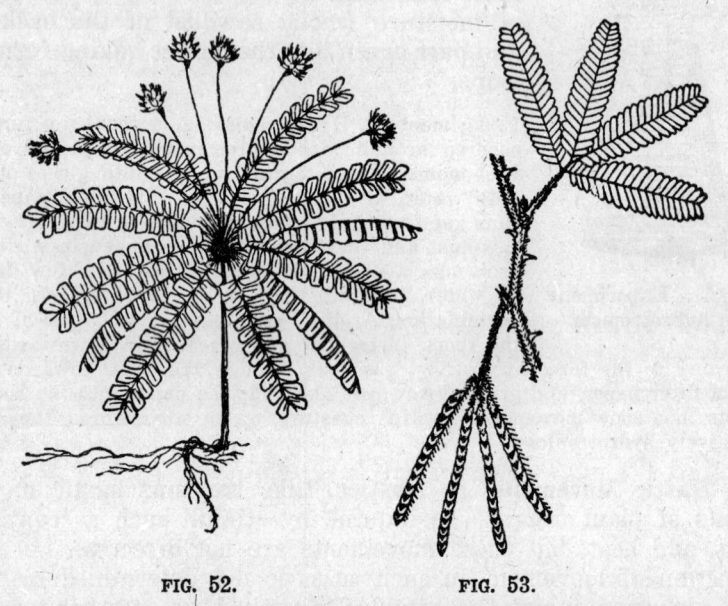

FIG. 52. FIG. 53.

FIG. 52. Sensitive wood-sorrel (*Biophytum sentitivum*).
FIG. 53. Sensitive plant (*Mimosa pudica*).

close up suddenly. In *Bignonia*, *Mimulus* and *Martynia* the two stigmatic lobes close up when touched or when pollen grains fall on them. In the sunflower family (*Compositae*) and China rose family (*Malvaceae*) it is seen that the stigmas bend down and touch the anthers to achieve self-pollination, in case cross-pollination fails. Sweet sultan (*Centaurea moschata*) is a very interesting case. Its stamens are very sensitive to contact with any foreign body. Thus when pollinating insects touch them they begin to twist and oscillate; the florets also move. This movement may be watched if the stamens are gently poked with any hard object. The stamens of barberry (*Berberis*), purslane (*Portulaca*), prickly pear (*Opuntia*), etc., are also sensitive to contact.

(2) **Photonasty.** The movement induced by changes in the intensity of light is called **photonasty**. Many flowers open when there is strong illumination and close up on darkening or when artificially shaded, e.g. noon flower (*Pentapetes*). Some flowers open in weak light in the morning but with the increasing intensity of light as the day advances they close up, e.g. garden purslane (*Portulaca grandiflora*). Some flowers open at night, that is, in darkness but close up at daybreak, e.g. night-blooming cacti (*Cereus* and *Phyllocactus*). Stomata open when light appears but close up again when light fails.

(3) **Thermonasty.** The movement induced by variations in the degree of temperature is called **thermonasty**. Many flowers open rather quickly when the temperature rises and close up when it falls. A similar effect is produced in the case of leaves of most *Leguminosae* and also of *Oxalis*. When the temperature rises high or falls low the leaves close up, and at optimum temperature they open.

(4) **Nyctinasty.** The movement induced by alternation of day and night is called **nyctinasty** (or nyctitropism) or **sleep movement**. Leaves and flowers, particularly the former, are markedly affected by nyctinasty. Nyctinasty is caused by both the factors (light and temperature) acting simultaneously but always more so by light. This kind of movement is most remarkably exhibited by *Leguminosae*. Leaflets of these plants close up and often the leaf as a whole droops in the evening when the light fails, and they open up again when the light appears in the morning. A few other plants like *Chenopodium*, carambola (*Averrhoa*), etc., also show the same phenomenon. Leaflets normally fold on their upper surface, and by doing so excessive radiation of heat is prevented during the night. Movement of this nature is due to difference in the turgidity of the cells of the pulvinus at the base of the leaf and the leaflets. Among the flowers showing nyctinasty mention may be made of *Gerbera* (a garden herb), *Portulaca* (wild or garden variety), etc.

The various kinds of movements that have already been described are scheduled below.

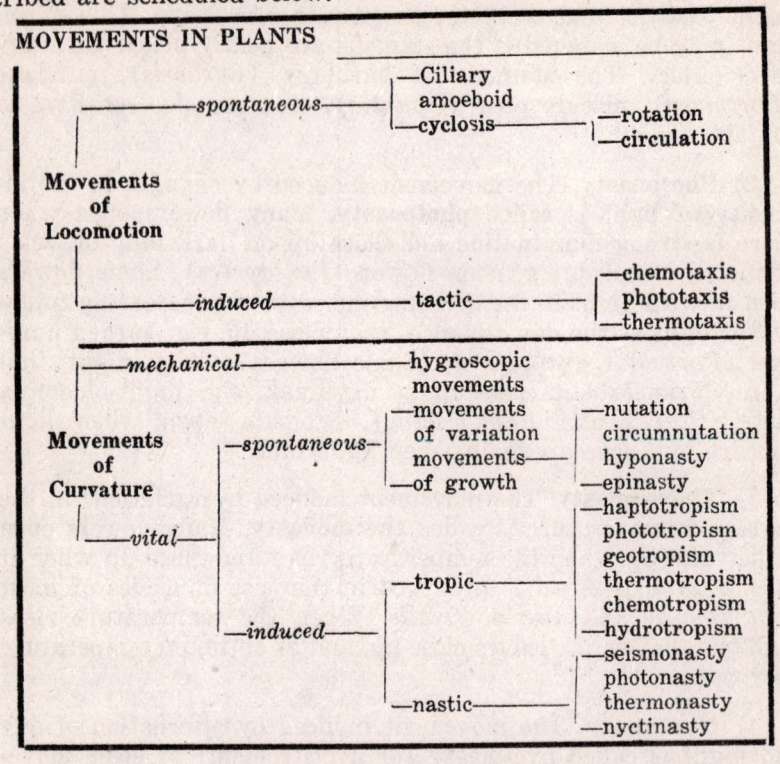

```
MOVEMENTS IN PLANTS
                                    ┌─ Ciliary
                    ┌─ spontaneous ──┼─ amoeboid
                    │                └─ cyclosis ──┬─ rotation
Movements           │                              └─ circulation
of
Locomotion
                    │                              ┌─ chemotaxis
                    └─ induced ─── tactic ─────────┼─ phototaxis
                                                   └─ thermotaxis

    ┌─ mechanical ─────────── hygroscopic movements
    │                       ┌─ movements      ┌─ nutation
    │       ┌─ spontaneous ─┤  of variation   │  circumnutation
Movements   │               │  movements      ├─ hyponasty
of          │               └─ of growth      └─ epinasty
Curvature   │                                 ┌─ haptotropism
    │                                         │  phototropism
    └─ vital ─┐                               │  geotropism
              │                ┌─ tropic ─────┤  thermotropism
              │                │              │  chemotropism
              └─ induced ──────┤              └─ hydrotropism
                               │              ┌─ seismonasty
                               │              │  photonasty
                               └─ nastic ─────┤  thermonasty
                                              └─ nyctinasty
```

Chapter 15 REPRODUCTION

Since the life of an individual plant is limited in duration, it has developed certain mechanisms by which it can reproduce itself in order to continue the perpetuation of the species and also to multiply in number. The following are the principal methods of reproduction: **vegetative, asexual** and **sexual**.

(1) Vegetative Reproduction

A. *NATURAL METHODS OF PROPAGATION*

In any of these methods a portion gets detached from the body of the mother plant, and this detached portion embarks on a new career under suitable conditions. Gradually it grows up into a new independent plant. The methods by which vegetative propagation takes place are many and varied.

(1) **Budding.** In the case of yeast (see FIG. II/25) one or more tiny outgrowths appear on one or more sides of the vegetative cell immersed in sugar solution. Soon these outgrowths get detached from the mother cell and form new individuals. This method of outgrowth formation is known as budding. Often budding continues one after the other so that finally one or more chains, sometimes sub-chains, of cells are formed. All the individual cells of the chain separate from one another and form new yeast plants.

(2) **Gemmae.** In some mosses and liverworts (e.g. *Marchantia*; see FIGS. V/118 and 120) special bodies known as gemmae develop on the leaf, branch or thallus for the purpose of vegetative propagation.

(3) **Leaf-tip.** There are certain ferns, commonly called 'walking' ferns (e.g. *Adiantum caudatum, A. lunulatum* and *Polypodium flagelliferum*), which propagate vegetatively by their leaf-tip (FIG. 54). As the leaf bows down to the ground the tip strikes roots and forms a bud. This bud grows into a new independent fern plant. Ferns normally, however, reproduce vegetatively by their rhizome.

FIG. 54. Walking fern (*Adiantum caudatum*).

In the 'flowering' plants the methods of vegetative propagation are diverse. The resulting offspring resemble the parent forms in almost all respects and, therefore, gardeners often use these methods for multiplying the number of individuals for their gardens.

(1) **Underground Stems.** Many 'flowering' plants reproduce themselves by means of the rhizome, (e.g. ginger), the tuber (e.g. potato), the bulb (e.g. onion) and the corm (e.g. *Gladiolus*). New buds are produced on these modified stems, which gradually grow up into new plants.

(2) **Sub-aerial Stems.** The runner, the stolon, the offset and the sucker are also made use of by plants like Indian pennywort (*Centella*; FIG. 55), taro (*Colocasia*), water lettuce (*Pistia*) and *Chrysanthemum* for vegetative propagation (see FIGS. I/32-5).

(3) Adventitious Buds. In sprout-leaf plant (*Bryophyllum pinnatum*; see FIG. I/16A), *Kalanchoe daigremontiana* (FIG. 56)

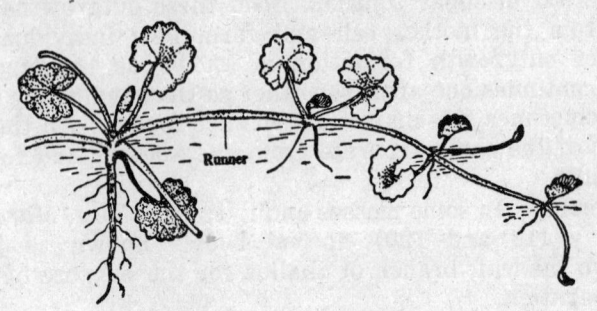

FIG. 55. Runner of Indian pennywort (*Centella*).

and *Crassula* rows of adventitious (foliar) buds are produced on the leaf-margin, each at the end of a vein. These buds may drop from the leaf and grow up into new plants, or they may drop

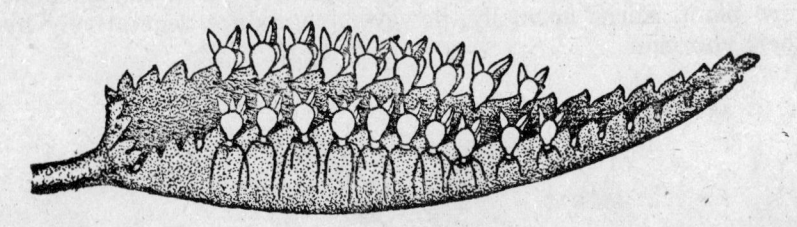

FIG. 56. A leaf of *Kalanchoe daigremontiana* with adventitious buds.

together with the leaf and then grow up. In *Kalanchoe verticillata* (FIG. 57) bud-formation is restricted to the apical part of the

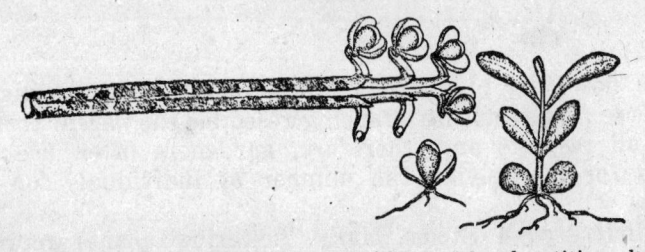

FIG. 57. A leaf of *Kalanchoe verticillata* with adventitious buds.

leaf. In elephant ear plant (*Begonia*; see FIG. I/16B) a few adventitious buds are produced on the surface of the leaf from the veins and also from the petiole, particularly when lightly incised. Similarly, roots of some plants may produce adventitious (radical) buds for the same purpose, as in sweet potato (see FIG. I/9A), *Trichosanthes* (B. PATAL; H. PARWAL), wood-apple (*Aegle*), ipecac (see FIG. I/10B), etc.

REPRODUCTION

(4) Bulbils. In *Globba bulbifera* (FIG. 58) and garlic (*Allium sativum*) some of the lower flowers of the inflorescence become modified into small multicellular bodies known as bulbils. They fall to the ground and grow up into new plants. Sometimes they are seen to grow to some extent on the plant itself. In American aloe (*Agave*; FIG. 60) and certain species of *Crassula* reproductive buds or bulbils often take the place of many flowers of the inflorescence. In *Marica northiana* 1 or 2 large leafy bulbils develop on the green flat scape near its apex. The scape soon bends down to the ground, and the bulbils strike roots and grow up into new plants. Bulbils, big or small, are also produced

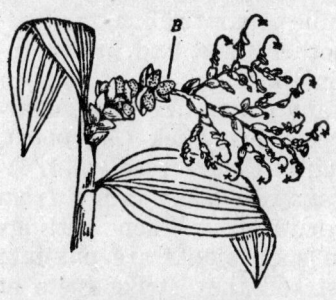

FIG. 58. *Globba bulbifera*. B, bulbil.

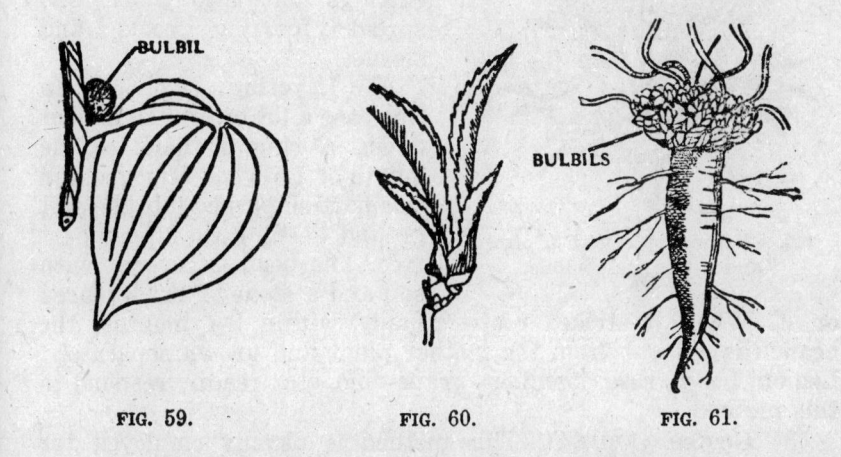

FIG. 59. FIG. 60. FIG. 61.

FIG. 59. *Dioscorea bulbifera*. FIG. 60. Bulbil of American aloe (*Agave*). FIG. 61. Wood-sorrel (*Oxalis*).

in the leaf-axil of wild yam (*Dioscorea bulbifera*; FIG. 59) and *Lilium bulbiferum*. In wood-sorrel (*Oxalis*; FIG. 61) a cluster of small buds (bulbils) may be seen on the top of the swollen tuberous root. These buds being brittle at the base easily fall off and grow up into new plants. In pineapple (*Ananas*) the inflorescence generally ends in a reproductive bud; but in some varieties of pineapple (FIG. 62) the inflorescence becomes surrounded at the base by a whorl of such buds and also crowned by a few of them.

B. ARTIFICIAL METHODS OF PROPAGATION

In any of these methods a portion can be separated out by a special method from the body of the mother plant and grown independently. There are several such methods.

(1) **Cuttings.** (*a*) *Stem-cuttings.* Many plants like rose, sugarcane, tapioca, garden croton, China rose, drumstick (*Moringa*), Duranta, Coleus (see FIG. I/6), etc., may be easily grown from stem-cuttings. When cuttings from such plants are put into moist soil they strike roots at the base and develop adventitious buds which grow up. (*b*) *Root-cuttings.* Sometimes, as in lemon, citron, ipecac (see FIG. I/10B), tamarind, etc., root-cuttings put into moist soil sprout forming roots and shoots.

(2) **Layering** (FIG. 63). In this case a lower branch is bent down, a ring of bark to the length of 2.5-5 cm. removed and this portion pushed into the soft ground keeping the upper part free. The bend is covered with soil and a stone or brick placed on it. When it strikes roots, usually within 2-4 months, the branch is cut out from the mother plant and grown separately. Lemon, *Ixora*, rose, jasmines, grape-vine, etc., readily respond to this method.

FIG. 62. Pineapple with a crown and a whorl of bulbils.

(3) **Gootee** (FIG. 64). This method is usually employed for propagating lemon, orange, pummelo, guava, litchi, *Magnolia*, etc. During early rains a healthy, somewhat woody branch is selected and a ring of bark, 2.5 to 5 cm. in length, sliced off from it. A sufficiently thick plaster of grafting clay[1] is applied all round the ringed portion which is then wrapped up with straw or rag and tied securely. It should be wetted with water every morning and afternoon. In drier climates an earthen pot with a hole at the bottom may be hung over the bandage in a convenient position, and the two connected by a long piece of cloth or soft

---

Grafting Clay. Clay (2 parts), cowdung (1 part) and some finely cut hay mixed with water.

cotton cord. As the pot is filled with water, the latter trickles down the cloth or cord and keeps the bandage constantly moist. Usually within 1-3 months the gootee is ready, as is indicated by its striking roots. It is then cut out below the bandage.

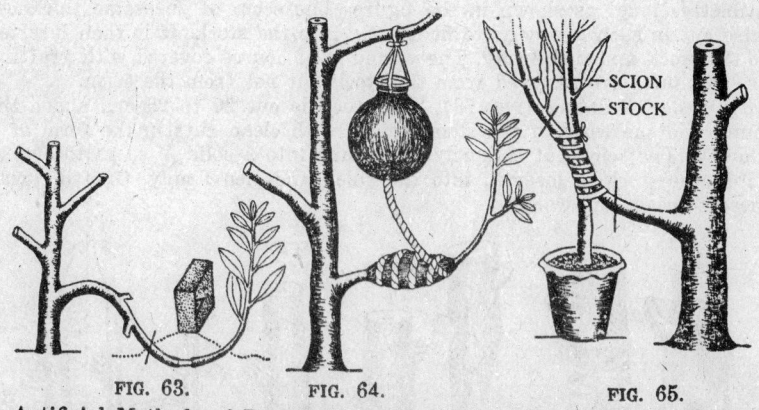

FIG. 63. FIG. 64. FIG. 65.
Artificial Methods of Propagation. FIG. 63. Layering. FIG. 64. Gootee. FIG. 65. Inarching or approach grafting.

(4) **Grafting.** This consists in inserting a small branch of a plant into a rooted plant of the same or allied species in such a way as to bring about an organic union (fusion of tissues) between the two and finally make them grow as one. The branch that is inserted is known as the *scion* or *graft*, and the plant that is rooted to the soil as the *stock*. The scion grows retaining all its qualities, while the stock which may be of inferior quality but physically sturdy supports it by supplying water and food material. Grafting thus ensures the production of particular desired characters in the scion or graft, originally exhibited by the scion-mother. Grafts are prepared for the purpose of propagation of certain fruit and ornamental shrubs and trees. Some of the common methods are as follows:

(*a*) **Inarching** or **Approach Grafting** (FIG. 65). By this method a branch (scion) of a plant is made to unite with a seedling (stock) or a small plant in a pot by firmly tying them together with a cord. Before doing this a small portion of the bark is sliced off from each to ensure a closer contact and a quicker union between the two. When proper fusion has taken place, usually within 2-3 months, the stock is cut above the joint and the scion below, thus leaving the scion standing on the stock. Some of the fruit trees like mango, litchi, guava, sapodilla plum, etc., readily respond to this method.

(*b*) **Bud Grafting** (FIG. 66). For this method a T-shaped incision is made in the bark of the stock, and a bud cut out clean from a selected plant is inserted into the T-shaped slit and properly bandaged. By this method it has been found possible to grow several varieties of roses on one rose stock, good varieties of orange, lime, lemon, citron, pummelo, etc., on inferior stocks, several varieties of *Hibiscus* on one, several varieties of cacti on one, and so on. Luther Burbank of California was able to grow by bud grafting several varieties of prune and allied species on stock.

(c) **Whip or Tongue Grafting** (FIG. 67). The stock, usually 1 to 1·5 cm. thick, is cut down above the ground. Sloping cuts are made in it a few centimetres long, as shown in the figure. The scion of the same thickness is also cut in such a way as to fit exactly into the stock. It is then inserted into the stock and tied firmly. The wound is of course covered with grafting wax.[2] All buds are removed from the stock but not from the scion.

(d) **Wedge Grafting** (FIG. 68). The stock is cut 20 to 25 cm. above the ground and the wood of the stem incised with clean cuts in the form of a hollow V. The scion cut obliquely downward into a solid V so as to closely fit into the stock is inserted into the stock and tied firmly. Grafting wax is used to cover the wound.

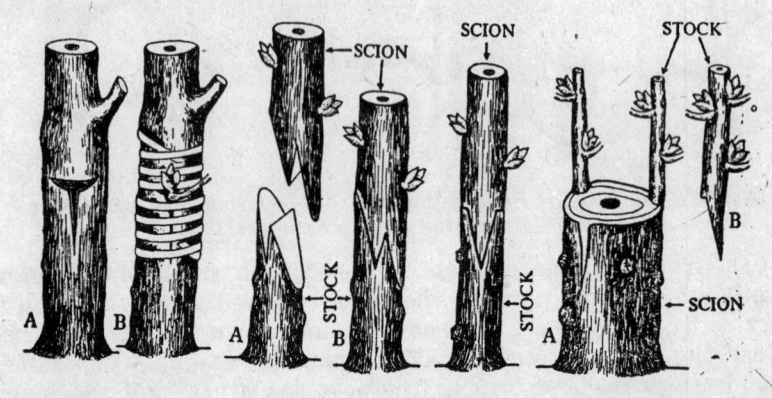

FIG. 66.   FIG. 67.   FIG. 68.   FIG. 69.

Artificial Methods of Propagation (*contd.*). FIG. 66. Bud grafting. FIG. 67. Whip or tongue grafting. FIG. 68. Wedge grafting. FIG. 69. Crown grafting.

(e) **Crown Grafting** (FIG. 69). An old tree may be rejuvenated by this method. The stem is cut across 20 to 25 cm. above the ground. The bark of the stock is cut through from the surface downward to a length of 12 to 15 cm. The bark is partially opened on either side. Prior to this a small branch cut out from another tree of the same species is incised at the base with a sloping cut and this is now inserted into the slit in the bark and tied firmly. The wound is of course covered with grafting wax.

Chimaera (or Chimera). Chimaera refers to a fabulous monster with a lion's head, goat's body and serpent's tail. It means that chimaera is an organism which is made up of two or more genetically distinct tissues. In plant the grafting sometimes produces chimaeras, otherwise called graft hybrids. Here genetically distinct tissues (cells) of the two (stock and scion) become associated, evidently producing a mingling of two sets of characters. As a result the new buds that appear on the junction of the two plants (stock and scion) possess mixtures of characters of the two. Finally branches

---

[2] **Grafting Wax.** A mixture to tallow (1 part), beeswax (1 part) and resin (4 parts), melted together and worked into a soft dough under water.

developing from such buds show such mixtures in leaves or flowers or in both. Sometimes chimaeras may be produced on normal plants by bud mutations.

(2) Asexual Reproduction

This commonly takes place by means of asexual reproductive units, called **spores**, produced by the mother plant, or sometimes by the division (fission) of the mother cell in the case of unicellular plants. Asexual reproduction thus takes place by two methods: fission and spore formation.

1. By Fission. In the simplest cases, as in many unicellular algae and fungi and in bacteria, the mother cell splits into two new cells. The new cells thus formed contain all the materials of the mother cell, and soon they grow to the size of the latter, becoming new independent plants. This method of reproduction by the division of the mother cell is called fission (see FIG. V/62A).

2. By Spore Formation. Spores are asexual reproductive units which can grow independently, i.e. without fusing with another unit, and are always unicellular and microscopic in size. They may be motile or non-motile.

(1) Ciliate motile spores, called **zoospores**, produced by many algae and fungi, swim about in water for some time with the help of their cilia, like minute aquatic animalcules, and then directly

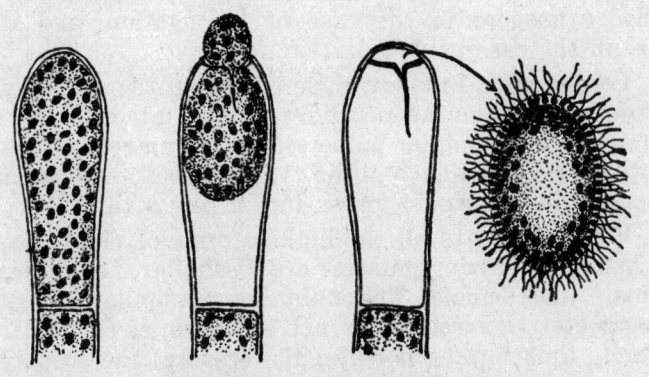

FIG. 70. Rejuvenescence in *Vaucheria*.

develop into new independent individuals. Zoospores are commonly formed in large numbers, as in *Ulothrix* (see FIG. V/20). In *Vaucheria* (FIG. 70), however, the whole mass of protoplasm escapes from the mother cell as a single large multiciliate and multinucleate zoospore. It swims in water for some time and then comes to rest. Almost immediately it germinates into a new *Vaucheria* filament. This is a case of **rejuvenescence**.

(2) **Non-ciliate non-motile spores** of various kinds are most common among terrestrial fungi. Such spores are light, dry and provided with a tough coat, and are well adapted for dispersal by wind; at the same time they are resistant to unfavourable atmospheric conditions.

(3) True spores are always borne by a sporophyte. Thus the sporogonium (sporophyte) of moss reproduces asexually by spores. Similarly ferns, *Lycopodium* and *Equisetum* bear spores and reproduce asexually by them. It is further to be noted that these plants are **homosporous**, i.e. they bear only one kind of spores. The more advanced types of plants, e.g. *Selaginella* and 'flowering' plants (both gymnosperms and angiosperms), are **heterosporous**, i.e. they bear two kinds of spores: microspores (male) and megaspores (female).

(3) Sexual Reproduction

This consists in the fusion of two sexual reproductive units, called **gametes**, which, like the spores, are also unicellular and microscopic. To reproduce sexually two similar or dissimilar gametes fuse together to give rise to a **zygote**; the zygote develops into a new plant. Sexual reproduction in which the pairing gametes are similar is known as **conjugation**, and that in which the pairing gametes are dissimilar is known as **fertilization**. The zygote is called **zygospore** in the case of conjugation, and is called **oospore** in the case of fertilization.

(1) **Conjugation.** In lower algae and fungi the pairing gametes are essentially similar, i.e. not differentiated into male and female. The union of such similar gametes, i.e. **isogametes** (*isos*, equal), is known as **conjugation**, and the zygote thus formed is called the **zygospore**; the zygospore grows into a new plant.

(2) **Fertilization.** In all the higher forms of plant life, on the other hand, the pairing gametes are dissimilar, i.e. differentiated into male and female. The union of dissimilar gametes, i.e. **heterogametes** (*heteros*, different), is known as **fertilization**, and the zygote thus formed is called the **oospore**; the oospore grows into a new plant. In Bryophyta and Pteridophyta and in many algae and fungi the male gametes are *ciliate, motile* and *active*, and are called **antherozoids** or **spermatozoids**; they are very minute in size. The female gamete in them on the other hand is much larger, *nonciliate, stationary* and *passive*, and is called **egg-cell, ovum** or **oosphere**. The corresponding male and female reproductive units in the 'flowering' plants are the **two gametes** of the pollen-tube, and the **egg-cell** or **ovum** of the embryo-sac within the ovule.

Gametes are always borne by the gametophyte. Moss is a gametophyte; it directly bears the gametes. In Pteridophyta the gametes are borne by a small green body, called the prothallus; so the prothallus is the gametophyte. In the 'flowering' plants the gametophyte is an extremely reduced body.

(4) Special Modes of Reproduction

1. **Apomixis.** Apomixis is an irregular mode of reproduction resulting in the development of an embryo without the act of fertilization. It may be (a) parthenogenesis, (b) apogamy and (c) sporophytic budding. (a) **Parthenogenesis** is the development of zygote from the egg-cell without the act of fertilization, as seen in many lower plants, e.g. *Spirogyra, Chara, Mucor, Saprolegnia*, certain ferns and certain species of *Selaginella* and *Marsilea*. In some 'flowering' plants, as in *Thalictrum, Alchemilla*, certain species of *Compositae* and *Solanaceae* the embryo also may develop by parthenogenesis, i.e. without fertilization. In them the embryo may develop from a haploid egg-cell or diploid egg-cell. In the former case although germination is more or less normal, the plant becomes small and sterile. In the latter case the plant is normal in all respects. Sometimes the ovary develops normally into a fruit without fertilization. This type of fruit development is called **parthenocarpy**. Parthenocarpic fruits are almost always seedless. Natural parthenocarpy is found in many plants, as in certain varieties of banana, pineapple, guava, grapes, apple, pear, papaw, etc. It is rather peculiar that sometimes fruit formation may be induced by artificial pollination with foreign pollen (pollen from another species or even genus) or even by the application of pollen extracts, but without subsequent fertilization. In such cases, as later found, auxins formed in the ovary appear to be involved in the setting of the fruit (Gustafson, 1939). Sometimes mere spraying with certain growth-promoting chemicals like indole-acetic acid and naphthalene-acetic acid results in the setting of fruits without fertilization (induced parthenocarpy; Gustafson, 1936). (b) **Apogamy** is the development of an embryo from any cell of the gametophyte (prothallus) other than the egg-cell, evidently without the intervention of gametes. The embryo so formed grows into the sporophyte. It is of common occurrence in ferns. In 'flowering' plants it is sometimes seen that one or more embryos may be formed from the synergids, as in onion (*Allium*), lily (*Lilium*), aconite (*Aconitum*) and *Iris*, or from an antipodal cell, as in onion (*Allium*). The synergid or antipodal cell may be haploid or diploid. (c) **Sporophytic budding** may sometimes occur resulting in the development of an embryo. This means that an

embryo may be formed from the diploid cells of the nucellus, as in orange, mango, prickly pear (*Opuntia*), etc., or even from those of the integument, as in onion (*Allium*). The embryo thus formed is pushed into the embryo-sac during the course of its development.

2. Apospory (*apo*, off or without). Apospory is the development of the gametophyte directly from vegetative cells of the sporophyte without the intervention of a spore. In ferns the prothallus may develop from certain vegetative cells of the leaf in place of spores, or it may develop from one or more cells of the sporangium other than a spore. Apospory has been found in *Pteris*, *Asplenium*, *Osmunda* and certain other ferns. The gametophyte that develops by this method is commonly diploid, and it may bear both antheridia and archegonia. Apospory is also sometimes seen in some mosses.

Polyembryony. The occurrence of more than one embryo in the seed is known as *polyembryony*. Many species of both dicotyledons and monocotyledons exhibit this phenomenon. Polyembryony is, however, particularly common among conifers (see FIG. VI/18H). These embryos may be formed in the seed in a variety of ways: (1) there may be more than one egg-cell in the embryo-sac or more than one embryo-sac in the ovule, and all the egg-cells may be fertilized; or (2) a number of embryos may develop simultaneously from different parts of the ovule. Thus they may be formed from the synergids and antipodal cells (vegetative apogamy), from the fertilized egg-cell or unfertilized egg-cell (parthenogenesis), and from the tissue of the nucellus and the integument by sporophytic budding (apospory). Examples have been found in onion, groundnut, mango, lime, lemon and orange (but not in shaddock and citron). Polyembryony is commonly found in conifers where there are many archegonia in the ovule. All the archegonia may be fertilized resulting in the formation of as many embryos as there are archegonia; but ultimately one embryo reaches maturity, while others die off. In addition, one or four embryos may be formed in *Pinus* from the fertilized egg-cell.

Part IV ECOLOGY

Chapter 1 PRELIMINARY CONSIDERATIONS

Ecology (*oikos*, home; *logos*, discourse or study) deals with the study of interrelationships between the living organisms (plants and animals) as they exist in their natural habitats, and the various factors of the environment surrounding them. Ecology, therefore, includes a detailed study of the flora and fauna (together called **biota** or biotic community) of a particular region and also the various conditions of the environment prevailing in that region. This is expressed as ecological complex or **ecosystem**. It includes an intensive study of all the principal aspects of life from birth to death, as influenced by the environment. It investigates the various structural peculiarities (external and internal), all vital functions including growth and reproduction, survival capacity of the offspring, adaptations for distribution, migration, colony formation, etc., of various species of plants and animals of a particular region, and also the various environmental factors influencing such aspects of their life. Ecology is thus a vast and intricate subject, requiring a good preliminary knowledge of morphology, taxonomy, anatomy, physiology, climatology, soil science, geography, etc., as a prerequisite to the study of this branch of science. Apart from the physical features of the environment the influence of plants and animals upon one another is also of considerable importance. It is known that all animals including human beings are directly or indirectly dependent upon plants, particularly for food and shelter; while the effects of animals and human communities on plants are also manifold. A study of ecology necessarily includes both plants and animals, and also the interactions between them.

Further, in different climatic regions of the earth, often covering vast areas, particular species of plants and animals live together under identical environmental conditions, and show a considerable amount of interdependencies between them in many respects. In such areas both plants and animals form distinct communities, major and minor. Major or climax communities consisting of one or more dominant species have permanently established themselves there, while the minor communities consisting of other species are still unstable. Such a complex of communities (major and minor) of plants and animals of a region existing under identical climatic conditions is called a **biome**. The major communities always constitute the main features of a biome, but they are associated with minor communities, whatever be their composition in different regions. Biomes may be aquatic, e.g. freshwater biome and marine (saltwater) biome, or terrestrial, e.g. evergreen forests, deciduous forests, coniferous forests, thorn forests, grasslands, scrubs, deserts, Arctic tundra, etc. The above biomes (with the exception of tundra) are well represented in India, while the coniferous forests are restricted to the hills, developing there in successive stages according to altitude.

Interdependence between Plants and Animals. (1) **Food.** Green plants manufacture food and store it in their body, while animals mostly depend on this stored food. (2) **Oxygen.** Green plants purify the atmosphere by absorbing carbon dioxide and giving out pure oxygen. (3) **Shelter.** Plants furnish shelter to a variety of animals, particularly birds and wild beasts and forest dwellers. (4) **Clothing.** Many plants furnish materials for clothing, while silk worms, MUGA worms and ENDI worms are reared on mulberry plant, SOM plant (*Machilus bombycina*) and castor plant respectively. (5) **Drugs.** Many plants have curative values in different illnesses and are extensively used as drugs. (6) **Vitamins.** These are almost exclusive products of plants required by human beings and also other animals, as also by plants themselves, for their healthy growth. (7) **Pollination.** Many animals, particularly insects, are very serviceable in this direction resulting in setting of seeds and fruits. (8) **Dispersal of Seeds.** Similarly, many animals are useful agents in disseminating seeds and fruits. (9) **Parasitism.** Certain animals lead a parasitic life upon plants, while many plants do likewise on animals. (10) **Disintegration of Soil.** Many lower organisms (the soil-dwellers) disintegrate and sometimes chemically alter the soil for its better utilization by higher plants. (11) **Civilization.** Plants have materially contributed to the growth of civilization.

Environment. Environment includes all the factors that affect the form and growth not only of individual plants, but also of plant associations. Environmental or ecological factors may be (1) climatic, (2) edaphic, (3) biotic, and (4) topographic. The climatic factor includes rainfall, temperature, light, wind and humidity. The edaphic factor includes physical and chemical properties of the soil, its water- and air-contents. The biotic factor includes bacteria, protozoa, etc., in the soil, and other living plants and animals including human beings, directly or indirectly affecting vegetation in a number of ways. The topographic factor includes undulation, elevation (altitude), slope, and exposure to sun, rain and wind. It must be noted that the vegetation of an area depends on the combined action of a number of factors.

PRELIMINARY CONSIDERATIONS

1. Climatic Factor. This includes all the conditions of the atmosphere such as rainfall, temperature, light, wind, etc., affecting primarily the shoot system of the plant.

(1) **Rainfall.** Water is the most important factor. It is responsible for various structural modifications of plants. Water is indispensable for all the vital functions of the plant. Protoplasm is saturated with water, and often we find that over 90% of the total weight of active tissues is water. The source of water for terrestrial plants is rain. Rainfall has a marked effect on the geographical distribution of plants. Thus depending on the amount of precipitation the vegetation may broadly be of the following types: evergreen forest (with abundant rainfall), deciduous forest (with moderate or low rainfall), grassland (with low rainfall) and thorn scrub (with scanty rainfall). Availability of rain-water depends on the water-retaining capacity of the soil, as also often on plants themselves. The soil may be *physically dry* or sometimes *physiologically dry* due to the cold state of the ground or the preponderance of salts in it. Topographical factors are also very important in this respect. Cherrapunji in the Khasi Hills, the rainiest spot in the world with an annual rainfall of over 12,700 mm., has a very luxuriant vegetation; while Rajputana with very little or no rain is extremely arid. It should also be remembered that abundance or scarcity of available water determines the life-cycle of the plant, the duration of plant growth and the time of reproduction in addition to certain well-marked features of the plant. Two extremes are hydrophytes and xerophytes.

(2) **Temperature.** A certain temperature is essential for all the vital functions of the plant. It markedly affects germination, growth, reproduction and movements. Temperature requirement is, however, different for plants growing in different climatic regions of the earth. Under certain conditions some organs of the plant are thermotropic; for instance, the opening and closing of flowers and of stomata, the drooping of leaves at night, etc., are caused partly by heat. In many cases temperature helps the dehiscence of fruits and thus the dissemination of seeds. Plants normally prefer a temperature varying from 20°C. to 40°C. Active tissues filled with water have much less ability than dry seeds and spores to withstand extremes of temperature. Most flowering plants are killed at a temperature below 0°C. and above 45°C; while seeds remain uninjured at a temperature far beyond these limits. Freezing tem-

perature or frost kills plants, but at high altitudes where frosts frequently occur plants become unusually resistant. Temperature has an important bearing on plant geography. We find a considerable difference between the flora of tropical, sub-tropical, temperate, arctic and alpine regions.

(3) **Light.** Physiologically light is a very important factor. It is responsible for the formation of chlorophyll and for carbon assimilation; it accelerates transpiration. Although strong light checks growth, it has a tonic effect on plants. Light induces certain kinds of movements like photonasty and phototropism. Relative length of day and night has a marked effect on the development of flowers and maturation of fruits. Of all parts of the plant the leaves undergo by far the greatest modifications under the action of light. Plants growing in shady places are called **sciophytes**, and they usually have large leaves which are thin in texture and sparsely distributed on the stem; the stem is thin with long internodes; both the stem and the leaves are glabrous; palisade tissue is poorly developed; the leaf consists largely or entirely of spongy tissues; epidermis often contains chlorophyll and the cuticle is thin; stomata may be present on both sides. Common examples are *Begonia,* aroids, wood-sorrel, ferns, mosses and liverworts. Plants which can only grow well in the light are called **heliophytes,** and they, on the other hand, have small leaves which are thicker and crowded together on the stem; the stem is stouter with short internodes; the stem and sometimes the leaves are very hairy; palisade tissue is well developed; epidermis is provided with a thick cuticle, but no chlorophyll; stomata are present on the lower side, often sunken or occluded. Aqueous tissue is often present. Most thick-leaved plants are heliophytes.

(4) **Wind.** Wind has usually a destructive action on vegetation. It enhances transpiration; very strong, dry wind is often fatal to plants, particularly to young seedlings. In forests it has been seen that some plants can resist the action of wind far better than others. On the seashore coconut-palm is exposed to strong gusts of wind. The plant can withstand them well because its leaves are cut into narrow segments with stout mid-rib; the cuticle also is very thick. Wind is useful in disseminating seeds and fruits, particularly those provided with some kind of appendages (see pp. 136-39). In deserts there are certain species which, when the air is dry, roll up into balls and are driven by wind from one place to another (see p. 424). They strike roots and grow when the conditions are favourable.

2. **Edaphic Factor: Soil.** Since most plants are fixed to the ground and draw their supply of essential food elements from

the soil, with the exception of carbon dioxide and a little nitrogen in some cases, which are obtained from the air, it is evident that there is an intimate relationship between plants and soils, particular types of the latter favouring the growth of particular species of the former. Thus the distribution of plants over different regions of the earth, and their structural peculiarities, are largely determined by edaphic factors. This being so, a study of soils in different aspects is as important as the study of plants themselves from ecological point of view. These aspects of soils have been dealt with in Part III, Chapter 2.

3. **Biotic Factor.** The term 'biotic factor' includes all influences exerted by living organisms to bring about any change in the vegetation of a locality. The most important in this respect are the human beings and the grazing animals. Man acts as a powerful agent, influencing profoundly the original vegetation of a locality in a number of ways: cultivation, terrace cultivation in the hills, reclamation of land, cutting down trees for wood and fuel, deforestation, afforestation, etc. Grazing by domesticated animals destroys many herbaceous plants in pastures and places round about villages of agriculturists. Grazing constantly interferes with the normal development of the vegetation of particular areas. The relation of a species to other living organisms is apparent in many cases. There is the competition with its neighbours for food, water and sunlight. It may be attacked and sometimes wholly destroyed by parasitic plants or animals, or it may be fed upon by animals; in some cases plants are of mutual help to each other, e.g. symbionts. Symbiosis between bacteria and leguminous plants may be recalled to mind in this connexion. Soil bacteria, protozoa, earth-worms, rats, etc., are useful agents in altering the soil. They also damage vegetation. This leads us to see how closely interwoven are the lives of plants and animals (see also p. 412).

4. **Topographic Factor.** This has an important bearing on vegetation, particularly in mountainous regions such as the Himalayas. Altitude is the dominant factor in this respect, determining zonal distribution of vegetation, e.g. tropical, sub-tropical, temperate, sub-temperate, and alpine. Within the zone rainfall plays an important part. Slope is another important factor, either retaining water in the soil after rainfall or draining it down. It thus influences the type of vegetation. Besides, exposure to sun, monsoon rain and wind also affect vegetation. Thus the southern slope of the Himalayas develops luxuriant deciduous and evergreen forests, whereas the northern side develops only dry forests.

Units of Vegetation. Vegetation is not uniform in any country. Depending on climatic and edaphic factors, the vegetation of an area, big or small, may be divided into a number of natural units whose composition is different and distinct. A large unit of natural vegetation in an area under identical climatic conditions is called a **plant formation.** Formations are controlled by their own climate, and may be of the following types: deciduous forest, evergreen forest, coniferous forest, arctic tundra, alpine tundra, grassland, etc. A formation is said to be a **climax** (or climax formation) when it is dominated by one or more species which are abundant in it; for example, in evergreen forest there may be certain dominant tall trees, in scrub vegetation certain dominant shrubs, and in grassland certain dominant grasses or sedges. A climax is the direct effect of the climate.

A major subdivision of a plant formation is called an **association.** Within the formation there may be a number of associations depending on sub-climates and nature of the soil. An association is similar in general outward appearance, ecological structure and floristic composition, e.g. marsh association, or hydrophytic, halophytic or xerophytic association. A plant association may have one or more dominant species. Within the association there may be **communities** of plants closely associated, e.g. *Acacia-Dalbergia* (KHAIR-SISSOO) along the banks of large streams, different species of *Quercus* or *Rhododendron* in the Himalayas. When a community invades a new area it is termed a **colony.**

The study of ecology from the community point of view (i.e. a mass of vegetation) is called **synecology** and that from the standpoint of component (individual) species is called **autecology.** The distribution of plants over the surface of the earth, including hills, rivers, lakes and seas, and the factors concerned in their distribution and migration are together known as plant geography or **phytogeography.**

Succession. The primitive vegetation of an area is never stable; it passes through a series of changes until finally a stable form or climax is reached. It may start initially in water or on barren rock or elsewhere.

(a) **Hydrosere.** The series of changes in the vegetation of a pond, lake, marsh or a stream are together known as hydrosere. The vegetation originally starts with some submerged plants rooted to the soil, e.g. *Ceratophyllum, Vallisneria, Hydrilla, Naias, Chara, Potamogeton,* etc. They increase in number and form a sort of underwater garden. As older plants and also animals die, a thin substratum is at first formed; this is further thickened by accumulation of soil particles as a result of erosion and transportation. Soon various floating plants, some rooted to the soil, e.g. water lily (*Nymphaea*), lotus (*Nelumbo*), *Euryale* (B. & H. MAKHNA), *Limnanthemum*, water chestnut (*Trapa*), etc., and some entirely floating, e.g. water lettuce (*Pistia*), water hyacinth (*Eichhornia*), duckweed (*Lemna*), bladderwort (*Utricularia*), *Salvinia*, *Ceratopteris* (a water fern), etc., invade the area. Some of them increase very rapidly and cover the water surface. Often they kill the submerged ones either by cutting off light or by competition. With the death of one set of plants the ground is prepared for the next set, i.e. succession continues. The pond in the meantime is being silted up. At this stage, still with sufficient water, plants like *Phragmites*, *Arundo*, *Typha*, *Scirpus*, etc., which are partly submerged, begin to invade the area and establish themselves. Besides, plants like *Sagittaria*, *Jussiaea*, *Alisma*, several sedges and rushes occupy the available space. They grow and eventually die, and further sedimentation takes place. The substratum rises and water-depth decreases. The next stage is the appearance of wet soil and then dry soil. By now hydrophytes disappear giving place to mesophytes. Species of shrubs and trees that can stand water-logging during the rainy season only make their appearance. Further invasion takes place. Finally a permanent forest stage is reached. This is the climax. The type of forest is determined by the amount of annual rainfall. If succession takes place in saline areas the hydrosere is called *halosere*.

(*b*) **Xerosere.** This is another interesting feature of succession. The series of changes in the vegetation of bare rocky beds, rocky hill-slopes, sand beds and other areas with extreme scarcity of water are together known as xerosere. The first invasion on rocks under such an extreme condition is that of crustose lichens which can only grow for a short period during wet weather and then undergo desiccation. They gradually spread and disintegrate rock into soil to some extent at least. Soon foliose lichens make their appearance. In due course humus accumulates and the soil becomes receptive for the next series. Some forms of hardy mosses now invade the area. Soil is in the process of formation, humus accumulates, and water is retained. The next series is the appearance of hardy short-lived annuals followed by biennials and perennials. They help further disintegration of rock into soil, accumulation of humus and retention of soil water. As herbaceous plants multiply in number, the lichens and mosses become scarce. The stage is now set for invasion of hardy shrubs which in due course form the dominant vegetation. The depth of the soil further increases, and soon becomes suitable for tree-growth. Trees at this stage are, however, stunted in growth and xeromorphic in habit, and are sparsely distributed. If conditions are favourable xeromorphic plants soon become replaced by mesophytes. Trees increase in number and new types of herbs grow in their shade. The invading trees soon become dominant forming a climax. If succession takes place in sandy areas the xerosere is called *psammosere*.

Ecesis. Ecesis means successful establishment of a vegetation in a new locality after migration from elsewhere for some reason or other. It refers to all the factors (climatic and edaphic) prevailing in the locality, and the adaptability of the plants to the new environment under new conditions. It includes, therefore, all the factors favouring seed-germination, plant-growth and reproduction, leading finally to their successful establishment. It is evident any one factor inhibiting any of the above processes may stand in the way of complete ecesis. Competition with neighbours and also among the invading plants may also be a setback in this direction. It is evident that ecesis is next in importance to migration of plants to a new locality because migration alone cannot give rise to a new vegetation without ecesis, i.e. without proper and favourable conditions for ecesis a vegetation cannot establish itself in a new environment.

Commensalism. When members of two species become closely associated with each other, temporarily or permanently, each being self-supporting so far as food and nutrition are concerned, the relationship between the two is expressed as commensalism (*com*, together; *mensa*, table). Literally it means 'eating together at the same table, without causing any harm or inconvenience to the other'. Commensalism involves growing together on the same substratum, or one on the body of the other for shelter or support, or even for transport, or often for facility of food manufacture. The epiphytes, total or partial, e.g. many orchids, Spanish moss (*Tillandsia usneoides*, an American plant, hanging in festoons from branches of trees and usually carried by wind from tree to tree), *Scindapsus*, *Pothos*, *Dischidia*, several ferns (e.g. *Drynaria*), *Lycopodium phlegmaria*, *Usnea* (a lichen), etc., using branches of trees for support and adequate light, are typical commensals. Besides, *Nostoc* living in the cavities of cycad root, certain bacteria (e.g. *Escherichia coli*) living in the animal intestine consuming undigested food, *Protoccoccus* living on tree trunks, etc., are other examples of commensalism.

Chapter 2 ECOLOGICAL GROUPS

Although plants sometimes occur as isolated individuals, more commonly we find that they become adapted to the same environment and are associated together in groups. The groups may include different plant species, belonging to different families, and differing in shape, size, form and relationship, but they live under the same environmental conditions of the soil, moisture, heat and light. Some of the common groups are described below.

1. Hydrophytes. Hydrophytes are plants that grow in water or in very wet places. They may be submerged or partly submerged, floating or amphibious. Their structural adaptations are mainly due to the high water content and the deficient supply of oxygen. The various adaptations met with in hydrophytes are as follows:

Adaptations. The main features of aquatic plants are the reduction of protective tissue (epidermis here is meant for absorption, and not for protection), supporting tissue (lack of sclerenchyma), conducting tissue (minimum development of vascular tissue) and absorbing tissue (roots mainly act as anchors, and root-hairs are lacking), and the special development of air-chambers for aeration of internal tissues.

The root system in hydrophytes is feebly developed and root-hairs and root-cap are absent. In some floating plants such as bladderwort (*Utricularia*), hornwort (*Ceratophyllum*), etc., no roots are developed at all, and in submerged plants such as *Vallisneria, Hydrilla, Naias*, etc., water, dissolved mineral salts and gases are absorbed by their whole surface. In plants like water lettuce (*Pistia*), water hyacinth (*Eichhornia*), duckweed (*Lemna*), etc., no root-cap develops, but an analogous structure called root-pocket is found instead.

The stem is soft and more or less spongy, owing to the development of a large number of air-cavities in it and also in the leaf,

FIG. 1. Giant water lily (*Victoria amazonica*=*V. regia*).

filled with gases: these air-cavities on the one hand give buoyancy to the plant for floating and on the other serve to store up air (oxygen and carbon dioxide). The carbon dioxide that is given off in respiration is stored in these cavities for photosynthesis, and again the oxygen that is given off in photosynthesis during the daytime is similarly stored in them for respiration. There is a minimum development of the mechanical and the vascular tissues. Xylem is reduced to only a few elements, while phloem is reduced to a few narrow sieve-tubes. The epidermis has no cuticle on it, but contains some chloroplasts. The stem and the leaf-stalk are in some cases provided with prickles and spines for defence against the attack of aquatic animals.

Aquatic plants may be fixed to the substratum or they may be floating freely. Leaves likewise may be submerged or floating. Submerged leaves are thin, and often become elongated owing to subdued light under water; they are generally ribbon-shaped, finely dissected or linear. Cuticle is absent and so usually are the stomata; the latter, if present, are functionless. Exchange of gases and absorption of water and mineral salts take place through the epidermis of the leaf. The mesophyll is thin and not differentiated into palisade tissue and spongy tissue, and the epidermis contains chloroplasts to utilize the weak light under water. Floating leaves are well developed, and have a thick cuticle and a large number of stomata on the upper surface; no stomata or only functionless ones are present on the lower surface. Exchange of gases takes place through the upper surface, and absorption of water through the lower. Many air-cavities develop in them for the purpose of aeration and necessary buoyancy. Amphibious plants are subjected to alternate flooding and drying. They usually grow at the edge of a pool of water with the lower leaves submerged and the upper ones above water. Many such plants often show **heterophylly** (*heteros*, different; *phylla*, leaves), i.e. they bear different kinds of leaves on the same plant (see pp. 58-9).

Some Common Aquatic Plants. (*a*) **Submerged**: *Vallisneria*, *Hydrilla*, *Naias*, *Ottelia*, *Potamogeton*, etc. (*b*) **Floating**: *Wolffia* (smallest plant with frond like a sand grain), *Hydrocharis*, bladderwort (*Utricularia*), hornwort (*Ceratophyllum*), duckweed (*Lemna*), water lettuce (*Pistia*), water hyacinth (*Eichhornia*), water chestnut (*Trapa*), *Neptunia*, *Azolla*, *Salvinia*, *Ceratopteris*, etc. (*c*) **Plants with floating leaves**: water lily, giant water lily (FIG. 1), *Euryale* (B. & H. MAKHNA), *Limnanthemum*, etc. In lotus (*Nelumbo*) the leaves stand above the water. (*d*) **Amphibious plants showing heterophylly**: water crowfoot (*Ranunculus aquatilis*), water plantain (*Alisma*), arrowhead (*Sagittaria*), *Limnophila heterophylla*, *Cardanthera triflora*, etc.

2. Hygrophytes. These are plants that grow in constantly moist situations. They occur in moist shady places, in forests, or in the moist soil near water-logged localities. The root system and the vascular system in hygrophytes are poorly developed. Plants are stunted in growth, their parts generally soft and spongy, and the stem usually an underground rhizome. There is a feeble development of mechanical tissues. Leaves on the whole are well developed, fully expanded, and with numerous stomata. They are smooth, shining and with a thin cuticle. The situation in which these plants grow being moist, transpiration is not active, but to get rid of the excess water, leaves are provided with hydathodes (see FIG. II/42) through which exudation of water takes place in liquid form. Common hygrophilous plants are aroids, ferns, begonias, some grasses, etc.

3. Mesophytes. These are plants that grow under average conditions of temperature and moisture; the soil in which they grow is neither saline nor is it water-logged, and the temperature of the air is neither too high nor too low. Mesophytes are, therefore, intermediate between hydrophytes and xerophytes.

Adaptations. The root system is well developed with the tap root and its branches in dicotyledons, and a cluster of fibrous roots in monocotyledons; root-hairs are luxuriantly produced for absorption of water from the soil. The stem is solid (and not spongy, as in water plants), erect, and normally branched. Thorns on the stem are absent or few. All the different kinds of tissues, particularly the mechanical and conducting tissues, have reached their full development in the mesophytes. The aerial parts of plants, such as the leaves and the branches, are provided with cuticle. In dorsiventral leaves the lower epidermis is provided with numerous stomata; there are few stomata or none at all on the upper surface. In erect leaves, as in monocotyledons, stomata are more or less equally distributed on both surfaces.

4. Epiphytes. Epiphytes are plants that grow perched on the trunk and branches of other plants, but only for the purpose of support, i.e. they do not suck the supporting plants as do parasites. They grow on almost all kinds of trees and shrubs, very rarely with any choice. They are normally green in colour and autotrophic in habit. Considering, however, the conditions under which they grow, they seem to lead a very hard life for want of adequate supply of their nutrients, more particularly water which is the first factor to sustain life. They have mainly to depend upon rain-water in tropical climates and on dew in temperate climates and in the hills. They normally draw their food material from the dead bark, wind-borne debris, humus and rain-water that collect in the network of their roots, and also absorb the water trickling down the hanging roots. Although they are seen to grow on isolated trees here and there, epiphytes show luxuriant growth in tropical rain-forests, often thickly covering the tree-trunk. As they have to depend on rain or dew, during the dry season they lead a precarious life, mostly going into a dormant, often defoliated, state to tide over the difficult period. Then again with a few showers of rain they begin to crop up. Considering the conditions under which they live the epiphytes commonly show the following adaptations: (a) they can quickly attach themselves to their support by the formation of clasping roots soon after the germination of the seeds, (b) many of them have xeromorphic characters so that they can stand droughts, and (c) in angiosperms further the mechanism for seed-dispersal is efficient; in orchids a large quantity of extremely minute seeds is produced in a single capsule. The xeromcrphic types may continue to grow even through the bad season without drying up. They may be leaf-succulent or stem-succulent; many orchids have succulent pseudo-bulbs. Besides, in many of such epiphytes the cuticle is thick, stomata are sunken, and leaves reduced in size and number. Some plants, particularly many orchids, pass their whole life on supporting plants; they are total epiphytes or holoepiphytes. Some plants, particularly the rootlet climbers, are at first terrestrial and later epiphytic having lost their connexion with the soil; they are partial epiphytes or homoepiphytes, e.g. *Cereus triangularis* (see FIG. VII/35A), *Scindapsus* (B. GAJPIPAL), etc.; conversely some plants are epiphytic at first and later terrestrial being rooted to the soil, e.g. banyan, peepul, etc.

Examples. Among angiosperms the orchids predominate as epiphytes, e.g. *Vanda* (see FIG. I/13), *Dendrobium* (see FIG. VII/79), *Cymbidium, Eria,* etc., and also certain plants of *Araceae,* e.g. *Scindapsus officinalis, Pothos scandens,* etc.; Bromeliaceae, e.g. *Tillandsia usneoides*—a peculiar American plant hanging from branches and looking like *Usnea* (a lichen; see FIG. V/109); Asclepiadaceae, e.g. *Dischidia rafflesiana* (see FIG. I/70) and *D. nummularia;* Cactaceae, e.g. *Cereus triangularis* (see FIG. VII/35A) —a large climbing epiphyte with triangular stem often reaching tree-tops, etc. Among Pteridophyta several ferns are epiphytes e.g. bird's nest fern (*Asplenium nidus*), *Drynaria quercifolia* with dimorphic leaves, many species of *Polypodium,* e.g. *P. fissum* —all growing in tufts on tree-trunks, and *Cyclophorus adnascen*. —creeping, branching and almost covering the whole tree, etc.; a few species of *Lycopodium,* e.g. *L. squarrosum* and *L. phlegmari* (see FIGS. V/150-51A), are also somewhat common epiphytes Among Bryophyta several forms of mosses often cover tree trunks as cushions; *Porella* (see FIG. V/128), one of the Junger manniales, is also fairly common. Among algae mention may be made of *Protococcus* and *Trentepohlia* growing on barks of trees, the latter on leaves also.

5. Xerophytes. These are plants that grow in deserts or in very dry places; they can withstand a prolonged period of drought uninjured; for this purpose they have certain peculiar adaptations. Xerophytes are really drought-resistant plants. It is not that they thrive under desert conditions. The property of drought-resistance is not attributed to the anatomical features of such plants but to the capacity of the protoplasm to endure a high degree of desiccation with practically no injury. Dominant factors in a desert or very dry region are: scarcity of moisture in the soil and the extreme atmospheric conditions such as intense light, high temperature, strong wind and aridity of the air. These being so, the xerophytic plants have to guard against excessive evaporation of water; this they do by reducing evaporating surfaces. They have also to adopt special mechanisms for absorbing moisture from the soil and for retaining it in their body. Xerophytic characters are also found in plants growing in cold regions such as temperate and sub-arctic zones and high altitude, rocky beds, sandy regions, dry places with scanty rainfall, and in salt marshes.

Adaptations. Plants produce a long tap root which goes deep into the subsoil in search of moisture; many of the desert plants which live for a short period produce a superficial root system to absorb moisture from the surface-soil after a passing shower of rain. To retain the water absorbed by the roots, the leaves and stems of some plants become very thick and fleshy, as in Indian aloe (*Aloe*) and American aloe or century plant (*Agave*); sometimes, roots also become fleshy, as in *Asparagus*. Aqueous tissue develops in them for storing up water; this is further facilitated by the abundance of mucilage contained in them. Multiple epidermis sometimes develops in the leaves for the same purpose, as in oleander (*Nerium*). Modification of the stem into phylloclade for storing water and food and at the same time performing functions of leaves is characteristic of many desert plants, e.g. most cacti and several species of *Euphorbia*.

Leaves and stems are provided with thick, sometimes very thick, cuticle; epidermal cells often become strongly cutinized to prevent excessive loss of water by transpiration. Tannins and gums are frequently found in the epidermal layer. In many cases the stem becomes reduced in size and provided with prickles, as in *Euphorbia splendens*. Leaves are also reduced in size minimizing their evaporating surfaces. Thus these may be divided into small segments, as in *Acacia*, or modified into spines, as in many cacti and spurges (*Euphorbia*), or sometimes reduced to small scales only, as in *Tamarix* and *Asparagus*. In some plants, as in *Gnaphalium* and *Aerua*, there is a dense coating of hairs. Stomata are fewer in number—usually 10-15 per sq. mm., and these remain sunken in grooves and occluded, sometimes covered by hairs. There is a strong development of sclerenchyma in most of the xerophytes. Modification of the leaf into phyllode, turning its edge in the vertical direction in strong sunlight to minimize transpiration, is characteristic of Australian *Acacia* (see FIG. I/67). Under conditions of extreme dryness leaves of most xerophytic grasses and also of many other plants roll up, considerably reducing their evaporating surfaces. In such cases stomata also become shut up.

Many of the xerophytic herbs lie prostrate on the ground, completing their life-history within a short time, e.g. *Solanum surattense*, *Tribulus terrestris*, *Trianthema monogyna* and *Suaeda fruticosa*; some are perennial in habit. Many xerophytes are elaborately armed with prickles and spines.

Two very peculiar cases showing special xerophytic adaptations may be mentioned here. These are certain species of *Selaginella* and *Anastatica*. When the dry season prevails in the desert, these plants curl up into a sort of ball, and are driven about by the wind. They are fixed only when they reach wet soil or the rains begin.

Some Common Xerophytic Plants. Many spurges (*Euphorbia*), e.g. *Euphorbia splendens*, *E. royleana*, *E. neriifolia*, *E. tirucalli* (see FIG. I/40A), etc., many cacti, e.g. prickly pear (*Opuntia*), *Cereus*, *Pereskia*, etc., dagger plant (*Yucca*; see FIG. I/81), Indian aloe (*Aloe*), American aloe (*Agave*), *Capparis aphylla*, *Tamarix*, prickly poppy (*Argemone*), jew's slipper (*Pedilanthus*), globe thistle (*Echinops echinata*), safflower (*Carthamus*), amaranth (*Amaranthus*), wild plum (*Zizyphus nummularia*), purslane (*Portulaca*), Indian spinach (*Basella*), saltwort (*Salsola*), seablite (*Suaeda*), *Asparagus*, gum tree (*Acacia nilotica*), *Prosopis spicigera* (B. & H. SHOMI), camel thorn (*Alhagi*), *Solanum surattense* (B. KANTIKARI; H. KATELI), *Tribulus terrestris* (B. GOKHRIKANTA; H. GOKHRU), *Gnaphalium*, *Aerua*, etc. Common xerophytic grasses are *Stipa*, *Sporobolus*, *Saccharum spontaneum* (B. KASH; H. KANS), *S. munja* (B. SAR; H. MUNJA), *Aristida*, etc.

6. Halophytes. These are special types of plants growing in saline soil or saline water with preponderance of salts in it. Hence halophytes show some special characteristics (or adaptations). It is to be noted that plant cells of ordinary land plants or freshwater plants maintain a low osmotic pressure which may be equal to or more often slightly higher than the osmotic pressure of the medium (soil or water) where the plants are growing, i.e. the two pressures inside and outside the plants are more or less in a state of equilibrium. But when the water medium or the

FIG. 2. Mangrove plants showing (a) pneumatophores for respiration, (b) stilt roots for support, and (c) viviparous germination for survival.

soil medium is definitely saline, as in the sea, seacoast or salt-lake, the water balance between the plants and their habitat becomes disturbed, much to the disadvantage of the former. The osmotic pressure of the saline water is very high. Evidently this condition tends to extract water from the plant tissues having lower osmotic pressure, rather than help them to absorb water from such a medium. The saline water or soil thus behaves as a *physiologically dry* medium for such plants. To meet this exigency the plant tissues must maintain a higher osmotic pressure than their saline medium. Therefore, only those plants can adapt themselves to such a situation, whose tissues have a high concentration of salts and consequently an exceptionally high osmotic pressure—as high as 35-40 atmospheres or even much higher. Such plants are called **halophytes,** and they show the following characteristics (or adaptations).

Xeromorphic Adaptations. The majority of halophytes show xeromorphic characters, i.e. the characters of xerophytes. The relationship between the two groups is not, however, clearly understood. Many of them have fleshy leaves, e.g. sea-blite (*Suaeda maritima*) and glasswort (*Salsola foetida*); leaves sometimes

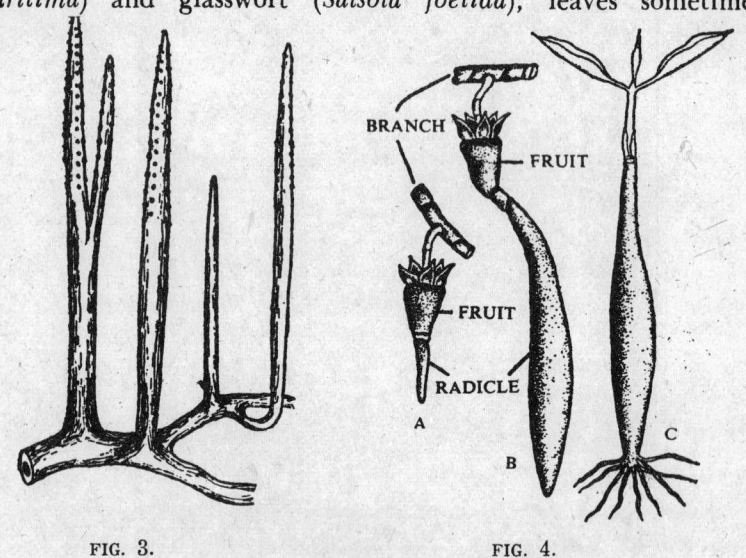

FIG. 3.　　　　　　　　FIG. 4.

FIG. 3. Pneumatophores growing vertically upwards from an underground root. FIG. 4. Viviparous germination.

altogether absent and the stem fleshy and jointed, e.g. saltwort (*Salicornia brachiata*) and *Arthrocnemum indicum*; sometimes stems fleshy and mucilaginous, e.g. Indian spinach (*Basella rubra*—wild or cultivated); epidermis often strongly cutinized and thickened, sometimes with hairs; spines and prickles present in several cases, e.g. *Acanthus ilicifolius* (B. HARGOZA; H. HARKUCH-KANTA), *Asteracantha longifilia* (=*Hygrophila spinosa;* B. KULE-KHARA; H. GOKULA-KANTA), *Solanum surattense* (= *S. xanthocarpum*; B. KANTIKARI; H. KATELI, KATITA), etc.; internal structure often like that of xerophytes. Other common examples of halophytes are *Pluchea indica, Blumea maritima, Azima tetracantha, Vigna luteola, Caesalpinia nuga, Flagellaria indica, Tamarix indica,* nipa-palm (*Nipa fruticans*). etc.

Special Adaptations. Halophytes growing on muddy swamps of tropical estuaries and sea-coasts inundated by tides almost daily form a special type of vegetation known as the **mangrove**. Mangrove vegetation is most extensively represented in the Sundarbans (West Bengal). Mangrove plants show some special adaptations. (1) They produce a large number of **stilt roots** (FIG. 2) from the main stem and the branches. (2) In several cases, in addition to the stilt roots special roots, called **respiratory roots** or **pneumatophores** (FIGS. 2-3), are also produced in large numbers. They develop from underground roots, and projecting beyond the water level they look like so many conical spikes distributed all round the trunk of the tree. In some places they grow so thickly that the passage through them by boat is difficult. They are provided with numerous pores or respiratory spaces in the upper part, through which exchange of gases for respiration takes place. (3) Mangrove species also show **vivipary** (FIG. 4), i.e. the seed germirates inside the fruit while it is still on the parent tree and is nourished by it. Germination is almost immediate without any period of rest. The radicle elongates to a certain length and swells at the lower part. Finally the seedling separates from the parent tree and falls vertically down. The radicle presses into the soft mud, keeping the plumule and cotyledons clear above the saline water. Lateral roots are quickly formed for proper anchorage. The advantage is that the fruit cannot be swept by tidal waves. Typical mangrove plants are *Rhizophora* (B. KHAMO), *Ceriops* (B. GORAN), *Kandelia* (B. GORIA), *Bruguiera* (B. KANKRA), *Sonneratia* (B. KEORA), *Heritiera* (B. SUNDRI), *Excoecaria* (B. GEO), *Avicennia* (B. BAEN or BINA), etc. (See also p. 433).

7. **Marsh Vegetation.** A marsh is a tract of low, wet land. During the greater part of the year it remains covered with water; for a shorter period it is dry. There is stagnant mud in a marsh with abundance of mineral salts. The water is alkaline or even neutral but not acid. The peculiarity of marsh vegetation is that the submerged parts show hydrophytic adaptations; while the aerial parts have the peculiarities of land plants. From the margin to the mid-marsh the vegetation shows often three zones with, of course, a certain amount of overlapping. The first or the outermost zone, lying nearest the dry land, is composed of plants whose roots or rhizomes only are in waterlogged areas. This zone exhibits to some extent the characteristics of hygrophytes. Leaves are usually well developed with numerous stomata but with thin cuticle, and the stem usually spongy in nature. There is dominance of low herbs in this zone, the common plants being water dropwort (*Oenanthe*), marsh pennywort (*Hydrocotyle*), various sedges (*Scirpus, Carex*, etc.), rush (*Juncus*), horsetail (*Equisetum*), etc.

The second or the intermediate zone lies between the first

zone and the mid-marsh. Here the roots or rhizomes are in the wet soil under water, parts of the stem are in water, and the leaves and flowers are either floating or raised above water. This zone is characterized by tall herbs of upright growth. Typical plants of this zone are reed (*Phragmites*), *Arundo donax*, elephant grass (*Typha*), etc. This zone is often called reed-swamp association. Low herbs or floating plants associated with them are water plantain (*Alisma*), water crowfoot (*Ranunculus*), water lilies (*Nymphaea*), lotus (*Nelumbo*), duckweed (*Lemna*), water lettuce (*Pistia*), arrowhead (*Sagittaria*), hornwort (*Ceratophyllum*), *Hydrilla*, *Vallisneria*, *Utricularia* as well as some floating grasses such as *Hygrorhiza*, *Vossia*, *Panicum proliferum*, etc., and some sedges such as *Scirpus*, *Cyperus*, etc.

The third or the innermost zone is submerged, and the vegetation typically hydrophytic. Common plants of this zone are *Potamogeton*, *Vallisneria*, *Hydrilla*, *Ceratophyllum*, etc. Besides many of the aquatic plants of the second zone extend into this zone.

Chapter 3   TYPES OF VEGETATION IN INDIA

Mainly depending on rainfall together with temperature, altitude and topography, the types of vegetation met with in India may be broadly classified as follows. It must, however, be noted that there is a certain amount of mixtures and overlappings of one type and the other both in the plains and the hills. It may be of further interest to note that *Orchidaceae* is the largest family in the Indian flora, being represented by about 1,700 species; while *Compositae* (the largest dicotyledonous family) is represented in India by only about 674 species. Further, according to Calder, the following families stand in their order of dominance in the Indian flora taken as a whole: *Orchidaceae, Leguminosae, Gramineae, Rubiaceae, Euphorbiaceae, Acanthaceae, Compositae, Cyperaceae, Labiatae* and *Urticaceae*. The proportion of dicotyledons to monocotyledons stands as 7:1.

1. **Tropical Evergreen or Rain Forests.** They occur in areas with a heavy annual rainfall exceeding 2,000 mm., and such areas are the western parts of the Western Ghats and the eastern parts of the sub-tropical Himalayas. Evergreen forests are composed of tall and medium-sized trees, many shrubs, several climbers and epiphytes, different types of bamboos and ferns, all in luxuriant growth. They do not shed their leaves annually, at least not all together. Such forests generally show a 3-storeyed appearance (see p. 434). Dominant families of such forests are: *Dipterocarpaceae, Guttiferae, Annonaceae, Meliaceae, Burseraceae, Sapotaceae, Euphorbiaceae* and *Palmae*. Families such as *Ebenaceae*, *Sapota*-

ceae, Capparidaceae, Rhamnaceae and *Myrtaceae* are also well represented. Further, *Tetrameles nudiflora, Dipterocarpus macrocarpus, Hopea odorata, Artocarpus chaplasha, Terminalia myriocarpa, Cedrela toona,* etc., are the common tallest trees of evergreen forests and are of considerable economic value. Other common trees are: ebony (*Diospyros*), ironwood tree (*Mesua*), *Mimusops,* rosewood (*Dalbergia*), *Chikrassia, Pterospermum, Calophyllum,* several species of *Terminalia* and *Cinnamomum, Amoora, Alstonia,* etc.

2. **Tropical Deciduous or Monsoon Forests.** Such forests (see p. 434) occur in areas with moderate or low annual rainfall of 1,000-1,500 mm., as in the eastern parts of the Western Ghats, Deccan Plateau, Madhya Pradesh, Uttar Pradesh, Bihar, West Bengal (western parts), and lower slopes of the Himalayas (forming tropical hill forests). SAL (*Shorea*) is dominant in certain areas, as in Assam and Central India extending to Orissa; while teak (*Tectona*) is dominant in Deccan Plateau; and sandalwood (*Santalum*) in Mysore and adjoining country. The common trees of such forests are: JARUL (*Lagerstroemia*), redwood (*Dalbergia*), GAMHAR (*Gmelina*), *Bombax, Terminalia, Sterculia, Dipterocarpus, Butea, Cassia, Bauhinia, Acacia, Albizzia,* padauk (*Pterocarpus macrocarpus*), red sandalwood (*P. santalinus*), *Adina, Odina* etc.

3. **Tropical Thorn Forests.** Such forests (see p. 434) are restricted to areas where the rainfall is very low (500-760 mm.), as in South Punjab, Rajputana, North Kutch and Deccan Plateau. Thorny shrubs and low trees are characteristic of these forests. Common plants are: SHOMI or JHAND (*Prosopis*), *Acacia nilotica Zizyphus, Capparis, Balanites, Euphorbia, Flacourtia, Tamarix,* etc., and in the saline soil, as in Punjab, thick-leaved species of *Salsola, Suaeda, Chenopodium,* etc. Chief families are: *Leguminosae, Capparidaceae, Salvadoraceae, Tamaricaceae* and *Rhamnaceae.*

4. **Mangrove or Littoral Forests.** These forests (see p. 433) are characteristic of the deltas of large rivers (Ganges, Mahanadi, Godavari, Krishna, Kaberi, and Indus), some areas of sea-coasts and marine islands.

5. **Riparian Forests.** They are found along banks of rivers (see p. 435), and consist mainly of KHAIR (*Acacia catechu*) and SISSOO (*Dalbergia sissoo*), mixed with *Bombax, Trewia, Barringtonia, Populus, Tamarix,* etc.

6. **Grasslands.** They are not common in India. Some grasslands, however, occur in the Gangetic plain and in the plains of Assam. Elsewhere they are sporadic. Hill grasslands are found in the Khasi Hills, South Indian Hills, and the tropical and sub-tropical Himalayas. Grasslands dotted with open forests of low trees and

shrubs, called **savannahs**, are found in Uttar Pradesh and Gujarat.

7. Temperate Evergreen or Coniferous Forests. They occur in the temperate eastern and western Himalayas and in the Nilgiri Hills at an altitude of 1,600-3,500 metres. Such forests are mainly composed of Coniferae such as pine (*Pinus*), deodar (*Cedrus*), spruce (*Picea*), silver fir (*Abies*), etc., and Cupuliferae (*Fagaceae* and *Betulaceae*) such as oak (*Quercus*), beech (*Fagus*), and birch (*Betula*) mixed with poplar (*Populus*), elm (*Ulmus*), *Rhododendron*, chestnut (*Castanopsis*), walnut (*Juglans*), maple (*Acer*), etc. Leaving out Coniferae and Cupuliferae the dominant families are: *Sapindaceae, Lauraceae, Magnoliaceae, Salicaceae* and *Urticaceae*. [see also p. 436].

8. Alpine Forests. They occur in the Alpine area of the Himalayas beyond the limit of tree growth (3,500 metres), and consist of dwarf shrubs of juniper (*Juniperus*), silver fir (*Abies*), *Betula* and *Rhododendron* (several species). At still higher altitude Alpine scrub of low herbs is the only type of vegetation found (see p. 437).

Chapter 4 PHYTOGEOGRAPHICAL REGIONS OF INDIA

The division of India into certain phytogeographical regions was first proposed by Clarke in 1898. It was revised by Hooker in 1904. Based on recent publications of provincial and local floras the following division was proposed by Calder[1] in 1937. On more information being available it was further modified by Chatterjee[2] in 1939. An outline of the above divisions is given below:

Clarke (1898)—1. West Himalaya. 2. India Deserta. 3. Malabaria. 4. Coromandalia. 5. Gangetic Plain. 6. East Himalaya. 7. Assam.

Hooker (1904)—1. Eastern Himalaya. 2. Western Himalaya. 3. Indus Plain. 4. Gangetic Plain. 5. Malabar. 6. Deccan.

Calder (1937)—1. Western Himalaya. 2. Eastern Himalaya. 3. Indus Plain. 4. Gangetic Plain. 5. Deccan. 6. Malabar.

Chatterji (1939)—1. Deccan. 2. Malabar. 3. Indus Plain. 4. Gangetic Plain. 5. Assam. 6. Eastern Himalaya. 7. Central Himalaya. 8. Western Himalaya.

The main factor in the distribution of species is the amount of annual rainfall in a particular region; within the region the type of soil is another important factor. In the mountainous regions such as the Himalayas, the altitude is the dominant factor, coupled with rainfall, determining zonal vegetation. Where rainfall is heavy, over 2,000 mm., sometimes very heavy, as in Assam, North Bengal and Malabar, luxuriant vegetation with **evergreen forests** and diversity of species is the rule; where rainfall is low

---

[1] *An Outline of the Vegetation of India* by C. C. Calder, 1937.
[2] *Studies on the Endemic Flora of India and Burma* by D. Chatterjee, 1939.

(usually 1,000-1,500 mm.), as in Deccan, Central India and parts of Punjab, sparse vegetation with **deciduous forests** and uniformity of species is the rule; and where rainfall is very low (500-760 mm.), sometimes much less, as in South Punjab, Rajputana and Sind (Pakistan) the vegetation is thin, sparse and xerophilous, with **thorny scrubs** predominating. Based on the above factors India may be divided into the following phytogeographical regions :

1. **Deccan** consisting of Deccan Plateau Tamil Nadu, Andhra and major part of Karnataka, a hilly tract of land, with an annual rain-fall of 840-1,000 mm., and Central India (Madhya Pradesh, Orissa and part of Gujarat) with an annual rainfall of 840-1,400 mm., shows the following types of vegetation : (a) dry deciduous forests predominating, with teak (*Tectona*), mahogany (*Swietenia*), *Acacia*, nux-vomica (*Strychnos*), *Lagerstroemia*, Indian rosewood (*Dalbergia*), *Dillenia*, *Cassia*, *Terminalia*, sandalwood (*Santalum*) in Karnataka, red sandalwood (*Pterocarpus*), *Butea*, *Bauhinia*, *Dillenia*, *Odina*, *Grewia*, *Buchanania*, *Chloroxylon*, *Chikrassia*, etc., and also many epiphytic orchids, etc.; (b) SAL (*Shorea*) forests in Central India extending to Orissa; (c) thorny scrubs mainly consisting of *Zizyphu*, *Acacia*, *Capparis*, *Balanites*, *Euphorbia*, *Flacourtia*, *Prosopis*, etc., are widespread; (d) an almost unbroken mass of coconut-palms along the coasts extending to Karnataka; (e) estuarian vegetation of mangrove type covering a small area. In the above region black soil covering a wide area is extensively used for the cultivation of cotton. On the whole, the following families are well represented in this area : Sterculiaceae, Meliaceae, Leguminosae, Combretaceae, Bignoniaceae, Urticaceae, Acanthaceae, Labiatae, Commelinaceae, Gramineae, etc.

2. **Malabar** consisting of the Western Ghats and extending from Gujarat to Travancore with an annual rainfall of over 2,500 mm. shows luxuriant vegetation of the following types : (a) tropical evergreen forests with luxuriant growth of trees, mainly *Hevea* and *Ficus elastica* (both rubber-yielding), *Dipterocarpus*, *Artocarpus*, sandalwood (*Santalum*), red sandalwood (*Pterocarpus*), nutmeg (*Myristica*), ebony (*Dyospyros*), toon (*Cedrela*), Alexandrian laurel (*Calophyllum*), *Michelia*, *Ternstroemia*, coconut-palm (*Cocos*), talipot palm (*Corypha*), *Mimusops*, *Hopea*, *Sterculia*, etc.; there is also a rich growth of shrubs, climbers and epiphytes; (b) deciduous forests occurring in strips with an annual rainfall 1,500-2,000 mm.; the vegetation is sparse and trees mostly deciduous; common trees are mountain ebony (*Bauhinia*), teak (*Tectona*), Indian redwood (*Dalbergia*), *Adina*, *Lagerstroemia*, *Terminalia*, *Grewia*, bamboo (*Bambusa*), etc. : (c) temperate

evergreen forests at higher elevations (to a maximum of 2,670 metres) with a rich vegetation of evergreen trees, mainly *Michelia, Eugenia, Ternstroemia,* etc. The following families have gained dominance in this area: *Guttiferae, Dipterocarpaceae, Palmae, Sterculiaceae, Anacardiaceae, Meliaceae, Myrtaceae, Melastomaceae, Scitamineae, Orchidaceae, Araceae,* etc.

3. **Indus Plain.** The average rainfall is very low—Punjab 630-760 mm., Rajputana 250-500 mm., and Sind (Pakistan) 200 mm.; north-eastern Punjab, however, is comparatively wet. A large tract of land in Rajputana is covered with THAR or Great Indian desert. The vegetation in its outskirts is typically xerophytic. Mean humidity of the plain is 50-52 or less, and is responsible for very hot summers and very cold winters. **Soils.** Tracts of land impregnated with salts, called KALLAR, are widely distributed. In Sind Sagar Doab, in southern districts of Punjab and in Rajputana the soil is very sandy and highly unstable, often forming dunes. Along the banks of larger rivers the soil, though sandy, has abundance of subsoil moisture. Black saline soil, called RAPAR, is devoid of any vegetation.

Vegetation. (a) **Desert Thorn Forest.** *Prosopis spicigera, Salvadora oleoides, Capparis aphylla,* etc., are the dominant species. They usually grow in clumps leaving the ground bare in between them. Such forests are commonly called RAKHS. Associated with the dominant species are commonly found *Tamarix articulata, Acacia leucoploea, A. nilotica, A. modesta, Albizzia lebbek, Morus alba, Zizyphus jujuba, Boswellia, Odina, Dalbergia, Salix,* etc. There are saline tracts over wide areas. *Sporobolus arabicus,* an Arabian grass, associated with saltwort (*Salsola foetida*), sea-blite (*Suaeda fruticosa*), *Chenopodium,* etc., is seen to remain confined to the saline soil. Plants like *Acacia nilotica, Tamarix articulata, Butea monosperma,* etc., also stand saline conditions well. Plants forming undergrowth are *Calotropis procera, Kochia indica, Chenopodium album, Zizyphus nummularia, Asparagus gracilis, Ephedra foliata,* etc. (b) **Dune Scrub.** In Sind, southern Punjab and Rajputana the vegetation is open, irregular and xerophytic. Stunted trees and bushes, thorny in character, form the only type of vegetation. Dominant species are *Leptadenia spartium, Euphorbia royleana, Acacia jacquemontia* (the only tree form), *Crotalaria burkia, Sericostoma pauciflorum, Calligonum polygonoides, Calotropic procera,* etc. (c) *Acacia-Dalbergia* (KHAIR-SISSOO) **Forest.** This is found along large rivers on sandy or gravelly alluvium. The dominant species are *Dalbergia sissoo, Acacia catechu,* poplar (*Populus euphratica*), etc. richly associated with *Tamarix dioica, Acacia farnesiana, A. nilotica, Saccharum spontaneum, S. munju,* etc.

4. **Gangetic Plain** with (a) an upper dry region extending from Punjab with an average annual rainfall of about 500 mm. to Allahabad with a rainfall of about 1,000 mm., (b) a lower humid region extending from Allahabad to West Bengal with a rainfall gradually increasing to 1,900-2,500 mm., and (c) Gangetic delta including the vast expanse of the Sundarbans. The Gangetic Plain is now mostly under cultivation and, therefore, the original flora is lost to a considerable extent. In the upper Gangetic Plain plants like *Peganum, Acacia, Moringa, Prosopis, Tecoma, Boswellia, Anogeissus, Rhus,* some palms, etc., still exist

In the alkali land *Salvadora* is common. Some grasslands or savannahs also occur here dotted with trees such as *Bombax, Zizyphus, Randia, Butea,* etc. Families such as *Leguminosae, Gramineae, Cyperaceae* and *Compositae* are dominant. In the lower Gangetic Plain there is a vast area of rice fields. In the uncultivated areas flourish groves of *Mangifera, Artocarpus, Ficus, Areca. Borassus, Phoenix,* etc. Several species of aroids are very common. Certain plants though introduced have a strong foothold here. These are *Lagerstroemia, Pterospermum, Bombax, Polyalthia, Casuarina,* etc. This area abounds in 'jheels' and large tanks where grow in luxuriance aquatic plants like *Nymphaea, Nelumbo, Euryale, Limnanthemum, Vallisneria, Hydrilla,* several grasses and sedges. Several families of 'flowering' plants are represented in this area.

Mangrove or Littoral Forests. The lower Gangetic plain nearest to the sea—a large tract of land, called the Sundarbans, with a network of water-channels, sea-creeks and swampy islands—is densely covered with extensive tidal swamp-forests of evergreen trees and shrubs of the Malayan type. The mangrove forest of the Sundarbans stretches over an area of about 6,000 sq. miles, and is the largest in the world. This tract of land spontaneously separates itself from the upper alluvial plain. The vegetation is dense and compact, and consists of species distinct from the rest of the Gangetic plain except its northern boundary with mixed types. The rivers are subject to tidal influence and are, therefore, saline. The numerous islets are mostly covered with savannahs composed of reed (*Phragmites karka*) associated with other tall grasses and tall sedges, and often dotted with trees. Throughout much of the Sundarbans the vegetation is of the mangrove type which is more pronounced towards the sea-face. The dominant feature of the mangrove is *Rhizophoraceae* consisting of the following typical mangrove trees: *Rhizophora mucronata* (BARRA-KHAMO), *R. conjugata* (KHAMO), *Ceriops roxburghiana* (GORAN), *Kandelia rheedei* (GORIA), *Bruguiera gymnorhiza* (KANKRA), etc. Associated with them are other mangrove species like *Heritiera minor* (SUNDRI—*Sterculiaceae*) which is very widespread in the area, *Excoecaria agallocha* (GEO—*Euphorbiaceae*) which is also plentiful, *Aegiceros majus* (KULSI—*Myrsinaceae*), *Sonneratia apetala* (KEORA—*Lythraceae*), *Avicennia officinalis* (BAEN or BINA—*Verbenaceae*) which is the largest tree in the Sundarbans, etc. There are 36 species of mangrove trees in the Sundarbans. Palms like *Nipa fruticans* (GOLPATA) and *Phoenix paludosa* (HITAL) are characteristic mangrove species. Coconut-palm (*Cocos nucifera*) and cane (*Calamus tenuis*) grow extensively, often planted. Along the edges of streams, canals, ponds and swamps elephant grass (*Typha angustata* and *T. elephantina*), *Alpinia allughas,* screwpine (*Pandanus fascicularis*), etc., are very common. *Panicum repens* and *Ipomoea pes-caprae* are two important sand- and mud-binding species. *Acanthus ilicifolius* (HARGOZA), *Asteracantha longifolia* (KULEKHARA), sea-blite (*Suaeda maritima*), *Salicornia brachiata, Arthrocnemum indicum, Tamarix gallica* (BAN-JHAU), *Allophylus cobbe, Crotalaria retusa, Derris sinuata,* etc., are some of the other common plants of the area. 13 species of *Orchidaceae,* 14 species of terrestrial and epiphytic ferns. *Lycopodium phlegmaria* (see FIG. V/151A), *Aldrovanda vesiculosa* (see FIG. III/29), *Psilotum* (in the eastern Sundarbans), etc., have been recorded from this area. It may be noted that Prain in his *Flora of the Sundarbans* has recorded 334 species belonging to 75 families. Some of these families are *Leguminosae* with 38 species, *Gramineae* with 29 species, *Cyperaceae* with 19 species, *Euphorbiaceae* with 16 species, *Orchidaceae* with 13 species, *Asclepiadaceae* with 12 species, and the rest of the families with less than 10 species. Apart from botanical interest the flora of the Sundarbans is economically very important, being the source of timber and firewood. Many medicinal plants also grow in this area.

5. **Assam**[1]. The climatic factors of North-East India are high humidity (80-90), frequent rainfall and moderate to mild temperature without extremes of heat or cold (generally 29°-19°C. in the plains, and mild to cold in the hills; and edaphic factor—high fertility of the soil. The average rainfall is heavy but varies —in lower Assam the average is 2,000 mm., in upper Assam 3,200 mm., in Cachar and Mizo Hills 2,900 mm. or more, in Garo Hills 2,700 mm., in Khasi and Jaintia Hills 5,800 mm. (in Shillong, however, 2,000 mm.), with the heaviest in Cherrapunji 12,700 mm. or more and its neighbourhood (Mawsynram) 15,200 mm. or more. This region consists of plains and hills, criss-crossed by several rivers. The Brahmaputra flows through the whole length of Assam valley with alluvial deposits on either side. Assam is a country of rivers, hills and plains, extraordinarily rich in vegetation; a region of the Khasi Hills above the pine zone (2,000 metres) is considered the richest, not only in India but perhaps in the whole world (see *Flora of Assam*, vol. I). The flora of Assam may be divided into the following types:

(*a*) **Evergreen Forests.** With abundant rainfall this type of forest extends in a more less continuous belt from the north-east corner of Arunachal (NEFA) to the Darrang district along the foothills of the Himalayas; such forests also occur in the Nowgong district, Cachar district and greater part of the Khasi Hills; elsewhere they are found to occur in isolated patches. These forests are composed of a very large number of species, and present a 3-storeyed appearance. The top storey consists of some isolated tall, evergreen or deciduous trees (some 46 metres high) towering above others, of which the following are common—*Dipterocarpus macrocarpus* (HOLLONG), *Artocarpus chaplasha* (CHAM), *Tetrameles nudiflora*, *Terminalia myriocarpa* (HOLLOK), etc. The middle storey consists of several medium-sized trees (up to about 23 metres) such as *Colophyllum*, *Mesua* (NAHOR), *Amoora* (AMARI), *Cinnamomum* (GONSEROI), *Phoebe* (BONSUM), *Machilus* (SOM), *Duabanga* (KHOKAN), *Ficus elastica* and other species, *Michelia*, *Magnolia*, *Schima*, etc. The lowest storey is made up of several shrubs. Climbers and lianes are very common, as also epiphytes (many orchids, some ferns and aroids). In the Khasi Hills pine forests are evergreen but without undergrowths and climbers. Here pine is associated with several species of oak (*Quercus*), *Pieris*, chestnut (*Castanopsis*), birch (*Betula*), etc., and in some places with yew (*Taxus*), *Cephalotaxus*, *Araucaria*, spruce (*Picea*), silver fir (*Abies*), deodar or cedar (*Cedrus*), *Tsuga*, *Cryptomeria*, cypress (*Cupressus*), juniper (*Juniperus*), etc. At high elevations in NEFA, particularly in Aka and Dafla Hills, the following conifers occur—*Podocarpus*, *Taxus*, *Abies*, *Pinus longifolia*, *P. excelsa*, *Tsuga* and *Cupressus*.

(*b*) **Deciduous Forests.** These forests occur mainly in the lower Assam Valley, Garo Hills and North Cachar Hills (particularly in the angle formed by the Mikir Hills and the Naga Hills), often mixed with some evergreen species. These tracts except the North Cachar Hills are mainly composed of *Shorea robusta* (SAL) forests and scrub forests, commonly associated with or surrounded by *Careya arborea* (KUMBHI), *Lagerstroemia* (AJAR), *Schima wallichii* (*evergreen*),

---

[1] This refers to flora of North-East India which includes Assam, **Nagaland**, Meghalaya, Arunachal (NEFA), Mizoram, Manipur and Tripura.

Dillenia pentagyna (AKSHI), *Kydia calycina* (KOTRA), *Terminalia belerica* and *T. chebula*, *Cassia fistula* (SONARU), *Albizzia* (KOROI and MOZ), *Gmelina* (GUMHAR), *Stereospermum* (PAROLI), *Alstonia*, walnut (*Juglans*), *Engelhardtia* (LEWA), *Dalbergia*, *Bombax*, *Sterculia* (ODAL), etc. *Shorea assamica* (MAKAI), often over 30 metres in height, forms pure forests in Lakhimpur. In the North Cachar Hills and the dry regions of Cachar, where SAL is absent, the forests are of mixed types and consist of *Dipterocarpus turbinatus* (GARJAN), *Bombax*, *Adina*, *Stephegyne*, several species of *Ficus*, *Cassia nodosa*, several grasses and bamboos (*Bambusa* and *Melocanna*). Further, *Coffea bengalensis* with white flowers, *Strobilanthes* with blue flowers, *Mussaenda* with a white or yellow leaf-like (modified) sepal, *Holmskioldia* with scarlet-red flowers, etc., adorn the deciduous forests as undershrubs and shrubs. Storeys, so characteristic of evergreen forests, are absent in deciduous forests.

(*c*) **Swamp Forests.** Cachar abounds in swamps; small swamps are not uncommon in Assam Valley. Various aquatic and semi-aquatic grasses, e.g. *Panicum* (many species), *Phragmites*, *Arundo*, *Erianthus*, *Vossia*, *Hygrorhiza*, etc., and sedges, e.g. species of *Scirpus* and *Cyperus*, etc., occur. Besides, *Ceratopteris* (an aquatic fern), *Azolla*, *Salvinia*, *Marsilea*, etc., and among 'flowering' plants *Euryale*, *Alpinia*, water lilies (*Nymphaea*), lotus (*Nelumbo*), etc., are quite common. At the edges of such swamps are commonly found *Barringtonia* (HIJAL), *Cephalanthus* (PANIKADAM), *Clinogyne* (SITAL-PATI), *Ficus heterophylla*, *Dracaena spicata*, etc.

(*d*) **Grasslands.** These are widespread in low-lying areas, riparian tracts, and dry lands with low rainfall. In the wet tracts characteristic genera are *Phragmites*, *Arundo*, *Erianthus*, *Saccharum*, etc. They often cover extensive areas along the banks of large rivers and some of them often grow to a height of 6 metres. In the dry lands the grasses are low and hardy, and are commonly *Imperata*, *Andropogon*, *Erianthus*, *Panicum*, *Apluda*, *Saccharum*, *Isachne*, etc.

(*e*) **Riparian Forests.** These forests extend along large streams at the foothills of the Bhutan Range from Goalpara district to Darrang district. *Acacia catechu* (KHAIR) and *Dalbergia sissoo* (SISSOO) are the two important species occurring in abundance in such forests. Such forests are generally known as *Acacia-Dalbergia* (KHAIR-SISSOO) forests. The two species occur mixed up with *Duabanga* (KHOKAN), *Bombax* (SIMUL), *Trewia* (PITULI), *Barringtonia* (HIJAL), *Salix tetrasperma* (PANI-HIJAL), *Anthocephalus* (KADAM), etc., which are ubiquitous.

Interesting Aspects of the Flora of North-East India (Assam and neighbouring States and Territories). Some of the many interesting plants occurring in this region deserve mention. Their morphological and physiological features and their economic importance have been described in the text. A general survey is given here. Many of them are rare found growing in this region only. Carnivorous plants—pitcher plant (*Nepenthes khasiana*) in Garo Hills and Khasi-Jaintia Hills, *Drosera burmanni* in Jorhat; and in Khasi and Garo Hills, *Drosera peltata* and *Pinguicula alpina* on Shillong Peak, *Aldrovanda vesiculosa* in tanks in Manipur, land *Utricularia* on hill-sides in Shillong. Saprophytes—Indian pipe (*Monotropa uniflora*) and *Burmannia* (4 sp.) in Khasi Hills. Root parasites—*Balanophora dioica* in Khasi Hills, and *Sapria himalayana* (akin to *Rafflesia*) in Aka, Dafla and Naga Hills. Orchids abound (estimated to be well over 400 species) and so do many beautiful ferns and some tree ferns. Other interesting plants—*Coptis teeta* (a medicinal plant) in Mishmi Hills; *Aquilaria agallocha* (AGARU) in Sibsagar; *Mussaenda* (9 sp.), lady's umbrella or Chinese hat (*Holmskioldia*) and *Cycas pectinata* adorning low hills; *Rhododendron arboreum* (a tree with deep red flowers) and a few other species in the Khasi Hills (alt. 1,624-1,830 m.) and several species in the Naga Hills (alt. 2,438-3,048 m.); *Gnetum gnemon* (a shrub) in Sibsagar and some other districts, *Gnetum montanum* (a climber) throughout the State; epiphytic lycopods—*Lycopodium phlegmaria* in plains districts, and *L. squarrosum* in the Khasi Hills; *Equisetum debile* (often 4 metres in length) among bushes besides hill-streams; timber trees (about 56 sp.) of which about

32 sp. have extensive uses; bamboos (*Bambusa, Melocanna, Arundinaria, Dendrocalamus*, etc.) about 20 sp., and canes (*Calamus, Zalacca, Daemonorops*, etc.) about 14 sp. *Aldrovanda* occurs in Manipur. This region is famous for oranges, pineapples, bananas, papaws, apples, plums, pears, etc.

6. **The Himalayas.** More or less sharply divisible into (a) the Eastern Himalayas extending from Darjeeling (West Bengal) to the Mishmi Hills (NEFA) including Sikkim and Bhutan, with an average rainfall of 3050 mm.; (b) the Central Himalayas—the region of Nepal, with much less rainfall; and (c) the Western Himalayas extending from the Kumaon Hills to Peshawar (Pakistan) with rainfall gradually decreasing from 1,100 mm. to about 500 mm.

The single dominant factor responsible for zonal vegetation of the Himalayas is the altitude (and, therefore, decreasing temperature), each zone having its characteristic flora with of course some or even considerable amount of overlapping, i.e. without any line of demarcation between one zone and the next. Three broad zones may, however, be more or less distinctly recognized, viz. (a) tropical zone with warm to mild climate; (b) temperate zone with mild to cold climate; and (c) alpine zone with intensely cold climate, merging into the snow-line. Within the zone rainfall and topography are the next important factors, the western wing of the Himalayas being drier and the eastern wing wetter. Within each zone often two sub-zones are distinguishable.

(a) **Tropical Zone** (500-1,600 metres). This forms the submontane belt of the Himalayas, called Terai, and is covered by dense forests. Depending on rainfall two types of forests develop in this zone—**evergreen** with abundant rainfall (over 2,500 mm.) in the eastern wing, and **deciduous** with scanty rainfall (1,100 mm. or much less) in the western wing. On the whole SAL (*Shorea robusta*) forests are dominant in this zone (uninfluenced by rains). Other common trees of this zone are *Lagerstroemia, Bauhinia, Bombax, Cassia fistula*, flame of the forest (*Butea*), *Sterculia, Schima, Mangifera, Anthocephalus, Dillenia, Michelia*, etc. A good number of timber trees occur in this zone. Several species of *Melastoma, Osbeckia, Strobilanthes, Mussaenda, Coffea bengalensis*, many grasses, plenty of ferns, etc., form the undergrowth. Climbers and lianes are abundant, and so also clumps of bamboos and wild bananas. Tree trunks are densely covered with mosses and lichens, and orchids are plentiful. Several families are represented in this zone, the dominant ones being *Orchidaceae, Gramineae, Cyperaceae, Leguminosae, Compositae, Scrophulariaceae, Urticaceae, Rosaceae, Rubiaceae* and *Euphorbiaceae*. Above 1,000 metres pines gradually become the dominant feature of vegetation (see next zone).

(b) **Temperate Zone** (1,600-3,500 metres). Coniferae and Cupuliferae (*Betulaceae* and *Fagaceae*) are dominant in this zone. This zone is otherwise called **coniferous zone** because of the predominance of conifers. In the lower temperate sub-zone pine forests predominate (*Pinus khasya* in the east, and *P. longifolia* in the west), associated with *Podocarpus* here and there. With ascent there is rapid change in the composition of flora. In the higher temperate sub-zone deodar (*Cedrus deodara*) predominates. Associated with pines and other conifers (see below) occur oak (*Quercus*), beech (*Fagus*), *Pieris, Castanopsis*, birch (*Betula*), alder (*Alnus*), poplar (*Populus*), willow (*Salix*), maple (*Acer*), elm (*Ulmus*), walnut (*Juglans*), *Magnolia*, etc., most of which ascend upwards to 3,500 metres or even higher. Orchids and ferns are abundant. *Bucklandia*, a handsome tree, is spotted here and there. *Rhododendron* (125 sp., mostly in the eastern Himalayas) deserves special mention. At lower elevations of this zone the species (e.g. *R. arboreum*) are trees or large shrubs; higher up the species are mostly shrubby (e.g. *R. companulatum, R. triflorum*, etc.); still higher up in the cold temperate sub-zone the species (e.g. *R. glaucum*) are further stunted in growth. The temperate zone as a whole is rich in vegetation and abounds in several species of herbs and shrubs. Special mention may be made of *Rosa, Rubus, Berberis, Pyrus, Sambucus, Hydrangea, Strobilanthes, Impatiens,* etc., among shrubs, and *Potentilla, Anaphalis, Senecio, Aster, Anemone, Aconitum, Ranunculus, Saxifraga, Sedum, Meconopsis, Primula, Digitalis, Podophyllum,* etc., among herbs. Mosses and lichens are abundant. Dominant families of this zone are: *Orchidaceae, Compositae, Leguminosae, Rubiaceae, Euphorbiaceae, Urticaceae, Ranunculaceae,* etc., but with ascent the following families predominate: *Rosaceae, Labiatae, Cruciferae, Ericaceae, Umbelliferae, Primulaceae* (e.g. *Primula* with 148 sp., mostly eastern, and *Androsace* with a few species, mostly western), and *Saxifragaceae* (e.g. *Saxifraga* and *Hydrangea*). The conifers occurring at higher elevations of this zone are: blue pine (*Pinus excelsa*), spruce (*Picea*), silver fir (*Abies*), deodar (*Cedrus*), cypress (*Cupressus*), yew (*Taxus*), *Cephalotaxus, Tsuga,* juniper (*Juniperus*), etc.

(c) **Alpine Zone** (3,500-4,500 metres). This zone extends beyond the limit of tree growth up to the perpetual snow-line, and is characterized by the dominance of small herbaceous dicotyledons spotted with low shrubs or undershrubs (mainly species of *Juniperus, Betula, Rhododendron, Ephedra, Berberis, Rosa, Lonicera,* etc.). Vegetation is sparse. This zone is under the influence of extreme conditions of life, viz. freezing temperature, snowcover for the greater part of the year, powerful sunlight, strong gales, low atmospheric pressure, etc.,—conditions which reduce absorption but enhance transpiration. This being so, the vegetation shows a tendency towards xeromorphism: short growth season, stunted habit, hairy covering, thick cuticle, reduced leaves, sometimes fleshy storing water, etc. Several shrubby and herbaceous species of *Rhododendron* (many gregarious in habit) appear in their exquisite beauty from 3,500 metres to the alpine meadows, particularly in the eastern Himalayas. As they ascend upwards to 4,500-4,800 metres their height diminishes to 0.6-0.3 metre or less. Some of the other common genera represented in this zone are *Meconopsis, Podophyllum, Rosa, Rheum, Potentilla, Primula, Saussurea, Saxifraga, Nardostachys, Anaphalis, Artemisia, Astragalus, Aster, Chrysanthemum, Erigeron, Inula, Senecio, Aconitum, Anemone, Caltha, Delphinium, Ranunculus, Thalictrum, Androsace, Sedum, Corydalis,* etc. Dominant families of this zone are: *Compositae, Scrophulariaceae, Primulaceae, Saxifragaceae, Rosaceae, Ranunculaceae,* and *Leguminosae.* Alpine mosses and lichens are plentiful. In such a region there is a preponderance of endemic species. Further it may be noted that *Polygonum viviparum* covers a wide range (1,500-5,400 metres); *Erigeron* shows a very wide range of distribution from tropical to temperate to alpine zone ascending up to the snow-line.

It may be of interest to note that the following species ascend right up to the snow-line : *Thalictrum alpinum, Ranunculus pulchellus, Caltha scapiosa, Delphinium caeruleum* and *D. viscosum, Potentilla microphylla, Saxifraga cernua* and *S. saginoides, Sedum himalense, Corydalis crassifolia, Rhododendron anthopogon, R. lepidotum,* edelweiss (*Leontopodium alpinum*) covered with woolly hairs, *Nardostachys jatamansi* (B. JATAMANSI), and a few species of *Aster, Erigeron, Saussurea,* etc.

The Eastern and the Western Himalayas. As already stated, the eastern side is wetter and the western side is drier. Obviously there is some difference in the floristic composition of the two wings of the Himalayas. There are, however, several genera and species common to both, and both the wings are divisible into the tropical, temperate and alpine zones. There is, however, preponderance of European and Siberian flora in the western Himalayas, and of Chinese and Malayan flora in the eastern Himalayas.

Eastern Region. This region, particularly. Sikkim, is very rich containing over 4,000 species of 'flowering' plants and over 250 species of ferns (including 8 tree ferns). The ratio of dicotyledons to monocotyledons may stand as 2.5:1. The coniferous plants of the east mainly are: *Pinus khasya, P. excelsa* and *P. longi-*

folia (the last two both eastern and western), *Cephalotaxus mannii* and *C. griffithii*, Himalayan silver fir (*Abies spectabilis = A. webbiana*), *Tsuga brunoniana* (both eastern and western), *Juniperus recurva, Taxus baccata, Podocarpus neriifolia* and *P. latifolia, Picea morinda* (both eastern and western). *Larix griffithii*, etc. Of the 'flowering' plants orchids, palms and bamboos are plentiful. Several species of *Rhododendron* appear in their brilliance. Epiphytes and climbers are in great profusion. Besides, the following flowering plants are abundantly represented in this area: oak (*Quercus*), *Castanopsis*, birch (*Betula alnoides*), alder (*Alnus nepalensis*), walnut (*Juglans*), *Magnolia, Michelia, Bucklandia, Primula* (several species), *Photinia, Eriobotrya*, Himalayan poppy (*Meconopsis*—6 sp.), *Berberis, Impatiens, Saxifraga*, etc. Of the many species of *Quercus* in the east some common ones are: *Q. lineata*, Q_H *glauca*, etc: The dominant families of 'flowering' plants are: Orchidaceae, Gramineae, Leguminosae, Compositae, Cyperaceae, Urticaceae, Scrophulariaceae, Rosaceae, Rubiaceae and Euphorbiaceae. Besides, several other families are also well represented in this area.

Western Region. Flora of this region is much poorer than that of the east. The ratio of dicotyledons to monocotyledons may stand as 3:1. The coniferous plants of the west mainly are: *Pinus longifolia, P. excelsa* and *P. gerardiana, Juniperus communis, J. recurva* and two other species, *Abies pindrow, Cupressus torulosa, Cedrus deodara, Picea morinda*, etc. Cycadaceae, however, disappears in the west. Of the 'flowering' plants orchids, palms and bamboos are fewer in number. A stemless palm *Nannorhops*, however, is notable here. Species of *Rhododendron* are fewer in number, but there are some common alpine species in both east and west, e.g. *R. campanulatum, R. lepidotum, R. anthopogon*, etc. Epiphytes and climbers diminish to a considerable extent. Grasses and leguminous plants are more numerous here than in the east. The following 'flowering' plants, however, are abundantly represented in this area: *Astragalus, Artemisia, Saussurea, Ranunculus, Rubus, Rosa, Prunus, Pyrus*, willow (*Salix*), poplar (*Populus*), elm (*Ulmus*), maple (*Acer*), barberry (*Berberis*), birch (*Betula utilis*), alder (*Alnus*), *Potentilla, Punica, Rhamnus, Polygonum, Nepeta, Pistacia*, etc. A few species of *Quercus* in the west are: *Q. incana, Q. dilatata, Q. semecarpifolia*, etc. The following families of 'flowering' plants gain dominance in the west: Gramineae, Compositae, Cyperaceae, Leguminosae, Labiatae, Ranunculaceae, Cruciferae, Orchidaceae, Rosaceae, and Scrophulariaceae. It may further be noted that some families of the east such as Dilleniaceae, Guttiferae, Passifloraceae, Burmanniaceae, Pandanaceae, etc., practically disappear in the west; while certain new families such as Salvadoraceae, Polemoniaceae, etc., make their appearance.

Endemism. Certain species and even genera are seen to remain confined to a small area, a small section of a country or an island from generation to generation; such species and genera are said to be **endemic**. They cannot migrate and spread out owing to certain natural barriers, e.g. high mountains, deserts and surrounding seas. In this respect oceanic islands have a high percentage of endemic species; for example, 82% in the Hawaii Islands and 72% in New Guinea. Although isolated, endemic species multiply and often freely give rise to new species there. There are two views regarding endemism. Some are of opinion that endemism represents the remnants of once flourishing flora, e.g. tree ferns, *Ginkgo biloba*, etc. Others believe that endemic flora is of recent origin, e.g. *Gentiana, Impatiens, Primula, Rhododendron*, etc. The latter view is more generally favoured. According to Chatterjee (*Studies on the Endemic Flora of India and Burma*) in India the Himalayas (temperate and alpine) and the Indian Peninsula show the largest number of endemic species, e.g. 3,169 species in the former region and 2,045 species in the latter region, while only 533 species in the general area. 'Wides' in India, i.e. those that have spread in from outside, may be approximately 4,000 species. Working on the above basis endemism in India comes to almost 59% (ranging, however, according to areas, from 50-70%, the Himalayas showing the highest percentage).

Part V CRYPTOGAMS

Chapter 1 DIVISIONS AND GENERAL DESCRIPTION

Cryptogams or 'flowerless' or 'seedless' plants are lower and more primitive plants. They form three main groups, viz. **Thallophyta, Bryophyta** and **Pteridophyta**. Thallophyta include algae, fungi, bacteria and lichens; Bryophyta include liverworts, horned liverworts and mosses; while Pteridophyta include ferns and their allies. All these have been further divided and subdivided into smaller groups. [See also Introduction, p. XV].

Reproduction. Of the three methods of reproduction, viz. vegetative, asexual and sexual, a particular plant may take to one or more methods. Vegetative reproduction commonly takes place by cell division or by fragmentation. Asexual reproduction takes place by fission or by spores of varied types in different groups of plants. Sexual reproduction takes place by the fusion of two gametes, and the degree of sexuality passes through progressive stages—isogamy to anisogamy to oogamy. In the primitive forms of plants there is fusion of two gametes similar in shape, size and behaviour (isogametes); the fusion of such gametes is called **isogamy**. In other forms the two gametes may be slightly different in size and behaviour (anisogametes); the fusion of such gametes is called **anisogamy**. In the advanced forms the gametes become differentiated into male (microgametes) and female (megagametes); their fusion is called **oogamy**. In oogamous forms the male gamete is small, motile, ciliate, active and initiative and is called an **antherozoid** (spermatozoid or simply sperm), while the female gamete is large, non-motile, non-ciliate, passive and receptive and is called an **egg** (egg-cell, ovum or oosphere). In isogamous and anisogamous forms both types of gametes may be discharged from the body of the plant and gametic union may take place outside the plant body; while in oogamous forms the antherozoids only are discharged, but the egg-cell is retained within the plant body and fertilization takes place in the oogonium within the body of the mother plant.

Alternation of Generations.[1] The life-history of many plants (higher algae, liverworts, mosses, ferns and their allies) is complete in two stages or generations, alternating with each other. These two generations differ not only in their morphological characters but also in their modes of reproduction. One genera-

---
[1] Hofmeister was the first to give a clear idea in 1851 about the alternation of generations in mosses and ferns. He also tried to extend the idea to gymnosperms and angiosperms.

tion reproduces by the asexual method, i.e. by spores, and the other by the sexual method, i.e. by gametes. The former is, therefore, called the **sporophytic** or asexual generation, and the latter the **gametophytic** or sexual generation. To complete the life-history of a particular plant one generation gives rise to the other—the gametophyte to the sporophyte and the sporophyte to the gametophyte, or to put it another way, the two generations regularly alternate with each other. This alternation of gametophyte with the sporophyte and *vice versa* is spoken of as alternation of generations.

Cytological Evidence of Alternation of Generations. In order to keep the chromosome number the same through successive generations plants reproducing sexually (fertilization), evidently with the doubling of chromosomes in the zygote as a result of fusion of two gametes, must have a counter-stage (meiosis) involving reduction of chromosomes. It is a fact that the gametophyte always possesses half as many chromosomes as the sporophyte, or in other words, if the sporophyte bears $2n$ or *diploid* chromosomes the gametophyte would bear n or *haploid* chromosomes (n signifying the number of chromosomes). At the time of reproduction the sporophyte bears spore mother cells (each with $2n$ chromosomes). These undergo meiosis or reduction division and the chromosome number is reduced to half in the spores, evidently with n or *haploid* chromosomes. The spore germinates and gives rise to the gametophyte. The spore, therefore, represents the beginning of the gametophytic generation. The gametophyte, evidently with n chromosomes, in due course bears gametes. When the two gametes (male and female, each with n chromosomes) fuse to form the zygote the chromosome number is doubled, i.e. it becomes $2n$. The zygote develops into the sporophyte with $2n$ chromosomes in all its cells. The zygote, therefore, represents the beginning of the sporophytic generation which continues right up to the spore mother cells.

We may summarize the matter, therefore, by saying that spores, gametophyte, sexual organs and gametes, all with n chromosomes (i.e. the phase interpolated between fertilization and meiosis) represent the gametophytic generation, and zygote, sporophyte, sporangium and spore mother cells, all with $2n$ chromosomes (i.e. the phase interpolated between meiosis and fertilization) represent the sporophytic generation, and that haploid (n) or gametophytic generation begins with the spore and ends in the gametes; while the diploid ($2n$) or sporophytic generation begins with the zygote and ends in the spore mother cells.

In most green algae and most fungi the $2n$ or diploid phase is represented only by the zygote and not by any definite structure developing from it, which may be regarded as a sporophyte. In them, therefore, there is no true alternation of generations. But in the higher green algae, some fungi, most brown algae and red algae, and more particularly in the higher cryptogams—liverworts, mosses, ferns and their allies—an alternation of generations is very regular. In them progressive stages in the development of the sporophyte and reduction of the gametophyte can be traced, culminating in the 'flowering' plants. In the latter the main plant body is always a sporophyte, and the gametophyte is represented only by a few cells (gymnosperms) or by a few nuclei (angiosperms).

Chapter 2   ALGAE

Differences between Algae and Fungi. (1) Algae are green thallophytes containing the green colouring matter *chlorophyll*. In many algae the green colour may be masked by other colours, but in all of them chlorophyll is always present; fungi, on the other hand, have no chlorophyll in them. (2) Algae are *autotrophic* plants, i.e. they manufacture their own food with the help of chlorophyll contained in them; whereas fungi are *heterotrophic* plants, i.e. their modes of nutrition are diverse; they have to depend on prepared food materials supplied to them. They are either parasitic or saprophytic in habit. (3) The body of an alga is composed of a *true parenchymatous tissue*, but the body of a fungus is composed of a *false tissue*, or pseudoparenchyma, which is an interwoven mass of fine delicate threads, known as *hyphae*. (4) The cell-wall of an alga is composed of true cellulose, and that of a fungus of fungus-cellulose or chitin mixed with cellulose, callose, pectose, etc., in different proportions. (5) Algae live in water or in wet substrata; whereas fungi live as parasites on other plants or on animals, or as saprophytes on decaying animal or vegetable matter. (6) Reserve carbohydrate in algae is usually starch; but in fungi it is glycogen.

In structure both the groups may be unicellular, multicellular, filamentous or thalloid, and reproduction in them may take place vegetatively by cell division or by detachment of a portion of the mother plant, or asexually by spores, or sexually by gametes.

Classification of Algae (20,000 sp.)

Class I   Myxophyceae or Cynophyceae or blue-green algae (1,500 sp.), e.g. *Gloeocapsa, Oscillatoria, Nostoc, Anabaena, Rivularia*, etc.

Class II   Euglenophyceae (350 sp.), e.g. *Euglena*.

Class III   Chlorophyceae or green algae (6,500 sp.). **Order 1.** Volvocales, e.g. *Chlamydomonas, Pandorina, Eudorina* and *Volvox*. **Order 2.** Chlorococcales, e.g. *Chlorococcum, Chlorella, Pediastrum, Protosiphon* and *Hydrodictyon*. **Order 3.** Ulotrichales, e.g. *Ulothrix* and *Ulva*. **Order 4.** Chaetophorales, e.g. *Chaetophora, Coleochaete* and *Protococcus*. **Order 5.** Conjugales (or Zygnematales), e.g. *Spirogyra, Zygnema* and desmids (e.g. *Cosmarium*). **Order 6.** Oedogoniales, e.g. *Oedogonium*. **Order 7.** Cladophorales, e.g. *Cladophora* and *Pithophora*. **Order 8.** Siphonales, e.g. *Vaucheria* and *Caulerpa*. **Order 9.** Charales, e.g. *Chara* and *Nitella*.

Class IV   Bacillariophyceae or diatoms (5,300 sp.).
Class V   Phaeophyceae or brown algae (about 1,000 sp.), e.g. *Ectocarpus, Laminaria, Fucus* and *Sargassum.*
Class VI   Rhodophyceae or red algae (about 3,000 sp.), e.g. *Polysiphonia* and *Batrachospermum.*

CLASS I CYANOPHYCEAE or MYXOPHYCEAE or blue-green algae   1,500 sp.

General Description. Cyanophyceae or Myxophyceae or blue-green algae are a small group of primitive algae characterized by the presence of a blue pigment *phycocyanin* in addition to chlorophyll (together making a blue-green colour), simple construction of the plant body, not clearly differentiated protoplast, and simple method of reproduction. Some species are truly unicellular, while in others the daughter cells after divisions adhere together to form a chain of cells (filament) or a flat or spherical colony. A great majority of them are freshwater dwellers and often abundantly found in almost every stagnant pool of water, wet ground and as road slime after rains. The cell structure is of primitive type. There is no definite nucleus nor any plastid, and the protoplast is differentiated into a peripheral coloured zone—the **chromoplasm**, and an inner colourless portion—the **central body** (see FIG. 2C). The cell-wall is made of cellulose and pectic compounds. Carbohydrate occurs in the form of glycogen, starch being altogether absent. A gelatinous sheath is a common feature in most of them. Some filamentous forms, particularly *Oscillatoria*, show a slow spontaneous movement. Blue-green algae never reproduce sexually nor do they bear any kind of ciliated bodies. The common methods of vegetative reproduction are cell division in unicellular forms, breaking up of the colony in colonial forms, and fragmentation of the filament into short pieces called **hormogonia** (see FIG. 2B) in filamentous forms. In some filamentous forms (except *Oscillatoria*) a vegetative cell may act as a resting spore, called **akinete** (see p. 447). In such forms again one or more enlarged vegetative cells with transparent contents and thickened walls may be seen; these are called **heterocysts** (see p. 446).

Origin and Relationship of Cyanophyceae. Cyanophyceae or blue-green algae are supposed to have originated from some non-ciliate unicellular ancestor. Cyanophyceae is a very primitive group and has not given rise to higher forms of plants, but has remained confined within its own group. Its primitive nature is evident from the simple construction of the plant body and lack of organized protoplast (cytoplasm, nucleus and plastids). Further evidence lies in the fact that sexual reproduction is altogether wanting and so also the asexual reproduction by ciliated bodies (zoospores). Cyanophyceae has, however, some resemblance to Rhodophyceae in sometimes possessing *phycoerythrin* (the red pigment of the latter) and in lacking ciliated motile cells at any stage of its life-history. Blue-green algae seem to be related to bacteria, both being ancient groups with some similarities in characters.

1. GLOEOCAPSA

Occurrence. *Gloeocapsa* (family *Chroococcaceae*; FIG. 1) represents a simple primitive form of unicellular blue-green algae. It is the commonest alga forming the road slime, and occurs abundantly on wet rocks, wet ground, in a pool of water, often in laboratory aquaria, forming small masses of jelly. It flourishes during the rainy season and dries up in winter. A small mass of jelly examined under a microscope reveals a large number of single cells or small colonies of cells, oval or spherical in shape, lying embedded in a mucilaginous matrix. *G. quarternata* is a very common species of the road slime.

Structure. A single cell, more or less spherical in shape, represents a *Gloeocapsa* plant. The protoplast of the cell is generally differentiated into two regions: a blue-green peripheral region with chlorophyll and phycocyanin diffused through it—the **chromoplasm**, and a central region with a mass of chromatin granules constituting an incipient nucleus—the **central body**. The plant is always unicellular but often 2 to 4, sometimes several, daughter cells are held together in a colony by the mucilaginous sheath which occurs in concentric layers surrounding the individual cells as well as the whole colony. Mucilage is always derived from the walls of the individual cells.

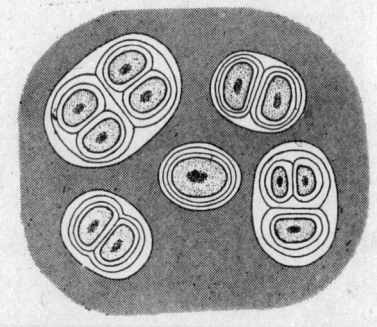

Gloeocapsa. FIG. 1. A cell and four colonies embedded in gelatinous matrix.

Reproduction. *Gloeocapsa* reproduces vegetatively only by the process of cell division. In this process the central chromatin matter divides first into two parts, and this is followed by the formation of a ring-like wall across the cell and its growth inward dividing the cells into two. Each daughter cell secretes its own mucilaginous sheath, grows and finally behaves as a new plant. Two or more such cells are frequently held together in a colony in a common mucilaginous matrix secreted by the individual cells. Sometimes a few species form thick-walled resting spores.

2. OSCILLATORIA (100 sp.)

Occurrence and Structure. *Oscillatoria* (family *Oscillatoriaceae*; FIG. 2A) is a dark blue-green alga. It consists of a slender, unbranched, cylindrical filament. It commonly occurs in ditches, in a shallow pool of water, wet rocks and walls, and in sewers. *O. tenuis, O. princeps*, etc., are some of the common species. Filaments of *Oscillatoria* float in water, either free or entangled in masses. Each filament is made up of numerous short cells. The individual cells are the *Oscillatoria* plants, and the filament is regarded as a colony. All the cells are alike except the end cell

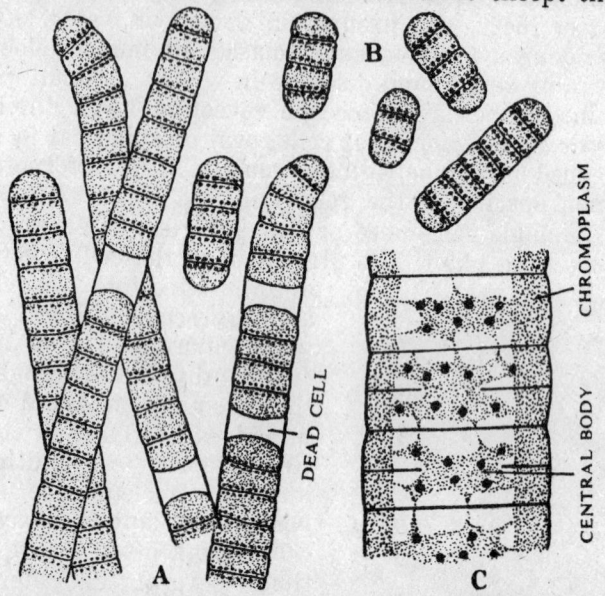

Oscillatoria. FIG. 2. *A*, filaments; *B*, hormogonia; and *C*, a portion of the filament magnified.

which is usually convex, and there is no differentiation of the filament into the base and the apex. Here and there some dead and empty cells occur in some of the filaments. The protoplast of each cell is differentiated into two regions: a coloured peripheral zone—the **chromoplasm,** and an inner colourless zone—the **central body** (FIG. 2C). The colour is due to the presence of chlorophyll and phycocyanin (a blue pigment) which diffuse through the chromoplasm and are not associated with any kind of plastids. Both regions are granular in nature with various spherical or irregular inclusions which are mainly food grains, particularly glycogen and proteins. There is no true nucleus. The central body, however, is regarded as an incipient nucleus with only some chromatin but without nuclear membrane and nucleolus. Cell division takes place in one direction only Each filament remains enveloped in a thin mucilaginous sheath. Under the microscope a slow swaying or oscillating movement of the filaments with ends tossing from side to side may be distinctly seen. The filaments may sometimes exhibit a twisting or rotating motion. This is a characteristic feature of *Oscillatoria*.

Reproduction. In the blue-green algae reproduction takes place vegetatively by cell division or by fragmentation of the filament, or asexually by spores. Gametes and zoospores are altogether absent in them. In *Oscillatoria* the filament breaks up into a number of fragments, called **hormogonia** (FIG. 2B). Each hormogonium consists of one or more cells and grows into a filament by cell divisions in one direction. The hormogonium has a capacity for locomotion.

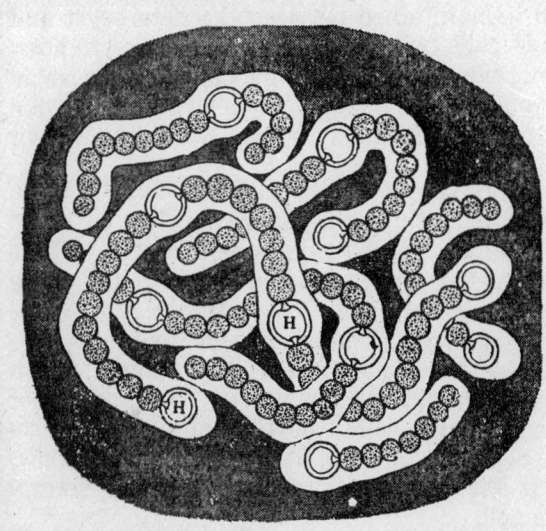

Nostoc.
FIG. 3.
Filaments embedded in gelatinous matrix. Note the heterocysts (H) with the polar nodules.

3. *NOSTOC* (29 sp.)

Occurrence. *Nostoc* (family *Nostocaceae*; FIG. 3) is a common blue-green alga of filamentous form. Species of *Nostoc* may be terrestrial or aquatic, and generally occur in ponds, ditches and other pools of water, and also in the damp soil as little, somewhat firm, masses of jelly. A few species are endophytic in habit occurring in the intercellular cavities of plants like *Anthoceros, Lemna*, root of *Cycas*, etc. Some lead a symbiotic life in association with a fungus forming a lichen.

Structure. Each gelatinous mass contains numerous slender filaments which under a microscope look very much like strings of beads. The filaments are interwoven forming an intricate mass. Each filament is unbranched and consists of a colony of cells like beads in a chain, and is often invested by a gelatinous sheath of its own, in addition to the gelatinous matrix in which the tangled mass of filaments remains embedded. The sheath may be colourless or slightly tinged. The filament minus the sheath is commonly called a **trichome**. Each cell of the filament is more or less spherical or oval (sometimes barrel-shaped or somewhat cylindrical) in shape and blue green in colour. The constitution of the cell is very much like that of *Gloeocapsa*. The filament increases in length by cell division in one plane only. The divided cells grow and round off, and the filament elongates and appears like a beaded chain. A characteristic feature of *Nostoc* is the presence of some enlarged vegetative cells, at frequent intervals, terminal and intercalary, with thickened walls and transparent contents; these are called **heterocysts**. At the two poles of each heterocyst there are two pores, one at each end, through which cytoplasmic connexion is maintained between the heterocyst and the adjacent vegetative cells. In the terminal heterocyst, however, only one such pore is formed. In any case each pore is later closed by a button-like thickening of the wall—the *polar nodule*. The function of the heterocyst is not definitely known (see below).

Reproduction. *Nostoc* reproduces vegetatively by fragmentation, and sometimes asexually by resting cells (spores) called akinetes.

Fragmentation. In vegetative reproduction the filament breaks up into a number of short segments, called **hormogonia**. Each hormogonium by repeated cell divisions in one direction gives rise to a long filament. A large number of such daughter filaments may be seen in the gelatinous matrix. The fragmentation of the filament into hormogonia takes place at the junction of the heterocyst and the adjoining cell. The function of the heterocyst is otherwise not definitely known. It may, however, sometimes act as a spore. It has also been suggested that the heterocyst is a food storage cell.

Akinetes. In a few species of *Nostoc* under unfavourable conditions, e.g. winter or drought periods, certain vegetative cells of the filament become enlarged and thick-walled containing reserve food; these are called resting cells (spores) or **akinetes** and may be produced singly or in a chain. An akinete is regarded as a modified vegetative cell meant to act as a resting spore having no wall of its own distinct from that of the mother cell. According to the species the akinetes may develop anywhere in the filament or in a specific position in it, either close to or away from the heterocyst. Later under favourable conditions each germinates and gives rise to a *Nostoc* filament.

4. *ANABAENA* (28 sp.)

Occurrence. *Anabaena* (family *Nostocaceae*; FIG. 4) is widely distributed, being commonly found in fresh water, floating freely or in thin mucous layers, and also sometimes in wet soil, rocks and tree trunks. Some species also grow as endophytes, e.g. *A. azollae* in the cavities of *Azolla* frond, and *A. cycadacearum* in *Cycas* roots.

Structure. *Anabaena* is a filamentous type of blue-green alga. The filament is a colony of individual cells (plants) formed in a chain, looking like a string of beads. In this respect *Anabaena* resembles *Nostoc* but differs from the latter in certain other respects (see p. 447). The filament of *Anabaena*, like *Nostoc*, is unbranched and consists of a row of oval or spherical cells, bead-like in appearance,

ALGAE

intermixed at frequent intervals with some enlarged thick-walled cells called **heterocysts**, and also another kind of much enlarged thick-walled cells called **akinetes** or resting spores. Akinetes may develop singly or in series, and are more or less cylindrical or oval in shape. They are considered important for diagnosis of species. The filament elongates by division of individual cells in one plane only, and their rapid enlargement. The protoplast of each cell is differen-

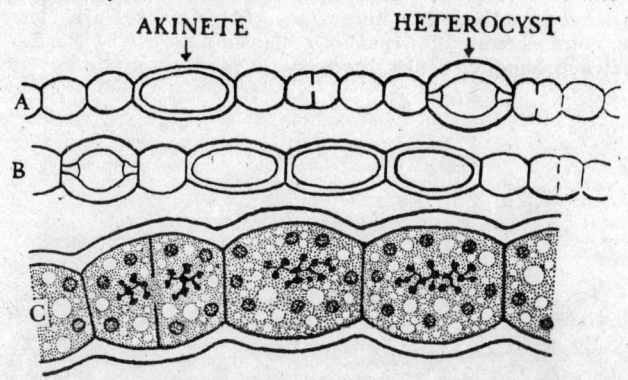

Anabaena. FIG. 4. *A-B,* two filaments; *C,* structure of cells (magnified); note the central body, cyanophycin granules, and the division of some of the cells of the filaments.

tiated into a central colourless region called **central body** which may correspond to an incipient nucleus, and a peripheral region called **chromoplasm** which corresponds to the cytoplasm. The central body is irregular in shape, radiates into arms and contains many chromatin granules. The peripheral region contains pigments—*chlorophyll* in minute granules and *phycocyanin* (the characteristic blue pigment of Cyanophyceae) in solution. The reserve food mainly occurs in the form of glycogen, fat globules, a variable number of cyanophycin granules which are protein in nature. Commonly they occur in the peripheral region. There are also several minute *gas vacuoles* or *pseudovacuoles* whose function is not definitely known. Possibly they give buoyancy to the filament. Each cell has a thin wall of cellulose and pectic compounds surrounded by a thin gelatinous sheath which helps to retain moisture for the plant. The heterocysts, sometimes many, are intercalary in position in *Anabaena*. Each heterocyst is an enlarged vegetative cell with a thick wall and colourless contents. It is connected with the adjacent cells of the filament by a pore at each end.

Reproduction. There is no sexual mode of reproduction in Cyanophyceae. The group also lacks in motile cells. *Anabaena* commonly reproduces vegetatively by fragmentation of the filament, and sometimes asexually by resting spores called **akinetes**. Fragmentation of the filament usually takes place at the junction of the heterocyst and the adjoining vegetative cell. The filament thus breaks up into short segments called **hormogonia** which by cell-divisions in one direction only again grow up into normal filaments. Under extreme conditions *Anabaena* may also reproduce by akinetes or resting spores which are a special type of reproductive units, very resistant to unfavourable conditions. They may ultimately separate from the filament and germinate.

Differences between Nostoc and Anabaena. Although both are characterized by their unbranched filamentous forms made of beaded cells and the presence of heterocysts, one differs from the other by the following characteristics. (1) *Nostoc* filaments are twisted and flexuous, forming a tangled mass, and have a thick (but not very firm) sheath; while *Anabaena* filaments are straight or slightly curved, more rigid, often free, and with a thin sheath. (2) *Nostoc* filaments often lie embedded in a more or less firm gelatinous matrix; while *Anabaena* filaments are often free or sometimes in a thin gelatin. (3) In *Nostoc* the heterocysts are intercalary as well as apical; while in *Anabaena* these are intercalary. (4) Akinetes are more frequent and much more elongated in *Anabaena* than in *Nostoc*. (5) Nostoc is equally terrestrial and aquatic; while *Anabaena* is mostly aquatic.

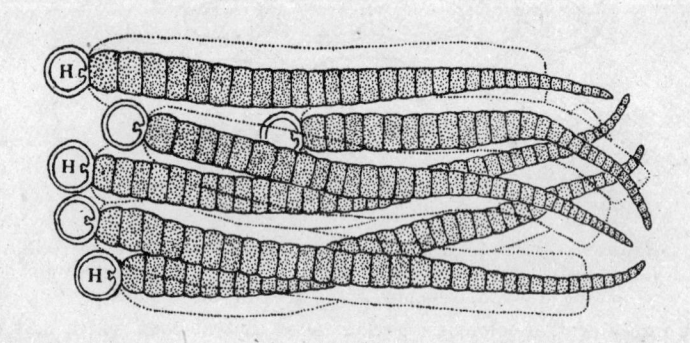

Rivularia. FIG. 5. A group of filaments; note the heterocysts (H) and the sheaths (dotted lines).

5. *RIVULARIA* (FIG. 5)

Rivularia (family *Rivulariaceae*; FIG. 5) is a filamentous type of blue-green alga. It commonly grows on submerged rocks, stones, aquatic plants, and sometimes on rocky cliffs. A few species are marine. The filaments form a colony which remains embedded in a gelatinous mass, sometimes incrusted with lime. In it the filaments are seen to occur in a radiating manner. Each filament has a distinct gelatinous sheath of its own. The filament, called **trichome**, is differentiated into (a) a basal large colourless cell—the **heterocyst**, (b) a narrow (attenuated) apical portion which is whip-or tail-like, consisting of a row of small cells, and (c) the main body consisting of a series of more or less rectangular cells in a row. *Rivularia* reproduces vegetatively by **hormogonia**, i.e. the filaments break up into short fragments called hormogonia. Each fragment or hormogonium then grows by cell divisions into a mature *Rivularia* filament.

CLASS II | EUGLENOPHYCEAE | (over 18 sp.)

EUGLENA (*over* 18 *sp.*)

Euglena (family *Euglenaceae*; FIG. 6) is a most simple unicellular organism from which evolution of the higher forms of plants has possibly started. It belongs to the group Flagellatae in which the organisms are difficult to refer to the kingdom of plants or of animals. It grows in large numbers in polluted water contain-

ALGAE

ing organic substances, and colours it bright green. It is a single-

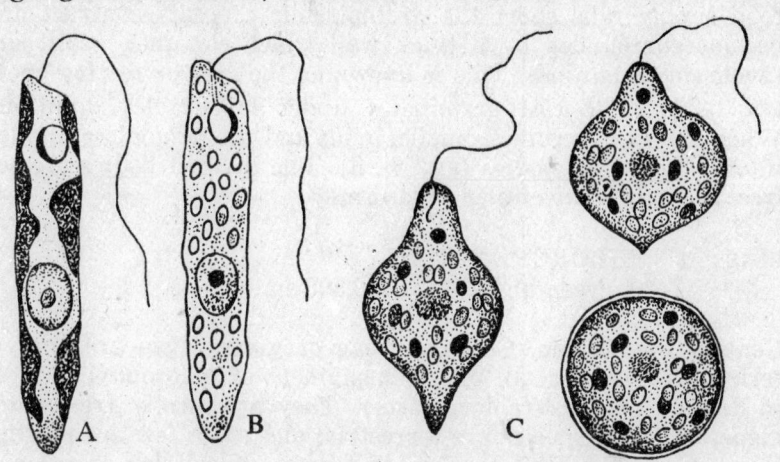

Euglena. FIG. 6-1. *A*, green form; *B*, colourless (saprophytic) form; *C*, various forms (three shown) assumed by a single cell.

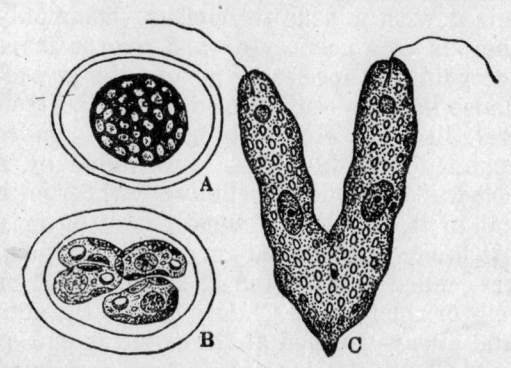

Euglena. FIG. 6-2. *A*, resting spore or cyst; *B*, four daughter cells formed by division of the cyst; *C*, longitudinal splitting of *Euglena* cell.

celled, naked, free-swimming organism, elongated in shape, with one end blunt and the other end tapering. It is provided with a single cilium, i.e. a long, slender, whip-like projection, which vibrates and helps the plant to swim in water. It can also crawl by changing its shape (FIG. 6-1C). The protoplast contains a central nucleus, several green plastids, a contractile vacuole contracting and expanding in a few seconds, and a red spot near the blunt end, called the *eye spot*. It feeds itself by photosynthesis and at the same time it ingests solid particles as food from the surrounding water. When grown in the dark it loses its colour and leads a saprophytic life, obtaining nourishment from the aquatic medium in which it grows (FIG. 6-1B). It does not contain starch as the product of carbon-assimilation.

Reproduction. There is no sexual mode of reproduction in this plant. It multiplies by dividing longitudinally into two, the nucleus taking the initiative in the process (FIG. 6-2C). When the food supply falls short the protoplasmic contents contract and become surrounded by a thick wall which can then resist unfavourable conditions. This is known as the cyst or resting spore (FIG. 6-2A). The cyst germinates under favourable conditions When the wall becomes mucilaginous and the protoplast divides into 2, 4 or more bodies (FIG. 6-2B). The divided bodies are set free as naked, unicellular organisms.

CLASS III CHLOROPHYCEAE
or green algae   6,500 sp.

General Description. Chlorophyceae or green algae are characterized by the presence of green pigment or chlorophyll located in definite plastids (chloroplasts). They are mostly freshwater algae, but some species are terrestrial and not a few are marine. Green algae exhibit a variety of forms—unicellular or colonial being motile or non-motile, multicellular being thalloid or filamentous; some species are coenocytic. The protoplast in all Chlorophyceae is well organized with a definite nucleus (commonly one in each cell or numerous in a coenocyte) and one or more distinct chloroplasts. According to species or genera the chloroplasts vary in shape and also in sizes—cup-shaped, plate-like, stellate, spiral, spherical, oval, discoidal, etc., containing one or more pyrenoids which are rounded protein bodies surrounded by a starchy envelope. The cell-wall is made of cellulose with often a layer of pectose external to it. Gelatinous sheath may or may not be present. Most unicellular and colonial forms are provided with whip-like structures, called *cilia*—often 2, sometimes 4 or many—for motility of cells or colonies. In Chlorophyceae the cilia are of uniform length and always formed at the anterior end of the cell. In higher forms of Chlorophyceae the cilia are restricted to the reproductive bodies only—zoospores and zoogametes. Primitive forms of Chlorophyceae have two or more contractile vacuoles and a small eye spot (see FIG. 7A).

Reproduction. Vegetative reproduction takes place commonly by cell division or by fragmentation. Asexual reproduction takes place by spores which are of varying types: a motile, ciliate spore is called a **zoospore**; a non-motile, non-ciliate spore with a distinct wall of its own but produced within a mother cell is called an **aplanospore** (abortive zoospore); and a vegetative cell acting as a spore having no wall of its own—the wall of the mother cell acting as the wall of the spore—is called an **akinete** (modified vegetative cell). Sexual reproduction takes place by **isogamy, anisogamy** or **oogamy** according to species (see p. 452). Whatever be the mode of sexual reproduction some species are **homothallic** (i.e. the pairing gametes come from the same parent) and others **heterothallic** (i.e. the pairing gametes come from two separate parents). In many green algae it has been observed that a gamete grows *parthenogenetically* (i.e. without fusion with another gamete) into a new plant; the gamete thus behaves as a spore and is called **parthenospore** or **azygospore**. Sometimes, as in *Spirogyra*, the gamete is without cilia and is called **aplanogamete**.

Origin and Evolution of Sexuality in Chlorophyceae. The vegetative method of reproduction is the most primitive method of multiplication of individual plants. Asexual reproduction by zoospores appeared later in the early (lower) Chlorophyceae, possibly as a means of rapid multiplication. Sexual reproduction appeared still later and continued right up to the highest division of the plant kingdom, evidently to achieve something not obtained by the other methods. This something is predominantly protection rather than reproduction as the thick-walled zygote—the result of the sexual act—is better fitted to withstand unfavourable conditions of the environment prior to its starting a new life. Sexual reproduction has other advantages too. It may be noted that when conditions are favourable for vegetative activity neither spores nor gametes are produced; when conditions are less favourable asexual cells or spores are produced; and when the plant is approaching the end of its life or when the conditions are very unfavourable sexual cells or gametes are produced. The mode of reproduction is thus greatly influenced by the changing environment and age of the plant.

The basic fact with regard to the origin of sexuality in Chlorophyceae is that it appeared as a modification of the older asexual method, and is directly correlated with the origin of the sexual cells or gametes for the first time from the asexual cells or spores (zoospores). The gametes because of their smallness in size due to repeated divisions have lost the power of functioning individually; they have thus developed some kind of mutual attraction and freely come together in pairs and fuse. This is the earliest indication of sexuality. It may then be rightly said that gametes are derived from spores (zoospores). It is also seen that spores and gametes are similar in several members of Chlorophyceae, e.g. *Chlamydomonas, Ulothrix, Oedogonium*, etc., excepting that the latter are smaller in size and more numerous.

Once sexuality appeared it established itself, and its evolution through isogamy to anisogamy to oogamy based on the differentiation of sexual cells and sexual organs proceeded towards a high degree of complexity, possibly towards a state of perfection through successive and progressive stages. In the simple and primitive forms of Chlorophyceae there is fusion of two gametes (zoogametes) similar in shape and size; this is called **isogamy**, as in *Chlamydomonas, Ulothrix,* etc. The next stage in the evolution of sexuality is what is called **anisogamy**, as in *Pandorina,* certain species of *Chlamydomonas,* etc.; here a slight difference is noticed in the size of the gametes or in their behaviour —the first indication of differentiation into male and female. A complete differentiation of gametes and gametangia into male and female is found in the advanced forms of Chlorophyceae. The union of such differentiated gametes is called **oogamy**, as in *Oedogonium, Vaucheria,* etc. Throughout Chlorophyceae the gametangia have, however, remained single-celled. It is also significant that there has been parallel progress in the origin and evolution of sexuality along the same evolutionary lines through the different orders of Chlorophyceae following divergent trends of evolution. Some representative types of Chlorophyceae may now be considered to illustrate the above.

Chlamydomonas (see FIGS. 7C & 8B). The simplest type of gamete-formation is found in *Chlamydomonas.* Here the vegetative cell divides and forms 2, 4 or 8 ciliate motile zoospores for asexual reproduction. In the same way the vegetative cell forms 16, 32 or 64 (sometimes more) ciliate motile gametes for sexual reproduction. The zoospores and gametes are similar in structure, but the latter are smaller and more numerous. This is suggestive of the fact that gametes are derived from zoospores and that sexuality has its origin in the transformation of the asexual zoospores into sexual gametes. This is a case of isogamy.

Ulothrix (see FIG. 20). *Ulothrix* is another case to illustrate the origin of sexual cells or gametes (C) from asexual cells or zoospores (C). It produces zoospores of different sizes—large with 4 cilia, medium with 2 or 4 cilia, and small with 2 cilia. The large zoospore germinates into the normal filament; the medium one into a slow-growing filament; while the small one germinates, if at all, into a short filament. The small ones, however, freely come together into pairs and fuse into a zygote which then germinates normally. Small spores not being able to function individually behave as gametes. Transition from spores to gametes and, therefore, from an asexual to a sexual condition is thus clear. A reproductive unit is thus a spore or a gamete according to its behaviour, or in other words, the origin of sex is correlated with the behaviour of the spore as a gamete.

Spirogyra and *Zygnema.* In them the special feature is that the gametes are without cilia and are called *aplanogametes.* In some species of *Zygnema* the gametes are truly isogamous, the two gametes meeting and fusing in the conjugation tube; while in other species of *Zygnema* (see FIG. 31B-C) and in all species of *Spirogyra* (see FIG. 25) the gametes are morphologically isogamous but physiologically anisogamous. One gamete (male) is actively amoeboid and the other gamete (female) is passive. This order (Conjugales) does not form spores (zoospores) for asexual reproduction. But sometimes when conjugation fails the protoplast (gamete) of a cell behaves as a spore and is called the *azygospore*.

Oedogonium (see FIGS. 38 & 39). This shows a high degree of sexual differentiation (oogamy). It reproduces asexually by a solitary zoospore and sexually by highly differentiated gametes —antherozoid and ovum—borne respectively by antheridium and oogonium which are only certain cells of the parent filament. Here the zoospores and the antherozoids, each provided with a ring of cilia, are similar in appearance excepting that the latter are smaller in size.

Vaucheria (see FIGS. 41 & 42). This reproduces, as in the previous case, by a solitary zoospore, and by highly differentiated male and female gametes and gametangia—oogamy. Here the antherozoids are quite distinct from the zoospores. The former are very minute, biciliate and produced in large numbers; while the latter are large, solitary and multiciliate. The antheridium and the oogonium bearing antherozoids and ovum respectively are produced laterally by the vegetative filament or they occur on special short branches borne by the parent filament.

Origin and Evolution of Chlorophyceae. It is presumed that Chlorophyceae or green algae have evolved from some motile (ciliate) unicellular ancestor of the type of *Chlamydomonas*. Several orders of Chlorophyceae seem to have evolved independently and followed different lines of development— progressive or retrogressive. Three such lines or trends of development (evolution) may be specially noted: one line has progressed towards the formation of a motile colony culminating in *Volvox*; a second line towards a non-motile colony-formation with multinucleate cells, e.g. *Chlorococcum, Protosiphon* and *Hydrodictyon*, with a further tendency to form multinucleate siphonaceous or tubular types, e.g. *Cladophora, Vaucheria* and *Caulerpa*; and a third line towards filament-formation—first unbranched and later branched, as represented by Ulotrichales, which may have given rise to the next higher group—the Bryophyta. The first and the second lines of evolution have ended blindly without giving rise to higher forms. It is noticeable that there has been a parallel development in the methods of reproduction through the same stages in all the above three lines: first by cell division, later by zoospores, and still later by the sexual method in progressive stages—isogamy to anisogamy to oogamy.

1. *CHLAMYDOMONAS* (43 *sp.*)

Occurrence. *Chlamydomonas* (family *Chlamydomonadaceae*; FIG. 7A) is a unicellular green alga found in ponds, ditches and other pools of stagnant water. A few species are found in snow in different regions forming blood-red patches due to the development of a red pigment by such species.

Structure. *Chlamydomonas* cells (FIG. 7A) are unicellular, usually spherical or oval in shape, with a thin wall. *Chlamydomonas* may be regarded as an intermediate form between the flagellate algae and the higher algae. The protoplasm at the anterior end of the cell is clear; it gives off two cilia and contains two contractile vacuoles which are pulsating in nature, undergoing alternate expansion and contraction. These may be respiratory or excretory in function. There is a lateral orange or red pigment spot, commonly called the *eye spot*. This is sensitive to different intensities of light. In the posterior region there is a single large cup-shaped chloroplast with a *pyrenoid* in it. The pyrenoid consists of a central protein body surrounded by numerous minute starch grains. There is a nucleus more or less centrally placed. By the lashing of the cilia the cells briskly swim about in water.

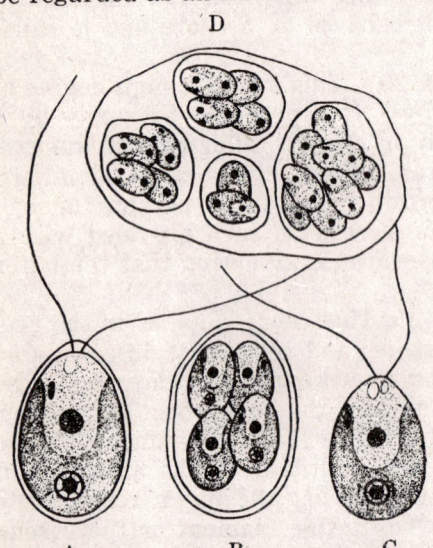

Chlamydomonas. FIG. 7. *A*, a mature cell—note the two contractile vacuoles, eye spot, nucleus, plastid and pyrenoid; *B*, four daughter cells (zoospores) formed by asexual method; *C*, a zoospore after escape; *D*, palmella stage.

Asexual Reproduction. *Chlamydomonas* reproduces asexually by zoospores. In the formation of the zoospores the cilia of each cell are withdrawn, and the contents divide into 2, 4 or 8 cells, seldom more (FIG. 7B). The daughter cells grow, develop two cilia each, and become motile zoospores. The wall of the mother cell dissolves and the zoospores are set free (FIG. 7C). Each zoospore enlarges and becomes a vegetative cell again.

Palmella Stage. Under unfavourable conditions the daughter cells instead of forming zoospores divide repeatedly into numerous cells. Their walls become gelatinous, and the cells are held together in colonies by the gelatinous envelope of the mother cell. Thus numerous colonies are seen to lie embedded in a gelatinous matrix. This is known as the palmella stage (FIG. 7D).

When the conditions are favourable the cells develop cilia, swim out of the gelatinous matrix, and become motile again.

Sexual Reproduction. Sexual reproduction takes place by motile ciliate gametes which are formed in the same way as the zoospores and are also alike but are somewhat smaller in size and

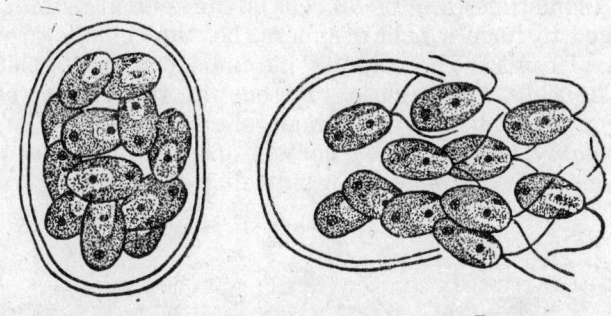

Chlamydomonas. FIG. 8. *A*, gametes formed; *B*, gametes escaping.

more numerous—16, 32 or 64, or even more (FIG. 8). In most species of *Chlamydomonas* all the gametes are similar and are called *isogametes*, and their fusion is known as *isogamy*. A few species, however, show *anisogamy*, i.e. slight differentiation of gametes. Gametes of different parents usually conjugate in pairs (FIG. 9A). A zygospore—the product of fusion of two similar gametes—is formed. Their ciliate ends conjugate first. Soon after fusion the cilia are withdrawn and the zygospore clothes itself with a thick wall (FIG. 9B). It undergoes a period of rest, and then its contents divide and form 2 or 4 motile daughter cells (FIG. 9B). They grow in size, escape from the mother cell, and become individual motile *Chlamydomonas* cells.

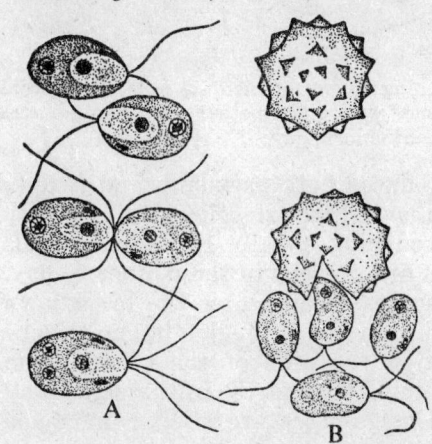

Chlamydomonas. FIG. 9. *A*, free swimming gametes and conjugation; *B*, (*top*) a resting zygote; (*bottom*) four cells formed from the zygote.

It may be noted that in *Chlamydomonas* gametes are mostly alike (isogametes), while there are cases showing slight differentiation of gametes (anisogametes). This is an early indication of sexual differentiation. Further, similarity of gametes and zoospores is suggestive of the origin of sexual cells (gametes) from asexual cells (zoospores) by transformation of the latter into the former.

2. *PANDORINA* (3 *sp.*)

Pandorina[1] (FIG. 10) forms small oval colonies of usually 16 biciliate cells (sometimes 4, 8 or 32 cells); the cells are similar and are arranged to form a hollow sphere, i.e. the cells lie around a small central cavity. The colonies lie embedded in a gelatinous matrix. The cells being close together in the colony become laterally compressed. The individual cells of the colony are like those of *Chlamydomonas*. The colony is propelled in water by the vibration of the two widely divergent cilia of each cell.

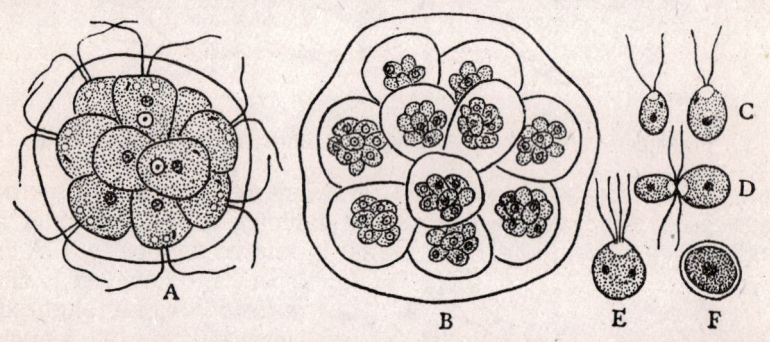

Pandorina. FIG. 10. *A*, a free-swimming colony of biciliate cells; *B*, asexual reproduction: daughter colonies formed within the mother colony; *C*, gametes (large and small); *D-E*, anisogamy; *F*, zygote.

Reproduction. *Pandorina* reproduces both asexually and sexually. In asexual reproduction the individual cells of the colony simultaneously divide, each producing usually 16 daughter cells in a colony or as many cells as are present in the mother colony. Later the new (daughter) colonies escape after the breakdown of the mother cells and swim away through the gelatinous envelope. In sexual reproduction the cells of the colony form biciliate gametes, 16 or 32 in number, exactly in the same way as in asexual reproduction. The gametes are similar in shape

---

[1] *Pandorina, Eudorina* and *Volvox* of the family *Volvocaceae* are motile, colonial, green algae growing sometimes abundantly in fresh water and floating about in it. The cells of the colony lie embedded in a more or less spherical, hollow, mucilaginous sphere. Each colony known as the coenobium consists commonly of 16 cells in the case of *Pandorina*, 32 cells in the case of *Eudorina*, and a few hundreds, often many thousands in the case of *Volvox*. *Volvocaceae* follows a distinct line of evolution. Originating perhaps from *Chlamydomonas* it passes through a simple type of colony to more complex types, and through isogamy to anisogamy to oogamy, ending finally in *Volvox*.

but dissimilar in size, some slightly larger and less active than others. They escape in groups but sooner or later they separate into free individual gametes and swim about. Soon the dissimilar gametes (anisogametes) fuse in pairs, the process being known as anisogamy. This anisogamy, it may be noted, becomes more pronounced in *Eudorina*. The result of fusion is a zygote. It soon settles down and forms a wall round itself. Later it divides and produces four zoospores, out of which one divides and forms a new colony; the remaining three degenerate.

3. *EUDORINA* (4 or 5 sp.)

Eudorina (FIG. 11) forms a spherical colony of usually 32 cells (sometimes 16 or 64 cells) which are also spherical and lie loosely towards the periphery of the mucilaginous sphere. The cells are arranged in five tiers—the three middle tiers with 8 cells each, and the anterior and posterior tiers with 4 cells each. The cells remain connected by fine protoplasmic strands, hardly visible except by proper staining. Each cell (12A) has a mucilaginous sheath and is provided with two cilia. There is a distinct reddish-brown *eye spot* which is sensitive to light. Besides, there are two contractile vacuoles at the base of the cilia, a nucleus somewhere in the middle, and a cup-shaped chloroplast with 1 or more pyrenoids.

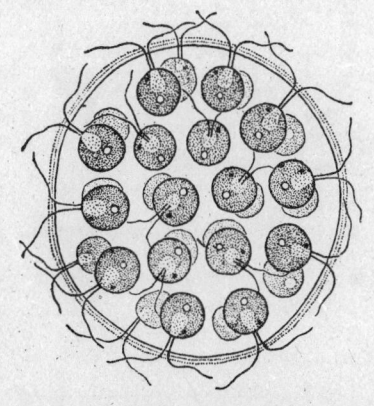

FIG. 11. *Eudorina* colony.

Asexual Reproduction. Each cell of the colony divides and forms its own small group of 32 cells. All the newly-formed small colonies separate from one another by the disintegration of the mother colony. Each colony grows and forms a new mature coenobium.

Sexual Reproduction. Sexual reproduction takes place by differentiated gametes. The coenobia are dioecious being differentiated into male and female. The vegetative cells of the female colony enlarge to some extent, lose their cilia and come to lie near the surface of the mucilaginous sphere (12E). Each such cell is a single female gamete. The cells, often not all, of the male colony divide successively giving rise to a packet of 64 male cells (12B). The packet of cells moves out of the colony and swims as a unit to the female colony (12C-D). The packet then splits up into individual male gametes or spermatozoids, as shown in (12E). The spermatozoids are more or less spindle-shaped and provided with two cilia. They swim into the female colony (12E) and each fuses with a female gamete to form an oospore. It is to be noted

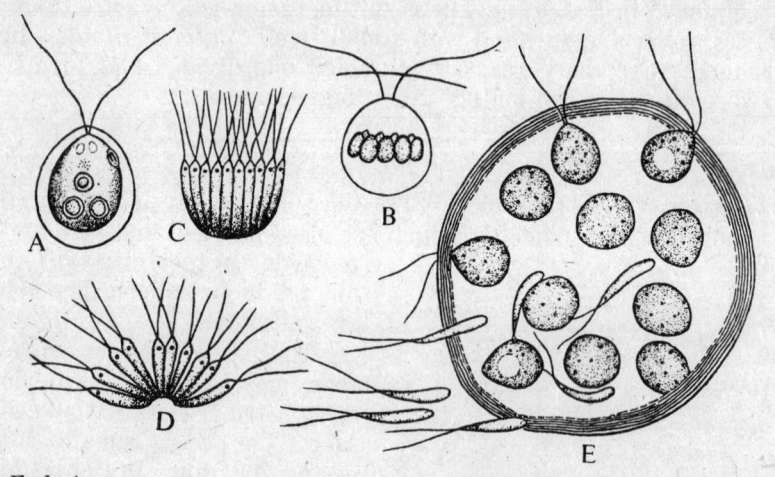

Eudorina. FIG. 12. *A*, a single cell of the colony (note the two contractile vacuoles, eye spot, nucleus and two pyrenoids); *B*, *C*, and *D*, stages in the formation of the male gametes; and *E*, a female colony with female gametes, and some free swimming male gametes.

that two dissimilar gametes (male and female) fuse in pairs; so this is a case of *oogamy*. Several such oospores are formed. They remain in the female colony until the decay of the latter has set in. In the germination of the oospore its contents form one zoospore together with 2 or 3 smaller hyaline bodies (possibly degenerated zoospores). The zoospore divides repeatedly and soon forms a colony.

4. *VOLVOX* (over 12 sp.)

Volvox (FIG. 13) is a freshwater, colony-forming, free-swimming, green alga occurring in ponds and other pools of water during and after rains. It often appears in abundance colouring the water green, particularly in the spring, and then abruptly disappears in the summer. During the rest of the year it lies dormant in the zygote stage. *V. globator* is a common species found floating in tank water.

Among the Volvocales *Volvox* has reached the highest degree of colony formation. As a matter of fact, each colony or **coenobium** (*koinos*, common; *bios*, life), as it is called, consists of a few hundreds to several thousands (500-40,000) of cells which are so arranged in a peripheral layer as to form a hollow sphere (FIG. 13A) containing water or a dilute solution of a gelatinous material. Each cell (FIG. 13B) of the colony has a gelatinous sheath of its own, and at the same time the cells are held together

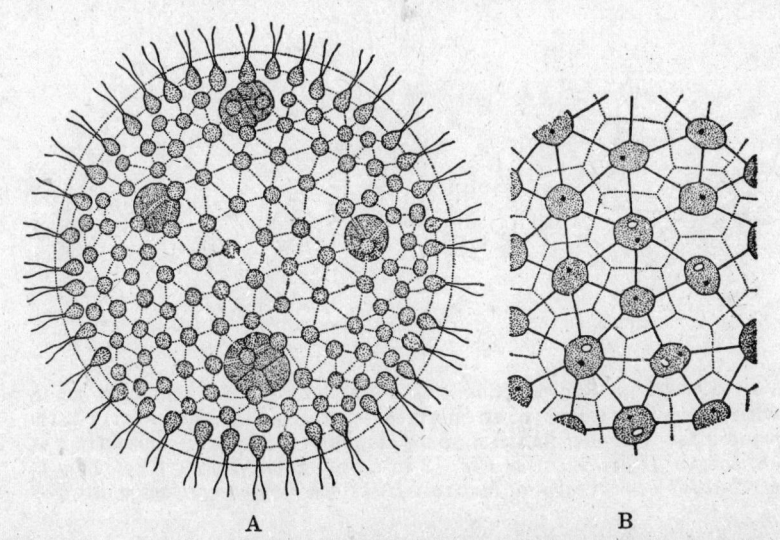

Volvox. FIG. 13. *A*, a colony showing vegetative cells connected by cytoplasmic strands, four colony-forming cells (including two daughter colonies) and outer sheath; *B*, a portion of a colony (magnified) showing vegetative cells connected by cytoplasmic strands (thick lines) and polygonal sheaths (dotted lines).

in a colony by the sheaths secreted by the individual cells. The cells are connected by delicate but distinct strands of cytoplasm. The colonies, approximately 1 mm. in diameter, sometimes up to 2 mm., swim about freely in water. Each *Volvox* cell is very much like that of *Chlamydomonas*.

A mature colony (FIG. 13A) shows two kinds of cells: numerous small vegetative cells and also a few (5-20) large cells among them. A vegetative cell has two cilia protruding outwards and vibrating, 2 to 5 contractile vacuoles, a central nucleus, a cup-shaped or plate-like chloroplast with one pyrenoid, and an eye spot. The vegetative cells do not divide. The larger cells of the colony are the reproductive cells. These cells may behave exclusively as asexual cells or as sexual cells. Normally they act as asexual cells in the beginning of the season and as sexual cells at the close of the season.

Asexual Reproduction (FIG. 14). The above enlarged cells (called gonidia) of the mother colony, after retracting their cilia and pushing back to the posterior side, divide and redivide in the longitudinal plane and give rise to a large number of cells in one plane, thus forming new young daughter colonies within the mother colony. When the cells cease dividing they turn round, develop cilia and form again hollow spheres. These are seen to

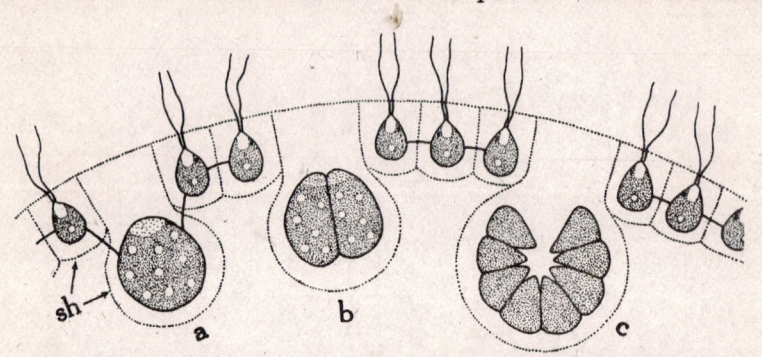

Volvox. Asexual Reproduction. FIG. 14. Formation of a daughter colony within a mother colony; *a*, an enlarged vegetative cell (gonidium); *b*, the same after first division; *c*, a young daughter colony developed from it; *sh*, sheath. *Redrawn after Fig.* 12 *in Cryptogamic Botany, Vol.* I *by G. M. Smith by permission of McGraw-Hill Book Company. Copyright* 1938.

float and slowly revolve within the much enlarged hollow portion of the mother colony. Soon they escape from their imprisoned state by a rupture of the membrane of the mother colony or through a pore in it, and swim away as independent colonies.

Sexual Reproduction (FIG. 15). Sexual reproduction is oogamous in *Volvox*. In the monoecious species both types of gametes (male and female) are borne by the same colony (*homothallic*), while in the dioecious species they are borne by separate colonies (*heterothallic*). These gametes are borne by certain enlarged cells called gametangia (gamete-bearing cells) which lie in the posterior side of the colony. Some of these cells are antheridia or male reproductive organs, the protoplast of which divides many times and produces a cluster of minute, biciliate, male gametes called the **antherozoids** or sperms (FIG. 15A); while other cells are **oogonia** or female reproductive organs, the protoplast of which forms a single, large, female gamete called **egg** or **ovum** (FIG. 15B). The egg is large, passive and non-motile; while the sperms are very minute, active and motile. The latter may be in a plate-like colony escaping from the mother colony as a unit, or they may be arranged to form a hollow sphere. In the former case the unit, as it approaches an egg, breaks up into individual sperms, and in the latter case the sperms are liberated singly. The mode of fertilization is oogamous. The sperms swim and enter through the gelatinous sheath into the oogonium lying in the mother colony, and one of them finally fuses with the egg (FIG. 15B*b*). Thus fertilization is effected.

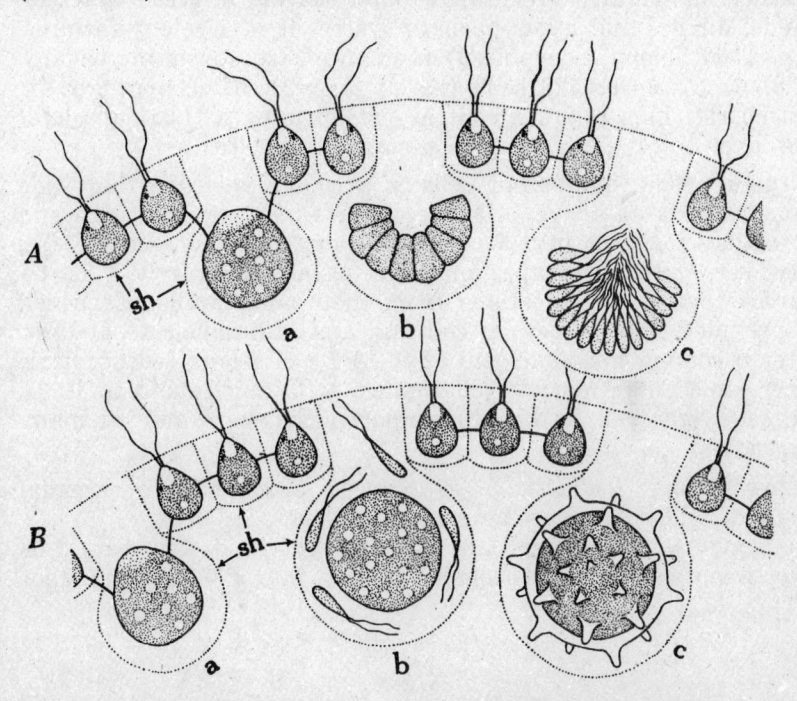

Volvox. Sexual Reproduction. FIG. 15. *A*, formation of antheridium and antherozoids; *a*, an antheridium; *b*, antherozoids being formed; *c*, the same already formed in a cluster; *sh*, sheath; *B*, formation of oogonium with an egg, fertilization, and zygote; *a*, a young oogonium; *b*, the egg about to be fertilized by one of the antherozoids; *c*, zygote; *sh*, sheath. *Redrawn after Fig. 13 in Cryptogamic Botany, Vol. I by G. M. Smith by permission of McGraw-Hill Book Company. Copyright 1938.*

Zygote. After fertilization the zygote clothes itself with a thick spiny wall and turns orange-red (FIG. 15B*c*). It is set free from the mother colony only after the decay or disintegration of the latter. The zygote sinks to the bottom of the pool of water, and then after a period of rest it germinates with the approach of the favourable season. The protoplast of the zygote undergoes reduction division prior to germination. In some species the protoplast of the zygote divides and forms a new colony directly; in others it forms a single biciliate zoospore which escapes by the rupture of the zygote wall and swims away. The free-swimming zoospore divides and forms a new colony.

5. *PEDIASTRUM* (*30 sp.*)

Occurrence. *Pediastrum* (family *Hydrodictyaceae*; FIG. 16A) is a widely distributed green alga found floating in fresh water of ponds, ditches and other pools of water. It is a colony-forming alga. Each colony (coenobium) is small in size, consisting usually of 8, 16, 32, 64 or 128 cells (rarely 4 or 2), often arranged in concentric rings round a central one, forming a thin flat plate, one layer thick. *P. simplex* is a common species.

Structure. The number of cells in a colony varies within wide limits, as stated above. All the cells are more or less alike and fit closely together into a compact colony. The peripheral cells, however, differ in shape, and bear usually 2 (rarely 1 or 3) narrow hair-like projections from their outer walls. Each cell is provided with a distinct cell-wall and is uninucleate at first, later becoming multinucleate (2 to 8) by divisions (coenocytic). There is a single parietal chloroplast with a pyrenoid in it. In old cells, however, with diffuse chloroplasts there may be many pyrenoids.

Reproduction. *Pediastrum* reproduces commonly by asexual method, and sometimes by sexual method.

- **Asexual Reproduction** (FIG. 16). The protoplast of certain cells (not many cells simultaneously) divides into a varying number

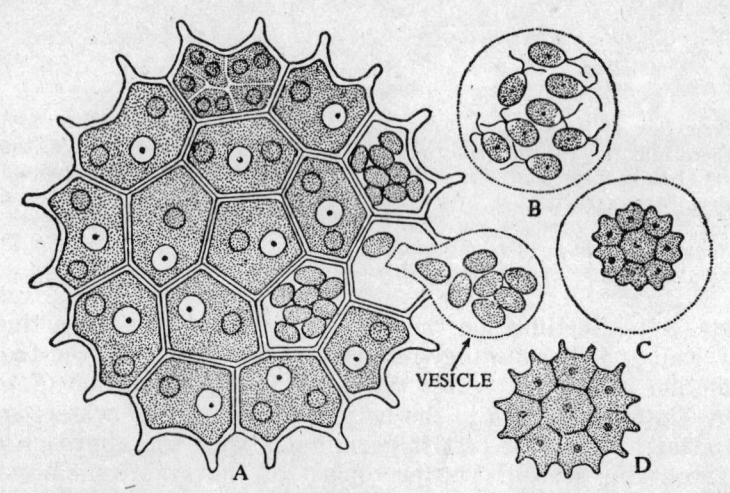

Pediastrum. FIG. 16. *A*, a mature 16-celled colony showing formation of zoospores, and escape of one group of them into a vesicle; *B*, zoospores swarming within the vesicle; *C*, zoospores rearranging themselves to form a daughter colony; *D*, daughter colony enlarging.

of segments, depending on the physiological condition of the cell. Thus it is seen that a 16-celled colony may produce 4, 8, 16, 32 or 64 zoospores in each parent cell. The division of the protoplast is accompanied by nuclear divisions. Each segment becomes converted into a biciliate, uninucleate and ovoid zoospore. The zoospores escape as a packet, usually in the morning, into a vesicle formed from the inner layer of the mother cell, and swarm within it for a while (*B*), Soon they rearrange themselves within the vesicle, grow and take the shape of the mother cells (*C*). Cell-walls appear round them almost immediately, and thus a new (daughter) colony is formed (*D*). The cells of the daughter colony vary in number, depending on the number of zoospores formed in the mother cell.

Sexual Reproduction (FIG. 17). This has been observed in a few species. In the same way as the zoospores are formed, the protoplast of a cell divides and forms a number of biciliate,

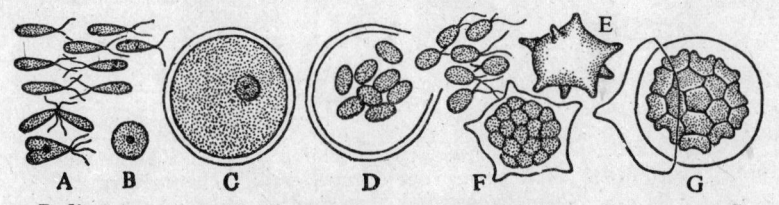

Pediastrum. FIG. 17. *A*, gametes and conjugation; *B*, zygote; *C*, zygote (magnified); *D*, zoospores escaping from zygote; *E*, a thick-walled polyhedron formed from a zoospore; *F*, zoospores within a polyhedron; *G*, zoospores forming a new colony.

spindle-shaped gametes (*A*). All the gametes are alike (isogametes) but are smaller than the zoospores. They escape from the mother cell, one at a time, into the surrounding water. Soon the gametes fuse in pairs (*A*); and the fusion product forms a smooth, spherical zygote (zygospore; *B-C*). The zygote grows in size and its protoplast divides into a large number of biciliate zoospores (*D*). They escape into the surrounding water, swim freely in it for a while, and come to rest. Each then develops into a thick-walled polyhedral cell (polyhedron; *E*). The latter increases in size, and its contents divide into a number of zoospores (*F*). As they escape, a vesicle is formed, and the zoospores stream into it. Within the vesicle the zoospores regroup themselves (*G*), as in asexual reproduction, grow and form a new colony.

6. *HYDRODICTYON* (2 *sp.*)

Hydrodictyon reticulatum (family *Hydrodictyaceae*; FIG. 18), commonly called water net (*hydro*, water; *diktyon*, net), is a common, freshwater, net-like, green alga. The other species *H. africanum* has been only recorded from South Africa. The net, a fairly big one, sometimes growing up to 25 cm. or so, and hollow within, freely floats on water. The plant body (net) consists of elongated cylindrical cells which anastomose to form a sac-like net. The cells are at first uninucleate and contain a single chloroplast with a pyrenoid but later they become multinucleate and contain a reticulate chloroplast with many pyrenoids. A mature cell has a large central vacuole and a lining layer of cytoplasm. *Hydrodictyon* has reached the highest degree of nonmotile colony-formation.

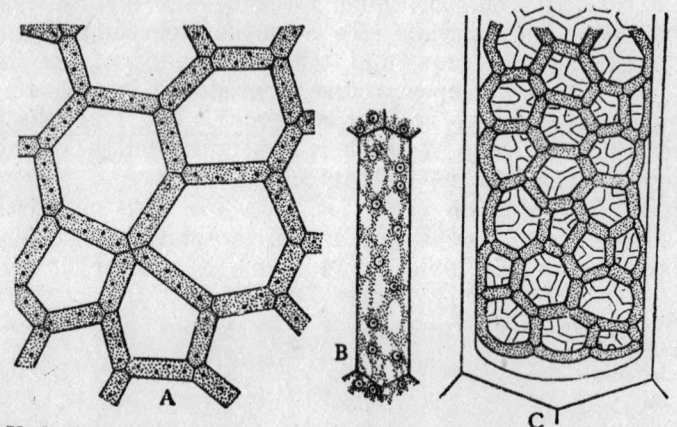

Hydrodictyon. FIG. 18. *A*, a portion of a net; *B*, a single cell (magnified); *C*, a young net formed within the mother cell by asexual reproduction.

Reproduction. Asexual reproduction (FIG. 18C) takes place through zoospores. The protoplast of a vegetative cell produces a large number of zoospores (7,000-20,000), which swim for a time within the mother cell. Soon they come together and form a new net within the mother cell. After the disintegration of the mother cell the new (daughter) net grows into an adult one. Sexual reproduction (FIG. 19) is isogamous. A vegetative cell forms

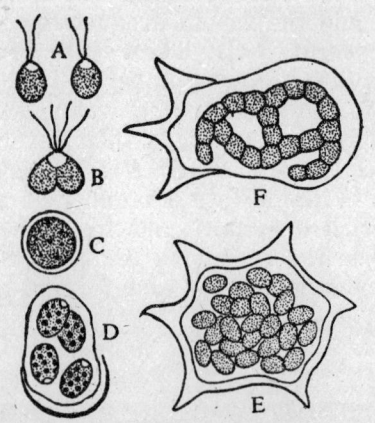

Hydrodictyon. FIG. 19.
A, isogametes;
B, conjugation;
C, formation of zygote;
D, zoospores escaping into a vesicle from the zygote;
E, a zoospore forming a thick-walled polyhedral body, the protoplast of which has divided into numerous zoospores;
F, a young net formed by rearrangement of zoospores.

30,000-100,000 minute, biciliate gametes (isogametes). They escape through an opening of the mother cell, and fuse in pairs. The zygote, thus formed, divides by meiosis to form four biciliate zoospores. Each zoospore at first grows into a thick-walled polyhedral case. Its protoplast undergoes rest for some months and then divides to form a large number of small zoospores which soon escape into a vesicle. Within the vesicle the zoospores rearrange themselves in such a way as to form a short and irregular net which soon grows up into the adult size.

7. *ULOTHRIX* (30 *sp.*)

Ulothrix (family *Ulotricaceae*; FIG. 20) is a green filamentous alga occurring in fresh water in ponds, ditches, water-reservoirs, horse- or cow-troughs, slow streams, etc., particularly in the spring; a few species grow in the sea. The filament of *Ulothrix* is unbranched and consists of a single row of more or less rectangular cells. It is fixed to the substratum or to any hard object in water by the basal elongated colourless cell called the *holdfast*. The filament, if detached, may float freely on water. Each cell of the filament contains a nucleus and a peripheral band-like chloroplast with entire or lobed margin. Usually there are many (sometimes one or few) *pyrenoids* lying embedded in the chloroplast These are rounded protein bodies with a starchy envelope.

Reproduction takes place asexually by zoospores, sexually by gametes, and vegetatively by fragmentation of the filament.

Asexual Reproduction (*B-E*). (1) Zoospores with 4 cilia (**megazoospores**) are produced for the process of asexual reproduction by division of the protoplast of any cell of the filament except the holdfast. They are larger than the gametes but produced in fewer numbers—2, 4, or 8 or sometimes even 1, rarely as many as 16 or even 32—in each cell. Each zoospore is more or less pear-shaped and contains a distinct red *eye spot* on one side, a pulsating vacuole close to the flagellate end, and a large chloroplast. The zoospores escape by an opening in the lateral wall of the cell and swim briskly about in water for some hours or even for a few days. Then they come to rest and attach themselves to any hard object in water by their colourless end. Cilia are withdrawn and a cell-wall is formed round each zoospore. Then it germinates directly into a new filament. (2) Sometimes smaller zoospores (but bigger than gametes), called **microzoospores**, are produced in the filament, and they possess either two cilia or four cilia. Either they germinate directly into new *Ulothrix* filaments like the megazoospores, or they fuse in pairs like the gametes. This indicates that the origin of gametes lies in zoospores. (3) Sometimes the whole protoplast of a cell may round itself off and form a thick-walled spore known as the **aplanospore** (non-ciliate, non-motile, modified zoospore).

Sexual Reproduction (*B-H*). Sexual reproduction is isogamous consisting in the fusion of two similar biciliate gametes (*isogametes*). The gametes may be formed in any cell of the filament except the holdfast. They are smaller than the zoospores, biciliate and may be 8, 16, 32 or 64 in number in each cell. Each gamete

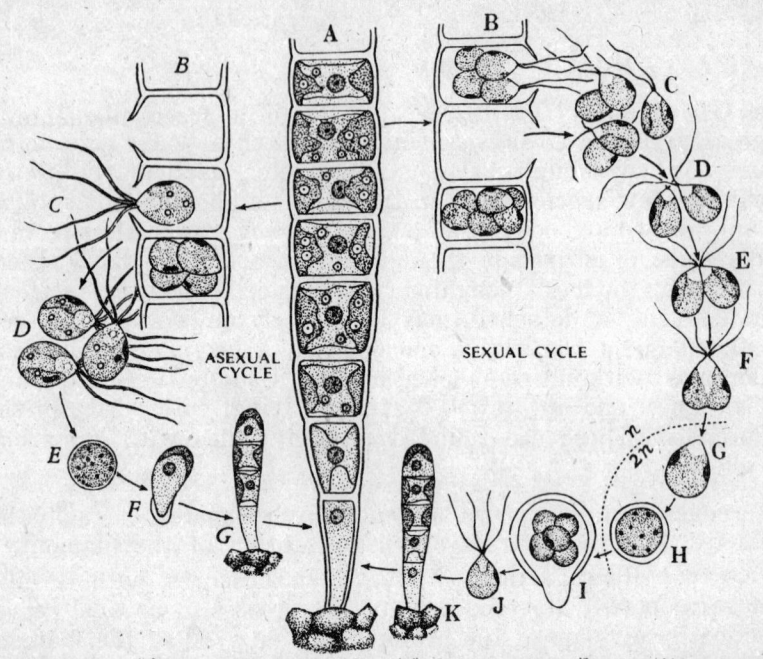

Ulothrix. FIG. 20. *Life-cycle*: **sexual reproduction**—A, vegetative filament; B, formation of gametes; C, gametes swimming; D-G, stages in the conjugation of gametes; H, zygospore; I, the germ plant with zoospores; J, a zoospore (quadriciliate); K, a young filament; **asexual reproduction**—B, a portion of the filament showing the formation of zoospores; C, a quadriciliate zoospore; D, zoospores swimming; E, a zoospore rounded off; F, zoospore germinating; and G, a young filament.

possesses a red *eye spot* and a chloroplast band. The gametes are set free from the cell exactly in the same way as the zoospores and they swim about in water with the help of their cilia for some time. Two gametes coming from two different filaments (*heterothallic*) get entangled by their cilia and gradually a complete fusion (conjugation) of the two takes place laterally. Cilia are withdrawn towards the close of the process, and the fusion product still moves for a while but soon comes to rest. It rounds itself off and clothes itself with a thick cell-wall, and forms into a **zygospore**. After a period of rest the zygospore germinates into a unicellular *germ plant* which produces zoospores or aplanospores—4 to 16 in number. They are quadriciliate (zoospores) or nonciliate (aplanospores) and each develops into a new plant. If fusion fails, each gamete may behave as a zoospore. It withdraws the cilia, rounds itself off and clothes itself with a cell-wall. After a dormant period it germinates directly into a new *Ulothrix* filament. The zygote nucleus has $2n$ or *diploid* chromosomes. It undergoes reduction division and the zoospores are provided with n or *haploid* chromosomes. This haploid number continues throughout the life of *Ulothrix* plant.

Vegetative Reproduction. This takes place by **fragmentation** of the filament into short pieces, each consisting of a few cells. Each piece or fragment grows into a long filament by transverse divisions of cells and their enlargement.

Note. In *Ulothrix* we get a very early indication of the sexual differentiation which becomes so pronounced in the higher plants. The behaviour of gametes or sexual cells and zoospores or asexual cells suggests that the former were originally derived from the latter. The gametes are similar in appearance, but not in their behaviour. The passive one may be regarded as the egg-cell or female gamete, and the active one as the male gamete. *Ulothrix* thus shows the beginning of sexual differentiation.

8. *CHAETOPHORA* (12 *sp.*)

Occurrence. *Chaetophora* (family *Chaetophoraceae*; FIG. 21) is a fresh-water green alga, growing attached to water plants, stones, pebbles, shells, etc., submerged in standing (rarely flowing) water, commonly at the edge of ponds, lakes, marshes and bogs. *C. elegans* is a common species.

Structure. The vegetative body consists of two parts: a part made of erect, profusely branching, deep green filaments, and another part made of a prostrate, sparingly branched, cushion-like thallus (usually 2 or 3 cm. in diameter). The branching filaments are hardly visible with naked eyes. The prostrate part remains attached to some object in water, and sometimes produces rhizoids from the older cells. The erect part develops from the prostrate part, and its filaments (branches) are held together within a tough gelatinous envelope. The cells are somewhat cylindrical or barrel-shaped and are *Ulothrix*-like in appear-

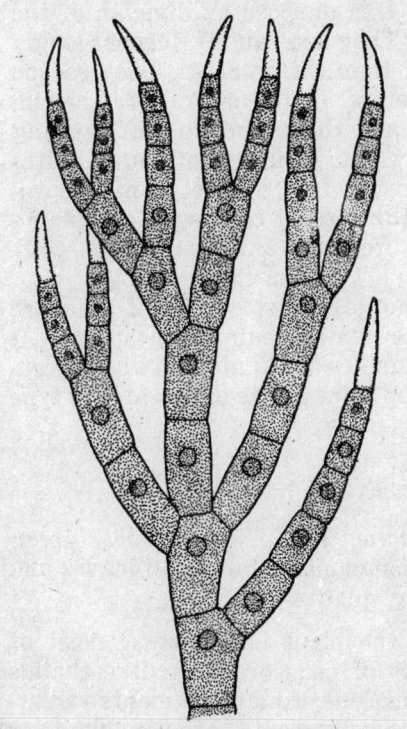

Chaetophora. FIG. 21. A branched thallus.

ance. The cells gradually become smaller in size towards the apex of the filament, and in several species of *Chaetophora* the filament ends in a long or short unicellular or multicellular, colourless hair (devoid of chloroplasts) which usually tapers to a point. Each cell of the filament contains a single plate-like, parietal chloroplast with usually 1 or 2 pyrenoids in it, and a single nucleus. *Chaetophora* follows the same trend of life-cycle as does *Ulothrix* but the structural differences between the two are conspicuous. The branching habit of *Chaetophora* indicates an evolutionary advance over *Ulothrix* and other related unbranched forms.

Reproduction. *Chaetophora* reproduces both asexually and sexually. Both the methods are much the same as those of *Ulothrix*. Vegetative reproduction sometimes occurs by fragmentation of the thallus but is not a common feature because of its toughness.

Asexual Reproduction. Large motile zoospores with 4 cilia (megazoospores) are formed for this purpose by division of the protoplast of any vegetative cell. They are usually formed singly in each cell, and are much larger than the gametes. They escape from the mother cells by an opening in the lateral wall, swim briskly in water for some time, and then come to rest, getting attached to some hard object in water. Cilia are of course withdrawn. Soon they germinate directly into a new plant. Reproduction may also be effected by the resting cells such as aplanospores (1 to 4 in each cell) and brownish akinetes (see p. 450) usually developing in the upper part of the filament.

Sexual Reproduction. Small motile gametes with 2 cilia are formed for this purpose. They are smaller than the zoospores but formed in larger numbers. The gametes are similar in all respects (isogametes); evidently sexual reproduction is of isogamous type (cf. *Ulothrix*).

9. *COLEOCHAETE* (10 sp.)

Coleochaete (family *Coleochaetaceae*; FIG. 22) is a small green alga. It is found in fresh water commonly attached to leaves and stems of some aquatic 'flowering' plants.

Structure. According to species the plant body may consist of branched filaments made of rows of cells or a disc-like thallus with lobed margin (*A*), the thallus consisting of filaments radiating from a common centre; the filaments are so close or adpressed together as to simulate a continuous disc. In certain species the filaments grow vertically from a cushion-like prostrate base. The disc-like thallus appears like a pinhead, varying in size from 2-5 mm., and is one layer of cells thick. Many of the cells give out long slender hair-like bristles, called *setae*, with a gelatinous sheathing base and a linear thread of cytoplasm within. The seta is an outgrowth of the wall formed at a pore in it (wall). Each cell of the thallus has a nucleus and a parietal chloroplast, the latter with 1 or 2 pyrenoids. The growth is apical or marginal.

Reproduction. *Coleochaete* reproduces both asexually and sexually.

Asexual reproduction takes place by **zoospores** (*B*). Any vegetative cell may act as a zoosporangium, producing a single relatively large ovoid biciliate zoospore. In the spring frequently all the vegetative cells form zoospores simultaneously. The zoospore has a single chloroplast and a nucleus. It escapes through a pore in the wall of the mother cell and swims in water for some time; then it comes to rest and begins to divide to produce a new thallus.

Sexual Reproduction. Sexual reproduction is of advanced type and *oogamous*, the reproductive organs being differentiated into **antheridium** and **oogonium**, with motile antherozoid and non-

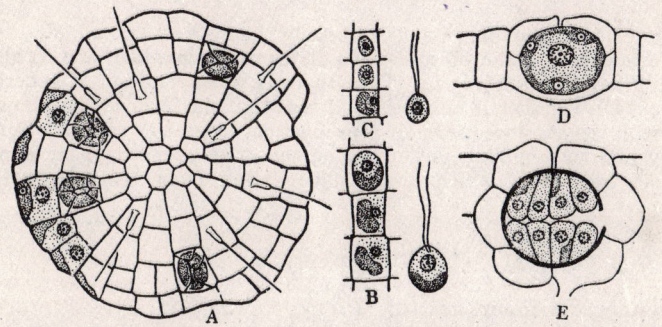

Coleochaete. FIG. 22. *A*, a disc consisting of numerous filaments growing close together (note the zygotes overgrown by the surrounding cells and some bristles with sheathing base); *B*, some vegetative cells showing the formation of zoospores; *right*, a biciliate zoospore; *C*, some antheridial cells showing the formation of antherozoids; *right*, a biciliate antherozoid; *D*, an oogonium within a thallus (in section); *E*, zygote (in section) showing a group of cells, each of which will develop into a zoospore.

motile egg respectively. The plant may be *homothallic* or *heterothallic*. Any vegetative cell may divide into a number of smaller cells, each cell being a male gametangium or antheridium (*C*), producing a single small motile biciliate antherozoid. In the formation of the oogonium (*D*) some of the marginal cells of the thallus enlarge and become oogonia. The protoplast of the oogonium is converted into a large passive (nonmotile) egg. The remaining marginal cells not involved in the formation of the oogonia continue to grow so that the oogonia soon come to lie away from the margin.

In the filamentous species the antheridia and the oogonia are formed terminally at the ends of filaments, either by the same plant or by two different plants. In certain species the oogonium is provided with a beak-like projection, called the *trichogyne*, which receives the antherozoids and allows them to pass through it into the oogonium.

Fertilization. The antherozoids swim to the oogonium and one of them fuses with the egg-nucleus. After fertilization the egg clothes itself with a heavy wall which soon turns brown; this is the **zygote** (oospore; E). The adjoining vegetative cells of the thallus overgrow the zygote and completely enclose it. After a period of rest the zygote, still within the encased oogonium, divides and gives rise to 16 or 32 daughter cells, each of which is converted into a biciliate zoospore. The zygote wall and the encasing layer break and the zoospores are liberated. Each then directly develops into a new thallus.

The zygote is diploid, but its first division being reductional all the zoospores formed as a result of successive divisions are haploid, and so also the plant body developed from each. Thus there is no alternation of generations in *Coleochaete*. Further it may be noted that thalloid form of the plant body coupled with the advanced form of reproduction (oogamy) raises *Coleochaete* to a very high systematic position among the green algae—a near approach to thalloid liverworts which might have arisen from the former in the remote past.

10. *PROTOCOCCUS* (14 sp.)

Occurrence. *Protococcus* (or *Pleurococcus;* family *Protococcaceae*) is a very common unicellular green alga (FIG. 23). It is

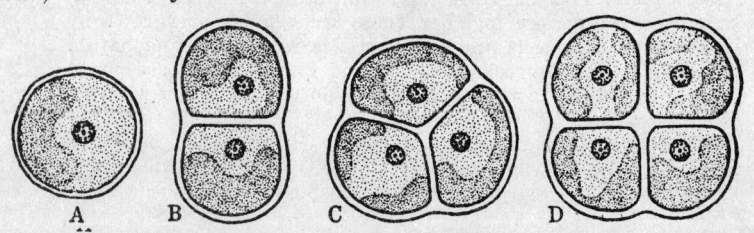

Protococcus. FIG. 23. *A*, a single cell; *B-D*, small colonies formed by divisions of the cell.

terrestrial in habit and is very widely distributed. Species of *Protococcus* (e.g. *P. viridis*) grow in moist shady places and form a green covering on tree-trunks, branches, posts, old damp bricks or brick-walls, flower-pots and other similar objects.

Structure. Each plant is represented by a single more or less globose cell, occurring either as isolated individuals or forming small groups (colonies) of 2, 3, 4 or sometimes more cells as a result of division of the solitary cell. The cells of a group become flattened on the contact side. Ciliate cells and gelatinous covering are conspicuous by their absence. Under conditions of excessive moisture and possibly under other conditions, *Protococcus* may divide in one direction and form a short filament consisting of a few cells, usually 3 or 4, sometimes many more. It is thus suggested that originally *Protococcus* was a filamentous form, but due to lack of vegetative growth together with loss of normal reproductive function it has become reduced to a unicellular stage. The cells have the remarkable power of resisting desiccation but their vital activity becomes very slow then. With the availability of rain-water, dew or moist air the cells become very active again. Individual cells are very small and spherical or oval in shape, filled with a dense cytoplasm and covered by a rather heavy cellulose wall. A single nucleus is present in each cell, and there is a large parietal chloroplast with lobed margin but no pyrenoid.

Reproduction. The only method of reproduction is vegetative cell division which is often very rapid. The first division of the cell is transverse (median) and the second division in one or both the daughter cells is at right angles to the first division; succeeding divisions, if any, are at right angles to the first one. The daughter cells may remain attached to one another in a small group or they may separate and form independent cells (plants). Each cell thus represents a plant. It begins to divide again under favourable conditions.

In the past *Protococcus* was regarded as one of the most primitive forms of green algae but now it is regarded as a reduced form of some filamentous type, possibly of Ulotrichales or Chaetophorales. The occasional branching, short and irregular though it is, lends support to this view.

11. *SPIROGYRA* (100 *sp.*)

Occurrence. *Spirogyra* (family *Zygnemataceae*) is a green filamentous alga (FIG. 24) occurring in a tangled mass which is seen to float about freely in water. It is a cosmopolitan plant of fresh water, and is found growing abundantly in ponds, ditches, springs, slow-running streams, etc. In some species growing in running water a short unicellular organ of attachment called *hapteron* is, however, formed. *S. maxima, S. longata, S. nitida,* etc., are some of the common species.

Structure. Each *Spirogyra* plant is an unbranched filament, a few to many cm. in length, consisting of a single row of cylindrical cells. The walls are made of cellulose and pectin. Pectin swells in water into a gelatinous sheath, and the *Spirogyra* filament becomes invested by this sheath. It is, therefore, slimy to touch. The filament shows no differentiation into the base and the apex. Each cell has a lining layer of cytoplasm in which one or usually more *spiral bands* of chloroplasts—the characteristic feature of *Spirogyra*—lie embedded. The nucleus is situated somewhere in the centre suspended by delicate strands of cytoplasm, and there is one large central vacuole. Chloroplasts vary in number from 1 to 14 in each cell, and run along the whole length of the cell. The margin of each chloroplast may be quite smooth or wavy or serrated. It includes in its body a number of small nodular protoplasmic bodies, known as the **pyrenoids**. Pyrenoids are connected by a sort of ridge which develops on the inner side of the chloroplast, and around them minute starch grains are deposited. There is usually one large nucleolus in each nucleus, but frequently more. Growth by elongation and cell division by mitosis usually take place at night. If the filament happens to be broken up into individual cells or into short pieces, the cells divide and give rise to new filaments (vegetative propagation).

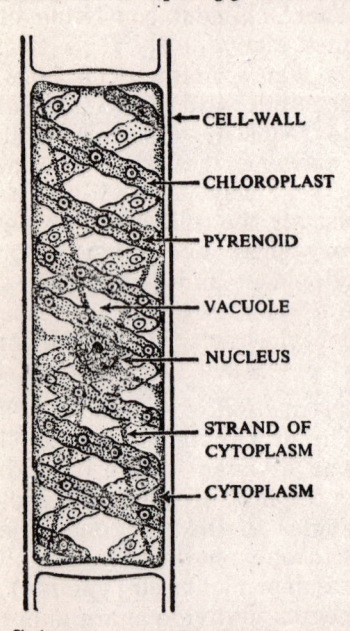

Spirogyra. FIG. 24. A cell of a filament (magnified).

Reproduction. This takes place in *Spirogyra* by the sexual method, and consists in the fusion (conjugation) of two similar reproductive units or gametes, i.e. isogametes. Conjugation usually takes place between the cells of two filaments or even three; this is called scalariform (or ladder-like) conjugation. Sometimes, however, conjugation takes place between the cells of the same filament; this is called lateral conjugation. **Scalariform conjugation** (FIG. 25). When two filaments come to lie in contact in the parallel direction they repel each other. As a result of this repulsion tubular outgrowths develop from the opposite or corresponding points of contact of the two filaments. These tubular outgrowths are called **conjugation tubes**, and when all or most of the cells of the two filaments have formed such tubes

ALGAE

the whole structure looks more or less like a ladder and hence the name scalariform or ladder-like conjugation. Their end- or

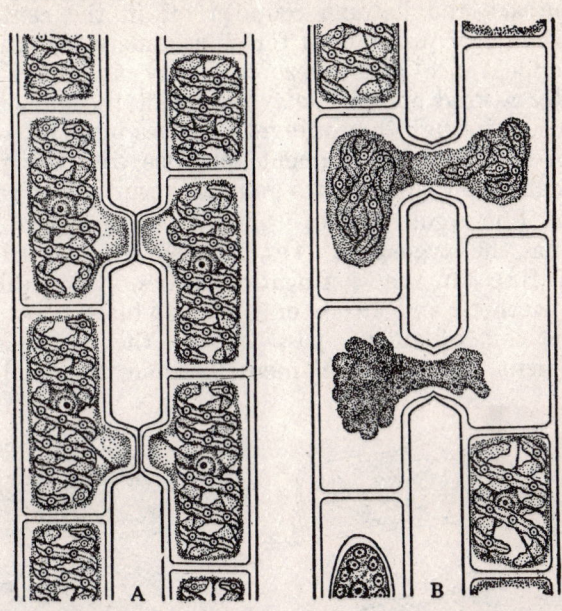

Spirogyra. FIG. 25. Scalariform conjugation. *A-B* are stages in the process.

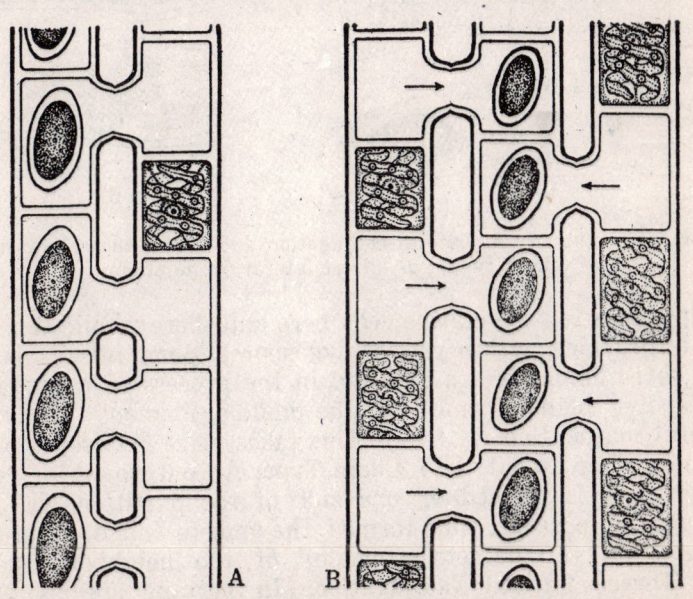

Spirogyra. FIG. 26. *A*, formation of zygospores after conjugation; *B*, scalariform conjugation between three filaments.

partition-walls dissolve and an open conjugation tube is formed. In the meantime the protoplasmic contents of each cell lose water, contract and become rounded off in the centre. Every contracted mass of protoplasm forms a **gamete**. All gametes are alike in appearance and, therefore, they are known as *isogametes*. By a kind of *amoeboid* movement the gametes of one filament creep through the conjugation tubes into the corresponding cells of the adjoining filament and fuse with the gametes of that filament. The fusion of two gametes results in the formation of a zygote. The zygote clothes itself with a thick dark wall, and is known as the **zygospore** (FIG. 26). Sometimes the gametes meet and fuse in the conjugation tubes. The wall of the zygospore is thick and black or brownish-black. Normally the gametes of one filament pass on to the gametes of the other filament. Thus one filament becomes practically empty

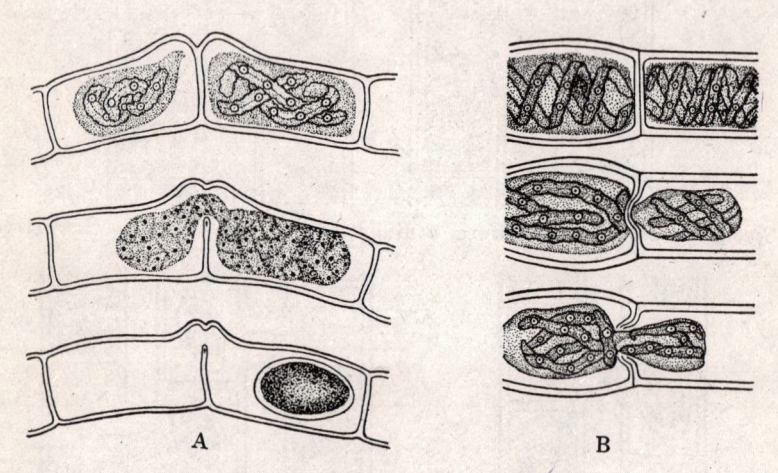

Spirogyra. FIG. 27. *A*, Lateral conjugation and formation of zygospore (chain type); *B*, direct lateral conjugation.

except for a few vegetative cells here and there; while the other one is provided with a row of zygospores. Sometimes it is seen that three filaments are involved in the process of conjugation, zygospores being formed in the middle filament (FIG. 26B). **Lateral conjugation** (FIG. 27). This takes place between the cells of the same filament. (*A*) **Chain Type.** An outgrowth or conjugation tube is formed on one side of the partition wall, and through the passage, thus formed, the gamete (male) of one cell passes into the gamete (female) of the neighbouring cell. (*B*) **Direct Lateral Conjugation.** In certain species as in *S. jogensis* (as reported by Iyengar, 1958) the male gamete passes into the female gamete by perforating the septum in its centre. The protoplast of one cell (male) tapers towards the next cell (female) which now swells considerably. The tapering end pushes and pierces the septum in between these two cells, and the whole protoplast of the male cell moves into the female cell through the perforation, and fuses with the female gamete. After fusion a zygospore is formed. It is believed that the perforation is effected by the secretion of an enzyme. In lateral conjugation the gametes of alternate cells only move to the neighbouring cells, and thus later on zygote-bearing cells are seen to alternate with empty cells in the same filament.

Sometimes it so happens that conjugation does not take place, and then the gametangia become converted into thick-walled bodies identical with zygospores; these bodies which are thus formed parthenogenetically (see p. 409) are called **azygospores** or **parthenospores.** They germinate like the zygospores.

Note. In *Spirogyra* there is no distinction between male and female gametes so far as their shape and structure are concerned, but there is some differ-

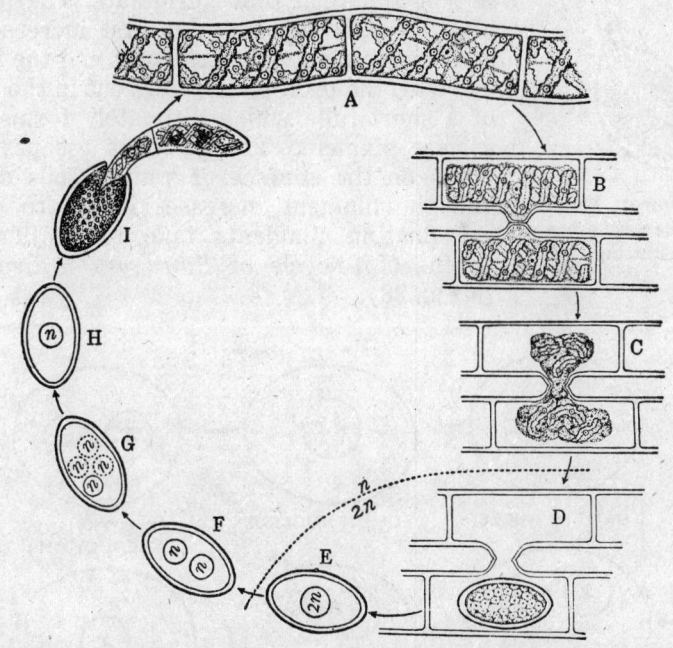

FIG. 28. Life-cycle of *Spirogyra*. *A*, vegetative filament (portion); *B-C*, stages in conjugation; *D*, zygospore formed; *E-H*, reduction division and nuclear changes within the zygospore (see also FIG. 30); and *I*, zygospore germinates.

ence in their behaviour; one is active, motile and initiative and may be regarded as male; while the other is passive, non-motile and receptive and may be regarded as **female**. Normally all the cells of one filament behave as male, and those of the other as female. It is also seen that the chloroplasts of the male gamete become disorganized; while those of the female gamete persist as broken filaments.

Reduction Division (FIG. 30). It is to be noted that the zygote is formed as a result of fusion of two gametes, each with n (or x) chromosomes and, therefore, the zygote-nucleus has $2n$ (or $2x$) chromosomes. The nucleus of the zygote at first undergoes a reduction division, the resulting nuclei divide again so as to form four nuclei, each with n (or x) chromosomes. Three of these nuclei degenerate so that the mature zygote contains a single nucleus with n (or x) chromosomes.

Germination of Zygospore

(FIG. 29). The zygospore is provided with a thick cellulose-wall, composed of three layers, of which the middle one contains some chitin. With the rapid decay of the parent filament all the zygospores are set free and they sink to the bottom of the pool of water. They undergo a period of rest till the next favourable season and then they germinate. The protoplast of each zygospore at first increases in size; then its outer layers burst and the inner one with the protoplast grows out in the form of a short tube which ultimately forms into a new filament. The filament escapes and floats on the surface of water. Cells divide and the filament increases in length. Soon the floating filaments take to conjugation again. (Life-cycle of *Spirogyra* is depicted in FIG. 28).

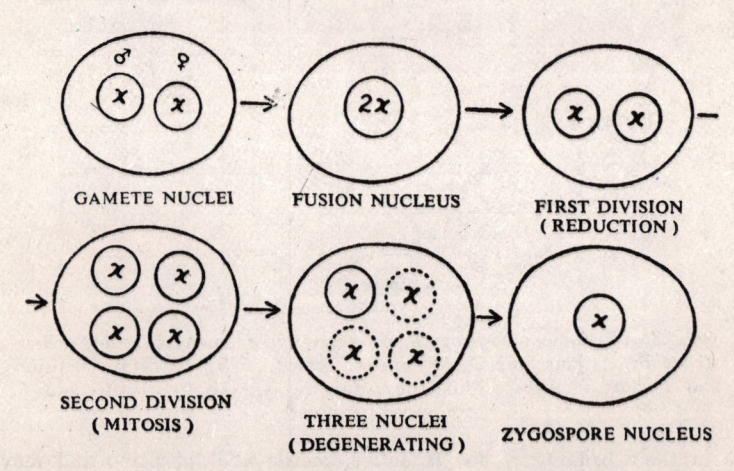

Spirogyra. FIG. 29. Zygospore germinating.

Spirogyra. FIG. 30. Reduction division of zygospore-nucleus.

12. *ZYGNEMA* (95 *sp.*)

Occurrence. *Zygnema* and *Spirogyra,* well-known members of *Zygnemataceae,* are widely distributed green algae occurring in almost every pool of fresh water and floating on it. Some species, however, produce rhizoid-like organs of attachment, called *haptera*.

Structure. Each plant of *Zygnema* (FIG. 31) is an unbranched filament without any differentiation into base and apex. The filament consists of a row of cylindrical cells. The protoplast of each cell is uninucleate, and the nucleus lies embedded in a broad strand of cytoplasm connecting two star-shaped chloroplasts

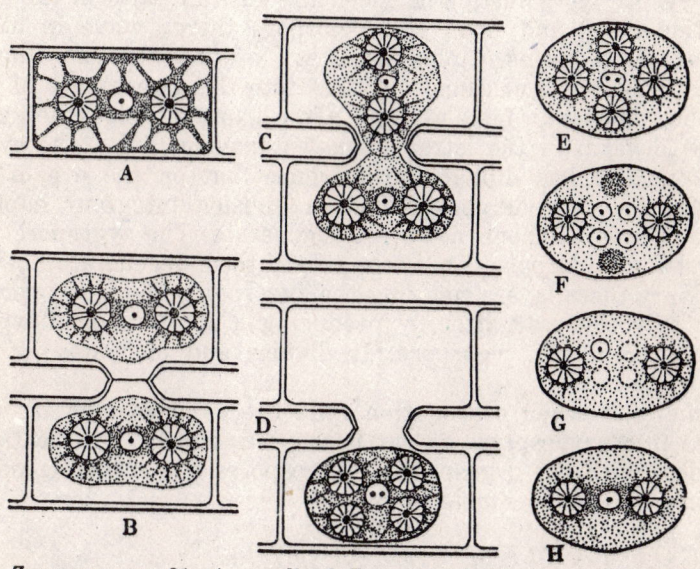

Zygnema. FIG. 31. *A*, a cell of *Zygnema* showing two star-shaped chloroplasts and a central nucleus; *B-C*, formation of gametes and mode of conjugation; *D*, formation of zygospore after conjugation; *E-H*, reduction division of zygospore-nucleus; note that three nuclei degenerate and so do two chloroplasts.

which lie in the axial direction (*A*). Each chloroplast has delicate strands of cytoplasm radiating outwards, often to the wall. The stellate chloroplast is characteristic of *Zygnema*. A single pyrenoid lies at the centre of each chloroplast and is surrounded by radiating starch grains. In the division of the cell the nucleus first undergoes mitosis and this is followed by the furrowing of the cytoplasm in the middle of the cell. Each of the two chloroplasts, thus separated, divides and so does the pyrenoid. The different organs then take their respective positions, as in the mother cell.

Reproduction. As in *Spirogyra*, reproduction in *Zygnema* takes place vegetatively and sexually. Both are conspicuous by the total absence of ciliate gametes or zoospores. **Vegetative reproduction** takes place only by the accidental breaking-off of the old filament into short fragments which grow by cell division and cell elongation.

Sexual Reproduction takes place by means of conjugation (*B-C*) usually in the spring. Both the methods of conjugation, viz., scalariform and lateral, are found in *Zygnema*, and show a marked similarity with those of *Spirogyra*. Scalariform conjugation is, however, the usual method. In some species the gametes (isogametes) of the conjugating filaments migrate from their respective gametangia, meet and fuse in pairs in the conjugation tube, and form the zygospore there; while in other species the gametes (aplanogametes) of one filament (male) move through the conjugation-tubes into the gametangia of the attached filament (female), and after gametic union the zygospores develop in the latter (female) filament (*D*). The zygospore-nucleus has diploid chromosomes but on the eve of its germination it undergoes reduction division into four haploid nuclei. Three of them, however, degenerate. The zygospore has at first four chloroplasts but soon two of them degenerate (*E-H*). There are three layers covering the zygospore. The latter germinates only the year after by producing a short filament which escapes from the zygospore. It divides and elongates into a mature filament.

Sometimes when conjugation fails the gametes may be converted into **azygospores** or **parthenospores** which are essentially identical with the zygospores. Azygospores are rather common in some species of *Zygnema*.

13. *COSMARIUM* (*over* 800 *sp.*)

The genus **Cosmarium** (FIG. 33) is a member of the family *Desmidiaceae*, commonly called **desmids**. Desmids are a big group of unicellular plants comprising about 2,500 species. They have a variety of peculiar, often exceedingly beautiful forms (FIG. 32), and are very widely distributed; they are found abundantly in almost every pool of fresh water. Sometimes the cells are held together in an unbranched filament or in an amorphous colony. Many of the desmids show a jerky movement which is due to the secretion of mucilage through the pores in the cell-wall. *Cosmarium* like other desmids is a freshwater, free-floating alga.

Structure. The body of *Cosmarium* (FIG. 33A) is represented by a single cell with a median constriction, called *sinus*, which divides the cell into two distinct symmetrical halves (*semicells*)

ALGAE

connected by a narrow zone, called *isthmus*. The cell-wall is smooth and consists of three concentric layers—(*a*) an innermost thin layer made of cellulose, (*b*) a somewhat thicker median layer made of cellulose and pectic compounds, and (*c*) an outermost layer of gelatinous sheath. The two inner walls are provided with pores. The protoplast contains a single nucleus which lies at the isthmus, and commonly a single chloroplast with 1 or 2 pyrenoids in each semicell; 2-4 chloroplasts, sometimes many, are not, however, uncommon. A large chloroplast has often many pyrenoids.

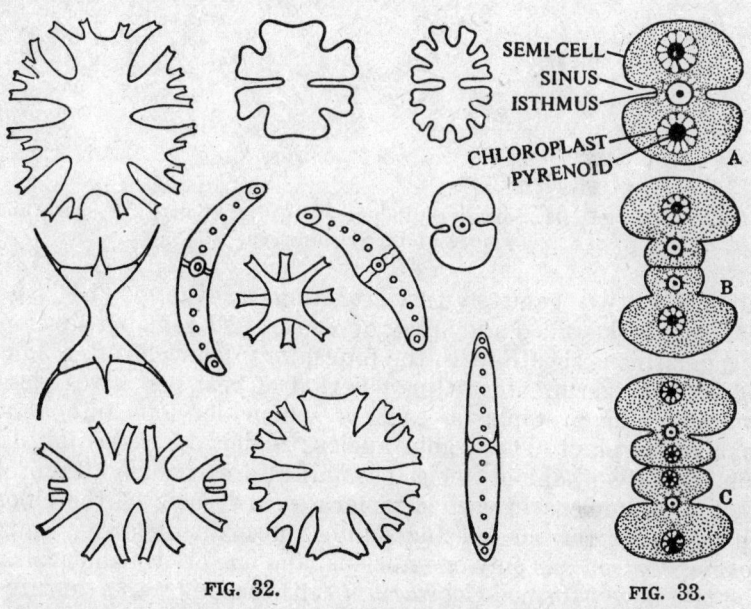

FIG. 32. FIG. 33.

Desmids. FIG. 32. Various forms of desmids including *Cosmarium*. FIG. 33. *Cosmarium*. *A*, a vegetative cell; *B-C*, mode of vegetative reproduction.

Reproduction. *Cosmarium* reproduces vegetatively and sexually.

Vegetative Reproduction (FIG. 33B-C). During vegetative reproduction, which is a process of cell division, the nucleus of each cell divides into two and the isthmus elongates. A partition wall is formed across the isthmus. Each elongated portion of the isthmus then enlarges and gives rise to a new semicell. The chloroplast also divides. The pyrenoids may divide or may be newly formed. The daughter cells separate, and thus two *Cosmarium* cells are formed. It is evident that each of the two newly-formed semicells is always younger than the one belonging to the parent cell. No zoospore is formed in desmids.

Sexual Reproduction (FIG. 34). Sexual reproduction, rare though it is, is isogamous. During conjugation two mature cells or sometimes two newly-formed daughter cells come in contact and are invested by a common gelatinous sheath. The protoplast of each cell forms a gamete. Each cell then breaks open at the isthmus and the two gametes escape and meet midway between the two. In some species a distinct conjugation tube may be formed. In such cases the zygote is formed in the conjugation

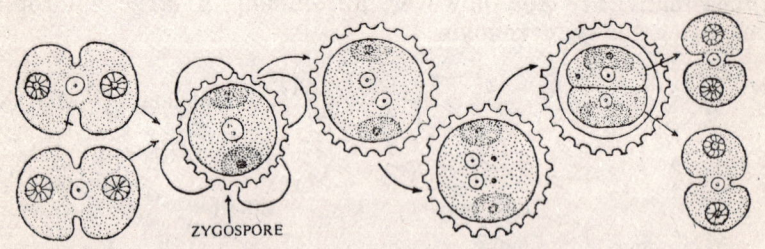

Cosmarium. FIG. 34. Sexual reproduction (conjugation), zygospore and mode of its germination.

tube. The two gametes fuse forming a zygote. The zygote becomes thick-walled and spiny or warty, and remains enveloped by a gelatinous sheath, with the four empty semicells often adhering to it. It germinates after a period of rest and gives rise to two daughter protoplasts, each of which develops into a new *Cosmarium* plant. The zygote nucleus undergoes reduction division, with two haploid nuclei passing into each daughter protoplast. One nucleus in each degenerates and some of the chloroplasts except for the specific number also disintegrate. Thus a *Cosmarium* cell has only one nucleus, and one or two chloroplasts. The two newly-formed *Cosmarium* cells escape by the rupture of the zygote wall.

14. *OEDOGONIUM* (300 sp.)

Occurrence. *Oedogonium* (family *Oedogoniaceae*) is a common green filamentous alga living in fresh water of ponds and pools. The filaments usually remain attached to any object in water by an irregularly lobed basal cell called the **holdfast**. Older filaments may, however, float freely on water.

Structure. Each plant is an unbranched filament consisting of a row of cylindrical cells (FIG. 35). The apex of the filament may be rounded or may end in a hair-like structure. Growth takes place by cell division which may be apical or intercalary. Cell-wall is somewhat thick and rigid, consisting of cellulose internally, pectose in the middle, and chitin externally. Each cell con-

ALGAE

tains a single nucleus placed in the peripheral layer of protoplasm, somewhere in the centre of the cell, and a single large peripheral chloroplast which takes the form of a cylindrical network, extending lengthwise from one wall to the other.

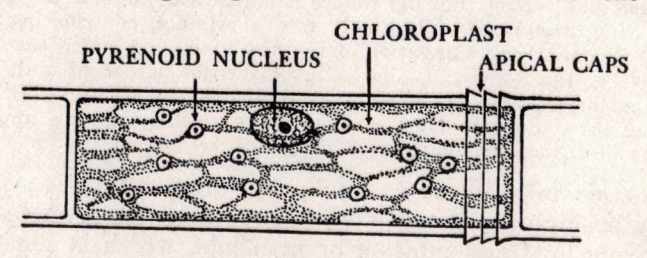

Oedogonium. FIG. 35. A vegetative cell.

Pyrenoids are often several in each chloroplast, occurring commonly in the intersections of the network. Each pyrenoid is a protein granule surrounded by a sheath of starch plates. There are certain special features of *Oedogonium*: peculiar mode of cell division, formation of apical caps, and development of dwarf males.

Method of Cell Division and Formation of Apical Caps. (FIG. 36). Only certain cells of the filament divide with the corresponding formation of apical

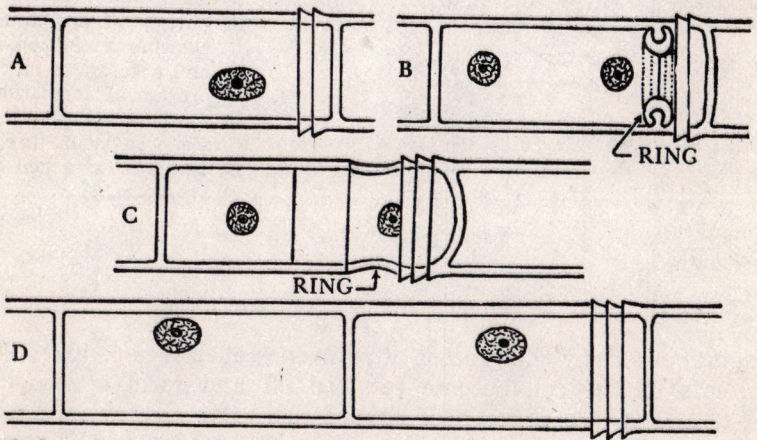

Oedogonium. FIG. 36. *A-D,* cell division and formation of apical caps.

caps at the distal end of each such cell. The division may be intercalary or apical (never basal). With the initiation of cell division the nucleus moves upward, elongates to some extent, and divides mitotically. As the nucleus begins to divide, a ring-like thickening of cellulose appears within the lateral wall of the cell at its distal end. A small groove is formed in this ring, completely encircling it. External to this groove a rent appears in the lateral wall. The result is that the wall splits here transversely. About this time or a little earlier mitosis comes to an end with the formation of a thin transverse plate (wall) immediately after cytokinesis, dividing the protoplasm into two. The lower cell now begins to elongate, rather rapidly, and within a very short time it reaches the size of the parent cell. With the elongation of the cell the transverse wall is pushed up to about the level of the ring, where it unites with the lateral wall. The cellulose ring stretches longitudinally into a cylinder, contributing to the new wall-formation and elongation of the new upper cell which soon grows to the size of the lower cell. The strip of the old persistent wall at the top of the upper cell now appears as a distinct apical cap. With each division of the cell a new cap is formed. Thus overlapping caps appear on successive divisions, and many such caps may be formed.

Reproduction takes place asexually by large solitary zoospores, sexually by highly differentiated male and female gametes, and vegetatively by fragmentation of filaments, which is sometimes common.

Asexual Reproduction (FIG. 37). This method of reproduction takes place by zoospores. Any cell of the filament may form a

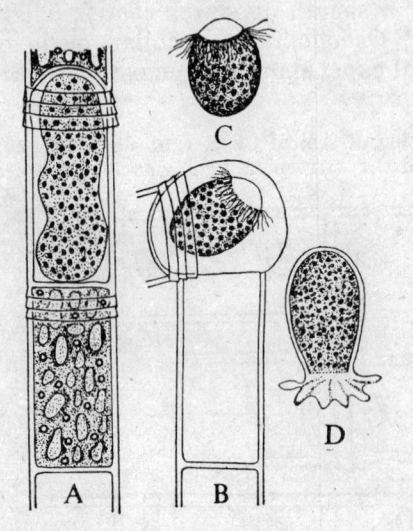

Oedogonium.
FIG. 37.
Asexual Reproduction.
A, portion of a filament showing a chloroplast and a zoospore in the process of formation;
B, zoospore escaping;
C, zoospore swimming;
D, zoospore attached to an object.

single zoospore by the process of **rejuvenescence** (see p. 407). The contents of the cell become rounded off and form a **zoospore** (*A-B*). This is a large pear-shaped body. Its narrower end is clear and bears a ring of cilia; while its broader end is green containing a chloroplast (*C*). The zoospore escapes by transverse splitting of the upper wall of the cell, swims in water for a while and then gets attached to some object (*D*). All the cilia are withdrawn; it then clothes itself with a cell-wall and eventually germinates into a filament.

Sexual Reproduction (FIG. 38). Sexual reproduction takes place by differentiated male and female gametes, known respectively as antherozoid and oosphere or egg-cell. Many species are monoecious (*homothallic*), and some dioecious (*heterothallic*). Antherozoids are borne in certain small cells, known as **antheridia** (A). Antheridia are produced in a series by repeated divisions of any cell of the filament. The protoplasmic contents of each antheridium divide once and produce a pair of **antherozoids**. Each antherozoid is furnished with a ring of cilia and is like the zoospore, excepting that it is smaller in size. Antherozoids are liberated by a transverse slit of the antheridial wall.

The egg-cell is borne in a large spherical cell known as the **oogonium** (B). Oogonia occur amongst the ordinary vegetative cells of the filament, either singly or 2 or more in a row. The protoplasmic contents of the oogonium separate from the cell-wall and become rounded off, forming a single, large, nonmotile **egg-cell** or oosphere. This egg-cell enlarges and becomes spherical or oval. There is a colourless receptive spot at one end of it, and close to this spot the oogonium opens by a pore or a transverse slit of the wall.

Fertilization. When the antherozoids are liberated they swim to the oogonium with the help of their cilia. Then one antherozoid enters through the slit in the oogonium wall and fuses with the

Oedogonium.
FIG. 38.
Sexual Reproduction.
A, portion of a filament showing antheridia (A) and antherozoids (B);
B, a filament showing oogonium (C) and egg-cell (D) with receptive spot (E);
C, an oospore;
D, formation of zoospores from it.

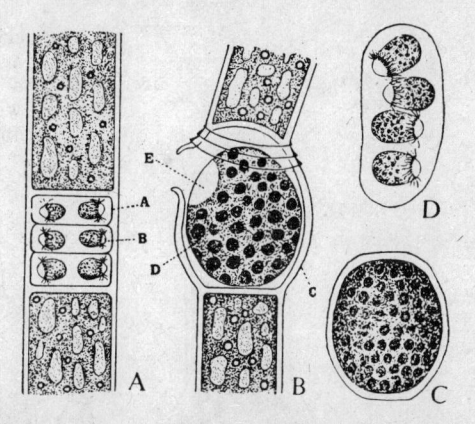

egg-nucleus at the receptive spot (B). The oosphere then covers itself with a thick cell-wall and becomes a reddish-brown **oospore** (C). The oospore sinks to the bottom, undergoes a period of rest and then germinates. The nucleus of the oospore has $2n$ chromosomes. It undergoes reduction division giving rise to four **zoospores** (D), each with n chromosomes. The zoospores escape and swim about for some time. Then they rest for a while, attach themselves to some object, and each germinates into a new filament.

In certain dioecious species of *Oedogonium* a complicated process of reproduction takes place. In them a special type of zoospore, called **androspore** (FIG. 39A), is produced by the same filament that bears the oogonia or by a distinct filament. Androspores are produced in special cells, called **androsporangia,** which are formed either singly or in a row like the antheridia by division of the ordinary vegetative cells of the filament. Each androsporangium produces a single androspore which like the antherozoid is provided with a crown of cilia and is motile. The androspore is intermediate in size between the zoospore and the antherozoid. When liberated, the androspore swims for a while and soon attaches itself direct to the oogonium or to a cell close to it. It then produces a short narrow filament, called the **dwarf male** or **nannandrium** (FIG. 39B), consisting of an elongated basal cell and a terminal cell or a row of cells usually (2 to 4). Each such cell is an antheridium. It bears a pair of small motile antherozoids crowned with cilia. The antheridium opens by a lid at the apex or it ruptures by the wall and the antherozoids are liberated. They swim to the oogonium and fertilization takes place in the usual way.

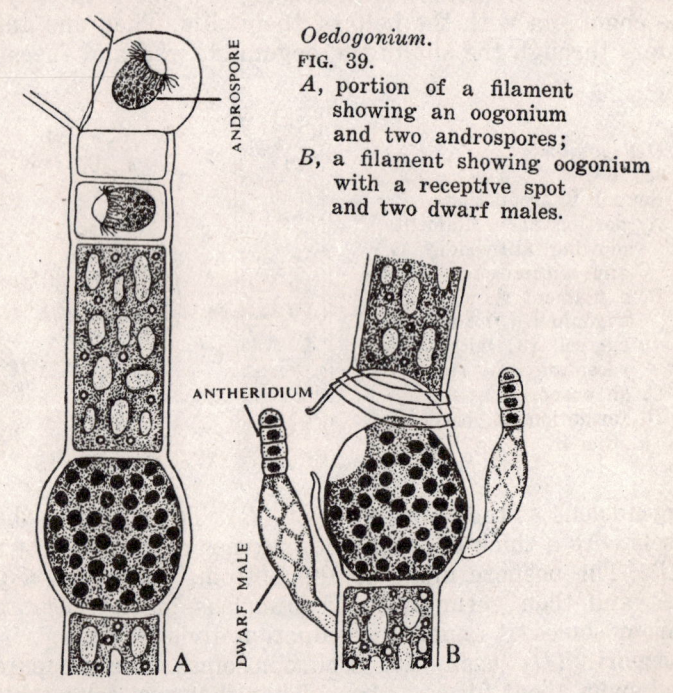

Oedogonium.
FIG. 39.
A, portion of a filament showing an oogonium and two androspores;
B, a filament showing oogonium with a receptive spot and two dwarf males.

15. *CLADOPHORA* (160 sp.)

Occurrence. *Cladophora* (family *Cladophoraceae*; FIG. 40) is a very widely distributed genus commonly found in freshwater ponds, lakes and streams, growing there attached to some objects by rhizoid-like branches. A few species are marine.

ALGAE

Structure. The plant body consists of freely branched filaments. Each filament is made of a row of cylindrical cells united end to end (*A-B*). The branching is lateral although sometimes it looks dichotomous. The cells are coenocytic in nature containing numerous nuclei. The chloroplast of the young cell is parietal encircling the cytoplasm and is reticulate enclosing many pyrenoids. In the older cell the chloroplast may break up into many discoidal ones, with pyrenoids in some of them. There is a large central vacuole. The cell-wall is made of three layers—the outer composed of chitin, the middle of pectic compounds and the inner of cellulose.

Reproduction. Vegetative reproduction takes place by certain cells, lying at the base of the plant, filled with a heavy store of food. After the filaments die back such cells grow up into new filaments in the following season. **Asexual reproduction** takes place by quadriciliate zoospores (*C*) which are formed in large numbers in the apical cell and in other cells close to it

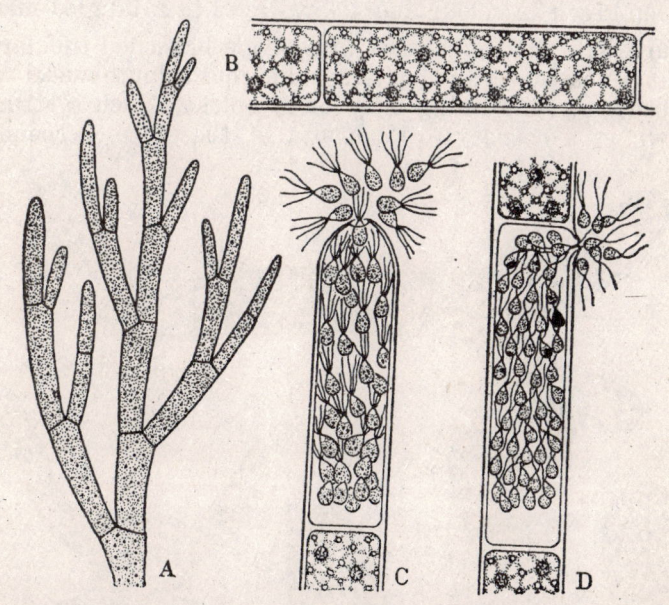

Cladophora. FIG. 40. *A*, a portion of the plant; *B*, a cell (magnified) showing peripheral reticulate chloroplast with many pyrenoids, and also several nuclei; *C*, quadriciliate zoospores escaping from the zoosporangium; *D*, biciliate gametes escaping from the gametangium.

(asexual plant). The zoospores are uninucleate. They escape one by one through a minute pore formed at or near the upper end of the cell-wall. After swarming for a while each secretes a wall, quickly elongates and develops into a new plant (sexual plant). **Sexual reproduction** takes place by **isogametes** borne by the sexual plant only in any of its vegetative cells. The gametes are biciliate (*D*), and are produced in large numbers like the zoospores. Gametes borne by two different plants fuse in pairs, *Cladophora* being *heterothallic*. The zygote formed as a result of fusion grows within a day or two directly into a new plant (asexual plant). It reproduces by zoospores again. It will be noted that the spore- (zoospore-) bearing plant,

i.e. the asexual plant, is diploid, and the gamete-bearing plant, i.e. the sexual plant, is haploid, but both are alike in appearance. Reduction division takes place in the formation of zoospores. *Cladophora* shows *isomorphic* alternation of generations. Stages in the life-history are shown below:
1. Asexual plant $(2n)$ —→zoospores (n) —→sexual plant (n).
2. Sexual plant (n) —→isogametes (n) —→zygote $(2n)$ —→asexual plant $(2n)$
   (fusion in pairs)

16. *VAUCHERIA* (35 *sp.*)

Occurrence. *Vaucheria* (family *Vaucheriaceae*; FIG. 41A) is a green freshwater alga, growing in ponds, ditches and also in the wet soil. It is not free-floating like *Spirogyra* but is mostly attached to a substratum by means of colourless rhizoids or holdfasts. It is deep-green in colour and always lives in a tangled mass.

Structure. The thallus consists of a single branched tubular filament. It is unseptate, and contains numerous minute nuclei which lie embedded in the lining layer of cytoplasm. Such a structure is known as a **coenocyte**. *Vaucheria* is, therefore, a coenocyte.

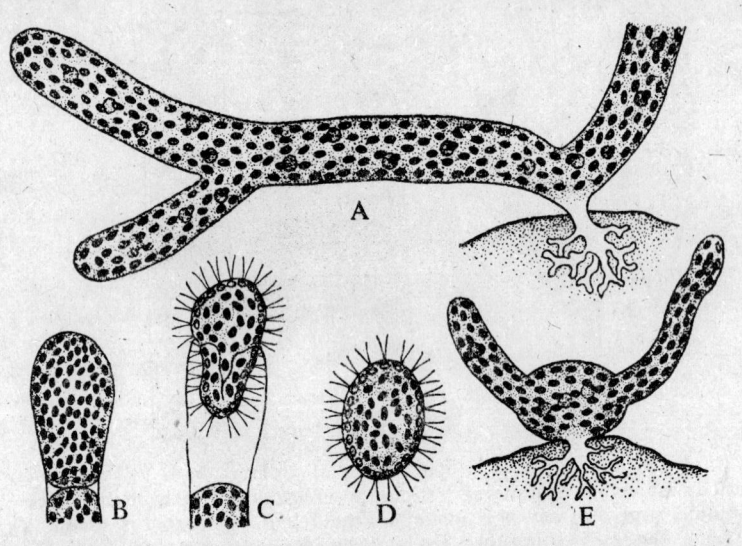

Vaucheria. FIG. 41. *A*, a *Vaucheria* filament; *B*, formation of zoosporangium; *C*, zoospore escaping; *D*, free-swimming zoospore; *E*, zoospore germinating.

There is a large central vacuole which runs the whole length of the coenocyte. Septa, however, normally appear in connexion with the reproductive organs. Injury also results in the production of septa cutting off the injured parts which then develop into new plants. Filaments increase in length by apical growth. Chloroplasts are numerous, very small and discoidal in shape.

ALGAE

They lie embedded in the lining layer of cytoplasm, and are without pyrenoids. Protoplasm contains abundant oil-globules, but lacks in starch.

Reproduction. This takes place asexually as well as sexually.

Asexual Reproduction. This takes place by a large solitary zoospore. During its development the apex of the filament swells up, becomes club-shaped and is partitioned off from the rest of the filament by a septum. This club-shaped body is known as the zoosporangium (FIG. 41B). Its protoplast becomes rounded off, forming a single zoospore. The wall of the zoosporangium ruptures at the apex, and the zoospore escapes through the apical pore (FIG. 41C) and begins to rotate. The zoospore (FIG. 41D) is an oval body of large size. The central part of it is occupied by a large vacuole, and in the surrounding zone of the cytoplasm there lie embedded numerous small chloroplasts, giving the zoospore an intensely deep-green colour. The whole surface of the naked (without cell-wall) zoospore is covered with numerous short cilia arranged in pairs and under each pair of cilia there lies a nucleus. For this reason the zoospore is regarded as a com-

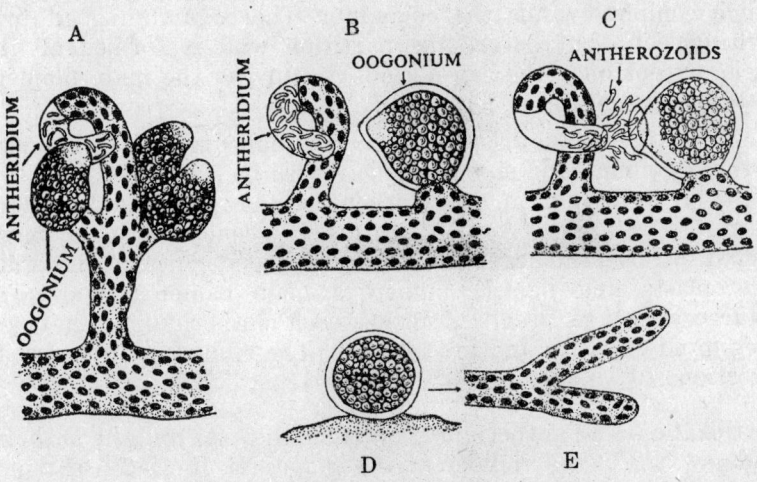

Vaucheria. FIG. 42. *A*, antheridium and oogonia on a short lateral branch; *B*, the same borne directly by the filament; *C*, mature antheridium and oogonium; antherozoids discharged and ovum about to be fertilized; *D*, oospore; *E*, a new filament developed from the oospore.

pound one. The zoospores generally escape in the morning. They swim about freely in water for a while (half an hour or less) by the vibration of their cilia and soon come to rest. The cilia are immediately withdrawn and a cell-wall is developed round each. After coming to rest the zoospore germinates (FIG. 41E) almost immediately by the protrusion of one or more tube-like filaments, one of which, at least, produces a colourless branched rhizoid, and attaches the plant to the substratum. The protoplasm leaves the old cell and rejuvenates, i.e. it becomes young and active again; this method of asexual reproduction is known as **rejuvenescence**.

Sexual Reproduction. This takes place by the method of **fertilization**, that is, by sharply differentiated male and female organs. Male organs are known as **antheridia** and female organs as **oogonia** and these are developed at scattered intervals as lateral outgrowths. Antheridia and oogonia commonly arise side by side on the same vegetative filament (FIG. 42B-C) or on short lateral branches of it (FIG. 42A). All fresh-water species of *Vaucheria* are homothallic.

The outgrowth that forms the **oogonium** swells out, assumes a more or less rounded form, and is cut off by a basal septum (FIG. 42A-C). The apex of the oogonium generally develops a *beak*, either towards the antheridium or away from it. The protoplasm of the oogonium contains much oil, numerous chloroplasts, but only one nucleus. There is a single **egg-cell** or oosphere which completely fills the oogonium. The oogonium is at first multinucleate, but before the partition wall is formed all the nuclei except one (the egg-nucleus) return to the main filament or they degenerate.

Each **antheridium** arises as a short tubular branch by the side of the oogonium, and simultaneously with it. Its terminal portion is cut off by a septum, and it then becomes the actual antheridium (FIG. 42A-C). As it matures it usually becomes much curved towards the oogonium. The protoplasm contains numerous chloroplasts and nuclei. Numerous male gametes, known as **antherozoids**, are produced inside each antheridium. They are very minute in size and are biciliate. The cilia point to opposite directions.

Fertilization. The antheridium bursts at the apex and the antherozoids swarm in the vicinity of the oogonium, the beak of which opens at about the same time (FIG 42C). Several antherozoids may enter through the beak but only one of them fuses with the egg-nucleus, while the rest perish. Thus fertilization is effected. *Vaucheria* is homothallic, and self-fertilization is the general rule. After fertilization the oosphere becomes invested with a thick cell-wall, and is known as the **oospore** (FIG. 42D). The oospore undergoes a period of rest and then it germinates directly into a new *Vaucheria* filament (FIG. 42E). Reduction division has not yet been observed in *Vaucheria*.

ALGAE

17. *CAULERPA* (60 sp.)

Occurrence. All the species of *Caulerpa* (family *Caulerpaceae*) are marine, growing in shallow or deep water in tropical seas. They grow attached to the mud- or sand-bottom, rock, coral-reef or to the roots of mangrove plants.

Structure. The plant body is represented by a single-celled thallus with a slender creeping rhizome which bears colourless root-like rhizoids on its lower surface, and erect leafy shoots on the upper surface. The plant body is a branched coenocyte (unicellular without any partition wall but multinucleate), usually 10-30 cm. in height. The cell-wall made of callose and pectic substances is comparatively thick with many ingrowths in the form of transverse and longitudinal rods, called the *trabeculae*, which give rigidity to the thallus to some extent. The cytoplasm forms a lining layer within the cell-wall and contains numerous nuclei and disc-shaped chloroplasts, the latter, however, without any pyrenoid. There is a large vacuole at the centre running lengthwise through the entire body of the plant.

Reproduction. Vegetative reproduction is effected by fragmentation of the thallus into several parts or by detachment of the leafy shoot from the rhizome. Asexual or sexual reproduction is only imperfectly known. **Asexual reproduction** takes place by means of zoospores. These are generally produced from any portion of the leafy shoot or sometimes from the rhizome. Numerous papilla-like outgrowths called the *extrusion papillae* develop on the surface of the thallus and through them zoospores are liberated, sometimes in a large quantity like a mass of fine particles. The zoospores are pear-shaped and biciliate. Each is provided with a single chloroplast without pyrenoid, and an eye spot. **Sexual reproduction** has not been observed in all species. In *Caulerpa clavifera* and some other species the mode of reproduction is *anisogamous* and the plants are dioecious (heterothallic). Gametes formed in the erect leafy shoot are biciliate. Female gametes are somewhat longer and broader than the male gametes. As the zygote matures the thallus disintegrates. But the germination of the zygote has not been observed.

18. *CHARA* (90 sp.)

Occurrence. *Chara* (family *Characeae*; FIG. 43), a common stonewort, grows submerged in fresh water of ponds, lakes, streams and other pools of standing water, remaining attached to the bottom by branched filamentous rhizoids, and is gregarious in habit. Some species grow in brackish water. The body is often encrusted with lime which sometimes forms a thick deposit at the bottom. *Chara* and *Nitella* are two very common genera of *Characeae*. They are very widely distributed. About 20 species of *Chara* and 30 species of *Nitella* have been recorded from India.

Structure. The plant body consists of an erect branched stem, usually 20-30 cm. in height, differentiated into distinct nodes and internodes. Branching is of the following types: (*a*) at each node of the stem and the long branches there is a whorl of short branches or branchlets or leaves (of limited growth) consisting usually of 3-8 nodes and internodes (FIG. 43A); (*b*) long branches (of unlimited growth) bearing whorls of short branches; and (*c*) the branchlets or leaves bearing still shorter branches or appendages having one internode and somewhat pointed end (with complex sexual organs; FIG. 43B), called *bracts and bracteoles*. Besides, there may be some unicellular outgrowths arising in 1 or 2 whorls from the basal node of each branchlet or leaf, called

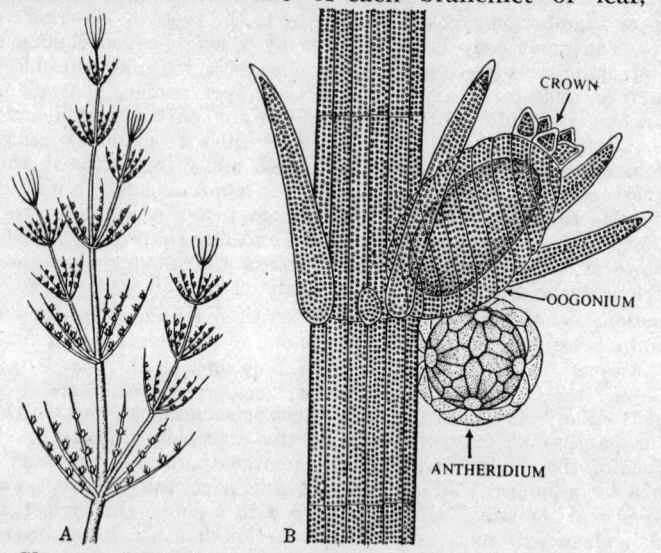

Chara. FIG. 43. *A*, a plant (upper portion) showing long and short branches; *B*, a branch showing an oogonium (or nucule) with sterile jacket and crown, and an antheridium (or globule) covered by shield cells.

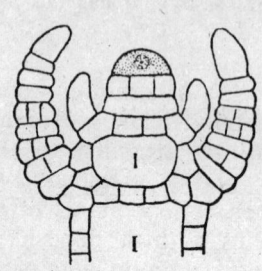

Chara. FIG. 44. Growing apical cell, nodes and internodes (*I*) in stem-apex.

stipulodes (or *stipuloids*). The node consists of two or a few central cells and a definite number of peripheral cells. The latter form a definite number of short branches or leaves (of limited growth). Their number is constant for a species. The central cells of the first node give rise to long branches (of unlimited growth). The internode consists of a single large cylindrical cell ensheathed by a layer of narrow vertical cells forming the cortex. The cortical cells arise partly from the upper node and partly from the lower node and meet somewhere in the middle of the internode. Growth in length of the stem takes place by a dome-shaped apical cell (FIG. 44). All the cells contain numerous small chloroplasts but no pyrenoids, and the reserve food occurs in the form of spindle-shaped starch. The rotatory movement of the protoplasm round a large central vacuole is very conspicuous in the internodal cells.

Reproduction. The plant normally reproduces sexually and sometimes vegetatively. Asexual reproduction is not known. Vegetative reproduction commonly takes place by (*a*) bulbils formed

ALGAE

on the stem at the node or on the rhizoid; they are small masses of cells, usually spherical or star-shaped; or, by (b) protonema-like outgrowths formed at the nodes.

Sexual Reproduction. All species of *Chara* normally reproduce sexually. The male and the female fructifications, respectively called **antheridium** (otherwise called *globule*) and **oogonium** (otherwise called *nucule*), are very complex structures with enveloping sheath. These are borne at the nodes of short branches together with the still shorter unicellular branches (FIG. 43B), and are visible to the naked eye. Most species of *Chara* are

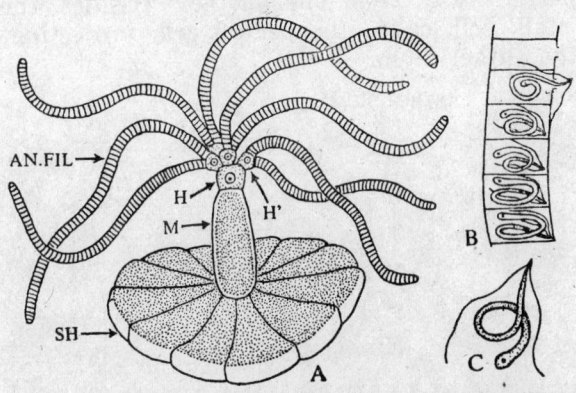

Chara. FIG. 45. *A*, a shield cell (*SH*) with manubrium (*M*), primary head cell or primary capitulum (*H*), secondary head cell or secondary capitulum (*H'*), and antheridial filaments (*AN, FIL*), each consisting of a row of minute antheridial cells; *B*, a few antheridial cells of a filament (magnified), each cell with a biciliate antherozoid; *C*, a free-swimming antherozoid after escape.

homothallic (monoecious), e.g. *C. zeylanica, C. fragilis,* etc., and in them the antheridium and the oogonium always occur in pairs, with the latter lying just above the former; only a few species are dioecious, e.g. *C. wallichii.*

The **antheridium** or globule (FIGS. 43B & 46A) is a very complex body. It is spherical in shape and turns red or orange when mature. Its wall (jacket) is made of eight curved plate-like cells, somewhat triangular in shape, called **shield cells**, with peculiar thickening on the surface and their walls folded with the joints fitting into one another. The shield cells expand rapidly, giving rise to a cavity within the antheridium. Besides, their outer walls form ingrowths, dividing the cells into a number of compartments. From the centre of each shield an elongated cylindrical cell called the **manubrium** or handle cell projects inwards. Evidently there are eight manubria. Each of them terminates inwardly in a roundish cell, called the *primary capitulum* or primary head cell. Each head cell cuts off six smaller cells known as the *secondary capitula* or secondary head cells. Each of them produces a pair of long slender **antheridial filaments** (FIGS. 45A & 46A), each consisting of 100-200 tiny cells called **antheridial cells**. The antheridial filaments form a tangled mass in the cavity of the antheridium. The protoplast of an antheridial cell forms a single, coiled, biciliate **antherozoid** (FIG. 45B-C). There may be as many as 20,000-50,000 antherozoids produced by a single antheridium. When the antheridium matures, the shields separate away from one another. Besides, there is an elongated stalk cell, called the **pedicel cell**, projecting into the cavity of the antheridium.

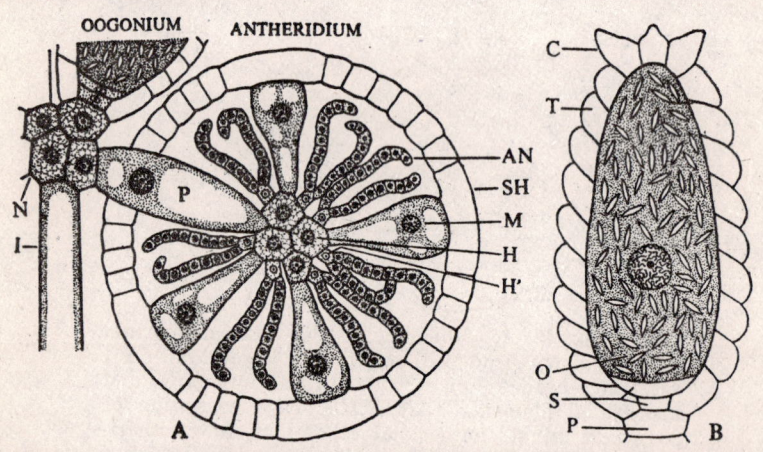

Chara. FIG. 46. *A*, antheridium (globule) in section. *AN*, antheridial filament; *SH*, shield cell; *M*, manubrium; *H*, primary head cell (or primary capitulum); *H'*, secondary head cell (or secondary capitulum); *P*, pedicel; *N*, node; *I*, internode; *B*, oogonium (nucule) in section. *C*, corona; *T*, tube cell (or turning cell); *O*, oogonium with an egg, and plenty of starch grains (spindle-shaped) and oil globules; *S*, stalk cells; *P*, pedicel cell.

The **oogonium** or nucule is an ovoid structure and much bigger than the antheridium. Closely encircling the oogonium there are five spiral bands of cells, called **jacket cells** (or tube cells or turning cells). These cells arise at the base and after completely surrounding the oogonium cut off at their tips five small cells which form the crown or corona. The oogonium contains a single large uninucleate **egg** and plenty of food material.

Fertilization. When the antheridium matures it gets ruptured at the junctions of shield cells which then fall apart, and the antheridial filaments become exposed. The antherozoids escape through slits or pores in the walls of the antheridial cells and swim about in water. The crown cells of the oogonium slightly separate at their base leaving five small slits through which the antherozoids enter. Finally one of them fuses with the egg-nucleus and thus fertilization is effected. The **zygote** (oospore) clothes itself with a thick wall, and the oogonium as a whole hardens. The zygote germinates after a period of rest. The zygote-nucleus undergoes reduction division forming four haploid nuclei, three of which degenerate. On germination the zygote gives rise to a rhizoid and a green filament (protonema) ; the shoot of the *Chara* plant arises from the protonema as a lateral bud.

Chara plant is haploid with 28 chromosomes in their vegetative cells, while the diploid phase is represented by the zygote (oospore).

Origin and Relationship of *Characeae*. *Characeae* or stoneworts hold a unique position among the green thallophytes. They are possibly related to Chlorophyceae but the degree of relationship cannot be ascertained. *Characeae* is an ancient group and may have arisen as an offshoot from some early Chlorophyceae. The bright green colour of the plants (chlorophyll being the only pigment present in them), storage of carbohydrates in the form of starch, gametophytic nature of the plant body, sporophytic stage represented only by the zygote, etc., as in all green algae, indicate some relationship of *Characeae* with Chlorophyceae. So the modern tendency is to place this group as an order of Chlorophyceae (see p. 442). But the erect vegetative body with nodes and internodes, and the complex reproductive organs with sterile sheath are distinctive features. The multicellular sex organs tend to raise this group to a higher level (bryophytic level), but the structure and development of these organs are very different in the two groups, and the sporophytic phase, so well developed in stages in bryophyta, is practically absent except the zygote. So some algologists still prefer to treat *Characeae* as a distinct and isolated group of green thallophytes, and they classify the green algae (or Chlorophyta) into two separate classes: Chlorophyceae and Charophyceae.

| CLASS IV | BACILLARIOPHYCEAE or diatoms | 5,300 sp. |

General Description. Bacillariophyceae, commonly called **diatoms** (FIG. 47), constitute a big isolated group of mostly one-celled algae which are of infinite varieties of forms and often of exquisite beauty; the single cells may occasionally form filaments and colonies. They are universally distributed in fresh water as well as in salt water and also in wet ground. In some parts of the ocean they occur in a vast assemblage as floating *plankton*. Sometimes they occur in huge numbers in a small space. Diatoms are mostly free-floating, while some remain attached by a gelatinous stalk. Many of the free-floating diatoms exhibit a jerky movement visible under a microscope. There are over 5,300 living species of diatoms. Fossil diatoms are seen to have formed huge deposits of siliceous or diatomaceous earth, often of considerable depth, in various parts of the world.

Structure. Diatoms (FIG. 47) may be boat-shaped, rod-shaped, disc-shaped, wedge-shaped, spindle-shaped, circular, oval, rectangular, etc. The wall of the diatom cell is made of two halves or valves, one (older) fitting closely over the other (younger)

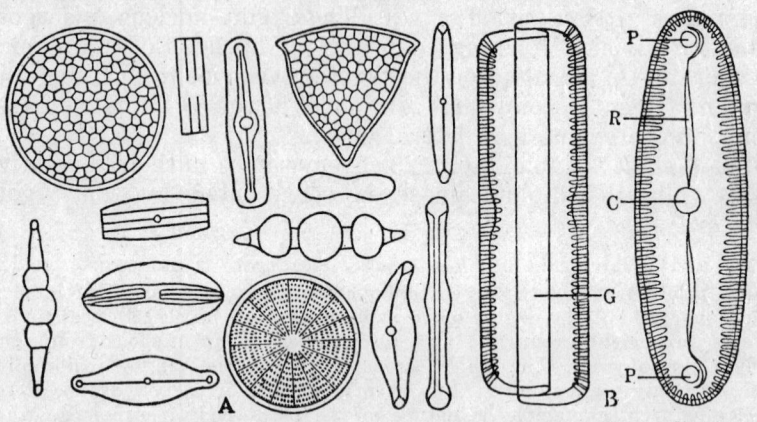

Diatoms. FIG. 47. *A*, various forms of diatoms; *B*, a diatom (*Pinnularia*): *left*, girdle view showing overlapping valves; *right*, valve view showing raphe; *G*, girdle; *P*, polar nodule; *R*, raphe; *C*, central nodule.

very much like a pill-box or soap-case, the outer valve being known as the *epitheca* and the inner one as the *hypotheca*. The valves are made of pectin impregnated with silica. The valves are ornamented with numerous fine lines which are really series of very fine *dots*. The ornamentation which is a special feature of diatom valves is radially symmetrical in the round or centric diatoms (called **Centrales**), so frequent in the ocean, and it is bilaterally symmetrical (in two series, one on each side of the valve) in the elongated or pinnate diatoms (called **Pennales**), so frequent in fresh water. In some genera there are *ingrowths* of the wall which, according to their position, are called central nodules or polar nodules. Extending from the central to the polar nodule there is in many diatoms a longitudinal line or slit consisting really of a series of extremely minute openings; this line is called *raphe*. The jerky movement seen only in those diatoms which possess a raphe may be due to the streaming movement of the cytoplasm along the raphe thrusting pseudopodia through the openings of it (raphe) and setting up a water current. Under the microscope commonly either the *valve side* is seen, i.e. the valve is uppermost, or the *girdle side* is seen, i.e. the connecting band is uppermost.

The protoplast consists of a thin peripheral layer of cytoplasm just within the cell-wall, one large or many small yellow to

ALGAE

golden-brown plastids of varied shape and size, a central nucleus suspended by distinct slender cytoplasmic threads or by a broad cytoplasmic band, and a conspicuous central vacuole. The colour is due to the presence of a golden-brown pigment called *diatomin* in addition to chlorophyll. Pyrenoids may or may not be present. If present, they are without the starchy envelope. The reserve foods are globules of fats and globules of an insoluble complex substance called *volutin*. No starch is formed in diatoms.

Reproduction. Diatoms may reproduce *vegetatively* by cell division generally at night, *asexually* by auxospore, and *sexually* by conjugation of gametes; but all these methods are *unique*. Vegetatively the protoplast grows and divides into two, resulting in the separation of the two valves. Each of the two half-cells forms a new valve against the old one fitting into it. Divisions and valve-formations continue one after the other. The result, as is evident, is that by this method one set of cells gradually become smaller and smaller. The reduction in size is not, however, true of all forms. It is seen that when a particular minimum size is reached, a reversion to original size takes place through the formation and activity of a special kind of cell called **auxospore**. The auxospore may be formed in a variety of ways. (1) In the first method the protoplast of a cell may escape from the valves after separation of the latter; it grows to its maximum size and then forms new valves. The protoplast acts as an auxospore. (2) In the second method the protoplast divides into two, each daughter protoplast (an auxospore) growing and forming new valves. (3) In the third method, which is a sexual one, the protoplasts of two cells escape and act as two gametes; they fuse to produce a zygote which behaves as an auxospore. (4) In the fourth method two contiguous diatom cells form two gametes each; they fuse in pairs forming two zygotes which act as auxospores. The auxospore grows and helps the diatom to return to its original size.

Uses. Diatomaceous earth has a variety of uses. It is used in metal polish, toothpaste, paints and plastics. It is extensively used as a filter in filtration of liquids, e.g. sugar refining. It is also used as a heat-insulator in boilers and furnaces which can then stand very high temperature.

CLASS V | PHAEOPHYCEAE or brown algae | 1,000 sp.

General Characters. Phaeophyceae or brown algae are a very interesting group of seaweeds with a variety of peculiar forms and sizes, and comprise about 1,000 species. They are widely distributed along sea coasts between tidal levels, predominantly along the coasts of temperate seas, and grow attached to rocks or some other substrata; in colder seas they seldom go beyond a depth of 20 m., while in warmer seas a few species may grow to the maximum depth of 90 m.; some also grow as epiphytes or endophytes on other algae; a few are free-floating. Their colour ranges from brown to olive-green due to the presence of a brown pigment *fucoxanthin* which is associated with the chloroplasts masking the chlorophyll. Pyrenoids are absent. The reserve food may be a kind of sugar (and not starch) or more commonly a complex carbohydrate called *laminarin*. Some like *Ectocarpus* are short filaments; some like *Fucus* and *Sargassum* are usually a few cm. to 1 m. in length; while some are massive sea-weeds, called giant kelps (*Laminaria*—2-9 m.; *Nereocystis*—45 m.; and *Macrocystis*—60-90 m.); they grow at or below the low tide level, extending far into the sea to a depth of about 90 m.; small kelps are only about a metre in length; unicellular brown algae are not known. The body of the kelp is differentiated into a basal root-like branched **holdfast**, a long or short stem called **stipe**, and one or more leaf-like blades called **fronds** which are provided with air-bladders for facility of floating. Fronds are of massive sizes in some species. Phaeophyceae (except *Fucus* and *Sargassum*) show a regular alternation of generations with different degrees of development of the sporophyte and the gametophyte, the two being similar (isomorphic, as in *Ectocarpus*) or dissimilar (heteromorphic, as in *Laminaria*) in external appearance, and their motile cells (zoospores and gametes or sperms) are laterally biciliate, the two cilia being of unequal lengths, in contrast with the apically ciliate cells of most algae.

Reproduction. Motile reproductive bodies (zoospores and gametes or sperms) are universally present throughout Phaeophyceae. Several species reproduce vegetatively by fragmentation of the thallus; most of them reproduce asexually by zoospores or aplanospores (except *Fucus* and *Sargassum*), and sexually by isogamy, anisogamy or oogamy. The zygote germinates without any period of rest.

Uses. Kelps are sources of iodine. Many of them are rich in potassium and other minerals, and are used as fertilizers along coastal areas. They contain sugar and many are rich in vitamins. Some are eaten as food by the coastal people of China and Japan. Most of the brown algae form an important food for fishes. Algin, another product obtained from kelps (particularly *Laminaria* and *Macrocystis*), is used in certain industries, as in the making of ice cream, in the rubber industry, baking industry, paint industry and in certain pharmaceutical preparations.

Origin and Relationship of Phaeophyceae. Phaeophyceae or brown algae are supposed to have originated from some brown ciliate unicellular ancestor. The universal presence of motile ciliate reproductive cells lends support to this view. The brown algae do not seem to be related to the green algae. Their vegetative body (thallus) is always multicellular as opposed to many unicellular green algae. Then again the presence of fucoxanthin and the formation of laminarin are exclusive to this group. But because of the missing links their origin and relationship with other algae cannot be traced. Like green algae, however, they show progressive stages of sexual reproduction—isogamy to anisogamy to oogamy. Phaeophyceae have remained confined to the sea and have not given rise to land forms. Within the group, however, there seem to be two divergent lines of evolution—in one, the two alternating generations (sporophyte and gametophyte) are similar in external appearance, while in the other, the two generations are dissimilar. The origin of Fucales and their relationship with other brown algae are further shrouded in mystery.

1. *ECTOCARPUS*

Occurrence. *Ectocarpus* (family *Ectocarpaceae*; FIG. 48A) is a simple brown alga. It is very widely distributed and found in plenty growing attached to other bigger algae or to rocks along the coasts of almost all seas.

Structure. *Ectocarpus* (FIG. 48A) occurs as tufts of branched filaments. Each filament is monosiphonous, i.e. it consists of a single row of cells. The plant body is differentiated into two portions: a prostrate portion irregularly branched, and an erect portion with much-branched filaments. The cells are uninucleate with a few band-shaped or many small disc-shaped brown plastids (chromatophores). Although all individual plants of a species look alike vegetatively investigations have shown that there are actually two different kinds—one kind (asexual plant) is the sporophyte with diploid ($2n$ chromosomes) and the other kind (sexual plant) is the gametophyte with haploid (n chromosomes).

Reproduction. The sporophytic (diploid) plant reproduces asexually by two kinds of zoospores: one kind formed in a unilocular zoosporangium is haploid (FIG. 48B), and the other kind borne in a plurilocular zoosporangium is diploid (FIG. 48C). The gametophytic plant on the other hand, though very much like a sporophyte and hardly distinguishable from it, reproduces by gametes borne in plurilocular gametangia. The gametangia with gametes and the plurilocular zoosporangium with zoospores being alike, a good deal of confusion arose as to whether the structure concerned is really a gametangium or a sporangium. Investigations on cytology and behaviour of such reproductive bodies clarified the position. It was found that such zoospores are diploid and each grows by itself; whereas the gametes are haploid and they fuse in pairs. Therefore, the zoospore-bearing organ is the sporangium and the gamete-bearing organ is the gametangium, and the corresponding plants, though alike, are sporophyte and gametophyte respectively.

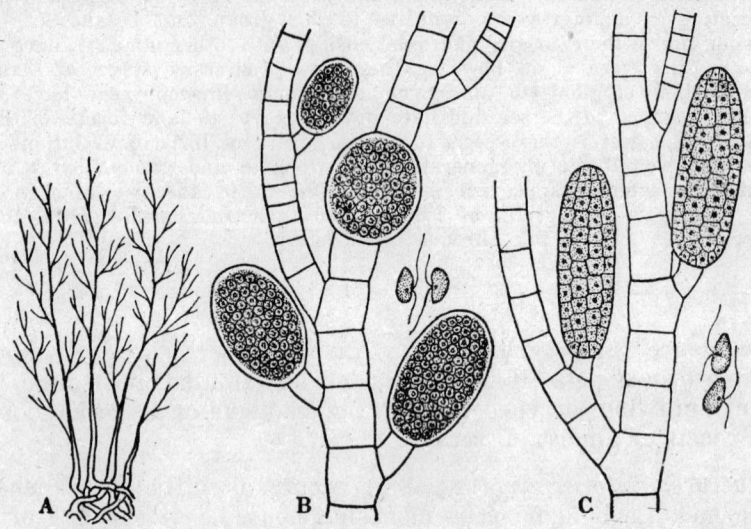

Ectocarpus. FIG. 48. *A*, habit of the plant; *B*, asexual plant (n) with unilocular sporangia; two zoospores (n) shown separately; *C*, asexual plant ($2n$) with plurilocular sporangia; two zoospores shown separately.

Asexual Reproduction. Cytologically there are two kinds of zoospores (borne by the sporophytic or asexual plant)—some haploid (with n chromosomes) and some diploid (with $2n$ chromosomes) borne respectively in a unilocular sporangium and a plurilocular sporangium. Both kinds of sporangia may be borne by the same plant or by two different plants. The unilocular sporangium may be a cell of the filament flattened to form it or it may develop at the end of a short lateral branch. In any case it bears 32 or 64 zoospores which, as said above, are haploid. These finally escape from the sporangium through a terminal opening and give rise to the gametophytes. The plurilocular sporangium on the other hand, borne likewise laterally, consists of several hundreds of small cells arranged in several (20-40) transverse tiers. Each cell of the sporangium develops a single zoospore. In both the cases the zoospores are laterally biciliate, with the cilia of different lengths. The zoospores of the plurilocular sporangium are diploid and on germination each zoospore gives rise to the diploid sporophyte again bearing plurilocular sporangia. *Ectocarpus* may indefinitely multiply by this method. Sometimes, depending on seasonal changes, the diploid zoospores borne in plurilocular sporangium develop plants which bear unilocular sporangia, with haploid zoospores. In this case reduction division takes place at an early stage of sporangium formation. Such zoospores grow into gametophytic plants

Sexual Reproduction. The gametophytic or sexual plant (FIG. 49) is like the sporophyte in appearance, and produces laterally long multicellular gametangia resembling plurilocular sporangia. Each cell of the gametangium produces a single laterally biciliate gamete. All gametes are alike in size (isogametes) but there is some difference in the behaviour (activity) of the pairing gametes. Gametes coming from different plants fuse in pairs, *Ectocarpus* being heterothallic. The gametes escape from the gametangium and swim in water. A particular gamete (female) soon comes to rest, and one of the many active gametes (male), which cluster round it, fuses with it. The zygote that is formed is diploid and it grows without any period of rest into a sporophyte bearing unilocular or plurilocular sporangia or both, as described above. Sometimes some of the gametes, particularly the female ones, develop by parthenogenesis, i.e. without fertilization, into new haploid plants (gametophytes). The gametes thus behave as haploid zoospores.

Ectocarpus. FIG. 49. Sexual plant (n) with plurilocular gametangia; two gametes (n), conjugation, and zygote ($2n$) shown separately.

Alternation of Generations. *Ectocarpus* shows *isomorphic* alternation of generations. The two generations are more or less equally developed. The spore- (zoospore-) bearing plant is the sporophyte (diploid), while the gamete-bearing plant is the gametophyte (haploid). The gametophyte on sexual reproduction through gametes produces a diploid zygote which gives rise to the diploid plant, i.e. the sporophyte. The sporophyte on the other hand reproduces by haploid zoospores borne in unilocular sporangium and gives rise to the gametophyte. Thus the two generations (diploid and haploid) alternate with each other. But there are some discrepancies in the life-cycle of *Ectocarpus*. Thus it is seen that the diploid zoospores borne in the plurilocular sporangium give rise to the sporophytes again which bear only plurilocular sporangia with diploid zoospores. In this case the gametophytic generation is omitted. Then again gametes may act as spores (zoospores) and develop parthenogenetically into gametophytes again. In this case the sporophytic generation is omitted. Stages in the life-history are shown below:

1. Asexual plant $(2n) \to (a)$ zoospores $(2n)$ in plurilocular sporangium $\to$ asexual plant $(2n)$. (b) zoospores (n) in unilocular sporangium $\to$ sexual plant (n).
2. Sexual plant $(n) \to$ isogametes (n) in plurilocular gametangium by fusion in pairs $\to$ zygote $(2n) \to$ asexual plant $(2n)$.

2. *LAMINARIA* (30 sp.)

Occurrence. *Laminaria* (family *Laminariaceae*; FIG. 50A) is widely distributed in cold waters along the shores of the North Pacific Ocean and the North Atlantic Ocean. Several species are found below the low-tide level, often extending far out into the sea to a depth of about 30 metres.

Structure. The plant body consists of a large blade (frond), a stipe and a holdfast (FIG. 50A). The blade represents the main part of the plant, varying in length usually from 2m. to 4m. In some species the thallus may be as big as 9m.; while in a few others it is only 0 5m. The blade may be simple with smooth or wavy margins, or it may be palmately divided (*L. digitata*; FIG. 50B). The stipe is a slender long flexuous stalk supporting the blade. The holdfast is a disc-like structure at the base, with root-like branches, sometimes provided with a dense mass of rhizoids; by the holdfast the plant attaches itself to some submerged rock. A meristematic zone lies at the base of the blade, which annually produces a new blade, the old one dying off (FIG. 50C).

Thallus in Section (FIG. 51). The thallus (stipe or blade) in a vertical section shows a differentiation into three distinct zones: (*a*) externally the **epidermis**, 1 or 2 layers thick, with many chromatophores, is assimilatory in nature; (*b*) a deep **cortex** made of vertically elongated cells arranged in radial rows but slightly separated by an anastomosing system of mucilage ducts except at certain points; and (*c*) centrally an axis or **medulla** consisting of (*i*) multicellular vertical filaments ('*hyphae*'), somewhat parallel to each other, (*ii*) some connecting filaments, also multicellular, running horizontally or diagonally through the medulla, and (*iii*) certain cells of the medulla modified into long filaments ('*trumpet hyphae*') with perforated transverse walls, as in sieve-plates; the perforations may be blocked by callus pads at a later stage. The trumpet hyphae may have spiral

ALGAE

bands of thickening like a spiral vessel, but made of cellulose. The medulla as a whole is conducting in nature.

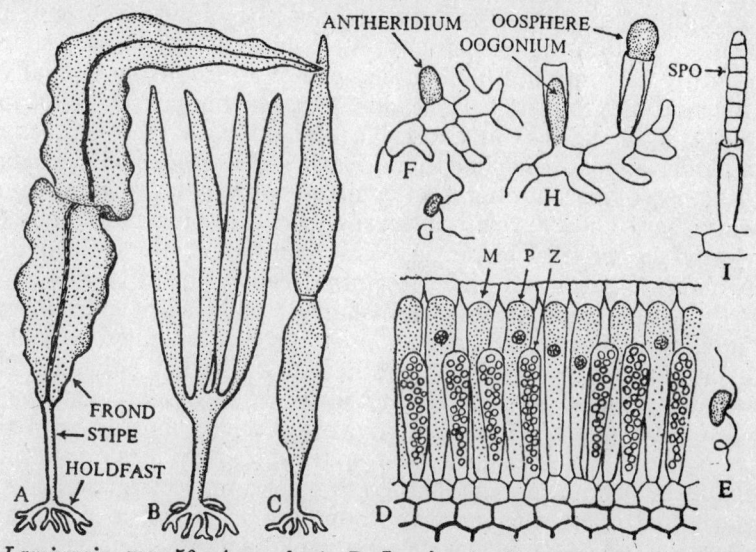

Laminaria. FIG. 50. *A*, a plant; *B, Laminaria digitata* with a palmate blade; *C*, a plant showing old blade (*top*), meristematic zone (*middle*), and new blade (*bottom*); *D*, a portion of a sorus in section (*M*, mucilage cap; *P*, paraphysis; *Z*, zoosporangium); *E*, a biciliate zoospore; *F*, a male gametophyte; *G*, a biciliate antherozoid; *H*, a female gametophyte; and *I*, a young sporophyte (*SPO*) on oogonium.

Reproduction. Most species of *Laminaria* are perennial. Although there are specific differences in external forms, the general life-

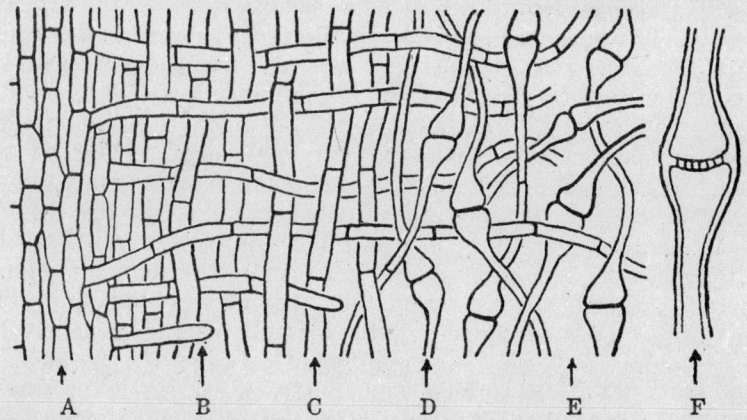

Laminaria. FIG. 51. Stipe in longi-section through the medulla. *A*, part of the cortex (note the mucilage ducts); *B*, transverse connecting filament; *C*, vertical filament ('hypha'); *D*, trumpet hypha; *E*, mucilage ducts; and *F*, a trumpet hypha showing perforations like a sieve-plate.

history including the modes of reproduction is remarkably the same in all. *Laminaria* shows distinct alternation of generations. The main plant which is large is the sporophyte reproducing asexually; while the gametophytes, male and female, are represented by very minute filamentous bodies reproducing sexually.

Asexual Reproduction. For this purpose numerous unilocular zoosporangia develop in extensive sori on both the surfaces of the blade, almost covering them. Each **sorus** (FIG. 50D) consists of a mass of zoosporangia intermixed with a large number of paraphyses. They develop from superficial (epidermal) cells on the surface of the blade. At first an epidermal cell divides into two—a basal cell and a terminal cell. The latter elongates and becomes more or less club-shaped; this is the **paraphysis**. It is capped by a mucilaginous layer. By the mucilaginous caps all the paraphyses are held together in a distinct palisade. The **zoosporangia** also develop in the same way, and are also more or less club-shaped. They are, however, much shorter than the paraphyses. The nucleus of the sporangium divides repeatedly into 32 or 64 nuclei and finally they become converted into a group of 32 or 64 zoospores, more or less of the same size. Each **zoospore** (FIG. 50E) is pear-shaped and biciliate. The two long cilia are placed laterally, and there may be an eye spot also. The zoospores escape in a mass of mucilage through the apex of the sporangium. The mucilage dissolves and the zoospores are set free. They are motile and briskly swim in water. Soon, however, they form a wall round each, settle down and germinate.

Sexual Reproduction. The zoospores germinate and produce two kinds of minute filamentous bodies—the gametophytes. At first a zoospore grows into a *germ tube* which soon enlarges at the tip. The zoospore nucleus divides into two, and the protoplast together with a nucleus migrates into the enlarged tip. The other nucleus degenerates. The enlarged cell then divides, giving rise to a short filament with a few short branches. The whole structure is the **gametophyte**, male or female. The zoospores of the sporangium produce male and female gametophytes (FIG. 50F & H) in equal numbers. All the gametophytes are very minute and microscopic in size. It is, however, seen that the male gametophyte has smaller cells than the female gametophyte. Antheridia and oogonia are borne respectively by the male and the female gametophytes in their lateral branches. The **antheridium** (FIG. 50F) is a small spherical cell with one antherozoid in it. The **antherozoid** (FIG. 50G) is extremely minute in size and has two long but unequal cilia placed laterally. The **oogonium** (FIG. 50H) is an enlarged structure with an oosphere in it. The oogonium is covered by a mucilaginous layer forming a sort of cup at its apex. The oosphere partly frees itself from the oogonium and rests on the mucilage cup. **Fertilization** is effected in the usual way. The antherozoids swim to the oogonium, and one of them fuses with the oosphere which then clothes itself with a wall and becomes known as the **oospore**. The latter soon divides and produces a short filament *in situ* (FIG. 50*I*). With the formation of rhizoids at the base the filament gets detached from the oogonium, and soon attaches itself to a submerged rock. Gradually it grows into a full-fledged *Laminaria* plant with a holdfast, stipe and blade.

Alternation of Generations. *Laminaria* shows a distinct heteromorphic alternation of generations—a large sporophyte and minute gametophytes. The main plant is the sporophyte which reproduces asexually by zoospores, giving rise to the gametophytes, male and female, in equal numbers. They are extremely reduced in size and bear antheridia and oogonia respectively, and reproduce sexually by the antherozoid and the oosphere, giving rise to the sporophyte again. The stages in the life-cycle are given below:

Laminaria plant $(2n)$ → sori $(2n)$ → zoosporangia $(2n)$ → zoospores (n)

↑ antherozoid (n) ← antheridium (n) ← male gametophyte (n) ←   ↓

oospore $(2n)$ ← × oosphere (n) ← oogonium (n) ← female gametophyte (n)

3. *FUCUS* (16 sp.)

Occurrence. *Fucus* (family *Fucaceae*) is a widely distributed marine alga, growing along the coast between the high tide level and the low tide level, being more abundant in temperate seas.

Structure. The *Fucus* plant (FIG. 52) consists of a dichotomously branched thallus, generally 0.3 to 1m. in length; some species growing below the low tide level may be as big as 4.5m. The plant body is differentiated into three distinct parts: (*a*) a basal disc-shaped **holdfast** by which it attaches itself to a rock, (*b*) a long or short **stipe**, i.e. the stem-like portion, and (*c*) leathery **fronds** or laminae, i.e. the flattened portions,

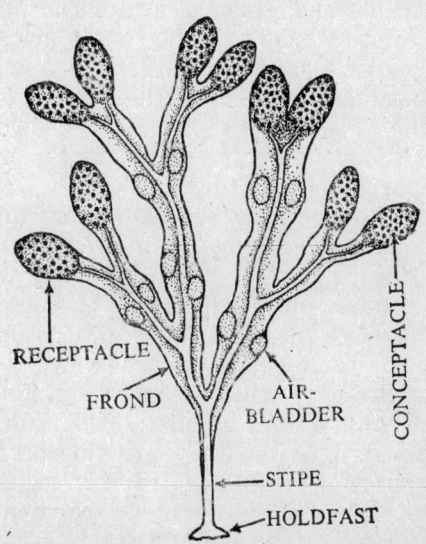

Fucus. FIG. 52. A *Fucus* plant. (Dark dots on the receptacle are conceptacles).

provided with a distinct mid-rib. The inflated tip of the frond is called the **receptacle**. It has some small openings, each leading into a cavity known as the **conceptacle**. The branches have some swellings here and there along the mid-rib, often close to dichotomy; these swellings are filled with air and are known as *air-bladders*. They give buoyancy to the plant helping it to float. The body of *Fucus* is composed of parenchymatous cells. Each cell contains a nucleus and a number of plastids which contain *fucoxanthin* in addition to chlorophyll, the former masking the chlorophyll and giving the plant a brown appearance. As a result of photosynthesis a carbohydrate, called *laminarin*, accumulates in the cells. No starch is formed.

The **thallus** in section shows a loose mass of elongated cells, collectively known as the **medulla**; it occupies the main bulk of the thallus. The interspaces in the medulla are filled with a gelatinous substance. Surrounding the medulla and lying towards the surfaces there is a compact mass of cells known as the **cortex**. Growth of the thallus takes place by a single pyramidal apical cell.

Reproduction. Asexual reproduction is absent in *Fucus*. Vegetative reproduction is by the fragmentation of the thallus; the detached parts then float on water and vegetate. *Fucus*, however, commonly reproduces sexually by heterogametes—antherozoid and egg—borne respectively by antheridium and oogonium.

Sexual Reproduction. Antheridia and oogonia are borne in special globose or flask-shaped cavities known as the **conceptacles** which lie embedded in the swollen tip or **receptacle** of the thallus. Externally the conceptacles appear as small scars. Both antheridia and oogonia may occur in the same conceptacle or in two separate conceptacles borne by the same plant (monoecious), or they may be borne by two separate plants (dioecious).

A conceptacle (FIG. 53A) of monoecious species shows in section the following: (*a*) numerous unbranched multicellular sterile hairs called **paraphyses**, (*b*) an apical opening known as the **ostiole**, (*c*) numerous oval sac-like **antheridia** borne on small much-branched *antheridial filaments* (FIG. 53B), and (*d*) a number of isolated oval **oogonia**, each borne on a very short stalk-cell. The antheridium produces numerous more or less pear-shaped **antherozoids**, generally 64 in number, each with two lateral cilia of unequal lengths. After they are liberated from the antheridium the antherozoids (FIG. 53C) freely swim about in water. Each oogonium (FIG. 53A) is a single large oval cell. Its nucleus divides three times to produce *eight* nuclei. With the cleavage of the cytoplasm each forms a large round **egg**. The oogonial wall is made of three layers. The outer layer bursts

ALGAE

and the oogonium, still surrounded by two inner layers, is pushed out of the conceptacle as a result of swelling of the gelatinous material within the conceptacle. Later the middle layer ruptures at the apex, and as the inner layer imbibes more water it swells

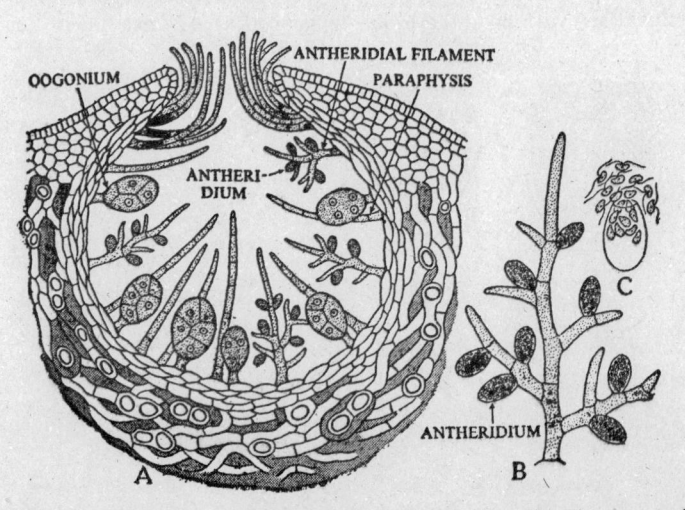

Fucus. FIG. 53. *A*, a conceptacle in longi-section showing oogonia in various stages of development, antheridial filaments with antheridia, and paraphyses; *B*, an antheridial filament (magnified) showing antheridia with antherozoids; *C*, an antheridium discharging biciliate antherozoids.

and the eggs become rounded off. The inner layer soon dissolves and the eggs are set free (FIG. 54*A-C*). Both antheridia and oogonia are liberated in packets through the ostiole in a quantity of mucilage.

Fertilization. Eggs are large and passive, while the antherozoids are minute, ciliate and active. A vast number of antherozoids swim to the oogonium being attracted by some chemical substance, and get attached to it by one cilium (FIG. 54*D*). Their vibration around the egg causes the egg to rotate. One or more antherozoids may enter the egg but finally only one fuses with the egg-nucleus. The fertilized egg or zygote clothes itself with a wall, and almost immediately (within a few hours) divides and grows into a new thallus.

Alternation of Generations. There is no alternation of generations in *Fucus*. The plant itself is diploid, i.e. sporophytic. Reduction division takes place in the formation of gametes (eggs and antherozoids); so the gametes are haploid, i.e. gametophytic. The diploid condition is restored in the zygote upon fertilization of the gametes. This diploid condition continues in the *Fucus* plant. Stages in life-cycle are shown below:

Fucus plant $(2n) \to$ conceptacle $(2n) \to$ $\begin{cases} \text{antheridium } (2n) \to \text{antherozoid } (n) \\ \text{oogonium } (2n) \to \text{egg } (n) \text{ by } fusion \end{cases}$
(fertilization) with an antherozoid$\to$zygote $(2n) \to$*Fucus* plant $(2n)$.

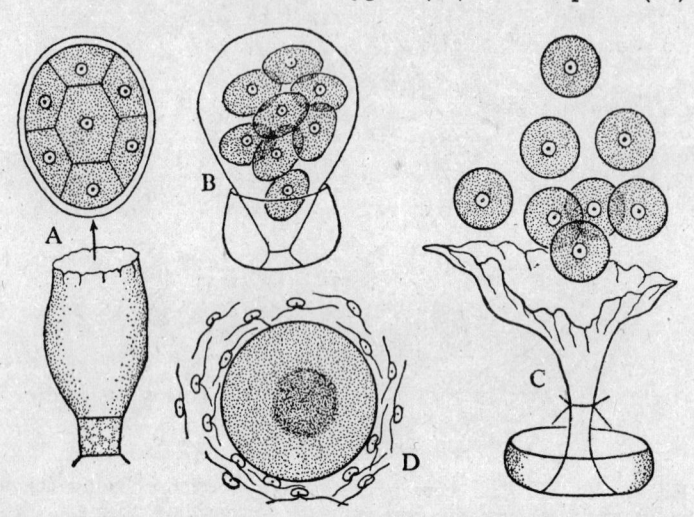

Fucus. FIG. 54. *A*, escape of oogonium with 8 (nearly mature) eggs from the outer layer; *B*, oogonium with 8 mature eggs escaping from the middle layer which has burst and slipped down: *C*, breaking of the inner layer setting free the 8 fully formed eggs; *D*, an egg surrounded by numerous biciliate antherozoids just prior to fertilization. *Redrawn partly after Fig. 171 in Plant Morphology by A. W. Haupt by permission of McGraw-Hill Book Company. Copyright 1953.*

Note that the *Fucus* plant being a sporophyte with diploid chromosomes the so-called antheridium is interpreted as a microsporangium producing small motile microspores which function as antherozoids, and the so-called oogonium as a megasporangium producing large immobile megaspores which function as eggs. *Fucus* is thus heterosporous.

4. *SARGASSUM* (250 *sp.*)

Occurrence. *Sargassum* (FIG. 55) is a common genus of *Sargassaceae* growing in abundance in tropical and sub-tropical seas along rocky coasts. The plant grows to a length of about 6m. or so and may remain attached to some rocks by a *holdfast* or may be commonly free-floating, often drifted to the coast or far out into the sea. There is a vast accumulation of this plant in the Sargasso Sea. Columbus was the first to sail through this sea in 1492 despite the fear of his sailors.

Structure. The plant body consists of an axis (stipe) which is much branched. Branching is always monopodial. The stipe and the lateral branches bear a large number of green flattened or cylindrical leafy fronds with wavy serrate margins and a mid-rib. Berry-like air-bladders develop in small clusters in the axil of the frond or close to its base; these give buoyancy to the plant to float. Species of *Sargassum* show ranges of forms. Growth takes place by a three-sided apical cell.

Reproduction. *Sargassum* is closely related to *Fucus* and follows the same life-cycle. Vegetative reproduction of this plant is most common and takes place by fragmentation of the thallus. There is no asexual mode of reproduction in *Sargassum*.

Sexual Reproduction. In the axil of the frond or a little beyond, often much-branched **receptacles** are borne. Sexual reproduction takes place by well-developed sex organs (antheridia and oogonia) which occur in hollow flask-shaped cavities called **conceptacles** borne by the receptacle at the ends of its branches. *Sargassum* species may be monoecious bearing the sex organs in two distinct conceptacles on the same receptacle or on two separate receptacles, or they may be dioecious with the sex organs being borne by two separate plants. The monoecious condition is, however, more usual than the other. Each conceptacle is provided with a minute opening at the apex, called the **ostiole**. Sterile conceptacles, called *cryptoblasts*, often occur scattered over the thallus. Several **oogonia** lie embedded in the wall of the female conceptacle. The nucleus of the oogonium divides thrice to form eight nuclei but only one of them is functional acting as the egg-nucleus; others degenerate. Nuclear divisions usually take place while the oogonium is still within the conceptacle, or sometimes only after the oogonium moves out of it. The oogonium wall is differentiated into three layers, as in *Fucus*. All the oogonia are discharged from the wall of the conceptacle but each still remains attached to it by a long slender mucilaginous thread. Numerous **antheridia** develop on branched antheridial filaments in the male conceptacle in much the same way as in *Fucus*. Each antheridium bears 64 antherozoids which are biciliate and more or less pear-shaped; the two cilia are of unequal lengths.

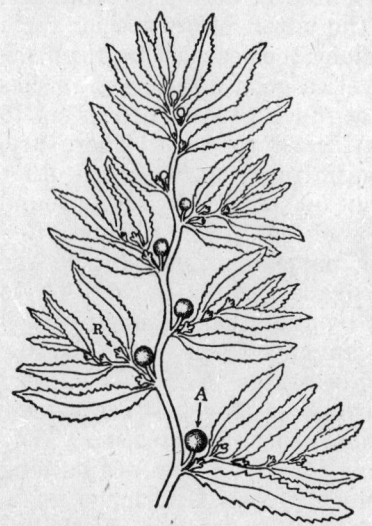

Sargassum. FIG. 55. A plant (portion) showing numerous green fronds, air-bladders (*A*) and receptacles (*R*).

Fertilization. This takes place outside the plant body, as in *Fucus*, and the process is much the same in both the cases. Several antherozoids swim to each oogonium but only one of them fuses with the egg-nucleus. This results in the formation of a zygote which immediately germinates without any period of rest.

Alternation of Generations. *Sargassum*, like *Fucus*, shows no alternation of generations. The plant itself is diploid and the gametes only are haploid. The zygote ($2n$) grows into the plant without any reduction division. Life-cycle is as follows:

Sargassum (2n) →receptacle (2n) → ♂ conceptacle (2n) →antheridium (2n)
↑   antherozoids (n) ←by meiosis
| → ♀ conceptacle (2n) →oogonium (2n)
zygote (2n) ← (by fusion with an antherozoid) egg (n) ←by meiosis

CLASS VI | RHODOPHYCEAE or red algae | about 3,000 sp.

General Characters. Rhodophyceae or red algae form a big group of highly specialized marine algae comprising about 3,000 species. They are very widely distributed in both temperate and tropical seas, more particularly in the latter. Many species form a characteristic belt of vegetation along the seacoast between the high tide level and the low tide level; a good number of species grow at depths of 60-90m; a few at much greater depths up to about 180m. They mostly grow attached to rocks. There are, however, some epiphytic and parasitic forms growing on other algae. Although mostly marine, about 50 species have been found to occur in fresh water, mostly restricted to hill streams. Red algae are characterized by red or purplish colour due to the presence of a red pigment called *phycoerythrin* in addition to chlorophyll which is often masked by the other pigment. Many of the red algae contain a small amount of *phycocyanin*, the blue pigment of Cyanophyceae. Freshwater species are green in colour. Red algae show a variety of forms—filamentous, ribbon-shaped, distinctly leaf-like marked with veins, etc. They are mostly a few to about 25 cm. in length; a few are as long as 1-1.3 m.; deep-water ones are much longer. Gelatinous material is abundant in red algae, either occurring within the thallus or forming a sheath in the filamentous forms. Some of the red algae are heavily incrusted with lime. The cells may be uninucleate or multinucleate with one or more plastids which may be with or without pyrenoid. As a result of photosynthesis a sugar or more commonly a special kind of starch called *floridean starch* accumulates in the cells. There is total absence of motile ciliate cells in red algae; zoospores are altogether absent, and the gametes are never ciliate. The peculiar mode of sexual reproduction also makes the life-history more complicated. Members of Rhodophyceae are either haploid, or they show a regular alternation of similar haploid and diploid stages, as in *Polysiphonia*.

Uses. Red algae are very important from economic point of view. Some of them (e.g. *Porphyra* and *Chondrus*) are used as food by the Chinese and the Japanese. In Europe their use for making soups and puddings is extensive. Many red algae are an important source of food for fish, and some of them are fed to cattle. An important commercial product known as **agar-agar** (or simply called agar) is a jelly-like material obtained from some red algae (e.g. *Gelidium* and *Gracilaria*). It has a variety of uses. It is universally used in the laboratories as a medium of culture for fungi and bacteria. It is also used in medicines, commonly as a laxative. It is widely used as a thickening material for soups, puddings and jellies. In textile industry it is extensively used as a sizing material, an emulsifying agent and also as a dyeing and printing material. The agar industry was confined to Japan for many years but has recently extended to U.S.A., Canada, Australia, New Zealand and South Africa. In recent years gelatine obtained from some red algae has been put to a variety of uses: as a base for shoe-polish, shaving cream, cosmetics, etc. It is further known that many red algae materially contribute to the formation of coral reefs.

Origin and Relationship of Rhodophyceae. Rhodophyceae are the most highly specialized group of all algae. The origin of red algae and their relationship with other algae have so far remained a mystery. This is indeed a unique group among the algae because of the characteristic type of female organ—the carpogonium, mode of zygotic germination, and absence of ciliate reproductive bodies. This group may have originated from some non-ciliate unicellular ancestor. The primitive red algae have, however, some resemblance with blue-green algae; both have blue pigment and red pigment, primitive type of nucleus, and both are wanting in ciliate reproductive bodies. They differ, however, in many other important characteristics, particularly in the mode of reproduction.

1. *POLYSIPHONIA* (150 *sp.*)

Polysiphonia (FIG. 56*A*) is a common marine red alga. Its thallus is cylindrical and much branched (filamentous), attaining a

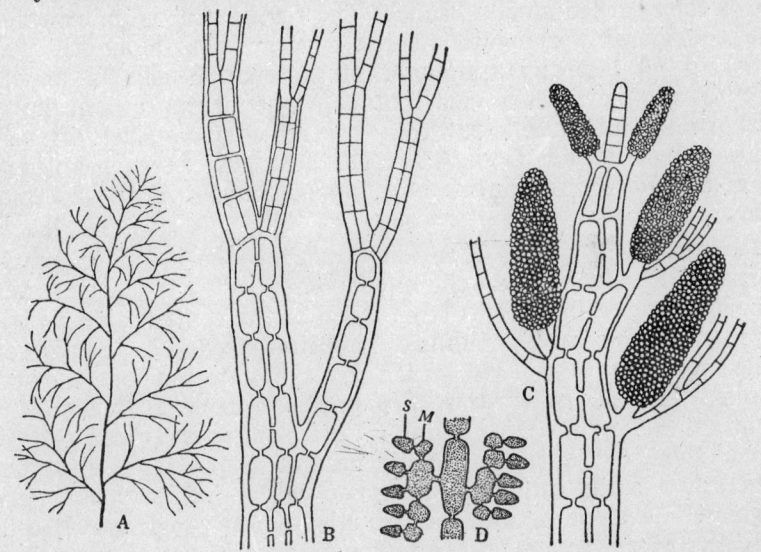

Polysiphonia. FIG. 56. *A*, a plant (portion); *B*, branching, and tubular or siphonic cells with pores maintaining protoplasmic connexions through them; *C*, a male plant (portion) with antheridial branches bearing clusters of antheridia; *D*, antheridial branching showing antheridial (spermatial) mother cells (*M*) and spermatia (*S*).

length of a few to about 25 cm. Many species commonly grow attached to the thallus of *Fucus* and other marine algae. The body is polysiphonous (FIG. 56*B*) being composed of a central row of elongated cells—the central or axial siphons, and encircling it another layer of more or less similar cells—the pericentral siphons, and hence the name *Polysiphonia*. The number of pericentral siphons varies from 4-24 according to the species. There are distinct pores and protoplasmic connexions through them between the adjoining cells. The cells are uninucleate and with many red discoidal plastids. Growth takes place by a single apical cell.

There are three distinct forms of *Polysiphonia* representing three stages in its life-history. One form reproduces by four spores formed in a tetrad, called the **tetraspores**; the second form is male and bears only **antheridia** (or spermatangia); and the third form is female and bears **carpogonia** and later carposporangia with carpospores. *Polysiphonia* is thus *heterothallic*. No motile ciliate cells develop at any stage. The three forms are alike in general appearance.

Male Plant (FIG. 56C). The antheridia borne near the thallus apex by the male plant occur in dense clusters on one arm (fertile) of a dichotomous branching, while the other arm grows into a long forked sterile branch. The fertile arm is unbranched and considerably elongated, being several cells in length. The cells cut off laterally a number of pericentral cells in an encircling manner. Then each cuts off one or more unicellular antheridial mother cells towards the free surface. Each of them bears 2-4 **antheridia** (FIG. 56D). The protoplast of each antheridium functions as a single male gamete known as the **spermatium**.

Female Plant (FIG. 57A). The procarp (the complex female organ of Rhodophyceae—the carpogonium with trichogyne and the associated cells) arises from a cell of the central axis a little behind the apex of the thallus. This first produces a large cell called the *supporting cell* which gives rise to a short, curved, 4-celled branch, called the *carpogonial branch*, of which the terminal cell is the **carpogonium** with a slender elongated projection at the apex, called the **trichogyne**. The supporting cell further cuts off a basal cell and an upper cell (auxiliary cell). An envelope develops as an outgrowth of the pericentral cells adjacent to the supporting cell. The nucleus of the carpogonium is the egg-nucleus.

Fertilization. The spermatium on liberation from the antheridium is carried by water current to the trichogyne. The tip of it dissolves and a spermatium passes through it into the carpogonium and finally unites with the corpogonium-nucleus (egg-nucleus). The trichogyne is cut off by a wall at its base, and the fusion-nucleus (diploid) passes into the auxiliary cell. The carpogonium and the branch cells break down, while the auxiliary cell, supporting cell and the other cells fuse into a large irregular cell—the placental cell. From this grow a number of large, elon-

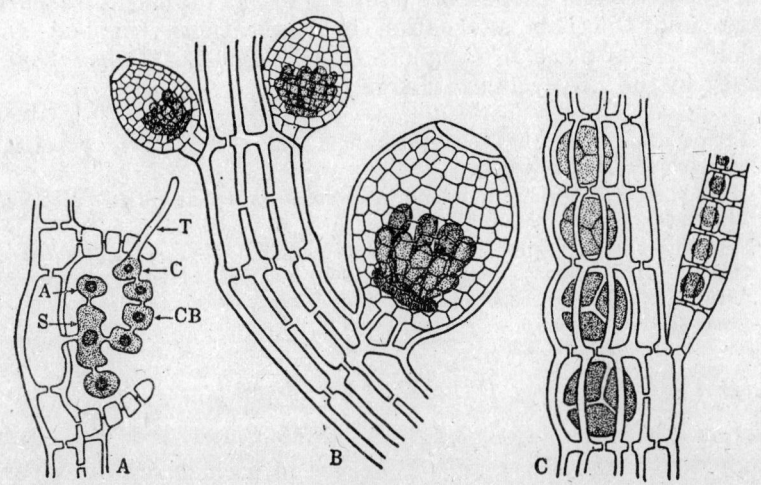

Polysiphonia. FIG. 57. *A*, a female plant (portion) with procarp (in section) showing carpogonial branch (*CB*), carpogonium (*C*) with trichogyne (*T*), supporting cell (*S*), auxiliary cell (*A*) and a sterile basal cell; note the envelope (pericarp) in the process of formation; B, cystocarps with a wall (pericarp) and carposporangia (carpospores) inside; *C*, tetrasporic plant (portion) bearing tetraspores (old and young).

gated, single-celled **carposporangia**. The contents of each constitute a **carpospore**, and the nucleus of it is diploid. In the meantime the envelope grows and surrounds the carposporangia as a large urn-shaped **pericarp** leaving a small opening or **ostiole** at the apex. The whole structure consisting of the carposporangia and the pericarp is known as the **cystocarp** (FIG. 57*B*).

Tetrasporic Plant (FIG. 57*C*). Carpospores liberated from the carposporangia germinate and develop into tetrasporic plants which are asexual. A fertile central cell (siphon) of a tier produces a short-stalked sporangium which lies between the central and the pericentral siphons. Its nucleus undergoes reduction division and four **tetraspores** are formed in a tetrad. In this way a series of tetraspores may be formed in the fertile filament. On liberation from the sporangium the tetraspores germinate, two of them developing into male plants and two into female plants.

Alternation of Generations. The tetrasporic plant is diploid (or sporophytic). Reduction division takes place in the formation of the tetraspores which evidently are haploid. They grow into haploid (or gametophytic) male and female plants. With the fusion of the gametes the diploid condition is restored in the zygote-nucleus. The diploid number continues into the carpospore. The diploid carpospore grows into the diploid tetrasporic plant, and thus an alternation of generations (haploid and diploid) takes place to complete the life-cycle of *Polysiphonia*. Stages in the life-cycle are shown below:

1. Tetrasporic plant $(2n)$ →tetrasporangia $(2n)$ →tetraspores (n) →male and female plants (n).
2. Male plant (n) →antheridia (n) in clusters→spermatia (n) →fuse with egg-nuclei (n) of female plant.
3. Female plant (n) →procarp (n) →carpogonium (n) →egg-nucleus (n) on fusion (fertilization) with spermatium (n) of male plant→zygote-nucleus $(2n)$ →carposporangium $(2n)$ →carpospore $(2n)$ →tetrasporic plant $(2n)$.

2. *BATRACHOSPERMUM* (40 sp.)

Batrachospermum (FIG. 58A-B) is a freshwater red alga commonly found in hill streams. It occurs in a clustered mass very close to a spring, from where the filaments are carried a long way down the stream. The colour is usually blue-green, but sometimes violet or brown. The cells are uninucleate with parietal lobed plastids, each with a pyrenoid. The thallus, a few cm. in length, is filamentous, and shows monopodial branching. Whorls of short (or dwarf), very delicate filaments arising from each node of the thallus give it a distinct beaded appearance (FIG. 58A) which is clear under a pocket lens or even to the naked eye. Each dwarf filament (FIG. 58B) is much branched and consists of a row of ellipsoidal or narrow and elongated cells, and commonly bears terminally some fine hairs. The thallus remains attached to some hard object by a prostrate shoot which sends off numerous primary thalli floating on water. The basal cells of the whorl of short filaments develop threads of cells which grow downwards over the cells of the thallus, forming a cortex.

Reproduction. *Batrachospermum* reproduces both sexually and asexually. There are two distinct stages or forms in the life-history of the plant—the mature form reproducing sexually and the juvenile form reproducing asexually. The juvenile form is a simple protonema-like filament (FIG. 58C).

Asexual Reproduction. Asexual reproduction of the juvenile form often takes place by the formation of **monospores** (FIG. 58C). The monospore is always formed singly within the monosporangium and is really the terminal cell of a short lateral branch growing from the juvenile filament. The latter may reproduce by monospore year after year and remain in this

ALGAE

stage, or under suitable conditions of illumination it may give off mature thalli of *Batrachospermum* by lateral branching.

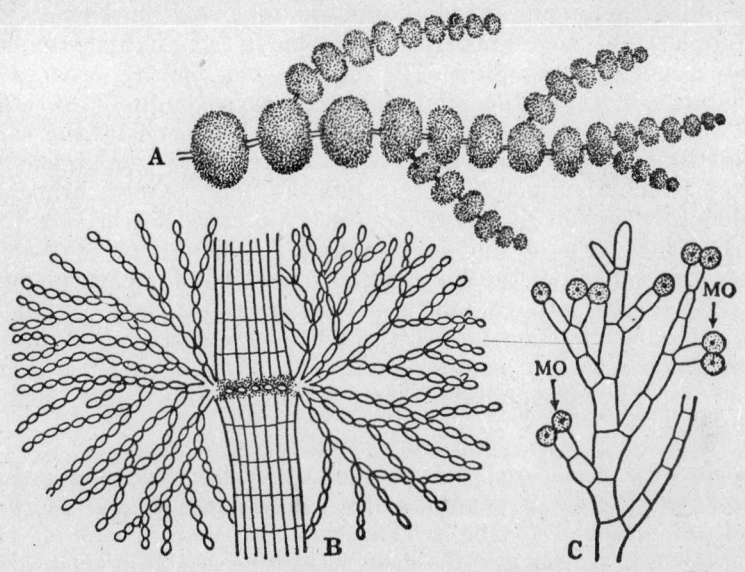

Batrachospermum. FIG. 58. *A*, a plant showing lateral branching and whorls of dwarf filaments; *B*, a whorl (magnified); *C*, the juvenile form producing monospores.

Sexual Reproduction. Atheridia or spermatangia, i.e. the male reproductive organs, develop in small groups at the end of a

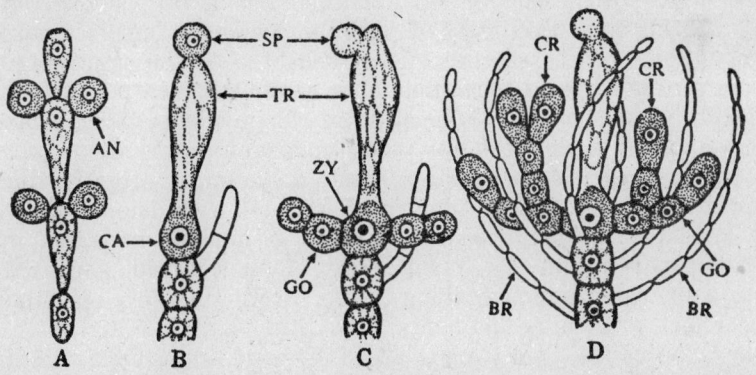

Batrachospermum. FIG. 59. *A*, an antheridial branch bearing antheridia; *B*, a mature carpogonium; *C*, fertilized carpogonium; *D*, cystocarp. *AN*, antheridium; *CA*, carpogonium; *TR*, trichogyne; *SP*, spermatium; *ZY*, zygote; *GO*, gonimoblasts; *BR*, enveloping branches (filaments); *CR*, carposporangium (with carpospore). *Redrawn after Fig. 76 in* Plant Morphology by *A. W. Haupt by permission of McGraw-Hill Book Company, Copyright* 1953.

lateral (short or dwarf) filament (FIG. 59A). The antheridium mother cell produces one or more antheridia terminally. They are uninucleate and usually globose or ovoid in shape. The protoplast of each antheridium forms a single male gamete known as the **spermatium**. The latter, when mature, escapes by a rupture of the antheridial wall. The **carpogonium** (FIG. 59B), i.e. the female reproductive organ, may be borne by the same plant that produces the antheridia (homothallic), or, it may be borne by another plant (heterothallic). Short lateral branches, called *carpogonial filaments*, consisting of a few cells, are produced from some of the short or dwarf filaments of the whorl. The terminal cell of the carpogonial filament is the **carpogonium**. The tip of the carpogonium is prolonged into a distinct inflated cell, called the *trichogyne*. All the cells of the carpogonial branch have a dense mass of protoplasm and are uninucleate. The protoplast of the carpogonium forms a single uninucleate female gamete.

Fertilization. After escape from the antheridium the spermatium is carried to the carpogonium by water current, and there it becomes attached to the trichogyne. The contact wall of the trichogyne dissolves and the spermatium passes through it down into the carpogonium and fuses with the female gamete. The trichogyne is cut off by the formation of a plug. It is, however, persistent in *Batrachospermum*. The basal part of the fertilized carpogonium (FIG. 59C), now partitioned off, forms the **zygote**. This produces a mass of short branching filaments called *gonimoblasts*. The terminal cell of a filament is a *carposporangium* (FIG. 59D). The protoplast of each carposporangium is a **carpospore**. Some of the sterile cells at the base of the carpogonium form an envelope of loose filaments around the carposporangia, and the whole structure consisting of the envelope, gonimoblasts, carposporangia and the fertilized carpogonium is then known as the **cystocarp** (FIG. 59D). Almost immediately after fertilization the zygote-nucleus undergoes reduction division, followed by repeated mitotic divisions. Finally a nucleus migrates into a carposporangium. The carpospore, when released, germinates into a pad of parenchymatous tissue. From this develop numerous short branching filaments (juvenile plants). This stage is otherwise called *Chantransia* stage (resembling an alga of this name). Later special branches arise from the *Chantransia* filaments, or commonly direct from the parenchymatous pad, and grow up into new *Batrachospermum* plants. It will be noted that the main plant including the juvenile forms is *haploid*, while the diploid phase is represented by the zygote only.

Chapter 3 BACTERIA

A Short Historical Account. Antoni von Leeuwenhoek (1632-1723) of Delft in Holland was the first to discover bacteria (1653-1673) with the help of the microscope considerably improved by himself (see also p. 143). Louis Pasteur (1822-1895), the famous French chemist and bacteriologist, thoroughly established the science of bacteriology. He carried out extensive work on fermentation and decay, and the cause of hydrophobia. About the year 1875 Pasteur made known to the world the importance of bacteria. He was the first to prepare vaccine and use it for the cure of the disease. He saved many Russians from hydrophobia by the use of this vaccine, and the Tsar of Russia in honour of his discovery sent him a diamond cross and also a hundred thousand francs to build a laboratory in Paris—now called the Pasteur Institute. About the same year Robert Koch of Germany proved that anthrax disease so common in cattle was caused by a kind of bacteria. He also showed in 1882 that tuberculosis and Asiatic cholera were caused by bacteria. From that time onwards many workers were attracted to this new science. Prominent among them were Emile Roux—helper of Pasteur, and Emil Behring—pupil of Koch, who discovered the diphtheria antitoxin about the year 1894. Elic Metchnikoff of South Russia in the eighties of the last century proved that phagocytes eat up germs and protect the body. Theobald Smith of America discovered in 1893 germs of Texas fever in the blood corpuscles of cows. It may be of interest to note in this connexion that in 1870 Joseph Lister (1827-1912), an English surgeon, developed the technique of antiseptic surgery, and in 1929 Alexander Fleming, an English bacteriologist, discovered penicillin which proved to be an effective remedy for many dreadful infectious diseases.

Shapes of Bacteria

1. Bacilli (sing. bacillus)—rod-shaped bacteria, e.g. *Bacillus* (=*Mycobacterium*) *tuberculosis*, *B.* (=*Clostridium*) *tetani*, *B. typhosus*, etc.
2. Cocci (sing. coccus)—spherical bacteria, e.g. *Staphylococcus*, *Streptococcus*, *Azotobacter*, etc.
3. Spirilla (sing. spirillum)—bacteria with spirally wound body, e.g. *Spirillum*, *Spirochaete*, etc.
4. Commas—slightly twisted like a comma, e.g. *Vibrio cholerae*.

General Description. Bacteria (class Schizomycetes) are the smallest and the most primitive cellular organisms, and number over 2,000 species. They are single-celled, usually spherical, rodlike or branched. In some forms the cells may be united into filaments or masses. They may be motile or nonmotile, and aerobic or anaerobic. There is no definite nucleus in the bacterial cell; chromatin granules representing an incipient nucleus are, however, often present. The cell-wall is made of chitin. Some forms of bacteria are provided with 1 or more cilia (or flagella). Chlorophyll is altogether wanting in them. Many of these microorganisms are as small as 1μ or even 0.5μ, particularly the spherical ones, while the rod-like forms are usually 2μ to 10μ in length (1 micron=1/1,000 mm.). Being so minute in size they are imperfectly seen even at the highest magnification of the microscope.

Occurrence. They occur almost everywhere—in water, air and soil, and in foodstuff, fruits and vegetables. Many of them float in the air on dust particles; many are abundant in water; and many are specially abundant in the soil, particularly to a depth

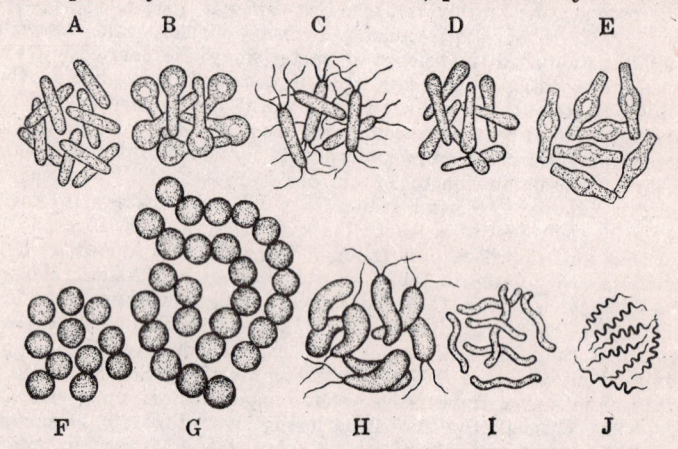

Bacteria. FIG. 60. **Bacilli:** A, *Bacillus* (=*Mycobacterium*) *tuberculosis;* B. *B.* (=*Clostridium*) *tetani;* C, *B. typhosus;* D, *B.* (=*Corynebacterium*) *diphtheriae;* E, *B. anthracis.* **Cocci:** F, *Staphylococcus;* G, *Streptococcus;* **Comma:** H, *Vibrio cholerae.* **Spirilla:** I, *Spirillum* (common in water); J, *Spirochaete.*

of 30 cm., and also in sewage. A few thousands may occur in 1 cc. of water, and a few millions in 1 gram of soil. Many live within and upon the bodies of living plants and animals. The intestines of all animals always contain a good number of different kinds of bacteria.

Structure. Electron microscope reveals the following structure in greater detail (FIG. 61). There is a distinct but complex cell-wall made of proteins and carbohydrates; chitin is often present but seldom any cellulose. Surrounding the cell-wall there is frequently a slime layer, often changed into a distinct **capsule** or **sheath.** The capsulated form is very resistant to adverse conditions and to treatments; such a form is commonly the cause of a disease. Several types of bacteria are provided with 1 or more slender, whip-like threads called *flagella* originating from the cytoplasm; such bacteria are motile. Internal to the cell-wall there is a thin plasma membrane formed by the cytoplasm. Cytoplasm spreads uniformly throughout the cell, and contains many small vacuoles, stored food granules such as volutin, glycogen, and fats, sometimes sulphur also, and an **incipient nucleus.** The active cell remains saturated with water, occurring to the extent of 90%. An organized nucleus, as found in higher plants, is absent in bacteria; nucleolus and nuclear membrane are absent. There is, however, a nuclear material (or chromatin) present in the bacterial cell in the form of 1 or 2 deeply staining bodies, possibly representing chromosomes. Chemically these bodies are

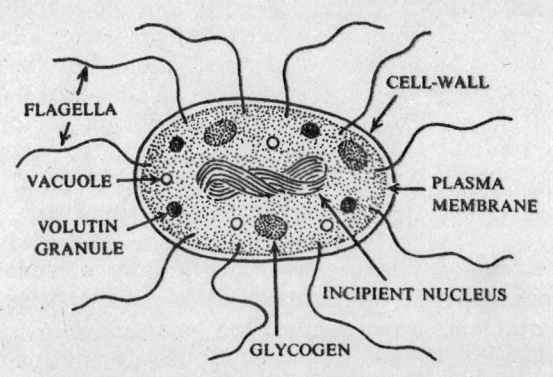

FIG. 61. A bacterial cell (highly magnified).

composed of twisted and folded strands of DNA (deoxyribonucleic acid) which is the genetic material of the chromosomes of living cells, responsible for transmission of hereditary characteristics. The DNA bodies divide prior to the division of the bacterial cell, and are distributed equally among the daughter cells.

Reproduction. Bacteria commonly reproduce by fission. Sexual reproduction in them is not known with any amount of certainty.

Fission (FIG. 62A). Bacteria commonly divide by the process of fission (see p. 407). Fission may take place in 1 plane or in 2 or 3 planes. By this method they may multiply rapidly in number. Hay bacillus (*Bacillus subtilis*), for instance, divides 2 or 3 times an hour under favourable conditions. At the minimum rate of division a single cell may give rise to over sixteen million (16, 777, 216) offspring at the end of 12 hours.

Sexual reproduction in bacteria has remained unknown for a long time. But in 1947 Lederberg and Tatum reported that they had observed a sort of conjugation between two strains of the common intestinal bacterium (*Escherichia coli*) evolved by them by exposure to ultraviolet light. They are of the view that during the short period of conjugation a part of the chromosome of one strain is injected into the other. The new bacterial cell thus produced is seen to be capable of living under conditions, not suiting any of the parent cells. Evidently it has received material from both the parent cells, and is better fitted in the struggle for existence. By this method, however, no multiplication of bacterial cells takes place. Beyond this, sexual method of reproduction is unknown in bacteria.

Spore Formation (FIG. 62B). Some bacteria, particularly some rod-shaped ones, form spores which are always 'resting' spores. A small mass of protoplasm within the mother cell clothes itself with a thick membrane, forming an **endospore**. The mother cell soon dissolves. The endospore may remain dormant for months or even several years resisting adverse conditions of life such as high temperature, freezing, extreme dryness, presence of many poisonous chemicals, etc. Then under favourable conditions of moisture, temperature and suitable food the endospore enlarges, and its contents form a full-fledged bacterial cell. Spore formation is not a method of multiplication.

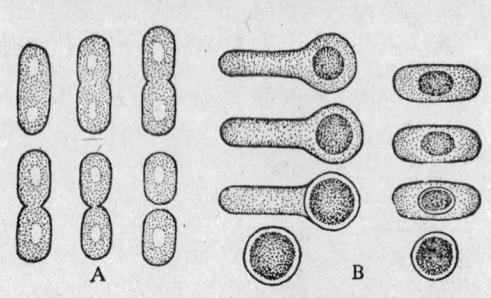

FIG. 62. A, fission of a bacterial cell; B, spore formation in two types of bacteria.

Nutrition of Bacteria. Bacteria like other living organisms require energy for their manifold activities. From the standpoint of their nutrition bacteria may be autotrophic (only a small number of them), or heterotrophic (majority of them). The former may be photosynthetic having a pigment closely related to chlorophyll, or chemosynthetic. The latter obtain their energy for synthetic processes by oxidation of some inorganic compounds such as sulphur, iron and nitrogen compounds (see pp. 335-37). Most of the bacteria are, however, heterotrophic in their mode of nutrition. They may be saprophytic or parasitic in habit. The former commonly live in water or soil containing organic compounds of plant or animal origin, in dead bodies of plants and animals, in vegetables and fruits in storage particularly, and in a variety of other media. They secrete enzymes to bring about the digestion of complex organic compounds into simpler ones which they absorb for their nutrition. Parasitic bacteria infect living plants and animals and absorb the food products from their hosts by the same process of enzyme-secretion and digestion.

Harmful Effects of Bacteria. Many parasitic (or pathogenic) bacteria infect living plants and animals, particularly the latter, and cause various and often serious diseases in them, sometimes in epidemic form. Normally they infect the host through wounds or they may be breathed in or taken in with food, water and milk. After infection of the body they produce a toxin (poison)

BACTERIA

which causes many of the serious diseases in human beings and domestic animals. Some of the common disease-producing bacteria are: *Bacillus typhosus* causing typhoid fever, *B. anthracis* causing anthrax, *B.* (=*Clostridium*) *tetani* causing tetanus, *Clostridium botulinum* causing a dangerous type of food-poisoning (called **ptomaine** poisoning), *B.* (=*Corynebacterium*) *diphtheriae* causing diphtheria, *B.* (=*Mycobacterium*) *tuberculosis* causing tuberculosis, *Mycobacterium leprae* causing leprosy, *B. dysenteriae* causing dysentery, *B.* (=*Diplococcus*) *pneumoniae* causing pneumonia, *Vibrio cholerae* causing cholera, etc. Some species of streptococci (the blood-poisoning bacteria) are possibly the deadliest enemy of mankind. They have the remarkable power of dissolving the red corpuscles of the human blood, and are responsible for erysipelas and extremely dangerous kinds of blood-poisoning.

Parasitic bacteria also attack plants and cause various diseases such as fire blight of apple and pear, ring disease of potato, black rot of cabbage, canker of *Citrus* (orange), etc. In plants, however, fungal diseases are far more common than bacterial diseases, while in animals the reverse is the case. Many bacteria are also responsible for decay of vegetables, fruits, meat, cooked food, etc., particularly during summer months.

Beneficial Effects of Bacteria. Although some bacteria (the disease-producing ones) are most harmful, a large number of them are most useful in various ways, particularly in agriculture and some industries. Many bacteria are nature's scavengers.

(1) **Agricultural.** (*a*) **Decay of Organic Substances.** But for the most useful work of many bacteria the dead bodies of plants and animals would remain unaltered covering possibly the whole or at least a very large part of the earth's surface leaving very little room for anything or anybody else; in addition, the organic compounds contained in such dead bodies would remain permanently locked up in them without any further use. Fortunately, however, bacteria play a most important role in this direction. They act on these bodies, reduce the various organic compounds to simple forms such as nitrates, sulphates, phosphates, and release them to the soil for utilization by green plants again in food manufacture. Carbon dioxide, water, oxides of nitrogen or even free nitrogen formed in the process, however, escape unused. (*b*) **Nitrification.** Proteins contained in the dead bodies of plants and animals are acted on by different kinds of bacteria and converted into ammonium compounds (ammonification) and then oxidized into nitrite and nitrate (nitrification) suitable for absorption by the green plants (see p. 292). (*c*) **Nitrogen Fixation.** Fixation of free nitrogen of the air by many soil

bacteria like *Azotobacter* and *Clostridium* directly in their own bodies, and *Rhizobium* (nodule bacteria) in association with the roots of leguminous plants is very important from an agricultural standpoint (see pp. 293-95). (*d*) **Fertilizers.** Conversion of cowdung, animal excreta into manures, and formation of humus or leaf-mould, as sources of nitrogen especially, are due to bacterial activities (see p. 285). Many chemical changes in the soil which make it fertile are mainly due to the activity of bacteria (and also of many other organisms in the soil). In fact the fertility of the soil may largely be attributed to the presence of bacteria in it. (*e*) **Plant Disease Control.** Some antibiotics obtained from bacteria have been successfully used to control many plant diseases.

(2) **Industrial.** From an industrial standpoint also many bacteria are most useful. Curing and ripening of tobacco leaves, fermentation of tea leaves, ripening of cheese, etc., for their characteristic flavours, retting of fibres, formation of vinegar from alcohol by acetic acid bacteria (*Acetobacter aceti*), fermentation of sugar into alcohol by yeast and a few bacteria, curdling of milk by lactic acid bacteria, conversion of hide into leather during the process of tanning, silage preparation (a fermented nutritious food for cattle) and such other cases of fermentation are specially important.

(3) **Medical.** (*a*) In the field of medicine valuable antibiotic drugs have been obtained from a number of bacteria and fungi. Thus, antibiotics like penicillin (isolated from *Penicillium notatum*, a mould), streptomycin (isolated from *Streptomyces griseus*, a bacterium), chloromycetin (isolated from *S. venezuelae*, another bacterium), etc., have been successfully used to control and cure some dreadful bacterial diseases such as pneumonia, diphtheria, wound infections, tuberculosis, typhoid, etc., saving the lives of millions of people throughout the world. (*b*) Bacteria always dwell in large numbers in the human system, particularly in the intestines. Although their activity is not clearly understood, some physiologists are of the view that these bacteria are in some way connected with digestive activities in the intestinal tracts. Possibly also they check the growth of putrefactive and pathogenic bacteria. Thus, lactic acid bacteria in fermented milk (curd) are supposed to cure or prevent dysentery. In any case bacterial flora in the human intestines is considered essential for the maintenance of normal health.

Hay bacillus (*Bacillus subtilis;* FIG. 63) is a common form of saprophytic bacteria growing in a decoction of hay. It can be easily grown in the laboratory by soaking hay in water and boiling it; the spores of hay bacillus withstand prolonged boiling. The decoction may then be kept in a warm place for a day or two. One or two drops may then be examined under a microscope at high magnification. Hay bacillus is unicellular and rod-shaped, provided with a number of flagella all over its body. The cells may be held together in chains. There is a granular vacuolated mass of protoplasm with chromatin granules but no definite nucleus. While growing in the fluid

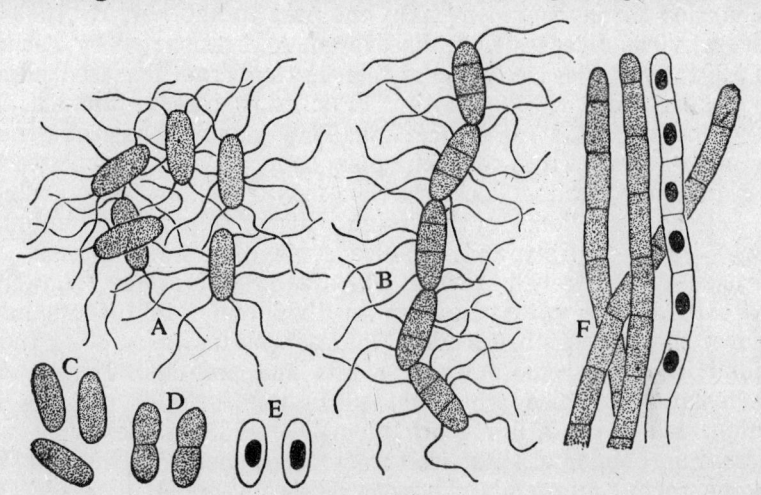

Hay bacillus (*Bacillus subtilis*), FIG. 63. *A*, motile forms; *B*, a chain of motile forms; *C*, nonmotile forms; *D*, fission; *E*, spore formation; and *F*, chains of nonmotile forms with spore formation in one chain.

hay bacillus reproduces by fission. The cell undergoes constriction in the transverse plane and splits up into two. The process of fission may be repeated several times and numerous cells formed within a short time. It is seen that the cells overcrowd the liquid and tend to come to the surface. They lose their cilia and become nonmotile. Their walls become gelatinous and the cells are held together in long chains. Several such chains form a mucilaginous mass, called a *zoogloea*, which floats as a thin film or scum on the surface of the liquid. When food is exhausted, the bacillus cells form 1 or 2 spores, called *endospores*, within the mother cell. The protoplasm withdraws from the wall and clothes itself with a fresh firm wall which can resist the action of high temperature and many poisonous substances. Later under favourable conditions the spore germinates in a suitable medium. The wall of the mother cell decays and the spore is liberated. The tough coat of the spore splits and the protoplasm escapes into the surrounding medium. Cilia are formed and the bacillus cell leads an active life.

Viruses. Viruses (*virus*, poison) are the smallest and possibly the most primitive living organisms yet known to science. They are very much smaller than bacteria, and cannot be detected even under the most powerful microscope. Their presence is revealed only when they produce certain diseased conditions in the plant or the animal. Mayer in 1886 first described the virus disease of tobacco and called it 'tobacco mosaic'. In 1892 Iwanowski, a Russian biologist, first demonstrated that the sap of the affected tobacco plant strained through a bacteria-proof filter could infect healthy tobacco plants. Beijerinck, a Dutch microbiologist, in 1899 showed that the dried sap containing virus retains the power of infection. According to him the virus is carried through phloem from one part to another. By 1933-34 several virus diseases were discovered. As first shown by Takami in 1901 virus diseases may be transmitted from diseased plants to healthy ones through biting or sucking insects and also by any kind of contact, as in grafting. They may also be seed-, leaf-, or stem-borne. Virus-containing sap may also be inoculated to non-infected plants. The infection may then spread from cell to cell by diffusion. In the living cells of plants and animals they grow, multiply and produce some disease symptoms. All viruses are entirely parasitic and are quite inert in their free state in air or water. Bawden in 1936 and later others have shown that viruses are made of nucleoproteins resembling those found in the chromosomes of plants and animals. Viruses can be isolated, purified and obtained in the form of crystals (a unique feature in a living organism). In 1935 Stanley, an American microbiologist, first isolated tobacco mosaic virus (TMV) in the form of crystals. These crystals dissolved in water and rubbed on tobacco leaves quickly produce disease symptoms. Since then many more have been obtained in such forms.

Recent investigations using electron microscope and X'ray photography have revealed some detailed facts about viruses. Although virus particles have no cellular structure they are still complex organisms having a genetic mechanism. They are of varying shapes and sizes ranging in lengths from 10 mμ to 200 mμ, sometimes up to 450 mμ (1 mμ=1/1,000μ). Further a virus is now known to contain a core of nucleic acid, mostly DNA (sometimes RNA) surrounded by a thin film of protein (protein shell). The protein shell is mainly a protective layer and is often very complex. The DNA is a genetic (hereditary) material, and is responsible for all biochemical activities. It has been possible to separate the protein from the DNA in different strains when it is found that the DNA of one strain can use the protein of another strain and produce new hybrids. Subsequently when they divide they make their own protein.

There are three kinds of viruses infecting plants, animals and bacteria. A few hundred plant diseases caused by viruses have been recorded so far, such as **mosaic diseases** in many cases as in apple, bean, beet, cabbage, cauliflower, cucumber, gourd, groundnut, mustard, pea, potato, radish, tapioca, tobacco, tomato, wheat, etc.; **black ring spot** of cabbage; **leaf roll** of potato; **chlorotic diseases** as in *Abutilon*, apple, pepper, rose, snapdragon, etc.; **leaf curl** as in bean, beet, cotton, gingelly, papaw, raspberry, soya-bean, tobacco, *Zinnia*, etc.; **spike disease** as in sandalwood;

BACTERIA

yellow diseases as in beet, carrot, marigold, peach, strawberry, sugarbeet, etc.; and **necrosis** (quick killing of affected tissues) as in potato and tomato. Some human diseases such as mumps, smallpox, chickenpox, measles, herpes, polio, yellow fever, scarlet fever, influenza, common cold, cancer, hydrophobia, etc., are supposed to be caused by viruses.

Bacteriophages. Some viruses also attack bacteria and destroy their nuclear material; they are called bacteriophages (*phagein*, to eat). Such viruses have a tail and a head, surrounded completely by a contractile protein sheath. The head contains DNA. When the protein sheath contracts, the tail end penetrates into

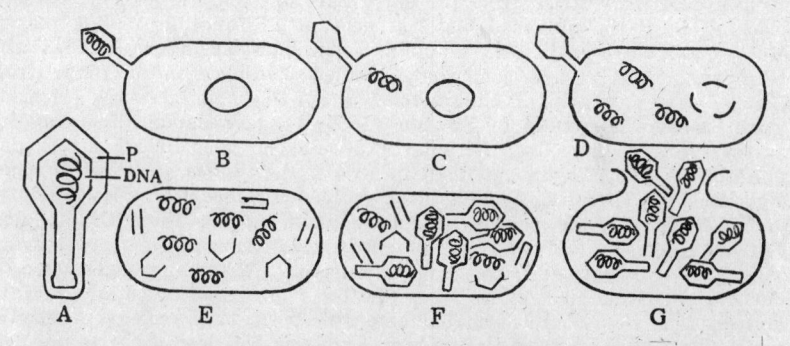

Bacteriophage. FIG. 64. *A*, a virus particle showing its structure and composition (*P*, protein; *DNA*, deoxyribonucleic acid); *B-G*, stages showing how a bacteriophage infects a bacterial cell, destroys the bacterial chromosome (shown here diagrammatically as an oval body), and replicates itself.

the bacterial cell. Then all of the virus DNA is injected into the latter, while the whole of the protein remains outside. The DNA of the virus now controls all the biochemical activities of the bacterial cell and, peculiarly enough, induces the latter to form more of virus DNA and protein. The result is the appearance, within about 20 minutes, of several new strains of virus particles, destruction of the bacterial chromosome (nuclear material) and finally bursting of the bacterial cell. No such effect is produced if the protein portion of the virus is injected into the bacterium. Evidently DNA is the genetic material (or gene).

Virus Reproduction. Outside the host body the virus is inactive. But inside it the virus DNA or RNA controls the biochemical activities of the infected cells which, peculiarly enough, now begin to make DNA or RNA and proteins characteristic of the invading virus. At the same time the virus particle replicates itself several times forming hundreds of particles like the original one with identical nucleic acids and proteins. The reproduction of virus has, however, been studied in greater details in connexion with the bacteriophage (see FIG. 64). Many types of viruses also undergo mutation producing new kinds of disease symptoms in host plants and animals.

Chapter 4 FUNGI

A Short Historical Account. Among the earliest workers mention may be made of **Gleditsch**, a German botanist, who attempted in 1753 a short classification of fungi. **Fontana**, an Italian scientist, worked assiduously for several years on rusts of cereals and published his account in 1767. Serious studies on fungi started from the early part of the 19th century. **Persoon** working in Paris published his first account of fungi in 1793, *Synopsis Methodica Fungorum* in 1801, edible and poisonous mushrooms in 1818, and European fungi and a somewhat detailed classification of fungi from 1822-28. **Fries'** system of classification (1821-32) and later his elaborate work on Hymenomycetes of Sweden (1857-63) gave impetus to many for further study of mycology. **Berkeley** (1803-89), a great British mycologist, published his *Cryptogamic Botany* in 1857 and *Outlines of British Fungology* in 1860. Besides, **Cooke's** (1825-1914) *Handbook of British Fungi* published in 1871, *Handbook of Australian Fungi* in 1892, Cryptogamic Journal *Grevillea* (1884-90), his extensive collections and elaborate drawings at Royal Botanic Gardens, Kew, are of outstanding merit. Special mention should be made of **De Bary** (1831-88), professor of botany at Halle and later at Strassburg, who did extensive work on mycology, mostly of a physiological and biological nature, and was the leading mycologist of the period. His new classification of fungi was published in 1884. His investigations on Uredinales and Ustilaginales cleared up many early misconceptions about them, and he was the first to establish heteroecism in rusts in 1864. He also introduced methods of culture of fungi. De Bary also discovered an association of algae and fungi in lichens. He actually laid the foundation of modern mycology. **Saccardo** (1845-1920), a famous Italian mycologist, published his monumental work on systematic mycology *Sylloge Fungorum* in 18 volumes (1892-1906); some supplementary volumes were added later. The technique of culture was further advanced by **Brefeld** (1839-1925) and **Van Tieghem** (1839-1914); the former introduced the agar method and the latter the glass-cell method. **Dangeard** about 1893 or so discovered sexuality in Uredinales and Ustilaginales, and later in Ascomycetes and Basidiomycetes. **Blackman** in 1904 discovered a kind of alternation of generations in rusts. **Blakeslee** in 1904 discovered heterothallism in *Mucor* and allied fungi, and he designated the two strains as + and −. Subsequently other investigators noticed the same phenomenon in some other fungi. The importance of mycology in agriculture and certain industries and also as an academic study has been realized and many specialists all over the world have undertaken investigations on fungi in the current century.

Classification of Fungi (90,000 sp.)
Class I   Myxomycetes or slime fungi (over 300 sp.).
Class II   Phycomycetes or alga-like fungi (1,500 sp.). Sub-class (i) Oomycetes (or Biflagellatae)—reproduction oogamous. **Order 1.** Saprolegniales, e.g. *Saprolegnia* (water mould). **Order 2.** Perenosporales, e.g. *Pythium, Phytophthora, Albugo* (=*Cystopus*) and *Pereno-*

FUNGI

spora. Sub-class (ii) Zygomycetes (or Aplanatae)—reproduction isogamous. **Order 3.** Mucorales, e.g. *Mucor* and *Rhizopus*.

Class III Ascomycetes or sac fungi (25,000 sp.). Sub-class (i) Protoascomycetes (no ascocarp; asci naked). **Order 1.** Saccharomycetales, e.g. yeast (*Saccharomyces*). Sub-class (ii) Euascomycetes (asci in ascocarp). Series (1) Plectomycetes (closed ascocarp—cleistothecium; no hymenium; asci scattered). **Order 2.** Aspergillales, e.g. *Penicillium* and *Aspergillus*. Series (2) Pyrenomycetes (ascocarp with an apical opening—perithecium; asci in hymenial layer). **Order 3.** *Erysiphales*, e.g. *Erysiphe* and *Uncinula*. **Order 4.** Sphaeriales, e.g. *Neurospora* and *Xylaria*. **Order 5.** Hypocreales, e.g. *Claviceps*. Series (3) Discomycetes (open ascocarp—apothecium; asci in hymenial layer). **Order 6.** Pezizales, e.g. *Peziza* and *Ascobolus*.

Class IV Basidiomycetes or club fungi (23,000 sp.). Sub-class (i) Hemibasidiomycetes (basidia septate or divided; teleutospore germinating into promycelium which bears basidiospores). **Order 1.** Ustilaginales or smuts (700 sp.), e.g. *Ustilago*. **Order 2.** Uredinales or rusts (4,600 sp.), e.g. *Puccinia*. Sub-class (ii) Holobasidiomycetes (basidium simple, club-shaped, directly bearing basidiospores; karyogamy and meiosis occur in the basidium). Series (1) Hymenomycetes (basidia on distinct hymenium). **Order 3.** Agaricales (7,000 sp.); it includes various forms of gill-fungi, e.g. mushrooms (*Agaricus, Amanita*, etc.), coral fungi (*Clavaria*), and various forms of pore fungi (*Polyporus, Polystichus, Fomes, Boletus*, etc.). Series (2) Gasteromycetes (hymenium indistinct; fruit body remains enclosed by peridium, i.e. distinct outer wall). **Order 4.** Lycoperdales, e.g. puff-balls (*Lycoperdon*). **Order 5.** Phallales, e.g. stinkhorns (*Phallus*). **Order 6.** Nidulariales, e.g. bird's nest fungi (*Nidularia* and *Cyathus*). Gasteromycetes are regarded as the highest of all fungi.

Class V Deuteromycetes or fungi imperfecti (over 24,000 sp.), i.e. fungi with imperfect life-history (sexual stages being unknown), e.g. *Helminthosporium* and *Fusarium*. They commonly reproduce by conidia. Many of them form important plant and animal diseases.

General Description. Fungi are a group of thallophytes lacking in chlorophyll. They may develop a variety of other pigments in the cell-wall or in the cell-cavity, and have an infinite variety of shapes and sizes, sometimes massive (fleshy or hard). Being non-green, they lead a heterotrophic mode of life, either as saprophytes or as parasites. The carbohydrate food is stored in the form of *glycogen* (and not starch). Saprophytic fungi grow in a variety of situations, and thrive under conditions of moisture, warmth and supply of organic food. Parasitic fungi on the other hand grow on living plants (wild or cultivated) and also on animals including human beings. They have the remarkable power of disintegrating or dissolving almost anything they attack by the secretion of suitable enzymes. Parasites that pass their entire life on living hosts are called *obligate parasites*, e.g. rusts; while those that normally live upon living hosts but may lead a saprophytic life, if need be, are called *facultative parasites*, e.g. some smuts. The plant body except the unicellular forms is commonly made of an interwoven mass of very fine and delicate threads called **hyphae**, collectively called **mycelium**. The wall of the hyphae may be made of chitin or pure cellulose.

Reproduction. *Vegetative reproduction* may take place by fragmentation of the body of the fungus or by detachment of a part of it or by budding or in some cases by special bodies called sclerotia. A **sclerotium** is a compact, often hard and rounded, mass of hyphae, with sometimes a dark firm outer covering layer but normally without any spore in it. It varies in size from a small pin-head to a few or sometimes several cm. in diameter, and gives rise to mycelium or fruit-body. *Asexual reproduction* takes place by means of spores of varied nature: (*a*) ciliate spores or **zoospores**, (*b*) ordinary **spores**, sometimes called gonidia, borne often in large numbers in a case called sporangium, (*c*) **conidia** formed singly or in groups or chains by specialized hyphae or conidiophores at their tips by the process of abstriction, (*d*) **chlamydospores** are thick-walled resting spores formed singly or in a chain by certain vegetative hyphae; these are transformed vegetative cells, (*e*) **oidia** are short segments of a vegetative hypha, functioning as spores, (*f*) **ascospores** are spores, usually 8 in number, formed in a sac called ascus, and (*g*) **basidiospores** are spores, usually 4 in number, formed externally by a club-shaped basidium on short slender stalks called sterigmata. There are also other kinds of specialized spores. *Sexual reproduction* in fungi shows three distinct phases: (*a*) **plasmogamy** which is the fusion of two protoplasts with or without nuclear fusion; by this process two haploid nuclei (called a **dikaryon**) of opposite sexes are brought together in one cell; dikaryotic condition may continue for a considerable length of time, as in higher fungi (e.g. (*Puccinia*), or the two nuclei may fuse almost immediately, as in lower fungi (e.g. *Mucor* and *Phythium*); (*b*) **karyogamy** which is the fusion of the two nuclei of a dikaryon resulting in a diploid (zygote) nucleus; this is sooner or later followed by meiosis to revert to the haploid condition; (*c*) with the development of gametangia and gametes sexual reproduction may be **isogamous, anisogamous** (rare) or **oogamous.**

Characteristics of the Main Groups

Phycomycetes. (1) Mycelium unseptate and coenocytic. (2) **Sporangium** with innumerable sporangiospores (zoospores or aplanospores) formed *endogenously*. (3) Sexual reproduction oogamous in Oomycetes, and isogamous in Zygomycetes. (4)

Biciliate motile cells (zoospores) produced by many species. (5) Zygote unicellular and simple.

Ascomycetes. (1) Mycelium septate; primary mycelium uni- or multinucleate; ascogenous hyphae always binucleate. (2) Conidia formation a common feature. (3) **Ascus** with usually 8 ascospores formed *endogenously*. (4) Sexual reproduction reduced to the fusion (karyogamy) of two compatible nuclei (+ and −) in the young ascus, with gradual suppression of sex organs (or gametangia); karyogamy immediately followed by meiosis, resulting in usually 8 ascospores within the ascus. (5) Motile cells absent. (6) **Ascocarp** (fruiting body) multicellular and complex, bearing the asci; open and cup- or saucer-shaped (called **apothecium**), oval- or flask-shaped with a small apical opening (called **perithecium**), or completely closed (called **cleistothecium**). (7) Hook or crosier common (see below).

Basidiomycetes. (1) Mycelium septate; primary mycelium multinucleate but secondary mycelium typically binucleate. (2) Conidia formation not a common feature. (3) **Basidium** with usually 4 basidiospores formed *exogenously*. (4) Sexual reproduction reduced to the fusion (karyogamy) of two compatible nuclei (+ and −) in the young basidium; sexual organs none (except spermatia and receptive hyphae in rusts); karyogamy immediately followed by meiosis, resulting in usually 4 basidiospores borne exogenously by the basidium. (5) Motile cells absent. (6) **Basidiocarp** (fruiting body) multicellular and complex, bearing the basidia; often open, sometimes closed (basidiocarp not formed in Uredinales and Ustilaginales). (7) Clamp connexion common (see p. 528).

Resemblances between Ascomycetes and Basidiomycetes. (1) The binucleate ascogenous hyphae of Ascomycetes may be homologous with the binucleate secondary hyphae of Basidiomycetes. (2) The asci may be homologous with the basidia; both are binucleate (dikaryotic) in their early stages. (3) Sexual reproduction is reduced to nuclear fusion (karyogamy) in the young ascus or the young basidium, as the case may be. (4) The ascospore may be homologous with the basidiospore; both are uninucleate, and in their formation stages like plasmogamy, karyogamy and meiosis are passed through, the last two processes taking place in the young ascus or in the young basidium. (5) No motile cells are formed in any of these two groups. (6) The hook or crosier method in Ascomycetes, and the clamp connexion in Basidiomycetes are hardly different in bringing together (dikaryotization) the reproductive nuclei of two opposite strains (+ and −).

Hook or Crosier Method (FIG. 65). The ascus in the majority of Ascomycetes is formed by hook or crosier (or crozier, meaning a crook or bend) method. In its formation the tip of the ascogenous hypha elongates and bends over, forming a sort of hook (*B*). The ascogenous hypha itself is formed from the ascogonium after pairing of nuclei in it (dikaryotic condition; *A*). The ascogenous hypha evidently contains pairs of nuclei received from the ascogonium. Of each pair one nucleus is male and the other is female. The tip of the hypha similarly contains a complementary pair of nuclei (dikaryon; see p. 526). These two nuclei undergo mitotic division simultaneously, with the two spindles standing parallel to each other in the vertical direction (*C*). It will be evident that one pair of the daughter nuclei (one male and one female) lies near the bend, close together in a pair; while one nucleus (either male or female) of one spindle lies towards the tip of the hook, and another nucleus (of opposite sex) of the other spindle lies towards the basal septum of the ascogenous hypha (*D*). Now

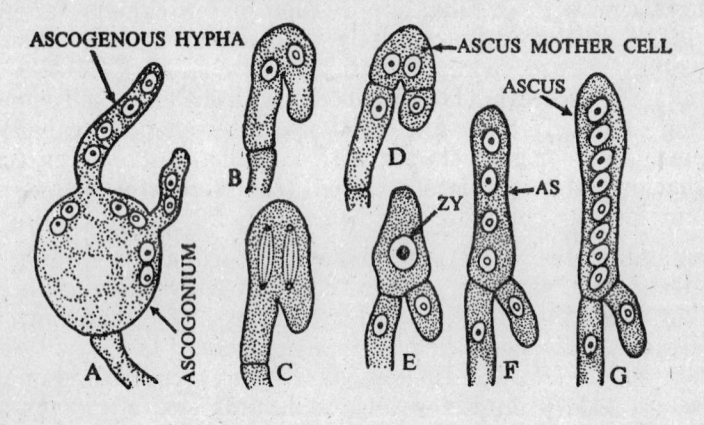

FIG. 65. Development of ascus by hook or crosier method. *A*, ascogonium producing ascogenous hyphae; *B*, hook or crosier with a pair of nuclei (dikaryon); *C*, conjugate nuclear division (mitosis); *D*, ascus mother cell with a dikaryon; *E*, zygote (ZY) with diploid (2*n*) nucleus; *F*, young ascus (*AS*) after meiosis; *G*, mature ascus with 8 ascospores.

two septa appear, resulting in three cells. It will be noted that the middle or the hook cell is binucleate (one male and one feemale); this is the **ascus mother cell**. The two other cells are uninucleate (one male and the other female). The binucleate cell (ascus mother cell) will soon develop into the **ascus**. Karyogamy (see p. 526), i.e. fusion of the two nuclei, takes place in the young ascus, and the zygote that is formed evidently contains a diploid nucleus (*E*). The zygote-nucleus undergoes divisions, the first one being meiotic (*F*). Usually the divisions cease when 8 nuclei are formed. Each nucleus clothes itself with a wall and becomes converted into an **ascospore**. The ascus simultaneously grows and elongates, containing the ascospores (*G*). Several asci may be formed in the above way from repeatedly branched ascogenous hyphae.

Clamp Connexion (FIG. 66). This is a common feature in most types of Basidiomycetes except the rusts. Clamp connexion is a special mechanism by which the sister nuclei of a dikaryon become separated into two daughter cells. This mechanism is found in the secondary mycelia which consist of typically binucleate cells. The binucleate condition has originated from the fusion of two uninucleate cells of two primary hyphae without, however, any karyogamy, i.e. without the fusion of nuclei. Clamp connexion is usually formed in the terminal cells of the hyphae. In some cases, however, it occurs in most of the cells of the secondary hyphae, sometimes only in some cells

FUNGI

here and there. Between the nuclei a short branch arises forming a sort of hook (clamp cell). One nucleus passes into the clamp cell, and now both the nuclei divide simultaneously. The clamp cell bends over and its end touches the wall. This bridge-like short connexion (branch) is called the clamp connexion. Now one nucleus of each pair approaches the other one, evidently passing through the bridge. After this migration a septum is formed at the base of the clamp cell and another septum below the bridge, thus separating a pair of nuclei, one from each parent cell.

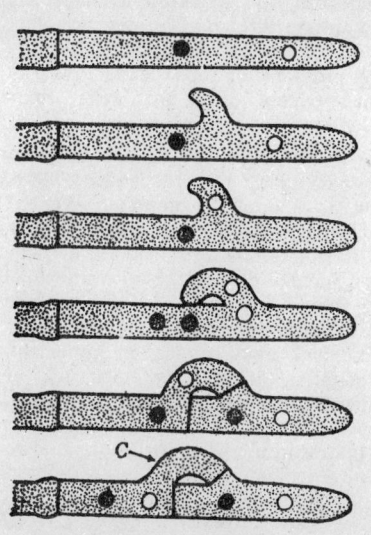

FIG. 66. Clamp connexion (*C*).

Origin of Fungi. Regarding the origin of fungi no definite statement can be made. There are two views in this respect. According to one view the fungi have been derived from algae. In this connexion it is said that Phycomycetes have arisen from Chlorophyceae, possibly Siphonales, by the loss of chlorophyll which evidently has brought about a change in the mode of nutrition but not in the mode of sexual reproduction, the latter showing parallelism in development in both the groups. The second view is that since primitive Phycomycetes (which are Chytridiales—unicellular) bear uniflagellate zoospores and gametes resembling protozoa (green algae are never uniflagellate) they may have been derived from some uniflagellate organisms like protozoa. If this is so, fungi must have followed an independent line of development (evolution) standing midway between animals and plants. Regarding the origin of Ascomycetes there are again two views. According to one view Ascomycetes may have been derived from some red algae since there are marked similarities in the sexual characters between these two groups. Another view which is more generally accepted is that Ascomycetes have arisen from oogamous Phycomycetes. Regarding Basidiomycetes there is general agreement among mycologists that these have been derived from Ascomycetes. There is a close resemblance between the two groups (see p. 527). Basidiomycetes are of course more advanced than Ascomycetes.

CLASS I | MYXOMYCETES or slime fungi | about 400 sp.

Myxomycetes, commonly called slime fungi, form a peculiar group of organisms which are animal-like in their vegetative stages and plant-like in their reproductive stages. Slime fungi are widely distributed growing in damp shady places in the soil rich in humus, damp old planks of wood, rotting logs of wood, decaying leaves, etc., always preferring moisture and darkness for their normal growth. They are saprophytic in habit and like other fungi are lacking in chlorophyll. They may be colourless or variously coloured—yellow, orange, brown, red, violet, etc.

Plant Body. The body of the Myxomycete consists of a naked slimy mass of protoplasm with many nuclei in it; such a body is otherwise known as

plasmodium. The plasmodium forms pseudopodia and shows amoeboid movement creeping over and within the substratum and growing in different directions, commonly towards moisture avoiding light as far as possible. The diameter of the body varies from a few to several centimetres according to species. The plasmodium engulfs solid particles of food; it also absorbs food material in solution. In the dry season it passes into a hard sclerotium-like body.

Reproduction. Vegetative reproduction may take place by the breaking up of the plasmodium into fragments, each growing as an independent body. Two fragments may also unite and grow as a single organism. Asexual reproduction. For this purpose the plasmodium comes to the surface of the substratum in which it grows, and forms different kinds of sporangia according to species. The sporangia and spores are formed under lighted conditions. The plasmodium may form one or more large sporangia on the surface of the substratum or a number of small separate sporangia on longer or shorter stalks. In shape they may be cylindrical, oval (ovate) or spherical. The sporangium is multinucleate and consists of a tough network of plasmodium strands, known as the **capillitium**, enclosing numerous spores in its meshes. The protoplasm of the sporangium excretes waste materials which are deposited on the surface of the sporangium as a wall-layer known as the peridium. The spores are uninucleate and surrounded by a definite cellulose wall. Capillitium shows hygroscopic movement and the sporangium bursts irregularly at the apex. The spores being liberated are scattered by the wind. *Germination of the Spore.* The protoplast of the spore escapes from it by amoeboid movement and forms a zoospore which develops a long cilium at its anterior end. Sometimes 2 or 4 zoospores may be formed. The zoospores swim for a while, retract their cilium and pass into an amoeboid condition or into a resting stage. The zoospores may divide repeatedly and multiply vegetatively. Sexual reproduction. Sooner or later the amoeboid zoospores behave as gametes and begin to fuse in pairs. The fusion product may be regarded as a zygote. It is amoeboid and as it develops into a plasmodium the fusion nucleus divides repeatedly. The young plasmodium seems to exert an attractive force upon the neighbouring zygotes which move towards it and fuse with it (no nuclear fusion takes place), contributing to the growth of the plasmodium body. The plasmodium evidently is multinucleate and coenocytic. The plasmodium creeps in different directions by the formation of pseudopodia and grows in size. The nuclei of the plasmodium are diploid, and reduction division takes place at some stage prior to the formation of the spores.

CLASS II | PHYCOMYCETES or alga-like fungi | 1,500 sp.

1. *SAPROLEGNIA* (25 sp.)

Occurrence. *Saprolegnia* (family *Saprolegniaceae*; FIG. 67*A*), commonly called water mould, grows in water as a saprophyte on dead bodies of insects, fishes and other animals, and on plant remains. Some species are parasitic on living fish and fish eggs causing diseases in them. They commonly occur in fish nurseries. Many species are also soil-dwellers. A dead insect placed in a pot of water shows within a few days a flocculent growth of this fungus.

FUNGI

Structure. The mycelium consists of much-branched, coenocytic (nonseptate and multinucleate) hyphae. Two types of hyphae are often distinguished: short rhizodal hyphae penetrating into

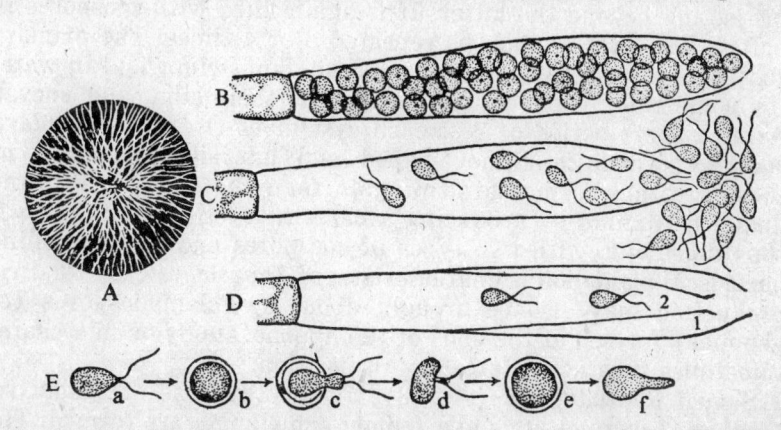

Saprolegnia. FIG. 67. *A*, growth of *Saprolegnia* on a dead insect; *B-E*, asexual reproduction; *B*, a mature zoosporangium with primary zoospores; *C*, a zoosporangium discharging primary zoospores; *D*, internal proliferation of zoosporangium (1, primary; 2, secondary); *E*, development of secondary zoospore from primary zoospore (*a*, primary zoospore; *b*, encysted zoospore; *c*, secondary zoospore being formed; *d*, secondary zoospore; *e*, encysted zoospore; *f*, and germinating zoospore).

the substratum to obtain food from it, and a mass of long profusely branched hyphae spreading over the substratum and finally bearing the reproductive organs. This mass is distinctly visible with unaided eyes. The hyphae vary considerably in width. Septa develop only in connexion with the reproductive organs.

Reproduction. *Saprolegnia* reproduces asexually as well as sexually.

Asexual Reproduction. (FIG. 67 *B-E*). For this purpose the end of a vegetative hypha swells and forms a **zoosporangium** (*B*). It is an elongated, tapering structure, separated from the vegetative hypha by a septum. When young, it is filled with a dense mass of protoplasm, and looks somewhat brownish under the microscope. Before the septum is formed a large number of nuclei flow into the zoosporangium. Soon by furrowing cleavage) of the protoplasm, starting from the wall, small uninucleate masses appear, and each mass becomes converted into a uninucleate **zoospore**. The zoosporangium is closely packed with such zoospores. These are the *primary zoospores*, and are pear-shaped and apically biciliate. They escape into the surrounding water through an apical opening of the sporangium (*C*). There is proliferation of the sporangium, and this means that after the zoospores escape a secondary sporangium grows from the base upwards within the primary sporangium (*D*), sometimes projecting beyond the latter. It is again filled with zoospores as before. The process may be repeated 3 or 4 times. The primary zoospores after escape from the sporangium swim about in water for some time, come to rest, withdraw their cilia, and encyst. After a short period of rest each cyst gives rise to a *secondary zoospore* which is kidney-shaped and laterally biciliate. The zoospores thus formed swarm again for some time, encyst, and finally germinate by producing a *germ tube* (*E*). *Saprolegnia* is thus **diplanetic**, with two types of zoospores and two swarming periods. **Diplanetism** is characteristic of *Saprolegnia*. Asexual reproduction may also be brought about by chlamydospores (or gemmae) formed at the ends of the hyphae, singly or in a chain; sometimes they are intercalary in position.

Sexual Reproduction (FIG. 68). For the purpose of sexual reproduction distinct male and female gametangia are formed. The female gametangium or **oogonium** may be the terminal cell of a vegetative hypha or it may be intercalary. It swells at the apex and becomes more or less spherical, sometimes oblong, in shape. Its contents form one, often more, spherical, uninucleate **oospheres** or **eggs**. One or more oil drops are seen to occupy a particular position in the eggs according to the species. The male

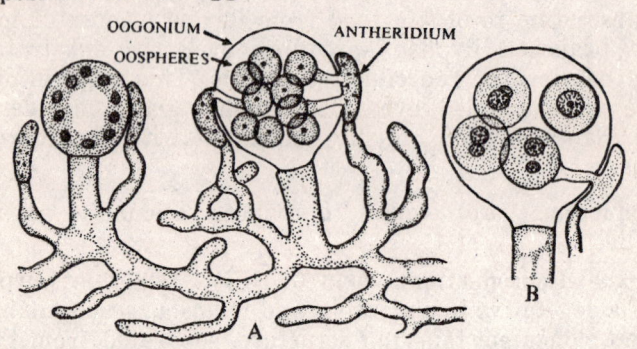

Saprolegnia. FIG. 68. Sexual reproduction; *A*, development of antheridia and oogonia; *B*, fertilization and formation of zygote.

gametangium or **antheridium** may arise from the same hypha that bears the oogonium, or from a different hypha close to the oogonium. A septum develops separating the antheridium from the rest of the hypha. The antheridium is a multinucleate, elongated, somewhat tubular body, much smaller than the oogonium. One or more antheridia may be attached to the oogonial wall. A *fertilization tube* develops from the antheridium, penetrates the

FUNGI

oogonial wall and reaches an oosphere (*A*). Within the oogonium the fertilization tube may be branched, each branch reaching an oosphere. Several nonmotile antheridial nuclei (male gametes) migrate through this tube into the oogonium, and one such male nucleus passes into an oosphere and fuses with the egg-nucleus. The fertilized egg-nucleus or zygote-nucleus is evidently diploid ($2n$). It clothes itself with a thick but smooth wall and becomes an **oospore** (*B*). After a prolonged period of rest the oogonial wall disintegrates and the oospore germinates by producing a *germ tube*. Reduction division of the zygote-nucleus takes place during the germination of the oospore. The oospore may also develop parthenogenetically, i.e. without fertilization, in many cases. In some species the antheridium is not formed at all.

2. *PYTHIUM* (40 sp.)

Pythium (family *Pythiaceae*) is a parasitic fungus, commonly attacking seedlings of cress and mustard at the base of their hypocotyl (FIG. 69*A*) under conditions of overcrowding and overwatering. It causes the disease commonly known as the 'damping-off' of seedlings. Members of *Cruciferae* are particularly susceptible to this disease; other crops like tobacco, ginger, etc., are also infected by this fungus. The most common

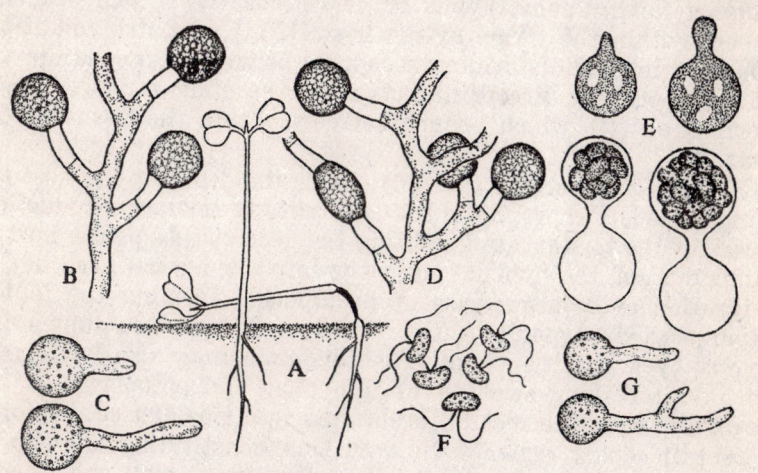

Pythium. FIG. 69. *A*, healthy and infected seedlings; *B*, hyphae (conidiophores) with conidia; *C*, conidium germinates by sending out a germ tube; *D*, hyphae with zoosporangia; *E*, zoosporangium germinating and finally forming a vesicle with zoospores; *F*, biciliate zoospores; and *G*, zoospore germinates by producing a germ tube.

species of *Pythium* causing the 'damping-off' is *P. debaryanum*. When attacked, the seedlings become weakened at their base and soon fall over. At first the fungus is a parasite, but later after the death of the host seedlings it thrives on them as a saprophyte. The mycelium ramifies in all directions through the intercellular spaces, penetrating here and there into the living cells. At a later stage white cottony threads (hyphae) may be seen on the surface of the seedlings. The long, slender, much-branched hyphae are unseptate and coenocytic enclosing many small nuclei (cf. *Mucor*).

Reproduction takes place both asexually and sexually.

Asexual Reproduction (FIG. 69 *B-G*). Here and there the mycelium sends out aerial hyphae through the stomata or the cuticle, which bear short lateral branches, each swelling into a spherical head. This head, which is partitioned off at the base by a septum, may act as a conidium (*B*) or as a zoosporangium (*D*) according to external conditions. (*a*) Under moist conditions it acts as a zoosporangium; it protrudes and bulges out into a bladder-like vesicle (*E*). The protoplasm migrates into it and divides to form a number of small naked uninucleate and biciliate zoospores. The zoospores, when set free, swim about in water for a short time (*F*). Soon they withdraw their cilia and clothe themselves with a wall. Eventually they infect a new seedling and germinate by putting forth a short hypha or *germ tube* (*G*) which branches freely within the tissue of the host. (*b*) Under dry conditions the head instead of forming zoospores behaves as a condium (or conidiospore). It directly infects a new seedling and produces a germ tube (*C*) which branches freely within the tissue of the host.

Sexual Reproduction (FIG. 70). After the death of the host the fungus leads a saprophytic life and takes to sexual reproduction, possibly due to starvation. Within the dead tissues of the host or outside them, the hyphal end swells into a spherical head due to migration of a large mass of protoplasm, forming the female organ, called **oogonium** (*A*). It is cut off from the supporting hypha by a partition wall. The oogonium may also be formed as an intercalary swelling of a hypha. The cytoplasm of the oogonium soon becomes differentiated into two distinct regions: a central denser region with a nucleus, constituting the **egg** or **oosphere**, and a peripheral region with many small nuclei, constituting the **periplasm**. From the same hypha or from another close to the oogonium a small branch arises. It swells and becomes more or less club-shaped, and is cut off by a septum at the base. This is the male organ called **antheridium**. Its protoplast becomes differentiated into a central male gamete which is uninucleate and an outer periplasm which is multinucleate. The nuclei of the periplasm soon degenerate. The antheridium bends

FUNGI

towards the oogonium and comes in contact with it. A short cylindrical tube, called the *fertilization tube* or *beak*, is produced from it, which pierces the oogonium. The male gamete passes into the oogonium through this tube and fuses with the egg-

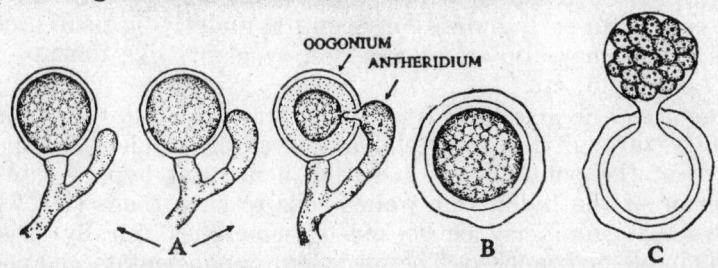

Pythium. FIG. 70. *A*, formation of sexual organs—oogonium with oosphere, and antheridium (note the beak); *B*, thick-walled oospore inside oogonium formed after fertilization; and *C*, oospore germinates and produces a vesicle (zoosporangium) with zoospores.

nucleus of the oosphere, and thus fertilization is effected. Fertilization in this case (with the antheridium and the oogonium lying side by side on the same stalk) is known as **paragynous**. The fertilized oospore (*B*) forms a thick wall round itself and rests for some months in the soil. Infection takes place in the way described above. The oospore germinates by putting forth a germ tube or it forms a zoosporangium (*C*), the protoplasmic contents of which divide to form a number of zoospores.

Control of the Disease. (1) Overcrowding the seedlings and overwatering the soil are to be avoided. (2) Steam sterilization of the soil. (3) Heat sterilization of the soil by burning wood in the field. (4) Sprinkling dilute formalin solution over the soil. (5) Spraying with Bordeaux mixture.

3. *PHYTOPHTHORA* (13 *sp*.)

Occurrence. Species of **Phytophthora** (family *Pythiaceae*) are widely distributed. Common species of the fungus occurring in India are: *P. infestans* attacking potato plant, *P. palmivora* attacking coconut-palm and palmyra-palm, *P. arecae* attacking betel-nut-palm, *P. colocasiae* attacking taro (*Colocasia*), *P. parasitica* attacking castor plant, etc. *Phytophthora infestans* is a notorious one causing a serious disease in potato plants, known as 'late blight' (early blight being caused by *Alternaria solani*). The disease is of a more serious nature in the hills than in the plains, sometimes appearing there in an epidemic form. The formation of black patches on the undersurfaces, less often on the upper, of the potato leaves indicates the diseased condition of the plant (FIG. 71*A*). The disease may spread to the entire leaf and other parts, extending down to the underground parts, particularly the tubers. All the neighbouring plants may be similarly affected if the weather is warm and humid. The fungus causes 'wilting' of the leaves and 'rotting' of the tubers, either in the field or in storage. In such tubers brownish markings may be noticed below the skin as an early indication. Soon the underlying tissues soften and rot. It may also attack and destroy plants like tomato, pepper, egg-plant, etc.

Structure. The mycelium is profusely branched, and the branches ramify through the intercellular spaces of the leaf, stem and tuber of the potato plant and send haustorial hyphae into the interior of the living thin-walled cells to suck them (FIG. 71C). The haustorium may be hooked or sometimes spirally twisted, and simple or branched. The mycelium is non-septate and coenocytic (cf. *Pythium*). Septa may, however, develop later in old mycelia.

Reproduction. The fungus commonly reproduces asexually. Sexual reproduction, though not common, was first observed in pure culture and later in the infected tuber.

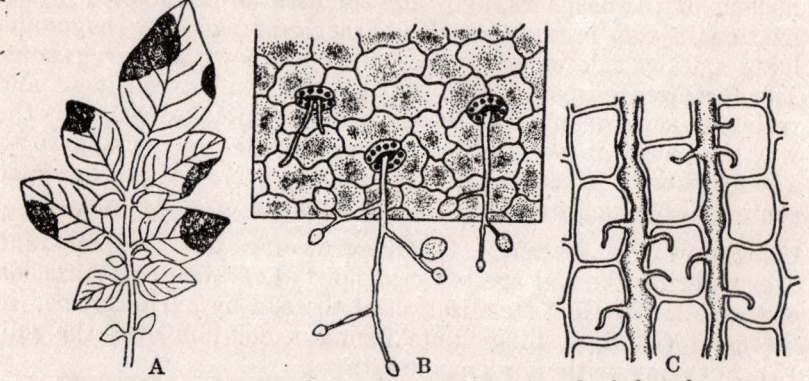

Phytophthora. FIG. 71. *A*, a potato leaf showing the infected areas; *B*, sporangiophores protruding through the stomata and bearing sporangia; *C*, haustorial hyphae.

Asexual Reproduction. At the time of reproduction long slender erect hyphae, called **sporangiophores**, grow out in small groups through the stomata on the lower surface of the leaf (FIG. 71B). On the tuber, however, they appear in large numbers. The sporangiophore becomes branched, and each branch bears at its apex a short-stalked multinucleate ovoid or lemon-shaped **sporangium** with a papilla-like tip. The tip of the sporangiophore continues to grow pushing the sporangium to one side, and the tip again forms a sporangium. Thus the sporangium, though terminal at first, soon becomes lateral on the sporangiophore which obviously shows a zig-zag growth (a sympodium). Further, a distinct nodular swelling of the branch appears just above the sporangial base. When mature, the sporangia are dispersed by wind or washed away by rain. Commonly at high temperature

FUNGI

the sporangium germinates directly on a potato plant by pushing out a *germ tube*. At low temperature, however, the contents of the sporangium divide into several uninucleate segments. Each segment then develops into a biciliate **zoospore**, with the two cilia attached laterally (FIG. 72*A-C*). The tip of the sporangium bursts and the zoospores

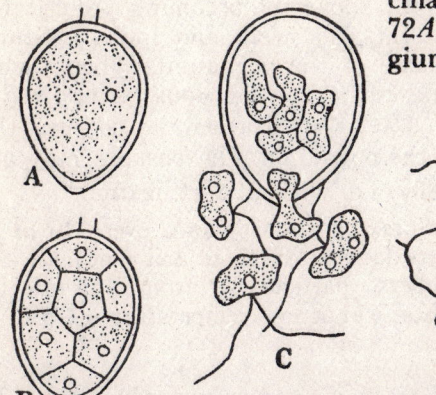

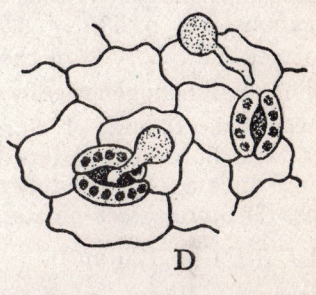

Phytophthora. FIG. 72. Asexual reproduction: *A*, sporangium; *B*, a sporangium becoming a zoosporangium; and *C*, zoospores escaping; *D*, zoospores germinating on the host plant (note the infection through the stomata).

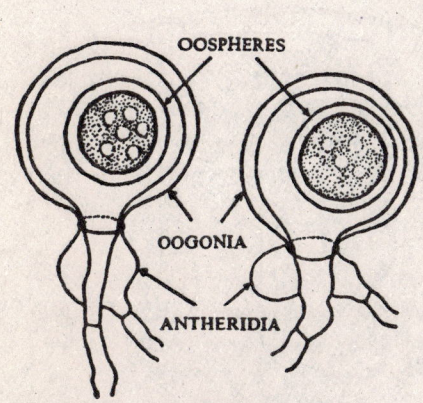

FIG. 73. Sexual reproduction in *Phytophthora*.

escape. They swim for a while, then come to rest, lose their cilia and form a wall round each. Under favourable conditions of temperature and moisture they germinate on the leaf by producing a *germ tube* which penetrates through a stoma into the tissue of the leaf (FIG. 72*D*). The disease thus spreads rapidly from plant to plant by this method. Sporangia and zoospores are short-lived, and the fungus hibernates in the tuber in the form of mycelia. Thus initial infection of the potato plant starts from this diseased tuber.

Sexual Reproduction. This takes place by means of oogonia (female) and antheridia (male) which develop from two neighbouring hyphae (FIG. 73). The **oogonium** is spherical or pear-shaped with a smooth reddish-brown wall. It contains a large oosphere or egg-cell, lying loose and free within it, surrounded by a scanty zone of protoplasm, called the **periplasm**. All the nuclei of the oogonium except one egg-nucleus of the oosphere degenerate.

The **antheridium** is broadly club-shaped, and develops before the oogonium. It contains many nuclei but finally only one male nucleus persists while others, originally present, degenerate. Peculiarly enough, the oogonium, as it grows, penetrates through the antheridium and swells above it, becoming spherical or pear-shaped. The antheridium also swells and forms a funnel-shaped collar around the base of the oogonium. When such is the case, fertilization is said to be **amphigynous** (as opposed to paragynous, see p. 535). After fertilization the oospore that is formed lies loosely in the oogonium. The oospore may also develop parthenogenetically, i.e. without fertilization.

Control of the Disease. (1) Spraying young plants, every 10 or 15 days, with Bordeaux mixture (copper sulphate and lime). (2) Selecting disease-free seed-tubers, particularly* from non infected areas. (3) Storing seed-potatoes at a low temperature - 4-5°C.

4. *ALBUGO* (25 sp.)

Occurrence. *Albugo candida* (= *Cystopus candidus;* FIG. 74), commonly known as 'white rust', is a parasitic fungus, attacking several plants of mustard family such as mustard, radish, cabbage, turnip, etc. White blisters on the stem and the leaves (*A*) indicate the diseased condition of the plants, finally affecting flowers and ovaries which become distorted in shape and size

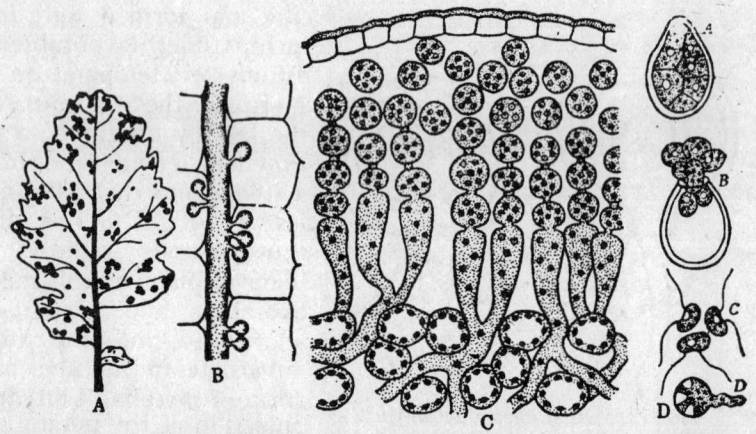

Albugo. FIG. 74. A, infected leaf of mustard; B, an intercellular hypha with button-like haustoria; C, an infected leaf in section showing chains of multinucleate sporangia under the epidermis (note the neck separating the sporangia); D, germination of a sporangium— *A*, sporangium dividing; *B*, zoospores escaping; *C*, biciliate zoospores swimming; and *D*, a zoospore germinating.

However, it is not a serious disease in India. *Albugo* belongs to the family *Albuginaceae*.

FUNGI

Structure. The mycelia ramify through the intercellular spaces of the host plant and branch profusely. The hyphae are unseptate and multinucleate. Here and there they send globular or button-like haustoria (B) into the living cells of the host to absorb food from them.

Reproduction. *Albugo* reproduces both asexually and sexually.

Asexual Reproduction (FIG. 74C). The hyphae grow luxuriantly at certain points below the epidermis. They form erect, branched or unbranched, club-shaped, multinucleate sporangiophores in clusters. The latter begin to cut off spherical multinucleate **sporangia** in chains from their tips. The sporangia are separated from one another by short necks made of gelatin. The epidermis soon gets ruptured as a result of internal pressure exerted by the increasing number of sporangia which then appear on the surface as a white powdery mass. The sporangia are now carried by the wind to other plants. The protoplasmic contents of each sporangium (FIG. 74D) divide to form a few (4 or 8 or more) kidney-shaped **zoospores**, each with two lateral cilia. The sporangium bursts and the zoospores escape. They swim about for some time in water that collects on the surface of the host plant. They soon lose their cilia, cover themselves with a wall and come to rest. Later they germinate by producing a *germ tube* which enters the host through a stoma. Sometimes a sporangium germinates directly without the intervention of the zoospores.

Sexual Reproduction (FIG. 75). Later in the season the hyphae produce separately male and female gametangia within the intercellular spaces of the host (A). To form the female gametangium or oogonium a hypha swells at the tip and becomes spherical. A septum appears at the base. The oogonium is multinucleate, and its cytoplasm becomes differentiated into two distinct zones: a central zone called the **ooplasm** which is the egg or oosphere, and a peripheral zone called the **periplasm**. The central zone is a dense mass of cytoplasm with an egg-nucleus in it (other nuclei of this zone usually degenerate); while the peripheral zone is

Albugo. FIG. 75. A, fertilization; B, zygote (oospore); C, germination of zygote—zoospores escaping into a vesicle; D, biciliate zoospores after escape.

lighter and multinucleate. To form the male gametangium or antheridium a hypha close to the oogonium swells at the tip and becomes more or less club-shaped. It is cut off at the base by a septum. The antheridium is multinucleate. Soon it comes in contact with the oogonium and produces a *fertilization tube* which penetrates through the oogonial wall and the periplasm and goes deep into the oosphere. One or more male nuclei are set free from the antheridium through this tube but only one of them fuses with the egg-nucleus. If more egg-nuclei be present all of them may be fertilized. Thus fertilization is effected. The zygote (oospore; *B*) formed thereby clothes itself with a thick wall, the periplasm taking part in its formation. The zygote is set free only after the decay of the host tissue. It undergoes a period of rest and then it produces numerous (over 100) **zoospores**. The zygote bursts and the zoospores escape into a vesicle (zoosporangium; *C*). The vesicle dissolves and the zoospores, each with two lateral cilia, are set free to swim about in water (*D*). The zoospore germinates under appropriate conditions by producing a *germ tube*.

5. *MUCOR* (50 *sp.*)

Occurrence. *Mucor* (family *Mucoraceae*), commonly called 'pinmould', is a saprophytic fungus (FIG. 76). It grows on animal-

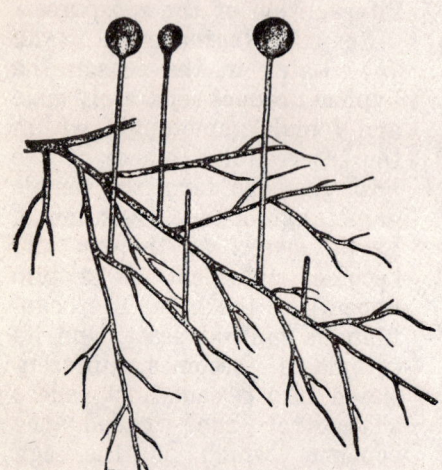

Mucor. FIG. 76. Ramifying mycelia with some sporangia (or gonidangia).

dung, wet shoes, stale moist bread, rotten fruits, decaying vegetables, shed flowers and other organic media, spreading like a cobweb. It can be easily grown in the laboratory in a piece of moist bread kept under a bell-jar in a warm place for 3 or 4 days.

FUNGI

Structure. The plant body is composed of a mass of white, delicate, cottony threads collectively known as the **mycelium** (FIG. 76). It is always very much branched, but is coenocytic, i.e. unseptate and multinucleate. Each individual thread of the mycelium is known as the **hypha** (pl. hyphae).

Reproduction. This takes place by two methods, viz. asexual and sexual. The asexual method is most common.

Asexual Method (FIG. 77). This method of reproduction takes place by means of **spores** (or gonidia) which develop in a case, called **sporangium** (or gonidangium), under favourable conditions of moisture and temperature. It is seen that mycelia give off here and there numerous slender *erect* hyphae, called sporangiophores, each ending in a spherical head—the sporangium (FIG. 76). In the formation of the sporangium the apical portion of each of these hyphae swells out into a spherical head (FIG. 77). As the hypha begins to swell the protoplasmic contents migrate to its tip, but accumulate more densely towards the periphery,

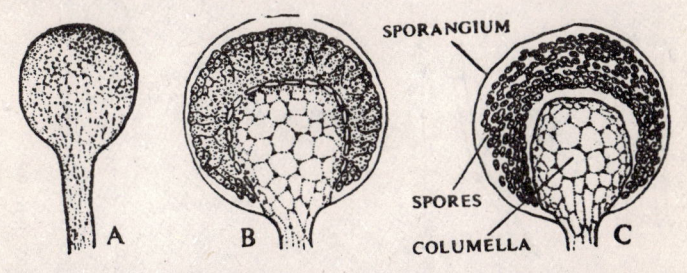

Mucor. FIG. 77. Development of sporangium, spore and columella. *A*, the end of the hypha swells; *B*, two regions—dense and light—are apparent with a layer of vacuoles between them; and *C*, mature sporangium (or gonidangium) with spores (or gonidia) and dome-shaped columella.

the central part remaining comparatively thin and vacuolate (*A*). Later a row of vacuoles appears around the central part and these soon fuse up (*B*). Consequently a cleft is formed between the outer denser portion and the inner thinner portion, thus separating them (*C*). The central portion which is dome-shaped and sterile, i.e. without spores, is called the **columella**. The peripheral protoplasm now gives rise to a number of small, multinucleate, angular masses by cleavage (*B*). Each multinucleate mass becomes rounded off and covered by a wall forming a **spore** (*C*). Its wall thickens and darkens. The wall of the sporangium is thin and brittle. Finally, as the columella swells due to accumulation of a quantity of fluid in it, it exerts a considerable pressure on the wall of the sporangium which as a consequence bursts, setting free the spores. The spores are blown about by the wind. The columella persists for some time after the bursting of the sporangium. The spores being very minute, light and dry, float about in the air, and under favourable conditions they germinate in suitable medium directly into the *Mucor* plant. Sometimes hyphae develop from the columella when the **sporangiophore** (i.e. the slender erect stalk of the sporangium) happens to fall over and these hyphae then bear the sporangia.

Sexual Method (FIG. 78). Sexual reproduction takes place by the method of **conjugation**, only under certain conditions, particularly when the food supply becomes exhausted. Conjugation consists in the fusion of *two similar* gametes, i.e. isogametes (cf. *Spirogyra*). The process is as follows: When two hyphae borne by two different plants of opposite sexes (called the + strain and the —strain) come close together, two short swollen protuberances, called the conjugating tubes or *progametes* (FIGS. 78 *A-B* & 80) develop forming a contact at their tips. As they elongate they push the parent hyphae apart from each other. Each pro-

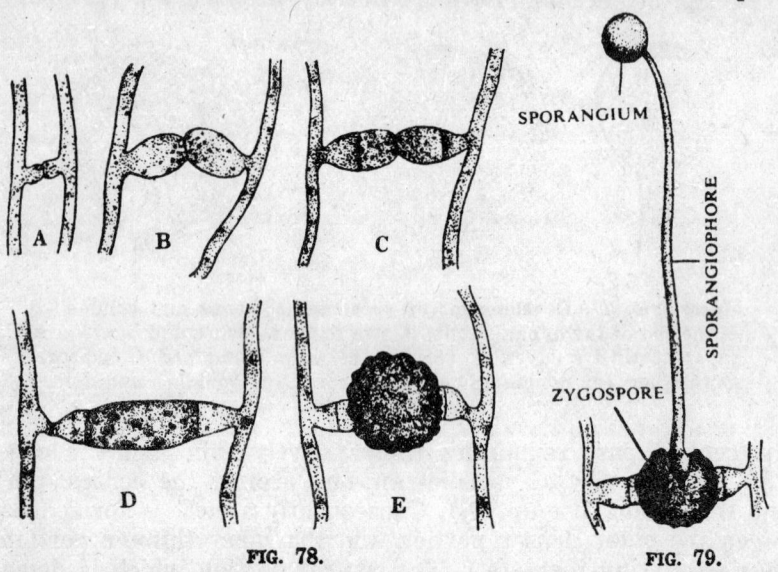

Mucor. FIG. 78. Conjugation: *A-E* are stages in the process; note the thick-walled zygospore at *E*. FIG. 79. Germination of zygospore.

gamete enlarges and becomes club-shaped. Soon it is divided by a partition wall into a basal **suspensor** and a terminal **gametangium** (FIGS. 78*C* & 80). The protoplasmic contents of each gametangium constitute a gamete. The gametes are multinucleate and are called *coenogametes*. The two gametes are identical in all respects. The end- (or common) walls of the two gametangia become dissolved, and the two gametes fuse together

FUNGI

(FIG. 78*D*) and form a zygospore (FIG. 78*E*). The zygospore swells into a rounded body, and its wall thickens, turns black in colour and becomes warted. It contains an abundance of food, particularly fat globules.

It has been observed that sometimes no sexual reproduction takes place even though the fungus grows under all favourable conditions. The investigations of the American botanist, Blakeslee (1904), have revealed the fact that there are two different strains or races of the fungus (the +strain and the —strain) and that sexual reproduction only takes place between the hyphae of these two different strains, apparently of opposite sexes. Evidently these two strains formed from separate spores—some giving rise to the *plus* strains and the others to the *minus* strains—must grow together (FIG. 80). Morphologically there is no difference between these two strains, excepting that the +strain (regarded as female) shows a little more vigorous growth than the —strain (regarded as male), but physiologically they are different behaving as two opposite sexes. Such species are said to be heterothallic, and the condition is designated as heterothallism. Heterothallism has also been found in certain Ascomycetes (e.g. *Ascobolus* and *Aspergillus*) and Basidiomycetes (e.g. rust fungi). There are, however, many species which form zygotes by the conjugation of the hyphae of the same mycelium. Such species are said to be homothallic.

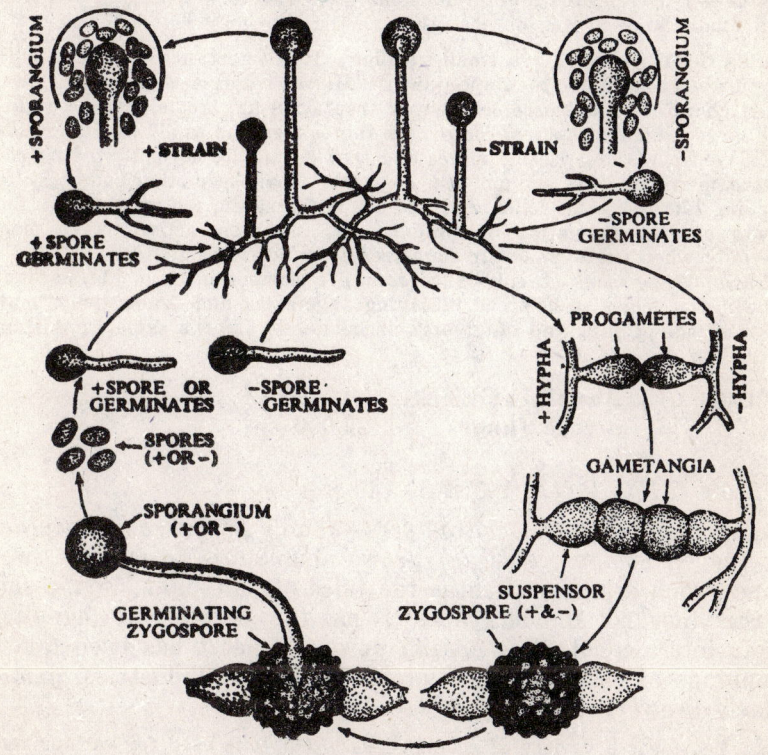

FIG. 80. Life-cycle of *Mucor*

Sometimes it so happens that although the conjugating hyphae meet, no fusion of gametes takes place. The gametangia then develop parthenogenetically (see p. 409) into thick-walled bodies called **azygospores** or **parthenospores** (cf. *Spirogyra*). The azygospore looks similar to the zygospore. Sometimes the free end of a hypha may produce a solitary azygospore. Germination of the azygospore in *Mucor* has not been followed.

Germination of Zygospore (FIG. 79). The zygospore undergoes a period of rest and then germinates. The outer wall bursts and the inner wall grows out into a tube, called the **sporangiophore** or **promycelium**, which ends in a single sporangium. The sporangiophore may be branched, each branch bearing a sporangium. The sporangium contains numerous small **spores** but no columella. The spore germinates, giving rise to the *Mucor* plant.

Blakeslee also found that in heterothallic species all the spores of a sporangium give rise to either +mycelia or to —mycelia, but not both. The zygote is diploid and contains material of both the +strain and the —strain. It undergoes reduction division at the initial stage of its germination, and of the four nuclei so formed three degenerate. Evidently the surviving one (+ or —) by repeated mitotic divisions gives rise to +spores, or —spores, and finally to +strains or —strains, as the case may be.

Torula Condition (*torula*, a small swelling). It is sometimes seen that under favourable conditions the mycelium of *Mucor* becomes segmented into a short chain of cells. These cells either then swell up, becoming thick-walled and large (chlamydospores), or remain thin-walled and small (oidium cells). Chlamydospores are resting spores and they germinate normally giving rise to the mycelium; but oidium cells separate from each other, multiply by budding like the yeast cells, and, like the latter again, set up alcoholic fermentation; the formation of oidium cells and their activity take place specially when the hyphae are immersed in a nutritive liquid.

Rhizopus nigricans (family *Mucoraceae*), a common black mould, is found growing on moist stale bread, decaying vegetables and fruits, jelly, male inflorescence of jack and other organic media. It has the same life-history as that of *Mucor*.

| CLASS III | ASCOMYCETES or sac fungi | 25,000 sp. |

1. *SACCHAROMYCES* (40 sp.)

Occurrence. Yeast (*Saccharomyces*—family *Saccharomycetaceae*, now called *Endomycetaceae*) grows abundantly in organic substances rich in sugar such as the juice of date-palm, in the soil of the vineyard, and in grapes; it has the property of changing sugar into alcohol. This special power of yeast has been taken advantage of in developing certain important industries, particularly brewery and bakery (see below).

Uses. Yeast has a variety of economic uses. (1) It is used for various fermentation processes such as the manufacture of beer from germinating

FUNGI

barley grains, of wine from grapes, of country liquor (toddy) from date-palm and palmyra-palm juices, etc. (2) Yeast is used in the preparation of industrial alcohol from different kinds of cereals and also from potato. (3) It is used in bread-making. The dough mixed with yeast gives sponginess and flavour to the bread. (4) It has a medicinal value since it is rich in vitamins and enzymes. (5) It has a nutritive value since it is rich in digestible compounds, specially proteins and fats.

Structure (FIG. 82). [Yeast was first microscopically examined by Leeuwenhoek in the year 1680. Its true nature was discovered

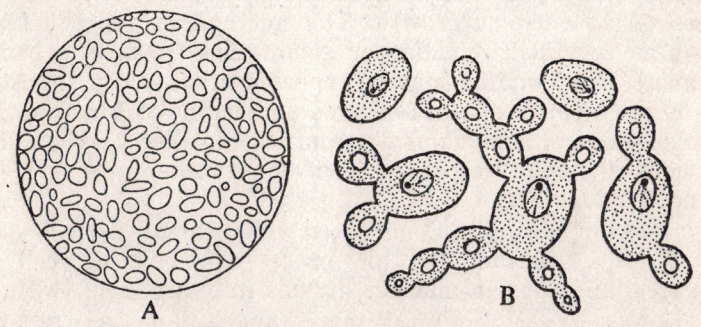

Yeast. FIG. 81. *A*, yeast cells as seen under the microscope; *B*, budding or gemmation.

by Schwann in Germany as late as 1836.] Its structure is very simple. A single cell represents the whole body of the plant. It is very minute in size and looks like a pinhead under the microscope (FIG. 81*A*). Each cell is colourless, oval or almost spherical or slightly elongated, and provided with a distinct cell-wall, possibly made of *chitin*, and contains a mass of cytoplasm and a single nucleus. The nucleus contains a large vacuole (FIG. 82), and this nuclear vacuole is a peculiarity in yeast. In the vacuole lies the nuclear reticulum with a nucleolus, and a centrosome on the side. Embedded in the cytoplasm there are granules of glycogen, protein and volutin and also several oil-globules. Mitochondria are common in the yeast cell.

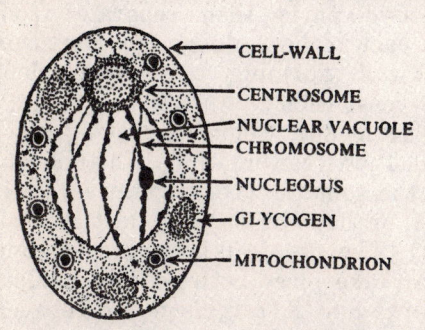

FIG. 82. One yeast cell (magnified) showing the nuclear vacuole.

Reproduction. This takes place commonly by budding, and sometimes by fission, and rarely sexually, as observed in a few species.

By Budding (FIG. 81*B*). The process of budding takes place under normal conditions when the yeast cells grow in sugar solution. As they grow there two changes are noticed—one in the yeast cells and the other in the sugar solution. The former change is the *budding* of yeast cells leading to vegetative reproduction, and the latter change is the *alcoholic fermentation* (see p. 548) leading to the breakdown of sugar into alcohol and carbon dioxide. In the process of budding each cell gives rise to one or more tiny outgrowths which gradually increase in size and are ultimately cut off from the mother cell; these then lead a separate existence. The nucleus divides amitotically and one passes on to each outgrowth. This method of reproduction is known as vegetative **budding** or **gemmation** (*gemma*, a bud; pl. *gemmae*). The budding may be repeated resulting in the formation of one or more chains and even sub-chains of bead-like cells; these chains and sub-chains are sometimes called *pseudomycelia*. The cells ultimately separate from one another and each leads an independent life.

By Fission (FIG. 83). Some yeast cells, called 'fission' yeasts, multiply by division. In this process the mother cell first elongates, and then its nucleus divides into two. The two nuclei move apart, and a transverse partition wall is formed somewhere in the middle of the mother cell, thus dividing it into two parts, each having a nucleus. The two parts then separate from each other along the partition wall, forming two independent yeast cells.

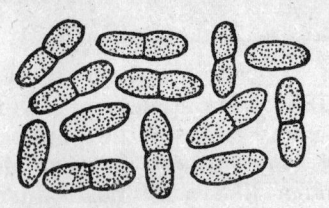

FIG. 83. Fission in yeast cells.

Sexual Reproduction (FIGS. 84 & 85). Some species of yeast also reproduce by the sexual method (conjugation). In this connexion it is to be noted that the somatic (i.e. vegetative) cells of yeast may be diploid ($2n$) or haploid (n), while the ascospores are always haploid (n). The zygote is of course diploid ($2n$). Sexual reproduction may take place between two haploid somatic cells, as in *S. octosporus* and *S. cerevisiae*, or between two ascospores, as in *S. ludwigii*, resulting in all cases in a zygote (diploid or $2n$). This zygote by budding may give rise to diploid somatic cells, as in the last two species, or by meiosis to ascospores, as in the first species. The ascospores are usually 8 or 4 in number according to species. The few species of yeast in which sexual reproduction has been observed so far go to show that the process may take any of the following three patterns. (Guilliermond, 1940).

(a) In certain species, as in *S. octosporus*, the diploid phase is very short, while the haploid phase is prolonged. Here two somatic cells (n) come in contact. At the point of contact they send out short, neck-like protuberances

FUNGI

(conjugating tubes) which unite by their tips. The two nuclei then pass on to the conjugating tubes. The partition is dissolved, and the two nuclei fuse at the neck. The conjugating tubes widen and the contents of the two cells unite, resulting in a zygote ($2n$). The zygote behaves as an ascus. Its nucleus now divides thrice, the first division being meiotic, and thus 8 nuclei are formed (each haploid or n). Each nucleus forms a wall round itself, enlarges

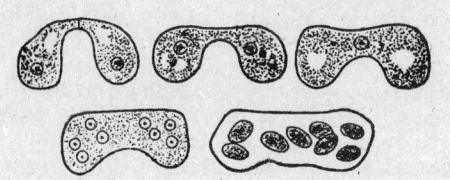

Yeast. FIG. 84. Conjugation of yeast cells and formation of ascospores.

and becomes as ascospore (n). The wall of the ascus breaks, and the ascospores are set free. Each ascospore enlarges and becomes a somatic cell (n). It divides by a transverse wall into two daughter cells (n). They may again take to conjugation under suitable conditions. Commonly, however, they multiply by budding. The sexual cycle may be represented as follows:

Somatic cells (n), by fusion in pairs ($n+n$) → zygote ($2n$), by meiosis → 8 ascospores (n) in ascus, by budding → somatic cells (n).

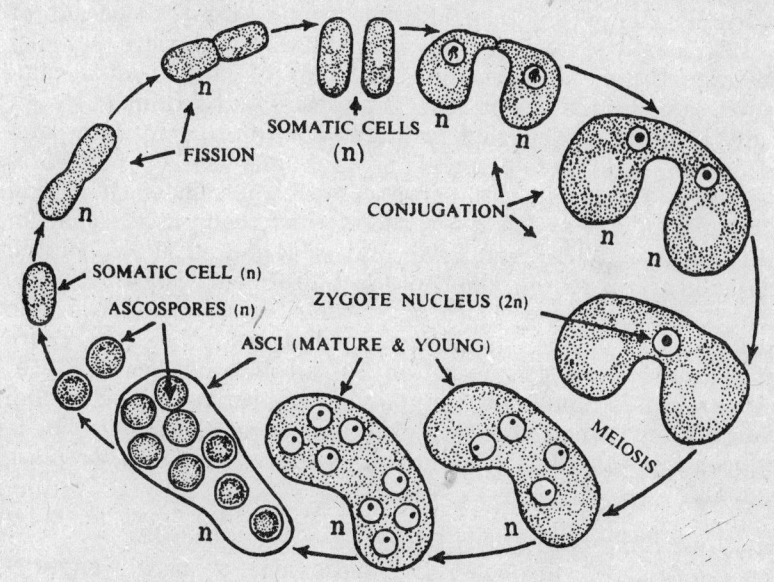

FIG. 85. Sexual cycle in a yeast (*Saccharomyces octosporus*).

(b) In certain other species, as in *S. ludwigii*, the diploid phase is prolonged, while the haploid phase is very short. Here the four ascospores (n) unite in pairs within the ascus, resulting in two zygotes (each $2n$). The zygote cell produces a *germ tube* ($2n$) which grows into a separate mycelium ($2n$) called *sprout mycelium*. Each cell of this mycelium produces somatic (yeast) cells ($2n$) by budding. The cells soon get detached. The somatic cells enlarge and become converted into asci, each with 4 ascospores

(n) produced by meiosis. They may again take to conjugation under suitable conditions. The sexual cycle may be represented as follows:

Somatic cells ($2n$), by meiosis → 4 ascospores (n) in ascus, by fusion in pairs → zygote ($2n$), by sprouting and budding → somatic cells ($2n$).

(c) In still other species, as in the common bread yeast (*S. cerevisiae*) both the phases (haploid and diploid) are more or less equally important since each phase may continue for an indefinite period by budding. Two such haploid somatic cells (+ and −) may fuse giving rise to a diploid cell, i.e. the zygote ($2n$). It continues to multiply by budding, giving rise to a large number of well-developed yeast cells ($2n$). Eventually they may behave as asci, each with 4 haploid ascospores formed by meiosis. After liberation from the ascus the ascospores continue to multiply by budding, giving rise to several haploid yeast cells (+ or −). They are somewhat smaller than the diploid cells. The sexual cycle may be represented as follows:

Somatic cells (n, + or −), by fusion in pairs → zygote ($2n$, + −), by budding → somatic cells ($2n$, + −), by meiosis → 4 ascospores (n, 2 + and 2 −) in ascus, by budding → somatic cells (n, + or −).

Alcoholic Fermentation. The process of fermentation was first elaborately studied by Louis Pasteur during the fifth decade of the 19th century, and many of his discoveries were actually applied to the making of different kinds of liquors on a commercial basis by the use of different species and strains of yeast. When the yeast cells grow in sugar solution, as in date-palm juice, palmyra-palm juice or grape juice, they set up fermentation (see p. 375) in it in the absence of oxygen under the action of the enzyme *zymase complex*, secreted by them, as first shown by Buchner in 1897. Sugar is decomposed, and ethyl alcohol and carbon dioxide are the chief products formed. Carbon dioxide escapes, and often gives rise to frothing on the surface of the solution. Ethyl alcohol is poisonous, and the yeast cells cease functioning if the concentration of alcohol reaches 14-18%. When oxygen is abundantly supplied comparatively little alcohol is formed, but when the supply of oxygen is cut off alcohol is produced more freely. The following chemical change takes place in sugar.

$$C_6H_{12}O_6 + Zymase = 2C_2H_5OH + 2CO_2 + Zymase + Energy$$
(sugar + zymase = alcohol + carbon dioxide + zymase + energy)

2. *PENICILLIUM* (137 sp.)

Penicillium and *Aspergillus* are two common and important genera of the family *Aspergillaceae*. They are common green and blue moulds. Both follow the same trend of life-history but their morphological structures are different. They can be readily distinguished by the nature of conidiophores that they abundantly produce. They belong to the order Aspergillales numbering about 800 species.

FUNGI

Occurrence. *Penicillium* (FIG. 86), a blue or green mould, is a very common and widely distributed fungus. Most species of *Penicillium* are saprophytic in habit, and they commonly grow on bread, vegetables, fruits, jams and other foodstuffs, and also on leather, shoes, fabrics, paper, books, etc. Spores of this fungus are present almost everywhere in the air and the soil, and are often sources of contamination of foodstuffs including fruits and vegetables, sometimes resulting in huge spoilage. A few species are parasitic on animals including human beings. *P. italicum*, a blue mould, and *P. digitatum*, a green mould, are common parasites of *Citrus* fruits. *P. notatum*, a blue-green mould, is the source of world famous **penicillin** (an antibiotic) first isolated from this fungus by the late Sir Alexander Fleming, a bacteriologist, in 1929. Certain species of *Penicillium* are used industrially in making various **organic acids, and in making special types of flavoured cheese.**

Structure. The mycelium consists of an interwoven mass of hyphae which spread on the surface of the substratum, pene-

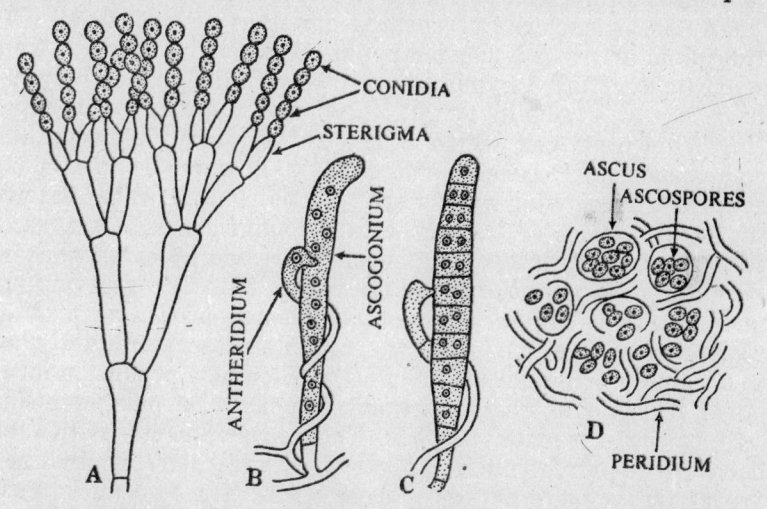

Penicillium. FIG. 86. *A*, branched conidiophore ending in sterigmata with conidia in chains (penicillus); *B*, sex organs: antheridium and ascogonium; *C*, ascogonium with binucleate cells; *D*, cleistothecium.

trating here and there deep into it. The hyphae branch freely and are septate and multinucleate.

Reproduction. *Penicillium* freely reproduces asexually by conidia. A few species take to sexual reproduction.

Asexual Reproduction. A number of hyphae stand erect from the undifferentiated vegetative mycelium; these are the **conidio-**

phores (*A*). They branch repeatedly near the apex, regularly or irregularly, in a broom-like fashion, and are septate. The slender ultimate branches known as the **sterigmata** (sing. sterigma) cut off chains of cells by the process of budding. These are the spores called the **conidia** which are formed in countless numbers. They are spherical or oval in shape, and commonly bluish or greenish in colour. The branched conidiophores including the sterigmata and the conidia are together called **penicillus** (a little brush). The conidia are easily dispersed by the wind, and under favourable conditions they germinate by producing a *germ tube*.

Sexual Reproduction. This has been observed only in a few cases, first by Dangeard in 1907. Perfect stages, however, are not known yet. In certain species the development of the antheridium (male) and the ascogonium (female) has been observed (*B*); in others the antheridia are absent or do not function. The ascogonium is a long straight and somewhat club-shaped body. It is at first uninucleate but later it becomes multinucleate (with as many as 64 nuclei) by repeated divisions of the nuclei. The antheridium arises as a slender hypha (**antheridial branch**) from a separate vegetative hypha and grows twining round the ascogonium up to a certain height. A septum appears in it, separating a terminal cell which is the antheridium. It swells and becomes more or less club-shaped, and is uninucleate. Its tip touches the ascogonial wall laterally. The contact walls dissolve and the protoplast of the antheridium migrates into the ascogonium. A doubt, however, has been expressed by some workers if the antheridium nucleus at all passes into the ascogonium. Commonly, however, the ascogonial nuclei approach each other in pairs. After this pairing of the nuclei the ascogonium divides into a number of binucleate cells (*C*) which now produce a large number of hyphae—the **ascogenous hyphae**. The pairing nuclei pass into the ascogenous hyphae. They develop septa, with each cell having a pair of nuclei (**dikaryon**), and their terminal cell develops into a more or less globose **ascus**. The two nuclei fuse in the young ascus (or ascus mother cell). The fusion (or zygote) nucleus ($2n$) divides thrice (the first division being meiotic) and finally gives rise to 8 **ascospores** (n) in each ascus.

In the meantime a closed 'fruiting' body or **ascocarp** (see (p. 527), called the **cleistothecium** (*D*), is formed from the surrounding vegetative hyphae, evidently enclosing a number of asci. It has a protective sheath, called the **peridium**, which is pseudoparenchymatous in nature. The inner layer of the peridium is nutritive in function. As the ascospores mature, the asci dissolve away leaving the ascospores free and scattered within

FUNGI

the cleistothecium. The peridium decays and the ascospores are liberated. Finally they are blown away by the wind.

3. ASPERGILLUS (78 sp.)

Occurrence. *Aspergillus* (=*Eurotium*; FIG. 87), commonly called blue mould, is a very widely distributed fungus like *Penicillium*, and its spores are abundant in the air and in the soil. Species of *Aspergillus* commonly grow on almost all kinds of foodstuffs including butter, bread, fruits, vegetables and jams, and on leather goods, fabrics and books, sometimes causing considerable damage to them, particularly during the rainy season. Air-borne spores (conidia) are often sources of contamination of foodstuffs and their decay. *A. niger*, a black mould, is such a nuisance. *Aspergillus* species are mostly saprophytic in habit, but a few (e.g. *A. fumigatus*) are parasitic on animals including human beings, causing diseases of the ear and the lungs.

Aspergillus can produce a large number of enzymes which enable them to grow on a variety of organic media. *Aspergillus* is economically an important fungus. Some species (e.g. *A. oryzae*) are used industrially in the manufac-

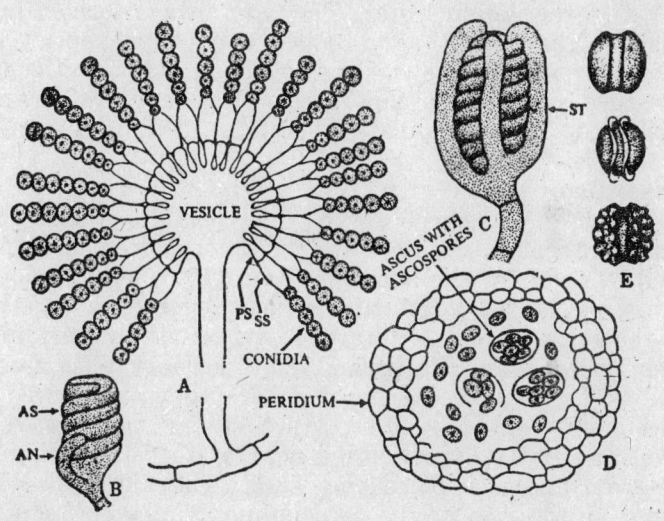

Aspergillus. FIG. 87. *A*, conidiophore with primary sterigmata (*PS*), secondary sterigmata (*SS*), conidia, vesicle and foot cell (at the bottom); *B*, sex organs: antheridium (*AN*) and ascogonium (*AS*); *C*, sterile hyphae (*ST*) enclosing the ascogonium; *D*, cleistothecium; *E*, ascospores of different species.

ture of alcohol from rice starch, and some species used in the manufacture of certain organic acids (e.g. citric, gluconic, etc.) on a commercial basis. Some enzyme preparations have also been made possible through their activity. Some species are sources of certain antibiotics.

Structure. The mycelium consists of an interwoven mass of hyphae which branch freely and spread through the substratum on its surface as well as into it. The hyphae are hyaline, septate, much branched and multinucleate.

Reproduction. *Aspergillus* freely reproduces asexually by conidia. Only a few species reproduce sexually.

Asexual Reproduction. Several hyphae stand erect from certain cells (called **foot cells**) of the vegetative hyphae; these are the **conidiophores** (*A*). They are long and erect but unseptate. Each conidiophore swells at the apex into a more or less spherical, multinucleate head called the **vesicle**. Over the entire surface of the vesicle, growing from it, there are innumerable cells in 1 or 2 layers according to the species. Of the two layers those of the first are short, and are called the *primary sterigmata*. The cells of the second layer are long and bottle-shaped, and are known as the *secondary sterigmata*. If only one layer is present the cells become bottle-shaped. Each such sterigma produces spores, called **conidia**, in a chain in basipetal order. The conidia are spherical or oval in shape, and commonly multinucleate (rarely uninucleate). They are always formed in great profusion. The colony as a whole may then appear bluish, greenish, blackish or brownish in colour according to the species. The mature conidia eventually become loose and are blown away by the wind. They germinate in suitable medium by producing a *germ tube*.

Sexual Reproduction In a few species of *Aspergillus* sexual reproduction has been observed but perfect stages are not known. Antheridia (male) and ascogonia (female) may be formed in them for the above purpose (*B*). They develop from separate specialized vegetative hyphae lying close together, or from the same hypha at different levels. Both antheridium and ascogonium are unicellular and multinucleate. The ascogonial hypha, as it grows, becomes soon differentiated into three parts: a unicellular, tightly coiled structure—the **ascogonium**, a terminal cell (which is the receptive neck of the ascogonium)—the **trichogyne**, and a multicellular **stalk**. The antheridial hypha, growing by the side of the ascogonium, climbs it. Soon it cuts off a terminal cell—the **antheridium**. It further climbs and its tip reaches the trichogyne. The contents of the antheridium migrate through the trichogyne into the ascogonium where pairing of nuclei (male and female) takes place. Their actual fusion has not been observed. Commonly, however, the sister nuclei of the ascogonium come together in pairs. The antheridium may also remain undeveloped or functionless, or it may not reach the trichogyne. After pairing of the nuclei within the ascogonium,

FUNGI

whatever be the method, the latter becomes septate forming a number of binucleate cells. These cells now produce several **ascogenous hyphae** which branch within the 'fruiting' body (see below), and are of different lengths. They become septate, with binucleate cells, and produce asci at their tips. The **asci** are spherical, oval or pear-shaped, and binucleate (dikaryotic). Fusion of the nuclei (karyogamy) takes place in the young ascus, and the zygote nucleus ($2n$) divides thrice (the first division being meiotic) to form 8 nuclei. Each nucleus clothes itself with a wall and becomes an **ascospore**. Thus there are 8 ascospores in each ascus. The ascospores are peculiar in shape being like pulley-wheels (E). In flat view, however, they are commonly spherical, oval or star-shaped.

In the meantime, soon after the formation of the sex organs, a number of sterile vegetative hyphae grow up the ascogonium (C), and enclose it as a 2-layered sheath called the **peridium**. It is pseudoparenchymatous in nature. Its outer layer is protective in function, while its inner layer is nutritive in function. The peridium enclosing the asci appears as a small globose body; this is the 'fruiting' body or ascocarp of the fungus, and being closed is otherwise known as the **cleistothecium** (D). The asci within the cleistothecium soon dissolve away, leaving the ascospores free and scattered within it. The peridium decays, and the ascospores are blown away by the wind. They germinate under suitable conditions by producing a *germ tube*.

4. *ERYSIPHE* (10 *sp.*)

Occurrence and Structure. *Erysiphe* (family *Erysiphaceae*; FIG. 88) is an obligate parasite causing a disease commonly

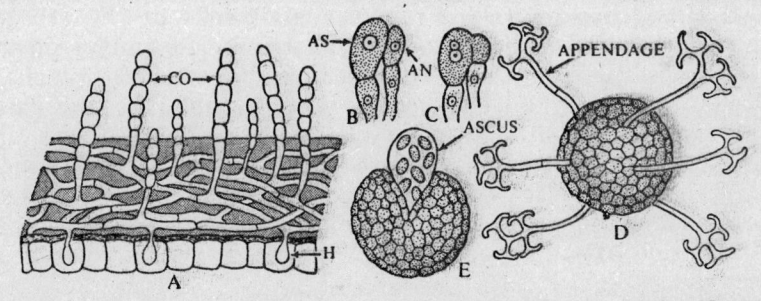

Erysiphe. FIG. 88. *A*, conidiophores and conidia (CO) on leaf-surface, and haustoria (H) in epidermis; *B*, sex organs: ascogonium (AS) and antheridium (AN); *C*, the same showing pairing of male and female nuclei; *D*, cleistothecium with dichotomously branched appendages; and *E*, the same bursting with the liberation of asci (appendages not shown).

called powdery mildew. Powdery mildews are mostly superficial parasites which first appear as small white spots on the leaf and then rapidly spread on the entire leaf-surface, usually the upper. The surface then appears to be coated with a fine white powder which can be easily brushed off. Common hosts of *Erysiphe* are cucurbits, cereals, some grasses, pea, bean, rose and numerous other plants. *E. polygoni* is almost omnivorous, growing on a wide range of hosts, wild or cultivated. In India this species is common on pea and some other leguminous plants, and *E. graminis* on barley and some other cereals, and also on many grasses. The mycelium of the fungus is superficial and consists of a matted mass of colourless septate hyphae (*A*) which here and there send haustoria into the epidermal cells of the host. The haustoria swell or branch off within the host cells. The hyphae are hyaline at first but later they change to greyish, brownish or reddish colour. On the surface of the leaf the mycelium produces a large number of simple erect conidiophores, each cutting off from the top downwards a chain of oval **conidia**, having the appearance of a powdery coating. The fungus multiplies rapidly by conidia.

Later with the cessation of conidia-formation, the fungus produces uninucleate sex organs—**ascogonium** (female) and **antheridium** (male). The latter is more slender than the former. They develop side by side at the tips of hyphae pressing together closely (*B*). The male nucleus enters the ascogonium through a perforation made in its wall and pairs with the female nucleus of the ascogonium (*C*). It is now generally agreed that karyogamy (see p. 526) takes place in the young ascus mother cell. After this the ascogonium forms into a row of cells; the penultimate cell of this row produces ascogenous hyphae which then form **asci**, usually 2-8, within a closed ascocarp (see p. 527), called **cleistothecium** (*D*). From the stalk-cells of the ascogonium and the antheridium sterile hyphae arise forming a sheath of the ascocarp. From the sheath-cells peculiar hooked, branched, bulbous or simple *appendages* develop around the ascocarp. These are important diagnostic characters. The ascocarps appear on the leaf-surface as minute, spherical, dark bodies, looking like so many black dots. The ascus commonly bears 8 ascospores, sometimes less (2-6) in some species. The ascospores are liberated after the bursting of the ascocarp and the ascus (*E*).

The source of infection may be the dormant mycelium in the seed, or the ascocarp remaining in the soil. *Control.* (*a*) Seeds may be soaked in hot water (50°C) for about 10 minutes; (*b*) the plants may be dusted with fine sulphur dust.

5. *UNCINULA* (20 *sp.*)

Occurrence. **Uncinula necator** (family *Erysiphaceae*; FIG. 89) is an obligate parasite like *Erysiphe,* causing a powdery mildew disease on grape vine (*Vitis vinifera*). The disease is prevalent in the vineyards of India, as at Nasik where grapes are extensively cultivated. Sometimes the disease breaks out in an epidemic and destructive form causing a heavy damage to the crop. The disease was first noticed in Kent (England) in 1845, and in 1851 almost all the vineyards in France were found to be seriously affected. The disease is also common in America and Australia.

Nature of the Disease. The vine may be attacked at any stage of its growth. The disease appears mainly on the leaves (A) and berries as white, diffuse, powdery or dusty patches which soon turn grey and finally dark. Microscopic examination shows that the dusty patches are made of mycelia and conidia. All parts of the plant body may be infected. Infected flowers do not produce fruits, while the infected young fruits drop off. Comparatively old fruits, if infected, become distorted in shape; their skin cracks and they seldom ripen. Diseased leaves curl up and become discoloured and deformed. On the stem black marks are left when the fungus is rubbed off. Diseased plants on the whole show stunted growth, and they may wilt. Moderately high temperature and cloudy sultry weather are conducive to the growth of the fungus.

Morphology of the Fungus. *Uncinula necator,* like other powdery mildews, is superficial (ectophytic). The mycelia together with

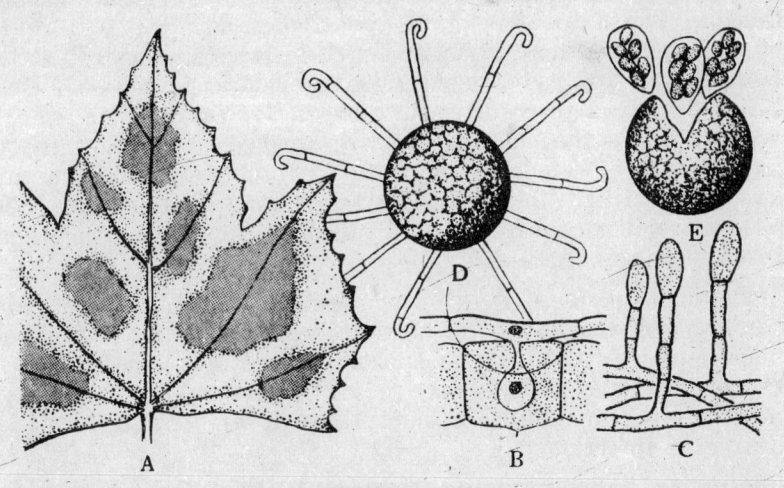

Uncinula. FIG. 89. *A,* an infected leaf of grape vine; *B,* a haustorium penetrating into an epidermal cell of the host; *O,* hyphae with conidiophores and conidia (on the top); *D,* a cleistothecium with hooked appendages; and *E,* a bursting cleistothecium showing the liberation of asci.

the conidiophores gradually spread on the surface of the affected parts. Here and there the hyphae send sucking organs or haustoria (B) into the epidermal and even sub-epidermal cells of the host plant. These enlarge and become pyriform in shape within the host cell. The hyphae are branched and septate, forming at first a white web on the surface, and darkening later.

Reproduction. The fungus commonly reproduces asexually by conidia, and sometimes sexually resulting in the formation of cleistothecia (closed fruiting body) enclosing asci. This, however, is not a common feature in India. Environmental factors may be responsible for this.

Asexual Reproduction. For this purpose **conidia** appear in abundance on leaves and fruits, forming a dense powdery film (A & D). Conidia, usually 3 or 4, are produced in a chain at the end of each simple erect conidiophore. They are oval in shape. When mature they are easily dispersed by wind or rains. They are resistant to cold and dry atmospheric conditions. Under warm moist conditions conidia readily germinate on the host (grape vine) and grow vigorously on it in shady position.

Sexual Reproduction. Stages in sexual reproduction are but imperfectly known although **cleistothecia** (D) have been abundantly found in some countries, though not in India. They lie embedded in the ectophytic mycelia on the surface of the leaf or shoot. They are hyaline when young but soon turn black, usually appearing in early winter or sometimes much earlier. They are almost spherical but somewhat flattened on the top. Each cleistothecium has 8 to 25 septate *appendages* developing from its surface cells. Each appendage is curled inwards or hooked at the tip. The fruiting body contains 4 to 8 ovoid **asci**, and each ascus bears usually 4 to 6 oval **ascospores** (E), rarely 8 or even 2. When mature the cleistothecia are washed down by rains or blown away by wind, either singly or in clusters. They rupture and the asci are liberated. The asci then burst and the ascospores are dispersed by the wind. Under suitable conditions they infect new hosts.

Mode of Infection. (1) Fresh infection normally takes place through conidia which have survived winter. (2) Cleistothecia may remain on the vines and leaves or on the ground. Later in the spring the ascospores may infect leaves and shoots. (3) Dormant mycelia remaining in buds through winter may also be a source of infection.

Control. (1) Sulphur dusting is regarded as a universal treatment against this vine mildew. Timely dusting (once when the shoots are still young and again when the flowers are about to appear; a third application some time later may also be necessary) effec-

FUNGI

tively controls the disease. (2) Pruning and removing all diseased parts. (3) Washing the vines with an acid solution of iron sulphate. (4) Coating the stems with a paste of sublimed sulphur and soft soap.

6. *NEUROSPORA* (3 sp.)

From academic standpoint *Neurospora* is considered very important since a considerable amount of genetical work, possibly next to yeast, has been done on this fungus, first by Dodge in 1927-28 and later, and subsequently by Lindegren in 1932 and later, Beedle in 1944, Beedle and Tatum in 1945, Tatum in 1950, and also by others. The three species of this genus are *N. sitophila, N. tetrasperma* and *N. crassa*. All of them have been thoroughly studied.

Occurrence and Structure. *Neurospora* (family *Fimetariaceae*; FIG. 90*A*), commonly called pink bread mould, is a saprophytic fungus, frequently occurring in bakeries and sometimes causing much damage. In the laboratory it often contaminates cultures of fungi and bacteria. The mycelium consists of a mass of branched, septate, somatic hyphae, spreading on the substratum. The aerial hyphae produce countless *pink* spores (conidia) and hence the fungus is easily recognizable. *Neurospora* is mostly heterothallic, being differentiated into +strain and —strain (cf. *Mucor*).

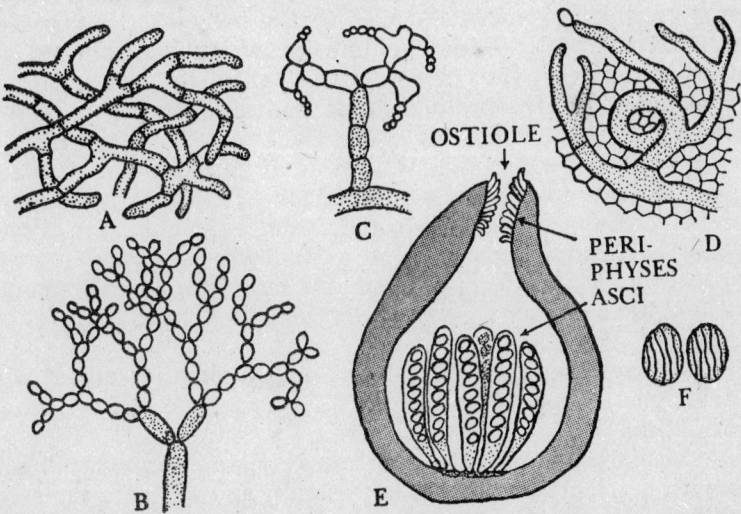

Neurospora. FIG. 90. *A*, branched and septate mycelia; *B*, a branched macroconidiophore with beaded macroconidia; *C*, a branched microconidiophore with microconidia; *D*, ascogonium (a coiled fertile branch); *E*, a mature perithecium with asci; and *F*, ascospores (2 shown) with ridges on the wall.

Reproduction. The most common method of reproduction is asexual, effected by conidia which are produced in immense numbers. Sexual reproduction was first discovered by Dodge in 1927-28. Since the discovery of this phenomenon much work has been done on the genetics of this fungus.

Asexual Reproduction. This takes place freely by **conidia** which are produced by the somatic aerial hyphae. Conidia are of two kinds: large (*macroconidia*) and small (*microconidia*). The former are borne in huge numbers in beaded chains on branched conidiophores, and are oval in shape and pink in colour (FIG. 90*B*). They are easily dispersed by the wind. The microconidia are minute in size and occur singly or in very short chains (each consisting of a few conidia) on special branched conidiophores (FIG. 90*C*). The fungus reproduces itself indefinitely by the conidia, generation after generation. Under favourable conditions they germinate and give rise to somatic hyphae (+ or −) again.

Sexual Reproduction. *Neurospora,* as stated before, is mostly heterothallic, and this being so, sexual reproduction occurs only when two opposite strains (+ and −) happen to come together. No antheridia or sperm cells have been found in *Neurospora.* The female gametangium, as in other Ascomycetes, is the ascogonium. It is seen that a loose mass of somatic hyphae forms into a flask-shaped body—the **perithecium.** The young perithecium is seen to contain a somewhat coiled fertile body—the **ascogonium** (FIG. 90*D*). From the ascogonium grow out certain curved hyphae—the **trichogynes** or receptive hyphae. Now plasmogamy, i.e. the union of two protoplasts of opposite strains, may take place in any of the following ways, leading to the production of asci and ascospores and the maturation of perithecia or 'fruiting' bodies. The mycelia of any strain (+ or −), though hermaphroditic in nature bearing both conidia and ascogonia, are mostly self-sterile. Plasmogamy may be as follows:

(1) Two hyphae of opposite strains may fuse (somatogamy).

(2) A hypha of one strain may come in contact with a trichogyne of opposite strain.

(3) A macroconidium of one strain may come in contact with a trichogyne of opposite strain and act as a male cell or gamete (spermatium).

(4) A microconidium of one strain may similarly come in contact with a trichogyne of opposite strain and act as a male cell or gamete (spermatium).

Perithecia, Asci and Ascospores. After plasmogamy the perithecia begin to mature, with several asci formed from the ascogonium within each perithecium. The young ascus is binucleate (one + and one −). But at an early stage of ascus formation the two nuclei within it fuse (karyogamy) and thus the ascus becomes uninucleate (+ −). This fused nucleus is the zygote which represents the only diploid ($2n$) phase of *Neurospora*. Then by meiosis followed by mitosis 8 ascospores (only 4 in *N. tetrasperma* by meiosis) are formed in each ascus. Sexual differentiation, i.e. differentiation into + strain and − strain, occurs mostly during the first division, sometimes second, of the ascus-nucleus in the young ascus (see p. 560). The mature perithecium (FIG. 90*E*) is pyriform, beaked, dark-coloured, and is provided with a thin wall and a long neck with an apical pore—the **ostiole**. The perithecium is superficial on the substratum, i.e. not on or within a stroma (see p. 562). It encloses several cylindrical asci, each opening by a minute apical pore. There may be hair-like growths (**periphyses**) forming an inner lining of the ostiole. Paraphyses are, however, absent in a mature perithecium. Each mature **ascus** contains 8 brownish or blackish, oval **ascospores**, of which 4 are of one sexual strain and 4 of the other strain. Dodge further found that the 4 ascospores formed in *N. tetraspora* are binucleate (+ and −) and, therefore, self-fertile, i.e. when such spores are germinated abundant perithecia are formed. The ascus has a definite apical pore, very minute though, through which the ascospores are expelled with some force. Within the perithecium the asci mature at different times. Each ascospore (FIG. 90*F*) is provided with distinct nerves or ridges on its wall, and hence the name *Neurospora* (*neuron*, a

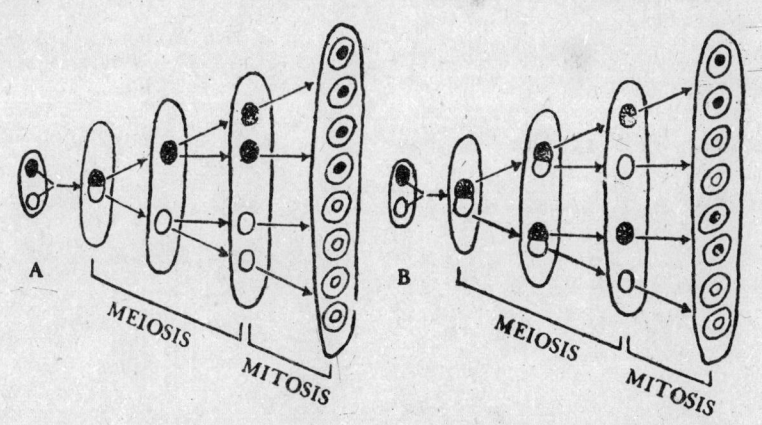

Neurospora. FIG. 90-1. Segregation of sex in ascospores. *A*, segregation of alleles (paternal and maternal) in the first meiotic division (but no crossing-over between two genes) — the result is 4:4 distribution of ascospores in the ascus; *B*, segregation of alleles in the second meiotic division (with crossing-over between two genes) — the result is 2:2:2:2 distribution of ascospores in the ascus.

nerve). The ascospores germinate easily in suitable medium into + mycelia or − mycelia, as the case may be.

Segregation of Strains. Cultures from ascospores of an ascus are seen to produce abundant perithecia; evidently the spores are of two strains (+ and −). By a special technique Dodge in 1928 was able to isolate the spores of an ascus one by one, and grow them separately into mycelia. He found that the first 4 spores at the bottom of an ascus gave rise to mycelia of one strain (+ or −), while the 4 spores at the other end gave rise to mycelia of the opposite strain (− or +), thus indicating that the segregation of sex factors took place in the *first division* (meiotic) of the ascus-nucleus in the young ascus (FIG. 90-1A). He had further shown that sex segregation might also take place to a certain extent in the *second division* (FIG. 90-1B). The above facts Dodge experimentally proved by growing the spores of specific locations in the ascus and observing to what extent mating of + strain and − strain took place, resulting in the formation of perithecia, i.e. whether both the strains were represented in each culture or only one strain. Dodge was also able to bring about crossings between different species of *Neurospora*. The experimental results of Dodge were corroborated by Lindegren and later by Beedle and Tatum, as stated in the beginning. Lindegren further found that the *first division* of the ascus-nucleus was meiotic in about 85% cases, while the *second division* was so in about 15% cases.

7. *CLAVICEPS* (12 sp.)

Claviceps (family *Clavicipitaceae*) belongs to the order Hypocreales which number over 800 species.

Claviceps purpurea (FIG. 91), commonly called ergot fungus, grows as a parasite on rye (*Secale cereale*) and other grasses.

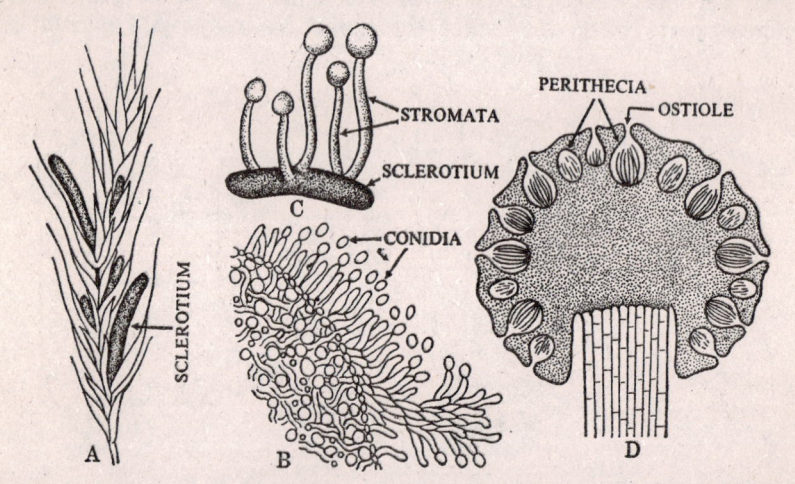

Claviceps. FIG. 91. *A*, sclerotia on spikelets of rye; *B*, conidia (*Sphacelia*) stage on the ovary; *C*, germinating sclerotium producing stromata; *D*, a stroma in vertical section showing numerous perithecia.

It is well known for the ergot disease it produces on the ovaries of rye, for its poisoning effect on many herbivorous animals and

FUNGI

human beings, and also for its universal use as a medicine known as ergot. In the spring if the ascospores of the fungus happen to fall on the flowers of rye, as they always do under natural conditions, they germinate into germ tubes which penetrate into the ovary. Thus the ovary becomes infected, and within it the mycelium ramifies and forms into a compact cottony mass of septate hyphae. The mass further grows, becomes hard and turns dark-purple or violet. The body thus formed is the **sclerotium** of the fungus (91A), and usually ranges in length from 1 to 3 cm. (see p. 526). It is pseudoparenchymatous in nature. The ovary gets completely destroyed, and the sclerotium takes its place. Finally it appears through the spikelet as an elongated, often somewhat curved and dark purplish body. In affected rye fields sclerotia are of common occurrence but so far it has not been possible to induce their formation in artificial culture.

Economic Importance. The sclerotium of rye is called the ergot. It contains seven alkaloids, of which **ergotamine** and **ergotoxine** are powerful. Ergot is distinctly poisonous, and its poisoning effect is known as ergotism. Grazing animals feeding upon diseased grains suffer badly; their hoofs, tails and horns are affected due to contraction and thickening of the blood-vessels supplying blood to these parts, finally leading to gangrenous condition. Besides, paralysis, nervous disorders and abortion are also the effect of ergot poisoning. Human beings taking contaminated bread or food equally suffer. Nevertheless, ergot is a universal drug. It is used, in medicinal doses of course, to stop haemorrhage and to induce uterine contraction after childbirth. Ergot is collected from naturally diseased rye in Russia, Spain, Portugal and Poland. A large quantity is exported to Britain and America.

Asexual Reproduction. Mycelium extends to the surface of the ovary and spreads over it as an interwoven mass of hyphae. Their tips behave as short slender **conidiophores** which cut off from their ends minute, oval or spherical **conidia** in great profusion (91*B*). Conidia formation takes place within two days after infection. A sweet yellow sticky fluid—'honeydew' as it is also called—emitting also an odour, is secreted by the infected flowers, bathing the conidia. This stage is otherwise known as the *Sphacelia* stage of the fungus, and indicates an early sign of infection. The liquid attracts insects which carry the conidia to other flowers of rye and also to other grasses, and infect them. In due course they may produce sclerotia again.

The **sclerotium** (FIG. 91*A*) is a hard compact mass of hyphae, dark outside and white inside. With the cessation of conidia formation the sclerotia begin to grow and take shape. When fully formed many of them fall to the ground or are harvested along with the grains. They contain plenty of food—proteins and fats—and pass into a dormant stage during the winter. In the next spring they grow and each gives rise to a number of mushroom-like bodies with long stalks and more or less globose heads which are yellowish-brown at first but turn pink or violet later; they are called **stromata** (91C). (A stroma is a compact mass of hyphae like the sclerotium but it bears perithecia or fruiting bodies). The stroma is also pseudoparenchymatous in nature like the sclerotium. A broken sclerotium may also grow normally like the intact one. Sunken in the stromatic head there are numerous flask-shaped fruiting bodies called **perithecia** (91D), lying in a single rink-like layer.

The perithecium (FIG. 92A) has a thin wall, and it opens outside by a pore called the **ostiole**. Within the perithecium develop many elongated, cylindrical **asci** interspersed with slender paraphyses. Each ascus contains 8 acicular (needle-like) **ascospores** (FIG. 92B). The perithecium grows along with the formation of the asci. When mature, the ascospores are forced out of the asci and finally dispersed by the wind. The ascospore germinates by putting forth *germ tubes* (92C).

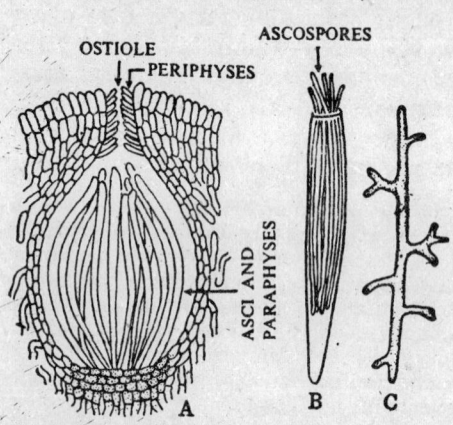

Claviceps. FIG. 92. A, a perithecium in vertical section (magnified); B, an ascus discharging ascospores; C, an ascospore germinating.

Sexual Reproduction. Asci must have been formed as a result of sexual union, as in other members of Ascomycetes, but details of their formation in *Claviceps* are not known yet. From certain hyphae at the base of the perithecium arise the sexual organs. These hyphae have much denser protoplasmic contents and are, therefore, easily distinguishable from the rest of the hyphae. They are branched, and from the terminal cell of a branch develop an ascogonium (female) and one or more antheridia (male) side by side. Both are multinucleate but the ascogonium is broader than the antheridium. The male nuclei of the antheridium migrate into the ascogonium, and plasmogamy (see p. 526) takes place; the nuclei are thus arranged in pairs, each pair being called a dikaryon (see p. 526). The ascogonium then produces ascogenous hyphae. Ascus formation takes place at the tips of such hyphae by hook or crozier method (see p. 527).

8. *PEZIZA* (150 *sp.*)

The family *Pezizaceae* is represented by a number of genera, of which *Peziza* and *Ascobolus* are described in the following pages. Both look alike with well-developed, fleshy, cup-shaped fruiting body but they differ mainly by the following points: in *Peziza* the ripe asci do not project beyond the level of the hymenium, and the ascospores are uniseriate in the ascus and colourless; while in *Ascobolus* the ripe asci elongate greatly and project

FUNGI

beyond the level of the hymenium, and the ascospores are biseriate or multiseriate in the ascus and often coloured (brown or violet).

Peziza (family *Pezizaceae*) belongs to the order **Pezizales** which has about 500 species.

Peziza (FIG. 93A) is a cup fungus. It has characteristically a fleshy, cup- or saucer-shaped, superficial fruiting body, commonly varying in sizes from 1-10 cm. in diameter; some may be much smaller or much bigger. The cups are commonly regular in form, sometimes irregular and often contorted. *Peziza* is saprophytic in habit. It is found growing abundantly during the rainy season on heaps of semi-decomposed cowdung, decaying wood and in soil heavily manured or rich in humus.

The hyphae are much branched and grow extensively penetrating into the substratum. The interwoven hyphae are massed together forming the aerial, fleshy, cup-shaped reproductive or fruiting body known as the **apothecium** (see p. 527). The apothecium (A) may be variously coloured, but often brownish, particularly on the inner surface, and is commonly sessile or short-stalked. It decays soon after the spores mature and shed. The inner wall of the apothecium is lined with a continuous layer (hymenium) consisting of countless numbers of **asci** intermixed with slender sterile hyphae—the paraphyses (B). Each ascus is a cylindrical body and contains 8 distinct, hyaline **ascospores** which are arranged obliquely in a row (uniseriate). The ascospores, when mature, are liberated from the ascus through a terminal pore. Under suitable conditions and in an appropriate medium they germinate producing new mycelia.

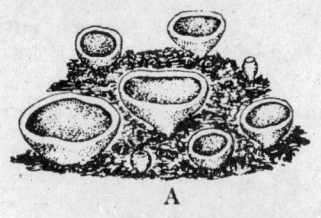

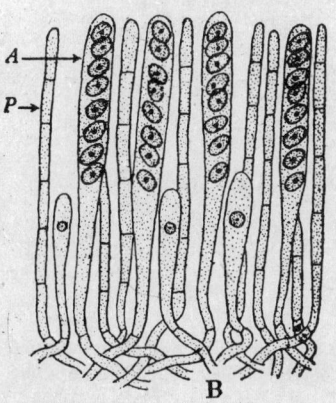

Peziza. FIG. 93. A, fruit bodies or cups (apothecia); B, a portion of apothecium (in section) showing asci (A) with ascospores and paraphyses (P).

Sexuality. There are no sexual organs in *Peziza* but in a few species a nuclear fusion of two vegetative hyphal cells has been observed. Thus prior to the formation of the apothecium nuclei of certain hyphal cells fuse in pairs indicating reduction of sexuality. These cells then give rise to the ascogenous hyphae from which asci are subsequently formed. If this is true the nucleus of the young ascus must be diploid. Then by three successive divisions, the first being reductional, 8 ascospores are formed in the ascus.

9. *ASCOBOLUS* (25 sp.)

Occurrence. *Ascobolus* (FIG. 94*A*) is also a cup fungus like *Peziza* and both belong to the same family, i.e. *Pezizaceae*. Most species of *Ascobolus* are coprophilous like *Peziza*, commonly growing on the dung of particular animals; for example, *A. magnificus* grows on horse dung. Some species grow in the soil rich in organic manures, and break up the organic compounds for the use of green plants. Certain species are also seen to grow on rotten wood, or on burnt ground with the deposit of carbon. All the species of *Ascobolus* are saprophytic in habit.

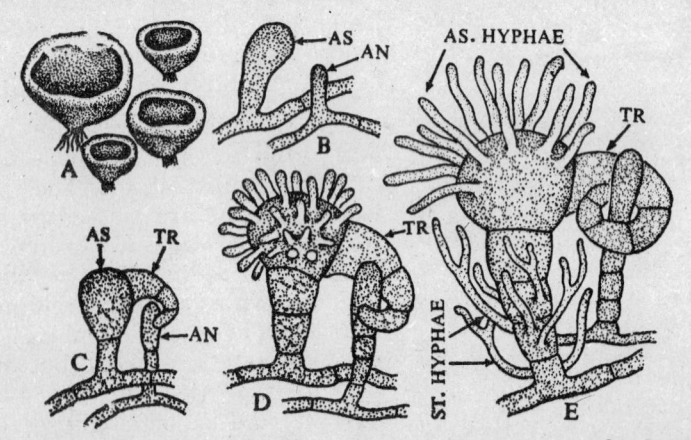

Ascobolus. FIG. 94. *A*, fruiting bodies; *B-E*, stages in the development of sexual organs; *AS*, ascogonium; *AN*, antheridium; *TR*, trichogyne; *AS. HYPHAE*, ascogenous hyphae; *ST. HYPHAE*, sterile hyphae.

Structure. The mycelium consists of much branched, hyaline hyphae which ramify through the substratum. The hyphae are septate. Within a few days after the germination of the spore a soft fleshy, cup-shaped body, closed at first, appears on the surface; this is the most conspicuous part of the fungus, and is its fruiting body called ascocarp (*A*). The ascocarps are small, usually varying in sizes from 5 mm. to 2.5 cm. in diameter, sometimes smaller or larger.

Sexual Reproduction. *A. magnificus* and also many other species are monoecious bearing both the sexual organs (male and female) but are self-sterile, as first shown by Dodge in 1920 and later by others. In them the ascocarps are produced only after two complementary strains (i.e. those of opposite sexes) are brought together. Morphologically, however, no difference exists between the two. When the two strains (*A* strain and *B* strain) grow mixed up together, sexual branches grow up and become differentiated into male (antheridium) and female (ascogonium) organs (*B*). Cross-fertilization takes place between these two organs borne by two separate strains. A few species of *Ascobolus*, however, are self-fertile. In certain species the antheridia may not develop at all. Then a single strain produces the fruiting body by fusion of hyphae. The antheridial branch (*C*) consists

FUNGI

of a multicellular stalk and a terminal **antheridium** which is cylindrical or somewhat club-shaped. It bears numerous nuclei, sometimes as many as 400. The female branch (C) consists of unicellular or multicellular stalk and an **ascogonium** (oogonium) which is more or less globose. The ascogonium is uninucleate at first, and forms terminally a receptive outgrowth called the **trichogyne** which consists usually of a few cells. As the female branch elongates, the trichogyne becomes twisted round the antheridium, and thus communication is established between the two sexual organs.

Fertilization. This, as said before, takes place between the antheridium and the ascogonium borne by two separate strains. The male nuclei pass into the trichogyne and move towards the ascogonium by perforating the septa of the trichogyne. Each male nucleus has two distinct chromatin bodies by which it can be distinguished from the female nucleus even at later stages. About this time the female nucleus of the ascogonium divides repeatedly by mitosis and the newly-formed nuclei spread themselves out near the periphery. The male nuclei enter into the ascogonium and now the sexual nuclei fuse in pairs (actually a close association). Each such pair, called a *dikaryon* (see p. 526), otherwise called *definitive nucleus*, is a functional unit ($2n$). It now undergoes repeated conjugate divisions mitotically, resulting in several nuclei (each $2n$). The ascogonium produces a number of filaments called **ascogenous hyphae** (D-E). The contents of the ascogonium pass into them. The ascogenous hyphae elongate and become septate, each cell having a pair of diploid nuclei. The ascogenous hypha produces terminally an **ascus** by the hook or crosier method (see p. 527), and a pair of nuclei move into it; the young ascus is always binucleate. It is here that actual nuclear fusion (second fusion) takes place, and the zygote nucleus becomes tetraploid ($4n$). Almost immediately the zygote nucleus undergoes meiosis, forming four nuclei (each $2n$). A second reduction called *brachymeiosis* follows. Thus finally 8 nuclei are produced (each n). The ascus simultaneously grows and becomes cylindrical. Cytoplasm collects round each nucleus and a wall is formed round each. Thus altogether 8 **ascospores** appear in each ascus. They are brown or violet.

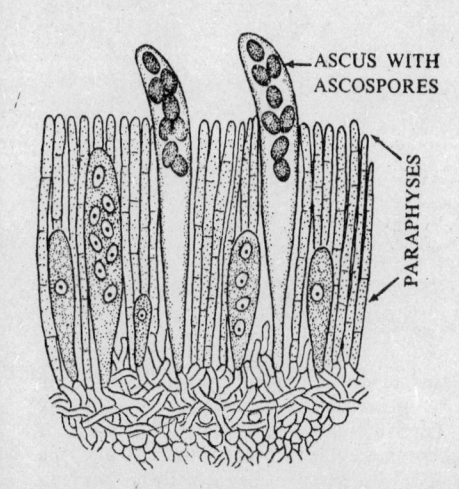

Ascobolus. FIG. 95. Section through apothecium showing asci (in several stages of development) and paraphyses.

Ascocarp. With the development of the asci a fruiting body called ascocarp (94A) is formed containing the asci. In *Ascobolus* the ascocarp is open and cup-shaped, and is called **apothecium**. It is formed of an interwoven mass of sterile vegetative hyphae growing around the sexual organs. The apothecium (FIG. 95) consists of a fleshy protective wall (pseudoparenchymatous in nature) known as the **peridium** and a **hymenium** on its inner surface. The hymenium bears a large number of **asci** and **paraphyses** (sterile hyphae) arranged in a parallel series. The asci elongate and protrude beyond the level of the hymenium. Their tips bend towards the source of light (positively phototropic). Eventually the asci burst and the **ascospores** are discharged. If the air is dry, 'puffing' of the spores may be noticed.

| CLASS IV | BASIDIOMYCETES or club fungi | 23,000 sp. |

1. *USTILAGO* (300 *sp.*)

Ustilago (family *Ustilaginaceae*) belongs to the order Ustilaginales or smuts which number about 700 species.

Ustilago or smut (FIG. 96) is a parasitic fungus commonly attacking members of *Gramineae*. It may also grow as a saprophyte in the soil rich in organic material. Some of the crops commonly attacked by *Ustilago* are maize, wheat, barley, oat, rice, sugarcane, etc. The damaging effect of corn-smut by *U. zeae* on maize (*A*), of oat-smut by *U. avenae* on oat, of loose-smut by *U. tritici* on wheat (*B*), etc., is often immense, and the total loss on this account is often quite heavy.

The plants may be infected in different ways. Infection usually takes place through teleutospores or basidiospores. Dormant mycelia may also exist in the seed or the plant body, becoming active afterwards. Types of infection, as given below, may be

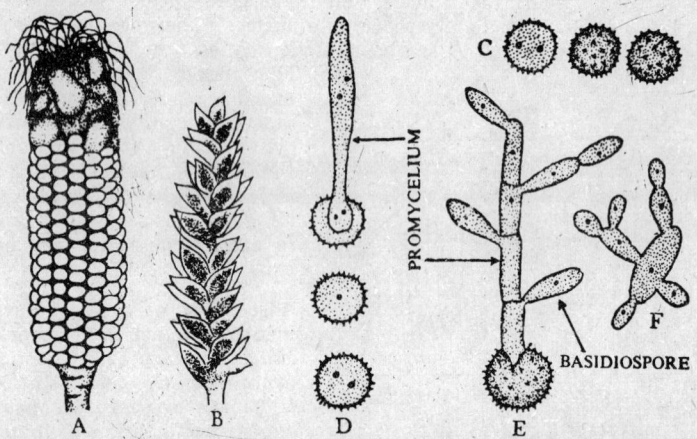

Ustilago. FIG. 96. *A*, maize cob infected; *B*, wheat spikelets infected; *C*, teleutospores; *D*, teleutospore germinating: *bottom*, teleutospore (binucleate); *middle*, the same (uninucleate and diploid on fusion of the two nuclei); *top*, the same germinating into promycelium (note the four nuclei on reduction division); *E*, mature promycelium bearing basidiospores; and *F*, basidiospore budding.

noted. (1) Falling on the stigma or feathery style the teleutospores may germinate and produce a promycelium (*D, top*). The latter forms infective hyphae (but no basidiospores) which penetrate into the ovary. Subsequently, as the ovary develops, mycelia remain dormant in the grain (seed), and as the latter germinates the mycelia grow and ramify. In such a case, as in *U. tritici* (on wheat), the disease is internally seed-borne. (2) The spores may also adhere to the grains externally. They are usually released after the threshing of the ears. As they get scattered they may infect new hosts through their young parts—roots, stems, leaves and more particularly spikelets, as in *U. zeae* (on maize) and *U. avenae* (on oat). (3) The basidiospores may produce uninucleate hyphae which may infect the host, or grow on the surface of the host body, but two such hyphae must fuse in the tissue of the host or just outside, producing binucleate mycelium. This then grows normally and causes the disease. Two basidiospores may also unite to give rise to a binucleate mycelium. (4) Binucleate mycelium may produce binucleate conidia on the surface of the host body. They are dispersed by the wind and become sources of infection. Uninucleate conidia formed by uninucleate hypha do not survive long. But they may unite with a hypha of opposite strain to produce binucleate mycelium first.

The binucleate mycelia may be local or widespread in the body of the host, and are sparsely septate but profusely branched. The hyphae ramify through the intercellular spaces, producing here and there some *haustoria* in some species, which penetrate into the living tissue of the host to absorb food material from it. As a consequence the host plant suffers, and it becomes stunted in growth. The hyphae tend to grow towards the meristematic regions, drying up from the older parts. When the flowering stage of the host plant is attained, the hyphae extend to all parts of the ears (spikelets) and cover them with black soot-like masses of spores, and hence the name 'smut'. All floral parts except the awns soon become replaced by the black powdery mass of spores. The spores are called **teleutospores** (or chlamydospores; *C*). They are formed in countless numbers and are somewhat thick-walled. The ovaries often become much swollen and usually distorted and destroyed.

The teleutospores are blown about by the wind. They may germinate in the soil or if they are carried over to the stigmas and styles of healthy plants they germinate there by producing a short *germ tube* called the **promycelium** (or basidium; *D, top*). The promycelium becomes septate (characteristic of *Ustilago*) and consists of 4 uninucleate cells (*E*). Young teleutospores are binucleate (*D, bottom*). When they mature, nuclear fusion (karyogamy) takes place within them (*D, middle*). This is the only *diploid condition* of the fungus. The diploid nucleus of the zygote divides meiotically while the promycelium is being formed (*D,* *top*). Each cell of the promycelium is provided with a haploid nucleus (*E*).

The next stage may take any of the following patterns according to the species. (1) From each cell of the promycelium one or more thin-walled elongated uninucleate (haploid) **basidiospores** (also called **sporidia**) may be formed by budding (*E*). Each such basidiospore may take to budding again before or after shedding from the promycelium, forming secondary basidiospores (*F*). The nuclei correspondingly divide mitotically to supply each spore with a nucleus. Two such basidiospores of opposite strains (+ and −) unite (but no karyogamy, i.e. nuclear fusion, takes place), and produce a binucleate mycelium (dikaryon, + and −). (2) A basidiospore may germinate into a uninucleate hyphae (+ or −). Two such hyphae unite and produce a binucleate mycelium (dikaryon). (3) A basidiospore may germinate into a uninucleate hypha (+ or −). Then a basidiospore of opposite strain may unite with this hypha, producing a binucleate mycelium (dikaryon). (4) The promycelium may directly produce infective hyphae without the intervention of basidiospores. Two such hyphae of opposite strains unite and produce binucleate mycelium (dikaryon).

It is to be noted that it is only after dikaryotic condition is attained by any of the above methods the binucleate mycelium grows vigorously and spreads through the body of the host, initiating the parasitic phase of the fungus and finally producing teleutospores—a distinct symptom of the disease.

Control. (1) Rotation of crops in the case of maize. (2) Hot water treatment of grains in the case of wheat, and then drying them under strong sun. (3) Spraying the grains with equal parts of commercial formalin and water in the case of oats, or dusting them with copper carbonate powder. (4) Cross-breeding with types resistant or immune to smut; this is possibly the best method.

---

2. *PUCCINIA* (700 *sp.*)

Puccinia (family *Pucciniaceae*) belongs to the order Uredinales or rusts which are destructive parasites and number about 4,600 species. *Puccinia* species may be *heteroecious* requiring two distinct hosts to complete their life-cycle, or *autoecious* requiring only a single host to complete the entire life-cycle. Species of *Puccinia* attack a variety of host plants, particularly members of Gramineae. *Puccinia graminis*, commonly known as the 'black rust' of wheat, is a *heteroecious* species attacking wheat plants and common barberry plants in rotation. This species is further

FUNGI

polymorphic bearing different kinds of spores and spore-structures on wheat and barberry.

Life-history of *Puccinia graminis*. This species is a virulent parasite. It attacks wheat plants and often very seriously diseases them. The disease sometimes breaks out in an epidemic form and causes very heavy damage to the crop, resulting in great economic loss. It also attacks barley, oats and rye.

Stages on Wheat Plant. (a) **Uredium and Uredospores** (FIG. 97). In late spring or early summer the spores (aeciospores; see FIG. 100) carried by the wind from barberry to wheat germinate on the latter, each producing a germ tube through a stoma (*A*).

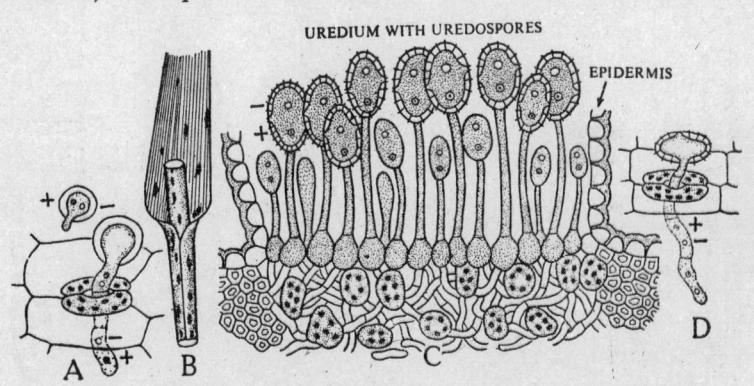

Puccinia. FIG. 97. *A*, germination of aeciospore (binucleate) on wheat plant—germ tube entering through a stoma; *B*, wheat leaf and leaf-sheath showing uredia; *C*, a uredium (in section) showing binucleate uredospores (+ and −) and also the infecting hyphae; *D*, germination of a uredospore on wheat plant again.

Within 10-12 days of infection reddish-brown streaks appear on the stem, leaf-sheath and leaf (*B*) indicating the diseased condition of the plant. A section through the infected part shows a mass of mycelia ramifying through the intercellular spaces, penetrating here and there into the living cells of the host to absorb food from them, and on the surface a number of spore-clusters known as **uredia** (or uredosori). The hyphae of the mycelium are septate and the cells binucleate (+ and −). The uredia, as they grow, break through the epidermis and appear on the surface as reddish-brown streaks (*C*). The uredium bears numerous slender hyphae projecting outwards, each ending in a one-celled, rough-walled, brownish or reddish, binucleate (+ and −) spore called the **uredospore**. This stage is known as the 'red rust' of wheat. The spores, when mature, are blown about by the wind over a wide area, and they directly infect other wheat plants (*D*). The disease may thus appear in an epidemic form, destroying the whole or major part of the crop. The uredospores may be produced successively throughout the summer infecting the wheat plants each time. Ordinarily these spores cannot stand a very severe winter.

(*b*) **Telium and Teliospores** (FIG. 98). Later, i.e. in late summer, the mycelia still existing in the wheat plant after the formation of the uredospores grow and mass together below the epidermis to give rise to black spots or streaks here and there on the stem (*A*), leaf-sheath and leaf. Each such spot or streak is a sorus called **telium**. The telium produces a large number of slender stalks, each ending in a black or dark-brown, elongated, two-

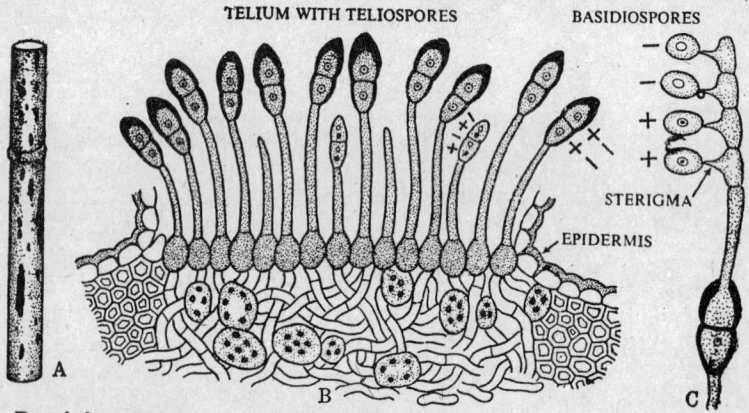

Puccinia. FIG. 98. *A*, a wheat stem infected showing telia; *B*, a telium (in section) showing young binucleate teliospores (+and−) and mature uninucleate teliospores (+−) and also the infecting hyphae; *C*, germination of a teliospore (on old dry wheat plant or in the soil) producing four basidiospores (uninucleate, +or−), each on a short sterigma.

celled, heavy-walled spore called the **teliospore** (*B*). This stage is the 'black rust' of wheat. The teliospores are resting spores and help the fungus to tide over the severe conditions of winter. Each cell of the young teliospore is binucleate (+ and −) but soon the two nuclei fuse together (a reduced form of the sexual act), and the mature teliospore has two uninucleate cells (+ −). The spores, evidently diploid, remain dormant on the wheat plant or in the soil till the following spring. They do not infect the wheat plant again.

(*c*) **Basidium and Basidiospores** (FIG. 98*C*). One or both cells

---

Nomenclature. Uredium, uredosorus or uredinium → uredospore, uredospore or urediniospore. Telium or teleutosorus → teliospore or teleutospore. Spermogonium, pycnium or pycnidium → spermatium, pycnospore or pycniospore. Aecium or aecidium → aeciospore or aecidiospore.

of the teliospore germinate independently, each producing a slender elongated hypha called the **basidium** which consists of four terminal cells. Each cell produces a short slender stalk called the **sterigma**. Its end dilates and forms a spore called the **basidiospore**. The diploid nucleus (+ −) of the teliospore undergoes reduction division so that the basidial cells and the basidiospores become haploid. They are of course uninucleate but are of opposite strains (two + and two −); *Puccinia* is thus *heterothallic* (see p. 543). The basidiospores do not infect the wheat plant. They are blown about by the wind and incidentally many of them are carried over to the barberry (*Berberis vulgaris*) bush where further stages appear in continuation of its life-cycle.

Stages on Barberry Plant. (d) **Spermogonium and Spermatia** (FIG. 99 *B-C*). The basidiospore germinates on the barberry leaf

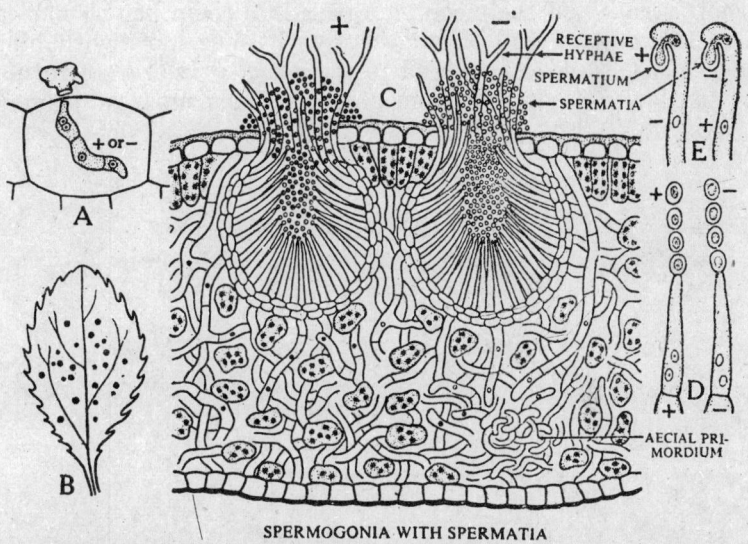

SPERMOGONIA WITH SPERMATIA

Puccinia. FIG. 99. *A*, germination of a basidiospore (uninucleate) on barberry leaf through the cuticle; *B*, barberry leaf infected showing spermogonia on the upper surface; *C*, a section through the infected part showing two spermogonia (one + and one −), a mass of infecting hyphae and some receptive hyphae which have become binucleate (+ and −) after the entry of spermatial nuclei (see *E*); the receptive hyphae are now extending towards the aecial primordium; *D*, spermatial hyphae cutting off spermatia (+ or −); *E*, spermatia and receptive hyphae of opposite strains uniting (nuclei, however, do not fuse).

by producing a germ tube which enters the leaf through the cuticle (*A*). The mycelia grow extensively in the leaf-tissue and soon mass together beneath the epidermis (usually upper) forming in about 7-10 days of infection slightly raised yellowish or reddish spots, called **spermogonia** (or pycnia), on the leaf-surface (*B*).

The cells of the mycelium as well as those of the spermogonium are uninucleate, either of + strain or of − strain, produced correspondingly from a + basidiospore or from a − basidiospore. In section the spermogonium is more or less flask-shaped (*C*). Its inner wall is lined with numerous fine fertile hyphae (spermatial hyphae) which successively cut off from their ends very minute uninucleate cells called **spermatia** (or pycnospores; *D*). Several sterile hyphae called **periphyses** also grow from the upper part, projecting outwards through the narrow pore or **ostiole**. Besides, there are certain special hyphae called **receptive hyphae**, protruding outwards through the ostiole. The spermatia are exuded through the ostiole in a drop of sweet fluid. The sweet fluid attracts insects which carry the spermatia from one spermogonium to another which may be of the opposite strain. The spermatia come in contact with the receptive hyphae of the opposite strain and their contents pass into them but no nuclear fusion takes place (*E*). The receptive hyphae, now with binucleate cells (+ and −), extend to the *aecial primordium* formed near the lower epidermis from an interwoven mass of primary

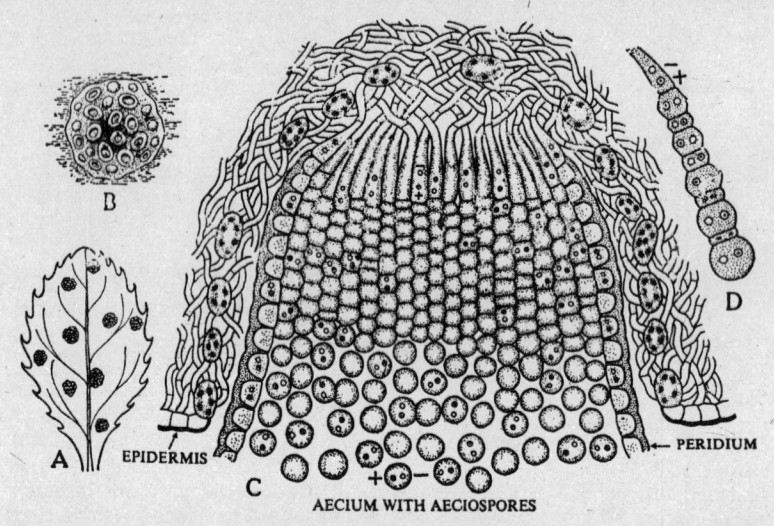

Puccinia. FIG. 100. *A*, barberry leaf showing clusters of aecia on the lower surface; *B*, a cluster (magnified); *C*, an aecium (in section) showing binucleate (+ and −) aecial cells, aeciospores and peridium, and a tangled mass of infecting hyphae; *D*, a chain of aeciospores and sterile cells produced in an alternating manner.

hyphae which have already penetrated the entire leaf-area. Some of the periphyses also may behave as receptive hyphae. There is also evidence that a + spermatium may unite with a − spermatium giving rise to hyphae with binucleate (+ and −) cells. The receptive hyphae with binucleate cells eventually form the basal cells of the aecium (FIG. 100C).

(e) **Aecium and Aeciospores** (FIG. 100). The elongated binucleate basal cells (+ and −) of the aecium now give rise to clusters of comparatively large cup-like blisters called **aecia** or cluster-cups on the lower surface of the leaf (A-B). As the aecium grows it breaks out of the epidermis (C). The basal cells begin to cut off from the bottom chains of binucleate cells (+ and −) which immediately divide producing large, orange or yellow binucleate cells (spores) called **aeciospores** (+ and −) and small sterile cells (also binucleate, + and −) in an alternating manner (D); the latter soon disintegrate. Aeciospores are the first *binucleate spores* to appear in the life-cycle of the fungus. A protective layer called the **peridium** also develops from the basal cells of the aecium. Soon the peridium bursts and the spores are liberated. These are shed in late spring and early summer, and are blown about by the wind. If they happen to fall upon the wheat plant they infect it (FIG. 97A), and the life-cycle is repeated.

Sexuality, Diplophase and Haplophase. Sexuality in *Puccinia* is reduced to the fusion of two nuclei in the young teliospore. Diplophase (binucleate condition) begins with the cells of the receptive hyphae after spermatization (FIG. 99E) and continues through aecium and aeciospores on barberry and later on wheat through infecting mycelia, uredium, uredospores and finally young teliospores (where the actual fusion of nuclei takes place). The haplophase (uninucleate condition) begins with the mature teliospores on wheat and continues on barberry through germinating basidiospores, infecting mycelia and finally spermogonium and spermatia. *Puccinia* is heterothallic, the basidiospores being distinctly of two opposite strains (+ and −). It should also be noted that in the whole life-cycle of *Puccinia* the diploid ($2n$) condition is represented by the mature teliospores only.

Control. No special method has yet been discovered to prevent or radically cure the disease. Certain methods have, however, been devised to control its intensity. (1) Eradication of the barberry bush near a wheat field is a good established practice. (2) By cross-breeding with rust-resistant varieties it has been possible to evolve new types of wheat which are immune to the rust for some years at least and for a particular locality. (3) Elimination of cultivation of wheat in the hills during summer may reduce the spread of the disease to the plains through uredospores. (4) Cultivation of rust-resistant varieties.

3. *AGARICUS* (about 70 sp.)

Agaricus (=***Psalliota***; family *Agaricaceae*), commonly called mushroom (when edible) or toadstool (when poisonous), is a fleshy saprophytic fungus. It grows on damp rotten logs of wood, trunks of trees, decaying organic matter, and in damp soil rich in organic substance. The family *Agaricaceae* has about 5,000 species. Other common genera of the family are *Amanita, Lepiota, Coprinus, Marasmius*, etc.

Edible and Poisonous Forms. There are about 200 species of fleshy fungi that are edible; many are non-edible, and above 12 species distinctly poisonous. All puff-balls are edible, particularly when they are young. Other common edible fungi are *Agaricus campestris* (=*Psalliota campestris*), *Morchella esculenta, Volvaria terastria, Lepiota mastoides*, etc. Certain species of *Amanita* which resemble edible *Agaricus* are extremely poisonous; these, however, are usually distinguished from the latter by their possession of a cup-like structure (called **volva**) at the base, which is wanting in *Agaricus*. Edible types cannot be easily distinguished from poisonous ones except by critical examination. Generally speaking, (1) most of the species having a bright colour are to be regarded as poisonous; (2) those bearing pink spores, and (3) those having a cup at the base are also poisonous; (4) a hot burning taste or acid flavour should as a rule be avoided; (5) those growing on the wood, and (6) those whose stem does not break easily, when touched, are non-edible; and (7) non-edible types do not generally grow in open sunny places.

Structure (FIG. 101). The mycelium consists of a mass of much-branched hyphae which unite (anastomose) at their points of contact and form a network in the substratum in which the fungus grows. The hyphae are very slender, hyaline and septate, mainly consisting of binucleate cells (+ and −). Frequently several hyphae are seen to be massed together here and there into thick twisted strands, called **rhizomorphs**, covered by a sheath. To start with, as the uninucleate basidiospore germinates it produces a primary mycelium (+ or −). Soon, however, it becomes multinucleate by repeated nuclear divisions. Septa appear between the nuclei, dividing the mycelium into a number of uninucleate cells. This stage is short. Soon, however, two such hyphae (one + and one −) come into contact and fuse. The fusion is in the nature of *plasmogamy* (i.e. fusion of two protoplasts without nuclear fusion or karyogamy; this takes place through a pore punctured in the hyphal wall. The new hypha, thus formed, is the secondary mycelium, and its cells are typically binucleate (+ and −). This is followed by *clamp con-*

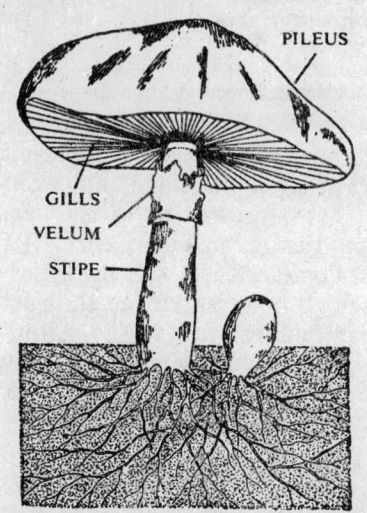

Agaricus. FIG. 101. Two plants, young and old, with ramifying mycelia.

FUNGI

nexion (see FIG. 66) in some cells, at least. The secondary mycelium spreads in all directions through the substratum, perennates from year to year, and in season produces the main fleshy aerial body which is the **fructification** or fruit body of the fungus, otherwise called **basidiocarp** (basidia forming body) or **sporophore** (spore-producing body).

Basidiocarp (FIG. 101). It consists of a fleshy stalk known as the **stipe** (a stem) and an umbrella-like head or cap known as the **pileus** (a cap or hat) on its top. The stalk and the head are composed of an interwoven mass of hyphae, and in sections they have the appearance of a tissue—a false tissue—known as *pseudoparenchyma*. The stipe is stout and cylindrical, while the pileus

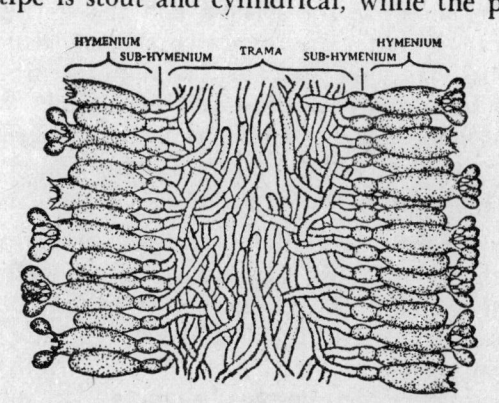

Agaricus.
FIG. 102.
A gill in section.

is expanded, roundish and convex. When young, the fructification is spherical in shape (button stage) and is completely enveloped by a thin membranous covering called the **veil** or **velum**. With the rapid growth of the fruit body, specially the pileus, the velum gets ruptured, while the lower part of it still remains attached to the stipe in the form of a ring (annulus). The pileus soon spreads in an umbrella-like fashion on the top of the stipe. On the undersurface of the pileus a large number

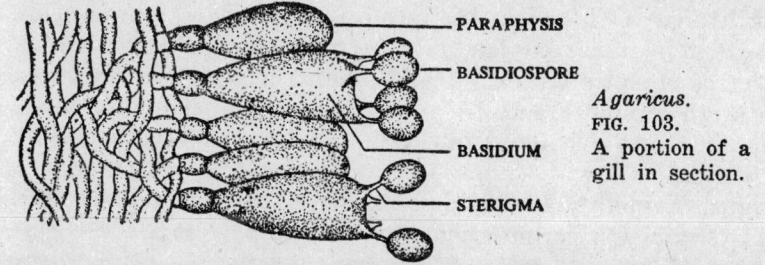

Agaricus.
FIG. 103.
A portion of a gill in section.

of thin vertical plate-like structures, extending radially from the stipe to the margin of the pileus, are seen; these are known as the **gills** or **lamellae**.

Gills. Gills occur in large numbers, usually varying from 300 to 600 for each fructification. Each gill bears innumerable spores (basidiospores) on both the surfaces. A gill (FIGS. 102-3) in section shows three distinct portions: trama, sub-hymenium and hymenium. The **trama** is the central portion of the gill and consists of an interwoven mass (false tissue or pseudoparenchyma) of long slender hyphae. The hyphal cells of the trama curve outwards on either side of the gill and terminate in a layer of small rounded or oval cells; this layer is the **sub-hymenium**. External to it lies the **hymenium** composed of a compact layer of club-shaped cells— basidia and paraphyses. The spore-bearing ones are the **basidia** and the sterile ones are the **paraphyses**; the latter are somewhat shorter than the basidia and are regarded by some mycologists as immature basidia. Each basidium bears four **basidiospores** (sometimes two, as in cultivated mushroom) on short slender stalks known as the **sterigmata** (sing. sterigma). The basidiospores, when mature, fall off and germinate under suitable conditions.

Reproduction. Asexual reproduction is not a regular feature in the life-cycle of *Agaricus*. Sometimes, however, it may take place through a kind of 'resting' spores called **chlamydospores** which are certain enlarged thick-walled vegetative cells of a hypha, formed singly or in chains. They germinate by producing a germ tube. Sometimes hyphae break up into small unicellular fragments called **oidia** (uninucleate or binucleate). Accordingly they grow up into primary or secondary mycelia. A uninucleate oidium may also directly fuse with a primary mycelium. In some species a basidiospore may give rise to a large number of **conidia** by budding. Each conidium then germinates into a mycelium. **Sexual Reproduction.** We have already seen how plasmogamy takes place between two primary hyphae of opposite strains ($+$ and $-$), leading to dikaryotic (binucleate) secondary hypha. A short but distinct sexual phase is, however, represented by the complete fusion (karyogamy) of two haploid nuclei of opposite strains ($+$ and $-$) in the young basidium to produce a diploid zygote ($+$ $-$). The zygote nucleus divides by meiosis to form 4 haploid nuclei ($2+$ and $2-$) in the basidium. Each nucleus ($+$ or $-$) pushes into a basidiospore through a sterigma. Evidently 2 basidiospores are of $+$ strain and the other 2 of $-$ strain. On germination the $+$ basidiospore produces $+$ primary mycelium, and the $-$ basidiospore produces $-$ primary mycelium, as described before. There are, however, no sex organs in *Agaricus*.

Development of Basidium and Basidiospore (FIG. 104). The basidium is at first binucleate (*A*). The two nuclei, each with *n* chromosomes, fuse to form the zygote nucleus (*B*). The latter, evidently provided with 2*n* chromosomes, undergoes reduction division giving rise to four daughter nuclei, each with *n* chromosomes (*C-D*). Slender projections or sterigmata—usually 4, sometimes 2 in number—are formed at the end of each basidium (*D*). Each sterigma swells at the end, and a nucleus migrates into it from the basidium (*E*). The swollen end-cell with a nucleus in it is the basidiospore (*F*). A small outgrowth, called hilum, is formed at the junction of the basidiospore and the sterigma. A drop of water accumulates on the hilum, and then the basidiospore together with the drop of water suddenly shoots off from the sterigma (*G*). This explosive mechanism is not, however, understood.

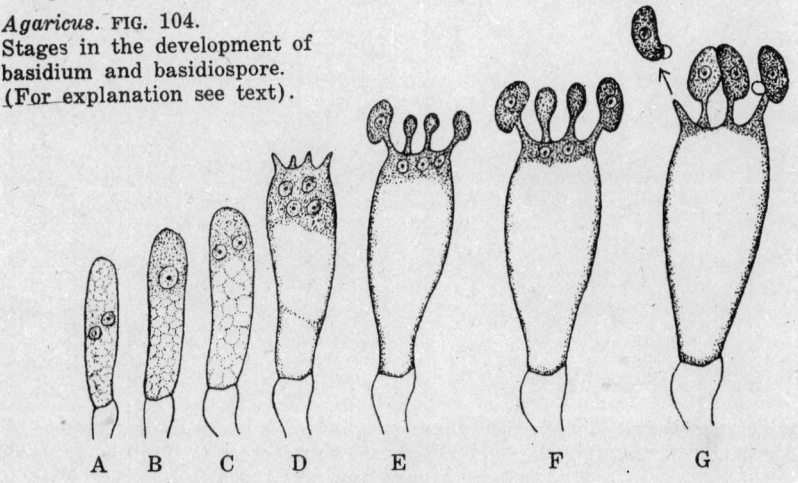

Agaricus. FIG. 104.
Stages in the development of basidium and basidiospore.
(For explanation see text).

A   B   C   D   E   F   G

4. *POLYPORUS* (about 500 sp.)

Polyporus (FIG. 105), a pore fungus, belongs to the family *Polyporaceae* which has over 1,000 species. Other common genera of this family are *Polystichus, Fomes, Daedalea* and *Lenzites*. Many species of *Polyporus* grow as bracket or shelf fungi, either singly or in groups, on many forest trees, stumps and logs of wood, and are responsible for wood decay, sometimes causing heavy damage. This has necessitated the use of wood preservatives, particularly in the case of timber. *Polyporus* is commonly annual. The mycelia develop within and below the bark, and sooner or later form on it a more or less flat fruit body (*A*) called **basidiocarp** (basidia-bearing body) or **sporophore** (spore-bearing body). The fruit body is leathery, corky or woody, and is whitish or slightly greyish or brownish in colour. The upper surface may be smooth, rough or warted, often undulating, and in some species distinctly striated, particularly towards the outer margin; while the lower surface is porous (*B*). The fruit body in section (*C*) is seen to consist of (*a*) a **context** which is the upper or outer fibrous part made of thick-walled hyphae; (*b*) a **trama** which is a loose mass of much-branched, septate and anastomosing hyphae; (*c*) a series of **pores** or **tubes** which extend from below the context to the lower surface; on the lower surface the pores appear as innumerable minute holes, practically covering it; and (*d*) a **hymenium** which is made of a distinct layer of basidia lining

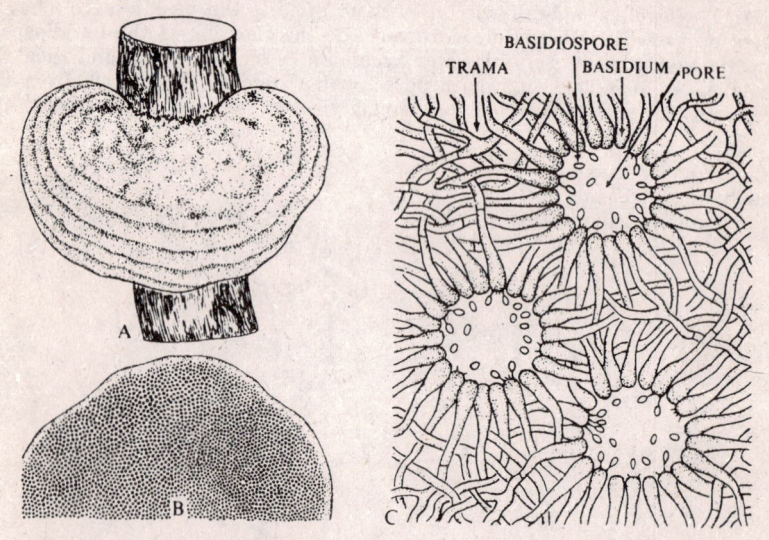

Polyporus. FIG. 105. *A*, the fungus growing on a dead branch; *B*, lower surface of the fungus (a portion) showing pores; *C*, fruit body (in transection through the pore-tubes).

each pore. The **basidia** are club-shaped and slightly project into the pore. Each basidium bears four short slender **sterigmata**. Each sterigma forms a **basidiospore** at its end by abstriction. The basidiospores are discharged continually for some weeks into the pores through which they escape freely and are blown about by wind. The basidiospore shoots from the sterigma exactly in the same way as in *Agaricus* (see FIG. 104). An enormous quantity of basidiospores is produced by a fruit body. They germinate under favourable conditions.

It may be noted that the primary hyphae formed from the basidiospores have uninucleate (*monokaryotic*, + or −) cells; while the secondary hyphae have binucleate (*dikaryotic*, + and −) cells. The basidiospores (+ or −) germinate close together, and the hyphae freely anastomose with the result that the + or − nucleus of one primary hypha passes into another primary hypha of the opposite strain. A dikaryon (see p. 526) is the result. The secondary hyphae that develop from dikaryotic hyphae have binucleate (*dikaryotic*, + and −) cells. The basidium which is the terminal cell of a

FUNGI

secondary hypha is similarly binucleate (+ and —). One nucleus goes to a basidiospore which evidently is either + or —, finally giving rise to a + hypha or a — hypha.

CLASS V | DEUTEROMYCETES or fungi imperfecti | over 24,000 sp.

1. *HELMINTHOSPORIUM* (175 *sp.*)

Helminthosporium (FIG. 106) belongs to the family *Dematiaceae*. Several diseases are caused by this fungus. The 'leaf spot' disease of rice (A) due to the attack of *H. oryzae* is fairly common in Assam and West Bengal but only occasionally appears in other areas. The fungus commonly attacks all parts of the rice plant, particularly the leaves, mostly on their lower surface, as brown

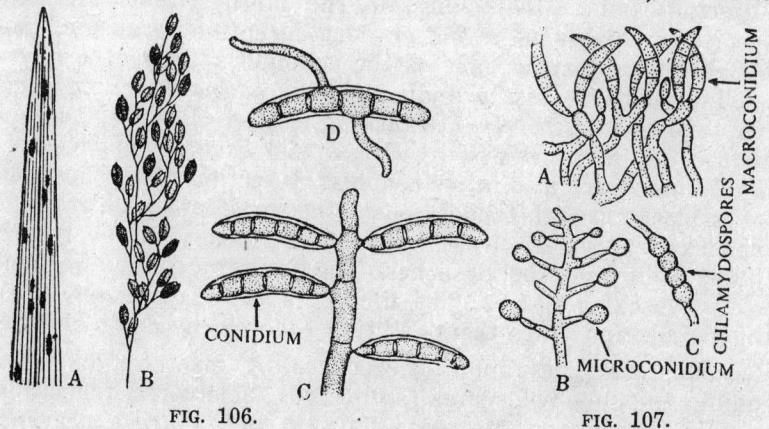

FIG. 106. FIG. 107.

Helminthosporium. FIG. 106. *A*, rice leaf infected; *B*, rice spikelets infected; *C*, conidiophore bearing conidia; *D*, a germinating conidium.
Fusarium. FIG. 107. *A*, sporodochium (a portion) with macroconidia; *B*, conidiophore with microconidia; *C*, hypha with chlamydospores.

to dark-brown spots with a yellowish halo around them. The spots rapidly increase in number. The ears may also be affected (*B*), which then become distorted and sterile. Consequently the crop suffers. The plant may be attacked at any stage of development. Seedlings, when attacked, die off soon. Other common diseases caused by the fungus are: 'foot rot' of barley due to the attack of *H. sativum* which also attacks wheat and many other grasses; 'leaf stripe' of barley caused by *H. gramineum*; 'leaf spot' of oats caused by *H. avenae*; and 'eye spot' of sugarcane caused by *H. sacchari*.

Structure. The mycelium consists of branched, septate hyphae which grow through the intercellular spaces and penetrate into the living cells of the host plant, i.e. the hyphae are intercellular as well as intracellular.

Reproduction. The only method of reproduction known so far is by **conidia** (C). Erect stout **conidiophores** emerge in groups, mainly through the stomata. They are unbranched; sometimes, however, branches appear at their base. The conidiophores bear conidia towards the top. The conidia are multiseptate, the septa varying in number from 5 to 10. They are dispersed by the wind. The disease spreads under conditions of rain and cloudy weather. It is commonly a seed-borne disease. Control of the disease has not proved successful yet.

2. *FUSARIUM* (65 sp. and several varieties)

Fusarium (FIG. 107) belongs to the family *Tuberculariaceae*. Species of this genus are often very deceptive because of great variability of forms. Some of the common species of *Fusarium* causing wilt-diseases in India are: *F. udum* attacking pigeon pea, and *F. vasinfectum* attacking cotton. Besides, the wilt-disease of linseed caused by *F. lini* and that of cabbage by *F. conglutinans* are also fairly common. Infection of the host plant takes place through tender roots. The mycelium penetrates into the vascular tissues. Black streaks are formed, which soon spread to the stem and the branches. The fungus remains restricted to the vascular tissues. The mycelium grows profusely within the vessels and plugs them. Wilting followed by death results.

Reproduction. The fungus reproduces by macroconidia, microconidia and chlamydospores (see p. 526). Sclerotium formation is not also uncommon. **Macroconidia** (FIG. 107A) are long, septate and crescent-shaped, and are formed on short conidiophores typically borne on a sporodochium on the surface of the host plant. The **sporodochium** is a cushion-shaped stroma, i.e. a compact mass of vegetative hyphae much like a mattress, covered with conidiophores on the surface of the host plant. **Microconidia** (FIG. 107B) are small, usually oval and unseptate or uniseptate. They are formed within the tissue, but on the surface of the plant they are held together in groups in a drop of liquid. **Chlamydospores** (FIG. 107C) are formed in chains within the tissue. The Conidia and the spores, when shed, remain in the soil and are viable for long periods. Control of the disease has not been successful yet by any direct treatment. Long periods of rotation have, however, been found effective.

Plant Pathology

A Short Historical Account. Work on plant pathology was initiated by Tillet whose experimental investigations on the bunt (stinking smut) of wheat elucidated the cause of the disease and its prevention. He published the results of his work in 1755, and was awarded the first prize offered by the Academy of Bordeaux. **Provost's** further work in Geneva in 1807 threw more light on the causal fungus (*Tilletia tritici*) and the effective treatment by copper sulphate. In the meantime **Andrew Knight's** work on wheat rust in 1804 established the fact that spores from barberry infect wheat plants. Plant pathology was actually founded as an important branch of study through the extensive work carried out by **Persoon** (1761-1836), **Fries** (1794-1878), **De Bary** (1831-88) and **Kuhne** (1825-1910). **Sir Edwin J. Butler** (1874-1943), Imperial Mycologist, Pusa Agricultural Institute, India (1905-19), and later Director of the Imperial Mycological Institute, Kew (1920-35), did very valuable work on fungal diseases of plants. His *Fungi and Diseases in Plants* published in 1918 was a notable contribution to mycology. A revised and enlarged edition of the book has been published in 1949 under the title *Plant Pathology* by Butler and Jones.

Plant Diseases Caused by Fungi.[1] *Symptoms, Causes* and *Effect.* Abnormal symptoms or outward signs are indications of the diseased conditions of plants. Symptoms appear in different forms such as leaf spots, leaf curl, discolouration, blights, rots, smuts, rusts, mildews, wilts, blisters, hypertrophies, cankers, damping off, etc. Causes are mainly attributable to many parasitic fungi, commonly called **pathogens**, which attack several field crops, cultivated plants, ornamental and useful plants, and also many wild plants, often causing serious diseases in them. The fungi plunder the food stored in the host plants, block the conducting tissues, destroy the affected cells and tissues, produce toxins (poisons) in them, and finally cause their decay and death. The spores of the fungi may be seed-borne, wind-borne or insect-borne, and a particular disease may spread from field to field, sometimes appearing in an epidemic form unless controlled in time. The annual loss in agricultural crops on this account alone is very heavy. In India the loss of foodgrains due to diseases, pests and rodents may roughly be 10%, sometimes much more, of the total annual production. It may be noted that a plant, as an animal, may suffer from one or more diseases at a time. Some of the above diseases are as follows:

A. **Late Blight of Potato.** This is a common and sometimes a serious disease of potato plants, often appearing in a severe form in the hills. The disease is caused by *Phytophthora infestans* (see p. 535). Black patches appear on the undersurface of the leaves, less often on the upper, indicating diseased conditions of plants. The disease may spread to the entire leaf and to all parts of the plant body including the tubers. If the weather is warm and humid, and the soil water-logged, the disease rapidly spreads

---

[1] References: *Plant Pathology* by Sir Edwin J. Butler and S. C. Jones, *Fungi and Plant Disease* by B. B. Mundkur, *Introduction to Plant Pathology* by F. D. Heald, and *Plant Pathology* by J. C. Walker.

to the neighbouring plants through the multinucleate sporangia of the fungus. The fungus finally causes 'wilting' of leaves and 'rotting' of tubers. In such tubers brownish markings may be noticed below the skin as an early indication. **Control.** (a) Spraying the young plants with Bordeaux mixture. (b) Selection of seed tubers from non-infected areas. (c) Storage of seed tubers at a low temperature—4-5°C.

B. Smuts. Smuts (soot-like diseases) are common and serious diseases of wheat, barley, maize, oats, and sugarcane, caused by different species of *Ustilago* The **loose smut of wheat** is caused by *Ustilago tritici* (see p. 566). The disease is common in the wheat-growing areas of India, as elsewhere. The fungus mainly attacks the stem, flowers and often the whole inflorescence. The infected parts turn black, and all the grains are often totally destroyed, thus causing a heavy loss to the cultivators. The disease becomes manifest only when a black sooty mass appears on the infected parts, particularly the 'ears'. The fungus is seed-borne, and as the seed (grain) germinates, the mycelia lying dormant in the grain grow and ramify through the intercellular spaces of the young plants, and pass on to the flowers, forming sooty masses of spores (teleutospores). The ovaries become swollen at first and then get distorted and destroyed. The final effect is that the entire spikelets become deformed and covered over with a black powdery mass of spores, replacing all the floral parts and the glumes (except the awns). When the spores are blown away by the wind or washed away by the rain, the stalk only stands bare of grains. The spores may thus spread over wide areas, and a fresh infection of flowers may take place. The spores germinate on the stigma and infect the grains of healthy plants. In this way the disease is carried over to the next generation. **Control.** (a) Varieties of wheat already immune and resistant to smut should be cultivated. (b) Cross-breeding with types immune or resistant to smut is possibly the best method. (c) Hot water treatment of wheat grains and then drying them under strong sun may reduce the intensity of the disease.

C. Rusts. Wheat suffers from a variety of rust diseases, caused by different species of *Puccinia*, viz. black or stem rust of wheat caused by *Puccinia graminis*, yellow rust caused by *P. glumarum*, and brown rust caused by *P. triticina*. All these rusts are common almost throughout India. *P. glumarum* appears on the leaves as yellow rusty spots or stripes on the leaves, and in case of severe attack the crop may dry up. *P. triticina* appears as brown or orange spots (not as stripes) on the leaves in clusters, often irregularly scattered. The loss on account of this disease seems to be insignificant. *P. graminis* (see p. 568), however, is a virulent type of parasite, and often causes a serious disease of the wheat plants both in the hills and the plains. The symptom of the disease is the appearance first of reddish-brown spots and streaks on the stem, leaf-sheath and leaf as a result of the formation of *uredia with uredospores,* and this stage is known as the 'red rust' of wheat. As the uredospores mature, the uredia burst and the brown oval uredospores are blown about by the wind over large areas. They may directly infect other healthy wheat plants, and thus the disease may appear in an epidemic form causing considerable damage to the crop, evidently resulting in a heavy economic loss. Later dark spots appear on the stem, leaf-sheath and leaf as a result of the formation of *telia with teliospores,* and this stage is known as the 'black or stem rust' of wheat as it is the stem that is most severely affected. The final effect of the disease is the weakening of the plants, reduction of grains in size and number, and their shrivelling up. The teliospores do not infect wheat plants again but attack barberry plants as their next hosts. **Control** (see p. 573).

D. **Mildews.** These diseases appear as whitish, yellowish or brownish spots on the leaves and also on other parts. There are two kinds of mildews: downy and powdery. The former are caused by *Cystopus, Plasmopara* and *Peronospora.* They are endophytic and, therefore, some damage is caused to the crops. Powdery mildews on the other hand are caused by *Erysiphe* and *Uncinula.* The fungus being ectophytic damage to the crops is not often heavy. Common mildews are the following: (*a*) **White rust of crucifers** (e.g. mustard, radish, cabbage, cauliflower, etc.) is caused by a downy mildew called *Cystopus candidus* (see p. 538). As a result of infection white or yellow blisters of variable shapes and sizes appear on the leaves (mainly), branches and even inflorescences. Frequently *Cystopus* is associated with *Peronospora*. In case of heavy attack it is seen that the nutrient cells collapse, soft parts disintegrate and the infected parts turn brown and dry up, and the flowers become deformed. The disease though common in India is not a serious one as it commonly appears only in mild form, and, therefore, no control measures are taken.

(*b*) **Powdery Mildew of Grape Vine** is caused by *Uncinula necator* The disease is prevalent in India as well as in some other countries. It appears mainly on leaves, flowers and berries as white powdery or dusty patches which soon turn grey and finally dark. When severely infected, the crop badly suffers (see p. 554).

(*c*) **Powdery Mildew of Cereals** (e.g. barley, oats, rye and wheat), and also several grasses is a very common but not serious disease caused by *Erysiphe graminis* (see p. 553). The mycelia and conidia form superficial growths on the upper surface of the leaves, stem and sometimes flowers, having a sort of powdery appearance which is white at first but reddish afterwards. The effect is that the plants become stunted in growth, and the leaves shed or they become deformed and twisted. The fungus being ectophytic much damage is not caused to any crop. The disease is not widespread in India either, and, therefore, no control measures are taken.

It may also be noted that many moulds damage vegetables, fruits and food, particularly in storage; similarly they also damage fabrics, paper, books, leather, shoes, etc., particularly during the rainy season.

Control. *Prevention and Check.* Considering the heavy economic loss due to various plant diseases it is imperative that proper and adequate ways and means should be devised to control them as far as practicable. Some of the common methods widely practised to destroy the causative fungi or keep them under check are as follows. (1) Spraying or dusting the affected plants with certain poisonous chemicals, called fungicides, e.g. copper sulphate, copper sulphate and lime (Bordeaux mixture), sulphur, sulphur-lime, quick-lime, mercury compounds, formaldehyde, etc., or a mixture of them. (2) Fumigation by sulphur dioxide gas. (3) Seed treatment—cautious application of hot water, formaldehyde or certain compounds of copper, sulphur or mercury. (4) Soil sterilization by burning wood in the field or by application of steam or some poisonous chemicals. (5) Selection of disease-free seeds and plants. (6) Eradication and destruction of diseased plants. (7) Destroying disease-carrying insects. (8) Breeding of disease-resistant varieties of plants. (9) Rotation of crops—growing some other crop in place of the existing one for one or more years.

Plant Diseases Caused by Bacteria. Some of the plant diseases are also caused by pathogenic bacteria as follows. (*a*) **Canker.** This appears as a dead area on the surface of the stem, leaves and fruits, having a crater-like depression in the centre, usually surrounded by a raised margin; this disease may be caused by bacteria as well as by certain fungi. Canker of *Citrus* (orange and lemon) is a common such disease caused by *Pseudomonas* (=*Xanthomonas*) *citri*. It occurs in most of the *Citrus* orchards in India, sometimes taking a serious turn. It appears on the leaves, branches and fruits under conditions of moderate temperature, adequate rainfall and humid weather. The cankerous spots soon become corky and turn brownish or sometimes pinkish, affecting shape, size, quality and appearance of the fruits. The disease may spread through the agency of wind, rain and insects, and also human beings. Canker of plum, peach and prune is caused by *Xanthomonas pruni*. (b) **Black rot of cabbage** is caused by *Pseudomonas campestris*. (c) **Wildfire of tobacco** is caused by *Pseudomonas tabaci*. (d) **Fire blight of apple and pear** is caused by *Bacterium* (=*Erwinia*) *amylovorum*. (e) **Ring disease of potato** is caused by *Bacterium* (=*Corynebacterium*) *solanacearum*. (f) **Soft rot of potato** is caused by various bacteria. There are many other such bacterial diseases. Fungal diseases, however, are more common than bacterial diseases.

Plant Diseases Caused by Viruses. A few hundred viral diseases of plants have been recorded so far (see p. 522).

Antibiotics

Antibiotics (*anti*, against; *bios*, life) are certain chemical substances, possibly enzymes, secreted by a good number of soil bacteria and soil fungi, which have been found to check the growth of particular types of infective bacteria (germs) and even destroy them. Antibiotics are the miracle drugs of the modern times. They act like magic bullets shooting down the germs which have invaded the human body and caused infectious diseases, often of a virulent nature, e.g. pneumonia, typhoid, diphtheria, tuberculosis, cholera, boils, abscesses, erysipelas, etc., and that too within a very short time. Within the last 20 years or so some 300 antibiotics have been isolated and studied at an almost incredible cost. Of these about 13 have an established therapeutic value in different bacterial diseases. The first antibiotic was an accidental discovery. For others which came in succession the persistent and painstaking labour, patience and perseverance, skill and cost involved in the examinations of several thousands of soil-samples, cultures and isolation of the bacteria present in them, study of their secretions and curative value, etc., are almost unimaginable. It is indeed a miracle that lay hidden in a spoonful of good earth for the benefit of mankind.

The first, best-known and most widely used antibiotic is **penicillin**[1] isolated by the late Sir Alexander Fleming, an English

---

[1] One morning in September, 1928, Fleming noticed some strange change in a culture of *Staphylococcus* (pus-forming bacteria) and found a blue-green mould in that plate of culture. It seemed to him that some deadly substance had been secreted by that mould which worked havoc on his fresh *Staphylococcus* culture. Fleming next cultured that mould and extracted a crude juice from it and injected it into mice after infecting them with some disease germs (*Staphylococcus, Streptococcus* and *Pneumococcus*) and found to his great amazement that the mice were cured. That was the age of sulpha drugs, and this wonderful discovery lay unheeded for many years, possibly

bacteriologist, in 1929 from a blue-green mould of the soil, called *Penicillium notatum*. It has a powerful antibacterial action and is amazingly effective against certain types of bacteria, called Gram-positive bacteria, which cause some virulent diseases like scarlet fever, tonsilitis, sore throat, rheumatic fever, erysipelas, wound infections, abscesses, carbuncles, tetanus, pneumonia, meningitis, etc. Penicillin really acts like a shotgun on a wide range of its objectives. It has come into general use since 1943-44 when mass production first got under way.

Another antibiotic, called **streptomycin**, was isolated in 1944 by Waksman, a microbiologist, from a species of soil-bacteria, called *Streptomyces grisesus*. It mainly attacks some of the Gram-positive germs, particularly tubercle bacilli, and has proved to be very valuable against tuberculosis. A vigorous search for more antibiotics went on at this time, and soon another antibiotic, named **chloromycetin**, was discovered by Burkholder, a microbiologist, in 1947. It was isolated from *Streptomyces venezuelae*. It has a powerful action on a wide range of infectious bacteria —both Gram-positive and Gram-negative, and is very effective in severe types of dysentery, intestinal infections and whooping coughs. It has proved to be a magic drug in the treatment of typhoid fever caused by typhus bacilli. Next in the chain of antibiotics were **aureomycin** isolated in 1948 by Duggar, a botanist and a world authority on mushrooms, from *Streptomyces aureofaciens*, and **terramycin** isolated in 1950 or a little earlier from *Streptomyces rimosus* under the auspices of the Pfizer Company of Brooklyn, the world's largest producers of antibiotics. Both are yellowish powders, and are very effective on a wide range of disease germs, particularly on penicillin-resistant cases, cholera and some virus diseases. **Erythromycin**, another antibiotic, discovered by McGuire in 1952 from *Streptomyces erythreus*, has proved to be particularly effective on drug-resistant *Staphylococcus*. It is also remarkably active against whooping cough, diphtheria, large viruses, etc. Its range of action is wide like penicillin.

Gradually some more new and hitherto unknown soil bacteria have been discovered, and antibiotics manufactured and released to the market to relieve human sufferings. These wonder drugs have already saved and are still saving millions of human lives.

---

forgotten. Fleming, however, continued his research, and in February, 1941, he gave his first injection to a human patient, a dying man unaffected by sulpha drugs. Amazingly enough, the patient showed immediate improvement but unfortunately for him only a teaspoonful could be prepared then which was too little to effect a radical cure. A second patient was similarly tried but again the stock was insufficient. The future of penicillin could, however, be foreseen, and soon some British and American pharmaceutical firms undertook to manufacture it on a large scale.

The latest, possibly not the last, in the series of antibiotics is **jawaharene** which was discovered in 1963 by Dr. D. K. Roy at the Institute of Biochemistry and Experimental Medicine in Calcutta, from a species of *Aspergillus* growing on rotten potato tuber. This antibiotic has proved to be very active against various diseases such as pox, poliomyelitis, influenza, etc., and against ameobic dysentery, leukaemia (or blood cancer—a deadly disease) and various forms of tumours.

Chapter 5 LICHENS

General Description. **Lichens** comprising over 15,000 species form a large peculiar and interesting group of plants, being associations of specific fungi and algae, the former constituting the greater part of the lichen body. They were first discovered by Tulasne in 1852, and a few years later De Bary studied in detail the two organisms in them. The associations of different fungi and algae give rise to distinct species. Lichens commonly occur as greyish-green, greenish-white or bright-coloured incrustations, one to several cm. in diameter, on stems and branches of shrubs and trees, wooden posts, logs of wood, rocks, stones, old walls and ground, or sometimes hanging in shaggy tufts, a few to several cm. in length, from branches of shrubs and trees. They have a variety of other colours: white, yellow, orange, brown, red or black. Many of them grow under extreme conditions of humidity and temperature, and may survive long periods of desiccation. They are very widely distributed over the earth, being specially common in cold, even extremely cold regions and high altitude up to the snowline as well as in tropical rain forests. In lichens fungi and algae live together in intimate relationship. The two organisms are of mutual help to each other and lead a symbiotic life. The fungi absorb water and mineral matter from the substratum and supply the same to the algae, while the latter in their turn prepare food and supply it to the fungi. Lichens are thus typical examples of symbiosis. If they are separated from their associations they lead a precarious life, more particularly the fungi. It has been possible to synthesize several lichens by bringing together appropriate fungi and algae (Bonnier, 1886).

Classification. Depending on the nature of the fungi the lichens have been classified into two main groups: (1) **Ascolichens** and (2) **Basidiolichens**. In Ascolichens the fungi are members of Ascomycetes, reproducing by ascospores. They may further be divided into (a) **Discolichens** when the fungi in them are members of Discomycetes (or cup fungi), producing open cup- or saucer-shaped apothecia (see FIG. 112A), and (b) **Pyrenolichens** when the fungi in them are members of Pyrenomycetes (or flask fungi),

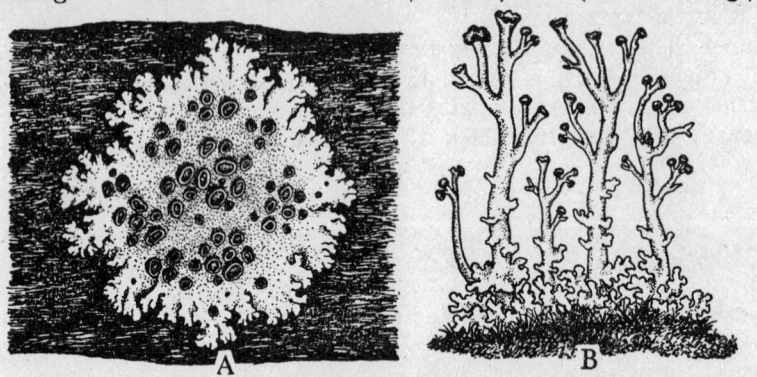

Lichens. FIG. 108. *A*, a foliose lichen (*Parmelia*); *B*, a fruticose lichen (*Cladonia*).

producing closed flask-shaped perithecia (see FIG. 92A) with an apical opening (ostiole). Ascolichens far outnumber Basidiolichens. In the latter the fungi are certain simple members of Agaricales, reproducing by basidiospores. There are only three genera of Basidiolichens, viz. *Cora, Corella* and *Dictyonema*.

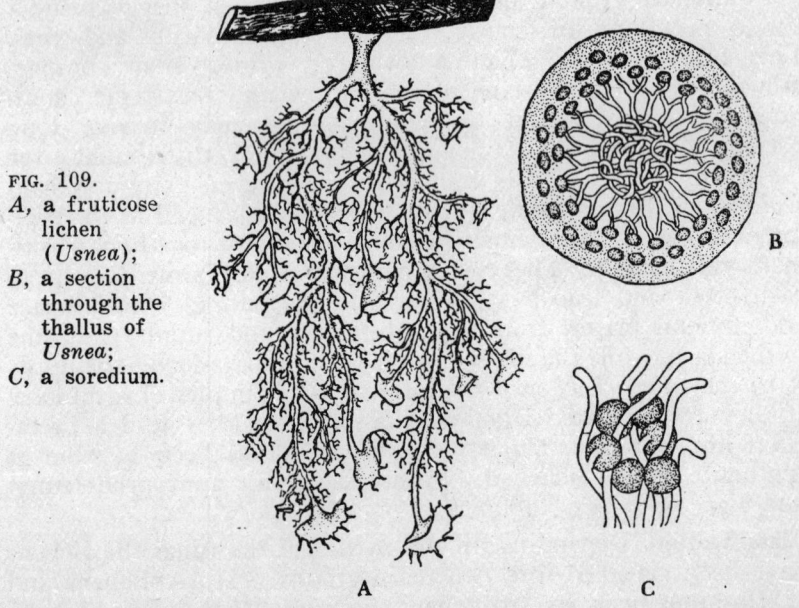

FIG. 109.
A, a fruticose lichen (*Usnea*);
B, a section through the thallus of *Usnea*;
C, a soredium.

The best known genus is *Cora* which is widely distributed in America. It is somewhat like a bracket fungus.

Thallus. Lichen thalli take three different patterns of growth in different genera, as follows: (*a*) **crustose lichens** forming hard granular crusts, very tenaciously adhering to rocks, barks of shrubs and trees and certain soils, e.g. *Graphis* and *Lecanora;* they show very little differentiation into the upper surface and the lower; (*b*) **foliose lichens** (FIG. 108*A*) forming definite flattened leaf-like thalli with lobed margin, adhering to walls, tree trunks, rocks and ground by delicate rhizoids (**rhizines**), e.g. *Parmelia* and *Physcia;* such lichens show distinct differentiation into the upper surface and the lower; and (*c*) **fruticose lichens** forming much-branched shrub-like bodies which remain attached by their narrow basal portion only; the branches may be flat and ribbon-like or slender and filamentous; such lichens may be standing erect, e.g. reindeer moss (*Cladonia;* FIG. 108*B*) or drooping from branches of shrubs and trees, e.g. old man's beard (*Usnea;* FIG. 109*A*). The main framework of the thallus is made of an interwoven mass of hyphae of a fungus, commonly as ascomycete (ascolichen) or in a few cases a basidiomycete (basidiolichen). Ascolichens enclose mostly unicellular or sometimes filamentous blue-green algae (e.g. *Chroococcus, Gloeocapsa,* etc.—unicellular, and *Nostoc, Rivularia, Scytonema,* etc.—filamentous) or certain green algae (e.g. *Chlorella, Pleurococcus,* and some other less known forms—unicellular, and commonly *Trentepohlia*—filamentous). Basidiolichens enclose similar blue-green algae. The type of a fungus and that of an alga associated together in a lichen are always constant. In some lichens the algae remain scattered in the thallus, while in others they occur in 1 or 2 layers.

A section through the thallus (FIG. 110) of a foliose lichen shows a loose mass of hyphae in the central region—the so-called **medulla,** a compact mass of hyphae in the peripheral region—the so-called **cortex,** and between these two regions usually lies the algal layer (commonly called the **gonidial layer**) with numerous algal cells (commonly called the **gonidia**) held together in the meshes of the hyphae. In *Usnea,* a fruticose lichen, the thallus (FIG. 109*B*) is seen to be differentiated into a central compact core of hyphae, a region of loosely interwoven hyphae, an algal region, and externally another compact region of hyphae.

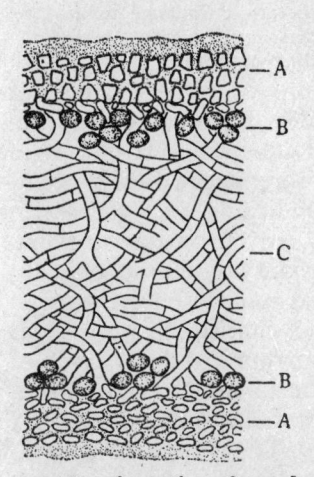

FIG. 110. A section through the thallus of a foliose lichen. *A*, cortex; *B*, gonidial layer; and *C*, medulla.

Reproduction. Lichens reproduce themselves in a variety of ways: (*A*) vegetative, (*B*) asexual and (*C*) sexual. But it must be noted that reproduction is predominantly fungal in character.

(*A*) **Vegetative Reproduction.** This may take place by various methods, the first three methods, as described below, being peculiar to lichens only. (*a*) **Soredia** (FIG. 109*C*). They are tiny granu-

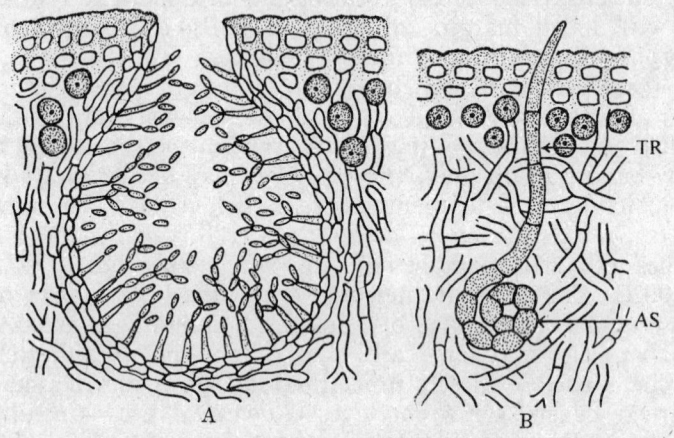

Lichen. FIG. 111. *A*, a pycnidium (or spermogonium) of *Physcia* in section showing pycnidiospores (or spermatia) formed in large numbers; *B*, a coiled ascogonium (*AS*) embedded in the thallus, with tube-like trichogyne (*TR*) protruding outwards.

lar bodies occurring in large numbers on the upper surface of the thallus as a greyish coating of powder. Each soredium consists of one to many algal cells wrapped up in a weft of fungal hyphae, as in *Usnea* and *Cladonia*. The soredia are blown about by the wind, and under appropriate conditions they germinate directly into lichen thalli, or sometimes they form new soredia. This is the commonest method of reproduction in lichens. (*b*) **Isidia.** In many lichens they are formed as minute outgrowths on the surface of the thallus. Each isidium consists of both algal cells and fungal hyphae, as usual, but surrounded by a layer of cortex. Isidia are primarily photosynthetic in function but sometimes they get detached from the parent thallus and then they develop into new thalli. (*c*) **Cephalodia.** They appear as dark-coloured swellings on the upper surface of the thallus, sometimes internally also. Each cephalodium consists of algal cells and fungal hyphae, as in the previous cases, but here the algal cells are different from those that normally occur in the thallus. Evidently they are foreign algae and may have been carried over to the lichen thallus, when young, ultimately forming such abnormal bodies. (*d*) **Oidia.** In a few lichens certain hyphae may break up into short segments, called oidia. An oidium germinates like a spore, producing normal hyphae. (*e*) **Fragmentation.**

In many lichens the thallus may be separated into long or short fragments. Each such fragment may grow up to the size of the parent thallus. In *Usnea* (FIG. 109A) the branches may be broken up by the wind into several fragments. Some of them at least get stuck up to some branches of trees and normally grow up.

(B) **Asexual Reproduction.** In Ascolichens this takes place by **spores** formed by the fungal partner. Each spore on germination sends out hyphae in different directions. If any of them happens to come in contact with the requisite alga it (the hypha) branches freely and covers up the algal cell. The combined body then grows up into a lichen thallus. Some workers have reported the formation of conidia in certain lichens. But this is disputed. Many lichens, e.g. *Physcia*, produce small spore-like bodies in large numbers within a flask-shaped cavity, called **pycnidium** (FIG. 111A), the spores then being called **pycnidiospores** (or pycnospores). Pycnidia appear in large numbers as small black dots on the surface of the thallus or as tiny protuberances on its margin. Pycnidiospores are known to germinate in certain species, producing a hypha. Coming in contact with an appropriate alga the combined body grows and forms into a lichen thallus. (It may be noted that in certain other species the so-called pycnidia behave as male organs, then called spermogonia, and the so-called pycnidiospores behave as male cells, then known as spermatia; see below). Basidiolichens. e.g. *Cora* (a common American genus), as stated before. reproduce by basidiospores, very much like *Agaricus*.

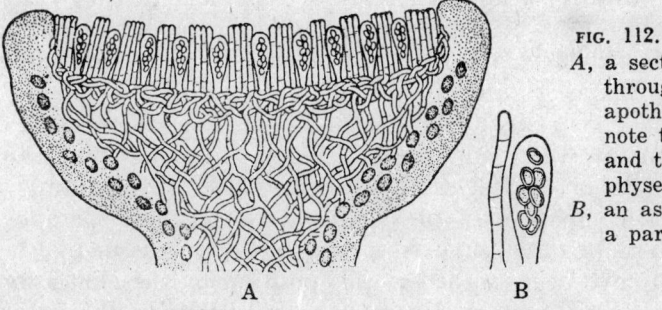

FIG. 112.
A, a section through an apothecium; note the asci and the paraphyses;
B, an ascus and a paraphysis.

(C) **Sexual Reproduction.** This has been observed in certain Ascolichens, as in *Collema*, the fungus alone taking part in the process. Sexual reproduction results in the formation of a fruiting body or fructification known as the **ascocarp** which is commonly an apothecium (FIG. 112A) or in some cases a perithecium (see FIG. 92A) with innumerable asci in each case. Sex organs are differentiated into male and female, and they develop in close proximity for facility of fertilization. The female organ is a multicellular stout filament of large cells and is known as the **carpogonium** (FIG. 111B). It consists of a coiled basal portion, called the **ascogonium**, lying within the thallus, and a tube-like upper portion, called the **trichogyne**, usually protruding beyond the thallus. The male organ is a flask-shaped chamber (FIG. 111A) with an apical opening (ostiole), and is known as the **spermogonium** (cf. pycnidium), and the minute non-motile male cells formed within it, known as the **spermatia**[1] (cf. pycnidiospores). Several spermogonia (pycnidia) may be formed in the thallus, with the inner wall of each lined with a layer of short slender hyphae. Spermatia are formed from them by abstriction like conidia, and occur in large numbers. Spermatia are very minute in size and cylindrical in form. They are liberated through the ostiole in slimy masses to float on the thallus.

Fertilization. This takes place when a spermatium comes in contact with the sticky protruding tip of a trichogyne. Its protoplast migrates into the trichogyne and apparently fuses with the ascogonium nucleus (though the process has not been observed yet). Thus fertilization is effected. Several **ascogenous hyphae** now begin to grow out from the basal portion of the ascogonium; these hyphae branch freely and develop an **ascus** always at the end of a branch. The ascogenous hyphae may also develop parthenogenetically in some cases. A mature ascus has usually 8 ascospores (FIG. 111B) but the number varies from 1 to 8. A complex fruiting body or fructification known as the ascocarp (see p. 527), bearing the asci, simultaneously develops. As stated before, the ascocarp commonly takes the shape of an open cup or saucer, called the **apothecium** (FIG. 112A), very much like that of *Peziza*. The apothecium consists of an investing tissue made of vegetative tissue and an interwoven mass of hyphae with a surface layer (hymenium) containing a compact mass of slender sterile hyphae (paraphyses) and club-shaped asci, occurring together in a palisade-like layer. The asci have been pushed to this position by the elongation of the ascogenous hyphae. In some cases, however, the ascocarp develops into a flask-shaped chamber, called the perithecium, with an apical opening (ostiole). The inner wall of the perithecium is lined with a layer of asci and paraphyses. The mature ascospores on liberation germinate producing hyphae, and those

---

[1] The nature of spermatia is a disputed point. Although in some cases spermatia act as male cells, in others they have been found to act as spores (pycnospores) germinating independently.

coming in contact with the right type of alga grow rapidly and eventually produce lichen thalli.

Uses. Lichens growing on rocks disintegrate them to form soils, thus preparing the ground first for mosses and subsequently for higher plants. Lichens have a variety of uses. Some of them are a valuable source of food for wild animals and cattle, e.g. reindeer moss (*Cladonia;* FIG. 108*B*) which grows in clumps to a height of about 30 cm. in the arctic tundra. Iceland moss (*Cetraria*) of the northern regions is used as food and medicine. In some countries certain lichens are fried for cattle feed and to some extent also for human food. Some types are used as medicines. Some yield beautiful dyes. Litmus is prepared from certain lichens. Some species are used in cosmetics, perfumes and soaps. Some lichens are used in brewing liquor, and some containing tannins used in tanning hide into leather.

Chapter 6   BRYOPHYTA

Classification of Bryophyta (23,725 sp.)
Class I   Hepaticae or liverworts (8,450 sp.). **Order 1.** Marchantiales (400 sp.), e.g. *Riccia* and *Marchantia* (thalloid liverworts). **Order 2.** Jungermanniales (8,050 sp.): (*a*) **anacrogynous**, e.g. *Pellia,* and (*b*) **acrogynous**, e.g. *Porella*; in the former the archegonia always develop behind the apical cell (never from the apical cell itself) and the gametophytes are thalloid; while in the latter the archegonia always develop from the segments of the apical cell and later from the apical cell itself and the gametophytes are leafy. The latter far outnumber the former.
Class II   Anthocerotae or horned liverworts (300 sp.). **Order 1.** Anthocerotales (only order), e.g. *Anthoceros* (gametophyte simple and thalloid but sporophyte complex).
Class III   Musci or mosses (14,975 sp). **Order 1.** Sphagnales or bog mosses (350 sp.), e.g. *Sphagnum.* **Order 2.** Andreaeales (125 sp.), e.g. *Andreaea.* **Order 3.** Bryales or true mosses (14,500 sp.), e.g. *Funaria, Polytrichum, Barbula* and *Dicranella* (gametophyte distinctly leafy and sporophyte very complex).

Origin and Evolution of Bryophyta. Bryophyta do, no doubt, occupy an intermediate position between the higher algae and the lower pteridophyta but their ancestors and descendants are not known. It is, however, most likely that bryophytes have evolved from some algal stock, possibly Ulotricales but because of the missing links between the higher algae and the lower bryophytes the actual position in regard to the origin of the latter cannot be ascertained. Having originated from some aquatic ancestor the bryophytes were pioneers in establishing themselves on land. This was a very important early event in the progressive evolution of plants under land conditions. Having established themselves on land they followed their own independent course of development (evolution), culminating in an individualistic group without giving rise to higher forms. Among bryophytes the liverworts are more primitive than the mosses, and the latter may have been derived from the former. According to another view the bryophytes have diverged from pteridophytes; later they followed an independent line of evolution ending in a blind alley without giving rise to higher forms. Bryophyta show an advance over algae by the development of archegonia, multicellular antheridia, and a distinct alternation of generations. By the same characters they are a near approach to the pteridophytes but the absence of vascular tissues and the dependence of the sporophyte upon the gametophyte have prevented them from coming to the pteridophytic level. The mode of fertilization by ciliate antherozoids through an aquatic medium has persisted from algae to bryophytes to pteridophytes and to certain lower gymnosperms; this may indicate a connecting link between these groups.

Development of the Sporophyte. The sporophyte is a distinct structure which has evolved from the zygote, and is diploid (with $2n$ chromosomes) and reproduces asexually by spores. It is the product of sexual reproduction and represents the stage between fertilization and subsequent meiosis (reduction division). In most green algae the zygote represents only a passing diploid ($2n$) phase without giving rise to a sporophyte. In them the zygote is only a resting body with protective covering to tide over unfavourable conditions. The true sporophyte developed in plants after they invaded the land from the sea in the remote past during their course of evolution and began to live under new adverse conditions of the environment. After its appearance it passed through successive developmental changes towards greater and greater complexities, and has in due course established itself as the main plant body; from ferns onwards all plants are sporophytes. Certain factors seem to be connected with the gradual development of the sporophyte: immediate germination of the zygote without any rest, prolongation of the vegetative period and delay in meiosis leading to the formation of spores, alternation of generations, land habit, facility of dispersion of spores by wind, sterilization of sporogenous tissue for other functions, and segregation of the spore-producing region from the vegetative region.

The sporophyte, to start with, is a simple structure lying embedded in the gametophytic thallus as a parasitic body, with nearly all its cells producing spores. This is the case of *Riccia*. A more complex type of sporophyte is formed in *Marchantia*. Here the sporophyte grows from the ray of the female receptacle and is already differentiated into a foot, seta and a capsule. In it some of the potentially sporogenous cells give rise to sterile cells—the elaters, and others to spores. Thus a partial sterilization is evident. Besides, the capsule of *Marchantia*, unlike that of *Riccia*, dehisces irregularly or by an apical lid to liberate the spores for their dispersal by wind. In *Anthoceros* the sporophyte has reached a high degree of complexity in many respects: relatively large size, continued growth in length by a basal meristem, extensive sterilization of the capsule leading to the development of the central axis—the columella, development of chlorophyllous cells and also other cellular differentiation in response to division of labour as a definite move towards an independent existence, partial sterilization of the sporogenous tissue leading to elaters and spores alternately, and all this on a simple primitive type of gametophyte, i.e. thallus. In it the mechanism of liberation of spores by dehiscence of the capsule into two valves is also very efficient. In moss the sporophyte has reached a high degree of specialization, on a highly developed gametophyte. In this plant the extensive amount of sterilization has resulted in an enlarged seta and a very complex capsule. The sporophyte of moss is not still quite an independent plant. *Anthoceros* and moss have, however, followed two independent lines of evolution. At the next higher stage, that is, in ferns and allied plants, the table is turned; the sporophyte in them has become the main or all-important body, and is quite independent of the gametophyte which has dwindled down to a simple structure—the prothallus. The sporophyte has elaborately developed roots, stem (often branched) and leaves (or sporophylls) with a distinct spore-producing region in them. Ferns and their relatives like *Equisetum* and *Lycopodium* are homosporous. At a higher level *Selaginella* is heterosporous. At the highest level the 'flowering' plants are all sporophytes and are heterosporous, reaching the highest degree of complexity; while the corresponding gametophytes have become reduced to a few cells or nuclei only. Thus it is seen that the sporophyte from a simple beginning several millions of years ago has at long last established itself as the highest form of plants—the angiosperms, which now dominate the vegetation of the earth; while the gametophyte, once the predominant feature of the early age, has now become reduced to almost nothing.

1. *RICCIA* (135 *sp.*)

Riccia (family *Ricciaceae;* FIG. 113) is a thalloid liverwort showing distinct dichotomous branching and taking on a rosette form (*A*). There are about 22 species of *Riccia* in India, of which *R. discolor* (=*R. himalayensis*) is a fairly common one. The thallus is a flattened structure showing dorsal (upper) surface with a longitudinal groove along the whole length of the mid-rib, and a ventral (lower) surface with usually a row of scales at the apex and a number of unicellular hairy structures known as the **rhizoids**. The rhizoids are of two kinds—smooth and tuberculate. The thallus is thicker in the middle and thinner at the two margins. The growth of the thallus takes place by a single wedge-shaped apical cell situated in an apical notch. The segments of the thallus are obcordate or linear with their margin sometimes ciliated. Species of *Riccia* are terrestrial growing as a green carpet on wet ground, old damp walls, old tree trunks and moist rocks; the only aquatic species is *Riccia fluitans*.

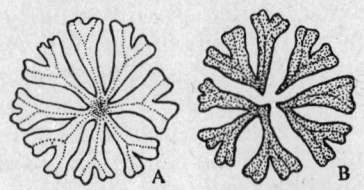

FIG. 113. *A*, a *Riccia* plant; *B*, vegetative propagation.

A cross-section of the thallus (FIG. 113C) shows the following structure: (*a*) a discontinuous upper epidermis, (*b*) an assimilatory tissue consisting of rows of cells with chloroplasts, leaving narrow irregular air-spaces between the rows for facility of diffusion of gases between the atmosphere and the thallus, (*c*) a lower colourless tissue consisting of fairly big thin-walled cells for storage of water and food, and (*d*) a lower epidermis with many rhizoids.

Vegetative Reproduction. The vegetative propagation of *Riccia* is common. It takes place by progressive decay of the older portion of the thallus, evidently at its base, and its separation into branches (FIG. 113B). These then grow into new thalli.

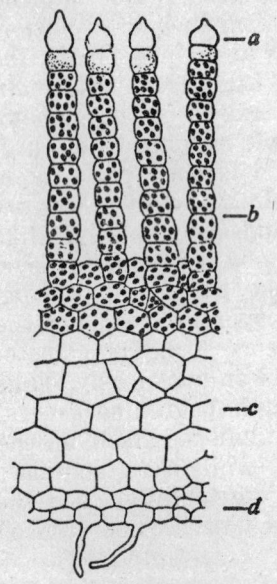

FIG. 113C. *Riccia* thallus in section (see text).

Sexual Reproduction. The plant is a **gametophyte**, i.e. it reproduces sexually by gametes. The two kinds of gametes—male and female—are borne in special structures known as the antheridia and the archegonia respectively. Some species (e.g. *R. robusta*) are *monoecious*, while others (e.g. *R. discolor*) are *dioecious*. In the monoecious species antheridia and archegonia develop together in the median groove on the dorsal (upper) side of the thallus. They grow in acropetal succession from the base of the thallus to its apex. Each **antheridium** (FIG. 114A), which is more or less pear-shaped, develops in a deep chamber formed by the overgrowth of the surrounding tissue of the thallus, and consists of a short stalk, a sterile wall and a compact mass of antherozoid mother cells. Each mother cell by a single division forms two cells, each of which is metamorphosed into a small twisted biciliate male gamete or **antherozoid** (FIG. 114B). Each **archegonium** (FIG. 114C-D) also lies sunken in a similar chamber. It is a short-stalked, flask-shaped body with a swollen basal portion known as the **venter** and a narrow tubular upper portion known as the **neck** which often projects beyond the epidermis and turns purplish. The neck contains a few neck canal

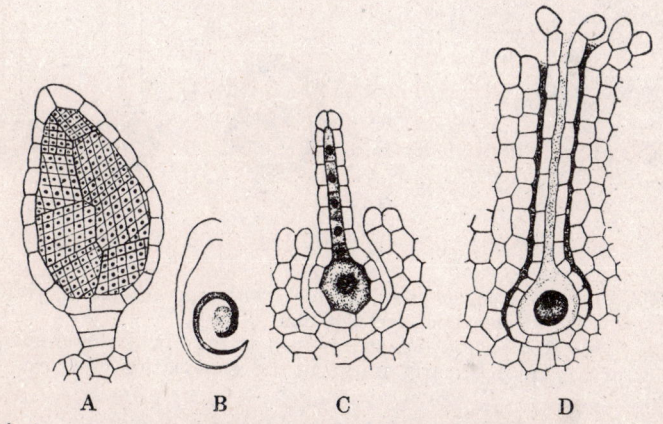

Riccia. FIG. 114. A, an antheridium; B, an antherozoid; C, a young archegonium; D, a mature archegonium.

cells (usually four) surrounded by six vertical rows of jacket cells, and the venter is occupied by a large cell—the **egg-cell** and a little higher up a small cell—the ventral canal cell. The egg-cell contains a distinct large nucleus which is the **egg-nucleus** (female gamete). As the archegonium matures, the neck canal cells and the ventral canal cell degenerate into mucilage. **Fertilization** takes place in the usual way. The antherozoids swim to the archegonium. The mucilage swells and forces out the cover cells of the neck canal (FIG. 114D). As the mucilage dissolves an open passage is established through the neck. The antherozoids enter the archegonium and one of them fuses with the egg-nucleus. After fertilization the ovum clothes itself with a wall and becomes the **oospore**.

Development and Structure of Sporophyte. The fertilized egg, i.e. the oospore, gives rise to the **sporophyte**. It reproduces asexually by spores. The sporophyte is a simple spherical body called the **capsule** (FIG. 115). It consists of a spore-sac and a wall surrounding it, the latter made of a single layer. The capsule develops *in situ* within the venter of the archegonium. With the growth of the capsule the venter also grows and invests it.

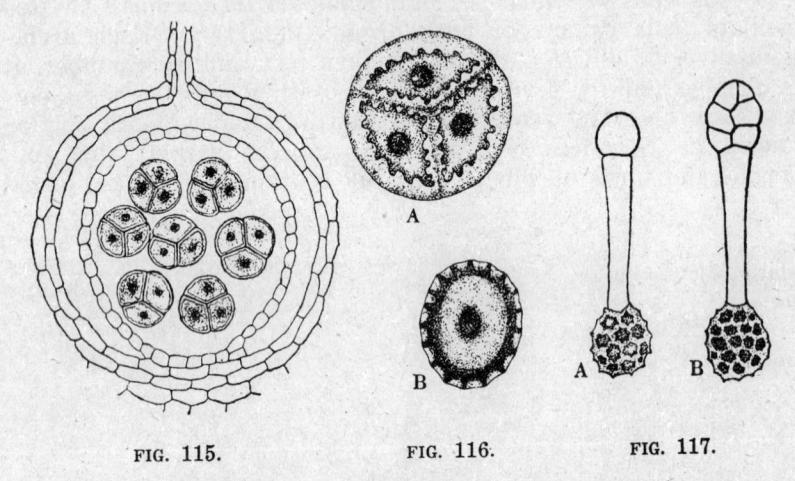

FIG. 115.   FIG. 116.   FIG. 117.

Riccia. FIG. 115. Sporophyte (capsule) with spore tetrads within enlarged archegonium.
FIG. 116. Spores—*A*, spores in a tetrad; *B*, a single spore.
FIG. 117. *A-B*, early stages in the germination of spore.

This investing structure is called the **calyptra**. The spore-sac contains a central mass of spore mother cells. Each mother cell undergoes reduction division and forms a *tetrad of spores* (FIG. 116*A*). Eventually by the rupture of the calyptra and the wall of the capsule the spores are set free. Each spore (FIG. 116*B*) is provided with a coat of two layers (three layers according to some authors). The outer layer is cutinized and the inner one made of pectose and callose. The whole coat is irregularly thickened and folded. In the germination of the spore the outer layer bursts and the inner one grows into a *germ tube* which gradually develops into *Riccia* thallus (FIG. 117). The sporophyte develops within the gametophyte and is wholly dependent upon it for its nutrition. In *Riccia* there is no special mechanism for spore dispersal.

Alternation of Generations. The plant passes through two successive generations—gametophytic with n chromosomes and sporophytic with $2n$ chromosomes—to complete its life-history. The gametophytic generation begins with the spore and ends in the gametes—antherozoid and ovum—prior to fertilization; while the sporophytic generation begins with the oospore and ends in the spore mother cells. The gametophyte gives rise to the sporophyte through sexual reproduction, and the sporophyte to the gametophyte through asexual reproduction. Thus there is a regular alternation of generations in *Riccia*. Life history showing alternation of generations is given below.

Riccia (gametophyte-n) → archegonium (n) → ovum (n) × → oospore ($2n$)
↑ → antheridium (n) → antherozoids (n) ↓
spore ← spore-tetrad ← spore mother cells ← capsule ← sporophyte
(n) (n) ($2n$) ($2n$) ($2n$)

2. *MARCHANTIA* (65 *sp.*)

Marchantia (FIG. 118) is a rosette type of thalloid liverwort showing conspicuous dichotomous branching, dorsiventral symmetry and a distinct mid-rib. It belongs to the family *Marchantiaceae*, of which there are about 11 species in India. *M. polymorpha* is a common and widespread species, and *M. palmata* is

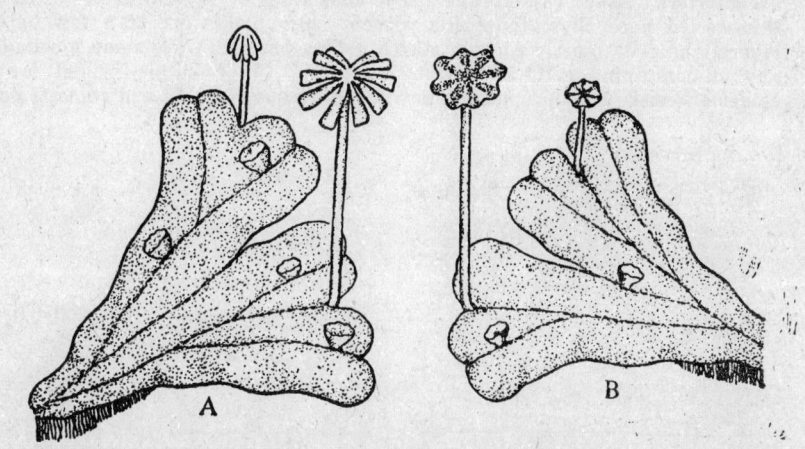

Marchantia. FIG. 118. *A*, female plant with archegoniphores and gemma-cups; *B*, male plant with antheridiophores and gemma-cups.

common in the Western Himalayas. *Marchantia* grows on damp ground or old walls and spreads rapidly during the rainy season forming a sort of green carpet. It occurs abundantly in the cold climate of the hills. The thallus bears on its undersurface (ventral) a number of unicellular rhizoids of two kinds—tuberculate and smooth-walled, a row of scales along the mid-rib, and 2 or 3 rows of them on each side of the mid-rib (the outer row being near the margin of the thallus). On the upper surface (dorsal), it bears a number of cup-like outgrowths, known as the **gemma-cups**, on the mid-rib. The thalli of some plants bear special *male* reproductive branches known as the **antheridiophores** (FIG. 118*B*) and those of other plants bear special *female* reproductive branches known as the **archegoniophores** (FIG. 118*A*). The two can be easily recognized—the former having a flat circular brownish lobed disc on the top, and the latter having a green smooth disc projected into distinct rays, at first bending downwards and later becoming horizontal. The growing point of the thallus lies in the notch of dichotomy and is represented by one or a few cells.

A section through the thallus (FIG. 119) shows: (1) a single-layered **upper epidermis** which is interspersed with numerous air-pores; the cells of the epidermis contain chloroplasts; (2) **air-chambers** lying below the epidermis and communicating with the exterior through a centrally placed air-pore; externally the chambers often appear as polygonal areas on the thallus; each air-pore is surrounded by a few tiers of cells; from the floor of the air-chamber arise short chains of cells, branched or unbranched, each cell containing several chloroplasts; these chains of green cells constitute the **assimilatory tissue**; (3) **storage tissue** consisting of several layers of large thin-walled parenchymatous cells without chloroplasts except a few upper layers; the cells mostly contain starch grains but there are some mucilage- and oil-containing cells here and there; and (4) a single-layered **lower epidermis** without chloroplasts but with numerous rhizoids and some scales.

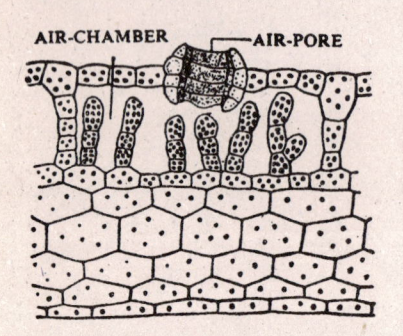

Marchantia. FIG. 119, Section through the thallus.

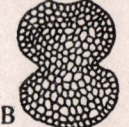

FIG. 120. *A*, gemma-cup; *B*, a gemma.

Reproduction. Vegetative reproduction may take place (*a*) by the decay of the old basal portion of the thallus, thus separating the branches, (*b*) by the formation of adventitious branches which get detached from the thallus, or (*c*) by **gemmae** (FIG. 120*B*) which develop in the **gemma-cup** or **cupule** (FIG. 120*A*). Each gemma is a small, more or less circular, flattened structure with a conspicuous depression on each side. The growing point lies in the depression. When the gemma gets detached from the gemma-cup it grows out into a dichotomously branched thallus. It is green in colour.

Sexual Reproduction. The thallus is the **gametophyte**, i.e. it reproduces sexually by gametes. *Marchantia* is **dioecious**, i.e. male and female plants are distinct and separate. The male plants bear antheridia or male reproductive organs on special erect branches called antheridiophores (FIG. 118*B*), and the female plants bear archegonia or female reproductive organs on almost similar branches called archegoniophores (FIG. 118*A*). The **antheridiophore** (FIG. 121) consists of an erect cylindrical **stalk** and a more or less circular, commonly 8-lobed disc or **receptacle** on the top. The stalk has two longitudinal channels on one side from which rhizoids and scales develop. The receptacle bears on its lower side a number of rhizoids and scales, and on the upper several

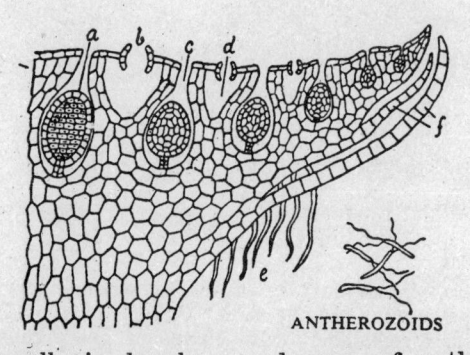

Marchantia.
FIG. 121.
Section through the antheridiophore.
 a, antheridium;
 b, air-pore;
 c, ostiole;
 d, air-chamber;
 e, hairs;
 f, scales.
Some antherozoids (*on the right*).

small air-chambers and rows of antheridia. Each air-chamber communicates with the exterior through a minute air-pore and has within it chains of green cells, as in the thallus. The antheridia are produced in acropetal order (the oldest towards the centre and the youngest towards the margin) from the segments of 8 growing points which are located at the tips of lobes. Each **antheridium** (FIG. 121) develops in a cavity lying embedded in the receptacle, and is more or less ovoid in shape; it communicates with the exterior by a narrow canal known as the **ostiole**. The antheridium is composed of a mass of small cubical cells (antherozoid mother cells) surrounded by a single-layered wall. Each antherozoid mother cell develops a minute biciliate spindle-shaped male gamete known as the **antherozoid** or **spermatozoid**. The **archegoniophore** (FIG. 122) similarly consists of a **stalk** (often longer than that of the antheridiophore) and an 8-lobed, star-shaped disc or **receptacle** with mostly 9 radiating **rays** or arms, somewhat like the ribs of an umbrella. The rays alternate with the lobes of the disc. The growing point is located at the tip of the lobe of the disc between two rays. Evidently there are 8 such growing points. A group of archegonia develop from the segments of each growing point in acropetal order, at first on the upper side. There are altogether 8 groups of archegonia alternating with the rays. By rapid elongation of the cells of the upper side of the disc the growing points are, however, pushed downwards and inwards with the result that the groups of archegonia come to lie underneath the disc. Each growing point and the youngest archegonium are brought close to the stalk, while the oldest archegonium lies near the margin. The stalk of the receptacle has two longitudinal channels on one side with rhizoids

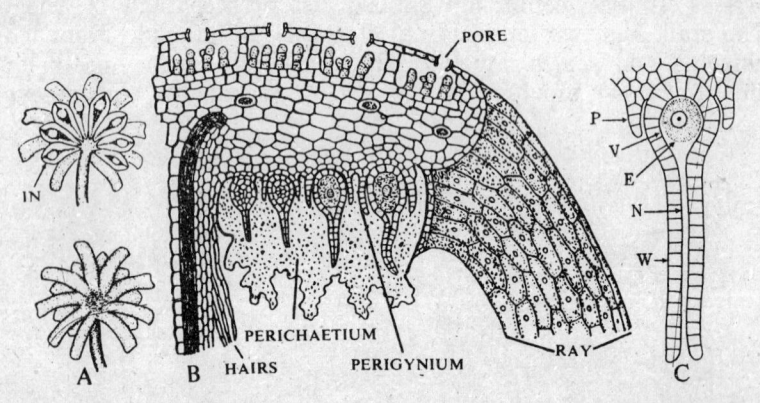

Marchantia. FIG. 122. *A*, (*top*) undersurface of the archegoniophore; *In*, involucre; (*bottom*) upper surface of the same; *B*, section through the archegoniophore showing air-chambers with chains of green cells and air-pores, ray, archegonia, etc.; *C*, an archegonium; *P*, perigynium or pseudo-perianth; *V*, venter; *E*, egg-cell; *N*, neck; and *W*, wall.

and scales, as in the male stalk. On the upper side the receptacle is provided with a number of air-chambers, as in the male receptacle, while groups of archegonia develop on the lower side hanging downwards. A membranous curtain-like outgrowth, known as the **involucre** (or perichaetium), fringed at the edges, is formed surrounding a group of archegonia as a protective covering (FIG. 122*A*-*B*). Moreover, at the base of each archegonium, ultimately surrounding it after fertilization, there is a cup-shaped outgrowth of it, known as the **pseudo-perianth** or **perigynium** (FIG. 122*B*-*C*). The **archegonium** (FIG. 122*C*) is a flask-shaped body consisting of a swollen basal portion—the **venter**, a narrow elongated portion—the **neck**, and a very short stout multicellular stalk. The neck of the archegonium, when young, is covered by a lid made of a few 'cap' or 'cover' cells. The venter contains a large cell—the **egg-cell** or ovum with a distinct large nucleus in it—the **egg-nucleus** (female gamete), and a small ventral canal cell, while the neck contains a row of usually 4-8 neck canal cells. The wall of the archegonium is made of six vertical rows of *jacket cells*. **Fertilization** takes place in the following way. When the antheridium bursts the ciliate antherozoids swim out of the antheridial chamber through the ostiole and frisk about in water lashing with their cilia. As the archegonium matures the neck canal cells and the ventral canal cell degenerate into mucilage; the mucilage swells up in contact with moisture, and the lid is forced open. A clear passage is thus formed. The mucilage contains some protein matter which attracts the antherozoids. They swim to the archegonium through the medium of dew or rainwater and many enter into the venter through the neck. Finally one of them fuses with the egg-nucleus in the venter. After fertilization the ovum develops a wall round itself and becomes the **oospore**.

Development and Structure of Sporogonium (FIGS. 123-24). The oospore germinates *in situ* and gives rise to the **sporogonium**.

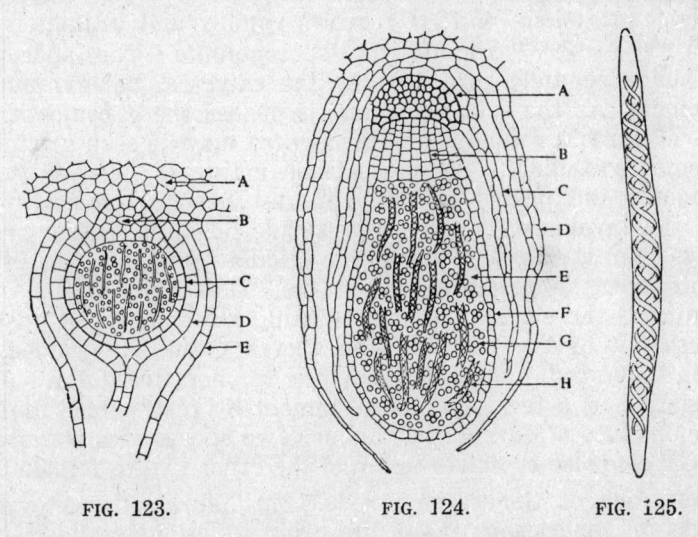

FIG. 123.   FIG. 124.   FIG. 125.

Marchantia. FIG. 123. A young sporogonium; *A*, tissue of the gametophyte; *B*, foot; *C*, capsule (wall); *D*, archegonium (wall); and *E*, perigynium or pseudo-perianth. FIG 124. A mature sporogonium; *A*, foot; *B*, seta; *C*, remnant of venter (calyptra); *D*, perigynium or pseudo-perianth; *E*, capsule; *F*, wall of the capsule; *G*, spore, and *H*, elater.
FIG. 125. An elater (enlarged).

The sporogonium is the **sporophyte**, i.e. it reproduces asexually by spores. The oospore divides into an upper cell and a lower.

The lower cell further divides and produces a **foot** and a short stalk called **seta** which elongates later. The foot penetrates into the tissue of the receptacle and absorbs nutritive materials from it. The upper cell divides and forms the **capsule**. The capsule consists of a single layer of wall-cells and a central mass of small cells (**archesporium**). Some of the archesporial cells grow up into elongated, spindle-shaped structures with internal spiral thickenings; these are known as the **elaters** (FIGS. 124-25). Other cells of the archesporium form **spore mother cells**. Each spore mother cell divides by meiosis to form *four* **spores** in a tetrad. After fertilization other parts of the archegonium also grow. Thus the wall of the venter grows forming the **calyptra** which surrounds the capsule (FIG. 124*C*); the neck withers and disappears. The **perigynium** (FIGS. 123*E* & 124*D*) grows rapidly and ultimately surrounds the sporogonium. The sporophyte is thus adequately protected by the calyptra, perigynium and involucre. As the seta elongates, it pushes the capsule through the calyptra; a remnant of the calyptra may be seen around the capsule (FIG. 124*C*). As the capsule matures, the seta further elongates and pushes it (the capsule) beyond the perigynium and the involucre. Finally the capsule dehisces, rather irregularly, from the apex to about the middle into a number of segments, and the spores are discharged (FIG. 126). Under humid conditions the elaters adopt a twisting movement and push the spores out of the capsule. The spores germinate immediately after they shed, and each gives rise to short irregular filament consisting of a few cells. This filament by further cell divisions develops into a *Marchantia* thallus. Two spores of a tetrad (FIG. 127*G*) give rise to male thalli and the other two to female thalli.

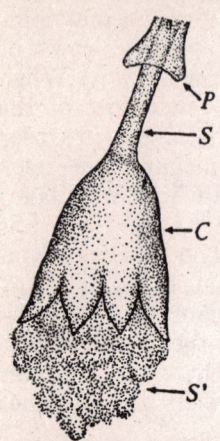

Marchantia. FIG. 126. Sporogonium dehiscing and discharging spores; *P*, perigynium; *S*, seta; *C*, capsule; and *S'*, spores.

Alternation of Generations. (FIG. 127). *Marchantia* shows two stages or generations in its life-history. The plant itself is the gametophyte having *haploid* or *n* chromosomes, and the sporogonium the sporophyte having *diploid* or 2*n* chromosomes. The gametophyte reproduces sexually by gametes and gives rise to the sporophyte, and the sporophyte reproduces asexually by spores and gives rise to the gametophyte. Thus the two generations regularly alternate with each other. All the stages from the oospore to the spore mother cells represent the sporophytic or asexual generation, and all the stages from the spores to the gametes—the ovum and the spermatozoid—represent the gametophytic or sexual generation. Life history showing alternation of generations is given below.

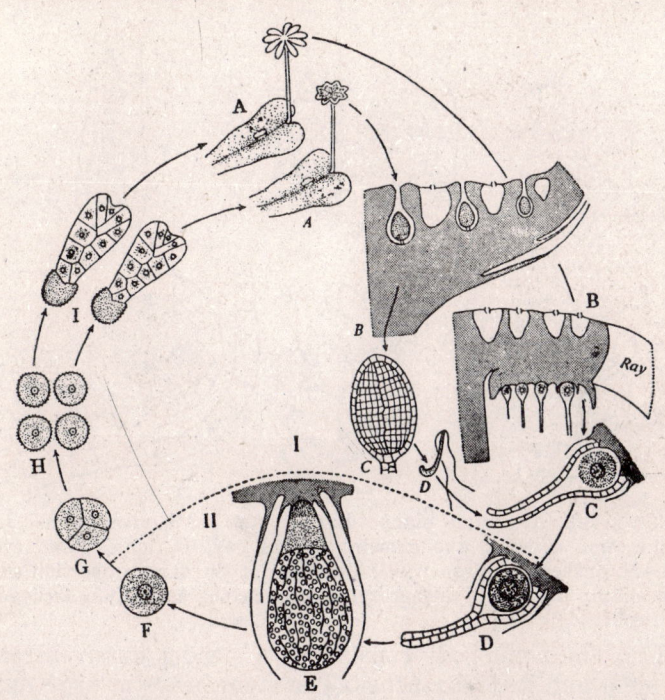

FIG. 127. Life cycle of *Marchantia*. *I*, gametophytic generation (haploid or *n*) and *II*, sporophytic generation (diploid or 2*n*). A, male thallus with antheridium; B, antheridiophore with antheridia (in section); C, an antheridium; D, an antherozoid; A, female thallus with archegoniophore; B, archegoniophore with archegonia (in section); C, an archegonium with egg-cell; D, zygote within archegonium; E, sporogonium; F, spore mother cell; G, spore-tetrad; H, spores, and I, young sporophytes (male and female).

Marchantia (♂) → ♂ receptacle (*n*) → antheridia (*n*) → antherozoids (*n*)
Marchantia (♀) → ♀ receptacle (*n*) → archegonia (*n*) → ovum (*n*) × ⎯
(gametophytes-*n*)

↑↑   ↓
spores (*n*) ← spore mother cells ← sporogonium (sporophyte, ← oospore
                         (2*n*)   (2*n*)   (2*n*)

3. *PORELLA* (180 *sp.*)

Occurrence. *Porella* (family *Porellaceae*; FIG. 128) is a common acrogynous (see p. 590) leafy liverwort. There are about 35 species in India, mostly in the Himalayas. *P. platyphylla* is a common species in the plains. *Porella* grows on moist rocks, tree trunks and old walls, and forms a compact greenish patch, practically covering the medium on which it grows.

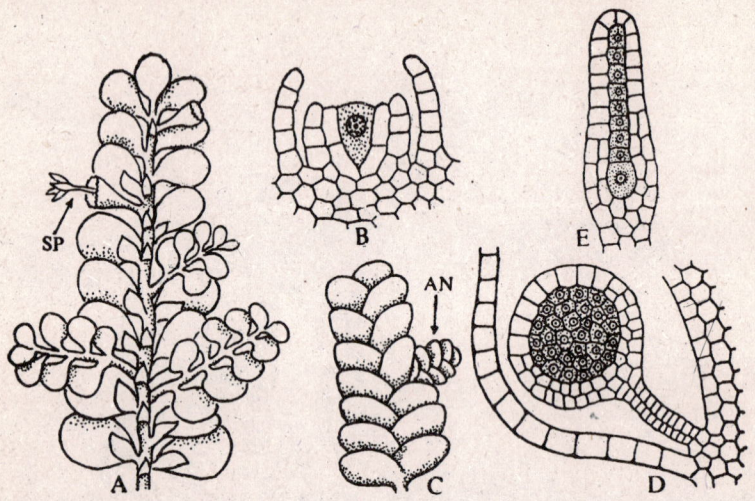

Porella. FIG. 128. *A*, female plant (dorsal view): *SP*, sporophyte; *B*, apex in longi-section showing the growing apical cell; *C*, male plant (ventral view); *AN*, antheridial branch; *D*, a portion of the same in section showing an antheridium filled with antherozoid mother cells; and *E*, an archegonium.

Structure. The plant body consists of a slender dorsiventral prostrate stem and leafy branches. The stem bears a large number of rhizoids from its lower side primarily for anchorage. Leaves are arranged in three rows: two rows of dorsal leaves and a row of ventral leaves much reduced in size, called *amphigastria*. Dorsal leaves are unequally bilobed and occur overlapping each other. The plant grows by means of an apical tetrahedral cell which evidently cuts off segments on three sides (FIG. 128*B*).

Vegetative reproduction may take place by the breaking-off of some of the branches, or by the formation of unicellular or multicellular gemmae on the margin or at the apex of the leaf.

Sexual Reproduction. The plant is a gametophyte and evidently reproduces sexually by gametes. *Porella* is dioecious. Male plants (FIG. 128*C*) are usually smaller in size and produce special short and lateral antheridial branches (FIG. 128*D*) which bear antheridia, each in the axil of a leaf (or bract). Paraphyses may be present. Female plants (FIG. 128*A*) are larger and produce lateral archegonial branches on which archegonia are always borne terminally either singly or in a group. Paraphyses may be present. Each antheridium (FIG. 128*D*) is a globular body surrounded by a wall (jacket) and provided with a long multicellular stalk, and is packed with antherozoid mother cells (androcytes), each giving rise to a minute biciliate antherozoid. Each **archegonium** (FIG. 128E) has a short multicellular stalk, a venter with an egg-cell and an egg-nucleus, a ventral canal cell, a long neck with 6-8 neck canal cells, and a wall. The neck is nearly as broad as the venter. Fertilization takes place in the usual way. The antherozoids, when liberated, swim in water to the archegonium. They enter through the apical opening of the archegonium, and finally one of them fuses with the egg-nucleus. The fertilized egg forms a zygote.

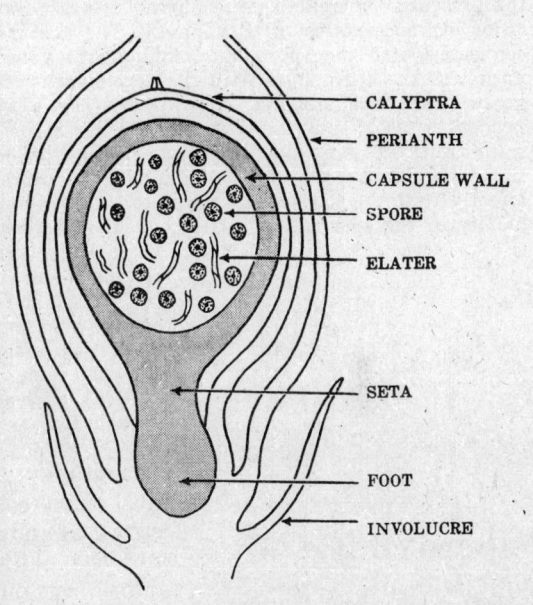

FIG. 129. Sporophyte of *Porella* in longi-section (semi-diagrammatic).

Sporophyte (FIG. 129). The zygote secretes a wall round it and soon increases in size. It divides and redivides and grows, and soon gives rise to the sporophyte. This consists of a foot, seta and capsule. The **capsule** is globose and is surrounded by a wall (jacket), 2 or 4 layers thick, and encloses short slender spirally thickened elaters and numerous spores. The sporophyte is surrounded by calyptra, perianth and involucre. **Calyptra** is the envelope developed from the venter. The other two envelopes are formed of united leaves (or bracts). When mature, the capsule dehisces by four valves, and the spores are liberated.
Germination of the spore. Under favourable conditions the spore germinates and gives rise to a small multicellular body—the **protonema.** Soon its apical cell becomes active and produces the shoot and leaves of a new *Porella* plant.

4. *ANTHOCEROS* (60 sp.)

Anthoceros (FIG. 130), commonly called horned liverwort, is a very interesting plant inasmuch as it shows certain special features in its life-history, particularly helping one to understand the course of evolution in the higher plants. While the plant shows a simple and primitive type of gametophyte, its sporophyte has already reached a high degree of development and complexity. The systematic position of *Anthoceros* is, therefore, uncertain. For the present it is treated as belonging to a separate class Anthocerotae and order Anthocerotales of Bryophyta. The features of interest are mostly connected with the sporophyte, and are as follows: (a) semi-independent nature of the sporophyte with the development of a considerable amount of green tissue and stomata, showing thereby a tendency towards an independent life; (b) a massive foot for greater absorption of water and mineral salts from the gametophyte; further with the decay of the gametophytic tissue the foot may touch the ground and absorb water and mineral salts directly from the soil—another step towards an independent career; (c) complexity of the sporophyte with a considerable development of sterile tissue in it being early indication of a more complex and quite independent sporophyte at a later stage in the evolution of higher plants; (d) establishment of the sterile axis (columella) representing the beginning of a conducting system; and (e) method of shedding spores comparable to that of ferns and allied plants.

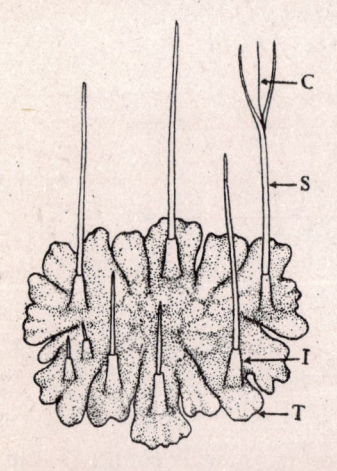

Anthoceros. FIG. 130. Thallus of *Anthoceros* with sporophytes; C, columella; S, sporophyte (capsule); I, involucre; and T, thallus (gametophyte).

Gametophyte. *Anthoceros* (FIG. 130) is a cosmopolitan plant and grows abundantly both in the hills and the plains in damp soils, hill sides, rotten tree trunks, damp walls, etc. There are about 25 species of *Anthoceros* in India, of which *A. punctatus* is a common one. The plant body of *Anthoceros* is a very simple type of gametophytic thallus, usually 2-3 cm. in diameter, with the reproductive organs lying sunken in it. Species of *Anthoceros* may be monoecious or dioecious. In the monoecious species both the male and female organs develop in the same thallus but separately, while in the dioecious species these organs develop in separate thalli. At a later stage the thalli bearing the female organs show a number of cylindrical deep-green sporophytes standing erect on them. The thallus is a small dark-green plate-like dorsiventral structure, often very irregularly lobed and without mid-rib. While there are many smooth-walled rhizoids developing on the ventral (lower) surface, scales are altogether absent. There are some intercellular mucilage-filled cavities opening to the ventral (lower) side of the thallus; these are occupied by colonies of *Nostoc*. The internal structure of the thallus is very simple consisting of a mass of thin-walled parenchyma without any differentiation of tissues. Each cell commonly contains a *single* large chloroplast with a large pyrenoid in it—a character not found in other Bryophyta. The pyrenoid consists of a mass of minute disc- or spindle-shaped bodies which are the rudiments of starch grains.

Reproduction. Vegetative reproduction may take place by the continued growth of the thallus and its separation into segments. In dry regions tubers may be formed on the margin, which may grow into new thalli under favourable conditions. **Sexual reproduction.** *Anthoceros* thallus is the gametophyte and it reproduces sexually by gametes. Species of *Anthoceros* may be monoecious (homothallic) bearing both antheridia and archegonia or dioecious (heterothallic) bearing either of the two. The sexual organs lie embedded in the dorsal (upper) surface of the thallus, antheridia appearing first in the monoecious species.

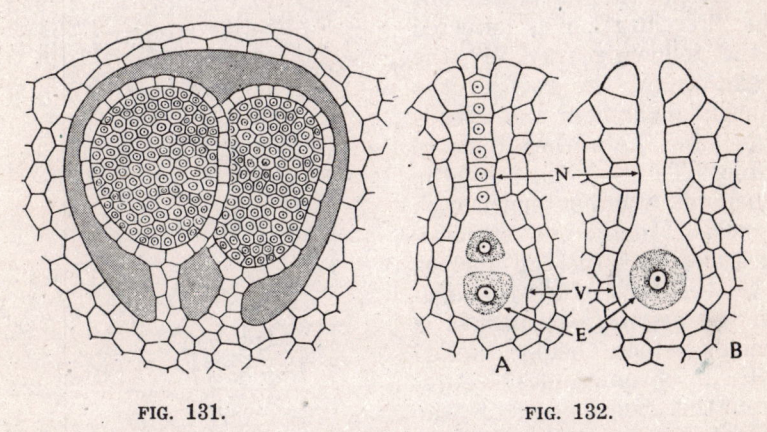

FIG. 131.   FIG. 132.

Anthoceros. FIG. 131. Two antheridia in an antheridial chamber (each with a stalk, wall and numerous antherozoid mother cells). FIG. 132. Two archegonia; *A*, an almost mature archegonium with egg-cell, ventral canal cell and neck canal cells; and *B*, a mature archegonium with egg-cell ready for fertilization; *N*, neck; *V*, venter; and *E*, egg-cell.

Antheridia (FIG. 131) grow in small groups (usually 2-4) within closed chambers, called antheridial chambers, which are filled with mucilage. Each chamber is covered over by a sort of roof made of 1 or 2 layers of cells. Each antheridium consists of a short multicellular stalk, a sterile outer layer (one or more cells thick) and a compact mass of antherozoid mother cells. Each mother cell gives rise to a single minute biciliate antherozoid. **Archegonia** (FIG. 132) develop singly and separately, lying partially embedded in the thallus. When fully developed, each archegonium consists of a venter and a neck. The neck consists of a vertical row of 4-6 neck canal cells, and the venter of a ventral canal cell and an egg-cell with a distinct egg-nucleus in it. At maturity the neck canal cells and the ventral canal cell get disorganized and become converted into mucilage. While the major part of the archegonium remains sunken in the thallus the upper end of the neck only protrudes out of it. When young the neck of the archegonium is covered by four 'cover' cells which separate out later. **Fertilization** is effected in the following way. By the breakdown of the roof of the antheridial chamber an outlet is formed for the antherozoids to escape. They swim to the archegonium and enter through its neck. Finally one antherozoid fuses with the egg-nucleus in the venter. After fertilization a zygote (oospore) is formed, which being diploid (with $2n$ chromosomes) represents the beginning of the sporophytic generation.

Anthoceros. FIG. 133. Longitudinal section of sporophyte (in three segments)—A, basal; B, middle and C, higher; *C*, columella; *S*, sporogenous tissue; *S'*, spore mother cell; *S"* spore-tetrad; *E*,' elater; *W*, wall of the capsule; *E*, epidermis; *I*, involucre or sheath; *M*, meristematic tissue; *F*, foot; and *G*, gametophytic tissue.

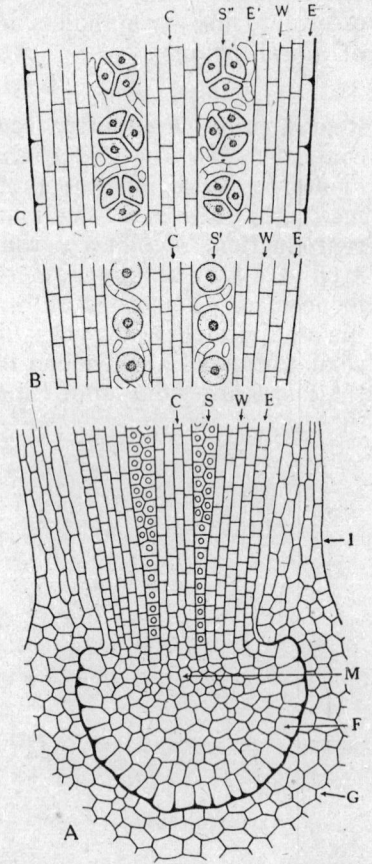

Sporophyte. The sporophyte develops from the zygote and consists of a **foot** and a **capsule**. It is surrounded at the base by a sheath or involucre formed by an upward growth of the archegonium for a time. Soon after fertilization the zygote grows and completely fills up the venter. It clothes itself with a wall and divides at first vertically into two cells; the second division which is transverse cuts them off into four cells; and the third division at right angles to the first one cuts them off into two tiers of four cells each. The lower tier by further divisions finally gives rise to a bulbous sterile structure at the base, thus increasing the absorbing surface; this is the foot which looks more or less like an inverted cap (FIG. 133A). The upper tier finally gives rise to the **capsule**. There is no seta (stalk) of the capsule in *Anthoceros*. The capsule (FIG. 130) is a slender cylindrical deep-green structure, usually 1-3 cm. long, sometimes much longer in some species. A longitudinal section through it shows the following regions (FIG. 133). (*a*) A **meristematic tissue** at the base of the capsule, by the activity of which the sporophyte continues to elongate and the sporocytes, i.e. the spore mother cells, also continue to be formed. (*b*) Centrally a sterile tissue—the **columella**, consisting of four rows of elongated cells each way, showing in a transverse section a solid square block of 16 rows of cells (FIG. 134); the sterile columella is an early indication of differentiation of the conducting system at a later stage in the higher plants. (*c*) Surrounding the columella there is a cylinder of **sporogenous tissue** (or archesporium). (*d*) Surrounding this there is the **capsule-wall** which is a jacket of green sterile tissue, 4-8 layers of cells thick, with 2 or sometimes more chloroplasts in

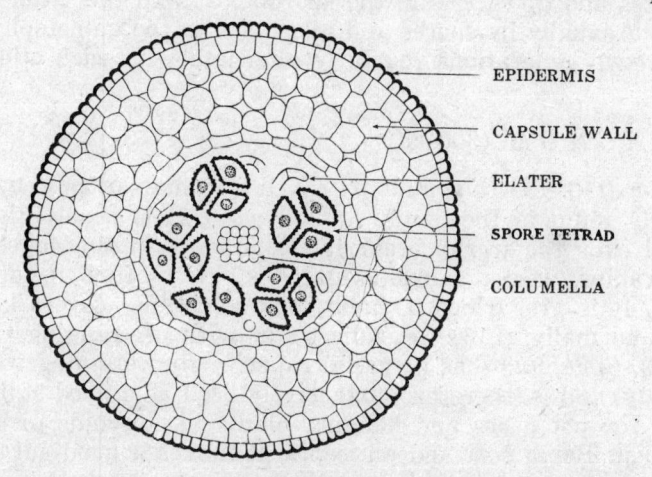

Anthoceros. FIG. 134. Transection of a sporophyte through the mature (upper) portion.

each cell; the outermost layer of this is the epidermis which is strongly thickened and cutinized, and provided with stomata. Because of the presence of the chloroplasts the sporophyte can manufacture most of its food and is only dependent on the gametophyte for water and mineral salts. The sporophyte is, therefore, a semi-independent body. The sporogenous tissue may extend down to the base of the capsule or only half-way down, and may be 1, 2, 3 or 4 layers of cells in thickness. The sporophyte matures from the apex downwards, i.e. the base is younger than the apex. The sporogenous cells develop into small groups of sterile cells called **elaters** and small groups of **spores** in an alternating manner. The elaters are mostly smooth-walled and rarely with spiral bands. Each spore mother cell undergoes reduction division and four spores are formed in a tetrad (FIG. 134).The sporophyte reproduces asexually by these spores. With the formation of the spores the gametophytic generation begins. The mature capsule dehisces from the apex downwards into two horn-like valves with the slender columella standing free in the centre (FIG. 130) and the spores are thrown out. The spore germinates and gives rise to an *Anthoceros* thallus.

Alternation of Generations. The life-history of *Anthoceros* is complete in two generations—gametophytic and sporophytic, which regularly alternate with each other. The *Anthoceros* thallus is the main gametophytic body with *haploid* or n chromosomes, and the capsule with the foot represents the sporophyte with *diploid* or $2n$ chromosomes. The gametophyte reproduces sexually by gametes and gives rise to the sporophyte, and the latter reproduces asexually by spores and gives rise to the gametophyte. Thus the two generations regularly alternate with each other.

5. *SPHAGNUM* (350 *sp.*)

Sphagnum (FIG. 135*A*), commonly called bog moss or peat moss, is the only genus of the family *Sphagnaceae*. It is widely distributed all over the world, occurring commonly at the edges of bogs, swamps, lakes, tanks, waterfalls, etc., often in dense cushions. It is gregarious in habit, and sometimes covers large areas. It normally grows in acidic water with pH ranging from 3.7 to 4.9. *Sphagnum* has a special capacity for retaining water in its body, and is, therefore, extensively used as a good stuffing material for pot herbs and hanging plants like orchids to keep them moist. Being soft and antiseptic it makes a good surgical dressing. It forms peat which may be used as a fuel. It is also added to alkaline soil to neutralize it. Altogether 17 species of *Sphagnum* have been recorded from the temperate Eastern Himalayas, mostly occurring in Sikkim, Bhutan, the Khasi Hills and the Assam Hill ranges. Some of them are: *S. khasianum, S. plumosum, S. papillosum, S. contortum*, etc.

Life-cycle. *Sphagnum* shows two distinct generations in its life-cycle. The main plant is the gametophyte which later bears on its top a stalked capsular body (varying, however, in number from 1 to 5) known as the sporogonium which is the sporophyte.

BRYOPHYTA

Sphagnum may be distinguished from mosses by the following characters: (1) the stem is without rhizoids; (2) the branching is copious and there are two kinds of branches—long and short; (3) the leaf consists of two kinds of cells—green and hyaline; (4) sex organs are borne on special lateral branches near the apex; they thus do not limit the indefinite growth of the main axis; further, antheridia are solitary, each in the axil of a leaf, while archegonia grow terminally, usually in a group of three; they are also without paraphyses; (5) the capsule is borne on a false stalk or pseudopodium (instead of a seta); (6) the capsule has no peristome; and (7) the prothallus is thalloid.

Gametophyte. The gametophyte (FIG. 135A) consists of a long or short slender erect axis (usually a few centimetres on land and sometimes about two metres in water), a profusion of slender branches, a dense mass of minute greenish leaves, and sex

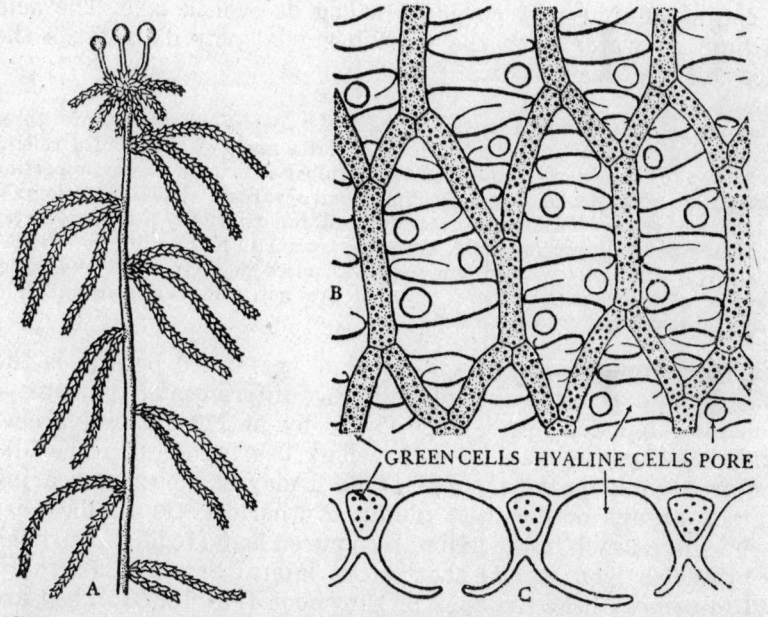

Sphagnum. FIG. 135. *A*, a stem with branches and leaves, and three terminal sporogonia; *B*, a mature leaf (surface view) showing narrow green cells, broad hyaline cells with pores, and spiral thickenings; *C*, the same in section (green cells may be triangular or rectangular).

organs on special short branches near the apex; this is called the **gametophore**. The branches are lateral, copiously formed, and are of two kinds: short ones (of limited growth) crowded near the apex and long ones (of unlimited growth) lower down in tufts. The plant is perennial in habit and continues to grow almost indefinitely by a tetrahedral apical cell, older parts always dying off from below. The leaves cover the whole plant but they are more closely set on the branches than on the main stem. They appear in three rows from the apical cell but later with the growth and twisting of the axis and the branches this arrangement becomes disturbed. The leaf is ovate or linear, and is composed of a single layer of cells and has no mid-rib. Under the microscope the leaf is seen to be composed of a network of elongated narrow green cells containing chloroplasts, interspersed with large broad hyaline dead cells filled with water (FIG. 135*B-C*). Such cells are spongy in nature (cf. velamen of orchid) absorbing and retaining water in enormous quantities. Some of the long strong branches often get detached from the older parts after their death and vegetate normally. The big mass of dead parts accumulating year after year forms peat which in course of time may cover up a bog or even a lake. The acid medium of water with the least bacterial activity retards the decay of the dead parts.

Anatomy of the Stem. Internally the stem is differentiated into three distinct regions: (*a*) a central pith or medulla made of thin-walled colourless cells; (*b*) a narrow cylinder of thick-walled cells acting as a supporting tissue; the cell-walls of this tissue may have various shades of colours—red, brown, yellow, blackish or greenish; and (*c*) externally a spongy cortex consisting of one layer (varying, however, from 1 to 5 according to species) of dead hyaline cells with circular or oval pores in their walls and sometimes spiral thickenings; these features are not, however, constant. The cortex absorbs and retains water.

Sexual Reproduction. *Sphagnum* plant, as stated before, is the gametophyte reproducing sexually by differentiated gametes—male and female, borne respectively by antheridia and archegonia. *Sphagnum* may be monoecious bearing both antheridia and archegonia on the same plant, or it may be dioecious bearing the sex organs on separate plants. Antheridia (FIG. 136*D*) are borne singly, each in the axil of a coloured leaf (reddish, purplish or brownish), on special short stout lateral branches (*antheridial branches*) near the apex of the shoot (FIG. 136*B*). They are produced in acropetal succession. Each **antheridium** is ovoid or spherical in shape and has a slender long stalk. It consists of a mass of antherozoid mother cells (androcytes), each enclosing a biciliate antherozoid, bounded by a jacket layer (wall). The two cilia are of equal length. Archegonia (FIG. 136*C*) grow in a group of three (varying, however, in number from 1 to 5) at the apex of very short branches (*archegonial branches*) just below the apex of the axis (FIG. 136*A*). Each **archegonium** consists of a swollen venter with an egg, and a long slightly twisted neck with neck canal cells, surrounded by a wall, and has a long multicellular stalk.

Fertilization. The antheridium bursts irregularly at the apex into valves and the antherozoids are liberated. They swim to the

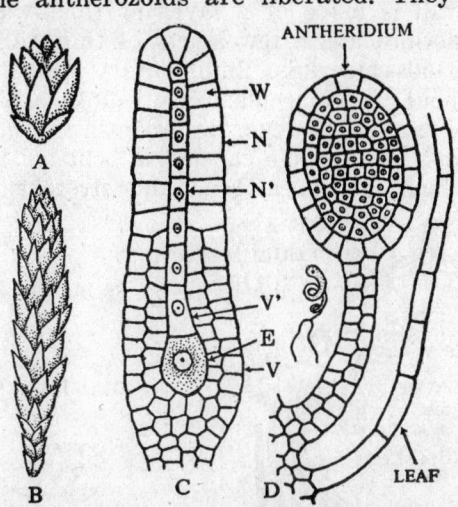

Sphagnum. FIG. 136. *A*, an archegonial branch; *B*, an antheridial branch; *C*, an archegonium (*V*, venter; *E*, egg; *V'*, ventral canal cell; *N'*, neck canal cell; *N*, neck; *W*, wall or jacket); *D*, an antheridium on a long stalk, developing from a branch, with antherozoid mother cells bounded by a jacket layer (wall); an antherozoid shown separately.

mature archegonium, and enter it through the open neck (neck canal cells dissolve into mucilage). One of the antherozoids then fuses with the egg-nucleus. Thus fertilization is effected. Sometimes the ventral canal cell may act as an oosphere, and fertilization takes place with it. The zygote, thus formed, first divides transversely into a short filament, 5-12 cells long. Longitudinal divisions then follow. Finally a spherical or ovoid spore-bearing body known as the **sporogonium** is formed on the top of the branch. Usually one zygote of an archegonial branch develops into a sporogonium.

Sporophyte. The sporogonium (FIG. 137*A*) is the sporophyte reproducing asexually by spores. It consists of a **capsule** developing from the upper part of the filament (see above), a very short neck-like stalk called the **seta** (often remaining undeveloped), a large bulbous **foot** developing from the lower part of the filament, and a pseudopodium (see p. 613). The **sporogonium** in longitudinal section (FIG. 137*A*) shows the following regions. Centrally there is a compact mass of colourless sterile cells, forming the **columella**. A dome-shaped **spore-sac** containing numerous spores formed in tetrads occurs on the top of the columella. There is a lid or cover on the top of the capsule known as the **operculum**;

it has a ring-like layer of thickened cells known as the **annulus**. The capsule wall is made of a layer of thick-walled, cutinized cells—the **epidermis**, and a few layers of thin-walled cells internal to it—the **sub-epidermis**. Rudimentary stomata (two guard cells only, without any chloroplast or opening) are present in the epidermis. The capsule, however, is greenish in colour containing some chloroplasts. The whole capsule is bounded by a loose cap or **calyptra** which is the enlarged and stretched archegonium

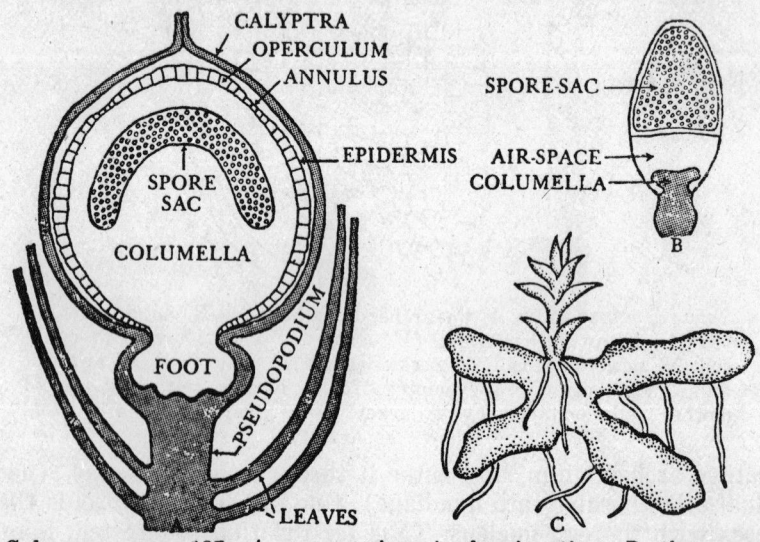

Sphagnum. FIG. 137. *A*, sporogonium in longi-section; *B*, the same showing the stage prior to bursting (diagrammatic); *C*, thalloid protonema with a young gametophore.

wall. Soon, however, it is torn off. The seta being very short (almost absent), a false stalk or **pseudopodium** develops from the stem at the base of the capsule. As the spores begin to mature this stalk elongates rapidly and pushes up the capsule. When the capsule is ripe the columella shrinks away from the spore-sac, leaving an air-cavity in between (FIG. 137*B*). The spore-sac now assumes a cylindrical form. The compressed air contained in the air-space then gives a heavy pressure to the spore-sac with the result that the capsule explodes, blowing off the lid and scattering the spores.

Germination of the Spore. The spore, tetrahedral in shape, on germination gives rise to a very short filament or *germ tube*. Its terminal cell divides in two directions, finally forming a green flat, irregularly lobed thallus called the **primary protonema** (FIG. 137*C*). It is a single layer of cells thick, and bears many septate rhizoids. Any marginal cell of the primary protonema may also divide and grow into a filament, finally becoming thalloid; this is the **secondary protonema**. A bud appears on the protonema from a tetrahedral cell on the margin of a lobe near its base. Leaves occur on the bud in three rows. The bud elongates rapidly and soon grows into an erect leafy much-branched gametophore. Usually one plant develops from a protonema. The latter soon disappears, and the mature plant having no rhizoids absorbs water by its whole surface.

Alternation of Generations. The foregoing description shows that the life-cycle of *Sphagnum* is completed in two stages: gametophytic and sporophytic. The main plant is the gametophyte (haploid or n) which reproduces sexually by gametes and gives rise to the sporophyte. The sporogonium is the sporophyte (diploid or $2n$) which reproduces asexually by spores and gives rise to the gametophyte again. Thus the two generations regularly alternate with each other to complete the life-cycle.

Vegetative Propagation. This is very common in *Sphagnum*, helping the plant to multiply rapidly and spread over large areas. The methods are: (a) separation of some of the long and strong branches after the death of the older parts; (b) development of secondary protonema by some of the short apical branches; (c) splitting of the protonema; and (d) formation of secondary protonema from the primary protonema.

6. *A MOSS*

Moss (FIG. 138A) occurs most commonly on old damp walls, trunks of trees, and on damp ground during the rainy season, while in winter it is seen to dry up. It is gregarious in habit; wherever it grows it forms a green patch or a soft velvet-like, green carpet. There are about 14,200 species of mosses and allies. Some of the common Indian mosses are *Funaria hygrometrica, Polytrichum commune, Barbula indica*, etc.

The moss plant is small, usually a few centimetres in height, and consists of a short axis with spirally arranged minute green leaves which are crowded towards the apex; true roots are absent; it bears a number of slender multicellular branching rhizoids which perform the functions of roots. The axis may be branched or unbranched.

Life-cycle. The life-cycle of moss is complete in two stages— gametophytic and sporophytic. The moss plant itself is the gametophyte and this is followed by another structure, called sporogonium, which grows dependent on the moss plant and is the sporophyte (FIG. 139C).

Gametophyte. The moss plant is a **gametophyte**, i.e. it bears gametes and reproduces by the sexual method. For this purpose highly differentiated male and female organs are developed at the apex of the shoot. The male organ is known as the **antheridium** and the female organ as the **archegonium**. These organs are sometimes intermixed with some multicellular hair-like structures known as the **paraphyses** (*para*, beside; *physis*, growth= offshoot). Antheridia and archegonia may occur together on the same branch or shoot or on two branches of the same plant (monoecious) or on two separate plants (dioecious).

The **antheridium** (FIG. 138*B-C*) is a multicellular, short-stalked, club-shaped body. It is packed with a mass of small cells known

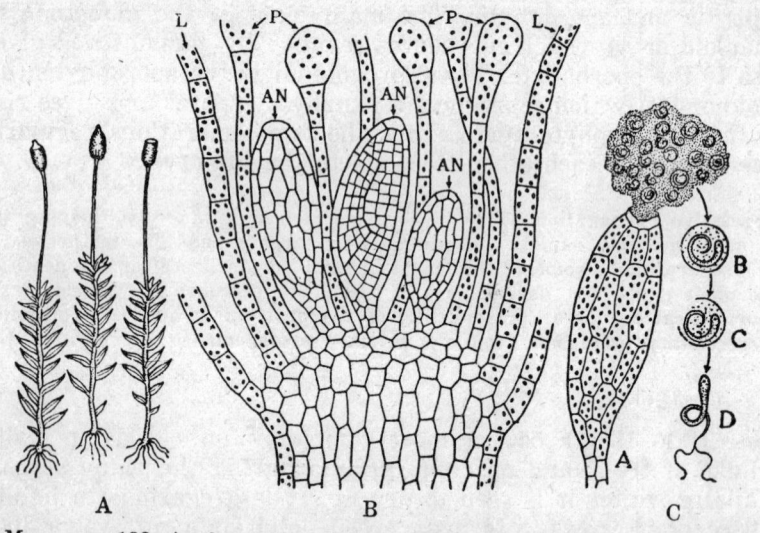

Moss. FIG. 138. A, three moss plants with capsule; B, apex of the male shoot in longi-section showing antheridia (*AN*), paraphyses (*P*) and leaves (*L*); C, a mature antheridium discharging antherozoid mother cells in a mass of mucilage (*A*), antherozoid mother cell (*B*), wall of the same getting dissolved (*C*), and biciliate antherozoid (*D*).

as the antherozoid mother cells (androcytes) and is surrounded by a single layer of cells known as the wall or jacket. The mother cells are regularly arranged in 5-15 segments (FIG. 138*B*), while the wall has a terminal lid or operculum consisting of one to many cells. Each mother cell develops a single antherozoid (or male gamete). As the antheridium matures the lid is forced open by the internal pressure of the contents, and the mother cells are liberated through the apical opening in a mass of mucilage (FIG. 138*C*). The mucilaginous walls of mother cells get dissolved in water and the **antherozoids** are set free. They are very minute in size, spirally coiled and biciliate; after liberation they swim in water that collects at the apex of the moss plant after rains.

The **archegonium** (FIG. 139 A-B) is also a multicellular body, but is flask-shaped in appearance. It is provided with a short multicellular stalk and consists of two portions: the lower swollen portion is known as the **venter** (belly) and the upper tube-like portion as the **neck**, surrounded by a wall or jacket which is

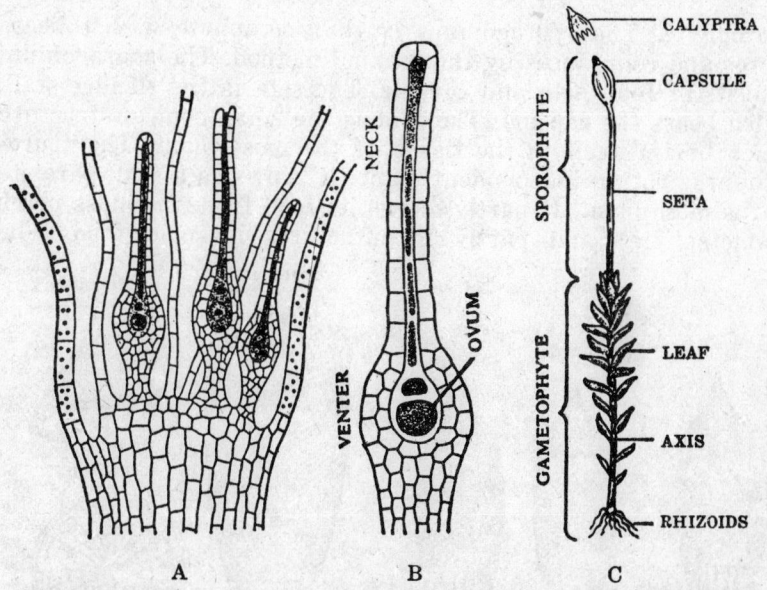

Moss. FIG. 139. *A*, apex of a female shoot in longi-section showing three archegonia, three paraphyses and two leaves; *B*, an archegonium; *C*, a moss plant showing the sporophyte growing on the gametophyte.

single-layered higher up and double-layered in the region of the venter. The neck is long, narrow and straight. Within the venter there lies a large cell—the **egg-cell** or ovum with a distinct nucleus in it—the **egg-nucleus** (female gamete); above this lies a small ventral canal cell, and higher up in the neck there are many neck canal cells. Except the ovum the other cells mentioned above are functionless and soon degenerate into mucilage. The neck at first remains closed at the apex by a sort of lid, but as the archegonium matures the lid opens as a result of internal pressure by mucilage and allows the antherozoids to enter and pass through it. **Fertilization** is effected through the medium of water—rain-water or dew—that collects on the moss plants. When the archegonium matures it secretes mucilage with cane-sugar. This attracts a swarm of antherozoids which enter through the neck canal and pass down into the venter; finally one of them fuses with the egg-nucleus and the rest die out. After fertilization the zygote clothes itself with a wall and is then known as the **oospore**. The latter germinates *in situ* and gives rise to the **sporogonium** on the moss plant (FIG. 139C). Although all the archegonia of a shoot or branch may be fertilized, ultimately one zygote (oospore) develops into the sporogonium.

Sporophyte. The sporogonium is the sporophyte, i.e. it bears spores and reproduces by the asexual method. The sporogonium consists of **foot, seta** and **capsule**. The seta is the slender stalk which bears the capsule. The foot is the small conical structure which buries itself in the tissue of the moss plant. The sporogonium is not an independent plant; it grows as a semi-parasite on the moss plant. It partly draws its food from the moss plant (gametophyte) and partly manufactures its own food. The

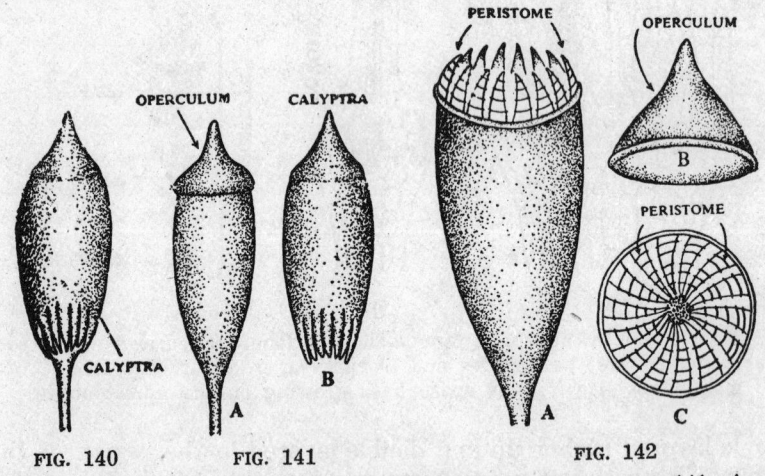

FIG. 140   FIG. 141   FIG. 142

Moss Capsule. FIG. 140. A capsule covered by calyptra. FIG. 141. *A*, a capsule without calyptra; *B*, detached calyptra. FIG. 142. *A*, a capsule showing peristome—open; *B*, opeculum; *C*, peristome—closed (surface view).

oospore divides into two cells—the upper and the lower; the lower cell by repeated divisions forms the seta with the foot, and the upper cell forms the multicellular complex body of the capsule. As the oospore grows into the sporogonium the archegonium gets ruptured somewhere in the middle. The upper half of the ruptured archegonium then forms a loose cap almost covering the capsule and is known as the **calyptra** (FIG. 140). It is afterwards blown away (FIGS. 141-42).

The **sporogonium** (FIG. 143) is a complex body, and more or less pear-shaped in appearance. A longitudinal section through it shows the following regions.

(1) **Operculum.** This is the lid of the capsule, and lies on the top of it. It is a few layers in thickness. When the capsule dehisces the operculum comes away as a circular, cup-shaped lid.

(2) **Annulus.** This is a special ring-like layer of epidermal cells, lying around the capsule at the base of the operculum. It is by the rupture of the annulus that the capsule dehisces.

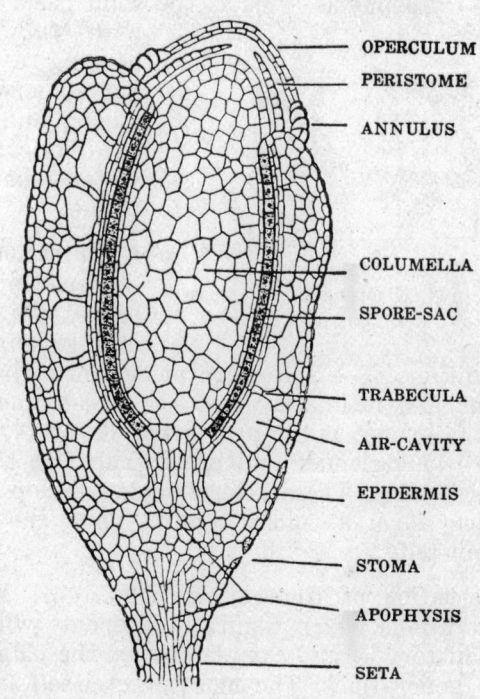

FIG. 143. Sporogonium of moss in longitudinal section.

OPERCULUM
PERISTOME
ANNULUS
COLUMELLA
SPORE-SAC
TRABECULA
AIR-CAVITY
EPIDERMIS
STOMA
APOPHYSIS
SETA

(3) **Peristome.** When the operculum falls off, the top of the capsule is seen to be furnished with one or two rows of thickened, tooth-like projections, constituting the peristome. These teeth are *hyrgroscopic*, and when they are dry they open out and help dispersion of spores (FIG. 142).

(4) **Columella.** This is the solid central column of the capsule. It is sterile, i.e. it contains no spores. Water and food material accumulate here for the developing spores.

(5) **Spore-sac.** This lies around the columella and contains numerous small cells. It is bounded externally by a few layers of cells, and internally by one layer. Each cell of the spore-sac is a spore mother cell. It soon undergoes reduction division to form four spores. The capsule dehisces at the annulus with the lid falling off. The capsule being seated on a long stalk is disturbed by wind, and the spores are thrown out of the spore-sac.

(6) **Air-cavity.** This lies as a cylindrical cavity surrounding the spore-sac, and is traversed by delicate strands of cells known as the **trabeculae** (sing. trabecula).

(7) **Capsule Wall.** This is composed of (a) a few layers of chloroplast-bearing cells just outside the air-cavity, (b) a few layers of bigger cells containing water—the sub-epidermis, and (c) externally a distinct layer—the epidermis.

(8) **Apophysis.** This is the solid basal portion of the capsule with (a) a distinct epidermis bearing a few stomata, (b) a sub-epidermis containing chloroplasts, and (c) a central region of elongated cells containing water—the water conducting tissue.

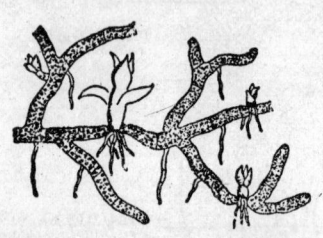

FIG. 144. Protonema of moss (note the buds and rhizoids).

Germination of the Spore. After dehiscence of the capsule the spores are scattered by the wind and germinate under favourable conditions. The spore grows out into a short tube which lengthens and ultimately forms a green, much-branched filament; this is known as the **protonema** (FIG. 144). It produces here and there long slender and brown rhizoids, and a number of small lateral buds. These lateral buds develop into new moss plants which form a colony again. Thus the life-cycle of moss is completed.

Alternation of Generations (FIG. 145). Moss plant shows two generations which regularly alternate with each other, and the life-history is only complete when the plant passes through these two generations. The moss plant itself is the gametophyte (or gamete-bearing plant), and the sporogonium is the sporophyte (or spore-bearing plant). Through sexual reproduction by gametes (antherozoid and ovum) the gametophyte gives rise to the sporophyte, and through asexual reproduction by spores the sporophyte gives rise to the gametophyte. In the life-history of moss the reduction of chromosomes to haploid or n takes place for the first time in the formation of spores from the spore mother cell. The spore is, therefore, the beginning of the sexual or gametophytic generation, and the various stages from the spore to the antherozoid and ovum represent the **gametophytic or sexual generation** because in all of them the chromosome number is n. The antherozoid and the ovum fuse, and the chromosome number is doubled, i.e. $2n$ is restored in the oospore. The oospore, therefore, represents the beginning of the asexual or sporophytic generation, and the oospore, sporogonium and spore mother

BRYOPHYTA

cells represent the **sporophytic** or **asexual generation** because in all of them the chromosome number is $2n$.

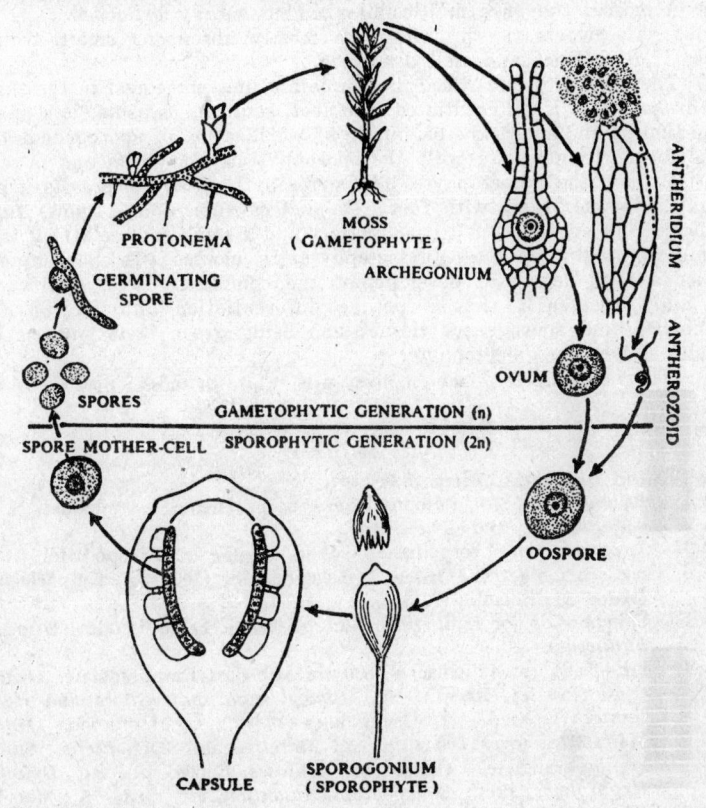

FIG. 145. Life-cycle of moss showing alternation of generations.

Vegetative Reproduction. The gametophytic plant reproduces vegetatively in a variety of ways: (1) by the formation of multicellular 'gemmae' which develop in groups usually at the apex of the leaf or at the apex of a comparatively long branch; they get detached and germinate in a moist soil, putting out a protonema; (2) by protonema which develops from any part of the plant—stem, leaf or even rhizoid; (3) by the production of resting buds on the protonema; they get detached from the protonema and develop into new moss plants; and (4) by separation of protonemal branches.

Any part of the sporogonium (foot, seta or capsule) may also develop protonema and give rise to moss plants. This is a case of **apospory**.

Comparison between Liverworts and Mosses. (1) In liverworts the gametophyte is mostly thalloid (except leafy Jungermanniales) and dorsiventral; whereas in mosses it is leafy and radial. In both, the gametophytes are the main plants and are green in colour performing photosynthesis, and they take to sexual reproduction through highly developed gametes and gametangia. Both show regular alternation of generations.

(2) Leaves in the leafy liverworts (e.g. *Porella* of Jungermanniales) have no mid-rib; while those in mosses have a mid-rib (with some exceptions).

(3) In liverworts the rhizoids are unicellular and commonly not branched; while in mosses they are multicellular and generally branched.

(4) In liverworts the protonema is mostly absent or small; while in mosses it is distinct and well developed.

(5) The sporophyte in *Riccia* is simple and lies embedded in the thallus. In *Marchantia* it is differentiated into foot, seta and capsule; it comes out of the thallus and shows the beginning of sterilization of sporogenous tissue in the form of elaters. In both, the sporophyte is non-green and is wholly parasitic upon the gametophyte. The sporophyte of *Anthoceros* is a green elongated complex body with foot, seta and capsule, and it shows further sterilization of sporogenous tissue (elaters and a small columella); it is only partially parasitic upon the gametophyte. In mosses the sporophyte has reached a high degree of development and complexity with distinct foot, seta and capsule; it shows further differentiation into sterile tissue (columella) and sporogenous tissue; and being green it is only partially parasitic upon the gametophyte.

(6) Elaters are mostly present in liverworts; while in mosses they are absent.

Chapter 7   PTERIDOPHYTA

Classification of Pteridophyta (9,000 sp.)

Class I   Psilotopspsida (or Psilophytinae)— 8 sp. **Order 1.** Psilotales (8 sp.), e.g. *Psilotum* and *Tmesipteris*.

Class II   Lycopsida (or Lycopodinae)—963 sp. **Order 1.** Lycopodiales (186 sp), e.g. *Lycopodium*. **Order 2.** Selaginellales (700 sp.), e.g. *Selaginella*. **Order 3.** Isoetales (77 sp.), e.g. *Isoetes*.

Class III   Sphenopsida (or Equisetinae)—25 sp. **Order 1.** Equisetales (25 sp.), e.g. *Equisetum*.

Class IV   Pteropsida (or Filicinae)—7,800 sp. Sub-class Eusporangiate. **Order 1.** Ophioglossales (70 sp.), e.g. *Ophioglossum*, *Botrychium* and *Helminthostachys*. **Order 2.** Osmundales (19 sp.), e.g. *Osmunda*. **Order 3.** Marattiales (over 200 sp.), e.g. *Marattia* and *Angiopteris*. Sub-class Leptosporangiate. **Order 4.** Filicales (7,600 sp.), e.g. *Dryopteris, Nephrolepis, Pteris, Polypodium, Adiantum*, etc. **Order 5.** Marsileales (67 sp.), e.g. *Marsilea*. **Order 6.** Salviniales (16 sp.), e.g. *Salvinia* and *Azolla*.

Leptosporangiate ferns are those in which the entire sporangium develops from a single superficial cell of the sporophyll; while the **eusporangiate ferns** are those in which a row or a group of superficial cells divide to form an outer and an inner layer of cells, the outer giving rise to the wall of the sporangium and the inner to the sporogenous tissue.

Origin and Evolution of Pteridophyta. The origin of Pteridophyta cannot be stated with any amount of certainty. It is, however, known that among the vascular plants the oldest and the most primitive group is Psilophytales which grew abundantly in the early Palaeozoic. The group soon declined, and before it disappeared (leaving only two genera as its living representatives) it gave rise to three independent lines of evolution as represented by Lycopodinae, Equisetinae and Filicinae. Each of them followed its own course of evolution. The origin of Psilophytales is again speculative; it may have been derived from algal ancestors or from some bryophytes. Of the three groups mentioned above Lycopodinae seems to be comparatively old although it has given rise to heterosporous condition (e.g. *Selaginella* and *Isoetes*). Equisetinae is more advanced than the former; while Filicinae is the most advanced group. Among the orders of this group the eusporangiate ferns are regarded as ancient and the leptosporangiate ferns as modern.

Types of Steles in Pteridophyta (FIG. 146). The vascular tissues as a whole together with the associated tissues making up the central column of the root, stem and leaf constitute a **stele** (*stele* means a column). The stele thus consists of all the tissues, mainly, however, the vascular, from the centre to the pericycle and is surrounded by the cortex (endodermis). Mainly depending on the relative position of xylem and phloem and on the presence or absence of pith different types of stele have made their appearance during the course of evolution. In this respect Pteridophyta are of special interest since they show, often with natural gradations, all the stelar types, as described below. Within each type a certain amount of variation or gradation is noticed. Sachs in 1875 gave an account of xylem and phloem (distribution and composition) and of cambium (origin and activity) in roots and shoots. But the theory of stele was formulated by Van Tieghem in 1886 who evidently laid the foundation for future study of this very important structure. He could not, however, show that different stelar types had arisen from a single original type. Jeffrey in 1897 and later modified Van Tieghem's concepts of stele and introduced the terms siphonostele and protostele. He concluded that the former, as a matter of fact all types of stele, had arisen from the latter.

(1) **Protostele** (*protos,* first). This is the most simple and primitive type of stele, in which the vascular tissues—xylem and phloem, particularly the former—have formed a solid central column *without any pith*. Protostele is characteristic of several members of Pteridophyta in stems generally but in roots almost universally. It has, however, certain types. In the simplest and earliest type xylem forms a solid central core, circular in outline, surrounded by a ring of phloem. Such a type is called **haplostele** (FIG. 146*A*), as in certain species of *Selaginella* and *Gleichenia*. In the latter, however, xylem occurs mixed up with some parenchyma. In the more advanced and complex types xylem appears as a star-shaped structure with phloem alternating with its rays. This is the radial type of stele, otherwise called **actinostele** (FIG. 146*B*), as in *Lycopodium serratum, L. phlegmaria* and *Psilotum*. In the most advanced type xylem and phloem occur in more or less parallel bands alternating with each other. This is the parallel-banded stele, otherwise called **plectostele** (FIG. 146*D*), as in *Lycopodium clavatum*. In *L. cernuum* xylem forms several small irregular groups which lie embedded in the ground mass of phloem. This is called **mixed protostele** (FIG. 146*C*). Protostele is also found in the pithless roots of many angiosperms and stems of some aquatic angiosperms.

With the development of the pith and the presence of leaf-gap; (see p. 625) the pattern of distribution of xylem and pholem changes, and the stele becomes differentiated into distinct types.

(2) **Siphonostele** (*siphon*, hollow or tube). *With the appearance of the pith the vascular tissues are pushed away from the centre, and it is seen that xylem forms a cylinder around the*

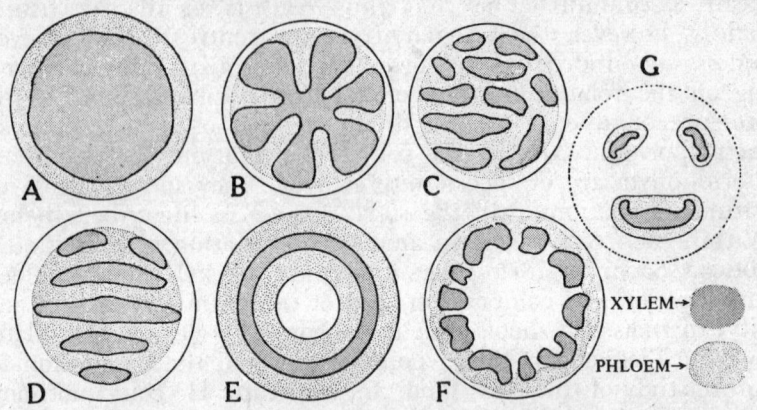

Types of Steles. FIG. 146. *A,* haplostele in *Gleichenia; B,* actinostele in *Lycopodium serratum; C,* mixed protostele in *L. cernuum; D,* plectostele in *L. clavatum; E,* amphiphloic siphonostele in *Marsilea; F,* ectophloic siphonostele in *Osmunda; G,* dictyostele in *Dryopteris.*

pith and phloem lies on both sides of it, often surrounding it, or phloem lies only on the outside of xylem. Siphonostele is thus a hollow or tubular stele. There are two types.

(a) **Amphiphloic Siphonostele** (*amphi*, on both sides). Here xylem forms a cylinder around the pith, and phloem (together with pericycle and endodermis) forms two cylinders, one on the outside of xylem and another on the inside of it (FIG. 146*E*). This type of stele is also called *solenostele*. This is seen in certain ferns, e.g. *Adiantum, Dicksonia,* etc. *Marsilea* typically shows this type of stele (see FIG. 190). Among the angiosperms it is found in *Cucurbitaceae*.

(b) **Ectophloic Siphonostele** (*ectos*, outside). Here also xylem forms a cylinder around the pith, as in the previous case, but only the external cylinder of phloem (together with only the external pericycle and endodermis) is present, the internal one being absent (FIG. 146*F*). This is found in some ferns, e.g. *Osmunda, Helminthostachys, Botrychium,* and also in *Equisetum.* Gymnosperms and dicotyledons show an advanced type of ectophloic siphonostele, with endarch collateral bundles. Here the leaf-gaps being large and indistinguishable from the interfascicular areas, and the pith and the cortex becoming connected by wide bands of parenchyma, the vascular system is broken up into separate vascular bundles.

Origin of Siphonostele. It is an admitted fact that siphonostele has been derived from protostele. But regarding the origin of pith there are two diametrically opposite views: *intrastelar origin* and *extrastelar origin*. According to Boodle and others supporting the 'intrastelar' view the origin of the pith lies in the gradual metamorphosis of inner elements (tracheids) of the protostele and consequent shifting of the vascular tissues outwards. According to them, therefore, the pith is intrastelar in origin, gradually expanding outwards in the form of medullary rays (expansion theory). This is the generally accepted view. According to Jeffrey, however, the pith is extrastelar in origin; first, by obliteration of the inner endodermis, and second, by the intrusion of the cortical tissue inwards, finally giving rise to the pith in the centre (invasion theory). This is not, however, a very convincing view.

(3) **Dictyostele** (*dictyo*, net). This is a much dissected type of stele derived from the siphonostele, i.e. the stele is broken up into a number of separate vascular strands, and is the most advanced type. The presence of numerous leaf-gaps caused by leaf-traces breaks up the stele into a network of separate strands, each constituting a concentric bundle (FIG. 146G). This is found characteristically in many species of *Polypodiaceae*, e.g. *Pteris, Pteridium, Polypodium, Aspidium, Dryopteris*, etc., and some species of *Selaginella*.

Leaf-traces and Leaf-gaps. It must be noted that the vascular cylinder is continuous through the whole plant body. At the node a strand of vascular tissue leaves the cylinder, passes out through the cortex and goes into the leaf; this strand of vascular tissue is called the **leaf-trace**. The leaf-trace while leaving the vascular cylinder of the stem causes small breaks or openings in it; each break in the vascular cylinder of the stem is called the **leaf-gap**. The leaf-gap occurs in the vascular cylinder just above the leaf-trace and is filled up with parenchyma. Above the leaf-gap the vascular cylinder again becomes continuous so that a transection at the node of the stem shows the break in the vascular cylinder, while a section at the internode above the leaf-gap shows a continuous ring-like cylinder. The presence of numerous leaf-gaps breaks up the vascular cylinder into a sort of network.

A reduction of the vascular tissues is noticed from gymnosperms and woody dicotyledons where the activity of the cambium is at its maximum, to herbaceous dicotyledons with much less or no activity of the cambium and sometimes absence of interfascicular cambium, finally to monocotyledons without cambium (except in a few cases). This reduction series indicates the trend of evolution from woody types (primitive condition) to herbaceous types (advanced condition).

1. *PSILOTUM* (2 *sp.*)

Psilotum (FIG. 147*A*) of the (family *Psilotaceae*) and *Tmesipteris* belong to the *living* order Psilotales which is closely related to the *extinct* order Psilophytales of the Devonian (see FIG. X/1). Psilophytales represented the earliest land plants and the oldest pteridophytes. The above two genera of Psilotales are regarded as the living representatives of the extinct order. *Tmesipteris* is confined to Australia and the neighbourhood, while *Psilotum* is somewhat common in the tropics, and in India *Psilotum nudum* (= P. *triquetrum*) is found in Pachmarhi (Madhya Pradesh), the Sundarbans and a few other places.

Psilotum, a slender tufted perennial herb, 15-60 cm. or more in length, grows as an epiphyte on tree trunks or on rocky slopes, often hanging downwards, or it grows on the ground or rocks or bases of trees, often standing erect. The plant shows conspicuous dichotomous branching. As a whole it consists of a slender, subterranean rhizome, and slender green aerial branches. The rhizome (*B*) is much branched, coralloid and brown in colour, and bears many rhizoids but no roots. The stem is unbranched at the base but higher up it becomes profusely branched in a dichotomous manner, bearing minute bifid scale-leaves singly at the nodes. The branches are somewhat angular in outline. The rhizome and the branch grow by a large pyramidal apical cell. Internally the rhizome is protostelic, with scanty development of phloem around. The stem on the other hand shows a siphonostelic structure, with star-shaped xylem around a central fibrous pith, and indistinct phloem in between the xylem arms. Xylem is always exarch. Vegetative reproduction is common and takes place by oval multicellular **gemmae** which develop in large numbers on the surface of the rhizome, and sometimes also on the prothallus.

Sporangia (FIG. 147*C-E*). In the axil of the scale-leaf a short stalk bears terminally a trilocular sporangium (or a union of three sporangia, i.e. a **synangium**). *Psilotum* is homosporous, i.e. it bears only one kind of spores. A section through the sporangium shows a central mass of sporogenous cells which produce minute oval spores with finely reticulated walls. The cells surrounding them disintegrate and offer nourishment to the developing spores. As the spores mature the sporangium bursts longitudinally from the apex to the base, and the spores are scattered.

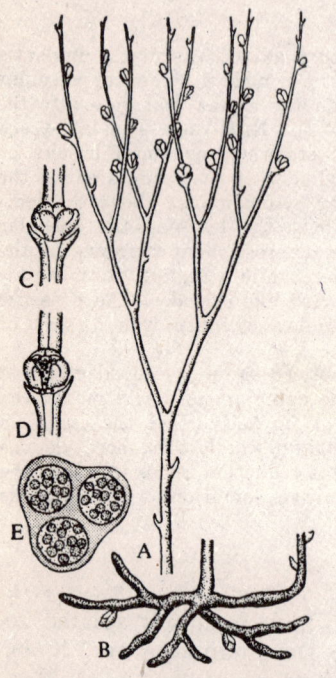

Psilotum. FIG. 147. *A,* a shoot with sporangia (synangia); *B,* much-branched rhizome; *C,* synangium closed; *D,* the same open; *E,* young synangium in section showing spore mother cells.

Prothallus. (FIG. 148*A*). Each spore germinates into a cylindrical branched prothallus. It is brownish in colour, and bears numerous rhizoids (but no roots) all over

PTERIDOPHYTA

its body. The presence of an endophytic mycorrhizal fungus is a common feature. Several antheridia and archegonia develop superficially, projecting outwards, on the prothallus. Each **antheridium** (*B*) is spherical and provided with a distinct wall, and produces numerous spiral and multiciliate **antherozoids** (*C*). The **archegonium** (*D*) is somewhat flask-shaped, with a neck and a venter. Fertilization takes place in the usual way. After fertilization the oospore that is formed grows into an embryo with a foot (buried in the prothallus) and an axis which shows dichotomy from the beginning. The axis grows into a branched rhizome from which aerial shoots soon develop.

Alternation of Generations. *Psilotum* shows a distinct alternation of generations but it is of *homomorphic* type, the prothallus and the sporophytic rhizome being somewhat similar in appearance.

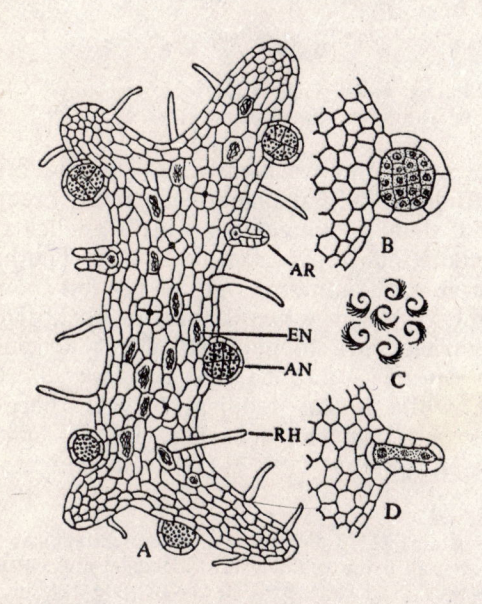

Psilotum. FIG. 148.
A, prothallus (gametophyte) with sex organs; *AR*, archegonium; *EN*, endophytic fungus; *AN*, antheridium; *RH*, rhizoid;
B, an antheridium;
C, antherozoids;
D, an archegonium.

Note. The primitive characters of *Psilotum* by which its close affinity with the ancient extinct group Psilophytales is indicated are: (*a*) axial nature of the plant body; (*b*) dichotomous branching; (*c*) complete absence of roots; (*d*) terminal sporangia; (*e*) homospory; (*f*) xylem made of annular and spiral tracheids only; (*g*) protostelic condition of the rhizome; (*h*) no cambium and, therefore, no secondary growth; and (*i*) homomorphic type of alternation of generations.

2. *LYCOPODIUM* (185 *sp.*)

Lycopodium (family *Lycopodiaceae*; FIGS. 149-51), commonly called club-moss, is a much branched herbaceous plant found

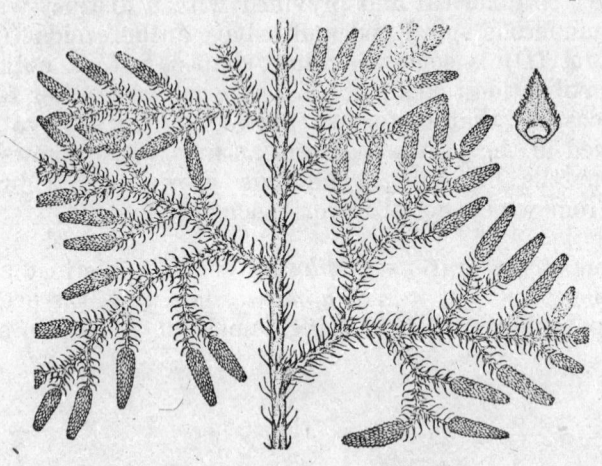

FIG. 149. *Lycopodium cernuum* (terrestrial); a sporophyll with a sporangium shown separately.

abundantly in the hills at a comparatively high altitude. There are about 8 species in India. The plant body consists of creeping rhizomes which give off slender elongated aerial branches from the upper side and adventitious roots from the lower. The branches are densely covered with numerous small narrow pointed leaves. *Lycopodium* mostly shows characteristic dichotomous branching, but some species are monopodial. Most species of *Lycopodium*, e.g. *L. cernuum* (FIG. 149) and *L. clavatum* (FIG. 151*B*) are terrestrial, while in the tropical forests there are some epiphytic species with pendent branches, e.g. *L. squarrosum* (FIG. 150) and *L. phlegmaria* (FIG. 151*A*), both common in North-East India.

Internal Structure of the Stem. (1) **Epidermis**—a single outermost layer of small cells with the outer and often radial walls thickened and cutinized; it is provided with numerous stomata. (2) **Cortex**—a wide region lying in between the epidermis and the stele. It varies considerably from species to species in its width, structure and composition. In *L. cernuum* (FIG.152) the cortex is differentiated into (*a*) inner cortex made of parenchyma, (*b*) middle cortex made of sclerenchyma, and (*c*) outer cortex made of one to a few layers of parenchyma or it may be absent altogether. In other species the inner cortex and the middle or outer cortex may be sclerenchymatous. (3) **Endodermis**—a single layer of small thin-walled cells surrounding the central stele; however, it is not well defined in all cases. (4) **Stele**—this is the central cylinder of the vascular system—xylem and phloem, surrounded by a few-layered, thin-walled parenchyma comprising the pericycle. The stele of the *Lycopodium* stem is a *protostele* with a central vascular cylinder but no pith. There are commonly three types of stele in different species of *Lycopodium*. In some species, as in *L. serratum* and *L. phlegmaria* xylem forms radiating arms or rays from the centre, with phloem alternating with them (radial arrangement); this is called **actinostele** (see FIG. 146B). In other species, as in *L. cernuum* (FIG. 152), xylem is broken up into isolated strands which lie embedded in the ground mass of phloem; this is called **mixed protostele** (FIG. 152). In still other species, as in *L. clavatum* and *L. complanatum*, xylem and phloem occur in alternating, more or less parallel bands; this is called **plectostele** (see FIG. 146D). In *Lycopodium* xylem is always exarch with protoxylem towards the pericycle, and metaxylem towards the centre. Protoxylem consists of a few narrow annular and spiral

FIG. 150. *Lycopodium squarrosum* (epiphytic).

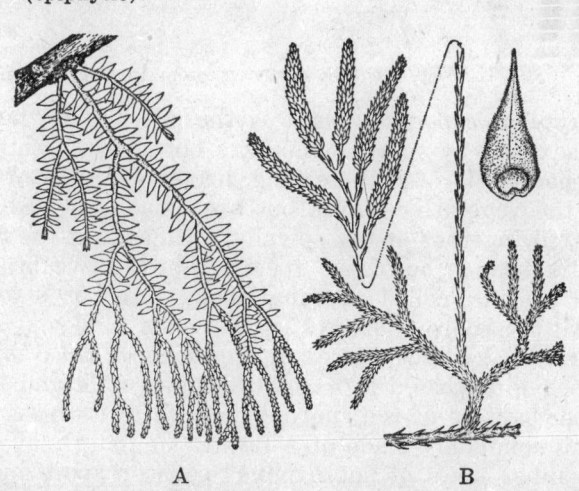

FIG. 151. A, *Lycopodium phlegmaria* (epiphytic); B, *L. clavatum* (terrestrial); a sporophyll with a sporangium shown separately.

tracheids, and metaxylem of large scalariform tracheids. Phloem consists of sieve-tubes and phloem parenchyma.

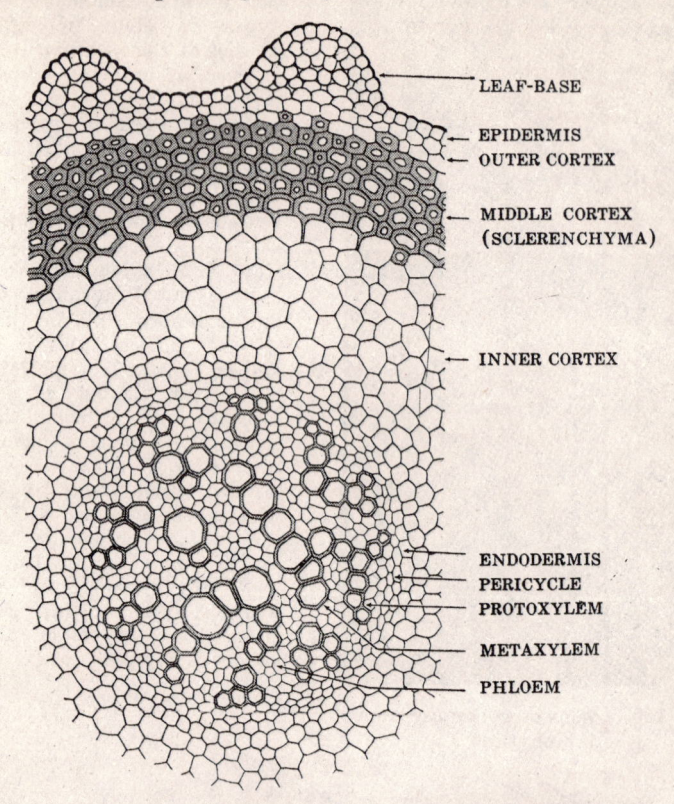

FIG. 152. *Lycopodium cernuum* stem in transection.

Sporophyte. *Lycopodium* plant is the sporophyte, i.e. it reproduces asexually by spores which are borne by specialized leaves called **sporophylls** (spore-bearing leaves). The sporophylls resemble the vegetative leaves but are smaller in size. They are aggregated together, being spirally arranged, at the apex of the vegetative branch or of the special reproductive branch in the form of a *cone*, called **sporangiferous spike** or **strobilus** (FIG. 153). All the sporophylls are of the same kind and so also are the sporangia and the spores. *Lycopodium is thus homosporous.* The sporangium has a short multicellular stalk and is borne on the upper surface of the sporophyll close to its base. It consists of a wall commonly made of a few (3 or more) layers of cells, and an inner mass of sporogenous cells or spore mother cells. By reduction division spores are formed in tetrads from these cells.

PTERIDOPHYTA

Gametophyte (FIG. 154*A-B*). After dehiscence of the sporangium the spores are scattered by the wind. The spores are remarkably

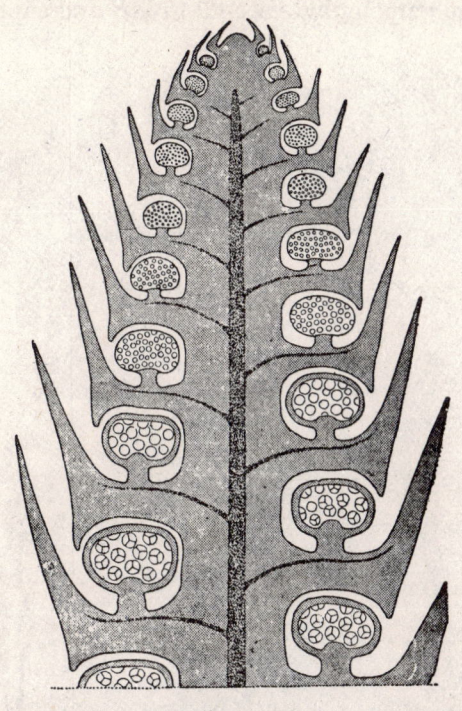

FIG. 153. Strobilus of *Lycopodium* in longitudinal section showing sporophylls, sporangia and spores (diagrammatic).

long-lived, often retaining their viability for a number of years. In many species the spores do not germinate for several months or sometimes even years after shedding. Even after the germination begins the rate of growth is slow. In *L. cernuum*, a common Indian species, the spores germinate within a few days and the growth of the gametophyte is completed in the same season. The spore, in any case, on germination gives rise to the gametophyte or prothallus. The gametophyte may be subterranean or sub-aerial, and vertical or partially horizontal. In the sub-aerial type the aerial portion turns green in colour and bears the sexual organs. In size it is about 2 or 3 mm. in length. The subterranean type is much bigger than the first type but non-green in colour. In shape the gametophyte may be a cylindrical or tuberous body with a lobed crown or it may be broad and irregularly cup-shaped. The crown bears the sexual organs. The tuberous portion shows a complicated internal structure, being differentiated into distinct regions, and is *always* associated with an endophytic (symbiotic) fungus which infects a definite region of it at an early stage of gametophyte-development. But for this fungal infection the growth of the gametophyte becomes arrested. Some rhizoids are produced from the epidermal layer. (For detailed structure of the gametophyte see FIG. 154B and caption).

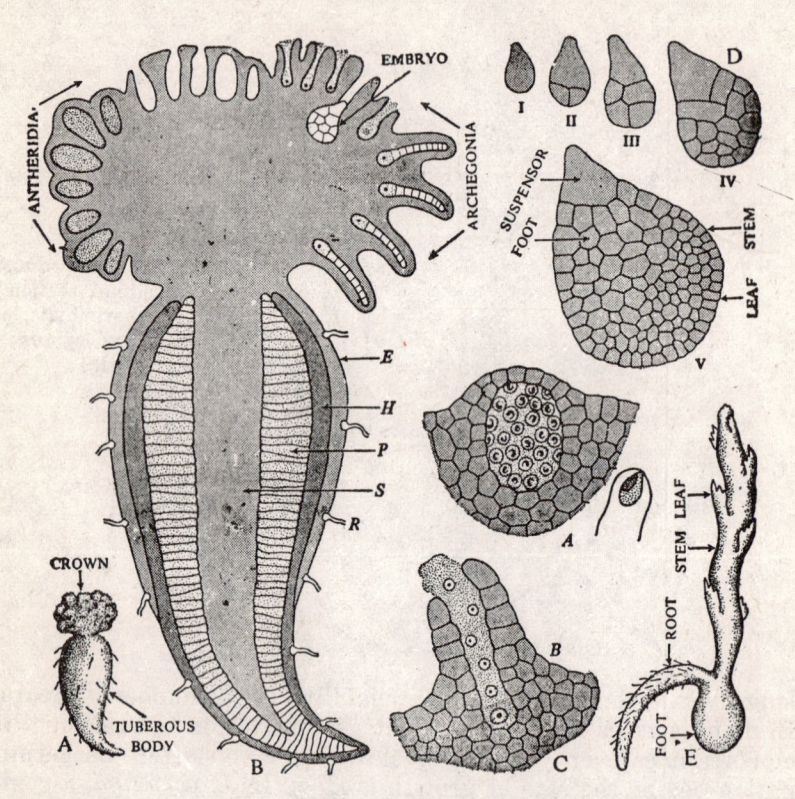

Lycopodium. FIG. 154. A, gametophyte; B, the same in longitudinal section; E, epidermis; H, hyphal tissue; P, palisade tissue; S, storage tissue; and R, rhizoid; C, reproductive organs—A, a mature antheridium with many antherozoid mother cells; an antherozoid on the right; B, a mature archegonium (open); D, stages in the development of the embryo (I-V); E, a young sporophyte (developing on the gametophyte). Redrawn after Fig. 182 and Fig. 186 in Plant Morphology by A. W. Haupt by permission of McGraw-Hill Book Company. Copyright 1953.

The gametophyte of *Lycopodium* is monoecious bearing both antheridia and archegonia. **Antheridia** and **Archegonia** (FIG. 154C-A & B). Numerous antheridia and archegonia are formed in the upper lobed portion (crown) of the gametophyte. The antheridium is more or less spherical consisting of a wall made of one layer of cells and a central mass of spermatogenous cells (antherozoid mother cells). It lies wholly or partially sunken in the tissue of the prothallus. The spermatozoids (antherozoids) are minute, broadly rounded at the base, slightly curved, and biciliate. The archegonium is a narrow elongated structure and lies almost wholly embedded in the prothallus except for the upper portion of the neck which projects beyond it. It consists of a narrow **venter** and a long **neck** with a variable number of neck canal cells (1-16, commonly 4-6). The venter contains a single large cell—the **egg-cell** with a distinct **egg** (egg-nucleus or female gamete), and a ventral canal cell. All the canal cells soon get disorganized.

Embryo (FIG. 154D). After fertilization, which takes place in the usual way, the fertilized egg or oospore divides into two cells—outer and inner. The outer cell is the suspensor cell which is elongated but not functional, while the inner one is the embryonal cell. The latter by successive divisions gives rise to two tiers of four cells each, of which the outer tier, i.e. the one next to the suspensor, produces the foot, and the inner tier produces the stem on one side and the leaf on the other side. The root develops later from the inner tier close to the leaf, and then the foot becomes disorganized (FIG. 154E).

Alternation of Generations. *Lycopodium* plant passes through two generations to complete its life-cycle. The main plant is the sporophyte (*diploid* or 2n). It bears one kind of spores (homosporous) and reproduces asexually by them to give rise to the next generation, i.e. the gametophyte (*haploid* or n). Each spore of the sporophyte on germination produces a minute body—the prothallus, which is the gametophyte. It bears sex organs and reproduces sexually by them to give rise to the sporophyte again. These two generations (the sporophyte including all the stages from the oospore to the spore mother cells, and the gametophyte including all the stages from the spores to the gametes) regularly alternate with each other.

3. *SELAGINELLA* (700 sp.)

Selaginella (family *Selaginellaceae;* FIG. 155) grows in damp places in the hills and in the plains. There are about 66 species in India, and some of the common ones are *S. caulescens* and *S. kraussiana* (commonly planted in gardens), *S. repanda* and *S. subdiaphana* (small and slender), *S. willdenovii* (very long and creeping), etc. In habit *Selaginella* may be prostrate, sub-erect, erect or rarely climbing; *S. lepidophylla* is xerophytic in habit (see p. 423). *Selaginella* plants are usually slender, much branched and creeping on the wall or on the ground. The slender stem bears four rows of leaves—two rows of small leaves on the upper surface and two rows of larger leaves at the two sides. A scaly structure, called **ligule** (FIG. 157), develops on the upper (ventral) surface of each leaf above its base. A long slender root-like organ is given off from the undersurface of the stem at the point of branching; this is known as the **rhizophore** (root-bearer). In some species the rhizophore bears small fibrous roots at the tip. The rhizophore resembles a true

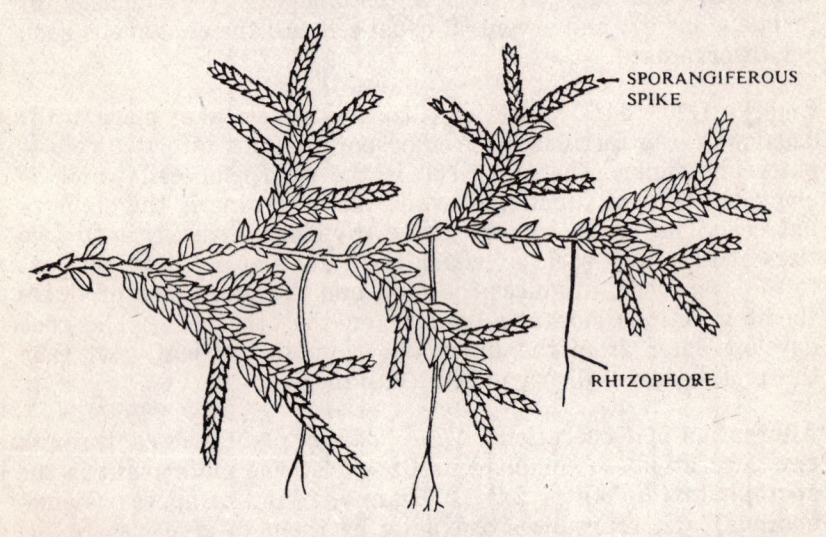

Selaginella. FIG. 155. A portion of a plant showing four rows of leaves, a number of spikes and three rhizophores.

root in its internal structure having a single protostele, in bearing no leaves, and in being positively geotropic; while it resembles a stem in having no root-cap and root-hairs, and in growing exogenously. It may be regarded as a leafless shoot behaving like a root in some respects. Sometimes it is seen to bear some scaly leaves and even cones.

Internal Structure of the Stem (FIG. 156). (1) **Epidermis**—a single layer with a cuticle. (2) **Sclerenchyma**—a few layers of sclerenchyma occur below the epidermis. (3) **Ground tissue**—a continuous mass of thin-walled, polygonal cells. (4) **Steles**—usually 2 or 3; each stele is surrounded by an air-space which is formed as a result of breaking-down of some of the inner layers of the cortex, and remains suspended in the air-space by delicate strands of cells, called **trabeculae** (sing. trabecula). Casparian strips are often present in the trabecular cells and appear as dots or thin walls. The stele, when young, is surrounded by a single-layered endodermis; later on the cells of the endodermis separate laterally and elongate considerably in the radial direction. These long radiating cells formed as a result of stretching of the endodermal cells are the trabeculae; in the mature stele they act as bridges across the air-space. Each stele, which is concentric in nature, consists of (a) pericycle, (b) phloem, and (c) xylem. Internal to the airspace there is a layer, sometimes two, of rather large but thin-walled cells—the **pericycle**. Phloem surrounds the central spindle-shaped **xylem**. Protoxylem lies at the two ends and metaxylem in the middle.

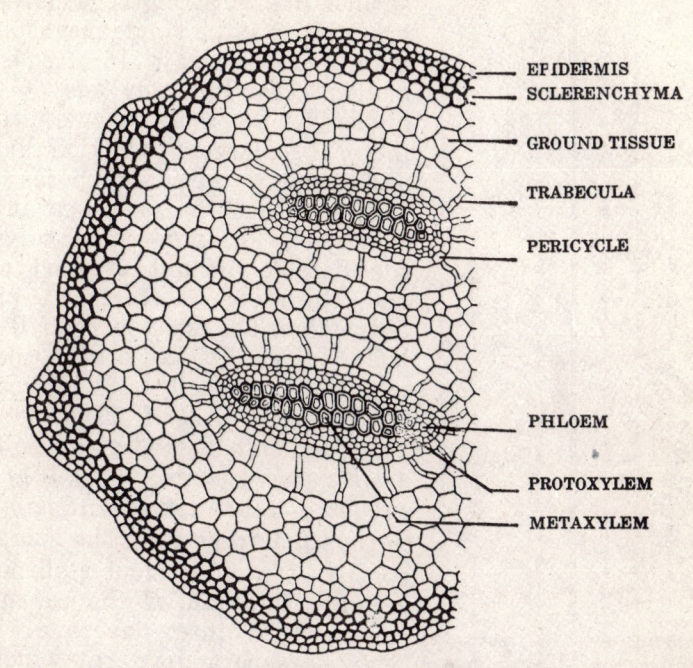

FIG. 156. *Selaginella* stem in transection.

Life-cycle. The life-cycle of *Selaginella* is complete in two stages—sporophytic and gametophytic—the former much more complicated and the latter much simpler than other pteridophytes. *Selaginella* plant is the sporophyte and this is followed by another *two* structures, called prothalli, which are the gametophytes (male and female).

Sporophyte. *Selaginella* plant is the sporophyte. It is **heterosporous** bearing *two* kinds of spores—microspores and megaspores—and reproduces asexually by them.

Sporophylls, Sporangia and spores. *Selaginella* bears two kinds of sporophylls—**microsporophylls** and **megasporophylls**. Both kinds of sporophylls may occur together in the same cone, or they may be borne in two separate cones either on the same plant (monoecious) or on two separate plants (dioecious). All the sporophylls are nearly of equal size and spirally arranged, usually in four rows, round the apex of the reproductive shoot, in the form of a more or less distinct four-angled *cone*, called the **sporangiferous spike** (FIGS. 155 & 157) or **strobilus.** The sporophylls are similar to the vegetative leaves in appearance, but are smaller in size. Each megasporophyll bears in its axil a single megasporangium with usually 16 megaspore mother cells. But only one of them divides, while others become disorganized. It undergoes reduction division and forms a tetrad of spores (FIG. 158A). Thus the megasporangium contains *four large* **megaspores.** A considerable amount of food material, chiefly oil, is stored up in the megaspore. The microsporophyll similarly bears in its axil a microsporangium with usually 16 microspore mother cells. All of them undergo reduction division and give rise to 64 *small* **microspores** (FIG. 161) in groups of four (tetrads). *Selaginella* is thus **heterosporous.** The sporangia consist of a short stout stalk and a capsule; the wall of the capsule is composed of three layers of cells. The megasporangia are somewhat larger than the microsporangia. It may be particularly noted that heterospory in *Selaginella* is an early important step leading to the evolution of the 'seed' in higher plants, i.e. gymnosperms and angiosperms (see Part VI, Chapter I).

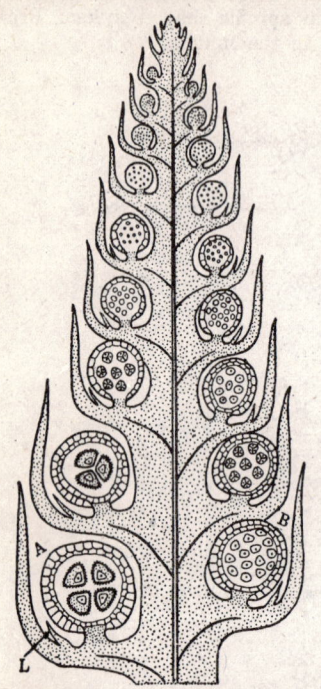

Selaginella. FIG. 157. A spike in longi-section; *A*, megasporophyll with megasporangium and megaspores; and *B*, microsporophyll with microsporangium and microspores; *L*, ligule.

Gametophytes. The two kinds of spores (micro- and mega-) germinate within their spore-coat and give rise to male and female prothalli respectively; the prothalli are the gametophytes, i.e. they bear male gametes or antherozoids, and female gamete or egg-cell, and reproduce sexually by the fusion of these two differentiated gametes (antherozoid and egg-cell).

Germination of the Megaspore: Female Prothallus (FIGS. 159-60). The megaspore-nucleus undergoes repeated free divisions, resulting in a number of free nuclei within the megaspore. Then walls appear round them, and thus a cellular mass of tissue is formed. This is the female prothallus, i.e. the female gametophyte. *The megaspore begins to grow before it is set free from the megasporangium,* but the formation of the female gametophyte is completed after the spore has fallen to the ground. At an early stage of gametophyte-formation a cavity appears at one end of it and is filled with reserve food, chiefly oil. This cavity subsequently becomes filled with cells. Further development of the gametophyte exerts pressure on the spore-wall which ruptures by a triradiate fissure, and the gametophyte becomes par-

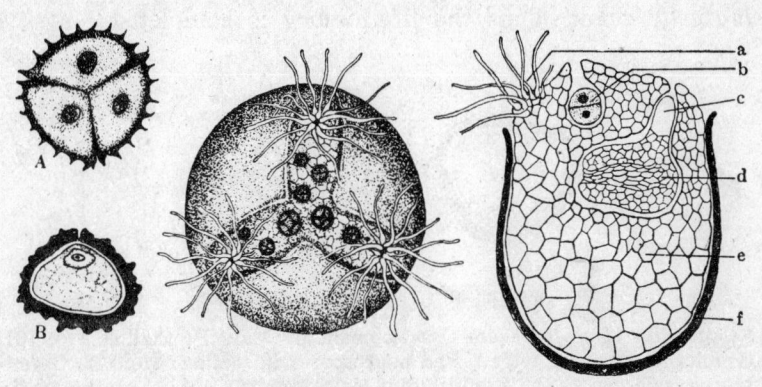

FIG. 158.   FIG. 159.   FIG. 160.

Selaginella. Megaspore and Development of Female Prothallus. FIG. 158. *A*, megaspores in a tetrad; *B*, a megaspore in section, FIG. 159. Germinating megaspore with female prothallus protruding through the triradiate fissure of the wall. FIG. 160. Female prothallus in longitudinal section; *a*, rhizoids; *b*, oospore after first division; *c*, suspensor; *d*, embryo; *e*, tissue of the prothallus; and *f*, wall of the megaspore.

tially exposed (FIGS. 159-60). *The gametophyte is partially endosporous. It is a much reduced structure* compared to that of fern and allied plants. It is also *not an independent structure* like that of fern and allied plants, being enclosed by the spore-coat and nourished by the food stored in the spore. A number of **archegonia** and some groups of rhizoids develop in the exposed green portion of the prothallus, while the inner larger non-green portion acts as a food reservoir. Each archegonium is also much reduced in size, consisting of a short neck with one neck canal cell, and a venter with an egg-cell and a ventral canal cell.

Germination of the Microspore: Male Prothallus (FIGS. 162-63). The microspore germinates and gives rise to the male prothallus, i.e. the male gametophyte. It begins to divide after it is set free from the microsporangium. A small cell is cut off at one end of the microspore; this is the **prothallus cell** representing an extremely reduced male gametophyte. The rest of the microspore forms a single **antheridial cell** which by a series of divisions forms a small mass of cells. These are differentiated into a layer of peripheral sterile cells—the **jacket cells**, enclosing about 128 or 256 **antherozoid mother cells**. Each mother cell encloses a single biciliate slightly twisted **antherozoid** or male gamete (FIG. 164).

Fertilization. After fertilization which is essentially the same as in fern and allied plants, the egg-cell becomes the **oospore**; this divides and forms the **embryo** which gradually develops into the *Selaginella* plant. Thus the life-history is completed.

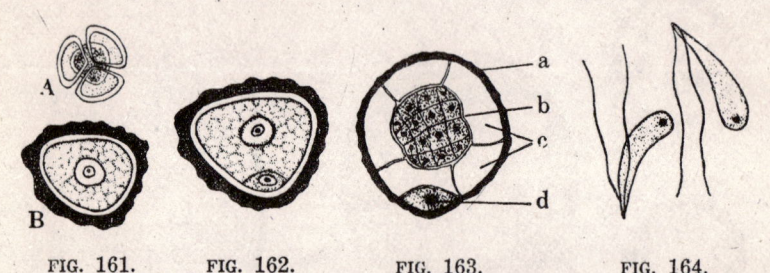

FIG. 161. FIG. 162. FIG. 163. FIG. 164.

Selaginella. Microspore and Development of Male Prothallus. FIG. 161. *A*, microspores in a tetrad; *B*, a microspores in section: FIG. 162. Germinating microspore with prothallus cell. FIG. 163. Male prothallus in section; *a*, wall of microspore; *b*, antherozoid mother cells; *c*, jacket cells; and *d*, prothallus cell. FIG. 164. Two antherozoids.

Alternation of Generations (FIG. 165). The life-history of *Selaginella* shows that the plant passes through two generations

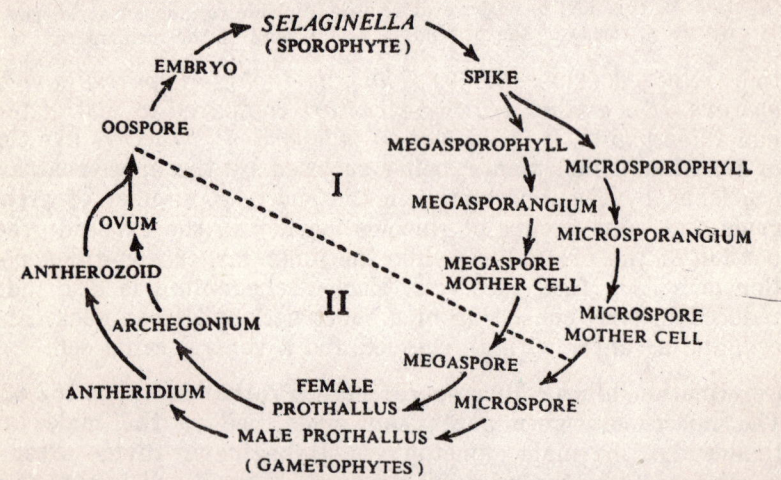

FIG. 165. Life-cycle of *Selaginella* (diagrammatic) showing alternation of generations. *I*, sporophytic generation (diploid or $2n$); and *II*, gametophytic generation (haploid or n).

which regularly alternate with each other. The plant itself is the sporophyte, and the two prothalli—male and female—are the gametophytes. The sporophyte reproduces asexually by two kinds of spores—microspore and megaspore—which give rise to the male prothallus and the female prothallus respectively. The male prothallus bears antherozoids in the antheridium, and the female prothallus bears an egg-cell in the archegonium, and the two prothalli (gametophytes) reproduce sexually by these two gametes, giving rise to the sporophyte again. The two gametes fuse together and give rise to the oospore. The oospore is the beginning of the sporophytic generation because $2n$ chromosomes are met with for the first time in the oospore, and all the stages from the oospore to the spore mother cells represent the sporophytic generation. Reduction division takes place in the formation of the spores; so all the stages from the spores (micro- and mega-) to gametes (antherozoid and egg-cell) with n chromosomes represent the gametophytic generation.

4. *ISOETES* (75 sp.)

Isoetes (FIG. 166A), commonly called quillwort, is the only **genus** of the family *Isoetaceae*. Species of this genus are mostly confined to temperate regions. There are about 6 species of *Isoetes* in India, of which *Isoetes coromandeliana* is fairly common. *Isoetes* commonly grows at the edge of a tank, ditch, stream or in a shallow pool of water, partly or completely submerged in water, and looks like a tuft of grass. Growth in length of the stem is due to the activity of the apical meristematic cells.

Sporophyte. *Isoetes* plant (FIG. 166A) is the sporophyte bearing two kinds of sporangia and spores (micro- and mega-) on the same plant, i.e. it is heterosporous and monoecious. It consists of a short 2- or 3- lobed corn-like stem, a rosette of linear leaves arising from the upper surface, and many slender often dichotomously branched roots developing from the lower sides. Each leaf is provided with a persistent **ligule** (FIG. 166B), as in *Selaginella*, and internally there are some longitudinal air-chambers (FIG. 168). Leaf-base is spoon-shaped and swollen owing to the development of a large sporangium on its inner concave surface just below the ligule (FIG. 166B). Most of the leaves except the central ones are potential sporophylls, the outer ones bearing megasporangia (FIG. 167A) and the inner ones microsporangia (FIG. 167B-C). Each sporangium is wholly or partially covered over by a membranous flap known as the **velum** which arises below the ligule (FIGS. 166B & 167) and grows downwards. The sporangium (micro- or mega-) is traversed by strands of cells, called the **trabeculae** (FIG. 167), which may be complete or incomplete. The two kinds of sporangia cannot be distinguished when they are young. When mature the microsporangium is seen to produce a very large number of minute microspores (150,000-1,000,000), and the megasporangium only a limited

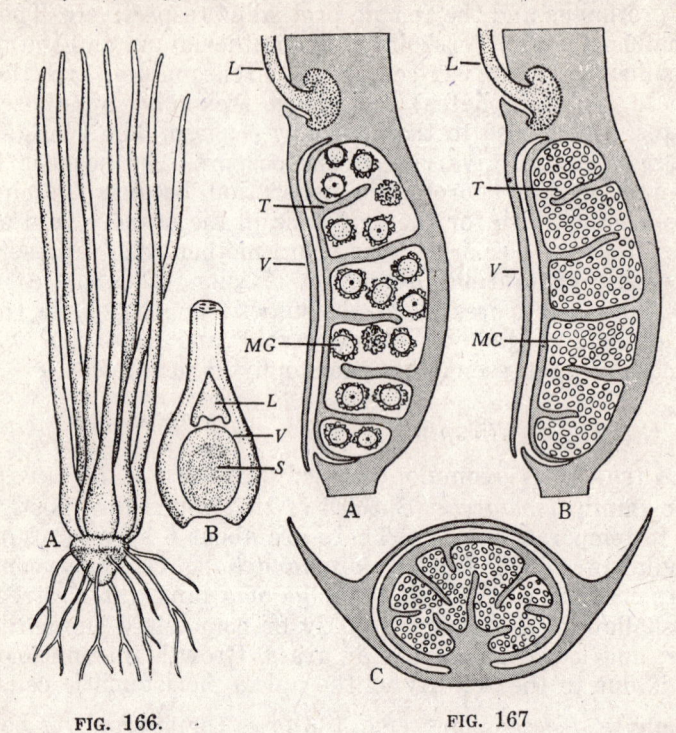

FIG. 166.

FIG. 167

Isoetes. FIG. 166. *A*, a plant; *B*, leaf-base (inner side). FIG. 167. *A*, megasporangium in longi-section with megaspores; *B*, microsporangium in longi-section with microspores; *C*, microsporangium in transection. *L*, ligule; *V*, velum; *S*, sporangium; *T*, trabecula; *MG*, megaspore; *MC*, microspores.

number of large megaspores (50-300). The megaspores are variously sculptured. Both microspores and megaspores are formed in tetrads, as usual.

Gametophytes. Spores are liberated after the decay of the sporangium wall. The microspore gives rise to the male gametophyte, and the megaspore to the female gametophyte. The **male gametophyte** (FIG. 169*A-D*) remains enclosed within the microspore coat, as in *Selaginella*, and consists of (*a*) a prothallus cell, (*b*) an antheridium of four antherozoid mother cells, and (*c*) a sterile jacket of four cells. Each antherozoid mother cell gives rise to a single large coiled multiciliate antherozoid (FIG. 169*E*). Thus altogether four antherozoids are produced by each male gametophyte, the lowest among all pteridophytes. The wall of

PTERIDOPHYTA

the microspore bursts and the antherozoids are liberated. The **female gametophyte** (FIG. 170A) develops inside the megaspore and does not protrude but, as in *Selaginella*, a triradiate fissure is formed in the megaspore wall exposing the rhizoids and the archegonia; the latter vary in number from one to many. The megaspore nucleus lying at one end begins to divide repeatedly, and gradually the whole of the female gametophyte becomes cellular. This is the female prothallus, and archegonia develop in it one after another in the first-formed region of the prothallus, if the previous one is not fertilized. The archegonium (FIG.170B) consists of (*a*) a neck with 3 or 4 tiers of cells, (*b*) a neck

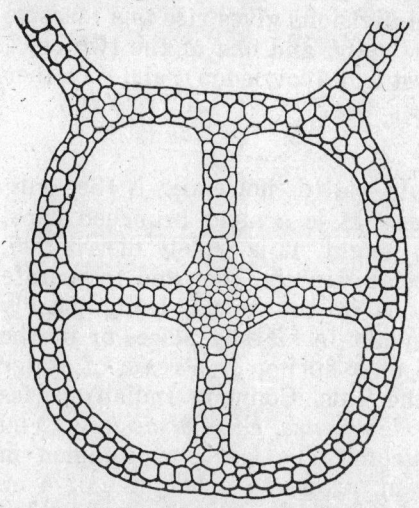

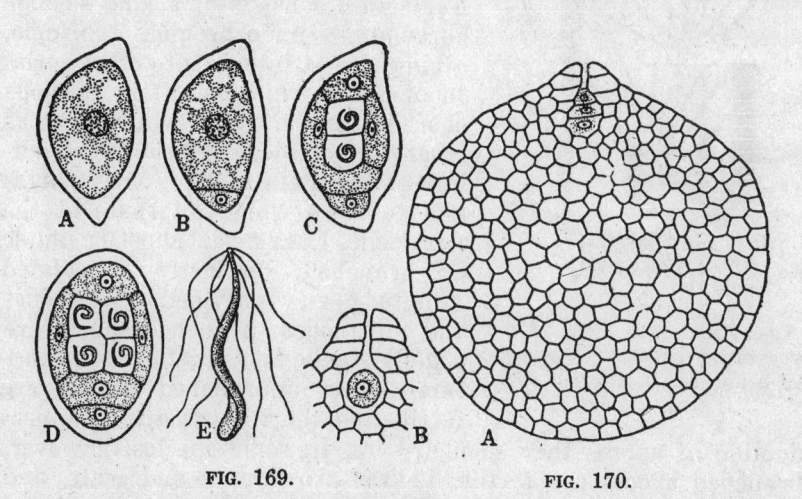

Isoetes. FIG. 168. Leaf in transection.

FIG. 169.   FIG. 170.

Isoetes. FIG. 169. *A-D*, male gametophyte in stages of development; *E*, an antherozoid. FIG. 170. *A*, female gametophyte; *B*, an archegonium. *Redrawn after Figs.* 203 & 204 *in* Plant Morphology *by A. W. Haupt by permission of McGraw-Hill Book Company. Copyright* 1953.

canal cell (sometimes binucleate), (*c*) a venter with a ventral canal cell, and (*d*) a conspicuous egg.

Embryo. The fertilized egg (oospore) divides transversely. The lower cell after further repeated divisions gives rise to a massive foot. The upper cell divides vertically, and one of the two cells, thus formed, gives rise to the first leaf (cotyledon), and the other cell to the root. The stem arises later.

5. *EQUISETUM* (25 sp.)

Equisetum (FIG. 171), commonly called horsetail, is the only genus of the family *Equisetaceae*. It is a much-branched herb, often not exceeding a metre in height. It is widely distributed, especially in the cool and temperate regions of the world, being usually abundant in marshy places or by the side of a spring or stream of water in the hills. Common Indian species are *E. arvense*, *E. ramosissimum* and *E. debile*. The last one, common in Assam, grows to a height of 3-4 m. through lofty bushes. It may be noted that an American species, *E. giganteum*, climbs neighbouring trees and grows to a length of about 12 m. *Equisetum* consists of a long slender horizontal underground rhizome, giving rise at intervals to erect aerial shoots. The rhizome often develops short tuber-like bodies which serve as reservoirs of food materials. The outline of the stem is wavy with ridges and furrows which alternate at the next node. Each aerial shoot is simple or branched, distinctly articulated (jointed) and provided with nodes and internodes. Branched shoots are usually sterile and vegetative in function, while unbranched shoots are fertile and short-lived; after the production of spores they soon dry up. In some species, however, branched shoots are fertile. Leaves are minute and scaly, and form a whorl at each node. These leaves are free and pointed at their tips but united below to form a sheath round the base of the internode. The lateral branches develop alternating with these leaves and grow upwards piercing the sheath. In consequence of reduction of leaf-laminae, the branches become green and perform photosynthesis. Roots are slender, adventitious and much branched developing from each node of the rhizome.

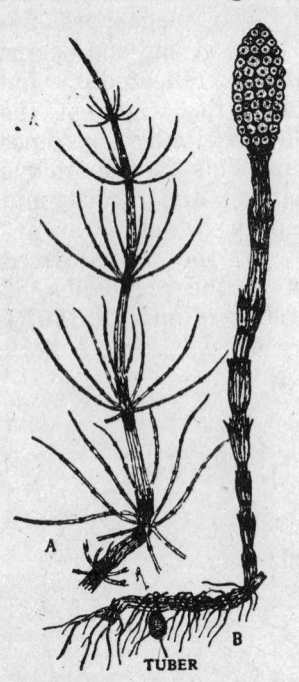

Equisetum. FIG. 171. *A*, a vegetative shoot with whorls of branches; *B*, a fertile shoot with a spike.

PTERIDOPHYTA

Internal Structure of the Stem (FIG. 172) The stem has distinct nodes and internodes, with longitudinal ridges and furrows. The internal structure is as follows: (1) **Epidermis**—a single outer layer of cells with a deposit of silica in their outer and lateral walls. It is wavy in outline and has stomata in two rows in the furrow. (2) **Sclerenchyma**—sclerenchyma develops, specially in the ridges, below the epidermis, interrupted in the furrows by the underlying cortex. (3) **Cortex**—this is many-layered, and in the middle of it large air-canals (*vallecular canals*), each corresponding to a groove, are

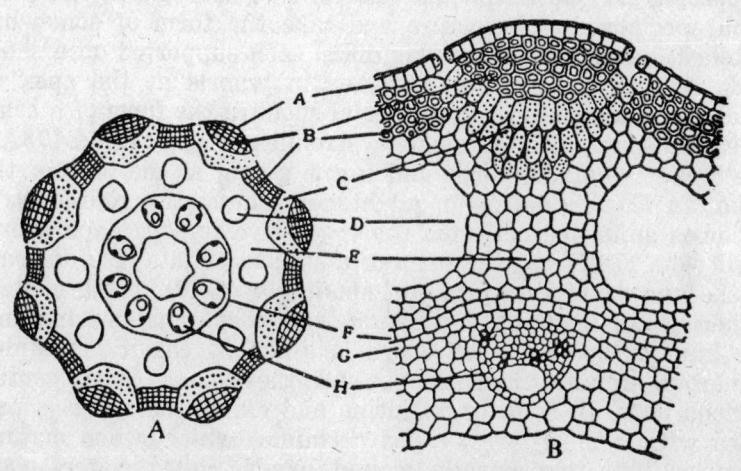

Equisetum. FIG. 172. A, section of stem (diagrammatic); B, a portion of the section (magnified). *A*, epidermis; *B*, hypodermis (sclerenchyma); *C*, outer cortex with chloroplasts (note the stomata); *D*, air-cavities; *E*, general cortex; *F*, endodermis; *G*, pericycle; *H*, vascular bundle (see text; note the pith and the pith cavity).

formed. Outer layers of the cortex contain chloroplasts; leaves being scaly, carbon assimilation is performed by the cortex (chlorenchyma) of the stem. The assimilating tissue extends up to the epidermis in the furrow where the stomata lie. (4) **Endodermis**—this is the innermost layer of the cortex, with often distinct Casparian strip in it. (5) **Pericycle**—this lies internal to the endodermis as a single layer. (6) **Vascular bundles**—these are closed, collateral, and arranged in a ring, each opposite to a ridge; each bundle is made of xylem and phloem, with some parenchyma; there is a water-containing cavity in it, called the *carinal cavity*, which has been formed lysigenously by the breaking down of some of the protoxylem elements (cf. maize stem). Xylem occurs in separate strands—the *protoxylem* (made of annular and spiral tracheids) lying in isolated strands against the carinal cavity, and the *metaxylem* (usually made of scalariform and reticulate tracheids) lying in two strands laterally outwards. Phloem lies between the two metaxylem strands, and consists of sieve-tubes and phloem parenchyma. (7) **Pith**—this lies on the inner side of the bundles but a major portion of it forms a large central cavity (pith cavity).

Life-cycle. The life-cycle of *Equisetum* is complete in two stages —sporophytic and gametophytic. The plant itself is the sporophyte (FIG. 171) and this is followed by another independent structure called prothallus (FIG. 174) which is the gametophyte.

Sporophyte. The *Equisetum* plant (FIG. 171) is the sporophyte, i.e. it reproduces asexually by spores which are borne by specialized leaves called sporophylls.

Sporophylls, Sporangia and Spores. The sporophylls are very much specialized in structure and take the form of somewhat flattened, hexagonal or circular discs, each supported on a short stalk, and are aggregated together in whorls at the apex of usually unbranched non-green aerial shoot in the form of a *cone*, called the **sporangiferous spike or strobilus** (FIGS. 171B & 173A). The lowest whorl is sterile and forms a **ring** at the base of the spike. In *Equisetum*, as in all higher plants, the reproductive region is quite distinct from the vegetative region. Each sporophyll (FIG. 173B) has the form of a stalked peltate disc. It bears on the undersurface a group of **sporangia** (5-10) which contain numerous small **spores**. *Equisetum* is *homosporous*, bearing only one kind of spores. Each spore is green in colour containing numerous minute chloroplasts, and there is a large central nucleus in it. In addition to intine and exine, the spore is provided with a third layer, called **perinium**, which, when mature, ruptures into two spirally wound bands, called **elaters** (FIG. 173-D-E), attached to the spores at their centre; the elaters appear as four distinct appendages. They are extremely hygroscopic; when the air is dry they unwind and stand out stiffly from the spore, and when the air is moist they roll up spirally round it. *Functions of Elaters*. The elaters expand and help the dehiscence of the sporangium. The spores become entangled by the elaters and are carried away in clusters by air-currents; this helps the germination of spores close together for facility of fertilization.

Gametophyte. The prothallus (FIG. 174A) is the gametophyte, i.e. it bears gametes—male (antherozoid) and female (ovum), and reproduces by the sexual method. Spores remain alive only for a few days. Under favourable conditions their germination begins within 10 or 12 hours after they are set free from the sporangium. Spores can be easily grown in culture in the laboratory. On germination they give rise to prothalli which are small in size, dull brownish-green in colour and much branched (lobed). The prothalli in most species are usually 3-6 mm. in diameter; in *E. debile*, however, they may be as big as 3 cm. in diameter. Prothalli, when they grow under favourable conditions, are normally monoecious bearing both antheridia and archegonia. If, however, they grow crowded together in the field or in culture they are dioecious, the smaller ones being male and the bigger ones female. It has been seen that the latter bear antheridia with age. This imperfect dioecism may be due to unfavourable conditions of growth. In *E. arvense*, however, about half of the spores give rise to male prothalli and the remaining half to female prothalli. If no fertilization takes place, the latter may produce antheridia. Prothalli usually live long, sometimes exceeding a period of two years. The prothallus is the **gametophyte** since it bears antheridia or archegonia or normally both. The sexual organs begin to appear in the gametophyte within 30-40 days of its growth, archegonia first and antheridia later. **Antheridia** develop at the apex of a branch (lobe) of the prothallus or on the margin of it. Each antheridium is more or less spherical in shape and contains numerous antherozoid mother cells (usually

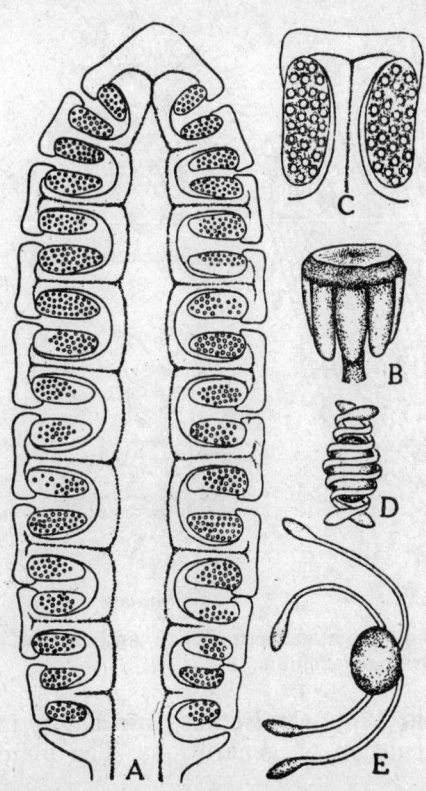

Equisetum. FIG. 173. *A*, a spike in longi-section showing sporophylls, each with sporangia, spores, stalk (sporangiophore) and peltate disc; *B*, a sporophyll with a whorl of sporangia; *C*, the same in longi-section; *D*, a spore with elaters coiled; and *E*, the same with elaters uncoiled.

256). In each mother cell one **antherozoid** or male gamete is produced. It is a large spirally coiled and multiciliate body (FIG. 174B). **Archegonia** always develop in the axial region of the prothallus and in the axil of a branch of it. Each archegonium (FIG. 174C) is flask-shaped with a swollen *venter* and a narrow *neck*, and encloses a large **egg-cell** with a distinct egg (egg-nucleus) in it, a small ventral canal cell and a narrow neck canal cell.

Fertilization. The method of fertilization is the same as that of ferns. After fertilization the **oospore** gives rise to an **embryo**

which develops into a branching rhizome. This then produces erect aerial shoots and a number of adventitious roots.

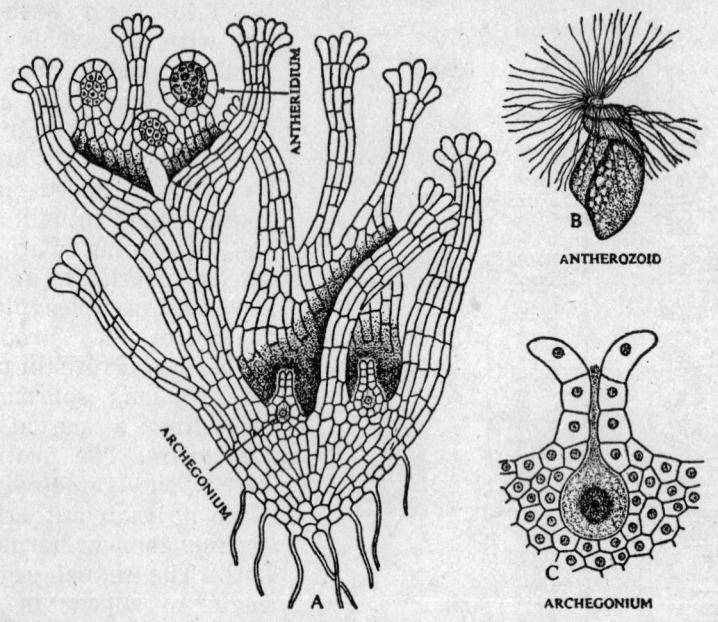

Equisetum. FIG. 174. *A*, prothallus (monoecious); *B*, an antherozoid; *C*, a mature archegonium.

Alternation of Generations (FIG. 175).

Equisetum plant shows in its life-history a regular alternation of generations. The plant

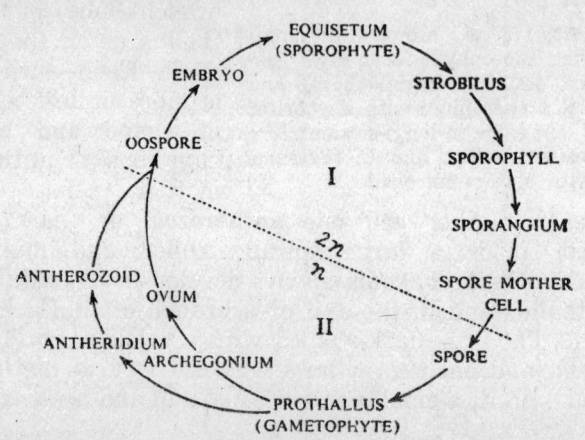

FIG. 175. Life-cycle of *Equisetum* (diagrammatic) showing alternation of generations. *I*, sporophytic generation (diploid or $2n$); and *II*, gametophytic generation (haploid or n).

itself is the sporophyte, and the prothallus—monoecious or dioecious—is the gametophyte. As in fern, the sporophyte or *Equisetum* plant reproduces asexually by spores and gives rise to the gametophyte or prothallus, and the prothallus reproduces sexually by gametes—antherozoid and ovum—and gives rise to the sporophyte. Thus the two generations regularly alternate with each other. The sporophytic generation begins with the oospore and ends in the spore mother cells because in all these stages $2n$ chromosomes are met with; while the gametophytic generation begins with the spores and ends in the gametes (antherozoid and ovum) because in all these stages there are only n chromosomes.

6. A FERN

Ferns (FIGS. 176-7) are a group of highly developed cryptogams

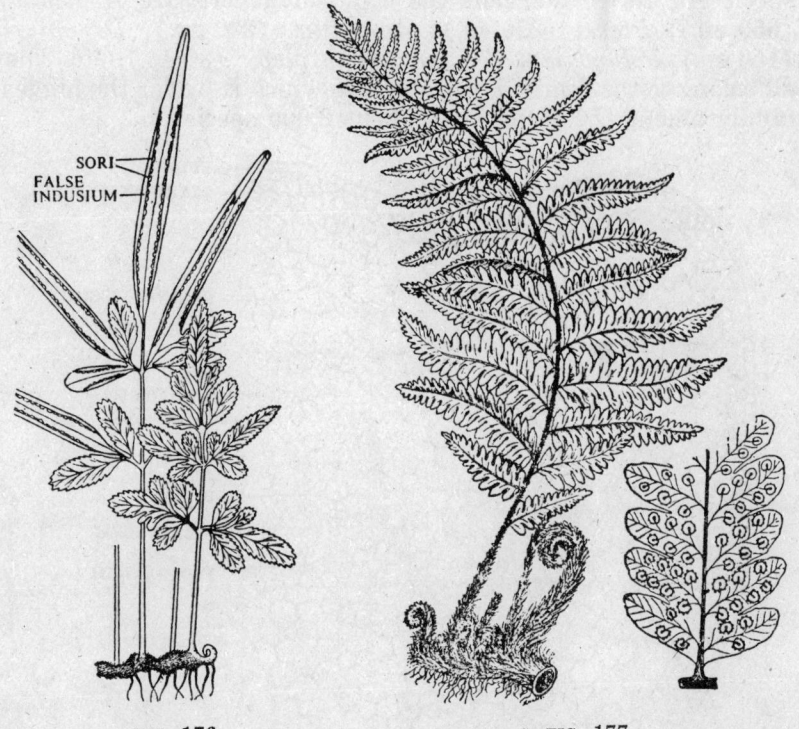

FIG. 176.

FIG. 177.

Fern Plants. FIG. 176. *Pteris* with continuous, linear sori and false indusium. FIG. 177. *Dryopteris* with sori on veins of the pinna and reniform indusium.

and are widely distributed all over the world. They are shade- and moisture-loving plants and, therefore, grow abundantly in

cool shady moist places, both in the hills and in the plains.
Ferns are mostly perennial herbs, with the stem often in the
form of a rhizome, by which they commonly reproduce vegetatively; in **tree ferns** (e.g. *Cyathea, Alsophila, Dicksonia*, etc.).
however, the stem is stout, erect and aerial. Roots are adventitious (fibrous) growing usually in clusters from the rhizome.
Leaves are commonly pinnately compound, and consist of two
parts: the frond (leafy portion) and the stipe (stalk). Young
fronds are **circinate**, i.e. coiled inwards on the upper surface
(FIG. 177), and are characterized by great apical growth. The
lateral leaflets borne by the axis or rachis are known as the
pinnae (sing. pinna); sometimes these are again more or less
deeply pinnately lobed, and then each lobe is known as the
pinnule. The stem and the petiole are covered with numerous
brownish scales known as the **ramenta**. There are about 7,600
species of ferns. Some of the common genera are *Asplenium*
(650 sp.), *Pteris* (250 sp.), *Adiantum* (200 sp.), *Dryopteris*
(150 sp.), *Polypodium* (75 sp.), *Nephrolepis* (30 sp.), etc. They
all belong to the family *Polypodiaceae* which is by far the biggest
family among the ferns, with about 3,000 species.

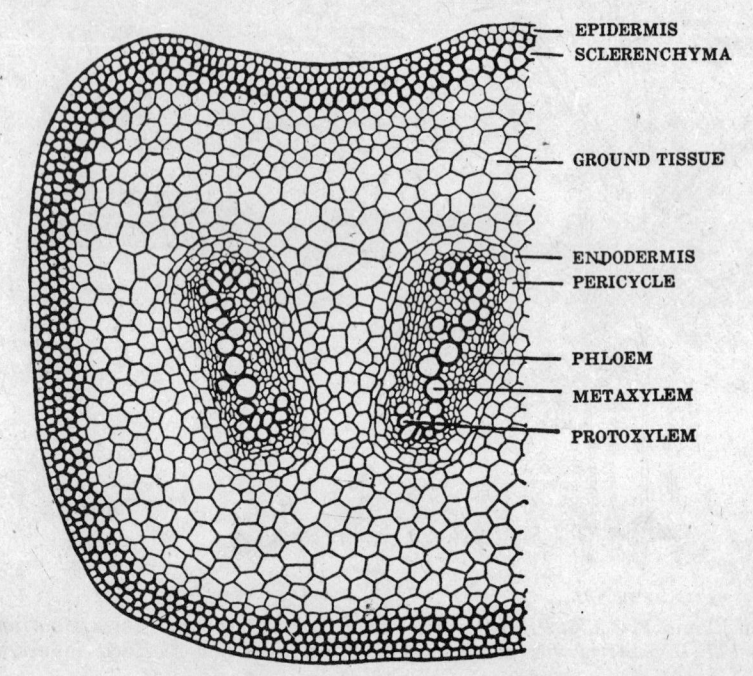

FIG. 178. Fern petiole in transection.

Internal Structure of the Fern Stem (or Petiole). (1) Epidermis—a
single layer of cells with the outer walls thickened and cutinized.

(2) **Sclerenchyma**—a few layers of sclerenchyma occur below the epidermis.
(3) **Ground tissue**—a continuous mass of polygonal parenchymatous cells.
(4) **Endodermis**—a single layer of narrow barrel-shaped cells surrounding each stele; this layer is often much thickened, particularly on the inner side. (5) **Steles**—in the young stem or petiole the stele is more or less horseshoe-shaped, but in the older part the stele is broken up usually into two or three smaller steles. Each stele consists of (*a*) **pericycle**, (*b*) **phloem** and (*c*) **xylem**. Pericycle surrounds the stele as a single layer (sometimes a double layer, particularly at the sides) and contains starch grains. Phloem surrounds the central xylem, the bundle being concentric, and consists of sieve tubes and phloem parenchyma. Xylem lies in the centre surrounded by phloem. It consists usually of two groups of protoxylem at the two ends, and metaxylem in the middle. Protoxylem is made of spiral tracheids, and metaxylem of scalariform tracheids.

Life-cycle. The life-cycle of fern is complete in two stages—sporophytic and gametophytic. The fern plant is the sporophyte and this is followed by another small green flat structure called the prothallus which is the gametophyte (FIG. 181).

Sporophyte. The fern plant, as stated above, is the sporophyte, i.e. it bears spores and reproduces by the asexual method.

Sporangia and Spores. On the undersurface of the fertile frond, i.e. the spore-bearing leaf or sporophyll (as it is called), a number of dark-brown structures, pale-green when young, may be seen; these are called **sori** (sing. sorus). In *Dryopteris* (and also in

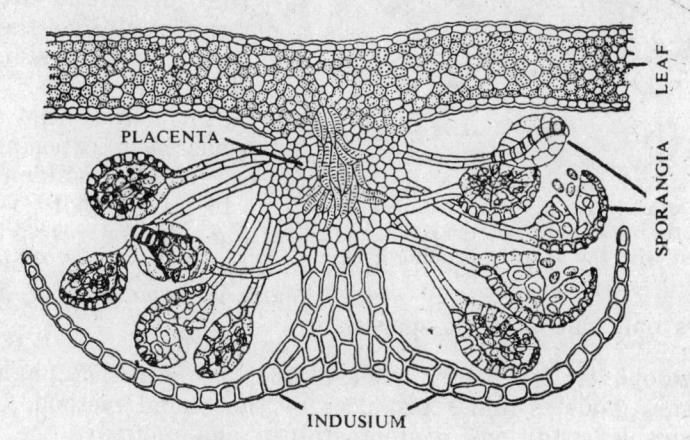

Fern. FIG. 179. Sorus in section

Polystichum, Nephrodium, etc.) the sori develop on the veins on the undersurface of the sporophyll, and are arranged in two rows in each leaflet or pinna (FIG. 177). Each sorus (FIG. 179) consists of a large number of **sporangia** which are covered over and protected by a reniform (i.e. kidney-shaped) or rounded shield, called the **indusium**. The sporangia and the indusium develop from a papilla-like outgrowth or **placenta** of the leaf.

In *Pteris* (FIG. 176) the sori are marginal, linear and continuous, with a thin membranous scaly indusium on the inner side of the sori. It is, however, seen that the reflexed margin of the pinna forms a continuous, overlapping indusium, called *false indusium*, which covers the sori. The indusium may be of other types; for example, in *Adiantum* the sori are distinct and separate on the margin of the pinna, usually in rounded groups, each covered by a *false indusium*; in *Cyathium* the indusium is cup-shaped; in *Asplenium* it is curved or horseshoe-shaped; and there are other types too. The indusium may also be absent, as in *Polypodium* and *Gleichenia*. The indusium is in fact an important character often used in the classification of ferns.

Each **sporangium** (FIG. 180) consists of a long slender multicellular **stalk** and a **capsule** which is biconvex. Inside the capsule lies a mass of very small grains; these are the **spores**. At first there are 16 spore mother cells in the capsule; these undergo reduction division into 32 daughter cells. The daughter cells then divide mitotically and 64 spores are formed. The wall of the capsule consists of a single layer of thin-walled cells, with a specially thickened and cutinized band or ring running round the margin of the capsule. This ring which is incomplete and thin-walled on one side is called the **annulus**, and its unthickened portion the **stomium**. When the spores mature they increase in size, and under dry conditions the capsule bursts at the stomium, liberating the spores. When the capsule bursts, the annulus bends back exposing the spores, and then suddenly returns to its original position, ejecting the spores with a jerk by this process. The fern plant is *homosporous*, i.e. it bears only one kind of spores.

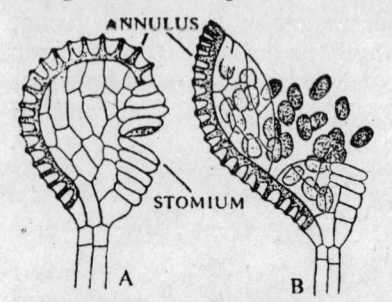

Fern. FIG. 180. Sporangia (capsule and stalk); *A*, capsule just opening at the stomium; *B*, capsule has burst with the annulus bending back.

Gametophyte. The prothallus (FIG. 181) is the gametophyte, i.e. it bears gametes and reproduces by the sexual method. Under favourable conditions of temperature and moisture the spore germinates. At first it gives rise to a short green filament (germ tube) resembling an alga or moss protonema. Subsequently by further divisions of cells it produces a small green flat heart-shaped body, about 8 mm. across; this is known as the **prothallus**. Its margin is very thin and single-layered, while its central part is comparatively thick and many-layered. Unicellular, hairy processes, called **rhizoids**, come out from the undersurface of the prothallus; these fix the prothallus to the soil and absorb water

PTERIDOPHYTA

and mineral salts. For reproduction highly specialized structures are produced on the undersurface of the prothallus; these are the **antheridia** or male organs and the **archegonia** or female organs. The former develop amongst the rhizoids and the latter near the groove.

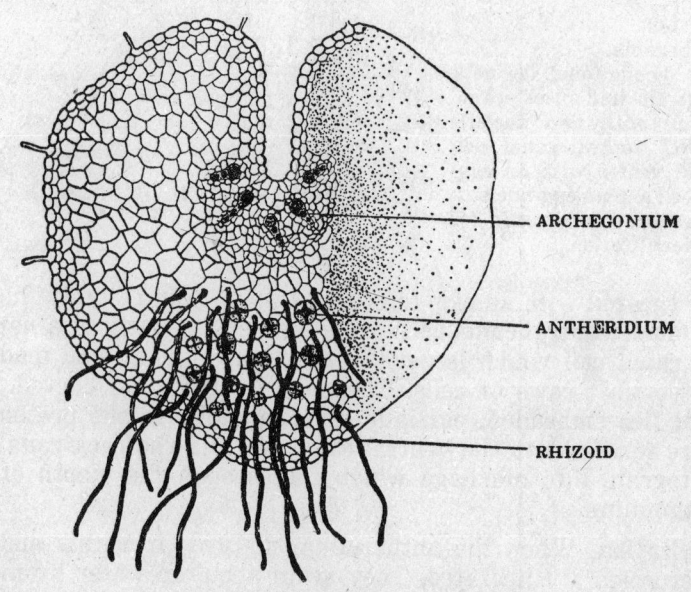

FIG. 181. Prothallus of fern.

The **antheridium** (FIG. 182A-B) is a spherical or oval body consisting of a wall or jacket with 1 or 2 caps or cover cells at the apex and a number of antherozoid mother cells (20-50). Each

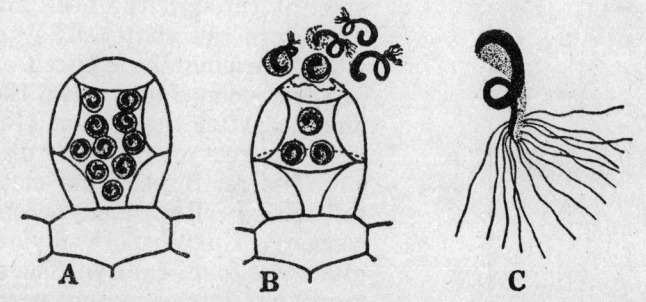

Fern. FIG. 182. Antheridium. *A*, a young one with antherozoid mother cells; *B*, a mature one after bursting; and *C*, an antherozoid.

mother cell develops a single antherozoid or male gamete which consists mostly of nuclear material. The antherozoid is comparatively large, spirally coiled and multiciliate (FIG. 182C).

The **archegonium** (FIG. 183) is a flask-shaped body. The swollen basal portion is known as the **venter**, and the slender tube-like upper portion as the **neck**. The venter encloses a single

Fern.
FIG. 183.
Archegonia.
A, a young one; N, neck (wall and neck canal cell with two nuclei); VC. ventral canal cell; V, venter with an egg-cell and an egg-nucleus; B, a mature one ready for fertilization.

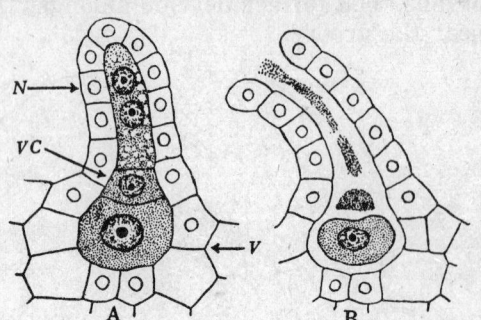

large **egg-cell** with an **egg** (egg-nucleus) in it and slightly higher up a small ventral canal cell, while the neck consists of a narrow neck canal cell which is usually binucleate and a wall made of four vertical rows of cells. The neck is short and curved. The venter lies embedded, partially or completely, in the prothallus. Before fertilization the ventral canal cell and the neck canal cell disintegrate into mucilage which forces open the mouth of the archegonium.

Fertilization. When the antheridium matures it bursts and the antherozoids are liberated. They swim about in water by means of their cilia. As the archegonium matures it secretes mucilage and malic acid. Attracted by these substances a large number of antherozoids swim to the archegonium, enter it through the neck and pass down into the venter. They quickly vibrate around the egg-cell and one of them soon fuses with the egg-nucleus. After this fusion (fertilization) the rest of the antherozoids die out. The fertilized ovum clothes itself with a cell-wall and becomes the **oospore**. The oospore divides and gives rise to an **embryo**. The embryo grows up into a young sporophyte (with a green leaf and a root) still attached to the prothallus (FIG. 184). With the penetration of the root into the soil and decay of the prothallus the young sporophyte develops into a fern plant as an independent body.

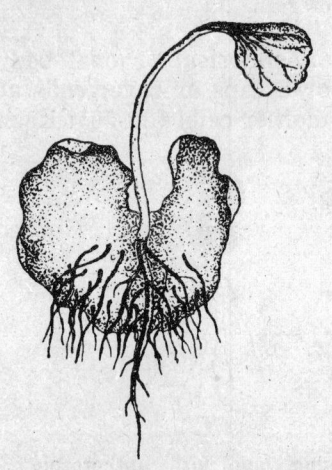

FIG. 184. Prothallus of fern with young sporophyte.

PTERIDOPHYTA

Alternation of Generations (FIG. 185). The fern plant, as its life-history shows, passes through two stages or generations. The plant itself is the sporophyte, and the prothallus the gametophyte. The sporophyte or the fern plant reproduces asexually by spores and gives rise to the gametophyte or the prothallus, and the latter reproduces sexually by gametes (antherozoid and ovum) and gives rise to the sporophyte or the fern plant. Thus

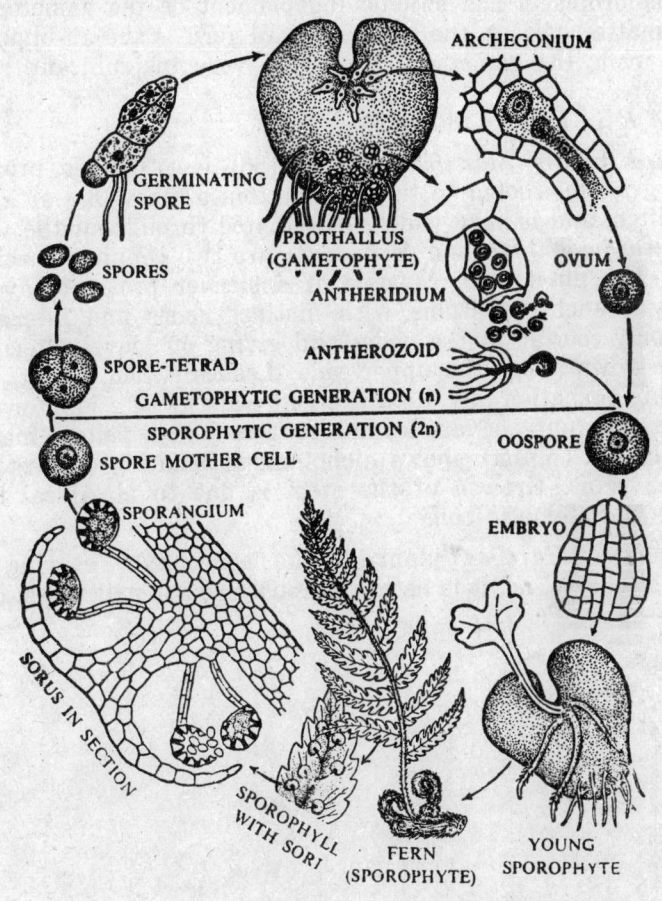

FIG. 185. Life-cycle of fern showing alternation of generations.

the two generations regularly alternate with each other. In the life-history of fern $2n$ chromosomes are met with for the first time in the oospore and, therefore, this is the beginning of the sporophytic generation, and all the stages from the oospore to the spore mother cells represent the sporophytic or asexual generation of fern. Reduction of chromosomes to n number takes place in the formation of the spores from the spore mother cells and, therefore, they (spores) represent the beginning of the gametophytic generation, and all the stages from the spores to the gametes (antherozoid and ovum) constitute the gametophytic or sexual generation. It is noticeable that the sporophyte (fern plant) has already reached a high degree of development and complexity, and with the formation of roots and green leaves with chloroplasts has become independent of the gametophyte. As a matter of fact, the sporophyte of fern is the all-important body, while the gametophyte of it is very insignificant.

7. *MARSILEA* (60 sp.)

Marsilea (family *Marsileaceae*; FIG. 186) is a slender prostrate herb, growing rooted to the mud-bottom at the edge of a tank or a ditch, and is very widely distributed throughout the world. *Marsilea quadrifolia* and *M. minuta* are two common species in India. The plant body consists of a slender prostrate dichotomously branched rhizome with distinct nodes and internodes, commonly rooting at the nodes and giving off leaves alternately in two rows along the upper side. Leaves, when young, show circinate vernation and a mature leaf consists of a long or short petiole and four obovate leaflets arranged in a peltate manner. Each leaflet (pinna) shows dichotomous venation connected by smaller veins. Growth of the stem is due to an apical tetrahedral meristematic cell.

Sporophyte. *Marsilea* plant is the sporophyte bearing two kinds of spores, i.e. it is heterosporous. Special structures, called

Marsilea. FIG. 186. A plant with sporocarps.

sporocarps, are seen to grow, in small groups of 2-5, sometimes singly, from the base of the petiole or a little above it as a segment of it (interpreted as modified fertile leaf-segment or entire leaf). Sporocarps develop only when the water recedes and the soil tends to dry up, and each is provided with a long or short stalk and has a very hard outer covering, when mature. In structure it is more or less bean-shaped (FIGS. 186-7) and is about 8×6 mm. in size. Each half of the sporocarp has distinct forked venation alternating with that of the other half. The sporocarp (FIG. 187B) contains within it usually 14-20 sori, each in a cavity, arranged in two rows on a receptacle; the sori develop in basipetal order. Each sorus is covered by a thin delicate layer known as the **indusium**. The sporangia at the apex of the receptacle are megasporangia and those lower down are microsporangia. The sporangium wall in each case is made of a single layer of cells. The sori are attached to a tissue which swells considerably in water and becomes gelatinous. In the early stages of development both kinds of sporangia form 8 or 16 sporocytes or spore mother cells which on reduction division produce 32 or 64 spores. But in the case of microsporangium all the microspores are functional, though very minute in size; while in the case of megasporangium only one megaspore grows larger in size and is functional; others degenerate.

With the development of the sporangia a stony layer is formed on the outer surface of the sporocarp. The resisting power of

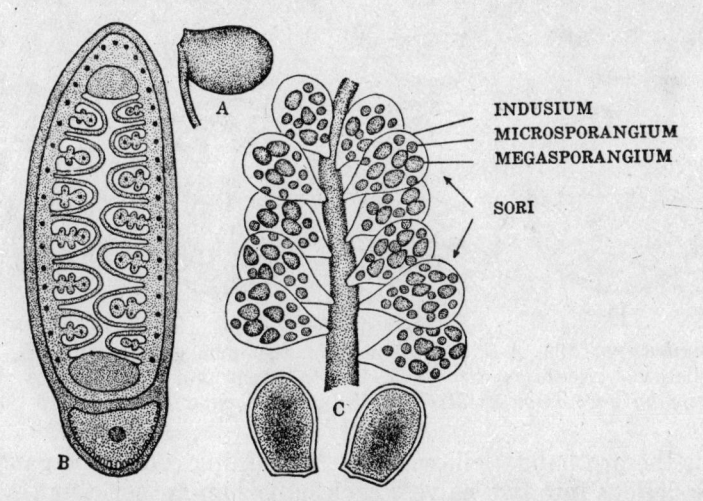

Marsilea. FIG. 187. *A*, a sporocarp; *B*, the same in longi-section showing young sori, in two rows, each sorus with megasporangium (terminal), microsporangia (lateral) and indusium (outer covering); *C*, part of the gelatinous ring carrying the sori; two empty valves of the sporocarp lying below.

the sporocarp and the longevity of the spores are remarkable; spores have been seen to germinate even after many years of desiccation. If the sporocarp be cracked at the edge and kept in water for half an hour or so it is seen that the gelatinized inner wall of the sporocarp pushes out of it in the form of a long gelatinous ring (**sorophore**) with the sori attached to it in an alternating manner[1] (FIG. 187C). In nature, however, the sporocarp takes at least 2 or 3 years for the decay of its stony layer. Thereafter the spores are liberated on the conversion of the indusium and the sporangium walls into mucilage. If the gelatinous ring be left in water the development of the male and female gametophytes may be seen on the following day or the day after.

Gametophytes. The spores germinate very quickly, the microspore giving rise to the male gametophyte, and the megaspore to the female gametophyte. Within 12-20 hours the development of the male gametophyte is complete, while the female gametophyte is slightly slower in development.

The male gametophyte (FIG. 188A-F) is endosporous developing within the microspore, as in *Selaginella*. A cell cut off on one

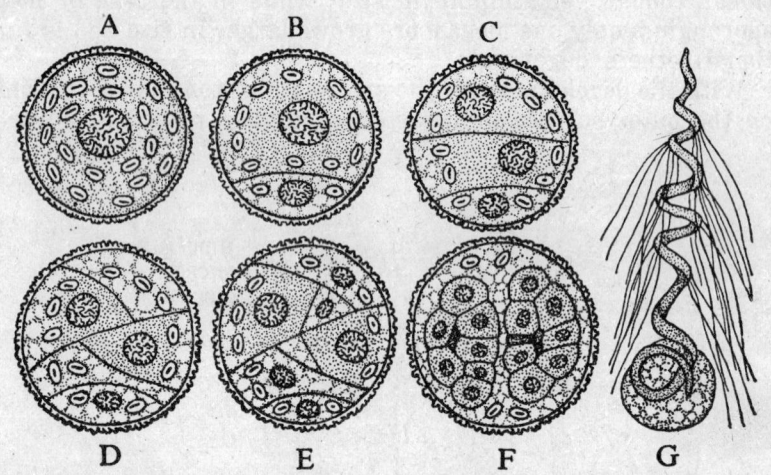

Marsilea. FIG. 188. *A-F*, development of the male gametophyte; *G*, an antherozoid. *Redrawn after Fig.* 252 *in* Plant Morphology *by A. W. Haupt by permission of McGraw-Hill Book Company. Copyright* 1953.

side is the prothallus cell, while the remaining cell of the gametophyte divides into two halves; each half is an antheridium. After further divisions two primary spermatogenous cells are formed

---

[1] The sporocarp may be split open with a sharp scalpel into two valves, the contents gently scooped out and spread on a slide in water, covered with a cover-glass, and lightly tapped.

surrounded by a jacket layer. Each cell then produces 16 *antherozoid mother cells*. Each mother cell is metamorphosed into an antherozoid (FIG. 188G). This is much coiled, corkscrew-like and multiciliate, with a mucilagious vesicle containing some food.

The female gametophyte (FIG. 189B) consists of an archegonium protruding out of the megaspore-coat at its apex, while the rest of the gametophyte without any cellular differentiation is a food reservoir. The archegonium consists of (a) a venter with a comparatively large egg and a small ventral canal cell, (b) a short neck with a neck canal cell, and (c) a sterile jacket of cells. Surrounding the female gametophyte there is a broad gelatinous envelope, which has a funnel-shaped portion converging to the archegonium.

Fertilization. Fertilization takes place almost immediately after the gametophytes are formed. Innumerable antherozoids swarm around the archegonium, many of them swim into the gelatinous envelope on the archegonium side, and some of them pass down the neck. Finally, however, one antherozoid fuses with the egg-nucleus.

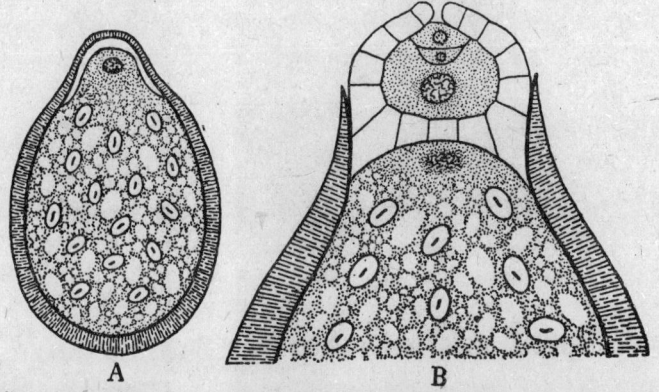

Marsilea. FIG. 189. *A*, a megaspore in longi-section; note the nucleus and several starch grains; *B*, female gametophyte with an archegonium projecting out of the spore-coat; note the large egg-cell with the egg-nucleus, ventral canal cell, neck canal cell and the sterile jacket.

Embryo. After fertilization the embryo is quickly formed. The oospore divides and redivides and the cells are arranged in four segments. The two segments, inner and outer, on one side give rise to the stem and the first leaf (cotyledon), while the remaining two segments, inner and outer, on the other side give rise to the foot and the root. The adjoining cells of the gametophyte form a cap-like structure or **calyptra** around the developing embryo. The embryo grows rapidly by bursting the calyptra.

Internal Structure of Rhizome (*Marsilea*; FIG. 190). (1) **Epidermis** lies externally as a single layer. (2) **Cortex** consists of usually two layers of thin-walled parenchyma internal to the epidermis, a ring of fairly big air-cavities traversed by trabeculae, usually 1 to 3 layers of thick-walled lignified cells, and internally several layers of thin-walled parenchyma containing starch grains. (3) **Stele** is an *amphiphloic siphonostele* with pith

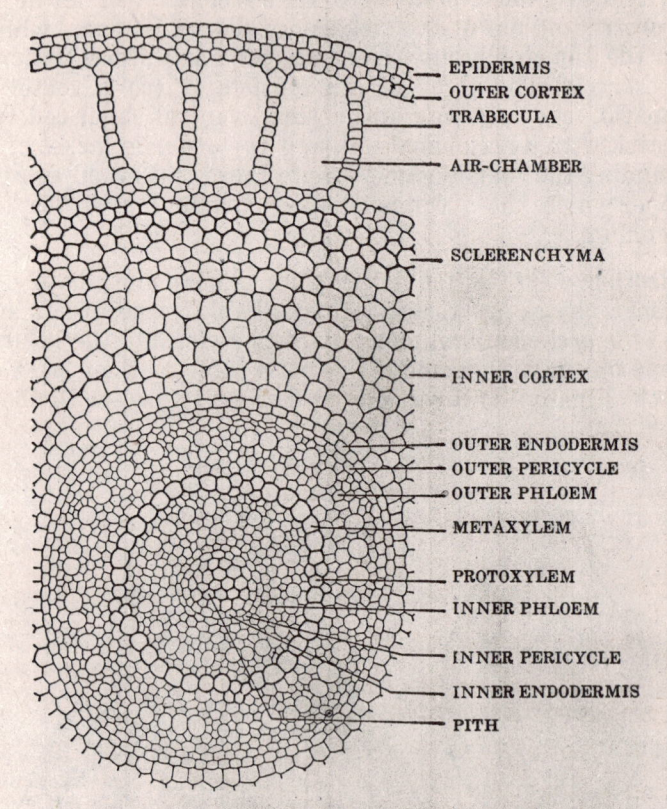

FIG. 190. Rhizome of *Marsilea* in transection (amphiphloic siphonostele).

in the centre (see p. 623). It is bounded both externally and internally by **phloem, pericycle** and **endodermis** which evidently occur twice. In the middle xylem occurs as a ring with more or less distinct protoxylem and metaxylem, and consists of thick-walled tracheids surrounded externally by outer phloem, outer pericycle and outer endodermis. Internal to xylem again occur inner phloem, inner pericycle and inner endodermis in the reverse order. (4) **Pith** occupies the central portion and is parenchymatous or sclerenchymatous (growing in water or on dry mud).

Part VI GYMNOSPERMS

Chapter 1 GENERAL DESCRIPTION

Spermatophytes (or 'seed' plants) or **phanerogams** (or 'flowering' plants) are divided into two sub-divisions—**angiosperms** (*angeion*, case; *sperma*, seed) and **gymnosperms** (gymnos, naked). Gymnosperms are closely related to the higher cryptogams on the one hand and to the angiosperms on the other, and thus they form an intermediate group between the two. It should be noted that lower gymnosperms like cycads have greater affinities with the higher cryptogams, while the higher gymnosperms—Coniferales and Gnetales—are closer to the angiosperms. Salient points of resemblances and differences are given below. Gymnosperms number about 700 species, of which a little over 50 species occur in India.

GYMNOSPERMS AND HIGHER CRYPTOGAMS

Resemblances. (1) Identical nature of general life-history in both with, of course, structural differences; both the groups show a regular alternation of generations with gradual reduction and loss of independence of gametophyte from higher cryptogams to gymnosperms, and gradual increase in the complexity of the sporophyte.

(2) Clear differentiation of the plant body into root, stem and leaves.

(3) Compound nature of leaves and circinate vernation, as in earlier gymnosperms (cycads).

(4) Xylem of the vascular bundle with only tracheids (and no vessels except in Gnetales) and phloem without companion cells.

(5) Gradual differentiation of sporophylls (micro- and mega-) and of spores—microspore and megaspore (heterosporous condition) in the higher cryptogams leading to the complete differentiation of the same in the gymnosperms, and the segregation of the sporophylls from the vegetative region and their arrangement in the form of cones (strobili).

(6) Development of the gametophyte within the spore-coat as a dependent body (partially endosporous in *Selaginella* but completely endosporous in gymnosperms) for advantages of food supply and adequate protection by the sporophyte.

(7) Development of ciliate spermatozoids in lower gymnosperms (cycads), as in cryptogams.

(8) Retention of the megaspore within the megasporangium (permanently in gymnosperms but for a short period only in *Selaginella*) and the development of the archegonia in the female gametophyte.

(9) Formation of the suspensor in gymnosperms during the stages of embryo development, as in *Selaginella*.

Differences. (1) Development of the ovule (and later the seed) in gymnosperms makes a fundamental difference between them and the cryptogams.

The ovule and the seed are characteristic of all gymnosperms, while they are altogether absent in cryptogams.

(2) Formation of the pollen-tube in gymnosperms is another special feature which is conspicuous by its absence in cryptogams. The pollen-tube carries the non-motile male gametes to the archegonium in most of the gymnosperms, while in all cryptogams the motile spermatozoids by themselves swim to the archegonium in a drop of water. In *Cycas*, however, the pollen-tube is a sucking organ (haustorium) and not a sperm-carrier.

(3) The megasporangium of the gymnosperm does not shed before fertilization and maturity; further it is provided with a new structure in the form of a coat—the integument. The megasporangium covered by the integument is the ovule. The ovule is altogether absent in cryptogams.

(4) The megaspore of the gymnosperm remains permanently enclosed within the megasporangium, and the female gametophyte is also completely endosporous (partially endosporous in *Selaginella*) and is consequently dependent on the mother sporophyte for its nourishment. The embryo also remains enclosed within the megasporangium and is nourished by it.

(5) Remaining enclosed the gametophytes of gymnosperms are not green.

(6) Archegonia are much simpler in construction in gymnosperms with shorter neck and no neck canal cells, unlike cryptogams. Final reduction of archegonia takes place in higher gymnosperms, as in *Gnetum*,—a condition more akin to angiosperms.

(7) Presence of the cambium in gymnosperms leading to secondary growth in thickness, and absence of it in cryptogams with no secondary growth. Besides, tracheids with circular pits and those with sclariform markings are characteristic of gymnosperms and higher cryptogams respectively.

GYMNOSPERMS AND ANGIOSPERMS

Resemblances. (1) The plant body is differentiated into distinct root and shoot, the latter with copious branches and leaves, however small they may be. In habit the plants of both groups may be shrubs and trees, with preponderance of angiospermic herbs.

(2) Vascular system is well developed in both, with xylem and phloem (note, however, their differences as given under item 1, next page).

(3) Flowers are developed in both cases for reproduction. But gymnospermic flowers are primitive in nature and simple in construction and are always unisexual and without perianth. Angiospermic flowers, however, are more advanced often with both sepals and petals, and are unisexual or bisexual. Gymnospermic flowers are pollinated through the agency of air-currents only, while angiospermic flowers are pollinated through various agencies, particularly insects.

(4) Microspore or pollen grain grows into a pollen-tube which carries the male gametes to a position close to the egg-cell or ovum for the purpose of fertilization. In *Cycas*, however, the pollen-tube is branched; the branches penetrate into the nucellus and act as sucking organs (haustoria).

(5) The megaspore remains permanently enclosed in the megasporangium (nucellus of the ovule) and it germinates into the female gametophyte or embryo-sac (reduced prothallus) within the megasporangium.

(6) The megasporangium remains enclosed by the integument (1 in gymnosperms and usually 2 in angiosperms) giving rise to a more compli-

cated structure—the **ovule** (and later the **seed**) for better protection of the embryo.

(7) The young sporophyte (embryo) develops at the expense of the food stored up in the parent sporophyte.

Differences. (1) In angiosperms xylem is mainly composed of vessels, and phloem contains companion cells; while in gymnosperms xylem is exclusively made of tracheids, and phloem contains no companion cells. In higher gymnosperms, however, as in *Gnetum*, there is a combination of gymnospermic tracheids and angiospermic vessels.

(2) Flowers are much simpler in gymnosperms; no calyx or corolla is present; also the flowers are always unisexual consisting of either microsporophylls (stamens) or megasporophylls (carpels) only, and the plants either monoecious or dioecious. In them again stamens and carpels are much simpler in construction than in angiosperms. The sporophylls in gymnosperms are borne in strobili (except the megasporophylls of cycads), whereas in angiosperms these are borne in flowers.

(3) For pollination in gymnosperms the only agency is air-current, while in angiosperms there are many different pollinating agents.

(4) Ovules in gymnosperms are borne freely exposed on the megasporophyll (carpel); while in angiosperms the ovules remain enclosed in the ovary, the carpel itself being differentiated into ovary, style and stigma.

(5) For pollination in gymnosperms pollen grains enter the micropyle and lodge on the nucellus; whereas in angiosperms they are deposited on the stigma.

(6) In angiosperms and higher gymnosperms the male gametes contained in the pollen-tube are two passive units, but in lower gymnosperms (*Cycas, Zamia* and *Ginkgo*) the male gametes are in the nature of ciliate spermatozoids.

(7) In gymnosperms the male gametophyte is represented by a few cells (usually 2 or 3)—a vestigial prothallus; while in angiosperms it is reduced to two nuclei only—the tube nucleus and the generative nucleus; in fact there is very little evidence of a male prothallus in angiosperms.

(8) The female gametophyte in gymnosperms is a relatively large structure *with distinct archegonia* embedded in it, each with an ovum; while in angiosperms the female gametophyte is a vestigial prothallus represented by an 8-nucleate embryo-sac and the ovum or egg-cell is free in it *without any archegonium*.

(9) The endosperm, when present in the angiosperm, is formed from the definitive nucleus only after fertilization and is triploid in nature; while in gymnosperms the endosperm is formed from the vegetative tissue of the female prothallus before fertilization (completed, however, after fertilization) and is haploid in nature. The seeds in all gymnosperms are endospermic.

(10) Cotyledons are 2 to 15 in gymnosperms, and 1 or 2 in angiosperms.

Homologous Structures in Cryptogams and Phanerogams

Megasporophyll	= carpel	Microsporophyll	= stamen
Megasporangium	= nucellus of the ovule	Microsporangium	= pollen-sac
		Microspore	= pollen grain
Megaspore	= embryo-sac mother cell	Male gametophyte	= germinating pollen grain (pollen-tube with the nuclei in it)
Female gametophyte	= embryo-sac		

Development of the Seed in Gymnosperms. The seed makes its appearance for the first time in gymnosperms. It develops from the ovule and the following factors are responsible for its development:

(1) Heterospory and differentiation of sporophylls and sporangia are the initial factors, as we find in *Selaginella*, leading towards seed production in the future.

(2) Retention of the megaspore within the megasporangium (nucellus of the ovule) and its germination into the female gametophyte (embryo-sac) within the megasporangium so that the female gametophyte becomes completely *endosporous*; it is partially endosporous in *Selaginella*.

(3) Enclosure of the megasporangium and the female gametophyte within a new structure, i.e. the *integument*—the whole complex body so formed being known as the *ovule*. This ovule after fertilization and maturity gives rise to the seed.

(4) Attachment of the megasporangium (and the ovule) to the megasporophyll till after its development into the seed.

(5) Development of the young sporophyte (embryo) within the tissue of the megasporangium (nucellus) which belongs to the mother sporophyte, ensuring better feeding and greater protection.

(6) Development of the pollen-tube for facility of fertilization under the new condition of ovule formation.

It is thus evident that the seed of the gymnosperm is a complex structure with three generations locked up in it: (1) the seed-coat or testa representing the parent sporophyte (old generation); (2) the female gametophyte representing the present generation; and (3) the embryo representing the new sporophyte (future generation).

Classification. Gymnosperms comprise 8 orders, of which 4 have become extinct, and the remaining 4 have living representatives. The orders are: (1) Cycadofilicales or Pteridospermales or seed ferns (extinct); (2) Bennettitales or Cycadeoideales (extinct); (3) Pentoxylales (extinct); (4) Cycadales (represented by the family *Cycadaceae* with 9 genera and 100 species); (5) Cordaitales (extinct); (6) Coniferales (largest order represented by 6 families with 41 genera and over 500 species); (7) Ginkgoales (represented by the family *Ginkgoaceae* with 1 genus and 1 species cultivated in China and Japan, sparingly cultivated in India); and (8) Gnetales (represented by three families with altogether 3 genera and about 71 species). An outline of classification is given in the following schedule:

GENERAL DESCRIPTION

Class A. Cycadopsida
  Order 1 Cycadofilicales
    (or Pteridospermales) —fossils only,   e.g. *Lyginopteris*
  Order 2 Bennettitales
    (or Cycadeoideales) —fossils only,   e.g. *Bennettites*
  Order 3 Pentoxylales —fossils only,   e.g. *Pentoxylon*
  Order 4 Cycadales —*Cycadaceae,*   e.g. *Cycas*

Class B. Coniferopsida
  Order 5 Cordaitales —fossils only,   e.g. *Cordaites*
  Order 6 Coniferales —*Abietaceae,*   e.g. *Pinus*
  Order 7 Ginkgoales —*Ginkgoaceae,*   e.g. *Ginkgo*

Class C. Gnetopsida
  Order 8 Gnetales —*Gnetaceae,*   e.g. *Gnetum*
    —*Ephedraceae,*   e.g. *Ephedra*
    —*Welwitschiaceae,*   e.g. *Welwitschia*

Origin and Evolution of Gymnosperms (see FIG. X/1). Evidence from fossil records goes to show that gymnosperms originated from the oldest and the most primitive group of pteridophytes—the Psilophytales—which flourished during the Devonian period of the Palaeozoic age. The two oldest groups of gymnosperms, now extinct, viz. the Cordaitales and the Cycadofilicates, which had a common but independent origin from this group of Palaeozoic pteridophytes, became quite abundant during the mid-carboniferous period. Towards the close of the Palaeozoic or very early Mesozoic both the groups became extinct. The four groups of Gymnosperms (leaving out Gnetales), viz. Bennettitales, Cycadales, Ginkgoales and Coniferales, which evolved independently from the two Palaeozoic extinct groups formed the dominant vegetation of the earth during the mid-mesozoic period. They, however, began to wane towards the late Mesozoic. One group, viz. Bennettitales, became quite extinct, while a few descendants only of the other three groups continued through the Cenozoic age as the present-day living forms. Of the four groups of Mesozoic gymnosperms, contemporaneous and widely distributed, the Bennettitales originated from the Cycadofilicales, flourished and died during the same age. The Cycadales which also originated from the Cycadofilicales, but independently of Bennettitales, have left some living representatives (about 100 species). Ginkgoales and Coniferales originated, independently of one another, from the Cordaitales; but the former has left only one living representative (*Ginko biloba*)—a large tree in western China (now widely cultivated), and the latter over 500 living representatives (the biggest group of living gymnosperms). The remaining group of gymnosperms, viz. the Gnetales, is regarded as the most recent and the most advanced group with a near approach to the angiosperms. It has 3 genera and about 71 species. The fossil records of the Gnetales are rare and fragmentary, and not found earlier than the Tertiary. Evidently the group is of recent origin, possibly as an offshoot of some Coniferales.

Chapter 3 CYCADALES

The family *Cycadaceae* comprises 9 genera with about 100 species. The genus *Cycas* is represented in India by a few species, of which *C. circinalis* (of Malabar and West Madras) and *C. revoluta* (a Japanese sp.) are widely cultivated in Indian gardens; both yield a kind of sago. *C. rumphii* (of Malacca) is also widely cultivated in India. *C. pectinata* grows in the low hills of Assam and also in Sikkim, and *C. beddomei* in the forests of Eastern Peninsula.

CYCAS (20 sp.)

The stem of **cycad** (*Cycas*; FIG. 1) is unbranched, erect, stout and palm-like, with a crown of pinnate leaves arranged spirally round the apex. Besides, there are small, dry, scale-like leaves alternating with the green pinnate leaves. Vernation (ptyxis) of the leaf is *circinate* like that of ferns. The plant has a long primary (tap) root.

Cycads are dioecious, i.e. male and female flowers are borne by two separate plants. The male flower is a cone (FIG. 2) borne at the apex of the stem which then grows by a lateral bud (the stem becomes a sympodium). In *Cycas pectinata* the male cone may be as big as half a meter in length. The male cone consists of a collection of **stamens** or **microsporophylls** arranged spirally round the axis. Each sporophyll (FIG. 3C) is in the form of a scale, narrowed below and broadened above. It bears on its undersurface several **pollen-sacs** or **microsporangia** grouped in sori. There are usually 2 to 6 pollen-sacs in each sorus. In each pollen-sac there are

FIG. 1. A female plant of *Cycas circinalis* with carpels.

CYCADALES

numerous **pollen grains** or **microspores**. The nucleus of each microspore divides once and produces an extremely reduced **prothallus cell** (male gametophyte) on one side and a large **antheridial cell** on the other side (FIG. 5A). The latter divides again and produces a **generative cell** and a **tube cell** (nucleus). The pollen grain sheds at this three-celled stage, and subsequent changes take place after pollination.

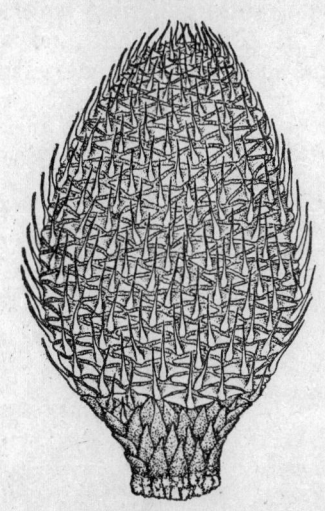

In *Cycas* there is no proper female flower; the plant bears near its apex a rosette of **carpels** or **megasporophylls** (FIG. 3A-B) which do not form a cone but are arranged alternating with the leaves. They are usually 15 to 30 cm. long, flattened or bent over like a hood, and often dilated above. In many species they are covered all over with soft brownish hairs. The margin may be entire, crenate or pectinate (pinnately divided). Ovules, usually 2-3 paris, sometimes up to 5 pairs, are borne in an alternate or opposite manner in notches on either side of

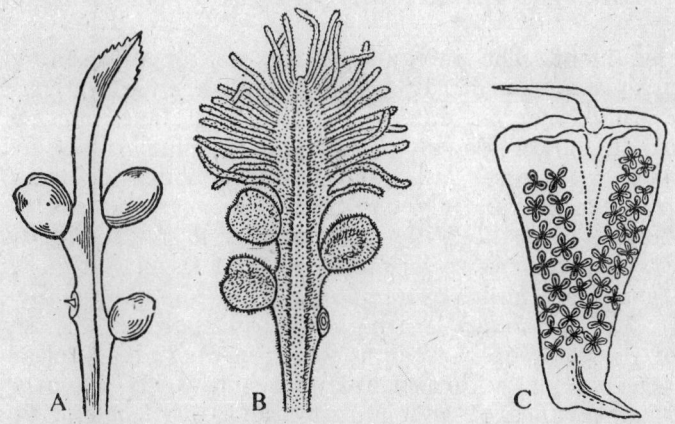

Cycas. FIG. 2. A male cone of *Cycas pectinata* consisting of innumerable microsporophylls arranged round the stout axis.

Cycas. Carpels and Stamen. FIG. 3. *A*, a carpel or megasporophyll of *Cycas circinalis*; *B*, the same of *Cycas revoluta*; *C*, a stamen or microsporophyll of *Cycas pectinata* with numerous pollen-sacs or microsporangia on the undersurface (slightly oblique view).

the stalk. These are commonly oval and large, sometimes very large, as in *C. circinalis* where they grow up to 6 cm. In gymnosperms the carpel is *always open,* i.e. it does not close up, as in angiosperms, to form the ovary, style and stigma, and the ovules are borne, freely exposed, on the two margins of the carpel. The ovule (FIG. 4A) grows considerably even before fertilization. It consists of a single thick integument and a nucellus or megasporangium fused practically throughout with

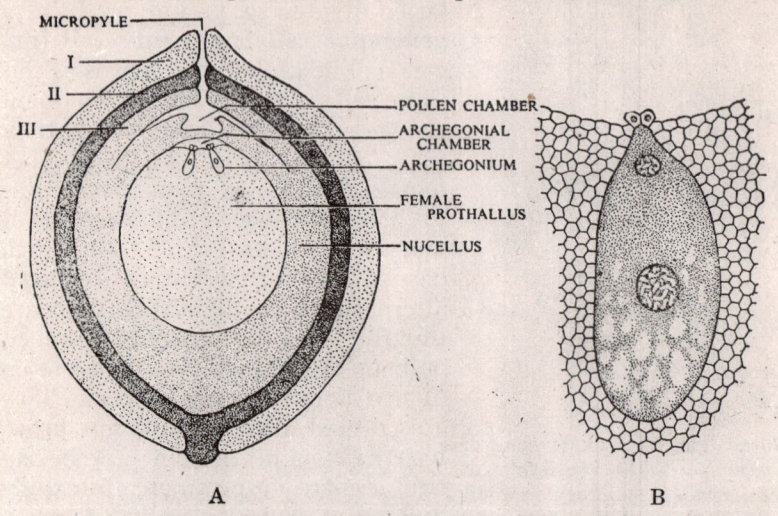

Cycas, FIG. 4. *A*, ovule in longitudinal section; *I, II* and *III*, outer, middle (stony) and inner layers of the integument; *B*, an archegonium; note the neck with two neck cells, a large venter (egg-cell) with a central egg-nucleus, and a small ventral canal nucleus (higher up).

the integument. The integument has an apical opening—the micropyle—and consists of three layers—a middle stony layer and a fleshy layer on either side. The inner layer merges into the nucellus lower down. Within the nucellus of the ovule a megaspore mother cell is produced which divides to form a row of four megaspores. Only one megaspore is functional; while the other three disintegrate. The functional megaspore divides rapidly and gives rise to a cellular mass of tissue within it; this is the **female prothallus** or **gametophyte** (designated as the endosperm). *The prothallus is completely endosporous.* The development of the gametophyte begins with repeated free nuclear divisions. The nuclei so formed are pushed towards the periphery by the appearance of an enlarging central vacuole. Cell-walls then develop round the nuclei from the peripheral region towards the centre. The fully developed gametophyte thus becomes cellular and large occupying the major part of the nucellus, with two distinct regions in it—a region of smaller cells towards the upper end, and a region of larger cells, nutritive in function, with sugar at first and starch later. The endosperm grows quickly after fertilization, invades the nucellus, and forms the major part of the seed. The small-celled region produces towards the micropyle a few (2-8) archegonia. Each **archegonium** (FIG. 4B) is extremely reduced in size, and consists of a very short neck with two small neck cells (but no neck canal cell) and a large venter with an egg-cell completely filling it. There is a distinct free **egg-nucleus** in it lying somewhere in the centre. The ventral canal cell is represented only by a nucleus which, however, soon becomes disorganized. The central cell is invested by a special layer of cells, called the **jacket,** which is pitted. Through the pits food materials enter the central cell. Just below the micropyle a chamber or cavity is formed due to the breaking down of some of the nucellus cells. This cavity is called the pollen chamber. Just below this another chamber is formed in the prothallus. This is the archegonial chamber.

Pollination and Fertilization. Pollen grains are carried by the wind. They fall on the micropyle and get bathed in the mucilage secreted by the latter; as the mucilage dries up the pollen grains are drawn into the pollen chamber. The pollen grain (FIG. 5A) divides and forms (a) a small prothallus cell (**male gameto-**

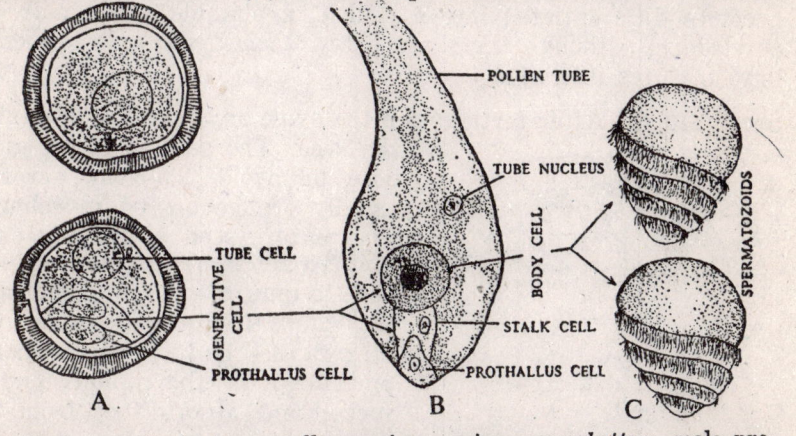

Cycas. FIG. 5. *A, top,* a pollen grain or microspore; *bottom,* male prothallus; *B,* pollen-tube (a portion); *C,* two spermatozoids.

phyte) on one side, (*b*) a generative cell, and (*c*) a tube cell. The tube cell elongates into a long branched pollen-tube (FIG. 5B) which enters into the archegonial chamber, the membrane between the two chambers having broken down. The pollen-tube of *Cycas* is a sucking organ (haustorium) absorbing food from the nucellus, rather than a sperm-carrier. The generative cell divides into two—the stalk cell and the body cell. The **stalk cell** is sterile and the **body cell** divides into two male gametes (**spermatozoids**; FIG. 5C). They are remarkably large (in fact the largest known, being about 300 μ in length), top-shaped and multiciliate (the cilia being arranged in a spiral band).

The pollen-tube bursts at the apex and the spermatozoids are set free. They swim to the archegonium by the vibration of their cilia and enter into the egg-cell. The male nucleus slips out of its cytoplasmic sheath, moves towards the egg-nucleus, and finally penetrates into it. Fertilization is thus effected. The time lapse between pollination and fertilization is about four months.

Development of the Embryo. After fertilization the egg-nucleus undergoes free nuclear divisions 8 or 10 times. As a result a large number of free nuclei (without cell-walls) appear in the enlarged egg. A central vacuole is formed which quickly enlarges, and as a consequence the free nuclei are pushed towards the periphery of the egg. Later cell-walls begin to be formed round the nuclei. A tissue thus appears which soon fills up the whole of the vacuole. The cellular mass at the base of the egg constitutes the **proembryo**, while the upper cells form a large food reservoir. The proembryo soon differentiates into a suspensor and a terminal embryo. The suspensor elongates quickly, becomes coiled and pushes the embryo deep into the nutritive tissue of the female gametophyte. The embryo, as it takes shape, becomes differentiated into a radicle, a plumule and a pair of cotyledons. Although there are a few archegonia, only one embryo matures in a seed.

Seed (FIG. 6). After fertilization the ovule as a whole grows into the seed. The seed is large, globose or ovoid, 2.5-5 cm. across, usually orange or red in colour. The mature seed bears only one embryo differentiated into a hypocotyl, a radicle, a plumule and *two* cotyledons (a constant feature of all cycads). The cotyledons remain embedded in the copious endosperm and absorb food from it. The seed-coat or integument surrounding the seed consists of an outer thick fleshy layer and an inner stony layer. The endosperm stores a considerable quantity of food for the use of the embryo while the seed germinates. Its germination is prompt, without any period of rest.

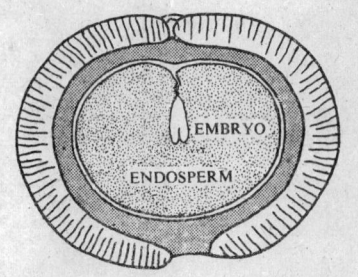

FIG. 6. *Cycas* seed in section.

Internal Structure of Cycad Leaf (FIG. 7). (1) **Epidermis**—the upper with very thick cuticle, while the lower with comparatively thin cuticle but with a large number of sunken stomata, each arched over by epidermal outgrowth. (2) **Hypodermis**—the upper is highly thickened, while the lower comparatively thin. (3) **Mesophyll** differentiated into upper distinct palisade parenchyma and lower spongy parenchyma. In between the two there is a tissue of long colourless cells forming the **transfusion tissue** or conducting channel which runs horizontally from the vascular bundle (mid-rib) to near the margin of the leaf. Veins being absent in cycad leaf food and water are conducted through the transfusion tissue. (4) **Vascular bundle** with xylem on the upper side and phloem on the lower. Xylem is mesarch, i.e. protoxylem lies in the middle of xylem; towards the upper side xylem is well developed and forms **centripetal xylem** (with distinct protoxylem and metaxylem), while towards the lower side, lying in the **parenchyma** of the bundle, small vascular strands detached from the centripetal xylem form **centrifugal xylem**. The vascular bundle is more or less completely surrounded by a sheath (**bundle sheath**). Some small thick-walled cells and some clusters of crystals may be seen here and there in the section.

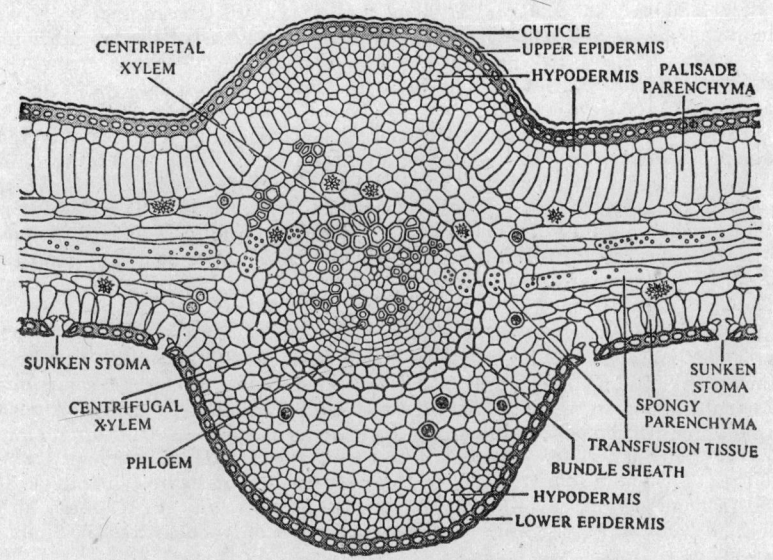

FIG. 7. *Cycas* leaf in transection.

Chapter 3 CONIFERALES

Coniferales (Coniferae or conifers) constitute the largest order among the gymnosperms and comprise six families, e.g. *Abietaceae* (=*Pinnaceae*), *Taxaceae*, *Cupressaceae*, etc., which are represented by 41 genera and over 500 species. They grow mostly in cold climates and at high altitudes in the hills (see below), and are evergreen trees or shrubs. *Abietaceae*, the largest family, is represented by 9 genera and about 230 species, of which pine (*Pinus*), juniper (*Juniperus*), fir (*Abies*), Himalayan deodar (*Cedrus deodara*), Atlantic cedar (*Cedrus atlantica*), spruce (*Picea*), etc., are some of the genera. *Pinus* is, of course, the typical genus of *Abietaceae*.

Distribution of Coniferales in India. There is a rich coniferous flora in the eastern and western Himalayas, occurring often in rich luxuriance between 1,200 m. and 3,300 m. Many of them also extend to higher elevations, and a few are found at lower heights. But investigations are far from complete yet. It will be noted that many exotic conifers have been introduced into Indian gardens and hill stations, and they have become naturalized. The distribution of conifers in India is as follows:

Abeitaceae (= *Pinaceae*). (1) *Pinus* (90 sp.): in India long-needle pine or CHIR pine (*P. longifolia*) grows in both eastern and western Himalayas at an altitude of 600-1,800 m.; in NEFA (Arunachal) at 1,200 m. and above; Khasi pine (*P. khasya*) in the Khasi Hills (Meghalaya) at 900-1,800 m.; blue pine (*P. excelsa*) in temperate Himalayas (both eastern and western) at 1,800-3,800 m.; in NEFA at 1,200 m. and above; *P. gerardiana* in N.W. Himalayas at 1,800-3,700 m.; *P. insularis* in the Khasi Hills at 1,800 m. (2) *Cedrus* (3 sp.): in India Himalayan deodar (*C. deodara*) grows in north-west Himalayas at 1,100-3,700 m. (planted in Shillong and other hill stations); Atlantic cedar (*C. atlantica*) cultivated in hill stations. (3) *Picea* (35 sp.): in India spruce (*P. morinda*) grows in eastern and western Himalayas at 1,800-3,700 m. (planted in Shillong and other hill stations). (4) *Abies* (24 sp.): in India Himalayan silver fir (*Abies spectabilis*=*A. webbiana*) grows in eastern and western Himalayas at 2,100-3,700 m.; in NEFA at 3,000 m. (5) *Tsuga* (9 sp.): in India *T. brunoniana* grows in eastern and western Himalayas at 2,400-3,300 m.; common in NEFA. (6) *Larix* (8 sp.): larch (*L. griffithii*) grows in eastern Himalayas at 2,400-3,700 m.

Cupressaceae. (1) *Cupressus* (12 sp.): in India *C. torulosa* grows in western Himalayas at 1,800-2,400 m. (planted in Shillong); *C. funebris* (planted in Shillong). (2) *Juniperus* (60 sp.): in India juniper (*J. communis*—a shrub) grows in western Himalayas at 1,700-4,300 m.; *J. recurva* (a bush or small tree) in temperate and alpine Himalayas at 2,300-4,600 m.; in Mishmi Hills at 2,700 m.; *J. macropoda* (a tree) in western Himalayas at 1,500-4,300 m.; *J. pseudo-sabina* (a tree) in western and eastern Himalayas at 2,700-4,600 m. (3) *Thuja* (=*Biota*; 6 sp.): *T. orientalis* (a Chinese sp.) and *T. occidentalis* (an American sp.) are commonly cultivated in Indian gardens.

Podocarpaceae. *Podocarpus* (70 sp.): in India *P. neriifolia* grows in the Khasi Hills and NEFA at 900-1,200 m.; *P. latifolia* in the Khasi Hills at 900-1,500 m.

Araucariaceae. (1) *Araucaria* (15 sp.): *A. excelsa* (Australian) and *A. cunninghamii* (South American) are commonly cultivated in Indian gardens; (2) *Agathis* (20 sp.): *A. australis* (a source of copal varnish) is planted in Indian gardens.

Taxaceae. (1) *Taxus* (8 sp.): in India *T. baccata* grows in temperate Himalayas at 1,800-3,300 m.; in the Khasi Hills and Mishmi Hills at 1,500 m.; (2) *Cephalotaxus* (6 sp.): *C. mannii* grows in the Khasi Hills and Naga Hills at 1,400-2,600 m.; *C. griffithii* in the Mishmi Hills and the Naga Hills at 1,800 m.

Taxodiaceae. *Cryptomeria* (1 sp.): Japanese cedar (*C. japonica*) is cultivated in Shillong and other hill stations.

PINUS (90 sp.)

Pine (*Pinus*; FIGS. 11-12) grows abundantly in the temperate regions of the eastern and western Himalayas at an altitude of 1,200 to 3,300 metres (see above). Common Himalayan species are: Khasi pine (*P. khasya*), long-needle pine or CHIR pine (*P. longifolia*), and blue pine (*P. excelsa*). *P. sylvestris*, a British pine, and a few others are exotic species.

Pine is a tall erect evergreen tree, sometimes growing to a height of 45 metres with a basal girth of 3 m. The plant has a well-developed tap root and numerous aerial branches with green needle-like leaves. The stem is rugged and covered with scale bark peeling off here and there in strips. Branches are of two kinds—long (of unlimited growth) in apparent whorls, developing from lateral buds in the spring, and dwarf (of limited growth; see FIG. 11). Leaves are also of two kinds—long green needle-like foliage leaves (commonly called needles) borne only on dwarf branches (or foliar spurs, as they are called) and small brown scaly leaves borne on both kinds of branches. The number of needles in a cluster varies from 1 to 5, 3 in *P. khasya*, *P. longifolia* and *P. gerardiana*, 5 in *P. excelsa*, 2 in *P. sylvestris*, and 4 or 1 in certain species.

Internal Structure of the Young Stem (FIG. 8). This resembles the dicotyledonous stem in many respects. The general arrangement of the various tissues from the circumference to the centre is the same. It differs, however, from the latter in having a large number of resin-ducts filled with brownish contents, i.e. resin. These ducts are distributed almost throughout the stem. The epidermis has an irregular outline. Endodermis and pericycle are not marked out in the stem. The vascular bundles are not wedge-shaped,

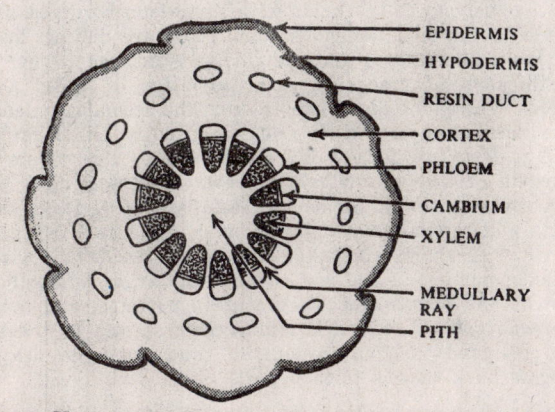

FIG. 8. Young pine stem in transection (diagrammatic).

as in the dicotyledons. Phloem consists of sieve-tubes and phloem parenchyma, but no companion cells. Protoxylem consists of annular and spiral tracheids which are irregularly disposed towards the centre. Metaxylem consists exclusively of tracheids with bordered pits. The tracheids are arranged in radial rows as seen in the transverse section of the stem. The pits of the coniferous wood are large and mostly restricted to the radial walls; while in angiosperms they are much smaller but more numerous, and are not confined to particular walls. There are no true vessels in the coniferous stem.

(1) **Epidermis.** A single layer with a very thick cuticle, and an irregular outline. (2) **Hypodermis.** A few layers of large, sometimes thick-walled and lignified cells lying below the epidermis. (3) **Cortex.** Many layers of more or less rounded cells, with conspicuous resin-ducts. (4) **Medullary Rays.** These run from the pith outwards between the vascular bundles. (5) **Pith.** There is a well-defined pith, consisting of a mass of parenchymatous cells. A few resin-ducts are also present in the pith. (6) **Vascular Bundles.** These are collateral and open, and arranged in a ring, as in the stems of dicotyledons. Each bundle consists of phloem, cambium and xylem. (a) **Phloem** consists of sieve-tubes and phloem parenchyma, but no companion cells. It lies on the outer side of the bundle. (b) **Cambium.** A few layers of thin-walled, rectangular cells lying in between xylem and phloem. (c) **Xylem** consists exclusively of tracheids; there are no true vessels in the wood. Resin-ducts are also present here. Protoxylem lies towards the centre and consists of a few annular and spiral tracheids which are not disposed in any regular order. Metaxylem lies towards the cambium, and consists of tracheids with bordered pits which develop on the radial walls. These tracheids are roughly four-sided and are arranged in definite rows.

Secondary Growth in Thickness of the Stem (FIG. 9). The secondary growth in a coniferous (pine) stem takes place in exactly the same way as in a dicotyledonous stem. The following points may, however, be mentioned here, showing slight differences between the two.

The coniferous (pine) stem is characterized by the presence of conspicuous resin-ducts which are distributed almost throughout the stem. The secondary wood consists exclusively of tracheids with numerous bordered pits on their radial walls. As in the dicotyledonous stem, there are distinct annual rings, consisting of the autumn wood and the spring wood; the former consists of narrow and thick-walled tracheids, and the latter of wider and thinner-walled tracheids. Vessels or fibres are absent. The secondary medullary rays are numerous running horizontally and are usually one layer of cells in thickness and a few (up to 12) layers in height. The xylem portion of each ray consists of (a) a few middle rows of rectangular, thick-walled cells (**ray parenchyma**) with numerous, comparatively large, simple pits, living contents and several starch grains, and (b) 1 to 3 marginal (upper and lower) rows of short tracheidal cells (**ray tracheids**) with bordered pits. The phloem portion of the ray consists of large, thin-walled cells extending upwards and downwards; they contain proteins, and are sometimes called *albuminous cells*. The **cork-cambium** soon arises in the cortex, and gives rise to cork and bark on the outside, and some parenchyma on the inside. The scale bark that is formed peels off in strips.

Internal Structure of the Pine Needle (FIG. 10). (1) **Epidermis**—a single layer of very thick-walled cells with a strong cuticle; the cell cavity is nearly obliterated. There are stomata sunken to some extent in this layer, each stoma opening internally into a respiratory cavity. (2) **Hypodermis** occurs internal to the epidermis in 1, 2 or 3 layers interrupted by the stomata. This tissue is deeper at the ridges. (3) **Mesophyll** consists of large thin-walled polygonal or irregular cells containing abundant chloroplasts and starch grains. There are peg-like projections of the cell-walls into the cell-cavity. In the mesophyll there are some **resin-ducts** here and there, adjoining the hypodermis, each surrounded by a layer (epithelium) of small, thin-walled cells. (4) **Endodermis** occurs as a conspicuous layer of large barrel-shaped cells. (5) **Pericycle** lies internal to the endodermis and is a many-layered

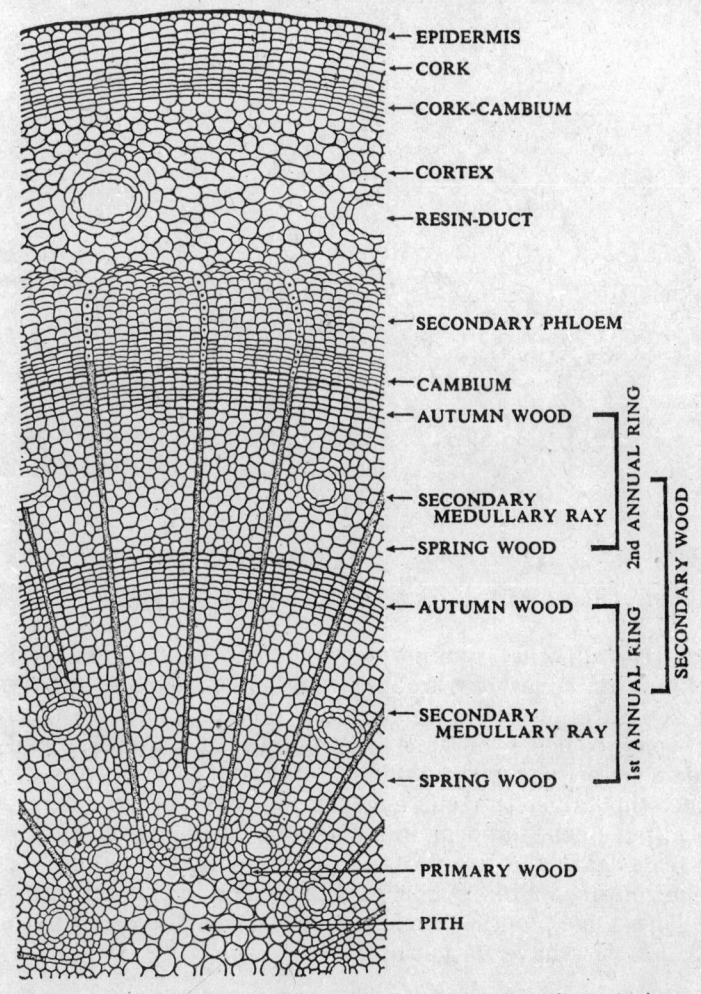

FIG. 9. A two-year old pine stem in transection (a sector).

tissue. It consists of some parenchyma, often some sclerenchyma, particularly on the phloem side, and transfusion tissue. **Transfusion tissue** consists of (*a*) *albuminous cells* which lie close to phloem and are parenchymatous in nature, living and rich in protein and starch, and (*b*) *tracheidal cells* which lie adjoining xylem and are thin-walled elongated dead cells, but provided with bordered pits like tracheids. Albuminous cells serve to conduct food from the mesophyll to the phloem, and tracheidal cells to conduct water and mineral salts from the xylem to the mesophyll. (6) **Vascular bundles**, 2 in number, lie embedded within the many-layered pericycle; they are collateral and closed. Xylem consisting of rows of

674　　　　　　　　　　　　　　　　　　　　BOTANY PART VI

tracheids lies towards the angular side of the leaf, and phloem consisting of sieve-tubes lies towards the convex side.

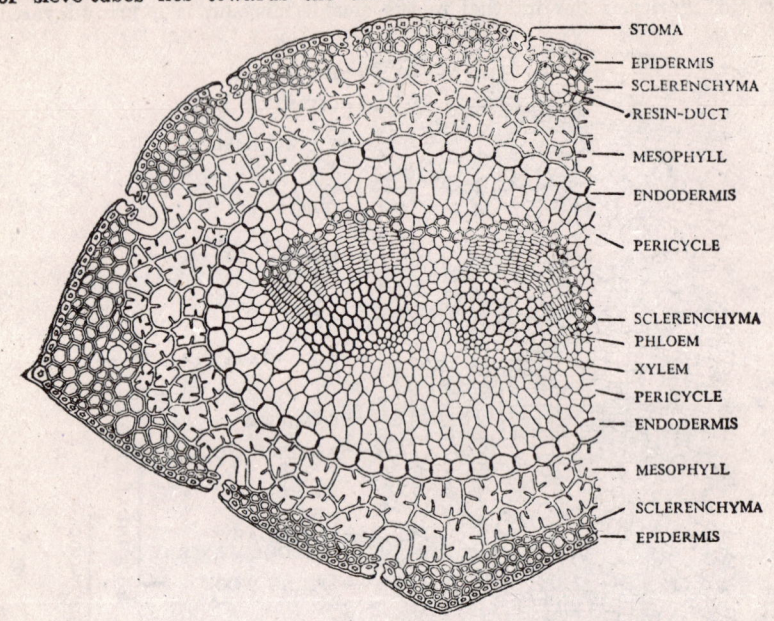

FIG. 10. Pine needle (leaf) in transection.

Pinus, like all other gymnosperms, is the **sporophyte**. It bears two kinds of cones or strobili (FIGS. 11-12)—male and female—on separate branches of the same plant (monoecious). The male cone consists of **microsporophylls** or **stamens**, and the female cone of **megasporophylls** or **carpels**. Cones always develop on the shoots of the current year a little below their apex. Several male cones appear in a cluster of spikes, each in the axil of a scale leaf, and are 1.5 to 2.5 cm. in length. Female cones may be solitary or in a whorl of 2 to 4, each in the axil of a scale leaf. The male cones develop much earlier than the female cones. Flowers have no perianth.

Male Cone (FIGS. 11 & 13*A*). This, as said before, is made of a number of **microsporophylls** or **stamens** arranged spirally round the axis (FIG. 13*A*). Each microsporophyll (FIG. 13*B-D*) is differentiated into a stalk (filament) and a terminal leafy expansion (anther), and its tip is bent upwards. It bears on its undersurface two pouch-like **microsporangia** or **pollen-sacs**. In some conifers these may be as many as 15. Each pollen-sac contains numerous microspore mother cells, each of which undergoes reduction division to produce a tetrad of **microspores** or **pollen grains**. Each pollen grain (FIG. 17*A*) has two coats—*exine*

CONIFERALES

(outer) and *intine* (inner). The exine forms two wings, one on each side. A huge quantity of pollen is produced for pollination to be brought about by wind.

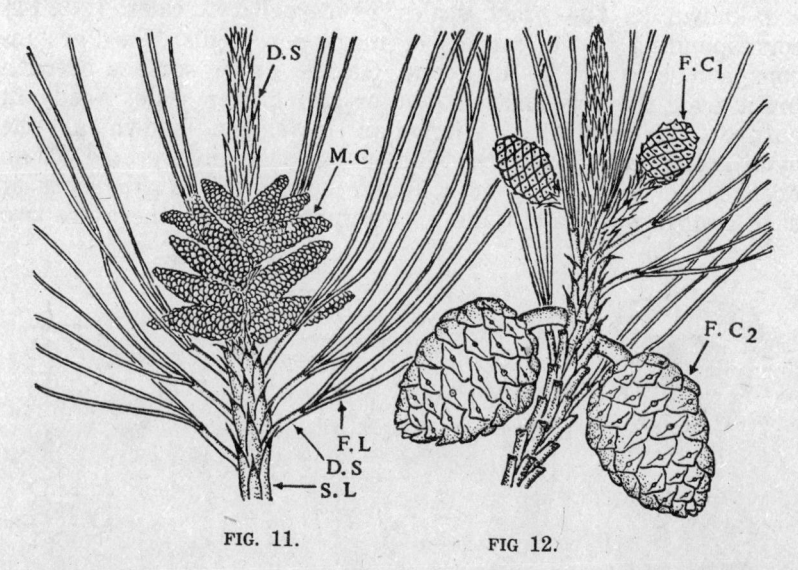

FIG. 11.   FIG 12.

Pinus. FIG. 11. A male shoot (branch) of unlimited growth showing male cones (*M.C.*), young dwarf shoots (*D.S.*) or shoots of limited growth (each in the axil of a scale), needle-like foliage leaves or needles (*F.L.*) borne only by dwarf shoots or foliar spurs (as these are called), and scale leaves (*S.L.*) spirally borne by both dwarf and long shoots. FIG. 12. A female shoot of unlimited growth showing two young female cones (*F.C₁*) of current year, two maturing female cones (*F.C₂*) of previous year (for the third year cone see FIG. 19A), and the rest as in FIG. 11.

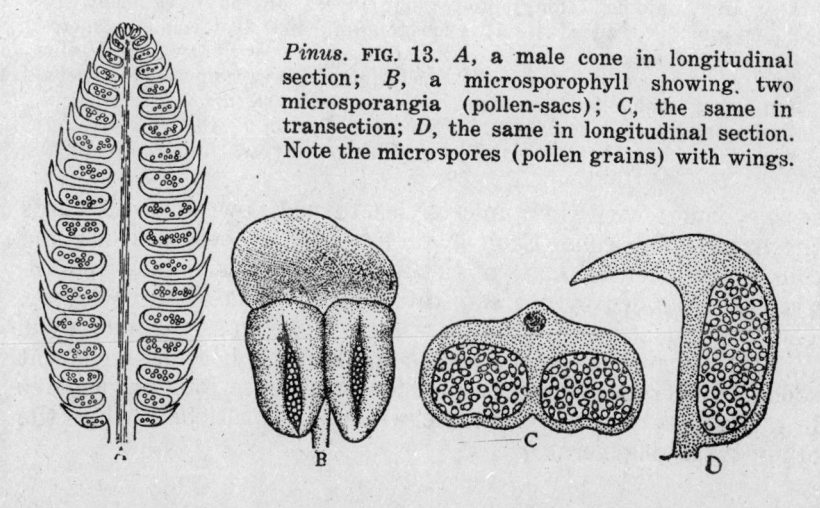

Pinus. FIG. 13. *A*, a male cone in longitudinal section; *B*, a microsporophyll showing two microsporangia (pollen-sacs); *C*, the same in transection; *D*, the same in longitudinal section. Note the microspores (pollen grains) with wings.

Female Cone (FIGS. 12 & 19A). This consists of a short axis round which a number of small thin dry brownish scales, slightly fringed at the upper part, are spirally arranged. These are known as the **bract scales** or **carpellary scales** (FIG. 14) corresponding to the carpels or megasporophylls. They are inconspicuous in the mature cone. On the upper surface of each bract scale there is another stouter and bigger scale, woody in nature and somewhat triangular in shape, known as the **ovuliferous scale** (FIG. 14). This is variously interpreted as an open carpel, a placenta or a ligular outgrowth. At the base of each ovuliferous scale, lying on its upper surface, there are two

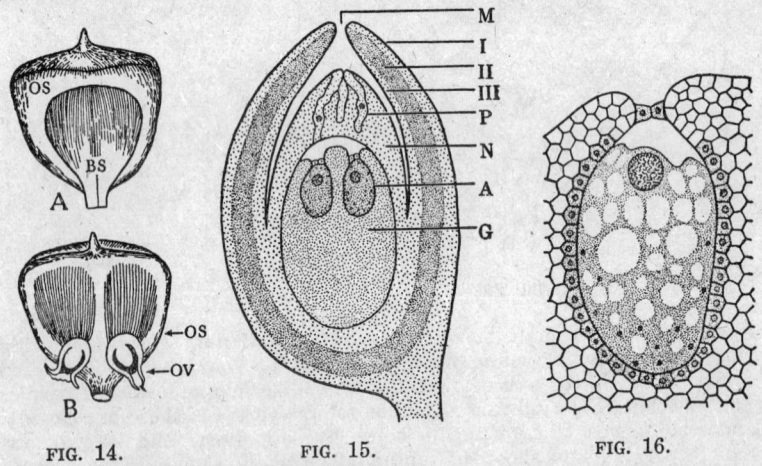

FIG. 14. FIG. 15. FIG. 16.

Pinus. FIG 14. Megasporophyll. A, lower surface; B, upper surface. *BS*, bract scale; *OS*, ovuliferous scale; *OV*, ovule with one integument. FIG. 15. Ovule in longi-section. M. micropyle; *I, II* and *III* are outer, middle (stony) and inner layers of the integument; *P*, pollen-tube; *N*, nucellus; *A*, archegonium; and *G*, female gametophyte (endosperm). FIG. 16. An archegonium with two neck-cells and a large venter (egg-cell) enclosing a conspicuous egg-nucleus and many small food particles. *Fig. 16 redrawn after Fig. 322 in Gymnosperms: Structure and Evolution by C. J. Chamberlain by permission of The University of Chicago Press. Copyright 1935.*

sessile **ovules** with their micropyles turned downwards towards the axis of the cone. Each ovule (FIG. 15) is orthotropous and consists of a central mass of tissue—the **nucellus** or **megasporangium**, surrounded by a *single* **integument** made of three layers. The integument leaves a rather wide gap known as the **micropyle**. Within the nucellus a megaspore mother cell becomes apparent soon. It undergoes reduction division to produce four **megaspores** in a linear tetrad. Only one megaspore is functional; while the other three degenerate.

Male Gametophyte (FIG. 17A). The microspore or pollen grain begins to divide before it is set free from the microsporangium or pollen-sac and gives rise to the extremely reduced male prothallus (male gametophyte) within the microspore coat. The prothallus consists of (a) 2 or 3 small cells (**prothallus cells**) at one end, which soon become disorganized, and (b) an **antheridial cell** which is the remaining large cell of the prothallus. The antheridial cell divides and forms a **generative cell** and a **tube cell**. The pollen grain sheds at this stage and subsequent changes take place after pollination.

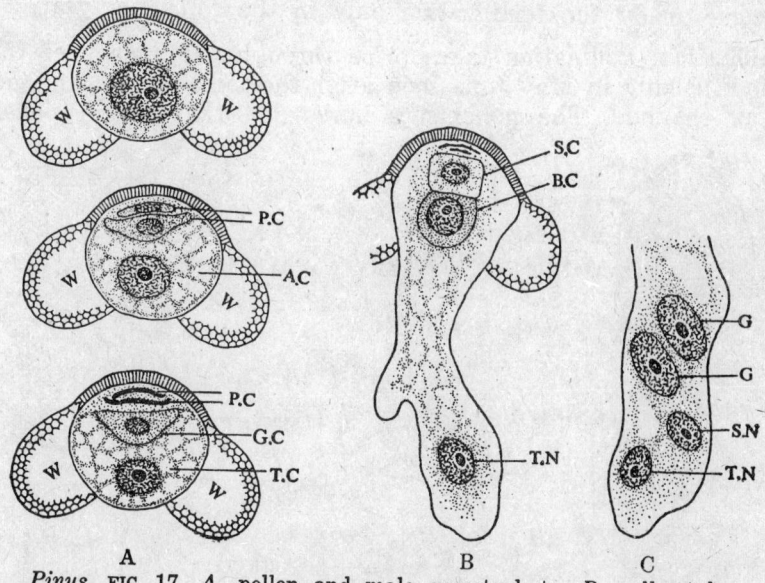

Pinus. FIG. 17. *A*, pollen and male gametophyte; *B*, pollen-tube; *C*, lower portion of the pollen-tube. *W*, wing; *P.C*, prothallus cells; *A.C*, antheridial cell; *G.C*, generative cell; *T.C*, tube cell; *S.C*, stalk cell; *B.C*, body cell; *T.N*, tube nucleus; *G*, male gamete; and *S.N*, stalk nucleus. *Redrawn after Fig. 311 in* Gymnosperm: Structure and Evolution *by C. J. Chamberlain by permission of The University of Chicago Press. Copyright 1935.*

Female Gametophyte (FIG. 15). The megaspore begins to divide after a period of rest, and gives rise to the female prothallus (female gametophyte) within the nucellus. It digests a big portion of the nucellus and enlarges considerably. Soon it takes to free nuclear divisions, finally forming a solid mass of tissue— the female gametophyte (otherwise designated as the endosperm). It is completely *endosporous,* i.e. it remains permanently enclosed within the nucellus. The endosperm grows quickly after fertilization, invades the nucellus, and surrounds the embryo in the seed (FIG. 19C). At the micropylar end of the prothallus, lying embedded in it, develop 2 to 5 **archegonia** (FIG. 15), much reduced in size as in *Selaginella*. As in the latter, each archegonium (FIG. 16) consists of a swollen venter and a short neck. The venter encloses a large **egg-cell** (almost filling the cavity) with a distinct **egg-nucleus** in it, and also a small ventral canal cell which, however, soon gets disorganized. The neck consists of 2 or more, usually 8, neck cells but no neck canal cell. The archegonial wall is made of a layer of small but distinct cells called the *jacket cells*. The archegonia, however, slowly mature and become ready for fertilization only in the following year.

Pollination. Pollination takes place through the agency of the wind, usually in May-June soon after the female cone emerges from the bud. The pollen-sacs burst and the winged pollen

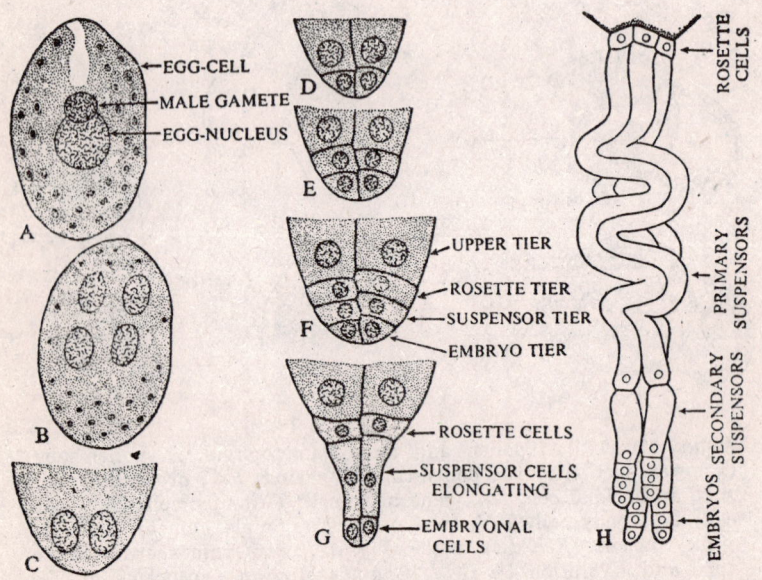

Pinus. FIG. 18. Fertilization and embryogeny. *A*, fertilization; *B-H*, development of the embryo by successive divisions of the fusion-nucleus; *B*, 4-nucleate stage; *C*, 2 of the 4 nuclei at the base; *D*, 8-celled proembryo; *E*, proembryo with 3 tiers of 4 cells each; *F*, proembryo with 4 tiers of 4 cells each; *G*, proembryo showing elongation of suspensor cells; *H*, 4 embryos (proembryos) separating, formation of secondary suspensors and rapid elongation of primary suspensors. At the next stage only one embryo matures (see FIG. 19C). *Redrawn after Fig. 300 in Plant Morphology by A. W. Haupt by permission of McGraw-Hill Book Company. Copyright 1953.*

grains are blown about by the wind like a yellow cloud of dust. Some of them happen to fall on the young female cone. A huge quantity is wasted, however. The pollen grains pass between the two slightly opened scales and are deposited at their base. A quantity of mucilage is secreted at the micropyle in which the microspores get entangled. After pollination the scales close up and so does the wide gaping integument. The mucilage is drawn in by the nucellus together with the microspores. The latter then lodge somewhere at the apex of the nucellus for about a year. Fertilization takes place in the following year at about the same time when the archegonia mature, and in the third year the cone fully develops.

Fertilization (FIG. 18A). The mode of fertilization in *Pinus* was first discovered in 1883 by Goroschankin who observed the entry of both the male gametes into the embryo-sac. Strasburger, however, discovered in 1884 the fusion of only one male gamete with the egg-nucleus, and the process is as follows. The outer coat (exine) of the pollen grain bursts and the inner coat (intine) grows out into a slender tube, i.e. the pollen-tube (FIG. 15) which pushes forward through the nucellus and finally reaches the neck of the archegonium. The pollen-tube is characteristic of all gymnosperms and angiosperms. The tube nucleus passes into the

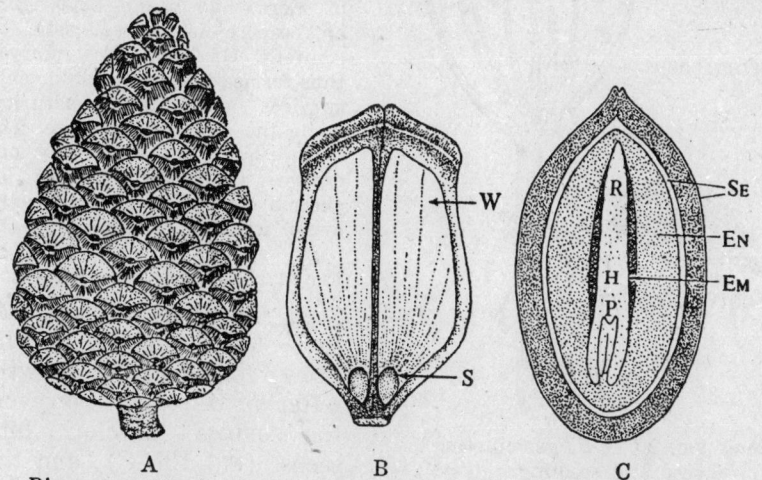

Pinus. FIG. 19. *A*, a mature female cone; *B*, a mature megasporophyll with 2 seeds (winged); *S*, seed; *W*, wing; *C*, a seed in longisection; *EM*, embryo (*P*, plumule with many cotyledons; *H*, hypocotyl; *R*, radicle); *EN*, endosperm; *SE*, seed coat (with outer and inner fleshy layers and middle stony layer).

pollen-tube. The generative cell divides and forms a **stalk cell** and a **body cell** (FIG. 17B). Both these cells migrate into the pollen-tube. The stalk cell is sterile while the body cell divides and produces two **male gametes** (FIG. 17C). The male gametes are not ciliate, as in cycads. The pollen-tube bursts at the apex and the two male gametes are liberated. The nucleus of the functioning male gamete slips out of the cytoplasm and passes on to the egg-nucleus.[1] The other male gamete, the stalk cell and the tube nucleus become disorganized.

Development of the Embryo (early stages; FIG. 18 B-H). By two successive divisions of the fusion-nucleus four free nuclei are formed within the egg-cell (B). These then move to the bottom of it in a horizontal plane (C). The four nuclei divide again and walls appear between them (D-E). Further divisions result in the formation of four tiers of four cells each; this 16-celled structure is called the **proembryo** (F). The **uppermost tier** of the proembryo with its four open cells, i.e. without cell-walls, merges into the general mass of cytoplasm of the egg-cell and acts as a part of the surrounding nutritive tissue. The next tier of four cells constitutes what is called the **'rosette tier'**; the rosette cells may develop short and abortive embryos; normally it transmits food to the suspensor and the embryonal cells. The third tier forms a four-celled **'suspensor tier'**. The lowest tier of four cells is the **'embryo tier'**. The suspensor cells begin to elongate rapidly (G) and soon become very big and tortuous (H) thrusting the embryos into the gametophytic tissue (endosperm). The embryonal cells (G) divide and give rise to *four* potential embryos and also secondary suspensor cells (H). This **polyembryony** (see p. 410), i.e. the development of more than one embryo from an oospore is characteristic of conifers. Of the four embryos thus formed the strongest one only survives and attains maturity, while the others degenerate. The mature seed has thus only one embryo (FIG. 19C). Besides, as there are a few archegonia in the ovule some more embryos may begin to develop but finally, as stated above, only one survives.

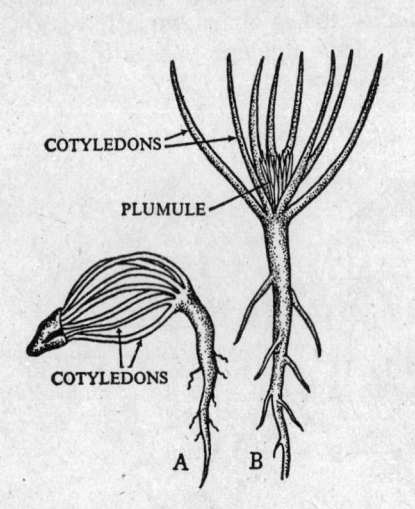

Pinus. FIG. 20. *A-B*, germinating seed and seedling.

Seed. Even before fertilization the megasporophylls show considerable growth. After fertilization, however, the ovules develop into seeds (FIG. 19*B-C*), and the whole female flower into a dry brown woody **cone** (FIG. 19*A*). The seed is albuminous in all gymnosperms, and in

---

[1] The male and the female gametes do not fuse in the resting stage but form two separate spindles with their respective chromosomes (12 in number) within the egg-nucleus. The two sets of chromosomes then orient themselves on a common spindle and their separate identity is lost. The chromosomes (24 in number) split longitudinally and the two sets move to the two opposite poles forming two daughter nuclei which immediately divide into four.

Pinus the seed-coat is provided with a membranous wing (FIG. 19B) which is formed from the ovuliferous scale. The seed-coat (integument) has three layers—the inner and the outer being thin and the middle stony. The fully developed embryo of the seed (FIGS. 19C & 20) consists of an axis with a hypocotyl, a radicle, and a tiny plumule with a number of cotyledons (2-15) surrounding it. Under suitable conditions the seed germinates (FIG. 20). The cotyledons are pushed upwards, germination being epigeal, and they turn green in colour, while the radicle grows downwards into a distinct tap root. The seedling thus becomes established.

Chapter 4 GNETALES

Among the gymnosperms the Gnetales have reached the summit of development (evolution) and bear a close resemblance to the angiosperms. Gnetales, however, cannot be regarded as the ancestors of the angiosperms. The living Gnetales are so much advanced and so highly specialized and the fossil forms so few that their origin from any particular group of gymnosperms cannot be traced. They may have originated from an extinct group of gymnosperms and have followed a parallel line of evolution with the angiosperms. The order Gnetales is represented by three families, each with a single genus: (a) *Gnetaceae*, e.g. *Gnetum*, (b) *Ephedraceae*, e.g. *Ephedra*, and (c) *Welwitschiaceae*, e.g. *Welwitschia*. *Gnetum* (35 sp.) is widely distributed in tropical Asia, Africa and South America; 2 sp. of *Gnetum*, viz. *G. gnemon* and *G. montanum*, are fairly common in Eastern India; *G. gnemon* grows in the Khasi Hills and Manipur Hills, while *G. montanum* in eastern tropical Himalayas, Assam and Deccan. *Ephedra* (35 sp.) is widely distributed in Mediterranean, Eastern Asia, and North and South America; in India *E. vulgaris* grows in temperate and alpine Himalayas at an elevation of 2,100-3,700 m.; (in Sikkim at 3,700-4,800 m.); *E. platyclada* in western Himalayas at or below 4,700 m.; *E. peduncularis* in the salt ranges of Punjab, Rajputana and Sind (Pakistan). *Welwitschia* (1 sp.), viz. *W. mirabilis*, has been found in South-West tropical Africa.

GNETUM (35 sp.)

Gymnospermic Characters of *Gnetum* (1) Ovules are naked, i.e. not enclosed within the ovary. (2) In pollination the wind-borne pollen grains are lodged directly on the ovule, there being no style, stigma or ovary. (3) Male and female strobili are of gymnospermic types although the flowers are more advanced with the development of the perianth. (4) The generative cell divides and produces a prothallus cell and a generative cell, the latter forming two male gametes. (5) Anatomically there is preponderance of gymnospermic tracheids with bordered pits. (6) Vascular bundles are in successive concentric rings, as in some cycads.

Resemblances with Angiosperms. *Gnetum* resembles angiosperms in many features—vegetative, anatomical and reproductive. (1) *Gnetum* bears well-developed broad evergreen leaves with distinct reticulate venation, hardly distinguishable from those of angiosperms (FIG. 21). (2) The general climbing habit of *Gnetum* is more angiospermic than gymnospermic. (3) There are true vessels (of angiospermic type) present in the secondary wood in addition to the tracheids (of gymnospermic type). (4) There is a perianth present in both the male and the female flowers. (5) There are two integuments surrounding the ovule. (6) Archegonia are altogether wanting, as in the angiosperms. (7) The female gametophyte with many free nuclei has a close resemblance to the embryo-sac of an angiosperm. (8) The stamen (microsporophyll) with a stalk (filament) and 1 or 2 anthers resembles that of an angiosperm. (9) The male gametophyte produces no prothallus cells, as in all angiosperms. (10) The endosperm is formed after fertilization. (11) The embryo has two cotyledons.

Life-history. *Gnetum* is found only in the tropics with 2 common species in India, i.e. *G. montanum* and *G. gnemon*, both being widespread in Assam. The species of *Gnetum* are mostly woody climbers (e.g. *G. montanum*) or shrubs or small trees (e.g. *G. gnemon*). Leaves are simple, decussate, broad, evergreen, leathery, lanceolate-ovate and distinctly net-veined (suggestive of dicotyledons). The primary stem often produces two kinds of shoots—the long and the short, the latter bearing one to a few pairs of leaves. Anatomically the primary vascular bundles are formed in a ring, and the secondary vascular bundles are formed in successive concentric rings by successive cortical cambia, the primary cambium being short-lived. The secondary xylem is made of true vessels (as in angiosperms) associated with gymnospermic tracheids with bordered pits. The peculiarity with the phloem is that the sieve-tube and the companion cell are formed from two separate cells (and not by the division of a single cell, as in angiosperms). Resin-ducts are absent unlike other gymnosperms.

Gnetum. FIG. 21. A branch with leaves (reticulate venation).

GENTALES

Species of *Gnetum* are dioecious, one plant bearing male inflorescences (strobili) and another female strobili. The strobili are commonly branched (compound), pendulous and catkin-like, bearing numerous male or female flowers, as the case may be. The strobili are commonly axillary, sometimes terminal, and grow to a length of two to a few cm.

Male Strobilus. The male strobilus (FIG. 22A-B) is a slender axis growing between two connate bracts at the base, and higher up at short intervals bracts appear in several whorls. The bracts of a whorl, however, fuse together at an early stage to form a cup-shaped structure called 'collar'. From a special tissue on the

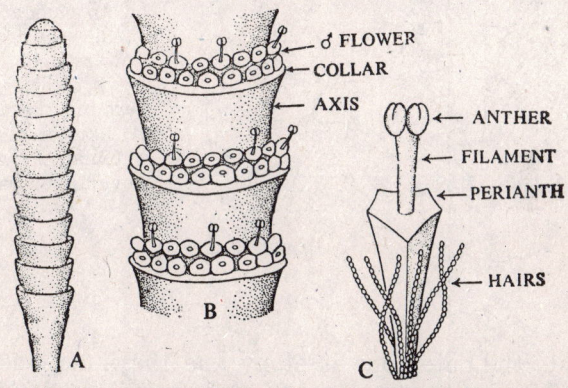

Gnetum. FIG. 22. *A*, a male strobilus; *B*, a portion of the same (magnified) showing whorls of male flowers; *C*, a male flower.

lower surface of the collar develop several minute male flowers (apparently axillary) in 2-5 whorls round the axis. There may be a whorl of sterile female flowers (ovules) interposed between the whorls of male flowers. Each male flower (FIG. 22C) consists of a sheathing perianth (two segments fused into a tube), often surrounded at the base by jointed hairs, and a single stamen or microsporophyll. The latter consists of a stalk (filament) and two unilocular anthers containing pollen grains or microspores. Each anther opens by a terminal slit. The presence of the perianth is an angiospermic character.

Female Strobilus (FIG. 23A-B). This is also a slender axis, branched or unbranched, growing between two bracts at the base. The female flowers, each represented by a single erect ovule, are arranged in successive whorls round the axis in the axils of connate cup-shaped bracts (two bracts are fused early into a cup). There are 4-10 female flowers or ovules in each whorl. Each ovule (FIG. 23C) is surrounded by a fleshy perianth (two outgrowths appearing from the base of the ovule get fused at a very early stage) forming the outer covering. The perianth finally turns orange-red in colour in the seed. The ovule (FIG. 25A) consists of two envelopes or integuments and a distinct nucellus or megasporangium, and is orthotropous in nature. Of the two

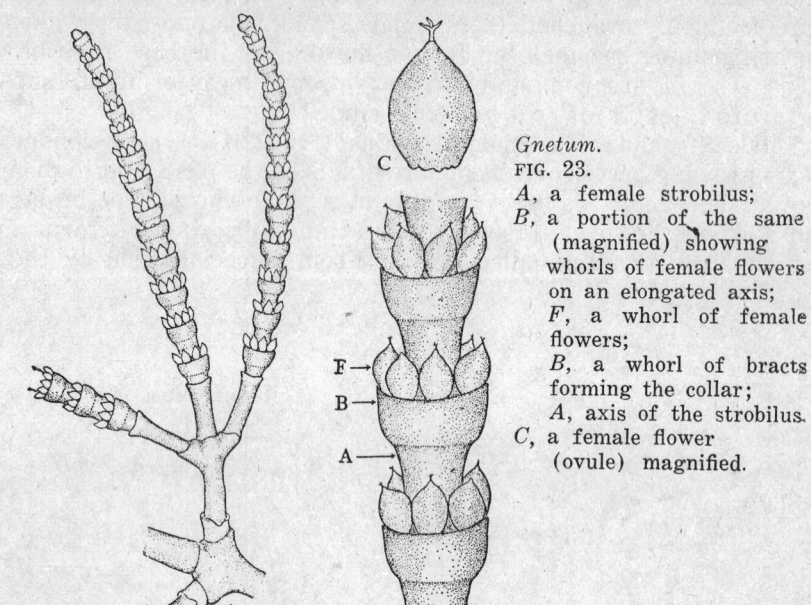

Gnetum.
FIG. 23.
A, a female strobilus;
B, a portion of the same (magnified) showing whorls of female flowers on an elongated axis;
F, a whorl of female flowers;
B, a whorl of bracts forming the collar;
A, axis of the strobilus.
C, a female flower (ovule) magnified.

integuments the outer one is stony and the inner one projects beyond the perianth into a sort of style or micropylar tube with its tip lobed or fimbriated. The embryo-sac or megaspore lies embedded in the nucellus towards the chalazal end. Beneath this a fan-shaped nutritive tissue, called the *pavement tissue,* formed of radiating rows of cells, develops in the nucellus. It, however, becomes disorganized after fertilization.

Male Gametophyte (FIG. 24). Pollen grains often germinate while still in the micropylar chamber. Some of them, however, directly reach the nucellus and germinate there. The pollen grain on germination gives rise to a very simple type of male gametophyte. The stages in the development of the male gametophyte, as worked out by Negi and Madhulata in 1957, are as follows (FIG. 24). The nucleus of the pollen grain or microspore (A) divides at first into two nuclei (B). The smaller one of these two is the **prothallus cell** (P), and the larger one the **generative cell** (G). The latter divides again into two nuclei (C), one of which is the vegetative nucleus or **tube nucleus** (T), while the other one is the generative cell (G). The former is without a wall, while the latter clothes itself with a distinct cell-wall. Thus the pollen grain at this stage shows the presence of a prothallus cell (P), a tube nucleus (T) and a generative cell (G). It will be noted that the earlier view regarding the presence of a stalk cell

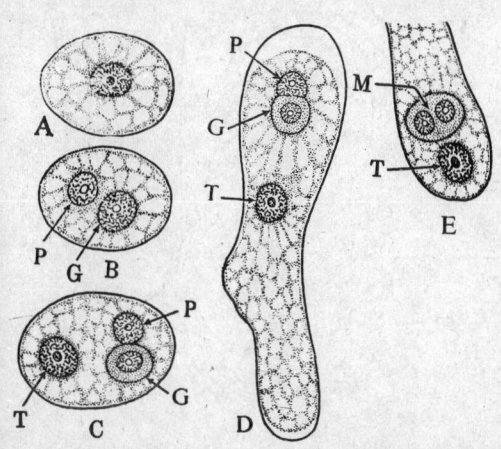

Gnetum.
FIG. 24.
Microspore and stages in the development of the male gametophyte.
A, a microspore;
B, the same after the first division;
C, the same after the second division;
D, pollen-tube formed;
E, lower end of the same;
P, prothallus cell;
G, generative cell;
T, tube nucleus;
M, male gametes.

Redrawn after FIG. 392 *in* Gymnosperms; Structure and Evolution by *C. J. Chamberlain by permission of The University of Chicago Press. Copyright* 1935.

and a body cell and the absence of the prothallial cell has since been discarded. The pollen grain sheds at this stage, and further development takes place after pollination. The pollen grains are carried over to the micropyle of the ovule by wind, and are drawn into the pollen chamber. Now the exine of the pollen grain bursts, and the intine grows into a pollen tube (*D*). The generative cell divides and produces two **male gametes** (*E*) which migrate to the tip of the pollen-tube.

Female Gametophyte (FIG. 25*B*). The megaspore mother cell divides and produces four megaspores, one or more of which may be functional. The megaspore germinates with repeated free nuclear divisions and gives rise to the embryo-sac or female gametophyte which lies deep-seated in the nucellus. One or more of the free nuclei, particularly at the micropylar end, may be organized with a mass of protoplasm around them into potential egg-nuclei, and may be fertilized. No archegonium is formed in *Gnetum* (cf. angiosperms.) In some species, as in *G. gnemon*, a multinucleate (later uninucleate) cellular tissue (prothallial tissue), interpreted as homologous with the antipodal cells of angiosperms, is formed at the basal part of the female gametophyte (FIG. 25*C*); after fertilization it fills up the whole of the gametophyte. In gymnosperms the prothallial tissue is designated as the endosperm. It is noticeable that the female gametophyte of *Gnetum* with many free nuclei and without any trace of archegonium has almost reached the angiosperm level.

Fertilization and Development of the Embryo. One or more pollen-tubes penetrate through the micropylar chamber and the

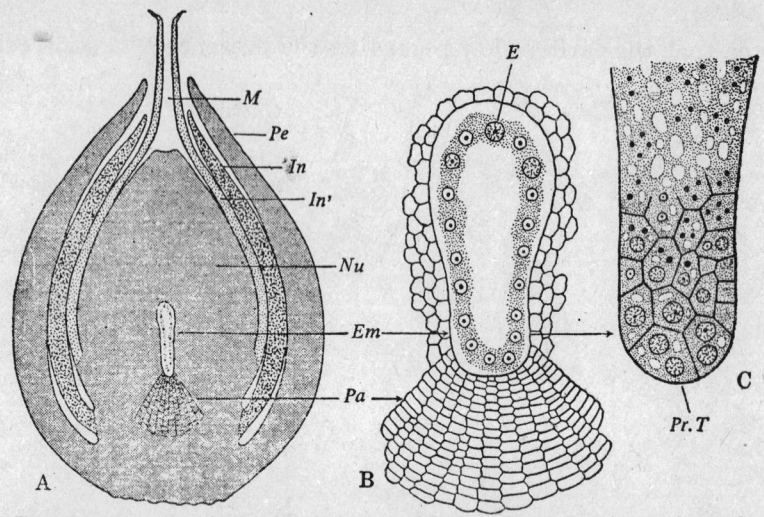

Gnetum. FIG. 25. *A*, ovule in longitudinal section showing micropylar tube (*M*), perianth (*Pe*), outer integument (*In*), inner integument (*In'*), nucellus (*Nu*), embryo-sac or female gametophyte (*Em*) and pavement tissue (*Pa*); *B*, female gametophyte (magnified) showing many free nuclei, a few eggs (*E*) and pavement tissue (*Pa*) beneath; *C*, development of prothallial tissue (*Pr. T*) at the base of the female gametophyte. *Redrawn after Figs.* 391*G*, 395 *and* 394 *in* Gymnosperms: Structure and Evolution *by C. J. Chamberlain by permission of The University of Chicago Press. Copyright* 1935.

nucellus. Finally the tubes reach the female gametophyte (FIG. 26A) and enter into it. The male gametes are discharged through a terminal pore in the pollen-tube. Both the male gametes are functional, and fertilization takes place by the fusion of any of them with any free egg-nucleus. Each fertilized egg-nucleus clothes itself with a wall and becomes the oospore (FIG. 26*B*). Some of the unfertilized nuclei divide and form a prothallial tissue (endosperm), while others become disorganized. The endosperm grows quickly and soon invades the whole of the nucellus space. In *Gnetum* most of the endosperm tissue is formed after fertilization. Many oospores or even embryos may be formed but ultimately one embryo comes to maturity. In the formation of the embryo the oospore divides and forms at first a two-celled **proembryo** (FIG. 26*C*). Each cell of the proembryo grows into a long slender tubular **suspensor** (FIG. 26*D*). Its nucleus divides into two; one of these two undergoes free nuclear divisions forming a group of four free nuclei at the end of the suspensor (FIG. 26*E*), while the other nucleus does not divide. These four nuclei undergo further divisions, and with the formation of walls round them a multicellular embryo (with uninucleate cells) is produced (FIG. 26*F*). In *Gnetum gnemon*, however, the embryo formation, as worked out in detail, shows some difference, as follows. In it the fertilized egg produces a single long tortuous

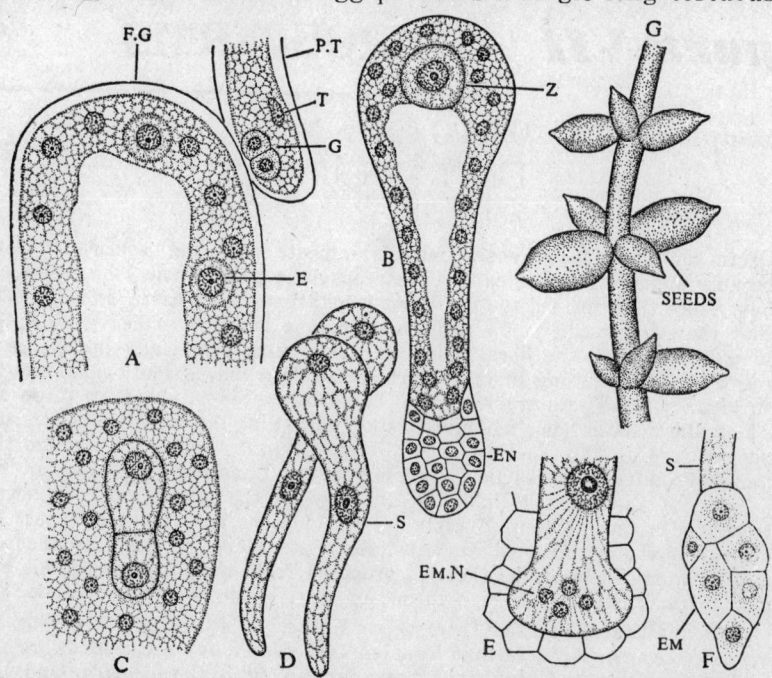

Gnetum. FIG. 26. *A*, pre-fertilization stage; female gametophyte (*F.G*) and pollen-tube (*P.T*); *E*, egg-nucleus; *T*, tube-nucleus; *G*, male gametes; *B*, female gametophyte after fertilization showing zygote (*Z*), free nuclei, and development of endosperm (*EN*); *C*, proembryo; *D*, two suspensors (*S*) developing from the 2-celled proembryo; *E*, end of a suspensor showing four embryonal nuclei (*EM.N*); *F*, embryo (*EM*) and suspensor (*S*); *G*, female strobilus with groups of seeds. FIG. 25*A-F* *redrawn after Fig. 395 in* Gymnosperms: Structure and Evolution *by C. J. Chamberlain by permission of The University of Chicago Press. Copyright* 1935.

suspensor in which free nuclear divisions in both earlier and later stages of embryo formation are more frequent. It is noted that at the tip of the suspensor a cell is cut off, containing a single nucleus associated with numerous starch grains; this is the **embryonal cell.** The nucleus in it undergoes free nuclear divisions, and then with the formation of walls round the free nuclei a multicellular embryo, as in other species, is developed.
Seed. Although there are many ovules, not many seeds are formed. The seed (FIG. 26*G*) is albuminous; its outer covering or perianth is fleshy and orange-red in colour; of the two integuments the outer one is hard and stony; and the embryo is with two cotyledons (dicotyledonous type).

part VII ANGIOSPERMS

Chapter 1 ORIGIN AND LIFE-HISTORY

Origin and Evolution. Fossil records indicate that angiosperms suddenly became abundant in the early Cretaceous in a great variety of forms and have since then formed the dominant vegetation of the earth. It is probable that they evolved in north temperate regions but due to heavy glaciation in a later period the flora is lost for ever with practically no chance of recovering the missing links. A part of the flora moved southward, and the survivors of the catastrophe, which could be traced in the form of fossils, e.g. willow, beech, elm, fig, laurel, etc., show that the angiosperms of that period were already well advanced. It may thus be taken for granted that they originated much earlier, probably in the Triassic or even earlier. But because of the missing links the ancestry of the angiosperms cannot be traced with any amount of certainty. Evidently views in this respect are divergent and speculative, the available data being meagre, fragmentary and isolated. Some of the theories proposed from time to time in this connexion, unsatisfactory though in one or more respects, are as follows:

1. **Bennettitales-Ranales Theory.** (*a*) Hallier's view (1906) regarding the origin of Angiosperms is that Ranales (e.g. *Magnolia*) seems to be related to Bennettitales and may have been derived from *Cycadeoidea*, and that the monocotyledons are an offshoot of the dicotyledons. The elongated floral axis of Ranales with spirally arranged male and female sporophylls and the cone of *Cycadeoidea* are definitely alike. Both the groups were also abundant in the Cretaceous. Ranales is regarded as the earliest stock from which the polycarpic families of dicotyledons have arisen, and also the monocotyledons as an offshoot. But the differences in the anatomical structure of the wood, types of sporophylls, nature and position of ovules, etc., in the two groups (Bennettitales and Ranales) have led to difficulties in accepting this view. It is more likely that both the groups have evolved from a common ancestry and developed in unrelated parallel lines. (*b*) Arber and Parkin (1907), while strongly supporting Hallier, proposed the existence of an imaginary group (Hemiangiospermae) having cycadeoid type of flowers, linking the above two groups. From this imaginary group the Ranalian type of flowers might have originated, and from this Ranalian type all other angiosperms had sprung up. The anatomical structure of the wood, however, does not support the Bennettitalean origin of the angiosperms. They also found no valid reason for the assumption that the monocotyledons had originated from the dicotyledons. As a matter of fact, they held the view that the earliest monocotyledons were more primitive than the dicotyledons. (*c*) Hutchinson (1925) considered the origin of angiosperms as *monophyletic*, and supported the views of Hallier, and Arber and Parkin. He believed in the Bennettitalean origin of angiosperms and laid stress on two parallel evolutionary lines for the primitive dicotyledons—a woody (arborescent) line, called **Lignosae**, starting from Magnoliales, and a herbaceous line, called **Herbaceae** starting from Ranales (see p. 701)

ORIGIN AND LIFE-HISTORY

He further held the view that the monocotyledons were derived from a primitive dicotyledonous order—the Ranales.

2. **Coniferae-Amentiferae Theory.** Engler and Prantl (1924) rejected the Cycadeoidean origin of angiosperms, as proposed by Hallier earlier. They held the view that dicotyledons and monocotyledons had arisen independently from a hypothetical group of extinct gymnosperms (allied to Coniferae) with unisexual strobilus, which developed in the Mesozoic. Thus, according to them, the angiosperms had had a *polyphyletic* origin, and evolution proceeded on several parallel lines from the beginning. They further considered the monocotyledons to be more primitive than the dicotyledons. The unisexual naked (without perianth) condition of the angiospermic flowers, as exemplified by Pandanales (monocotyledons) and Amentiferae (catkin-bearing dicotyledons, e.g. *Casuarina, Salix, Betula*, etc.) was most primitive. But according to the latest classification these orders are regarded as definitely more advanced.

3. **Gnetales-Casuarinales Theory.** Wettstein (1935) held the view that angiosperms of *Casuarina* type evolved from Gnetales (particularly *Ephedra*), a highly advanced group of gymnosperms, which branched off from the main gymnospermic line. There are, no doubt, some angiospermic features in Gnetales (see *Gnetum*, p. 681) but this group, as far as is known from meagre fossil records, has not gone earlier than the Tertiary. Fagerland (1947) was, however, of the view that both Gnetales and Proangiosperms had had a common ancestor and that the modern angiosperms evolved from the latter group in polyphyletic lines.

4. **Caytoniales-Angiosperm Theory.** Thomas (1925) suggested that Caytoniales, a small group of angiosperm-like plants, discovered in the Jurassic rocks of Yorkshire, might be the ancestor of angiosperms. Harris (1932-33), another investigator on Caytoniales, was, however, opposed to this view. Knowlton (1927) in his book *Plants of the Past* expressed the view that both Caytoniales and angiosperms evolved from the large extinct Palaeozoic group of pteridosperms. Arnold (1948) in his review of gymnosperms expressed the view that Caytoniales were definitely allied to the pteridosperms rather than to the angiosperms. In fact, according to him, they were the Mesozoic remnants of the Palaeozoic pteridosperms.

5. **Pteridosperm-Angiosperm Theory.** In reviewing the 'ancient plants' Andrews (1947) opined that 'seed-ferns' or pteridosperms, an ancient group of the Palaeozoic, might be the starting point for the angiospermic plants. Arnold in the same year expressed a similar view. But it must be noted that the absence or scarcity of angiospermic fossils earlier than the Cretaceous, connecting them with the pteridosperms, has led to a mere speculation rather than to any convincing proof. Pteridosperms might be the ancestors of cycads and true ferns but it is doubtful if they could represent the ancestry of the angiosperms.

There are other theories too on the subject but we are still far from any solution of this baffling problem. The consensus of opinion is that the angiosperms have evolved from some extinct group which has yet to be traced.

It must have been noted from the above that the question whether angiosperms are **monophyletic** (i.e. derived from a single ancestral stock) or **polyphyletic** (i.e. derived from several ancestral stocks) could not be satisfactorily solved yet. Generally speaking, the diversities of forms of the primitive families of both dicotyledons and monocotyledons and their early wide distribution suggest polyphyletic origin of the early angiosperms. On the other hand, certain common and constant characters of early angiosperms, such as 8-nucleate embryo-sac, formation of endosperm after fertilization as a result of triple fusion, types of flowers, stamens, carpels, ovules, etc., suggest monophyletic origin.

Life-cycle (FIG. 1) The life-cycle of a plant is the series of progressive changes it undergoes, usually from the stage of zygote to the same stage (zygote) again. In the life-cycle of an angiosperm there are certain special features, as compared with cryptogams. Development of the pollen-tube, the seed and the $3n$ endosperm, absence of antheridia and archegonia, double fertilization, complexity of the sporophyte and extreme reduction and loss of independence of the gametophyte are some such features. To start with, the zygote quickly grows into an embryo, and the latter gradually into a full-fledged plant with roots, stem, leaves and flowers. The flower bears stamen or microsporophyll and carpel or megasporophyll. The stamen bears pollen-sac or microsporangium and the carpel bears nucellus or megasporangium within the ovule which again develops within the ovary. The pollen-sac and the nucellus in their turn produce the pollen (or microspore) mother cell and the embryo-sac (or megaspore) mother cell. By reduction division the pollen mother cell gives rise to the pollen grains or microspores, and the embryo-sac mother cell to the megaspore (of the four megaspores formed three degenerate). The pollen grain germinates on the stigma of the carpel (pistil) and gives rise to the male gametophyte (the germi-

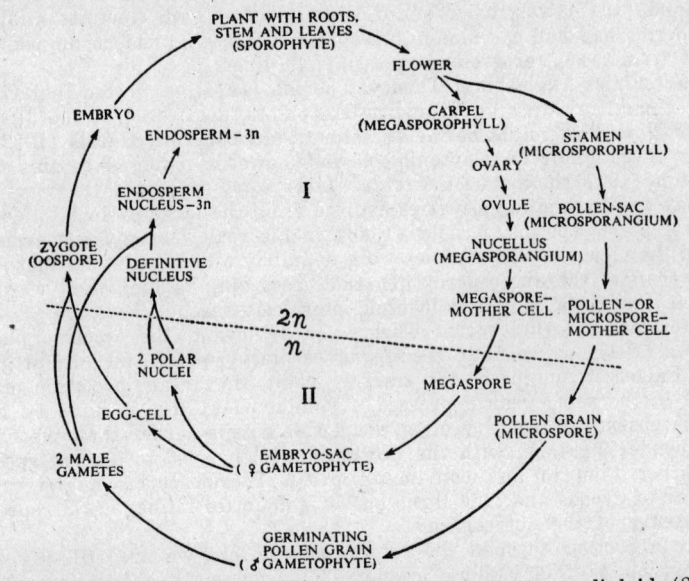

FIG. 1. Life-cycle of an angiosperm. *I*, sporophytic or diploid ($2n$) generation; *II*, gametophytic or haploid (n) generation.

nating pollen grain—the pollen-tube with the three nuclei in it, i.e. the tube-nucleus and the two male gametes), and the megaspore to the female gametophyte (the embryo-sac with the eight nuclei). Both the gametophytes are extremely reduced in size. The female gametophyte is completely *endosporous* and is entirely dependent on the mother sporophyte for protection and nutrition. Antheridia and archegonia are altogether absent in the gametophytes. The male gametophyte bears two male gametes and the female gametophyte bears a female gamete or egg-cell. Pollination mainly takes place through the agency of insects, wind, etc., and this is followed by fertilization. One of the male gametes of the pollen-tube fuses with the egg-cell of the embryo-sac and the other male gamete fuses with the two polar nuclei or their fusion product, i.e. the definitive nucleus with $2n$ chromosomes. As a result of this *double fertilization* rapid changes take place in the ovule and the ovary. The fertilized egg-cell becomes the zygote (oospore) which quickly grows into the embryo, the *triple-fusion* nucleus (two polar nuclei and one male gamete) into the endosperm (with $3n$ chromosomes), the ovule and the ovary into the seed and the fruit respectively. It will be noted that the zygote marks the end of the life-cycle. Sooner or later the next cycle begins from it (the zygote) and again passes through the same series of changes.

Alternation of Generations. In angiosperms the sporophyte has reached a high degree of complexity, while the gametophytes have become very small, simple and inconspicuous. The angiospermic plant is the sporophyte with $2n$ or *diploid* chromosomes. The embryo-sac with the eight nuclei and the germinating pollen grain, i.e. the pollen-tube with the three nuclei in it are the female and the male gametophytes respectively with n or *haploid* chromosomes. The endosperm with $3n$ chromosomes formed as a result of triple fusion is a unique structure in the angiosperm. All the stages from the zygote to the spore mother cells (mega- and micro-) represent the sporophytic generation (well advanced), and all the stages from the spores (mega- and micro-) to the gametes (male and female) represent the gametophytic generation (extremely reduced). These two generations (sporophytic and gametophytic) regularly alternate with each other to complete the life-cycle of an angiosperm.

Chapter 2 PRINCIPLES & SYSTEMS OF CLASSIFICATION

Systematic Botany or **Taxonomy.** This deals with the description, identification and naming of plants, and their classification into different groups according to their resemblances and differences, mainly in their morphological characteristics. So far as angiosperms or higher 'flowering' plants are concerned, it has been estimated that over 199,000 species (dicotyledons—159,000 and monocotyledons—40,000) are already known to us, and that many thousands more have yet to be discovered and recorded. Thus plants are not only numerous but are of varied types, and it is not possible to study them unless they are arranged in some orderly system. The object of systematic botany or taxonomy is to describe, name and classify plants in such a manner that their relationship with regard to their descent from a common ancestry may be brought out. The ultimate object of classification is to arrange plants in such a way as to give us an idea about their phylogenetic relationships, i.e. the sequence of their origin and evolution from simpler, earlier and more primitive types to more complex, more recent and more advanced types in different periods of the earth. The earlier classifications of plants were based on their economic uses, e.g. cereals, medicinal plants, fibre-yielding plants, oil-yielding plants, etc., or on gross structural resemblances, e.g. herbs, shrubs, trees, climbers, etc. These classifications were incomplete and fragmentary as plants that did not fit into such classifications or were of no economic value were usually ignored. An ideal system of classification should, therefore, not only indicate the actual genetic relationship but should also be within a reasonable limit of convenience for practical purposes.

Units of Classification

Species. A species is a group of individuals (plants or animals) having a very close resemblance with one another structurally and functionally. The individuals of a species interbreed in nature freely and successfully among themselves so that they may be taken as having descended from the same parent stock, and they also breed true to type, i.e. give rise to the progeny of the same kind. They have also normally the same number of chromosomes in their cells—$2n$ in somatic cells and n in reproductive cells. Thus all pea plants together constitute a species. Similarly all banyan plants, all peepul plants, and all mango plants constitute different and distinct species. Species, however, often show a wide range of variations, with many intermediate forms or gradations, serving as connecting links between two allied species. It has, therefore, not been possible to draw a line of demarcation between one species and another allied one, nor to delimit the boundary of any one particular species. This being so, it becomes very difficult to give an accurate and rigid definition of a species, and thus taxonomists have often differed regarding the latitude of a species. Nevertheless, the importance of species, however vaguely and arbitrarily defined, as fundamental units of classification, has never been ignored by the taxonomists. It will further be noted that all the characteristics of a species are attributable to the behaviour of DNA in the chromosomes. DNA carries the same genetic code from generation to generation, controls the development of the same set of parental characteristics in the offspring and thus normally limits the boundary of a species.

Sometimes under the influence of external conditions certain individuals of a species may show a marked degree of variation in form, size, shape, colour and other minor characteristics. Such plants are said to form **varieties**. A species may consist of one or more varieties or none at all. We have thus different varieties of common garden pea, rice, potato, mango, etc. Varieties, however, are not permanent. They tend to revert to the original species from which they were derived.

Genus. A genus is a collection of species which bear a close resemblance to one another in the morphological characters of the floral or reproductive parts. For example, banyan, peepul and fig are different species because they differ from one another in their vegetative characters such as the habit of the plant, the shape, size and surface of the leaf, etc. But these three species are allied because they resemble one another in their reproductive organs, namely, inflorescence, flower, fruit and seed. Therefore, banyan, peepul and fig come under the same genus and that is *Ficus*.

Nomenclature. A plant name has two parts. The first refers to the genus and the second to the species. This system of designating every type of plant with a *binomial*, i.e. a name consisting of two parts, is known as binomial nomenclature, and was first established by Linnaeus and finally settled by the International Botanical Congress[1] held at Amsterdam in 1935. Thus pea has received the name of *pisum sativum*, rice *Oryza sativa*, mango *Mangifera indica*, banyan *Ficus bengalensis*, peepul *Ficus religiosa*, fig *Ficus carica*, and india-rubber plant *Ficus elastica*. Referring to cottons, we find that they all belong to the same genus *Gossypium* which consists of about 20 or more species such as BANI cotton of India (*Gossypium indicum*), KUMPTA cotton of the southern Maratha country (*G. herbaceum*), American cotton (*G. barbadense*), KIL cotton of Assam (*G. cernuum*), BURI cotton—the upland American cotton naturalized in India (*G. hirsutum*) and so on. The name of the author who first described a species is also written in an abbreviated form after the name of the species, e.g. *Mangifera indica* Linn. Here Linn refers to the author 'Linne' or 'Linnaeus' who first described the plant.

Family. A family is a group of genera which show general structural resemblances to one another mainly in their floral organs. Thus in the genera *Gossypium, Hibiscus, Thespesia, Sida, Abuti-*

---

[1]International Rules of Botanical Nomenclature formulated by the International Botanical Congress at Vienna 1905, Brussels 1910, Cambridge 1930, Amsterdam 1935, and further revised by the American Society of Plant Taxonomists 1946-47, Stockholm 1950, Paris 1954 to synchronize with the centenary of the Botanical Society of France (1854-1954), Montreal 1959, Edinburgh 1964, Seattle 1969, Leningrad 1975, and Canberra 1980.

lon, Malva, etc., we find free lateral stipules, epicalyx, twisted aestivation of corolla, monadelphous stamens, unilocular anthers, axile placentation, etc. So all the above-mentioned genera belong to the same family, and that is *Malvaceae*.

Systems of Classification

Systems of classification may be artificial, natural or phylogenetic. Systems have developed in a stepwise manner from the earliest to the present times, and four periods depending on criteria used in classification may be recognized. *1st period*: this extends over several centuries from about 300 B.C. to the end of the 17th century; during this period systems were based on the **habit** of plants. *2nd period*: this extends from the thirties to the end of the 18th century; this period records a definite change in the pattern of systems; the habit has given place to **sex organs** (stamens and carpels) of plants, and Linnaeus' sexual system of classification (1735) is specially remarkable. *3rd period*: this extends from the beginning to the eighties of the 19th century; there are several famous botanists of this period; systems propounded during this period have been based on **natural relationships** of forms; Bentham and Hooker's natural system of classification (1862-83) is of outstanding merit. *4th period*: this period extends from 1875 (possibly 1883) to the present times; systems of this period have been based on **phylogenetic relationships** of plants.

Artificial System. In the artificial system only one or at most a few characters are selected arbitrarily and plants are arranged into groups according to such characters; as a result closely related plants are often placed in different groups, while unrelated plants are often placed in the same group because of the presence or absence of a particular character. This system enables us to determine readily the names of plants but does not indicate the natural relationship that exists among the individuals forming a group. It may thus be compared to the manner of arrangement of words in a dictionary in which, except for the alphabetical order, adjacent words have no necessary agreements with one another. Thus the defect in the artificial system of classification is that plants resembling one another closely, instead of being grouped together, become widely separated. Nevertheless, the artificial system is of very great advantage in view of the fact that by following this system of classification one can without much difficulty get to the name of an unknown plant; or in other words, the identification of an unknown plant is rendered much easier by this system.

Linnaean System (1735). The best-known artificial system is that promulgated by Linnaeus (1707-78) and published by him in the year 1735, further revised in his *Genera Plantarum* (1737) and *Species Plantarum* (1753). By 1760 his system became popular in Holland and Germany and partly in England. Linnaeus classified plants according to the characters of reproductive organs, viz. stamens and carpels. Since these are regarded as the sexual organs of plants, this artificial system of Linnaeus is commonly called the 'Sexual System'. According to this system plants are mainly divided into 24 classes including 23 of phanerogams and one of cryptogams. Phanerogams were further subdivided into groups with unisexual or bisexual flowers. Plants with unisexual flowers were again divided according to whether they were monoecious or dioecious. Further classification was based on the number of stamens. Plants with bisexual flowers were classified according to whether the stamens were united with the carpels, or were free from them. The next consideration was whether the stamens were free or united. Then the number of stamens, their length, and ultimately the number of carpels were taken into account.

Natural System. In the natural system all the important characters are taken into consideration, and plants are classified according to their related characters. Thus, according to their resemblances and differences, mostly in their important morphological characters, plants are first classified into a few big groups. These are further divided and subdivided into smaller and smaller groups until the smallest division is reached and that is a species. All modern systems of classification are natural and supersede the artificial ones by the fact that on the one hand they give us a true idea of the natural relationship existing between different plants based on a detailed description of characters and also of the sequence of their evolution from simpler to more complex types in different periods of the earth, and on the other they meet, like the artificial system, the practical need of identification of unknown plants. Plants as arranged or grouped together according to these systems are further seen to possess in most cases the same or identical properties.

According to the natural system the plant kingdom has been divided into two *divisions*, viz. **cryptogams** or 'flowerless' plants (see part V) and **phanerogams** or 'flowering' plants. Phanerogams have again been divided into two *sub-divisions*, viz. **gymnosperms** or naked-seeded plants (see part VI) and **angiosperms** or closed-seeded plants. Angiosperms have further been divided into two *classes*, viz. **dicotyledons** and **monocotyledons**. The classes have been divided into *orders*; the orders into *families*;

families again into *genera* and genera into *species*; and sometimes species into *varieties*. If a greater number of intermediate categories are required, the prefix *sub* is added to the particular term.

Bentham and Hooker's System (1862-83). Bentham (1800-84) and Hooker (1817-1911), two English botanists, contemporaries of Darwin, elaborated a natural system of classification of 'flowering' plants (dicotyledons, gymnosperms and monocotyledons) giving more or less equal status to all the three groups. Their system is based on (virtually an extension of) de Jussieu's system (1789) and de Candolle's system (1819). Their system was published in *Genera Plantarum* in which they discussed dicotyledons first, then gymnosperms and finally monocotyledons. They divided dicotyledons into 3 sub-classes and 14 series; series into cohorts (equivalent to orders), and cohorts into orders (equivalent to families). There are altogether 202 orders of angiosperms—165 orders of dicotyledons and 37 orders of monocotyledons. They started with the family *Ranunculaceae* (with polypetalous corolla and indefinite number of free stamens and carpels) and ended in *Labiatae* (with gamopetalous corolla and a definite number of stamens and carpels), leaving out Incompletae and an anomalous series. They divided monocotyledons into 7 series, series directly into orders, and started with orders having epigynous flowers, e.g. *Orchidaceae* and *Scitamineae*, passed through orders with petaloid hypogynous flowers, e.g. *Liliaceae*, and then through orders with flowers which have lost their petaloid character, e.g. *Palmae* and *Araceae*, and finally to *Gramineae* and *Cyperaceae* with simple construction of flowers and presence of glumes. An outline of this system of classification is as follows.

CLASS I. DICOTYLEDONES

Sub-class I. **Polypetalae.** Flowers with both calyx and corolla; corolla polypetalous; both stamens and carpels present. Within the sub-class the progress is indicated through polysepalous calyx to gamosepalous calyx, through indefinite number of stamens to definite number, and through hypogyny, perigyny and epigyny. It is divided into 3 series.

Series (*i*) **Thalamiflorae.** Calyx polysepalous, free from the ovary; stamens inserted on the thalamus, hypogynous; ovary superior. It has 11 cohorts and 57 orders, beginning with Ranales and ending in Malvales.

Series (*ii*) **Disciflorae.** Calyx poly- or gamosepalous, free from or adnate to the ovary; disc usually conspicuous; stamens usually definite, inserted on or around the disc; ovary superior. It has 5 cohorts and 21 orders (+2 anomalous orders), beginning with Geraniales and ending in Sapindales.

Series (*iii*) **Calyciflorae.** Calyx gamosepalous, often adnate to the ovary; stamens perigynous or epigynous, inserted on the disc; ovary superior or inferior. It has 5 cohorts and 27 orders, beginning with Rosales and ending in Umbellales.

Sub-class 2. **Gamopetalae.** Flowers with both calyx and corolla; corolla gamopetalous; stamens almost always definite and epipetalous; carpels usually two or sometimes more, often united; ovary superior or inferior. This sub-class is also called Corolliflorae. It is divided into 3 series.

Series (*i*) **Inferae.** Ovary inferior; stamens usually as many as the corolla-lobes. It has 3 cohorts and 9 orders, beginning with Rubiales and ending in Campanales.

Series (*ii*) **Heteromerae.** Ovary usually superior; stamens as many or twice as many as the corolla-lobes; carpels more than two. It has 3 cohorts and 12 orders, beginning with Ericales and ending in Ebenales.

Series (*iii*) **Bicarpellatae.** Ovary usually superior; stamens as many as or fewer than the corolla-lobes; carpels two. It has 5 cohorts and 24 orders, beginning with Gentianales and ending in Lamiales.

Sub-class 3. **Monochlamydeae** or Incompletae or Apetalae. Flowers incomplete; either calyx or corolla absent or sometimes both absent; flowers bisexual or unisexual. It includes the orders (families) which do not fall under any of the above two sub-classes. It is divided into 8 series and 36 orders, beginning with *Nyctagineae, Amarantaceae*, etc., and ending in *Casuarineae, Cupuliferae, Salicaceae* and *Ceratophylleae*.

CLASS II. MONOCOTYLEDONES

Series (*i*) **Microspermae** (inner perianth petaloid; ovary inferior, with 3 parietal placentae; seeds very minute, exalbuminous) with 3 orders—*Hydrocharitaceae, Burmanniaceae* and *Orchidaceae*, etc.

Series (ii) **Epigynae** (perianth petaloid, partly at least; ovary often inferior; seeds with copious endosperm) with 7 orders—*Scitamineae, Irideae, Amaryllideae*, etc.

Series (*iii*) **Coronarieae** (perianth partly petaloid; ovary superior; seeds with copious endosperm) with 8 orders—*Liliaceae, Commelinaceae*, etc.

Series (*iv*) **Calycinae** (perianth sepaloid; ovary superior; endosperm copious) with 3 orders—*Palmae*, etc.

Series (*v*) **Nudiflorae** (perianth absent or represented by scales; ovary superior; carpels 1-∞, syncarpous; endosperm usually present) with 5 orders—*Pandaneae, Typhaceae, Aroideae, Lemnaceae*, etc.

Series (*vi*) **Apocarpae** (perianth in 1 or 2 whorls or absent; ovary superior, apocarpous; endosperm absent) with 3 orders—*Alismaceae, Naiadaceae*, etc.

Series (*vii*) **Glumaceae** (flowers solitary, sessile, in the axils of bracts or glumes, in spikelets; perianth absent or modified into scales; ovary 1-locular, 1-ovuled; endosperm copious) with 5 orders—*Cyperaceae, Gramineae*, etc.

Merits. (1) Bentham and Hooker's system is a masterpiece of work on systematic botany. Some even go so far as to say that this system should have been further elaborated rather than replaced. (2) The placing of Ranales at the beginning of the system is very reasonable. This is also Hutchinson's view. Engler, however, holds a different view. (3) It is a natural system, and for its practical utility it is widely followed. (4) Monocotyledons are placed after dicotyledons, but the interpolation of gymnosperms in between them is an anomaly. (5) Position of many cohorts (orders), though not the series, of Monochlamydeae is natural, e.g. Cactales is regarded by them as well as by Hutchinson as related to Passiflorales, while Engler places the group near Myrtiflorae. (6) The system ends in *Verbenaceae* and *Labiatae* (leaving out Monochlamydeae). This view is shared by Hutchinson also, but not by Engler.

Demerits. (1) The greatest drawback of the system is the unfortunate introduction of the group Monochlamydeae. It is as a whole, except its first series Curvembryeae, regarded as an artificial group. In fact it has received the refuse of the other two sub-classes. Among achlamydeous families there are many unrelated types. Achlamydeous condition is now regarded as one of reduction rather than of progress. It is largely upon this that the system has been criticized by later systematists. So this group has been broken up, and the families redistributed. (2) In this system often characters have been selected arbitrarily so that the related orders have become separated, often widely. (3) Microspermae (*Burmanniaceae* and *Orchideae*) is placed at the beginning of monocotyledonous orders; while Engler and Hutchinson have shifted them to the end. (4) The distinction between *Liliaceae* (with superior ovary) on the one hand and *Irideae* and *Amaryllideae* (with inferior ovary) on the other is no longer held valid on the basis of one character. Engler has brought them together under *Liliflorae*. Hutchinson has, however, split it up into a number of smaller orders. (5) *Scitamineae* precedes *Liliaceae* and is placed together with *Bromeliaceae, Irideae* and *Amaryllideae*; while Engler raises it to the rank of an order with four families and places it after *Liliaceae* from which it may have been derived. Hutchinson calls the order Zingiberales and includes a number of small families in it, and places it back to its former position.

Phylogenetic System. This is based on phylogenetic relationship of plants bearing on the concepts of evolution. The systems of classification proposed by Engler, a German botanist, in 1886, by Hutchinson, an English botanist, in 1926, and by Tippo, an American botanist, in 1942 are phylogenetic.

Engler's System. Engler's system is based on Eichler's system (1883), the major categories of that system having been accepted by Engler and his associates. Adolf Engler (1844-1930), Professor of Botany at the University of Berlin, first proposed his system in 1886 as a guide to the botanical garden at Breslau. The system was published in an expanded and elaborated form in *Die Naturlichen Pflanzenfamilien* (1887-1909) in 23 volumes covering the whole range of the plant kingdom, under the editorship of Engler and Prantl. Engler and Gilg's *Syllabus der Pflanzenfamilien* published in 1892 gives a comprehensive idea about the systematic classification of 'flowering' plants and cryptogams, and is a very useful publication. According to Engler's system the plant kingdom has been divided into 13 divisions, of which the seed-bearing plants (Spermatophyta) form the last division named 'Embryophyta Siphonogama'. This has been further divided into two sub-divisions—(1) Gymnospermae and (2) Angiospermae; the latter further divided into two classes—(*a*) Monocotyledoneae and (*b*) Dicotyledoneae. Monocotyledoneae has been directly divided into 11 orders with 45 families, while Dicotyledoneae into two sub-classes— (*i*) **Archichlamydeae** containing 30 orders with 190 families representing the lower dicotyledons, and (*ii*) **Sympetalae** (or Metachlamydeae) containing 10 orders with 53 families representing the higher dicotyledons. The orders have been further subdivided into sub-orders, families and genera. There are altogether 288 families. Monocotyledoneae begins with *Typhaceae* and ends in *Orchidaceae*, while Dicotyledoneae begins with *Casuarinaceae* and ends in *Compositae*.

The principle involved in this system of classification is tracing the increasing complexity of flowers, particularly their accessory whorls, viz. achlamydeous flowers (no perianth), haplochlamydeous flowers (one whorl of perianth) and diplochlamydeous flowers (two whorls of perianth). Those with no perianth or with only one whorl or with polypetalous corolla form the sub-class **Archichlamydeae**; gamopetalous condition with 1 or 2 whorls of perianth comes under the sub-class **Sympetalae**. The latter represents more advanced groups of dicotyledons. Within the sub-class progress is indicated through hypogyny, perigyny and epigyny, and from a variable number of stamens and carpels to a definite number. Among dicotyledons Engler starts with the family *Casuarinaceae* on the assumption that woody plants with unisexual apetalous flowers borne in catkins are the most primitive.

Merits. (1) Classification of the whole plant kingdom with necessary sketches, records of numbers of species, and keys for identification of all known genera of plants. (2) A natural system based on relationship and compatible with evolutionary principles. (3) A system where groups have been built up step by step to form a generally progressional morphological series. (4) Phylogenetic arrangement of many groups. (5) Abolition of Apetalae (Monochlamydeae) considering it an artificial group, and amalgamation of Polypetalae and Apetalae into a new group, i.e. sub-class Archichlamydeae. (6) *Compositae* and *Orchidaceae* are considered, very reasonably, the highest families of dicotyledons and monocotyledons respectively in view of their most highly evolved characters. Because of its merits Engler's system was accepted by American and British botanists.

Demerits. (1) Precedence of monocotyledons over dicotyledons. Higher dicotyledons and the monocotyledons are believed to have developed from some primitive dicotyledonous stock. (2) *Casuarinaceae* is considered most primitive among dicotyledons and placed at the beginning of his system; this view is no longer acceptable. (3) Amentiferae, i.e. the catkin-bearing families with unisexual flowers, e.g. *Fagaceae, Salicaceae, Betulaceae*, etc., are considered primitive among dicotyledons and placed early in his system. This group is now considered as advanced rather than primitive. Ranales like *Ranunculaceae, Nymphaeaceae*, etc., can never be derived from them. The primitive nature of Ranalean families is now almost universally accepted by the taxonomists. Centrospermae, e.g. *Amarantaceae, Caryophyllaceae*, etc., cannot precede Ranales. (4) Derivation of dicotyledons from gymnosperms with unisexual strobili. (5) The system as a whole cannot be regarded as phylogenetic. (6) Position of Helobiae, e.g. *Potamogetonaceae, Alismaceae Hydrocharitaceae*, etc., between Pandanales and Glumiflorae is very unsatisfactory; both Pandanales and Glumiflorae are advanced groups. (7) Similarly, *Araceae, Lemnaceae* and *Typhaceae* are supposed to be derived from *Liliaceae*, and, therefore, cannot precede it.

Hutchinson's System. John Hutchinson, formerly Director of Royal Botanic Gardens at Kew, England, is a leading exponent of a phylogenetic system of classification. He classified only the angiosperms in his famous work *The Families of Flowering Plants*, vol. I (dicotyledons) which appeared in 1926, and vol. II (monocotyledons) in 1934. The system has been revised in his *British Flowering Plants* published in 1948, and in the second edition of *The Families of Flowering Plants* published in 1959. His system differs from all other previous ones in several fundamental facts. It is, however, nearer to Bentham and Hooker's or Bessey's in some respects rather than to Engler's. The main features of his system are as follows.

The system is based on the logical interpretation of the theory that the parts of an angiospermic flower are modified leaves. He held the view that the angiosperms having strobilus-like, hypogynous, bisexual and polycarpellary flowers had had their origin in an extinct Mesozoic group of gymnosperms—the Bennettitales (e.g. *Bennettites=Cycadeoidea*). According to Hutchinson two primitive orders of dicotyledons had arisen independently from this ancestral stock through an imaginary group of proangiosperms, and developed in nature simultaneously on two distinct parallel lines (see p. 688). He thus considered the origin of angiosperms as *monophyletic*, and supported Hallier, and Arber and Parkin. Hutchinson emphasized the old view of woody (arborescent) and herbaceous types of primitive dicotyledons, of course, together with other related characters. Accordingly he recognized two distinct divisions proceeding on two phyletic lines—woody and herbaceous. On one line evolved the orders, fundamentally and predominantly *woody*, which he called **Lignosae**, and on the other line evolved the orders, fundamentally and predominantly *herbaceous*, which he called **Herbaceae**. The woody line started from the primitive order Magnoliales (family *Magnoliaceae*), passed through *Annonaceae, Lauraceae, Dilleniaceae, Rosaceae,* etc., and ended in families like *Apocynaceae, Rubiaceae, Verbenaceae,* etc. The hrbaceous line started from the primitive order Ranales (family *Ranunculaceae*), passed through *Nymphaeaceae, Cruciferae, Caryophyllaceae,* etc., and ended in families like *Umbelliferae, Compositae, Solanaceae, Scrophulariaceae, Labiatae,* etc. Hutchinson recognized several climax orders in the phylogeny of the dicotyledons and rearranged them in proper sequences. Orders having both woody and herbaceous forms were considered polyphyletic, i.e. members originating from different ancestors, as found in Apetalae (without perianth), Urticales, Umbellales, Euphorbiales, etc. He further held the view that the monocotyledons were derived from a primitive dicotyledonous stock at a very early stage of evolution, the point of origin being the Ranales. His reshuffling of the monocotyledonous orders and families with elaborate description of each has led to a better understanding of the group as a whole. Further, the splitting up of some of the families, particularly *Liliaceae* on the basis of inflorescence characters, is very logical and sound. Monocotyledons began with the family *Butomaceae* and ended in the family *Gramineae,* with several climax orders, as in dicotyledons.

Primitive and Advanced Characters which formed the basis of his classification are: polypetalous condition is more primitive than gamopetalous condition; unisexual flowers are more advanced than bisexual flowers; apetalous forms have originated from petaliferous stock, and represent reduction and advance; free parts on the whole are considered more primitive than connate or adnate parts; thus numerous free stamens are earlier than few or connate stamens; similarly apocarpous pistil is more primitive than syncarpous pistil; spiral arrangement of parts is more primitive than cyclic, etc.

Hutchinson's revised scheme of classification, (1959) is as follows. He has divided the 'flowering' plants into two phyla: phylum I—**Gymnospermae** (not elaborated by him) and phylum II—**Angiospermae**; the latter into two sub-phyla: sub-phylum I

—**Dicotyledones** and sub-phylum II—**Monocotyledones**. Dicotyledones were divided into two big divisions: division I—**Lignosae** (with 54 orders and 246 families) and division II—**Herbaceae** (with 28 orders and 96 families). Monocotyledones were divided into three divisions: division I—**Calyciferae** (with 12 orders and 29 families) beginning with *Butomaceae* and ending in Zingiberales (*Scitamineae* of Bentham and Hooker), division II—**Corolliferae** (with 14 orders and 34 families) beginning with *Liliaceae* and ending in *Orchidaceae;* and division III—**Glumiflorae** (with 3 orders and 6 families) beginning with *Juncaceae* and ending in *Gramineae*. Thus there are altogether 411 families of Angiospermae—342 of Dicotyledones and 69 of Monocotyledones. In his scheme of classification Hutchinson placed the Gymnospermae first, then the Dicotyledones, and finally the Monocotyledones.

Merits. (1) Most taxonomists believe that this system has given a much better idea of phylogenetic conception and has stimulated phyletic rethinking to a greater extent since its publication. (2) Primitive or basic orders are Magnoliales representing arborescent families, and Ranales representing herbaceous families, giving rise to woody and herbaceous forms respectively on parallel lines. (3) Bisexual and polypetalous flowers precede unisexual and gamopetalous flowers. (4) Amentiferae (catkin-bearing families with unisexual, apetalous flowers), e.g. *Fagaceae*—oak, *Betulaceae*—birch, *Juglandaceae*—walnut, *Salicaceae*—willow and poplar, etc., is regarded as advanced (and not primitive as considered by Engler) and thus transferred (including Urticales) to a new phyletic position close to Rosales and Leguminosae; their apparent simplicity represents reduction and specialization (and not primitiveness). (5) *Casuarinaceae* has been assigned an advanced position and placed at the top of the Amentiferae. (6) Several big orders have been split up into distinct small families, thus simplifying matters, e.g. Rosales, Perietales, Malvales, etc. (7) Many families have been raised to the rank of orders, e.g. *Leguminosae* (now to an order), *Saxifragaceae* to Saxifragales, *Podostemaceae* to Podostemales, etc. (8) Reshuffling of several orders and families, e.g. Cactales (Opuntiales of Engler) placed close to Cucurbitales and Passiflorales. (9) Origin of monocotyledons from dicotyledons at an early stage of evolution, the point of origin being the Ranales. (10) Splitting up of Helobiae (of Engler) into separate orders and considering the order Butomales (*Butomaceae* and *Hydrocharitaceae*) as the starting point of monocotyledons. (11) Reshuffling of genera of *Liliaceae* and *Amaryllidaceae* on the basis of inflorescence characters (and not on ovary position). (12) Rearrangement of several orders, finally ending in Cyperales and Graminales.

Demerits. (1) Many doubt the wisdom of his rigid bifurcation of dicotyledons into Lignosae (woody types) and Herbaceae (herbaceous types), (2) Many hold different views regarding the relationship between various orders and families. (3) Monophyletic view regarding the origin of angiosperms is not universally accepted. (4) Monophyletic origin of monocotyledons from the Ranales, as against the polyphyletic (diphyletic) views of Lotsy (1911) and Hallier (1912). (5) Urticales, Umbellales Euphorbiales, etc., as originating from different ancestors.

Dicotyledons and Monocotyledons. The division of angiosperms into the two great classes is based on the following characters: (1) In dicotyledons the embryo bears *two cotyledons*; whereas in monocotyledons it bears *only one*. (2) In dicotyledons the primary root persists and gives rise to the *tap root*; while in monocotyledons the primary root soon perishes and is replaced by a cluster of *adventitious (fibrous) roots*. (3) As a rule venation is *reticulate* in dicotyledons and *parallel* in monocotyledons; among monocotyledons aroids, sarsaparilla (*Smilax*) and yams (*Dioscorea*), however, show reticulate venation, and among dicotyledons Alexandrian laurel (*Calophyllum*) shows parallel venation. Further, in dicotyledons the veinlets end freely in the mesophyll of the leaf; whereas in monocotyledons there is no free ending of veins or veinlets. (4) The dicotyledonous flower has commonly a *pentamerous* symmetry, sometimes *tetramerous* (as in *Cruciferae, Rubiaceae,* etc.); while the monocotyledonous one has a *trimerous* symmetry. (5) In the dicotyledonous stem the vascular bundles are *arranged in a ring* and are *collateral* and *open*, that is, they contain a strip of cambium which gives rise to the secondary growth; while in the monocotyledonous stem the bundles are *scattered* in the ground tissue and are *collateral* and *closed*, and hence there is no secondary growth (with but few exceptions); also the bundles are more numerous in monocotyledons than in dicotyledons; further, the bundles are more or less oval in monocotyledons and wedge-shaped in dicotyledons. (6) In the dicotyledonous root the number of xylem bundles varies from 2 to 6, seldom more, but in the monocotyledonous root they are *numerous*, seldom a limited number (5 to 8). It may also be noted that the cambium soon makes its appearance in the dicotyledonous root as a secondary meristem and gives rise to the secondary growth, but in the monocotyledonous root cambium is absent (leaving out the exceptions) and hence there is no secondary growth.

Floral Diagram. The number of parts of a flower, their general structure, arrangement and the relation they bear to one another (aestivation), adhesion, cohesion, and position with respect to the mother axis may be represented by a diagram known as the **floral diagram**. The floral diagram is the ground plan of a flower. In the diagram the calyx lies outermost, the corolla internal to the calyx, the androecium in the middle, and the gynoecium in the centre. Adhesion and cohesion (see p. 89) of members of different whorls may also be shown clearly by connecting the respective parts with lines; as, for example, FIG. 2 shows that there are altogether ten stamens, of which nine are united into one bundle (cohesion) and the remaining one is free; while FIG. 40 shows that petals and stamens are united (adhesion). The black dot on the top represents the position of the mother axis (not the pedicel) which bears the flower. The axis lies behind the flower and, therefore, the side of the flower nearest to the axis is

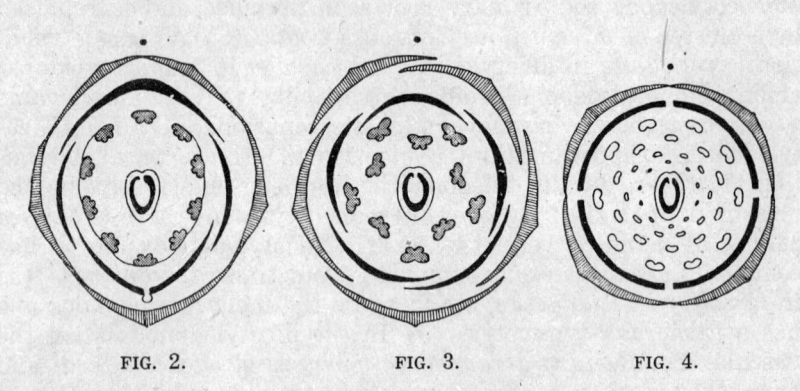

FIG. 2. FIG. 3. FIG. 4.

Floral Diagrams (three types). FIG. 2. *Papilionaceae*; FIG. 3. *Caesalpinieae*; FIG. 4. *Mimoseae*.

called the *posterior* side, and the other side away from the axis the *anterior* side. The floral characters of a species may be well represented by a floral diagram, while to represent a genus or family more than one diagram may be necessary.

Floral Formula. The different whorls of a flower, their number, cohesion and adhesion may be represented by a formula known as the **floral formula**. In the floral formula K stands for calyx, C for corolla, P for perianth, A for androecium and G for gynoecium. The figures following the letters K, C, P, A and G indicate the number of parts of those whorls. Cohesion of a whorl is shown by enclosing the figure within brackets, and the adhesion is shown by a line drawn on the top of the two whorls concerned. In the case of the gynoecium the position of the ovary is shown by a line drawn above or below G or the figure. If the ovary is superior the line should be below it, and if it is inferior the line should be on the top. Thus all the parts of a flower are represented in a general way by a floral formula; the floral characters of a family may also be represented by one or more formulae, as follows. Besides, some symbols are used to represent certain features of flowers; thus ♂ represents male, ♀ female, ⚥ hermaphrodite, ♂ ♀ dioecious, ♂ - ♀ monoecious, ♂ ♀ ⚥ polygamous, ⊕ actinomorphic, ⊹ zygomorphic, ∞ indefinite number of parts, etc.

Ranunculaceae: $\quad \oplus \male K_5 C_5 A_\infty \underline{G}_\infty \quad$ *Cucurbitaceae:* $\oplus \male \female \text{or} \male\text{-}\female$

Annonaceae: $\quad \oplus \male K_3 C_{3+3} A_\infty \underline{G}_\infty \qquad\qquad \male K_{(5)} C_{(5)} A_{3 \text{ or } 5}$

Nymphaeaceae: $\oplus \male K_4 C_\infty A_\infty \underline{G}_{(\infty) \text{or} \infty} \qquad \female K_{(5)} C_{(5)} \bar{G}_{(3)}$

Cruciferae: $\quad \oplus \male K_{2+2} C_4 A_{2+4} \underline{G}_{(2)}$ *Solanaceae:* $\oplus \male K_{(5)} \overline{C_{(5)}} A_5 \underline{G}_{(2)}$

Malvaceae: $\quad \oplus \male K_{(5)} \overline{C_5} A_{(\infty)} \underline{G}_{(5-\infty)}$ *Labiatae* $\quad \cdot |\cdot \male K_{(5)} \overline{C_{(5)}} A_4 \underline{G}_{(2)}$

Features used to describe an Angiospermic Plant

Habitat: natural abode of the plant.

Habit: herb (erect, prostrate, decumbent, diffuse, trailing, twining or climbing), shrub (erect, straggling, twining or climbing) or tree or any other peculiarity in the habit.

Root: nature of the root; any special form.

Stem: kind of stem— herbaceous or woody; cylindrical or angular; hairy or smooth; jointed or not; hollow or solid; erect, prostrate, twining or climbing; nature of modification, if any.

Leaf: arrangement—whether alternate, opposite (superposed or decussate) or whorled; stipulate or exstipulate; nature of the stipules, if present, simple or compound; nature of the compound leaf and the number of leaflets; shape and size; hairy or smooth; deciduous or persistent; venation; margin; apex; and petiole.

Inflorescence: type of inflorescence (to be explained).

Flower: sessile or stalked; complete or incomplete; unisexual or bisexual; regular, zygomorphic, or irregular; hypogynous, epigynous or perigynous; bracteate or ebracteate; nature of bracts and bracteoles, if present; shape of the flower, its colour and size.

Calyx: polysepalous or gamosepalous; number of sepals or of lobes; superior or inferior; aestivation; shape, size and colour.

Corolla: polypetalous or gamopetalous; number of petals or of lobes; superior or inferior; aestivation; shape, size, colour and scent; corona or any special feature.

(When there is not much difference between the calyx and the corolla the term **perianth** should be used; it may be sepaloid or petaloid; polyphyllous or gamophyllous, free or epiphyllous).

Androecium: number of stamens—definite (ten or less) or indefinite (more than ten); free or united; nature of cohesion—monadelphous, diadelphous, polyadelphous, syngenesious or synandrous; nature of adhesion—epipetalous or gynandrous, or any special feature; whether alternating with the petals (or corolla-lobes) or opposite them. Length of stamens—general length; inserted or exerted; didynamous or tetradynamous; position of stamens—hypogynous, perigynous or epigynous; attachment of the anther and its dehiscence; anther-lobes or appendages, if any.

Gynoecium or Pistil: number of carpels; syncarpous or apocarpous; nature of style—long or short; stigmas—simple, lobed or branched; their number and nature—smooth or papillose; ovary—superior or inferior; number of lobes; number of chambers (loculi); nature of placentation; number and form of ovules in each loculus of the ovary.

Fruit: kind of fruit (to be explained).

Seeds: number of seeds in the fruit; shape and size; albuminous or exalbuminous; nature of endosperm, if present.

Chapter 3 SELECTED FAMILIES OF DICOTYLEDONS[1]

SUB-CLASS I POLYPETALAE

Family 1 *Ranunculaceae* (*over* 1,200 *sp.*—157 *sp. in India*)

Habit—annual or perennial herbs or climbing shrubs, usually with an acrid juice. **Leaves** simple or compound, alternate or rarely opposite, radical and cauline, usually with sheathing base. **Inflorescence** cymose. **Flowers** mostly regular (actinomorphic), sometimes zygomorphic, as in larkspur (*Delphinium*) and monk's hood (*Aconitum*), bisexual and hypogynous, often showy; sepals and petals in whorls; stamens and carpels typically spiral on the elongated thalamus. **Calyx**—sepals usually 5, sometimes more, free, sometimes brightly coloured. **Corolla**—petals 5 or more, free, sometimes absent, often with nectaries, imbricate; perianth leaves (when calyx and corolla not distinguishable) free and petaloid. **Androecium**—stamens numerous, free, usually spiral. **Gynoecium**—carpels usually numerous, sometimes few or even 1, free (apocarpous), usually spiral, with one or more ovules in each; in *Nigella* carpels are united at the base. **Fruit** an etaerio of achenes or follicles, rarely a berry or capsule. **Seeds** albuminous. *Floral formula*— $\oplus \, \male \, K_5 C_5 A_\infty \underline{G}_\infty$.

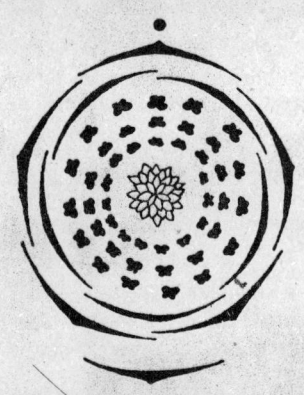

FIG. 5. Floral diagram of *Ranunculaceae*.

Ranunculaceae, according to Hutchinson, is a most primitive family having originated from some gymnospermic stock and showing parallel development with *Magnoliaceae* (see pp. 700-01). The former shows evolutionary progress through herbaceous families like *Nymphaeaceae, Papaveraceae, Capparidaceae, Cruciferae*, etc. Hutchinson is further of the opinion that monocotyledonous families like *Alismaceae, Hydrocharitaceae*, etc., have evolved from Ranalean stock.

Examples. [The larger genera are: *Ranunculus* (300 sp.), *Clematis* (250 sp.), *Delphinium* (250 sp.), *Anemone* (120 sp.), *Aconitum* (over 100 sp.), *Thalictrum* (over 100 sp.), *Aquilegia* (100 sp.)]. **Useful plants:** monk's hood (*Aconitum ferox;* B.

---

[1] In India dicotyledons are represented by 173 families and 11,124 species approximately.

SELECTED FAMILIES OF DICOTYLEDONS

KATBISH; H. BISH)—medicinal, tuberous roots containing a very poisonous alkaloid, black cumin (*Nigella sativa*; B. & H. KALA-JIRA)—seeds used as a condiment, *Thalictrum foliolosum*—a tall herb, root powder used as SURMA in eye diseases, and *Coptis teeta* (B. MISHMI-TEETA)—a stemless herb in the Mishmi Hills, roots medicinal; **ornamental:** larkspur (*Delphinium ajacis*), wind flower (*Anemone elongata*)—a small tuberous plant with

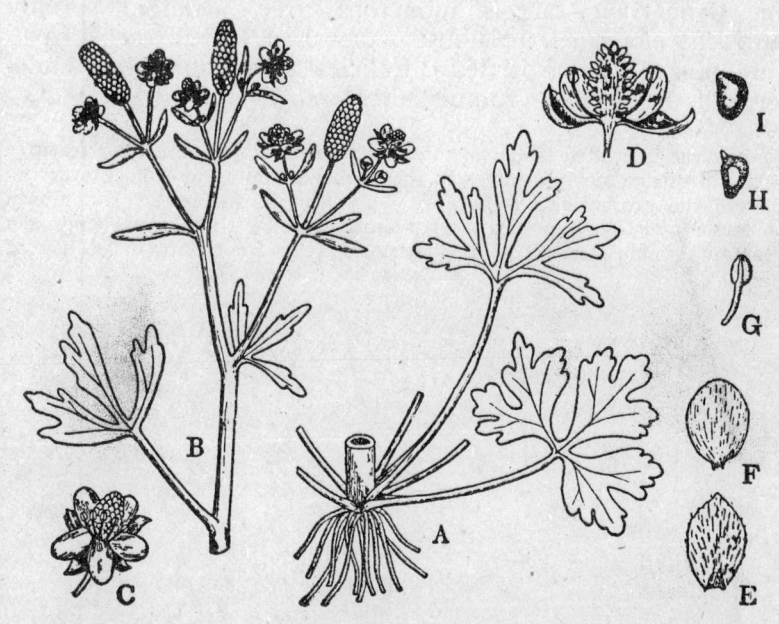

Ranunculaceae. FIG. 6. *Ranunculus sceleratus.* A, basal portion of the plant with leaves and roots; B, upper portion of the same with inflorescence; C, flower; D, flower cut longitudinally; E, a sepal; F, a petal; G, a stamen; H, a carpel; and I, a fruit (achene).

woolly achenes for wind-dispersal, virgin's bower (*Clematis*), e.g. *C. gouriana, C. cadmia,* etc.,—climbing shrubs, columbine (*Aquilegia vulgaris*), etc.; **other common plants:** *Ranunculus,* e.g. Indian buttercup (*R. sceleratus*)—usually growing in river- and marsh-banks, water crowfoot (*R. aquatilis*)—growing in water and showing heterophylly, etc., traveller's joy (*Naravelia zeylanica*)—a climbing shrub, *Thalictrum javanicum*—a perennial herb, etc.

Family 2 Magnoliaceae (250 *sp.*—30 *sp. in India*)

Habit—mostly ornamental evergreen trees and shrubs, some woody climbers. **Leaves** simple, alternate, often with large stipules covering young leaves. **Flowers** solitary, terminal (as in *Magnolia*) or axillary (as in *Michelia*), often large, showy and fragrant; regular, bisexual and hypogynous. **Perianth leaves** all alike, petaloid, deciduous; either cyclic being arranged in whorls of 3 (trimerous) or acyclic (spiral); sometimes the outer whorl sepaloid. **Androecium**—stamens numerous, free; filament short or absent; anther-lobes linear, 4, with prolonged connective. **Gynoecium**—carpels numerous, free, arranged spirally round the elongated thalamus; ovules 1 or few in each carpel. **Fruit** an aggregate of berries or follicles. **Seed** albuminous, Endosperm of the seed non-ruminated. *Floral formula*—$\oplus \; \male \; P_\infty A_\infty \underline{G}_\infty$

Magnoliaceae is related to *Annonaceae* but the main distinguishing features are: the presence of big stipules in the former standing as a hood over the bud, and the presence of ruminated endosperm in the latter. Hutchinson considers *Magnoliaceae* as the most primitive family with a near approach to certain gymnosperms like Bennettitales (e.g. *Bennettites*) because of

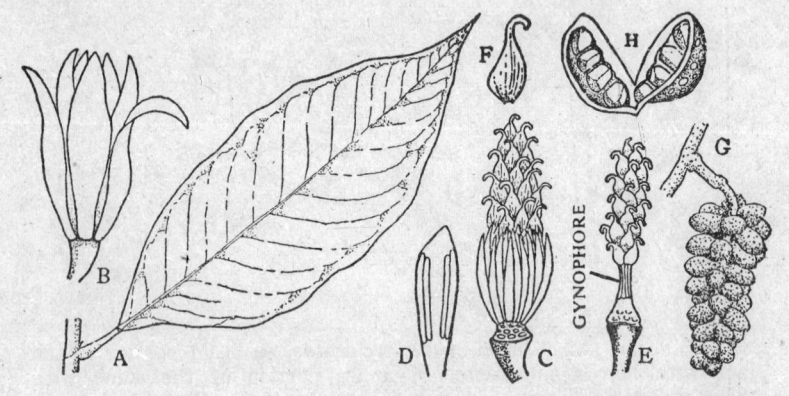

Magnoliaceae. FIG. 7. *Michelia champaca.* A, a leaf; B, a flower; C, stamens and carpels spirally arranged on the thalamus; D, a stamen with four anther-lobes; E, carpels (free); F, a carpel; G, aggregate fruits (follicles); and H, a follicle dehiscing.

spiral arrangement of free stamens and free carpels, and also the presence of tracheids with bordered pits and unisexual flowers in *Drimys*. *Magnoliaceae* shows evolutionary progress through woody families like *Annonaceae, Lauraceae,* etc.

Examples. *Magnolia* (80 sp.), e.g. *M. grandiflora*—a small American tree with large white flowers, *M. pterocarpa* (B. & H. DULEE-CHAMPA)—a tree with large fleshy greenish-white flowers, *M. griffithii* (B. GOURI-CHAMPA)—a tree with large pale-white flowers (about 15 cm. across), *M. fuscata* (B. CHINI-CHAMPA)—a shrub and *M. pumila* (B. JAHURE-CHAMPA)—also a shrub, are commonly planted in gardens, *M. globosa*—a small tree and *M. campbellii*—a shrub grown at high altitude (the latter planted on Kurseong-Darjeeling Road), etc., *Michelia* (50 sp.). e.g. *M. champaca* (B. SWARNA-CHAMPA; H. CHAMPA)—a tree with golden-yellow flowers, wood light but durable and used for cabinet work,

M. alba (B. & H. CHINA-CHAMPA)—a tree with white flowers, *M. oblonga*—a very tall tree (46 m.) with white flowers, wood much used for planking and cabinet work, etc., *Talauma hodgsoni*—a small tree with large red flowers, *T. phellocarpa* (B. TITA-CHAMPA)—a large tree with pale-white flowers, wood much used for cabinet work, etc., a few species of *Manglietia*, tulip-tree (*Liriodendron tulipifera*) grown in some gardens as an ornamental tree, *Kadsura roxburghiana* and *Schizandra elongata* are woody climbers. The family is well-represented in Assam and Meghalaya.

Family 3 Annonaceae (850 *sp.*—100 *sp. in India*)

Habit—shrubs and trees, sometimes climbers. **Leaves** simple, alternate, distichous and exstipulate. **Flowers** regular, bisexual, and hypogynous; often aromatic. **Perianth** usually in three whorls of three members each; sepals 3 and petals 6 in two whorls. **Androecium**—stamens numerous, free, arranged spirally round the slightly elongated thalamus; filament short or absent; anther-lobes linear, 4, extrorse, with an outgrowth of the connective. **Gynoecium**—carpels numerous, free or connate, each with a prolonged appendage; ovules one to many in each carpel. **Fruit** an aggregate of berries. **Seed** with the endosperm distinctly ruminated, i.e. marked by irregular wavy lines. *Floral formula*— $\oplus \; \male \; K_3 C_{3+3} A_\infty G_\infty$.

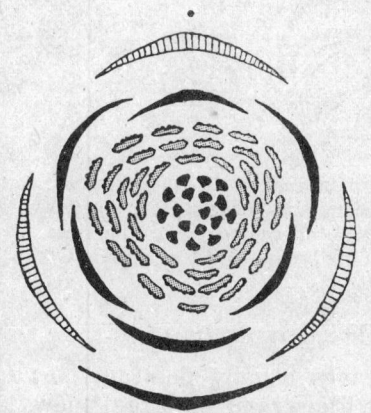

FIG. 8. Floral diagram of Annonaceae (*Artabotrys*).

Annonaceae is allied to *Magnoliaceae* but is distinguished from it by the presence of deeply ruminated endosperm.

Examples. [The larger genera are: *Uvaria* (150 sp.), *Annona* (120 sp.), *Polyalthia* (100 sp.), *Xylopia* (100 sp.), and *Artabotrys* (over 50 sp.)]. Custard-apple (*Annona squamosa*)—fruit edible, bullock's heart (*A. reticulata*)—fruit edible, sour sop (*A. muricata*)—fruit edible, *Artabotrys hexapetatus* (=*A. odoratissimus*)—flowers very fragrant, *Unona discolor* (B. LAVENDER-CHAMPA)—flowers very fragrant, mast tree (*Polyalthia longifolia*)—an evergreen tall tree, leaves used for decoration, *P. suberosa*—a handsome shrub, *Uvaria hamiltoni* and *Melodorum bicolor* are large woody climbers, *Cananga odorata* —a tall tree, flowers yield Macassar oil—a perfume, Negro pepper (*Xylopia aromatica*)—an African shrub, fruits (about 5 cm. long) used as a spice, etc.

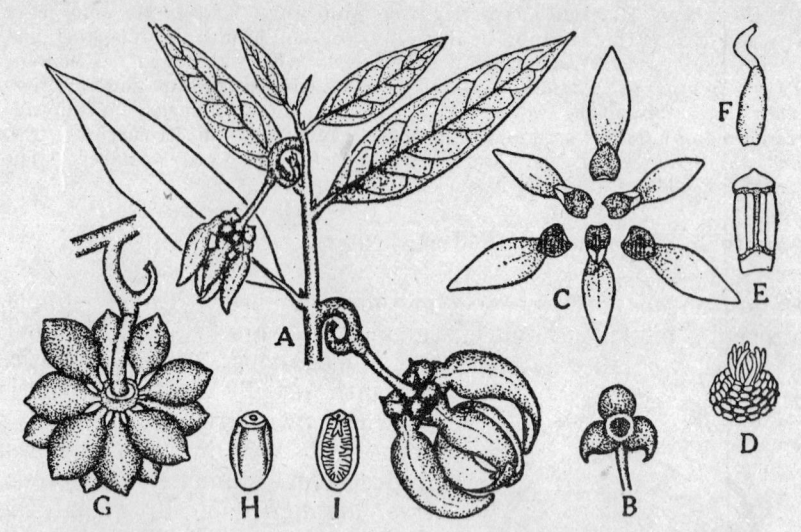

Annonaceae. FIG. 9. *Artabotrys.* A, a branch with two flowers; B, calyx; C, petals spread out; D, stamens and carpels; E, a stamen with four anther-lobes; F, a carpel; G, an aggregate of berries; H, a seed; and I, the seed cut longitudinally showing the ruminated endosperm.

Family 4. Nymphaeaceae (100 sp.—11 sp. in India)

Habit—aquatic perennial herbs. **Leaves** usually floating, borne on a long petiole, cordate or peltate. **Flowers** often large, showy, solitary, on a long pedicel, usually floating; bisexual, regular and usually perigynous, sometimes hypogynous or even epigynous; thalamus fleshy and goblet-shaped. **Perianth** leaves several, free; sepals usually 4, gradually merging into petals; petals numerous, gradually merging into stamens. **Androecium**—stamens numerous, free, usually perigynous, adnate to the fleshy thalamus that envelops the carpels. **Gynoecium**—carpels several, either free on the fleshy thalamus, as in lotus, or syncarpous lying embedded in the thalamus and surrounded by it; ovary unilocular with one ovule or multilocular with many ovules on superficial placentation; stigmas sessile, free or united, radiating, often with horn-like appendages. **Fruit** a berry. **Seeds**

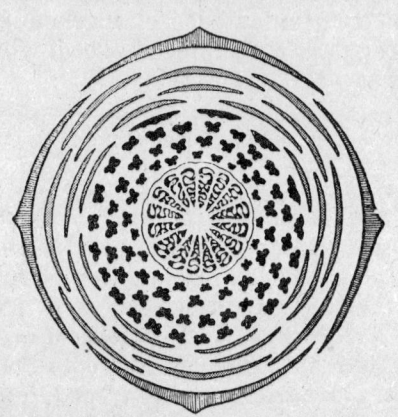

FIG. 10. Floral diagram of *Nymphaeaceae* (*Nymphaea*).

SELECTED FAMILIES OF DICOTYLEDONS

solitary and exalbuminous, or many with both perisperm and endosperm; spongy aril is often present and helps the seed to float. *Floral formula*—
$\oplus \, \male \, K_4 \, C_\infty \, A_\infty \, \underline{G}(\infty) \text{ or } \infty.$

Nymphaeaceae forms a link with *Ranunculaceae* and *Magnoliaceae* by spiro-cyclic nature of floral whorls, and sometimes apocarpous pistil, as in *Cabomba*. The family also bears an affinity with monocotyledons (e.g. *Alismaceae*) through *Cabomba* which has a typical trimerous symmetry ($P_{3+3} A_{3-6} \underline{G}_3$), apocarpous pistil, and scattered closed vascular bundles. It also shows affinity with *Papaveraceae* by its superficial placentation and radiating stigmas.

Examples. Plants often cultivated for pond decoration—water lilies (*Nymphaea* with 50 sp.), e.g. *N. lotus*, *N. rubra*, *N. stellata*, *N. alba*, etc., *Euryale ferox* (B. & H. MAKHNA), lotus (*Nelumbo nucifera*= *Nelumbium specisum*)—rhizome eaten, and giant water lily (*Victoria amazonica*=*V. regia*; see FIG. IV/I)—it bears huge tray-like leaves, each measuring 1-2 metres in diameter; although a native of South America this plant may be seen in luxuriant growth about October in the Indian Botanical Gardens near Calcutta. Lotus has some distinctive characteristics of its own: (1) leaves and flowers are raised above the surface of water; (2) flowers hypogynous; (3) carpels several, free (apocarpous), and embedded in the upper surface of the top-shaped thalamus; (4) ovary unilocular with one ovule; (5) stigmas sessile, solitary; and (6) seeds exalbuminous.

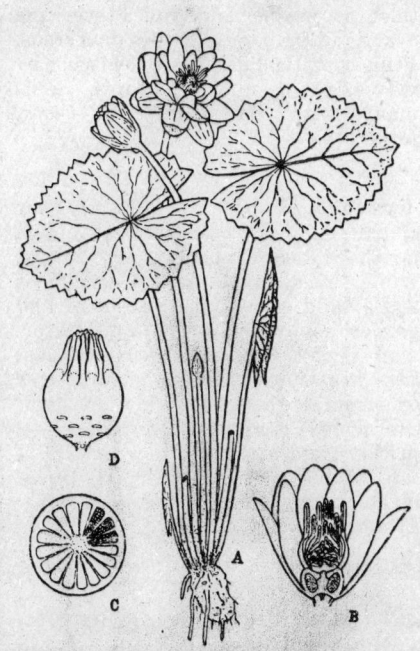

Nymphaeaceae. FIG. 11. Water lily (*Nymphaea lotus*). *A*, an entire plant; *B*, a flower cut longitudinally (see also FIGS. I/104-5); *C*, transverse section of the ovary; and *D*, a young fruit.

Family 5 Papaveraceae (250 *sp.*—40 *sp. in India*)

Habit—mostly herbs with milky or yellowish latex. Leaves radical and cauline, simple and alternate, often lobed. Flowers solitary, often showy, regular, bisexual and hypogynous. **Calyx**—sepals typically 2, sometimes 3, free, caducous. **Corolla**—petals 2+2 or 3+3, rarely more, in 2 whorls (rarely 3), large, free, rolled or crumpled in the bud, caducous; imbricate, **Androecium**—stamens ∞, sometimes 2 or 4, free. **Gynoecium**—carpels (2-∞), (4-6) in *Argemone*, syncarpous; ovary superior, 1-chambered, or spuriously

2- to 4-chambered, with 2-∞ parietal placentae which may project inwards, as in poppy (*Papaver*); stigmas distinct or sessile and rayed over the ovary, as in poppy; ovules numerous. Fruit a septicidal capsule dehiscing by valves, or opening by pores. Seeds many, with oily endosperm. Floral formula—$\oplus \; \male \; K_{2 \text{ or } 3} \; C_{2+2 \text{ or } 3+3} \; A_{\infty} \; \underline{G}_{(2-\infty)}$.

Examples. *Papaver* (110 sp.), e.g. opium poppy (*P. somniferum*)—opium is the latex obtained from unripe fruits, seeds used as a condiment POSTO and also yield an oil, and garden poppy (*P. orientale*), Californian poppy (*Eschscholzia*)—the last two are cultivated in gardens for ornamental flowers, prickly or Mexican poppy (*Argemone mexicana*)—a prickly weed bearing yellow flowers in late winter, seeds yield an oil, Himalayan poppy (*Meconopsis*; 45 sp.)—mainly occurring in Nepal and the Eastern Himalayas, etc.

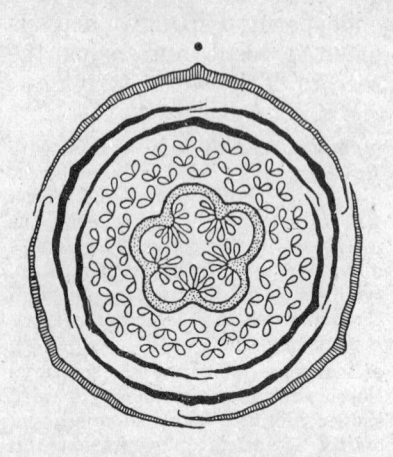

FIG. 12. Floral diagram of *Papaveraceae* (*Argemone*).

Family 6 *Cruciferae* (over 3,000 sp.—174 sp. in India)

Habit—annual herbs. **Leaves** radical and cauline, simple, alternate, often lobed, or rarely pinnately compound. **Inflorescence** a raceme (corymbose towards the top). **Flowers** regular and cruciform, bisexual and complete, hypogynous. **Calyx**—sepals 2+2, free, in two whorls. **Corolla**—petals 4, free, in one whorl, alternating with sepals, cruciform, each petal with distinct limb and claw. **Androecium**—stamens 6, in two whorls, 2 outer short and 4 inner long (tetradynamous). **Gynoecium**—carpels (2), syncarpous; ovary superior; at first 1-celled, but later 2-celled owing to the development of a *false septum*; ovules often many in each cell, sometimes only 2, anatropous or campylotropous; placentation parietal. **Fruit** a long narrow siliqua or a short broad silicula. **Seeds** exalbuminous; embryo curved; and remain attached to a wiry framework, called *replum* (see p. 129), which surrounds the fruit. Floral formula—$\oplus \; \male \; K_{2+2} \; C_4 \; A_{2+4} \; \underline{G}_{(2)}$.

FIG. 13. Floral diagram of *Cruciferae*.

Cruciferae is allied to *Capparidaceae* by having tetramerous perianth, bicarpellary ovary, parietal placentation, folding of the embryo in various ways, exalbuminous seeds, and often campylotropous ovules. It will be noted, however, that tetradynamous condition is characteristic of *Cruciferae*, while it is very rare in *Capparidaceae*; gynophore (sometimes also androphore) commonly present in *Capparidaceae*, but altogether absent in *Cruciferae*. The two families may have arisen from a common ancestor, or *Cruciferae* may have evolved from some primitive member of *Capparidaceae*.

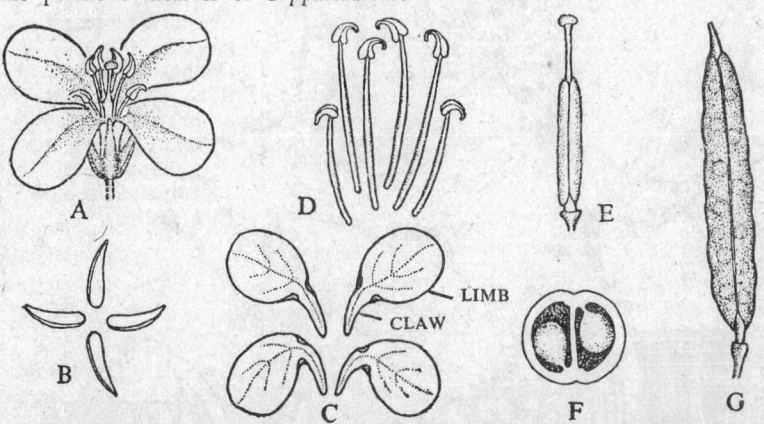

Cruciferae. FIG. 14. Mustard (*Brassica campestris*) flower. *A*, a flower —cruciform; *B*, calyx opened out; *C*, corolla opened out; *D*, androecium showing tetradynamous stamens; *E*, gynoecium showing two carpels united; *F*, ovary in transection showing parietal placentation and replum; and *G*, a fruit—siliqua. (See also FIG. 1 in Introduction).

Examples. [The larger genera are: *Draba* (about 300 sp. in north temperate and arctic), *Cardamine* (130 sp.), *Lepidium* (130 sp.), *Alyssum* (130 sp.), *Rorippa* (70 sp.), and *Brassica* (50 sp.)]. **Useful plants: oils and condiments:** mustard (*Brassica campestris*), Indian mustard (*B. juncea*), Indian rape (*B. napus*), white mustard (*B. alba*), black mustard (*B. nigra*), etc.; **vegetables:** radish (*Raphanus sativus*), cabbage (*Brassica oleracea* var. *capitata*), cauliflower (*B. oleracea* var. *botrytis*), turnip (*B. rapa*), kohl-rabi or knol-kohl (*B. caulorapa*), *B. rugosa* (B. LAI-SAK), garden cress (*Lepidium sativum*; B. HALIM-SAK; H. HALIM), water cress (*Nasturtium officinale*), etc.; **ornamental:** candytuft (*Iberis amara*; H. CHANDNI)— fruit a silicula, alison (*Alyssum maritimum*)—fruit a silicula, wallflower (*Cheiranthus cheiri*), etc.; **other common plants:** *Rorippa indica* (=*Nasturtium indicum*), *Eruca sativa*, bitter cress (*Cardamine debilis*), shepherd's purse (*Capsella bursa-pastoris*)—fruit a silicula, etc.

Family 7 *Capparidaceae* (650 *sp.*—53 *sp. in India*)

Habit—herbs, climbing shrubs, and trees. **Leaves** mostly alternate, rarely opposite, simple or palmately compound; stipules, if present, minute or spiny. **Flowers** regular (actinomorphic), sometimes zygomorphic, hypogynous or perigynous, and bisexual; thalamus in some cases elongated (gynophore) between the stamens and the pistil, sometimes both androphore and gynophore develop. **Calyx**—sepals 2+2, free. **Corolla**—petals 4, free. **Androecium**—stamens usually many, sometimes 6 (not tetradynamous), free. There

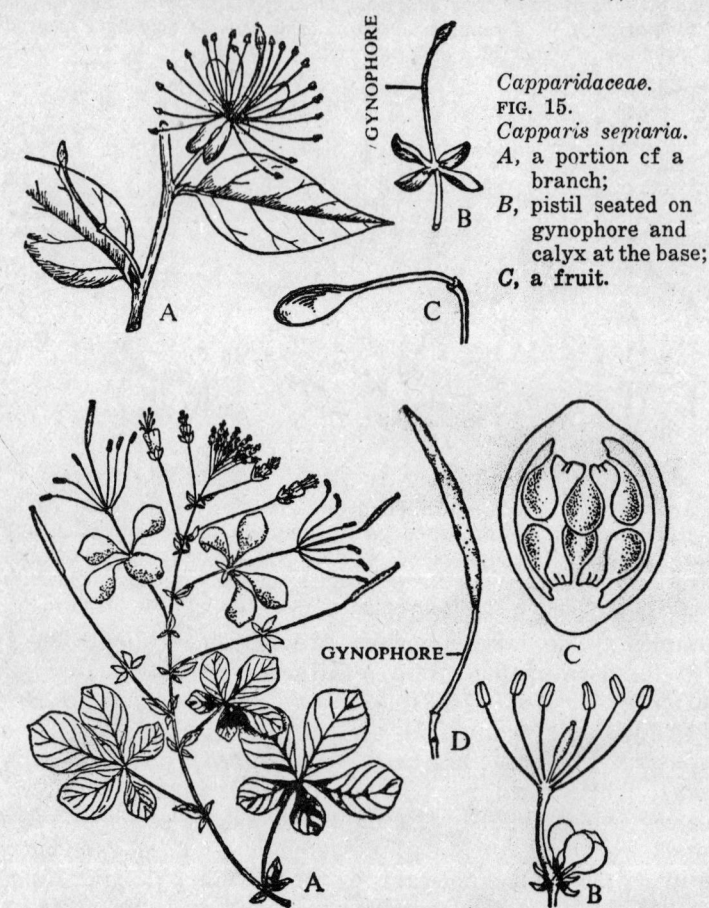

Capparidaceae.
FIG. 15.
Capparis sepiaria.
A, a portion of a branch;
B, pistil seated on gynophore and calyx at the base;
C, a fruit.

Capparidaceae. FIG. 16. *Gynandropsis gynandra*. A, a branch with leaves and flowers; B, a flower; C, section of ovary showing parietal placentation; and D, a fruit.

may be a disc between the perianth and the androecium. **Gynoecium**—carpels typically (2), rarely many, syncarpous; gynophore often present; ovary superior, 1-celled, or 2-celled due to a false partition wall; replum present or absent; placentation parietal; ovules many, campylotropous. **Fruit** an elongated capsule or a berry. **Seed** exalbuminous; embryo curved in various ways. *Floral formula*—$\oplus \; \male \; K_{2+2} C_4 A_{\infty \text{ or } 6} \underline{G}(_2)$.

From the structure of the flower it appears *Capparidaceae* stands midway between *Papaveraceae* and *Cruciferae*. It resembles *Cruciferae* in perianth and pistil (see p. 712), while it may resemble *Papaveraceae* in androecium.

Examples. *Polanisia icosandra* (=*Cleome viscosa*), *Gynandropsis gynandra*, *Capparis* (200 sp.), usually with spinous stipules, e.g. *C. sepiaria*, *C. horrida*,

C. aphylla (the whole leaf is modified into a tendril), caperbush (*C. spinosa*) —pickled flower-bud of it is called caper, etc., *Crataeva nurvala* (=*C. religiosa*)—a large tree, and *Roydsia suaveolens*—flowers delightfully scented.

Family 8 Caryophyllaceae ((2,000 sp.—106 sp. in India)

Habit—annual or perennial herbs, with swollen nodes. **Leaves** simple, opposite, sessile, sometimes connate, often stipulate. **Inflorescence** a cyme (usually dichasial). **Flowers** regular, bisexual, hypogynous, caryophyllaceous. **Calyx**—sepals usually 5, free or slightly connate. **Corolla**—petals commonly 5, free, usually clawed; sometimes androphore develops. **Androecium**—stamens usually 10, sometimes 8, free or sometimes united at the base, in two whorls, the outer opposite the petals and the inner opposite the sepals. **Gynoecium**—carpels (5) or (3); ovary 1-celled due to the breaking down of the septa at an early stage; placentation central; ovules usually many; styles 5 or 3, free; stigma along the inner surface of the style. **Fruit** a capsule dehiscing by valves. **Seeds** usually many, albuminous, with curved embryo.

Floral formula— $\oplus \; \male \; K_5 C_5 A_{5+5} \underline{G}_{(3-5)}$.

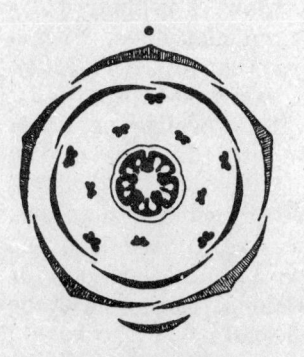

FIG. 17. Floral diagram of *Caryophyllaceae*.

Caryophyllaceae is related to *Portulacaceae*. According to Hutchinson *Amaranthaceae* and *Chenopodiaceae* are derived from *Caryophyllaceae* by way of reduction of floral members, while Engler holds the reverse view.

Examples. [The larger genera are: *Silene* (500 sp.)—a Mediterranean genus, *Dianthus* (over 300 sp.), *Arenaria* (over 200 sp.)—in north temperate regions, *Gypsophila* (125 sp.) and *Stellaria* (over 100 sp.)] *Dianthus*— many are cultivated for ornamental flowers, e.g. pink (*D. chinensis*), carnation (*D. caryophyllus*), sweet william (*D. barbatus*), etc., *Drymaria cordata* —a common diffuse weed, soapwort (*Saponaria vaccaria*; B. SABUNI; H. MUSNA)—a weed, *Polycarpon loeflingiae* (B. GIMA; H. SURETA)—a bitter diffuse herb, *Stellaria media*—a weed, *Spergula arvensis*—a weed, *Gypsophila elegans*—flowers in loose sprays used for table decoration, *Silene* (some species in temperate Western Himalayas), etc.

Family 9 Dipterocarpaceae (380 sp.—34 sp. in India)

Habit—mostly large resinous trees. **Leaves** simple, usually alternate, entire, leathery, evergreen or deciduous; stipules often prominent. **Inflorescence** an axillary or terminal panicle. **Flowers** regular, bisexual, hypogynous or perigynous, pentamerous. **Calyx**—sepals 5, polysepalous, persistent, 2 or more developing into wings. **Corolla**—petals 5, free or connate at the base; aestivation twisted. **Androecium**—stamens 5, 10, 15 or ∞, free or connate at the base; anther 2-celled, connective with an appendage. **Gynoecium**— carpels (3), syncarpous; ovary superior, 3-locular, with 2-∞ pendulous ovules in each. **Fruit** a 1-seeded nut, usually winged. **Seed** exalbuminous.

Floral formula— $\oplus \; \male \; K_5 C_5 A_{5,\;10,\;15 \text{ or } \infty} \underline{G}_{(3)}$.

Examples. [The larger genera are: *Vatica* (over 90 sp.), *Hopea* (90 sp.), *Shorea* (90 sp.), and *Dipterocarpus* (70 sp.)]. Wood-oil tree (*Dipterocarpus* *turbinatus*; B. & H. GARJAN; see FIG. I/177A)—a moderately hard timber tree; stem yields GARJAN oil and resin, *D. macrocarpus*—a very tall timber tree (46 metres); wood moderately hard, *Shorea robusta* (B. & H. SAL; see FIG. I/177C)—a very valuable timber tree, *S. assamica*—wood somewhat softer, *Vatica lancaefolia*—used as firewood and makes good charcoal, *Vateria indica*—yields a gum-resin used for Indian copal varnish, *Hopea odorata* (see FIG. I/176C)—a tall tree, etc.

Family 10 Malvaceae[1] (1,000 *sp.*—105 *sp. in India*)

Habit—herbs, shrubs and trees. **Leaves**—simple, alternate and palmately-veined; stipules 2, free lateral. **Flowers** regular, polypetalous, bisexual, hypogynous, copiously mucilaginous, with a whorl of bracteoles known as the *epicalyx* (except in *Abutilon* and *Sida*). **Calyx**—sepals (5), united, valvate. **Corolla**—petals 5, free, attached to the base of the staminal tube; aestivation twisted. **Androecium**—stamens numerous, monadelphous, i.e. united into one bundle called staminal column or tube, epipetalous (staminal tube adnate to the petals at the base); anthers reniform, unilocular; pollen grains large and spiny. **Gynoecium**—carpels commonly (5 to ∞), syncarpous, (2-3) in *Kydia*; ovary superior, 5- to ∞-locular, usually 5-locular, with 1 to many ovules in each loculus; placentation axile; style passes through the staminal tube; stigmas free, as many as the carpels. **Fruit** a capsule or sometimes a schizocarp. **Seed** endospermic.

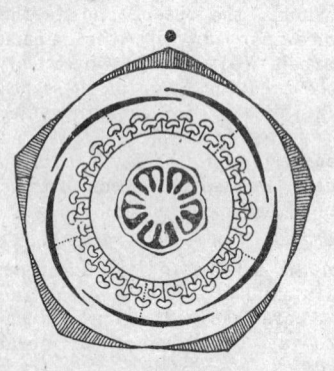

FIG. 18. Floral diagram of *Malvaceae*.

Floral formula—$\oplus \ \male \ K_{(5)} \overline{C_5 A_{(\infty)}} \underline{G}_{(5-\infty)}$.

Malvales may be allied to Guttiferales by various degrees of union of stamens, 5-merous calyx and corolla, and hypogynous flowers, and may have a common origin.

Examples. [The larger genera are: *Hibiscus* (over 200 sp.), *Sida* (200 sp.), *Abutilon* (100 sp.), and *Malva* (40 sp.)]. **Useful plants:** *Gossypium* (20 sp.) yields commercial textile cotton, rozelle (*Hibiscus sabdariffa*)—fruits used for a sour jelly, Madras or Deccan hemp (*H. cannabinus*)—a source of strong fibres, musk mallow (*H. abelmoschus*; B. MUSHAKDANA; H.

---

[1] Malvales of Bentham and Hooker—*Malvaceae, Tiliaceae* and *Sterculiaceae*; the same order of Engler—*Tiliaceae, Malvaceae, Bombacaceae, Sterculiaceae*, and a few more; Hutchinson has split the order into two: Malvales—*Malvaceae*, and Tiliales—*Tiliaceae, Sterculiaceae, Bombacaceae* and a few more.

MUSHAKDANA)—seeds smell like musk and are used as a flavouring agent and as a medicine, mallow (*Malva verticillata*)—cultivated as a winter vegetable, lady's finger (*Abelmoschus esculentus*)—green fruits used as a vegetable; **ornamental:** several species of

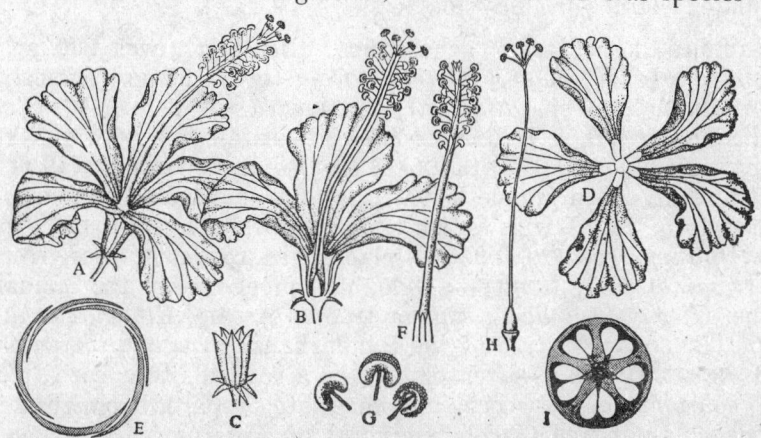

Malvaceae. FIG. 19. China rose (*Hibiscus rosa-sinensis*) flower. *A*, an entire flower; *B*, the same split open longitudinally showing the four whorls, more particularly the staminal column with the style passing through it; *C*, calyx with epicalyx; *D*, corolla opened out; *E*, twisted aestivation of corolla; *F*, androecium showing monadelphous stamens; *G*, one-celled anthers—young and mature (dehiscing); *H*, gynoecium showing five carpels united; and *I*, ovary in transection showing axile placentation.

Hibiscus, e.g. shoe-flower or China rose (*H. rosa-sinensis*), *H. mutabilis* (B. STHALPADMA; H. GUL-AJAIB), *H. radiatus,* etc., Chinese lantern (*Achania malvaviscus*; B. LANKAJABA), and hollyhock (*Althaea rosea*); **shade tree:** Portia tree (*Thespesia populnea*); **other common plants:** *Sida cordifolia, S. rhomboidea, Urena lobata, Hibiscus vitifolius* (B. & H. BAN-KAPAS), Indian mallow (*Abutilon indicum*; B. PETARI; H. KANGHI), *Malachra capitata* (B. & H. BAN-BHINDI) and *Malvastrum spicatum*—common weeds of waste places.

Family 11 Sterculiaceae (700 *sp.*—75 *sp. in India*)

Habit—shrubs or trees, rarely herbs. **Leaves**—leaves and stipules are like those of *Malvaceae*. **Inflorescence** cymose, often complex. **Flowers** (see FIG. I/95C) regular, sometimes zygomorphic, bisexual, rarely unisexual (as in *Sterculia*), hypogynous. **Calyx and Corolla** as in *Malvaceae*, sometimes corolla absent; no epicalyx. **Androecium**—stamens usually ∞ (but varying from 5-25), typically in two whorls, the outer whorl opposite to sepals and often reduced to staminodes or absent, while the inner whorl opposite to petals, fertile and often branched; all stamens more or less united below into a tube; sometimes on gonophore; anthers 2-locular. **Gynoecium**—carpels (5-2), often (5), syncarpous; ovary superior, 5- to 2-locular; with 2-∞ anatropous ovules in each; style simple; stigma lobed. **Fruit** varying, dry or fleshy, often a schizocarp. **Seed** with fleshy endosperm, sometimes arillate. *Floral formula*—⊕ ♂K($_5$)C$_5$A($_\alpha$)G̲($_{5-2}$).

Examples. [The larger genera are: *Sterculia* (over 200 sp.), *Dombeya* (over 100 sp.; African), *Cola* (over 100 sp.; African), *Hermannia* (150 sp.; African), and *Vatica* (over 70 sp.)]. *Sterculia foetida* (B. & H. JANGLI-BADAM)—a tall tree with digitate leaves, *S. villosa* (B. & H. UDAL, ODAL)—a moderately tall tree with deeply-lobed simple leaves, *S. alata*—a very tall handsome tree (46 m. high) with entire simple leaves, etc., *Pterospermum acerifolium* (see FIG. I/95C)—planted as roadside trees, *Heritiera minor* (B. SUNDRI)—a valuable timber tree of the Sundarbans, *H. macrophylla*—a timber tree of Assam, *Kleinhovia hospita* (B. BOLA)—a tree of the Sundarbans, *Guazuma tomentosa* (B. NEPAL TUNTH)—often planted as a roadside tree, cocoa tree (*Theobroma cacao*)—cocoa and chocolate prepared from roasted seeds, cola-nut tree (*Cola acuminata*) of West Africa—nuts contain much caffeine, devil's cotton (*Abroma augusta*; B. & H. ULATKAMBAL)—a shrub with fruits standing erect on the branches, *Helicteres isora*—a shrub with crimson flowers, *Dombeya mastersii*—an ornamental large shrub (planted in Shillong), *D. angulata*—a much-branched shrub with clusters of pink flowers, noon flower (*Pentapetes phoenicea*)—a tall herb with red flowers, *Melochia corchorifolia* and *Waltheria indica*—common weeds, etc.

Family 12 *Tiliaceae* (450 *sp.*—72 *sp.* in *India*)

Habit—generally trees or shrubs, sometimes herbs. **Leaves**—leaves and stipules, as in *Malvaceae*. **Inflorescence** cymose, often complex. **Flowers** regular, usually bisexual, hypogynous, pentamerous. **Calyx**—sepals (5) or 5, aestivation valvate; no epicalyx. **Corolla**—petals 5, polypetalous, rarely absent, imbricate in bud. **Androecium** — stamens usually many, free or polyadelphous, inserted at the base of the petals or on gonophore, as in *Grewia*; anther 2-locular. **Gynoecium**—carpels (5-2), syncarpous; ovary superior, 10- to 2-locular, with 1-∞ anatropous ovules in each; style simple; stigma capitate or

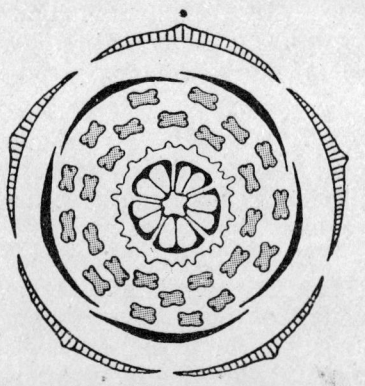

FIG. 20. Floral diagram of *Tiliaceae* (*Corchorus*).

SELECTED FAMILIES OF DICOTYLEDONS

lobed. **Fruit** varying, commonly a capsule, drupe or berry. **Seed** with fleshy endosperm. *Floral formula*— $\oplus \; \male \; K_{(5) \, or \, 5} \; C_5 \, A_\infty \, \underline{G}_{(5-2)}$.

Examples. [The larger genera are: *Grewia* (150 sp.), *Triumfetta* (over 100 sp.), and *Corchorus* (40 sp.)]. Jute (*Corchorus capsularis* and *C. olitorius*)—bast fibres form commercial jute, *C. aestuans* (B. TITAPAT)—a common weed, *Grewia asiatica* (B. & H. PHALSA)—a tree bearing edible fruits, *G. multiflora*—often grown as a hedge plant, *G. hirsuta*—a shrub, *Brownlowia lanceolata* (B. BOLA-SUNDRI)—a middle-sized tree of the Sundarbans—wood used as a fuel, *Triumfetta rhomboidea*—an undershrub with hooked fruits, *Tilia europea*—in temperate Himalayas, etc.

Distinguishing Characters of Malvales

Malvaceae	*Sterculiaceae*	*Tiliaceae*
Herbs, shrubs or trees	trees or shrubs, a few herbs	trees or shrubs, a few herbs
Leaves simple, often palmately lobed	simple or palmately compound	simple, entire or dentate
Flowers regular, bisexual, with epicalyx	regular, sometimes zygomorphic, bisexual, rarely unisexual, rarely corolla absent, with no epicalyx	regular, bisexual, rarely unisexual, with no epicalyx
Stamens (∞), monadelphous, epipetalous anther 1-locular	often many, typically in 2 whorls, the outer staminodial or 0, the inner fertile and branched, connate below, sometimes on gonophore; anther 2-locular	usually many, sometimes 10, free or connate at the base only, developing at the base of the petals or on gonophore, anther 2-locular
Carpels usually (5-∞), ovary 5- to ∞-locular, with 1-∞ anatropous ovules in each chamber	(5-2), usually (5), ovary 5- to 2-locular, with 2-∞ anatropous ovules in each chamber	(5-2), ovary 10- to 2-locular, with 1-∞ anatropous ovules in each chamber
Fruit capsular or schizocarpic; seed with scanty endosperm	capsular or schizocarpic, seed with fleshy endosperm	capsule or berry-like; seed with fleshy endosperm
Embryo with folded cotyledons	with flat or folded cotyledons	with large leafy cotyledons

Family 13 *Bombacaceae* (180 *sp.*)

Habit—large trees. **Leaves** simple or digitately compound, with deciduous stipules. **Flowers** regular, large, bisexual, hypogynous. **Calyx**—sepals (5), gamosepalous, valvate, often with epicalyx. **Corolla**—petals 5, polypetalous, imbricate. **Androecium**—stamens 5-∞ free or polyadelphous; anthers 2- or sometimes more-celled; pollen grains smooth; staminodes often present.

Gynoecium—carpels (2-5), syncarpous, when 5 they are opposite to the petals; ovary superior, multilocular, with 2-∞ ovules in each loculus. **Fruit** a capsule. Seeds smooth, often very hairy, with scanty or no endosperm. *Floral Formula*— $\oplus\ \male\female\ K_{(5)}C_5A_{5-\infty}\underline{G}_{(2-5)}$.

Examples. *Bombax* (60 sp.), e.g. red or silk cotton tree (*Bombax ceiba = Salmalia malabarica*) and *Ceiba* (20 sp.), e.g. white cotton tree or kapok (*Ceiba pentandra=Eriodendron anfractuosum*)—cotton in both used for stuffing pillows and cushions, and wood used for making tea- and matchboxes and matchsticks, baobab tree (*Adansonia digitata*)—trunk often reaching a diameter of 9 metres, balsa (*Ochroma pyramidale*)—a South American plant with very light wood, used for making models of boats, ships, etc.

Family 14 *Rutaceae* (900 *sp.*—66 *sp.* in India)

Habit—shrubs and trees (rarely herbs). **Leaves** simple or compound, alternate or rarely opposite, *gland-dotted*. **Flowers** regular, bisexual and hypogynous; disc below the ovary prominent—ring- or cup-like. **Calyx**—sepals 4 or 5, free or connate below, imbricate. **Corolla**—petals 4 or 5, free, imbricate. **Androecium**—stamens variable in number, as many or more generally twice as many as petals (obdiplostemonous), or numerous in *Citrus* and *Aegle*, free or united in irregular bundles (polyadelphous), inserted on the disc. **Gynoecium**—carpels generally (4) or (5), or ∞ in *Citrus*, syncarpous or free at the base and united above, either sessile or seated on the disc; ovary generally 4- or 5-locular, or multilocular in *Citrus*, with axile placentation (parietal in *Limonia* only); ovules 2-∞ (rarely 1) in each loculus, in two rows. **Fruit** a berry, capsule or hesperidium (see p. 133). **Seeds** with or without endosperm. Polyembryony is frequent in *Citrus*, e.g. lemon and orange (but not pummelo and citron). *Floral formula*— $\oplus\ \male\female\ K_{4-5}C_{4-5}A_{8,\ 10\ or\ \infty}\underline{G}_{(4,\ 5\ or\ \infty)}$.

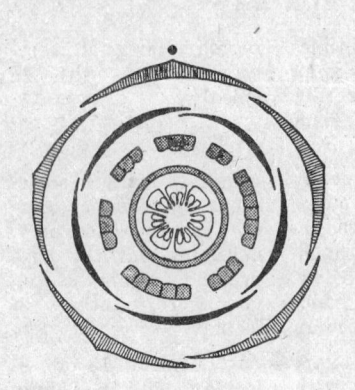

FIG. 21. Floral diagram of *Rutaceae*.

Rutaceae is allied to *Meliaceae* by obdiplostemonous stamens (i.e. stamens in 2 whorls, the outer opposite to the petals), presence of disc, carpels often (5), anatropous ovule with ventral raphe, and types of fruits. Hutchinson has separated the above two families from Geraniales and placed them under two separate orders Rutales and Meliales respectively. Rutales is also related to Sapindales (*Sapindaceae, Anacardiaceae*, etc.) but in the latter order leaves are not gland-dotted.

Examples. [The larger genera are: *Fagara* (over 200 sp.), *Ruta* (60 sp.), *Glycosmis* (60 sp.), and *Evodia* (45 sp.)]. **Useful**

SELECTED FAMILIES OF DICOTYLEDONS

Plants: *Citrus*[1] (e.g. lime, lemon, orange, citron and pummelo), wood-apple (*Aegle marmelos*)—the pulp containing *marmelosin* is an effective remedy for chronic dysentery and constipation, elephant-apple (*Limonia acidissima*), Chinese box (*Murraya paniculata*)—wood hard and useful, curry leaf plant (*M. koenigii*)—

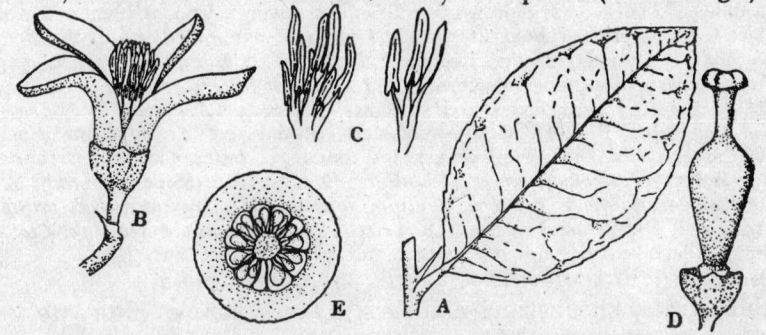

Rutaceae. FIG. 22. Sour lime (*Citrus aurantifolia*). *A*, a leaf; *B*, a flower; *C*, stamens (polyadelphous); *D*, pistil on a disc, with calyx at the base; and *E*, section of ovary showing axile placentation.

leaves used for flavouring curries, *Micromelum pubescens*—an evergreen tree, etc.; **other common plants:** rue (*Ruta graveolens*; B. ERMUL; H. SADAB)—a strongly smelling small herb, *Glycosmis arborea, Clausena heptaphylla, C. pentaphylla* (B. & H. PAN-KARPUR), *Luvunga scandens*—a large thorny climbing shrub, *Toddalia aculeata* (B. TODALI; H. DAHAN)—a large thorny climbing shrub, *Zanthoxylum budrunga* (B. BAZINALI; H. BADRANG) —a prickly tree, *Z. alatum* (B. NEPALI-DHANIYA, H. TEJPHAL or TUMRU)—a small aromatic tree, branches used as toothbrushes and fruits as a condiment like coriander, *Z. piperita*—pungent fruits yield Japanese pepper, *Evodia roxburghiana*—a tree, etc.

Family 15 *Meliaceae* (1,400 *sp.*—58 *sp. in India*)

Habit—mostly trees, rarely shrubs. **Leaves** pinnately compound, leaflets oblique. **Inflorescence** an axillary panicle. **Flowers** regular, often bisexual, sometimes polygamous (as in *Amoora*), hypogynous. **Calyx**—sepals (4-5), gamosepalous. **Corolla**—petals 4-5, usually polypetalous, imbricate. **Androe-**

---

[1] Common species of *Citrus*: sour lime (*C. aurantifolia*; B. PATI- or KAGZI-NEBU; H. NIMBOO), sweet lime (*C. limetta*; B. MITHA-NEBU or -KAGZI), lemon (*C. limon*; B. NEBU; H. KHATTI), rough lemon (*C. jambhiri*; B. JAMIR; H. JHAMBHIRI), *C. assamensis* (B. ADA-JAMIRI)—used for garnishing curries, citron (*C. medica*; B. BARA-NEBU; H. BARA-NIMBOO), pummelo or shaddock (*C. grandis*; B. BATABI-NEBU; H. CHAKOTRA), Mandarin orange (*C. reticulata*; B. KAMALA; H. SANGTRA)—loose skinned commercial orange, sour or bitter orange (*C. aurantium*)—used in preparing marmalade, sweet orange (*C. sinensis*—Malta, Mosambi or Mozambique, and Valentia are varieties of it)—tight-skinned, king orange (*C. nobilis*), wild orange (*C. indica*)—growing wild in Assam, bergamot orange (*C. bergamia*)—bergamot oil is prepared from it, and grapefruit (*C. paradisi*).

cium—stamens 8-10, generally united into a long or short staminal tube. **Gynoecium**—carpels (2-5), syncarpous; ovary superior, 2- to 5-locular, rarely 1-locular, with 1 or 2 ovules in each, seldom more; disc annular surrounding the ovary. **Fruit** a capsule, berry or drupe. **Seed** often winged, albuminous. *Floral formula*— $\oplus \ \male \ K_{(4-5)} C_{4-5} A_{(8-10)} \underline{G}_{(2-5)}$.

Examples. [The larger genera are: *Trichilia* (over 250 sp. in America and Africa), *Dysoxylum* (about 200 sp.)]. Timber trees: mahogany (*Swietenia mahagoni*), toon (*Cedrella toona*), satinwood (*Chloroxylon swietenia*), *Amoora rohituka*—timber moderately hard, *A. wallichii*—timber hard and suitable for furniture, doors and windows, margosa (*Azadirachta indica*= *Melia azadirachta*; B. & H. NEEM)—wood dark-red and very hard, also medicinal (NEEM oil extracted from fruits and leaves contains a bitter alkaloid margosine which is very efficacious in sores and ulcers), Persian lilac (*Melia azedarach*; B. GHORA-NEEM)—also yields firewood. *Walsura robusta* (B. LALI)—heartwood brown or light red, *Dysoxylum procerum* (B. LALI)— heartwood bright red, and *Chikrassia tabularis*—wood hard and suitable for planking and furniture.

Family 16 Rhamnaceae (900 *sp.*—51 *sp. in India*)

Habit—trees, shrubs and climbers. **Leaves** simple, alternate or rarely opposite, stipulate (somtimes spinous). **Inflorescence** an axillary cyme (often paniculate). **Flowers** small and inconspicuous, regular, bisexual or sometimes unisexual, usually pentamerous, sometimes tetramerous, perigynous (with the receptacle cup-shaped) to epigynous (with the receptacle united with the ovary); disc well developed (intrastaminal). **Calyx**—sepals (5-4), gamosepalous, valvate. **Corolla**—petals 5-4, polypetalous, often very small, clawed at the base and hooded above, sometimes absent. **Androecium**—stamens 5-4, opposite to and often enclosed by the petals. **Gynoecium**—carpels (3), syncarpous; ovary superior, free or immersed in fleshy disc, 3-locular, sometimes 2- or 1-locular, with 1 basal ovule in each. **Fruit** varying, a drupe or a nut or a dry one splitting into mericarps. **Seed** with thin endosperm. *Floral formula*— $\oplus \ \male \ K_{(5-4)} C_{5-4} A_{5-4} \underline{G}_{(3)}$.

Rhamnaceae is closely related to *Vitaceae*; the chief characters of *Rhamnaceae* distinguishing it from *Vitaceae* are: simple leaves (sometimes with spines), very small petals, structure of receptacle (free from or united with the ovary), ovary often sunken in receptacle or disc, fruit sometimes drupaceous, etc.

Examples. [The larger genera are: *Rhamnus* (over 100 sp.), *Zizyphus* (about 100 sp.), *Gouania* (about 70 sp.), and *Ventilago* (37 sp.)]. Indian plum or jujube (*Zizyphus mauritiana*=*Z. jujuba*)—a tree with edible fruits, *Z. oenoplia*—a sturdy shrub or undershrub, *Z. nummularia* and *Z. vulgaris*—shrubs, *Gouania leptostachya*—a strong climber with watch-spring-like tendrils, *Rhamnus nepalensis*—a rambling shrub, *Ventilago maderaspatana*—a strong hook-climber, etc.

Family 17 Sapindaceae (1,100 *sp.*—46 *sp. in India*)

Habit—trees, shrubs or lianes climbing by axillary tendrils which are often closely coiled. Balloon vine (*Cardiospermum*), however, is a slender tendril-climber (see FIG. I/37). The family shows anomalous secondary growth. Leaves alternate, usually pinnately compound, rarely simple. Inflorescence racemose or cymose. Flowers very small, regular or slightly zygomorphic, unisexual, monoecious, or bisexual, generally pentamerous, sometimes tetramerous; male flower often with rudimentary ovary, and female flower often with staminodes. Calyx—sepals 5, sometimes 4, usually polysepalous, imbricate. Corolla—petals 5, often 4 in regular flowers by the suppression of 1, free, imbricate, petals often with scales or tufts of hairs; an annular disc often present between the corolla and the androecium. Androecium—stamens often 8 by the suppression of 2, sometimes fewer in number, often inserted within the disc around the ovary. Gynoecium—carpels (3), syncarpous; ovary 3-locular, superior, with one ovule in each chamber. Fruit—dry (capsule or nut) or fleshy (berry or drupe), sometimes a schizocarp. Seed often arillate, exalbuminous; embryo curved.

Floral formula— $\oplus$ or $\cdot |\cdot \male - \female\, K_{5-4} C_{5-4} A_{4+4} \underline{G}_{(3)}$.

Examples. [The larger genera are: *Serjania* (over 300 sp., American), *Paullinia* (over 200 sp., American)—both are lianes with watch-spring-like tendrils, *Allophylus* (190 sp., tropical), and *Dodonaea* (50 sp.)]. Litchi (*Litchi chinensis*)—the edible part is the fleshy aril, longan (*Euphoria longana*)—aril edible, soap-nut (*Sapindus trifoliatus* and *S. mukorossi*)—fruits contain saponin which makes a lather with water and is used for washing silk and woollen fabrics, balloon vine (*Cardiospermum halicacabum*)—a common weed (see FIG. I/37), *Allophylus cobbe*—a shrub or small tree, *Aphania danura*—a small tree common in village shrubberies, *Dodônaea viscosa*—commonly grown as a hedge plant, *Schleichera trijuga*—a deciduous tree, furnishes the best Mirzapur lac and is also a valuable timber tree, etc.

Family 18 *Anacardiaceae* (600 sp.—58 sp. in India)

Habit—shrubs or trees, with often conspicuous resin. Leaves simple or pinnately compound, alternate, exstipulate. Inflorescence a panicle of many small flowers. Flowers small, regular, bisexual, sometimes polygamous, hypogynous to epigynous, typically pentamerous; disc present. Calyx—sepals usually 5, sometimes varying from 3 to 7, free or united. Corolla—petals as many as sepals, sometimes absent, free or connate. Androecium—stamens 10-5 (fertile stamens in many cases varying in number, as in *Anacardium* and *Mangifera* where only one stamen is fertile), free, inserted on an annular disc. Gynoecium—carpels commonly (3-1), rarely (5), syncarpous; ovary superior or sometimes inferior, often 1-celled, rarely 2- to

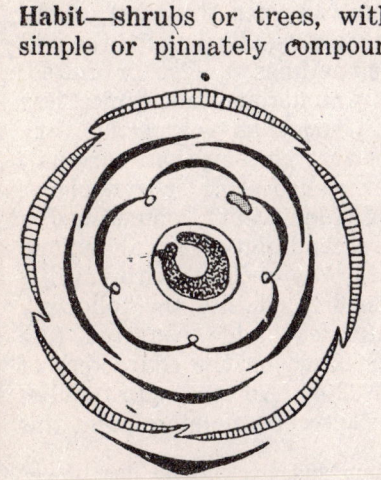

FIG. 23. Floral diagram of *Anacardiaceae* (*Mangifera*).

5-celled, with one ovule in each, often only one ovule matures into seed. Fruit commonly a 1-celled and 1-seeded drupe. **Seed** exalbuminous, with a large curved embryo. *Floral formula—* $\oplus \, \male \, K_5 C_5 A_{10-5} \underline{G}_{(3-1)}$.

Examples. [The larger genera are: *Rhus* (over 300 sp.), and *Semecarpus* (50 sp.)] Mango (*Mangifera indica*), cashew-nut (*Anacardium occidentale*), Indian hogplum (*Spondias pinnata= S. mangifera*), hogplum (*S. cytherea*), pistachio (*Pistacia vera*), marking-nut (*Semecarpus anacardium*; B. BHELA; H. BHILAWA), *Odina wodier* (B. JIYAL; H. JHINGAN), *Buchanania latifolia* (B. & H. PIYAL), *Tapiria hirsuta*, *Rhus* (some species are useful while others poisonous), e.g. *Rhus parviflora*—fruits sour, *R. semialata*—a common tree of the Khasi Hills, yields extremely sour fruits (sold in the Shillong market as NAGA-TENGA), *R. vernicifera*—the laquer tree of Japan, etc.

Family 19 *Leguminosae*[1] (12,000 *sp.*—951 *sp. in India*)

Habit—herbs, shrubs, trees, twiners and climbers. **Roots** of many species, particularly of *Papilionaceae*, have tubercles (see FIG. III/5). **Leaves** alternate, pinnately compound, rarely simple, as in rattlewort (*Crotalaria sericea*), camel's foot tree (*Bauhinia*) and some species of *Desmodium*, e.g. *D. gangeticum*, with a swollen leaf-base known as the pulvinus; stipules 2, usually free. **Flowers** bisexual and complete, regular or zygomorphic or irregular, hypogynous or slightly perigynous. **Calyx**—sepals usually (5) or 5, with the odd one anterior (away from the axis), sometimes (4), united or free. **Corolla**—petals usually 5, with the odd one posterior (towards the axis), sometimes 4, free or united. **Androecium**—stamens usually 10 or numerous, sometimes less than 10 by reduction, free or united. **Gynoecium**—carpel 1; ovary 1-celled, with 1 to many ovules, superior; placentation marginal; ovary often on a long or short stalk, called stipe or gynophore. **Fruit** commonly a legume or pod (dehiscent), sometimes a lomentum (indehiscent). This is the second biggest family among the dicotyledons (being second only to *Compositae*), with varying characters, and as such it has been divided into the following sub-families: *Papilionaceae*, *Caesalpinieae* and *Mimoseae* (see footnote). The division is primarily based on the characters of the corolla and the stamens (see FIGS. 2-4). All these sub-families are well represented in India. From an economic standpoint this

---

[1] The order Rosales according to Bentham and Hooker and also Engler includes both *Rosaceae* and *Leguminosae*; while Hutchinson has separated *Leguminosae* from Rosales and raised it to the rank of an order with three families—*Caesalpiniaceae*, *Mimosaceae* and *Papilionaceae*. It may also be noted that *Leguminosae* is the biggest family in India.

SELECTED FAMILIES OF DICOTYLEDONS

is one of the most important families; probably it ranks second to *Gramineae* in order of importance.

Distinguishing Characters of Leguminosae

	Papilionaceae	*Caesalpinieae*	*Mimoseae*
Leaves	usually 1-pinnate, rarely simple, stipels often present	1- or 2-pinnate, rarely simple, stipels absent	bipinnate, stipels present or absent
Flowers Inflor. Calyx	papilionaceous racemose gamosepalous, imbricate	zygomorphic racemose polysepalous, sometimes gamosepalous, usually imbricate	regular, small spherical head gamosepalous, usually valvate
Corolla	polypetalous, posterior petal largest and outermost, aestivation vexillary	polypetalous, posterior petal smallest and innermost, aestivation imbricate	gamopetalous, all petals equal, aestivation valvate
Androecium	stamens ten, (9) +1, rarely (10) or 10, pollen grains simple	ten or fewer, free, pollen grains, simple	often many or 10, rarely 4 or 8, free, pollen often compound
Floral formula—	$\cdot\!\!\mid\,\, \male K_{(5)} C_5 A_{(9)+1} \underline{G}_1$	$\cdot\!\!\mid\,\, \male K_5 C_5 A_{10} \underline{G}_1$	$\oplus \male K_{(5-4)} C_{(5-4)} A_{\infty \text{ or } 10} \underline{G}_1$

(For floral diagrams see p. 704)

(1) **Papilionaceae** (754 sp. in India). Herbs, shrubs, trees and climbers. **Leaves** unipinnate, sometimes trifoliate, rarely simple; stipels often present. **Inflorescence** usually a raceme. **Flowers** zygomorphic, polypetalous and papilionaceous. **Calyx**—sepals usually (5), gamosepalous, often imbricate, sometimes valvate. **Corolla**—petals usually 5, free, of very unequal sizes, the posterior largest one being the vexillum or standard, the two lateral ones being the wings or alae, and the two innermost ones (apparently united) forming the keel or carina; aestivation vexillary. **Androecium**—stamens ten, diadelphous—(9)+1, rarely 10, free, as in coral tree (*Erythrina*), or (10), connate, as in rattlewort (*Crotalaria*). Floral formula— $\cdot\!\!\mid\,\, \male K_{(5)} C_5 A_{(9)+1} \underline{G}_1$.

(2) **Caesalpinieae** (110 sp. in India). Shrubs and trees, rarely climbers or herbs. **Leaves** unipinnate or bipinnate, rarely simple, as in camel's foot tree (*Bauhinia*); stipels absent. **Inflorescence** commonly a raceme. **Flowers** zygomorphic or irregular and polypetalous. **Calyx**—sepals usually 5, polysepalous (sometimes

gamosepalous), imbricate. **Corolla**—petals usually 5, free, subequal or unequal, the odd or posterior one (sometimes very small)

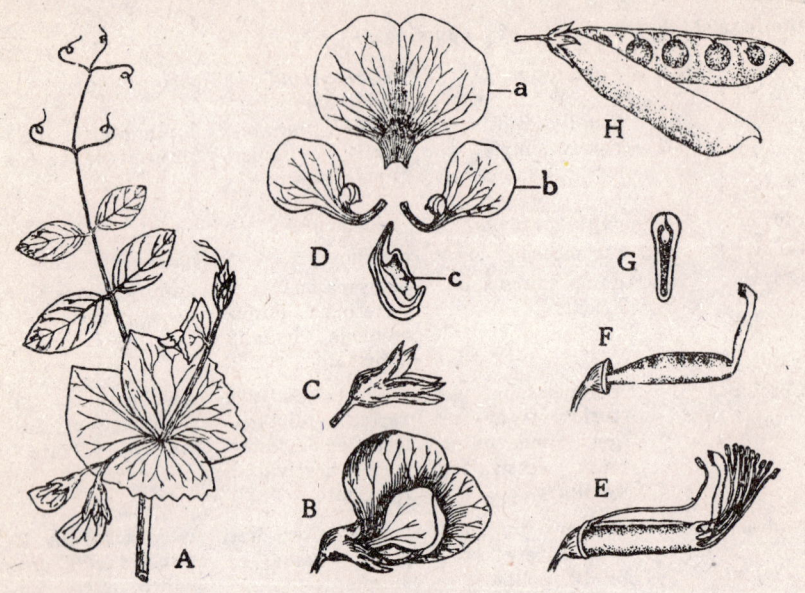

Papilionaceae. FIG. 24. Pea (*Pisum sativum*). *A*, a branch; *B*, a flower—papilionaceous (see also FIG. I/108); *C*, calyx; *D*, corolla—petals opened out (*a*, vexillum, *b*, wing; *c*, keel); *E*, stamens—(9)+1 and pistil; *F*, pistil—1 carpel (note the ovary, style and stigma); *G*, ovary in transection showing marginal placentation; and *H*, a fruit—legume.

always innermost; aestivation imbricate. **Androecium**—stamens 10 or less by reduction, free. *Floral formula*— ⊹ ⚥ $K_5 C_5 A_{10} \underline{G}_1$.

(3) *Mimoseae* (87 sp. in India). Shrubs and trees, sometimes herbs or woody climbers. **Leaves** bipinnate; stipels present or absent. **Inflorescence** a head or a spike. **Flowers** regular, often small and aggregated in spherical heads. **Calyx**—sepals (5) or (4), generally gamosepalous, valvate. **Corolla**—petals (5) or (4), mostly gamopetalous; aestivation valvate. **Androecium**—stamens usually ∞ sometimes 10 (as in *Entada, Neptunia, Prosopis* and *Parkia*), free, sometimes united at the base; pollen often united in small masses. *Floral formula*—⊕ ⚥ $K_{(5-4)} C_{(5-4)} A_{\infty \text{ or } 10} \underline{G}_1$.

Examples of Papilionaceae. [The larger genera are: *Astragalus* (over 1,600 sp.; xerophytic), *Crotalaria* (over 600 sp.). *Desmodium* (over 400 sp.), *Indigofera* (350 sp.), *Trifolium* (300 sp.),

Tephrosia (300 sp.), *Dalbergia* (over 250 sp.), *Phaseolus* (over 200 sp.), *Lupinus* (about 200 sp.), *Lathyrus* (130 sp.) and *Aeschy-*

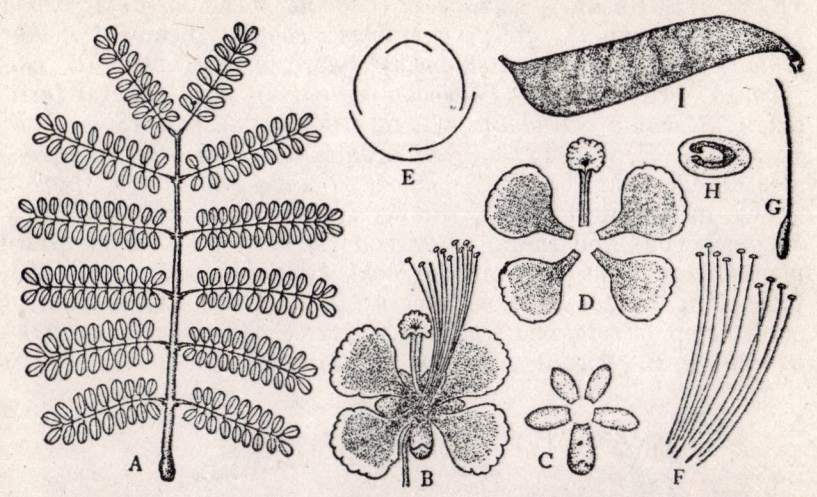

Caesalpinieae. FIG. 25. Dwarf gold mohur (*Poinciana pulcherrima*). A, a pinnately compound leaf; B, a flower; C, calyx; D, corolla—petals dissected out; E, aestivation (imbricate); F, stamens; G, pistil (one carpel); H, ovary in transection showing marginal placentation; and I, a fruit.

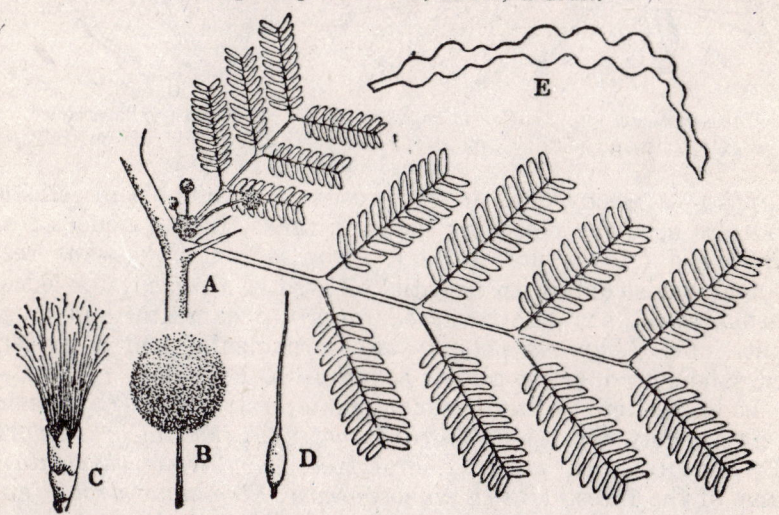

Mimoseae. FIG. 26. Gum tree (*Acacia nilotica*). A, a branch with bipinnate compound leaves; B, an inflorescence (head); C, a flower; D, pistil (one carpel); and E, a fruit (lomentum).

nomene (over 100 sp.)]. **Useful plants: pulses** (rich in proteins): Bengal gram (*Cicer arietinum*), lentil (*Lens culinaris*), pigeon pea or red gram (*Cajanus cajan*), pea (*Pisum sativum*), green gram (*Phaseolus aureus*), black gram (*P. mungo*), *Lathyrus sativus* (B. &. H. KHESARI), soya-bean (*Glycine max=G. soja*), broad bean (*Vicia faba*), etc.; **vegetables**: country bean (*Dolichos lablab*), cow pea (*Vigna sinensis*), sword bean (*Canavalia ensiformis*), French bean (*Phaseolus vulgaris*), etc.; **natural fertilizers**: *Sesbania cannabina* (B. DHAINCHA), sesban (*S. sesban*; B. JAINTI; H. JAINT), lucerne or alfalfa (*Medicago sativa*)—also an excellent fodder, *Tephrosia candida* and *Derris robusta* grown in tea gardens, etc.; **timber trees**: Indian redwood (*Dalbergia sissoo*) and Indian rosewood (*D. latifolia*); **other useful plants**: groundnut or peanut (*Arachis hypogaea*; see FIG. III/49), pith plant (*Aeschynomene aspera*; B. & H. SHOLA), Indian or sunn hemp (*Crotalaria juncea*), fenugreek (*Trigonella foenum-graecum*; B. METHI), indigo (*Indigofera tinctoria*), *Derris*

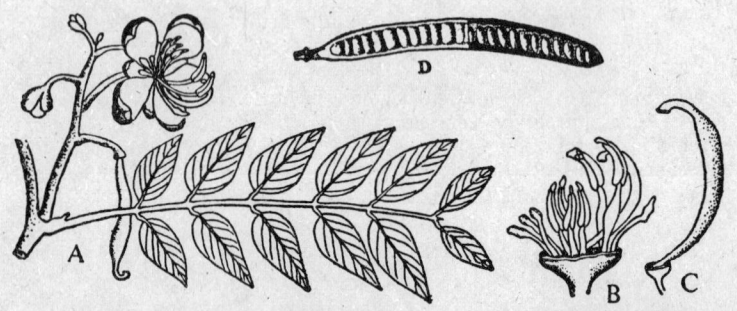

Caesalpinieae. FIG. 27. *Cassia sophera. A,* a branch with inflorescence; *B,* stamens and pistil; *C,* pistil (one carpel); and *D,* a fruit partially opened up.

elliptica—a woody climber, roots used as a valuable insecticide and also used for poisoning fishes in tanks, Indian liquorice or crab's eye (*Abrus precatorius*), *Psoralea corylifolia*—an erect annual, seeds contain an essential oil used as a remedy for leucoderma, sweet pea (*Lathyrus odoratus*)—ornamental and fragrant, lupin (*Lupinus polyphyllus*)—ornamental and a fodder, red sandalwood (*Pterocarpus santalinus*), *Pongamia pinnata*—a shade tree, etc.; **other common plants**: rattlewort (*Crotalaria sericea*), butterfly pea (*Clitoria ternatea*), *Sesbania grandiflora* (B. BAKPHUL; H. AGAST), coral tree (*Erythrina variegata*), flame of the forest (*Butea monosperma*), *Flemingia strobilifera*—a shrub with simple leaves and copious bracts, Indian telegraph plant (*Desmodium gyrans*), *D. gangeticum*, cowage (*Mucuna prurita*)—fruits with stinging hairs, wild pea (*Lathyrus aphaca*), wild indigo (*Tephrosia purpurea*), white clover (*Trifolium repens*)—a common prostrate weed bearing clusters of white flowers (in Shillong), etc.

SELECTED FAMILIES OF DICOTYLEDONS

Examples of *Caesalpinieae*. [The larger genera are: *Cassia* (450 sp.), *Bauhinia* (250 sp.), and *Caesalpinia* (100 sp.)]. **Useful plants**: tamarind (*Tamarindus indica*)—fruits widely used for sour preparations, Indian laburnum (*Cassia fistula*)—heartwood very hard and durable, flowers golden yellow, etc.; **medicinal**: Indian senna (*Cassia angustifolia*; B. SONAPAT or SONAMUKHI; H. SANAKKAPAT), *Saraca indica*, fever nut (*Caesalpinia crista= C. bonducella*), etc.; **dyes**: sappan or Brazil wood (*Caesalpinia sappan*; B. & H. BAKAM)—wood yields a valuable red dye used extensively for dyeing silk and wool, starch coloured with this dye forms 'ABIR' used in 'HOLI' festival, and pods yield a high percentage of tannin, logwood (*Haematoxylon*), an American plant—wood yields the dye haematoxylin; **ornamental**: camel's foot tree (*Bauhinia purpurea* and *B. variegata*), gold mohur (*Delonix regia*; see FIG. I/94), dwarf gold mohur or peacock flower (*Poinciana pulcherrima*; FIG. 25), Jerusalem thorn (*Parkinsonia aculeata*) and *Peltophorum ferrugineum*; **other common plants**: *Cassia sophera*, *C. occidentalis*, ringworm shrub (*C. alata*), *C. tora*, *C. auriculata*, etc.

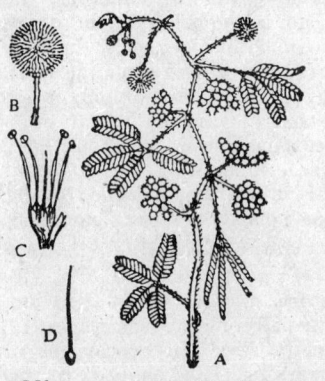

Mimoseae. FIG. 28. Sensitive plant (*Mimosa pudica*). A, a branch; B, an inflorescence; C, a flower; and D, pistil (one carpel).

Examples of *Mimoseae*. [The larger genera are: *Acacia* (780 sp.), *Mimosa* (over 400 sp.), *Inga* (200 sp.), *Pithecolobium* (120 sp.), and *Albizzia* (100 sp.)]. **Useful plants**: catechu (*Acacia catechu*)—catechu, a kind of tannin, is obtained by boiling chips of heartwood, *A. nilotica* (=*A. arabica*) and *A. senegal* yield gum, *A. dealbata*—an evergreen tree common in Shillong, its bark is rich in tannin, *A. farnesiana*—flowers very fragrant, used in perfumery, *A. pennata*—bark used for poisoning fish, many species of *Acacia* are sources of fuel and tannin, *Albizzia lebbek* (B. & H. SIRISH)—a timber tree, *A. procera*—wood suitable for tea boxes, many species of *Albizzia* are sources of fuel, rain tree (*Pithecolobium saman*)—planted as a shade tree, and *Parkia*—a handsome quick-growing lofty tree; **other common plants**: sensitive plant (*Mimosa pudica*)—a straggling weed, *M. himalayana* and *M. hamata*—shrubs, etc., *Neptunia oleracea* (B. & H. PANI-LAJUK), *Pithecolobium dulce* (B. & H. DEKANI-BABUL), nicker bean (*Entada gigas*=*E. phaseoloides*; B. & H. GILA)—pod up to a metre in length, and *Prosopis spicigera* (B. & H. SHOMI).

Family 20 *Rosaceae* (2,000 *sp.*—244 *sp. in India*)

Habit—herbs, shrubs, trees and climbers. **Leaves** simple or compound, alternate; stipules 2, often adnate to the petiole. **Inflorescence**—flowers solitary or in terminal cymes or racemes. **Flowers** (see FIG. I/125C) regular, bisexual, rosaceous, typically perigynous with the receptacle hollowed and cup-shaped, rarely epigynous (as in apple and pear). Disc often present in the form of a ring. **Calyx**—sepals 5, adnate to the receptacle, lobes free, sometimes with *epicalyx*. **Corolla**—petals 5 (many in cultivated roses), free, usually imbricate, alternating with the sepals, usually white or pink. **Androecium**—stamens numerous, incurved in the bud, arranged in cyclic order, rarely few. **Gynoecium**—carpels usually numerous, free (as in rose) or sometimes (5), united (as in apple and pear) or only 1 (as in plum and peach); ovary unilocular or 5-locular in syncarpous pistil, with usually 2 ovules in each loculus; ovules anatropous and pendulous. **Fruit** varying—drupe, follicle, berry, achene or pome. **Seeds** exalbuminous.
Floral formula— $\oplus \; \male \; K_5 C_5 A_\alpha G_\alpha$ or $(_5)$ or $_1$.

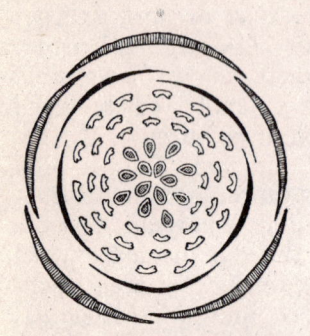

FIG. 29. Floral diagram of *Rosaceae*.

Rosaceae shows a wide range of floral structure: flowers bisexual to unisexual; petals 5, many or absent; stamens usually many, sometimes 10; carpels many, sometimes 5-10 or even 1 (as in *Prunus*), apocarpous or syncarpous; ovary superior or inferior; fruit varying. The family is related to *Leguminosae* by perigyny, and sometimes monadelphy, zygomorphy and simple pistil in some of its members. *Rosaceae* is also related to *Myrtaceae* through *Pyrus* having (2-5) carpels.

Economically this is an important family. Otto-of-rose (ATTAR) is mostly obtained from *Rosa damascena* and *R. centifolia*; Bulgaria, most famous for its roses, is the world's biggest centre of distillation of this essential oil; there are many fleshy edible fruits, e.g. plum, peach, prune, apricot, strawberry, apple, pear, etc.; and several varieties of rose are ornamental garden plants, and so also are many species of *Spiraea* (grown in hill stations).

Examples. [The larger genera are: *Potentilla* (over 300 sp.), *Rubus* (250 sp.), *Rosa* (over 200 sp.), *Prunus* (150 sp.), *Spiraea* (100 sp.), and *Pyrus* (50 sp.).] Dog rose (*R. canina*), wild rose (*R. gigantea*), wild rose of Bengal (*R. involucrata*), Damask or Bussora rose (*R. damascena* and *R. centifolia*), musk rose (*R. moschata*), *R. indica*, *R. alba*, etc., and several hybrids, *Spiraea cantoniensis*—flowers white, in clusters (grown as a hedge plant in Shillong), loquat (*Eriobotrya japonica*), plum (*Prunus domestica*)—prune (ALUBUKHRA) is the dried plum, peach (*P. persica*), apricot (*P. armeniaca*), almond (*P. amygdalus*), cherry (*P. avium*), etc., quince (*Cydonia oblonga*), strawberry (*Fragaria vesca*), wild strawberry (*F. indica*), apple (*Malus sylvestris*), pear (*Pyrus communis* and *P. pyrifolia*), silverweed (*Potentilla fulgens*)—common in the hills, raspberry (*Rubus idaeus*), wild raspberry (*R. moluccanus*), and many other wild species in the hills, and *Photinia notoniana*—a tree in the Khasi Hills.

Family 21 *Myrtaceae* (3,000 sp.—112 sp. in India)

Habit—shrubs and trees, rarely herbs; bicollateral bundles or internal phloem often present. **Leaves** simple, opposite, gland-dotted. **Inflorescence** cymose. **Flowers** regular, epigynous, bisexual; disc lining the calyx-tube. **Calyx**—sepals 4-5, free, or (4-5), connate, persistent or deciduous, valvate or imbricate. **Corolla**—petals 4-5, free, imbricate. **Androecium**—stamens numerous (rarely few), free, sometimes polyadelphous, epigynous. **Gynoecium**—carpels (2-5) or (∞), syncarpous; ovary crowned by a disc, inferior (or sometimes half-inferior), 1- to 2-locular, sometimes multilocular, with 2 to many ovules in each loculus; placentation axile (rarely parietal). **Fruit** a berry or capsule, inferior, usually with persistent calyx. **Seed** exalbuminous. *Floral formula*—

$\oplus \; \male \; K_{4-5} \text{ or } _{(4-5)} C_{4-5} A_\infty \bar{G}_{(2-5) \text{ or } (\infty)}$

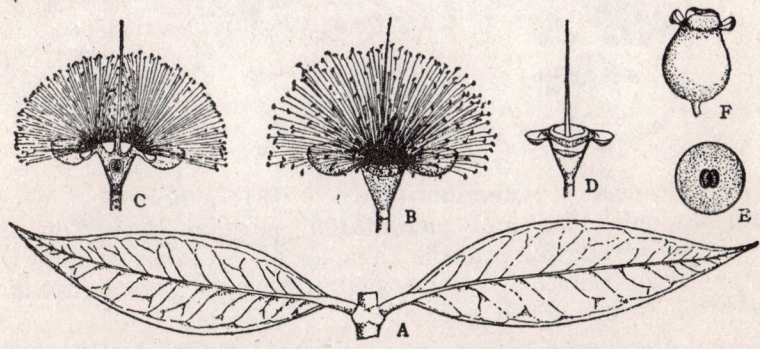

Myrtaceae. FIG: 30. Rose-apple (*Syzygium jambos*). *A*, opposite leaves; *B*, a flower; *C*, a flower cut longitudinally; *D*, pistil; *E*, section of ovary showing axile placentation; and *F*, a fruit.

Examples. [The larger genera are: *Eucalyptus* (about 600 sp.), *Syzygium* (500 sp.), *Psidium* (over 100 sp.), *Melaleuca* (100 sp.), and *Myrtus* (about 100 sp.)]. **Useful plants:** *Eucalyptus*—leaves yield eucalyptus oil, clove (*Syzygium aromaticum*), blackberry (*S. cuminii*), wild berry (*S. fruticosum*), rose-apple (*S. jambos*), Malay apple (*S. malaccense*), guava (*Psidium guayava*), allspice or pimento (*Pimenta officinalis*)—dried unripe fruits form allspice which combines the flavour of cloves, nutmeg and cinnamon, cajeput (*Melaleuca leucadendron*)—leaves yield cajeput oil, and timber useful, etc.; **other common plants:** *Barringtonia acutangula* (B. & H. HIJAL), myrtle (*Myrtus communis*)—an ornamental shrub, bottlebrush tree (*Callistemon lanceolatus*), etc.

Family 22 *Cucurbitaceae* (750 sp.—84 sp. in India)

Habit—tendril climbers; tendrils extra-axillary, simple or branched. **Leaves** simple, alternate and palmately veined. **Flowers** regular, unisexual, epigynous and monoecious or dioecious. **Calyx**—sepals (5), united, often deeply 5-lobed. **Corolla**—petals (5), united, often deeply 5-lobed, imbricate; inserted on the calyx-tube. *Male Flowers*: **androecium**—stamens usually 3, sometimes 5, varying in character; sometimes they are free but more commonly they are united in a pair (or in 2 pairs when stamens 5) throughout their whole length (*synandrous*), the odd one remaining free; in some cases the anthers only are united (*syngenesious*); each anther 1-lobed or 2-lobed; paired stamens have either 2-lobed or 4-lobed anthers; anther-lobes variously folded, or *sinuous*, i.e. twisted like a transverse ∽. Rudiments of the pistil sometimes present.

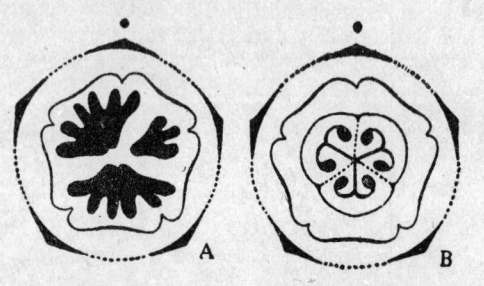

FIG. 31. Floral diagrams of *Cucurbitaceae*. A, male flower; B, female flower.

Female Flowers: **gynoecium**—carpels (3), syncarpous; ovary inferior, unilocular and placentation parietal but often the placentae intrude far into the chamber of the ovary making the latter falsely trilocular; ovules many; style 1; stigmas 3 which are often forked. **Fruit a pepo.**

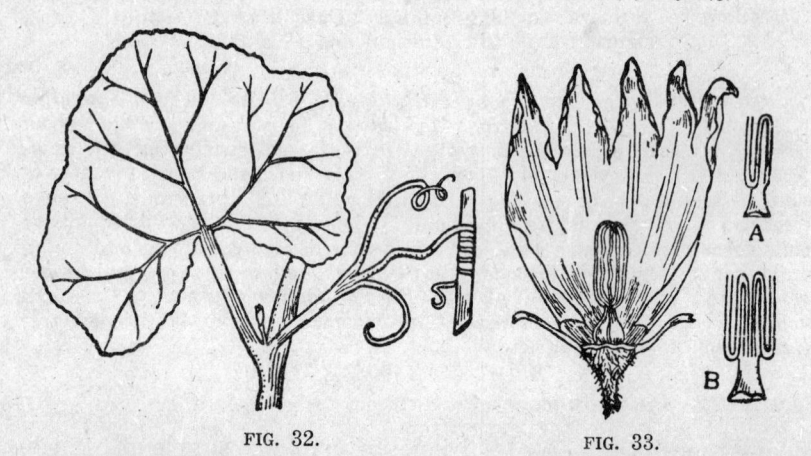

Floral formulae— $\oplus \male \female$ or $\male - \female \, K(_5)C(_5)A_{3 \text{ or } 5} \underline{G}_0 \mid A_0 \bar{G}(_3)$.

FIG. 32. FIG. 33.
Cucurbitaceae. FIG. 32. Gourd (*Cucurbita moschata*). Portion of a branch with a leaf and a tendril. FIG. 33. Male flower of the same. A, one stamen; B, two stamens united together.

The systematic position of *Cucurbitaceae* is disputed. There is controversy also regarding the nature of placentation (axile or parietal) and of tendrils. Bentham and Hooker have placed *Cucurbitaceae* and *Passifloreae* together in the same cohort (order), Passiflorales among polypetalous orders, not considering them as advanced, on the basis of the following characters: tendrils, regular flowers (gamopetalous or polypetalous), syncarpous pistil, 1-locular ovary, parietal placentation, etc. *Passifloreae*, however, is distinguished from *Cucurbitaceae* by its stipulate leaves, axillary tendrils, mostly bisexual flowers, often androgynophore and corona, free stamens, superior ovary, etc. Engler, however, considers *Cucurbitaceae* as a much advanced family and places it in Sympetalae as an only family of Cucurbitales prior to *Campanulaceae* and close to *Compositae*. Hutchinson has likewise separated the two families into two orders, Cucurbitales and Passiflorales, but regarding their systematic position he seems to have shared the same view as that of Bentham and Hooker. There is no doubt that *Cucurbitaceae* is closely related to *Campanulaceae*, the relationship being based on pentamerous symmetry, gamopetaly, epigyny, reduction in the number of stamens and carpels and syngenesious stamens (sometimes also found in *Campanulaceae*).

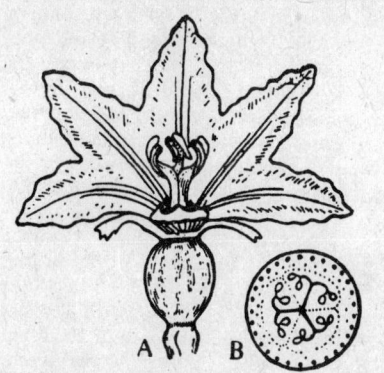

FIG. 34. *A*, female flower of *Cucurbita moschata*; *B*, ovary in transection showing placentation.

Plants of this family are mostly used as vegetables, a few yield delicious summer fruits, and a few are medicinal.

Examples. [The larger genera are: *Momordica* (45 sp.), *Cyclanthera* (40 sp.; American), *Cucumis* (25 sp.), *Trichosanthes* (25 sp.), and *Cucurbita* (15 sp.)]. Sweet gourd or musk melon (*Cucurbita moschata*), pumpkin or vegetable marrow (*C. pepo*; B. KUMRA; H. HALWAKADDU)—squash is a variety of it, giant pumpkin or giant gourd (*C. maxima*), bottle gourd (*Lagenaria siceraria*), cho-cho or chayote (*Sechium edule*)—commonly grown in hill stations, fruits (each with a large seed) are popular as a vegetable, snake gourd (*Trichosanthes anguina*), *T. dioica* (B. PATAL; H. PARWAL), bitter gourd (*Momordica charantia*; B. UCHCHE and KARALA; H. KARELI), *M. cochinchinensis* (B KAKROL; H. CHATTHAI), ash or wax gourd (*Benincasa hispida*), ribbed gourd (*Luffa acutangula*), bath sponge or loofah (*L. cylindrica*), *Coccinia indica* (B. TELAKUCHA; H. KUNDARU), cucumber (*Cucumis sativus*), melon (*C. melo*), water melon (*Citrullus lanatus*), colocynth (*C. colocynthis*; B. MAKAL: H INDRAYAN)—medicinal, and *Bryonia*—medicinal.

Family 23 Cactaceae   (about 2,000 *sp.*—6 *sp. in India*)

Habit—mostly succulent herbs, some are shrubs or climbers, strongly xerophytic in habit; it is predominantly a tropical American family. **Stem** often fleshy, exhibiting a variety of peculiar forms, often modified into phylloclade.

Leaves often modified into scales; axillary spines and sometimes a cluster of bristles present; spines are borne on the *tubercle* often mixed with hairs, usually from a depressed area at its tip known as the *areole*. The morphology of the tubercle and the spines is variously interpreted. *Pereskia* is, however, a much-branched shrub with flat green leaves and strong thorns. Roots often very deep. (Anatomically abundant water-storing parenchyma, mucilage-containing cells, thick cuticle, mechanical tissues in the ridges, sunken stomata, etc., are characteristic of this family). Flowers often solitary, sometimes very large, often brightly coloured, but white in night-blooming cacti, regular, bisexual, epigynous; they arise from the areole or from the axil of a tubercle, as in *Mammillaria*. **Perianth**—tepals many, often united to form a tube, spirally arranged, showing a gradual transition from sepaloid to petaloid stages. **Androecium**—stamens numerous, often epiphyllous. **Gynoecium**—carpels $(3-\infty)$, united; ovary inferior, 1-chambered, with $3-\infty$ parietal placentae bearing numerous anatropous ovules; style simple; stigmas correspond to the number of placentae. **Fruit** a many-seeded berry, sometimes edible. **Seed** with or without endosperm. *Floral formula*- $\oplus \; \male \; P_{(\alpha)} A_\alpha \bar{G}_{(3-\alpha)}$.

Cactaceae. FIG. 35. *A, Cereus triangularis; B, Phyllocactus latifrons* (see also FIG. I/40B).

The affinity of *Cactaceae* is difficult to trace. Bentham and Hooker have placed *Ficoideae* (*Aizoaceae* of Engler) and *Cactaceae* in one cohort. There may be a distant affinity of *Cactaceae* with *Aizoaceae*. Engler has separated the two families, and placed *Cactaceae* under the order Opuntiales close to Myrtiflorae with which the former has some affinity in floral structure, particularly stamens and inferior ovary. Some cacti bear external resemblances to certain species of *Euphorbia* in general habit but are readily distinguished from them by having no latex.

Examples: [The larger genera are: *Opuntia* (over 250 sp.), *Rhipsalis* (60 sp.), *Cereus* (50 sp.), *Echinopsis* (35 sp.), and *Melocactus* (30 sp.)]. Prickly pear (*Opuntia dillenii*; FIG. 36)—a troublesome weed, sometimes bearing edible fruit, *Rhipsalis cassytha*—a small fleshy shrub in Madhya Pradesh, *Nopalea*—similar to *Opuntia* but smaller in size, night-blooming cacti (*Cereus, Phyllocactus*, etc.), *Cereus grandiflorus*—flowers sweet-scented, *C. giganteus*—largest of all cacti, *C. hexagonus, C. multangularis* and *C. tetragonus*—commonly grown as hedge plants, *C. triangularis* (FIG. 35A)—a climber, etc., *Phyllocactus latifrons*—flat-stemmed (FIG. 35B); (see also FIG. I/40B), *Echinocactus, Melocactus*, Christmas cactus (*Epiphyllum* *truncatum*; see FIG. 1/39C), *Pereskia bleo*—a large thorny shrub with flat green leaves. Indigenous cacti are very few but a good number of exotic species are grown in Indian gardens as ornamental plants.

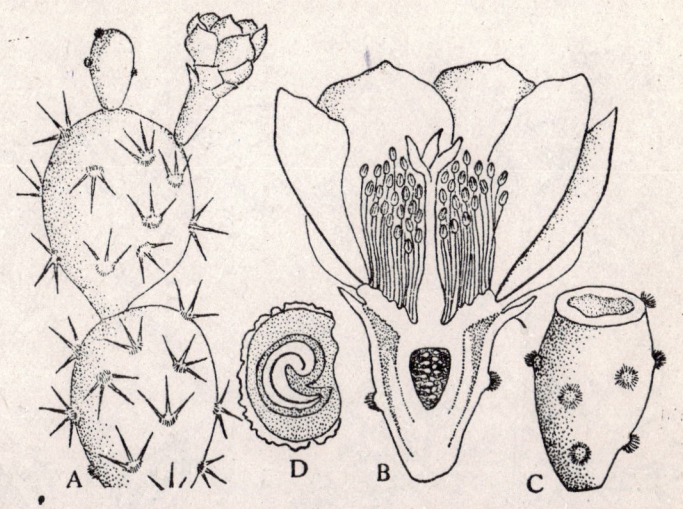

Cactaceae. FIG. 36. Prickly pear (*Opuntia dillenii*). *A*, plant (portion) with flower, thorns and bristles; *B*, flower cut longitudinally; *C*, fruit; *D*, seed cut longitudinally showing curved embryo.

Family 24 Umbelliferae (2,900 *sp.*—176 *sp. in India*)

Habit—herbs (rarely shrubs); stem usually fistular. **Leaves** alternate, simple, often much divided, sometimes decompound; petiole usually sheathing at the base. **Inflorescence** an umbel, usually compound or in a few cases simple as in *Centella*. **Flowers** regular (actinomorphic) or sometimes zygomorphic, epigynous, bisexual or polygamous, outer flowers sometimes rayed; mostly protandrous; bracts in the form of an involucre. **Calyx**—sepals 5, free, adnate to the ovary, often considerably reduced in size. **Corolla**—petals 5, rarely absent, free, adnate to the ovary, sometimes unequal, margin often incurved, valvate or imbricate. **Androecium**—stamens 5, free, alternating with the petals, epigynous; filaments bent inwards in the bud; anthers introrse. **Gynoecium**—carpels (2), syncarpous; ovary inferior, 2-celled, antero-posterior, crowned by a 2-lobed epigynous disc

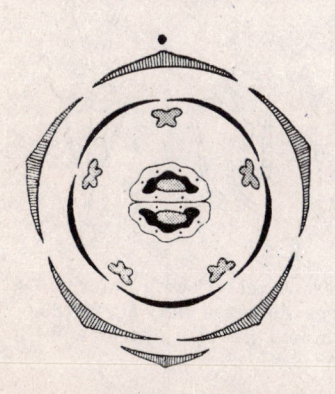

FIG. 37. Floral diagram of *Umbelliferae*.

(stylopodium) with two free styles arising from it; stigmas capitate; ovules 2, solitary in each cell, pendulous. **Fruit a cremocarp consisting of two indehiscent carpels laterally or**

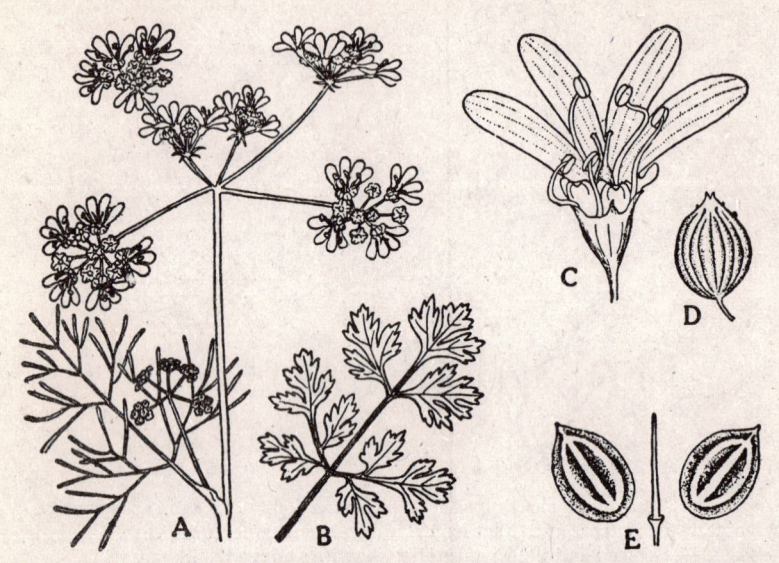

Umbelliferae. FIG. 38. Coriander (*Coriandrum sativum*). A, a branch with leaf and compound umbels; B, a lower leaf; C, a flower; D, a fruit; E, a fruit split into two mericarps, and the carpophore (central axis).

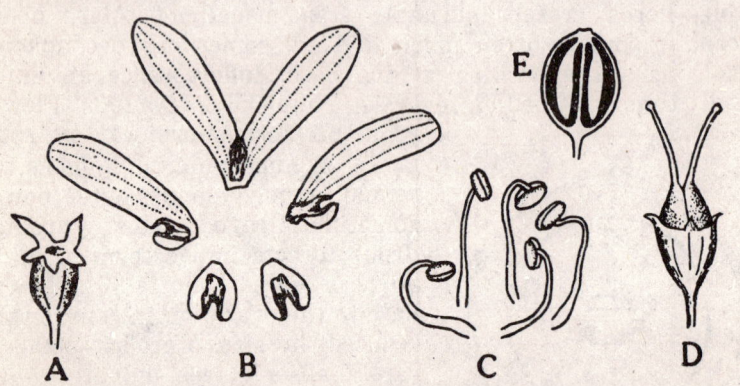

Umbelliferae (contd.). FIG. 39. Coriander. A, calyx with inferior ovary; B, petals dissected out; C, stamens dissected out; D, pistil with two styles, bilobed disc, calyx-teeth and inferior ovary; and E, ovary in longitudinal section.

dorsally compressed, breaking up into two parts, called *mericarps*, which remain attached to a slender, often forked, axis (*carpophore*); each mericarp usually shows five longitudinal ridges and oil-canals (*vittae*) in the furrows. **Seeds** 2, solitary in each mericarp, albuminous.

Floral formula— ⊕ or ·|· ⚥ $K_5 C_5 A_5 \bar{G}_{(2)}$.

Floral Range in *Umbelliferae*. Flowers commonly in a compound umbel, sometimes in a compact mass (head) as in *Eryngium*, or sometimes in a simple umbel or even single, as in some species of *Hydrocotyle* (=*Centella*). Flowers regular or zygomorphic, bisexual or polygamous but always epigynous. Flower development is, however, peculiar: stamens appear first, then petals followed by sepals, last of all the two carpels—at first separate but later united; by the rapid ingrowth of the ovary it soon becomes inferior. Disc variously 2-lobed. Sepals adnate to ovary, 5-toothed or entire or reduced to a few scales or to a narrow circular ridge. Petals 5, regular or zygomorphic, some petals of outer flowers often rayed; aestivation imbricate or valvate. Stamens and carpels more or less constant. Fruit—mericarps show a variety of forms and help identification of genera and species. *Umbelliferae* maintains a connexion with *Compositae* by the following characters: herbaceous nature of plants, reduction of calyx, presence of involucre, dense inflorescence (a near approach to capitulum of *Compositae*), outer flowers often sterile and rayed, bicarpellate pistil with two distinct styles, inferior ovary, solitary ovule in the ovarian chamber, etc. The fruit characters, however, readily distinguish the two families.

Examples. [The larger genera are: *Eryngium* (200 sp.), *Pimpinella* (200 sp.), *Ferula* (over 100 sp.), *Peucedanum* (over 100 sp.), *Hydrocotyle* (100 sp.), *Daucus* (60 sp.), *Centella* (40 sp.), *Oenanthe* (40 sp.), and *Carum* (30 sp.)]. **Useful plants: condiments and spices:** coriander (*Coriandrum sativum*), fennel (*Foeniculum vulgare*; B. PAN-MOURI; H. SAUNF), ajwan or ajowan (*Trachyspermum ammi*=*Carum copticum*; B. JOWAN; H. AJOWAN), *Carum roxburghianum* (B. RANDHANI), caraway (*C. carvi*; B. & H. SHIAJIRA), cumin (*Cuminum cyminum*; B. JIRA; H. SAFED-JIRA), dill (*Anethum graveolens*; B. SULPA; H. SOWA), parsley (*Petroselinum crispum*), anise (*Pimpinella anisum*), etc.; **vegetables:** carrot (*Daucus carota*), parsnip (*Pastinaca sativa*) and celery (*Apium graveolens*); **medicinal:** asafoetida (*Ferula assa-foetida*; B. & H. HING)—asafoetida (HING) of commerce is obtained from the roots, and Indian pennywort (*Centella asiatica*; B. THULKURI; H. BRAHMI)— leaves used as a remedy for dysenteric troubles in children; **other common plants:** wild coriander (*Eryngium foetidum*), *Centella rotundifolia*, dropwort (*Oenanthe*), e.g. *O. bengalensis* and *O. stolonifera*—common weeds of wet places, *Seseli indicum* (B. BAN-JOWAN)—a common much-branched weed, etc.

SUB-CLASS II GAMOPETALAE

Family 25 *Rubiaceae* (6,000 *sp.*—489 *sp. in India*)
Habit—herbs (erect or prostrate), shrubs, trees and climbers, sometimes thorny. **Leaves** simple, entire, opposite (decussate)

or whorled, with interpetiolar (sometimes intrapetiolar) stipules. **Inflorescence**—typically cymose, frequently dichasial and much branched, sometimes in globose heads. **Flowers** regular, bisexual, epigynous, sometimes dimorphic, as in some species of *Randia*

FIG. 40. Floral diagrams of *Rubiaceae*.

and *Oldenlandia*. **Calyx**—sepals usually (4), sometimes (5), gamosepalous, calyx-tube adnate to the ovary. **Corolla**—petals usually (4), sometimes (5), gamopetalous, generally rotate; aestivation valvate, imbricate or twisted. **Androecium**—stamens as many as petals, epipetalous, inserted within or at the mouth of the corolla-tube, alternating with the corolla-lobes. **Gynoecium**—carpels (2), syncarpous; ovary inferior, commonly 2-locular, with 1-∞ ovules in each; disc usually annular, at the base of the style. **Fruit** a berry, drupe or capsule. **Seed** with fleshy or horny endosperm.

Floral formula—$\oplus \male K_{(4-5)} \overline{C_{(4-5)}} A_{4-5} \bar{G}_{(2)}$.

Rubiaceae is related to *Caprifoliaceae* but in the latter the interpetiolar stipules are wanting and the carpels are (5-3). *Rubiaceae* is also distantly related to *Compositae* by head or captulum, as in *Anthocephalus*, *Uncaria*, *Nauclea*, *Adina*, etc.

Examples. [The larger genera are: *Psychotria* (over 600 sp.), *Ixora* (over 300 sp.), *Pavetta* (over 300 sp.), *Galium* (300 sp.), *Gardenia* (250 sp.), *Randia* (over 200 sp.), *Oldenlandia* (over 200 sp.), *Mussaenda* (200 sp.)]. **Useful plants: medicinal:** *Cinchona* yields quinine which is extracted from root- and stem-bark, ipecac (*Psychotria ipecacuanha*=*Cephaelis ipecacuanha*) yields emetine, *Paederia foetida*—leaves form a good stomachic, *Oldenlandia corymbosa*—entire plant used as a remedy for jaundice, liver disorders and remittent fever, etc.; **ornamental:** *Ixora coccinea*, *I. parviflora*, *Pavetta indica*, *Gardenia jasminoides* (=*G. florida*), *Anthocephalus indicus* (B. & H. KADAM), *Adina cordifolia* (B. KELI-KADAM; H. HALDU), *Cephalanthus occidentalis* (B. & H. PANI-KADAM), *Randia fasciculata*, *Hamelia patens*, *Mussaenda*—flowers usually yellow or orange and one of the sepals much enlarged (see FIG. I/103), the latter being white in *M. frondosa* and *M. roxburghii*, cream or yellow in

M. incana, and bright scarlet in *M. erythrophylla*, etc.; **dyes**: madder (*Rubia cordifolia*) and *Morinda tinctoria*; **beverage**: coffee (*Coffea arabica* and *C. robusta*)—seeds are the source of coffee powder; **other common plants**: *Coffea bengalensis, Oldenlandia diffusa, Dentella repens*—all growing wild, *Vangueria spinosa* (B. & H. MOYNA)—a thorny shrub, *Uncaria macrophylla*—a large woody hook-climber (see FIG. I/18B), *Galium rotundifolium*—a diffuse herb common in the Khasi Hills, etc.

Family 26 *Compositae* or *Asteraceae* (14,000 sp.—674 sp. in India)

Habit—herbs and shrubs, rarely a twiner, e.g. *Mikania scandens*, or a tree, e.g. *Vernonia arborea*, sometimes with internal phloem; some genera with latex, e.g. *Sonchus, Crepis, Lactuca, Picris*, etc. **Leaves** simple, alternate or opposite, rarely compound. **Inflorescence** a head (or capitulum), with an involucre of bracts. **Flowers** (florets) are of two kinds—the central ones (called *disc florets*) are tubular, and the marginal ones (called *ray florets*) are ligulate; sometimes all florets are of one kind, either tubular or ligulate.

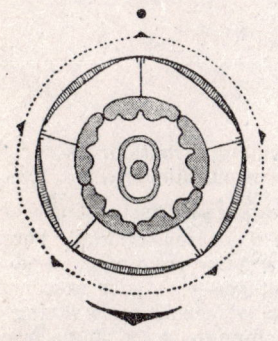

FIG. 41. Floral diagram of *Compositae* (disc floret).

Disc Florets: regular, tubular, bisexual and epigynous, each usually in the axil of a bracteole. **Calyx** often modified into a cluster of hairs called pappus, as in *Tridax* and *Ageratum*, or into scales, as in sunflower and *Eclipta*, or absent, as in water cress (*Enhydra*). **Corolla**—petals (5), gamopetalous, tubular. **Androecium**—stamens 5, epipetalous, filaments free but anthers united (syngenesious). **Gynoecium**—carpels (2), syncarpous; ovary inferior, 1-celled with one basal, anatropous ovule; style 1; stigma bifid. **Fruit**—a cypsela. *Floral formula*— $\oplus\ \male\female$ Kpappus or o $\overline{C_{(5)}A}_{(5)}\bar{G}_{(2)}$.

Ray Florets: zygomorphic, ligulate, unisexual (female) or sometimes neuter, as in sunflower, and epigynous, each usually in the axil of a bracteole. **Calyx** usually modified into pappus, sometimes it is scaly or absent. **Corolla**—petals (5), gamopetalous, ligulate (strap-shaped). **Gynoecium**, as in the disc florets. **Fruit** the same. *Floral formula*—$\cdot|\cdot\ \female$ Kpappus or o $C_{(5)}\bar{G}_{(2)}$.

Systematic Position of *Compositae*. For its many special characters *Compositae* is assigned an advanced position, the highest according to Engler, in systematic botany. This means that the family is of recent origin; fossils of it have been traced down to the Oligocene only (and not further back). Fossil records indicate that the genus *Senecio* came into existence first and other genera developed from it in due course. It is likely *Compositae* and *Rubiaceae* have arisen from a common ancestry. The former also maintains a phylogenetic connexion with *Umbelliferae* through inflorescence and floral mechanism. *Compositae* is remarkable in many respects: it has the maximum number of species among dicotyledons, and some genera with very

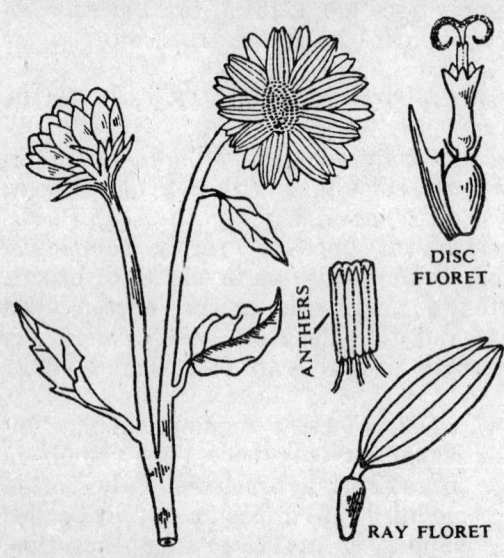

Compositae.
FIG. 42. Sunflower (*Helianthus annuus*). Note the branch with inflorescences (heads); disc floret (bisexual), with a bracteole at the base; anthers (syngenesious), and ray floret (neuter or female).

large number of species (see below); its worldwide distribution; its variety of forms; and its very effective mechanism for cross-pollination (see p. 102).

Special Features of *Compositae*. The special features characterizing *Compositae* as an advanced family are as follows: (1) predominantly herbaceous forms; (2) flowers massed together in a head (a perfect type of inflorescence) with the following decided advantages—(*a*) greater conspicuousness to attract insects for cross-pollination; (*b*) considerable saving of corolla-material; and (*c*) achievement of cross-pollination by a single insect within a very short time; (3) very simple but effective type of floral mechanism to achieve cross-pollination without at the same time losing a chance for self-pollination if the former method fails; (4) easy access of insects to the nectary which lies at the base of the style and is protected from rain; (5) flowers typically 5-merous, gamopetalous, bisexual, epigynous with definite number of stamens and carpels—an almost similar type of floral construction throughout the whole family; (6) efficient protection of floral buds by the involucral bracts; (7) very effective mechanism for seed- (fruit-) dispersal by parachute-like pappus (calyx) or by hooks or glands developing on the fruit.

Examples. [The larger genera are: *Senecio* (2,500 sp.), *Eupatorium*, an American genus (over 1,000 sp.), *Vernonia* (about 1,000 sp.), *Centaurea* (600 sp.), *Aster* (500 sp.), *Artemisia* (over 300 sp.), *Mikania*, predominantly American (over 200 sp.), *Gnaphalium* (200 sp.), *Bidens* (about 200 sp.), and *Erigeron* (about 200 sp.)]. **Useful plants: medicinal:** Indian wormwood

SELECTED FAMILIES OF DICOTYLEDONS

(*Artemisia nilagirica*=*A. vulgaris*; B. NAGDONA; H. NAG-DAMAN), santonin (*A. cina*), *Vernonia anthelmintica* (B. SOMRAJ; H. KALIZIRI), *Eupatorium ayapana* (B. & H. AYAPANA), *Wedelia calendulacea, Eclipta alba*, etc.; **vegetables**: chicory (*Cichorium intybus*; B. & H. KASNI)—its carrot-like roots are also roasted, ground and mixed with coffee, endive (*C. endivia*), lettuce (*Lactuca sativa*), *Enhydra fluctuans* (B. HALENCHA; H.

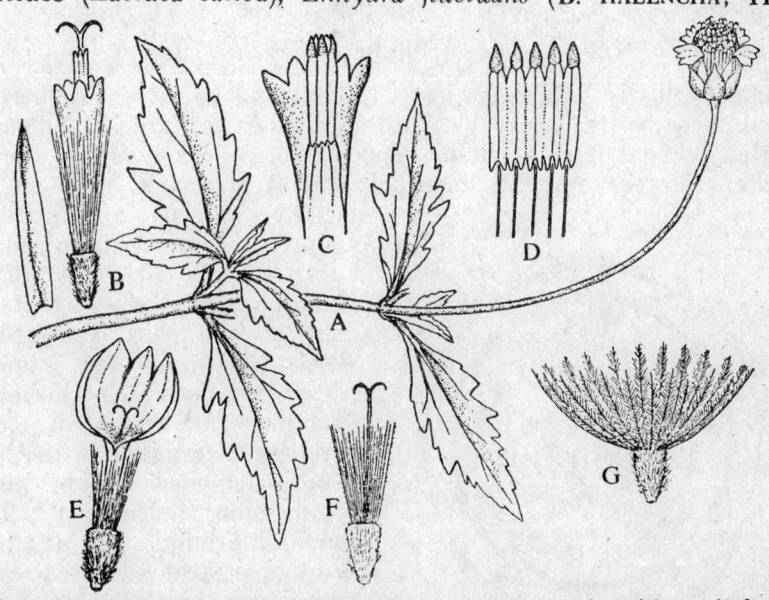

Compositae. FIG. 43. *Tridax procumbens. A*, a branch with capitulum; *B*, a disc floret with a bracteole; *C*, corolla (split open) with epipetalous stamens; *D*, syngenesious stamens (split open); *E*, a ray floret; *F*, pistil and pappus; and *G*, a fruit (cypsela) with pappus (parachute mechanism).

HARKUCH), Jerusalem artichoke (*Helianthus tuberosus*; B. & H. HATICHOKE), etc.; **oils:** safflower (*Carthamus tinctorius*)—seeds yield a good quality edible oil (about 25%) used for cooking, paint- and soap-making while the flowers yield a red dye, sunflower (*Helianthus annuus*; FIG. 42)—seeds yield about 30% of cooking oil, etc.; **insecticides:** a few species of *Chrysanthemum* (*Pyrethrum*), e.g. *C. cinerariefolium* yielding more or less 1% pyrethrin; **ornamental**: sunflower (*Helianthus annuus*), Mexican sunflower (*Tithonia tagetiflora*), *Zinnia, Cosmos, Dahlia*, daisy (*Bellis*), tree dahlia (*Montanoa*; in Shillong), *Calendula, Aster, Chrysanthemum, Gerbera*, golden rod or crest (*Solidago virgaurea*), marigold (*Tagetes patula*), everlasting flower (*Helichrysum*), *Centaurea*, e.g. sweet sultan (*C. moschata*), cornflower (*C. cyanus*), etc., *Erigeron mucronatus*—a common small herb in hill slopes (in Shillong), etc.; **other common plants:** goat-weed (*Ageratum conyzoides*) with purplish heads, *Blumea* *indicum*), *G. luteo-album* and *Anaphalis contorta*—common weeds in hill stations, *Eupatorium odoratum*—a common scandent shrub, elephant's foot (*Elephantopus scaber*), *Sonchus asper* and *Launea asplenifolia*—annual weeds with latex, *Tridax procumbens* (FIG. 43), cockle-bur (*Xanthium strumarium*), *Mikania scandens*—a large twiner, *Vernonia cineria* (B. KUKSHIM; H. SAHADEVI)—a herb, and *V. arborea*—a tree.

Family 27 **Apocynaceae** (1,400 *sp.*—67 *sp. in India*)

Habit—mostly twining or erect shrubs and lianes, a few herbs and trees; with latex; bicollateral bundles or internal phloem often present. **Leaves** simple, opposite or whorled, rarely alternate. **Flowers** regular, bisexual and hypogynous, in cymes, usually salver- or funnel-shaped, often with corona. **Calyx**—sepals (5), rarely (4), gamosepalous, often united only at the base. **Corolla**—petals (5), rarely (4), gamopetalous, twisted. **Androecium**—stamens 5, rarely 4, epipetalous, alternating with the petals, included within the corolla-tube; anthers usually connate around the stigma and apparently adnate to it. Disc present, ring-like or glandular. **Gynoecium**—carpels 2 or (2), apocarpous or syncarpous, superior. When apocarpous each ovary is 1-celled with marginal placentation, and when syncarpous the ovary may be 1-celled with parietal placentation, or 2-celled with axile placentation; ovules 2-∞ in each. **Fruit** a pair of follicles or berries or drupes. **Seeds** often with a crown of long silky hairs; mostly with endosperm. *Floral formula*— $\oplus \ \male \ K_{(5)} \overline{C_{(5)}} A_5 \underline{G}_{2 \text{ or } (2)}$.

FIG. 44. Floral diagram of *Apocynaceae*.

Apocynaceae is related to *Asclepiadaceae* in general habit, bicollateral vascular bundles, latex tubes, and general floral and fruit characters. The two families, however, can be easily distinguished from each other by the characters of androecium and gynoecium. In *Apocynaceae* stamens distinct (not united with the stigma), corona (staminal) absent, pollen grains distinct (not in pollinia), and style one (not two).

Examples. [The larger genera are: *Tabernaemontana* (110 sp.), *Rauwolfia* (about 100 sp.), *Ervatamia* (80 sp.), *Alstonia* (50 sp.), and *Landolphia* (50 sp.)]. **Useful plants: medicinal:** shrubs—*Rauwolfia serpentina* (B. SARPAGANDHA; H. SARPGAND),

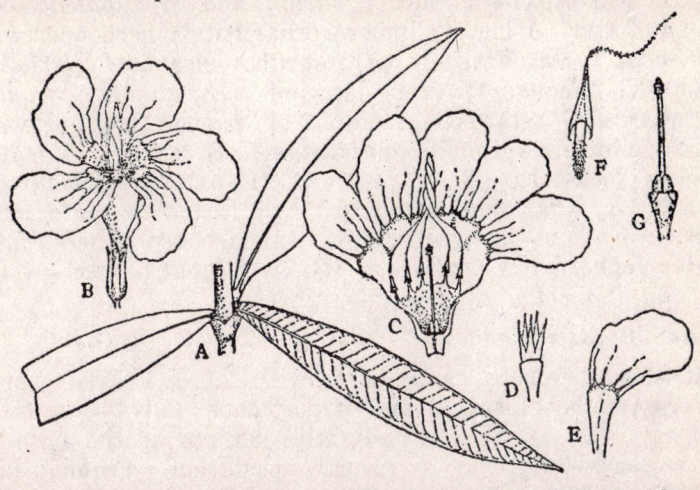

Apocynaceae. FIG. 45. Oleander (*Nerium indicum*). *A*, a whorl of leaves; *B*, a flower; *C*, a flower opened out; *D*, calyx; *E*, a petal; *F*, a stamen (connective with hairy appendage); and *G*, pistil.

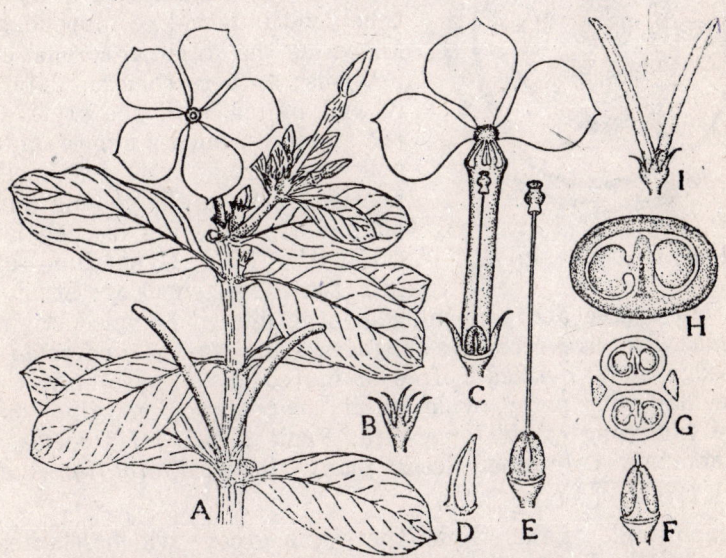

Apocynaceae. FIG. 46. Periwinkle (*Lochnera rosea*=*Vinca rosea*). *A*, a branch; *B*, calyx; *C*, a flower split longitudinally; *D*, a stamen; *E*, pistil; *F*, ovaries with disk; *G*, ovaries with disk (of two glands) in section; *H*, one ovary in section; and *I*, a pair of follicles.

yellow oleander (*Thevetia peruviana*)—seeds very poisonous; trees—*Holarrhena antidysenterica* (B. KURCHI; H. KARCHI),

Wrightia tomentosa (B. DUDHI-KHOROI; H. DUDHI), devil tree (*Alstonia scholaris*), etc.; **fruits**: *Carissa carandas* (B. KARANJA; H. KARONDA)—a thorny shrub, and *Willughbeia edulis* (B. LATA-AM)—a large climber; **ornamental**: herb—periwinkle (*Lochnera rosea=Vinca rosea*); shrubs—oleander (*Nerium indicum=N. oleander*), crepe-jasmine (*Ervatamia coronaria*; B. TAGAR; H. CHANDNI), temple or pagoda tree (*Plumeria rubra*); climbers—*Aganosma dichotoma* (B. MALATI; H. MALTI); *Vallaris solanacea* (B. HAPARMALI; H. RAMSAR), *Beaumontia grandiflora, Allamanda cathartica* and *Roupellia grata*; tree—*Cerbera odollam*—a tree of the Sundarbans; **other common plants**: *Ichnocarpus frutescens* (B. & H. DUDHI-LATA)—a climber, and *Rauwolfia canescens*—a shrub.

Family 28 *Asclepiadaceae* (1,800 *sp.*—213 *sp.* in India)

Habit—herbs, shrubs or twiners; with latex. **Leaves** opposite. **Flowers** regular, bisexual and hypogynous. **Calyx**—sepals (5), slightly connate at the base, odd sepal posterior. **Corolla**—petals (5), connate; aestivation imbricate, twisted, or valvate. **Androecium**—stamens (5), connate in a hollow tube, with horn-like appendages known as the staminal corona, epipetalous; anthers coherent laterally and united with the style and the stigma forming a gynostegium; pollen cohering into two pollen masses known as the pollinia (sing. pollinium), one lying in each lateral anther-lobe. **Gynoecium**—carpels two, free, superior; styles 2, free but united above forming a large dilated 5-angled stigma; stigma with five receptive surfaces lying on the underside or the edge of it; ovaries 2, free or united at the base only, each unilocular with many ovules in it; placentation marginal on a large intruding ventral placenta. **Fruit** a pair of follicles, or by abortion only one. **Seeds** many, hairy. *Floral formula*—
$\oplus \male K_{(5)} \overline{C_{(5)} [A_{(5)}} \underline{G}_2]$.

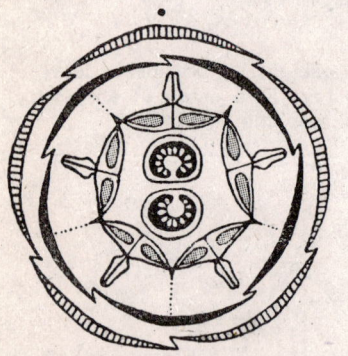

FIG. 47. Floral diagram of *Asclepiadaceae* (*Calotropis*).

At each angle of the stigma there is a groove which secretes a sticky body called the corpusculum. The sticky secretion extends on either side into a connecting thread (retinaculum) to which each pollinium becomes attached.

Examples. [The larger genera are: *Hoya* (about 200 sp.), *Asclepias* (120 sp.), *Ceropegia* (over 100 sp.), *Stapelia* (80 sp.),

SELECTED FAMILIES OF DICOTYLEDONS 745

Dischidia (70 sp.), and *Marsdenia* (70 sp.)]. **Useful plants**: **medicinal**: Indian sarsaparilla (*Hemidesmus indicus*; B. & H.

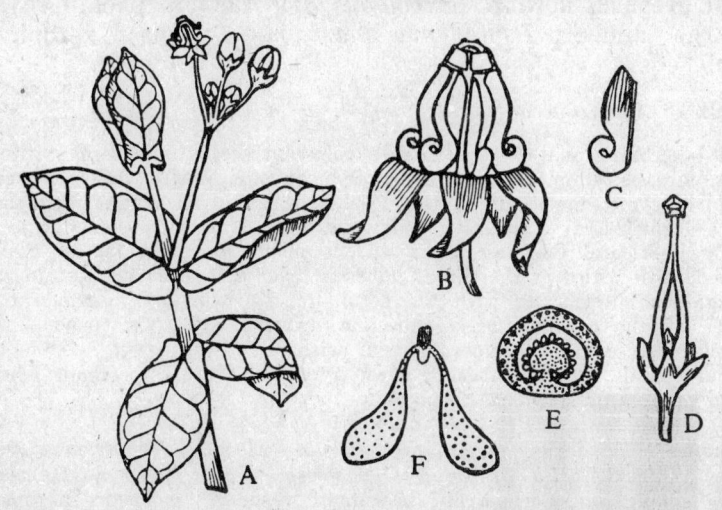

Asclepiadaceae. FIG. 48. Madar (*Calotropis gigantea*). *A*, a branch; *B*, a flower; *C*, staminal corona with the filament (inner side); *D*, apocarpous pistil with the calyx at the base; *E*, section of ovary showing marginal placentation; and *F*, a pair of pollinia (see also FIG. I/118).

ANANTAMUL)—a twining shrub. *Tylophora asthmatica*—a twining herb, madar (*Calotropis gigantea* and *C. procera*)—also floss from seeds used for stuffing pillows and cushions; **ornamental**: *Stephanotis floribunda*—a large climber with white fragrant flowers, *Cryptostegia grandiflora*—a woody climber with large magenta flowers, *Oxystelma esculentum* (B. & H. DUDHIALATA)—a slender perennial twiner with large showy flowers, *Pergularia pallida*—an extensive twiner with fragrant flowers, *Stapelia grandiflora*—a cactus-like pot herb with large red flowers, etc.; **other common plants**: milkweed or blood-flower (*Asclepias curassavica*; B. DUDHIA; H. KAKATUNDI) —an erect herb with orange-red flowers, *Sarcostemma brevistigma* (B. & H. SOMA-LATA)—a leafless shrub with jointed stem and pendulous branches, wax plant (*Hoya parasitica*)—a thickleaved epiphytic climber, *Dischidia rafflesiana* (see FIG. I/70)—

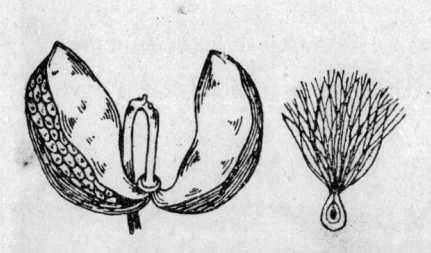

FIG. 49. *Left*, a pair of follicles of madar; *right*, a hairy seed of the same.

a stout epiphytic climber with pitchers, *Daemia extensa*—a foetid climbing undershrub. *Dregea volubilis*—a woody climber with small greenish flowers, *Sarcolobus globosus* (B. BAOLI-LATA)—a large climber, *Leptadenia spartium*—a leafless xerophytic shrub, etc.

Family 29 *Boraginaceae* (1,800 sp.—141 sp. in India)

Habit—herbs (annual or perennial), shrubs or trees, often covered with stiff hairs. **Leaves** simple, alternate, rarely opposite, entire. **Inflorescence** a scorpioid cyme, sometimes coiled. **Flowers** regular, bisexual, hypogynous. **Calyx**—sepals 5, free or united below into a tube, usually persistent, commonly imbricate. **Corolla**—petals (5), gamopetalous, tubular or funnel-shaped, with scales at the throat, lobes imbricate. **Androecium**—stamens 5, epipetalous, alternating with the corolla-lobes, commonly inserted at the mouth of the corolla-tube. **Gynoecium**—carpels (2), syncarpous, on an annular disc; ovary superior, 2-locular with 2 ovules in each, or commonly 4-locular with 1 ovule in each; style gynobasic, rarely terminal. **Fruit** a group of 4 nutlets. *Floral formula*— $\oplus \; \male \; K_5 \, C_{(5)} \, A_5 \, \underline{G}_{(2)}$.

Examples. [The larger genera are: *Cordia* (250 sp.), *Heliotropium* (over 200 sp.), *Tournefortia* (120 sp.), *Cynoglossum* (over 50 sp.)]. Heliotrope (*Heliotropium indicum*)—a diffuse annual weed, *H. strigosum*—a procumbent annual weed, hound's tongue (*Cynoglossum lanceolatum*)—an erect annual weed, *Trichodesma indicum*—a diffuse annual weed, *Coldenia procumbens*—a prostrate weed, forget-me-not (*Myosotis pallustris*) and *Onosma echioides*—an ornamental herb with blue or pink flowers, *Tournefortia roxburghii*—a rambling shrub, *Cordia myxa* (B. & H. LASORA)—a large shrub with small white flowers, *C. sebestena*—a small tree (planted in gardens) with large orange-coloured flowers, *Ehretia acuminata*—a tree with flowers in dense terminal panicles, etc.

Family 30 *Convolvulaceae* (over 1,600 sp.—157 sp. in India)

Habit—mostly twiners, often with latex and bicollateral vascular

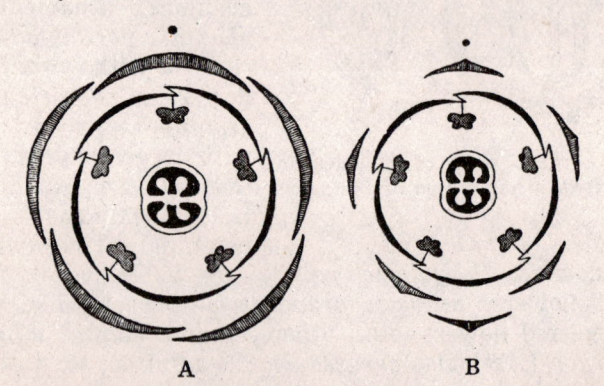

FIG. 50. Floral diagrams of *Convolvulaceae*. A, *Ipomoea*; B, dodder (*Cuscuta*).

bundles or internal phloem. **Leaves** simple, alternate and exstipulate. **Inflorescence** cymose. **Flowers** regular, bisexual, hypogynous, often large and showy. **Calyx**—sepals 5, usually free, odd one posterior, imbricate and persistent. **Corolla**—petals (5), gamopetalous, funnel-shaped, twisted in bud, sometimes imbricate. **Androecium**—stamens 5, epipetalous, alternating with the petals. **Gynoecium**—carpels (2), rarely more, connate; ovary superior, with a disc at the base, 2-celled, with 2 ovules in each cell, or sometimes 4-celled with 1 ovule in each cell; placentation axile. **Fruit** a berry or a capsule.

Floral formula— $\oplus \; \male \; K_5 \; C_{(5)} \; A_5 \; \underline{G}_{(2)}$.

Convolvulaceae is related to *Solanaceae* by its persistent calyx, regular gamopetalous corolla, 5 epipetalous stamens, often false septum in the ovary, bicollateral vascular bundles, etc.; but it is distinguished from *Solanaceae* by definite number (1 or 2) of ovules in each chamber of the ovary, micropyle pointing downwards, median carpels, etc.

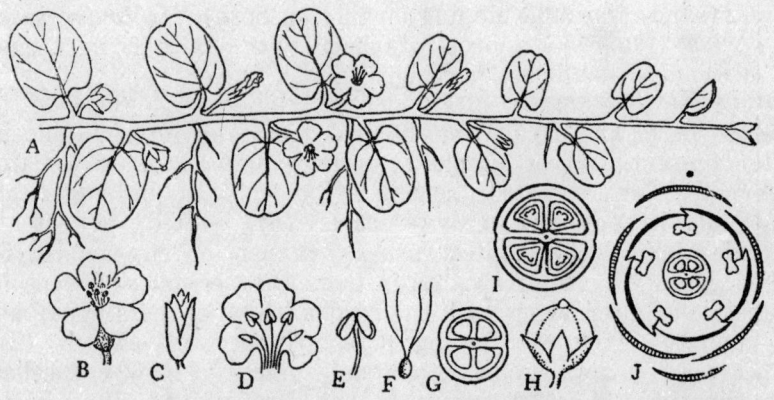

Convolvulaceae. FIG. 51. *Evolvulus nummularius.* A, a branch (prostrate); B, a flower; C, calyx; D, corolla (with epipetalous stamens) opened out; E, a stamen; F, gynoecium; G, ovary in transection with 4 ovules (sometimes 2 or 3); H, a fruit; I, the same in transection; and J, floral diagram.

Examples. [The larger genera are: *Ipomoea* (=*Pharbitis*; over 400 sp.), *Convolvulus* (200 sp.), *Jacquemontia* (120 sp.), *Cuscuta* (120 sp.), *Evolvulus* (about 100 sp.), *Argyreia* (90 sp.), and *Merremia* (80 sp.)]. **Vegetables:** sweet potato (*Batatas edulis*=*Ipomoea batatas*), water bindweed (*I. aquatica*=*I. reptans*; B. & H. KALMISAK), *Argyreia nervosa* (in the Khasi Hills), etc.; **medicinal:** *Ipomoea paniculata* (B. & H. BHUIKUMRA), *I. hederacea* (B. & H. NIL-KALMI—seeds sold as KALADANA are used as a purgative, Indian jalap or turpeth (*I. turpethum*=*Operculina turpethum*; B. TEORI or DUDH-KALMI; H. PITOHARI)—roots used as a mild cathartic, jalap (*I. purga*)—turnip-like rhizome used as a strong purgative, elephant climber (*Argyreia speciosa*; B. SAMUDRASOK; H. SAMUNDERPHEN)—roots used in rheumatic affections, and leaves used in skin diseases and wounds, etc.; **ornamental**: *Ipomoea* (several species), e.g. morning glory (*I. purpurea*)—flowers white fading to purple, *I. hederacea*—flowers blue or purple, railway creeper (*I. palmata*)—flowers dull violet, moon flower (*I. bona-nox*)—flowers white, opening at dusk, cypress vine (*I. quamoclit=Quamoclit pinnata*; B. KUNJALATA)—a slender twiner with scarlet flowers, *I. learii* and *I. nil*—flowers large, very showy, *I. versicolor*—an elegant twiner (in Shillong), *I. carnea*—a much-branched scandent shrub with dull violet flowers, woodrose (*I. tuberosa*)—a liane (in Assam), fruits used for table decoration, etc.; *Convolvulus* (mostly temperate), e.g. field bindweed (*C. scammonia*), *C. major*, *C. minor*, etc.; and *Jacquemontia coerulea*—a garden twiner; **other common plants**: dodder (*Cuscuta reflexa*, see FIG. I/22), bridal creeper (*Porana paniculata*), *Ipomoea pescaprae* (=*I. biloba*)—a creeping perennial plant (sand- and mud-binding on the seashore), *Evolvulus nummularius* (FIG. 51)—a prostrate herb with white flowers, and *E. alsinoides*—a diffuse herb with bluish flowers, etc.

Family 31 *Solanaceae* (over 2,000 *sp.*—58 *sp. in India*)

Habit—herbs and shrubs; bicollateral bundles or internal phloem often present. **Leaves** simple, sometimes pinnate, as in tomato, alternate. **Flowers** regular, seldom zygomorphic, as in *Brunfelsia*, bisexual, hypogynous. **Calyx**—sepals (5), united, persistent. **Corolla**—petals (5), united, usually funnel- or cup-shaped, 5-lobed, lobes valvate or twisted in bud. **Androecium**—stamens 5, epipetalous, alternating with the corolla-lobes; anthers apparently connate, often opening by pores. **Gynoecium**—carpels (2), syncarpous; ovary superior, obliquely placed (FIG. 52), 2-celled or sometimes 4-celled owing to the development of a false septum, as in tomato and thorn-apple, with many ovules in each; placentation axile. **Fruit** a berry or capsule with many seeds.

Floral formula—$\oplus \; \male \; K_{(5)} C_{(5)} A_5 \underline{G}_{(2)}$.

Solanaceae is related to *Convolvulaceae* (see p. 747). It is closely related to *Scrophulariaceae* through *Brunfelsia* and *Schizanthus* which have a zygomorphic corolla and 4 or 2 stamens. *Solanaceae* is, however, generally distinguished from the latter family by having regular corolla, twisted aestivation, five stamens, obliquely placed carpels, often bicollateral vascular bundles in the stem, etc.

Examples. [The larger genera are: *Solanum* (1,500 sp.), *Cestrum* (250 sp., mostly American), *Physalis* (100 sp.), *Nicotiana* (100 sp.), and *Capsicum* (50 sp.)]. **Useful plants**: vegetables: potato (*Solanum tuberosum*), brinjal (*S. melongena*),

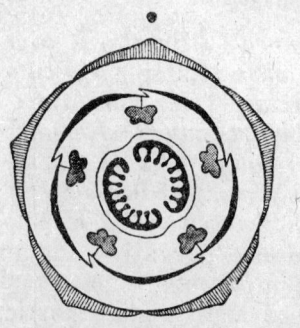

FIG. 52. Floral diagram of *Solanaceae*.

SELECTED FAMILIES OF DICOTYLEDONS

chilli or red pepper (*Capsicum annuum*)—fruits pungent, mainly used as a condiment, bell-pepper (*C. grossum*)—fruits not pungent, used as a vegetable, and tomato (*Lycopersicum esculentum*); **medicinal**: deadly nightshade (*Atropa belladonna*), thorn-apple (*Datura stramonium*) and *Datura metel* (=*D.*

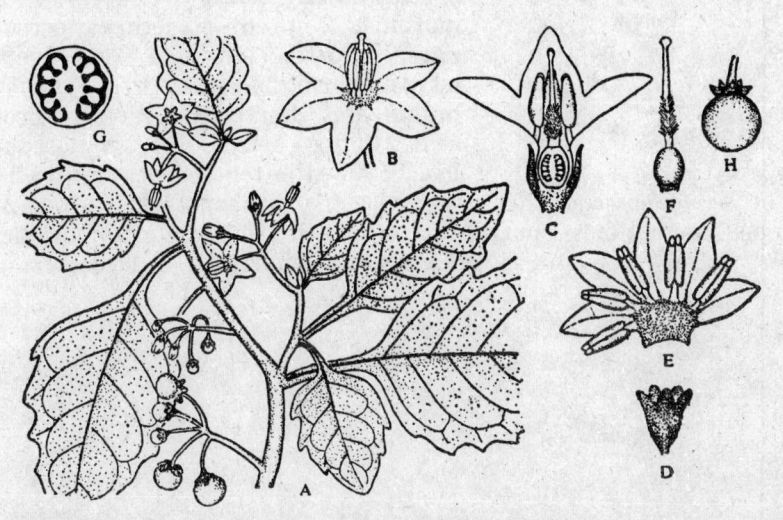

Solanaceae. FIG. 53. Black nightshade (*Solanum nigrum*). *A*, a branch; *B*, a flower; *C*, a flower cut longitudinally; *D*, calyx; *E*, corolla with epipetalous stamens; *F*, pistil; *G*, ovary in transection showing axile placentation; and *H*, a fruit (berry).

fastuosa)—seeds narcotic and very poisonous, henbane (*Hyoscyamus niger*), bittersweet (*Solanum dulcamara*; B. &. H. MITHA-BISH), *S. indicum* (B. BRIHATI; H. BIRHATTA), *S. surattense* (=*S. xanthocarpum*; B. KANTIKARI; H. KATELI), and *Withania somnifera* (B. ASWAGANDHA; H. ASGAND); **narcotic**: tobacco (*Nicotiana tabacum*)—tobacco of commerce and also a source of nicotine—an insecticide; **fruit**: gooseberry (*Physalis peruviana*); **ornamental**: *Petunia hybrida*, queen of the night (*Gestrum nocturnum*; B. HAS-NA-HANA; H. RAT-KI-RANI) and *C. parqui*—both sweet-scented, *Brunfelsia hopeana* (=*Franciscea bicolor*)—flowers white changing to blue, sweet-scented, *Schizanthus pinnatus*—a beautiful garden herb in Shillong, etc.; **other common plants**: black nightshade (*Solanum nigrum*; FIG. 53), egg-plant (*S. ferox*), wild gooseberry (*Physalis minima*) and wild tobacco (*Nicotiana plumbaginifolia*).

Family 32 *Scrophulariaceae* (3,000 *sp.*—258 *sp. in India*)

Habit—mostly herbs and undershrubs. **Leaves** simple, alternate,

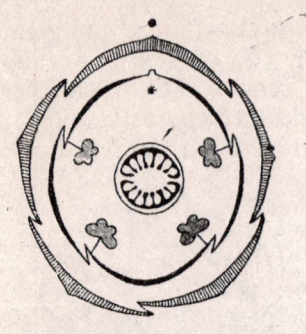

FIG. 54. Floral diagram of *Scrophulariaceae*.

opposite or whorled, exstipulate, sometimes showing heterophylly. **Inflorescence** commonly racemose (raceme or spike), sometimes cymose (dichasium), axillary or terminal; in some species flowers solitary. **Flowers** zygomorphic, 2-lipped, sometimes personate, but often showing a great diversity in form, bisexual, hypogynous; bracts and bracteoles generally present. **Calyx**—sepals (5), gamosepalous, 5-lobed, often imbricate. **Corolla** —petals (5), gamopetalous, often 2-lipped, sometimes spurred or saccate, medianly zygomorphic, very rarely regular as in *Scoparia*, imbricate. **Androecium**—

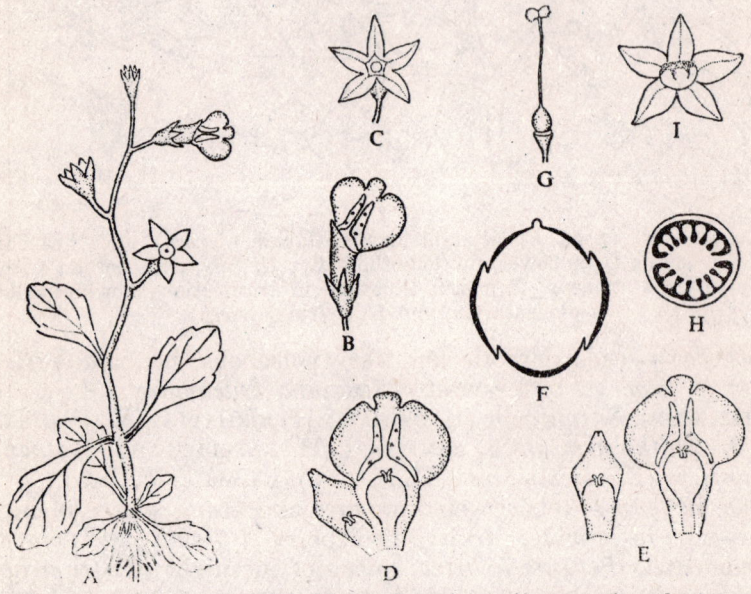

Scrophulariaceae. FIG. 55. *Mazus japonicus.* *A*, an entire plant—a small herb; *B*, a flower—bilabiate; *C*, calyx—gamosepalous and persistent; *D*, corolla split open with epipetalous, didynamous stamens arching over in pairs; *E*, corolla—the two lips separated, each with a pair of epipetalous stamens arching over; *F*, aestivation of corolla —imbricate; *H*, ovary in transection showing axile placentation; and *I*, fruit—capsule (dehiscing) with persistent calyx.

stamens 4, didynamous, sometimes 2, arching over in pairs, posterior stamen absent or a staminode; anthers divaricate. **Gynoecium**—carpels (2), syncarpous; ovary superior, bilocular, antero-posterior (and not oblique as in *Solanaceae*); placenta-

SELECTED FAMILIES OF DICOTYLEDONS

tion axile; stigma simple or bilobed; ovules usually numerous, sometimes few; disc ring-like around the base of the ovary, sometimes unilateral. **Fruit**—commonly a capsule, sometimes a berry. **Seeds**—usually numerous, minute, endospermic.

Floral formula—·|· ⚥ $K_{(5)} \overline{C_{(5)} A_{4 \text{ or } 2} \underline{G}_{(2)}}$.

Scrophulariaceae is closely related to *Solanaceae* but is distinguished from it by its simple collateral bundles in the stem, zygomorphic corolla, imbricate aestivation, stamens 4 (didynamous) or 2, median position of the ovary, etc. It is distinguished from *Labiatae* and *Verbenaceae* by its inflorescence and fruit. It is also related to *Acanthaceae* (see p. 752).

Examples. [The larger genera are: *Pedicularis* (over 500 *sp.*), *Veronica* (300 *sp.*), *Scrophularia* (300 *sp.*), *Verbascum* (300 *sp.*) and *Linaria* (150 *sp.*), and they mostly occur in north temperate regions; while *Calceolaria* (about 500 *sp.*) occurs in South America and *Mimulus* (150 *sp.*) in North America]. **Useful plants**: **medicinal**: foxglove (*Digitalis purpurea*)—leaves contain a bitter glycoside *digitalin* which acts on the heart and the circulatory system, *Bacopa* (=*Herpestis*) *monnieria* (B. & H. BRAHMI)—a brain and nerve tonic; **ornamental**: snapdragon (*Antirrhinum majus*), *Torenia fournieri*, fountain plant (*Russelia equisetiformis*=*R. juncea*), *Linaria cymbalaria*, monkey flower (*Mimulus maculosus*), *Veronica speciosa*, *Angelonia salicariaefolia*, *Scrophularia elatior*, etc.; **other common plants**: weeds: *Mazus japonicus* (=*M. rugosus*), *Torenia* (=*Vandellia*) *crustacea*, *Lindernia* (=*Vandellia*) *multiflora*, *L. ciliata* (=*Bonnaya brachiata*), *Limnophila heterophylla* showing heterophylly, *Hemiphragma heterophyllum* (in Shillong and Darjeeling) showing heterophylly, *Lindenbergia indica*—a common weed growing on old walls, toad-flax (*Linaria ramosissima*), *Mecardonia dianthera*, *Scoparia dulcis*, *Veronica anagallis*, *Celsia coromandeliana*, *Striga lutea* and a few other species—partial root-parasites, etc.

Family 33 *Bignoniaceae* (750 *sp.*—25 *sp.* in India)

Habit—trees, shrubs or climbers (rootlet- or tendril-climbers). **Leaves** usually pinnately compound, opposite, exstipulate; in *Bignonia* terminal leaflet modified into tendril (branched or unbranched) or into hooks. **Inflorescence** often a dichasial cyme. **Flowers** bisexual, zygomorphic, hypogynous; bracts and bracteoles present. **Calyx**—sepals (5), gamosepalous. **Corolla**—petals (5), gamopetalous, usually obliquely bell- or funnel-shaped; aestivation imbricate. **Androecium**—stamens 4. epipetalous, didynamous; anthers 2-lobed, lobes divaricate. **Gynoecium**—carpels (2), syncarpous, on a hypogynous disc; ovary superior, usually 2-locular, with many ovules in each; placentation axile. **Fruit** a 2-valved capsule, sometimes a berry. **Seeds** exalbuminous, usually flattened, with membranous wing. *Floral formula*—·|· ⚥ $K_{(5)} \overline{C_{(5)} A_4 \underline{G}_{(2)}}$.

Examples. [The larger genera are *Bignonia* (150 sp.; now split up into several genera), *Tecoma* (90 sp.; now split into several genera), and *Jacaranda* (50 sp.)]. *Bignonia*—usually tendril-climbers, often with showy flowers, e.g. *B. venusta, B. unguis-cati, B. magnifica,* etc., *Tecoma grandiflora*—a common garden climber, *T. stans*—a garden shrub, Indian cork tree (*Millingtonia hortensis*)—a tall robust tree with sweet-scented flowers, *Stereospermum chelonoides* (=*S. personatum*)—a large tree, *Oroxylum indicum* (see FIG. I/174A)—a small tree, *Spathodea campanulata*—a medium-sized tree with large red flowers, *Jacaranda mimosaefolia* (having (16-20 pairs of pinnae) and *J. acutifolia* (having 6-8 pairs of pinnae)—both planted as roadside or garden trees, with purplish flowers in panicles, calabash (*Crescentia cujete*)—fruit large, gourd-like, *Tecomella undulata*—a small timber tree (commonly called 'desert teak') of Rajasthan and Punjab, candle tree (*Parmentiera cerifera*)—candle-like yellow fruits (35-50 cm. long) hang from the stem and the branches in large numbers, etc.

Family 34 *Acanthaceae* (2,200 *sp.*—409 *sp. in India*)

Habit—herbs, shrubs and a few climbers; cystoliths often present in stem and leaf. **Leaves** simple, opposite, exstipulate. **Inflorescence** a spike or a cyme or sometimes a raceme; flowers in some species in axillary clusters, rarely solitary. **Flowers** zygomorphic,

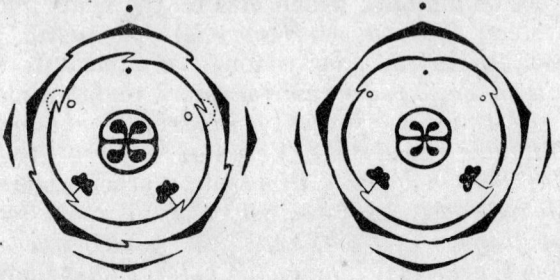

FIG. 56. Floral diagrams (two types) of *Acanthaceae*.

bilabiate or oblique, bisexual and hypogynous, often with conspicuous bracts and bracteoles, the latter often large, sometimes spiny. **Calyx**—sepals (5), rarely (4), united. **Corolla**—petals (5), connate in a two-lipped or oblique corolla, twisted or imbricate in bud. **Androecium**—stamens 2 or 4, if 4 didynamous, epipetalous; disc often conspicuous. **Gynoecium**—carpels (2), syncarpous; ovary 2-celled, superior, with 2 to many ovules in each cell; placentation axile; stigmas 2. **Fruit** a 2-valved capsule (FIG. 57*G*). **Seeds** are in most cases supported on curved hooks (jaculators); these press the fruit from inside, which bursts with a sudden jerk and scatters the seeds (see FIG. I/182).

Floral formula— $\cdot \vert \cdot \; \male \; K_{(5)} \; \overline{C_{(5)} \, A_{2 \, \text{or} \, 4}} \; \underline{G}_{(2)}$.

Acanthaceae is related to *Scrophulariaceae* but is distinguished from it by the presence of copious bracts and bracteoles, often unequal posterior sepal, loculicidal capsule dehiscing to the very base, seeds often with jaculators, absence of endosperm, frequent presence of cystolith, etc. It is also related to *Labiatae* (see p. 756), and to *Verbenaceae* (see p. 755).

Examples. [The larger genera are: *Justicia* (250 sp.), *Strobilanthes* (over 200 sp.), *Barleria* (over 200 sp.), *Thunbergia*

(about 200 sp.), *Dicliptera* (150 sp.), and *Hygrophila* (80 sp.)].
Useful plants: medicinal: *Andrographis paniculata* (B. KAL-
MEGH; H. MAHATITA)—an effective remedy for liver complaints in
children, and *Adhatoda vasica* (B. BASAK; H. ADALSA)—an excellent remedy for cough; **ornamental:** *Barleria*, e.g. *B. prionitis*
(B. KANTA-JHANTI; H. KATSAREYA)—spinous, flowers yellow, *B. cristata* (B. JHANTI; H. JHINTI)—flowers white or rose-coloured, *B.
strigosa*—flowers blue, *Meyenia erecta*—a pretty shrub with deep
blue flowers, *Crossandra undulaefolia*—an undershrub with
orange-coloured flowers, *Strobilanthes*, e.g. *S. scaber* (with yellow
flowers), *S. auriculatus* (with purplish flowers), and many other
species common in the Khasi Hills, *Eranthemum* (=*Daedala-*

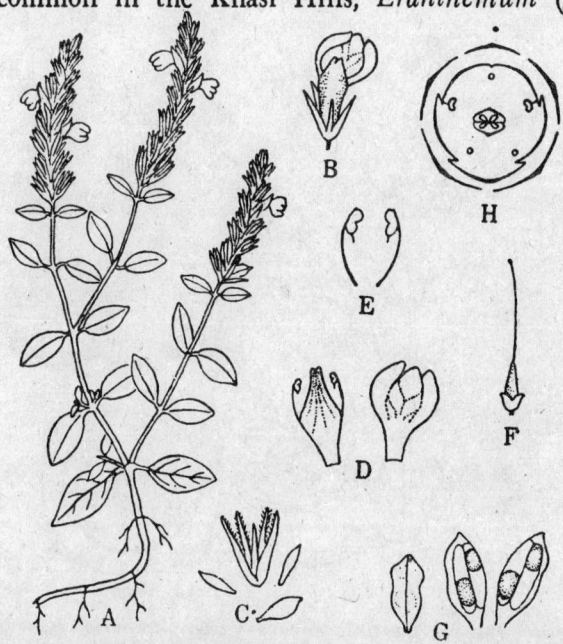

Acanthaceae. FIG. 57. *Justicia simplex.* A, a plant with spikes;
B, a flower (bilabiate); C, calyx with bracts and bracteoles; D, corolla
split open, upper lip with 2 epipetalous stamens; E, stamens—2;
F, gynoecium with glandular disc at the base; G, fruits—young, and
dehiscent (loculicidal capsule); and H, floral diagram.

canthus) *nervosus*—flowers bright blue, etc.; **other common
plants: herbs:** *Cardanthera triflora*—showing heterophylly (see
FIG. I/79A), *Hygrophila polysperma*—a common weed, *Astera-
cantha longifolia* (=*Hygrophila spinosa*; B. KULEKHARA; H.
TALMAKHANA)—an erect spinous herb, *Ruellia tuberosa* (see
FIG. I/182), *R. prostrata*, *Phaylopsis imbricata* (=*P. parviflora*),
Rungia parviflora, *R. elegans*—bracts with white margin, *Hypo-
estis triflora* (in Shillong), *Dicliptera roxburghiana*, *Justicia*

simplex (FIG. 57) and *Ecbolium linneanum* (B. NILKANTHA; H. UDAJATI); **climbers**: *Thunbergia alata* (flowers yellow) and *T. grandiflora* (flowers blue); **shrubs or undershrubs**: *Justicia gendarussa* (B. JAGAT-MADAN; H. NILI-NARGANDI), *Acanthus ilicifolius* (B. HARGOZA)—a mangrove shrub, *Phlogacanthus curviflorus*—flowers pink, *P. thyrsiflorus*—flowers orange, eaten cooked as a vegetable, *P. tubiflorus*—flowers red, eaten cooked as a vegetable, leaves rubbed in water give a lather, etc.

Family 35 *Verbenaceae* (over 2,600 *sp.*—107 *sp. in India*)

Habit—herbs, shrubs, trees or climbers, sometimes prickly, some xerophytic in habit, commonly strong-smelling; stem sometimes

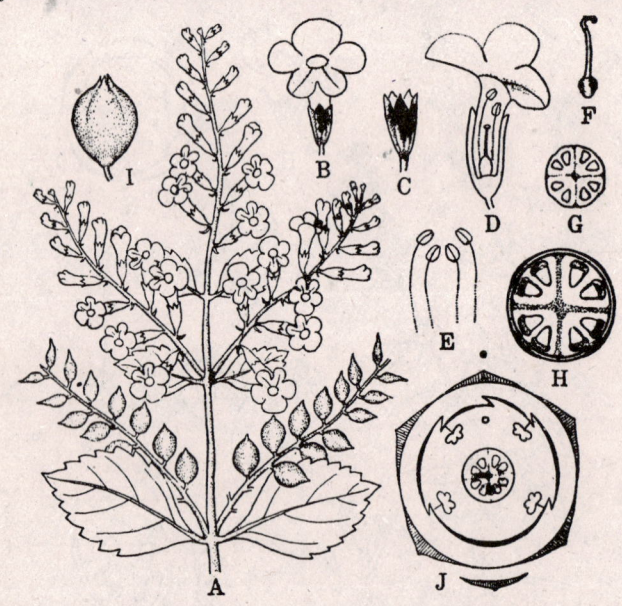

Verbenaceae. FIG. 58. *Duranta repens* (=*D. plumieri*). *A*, a branch with inflorescence; *B*, a flower; *C*, calyx; *D*, a flower split lengthwise; *E*, stamens (didynamous); *F*, gynoecium—carpels (4); *G*, ovary in transection (8 chambers, each with 1 ovule); *H*, young fruit in transection showing 4 pyrenes (being formed), each 2-chambered with 1 ovule in each; *I*, fruit—a fleshy drupe (the outer envelope is the fleshy persistent calyx); and *J*, floral diagram.

4-angled. **Leaves** simple, opposite or whorled, sometimes pinnately or palmately compound. **Inflorescence** a raceme, panicle or spike (long or condensed) or a dichasial cyme. **Flowers** bisexual, medianly zygomorphic, hypogynous, pentamerous; bracts sometimes in the form of an involucre, as in *Lantana*. **Calyx**—sepals usually (5), rarely (4) or more, gamosepalous, persistent. **Corolla**—petals usually (5), gamopetalous, at first 2-lipped and later 5-lobed, tube long or short and limb oblique; aestivation imbricate. **Androecium**—stamens 4, didynamous, epipetalous, very rarely 2 as in *Stachytarpheta*, or 5 as in teak (*Tectona*), often inserted, sometimes exserted (even far exserted, as in *Clerodendron*), alternating with corolla-lobes. **Gynoecium**—carpels commonly (2), rarely (4) as in *Duranta*, syncarpous; ovary superior, entire or lobed, 2-locular with 1 or 2 ovules in each or 4-locular with 1 ovule in each (8-locular in *Duranta*; FIG. 58G); style terminal. **Fruit** a drupe (consisting of 2 or 4 pyrenes), rarely a capsule. **Seed** exalbuminous.

Floral formula— ⋅⊦⋅ ⚥ $K_{(5)} \overline{C_{(5)}} A_4 \underline{G}_{(2)}$.

Verbenaceae is closely related to *Labiatae* both showing zygomorphy in the corolla, often extending to calyx in the latter. It is, however, distinguished from *Labiatae* by the following characters: in the former the inflorescence variously formed; ovary 2-locular with 1 or 2 ovules in each or *later* divided into 4 loculi with 1 ovule in each; style terminal; fruit usually drupaceous consisting of 2 or 4 pyrenes; whereas in the latter inflorescence a verticillaster or a cyme; 2-locular ovary *early* divided into 4 loculi with 1 ovule in each; style gynobasic; fruit consisting of 4 one-seeded nutlets. *Verbenaceae* is distinguished from *Acanthaceae*, another related family, by having 4-chambered ovary with 1 ovule in each or 2-chambered ovary with 1 or 2 ovules in each. The spike of *Acanthaceae* with often conspicuous bracts and bracteoles is another distinguishing feature. The fruit character also often distinguishes the two families.

Examples. [The larger genera are: *Clerodendrum* (about 400 sp.), *Vitex* (over 250 sp.), *Verbena* (230 sp.), *Lippia* (over 200 sp.), *Premna* (200 sp.), *Lantana* (155 sp.), *Callicarpa* (140 sp.), and *Stachytarpheta* (about 100 sp.)]. Teak (*Tectona grandis*)—a very valuable timber tree, *Gmelina arborea* (B. GAMHAR; H. GAMARI)—a very good timber tree, *Duranta repens* (=*D. plumieri*; FIG 58)—commonly grown as a hedge plant, *Lantana aculeata* (=*L. camara*)—a strong-smelling straggling shrub, stem prickly, *L. indica*—a strong-smelling erect shrub, stem not prickly, *Clerodendrum infortunatum* (B. & H. BHANT), *C. siphonanthus*, *C. inerme*, *C. thomsonae*—a garden climber with white calyx and red corolla (see FIG. I/143), etc., *Petrea volubilis*—a garden climber with profuse violet flowers in racemes, lady's umbrella or Chinese hat (*Holmskioldia sanguinea*)—a shrub bearing beautiful scarlet flowers (common in the low hills of Assam), *Lippia nodiflora* (=*Phyla nodiflora*)—a prostrate herb in wet places, *L. javanica* (=*L. geminata*)—a scandent shrub on the banks of tanks and canals, *Verbena officinalis*—a small erect weed, *Stachytarpheta indica*—a perennial herb with bluish flowers, often cultivated, *Avicennia officinalis*—a mangrove tree of the Sundarbans, *Premna esculenta*—a shrub, *P. bengalensis* —a tree, etc., *Vitex negundo* (B. NISHINDA; H. SHAMALU)—a shrub, *V. trifolia*—a tree, *Callicarpa arborea*—an evergreen tree, *C. macrophylla*—a tomentose shrub, etc.

Family 36 Labiatae or **Lamiaceae** (over 3,000 *sp.*—391 *sp.* in India)

Habit—herbs and undershrubs with square stem. **Leaves** simple, opposite or whorled, exstipulate, with oil-glands. **Flowers** zygomorphic, bilabiate, hypogynous and bisexual. **Inflorescence** verticillaster (see p. 71); sometimes reduced to true cyme, as in sacred basil (*Ocimum*: B. & H. TULSI). **Calyx**—sepals (5), gamosepalous, unequally 5-lobed or 2-lipped, persistent. **Corolla**—petals (5), gamopetalous, bilabiate, i.e. 2-lipped; aestivation imbricate. **Androecium**—stamens 4, didynamous, sometimes only 2, as in sage (*Salvia*; see FIG. I/139), epipetalous. **Gynoecium**—carpels (2), syncarpous; disc prominent; ovary 4-lobed and 4-celled, with one ovule in each cell, ascending from the base of the ovary; style gynobasic (FIG. 60 E-F), i.e. develops from the depressed centre of the lobed ovary; stigma bifid. **Fruit** a group of four nutlets, each with one seed; seed with scanty or no endosperm.

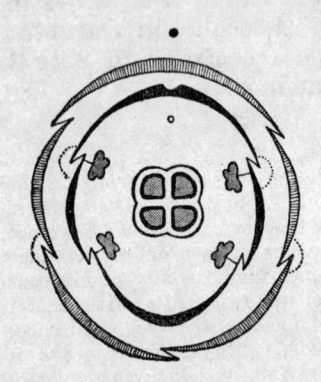

FIG. 59. Floral diagram of Labiatae.

Floral formula— ⋅|⋅ ♂ $K_{(5)} \overline{C_{(5)}} A_4 G_{(2)}$.

Labiatae is closely related to *Verbenaceae* (see p. 755). It may be related to *Boraginaceae* by its fruit character (4 nutlets) but is readily distinguished from it by its inflorescence. *Labiatae* is distinguished from *Acanthaceae* and *Scrophulariaceae* by its inflorescence, lobed ovary and fruit structure.

Labiatae abounds in volatile, aromatic oils which are used in perfumery and also as stimulants. Many of them possess a bitter astringent property.

Examples. [The larger genera are: *Salvia* (over 500 sp.), *Nepeta* (250 sp.), *Stachys* (over 200 sp.), *Teucrium* (over 200 sp.), *Scutellaria* (over 200 sp.), *Ocimum* (150 sp.), and *Coleus* (150 sp.)]. **Useful plants**: medicinal: sacred basil (*Ocimum sanctum*), mint (*Mentha*), e.g. spearmint or garden mint (*M. viridis*; B. PUDINA; H. PODINA)—commonly cultivated and used as a salad, peppermint (*M. piperita*)—yields peppermint oil from which menthol is obtained, Japanese peppermint (*M. arvensis*), garden thyme (*Thymus vulgaris*)—leaves used for flavouring soups and curries, thyme (*T. serpyllum*)—yields thyme oil from which thymol is obtained, patchouli (*Pogostemon heyneanus*)—

SELECTED FAMILIES OF DICOTYLEDONS

yields patchouli oil, lavender (*Lavandula vera*)—yields lavender oil, and rosemary (*Rosmarinus officinalis*)—yields oil of rosemary; **ornamental**: sage (*Salvia*), e.g. *S. coccinea* and *S. splendens*—flowers scarlet, cultivated garden herbs or undershrubs, *S. farinacea*—flowers lavender-blue, cultivated, *S. leucantha*—flowers purplish-blue, a scandent shrub, wild and cultivated in the hills, *S. plebeja*—an annual weed, etc., country

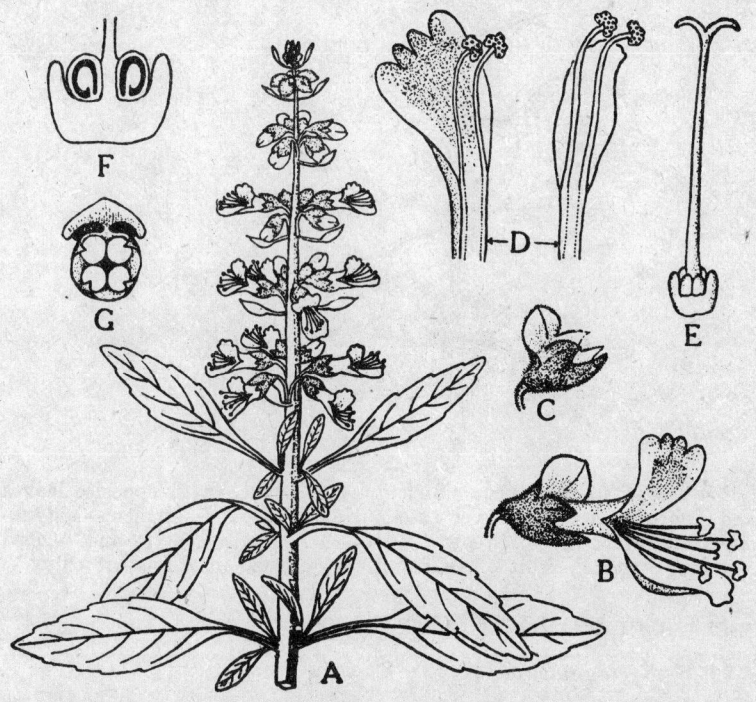

Labiatae. FIG. 60. Basil (*Ocimum basilicum*). *A*, a branch with inflorescences; *B*, a flower—bilabiate (note the didynamous stamens); *C*, calyx; *D*, corolla split open with epipetalous stamens; *E*, pistil (note the gynobasic style); *F*, ovary with the disc (in longi-section); and *G*, fruit of four nutlets enclosed in the persistent calyx.

borage (*Coleus aromaticus*—see FIG. I/6) and marjoram (*Origanum vulgare*)—cultivated for its scented leaves; **other common plants**: *Ocimum gratissimum* (B. & H. RAM-TULSI), basil (*O. basilicum*; B. & H. BABUI-TULSI), wild basil (*O. canum*; B. & H. BAN-TULSI), *Anisomeles indica*, *Leonurus sibiricus* (FIG. 61), *Leucas lavandulaefolia* (=*L. linifolia*), *L. aspera*, *L. cephalotes*, *L. lanata*—a whitish pubescent herb, *Dysophylla verticillata*—a marsh herb, *Pogostemon plectranthoides*—a common bushy undershrub, *Plectranthus ternifolius*—a bushy, tomentose undershrub, etc.

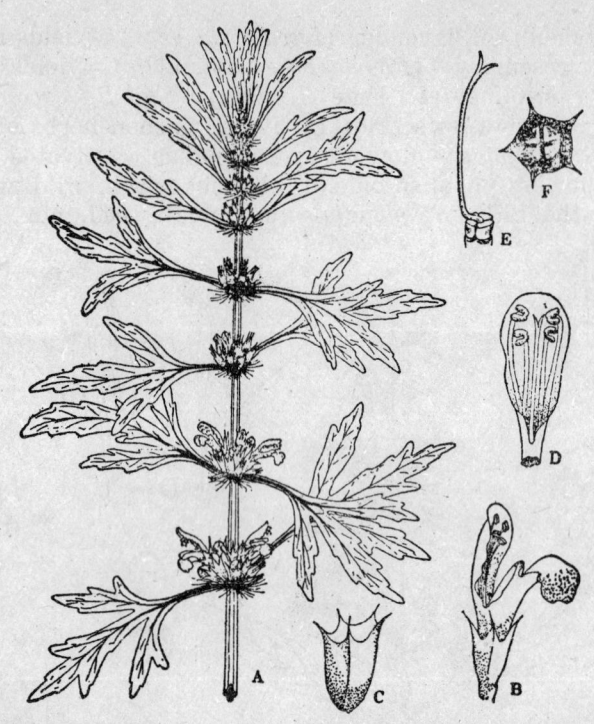

Labiatae FIG. 61. *Leonurus sibiricus*. *A*, a branch with opposite leaves and inflorescences; *B*, a flower (bilabiate); *C*, calyx; *D*, stamens—didynamous and epipetalous; *E*, pistil (note the gynobasic style and 4-lobed ovary); and *F*, fruit of four nutlets enclosed in persistent calyx.

SUB-CLASS III MONOCHLAMYDEAE

Family 37 *Nyctaginaceae* (300 *sp*.—8 *sp. in India*)

Habit—herbs, shrubs or climbers, showing anomalous secondary growth. **Leaves** simple, opposite, leaves of a pair often unequal, exstipulate. **Inflorescence** cymose. **Flowers** regular, bisexual or sometimes unisexual (as in *Pisonia*), hypogynous; bracts usually large and coloured, varying in number (3 petaloid bracts in *Bougainvillea* surrounding a group of 3 flowers; 5 sepal-like bracts in the form of an involucre in *Mirabilis*; in *Boerhaavia* the involucral bracts are reduced to scales). **Perianth** leaves united, tubular or funnel-shaped, 5-lobed, petaloid, lower part of the perianth is persistent in the fruit enveloping the latter like the pericarp (and known as the anthocarp). **Androecium**—stamens 5, alternating with the perianth-lobes but often very variable—1 or few or several (up to 30) by branching; filaments often unequal. **Gynoecium**—carpel 1; ovary superior, 1-locular, with a basal anatropous or campylotropous ovule; style long. Fruit an achene, 1-seeded and enclosed by the perianth-base. **Seed** with mealy perisperm; embryo straight, folded or curved. *Floral formula*— $\oplus \; \male \; P_{(5)} A_5 \underline{G}_1$.

Examples. [The larger genera are: *Mirabilis* (60 sp.), and *Pisonia* (50 sp.)]. Four o'clock plant (*Mirabilis jalapa*)—a garden herb with variously coloured flowers, glory of the garden (*Bougainvillea spectabilis*)—a large climber with petaloid bracts, *Pisonia aculeata*—a large climber elaborately armed with recurved spines, hogweed (*Boerhaavia diffusa*)—a diffuse prostrate herb (medicinal), and *Abronia umbellata*—an annual creeper with pink flowers.

Family 38 *Amaranthaceae* (850 sp.—46 sp. in India)

Habit—mostly herbs, sometimes climbing. **Leaves** simple, opposite or alternate, entire, exstipulate. **Inflorescence** an axillary cyme, a simple or branched spike, or a raceme. **Flowers** small, regular, bisexual, rarely unisexual, pentamerous, often with scarious bracts and bracteoles. **Perianth**—tepals usually 4-5, free or united, membranous. **Androecium**—stamens 5 (often some reduced to staminodes), opposite the perianth leaves, free, or united to the perianth, or to one another into a membranous tube, sometimes petaloid outgrowths are present between the stamens; anthers 2- or 4-locular. **Gynoecium**—carpels (2-3), syncarpous; ovary superior, unilocular, commonly with one campylotropous ovule (sometimes, as in cock's comb, several ovules are present). **Fruit** a utricle (1-seeded small fruit with loose perianth), or berry or nut or dehiscent (capsular). **Seed** endospermic. *Floral formula*—

$$\oplus \; \male\female \; P_{4-5 \text{ or } (4-5)} A_5 \underline{G}_{(2-3)}.$$

Examples. [The larger genera are: *Amaranthus* (60 sp.) and *Celosia* (60 sp.)]. **Leafy vegetables**: *Amaranthus caudatus* (B. NATE-SAK; H. CHAULAI), *A. gangeticus* (=*A. tristis*; B. LAL-SAK; H. LAL-SAG), *A. blitum* (=*A. oleracea*; B. SADA-NATE; H. CHAULAI)—a tall annual, widely cultivated, *A. polygamous* (B. CHAMFA-NATE), etc.; **ornamental**: *A. tricolor* bearing showy coloured leaves, *A. paniculatus*—a tall plant bearing long crimson or golden yellow pendulous spikes, cock's comb (*Celosia cristata*) bearing red fasciated inflorescence with red flowers, *C. argentea* bearing white flowers, *C. plumosa* bearing yellow flowers, button flower or globe amaranth (*Gomphrena globosa*), *Allmania nodiflora*—commonly grown as a garden border, *Deeringia celosioides*—a rambling climber bearing small globose scarlet fruits, etc.; **weeds**: prickly amaranth (*Amaranthus spinosus*; B. KANTA-NATE; H. CHAULAI)—a common spinous weed, *A. viridis*, *A. mangostanus*, *A. tenuifolius*, etc., chaff-flower (*Achyranthes aspera*; B. APANG; H. LATJIRA)—a common weed with long spinous spikes, *Cyathula prostrata*—a slender erect weed, *C. tomentosa*—a densely tomentose undershrub, *Digera arvensis*—a common weed of fields and roadsides, *Alternanthera amoena*—a very common prostrate weed, *Pupalia atropurpurea*—a climber with hooked fruits, *Aerua scandens*—leaves and branches reddish, *A. lanata* and *A. tomentosa* having a dense coating of hairs, *A. monsonia*—a much-branched slender herb, etc.

Family 39 *Chenopodiaceae* (1,400 sp.—40 sp. in India)

Habit—mostly herbs, rarely shrubs, often fleshy, sometimes covered with hairs; stem jointed in *Salicornia*; mostly xerophytic and halophytic. **Leaves** simple, usually alternate, rarely opposite, often fleshy, sometimes with hairs all over; in some species leaves remain undeveloped. **Inflorescence**—racemose with branches cymose. **Flowers** small, greenish, regular, bisexual or sometimes unisexual (as in *Spinacea* and *Atriplex*), monoecious or dioecious, hypogynous (except in *Beta*). **Perianth**—tepals 3-5, simple, sepaloid, free, imbricate. **Androecium**—stamens as many as tepals and opposite to them, free, or united at the base; disc sometimes present. **Gynoecium**—carpels usually (2-3), syncarpous; ovary superior (semi-inferior in *Beta*), unilocular, with one basal campylotropous ovule. **Fruit** a small nut, achene or berry. **Seed** often with mealy endosperm; embryo curved or rolled. *Floral formula*— $\oplus \, \male \, P_{3-5} A_{3-5} \underline{G}_{(2-3)}$.

Examples. [The larger genera are: *Atriplex* (about 200 sp.), and *Chenopodium* (over 100 sp.)]. **Vegetables:** spinach (*Spinacia oleracea*; B. PALANG; H. PALAK)—a fleshy herb, beet (*Beta vulgaris*)—the sugar-beet yielding about 18% of sugar is cultivated in Europe, Russia and America, while the garden-beet is extensively used in India as a vegetable, Indian spinach (*Basella rubra*; B. PUIN; H. POI)—a large twining fleshy herb, *Atriplex hortensis*—a succulent herb, goosefoot (*Chenopodium album*; B. & H. BATHUA), etc.; **weeds:** worm-seed (*Chenopodium anthelminticum*)—a roadside weed; essential oil is used as a vermifuge, *C. ambrosioides*—a much-branched weed; **halophytes:** glasswort (*Salsola foetida*)—a fleshy plant with spinous leaf-apex, saltwort (*Salicornia brachiata*)—a leafless succulent herb with jointed stems; yields barilla (an impure sodium carbonate), *Arthrocnemum indicum* (B. JADU-PALANG)—a leafless undershrub with jointed fleshy stem, growing along with *Salicornia*, sea-blite (*Suaeda maritima* and *S. fruticosa*)—herbs with fleshy leaves, etc.; **ornamental:** *Kochia* (90 sp.), e.g. *K. tricophylla*—an ornamental bushy herb, etc.

Family 40 Polygonaceae   (800 *sp.*—109 *sp. in India*)
Habit—mostly herbs, sometimes climbing. **Leaves** simple, entire, alternate, opposite or whorled, with distinct ochreate stipules (a characteristic feature of the family). **Inflorescence** a raceme or spike, with lateral cymes. **Flowers** small, regular, bisexual, usually hypogynous, trimerous (rarely dimerous), cyclic or acyclic, sometimes medianly zygomorphic; several species are dimorphic. **Perianth**—tepals 3 or 6 (in two whorls), sometimes 5, generally uniform, often persistent. **Androecium**—stamens varying, commonly 5–8 in two series. **Gynoecium**—carpels (3) or sometimes (2), syncarpous; ovary superior, unilocular containing a single erect (orthotropous) ovule; styles 3 or 2. **Fruit** a small hard triangular nut. **Seed** albuminous, sometimes ruminated; embryo curved or straight.

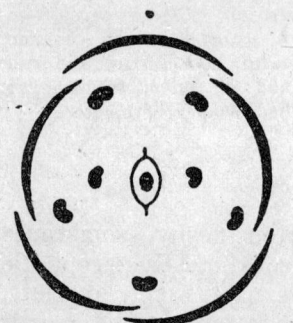

FIG. 62. Floral diagram of *Polygonaceae* (*Polygonum*).

Floral formula— $\oplus \, \male \, P_{3+3, \text{ or } 5} A_{5-8} \underline{G}_{(3)}$.

SELECTED FAMILIES OF DICOTYLEDONS

Polygonaceae is related to *Urticaceae* by having unilocular ovary with a single orthotropous ovule. According to Hutchinson it is derived from *Caryophyllaceae*.

Examples. [The larger genera are: *Polygonum* (275 sp.), *Cocoloba* (150 sp. in America), *Eriogonum* (150 sp. in America), *Rumex* (over 100 sp.), and *Rheum* (50 sp.)]. *Polygonum* (88 sp. in India), e.g. *P. plebejum*, *P. orientale*, *P. glabrum*, *P. hydropiper*, *P. capitatum* (common in hill slopes in Shillong), etc., dock or sorrel (*Rumex vesicarius*)—cultivated for its sour leaves, *R. maritimus*—a common weed, buckwheat (*Fagopyrum esculentum*)—cultivated in the hills for grains used for making bread, rhubarb (*Rheum*; 7 sp. in alpine and sub-alpine Himalayas), e.g. *Rheum emodi* (Indian rhubarb)—a medicinal (cathartic) herb, and *R. rhaponticum*—cultivated as a vegetable in the Khasi Hills, cocoloba (*Muehlenbeckia platyclados*) showing phylloclades (see FIG. I/39B), Sandwich Island climber (*Corculum leptopus* =*Antigonon leptopus*; see FIG. I/36B)—a common garden climber with pink or white flowers, *Calligonum polygonoides*—an almost leafless shrub of Rajasthan and Punjab—flowers and buds eaten and relished, etc.

Family 41 Loranthaceae (1,100 sp.—64 sp. in India)

Habit—mostly semi-parasitic shrubs or undershrubs, growing on tree branches and developing sucking roots or haustoria. **Leaves** simple, commonly opposite, thick and leathery, exstipulate, sometimes reduced to scales. **Inflorescence** racemose, spicate or cymose, sometimes flowers in fascicles. **Flowers** unisexual or bisexual, regular or slightly zygomorphic, greenish or brightly coloured, epigynous. **Perianth** grows from the margin of the cup-shaped receptacle, in two whorls, sepaloid (in *Viscum*) or petaloid (in *Loranthus*); perianth leaves free or united into a tubular structure, usually 3- to 6-lobed; in *Loranthus* a small outgrowth (called **calyculus**) of the axis is present below the perianth; the calyculus is regarded by some as a calyx. **Stamens** opposite to perianth-segments and equal in number, united with the latter. **Carpels** (3) or (4); ovary unilocular, inferior, sunken in the receptacle; ovule solitary; placenta not differentiated from the ovule. **Fruit** a drupaceous or berry-like pseudocarp, with often a very sticky substance (viscin) round the seed. **Seeds** commonly 1, sometimes 2 or 3, albuminous; on germination the hypocotyl first forms a swollen sucker fixing the embryo to the branch of the host plant.

Examples. With unisexual flowers: *Viscum* (60 sp.), e.g. mistletoe (*V. album*) showing dichasial branching (see FIG. I/24). With mostly bisexual flowers: *Loranthus* (over 500 sp.), now split up into a number of genera, e.g. *Dendropthe*, *Taxillus*, *Scurrula*, etc., as follows—*Dendrophoe falcata* (=*L. longiflorus*), *Taxillus vestitus* (=*L. vestitus*), *Scurrula parasitica* (=*L. scurrula*), *Tolypanthus involucratus* (=*L. involucratus*), *Helixanthera parasitica* (=*L. pentapetalus*), *H. coccinea* (=*L. coccinea*), *Macrosolen cochinchinensis* (=*L. globosus*), etc. *Phoradendron*, an American genus, has about 190 sp.

Family 42 *Euphorbiaceae* (7,000 *sp.*—374 *sp. in India*)

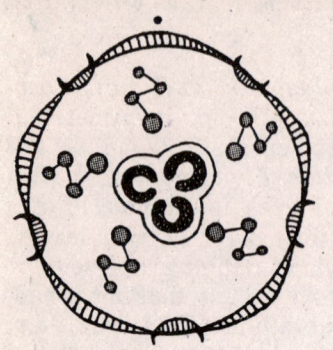

FIG. 63. Floral diagram of *Euphorbiaceae* (*Euphorbia*)

Habit—herbs, shrubs and trees, with acrid milky juice, many xerophytic. **Leaves** simple, usually alternate; stipules usually present. **Inflorescence** varying—racemose or cymose, or mixed, or a cyathium (see p. 70), as in spurge (*Euphorbia*) and jew's slipper (*Pedilanthus*). **Flowers** small, bracteate, regular, hypogynous, always unisexual, monoecious or dioecious; rudiments of the other sex are sometimes present. **Perianth** in 1 or 2 whorls, sometimes absent, dissimilar in male and female flowers.

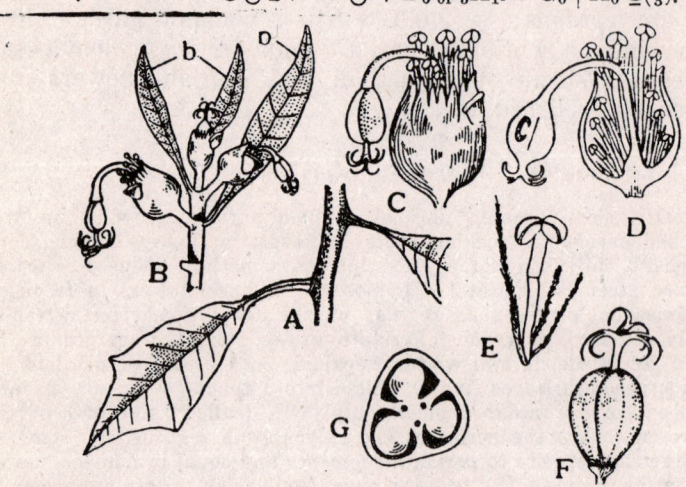

Floral formulae—$\oplus \male\ \female$ or $\male\ \female\ P_{0\text{ or }5} A_{1\cdot\infty}\ G_0\ |\ A_0\ \underline{G}_{(3)}$.

Euphorbiaceae. FIG. 64. Poinsettia (*Euphorbia pulcherrima = Poinsettia pulcherrima*). *A*, a portion of a branch; *B*, a branch with three inflorescences; *b*, petaloid bracts; *C*, an inflorescence (cyathium); *D*, cyathium cut longitudinally showing the centrally placed female flower surrounded by numerous male flowers; *E*, a male flower (reduced to a stamen only) with bract and bracteoles at the base; *F*, female flower (reduced to a pistil only); *G*, ovary in transection.

Male Flowers: in spurges (*Euphorbia*) and jew's slipper (*Pedilanthus*) flowers are reduced to solitary stamens without any perianth (see p. 70); in other cases stamens usually many or sometimes few; filaments either free or connate in 1 to many bundles. *Female Flowers*: **carpels** (3), syncarpous; ovary 3-celled, 3-lobed, superior, with 1 or 2 ovules in each loculus.

pendulous; styles 3, each bifid; stigmas 6. **Fruit** mostly a capsule or a regma. Seed albuminous.

Euphorbiaceae is closely related to *Sterculiaceae* of Malvales by various degrees of union of stamens and the presence of pistillode and staminode in male and female flowers respectively, and sometimes also androphore and gynophore, and may have originated from a common ancestral stock of Malvales, but separated from the latter by reduction of floral parts. It is also very closely related to Geraniales by the structure of the gynoecium.

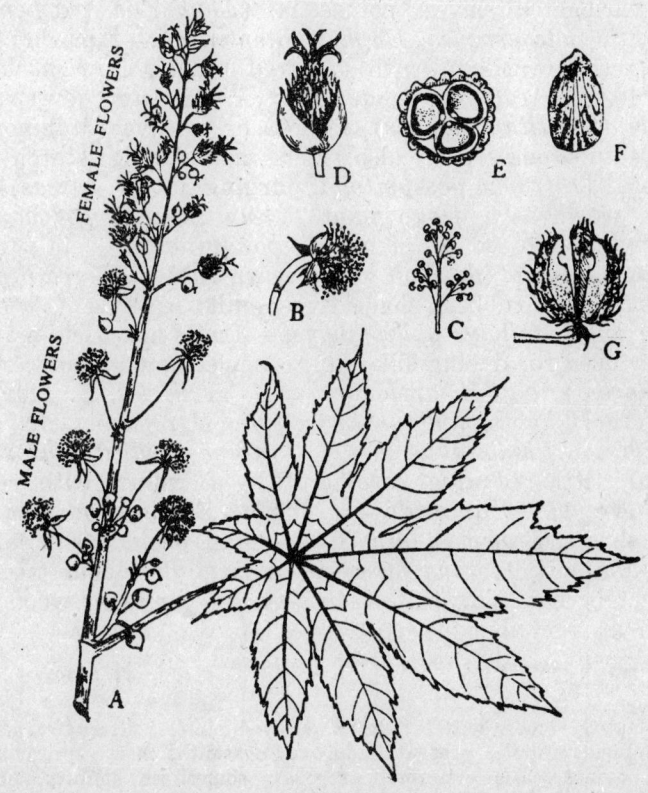

Euphorbiaceae. FIG. 65. Castor (*Ricinus communis*). *A*, a branch with a leaf and an inflorescence; *B*, a male flower; *C*, branched stamens; *D*, a female flower; *E*, ovary in transection; *F*, a seed with caruncle; and *G*, a fruit (regma) splitting.

Examples. [The larger genera are: *Euphorbia* (about 2,000 sp.), *Croton* (700 sp.), *Phyllanthus* (500 sp.), *Acalypha* (430 sp.), *Jatropha* (175 sp.), *Manihot* (170 sp.), etc.]. **Useful plants: oils:** castor (*Ricinus communis*; FIG. 65)—seeds yield castor oil. *Croton tiglium* (B. JAIPAL; H. JAMOLGOTA)—seeds yield croton oil used as a drastic purgative, Indian walnut (*Aleurites moluc-*

cana)—seeds yield a drying oil, sometimes used as candles, *A. cordata* and a few other species—nuts yield 35-40% of TUNG oil used for lacquering, varnishing, water proofing and manufacturing oil cloth, etc.; **fruits**: emblic myrobalan (*Emblica officinalis*=*Phyllanthus emblica*)—fruits rich in vitamins, medicinal and also used for tanning, *Phyllanthus acidus* (=*Cicca acidus*; B. NOAR; H. CHALMERI)—fruits sour but tasteful, *Baccaurea sapida* (B. LATKAN; H. LUTKO)—aril, pulpy and edible, etc.; **ornamental**: garden croton (*Codiaeum variegatum*) with variegated leaves, poinsettia (*Euphorbia pulcherrima*= *Poinsettia pulcherrima*), *Euphorbia splendens*—a prickly undershrub, each cyathium with two red bracts, *E. tirucalli* (see FIG. I/40A), *Acalypha sanderiana*, *A. tricolor*, jew's slipper (*Pedilanthus tithymaloides*), tapioca or cassava (*Manihot esculenta*)—tuberous roots also yield a valuable starchy food (tapioca), *Jatropha gossypifolia*, purging nut (*J. curcas*)—commonly grown as a hedge plant, *J. multifida*, *J. podogarica*—a xerophyte with gouty stem, *J. panduraefolia*, etc.; **rubber-yielding**: *Manihot glaziovii* yields ceara rubber, *Hevea brasiliensis* yields para rubber; **shade trees**: child life tree (*Putranjiva roxburghii*), *Mallotus philippinensis*—fruits also yield a crimson powder used for dyeing silk, etc.; **other common plants**: spurges (*Euphorbia*), e.g. *E. antiquorum*, *E. neriifolia*, *E. nivulia*, *E. pilulifera*, *E. heterophylla*, *E. thymifolia*, *E. royleana*, etc., *Croton sparsiflorus*, *Phyllanthus niruri*, *Acalypha indica*, *Chrozophora plicata*, nettle (*Tragia involucrata*)—a twiner with stinging hairs, *Breynia rhamnoides*—a shrub, *Fluggea microcarpa*—a large shrub, *Trewia nudiflora*—a small deciduous tree, wood white and soft, *Biscofia javanica*—a large deciduous tree, wood moderately hard, *Bridelia retusa*—a timber tree, wood cream or brown, very durable, etc.

Family 43 *Ulmaceae* (180 *sp*.—14 *sp. in India*)

Habit—mostly trees, with no latex. **Leaves** simple, alternate, distichous, often oblique; stipules present, caducous; cystoliths mostly present. **Inflorescence** cymose. **Flowers** small, regular, sometimes solitary, unisexual, monoecious, rarely bisexual, hypogynous. **Perianth**—tepals 4-5, free or united, sepaloid, imbricate. **Androecium**—stamens equal in number and opposite to the tepals, sometimes fewer; filaments straight in bud; anther splits longitudinally; rudimentary carpels often present. **Gynoecium**—carpels (2), syncarpous; ovary 1-locular, sometimes 2-locular; ovule solitary, anatropous or amphitropous. **Fruit** a nut, samara or drupe. **Seed** commonly exalbuminous; embryo straight or curved.

Floral formulae— $\oplus \male - \female \; P_{4\text{-}5 \text{ or } (4\text{-}5)} \, A_{4\text{-}5} G_0 \mid A_0 \underline{G}_{(2)}$.

Examples. *Celtis* (60 sp.), e.g. *C. australis*—a middle-sized deciduous tree, *Ulmus* (45 sp.), e.g. *U. lancifolia*—a large deciduous tree, *Trema* (30 sp.)

e.g. *T. orientalis*—a fast-growing tree, the fibrous bark of which is beaten into a coarse mattress by the Garos.

Family 44 Moraceae (over 1,000 *sp.*—106 *sp. in India*)

Habit—mostly trees, a few shrubs or herbs, with latex. **Leaves** simple, alternate, entire or lobed; stipules large, caducous; cystoliths present in some genera, e.g. *Ficus* and *Morus*, while absent in others, e.g. *Artocarpus*. **Inflorescence** cymose, usually in the form of raceme, spike, umbel or head; hypanthodium in *Ficus*. **Flowers** small, regular, unisexual, monoecious or dioecious, hypogynous. **Perianth**— tepals 4 or (4), free or united, often persistent in fruit. **Androecium** (in ♂ flowers)—stamens equal in number and opposite to the tepals, sometimes reduced to 1 or 2; filaments incurved or straight in bud; anther dehiscing. **Gynoecium** (in ♀ flowers)—carpels (2), syncarpous, 1-locular (one carpel usually abortive), superior to inferior; ovule solitary, campylotropous. **Fruit** a drupe, nut or achene; the whole inflorescence sometimes develops into a multiple fruit (sorosis or syconus). **Seed** with or without endosperm; embryo curved.

Floral formulae—$\oplus \male \female$ or $\male \female$ $P_{4 \text{ or } (4)} A_{4-1} G_0 \mid A_0 \underline{G}(_2) \text{ or } \bar{G}(_2)$.

Examples. Mulberry (*Morus alba*—with short spikes and *M. nigra*—with long spikes)—fruits eaten; wood very valuable, particularly of the latter; silk worms are reared on mulberry, *Ficus* (over 600 sp.) e.g. fig (*F. carica*; B. DUMUR; H. ANJIR), *F. glomerata* (B. JAJNA-DUMUR; H. GULAR), *F. virens* (=*F. infectoria*; B. PAKUR; H. PAKAR), *F. hispida* (B. KAK-DUMUR; H. KONEA-DUMBAR), *F. rumphii* (B. GAI-ASWATTHA; H. KHABAR), banyan (*F. bengalensis*), peepul or bo-tree (*F. religiosa*), indiarubber plant (*F. elastica*), Indian ivy (*F. pumila*; see FIG. I/17), *F. pomifera*—large reddish edible fruits of upper Assam, *F. lanceolato*—large purplish fruits of Khasi Hills, *Artocarpus* (40 sp.), e.g. jack (*A. heterophyllus*)—large summer fruit (sorosis), sweet and edible; wood used for furniture, monkey jack (*A. lakoocha*; B. & H. DEOPHAL)—fruit eaten, wood useful, bread-fruit (*A. incisa*)—fruit sliced, roasted and eaten like bread, chaplash (*A. chaplasha*; see FIG. I/79B)—wood valuable, cow-tree or milk-tree (*Brosimum galactodendron*) of Venezuela—its profuse milky latex is used as a substitute for milk, being sweet, tasteful and nutritious, *Streblus asper* (B. SHAORA)—a rigid evergreen tree, paper-mulberry (*Broussonetia papyrifera*) —wood very soft and light, bark used for making paper in Japan, *Cudrania javanensis*—a rambling shrub, *Conocephalus suaveolens*—a large evergreen woody climber, *Castilloa elastica* —source of panama-rubber, etc.

Distinguishing Characters of Urticales

Urticaceae	Moraceae	Ulmaceae
mostly herbs	trees or shrubs	mostly trees
Latex absent; cystoliths abundant	latex present; cystoliths present or absent	latex absent; cystoliths mostly present
Flowers unisexual, hypogynous	unisexual, hypogynous to epigynous	unisexual, rarely bisexual, hypogynous
Stamens 4-5, anther exploding, filaments incurred in bud	4, often reduced to 1 or 2, anther not exploding, filaments incurred or straight in bud	4-5, anther splitting longitudinally, filaments straight in bud
Carpel 1, style 1	(2), usually 1 aborted, styles 1 or 2	(2), styles 2
Ovary 1-locular, superior	1-locular, superior to inferior	2-locular or 1-locular, superior
Ovule orthotropous	campylotropous	anatropous or amphitropous
Endosperm oily	fleshy or absent	usually not present
Embryo straight	curved	straight or curved

Family 45 *Urticaceae*[1] (600 *sp.*—104 *sp. in India*)

Habit—mostly herbs, sometimes shrubs, with no latex. **Leaves** simple, alternate or opposite with three basal nerves, with or without stinging hair; cystoliths abundant and of various forms; stipules membranous. **Inflorescence** cymose (often condensed). **Flowers** small, regular, unisexual, monoecious or dioecious, **Perianth**—tepals usually 4, sometimes 5, free or connate, sepaloid. **Androecium**—stamens as many as the tepals and opposite to them; filaments incurved in bud; anthers exploding when mature. **Gynoecium**—carpel 1; ovary superior, 1-locular, with 1 basal orthotropous ovule. **Fruit** a small achene, nut or drupe. **Seed** usually with oily endosperm; embryo straight.

Floral formulae— $\oplus \male - \female \; or \male \; \female \; P_{4 \; or \; (4)} \; A_{4 \; or \; 5} G_0 \mid A_0 G_1$.

Examples. With stinging hairs—nettles: *Urtica* (50 sp.), e.g. *U. dioica* (also fibre-yielding), *Laportea* (=*Fleurya*) *interrupta*, *Girardinia zeylanica* (also fibre-yielding), devil or fever nettle (*Laportea crenulata*), etc. **Without stinging hairs:** *Pilea* (over 200 sp.), e.g. gunpowder plant (*P. microphylla*), *Boehmeria* (about 100 sp.), e.g. rhea or ramie (*B. nivea*)—cultivated for best fibres (longest, toughest and silkiest), *Pouzolzia indica*—common on roadsides and waste places, *Elatostema*—a herb or undershrub common on hill-slopes.

---

[1] *Urticaceae* of Bentham and Hooker has been split up by Engler into three families—*Ulmaceae*, *Moraceae* and *Urticaceae* under the order Urticales. Hutchinson further split up the order into four families—*Ulmaceae*, *Moraceae*, *Urticaceae* and *Cannabinaceae*; the last one has been isolated from *Moraceae*.

Family 46 *Cannabinaceae* (3 sp.—2 sp. *in India*)

Aromatic herbs. Leaves palmi-nerved and palmately divided; cystoliths present; stipules present and persistent; no latex. Flowers unisexual, dioecious, borne in cymes. *In male flowers* perianth leaves 5 and stamens 5 (opposite to perianth leaves), while *in female flowers* the perianth is entire and cup-shaped, and carpels (2); ovary 1-celled with 1 pendulous ovule. Fruit a nut or achene. Seed albuminous or exalbuminous; embryo curved or spiral.

Examples. Only 2 genera—*Cannabis* (1 sp.) and *Humulus* (2 sp.). Hemp (*Cannabis sativa*)—yields valuable bast fibres and is the source of a narcotic resin in three forms: GANJA (resinous flowering shoots of cultivated female plants), CHARAS (resinous exudation from twigs of plants grown in cold climates, particularly in Nepal) and BHANG (mature leaves and flowering shoots with resinous contents of plants growing wild), and hop (*Humulus lupulus*)—fruits are used in brewing.

Family 47 *Casuarinaceae* (40 sp—1 sp. *in India*)

Habit—trees with xerophytic habit; branches jointed, and internodes furrowed. Leaves—alternating whorls of 4-12 minute scale-like leaves united at the base to form a sheath. Flowers—extremely simple, unisexual, monoecious, with a bract and two bracteoles. *Male flowers* borne in terminal catkin-like spike at the end of a branch, each consisting of a single stamen, two small perianth leaves, a bract and a pair of lateral bracteoles; flowers in whorls, and bracts in a sheath round each whorl. *Female flowers* borne in a more or less spherical head on a short lateral branch, each consisting of two carpels, a bract and 2 bracteoles which harden in the fruit; pistil syncarpous; ovary 1-celled by the suppression of the posterior cell; ovules generally 2, orthotropous, ascending, but 1 matures; stigmas 2, long protruding. Pollination by wind; fertilization chalazogamic. Fruit 1-seeded winged nut (the whole head, however, becomes woody and cone-like). Seed exalbuminous, winged.

Examples. *Casuarina*, commonly called beef-wood or she-oak, is the only genus of the family (mainly Australian). *C. equisetifolia* is widely grown in India as a roadside or avenue tree.

Chapter 4 SELECTED FAMILIES OF MONOCOTYLEDONS

Family 1 *Hydrocharitaceae* (80 sp.)

Habit—aquatic herbs, usually submerged, sometimes floating; some marine. Leaves simple, radical and clustered, or cauline, entire, ribbon-shaped or flattened. Flowers often regular. trimerous. commonly unisexual, dioecious or monoecious, rarely bisexual, enclosed in a spathe of 2 or more fused bracts; female flowers solitary, sometimes each on a long stalk; male flowers in a cluster. Perianth segments 3+3, free, in 2 whorls (1 whorl in *Vallisneria*), the outer (calyx) often green and the inner (corolla) petaloid. In ♂ flowers **stamens** 3-12 in 1-4 trimerous series, the innermost whorl often reduced to staminodes. In ♀ flowers **carpels** (2-15), commonly (3) or (6-15), connate; ovary inferior, unilocular; placentation parietal, placentae 3-6, sometimes intruding into the loculus, almost reaching the axis; ovules ∞ on each placenta; style 1, usually divided into 3-12 stigmas. **Fruit** ovoid or oblong, dry or pulpy. **Seeds** few to ∞, exalbuminous; embryo large. *Floral formula*— $\oplus \male \female P_{3+3} A_{3-12} \bar{G}_{(2-15)}$

Examples. Dioecious: *Vallisneria spiralis*—a stoloniferous submerged herb (see FIG. I/141), *Hydrilla verticillata*—a submerged leafy herb, *Lagarosiphon roxburghii*—a submerged leafy herb, *Blyxa roxburghii*—a submerged tufted herb; **monoecious:** *Hydrocharis cellulosa*—a floating herb; **bisexual:** *Blyxa griffithii*—a submerged tufted herb, and *Ottelia alismoides*—a partly submerged herb.

Family 2 *Liliaceae* (over 3,000 *sp.*)

Habit—herbs and climbers, rarely shrubs or trees, e.g. *Dracaena* and *Yucca*, with a bulb or rhizome or with fibrous roots. **Leaves** simple, radical or cauline or both. **Inflorescence** a spike, raceme, panicle or umbel (solitary flowers in *Tulipa*), often on a scape. **Flowers** regular, bisexual (rarely unisexual, dioecious, as in *Smilax*), 3-merous and hypogynous; bracts usually small, scarious (thin, dry and membranous). **Perianth**—tepals petaloid, usually 6 in two whorls, 3 + 3, free (polyphyllous) or (3+3), united (gamophyllous). **Androecium**—stamens 6, in two whorls, 3+3, rarely 3, free, or united with the perianth (epiphyllous) at the base; anthers often dorsifixed. **Gynoecium**—carpels (3), syncarpous; ovary superior, 3-celled; ovules usually ∞ in two rows in each loculus; placentation axile; styles (3) or 3. **Fruit** a berry or capsule. **Seeds** albuminous.

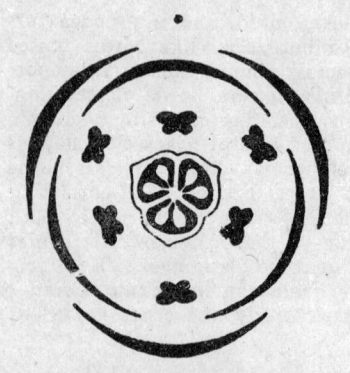

FIG. 66. Floral diagram of *Liliaceae*.

Floral formula— $\oplus \female P_{3+3} A_{3+3} \underline{G}_{(3)}$ or $P_{(3+3)} A_{3+3} \underline{G}_{(3)}$.

Examples. [The larger genera are: *Allium* (over 400 sp.), *Smilax* (300 sp.), *Asparagus* (300 sp.), *Aloe* (over 200 sp.), *Dracaena* (150 sp.), *Lilium* (70 sp.), *Colchicum* (65 sp.), and *Sansevieria* (60 sp.)]. **Useful plants—vegetables:** onion (*Allium cepa*), garlic (*A. sativum*)—also medicinal, shallot (*A. ascalonicum*), leek (*A. porrum*), etc.; **medicinal:** *Asparagus racemosus*—root is tonic and astringent, sarsaparilla (*Smilax zeylanica* and other species; see FIG. I/48)—dried roots form sarsaparilla (a blood purifier), Indian aloe (*Aloe vera*), meadow saffron (*Colchicum autumnale*), etc.; **ornamental:** lily (*Lilium*), e.g. *L. candidum*—

SELECTED FAMILIES OF MONOCOTYLEDONS

flowers white, *L. bulbiferum* reproducing by bulbils, etc., glory lily (*Gloriosa superba*; see FIG. I/63C), day lily (*Hemerocallis fulva*), dagger plant or Adam's needle (*Yucca gloriosa*; see FIG. I/81), dragon plant (*Dracaena*), *Sansevieria laurentii*—green foliage with yellow or white margin, *Cordyline australis*—stem erect, cylindrical and *Dracaena-like*, butcher's broom (*Ruscus aculeatus*; see FIG. I/41A), asphodel (*Asphodelus tenuifolium*), *Scilla indica* propagating by leaf-tips, *Asparagus plumosus*, *A. sprengeri*, etc.; **fibre-yielding**: *Phormium tenax* yielding New Zealand flax, and bowstring hemp (*Sansevieria roxburghii*)—leaves flat and spotted, *S. cylindrica*—leaves cylindrical, *S. zeylanica*—leaves furrowed, etc.

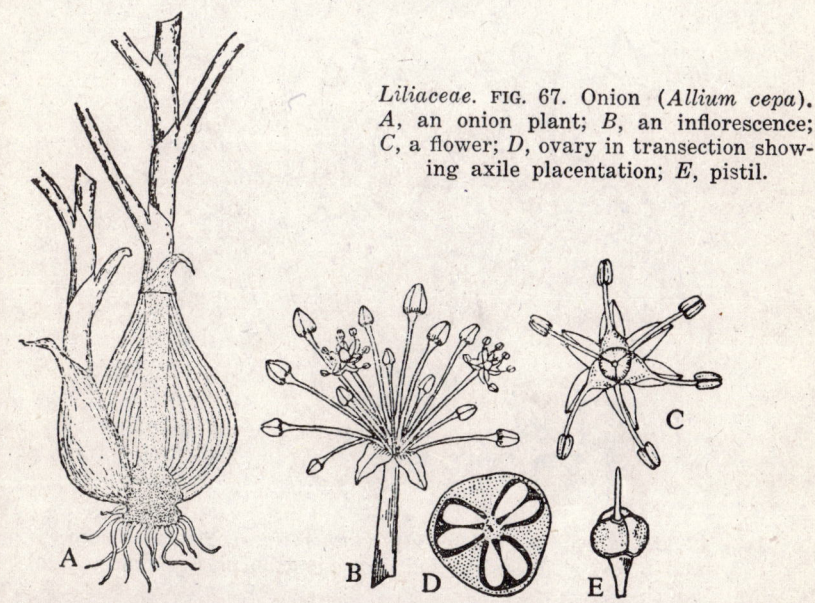

Liliaceae. FIG. 67. Onion (*Allium cepa*). *A*, an onion plant; *B*, an inflorescence; *C*, a flower; *D*, ovary in transection showing axile placentation; *E*, pistil.

Family 3 *Amaryllidaceae* (1,100 *sp.*)

Habit—often bulbous, sometimes rhizomatous, perennial herbs. **Leaves** radical. **Inflorescence** commonly *umbellate* on a scape, or a solitary flower on a scape, with *spathaceous involucrate* bracts. **Flowers** regular, bisexual, epigynous, often showy. **Perianth** petaloid, segments 3+3 or (3+3), biseriate, sometimes brightly coloured. **Androecium**—stamens 3+3, in two whorls; staminodes sometimes present; filaments often dilated at the base; anthers erect or versatile; corona prominent in some cases, as in *Pancratium*, *Narcissus*, *Eucharis*, etc. **Gynoecium**—carpels (3), syncarpous, inferior, rarely semi-inferior or even superior, trilocular; placentation axile,

---

It may be noted that *Dracaena*, *Cordyline*, *Yucca*, *Sansevieria* and *Phormium* of Liliaceae, and *Agave* and *Polianthes* of Amaryllidaceae have now been taken over to a new family *Agavaceae*. Further, *Allium* having spathaceous involucrate bracts and umbellate inflorescence has been transferred to Amaryllidaceae.

with often numerous anatropous ovules. **Fruit** a capsule, or sometimes a berry.

Floral formula— $\oplus \; ⚥ \; P_{3+3} A_{3+3} \bar{G}_{(3)}$ or $\overline{P_{(3+3)}} A_{3+3} \bar{G}_{(3)}$

Examples. [The larger genera are: *Agave* (300 sp., American), *Crinum* (over 100 sp.), and *Zephyranthes* (60 sp., American)]. Mostly ornamental, e.g. Easter lily (*Amaryllis bella-donna*), spider lily (*Pancratium gloriosa*; FIG. 68), *Crinum asiaticum, C. latifolium*, zephyr lilies, e.g. *Zephyranthes tubispatha*—flowers white, *Z. carinata*—flowers bright rose, *Z. andersoni*—flowers yellow, African lily (*Agapanthus umbellatus*)—flowers large and bright blue, pin-cushion lily (*Haemanthus multiflorus*), eucharis or Amazon lily (*Eucharis grandiflora*), daffodil (*Narcissus*), tuberose (*Polianthes tuberosa*), American aloe or century plant (*Agave americana*), *Curculigo orchioides* (B. TALMULI; H. MUSLIKAND), etc.

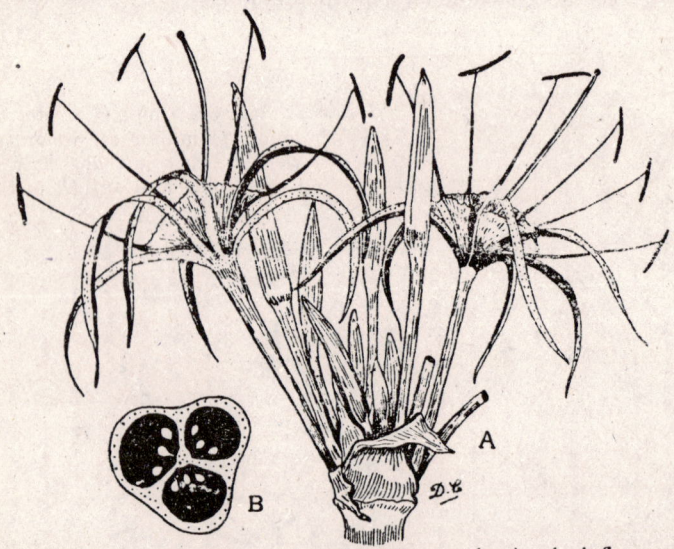

Amaryllidaceae. FIG. 68. Spider lily (*Pancratium*). *A*, inflorescence; and *B*, ovary in transection showing axile placentation.

Family 4 *Commelinaceae* (500 *sp.*)

Habit—annual or perennial herbs, with jointed stem. **Leaves** simple, alternate, with sheathing base. **Flowers** more or less regular, commonly blue, sometimes whitish or pinkish, bisexual, hypogynous, in monochasial cyme, enclosed in a distinct spathe. **Perianth**—6 segments in two series, distinguishable into the outer sepaloid calyx and the inner petaloid corolla, sepals and petals generally free. **Androecium**—stamens 6 in two whorls, either all perfect or some reduced to staminodes or absent; filaments often bearded with hairs. **Gynoecium**—carpels (3), syncarpous; ovary superior, 3-celled; placentation axile; ovules solitary or few, orthotropous. **Fruit** a dehiscent capsule or indehiscent. **Seeds** albuminous.

Floral formula— $\oplus \; ⚥ \; K_3 C_3 A_{3+3} \underline{G}_{(3)}$

Examples. *Commelina* (over 200 sp.), e.g. *C. bengalensis* (see FIG. I/136), *C. salicifolia, C. obliqua, C. nudiflora*, etc., *Aneilema* (100 sp.), e.g. *A. nudi-*

florum, A. spiratum, etc., *Cyanotis* (50 sp.), e.g. *C. axillaris, C. cristata,* etc., *Floscopa scandens, Tradescantia virginiana, Rhoeo discolor*, etc.

Family 5 **Scitamineae**[1] (over 1,200 *sp.*)

Habit—herbs, rarely woody and tree-like, e.g. traveller's tree (*Ravenala*), underground stem usually in the form of a slender or stout rhizome; aerial stem distinct or 'false' made of sheathing leaf-bases, and the flowering stem or scape pushing out through the 'false' stem and ending in an inflorescence. Leaves spiral or distichous, and sheathing. **Inflorescence** a raceme, spike or spadix with often large spathes, either terminal or axillary. **Flowers** zygomorphic, mostly bisexual and epigynous; bracts often spathaceous. **Perianth** of six segments in two whorls. **Androecium**—stamens 1 or 5 (-6) (see sub-families). **Gynoecium**—carpels (3), syncarpous; ovary inferior and trilocular; placentation axile; ovules usually many. **Fruit** a berry or capsule. **Seeds** with perisperm, often arillate. *Scitamineae* has been divided into the following four sub-families (later raised to the rank of families), mainly on the basis of the number of stamens.

(1) **Musaceae** (150 sp.). Leaves spiral, rarely distichous, sheathing but with no ligule. **Perianth** petaloid in two series—

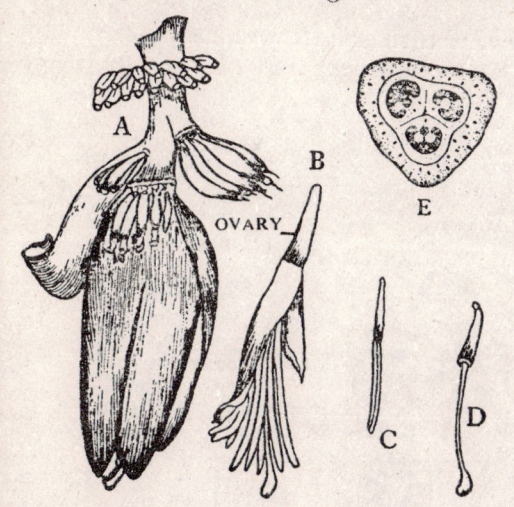

Musaceae.
FIG. 69.
Banana (*Musa paradisiaca*).
A, spadix;
B, a flower;
C, a stamen;
D, pistil; and
E, ovary in transection showing axile placentation (section taken from a wild variety).

one with 5 limbs united and another solitary and free. **Stamens** in 2 whorls, 5 perfect and the 6th one sterile or absent. In *Ravenala*, however, there are six stamens and all of them are fertile.

Floral formula— ⋅↓⋅ ⚥ $P_{(5)+1} A_{3+2} \bar{G}_{(3)}$

---

[1] *Scitamineae* of Bentham and Hooker has been raised by Engler to the rank of an order with four families—*Musaceae, Zingiberaceae, Cannaceae* and *Marantaceae*. Hutchinson renamed the order as *Zingiberales* and divided it into six families.

Examples. *Mussa* (35 sp.), e.g. banana (*M. paradisiaca*)—a dessert fruit, plantain (*M. sapientum*)—green fruit used as a vegetable, dwarf plantain of Assam (*M. sanguinea*)—fruit not edible, *M. superba* and *M. nepalensis*—ornamental, fruits not edible, *M. textilis* yielding commercial Manila hemp or abaca, *Heliconia* (80 sp.-American), traveller's tree (*Ravenala madagascariensis*; B. PANTHAPADAP—see FIG. I/74),—ornamental, and bird of Paradise (*Strelitzia reginae*)—very ornamental. [It may be noted that *Ravenala* and *Strelitzia* having distichous leaves and flowers in cincinnus have been separated from *Musaceae* to a new family *Strelitziaceae* by Hutchinson.]

(2) ***Zingiberaceae*** (700 sp.). Leaves distichous, sheathing, and with a distinct ligule. **Perianth** of 6 segments in 2 whorls, generally distinguishable into calyx and corolla. **Stamens** in 2 whorls—only 1 perfect, adnate to corolla-throat (this is the posterior one of the inner whorl), the other 2 stamens of this whorl are united to form a 2-lipped labellum; the anterior stamen of the outer whorl is absent and the remaining 2 modified into petaloid staminodes or absent. **Style** slender, passing through the two anther-lobes. *Floral formula*—·|· $\male \female K_{(3)} C_3 A_1 \bar{G}_{(3)}$

Examples. *Zingiber* (80 sp.), e.g. ginger (*Zingiber officinale*), wild ginger (*Z. casumunar*), turmeric (*Curcuma longa*), wild turmeric (*C. aromatica*), mango ginger (*C. amada*), butterfly

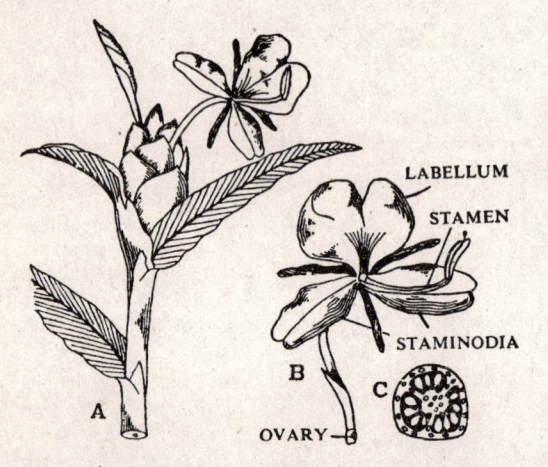

Zingiberaceae.
FIG. 70.
Butterfly lily (*Hedychium coronarium*).
A, a branch with inflorescence;
B, a flower; and
C, ovary in transection showing axile placentation.

lily (*Hedychium coronarium*; B. DULAL-CHAMPA), *Kaempferia* (70 sp.), e.g. *K. rotunda* (B. BHUI-CHAMPA), *Costus* (150 sp.), e.g. *Costus speciosus* (B. KUST; H. KEU), *Alpinia* (225 sp.), e.g. *A. allughas* (B. TARA), *A. galanga*—medicinal, *Globba* (100 sp.) e.g. *G. bulbifera* (see FIG. III/58), cardamom (*Elettaria cardamomum*), *Amomum* (150 sp.), e.g. *A. subulatum* (B. BARA-ELAICH), *A. aromaticum* (B. MORAN-HANCHI), etc.

(3) **Cannaceae** (over 40 sp.). Leaves spiral and sheathing but with no ligule. **Perianth** in 2 whorls of 3 members each—the outer 3 (sepals) free and the inner 3 (petals) united. **Stamens** in 2 whorls—only 1 anther-lobe of 1 stamen (the posterior one of the inner whorl) fertile, the other anther-lobe together with the filament becoming petaloid; one stamen is suppressed and the other stamens modified into petaloid staminodes, one of which forms the labellum covering the style. All petaloid staminodes together with the petaloid anther-lobe are united below with the corolla into a cylindrical tube. **Style** petaloid and flattened.

Floral formula— $\cdot\vert\cdot\ \male\female\ K_3 C_{(3)} A_{\frac{1}{2}} \bar{G}_{(3)}$

Examples. Only genus is *Canna* in tropical America. Indian shot (*Canna orientalis = C. indica*) with many varieties and hybrids is grown widely in Indian gardens. *C. edulis*—the starchy rhizome is eaten as a vegetable or ground into flour in the West Indies.

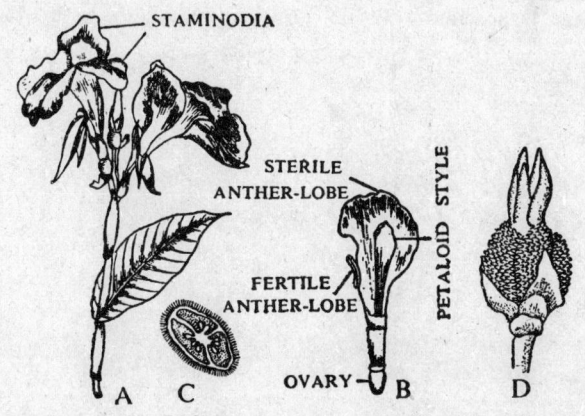

Cannaceae. FIG. 71. Indian shot (*Canna indica*). *A*, a branch; *B*, a flower (perianth and staminodia cut out); *C*, ovary in transection showing axile placentation; and *D*, a fruit.

(4) **Marantaceae** (over 350 sp.). Leaves distichous, sheathing, and with ligule. Floral characters like those of *Cannaceae*. A distinguishing feature of this family is the presence of a joint or swollen pulvinus at the junction of the petiole and the leaf-blade. It is chiefly a tropical American family.

Floral formula— $\cdot\vert\cdot\ \male\female\ P_{3+3} A_{\frac{1}{2}} \bar{G}_{(3)}$

Examples. *Maranta* (23 sp.), e.g. arrowroot (*Maranta arundinacea*), some ornamental species are *Maranta sanderiana*, *M. zebrina*, *M. bicolor*, *M. viridis*, etc., *Calathea* (about 150 sp.)—several species ornamental, *C. allouia*—small potato-like tubers are eaten as a vegetable in the West Indies, *Clinogyne dichotoma* (B. SITALPATI), and *Phrynium* (30 sp.), e.g. *P. variegatum*—ornamental.

Family 6 *Araceae* (over 1,500 sp.)

Habit—herbs, or occasionally climbing shrubs with two kinds of aerial roots—clinging roots fixing the plant to its support and hanging roots ultimately growing down into the soil; with an acrid juice or latex which is in some cases poisonous. Underground stem in the form of sympodial rhizome or corm or erect rootstock; in climbing shrubs branches are sympodial. Plants of this family grow in moist, shady places. **Leaves** are alternate, radical, simple, frequently pinnately or palmately divided, often broad, long-petioled and net-veined; in climbing species cauline leaves distinctly alternate, often broad, long-petioled and palmately veined (reticulate); petiole with sheathing base. **Inflorescence** in the form of spadix subtended by a large, often brightly coloured spathe; usually monoecious with female flowers at the base, higher up some neuter flowers in some species and above them a number of closely-packed male flowers; the axis of the spadix is prolonged into a sterile appendix. **Flowers** sessile, small, naked, unisexual (monoecious or dioecious) or sometimes bisexual, trimerous or dimerous. **Perianth** absent or of 6 or 4 scales. **Androecium**—stamens (6) in 1 or 2 whorls, or reduced in number, even to 1; filaments very short, often united at the base or into a *synandrium*; anthers 2-celled, often dehiscing by terminal pore. **Gynoecium**—carpels (1-3), connate; ovary 1- to 3-celled,

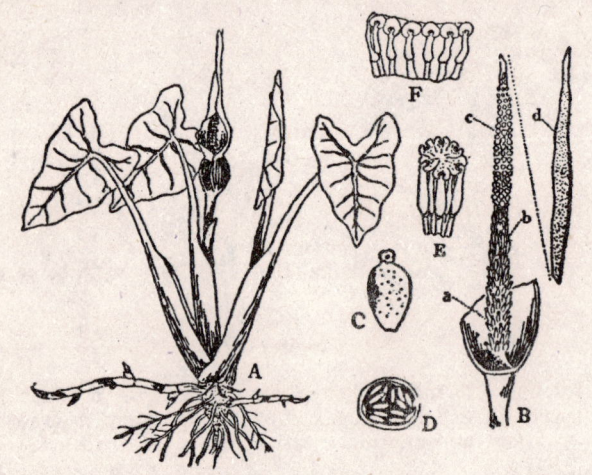

Araceae. FIG. 72. Taro (*Colocasia esculenta*). *A*, a plant; *B*, an inflorescence (spadix)—*a*, female flowers; *b*, neuter flowers; *c*, male flowers; and *d*, appendix; *C*. pistil; *D*, section of ovary; *E*, synandrium; and *F*, synandrium opened out.

superior; each cell with 1 or more ovules; placentation often parietal. **Fruit** a berry. Seed with or without endosperm. Flowers are protogynous and pollination is mainly effected by insects. Appendix in some species with strong, offensive smell, as in *Arum* and *Amorphophallus* (see p. 101).

Floral formulae— ♂-♀ or ♂♀ $P_0 A_{(6-1)} G_0 \mid A_0 \underline{G}_{(1-3)}$

Examples. [The larger genera are: *Anthurium* (over 500 sp.), *Philodendron* (over 250 sp.), *Arisaema* (over 100 sp.), *Amorphophallus* (about 100 sp.), *Pothos* (75 sp.), *Alocasia* (70 sp.), and *Monstera* (50 sp.)]. Taro (*Colocasia esculenta*; FIG. 72), *Alocasia indica* (B. MANKACHU; H. MANKANDA),

SELECTED FAMILIES OF MONOCOTYLEDONS

Typhonium trilobatum (B. GHETKACHU—see FIG. I/85), trumpet- or arum-lily (*Richardia africana*), *Amorphophallus campanulatus* (B. OL; H. ZAMIKAND—see FIG. I/137), Portland arrowroot (*Arum maculatum*), snake or cobra plants (*Arisaema consanguineum*; see FIG. I/83; and *A. tortuosum*), sweet flag (*Acorus calamus*; B. BOCH; H. WACH), water lettuce (*Pistia stratiotes*; see FIG. I/34). *Caladium*—ornamental herbs, *Pothos scandens*—a climber, *Scindapsus officinalis* (B. & H. GAJPIPAL)—an epiphytic climber, *Monstera* and *Philodendron*—ornamental climbing shrubs, *Anthurium*—ornamental herbs, dumb-cane or mother-in-law plant (*Dieffenbachia*)—stem fleshy and erect, leaves blotched, juice acrid and poisonous, and causes temporary paralysis and dumbness in contact with the tongue, etc.

Family 7 Palmae (over 2,500 *sp.*)

Habit—shrubs or trees, sometimes climbing, e.g. cane (*Calamus*). **Stem** (caudex) erect, unbranched and woody, marked with scars of shed leaves, rarely branched, e.g. *Hyphaene*. Some palms attain a height of 45 metres or even more. Some canes grow to a length of over 150 metres. A few palms are short-stemmed, e.g. *Phytelephas*, or even apparently stemless, e.g. *Nipa*, *Phoenix acaulis*, etc. While most palms are polycarpic usually flowering year after year, a few are monocarpic, flowering once only before dying down, e.g. talipot-palm, sago-palm, etc. **Roots** numerous, adventitious (fibrous) developing from the base of the stem. **Leaves** usually forming a crown, plaited in bud, sometimes very large (in some species 15 m. long and 2.5 m. wide); they are of two types—palmately cut or divided (**fan palms**) or pinnately cut or divided (**feather palms**); petiole often with stout sheathing base. **Inflorescence** a spadix, spike or panicle enclosed in one or more large sheathing spathes, often much branched, sometimes very large (as in *Corypha*), often axillary, sometimes terminal. **Flowers** sessile, small

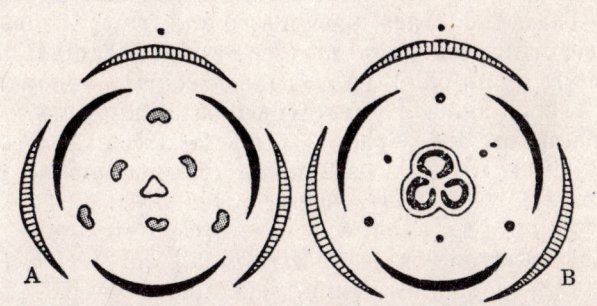

FIG. 73. Floral diagrams of *Palmae*. *A*, male flower; *B*, female flower.

and inconspicuous but sometimes produced in huge numbers presenting an imposing appearance, regular, hypogynous, unisexual (rarely bisexual); male and female flowers in the same inflorescence or in two, either monoecious or dioecious. **Perianth** in two whorls, 3+3, the outer being often smaller, usually free or sometimes connate, imbricate or valvate, persistent in the female flower. **Androecium**—stamens usually in two whorls, 3+3; filaments free or connate; anthers versatile, 2-celled. Rudiment of the pistil sometimes present in the male flower. **Gynoecium**—carpels (3), syncarpous or sometimes 3, apocarpous; ovary superior, unilocular or trilocular, with 1 or 3 ovules. **Fruit** a drupe, berry or nut. **Seed** albuminous. Pollination by wind; pollen produced in huge quantities. Some palms are pollinated by insects also. Flowers protandrous, and hence self-pollination is prevented.

Floral formulae— $\oplus \male\textrm{-}\female$ or $\male\female P_{3+3} A_{3+3} G_0 \mid A_0 \underline{G}(_3)_{\textrm{ or } 3}$

Economically this is one of the most important families since many useful products are obtained from several species of it. Many palms such as palmyra-palm, toddy-palm, date-palm, coconut-palm, etc., are tapped for toddy (fermented country liquor) or for sweet juice from which jaggery or sugar is made. Coconut-palm, date-palm, palmyra-palm, etc., yield edible fruits. Coir fibres of coconut-palm are used for making mats, mattresses and brushes, and also for stuffing cushions. Leaves of many palms are woven into mats, hats and baskets and also used for thatching. Some palms yield oil, e.g. coconut-palm, oil-palm, etc. Sago-palms (*Metroxylon* and *Caryota*) yield sago, which is obtained by crushing the pith. Betelnut is used for masticating with betel leaf. Endosperm of vegetable ivory-palm is very hard and made into billiard balls. Cane is used for making chairs, sofas, tables and baskets and for a variety of other purposes. Many palms are ornamental, sometimes called 'Princes of the Vegetable Kingdom', e.g. fan-palms: *Livistonia chinensis*, *Licuala elegans*, *Sabal palmetto*, *Pritchardia grandis*, etc.; and feather palms: cabbage-palm (*Areca oleracea*), sugar-palm (*Arenga saccharifera*), bottle-palm or royal-palm (*Oreodoxa regia*), *Pinanga spectabilis*, dwarf cane (*Calamus ciliaris*), etc.

Examples. Fan-palms: palmyra-palm (*Borassus flabellifer*), talipot-palm (*Corypha umbraculifera*)—grows to a height of about 24 m., flowers once after about 70 years and then dies, double coconut-palm (*Lodoicea maldivica*), a native of Seychelles Islands—bears the largest known seed and fruit; the latter sometimes measuring well over 1 m. (see FIG. I/181), oil-palm (*Elaeis guineensis*), doum-palm (*Hyphaene thebaica*)—showing dichotomous branching, etc. **Feather-palms:** Indian sago-palm or fishtail-palm or toddy-palm (*Caryota urens*), coconut-palm (*Cocos nucifera*), edible date-palm (*Phoenix dactylifera*), wild date-palm (*P. sylvestris*), *Areca* (over 50 sp.), e.g. areca- or betel-nut-palm (*A. catechu*), cane (*Calamus*—over 350 sp.), sago-palm (*Metroxylon rumphii*), nipa-palm (*Nipa fruticans*; B. GOLPATA)—a stemless palm, vegetable ivory-palm (*Phytelephas macrocarpa*), and *Zalacca beccarii*—its large cane-like fruits are sold in Shillong market.

Family 8 *Gramineae* **or** *Poaceae* **(10,000** *sp.*)

Habit—herbs, rarely woody, as bamboos. They are very widely distributed all over the earth. **Stem** cylindrical with distinct nodes and internodes (sometimes hollow), called *culm*. **Leaves** simple, alternate, distichous, with sheathing leaf-base which is split open on the side opposite to the leaf-blade; there is a hairy structure at the base of the leaf-blade, called the *ligule*. **Inflorescence** usually a spike or panicle of spikelets (FIG. 74); each spikelet consists of one or few flowers (not exceeding 5), and bears at the base two empty bracts or *glumes*, (G_1, G_{11}), one placed a little above and opposite the other; a third glume called *lemma* or flowering glume stands opposite glume II; the lemma encloses a flower in its axil; it may have a bristle-like appendage, long or short, known as the *awn*; opposite the flowering glume or lemma there is a somewhat smaller, 2-nerved glume called *palea*. The spikelet may be sessile or stalked. **Flowers** usually bisexual, sometimes unisexual, monoecious. **Perianth** represented by 2 or 3 minute scales at the base of the flower, called the *lodicules*; these are regarded as forming the rudimentary perianth. **Androecium**—stamens 3, sometimes 6 as in rice and bamboo; anthers versatile and pendulous. **Gynoecium**—carpels generally considered as (3), reduced to 1 (according to some authors) by their fusion or by suppression of 2; ovary superior, 1-celled, with 1 ovule; styles usually 2 (but 3 in bamboos, and 2 fused into 1 in maize, rarely 1), terminal or lateral; stigmas feathery. **Fruit** a caryopsis. **Seed** albuminous. Pollination by wind is most common; self-pollination in a few cases, as in wheat.

Floral formula— ⚥ P$_{\text{Lodicules 2 or 3}}$ A$_{\text{3 or 6}}$ G$_{(3) \text{ or } 1}$

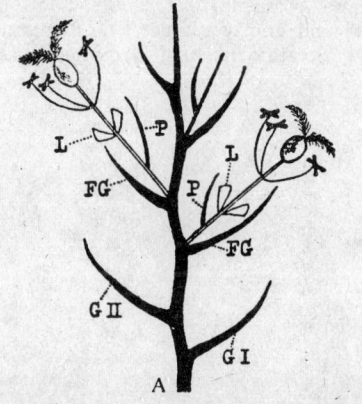

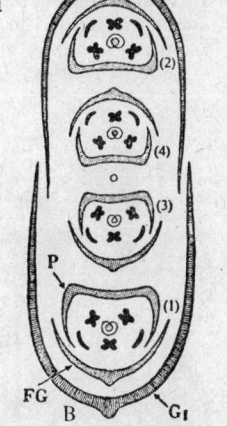

Gramineae. FIG. 74. *A*, spikelet of a grass; *B*, floral diagram of the same. G_1, first empty glume; G_{11}, second empty glume; *FG*, flowering glume; *P*, palea; *L*, lodicule; stamens and carpels of the florets are apparent.

From an economic standpoint *Gramineae* is regarded as the most important family as cereals and millets which constitute the chief foodstuff of mankind belong to it. Most of the fodder crops which are equally important to domestic animals also belong to this family. The importance of bamboo, thatch grass and reed as building materials and of sugarcane as a source of sugar and jaggery is well known. The importance of sabai grass and bamboo as a source of paper pulp cannot be over-emphasized. Common decorative grasses are: ribbon grass (*Phalaris arundinacea*), dwarf bamboo (*Bambusa nana*), certain species of *Panicum*, *Paspalum*, etc. Common lawn grasses are: dog grass (*Cynodon dactylon*), love thorn (*Chrysopogon aciculatus*), etc.

Examples. [The larger genera are: *Panicum* (over 500 sp.), *Digitaria* (over 350 sp.), *Aristida* (over 300 sp.), *Eragrostis* (300 sp.), *Paspalum* (250 sp.), *Poa* (over 200 sp.) *Stipa* (over 200 sp.), *Andropogon* (200 sp.), *Agrostis* (over 150 sp), *Sporobolus* (150 sp.),

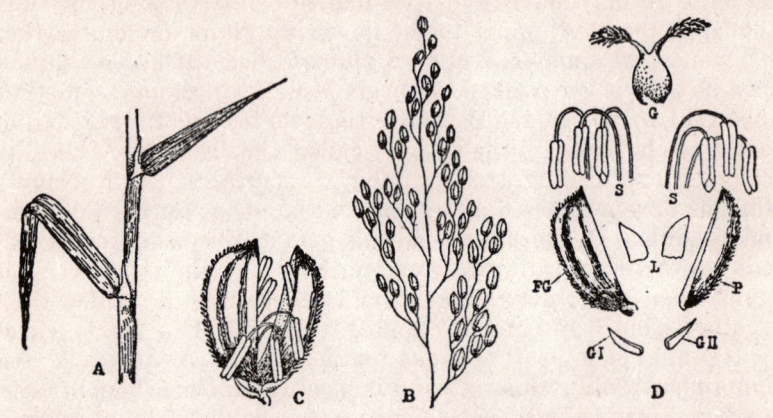

Gramineae. FIG. 75. Rice (*Oryza sativa*). *A*, portion of a branch with sheathing leaves and ligules; *B*, a panicle of spikelets; *C*, 1-flowered spikelet (note the glumes and stamens); *D*, spikelet dissected out *GI*, first empty glume; *GII*, second empty glume; *FG*, flowering glume; *P*, palea; *L*, lodicules; *S*, stamens; and *G*, gynoecium.

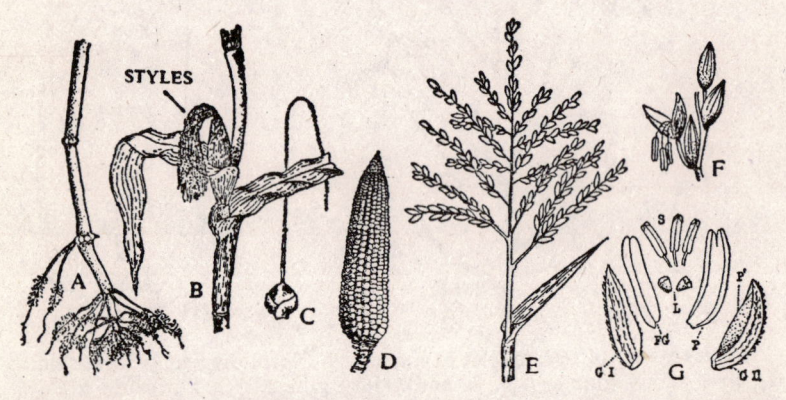

Gramineae. FIG. 76. Maize or Indian corn (*Zea mays*). *A*, adventitious roots; *B*, female spadix in the axil of a leaf; *C*, female spikelet; *D*, ripe cob; *E*, a panicle of male spikelets; *F*, two pairs of male spikelets; and *G*, a male spikelet dissected out—*GI*, first empty glume; *GII*, second empty glume; *P'*, palea of the lower flower, *FG*, flowering glume; *P*, palea of the upper flower; *L*, lodicules; and *S*, three stamens of the upper flower.

and *Pennisetum* (130 sp.)]. **Cereals** such as rice (*Oryza sativa*—FIG. 75), maize or Indian corn (*Zea mays*—FIG. 76), wheat (*Triticum aestivum*), barley (*Hordeum vulgare*), oat (*Avena* *sterilis*), etc.; **millets** such as great millet (*Sorghum vulgare*; B. & H. JUAR), Italian millet (*Setaria italica*; B. KAUN), Indian millet (*Panicum miliaceum*; B. & H. CHEENA), little millet (*P. miliare*), pearl millet (*Pennisetum typhoideum;* B. & H. BAJRA), *Eleusine coracana* (B. & H. MARUA), etc.; job's tears (*Coix lachryma-jobi*)—grains used as an important article of food by the poor hill tribes and also used as beads for ornamental purposes, sugarcane (*Saccharum officinarum*), thatch grass (*S. spontaneum*; B. KASH; H. KANS), reed (*Phragmites karka*; B. NAL; H. NUDA-NAR), giant reed (*Arundo donax*; B. GABNAL; H. NAL-DURA), bamboo (*Bambusa*), giant bamboo (*Dendrocalamus*), *Melocanna* (B. MULI-BANS)—ripe berries edible, guinea grass (*Panicum maximum*)—a fodder plant, *P. repens*—a mud-binding plant, lemon grass (*Cymbopogon citratus*)—yields lemon oil, *C. nardus*—yields citronella oil, *C. martini*—yields geranium-oil, *Vetiveria zizanioides* (= *Andropogon squarrosus;* B. & H. KHUS-KHUS)—fragrant fibrous roots are woven into summer screen (KHUS-KHUS), sabai grass (*Ischaemum angustifolium*)—paper is manufactured from this grass, etc.; **other common plants**: dog grass (*Cynodon dactylon*), love thorn (*Chrysopogon aciculatus*), *Imperata cylindrica* (B. ULU), several species of *Panicum*, e.g. *P. crus-galli* (B. SHYAMA), and a few species of *Paspalum*, e.g. *P. scrobiculatum*—common annual grasses.

Family 9 *Cyperaceae* (3,500 *sp.*)

The general habit of *Cyperaceae* is similar to that of *Gramineae* but the following points should be noted.

Habit—herbs, growing mostly in marshy places, pastures and sand beds. Stem solid, usually triangular. **Leaves** simple, alternate, tristichous; ligule absent; sheath closed. The total **inflorescence** may be a spike or panicle or globose head, but the unit of inflorescence is a spikelet. In each spikelet there may be one or more flowers, but each is borne in the axil of a glume and is minute in size, unisexual or bisexual. **Perianth** usually represented by 3 or 6 bristles or scales, or absent. **Androecium**—stamens usually 3; anthers basifixed, linear. **Gynoecium**—carpels (3) or (2), syncarpous; ovary 1-celled, with 1 ovule, superior; stigmas 3 or 2, long and feathery or papillose. **Fruit** a small nut or nutlet. **Seed** albuminous. Pollination is brought about by wind.

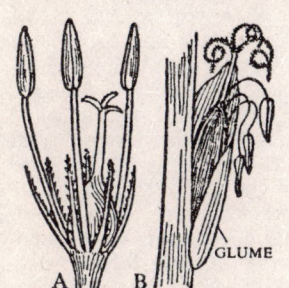

FIG. 77. *A*, flower of *Scirpus*; *B*, the same of *Cyperus*.

Examples. [The larger genera are: *Carex* (over 1,200 sp.), *Cyperus* (700 sp.), *Scirpus* (300 sp.), *Fimbristylis* (over 200 sp.), *Scleria* (200 sp.), *Eleocharis* (150 sp.), *Pycreus* (100 sp.), *Bulbostylis* (100 sp.), *Kyllinga* (60 sp.)]. Sedge (*Cyperus rotundus*; B. & H. MUTHA), *C. tegetum* (B. MADUR-KATI), *C. difformis*—a weed of rice fields, *C. nutans*—a large sedge, *C. cephalotes*—a floating sedge, umbrella grass (*Cyperus alternifolius*) —ornamental, etc., *Scirpus grossus* (B. KESHUR)—a tall rush, club rush

(*S. littoralis*)—a stout rush of the Sundarbans, *S. articulatus* (B. PATPATI)—a small herb in wet places, etc., *Kyllinga*—common weeds with white globose heads, e.g. *K. monocephala*, *Carex indica*—a short herb, *Juncellus inundatus* (B. PATI)—a stout herb in wet places, *Fimbristylis monostachya* (B. MARMARI) and *F. dichotoma*—small tufted herbs, *Bulbostylis barbata*—a small tufted herb, *Scleria elata*—a tall robust herb, *Eleocharis fistulosa*—a rush-like sedge in ponds, *Pycreus nitens*—a weed of wet places, *Cladium riparium*—a tall stout herb in the Sundarbans, etc. The family is of little economic importance.

Distinctions between Gramineae and Cyperaceae

	Gramineae	*Cyperaceae*
Stem	cylindrical; solid or hollow	triangular; solid
Leaf	distichous, with ligule; sheath split open	tristichous, without ligule; sheath closed
Inflorescence	spike or panicle of spikelets	spike or panicle or head of spikelets
Glumes	glumes 3, palea 1; lower 2 empty and 3rd one flowering	a flower in the axil of a glume
Perianth	represented by 2 or 3 lodicules	represented by 3 or 6 bristles or scales or absent
Stamens	usually 3; sometimes 6; anthers versatile	usually 3; rarely 6, anthers basifixed
Carpels	(3) or 1, styles usually 2	(3) or (2), styles 3 or 2
Fruit	caryopsis	nutlet

Family 10 Orchidaceae (about 20,000 *sp.*)

Habit—perennial herbs, predominantly epiphytic, some terrestrial, a few saprophytic. Orchids are widely distributed but they are abundant in the Tropics. In India over 1,700 species have been recorded so far, and about 400 species in N-E India; Sikkim is also rich in orchids. Mycorrhiza (see p. 21) is common. Epiphytic orchids usually with clinging roots, absorbing roots and hanging roots with velamen (see FIG. 1/26). Stems sometimes with pseudobulbs. **Leaves** simple, entire, usually thick, frequently sheathing at the base, and variously mottled in a few species. **Inflorescence** a raceme, panicle or commonly a spike (long or short). **Flowers** bisexual, epigynous, medianly zygomorphic, often very showy, sweet-scented in some species, with an endless variety of colours and forms, for which they are highly valued and largely cultivated; structure sometimes very complex. **Perianth** in two trimerous whorls, epigynous, petaloid and showy, segments of the outer whorl almost equal and so also the two lateral segments of the inner whorl, but the posterior one of this whorl is the largest and most conspicuous, being often folded, variously shaped and spotted, enclosing the column, and is known as the *labellum*. Owing to the twisting of the ovary the labellum, normally posterior, comes round to the anterior side. The labellum is often provided with a *spur*, long or short, which secretes nectar. **Androecium**—stamens in two trimerous whorls but undergo a considerable amount of suppression or modification into staminodes. Majority of orchids bear 1 stamen (anterior one of the outer whorl)—**Monandrae**; while some bear 2 stamens (two lateral ones of the inner whorl)—**Diandrae**. Filament and

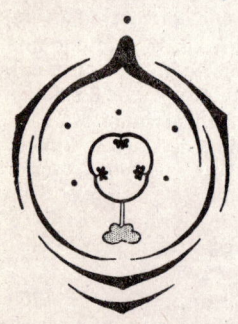

FIG. 78. Floral diagram of an orchid.

SELECTED FAMILIES OF MONOCOTYLEDONS

style united together (gynandrous) and they occur, together with a special organ—the rostellum—on a central structure called the *column* (or gymnostemium). The column is an extension of the floral axis, and the rostellum is a sterile stigma which aids pollination (see below).

The column with the anther(s) and stigmas stands opposite to and facing the labellum. The anther lies on the top of the rostellum, and is attached to the column at its back by a short filament. It is commonly 2-chambered with a pair of pollinia, sometimes 4- or more-chambered with as many pollinia; pollen sometimes granular. The base of the pollinium often extends into a

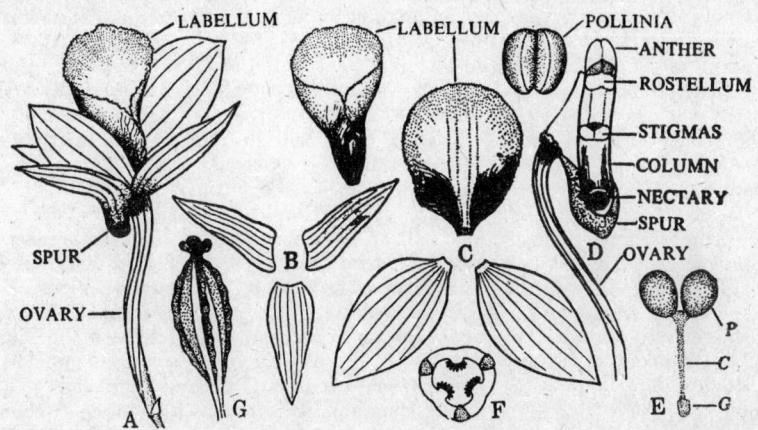

Orchidaceae. FIG. 79. *Dendrobium pierardi. A,* a flower; *B,* outer whorl of perianth; *C,* inner whorl of the same; *D,* column (or gynostemium), ovary and anther; *E,* a pair of pollinia (of *Vanda*)—*P,* pollinium; *C,* caudicle; *G,* gland; *F,* ovary in transection (semi-diagrammatic); *G,* fruit (of *Vanda*)—a capsule dehiscing into 6 valves.

slender stalk called the *caudicle* which may again end in a sticky gland attached to the rostellum. **Gynoecium**—carpels (3), syncarpous; stigmas 3 —the lateral two are fertile and very sticky and the third one sterile (the rostellum); in Diandrae three stigmas are fertile (no rostellum); ovary inferior, commonly cylindrical, 3-valved with 3 ridges (representing midribs of the three carpels), often twisted, unilocular (sometimes falsely trilocular); placentae and numerous extremely minute ovules develop only some time after pollination as a result of stimulus by the process. **Fruit** a loculicidal capsule, mostly cylindrical, dehiscing by valves and ridges; fruits and seeds take several months, often a year, to develop and mature; hygroscopic hairs often develop from the inner wall of the valves, helping dispersion of seeds. Seeds very numerous, extremely minute and powdery, **exalbuminous**, and with a microscopic embryo.

Floral formula— ·|· ⚥ $P_{3+3}[A_{1 \text{ or } 2} \bar{G}(_3)]$

Pollination. Orchids are adapted for cross-pollination by insects in a variety of ways. A common one is as follows. The insect is attracted by the bright colour of the flower, its peculiar form, and sometimes a sweet smell. As it visits the flower it sits on the labellum. It moves into the flower, drills its way to get to the nectar which is concealed in the spur at the base. In its effort to get in, it disorganizes a part of the rostellum from which a sticky substance adheres to the head (or proboscis) of the insect. After collecting the nectar as the insect crawls out of the flower it carries the pollinia with it. The pollinia easily fall off from the anther and stick to its head or become attached to it by the caudicle. When the insect visits a second flower the pollinia on its head point, sometimes by a hygroscopic movement of the caudicle, towards the receptive stigmas; the stigmas being very sticky pull away the pollinia from the body of the insect. Cross-pollination is thus effected. In the same way the insect carries the pollinia from this flower and pollinates a third one. The relative position of the stigmas and the anther, sometimes lying wide apart, often excludes the possibility of self-pollination. Some orchids are also self-sterile. Even then self-pollination in *Orchidaceae* is not uncommon. A few species have cleistogamous flowers. (For details read Darwin's *Fertilization of Orchids*.)

Examples. [The larger genera are: *Pleurothallis* (1,000 sp. in South America), *Dendrobium* (900 sp. in Asia and Australia), *Bulbophyllum* (900 sp. in the Tropics), *Habenaria* (700 sp. in the Tropics), *Epidendrum* (400 sp. in Tropical America), *Eria* (over 350 sp. in Asia and Australia), *Oncidium* (about 350 sp. in South America), *Oberonia* (over 300 sp. in the Tropics), *Microstylis* (=*Malaxis*; 300 sp. cosmopolitan), *Masdevallia* (275 sp. in South America), *Stelis* (270 sp. in Tropical America), *Liparis* (250 sp. cosmopolitan), *Angraecum* (220 sp. in South Africa and Madagascar), *Eulophia* (200 sp. in the Tropics), *Platanthera* (200 sp. widely distributed). There are several other genera with over 200 sp.]. **Epiphytic:** *Vanda* (60 sp.), e.g. *V. roxburghii* (B. & H. RASNA), *V. teres*—branches cylindrical and erect, blue vanda (*V. cerulaea*)—a rare beautiful orchid of the Khasi Hills, *Dendrobium* (many species in India), e.g. *D. pierardi* (FIG. 79)—pendulous, flowers pink, *D. densiflorum*—pendulous, flowers large and golden yellow, *D. moschatum*—erect, flowers pink, *D. formosum* and *D. nobile*—flowers large, white and showy, *D. multiforum*—flowers red, *D. clavatum*—flowers deep red, *D. chrysanthum*—flowers yellow, profuse, *D. nathanielis*—jointed stem, etc., lady's slipper (*Cypripedium*), *Aerides multiflorum* and *A. odoratum* (flowers scented)—showy epiphytes, *Bulbophyllum* and *Coelogyne*—several species, fox-tail orchid (*Rhynchostylis retusa*)—flowers pink, *Cymbidium aloifolium*—tufted orchid, chain orchid (*Pholidota imbricata*), jewel orchid (*Goodyera*), *Pleione, Calanthe, Eria*—several species, *Cattleya* (an American genus)—very ornamental, cultivated, etc. **Terrestrial:** *Arundina graminifolia*, butterfly orchid (*Havenaria*)—several species, *Eulophia, Zeuxine* (in grassy fields), *Phaius, Orchis, Pogonia, Geodorum, Lipparis*, etc. **Saprophytes:** bird's nest orchid (*Neottia nidus-avis*), coral-root orchid (*Corallorhiza*), *Galeola falconeri*—a curious leafless tall orchid of Arunachal, with stout rhizome and upright stout inflorescence axis (3 metres in height), etc. It may be noted that *Vanilla* (90 sp.)—climbing orchids, is the only genus of economic importance; pods of *Vanilla planifolia* yield the essence *vanillin* used in perfumery and in flavouring confectioneries, and in scenting tobacco. It is cultivated in Mexico and Madagascar and to some extent in Ceylon.

Part VIII Evolution & Genetics

Chapter 1 ORGANIC EVOLUTION

It is now an established fact that higher and more complex forms of plants and animals have evolved from earlier and simpler forms. At one time, however, it was believed that different species of plants and animals were created in the forms in which they exist today. There was also the belief that minute organisms like bacteria develop spontaneously in putrefying material. Louis Pasteur, a French scientist and founder of the science of bacteriology, proved definitely in 1864 that spontaneous generation is absolutely impossible. Bacteria are present everywhere, and whenever they get a suitable medium and favourable conditions they grow. The old view, therefore, is no longer tenable. Moreover, there is evidence to show that plants and animals have evolved from pre-existing forms, and have not been created. Fossil records are particularly instructive in this respect (see Part X). It has been possible to trace the gradual changes in the types of plants and animals through the successive periods of the earth. To go back to the earliest stage the question pertinently arises 'What is the origin of life?' or 'How did the protoplasm, the physical basis of life, first come into being?' But protoplasm once formed has become continuous from the earliest form of life to the present form through many millions of years.

According to geologists, and also astronomers, the age of the earth may be either 3,440 or 4,550 millions of years, while according to the *Puranas* the earth is 4,320 million years old. At some early age it was only a molten mass at an excessively high temperature. Gradually the earth cooled, and there was spontaneous generation of life—how and exactly when we are not in a position to say with any amount of certainty. According to fossil records life seems to be already somewhat advanced about 600 million years ago in the form of bacteria, fungi, blue-green algae and some primitive invertebrates. There is still evidence of life in the sedimentary rocks formed about 2,000 millions of years ago. Life must have originated much earlier. As time rolled by, newer forms of plants and animals were evolved through successive stages in the earth's history. It is believed that the primitive earth was surrounded by a gaseous atmosphere made of methane (CH_4), ammonia (NH_3), water vapour (H_2O) and hydrogen (H_2). All the elements necessary for the formation of organic molecules were thus present—C,H,O & N. Under special conditions (both chemical and physical) prevailing then the energy required for their formation was possibly obtained from ultraviolet rays of the sun, radioactivity, electric discharges, etc., and a soup of organic molecules was possibly formed in the ocean. Some or many of these molecules might have combined to form complex molecules (particularly nucleic acids as the basis of life), and some such molecules acquired the characteristics of life once for all. Evidently life originated in water (in the ocean), possibly in the form of some aquatic bacteria which could utilize energy by oxidation of iron compounds, sulphur compounds, etc. (and not directly from sunlight) for synthesis of organic compounds. The next phase in evolution was possibly the appearance of blue-green algae. Primitive unicellular animals might have originated at this stage, and they formed another line of evolution. Later with the evolution of green algae, evidently with the *development of chlorophyll* in organized chloroplasts, which could directly utilize sunlight as a source of energy for manufacture of food the trend of evolution leading to higher plants became established. This was the **first major event** in the early history of evolution of plants and animals, following separate lines of descent. For some millions of years primitive algae and primitive animals formed the dominant feature of the sea. At the next early stage green plants invaded land. This *migration of plants* from water (sea) to land (Psilophytales as the earliest land plants) was the **second major event** in the early history of evolution. The green plants invading land rapidly gave rise to a diversity of forms such as giant lycopods and horsetails, ferns and two groups of gymnosperms (seed-ferns or Cycadofilicales and Cordaitales) in the early geological age (see part X, chapter 2). Colonization of land by green plants soon provided for invasion of land by animals so far as food and shelter for them were concerned.

EVIDENCES OF ORGANIC EVOLUTION

1. Geological Evidence. Actual petrified remains of ancient plants and animals, or impressions left by them in rocks, are called **fossils**. They bear sound evidence regarding the existence of different types of plants and animals in different geological ages and periods of the earth, and help us to trace the facts correlated with their origin (first appearance) and evolution, climax of development, relationship, increasing complexity and specialization, extinction of certain groups, etc. The surface of the earth consists of layers or strata of rocks formed in different ways in different periods. These strata, thus formed successively, have been found to bear fossils of particular types of plants and animals in their increasing complexity. Thus earlier-formed rocks show fossils of simpler types of plants and animals, while later-formed rocks show fossils of more complex and advanced types (see Part X), maintaining in many cases a definite relationship between different groups of plants or of animals which appeared through the successive stages of the earth. Fossil records, therefore, are of considerable importance in elucidating the problem of evolution. These records are, however, incomplete for various reasons and as such they reveal several wide gaps in the evolutionary history of plants and animals. Because of the missing links the origin of several groups has still remained a mystery.

2. **Taxonomic Evidence.** According to resemblances and differences we classify plants and animals into certain well-marked groups, members of each group resembling each other more closely. It is difficult to conceive of these similarities in forms without having recourse to evolution. In addition, it is seen that between two or more species of a particular genus there are intermediate forms linking such species (*intergrading species*). If species were constant the occurrence of such forms could not be accounted for.

3. **Morphological and Anatomical Evidence.** The structural similarities of roots, stems, leaves, flowers and other morphological characters among certain groups of plants, and of bones and other organs among certain groups of animals, and the successive stages in the development of such organs from simpler to more complex forms, evidently show the evolutionary tendencies of plants and animals towards perfection. Morphological similarities in the type of venation, shape of corolla, cohesion or adhesion of stamens and other similar morphological traits among a group of plants are very significant in the problem of evolution. Similarly, the study of types of wood or xylem, development of stele, nature of tracheids and vessels among the higher cryptogams, gymnosperms and angiosperms, and of the development of tissues and nerves among animals lends additional support to the theory of evolution. Sometimes instead of progressive evolution some parts of plants show a reversion to an ancestral type.

4. **Embryological Evidence.** The study of the nature and development of the embryo reveals a great resemblance among certain groups of plants and of animals. The development indicates the evolutionary change that has taken place through successive stages. In all cases one fact at least is common, i.e., the embryo develops from the egg-cell or ovum. Sometimes some organs of plants or animals show a striking resemblance to certain forms

from which they have possibly been derived. Thus when a fern spore germinates it resembles a filamentous alga; it then assumes a thalloid form resembling a liverwort; and finally it grows into a fern plant. The frog passes through a tadpole stage resembling a fish which is supposed to be its ancestor. Seedlings sometimes show their resemblance to plants which may be their ancestors. Thus in Australian *Acacia* the seedling shows bipinnate compound leaves like other species of Acacia, although adult Australian *Acacia* has a winged petiole or rachis (phyllode) with or without the compound leaf (see p. 50).

5. Evidence from Geographical Distribution. It has been seen that many allied species of plants in their wild state remain confined to a particular area. The explanation is that they sprang up from a common ancestor in that region and could not migrate owing to some barriers such as high mountains, seas and deserts. Thus we find that double coconut-palm (*Lodoicea*) originated in the Seychelles, traveller's tree (*Ravenala*) in Madagascar, *Eucalyptus* in Australia, cacti in the dry regions of tropical America, cactus-like spurges (*Euphorbia*) in the deserts of Africa, etc., often with allied species near together, showing thereby that all the allied species have evolved from the same ancestral species.

6. Vestigial Structures. It is commonly observed that there are certain degenerated parts of plants and animals, which do not serve any useful purpose. Such parts are known as vestigial structures. It is assumed that in the early ancestry of the organisms concerned, such parts were functional and in course of evolution these parts have degenerated after ceasing to function. Thus in *Asparagus*, with the development of the green branches and the cladodes, the leaves have been reduced to scales. In parasites like broomrape (*Orobanche*) and dodder (*Cuscuta*) the leaves have been reduced to scales in the absence of their normal function of photosynthesis. There are many instances of reduction of floral organs, particularly stamens, in many families. The presence of vestigial structures is explained on the basis of the early history of the race and the course of evolution.

MECHANISM OF ORGANIC EVOLUTION

Variation. Variation is the rule in nature. No two forms, belonging even to the same species, are exactly alike. The differences between the individuals of a species are spoken of as **variations**. Variations are the basis on which evolution works. Variations may be of four types. (*a*) *Variation due to change in environment.* Variation in certain organs of plants or animals in certain directions due to this cause is not, however, believed to be inherited by the offspring. (*b*) *Slow but continuous variation* from generation to generation, however, according to Darwin, is the basis of organic evolution. (c) *Discontinuous variation* or *mutation*, on the other hand, means sudden and sharp variation of one or more individuals of the species in respect of one or more characters. The individuals show no gradations, as in the previous case, but at once assume new forms. The sharp variation of this type is directly inherited by the offspring, and is, according to De Vries, due to change or mutation in genes (see p. 804). As mutation occurs suddenly and spontaneously there is no knowing when a new form will appear by this process. There are many cases of mutation on record. Darwin also observed several cases of wide divergences which he called 'sports'. He thought that sports played only a very minor role in evolution. (d) *Variation due to hybridization.* In this process a mingling of two sets of contrasting characters (paternal and maternal) takes place and as a result some of the progeny at least show wider variation.

Adaptation. Adjustment of plants and animals to their environment by means of special structures or of functions is spoken of as **adaptation**. In ecology numerous instances of adaptation are met with. Adaptation may have an important bearing on evolution. Plants have an inherent capacity to adapt themselves to their environment. Many of them are plastic in nature, and consequently are in a position to adapt themselves to changed conditions according to their needs. This is even more true of animals. It is Lamarck's view that adapted structures are fixed and inherited by the offspring, so far at least as the same environment continues (inheritance of acquired characters). According to this view individuals of the species that invade two or more situations will give rise to a corresponding number of new forms.

Heredity. Heredity means transmission of characteristics and qualities of parent forms to their offspring (the term heredity means 'like begets like'). This is evident from the fact that a particular species on reproduction gives rise to the same species and to no other. Although no two forms are exactly alike, offspring still bear the closest resemblance to their parental forms, and they also resemble one another most closely with, of course, individual variations. Heredity tends to keep the individuals of a species within specific limits; while variation tends to separate them from one another by differences between them, however minute these may be. Variation, no doubt, is responsible for evolution; while heredity is a check on uncontrolled variation and evolution which would have otherwise given rise to a multitude of peculiar forms without any clear distinction between one species and another. When, however, heredity is due to cross-pollination, the chance of a certain amount of variation is inevitable.

Inheritance of Characters. In the process of reproduction we find that two reproductive nuclei (i.e. gametes—male gamete of the pollen-tube and egg-cell of the embryo-sac) of opposite sexes, each with n chromosomes, fuse to give rise to the oospore, the embryo and ultimately the mature plant, each with $2n$ chromosomes. Thus inheritance of characters takes place through these nuclei. In 1884 Strasburger and Hertwig established the fact that it is through the chromosomes that characters are transmitted from generation to generation. It is obvious, however, that any particular character of the parent (e.g. colour of flower) cannot be found in the chromosomes; but it may be safely assumed that something representing that particular character must be present in them. That 'something', obscure though it is, is called the **factor** or **determiner** or **gene** for that particular character, and the genes located in the chromosomes of the gametes or reproductive nuclei are responsible for all the characteristics of the parent plants and their transmission to the offspring. The theory of genes in the chromosomes was introduced by Morgan in 1926. Genes are extremely minute bodies, possibly made of a single protein molecule or at most very small groups of protein molecules. [For details see pp. 802-03].

THEORIES OF ORGANIC EVOLUTION

Pre-Darwinian Ideas of Evolution. The idea of evolution dates back to the earliest period of human civilization. The oldest theories discussed much about the origin of life, but they were mere speculations rather than facts. These may be summarized under three heads: (1) theory of eternity of the present conditions, (2) theory of special creation, and (3) theory of catastrophism. The believers of the first theory argued that there was neither beginning nor end to the universe. The life forms which existed many millions of years ago have remained unchanged till the present day and would continue to be the same throughout eternity. The theory of special creation was preached for many centuries by the Christian Church. The basis of the belief was the account of the creation of the world by God, and everything in this world including animals, plants and man. The theory of catastrophism was introduced by Cuvier, a palaeontologist, who carried out research on fossil fauna for a long time in Paris. He believed that at one time world-wide catastrophes brought about the death of the old fauna; their extinction caused the creation of a new fauna which, according to him, took a long

ORGANIC EVOLUTION

time covering millions of years due to the changed conditions of the environment.

The idea of organic evolution, although believed to be a modern one, can be traced many centuries back. The idea of evolution emanated from the researches of many Greek philosophers of whom Empedocles was the first. He believed that organisms did not improve through successive generations but nature tried to produce perfect organisms many times, and during this period unfit forms were eliminated. Then came the great philosopher Aristotle. He believed that there was an inherent tendency among the organisms in nature to attain greater and greater perfection according to the changes in the environment, and that was the reason why there was a perfect gradation from the lowest to the highest evolved in nature. After this period there was a halt in the speculation of organic evolution until the coming of such evolutionists as Linnaeus, Buffon, Erasmus Darwin, Lamarck, Charles Darwin and others.

Lamarck's Theory : Inheritance of Acquired Characters. The first modern theory of evolution was put forward in 1809 by the French biologist Lamarck (1744-1829). His theory resolves itself into three factors—(a) influence of the environment, (b) use and disuse of parts; and (c) inheritance of acquired characters. Lamarck held the view that environment plays the principal part in the evolution of living organisms. He noted many instances where individuals of the same species grown under different environmental conditions showed marked differences. Plants grown in the shade develop larger leaves than those grown in the open. In dry soil the root system becomes more extensive than in wet soil. In darkness leaves do not develop chlorophyll and the stem becomes weak and drawn out (etiolated). Many plants leading an amphibious life show heterophylly. From such observations Lamarck concluded that plants react to external conditions, and that as a result of cumulative effects produced by the changed conditions through successive generations new species make their appearance. In the case of plants, according to Lamarck, changes in characters (or adaptations) are brought about by the direct action of the environment, and in the case of animals they are brought about by the use and disuse of parts. The use or exercise of certain parts results in the development of those parts; while disuse or want of exercise results in the degeneration of the parts. He further believed that new characters, however minute, acquired in each generation under changing conditions of the environment, are preserved and transmitted to the offspring (inheritance of acquired characters). The classic example cited in this connexion is that of the giraffe. Lamarck's view was that horse-like ancestors of these animals living in the arid region in the interior of Africa had to feed on the leaves of trees. They had necessarily to stretch their limbs to reach up to the leaves. This use or exercise resulted in the lengthening of the neck and the front legs, and thus a new type of animal made its appearance from a horse-like ancestor. His theory is open to certain objections. One objection is that adaptations due to the influence of the environmont are very slight and superficial. Another objection is that the inheritance of acquired characters has not been proved yet. In fact, if seeds collected from plants growing elsewhere for many years under a new environment and acquiring new characters be brought back and grown in their original habitat, the plants are seen to revert to their original forms.

Darwin's Theory : Natural Selection. The next theory of evolution was put forward in 1859 by the English biologist Charles Darwin (1809-82) and published in his *Origin of Species by Means of Natural Selection*. His theory based on a mass of accurate observations and prolonged experiments led the whole scientific world to believe in the doctrine of evolution. His theory, called the *theory of natural selection*, is based on three important factors: (a) over-production of offspring and a consequent struggle for existence, (b) variations and their inheritance, and (c) elimination of unfavourable variations (survival of the fittest).

Struggle for Existence. If all the seeds of any particular plant were to germinate and all seedlings to grow up into full-sized plants, a very wide area would soon be covered by them in course of a few years. If other plants (and also animals) were to increase at this rate, a keen competition or, in other words, a struggle for existence, would evidently be set up among them because availability of food, water and space would fall far short of the demand. This struggle would soon result in the destruction of large numbers of individuals.

Variations and their Inheritance. It is known that no two individuals, even coming out of the same parent stalk, are exactly alike. There are always some variations, however minute they may be, from one individual to another. Some variations are suited to the conditions of the environment, while others are not. According to Darwin these minute variations are preserved and transmitted to the offspring, although no cause for these variations was assigned by him.

Survival of the Fittest. In the struggle for existence the individuals showing variations in the right directions survive, and these variations are transmitted to the offspring; others with unfavourable variations perish. This is what is called by him 'survival of the fittest'. The survivors gradually and steadily change from one generation to another, and ultimately give rise to new forms. These new forms are better adapted to the surrounding conditions.

Darwin's observations on the variations of domestic animals and cultivated plants served him as a clue to the elucidation of his theory of natural selection. Sometimes such extensive changes are found in course of several generations that it becomes difficult to believe that the first form has given rise to the last. Further, for the purpose of having a desired type, breeders and florists take note of certain variations among individuals, select them for future generations, rejecting and destroying the rest. They grow the selected types, generation after generation, until the desired result is obtained. New types are seen to appear by this process, called *artificial selection*. Many cultivated flowers and vegetables often show a number of varieties, and in course of time these variations become well marked.

Natural Selection. Now Darwin's explanation of natural selection is this: animals and plants are multiplying at an enormous rate. As we know no two individuals are exactly alike, the new forms naturally show certain variations. Some variations are favourable or advantageous so far as their adaptation to the conditions of the environment is concerned, and others are not so. Owing to an excessive number crowding together a keen struggle for existence ensues. And in this struggle those that have favourable variations and are, therefore, better fitted naturally survive, and the rest perish. Through this survival of the fittest the species change steadily owing to preservation and transmission of minute variations, and gradually give rise to newer forms. Darwin called this process 'natural selection' from analogy to artificial selection. It is the environment that selects and preserves the better types and destroys the unsuitable forms.

Pangenesis. This is Darwin's theory by which he assumes that every cell of the plant body or the animal body produces imaginary particles or units, called *pangenes*, which carry in their body not only the normal parental characters but also those acquired during the life-time of an individual plant or animal. Pangenes are formed in all parts of the body, and finally they collect together to form the germ cells. Through these cells all the characters, normal and acquired, are ultimately transmitted from one generation to another. By his pangenesis theory Darwin tried to explain how the characters are carried forward from the parents to the offspring, assuming that somatic cells at a certain stage produce germ cells which in turn produce somatic cells in the next generation. The above doctrine of Darwin has, however, been discarded by the modern scientists since, as now definitely known, it is only the germ cells that are the true bearers of hereditary characters, remaining practically unchanged and unaffected by the environment and thus passing down intact from generation to generation.

Although Darwin receives the fullest credit for bringing about the final acceptance of the doctrine of evolution, his theory is open to certain doubts. It is true that natural selection is operative in the preservation of certain forms and destruction of others. Yet some doubts have been expressed regarding the process being the cause of evolution of new species. Some of the reasons for these doubts are as follows. (a) Are slight variations of any decided advantage in the struggle for existence? It is only the perfected organs that are helpful to organisms, and not the organs during the process of perfection. (b) It is doubtful if slight variations can help the individuals to go beyond the boundary of the species; this has never been found possible by artificial selection in breeding experiments. (c) There are many organs which are not of any apparent use to the organisms. (d) If only the fittest survive, how is it that many unfit ones still exist? (e) If nature selects suitable forms and features, why were not the rest swept out of existence?

Weismann's Theory : Continuity of Germplasm. An ingenious theory explaining the cause of variation and evolution was put forward in 1895 by the German scientist Weismann (1834-1914), a disciple of Darwin. He divided the protoplasm of the animal or plant body into *somatoplasm* which gives rise only to somatic or body cells and *germplasm* which produces the reproductive cells, and they flow as two separate streams through the body of the plant or the animal. Somatoplasm is responsible for differentiation of tissues, development and growth of the individual plant or animal body, and is exhausted and lost at the end of the life-cycle, i.e. it is discontinuous, whereas the germplasm is ever-young and immortal, and is continuous from one generation to another, and is actually the bearer of hereditary characters. Somatoplasm may be influenced by the environment and new characters acquired during the life-time of an individual, but germplasm is not so affected, and, therefore, as is almost universally believed, inheritance of acquired characters is not a possibility. Weismann did not believe in Darwin's pangensis (see p. 791). During reproduction the fertilized egg gets the paternal germplasm from the sperm and the egg-cell respectively. In the nuclei of both somatic and germ cells there are certain factors which determine the character of the cell. It is believed that each somatic cell has a single factor, whereas a germ cell contains all the factors that are found in the somatic cells of the adult plant or animal. The inheritance of characters by the offspring depends upon these factors of the germ cells only. There is always a struggle for existence among these factors, and this results in a *germinal selection*. The stronger factors survive and are readily transmitted from one generation to another. Hence any mutation in the germplasm or any variation resulting from the struggle among the factors in the germ cells can only be handed down from generation to generation. Weismann's theory supports Darwin's theory of natural selection, but it has been criticized by many scientists as purely speculative Also, it is not a fact, as assumed by Weismann, that germplasm is permanently curtained off from the somatoplasm. With advance in knowledge it has been revealed that chromosomes come in direct contact with somatoplasm during nuclear divisions.

De Vries' Theory : Mutation. Another theory of evolution was advanced in 1901 by the Dutch botanist Hugo De Vries (1848-1935). He held that small variations, which Darwin regarded as the most important from the standpoint of evolution, are only fluctuations around the specific type. These variations are not inheritable. De Vries held that large variations appearing suddenly and spontaneously in the offspring in one generation are the cause of evolution. These variations De Vries called 'mutations'. That large discontinuous variations are the causes of organic evolution was first advocated by Bateson in 1894. De Vries strongly supported this view on the basis of his extensive observations. Thus, he observed an evening primrose (*Oenothera lamarckiana*), introduced from America, growing in a field in Holland. Among numerous plants he found two types quite distinct from the rest. These new types were not described before, and having bred true he regarded them as distinct species—*O. brevistylis* and *O. laevifolia*, as he named them. *Oenothera lamarckiana* and the new species were removed to his garden at Amsterdam, and cultivated through many generations. It was found that among thousands of seedlings raised a few (7 new species) appeared that were different from the rest. These when raised, generation after generation, always came true to types. These new forms are known as *mutants*. He propounded a mutation theory as his explanation of evolution. While De Vries agreed with Darwin's view, regarding natural selection weeding out unsuitable forms, he held the view that new species are not formed, as Darwin said, by the slow process of continuous variations. Since then several instances of plant mutations (as well as animal mutations) have been found in nature. Mutation is now known to be due to changes in genes—loss, degeneration, addition, recombination, etc.,—occurring in the gametes, zygote, or somatic cells, ultimately affecting the nature of the mature plant. This is 'gene mutation' (see p. 804) which may also be artificially brought about by treatment with X-rays. The mutation theory of De Vries has been widely accepted.

Chapter 2   GENETICS

A considerable amount of experimental work on hybridization of plants was carried out long before Mendel evincing interesting and valuable results. But it was Mendel who could for the first time elucidate and formulate the laws involved in the inheritance of parental characters by the offspring. In this connexion the following workers of the pre-Mendelian period deserve special mention. The first authentic work on artificial hybridization of plants was done by a German botanist, **Joseph Kolreuter** by name, in 1760. He hybridized two species of tobacco plants (*Nicotiana paniculata* and *N. rustica*). The progeny were intermediate between the two parents in respect of many characters. This was conclusive evidence to show for the first time that the pollen (male) parent also influences the characters of the progeny. **John Goss** (1820) in England hybridized two types of pea plants —one with bluish seeds and the other with yellowish-white seeds. In the first generation the seeds were all yellowish-white but in the second generation both bluish and yellowish-white seeds appeared. In 1854 **Naudin** was awarded a prize by the Paris Academy for his valuable work on plant hybridization. He showed that the parental characters did not actually blend in the offspring of the first generation; he further showed that the characters reappeared separately in parental forms in the second generation of the cross. He also made reciprocal crosses and proved the identity of the first generation. He almost hit upon the laws of heredity but failed to enunciate them because he did not count the number of progeny.

Genetics is the modern experimental study of the laws of inheritance (variation and heredity). The name 'genetics' was proposed by Bateson in 1906. Cytology, dealing with the structure, number, behaviour, etc., of chromosomes, is of immense value in understanding many intricate facts connected with genetics since chromosomes are the bearers of heritable characters. It is also to be remembered that in addition to their characteristic form and individuality their number is always constant for each species. First scientific studies in genetics were carried out by Gregor Mendel (1822-84), an Austrian monk. He entered a monastery in Brunn (Austria), where he performed his scientific investigations on hybridization of plants, particularly garden pea. The results of his eight years' breeding experiments were read before the Natural History Society of Brunn in 1865, and the following year they were published in the transactions of that Society, a publication with limited circulation. Consequently his work remained unnoticed until 1900, when three distinguished botanists, Hugo De Vries in Holland, Tschermak in Austria and Correns in Germany, working independently in the same line, discovered its significance and importance. Since then Mendel's work has formed the basis of studies in genetics, and it has been called Mendelism as a mark of honour to Mendel. Mendel died in 1884 before he could see his work accepted and

GENETICS

appreciated. The reason is not far to seek. The internal mechanism of the cell leading to heredity was still unknown; chromosomes and genes directly concerned in the transmission of hereditary characters remained undiscovered. Moreover, Mendel's work was overlooked in the excitement caused by the publication of the controversial *The Origin of Species by Natural Selection* by Charles Darwin in 1859.

Plant Breeding. The subject of plant breeding although developed in recent times on modern scientific lines after Mendel's discoveries was known in early times to the Egyptians and Assyrians. During the eighteenth century several crosses were made by many workers and interesting results obtained. New varieties were produced by them by such crosses, but the actual mechanism of fertilization remained undiscovered. About 1830 the development of the pollen-tube and its approach to the ovule were observed. In 1846 Robert Brown, an English botanist, did considerable work on this problem. But it was in 1884 that Strasburger clarified the whole process of fertilization and the transmission of hereditary characters through the reproductive nuclei, i.e. male gamete and egg-cell, which are directly concerned in fertilization. In the nineteenth century Gartner in Germany made extensive crosses—thousands of them—involving nearly 700 species and obtained about 250 hybrids. His work was published in 1849.

Plant breeding consists in producing offspring by artificially pollenizing the stigma of another flower according to certain principles. Two varieties or species or even genera differing from each other in one or more characters may thus be crossed and the results studied. The offspring or hybrids, as they are called, will usually show in the first generation some of the characters of each parent or will be intermediate between the two parent forms. In the subsequent generation the dormant characters are seen to appear in the offspring. The offspring resulting from the crossing of red-flowered and white-flowered plants, of tall and dwarf plants, and of other plants with contrasting characters are said to be **hybrids**. It is often seen that the hybrids are more vigorous than their parents. This phenomenon is known as **hybrid vigour** or **heterosis**—the term suggested by Shull in 1914. Since the characters of parent forms can be combined in the offspring by artificial breeding or **hybridization,** as it is called, it is possible to obtain new varieties with desired characters by selecting parent plants with suitable characters. Thus new varieties of economic and ornamental plants can be produced by hybridization. Herein lies the importance of plant breeding (see pp. 807-09).

Mendel's Experiments. Mendel selected for his work the common garden pea. In pea he found a number of contrasting characters—flowers purple, red or white; plants tall or dwarf; and seeds yellow or green, smooth or wrinkled. He concentrated his attention on only one pair of characters at a time, and traced them carefully through many successive generations. In one series of experiments he selected tallness and dwarfness of plants. The results he achieved in these experiments were the same in all cases. It did not matter whether he took the dwarf plant as the male one and the tall plant as the female one or vice-versa (reciprocal crosses).

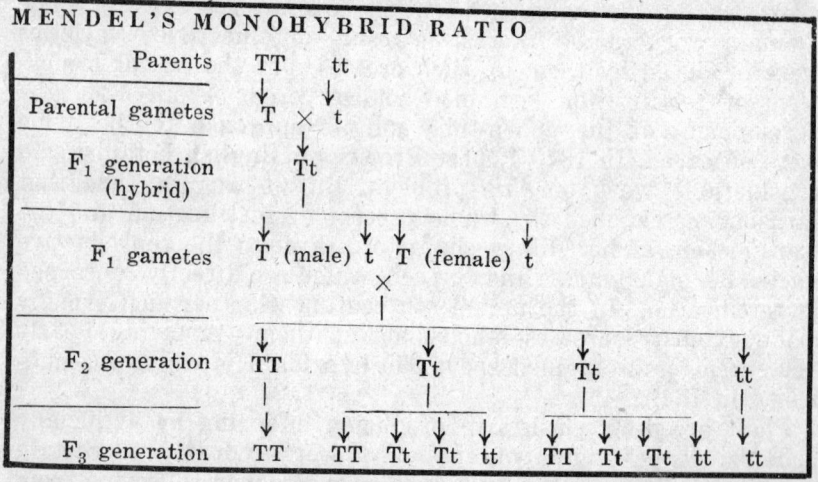

MENDEL'S MONOHYBRID RATIO

Parents	TT tt
Parental gametes	T × t
F_1 generation (hybrid)	Tt
F_1 gametes	T (male) t T (female) t
F_2 generation	TT Tt Tt tt
F_3 generation	TT TT Tt Tt tt TT Tt Tt tt tt

Monohybrid Cross. Mendel selected a pea plant, 2 metres in height, and another plant, 0.5 metre in height. He brought about artificial crossings between the two. In due course seeds collected from the crossed plants were sown next year. The hybrid progeny that came up were all **tall** (with dwarf character remaining latent in them and, therefore, impure tall). This generation was called the first filial generation or F_1 generation. Of the two contrasting characters the one that expressed itself in the F_1 generation was called **dominant** by Mendel, while the other character that remained suppressed (but not absent in the hybrids) was called **recessive** by him. All the F_1 plants were inbred, and so also those of the successive generations. Seeds were collected from them and sown next year. It was seen that they gave rise to a mixed generation of talls—787 plants and dwarfs—277 plants (but no intermediate). This gave an approximate ratio of 3:1, i.e. three-fourths talls and one-fourth dwarfs. This generation was called the second filial generation or F_2 generation. All dwarfs of the F_2 generation bred true producing dwarfs only in the third and subsequent generations. Seeds were collected separately from the F_2 tall plants and sown separately. They gave rise to the next generation or F_3 generation. It was seen that one-third of the talls of the F_2 generation bred true to type producing talls only, while the other two-thirds of the talls again split up in the same ratio of 3:1. The F_2 ratio may then be expressed as 1:2:1, i.e. one-fourth pure talls, half mixed talls, and one-fourth pure dwarfs. This scheme of inheritance is expressed symbolically in the table on p. 796. It is now the custom to use a capital letter to denote the factor for the dominant character—T in this case, and the corresponding small letter for the recessive character—t in this case.

Mendel's Laws of Inheritance. From the results of his experiments on crossings Mendel formulated certain laws to explain the inheritance of characters, as follows:

(1) **Law of Unit Characters.** This means that all characters of the plant are units by themselves, being independent of one another so far as their inheritance is concerned. There are certain factors or determiners (now called *genes*) of unit characters, which control the expression of these characters during the development of the plants. The factors occur in pairs; this is evident from the fact that F_1 generation splits in the F_2 generation into tall and dwarf individuals.

(2) **Law of Dominance.** As already mentioned, of the two contrasting characters the one that expresses itself in the F_1 generation is called *dominant*, while the other character that remains suppressed in the F_1 generation is called *recessive*. The latter, however, is always present in the F_1 individuals. Thus in the previous experiment tallness is the dominant character and suppressed dwarfness the recessive character. Mendel reasoned that there must be two factors separately responsible for each pair of contrasting characters, i.e. tallness and dwarfness, as in the previous experiment, and that these factors occur in pairs (now known to be arranged in a linear fashion in the chromosome). In the F_1 generation one factor masks the expression of the other factor, and, therefore, one character becomes dominant and the other suppressed or recessive. Later work has shown, however, that dominance does not hold good in all cases. The contrasting pairs of characters are called **allelomorphs** (or simply **alleles**). Thus tallness and dwarfness are allelomorphs.

(3) **Law of Segregation.** It is evident that the F_1 zygote contains factors for both the alternative characters, namely, tallness and dwarfness, although tallness has expressed itself in the F_1 generation. These factors remain associated in pairs in the somatic cells of the F_1 individuals throughout their whole life

Later in the life-history when spores—pollen grains and megaspores (and subsequently gametes)—are formed as a result of reduction division, the factors located in homologous chromosomes become separated out, and each of the four spores (and gametes) will have only one factor (tallness or dwarfness) of the pair but not both, i.e. a gamete becomes *pure* for a particular character. This law is also otherwise called the law of *purity of gametes*.

Phenotype and Genotype. When two individuals are similar in their external appearance (evident from morphological study) but differ in their genetic make-up, they are referred to as phenotypes, and when their genetic composition is the same (evident from cytological study), they are referred to as genotypes. Thus in Mendel's previous experiment TT and Tt individuals of the F_2 generation are alike externally although they differ from each other with respect to their genes; these individuals, therefore, belong to the same phenotype. But because of their differences in genetic composition they are said to belong to different genotypes—TT to one and Tt to another. It may further be noted that when the individuals have similar gene pairs they are said to be *homozygous,* and when dissimilar they are *heterozygous*. Thus TT and tt individuals are homozygous and Tt individuals are heterozygous.

Mendel also experimented on other pairs of alternative characters, and he found that in every case the characters followed the same scheme of inheritance. Thus in garden pea he discovered that a coloured flower was dominant over a white flower; yellow seed over green seed; and smooth seed over wrinkled seed.

Back Cross. Crossing back F_1 plants (hybrids) to either of the two parental types, normally a recessive type, is called a **back cross**. It is otherwise called **test cross** because by this method it has been found possible to test the purity of a particular race of plants and also the gametic proportion of F_1 hybrids. Back crosses are now extensively employed in experimental plant breeding. It is evident that only two alternative back crosses are possible either with the dominant parental type or with the recessive type. In the former case all the progeny would show only the dominant character; the latter, however, is much more important. In such a back cross (i.e. crossing with a recessive type) the dominant character and the recessive character are seen to appear in the ratio of 1:1, i.e. one-half of the total population dominant and one-half recessive. To take a concrete case, if a tall hybrid (Tt) of F_1, as in Mendel's experiment with tall and dwarf plants, be crossed back to a parental dwarf (recessive) type (tt), one-half of the progeny will be impure tall

(Tt) and one-half dwarf (tt), i.e. the ratio of tall (dominant) to dwarf (recessive) is 1:1. The explanation is: gametes of F_1 hybrid are T and t in equal numbers and those of the dwarf parent are t and t in equal numbers. Therefore, in such a back cross only four combinations are possible, viz. Tt, Tt, tt and tt, i.e. the ratio of tall (impure Tt) to dwarf (pure tt) is 1:1. This ratio has been verified by thousands of back crosses.

Xenia. The term xenia, first introduced by Focke in 1881, is used to indicate the direct influence of foreign pollen on the endosperm and other seed characters of the crossed plant in the same generation, i.e. on the same mother plant in the same year. Such influence has also been noted on certain maternal tissues outside the embryo and the endosperm, as on the size, colour, flavour of fruits like date-palm, apple, orange, etc.; such a phenomenon has been called *metaxenia* by Swingle in 1926. Swingle further suggested in 1928 that this effect might be due to the secretion of hormones by the embryo and the endosperm. Thus certain gametic differences have been noticed in some seeds developing directly on the mother plant. In such cases the dominant factors of the pollen are believed to have been directly introduced through crossing—natural or artificial, and corresponding characters have expressed themselves in the seeds directly borne by the mother plant. For instance, in maize the cob with wrinkled grains is seen to produce some smooth grains on it here and there; these are hybrids which later breed in the Mendelian ratio of 3:1. Similarly, coloured grains are occasionally found on white or golden-grained cob. There are several instances of xenia on record.

Dihybrid Cross. For the dihybrid cross two pairs of contrasting characters are taken into consideration at a time. Mendel selected a tall plant with red flowers—TRTR and a dwarf one with white flowers—trtr, their respective gametes being TR and tr. Four unit characters are, therefore, concerned in the dihybrid ratio. Factors for tallness or dwarfness and for red flowers or white are independently inherited. Artificial crossing was brought about between these two plants. In the F_1 generation all individuals were tall with red flowers—TRtr because tallness is dominant over dwarfness, and coloured flowers dominant over white, subsequently their gametes bearing factors TR (tall-red), Tr (tall white), tR (dwarf-red) and tr (dwarf-white). When the seeds from the F_1 generation were grown, a segregation of characters showing all possible combinations took place in the following proportions: 9 red talls, 3 white talls, 3 red dwarfs, and 1 white dwarf. This 9:3:3:1 is the dihybrid ratio.

Nos. 1, 2, 3, 4, 5, 7, 9, 10, 13 are tall-red = 9
Nos. 6, 8, 14.............are tall-white = 3
Nos. 11, 12, 15...........are dwarf-red = 3
No. 16..................is dwarf-white = 1

It will further be noticed that Nos. 1, 6, 11 and 16 are *homozygous* (i.e. they have two similar gametes), breeding true;

MENDEL'S DIHYBRID RATIO

Parents	TRTR		trtr
Parental gametes	TR	×	tr
F_1 generation (hybrid)		TRtr	
F_1 gametes (See below for the next generation)	TR Tr tR tr		

		Male gametes of F_1			
		TR	Tr	tR	tr
Female gametes of F_1	TR	TRTR (tall-red) [1]	TRTr (tall-red) [2]	TRtR (tall-red) [3]	TRtr (tall-red) [4]
	Tr	TrTR (tall-red) [5]	TrTr (tall-white) [6]	TrtR (tall-red) [7]	Trtr (tall-white) [8]
	tR	tRTR (tall-red) [9]	tRTr (tall-red) [10]	tRtR (dwarf-red) [11]	tRtr (dwarf-red) [12]
	tr	trTR (tall-red) [13]	trTr (tall-white) [14]	trtR (dwarf-red) [15]	trtr (dwarf-white) [16]

(F_2 generation)

while the rest are *heterozygous* (i.e. they have two dissimilar gametes), segregating in the next generation.

    No. 1 (TRTR) will breed true for tall-red
    No. 6 (TrTr)..................tall-white
    No. 11 (tRtR)..................dwarf-red
    No. 16 (trtr)..................dwarf-white

Polyhybrid Cross. In this way Mendel extended the number of characters; for example, when three pairs of contrasting characters were taken—tall and dwarf, red flower and white, and smooth seed and wrinkled—it was found that in the F_1 generation all individuals were tall with red flowers and smooth seeds with the factors TRStrs—three dominant and three recessive. They evidently produced eight kinds of gametes with triple factors each, e.g. TRS, TRs, TrS, Trs, tRS, tRs, trS, trs. When inbred they formed a possible range of 64 (8×8) types of plants in the F_2 generation. When analysed they were found to have split up in the following proportions—27 (tall-red-smooth) : 9 (tall-red-wrinkled) : 9 (tall-white-smooth) : 3 (tall-white-wrinkled) :9 (dwarf-red-smooth) :3 (dwarf-red-wrinkled) :3 (dwarf-white-smooth) : 1 (dwarf-white-wrinkled).

Linkage. Chromosomes are regarded as the bearers of hereditary characters. Each of the chromosomes is made up of several *genes,* each being generally responsible for a single character. These genes are arranged in a linear fashion in the chromosome. At a certain stage in meiosis we find that homologous chromosomes come in close association in two's and interchange their parts, and while doing so some of the genes of one chromosome go over to the other chromosome. But some genes of a particular chromosome, whether paternal or maternal, tend to remain together from generation to generation, i.e. these genes are linked together in inheritance even after crossing over (see p. 804). Such a phenomenon where genes tend to remain together in inheritance is known as **linkage,** as first noted by Morgan in 1910. For example, in sweet peas, as first noted by Bateson and Punnett in 1906, there are two flower-colours—*purple* and *red,* and two types of pollen—*long* and *round.* Purple colour and long pollen behave as dominants to red and round respectively, each character being represented by a single gene in inheritance. In a cross between purple long and red round the F_1 individuals are all purple long. In the F_2 generation it is found that the parental combinations, i.e. purple long and red round, are much more numerous than the expected Mendelian ratio of 9:3:3:1. The unexpected number is explained as due to the fact that the genes for purple long as well as red round remain together, i.e. linked in inheritance, and hence more of these parental combinations are produced. In tomatoes it has been found that there are two factor pairs Rr and Yy governing the fruit-colour. R factor produces red flesh and is dominant over factor r which produces yellow flesh. The dominant factor Y produces yellow colour and the recessive factor y represents colourless fruit. It has been found that the factor pair Yy is linked with the size of the fruit. There is a close association of y genes with large fruit. In maize or Indian corn ten linkage groups have been noted, e.g. coloured and full endosperm, colourless and shrunken endosperm, etc. Generally the number of linkage groups in any plant or animal is the same as the number of chromosome pairs. Linkage is not, however, always very complete, i.e. a few linked genes break apart and recombinations take place.

Genes[1]. Genes, invisible as they are, cannot be properly defined. But from their action they may be regarded as *functional* units of the chromosome. Genes are highly stable (except when they mutate) and continue as such from generation to generation.

---

[1] References: *Elements of Cytology* by N. S. Cohn, the *Scientific American* —54:1957, 60 & 104:1958, 68:1959, 92:1961, 119 & 123:1962 and 153:1963.

There are several hundreds or thousands of genes in each chromosome, and they are strung together in a linear order. A gene or a group of genes responsible for a particular character is located in a definite position of the chromosome. Genes divide during mitosis, and the two cells (subsequently other cells) thus formed receive an identical set of genes to function normally. Gene is a hereditary material in both plants and animals, as known for quite some time.

Chemical Basis of Gene. Within the last three decades or so intensive work on the chemistry of chromosomes almost throughout the world with the help of electron micrographs, X-ray photographs and radioactive isotopes (particularly S and P) has made many new interesting revelations about them. Each chromosome is now known to have a backbone of DNA (deoxyribonucleic acid) molecule (its short segments being regarded as genes) and some specific proteins. DNA thus consists of a large number of genes of different kinds, responsible for transmission of hereditary characters. Much is now known about the function and structure of DNA through the brilliant experimental

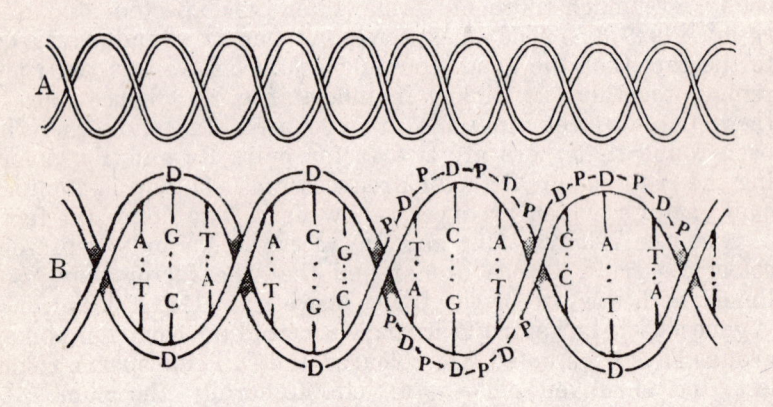

FIG. 1. *A*, Watson-Crick model of DNA molecule—long, double-stranded, wound about each other in a helical fashion; *B*, a portion of the same magnified, showing the distribution of deoxyribose sugar (*D*) and phosphate (*P*) in each strand, and cross-links made of thymine-adenine (*T-A*) and guanine-cytosine (*G-C*) in infinite sequences; dotted lines indicate linking by hydrogen bonds.

works of several investigators (see p. 152). Their works have revealed that the DNA of the chromosome (and not its proteins) acts in two principal ways: it is the chemical basis of gene and is the *essential genetic* material, and it controls all the *biosynthetic* processes of the cell, including protein synthesis. It may be noted that most of the work on DNA has been done on viruses and bacteria (see bacteriophage, p. 523) since with them the results of experiments may be observed within 12 to 20 minutes.

DNA Molecule. The model of a DNA molecule (Watson-Crick model; FIG. 1), as first worked out by Watson and Crick in 1953, has been described in detail on p. 153. A summary is given here. The two-stranded molecule occurs as a twisted ladder, and its cross-links—2 purines (adenine and guanine) and 2 pyrimidines (thymine and cytosine)—occur as rungs of the ladder. It is the rule that a specific purine always pairs with a specific pyrimidine as alleles (i.e. complementary pairs) —T-A and G-C, loosely linked by hydrogen bonds.

Action of Gene. Gene or DNA, as established on three decades' intensive research, is the 'brain' or 'control' centre of the living cell. It controls all the life processes of the cell by sending chemical messages or codes to all its parts. The bases, as named above, occur in infinite orders or sequences in a DNA molecule, and are used by it to coin an infinite number of codes (messages or informations) which are then transmitted to all parts of the cell for necessary biosynthetic processes. Thus DNA sends message to ribosomes in the cytoplasm (see FIG. II/6) through **messenger RNA** and directs the RNA of the ribosomes to make proteins of particular kinds and in particular quantities (see pp. 340-41). In stepwise reactions of such a synthesis (aminoacids to proteins) a specific enzyme is controlled by each gene. This is the familiar 'one gene—one enzyme theory' of Beadle and Tatum (Nobel Prize winners, 1958). DNA controls nuclear divisions and chromosomal changes, and final distribution of DNA among the newly-formed cells, both qualitatively and quantitatively. The importance of DNA or gene as an essential genetic material, transmitted as such, has been already mentioned.

Duplication of DNA. DNA molecule duplicates itself. As stated before, the DNA molecule occurs in two long strands entwined about each other. The two strands unwind, separate lengthwise little by little, and each constructs a new partner. In 1958 Herbert Taylor of Columbia University has elaborated this point. Briefly speaking, when the two strands are separated, each serves as a template (or mold), forming a new complementary strand or its partner. The new materials required for this purpose are obtained from the surrounding cytoplasm. Thus two molecules appear in place of one, and they are exact replicas of the original one. This is called **template hypothesis**. When the DNA molecule thus divides, the bases also form *alleles*. A special enzyme *DNA-polymerase*, as first discovered in 1956 by Arthur Kornberg (a Nobel Prize winner), is responsible for the duplication of DNA.

Gene Mutation. Genes are the hereditary units present in the chromosomes. They are responsible for various characters that are externally manifested by plants and animals. A single gene may affect one or more characters or a single character may be due to the interaction of several genes. Any change in the gene brings about a change in the character. Although genes are highly stable one or more of them sometimes (rarely though) mutate. This may occur in somatic cells, little affecting the species. Mutation taking place in gametic cells is, however, of great significance so far as variability among living organisms and evolution are concerned. Many mutations are, however, harmful, e.g. mutant genes carrying diseases may be passed on to the progeny. In many instances it has been seen that red-eyed species of *Drosophila* produced white-eyed individuals, or normal *Oenothera* plants produced dwarf species, or plants with short styles, and sometimes a red- or blue-flowered species gave rise suddenly to white-flowered species. It has also been found that these individuals with new characters bred true for many generations. It is assumed that in all these cases gene-changes occurred in the parents, evidently causing the appearance of new characters in the offspring. These changes are spoken of as **gene mutations**. The cause of these changes is yet unknown. On the basis of genetical work on *Neurospora* by Beadle and Tatum (1959) and further work on nucleic acids (DNA and RNA) it may be stated that a gene mutant fails to synthesize a particular enzyme (single gene—single enzyme theory of Beadle and Tatum). Evidently this disturbs the sequence of amino-acids, and leads to the production of altered types of proteins and, therefore, a new type of plant since each plant has its own specific types of proteins generation after generation through the work of DNA, a genetic material. Thus gene mutation produces a new strain with new proteins. It is also asserted that evolution itself is primarily due to changes in proteins. Gene mutations can be induced artificially by treatment with X-ray (as shown in 1927 by Hermann J. Muller—a 1946 Nobel Prize winner), exposure to radiation and high temperature. Gene mutations may be due to the complete loss of old genes or appearance of entirely new genes or recombination of genes. Direct observation of gene-changes is not possible. There are several instances of gene mutations on record.

Crossing Over. Each chromosome consists of a pair of chromatids. At pachytene stage of meiosis the homologous chromosomes pair but subsequently at diplotene stage they separate again remaining attached at one or few points known as chiasmata. At these points a break may occur due to the twisting of the chromatids about each other; as a result a part of a chromatid of one chromosome goes over to a chromatid of the other chromosome, and this interchange of the parts of the chromatids

GENETICS

of a pair of chromosomes is known as **crossing over**. Crossing over brings about parental combinations of linked genes. Linkage is not, however, always very complete, thus resulting in separation of a few linked genes and appearance of new combinations. It is to be noted that normally crossing over takes place between only two chromatids of the two homologous chromosomes, while the other two chromatids of the chromosome pair

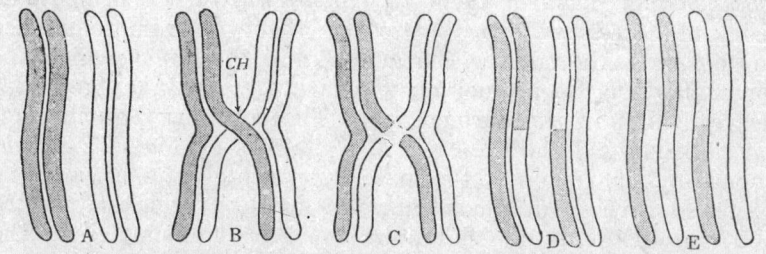

FIG. 2. Diagrammatic representation of crossing over (with one pair of chromosomes). *A*, two homologous chromosomes, each with two chromatids (at pachytene); *B*, chiasma (*CH*) formation (at diplotene); *C*, break of chromatids at chiasma and crossing over (at diakinesis or at metaphase), *D*, dissociation of two newly-constituted chromosomes (at anaphase). *E*, separation of four chromatids (at the end of second division), each entering into a daughter nucleus (spore or gamete).

preserve their own identity and pass on to the gametes intact. Temperature, X-ray and radium treatment have a profound influence on the frequency of crossing over.

Sex Chromosomes. Most animals and a good number of plants are dioecious. In them sexuality is controlled and sex determined by a definite pair of chromosomes called sex chromosomes. It has been observed that in most cases both the chromosomes of this pair are identical in the female, whereas in the male one is different in shape from the other. When the pair is similar, the chromosomes are known as X-chromosomes. The other dissimilar chromosome found in the male is known as Y-chromosome. Except for this pair all the other chromosomes are identical in both the male and the female. These chromosomes are called *autosomes*, while the sex chromosomes are otherwise called *allosomes*. The chromosome complements in this type may thus be expressed as follows: male—$2n+XY$, and female—$2n+XX$. This type is very widespread in both animals and plants. Subsequently the female yields gametes of one kind (homogametic), each carrying X, while the male yields gametes of two kinds (heterogametic) in equal numbers, i.e. one-half carrying X and one-half Y. When fertilization takes place two combinations become possible: XX (female offspring) and XY (male offspring). Now a question pertinently arises. Is it possible by any means to bring about any desired combination at will? The answer is a definite 'no' at the present state of our knowledge. In addition to the common type mentioned above there are other types also. For example, in birds the female has XY, while the male has XX, i.e. the order is seen to be reversed. In liverworts, mosses and some algae there is an X in the female, and a Y in the male. They combine into XY in the zygote and the succeeding sporophyte. There is another type in which the female has a pair of X-chromosomes, while the male has only one X-chromosome. Thus one sex is XX (female), while the other is XO (male). Such a type is known as the XO type.

Chromosome Mechanism in Heredity and Evolution. Recent work of Morgan and his colleagues on a fruit-fly (*Drosophila melanogaster*) has led them to put forward the chromosome theory of heredity. According to this theory, chromosomes are the bearers of hereditary characters, and the *genes* which are responsible for the production of characters are arranged in a linear fashion in the chromosomes. The continuity of chromosomes throughout the life-cycle of a plant is evident. Plants reproduce by the fusion of the male and female gametes, and the zygote so formed develops into the seedling and the mature plant. So the parental characters must be transmitted to the offspring through these gametes. The somatic (body) cells of the sporophyte have $2n$ chromosomes, of which n chromosomes are of paternal origin and n of maternal origin. When the gametes are produced the sporophytic number ($2n$) is reduced to half (n) by meiosis so that each gamete has n chromosomes. As soon as the gametes fuse to form the zygote, $2n$ is regained. The inheritance of the parental characters by the offspring is carried on by the **genes** which are ultra-microscopic particles occurring in pairs (one paternal and one maternal) in linear series in the chromosomes. When the chromosome splits, the genes become equally apportioned to the two chromatids. A particular plant or animal has a large number of gene pairs but comparatively few chromosomes and, therefore, each chromosome must carry several genes in its body. A gene (or a pair) is a factor or determiner responsible for the production of a particular character, e.g. colour of flower, shape of leaf, size of fruit or any other character. The several characters of a plant or an animal are, however, always influenced by the interaction of a number of genes. Although some investigators speak of transmission of hereditary characters through the cytoplasm of the egg or the sperm, yet it is regarded as of secondary importance at present. Whether the transmission of characters is chromosomal or cytoplasmic, the characters of mature plants and animals are modified to a great extent by environmental influences.

Chromosomes play an important role in evolution. New species of plants may arise due to changes in or loss or degeneration of genes or even the appearance of new genes or due to changes in the proportion of genes, known in all cases as 'gene' mutations. The cause of these changes is as yet unknown. The simple case of 'gene' mutation was observed by De Vries in evening primrose (*Oenothera lamarckiana*) which led him to put forward the 'mutation theory' of evolution. It was seen that now and then dwarf plants of evening primrose suddenly appeared in the field from normal plants, and these subsequently bred true. These were new species which, according to De Vries, had evolved only by 'gene' mutation. Sometimes a red- or blue-flowered species is seen to give rise suddenly to white-flowered species. The old gene, it is assumed, has changed and a new character has developed. New species of plants are also produced by a sudden increase in the number of chromosomes. Thus there may be $2n$, $3n$, $4n$, $5n$, $6n$, $7n$, $8n$, $9n$, or $10n$ or more or even a variable number. When this increased number is a multiple of haploid, it is expressed as **polyploid** (see pp. 188-92).

Economic Importance of Plant-breeding. Plant-breeding has been scientifically developed to such an extent in recent years that it is now recognized as the best practical method for the improvement of various crops—foodgrains, vegetables, pulses, oil-seeds, fruits, industrial plants, fodder crops and many other plants of economic importance and of aesthetic value as regards their productiveness, quality, colour, size, total yield and other useful characters. It has been found possible by the application of this method to combine the desireed characters of parent forms and evolve new types far better than original types in many respects. Babcock, a leading geneticist, even goes so far as to say that plants can almost be made to order. This being so, the subject is considered to be of utmost importance in agriculture and to a great extent in horticulture. By following practical methods of plant-breeding considerable improvements have already been achieved within the last few decades; for example, hardy rust-resistant and high-yielding types of wheat plants have been evolved, and the milling and bread-making qualities of wheat grains considerably improved. Several new varieties of rice with higher yield and better quality have been brought into existence. In tobacco the number of leaves per plant and their size and quality have been enhanced. Pea, bean, many vegetables, pulses, oil-seeds, etc., have been considerably improved. Fruits like apple, grapes, peach, pear, plum, orange, strawberry, etc., have similarly been much improved. In fibre-yielding plants like jute, cotton, flax, etc., better quality of the fibre (length, strength and fineness) and also higher yield have already been achieved. New types of maize, potato and tomato —high-yielding and rust-resistant—evolved recently in America are also some of the numerous achievements in this direction. In Russia new varieties of summer and winter wheats and 'perennial' wheat (results of crosses between varieties of wheat and couch-grass), wheat-*Elymus* (result of cross between wheat and

Elymus—a kind of grass), and barley-*Elymus* (result of cross between barley and *Elymus*) are some of the outstanding features of work in this line. A new hybrid called Triticale—result of cross between wheat (*Triticum*) and rye (*Secale*)—has shown higher production and better nutritional value). In India also a considerable amount of work has been done in this line on rice, wheat, millets, maize, sugarcane, pulses, oil-seeds, cotton, tobacco, jute, flax, hemp etc., and improved strains combining high yield, good quality and resistance to pests and diseases evolved by breeding. Pusa wheats deserve special mention in this connexion. Several new Pusa varieties tolerant to rust and resistant to smut and at the same time very high-yielding evolved at the Indian Agricultural Research Institute[1] are outstanding successes. A wheat, New Pusa 4 (NP 4), evolved at Pusa, was awarded the first prize in several international exhibitions in America, Australia and Africa. An outcome of several years' work at ICAR is the production of new types of bread wheat —NP 800, NP 200, NP 400, etc. A variety of rice—GEB 24— evolved at Coimbatore by Ramiah, a pioneer in rice-breeding in India, still remains unsurpassed in quality, productivity and wide adaptability. At the Central Rice Research Institute at Cuttack new varieties of rice resistant to disease have been evolved. Pusa 33, a hybrid BASMATI rice (a cross between BASMATI 370 and a dwarf type) recently evolved by IARI in New Delhi is very high-yielding (5,500 kg. per hectare), pleasantly aromatic, early-maturing and easy-cooking. An important achievement of the Institute is the development in recent years of hybrid millets and maize strains yielding 50% more than the common varieties. Some new strains of wheat have also been evolved by this Institute to suit different climatic regions of India. A hybrid maize, Texas 26, evolved by the same Institute, yields over 2,800 kg. per hectare per year. In addition the Institute has developed rust-resistant linseeds, wilt-resistant pigeon peas and late blight-resistant potatoes. Besides, highly improved types of chilli, gram, linseed, mustard, etc., evolved through hybridization are some of the special achievements of this Institute. A recent success is the production of a sweet-flavoured tomato with a high vitamin

---

[1] The Indian Agricultural Research Institute (formerly known as Pusa Institute) was established in 1905 at Pusa in Darbhanga (Bihar) on the liberal donation of £30,000 by an American philanthropist, Mr. Henry Phipps. At this Institute early work on the improvement of wheat by selection and breeding was carried out by late Sir Alfred Howard with considerable success. The laboratory buildings having been irreparably damaged by the great Bihar earthquake in 1934 the Institute was shifted to New Delhi in 1936. At present it has six sub-stations at Pusa (Bihar), Karnal (East Punjab), Simla (Himachal Pradesh), Pune (Maharashtra), Indore (Madhya Pradesh) and Willington (Madras).

content (result of a cross between a cultivated tomato and a wild South American species). New types of sugarcane evolved at Coimbatore in South India have already become world famous. These sugarcanes are mainly responsible for the improvement of the sugar industry in India. Although much has already been achieved there is still ample scope for improvement of several crops for food and industry.

It is evident from the foregoing that the subject of plant-breeding has grown so much in economic importance that the agricultural and plant-breeding stations all over the world have embarked on programmes of artificial plant-breeding for enhancing the quality and yield of particular crops, and the power of resistance to pests and diseases.

Part IX ECONOMIC BOTANY

Chapter 1   GENERAL DESCRIPTION AND ECONOMIC PLANTS

Economic Botany deals with the various uses of plants and plant products for the well-being of mankind. It also includes the various practical methods adopted for their improvement. Economic uses of plants are varied and the scope for improvement is immense to meet man's ever-increasing need. The primary needs of mankind are of course food, clothing and shelter which in their basic forms are supplied by nature and subsequently improved upon by man by the application of his scientific knowledge. The gifts of nature are almost unlimited and a variety of useful products are obtained from the plant kingdom. **Methods of Improvement.** The methods commonly employed for improvement of crops with regard to their quality, yield, etc., are: (1) pure line selection, (2) breeding (see pp. 795 & 807), (3) improved methods of cultivation, (4) selection and use of 'quality' seeds, (5) proper use of adequate amount of chemical fertilizers and manures (see p. 285), (6) judicious selection of crops for a particular locality, (7) introduction of high-yielding and disease-resistant varieties; (8) intensive and extensive cultivation; (9) introduction of short-duration and early-maturing crops to avoid flood or drought; (10) protection against diseases and pests (insects, moulds, rats, etc.), and against destruction by natural calamities (floods, droughts, hail-storms, etc.); and (11) proper irrigation—*major* by dams and barrages across perennial rivers diverting the water through canals, and *minor* by wells, tube-wells and tanks with the help of Persian wheels, MOTS and pumps. It may be noted that the loss of crops in India on account of diseases and pests is estimated to be 10-18.6% every year. The use of pesticides in India is only about 310 gms. per hectare per year, while it is 6 kg. in Japan. Further, world's rats devour about 40 million tonnes a year, enough to feed 250 million people.

Economic plants are numerous and have a variety of uses. Many of them occur in a natural state, particularly in forests; while a good number of them are cultivated for food and industry. Such plants may be classified under the following heads: (A) food—cereals and millets, (B) pulses, (C) vegetables, (D) oil-seeds, (E) fruits, (F) sugar, (G) spices and condiments, (H) medicinal (drug) plants, (I) beverages, (J) timber, (K) fibres, (L) rubber, and (M) paper. It may be of interest to note in this connexion that India is the largest producer of tea, sugarcane, groundnuts and jute; China of rice; U.S.A. of corn and cotton; Brazil of coffee; Ghana of cocoa; and U.S.S.R. of beet sugar. It may also be noted that the most important exchange earners of India are jute goods, tea and cotton cloth in their order of importance. These three together represent in value nearly half of India's total exports.

A. Food. Food may be classified into the following groups: (a) heat- or energy-producing food having high calorific values such as carbohydrates and fats; (b) body-building food such as proteins; (c) protective food such as vitamins and some minerals; and (d) luxury food such as confectioneries, etc. The energy value of food is expressed in terms of calories. A calorie is the amount of heat needed to raise 1 kilogram of water through 1°C. It may be noted that 1 gm. of carbohydrate yields about 4 calories, and 1 gm. of fat about 9 calories. The daily requirement of a man of average weight, doing moderate work, is about 3,000 calories which must be obtained from the food he eats. Average Indian diet, it may be noted, produces only about 1,620 calories. It is evident that food plants must contain sufficiently high percentages of carbohydrates, proteins, and fats and oils together with vitamins and essential minerals. All cereals and millets are rich in starch and contain vitamins A, B and C. They belong to *Gramineae* and are cultivated as annual crops. Cereals constitute the main foodstuff of human beings all over the world. Major cereals are rice, wheat, and maize, and major millets (smaller-grained ones) are JUAR or CHOLAM (*Sorghum*), RAGI (*Eleusine*) and BAJRA (*Pennisetum*). For proper nutrition of the human body, however, a balanced diet consisting of cereals, vegetables, pulses, vegetable oils, sugar, fruits, and also milk and milk products, and according to habit and custom fish, meat, eggs, etc., is indispensable. Among the important food crops of India cereals occupy about 60% of the total area under cultivation, pulses about 18% and oil-yielding plants about 8%. After many years of shortages, mainly due to heavy ever-increasing population, India has now attained self-sufficiency in food grains with an annual production of over 125 million tonnes on the basis of the steps taken in the following directions: improved method of cultivation, increased use of fertilizers (see p. 285), cultivation of some prolific varieties, arrangement for proper irrigation, and intensive and extensive cultivation of crops.

1. Rice (*Oryza sativa*) is the major agricultural crop in India, occupying about 37% of the total area under cereals. It covers a total area of 37.4 million hectares, which is the world's largest rice area. The total yield of this vast area although poor some years back has now gone up to 52.7 million tonnes a year with the cultivation of some special prolific varieties. Asia as a whole, however, accounts for 90% of the world's total rice production.

Rice is the staple food and the principal article of diet in India and tropical Asia, and feeds over 60% of the world's population. It contains 70-80% of starch, 7% of proteins, 1.5% of oils, some vitamins (commonly A, B and C) and some essential minerals in the pericarp and the embryo. In polished rice, however, the pericarp is destroyed with the loss of some precious nutrients (proteins, vitamins and minerals). Throwing away the water (gruel) after cooking results in further loss. Finally, starch mainly remains as food. Rice has been under cultivation in India and China from time immemorial. There are about 7,000 varieties of rice, of which about 4,000 varieties occur in India alone, derived from a few old wild forms which are still found. Although many new better varieties of rice—finer, higher-yielding and disease-resistant—have been evolved in India by the Agricultural Departments and the Central Rice Research Institute at Cuttack (Orissa) by means of selection and hybridization the annual average yield of the common varieties is very poor. Rice is widely cultivated in India except in northwest India where wheat is the main crop. The crop thrives under conditions of moderately high temperature and plenty of rainfall or proper irrigation. Under heavy manuring with cowdung, ashes and tank-earth, and by the use of chemical fertilizers like sulphate of ammonia and bonemeal the yield increases much above the average. The average yield of rice in India is more or less 1,450 kg. per hectare per year (Tamil Nadu leading with 1,974 kg.) much more under the Japanese method of cultivation and also under some special conditions, as against 3,117 kg. and 3,406 kg. in Egypt and Japan respectively, where fields are heavily manured. Rice straw is a fodder for cattle. Rice starch is used in preparing alcoholic beverages.

Cultivation. 2 or 3 croppings are practised in India according to the conditions of the soil and the climate. Indian rice fields generally remain undermanured. Common seasonal varieties are AUS or summer rice, AMAN or winter rice and BORO rice in between, with many sub-varieties of each. Some special varieties such a Taichung Native 1, Tainan 3, Kalimpong 1, Jaya, Padma, IR 8 and IR 24,—all high-yielding and early-maturing varieties— may yield on the same land, BORO paddy in April, AUS paddy in July, and AMAN paddy in October or even earlier in each case. ADT 27, now under cultivation in parts of Madras (Tamil Nadu), yields about 4,257 kg. per hectare per year. Further two new fine dwarf high-yielding varieties. IET 1919 and IET 1039, have shown great promise in several parts of Tamil Nadu.

1. *AUS* or **summer rice** is coarse, difficult to digest and low-yielding, as compared to AMAN. AUS paddy prefers high land without waterlogging. Sandy banks of rivers with silt deposits are ideal for its cultivation. After ploughing and cross-ploughing the fields are made ready. Paddy grains are sown in seed-beds or in the fields in early May and watered regularly. When the seedlings are about 23 cm. high they are transplanted to watersoaked fields after a few showers of rain. The seedlings are commonly

planted 15-23 cm. apart in small bunches of 3-5, or even singly, sometimes, however, in large bunches of 15 or so in flooded areas. AUS paddy is also broadcast, and transplanting is avoided; the yield, however, is low in this case. The time for harvesting is August-September before the grains fully ripen as they shed very easily. Paddy is threshed out of the stalk by beating or by trampling by cattle. The yield of AUS paddy varies from 1,075-2,240 kg. per hectare per year.

2. *AMAN* or **winter rice** is better in quality, finer and more easily digestible. AMAN paddy prefers a low-lying land with clay soil. The general mode of cultivation is almost the same as that of AUS. Paddy grains are sown in prepared seed-beds (nurseries) in May-June or sometimes broadcast. Sufficient rain-water being available transplantation is done in June-July. If necessary, irrigation is resorted to. If the season continues to be dry it is better to grow AUS, maize and millets. AMAN paddy ordinarily requires no manuring but addition of manures and fertilizers, as in the case of AUS, improves yield. AMAN paddy is harvested in November-December. It can be stored for months before threshing. Its yield is much heavier, being about 2,912-4,480 kg. per hectare per year.

3. *BORO* **paddy** is a minor crop. But two croppings may be practised: (a) KHARIF or rain crop sown in June-July and harvested in September-October; and (b) RABI or winter crop sown in October-November and harvested in March-April. The grains may be broadcast and seedlings later transplanted. The yield is 1,790-2,240 kg. per hectare per year.

4. **Deep-water paddy** is grown in low-lying areas during the rains, but not widely. It can stand water to a depth of 1·5-3 metres or even more. The peculiarity is that with the rise of water level the plant grows quickly—20-30 cm. a day but if it gets completely submerged it rots. The yield is low.

JHUM **Cultivation.** It is the practice with some hill tribes to select the best fertile areas on hill slopes, cut down and burn forest growths, and then raise crops in these areas for a year or two. After this period fresh areas are selected and the same practice repeated. Because of the very evil effects of JHUMMING or shifting cultivation, viz. deforestation, loss of soil-fertility, erosion of land, occurrence of floods, etc., this practice is being gradually replaced by ordinary cultivation.

2. **Wheat** (*Triticum aestivum*) is the second staple food of people in India, and the principal article of diet in Western countries. There are several varieties of wheat. They may be broadly classified into *hard* and *soft*. The former varieties are adapted for making SUJI and ATTA, while the latter varieties are used for making fine flour (MAIDA). Wheat is extensively cultivated in Uttar Pradesh, Madhya Pradesh, Gujarat, Rajasthan, Haryana and Punjab. Soft wheats are grown in the basins of the Ganges and the Indus, and harder varieties elsewhere. Grains are sown in October-December and the crop harvested in March-May. Wheat is a universal crop, i.e. it can be successfully grown in both temperate and tropical countries. The inflorescence usually consists of 15-20 sessile spikelets; each spikelet bears 1-5 flowers but all of them are not fertile. The average yield of wheat in India is poor—about 900 kg. per hectare per year, as against 1,396 kg. in Canada, 1,708 kg. in Japan, and 2,468 kg. in Great Britain. Several new varieties of wheat (e.g. SONA 227 evolved by IARI)—hardy, disease-resistant, high-yielding, with

better milling and bread-making qualities—have been evolved in India (see p. 808), with phenomenal increase in yield. Further with the cultivation of some dwarf Mexican varieties, e.g. Mayo 64, Sonara 63 and 64, Lerma Rojo 64A, etc., particularly in the Gangetic and Indus plains, the total production is now 31.33 million tonnes a year. Some of the best wheats are grown in Australia, America and Russia. Average chemical composition is: starch—66-70%, proteins 12% and oils 1.5%. Wheat starch is used in textile industry and in making alcohol. Wheat straw is a fodder for cattle. Wheat prefers a clay loam or sandy loam and a moderately dry climate, and thrives under irrigation. Saltpetre is the best manure for wheat.

3. **Maize** or **Indian Corn** (*Zea mays*) is an important cereal food for poor people. It is, however, mainly used for feeding livestock. Originally it is a Mexican plant but now grown all over the world. It was introduced into India by the Portuguese in the early part of the 17th century. Maize is now cultivated all over India both in the hills and the plains, predominantly in Uttar Pradesh, Madhya Pradesh, Punjab, Rajasthan and Bihar, and does well both in hot and cold climates. There are several varieties and hybrids. The usual sowing season is April-May and the harvesting season July-August. Each plant commonly bears one cob, sometimes two, and the average annual yield per hectare is 986 kg.; with some good varieties, e.g. dwarf Mexican variety, the yield has gone up to 1,570 kg. Some new hybrids of maize have been evolved in India, which are very high-yielding, disease-resistant, sweetish and nutritious (see p. 808). The total production of maize in India has now gone up to 7.41 million tonnes a year. Maize cobs may be 15-25 cm. in length, and the grain golden-yellow, dull-yellow, red, white, purplish, etc., in colour. Maize grains are taken as a substitute for other cereal grains and prepared by boiling. They are also often fried; commonly they are ground into fine flour called corn-flour; they are also powdered into starch. Maize starch is largely used in making alcoholic beverages. The young tender grains are nutritious and may be taken raw, roasted or boiled in milk. Of all cereals maize contains the largest amount of oils. The average chemical composition is: starch—68-70%, proteins—10% and oils—3.6-5%. In addition the grains contain an appreciable quantity of Ca and Fe. Maize prefers a high open land for its cultivation, and requires manuring as it exhausts the soil. Leaves, spathes and stems form a good fodder and the grains a nutritious food for farm animals.

4. **Sorghum vulgare** (= *Andropogon sorghum*; great millet—JUAR or CHOLAM) is the best of all millets. It affords nutritious food, nearly as good as wheat. It is widely cultivated in

India, particularly in South India, and also in Maharashtra and Gujarat. The plants are as tall as 3-4 metres, the stem stout and often sweetish, and the panicle much branched. There are several varieties and hybrids. The average annual yield per hectare is 785-896 kg., often double this quantity in properly irrigated black soil. The grains are made into flour, often mixed with wheat, forming a nutritious food. The average chemical composition is: starch—72%, proteins—9% and oils—2%. Grains are commonly sown in June-July, and the crop harvested in October-November. It takes 4-5 months to mature, and two croppings are generally practised. JUAR is a good fodder crop.

5. *Eleusine coracana* (African millet—RAGI or MARUA) is an important food crop of Karnataka, and is extensively cultivated in Karnataka, Tamil Nadu and Andhra, and to some extent in Maharashtra, Punjab, Uttar Pradesh and Bihar. RAGI being a short duration crop 2 or 3 croppings are practised a year. Commonly grains are sown in June-July, and the crop harvested in August-September. The plants are dwarf, being 0.6-1 metre in height, but very hardy. The spikes are short and occur in a whorl. RAGI is a dryland crop. The annual yield per hectare is 785—1,120 kg., sometimes well over 2,200 kg. in properly irrigated red soil. The grains can be stored for several years without injury. The average chemical composition is: starch—73%, proteins—over 7% and oils—1.5%. The grains are difficult to digest. The straw is a nutritious fodder for cattle.

6. *Pennisetum typhoideum* (pearl millet—BAJRA) is another important millet, and cultivated almost throughout India. The plants are 1-2 metres in height, and the dark-brown spikes, 15-23 cm. in length, occur in clusters. The crop takes three months to mature. BAJRA grows in regions with low rainfall on both red and black soils. Generally two croppings are practised a year. Commonly the grains are sown in May, and the crop harvested in July-August. The average annual yield per hectare is 560-670 kg. or a little more under irrigation, the total yield a year being over 8 million tonnes. The grains often require threshing and husking like paddy. The average chemical composition is: starch—71%, proteins—10% and oils—over 3%. The grains are cooked like rice or ground into flour. The straw is not commonly used as a fodder for cattle.

B. **Pulses.** As foodgrains, pulses (family *Papilionaceae*) stand next to cereals. They are cultivated extensively in India as winter crops in rotation with cereals. Pulses occupy about 18% of the total area under cultivation. They are valued as food because of their high protein content averaging 22-25% (in soya-bean as high as 35-42%), starch content about 58% and oil content 2% or more; in gram, however, the oil content may be as high as 5%. They contain vitamins A, B and C, particularly in sprouted seeds. The pulses commonly used in India are Bengal gram, black gram, pea, red gram or pigeon pea and lentil. With the exception of pigeon pea (which is a shrub) the rest are annuals and form short-duration crops. Pulses are widely used in various culinary preparations, particularly DAL (a kind of soup). The plants form good fodder. Having root-nodules for nitrogen-fixation they form excellent green manures.

1. **Bengal Gram** (*Cicer arietium*) is cultivated extensively all over India as an important pulse. The crop matures in three months. Gram is a nutritious food but somewhat difficult to digest. Split seeds are commonly used as a DAL. Grams are also boiled, fried or roasted. Soaked seeds are fed to horses for strength and working capacity. Dried seeds ground into flour are sometimes relished by many, and widely used in confectionery. Green seeds are often eaten raw. Apart from its high protein content it is also rich in oil, and is a good source of vitamin A. Because of such contents the germinating seeds are specially good for the growing children and also for the adults. Average chemical composition is: starch—58%, proteins—23.5% and oils—5%. The average annual yield per hectare varies from 680-1,130 kg. The straw forms a good fodder.

2. **Black Gram** (*Phaseolus mungo*) is one of the best pulses grown throughout India. The plants have a trailing habit. The seeds are usually dark-brown in colour. The average annual yield of seeds per hectare is 450-550 kg. The average chemical composition is: starch—54%, proteins—22% and oils—1%. Apart from their high protein and vitamin contents the seeds are rich in phosphoric acid. The proteins of black gram are more easily digestible and are almost as good as meat. Apart from its use as DAL black gram is used for preparation of various forms of pies. PAPPAD (wafer) is made of this pulse. Germinating seeds taken raw are considered nutritious. Husked pods, seeds and straw form a valuable cattle feed.

3. **Pea** (*Pisum sativum*) is another common pulse but not very extensively used. It is relished as an occasional substitute for other pulses. The dried split seeds of field pea are used in the preparation of DAL, while the green seeds of garden pea are eaten raw or in stews and commonly used as a vegetable; the latter are sweetish in taste. A large quantity is used for canning, particularly in America. In India dehydrated peas are coming into use as off-season vegetable. The average chemical composition is: starch—55%, proteins—28% and oils—1.5%. Average annual yield per hectare is 280 kg. It is cultivated mainly in Northern and Eastern India. Leaves and stems are used as a valuable fodder.

4. **Pigeon Pea** or **red gram** (*Cajanus cajan*) is a perennial shrub

grown as a pure crop or mixed crop, and widely cultivated in India. Sowing is done in May-July, and within six months it bears pods. The crop is harvested in December-March. Commonly two varieties are distinguished—one with longer pods and the other with shorter pods. The average annual yield per hectare is 450-900 kg. Average chemical composition is: starch—57%, proteins—22% and oils—about 2%. It is widely used in India as DAL, and is one of the important items of a vegetarian diet, particularly in South India. Various sweet cakes are also made of this pulse. Sometimes tender green pods are used as a vegetable. At one time it was a favourite food with the sailors. The leaves form a valuable fodder.

5. **Lentil** (*Lens culinaris*) is an important pulse of Northern and Eastern India. The crop takes only three months to mature and is generally harvested in February-April. The average annual yield per hectare is 450-670 kg., going up to 900 kg. under irrigation. In Eastern India, this pulse is in almost daily use and is nutritious. Average chemical composition is: starch—58%, proteins—23-25% and oils—1-1.5%. Dry leaves and stalks are used as a fodder.

6. **Soya-bean** (*Glycine max*=*G. soja*), a native of China and Japan, is an annual sub-erect herb. It was introduced into India late in the 19th century but so long mostly grown as a garden crop. Only recently, because of its high food value, an extension of cultivation of this very important crop in many States has been under serious consideration of the Government. The seeds, 3 or 4 in each pod, are very rich in proteins (35% or more), oils (19%) and also minerals, particularly Ca and Fe, but low starch and sugar contents (26%). Evidently they afford very nutritious food for all, being specially suitable for diabetics. As a matter of fact, soya-bean is considered to be the richest vegetable food. In India the main preparation, though not common, is DAL or flour. It makes, however, a good soup. Soya-bean milk is considered a good substitute for cow's milk, quite suitable for children and invalids, and is much cheaper too. Casein and curd made out of this milk are of superior quality. In China and Japan, soya-bean forms a standard article of diet, and preparations are of varied nature. A rather common preparation is a sort of cheese (casein) or paste. Seeds are also ground into flour, and biscuits and breads are made. They yield an edible drying oil which is used for margarine, and also as a cooking oil after refining. Its industrial uses are numerous: soap-making, paints, lacquer, varnish, candle, greese, disinfectant, insecticide, etc. A large quantity of seeds and oil is annually exported to Europe from Manchuria. Green plants and oil-cakes (with 40-48% of proteins) form an extremely rich cattle food. The crop becomes ready within three months and is harvested in Nov.-Dec. The yield of seeds per hectare per year is about 970 kg. There are several varieties of soya-bean. At present the annual production of soya-bean in India is near about 1,000 tonnes, while in China it is 10,000 tonnes and in America over 13,000 tonnes. Since 1967 American varieties are being grown with success as a commercial crop in Mysore, Uttar Pradesh, Madhya Pradesh and Rajasthan.

C. Vegetables. (*a*) Leafy vegetables like cabbage, lettuce, spinach, etc., are rich in vitamins, usually A, B, C and E, and should, therefore, be included in the daily diet; (*b*) several fruits (leguminous and non-leguminous) are also used as vegetables; and (*c*) tuber crops are fleshy underground roots or stems laden with a heavy deposit of food materials. Some of the common ones are as follows:

1. **Potato** (*Solanum tuberosum*—family *Solanaceae*), a native of Peru (South America), is a much-branched annual herb. It was first introduced into India by the Portuguese in the early part of the 17th century, and into Europe by Sir Francis Drake late in the 16th century, but it was grown as a commercial crop late in the 18th century. Over 70% of the world's potatoes are grown in Europe and the USSR. Potato is an underground stem-tuber, and is extensively cultivated in India in the hills as well as in the plains, usually during cold months. But it can be grown both as a summer crop and a winter crop. Potato is a universal article of diet all over the world. It is an excellent food with easily digestible starch, some essential amino-acids (e.g. lysine), certain useful minerals (e.g. K, Mg and P) and sufficient vitamin C, and is used in a variety of culinary preparations. The average yield of potato in India is on the whole very low, being in the neighbourhood of 7-8 tonnes (70-80 quintals) per hectare per year, sometimes double this quantity under suitable conditions. The Central Potato Research Institute in Simla has recently evolved some new hybrids and varieties which are very high-yielding and disease-resistant. The *per capita* consumption of potato in India is extremely low, being about 4 kg., as against 117-174 kg. in European countries. The average yield of tubers per plant in India is 500 gm., varying from 200 gm. to 800 gm. according to varieties. There are a few hundred varieties which may be broadly classified into two—*waxy* and *mealy*. The average chemical composition is: starch—18-20%, proteins—2% and oils—0.1%. Potato starch has a variety of uses, particularly in laundry and in the preparation of alcohol (see p. 169). Sandy loam is the best soil for cultivation of potato. Water-logging is very injurious, while irrigation with proper drainage is very beneficial to the crop. Potato requires 3-4 months to mature, and when the leaves have completely withered it is ready to be lifted, usually in February-March. Generally the yield is tenfold of the seed-potato sown. Late blight of potato (caused by *Phytophthora infestans*) is a serious disease of the crop, particularly in the hills.

2. **Sweet Potato** (*Ipomoea batatas*—family *Convolvulaceae*) is an underground tuberous root. The plant, a native of tropical America, is a perennial twiner, trailing on the ground. It is grown widely in India, always preferring a moist sandy soil. The plant is commonly propagated by branch-cuttings or by tuberous roots. There are two common varieties—one with white skin and the other with red skin. Sweet potato is tasteful and nutritious, and may be taken raw, boiled or fried, or in curries. It also makes tasteful CHUTNEY, sweet or sour. Average chemical composition is: starch and sugar—29%, proteins—2% and oils 0.7%.

3. **Tapioca** or **Cassava** (*Manihot esculentus*—family *Euphorbiaceae*) is the large fleshy root of the plant, a perennial shrub, cultivated mainly in Kerala, and to some extent only in other States. The numerous varieties may be broadly classified into *bitter* and *sweet*; the former contains some amount of poisonous hydrocyanic acid which, however, disappears on boiling or roasting. Tapioca makes tasteful curries and is very nutritious. Tapioca flour is used in making CHAPATIS, HALWA, puddings and biscuits. Granulated tapioca (called *pearl tapioca*) is sold in the market as a substitute for sago. Tapioca meal may be used as a substitute for arrowroot. Tapioca is a good anti-famine food. The plant is propagated by stem-cuttings. The yield of raw roots may be over 11 tonnes per hectare per year. A single plant may

produce 11-25 kg. of roots. On drying they lose about 75% of their weight. Tapioca is widely used in America.

4. **Yams** (*Dioscorea*—family *Dioscoreaceae*) are the large underground storage tubers (root tubers or stem tubers according to species) of *Dioscorea*, particularly *D. alata*. Yams sometimes weigh as much as 12 kg. or even more. There are also edible bulbils borne in the axils of leaves. Yams are cultivated in all tropical countries (particularly in tropical America) to a greater or less extent. Good varieties, when cooked, are palatable and nutritious. They also form a good feed to livestock. They are also ground into flour. *Dioscorea* plants are large twiners, and are easily propagated by bulbils or portions of tubers.

D. Oil-seeds. There are several species of plants yielding oils— edible and industrial—in high percentages. In oil-seeds India holds a prominent position in the world market. Oil-yielding plants occupy about 8% of the total cropped area in India, and the production is on the increase from year to year.

1. **Groundnut oil** is obtained from the seeds of *Arachis hypogaea*—family *Papilionaceae*, the yield of oil being 43-46%. Pods develop underground (see FIG. III/49). Groundnut cultivation is now a major agricultural operation in India, and occupies the largest area in the world. Tamil Nadu, Andhra, Gujarat (leading in production), Maharashtra, Madhya Pradesh, Punjab and Rajasthan are the principal areas of its cultivation. The average annual yield of pods per hectare is 900-1,100 kg., more with some varieties, and the total production is about 7.5 million tonnes. Nuts are nutritious, containing 31% of proteins. Groundnut proteins are easily assimilated and agreeable in taste. They may supplement milk. The nuts also contain Ca, P and vitamin B. They are eaten raw, fried or roasted, and used in some confectioneries, but groundnut is mainly cultivated for its edible oil which is used extensively for cooking. It is the principal commercial oil of the Vanaspati industry, often mixed with soya-bean oil to the extent of 25-30%, and lately with sunflower oil. In Europe 'margarine'— an imitation butter—is manufactured from this oil. Other uses of the oil are: soap-making, illuminant, lubricant, paints, varnishes, tanning, etc. The oil-cake is a good feed for cattle. There is a big export of groundnuts, oil and oil-cake to France, England and Germany. (2) **Gingelly or sesame oil** is obtained from the seeds of *Sesamum indicum* (family *Pedaliaceae*), an annual herb, 1 metre or so in height, mostly grown in Uttar Pradesh, Tamil Nadu and Karnataka, and to some extent in other States. India is the largest producer of this crop. It takes $3\frac{1}{2}$ to $4\frac{1}{2}$ months to mature. The seeds yield 45-50% or even more of non-drying edible oil. The superior quality of gingelly oil is used for cooking in South India, and also for annointing the body before bath. It is widely used as a

cooling hair-oil, often perfumed. The inferior quality is commonly used for lighting in villages, and also for soap-making. The average annual yield of seeds per hectare is 340-460 kg., sometimes up to 580 kg. The oil-cake makes a good cattle feed, particularly for milch cattle. (3) **Mustard oil** is obtained from the seeds of *Brassica campestris*—family *Cruciferae* and of a few other species of *Brassica*. Mustard is extensively grown in Uttar Pradesh, Rajasthan, Punjab, Madhya Pradesh, and to some extent in Bihar, Orissa, West Bengal and Assam. It is a winter crop of 5-6 months' duration, commonly harvested in March. The average annual yield of seeds per hectare may be taken as 400-560 kg., varying according to the species. The oil content of mustard seeds average 35% or a little more. It is a non-drying edible oil, chiefly used for cooking purposes in Northern India. It is also largely used as a toilet oil. Mustard powder is widely used as a condiment. (4) **Coconut oil** is obtained from the dry kernel (copra) of the seed of *Cocos nucifera* (family *Palmae*), the yield of oil being 50-65% or even more. It is a very valuable oil used for cooking, lighting, and several toilet preparations such as good quality soap, shampoo, hair-oil, cosmetics, shaving cream, etc. It is also extensively used for making 'margarine'. The oil-cake is a valuable fattening food for cattle. Coconut trees grow luxuriantly along the sea-coasts extending further in, and on marine islands. A healthy tree bears 60-80 coconuts a year, sometimes even 100, fruiting all the year round. The tree attains a height of 18-20 metres or even more. There are several varieties including some dwarf ones. Kerala leads in the production of coconuts, coconut oil, coir fibres and excellent coid goods. (5) **Castor oil** is obtained from the seeds of *Ricinus communis* (family *Euphorbiaceae*), a quick-growing perennial shrub. It is widely cultivated as an annual crop in Tamil Nadu and Karnataka, and so some extent in Bihar and West Bengal. The crop takes 5-8 months to mature. There are two distinct varieties—one with larger seeds and the other with smaller seeds. The former yields an inferior quality of oil, usually 25-30%, while the latter yields a superior quality of oil, usually 36-40% or even more. It is a non-drying oil. The yield of seeds per hectare per year varies from 450-560 kg., sometimes up to 1,000 kg. The plant grows well in sandy or clay loam and also in red soil. Castor oil has a variety of uses. The oil expressed from the smaller seeds and purified is used as a medicine, particularly as a safe purgative, as known from time immemorial, possibly because it contains *ricinolic acid*. The oil expressed from the larger seeds is widely used for lubricating machinery, particularly the rolling parts of railway carriages. It is also widely used for lighting in villages; it gives a bright light without soot. It is the main ingredient of copal varnish. It is also used for dressing leather and skin in tanning industry. Its other uses are soap-making and candle-making. Refined castor oil, often perfumed, is a good (light and non-sticky) cooling and refreshing hair-oil, widely used. Castor cake is a valuable manure but not a cattle feed since it contains *ricin*, a poisonous alkaloid. The oil, however, is free from it. ERI silk-worms are reared on castor plants. India is the largest exporter of castor seeds and castor oil. U.S.A., however, imports castor oil mostly from Brazil and Japan, and to some extent only from India.

E. Fruits. India abounds in some excellent dessert fruits. Apart from their food value they always contain some vitamins. There is a fairly heavy export of Indian fruits, particularly bananas and mangoes. Although cold storage methods have not been developed extensively yet in our country for the preservation of fruits in fresh condition, many of them are available out of the season in the form of various preserves, either as slices or as jams, jellies, pickles, marmalades, chutneys (hot or sweet), etc. Some of the common fruits are as follows:

1. **Mango** (*Mangifera indica*—family *Anacardiaceae*) is regarded as the 'king' of fruits and its edible part is the mesocarp. India is by far the largest producer of mango (90%) in the world, covering an estimated area of 593,520 hectares. It is a mid-summer fruit (drupe) having over 1,000 known varieties, of which about 500 are common, and several of them are good table varieties. The superior ones range in weight from 200-600 gm. or sometimes even more. Mango is possibly the best dessert fruit, specially noted for its very pleasant taste, pulpy flesh and fine flavour, but each variety lasts only for a short time. It makes palatable chutney, jelly, pickle and tarts. Mango slices and expressed juice spread on plates are also dried in the sun for off-season use. The keeping quality of ripe mangoes is very poor. They are rich in vitamin A, and also contain a little B and C. Mango is considered to be fattening, energising and laxative. Good varieties are propagated by 'inarching' (see FIG. III/65).

Some of the famous varieties are: LANGRA, DASHERI, and SAFEDA of Uttar Pradesh; LANGRA, BOMBAI, GULABKHAS, SEPIA and ZARDALU of Bihar; HIMSAGAR, MALDA, FAZLI, KOHINOOR, MOHANBHOG GOPALBHOG (possibly the best variety) and KISHENBHOG of West Bengal; ALFONSO and PAIRI of Maharashtra; FERNANDIX of Goa; ALFONSO and KESAR of Gujarat; GULABKHAS, HIMSAGAR, KISHENBHOG, DASHERI and SIROLI of North India; SIROLI and TAIMUNSO of Punjab; BANGALORA (or TOTAPURI), BANGANAPALLI, SUBARNAREKHA, JEHANGIR and MALGOA of South India. It may be noted that Uttar Pradesh is the leading mango-producing State in India accounting for nearly one-third of the total production, and Bihar ranks second to this State.

2. Pineapple (*Ananas comosus*—family *Bromeliaceae*) is the fruit (sorosis) with a very delicious taste and flavour. There are several varieties; common good ones may weigh 1 to 2 kg. and specially large ones (e.g. Singapore variety) may weigh up to 6 kg., sometimes even more. Pineapple slices and juice are canned on a large scale. The juice contains 8-15% of sugar, vitamins A and C, certain minerals—Ca, P and Fe, some fruit acids, and a digestive enzyme *bromelin*. The plant is a native of South America, and from there it has spread to all parts of the world. It is a stout perennial herb, and grows in the plains as well as in the low hills. It prefers sandy loam with humus and some amount of lime in it, a little shade and proper drainage, and bears fruits during the rainy season. The plant is propagated by suckers and crowns (see FIG. III/62).

3. Banana (*Musa paradisiaca*—family *Musaceae*) is one of the best dessert fruits and is a berry. The plant is a tall robust herb cultivated in most parts of India, covering an estimated area of 194,370 hectares. The yield of fruits (good varieties) per hectare per year may be 70-100 tonnes. It bears a single large spadix. The stout scape is cooked and eaten. Ripe fruits are sweet, soft, pleasantly flavoured, palatable, nutritious, easily digestible and laxative. Per-capita consumption of this highly nutritious fruit in India is, however, only 58 kg., as compared to 275 kg. or more in other tropical countries. There are 200 varieties, each having its own characteristic taste, size and flavour. Bananas are available throughout the year, specially during the rainy season. They are also used for making various sweet preparations. The fruit contains about 20% of sugar (but no starch), about 4.7% of proteins, and vitamins A, B, C and and also D and E. It is also rich in K, Ca, Fe and P. Green fruits of plantain (*M. sapientum*) are used only as a vegetable. Musa plants are propagated by suckers. There is a small export of banana from India but Latin America and Jamaica have practically monopolised the European markets.

4. Orange (*Citrus reticulata*—family *Rutaceae*) is a winter fruit (hesperidium)—very juicy, tasty and stomachic. The juice contains citric acid and is rich in vitamin C. The plant is a much-branched large shrub. It begins to bear fruits within five years of planting, and a healthy plant, ten years or so old, often bears 300 to 400 fruits, sometimes more, presenting a spectacular sight. The keeping quality of ripe fruits is very poor, and a good quantity sheds and is wasted. Orange plantations thrive at low elevations on the hill-slopes with plenty of lime and phosphate in the soil. Climate is also an important factor. Assam, West Bengal, Madhya Pradesh, Delhi, Punjab, Madras, Coorg and Hyderabad are important centres of orange cultivation in India. Rajasthan is famous for cultivation of sweet orange (*C. sinensis* —MALTA). There are many species and varieties (see p. 721). Good varieties are commonly propagated by bud grafting (see FIG. III/66) on a hardy variety or by gootee (see FIG. III/64). Orange juice contains citric acid (1-2%), sugar (5-10%), and is rich in vitamin C. Besides, an essential oil is obtained from the peel, bergamot oil (a perfume) from the flowers and the peel; and marmalade (a jam) is prepared from the skin and pulp of sour varieties.

5. **Papaw** (*Carica papaya*—family *Caricaceae*) is a pulpy fruit (berry). It is an excellent dessert fruit—palatable, cooling, refreshing, digestive and laxative; it contains vitamins A and C. The green fruit is used as a vegetable; the latex obtained from it contains *papain* which is a digestive enzyme. Fruits, ripe or green, are available throughout the year, particularly during the rainy season. Some good varieties (e.g. Ranchi variety) may bear fruits weighing up to 3 to 4 kg., particularly when some of the young green fruits are removed early. Each plant may bear 40 to 60 fruits. The plant is monoecious and, therefore, a few male plants in a plantation help pollination and development of large fruits. Proper manuring and watering are also a prerequisite to this factor. The plant, originally a native of America, is now extensively grown in all warm countries, including India.

F. **Sugar.** Cane-sugar or sucrose ($C_{12}H_{22}O_{11}$) is the main commercial sugar used universally for sweetening various food preparations. Apart from its taste cane-sugar is one of the best sources of energy available to man. An average acre of sugarcane yields more calories of energy than any other field crop covering the same area. But in India the *per capita* consumption of sugar is the lowest in the world, being only about 2.7 kg., as against 50.8 kg. in the U.K., 51.7 kg.. in Australia, and 58 kg. in Denmark. The *per capita* consumption of GUR (jaggery) in India is about 10 kg. The sugar industry is India's second largest industry next only to textiles. There are two main sources of supply of cane-sugar in the world—sugarcane in tropical countries and sugar-beet in temperate countries, and to a small extent maple (*Acer saccharum*) in the U.S.A. In India sugar is also obtained from various sugar-palms, mainly in the form of GUR (jaggery).

1. Sugarcane (*Saccharum officinarum*—family *Gramineae*) is a tall reed-like grass—2.5 to 4 metres or sometimes more in height. It has been in cultivation in India since 300 B.C., possibly much earlier. There are several varieties grown all over India. Main sugarcane-producing States in India are, however, Karnataka, Tamil Nadu, Bihar, Uttar Pradesh and Maharashtra. The plant takes 12 to 20 months to mature. It is propagated by stem-cuttings, and the rootstock continues to grow again after this operation. The majority of sugar factories now numbering 270 are located in Uttar Pradesh and Bihar. India is the largest sugar-producing country in the world. Her annual total production of sugar has now reached a high peak of about 6 million tonnes. 1969-70, but it declined to 4 million tonnes or a little more since then. Some improved varieties of sugarcane evolved by the Sugarcane Research Station at Coimbatore are now being cultivated widely in India to feed her numerous sugar mills. Nearly 55% of sugarcanes are used for making GUR (jaggery) and KHANDSARI (unrefined sugar) as cottage industries, while approximately 25% or a little more are used in mills for manufacture of white sugar. A small percentage is used for chewing. Although India is the largest cane-growing country in the world, the varieties of canes commonly grown and their yield are the poorest. The average yield of sugarcanes per hectare per year in India is nearly 50 tonnes, which is much less than that of other good cane-growing countries (Java, Hawaii, etc.). The highest yield of nearly 88 tonnes has gone to the credit of Karnataka. Proper irrigation and the use of fertilizers are important factors in the cultivation of sugarcanes. The plants are susceptible to many diseases, particularly 'red rot' caused by *Colletotrichum*, and to the attack of borers. Although sugarcane contains 10-15% of sugar (18-20% in rich canes) with an average of about 13%, the actual recovery of sugar from the cane in Indian sugar mills is only about 10%. On an average 9.5 to 10 tonnes of Indian sugarcane yield one tonne of sugar, whereas in other countries 8 to 9 tonnes yield one tonne. The average production of sugar in an Indian mill is much less than in other countries (Cuba, Egypt, etc.). Evidently there are many uneconomic sugar mills in India.

Manufacturing Process of Cane-sugar. Sugar is not really manufactured in the mills, but is extracted from the juicy pith of the stem and crystallized in the mills, and refined in the refineries. The process consists of the following stages. (1) **Juice Extraction: Milling.** The cane cut into short lengths by revolving knives is passed through a series of rollers for crushing and releasing the juice. It is then strained through perforated metal-sheet strainers to remove *bagasse* (crushed stalks of sugarcane). The juice thus obtained is turbid containing many suspended particles, and is acid (pH 5.1-5.7) in reaction. It is then sent to the boiling house where the extracted juice is first boiled. (2) **Clarification.** Lime (and sometimes also phosphoric acid) is added continuously to the juice in amounts sufficient to raise the pH to 7 or slightly above. The juice is then pumped through high-velocity juice-heaters. Impurities are mostly precipitated, and are removed in 'clarifiers'. The clarified juice is of bright yellow colour. The sediment called *press mud* is used as a fertilizer. (3) **Concentration.** The clarified juice is pumped continuously into a multiple-effect evaporator, where it is concentrated into a clear pale-yellow syrup with soluble solid contents to the extent of 55-64%. It is then boiled in a single-effect vacuum pan and further concentrated. This operation is carried out in batches. (4) **Crystallization.** The semi-solid mass in the pan is called *massecuite*. When the pan is full, the massecuite, still hot, is fed into high-speed centrifuges (1,500-1,800 r.p.m.). The centrifugal baskets have closely woven meshes. When the contrifuge is worked the massecuite is separated into molasses and sugar crystals. The formation of a regular crop of crystals is controlled by an ingenious method. Under centrifugal force the molasses passes out through the meshes and is drained off and collected. The sugar thus obtained is still coated with a thin film of molasses, and is called brown sugar. It has a wide use in confectionery because of its special flavour. To obtain white sugar water is added to the surface of the sugar in the form of a spray while the centrifuge is still revolving. This washes out the brown film, and white sugar is obtained. (5) **Refining.** The sugar thus formed contains 96-97.5% of sucrose. The refineries then take over. The sugar is washed and melted, and mixed with lime when the pH goes up to 10. The melted sugar is saturated with CO_2 and the pH brought down to 7. It is then heated and passed through charcoal filters. 'Continuous rotary pressure filters' are also used for the purpose. The previous process, as described under brown sugar, is repeated and finally the crystals are dried in a rotary drier.

Utilization of By-products. (1) *Bagasse* is mostly used as a fuel, and also used in the manufacture of wrapping paper and cardboard. (2) *Press mud* is used on a limited scale as a manure. A good quality wax obtained from it is used as a shoe polish. (3) *Molasses* has a variety of uses such as manufacture of alcohol, cattle feed, manure, fertilizer, curing tobacco, etc.

2. **Sugar-beet** (*Beta vulgaris*—family *Chenopodiaceae*) is the source of sugar in cold countries (Europe, Russia, Canada and U.S.A.). Sugar is extracted from the fleshy roots which contain 10-20% of sucrose with an average of 13-14%. Russia is the biggest producer of sugar-beet.

G. Spices and Condiments. Spices are certain aromatic and pungent plant products used for seasoning and flavouring food and various fruit and vegetable preserves, and enhancing their taste. They are used extensively in cookery and confectionery, hot or sweet chutney, beverages and for chewing alone or with betel leaf. They are also used in medicines. India is the only country in the world that produces and exports almost all kinds of spices, particularly pepper, ginger, cardamom and turmeric, earning a foreign exchange of Rupees 70 crores or so annually. Some of the common ones are as follows:

(1) **Cardamoms** are the dried fruits (capsules) of *Elettaria cardamomum* (family *Zingiberaceae*). The plants are tall perennial herbs growing in clumps from the rhizome. Cardamom cultivation is confined to the low hills of South India: the Western Ghats Range, Karnataka (Shimoga, Kanara, Hassan and Coorg), Kerala (the Cardamom Hills and the Annamalai Hills), and some southern districts of Tamil Nadu. The main concentration is in the Cardamom Hills. Heavy rainfall and red laterite soil favour their luxuriant growth. The main picking season is September to January. After picking the fruits are dried

in the sun or by different artificial methods. They have a yellowish skin and enclose 15-20 black seeds. The latter contain an aromatic volatile oil—usually 4-6%. Seeds are universally used as an important spice. They have also medicinal use as carminative. The average annual yield of capsules is 168-280 kg. per hectare. India is the largest producer of cardamoms, followed by Ceylon and Indo-China.

(2) **Pepper** is the dried berry of pepper vine (*Piper nigrum*—family *Piperaceae*). The dried berry forms commercial black pepper and the seeds form commercial white pepper. Kerala is the principal centre of cultivation. It is also grown in Karnataka, Madras, Maharashtra, West Bengal and Assam. It is propagated by stem-cuttings. More or less 1 kg. of cured pepper is obtained from each pepper vine.

(3) **Red Pepper** or Chilli is the red pod-like fruit (berry) of *Capsicum annuum*—family *Solanaceae*. Chilli is a native of tropical America and West Indies, and was introduced into India by the Portuguese in the 17th century. The plant is a herb or undershrub and extensively cultivated in all tropical countries, and in many States of India, Maharashtra supplying the largest quantity. Chillies are pungent, stimulant, stomachic and carminative. In small doses they help secretion of saliva and gastric juice, and also induce peristaltic movements. The active principle responsible for pungency is 'capsicin' contained in the skin of the fruit. Chillies have wide uses all over the world as a condiment in raw, ripe or dried form for flavouring and giving taste to curries, chutneys, salads, etc. Dried fruits are ground into a fine powder and sold as Cayenne pepper. Extracts from chillies have many pharmaceutical uses.

(4) **Camphor** is obtained from the wood of *Cinnamomum camphora* (family *Lauraceae*), a tall tree of China, Japan and Formosa (Taiwan); planted in some gardens in India. Camphor is extracted by steam distillation from old wood cut into chips. Sometimes, as in Florida and Ceylon, young twigs and leaves are used in the same way for extraction of camphor. It has a characteristic strong but agreeable odour and is widely used in very small quantities in various food preparations, perfumery and medicines. Camphor industry is practically a monopoly of Japan. Camphor is very slightly soluble in water but readily so in alcohol and ether. It volatilizes very slowly. Synthetic (artificial) camphor is also in wide use now. It is made from pinene, a derivative of turpentine.

(5) **Cinnamon** is the dried brown or dark-brown bark peeled off from the twigs of *Cinnamomum zeylanicum*, a small tree of Ceylon. It is grown in Kanara, Mysore, Travancore and Assam in the plains as well as in the hills. Cinnamon bark contains a volatile oil, tannin, sugar and gum. It is aromatic and tastes sweet. It is extensively used for flavouring foods, various fruit and vegetable preserves and also some sweet preparations. Cinnamon oil is extracted from the bark and the leaf. It is used in combination with certain drugs as an intestinal antiseptic.

(6) **Bay Leaf** is the dried leaf of *Cinnamomum tamala*. It is a medium-sized tree growing in many parts of India, abundantly in the Khasi Hills. The leaves are widely used as a spice for flavouring various kinds of curries, some sweet preparations, chutney, fruit and vegetable preserves, and often tea infusion.

(7) **Nutmeg** (*Myristica fragrans*) is a big evergreen tree of the Moluccas; in India it grows abundantly in the Western Ghats Range. Each female tree bears a few thousand oval fruits, each measuring more or less 3 cm. in length. The fruit has a hard shell which breaks into two pieces, exposing the seed. The aril of the seed is bright-red and deeply lobed, and is the

mace of commerce, while the kernel of the seed is the nutmeg of commerce. The mace is rich in a volatile oil and is aromatic. The mace as well as the kernel is an important spice used all over India for seasoning and flavouring various curries and confectioneries. It is also used in medicine. Usually 100 seeds produce about 85 gms. of dried mace. Nutmeg contains a yellowish fat (a fixed oil) called 'nutmeg butter'. Because of its agreeable aroma it is used in perfumes, hair lotions and ointments.

(8) **Cloves** are the dried flower-buds of *Syzygium aromaticum* (family *Myrtaceae*). It is a small tree of the Moluccas, spreading from there to most tropical countries. The green colour of the flower changes to dark-brown on drying. Cloves are very aromatic and have extensive uses in curries, preserves and medicines. Clove oil is extracted from unripe fruits and leaves, and is used in certain medicines, toothache and certain toilet products. In histological work its use as a clearing reagent is universal. Cloves are grown in the Western Ghats and in Kerala but the production is far short of the demand. The main source of supply is Zanzibar, popularly called the 'Island of Cloves', off the east coast of Africa. The yield of dried cloves per plant (after about 20 years of growth) is approximately 2-2.5 kg. per year.

(9) **Ginger** is the rhizome of *Zingiber officinale* (family *Zingiberaceae*). The plant is a small erect perennial herb. Ginger is considered to be the most important of all spices and condiments, and is used all over the world, largely as a condiment. It is commonly used to give aromatic hot taste to curries and various fruit and vegetable preserves. It contains an essential oil which is responsible for the aromatic odour, and an oleoresin (called *gingerin*) which is responsible for its pungent taste. Medicinally it is stomachic, digestive and carminative.

(10) **Garlic** (*Allium sativum*—family *Liliaceae*) is a strong-smelling, whitish bulb, the smell being due to the presence of a sulphur-containing volatile oil present in all parts of the plant. The plant is a small perennial herb cultivated throughout India. Garlic is used as a condiment, particularly in fish and meat preparations, and in various fruit and vegetable preserves. It has some important medicinal properties. It is an effective remedy for high blood pressure, rheumatic and muscular pain, and for giddiness and sore eyes. It is digestive and carminative, and removes pain in the bowels. It heals intestinal and stomach ulcers, and is in fact regarded as Nature's best antiseptic for the alimentary canal. It is highly efficacious in torpid liver and dyspepsia. It is a good tonic for the lungs.

H. Medicinal (Drug) Plants. India's forests abound in medicinal herbs, shrubs and trees. It is estimated that they number over 4,000 species. Of them, 2,500 to 3,000 species are in general use in some form or other. The Eastern and Western Himalayas and the Nilgiris are known to be the natural abodes of many such plants. A good number of them are now cultivated in different States on an experimental as well as commercial basis. The Central Drug Research Institute in Lucknow is carrying on research work on indigenous medicinal plants. The Tropical School of Medicine in Calcutta has also done a good amount of work in this direction, and so also some of the big pharmaceutical works. The Medicinal Plants Committee of the Government of West Bengal have already started experimental cultivation of certain very valuable medicinal plants at Rongpo in Darjeeling district with very encouraging results so far. Some of these plants are: ipecac (*Psychotria*), *Digitalis*, *Rauwolfia*, *Erythroxylum* (cocaine-yielding), etc. Some rare species of *Podophyllum* (for cancer), *Securinega* (for poliomyelitis), *Securigera* (for general debility), *Aralia* (for vigour), etc., have also been introduced.

1. **Ipecac** (*Psychotria ipecacuanha*—family *Rubiaceae*) is a herb with closely annulated roots, cultivated at Mungpo (Darjeeling) and in the Khasi Hills. The roots are bitter in taste, and contains 3 or 4 alkaloids to the extent of 2-3%, of which *emetine* is the most important one. Formerly powdered ipecac root was used in small doses to treat amoebic dysentery. It stimulates the liver helping secretion of gastric juice but it produces local irritation. Further, emetine makes the heart weak and slow, and lowers the blood pressure. Now emetine hydrochloride is injected for the treatment of amoebic dysentery. Emetine is a highly poisonous drug.

2. **Rauwolfia** (*Rauwolfia serpentina*—family *Apocynaceae*) is an evergreen, perennial undershrub found growing extensively in the tropical Himalayas, Assam and the Western Ghats, and to some extent in West Bengal, Orissa and Bihar. The plant has been named after Leonard Rauwolf, a German botanist of the 16th century. Rauwolfia drug has extensive uses all over the world for its hypnotic and sedative properties, and is used in the treatment of insomnia, mental imbalance and insanity. It has also the property of reducing high blood pressure. The roots are cleaned, dried in the sun, and then powdered. The powder is strained through a piece of muslin, and administered in appropriate doses. It has a bitter taste. It is now known that the dried roots of *Rauwolfia* contain five alkaloids to the extent of 0.5%. Of them *serpentine, serpentinine* and *rauwolfine* are most powerful. It is on record that the properties of *Rauwolfia* were known to Ayurvedic physicians in ancient India nearly 3,000 years ago.

3. **Nux-vomica** (*Strychnos nux-vomica*—family *Loganiaceae*) is a very valuable drug. The plant is a handsome tree growing almost throughout tropical India, specially in Deccan and West Coast. The seeds, intensely bitter in taste and extremely poisonous, contain a very important alkaloid *strychnine* (and also another alkaloid *brucine*) to the extent of 0.5-1.2%. Strychnine is an effective stomachic, a tonic and a stimulant in, of course, very minute doses. It increases secretion of gastric juices, sharpens appetite and promotes digestion. It helps peristaltic movement of the intestines. It is also used in the treatment of nervous disorders. It increases mental alertness, field of vision and capacity for muscular work. It is also effective in the treatment of paralysis.

4. **Cinchona** is the famous quinine-yielding plant. The bark of a plant (unknown for a long time) was in use as a febrifuge in South America till the late 19th century. The plant yielding this bark was discovered in Peru as late as 1739 by La Condamine, and in 1742 Linnaeus named it *Cinchona* after the Countess of Chinchon, wife of a Spanish Viceroy of Peru, who was cured of an attack of malarial fever by the use of its bark in 1638. Thereafter the Jesuits popularized its use. From then on it was popularly known as 'Countess bark' or 'Jesuit's bark' and also 'Peruvian bark'. In 1858 Markham was deputed by the Secretary of State for India to explore the forests of Peru in search of this plant. After a long, laborious and hazardous journey through dense forests, Markham collected as many as 529 living plants and a quantity of seeds. The small stock that could stand the journey was sent to Kew Gardens, London, in 1861; some seeds were brought to India. In the meantime, Dr. Anderson, the then Superintendent of the Royal Botanical Gardens, Calcutta, was sent to Java and he brought back with him 412 plants and a quantity of seeds. The whole lot was sent to Ootacamund in the Nilgiri Hills in 1861 where the plants showed luxuriant growth. He brought 193 plants from Ootacamund to Calcutta in 1862 but the plantation did not prove to be a success in the plains; so the whole stock was removed to Darjeeling district where finally the plantation was established in 1864. Dr. Anderson unfortunately died of malarial fever, and the work was entrusted to his successor, Dr. King, who made the plantation and the manufacture of quinine a commercial success.

Species. Of about 40 species of *Cinchona* (family *Rubiaceae*), the quinine-yielding ones commonly cultivated are *C. ledgeriana* (yellow bark), *C. officinalis* (brown bark), *C. cordifolia*, *C. succirubra* (red bark) and several varieties and hybrids. Cinchona plants are low trees (6-15 metres high) and prefer a cool climate and well-drained soil, as in hill-slopes.

Collection of Bark. There are various methods of harvesting the bark—thinning of plantation, uprooting, lopping of branches, coppicing and shaving. Quinine and other alkaloids are formed richly in the bark of the root and the stem in the region 30 cm. below and above the ground, more in the former than in the latter, gradually decreasing and disappearing downward and upward. The bark is stripped off and dried in the sun or in a drying shed during the rainy season.

Chemistry of Bark. It was in 1820 that quinine was first extracted. About the middle of the century several alkaloids, particularly quinine, quinidine, cinchonine, cinchonidine and an amorphous alkaloid, were recognized in the bark. In 1888 a quantity of about 136 kg. of quinine was manufactured for the first time. Prior to this a crude preparation of the bark powder was used under the name 'cinchona febrifuge'. Java bark is richest in alkaloid contents, and over 90% of the world output comes from Java plantations. Bengal bark and Madras bark are rather poor in this respect. An analysis of Bengal bark gives the following figures:

	Quinine	Other alkaloids	Quinine sulphate
C. ledgeriana	5.49%	2.03%	7.38%
C. officinalis	2.77%	2.62%	3.72%
C. succirubra	1.92%	4.13%	2.75%

All the bark collected in India used to be shipped to Europe for extraction of quinine, and the manufactured product imported into this country. About the year 1875 two factories were established in India—one at Mungpo in the Darjeeling district and another at Naduvattam in the Nilgiris.

Manufacture of Quinine. (1) **Grinding.** This is done by a disintegrator (a heavy circular iron casing) fitted with heavy iron bars which are made to spin with a tremendous force—2,500 revolutions per minute. The powdered bark is then strained through a fine piece of silk in contact with a spirally wound brush. (2) **Extraction.** 136-227 kg. of powdered bark are put into each of the several rows of cylindrical iron vats, called *digestors*, fitted with a spirally-coiled steam-pipe, an oil-pipe and a water-pipe, and a mechanical stirrer. To each vat is added about 900 litres of 20% caustic soda. After continued stirring and heating for about 3½ hours, the oil released is seen to float taking up all the alkaloids, and is then skimmed off. To the alkaloid-bearing oil dilute sulphuric acid is added and the mixture agitated by jets of steam. This treatment induces the oil to give up the alkaloids to the acid solution. (3) **Purification.** The acid liquor is neutralized with caustic soda solution, and poured into long troughs where the quinine sulphate partially crystallizes out in two days' time as dirty-looking greyish pulp. This is poured into centrifugal separator which is a cylindrical copper-gauze basket lined with a piece of calico. The basket is made to revolve at a speed of 1,200 revolutions per minute. All the liquid is strained out. A greyish cake of crystals is left in the basket as crude quinine sulphate with about 10% of other alkaloids. This is dissolved in boiling water. A precipitate settles down and the supernatant liquor now contains practically pure quinine sulphate. It is collected, and finally crystallized. The crystals are dried in a drying room. The precipitate forms 'cinchona febrifuge' which nowadays is not much in demand.

5. Aconite

(*Aconitum*—family *Ranunculaceae*), *Aconitum napellus* is the European aconite, while *A. ferox* is the Indian aconite. *Aconitum ferox* is an erect perennial herb growing wild in the sub-alpine Eastern Himalayas. The plant is highly poisonous. Its tuberous roots contain a few alkaloids, of which *aconitine* is the chief active one. Aconite relieves pain due to sciatica, neuralgia, rheumatism and inflamed joints. Aconite is used as a tonic, febrifuge, antiperiodic and sedative. It is a highly poisonous drug.

6. **Deadly Nightshade** (*Atropa belladonna*—family *Solanaceae*) is a short erect herb growing wild in the temperate Western Himalayas. It is also cultivated as a medicinal plant. The leaves are collected when the plant is in the flowering stage, and dried. The dried leaves contain *atropine* and two other alkaloids to the extent of 0.3%. All parts of the plant, however, are narcotic and poisonous. Atropine is the basis of the drug *belladonna* which is used to relieve pain in neuralgia and inflammation of muscles, palpitation of the heart and pain of the cardiac muscles, and has proved to be very useful in spasms due to bronchitis, asthma and whooping cough, and in spasms of the involuntary muscles. Atropine gives great relief in all such cases. It is an excellant remedy for night sweats, as in phthisis, and is an antidote to certain types of poisoning. Atropine is used to dilate the pupil of the eye and thus to facilitate its examination.

7. **Poppy** (*Papaver somniferum*—family *Papaveraceae*) grows wild in the Himalayas, and is also cultivated in Bihar and Uttar Pradesh under official control. It yields a narcotic drug which is the latex obtained by incising unripe capsules. The latex is dried and made into balls or flattened masses which form the commercial *opium*. Opium contains about 9.5% of *morphine* and also a number of other alkaloids. Opium and morphine relieve pain and induce sleep. Opium relieves intestinal pain and cures diarrhoea and dysentery, but it lessens appetite and retards digestion. It removes the sensation of hunger, coughing, fatigue, etc. In large doses the central nervous system becomes depressed. Opium is sedative, astringent and anodyne. The action of morphine is always more definite and less harmful than opium. The use of opium in regulated doses is regarded as a panacea for all physical ailments in old age. China is the world's biggest producer of poppy.

I. **Beverages.** Beverages are mild, agreeable and stimulating liquors meant for drinking. Tea, coffee and cocoa are examples of non-alcoholic beverages.

1. **Tea.** Tea is the dried and prepared leaves and buds of *Thea* (=*Camellia*) *assamica* and *T. sinensis* and several hybrids of the family *Theaceae*. Tea infusion is now a universal drink. But from the tea garden to the tea cup there is a long history. Tea plants are kept bushy by regular pruning. If left to themselves they grow to the size of small trees, and may live for 80-100 years.

A Short History. The plant growing wild and uncultivated was first discovered in North-East Assam by Charles Alexander Bruce, an army man, in 1826. He collected seeds and plants and sent them to the Royal (now Indian) Botanical Gardens near Calcutta. They were grown with success and found to be real tea plants although distinct from the Chinese species. A Tea Committee was appointed by Lord William Bentinck in 1834 to enquire into the possibility of profitable cultivation of tea plants in India. The Committee considered the discovery of tea plants in Assam of far-reaching importance, and strongly recommended their cultivation in Assam and also elsewhere in India, particularly because China had so far monopolized the world tea trade. Tea nurseries were immediately established by the Government in Upper Assam, and seeds and plants were also sent to Darjeeling, Dehra Dun, Kangra Valley and the Nilgiri Hills. In 1835 seeds and plants were also obtained from China, and in 1836 the first tea garden was started in Upper Assam by Bruce. By 1838 a marketable quantity of tea was produced in Assam and this was sent to England where it sold for 16-36 shillings per pound. In 1839 the Assam (Tea) Company was floated in England with a capital of £500,000. By 1859 about ten estates were actually growing tea in an area of 1,700 acres. By this time tea cultivation became established in Assam, North Bengal and South India (Kerala, Mysore and Nilgiri Hills). The first public auction of tea was started in 1861. By 1874 there were 11,680 acres of land under tea cultivation. In 1881 the Indian Tea Association was formed in Calcutta, which later in 1911 had its headquarters at Tocklai Experimental Station at Jorhat (Assam) for better facilities of research work on various aspects of the culture and manufacture of tea. India is the world's foremost tea-growing country with over 10,000 gardens, big and small. The principal tea-growing States are Assam, West Bengal, Tamil Nadu and Kerala. Karnataka is a promising one. India continues to be the largest producer and exporter of tea in the world. Her annual production has now gone up to 560 m. kg., and export over 220 m. kg., earning a foreign exchange of Rs 270 crores, next only to jute. The production of tea is on the increase from year to year. More than half of the world's output of tea is produced in India. Assam monopolises over 50% of India's output, and North Bengal a quarter. In Assam the yield of tea per hectare per year is about 1,680 kg. (the maximum going up to 2,800 kg.), as against the all-India average of 1,360 kg. There are 750 tea gardens (excluding many small ones) in Assam (mostly in Upper Assam) and 350 in West Bengal. The largest buyers of Indian tea are the United Kingdom and Russia. Other countries buying Indian tea in large quantities are Netherlands, UAR, West Germany, France, Afghanistan, Iran, Iraq, Australia, Japan and Canada. Internal consumption of tea has now increased to 285 m. kg. annually. Recently Sri Lanka and East Africa are making a rapid headway towards production and export.

Tea Leaves. There are different grades of manufactured tea. The terminal bud with *two* leaves just next to it forms *fine* tea; the same with *three* leaves forms *medium* tea; and the same with *four* leaves forms *coarse* tea. The average yield of manufactured tea in India has now gone upto 1,360 kg. per hectare per year, and 1.8 kg. of green leaves usually make 0.45 kg. of cured tea. The yield of green leaves per plant is more or less 0.9 kg. Tea plants grow in the plains and in the hills up to an altitude of over 2,100 metres (better tea always at a high elevation) and flourish in localities with abundant rainfall. Darjeeling tea is considered to be the world's best quality tea for its very agreeable taste and flavour, and fetches a record price from both foreign and inland markets (recently a maximum of Rs. 450 per kg. on auction for export). But the production of Darjeeling tea per hectare and as a whole is very low. Annual pruning after plucking (and the same operation even after the second year of planting) is a very important practice helping the plant to 'flush' profusely. Regular picking of leaves usually begins from the 7th year, sometimes earlier. Economic life of tea bushes continues for 50-70 years. Tea bushes require proper irrigation and adequate manuring for increased yield. Chemical fertilizer like sulphate of ammonia at the rate of 110 kg. per hectare, green manure (growing certain leguminous plants like *Tephrosia, Derris, Sesbania, Cajanus,* etc.), cattle manure and leaf compost are very beneficial to them.

Chemistry of Tea. Manufactured tea contains 4-5% of tannins (catechins) which are responsible for colour and strength of the infusion, 3.3-4.7% of caffeine which is a stimulant for the heart, a little volatile oil to which the aroma of the tea is due, about 8% of resinous matter which gives the reddish-brown colour to tea infusion, etc. Green tea leaves contain 13-18% of tannins but greater portions of them are converted into sugar and gallic acid during the process of manufacture. Starch is also converted into sugar during the same process. Caffeine distribution in tea leaves is as follows: 1st leaf and bud—4.7%, 2nd leaf—4.5%, 3rd leaf—3.7%, 4th leaf—3.3%. Caffeine contents are not changed during the process of manufacture. The production of caffeine from tea waste and coffee seeds at the Regional Research Laboratory at Jorhat (Assam) by a new process is under way.

Manufacturing Process. (1) **Plucking.** The quality of cured tea depends to a great extent on the standard of plucking, the proportion of desirable constituents gradually diminishing from the bud to the lower leaves. Plucking season is March to mid-December. (2) **Withering.** After weighing the leaves are taken to the withering house and spread out thinly on bamboo racks for about 18 hours (by employing heated air it has been possible to reduce the period to 2 to 3 hours). (3) **Rolling.** The withered leaves, sufficiently flaccid in the previous process, are passed through the rolling machine for half an hour. Rolling imparts twists to the leaves. Major chemical changes also take place in this process, and fermentation begins here. (4) **Sifting.** After the first rolling the sifting is done. Finer meal is taken to the fermentation room, while a second rolling is done with the coarser portion. (5) **Fermentation.** Sifted leaves are then spread on the floor of the fermentation room where the temperature is maintained at more or less 27°C. for 3 or 4 hours. During fermentation (actually enzymic oxidation, not bacterial) the flavour and colour develop. (6) **Firing.** This takes place usually in two stages—the first one at 93°C. and the second one at 82°C. Firing reduces moisture contents to 3-4%, thus ensuring better keeping quality. Firing also arrests further process of fermentation, and in this process the quality of tea further improves. This is *black tea.* *Green tea* (largely used in China and Japan) is unfermented tea made by steaming the leaves and then rolling and drying them immediately. It tastes bitter but its use is on the increase because it is definitely known to lower chloesterol level and prevent heart attacks. In India the best green tea is manufactured in Upper Assam and North Bengal, and to some extent in Himachal Pradesh (Kangra Valley) and Uttar Pradesh, and mostly exported to Afganistan and some other countries. Steps are being taken to produce more green tea in India for bigger export.

(7) **Sorting and Grading.** The fired tea is then sorted into grades by automatic devices. Finally it is packed into plywood chests (specially made for the purpose) for marketing.

2. **Coffee.** Coffee is a favourite drink in South India, and it is gradually becoming popular all over India. It is regarded as a wholesome and refreshing drink. Seeds of *Coffea arabica* and *C. robusta* (family *Rubiaceae*), particularly the former, are the sources of coffee. Over 70% of coffee plantations in India consist of *C. arabica*. It is noted for its quality and has already become popular, while *C. robusta* yields more and is resistant to pests and diseases. The seeds are roasted to a desired brown colour and then powdered. The aroma of coffee powder develops on proper and skilful roasting. Certain chemicals are also added for this purpose. Roasted coffee seeds contain 0.75-1.5% of caffeine, several vitamins, and a little volatile oil. A coffee bush, 5-6 m. high, bears abundant red fruits, each with two seeds, and usually yields 0.45-0.9 kg. of cured coffee. It prefers hill slopes with abundant rainfall. Main coffee plantations are in the low hills of South India—Karnataka (particularly Coorg), Kerala and Tamil Nadu It is also cultivated on a smaller scale in Bihar and Orissa. In Assam coffee plantation started in 1954 in the Mikir Hills and North Cachar Hills has proved to be a success, and extension is under way. India's coffee production has now gone up to about one lakh tonnes annually. More than half of this quantity is exported, while the balance is consumed internally. India ranks third in the world in her coffee production. The coffee plant, a native of Abyssinia, was first introduced into India by a Muslim pilgrim more than 250 years ago. Regular cultivation of coffee, however, dates from 1830. Brazil and Kenya are the world's largest suppliers of coffee. America consumes the largest quantity of coffee, well over half the world's supply.

3. **Cocoa.** Cocoa makes a refreshing and nourishing drink. It is prepared from the seeds of *Theobroma cacao* (family *Sterculiaceae*), a native of tropical America. It is a small tree cultivated more or less extensively in tropical America, West Indies, Brazil, Ghana and Kenya. The world's supply comes mainly from Brazil, Kenya and Ghana (Ghana supplying the largest quantity). Cocoa is also cultivated in Java and Ceylon. Each tree bears commonly 70 to 80 fruits, each measuring 15-22×7-10 cm. and bearing numerous seeds. Fruits are cut or broken open, and the seeds dried, roasted and powdered. In addition to its use as a drink, cocoa powder is used in the making of chocolate with certain ingredients such as sugar, spices and sometimes milk (milk chocolate) mixed with it. With protein and fat contents cocoa is also a food. Cocoa seeds contain theobromine and caffeine (1% or less), proteins (15%), starch (15%) and fatty oil

(30-50%). On an average 50 pods, each having about 30 good seeds, yield over 1 kg. of cured cocoa. Cocoa-butter, an opaque solid oil expressed from warmed seeds, is used in medicine.

J. Timber Trees. Timber is the wood (heartwood) used for various building purposes: houses, boats, bridges, ships, etc.; for making furniture, packing boxes, matchsticks and boxes; for making plywood tea chests, flush doors, partitions, walls, ceiling, shelves, cabinets, prefabricated houses, commercial boards, etc.; and for railway sleepers. In addition, wood chips and shavings are used for making compressed wood which is in demand for panelled doors, table tops, room partitions, hard blocks, etc. Timber and firewood (fuel) together with many useful forest products constitute the forest wealth of a country. To be self-sufficient in them a country should normally have about one-third of the total land area under forests. In this respect India having only 22.7% lags behind other countries. In the face of this we are consuming every year about 240 million cubic metres of wood as fuel, and over 200 million cubic metres as timber. According to an estimate India has already lost nearly 5 million hectares of virgin forests over the past 25 years. The present rate of deforestation is a warning to the future need of wood in the country. Systematic afforestation is the only means of maintaining a balance between loss and gain. In India Madhya Pradesh has now the largest forest area, while Assam occupies the second place. The timber trees of Indian forests number over 75 species. The quality of timber depends on its hardness, strength, weight, presence of natural preservatives like tannin, resin, etc., durability against heat, moisture and insect attack, workability, grains, colour, porosity and capacity to take polish and varnish.

1. **Teak** (*Tectona grandis*—family *Verbenaceae*) is the famous timber of the Deccan Plateau, Madras, Kerala, Maharashtra, Bihar and Orissa. It is also being successfully grown in Assam. The tree is about 36 metres in height. Teak yields a very valuable timber with straight grains and light golden-brown colour. The wood is hard, strong, moderately heavy (720 kg. per cu. m.) and extremely durable, being immune to insect and fungal attacks because of the presence of an oil and a resinous matter in it. It does not warp, shrink or expand. This timber is used for making handsome furniture of various designs. It is also used extensively for doors, windows, beams, rafters, staircases, etc. It is also used for carving. Teak is, however, a very costly wood. Burma teak is the best. The Deccan Plateau produces the best Indian teak.

2. **Indian Redwood**. (*Dalbergia sissoo*—family *Papilionaceae*) is a tree of sub-Himalayan forests extending from Assam to Punjab. It is a very valuable timber with fine to medium grains

and golden-brown to dark-brown colour. It is hard (harder than teak), strong, very durable and moderately heavy (800-850 kg. per cu. m). It makes handsome furniture. It is easy to work and takes a good polish, and is least susceptible to white ants and borers. The timber is also used for posts, rafters and boards. It makes durable carts, coaches and boats. The wood is widely used for carving.

3. **Sal** (*Shorea robusta*—family *Dipterocarpaceae*) is a very valuable timber tree, growing to a height of 20-25 metres, sometimes 30 metres or even more. The SAL tract stretches along the sub-Himalayan region from Assam to Punjab. Tracts of SAL also occur in Madhya Pradesh, Bihar, Orissa and Andhra. The timber is very hard, strong, heavy (900 kg. per cu. m.) and very durable even under water. It is used extensively for railway sleepers and for house-building as posts, rafters, planks, door- and window-frames, etc., and for construction of bridges and piles. It is also used for hubs of wheels and bottoms of carts and carriages. It stands heat and water well. *Shorea assamica* of Upper Assam and Nagaland is not very durable as a railway sleeper but it has variety of other uses. It is suitable for doors, windows, planking and tea chests, and makes good plywood.

4. **Jarul** (*Lagerstroemia speciosa* = *L. flos-reginae*—family *Lythraceae*) is a good timber tree common throughout India, particularly in Assam and the West Coast. It grows mainly along banks of rivers and is also cultivated for its ornamental mauve-purple flowers. The timber is hard, durable (even under water), straight-grained, and takes a good polish. The wood is light to moderately heavy (640 kg. per cu. m.) and pale red. It is used for house-building, boat-building, furniture- and cart-making, posts, bridge-piles and bridges. It is also good for general uses.

5. **Mahogany** (*Swietenia mahagoni*—family *Meliaceae*) is an evergreen tall tree (18-21 metres high, sometimes even 30 metres). It is a valuable timber tree. Mahogany is an indigenous tree of Central America. In India it is found to grow in the Deccan Plateau; it is also planted in gardens and forests, and alongside roads sporadically. The wood is very hard and durable, coarse-grained and takes a good polish. It is dark reddish-brown in colour, and is used for making furniture, boats, ships, caskets and the body of some musical instruments.

6. **Pines** (*Pinus longifolia* and *P. khasya*—family *Abietaceae*) are evergreen, tall, straight, coniferous trees. They attain a height of 30-45 metres, and grow abundantly and gregariously at an altitude of 900 to 1,900 metres or even higher. *Pinus khasya* grows in the Khasi Hills and *P. longifolia* in the Western Himalayas. The wood is light (530-610 kg. per cu. m.), moderately hard, easy to work, white to pale brown in colour, with straight

but uneven grains, numerous dark-coloured resin-ducts and large knots, and is odorous. It seasons well and takes a fairly good polish, and is durable if not exposed. Pine is used extensively in the hills for house-building and furniture-making, and for packing cases. The timber is not susceptible to the attack of white ants.

7. **Deodar** (*Cedrus deodara*—family *Abietaceae*) is an evergreen, elegant-looking, cone-shaped, coniferous tree attaining a height of 30-60 metres. It grows at an altitude of 1,800 to 2,500 metres. It is a well-known timber tree of the Western Himalayas and grows plentifully in Kashmir. The wood is light (560 kg. per cu. m.), moderately hard, extremely durable and seasons well. It is yellowish-brown in colour and odorous. It is easy to work and finishes well but is not suitable for fine work because of continuous oozing of resin. This timber is used for sleepers after proper treatment, for house-building as beams, rafters and flooring, bridges and light furniture. It is almost immune to white ants. The wood is a source of deodar oil.

Some Common and Useful Timber Trees of Assam. In addition to SAL, AJAR or JARUL, SISSOO and teak, as already described, the following may be mentioned: (1) BONSUM (*Phoebe attenuata* and *P. goalparensis*—family *Lauraceae*)—timber very valuable, used extensively for furniture, planks, doors and windows; (2) GAMHAR (*Gmelina arborea*—family *Verbenaceae*)—wood strong and durable, a good timber for furniture, doors and windows; (3) POMA or toon (*Cedrela toona*—family *Meliaceae*)—wood soft but much used for furniture, doors and windows, carriage-building, tea chests and panelling; (4) BOGA-POMA (*Chickrassia tabularis*—family *Meliaceae*)—wood hard, suitable for planking and furniture; (5) AMARI (*Amoora wallichii*—family *Meliaceae*)—wood hard, used for furniture, doors and windows; (6) LALI (*Dysoxylum procerum*—family *Meliaceae*)—wood bright red, moderately hard, much used for doors and windows; (7) GONSOROI (*Cinnamomum glanduliferum*—family *Lauraceae*)—wood soft but durable, scented; it makes fairly strong furniture, cupboards and boxes and is somewhat better than POMA; (8) KHOKAN (*Duabanga sonneratioides*—family *Lythraceae*)—wood soft, used for cheap furniture, suitable for plywood; (9) TITASOPA (*Michelia champaca* and *Talauma phellocarpa*—family *Magnoliaceae*)—wood light but durable, much used for furniture; (10) NAHOR (*Mesua ferrea*—family *Guttiferae*)—wood very hard and heavy, used for posts, beams, bridge-piles and railway sleepers.

K. Fibres.

Fibres are thread-like tissues obtained from different parts of the plant body. They are mostly made of sclerenchymatous cells, strongly lignified and thickened. Cotton fibres are, however, made of cellulose. Commercial vegetable fibres may be classified as (*a*) floss fibres or lint which are the hairy outgrowths of the seed, e.g. cotton, silk-cotton and madar; (*b*) bast fibres which are the sclerenchymatous tissues of the secondary phloem or bast, e.g. jute, hemp and rhea; (*c*) coir fibres which are the fibrous husk of coconut fruits; and (*d*) leaf fibres which are the sclerenchymatous tissues of the leaf, e.g. bowstring hemp

(*Sansevieria*) and American aloe (*Agave*). The quality of fibres depends on their length, strength, fineness, lustre, reaction to high temperature and water, etc. Some of the important commercial fibres are as follows:

1. **Cotton** (*Gossypium sp.*; see p. 693; family *Malvaceae*) is the most important commercial textile fibres, spun into yarn and woven into various kinds of garments, screens, sheets, canopies, sails, carpets, etc. Fibres are also used for making ropes, twines and threads. Raw cotton is used for stuffing pillows and cushions. Mercerized cotton (treated with caustic soda) is used for finer garments. Pure cotton, properly treated, is used for surgical bandages. Cotton thread is universally used for sewing and stitching. Cotton cultivation and weaving cloth date as far back as 1,800 B.C. The quality of cotton fibres is judged by their length, strength, fineness and silkiness. Indian cottons are poor in respect of length, having a short staple—12.7-25.4 mm.; *G. indicum* and Upland American cotton (*G. hirsutum*) have a lint length of 25.4 mm.; while Egyptian cotton (*G. peruvianum*) 31.7-38 mm. and American cotton (sea island cotton—*G. barbadense*) 38-50.8 mm. Upland American cotton, naturalized in India, is extensively cultivated in India. *G. arboreum*, a perennial tree cotton, is grown sporadically in India; while *G. herbaceum*, an annual shrub, has been grown in India from time immemorial. Of all Indian cottons Broach cotton (*G. herbaceum*) of Gujarat is the finest. The cultivation of long-staple foreign cottons has not yet proved to be a success in India. Recently Punjab Agricultural University has evolved a new strain of cotton, G 27, which yields 50% more than the existing varieties. Although the total area under cotton in India is the largest in the world, her total output is far below that of other cotton-producing countries, and she is still below her expected or target production. The average annual yield of cotton lint (and not seed cotton) in India is only about 97 kg. per hectare; while in U.S.A. it is 177 kg., in Japan 203 kg., in Egypt 416 kg., and in U.S.S.R. about 272 kg. The percentage of lint to seed cotton usually varies from 25-30. Of the total output (6.5 million bales) of cotton in India long-staple cotton comprises only about 7%. Thus there is ample scope for improvement of cotton in India. Maharashtra, Gujarat, Madhya Pradesh, Tamil Nadu, Karnataka, Andhra, Uttar Pradesh, Punjab and Rajasthan are the important cotton-growing States of India. The indigenous cottons of India like *G. herbaceum* yield 3-10 quintals of seed-cotton per hectare per year; while Upland American cotton (*G. hirsutum*) and hybrid cottons, when properly irrigated, yield 25-30 quintals, sometimes up to 40 quintals. Black soil is most suitable for cotton cultivation. It may be noted that India's first cotton mill

was established at Howrah (West Bengal) in 1832.

2. **Jute** (*Corchorus capsularis* and *C. olitorius*—family *Tiliaceae*) is a very valuable bast fibre obtained almost exclusively from the above two species. Jute is cultivated widely in the low-lying areas of West Bengal (mainly), Assam, Bihar and Orissa, and to some extent only in Uttar Pradesh, Meghalaya and Tripura. The cultivation of jute in Assam is rapidly on the increase. It is extensively cultivated in Bangladesh. Production of jute in Assam has increased immensely so much so that it has now taken a leading position. There is record of jute cultivation in West Bengal and Bangladesh in the early 19th century and of the use of fibres in making gunny bags and coarse cloth. With increasing demand for gunny bags for packing foodgrains, etc., the cultivation of jute was rapidly extended, and from the middle of the 19th century several jute mills sprang up in and around Calcutta. The plant thrives under conditions of plenty of rain and flooding at a later stage. The sowing season is March or a little later, and the harvesting season is July-September. Fibres mature with the ripening of fruits. After harvesting the jute plants are retted in water for 10 to 15 days, sometimes more, and then the fibres are stripped off the stalks by hand. The fibres are then washed, dried in the sun and finally baled. The annual yield in India usually varies from 823 to 1,646 kg. per hectare. With the extension of cultivation the annual production of jute has gone up to 64.54 lakh bales of 180 kg. each (West Bengal—41.23 lakh bales, Assam—8.99, Bihar—8.31, Orissa—4.39, and the rest—1.62. Recently the Jute Research Institute at Barrackpore (near Calcutta) has evolved a new type of jute plant—J.R. 524—which yields 25-32 quintals of fibres per hectare per year. Jute is the major foreign exchange earner for India. Jute fibres are used extensively for making gunny bags, cheap rugs, carpets, cordage, hessian (coarse cloth), curtains, etc., but they are much less strong than hemp. India's first jute mill to manufacture the above goods was established at Rishra (West Bengal) in 1854. There are now 69 jute mills in and around Calcutta and 1 in Assam (recently established).

3. **Hemp** obtained from GANJA plant (*Cannabis sativa*—family *Cannabinaceae*) is the true hemp. But this is an excisable plant and its cultivation is restricted. Its fibres are, however, very strong and durable. Commercial hemp is the sunn hemp or Indian hemp (*Crotalaria juncea*—family *Papilionaceae*). Sunn hemp yields very strong fibres used for various kinds of cordage—ropes, twine, fishing nets, etc., coarse sheets, tents, screens, sacks, cigarette paper, tissue paper, etc. Sunn hemp is much

stronger than jute and stands water well. It is also used for making strong paper. It is cultivated on a large scale in Uttar Pradesh and also in Bihar and Central India. Elsewhere it is cultivated as a green manure crop. It is a monsoon crop requiring about four months to mature and is harvested in August-September before pod-formation when the plants usually grow to a height of about 2.5 metres. The plants are cut or pulled up and steeped in water for a week or so for complete retting. Fibres are then pulled out in strips, beaten and washed in water when clean fibres are obtained. Sunn hemp is always grown thickly to ensure tallness free from branches. The average yield of fibres is a little over 448 kg. per hectare per year.

4. **Rhea** or **Ramie** (*Boehmeria nivea*—family *Urticaceae*), a native of China and Japan, is a perennial shrub yielding very good fibres. In India its cultivation (since the middle of the 19th century or earlier) has remained mostly restricted to small plots by fishermen. Recently, however, because of its importance as a fibre crop, an extension of cultivation is under way in Assam, North Bengal and Bihar. The fibres of this plant are known to be the longest, strongest, silkiest and most durable. But they do not yield to dyeing easily. Commonly the fibres are spun into threads, strings, cords, ropes, belts, nets, and more particularly fishing lines and fishing nets. They are also woven into cloths, sails, laces, sheetings, parachutes, hosepipes, mosquito nets, banknote paper, etc. Fibres adhere to the inner bark of the stem and, therefore, their extraction is a troublesome job. The plant stands 3 or 4 cuttings a year from the second year only, each shoot being 1.5 m. in length. The yield of dry fibres from the stem is only 2.5%. The yield per hectare per year in India is 150-160 kg., which is very low compared to jute. Under suitable conditions the yield may be much higher. In Malay plantation the yield is 320 kg., and in China it is over 222 kg. The plant prefers a rich sandy loam in highlands. It is propagated by root-cuttings and stem-cuttings, and also from seeds.

5. **Flax** is the fibre of linseed plant (*Linum usitatissimum*—family *Linaceae*). The fibres are fine, very strong and silky but they are rather short. They are woven into various kinds of valuable fine textiles or linen cloths, mixed or unmixed with cotton. The stem yields fibres, while the seeds yield linseed oil. The same plant cannot, however, be used for both the purposes. If required for fibres the seeds are sown thickly so that the plants may grow erect and unbranched, and the crop is also cut earlier. Fibres are extracted from the stems by retting them in water for 3 or 4 days, and then beating them on a board. This crop is raised in India mainly for oil, and the main areas of its cultiva-

tion are in Madhya Pradesh, Uttar Pradesh, Maharashtra, Bihar and West Bengal. The stems yield about 15% of fibres.

6. **Coir** is the husk fibre obtained from dry fruits of coconut (*Cocos nucifera*—family *Palmae*). The fibres are short, coarse and rough but very durable and resistant to water, and used for making door mats, mattings (to cover floors), mattresses, carpets, rugs, etc. They are also used for making coarse brushes, cords and ropes, and also for stuffing sofas and carriage seats. Kerala leads in the production of coconuts and in the manufacture of coir goods in India, being only second to the Philippines in this respect. Kerala contributes the maximum quantity for export to foreign countries. Other States like Mysore, Madras, Andhra, Orissa and Maharashtra have also developed this industry. The annual production of coir fibres in India is estimated to be 130,000 tonnes. Ceylon is another big centre of fibre production.

L. Rubber is obtained from the latex of *Hevea brasiliensis* (family *Euphorbiaceae*), a big tree, which is the main source of commercial rubber. The tree comes into production 7 or 8 years after planting; maximum production is usually obtained in about the 15th year. The latex is collected by tapping the bark. It is then allowed to coagulate with the addition of water and a little acetic acid. The coagulated mass (rubber) is then separated from the liquid portion, washed and dried in the smoke-house. It is then passed through rollers and pressed into blocks, sheets, crepe, etc. The use of rubber as tyres and tubes on the wheels of various types of vehicles, crepe soles, rubber shoes, rubber sheets, rubber-tubings, beltings, insulation of electric wires and various other goods of commercial importance is of course well known. Indian rubber is mostly consumed internally in the country. The majority of rubber plantations are in Kerala; the rest are in Tamil Nadu and Karnataka. The experimental rubber plantation in the Mikir Hills (Assam) and in the Garo Hills (Meghalaya) has yielded encouraging results. Kerala accounts for over 90% of the total Indian output which has gone up to 81,953 tonnes per year. But the demand is on the increase. The average yield of rubber in India is rather low, being only about 336 kg. per hectare per year. The rubber obtained from *Hevea brasiliensis* is called **para-rubber**, that from *Manihot glaziovii* **ceara-rubber**, that from *Castilloa elastica* **panama-rubber**, that from *Ficus elastica* **india-rubber** Synthetic rubber is gradually coming into use, with an annual production of over 30,000 tonnes by Synthetics and Chemicals Limited at Bareilly—the only unit in India.

M. Paper. The importance of paper for various essential purposes cannot be overestimated. Printing paper, writing paper, newsprint, wrapping paper, cardboard, poster paper, etc., are some of the items requiring an enormous quantity of different grades of paper. In India the total production of paper in various Indian paper mills cannot, however, meet her demand. Paper mills and pulp mills may be separate or both may be integrated.

An Early History. More than 5,000 years ago the Egyptians first produced a kind of paper from paper-reed (*Cyperus papyrus*—family *Cyperaceae*), a riverside plant abundant on the banks of the Nile. The stem of this plant split into thin strips was pressed into stiff sheets which were then used as a writing material. Much later, about 2,000 years ago, in Asia Minor animal skins were specially treated to make a sort of writing paper (parchment paper). Possibly about this time the Chinese began to make paper by boiling rags, rice, straw and bark of paper-mulberry (*Broussonetia papyrifera*) and stems of certain plants into pulp and finally beating it into sheets. This was the beginning of manufacture of the modern type of paper. In 751 A.D. the Arabs got the secret of paper-making from some Chinese prisoners who were skilled paper-makers, and founded a paper mill at Baghdad. In the 9th century paper was largely used for writing Arabic manuscripts. Within a few hundred years the art of paper-making spread to Europe—to Spain in the 12th century, to France in the 14th century, and to England in the 15th century. Paper was, however, made available to the world some time in the middle of the 18th century. In ancient India the foliated bark of BHURJJAPATRA (*Betula utilis*—family *Betulaceae*), which is easily separable into thin large white sheets, was used as a writing material. The plant grows at an altitude of 3,350-3,960 metres both in the Western and the Eastern Himalayas. The first paper mill was established in India in the year 1820 (see below).

Raw Material. Cellulose is the basic constituent of paper, and the various raw materials used for paper pulp are: wood of coniferous and other trees, different kinds of bamboo (*Bambusa, Melocanna, Dendrocalamus*, etc.), various grasses such as saboi (*Ischaemum*), *Imperata, Erianthus, Phragmites*, etc. In addition, waste paper, cotton and linen rags, straws, etc., are also used extensively. Bagasse (see p. 823) is also used widely. Fir and spruce of the Himalayas are considered very suitable for quality newsprint. Lignin and other non-cellulose components of raw materials are removed by cooking and bleaching. Finally it is the cellulose that makes paper.

Paper Mills and Production. In India the first paper mill was started at Serampore by Dr William Carey, a missionary, in the year 1820 but this venture ended in failure. Between 1867 and 1891, 5 paper mills were established in India—3 in West Bengal, 1 in Lucknow, and 1 in Pune. Recently, however, with the establishment of 56 paper mills a considerable progress has been made in the production of writing and printing paper. Still India's total annual production of 1,000,000 tonnes falls far short of her estimated demand for 1,150,000 tonnes. The position of newsprint is also far from satisfactory with the total production of about 40,000 tonnes a year (or about 109 tonnes daily) by the only mill in India at Nepanagar in Madhya Pradesh, while the country's minimum requirement at present is more or less 225,000 tonnes annually (or 616 tonnes daily). Evidently a huge quantity of newsprint has to be imported from foreign countries at an enormous cost of foreign exchange. Self-sufficiency in this commodity is thus the need of the day. To meet it the production at NEPA factory is being stepped up to 75,000 tonnes. Besides, four big new projects have been taken up for the exclusive production of newsprint. It may be noted that the average *per capita* consumption of paper in India is only about 0.64 kg., while it is 68 kg., in the United Kingdom, 79 kg. in Canada, and over 136 kg. in the U.S.A.

Pulp-Making. Raw material cut into small chips is passed through a series of screens of various meshes to obtain uniformity of size. It is then cooked in a huge quantity of water. For chemical pulp required for different grades of paper three processes are adopted: (*a*) *sulphate process*—sodium sulphate is used for this purpose; this method is increasingly used; (*b*) *soda process*—caustic soda is used in this process; and (*c*) *sulphite process*—calcium sulphite is used. Mechanical pulping is done by grinding the raw material with the addition of sufficient water. Screening the pulp to remove indigested particles and bleaching it to get white paper, usually with hypochlorites, liquid chlorine, milk of lime, sodium peroxide, etc., are important processes. Rags and waste papers, when used, are first boiled with lime and caustic soda. Different grades of paper are finally made according to choice and need by mixtures of different pulps in particular proportions. Different chemicals such as rosin, paraffin, wax, alum, sodium aluminate, etc., are used to give the paper a particular finish and to make it non-absorbent to liquids. The pulp is concentrated to 70% solids or even more before sending it to the mill.

Paper-Making. The dewatered pulp is passed through a series of roll-type presses. At the outgoing end the sheet is passed through 'driers' in the form of heated cylinders. The dried sheet is then passed through highly polished rolls known as 'calenders' to give the paper a polished surface. To prepare good quality printing paper preventing 'show-through' or 'strike-through' and to improve whiteness, smoothness and fineness certain materials like China clay, precipitated chalk, titanium dioxide, zinc sulphate, talc, barium sulphate, etc., called 'fillers' are used. The forward-moving sheet which is continuous is wound into large rolls, and finally cut into sizes.

Part X PALAEOBOTANY

Chapter 1 GENERAL DESCRIPTION

Palaeobotany deals with the study of fossil plants preserved in rocks of various geological periods. A fossil (*fossilis*, dug out) is any relic or trace of past life (plant or animal) preserved in the earth's crust during its formation in different ages and periods. The term was formerly used to refer to anything dug out of the earth, but is now used to designate any tangible evidence of former life embedded in the earth and preserved in some form or other. During the period between the cooling of the earth and the present day, the earth's crust has experienced several revolutions involving widespread changes in its topography, viz. redistribution of land and water, elevation of submerged land, submergence of elevated land, sedimentation of fragmentary materials and organic remains at the bottom of lakes and oceans, etc. The sediments gradually became transformed into rocks (sedimentary rocks) with plant and animal remains in them preserved in the form of fossils. These sedimentary rocks have been divided into different geological periods on the basis of their fossil contents (see table on p. 853). Only certain parts of plants are resistant to decay and these, when properly buried in muds and sands, become transformed into fossils in consolidated sediments such as shales and sandstones. Pteridophytes and gymnosperms have been found in large numbers in a fossil state, while bryophytes, algae and fungi having delicate parts are seldom encountered as fossils. Palaeontology dealing with plant and animal fossils gives us a glimpse of the occurrence and nature of ancient life—flora and fauna—as they existed in past geological ages. It precisely tells us about the period of the earth's history when particular types of plants and animals came into existence, flourished and became extinct, and also their geographical extent. The study of fossils is thus of utmost importance in tracing the evolutionary sequence

---

It may be noted that Birbal Sahni (1891-1949), a student of A. C. Seward, a world-renowned palaeobotanist of Cambridge, was the fountain-head of palaeobotanical work in India. Great advances were made by him in this field of research during the years 1932-48. The Birbal Sahni Institute of Palaeobotany at Lucknow established by him in 1946 is the centre of palaeobotanical research in this country. Sahni died in 1949 leaving a number of enthusiastic workers to continue the work at this Institute. It may further be noted in this connexion that important fossiliferous regions of India lie in Assam, Bihar, Gujarat, South India and Kashmir, which evidently require further exploration.

of flora and fauna—appearing, disappearing, and giving rise to more organized forms in successive stages. Palaeontology is directly correlated with the stratigraphy of the earth, i.e. the formation of the earth's strata in different periods. It has, therefore, been possible to determine the age of the particular strata from the occurrence of fossils in them. The carbon-dating method of fossils for fixing their ages is fast developing. Fossils also tell us about the extent of lands, lakes and seas of the past ages. It is thus known that a vast sea called the Tethys existed in the region of the present-day Himalayas[1], possibly with some land bridges across this sea. Palaeontology has its economic application in the exploration of minerals, specially coal, occurring in freshwater sedimentary formations (in the Carboniferous), and oil (petroleum) possibly derived from marine planktonic flora and fauna (in the Eocene). It is known that an ancient (Carboniferous) flora played an important part in the formation of coal. In India the coal seams of Raniganj and Jharia are of the Permian age. Palynology dealing with fossil spores and pollen grains is of great value in determining and correlating coal seams and sedimentary beds of both freshwater and marine origin.

Formation of Fossils. Two major factors are involved in the mode of preservation of plant and animal bodies in the form of fossils: rapidity of burial and prevention of normal decay. A combination of these two factors frequently occurs such as burial in stagnant water, complete burial under fine-grained sediment, or rapid infiltration of mineral substances into the cell-walls, in any of which the quantity of available oxygen is diminished.

Kinds of Fossil Plants. (1) **Petrifaction** (*petra*, rock: *facere*, to make). This means fossilization by cell to cell replacement of certain plant parts by a good number of mineral substances, of which carbonates of calcium and magnesium, iron sulphide, and silica are most common. Petrified fossils have shown the external form, internal structure, and sometimes substance of the original plant, often in great detail. It happened that before vertical pressure came into play, plant fragments became saturated with water containing mineral substances in solution. The mineral substances infiltrated into the plant body, and gradually separated out from the solution. In due course the water was also expelled. Finally the tissues and cells had a complete filling of solid material, and the whole formed a solid, incompressible, hard

---

[1] It may be noted that the upheaval of the Himalayas in the region of the Tethys took place somewhat later in two phases—one in the Oligocene (about 35 million years ago) and another in the Pliocene (about 12 million years ago)—evidently very young in age in geological time scale.

mass. Coal balls and silicified wood are the best examples of petrification. Coal balls remain embedded in the coal and are of varying sizes, usually about the size of potatoes. They are often very rich in calcified remains of plant materials. Silicified stumps of wood have often been so well preserved that it has been possible to prepare thin sections of them for microscopic examination. They often reveal minute structures in extraordinary detail. (2) **Incrustation** or **Cast**. This is a fossil with the external form as a cast. The internal structure is not preserved. Here the plant substances have disappeared and a cavity has been left; this cavity is subsequently filled up with mineral matter which thus forms a *cast* of the original plant. The surrounding material, the mould, forms the *incrustation*. Casts of pith cavities of hollow stems are sometimes found which have resulted from the entry of fine sand or mud into the hollow stems. In course of time the filling material is converted into an internal mould of the hollow stem, e.g. the pith cast of *Calamites*. (3) **Compression**. The external form of the plant is modified by vertical pressure of the sediment in which the plant material lies embedded. When a plant is subjected to compression, some of its parts—leaves, seeds, fruits, trunks, etc.,—leave impressions on the rock surface. The compression shows the outline of parts of plants. (4) **Compactions**. These are plants or plant fragments compressed by vertical pressure. Masses of plant fragments without intervening matrix such as are found in peat and coal are large-scale compactions. (5) **Impressions**. The forms impressed on a matrix, as on coal and shale, which harden afterwards, are usually termed impressions. The external features of plant parts are thus preserved.

Chapter 2   FOSSIL PLANTS

CLASSIFICATION OF FOSSILS
PTERIDOPHYTES AND GYMNOSPERMS

Pteridophytes

(A) Psilophytopsida (or Psilophytinae). **Order 1.** Psilophytales, e.g. *Psilophyton, Horneophyton, Rhynia*, etc.

(B) Lycopsida (or Lycopodinae). **Order 2.** Lepidodendrales, e.g. *Lepidodendron, Sigillaria*, etc.

(C) Sphenopsida (or Equisetinae). **Order 3.** Hyeniales, e.g. *Hyenia, Calamophyton*, etc. **Order 4.** Sphenophyllales, e.g. *Sphenophyllum.* **Order 5.** Equisetales, e.g. *Equisetites.* **Order 6.** Calamitales, e.g. *Calamites.*

FOSSIL PLANTS

(D) Pteropsida or Filicopsida (or Filicinae). **Order 7.** Coenopteridales, e.g. *Etapteris* and *Botryopteris*. **Order 8.** Marattiales, e.g. *Danaeopsis, Marattia*, and also some other orders.

Gymnosperms

Order 1. Cycadofilicales (or Pteridospermales), e.g. *Glossopteris, Lyginopteris,* etc.

Order 2. Cordaitales, e.g. *Cordaites*.

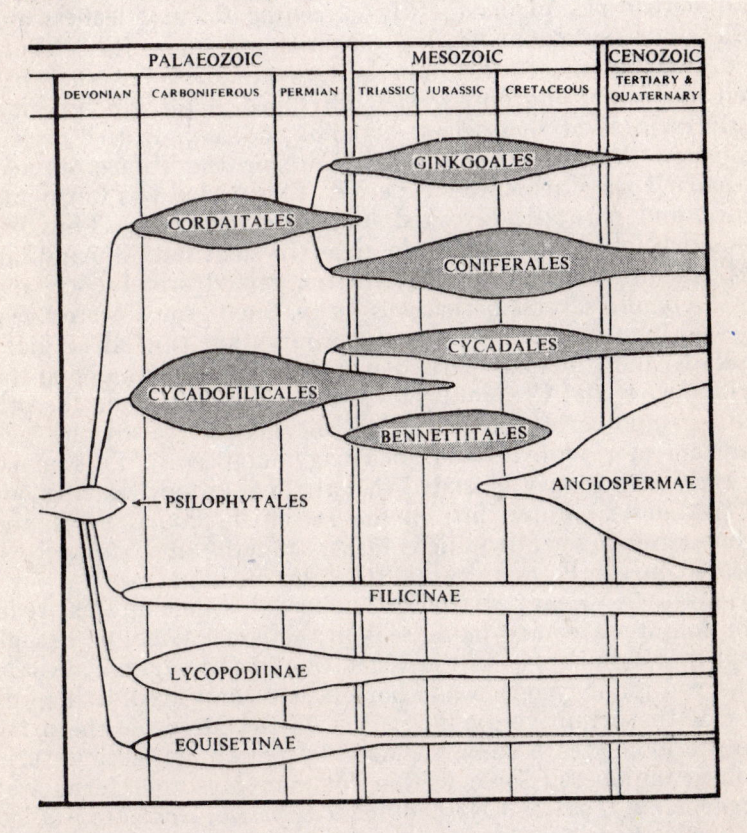

FIG. 1. Diagram showing possible origin, evolution and distribution of pteridophytes, gymnosperms and angiosperms in different geological periods. For time scale see table on p. 853.

Order 3. Bennettitales (or Cycadeoideales), e.g. *Williamsonia, Cycadeoidea* (=*Bennettites*), etc.

Order 4. Ginkgoales, e.g. *Baiera* (fossil) and *Ginkgo* (living).

Order 5. Cycadales, e.g. cycads (living and fossil).

Order 6. Coniferales, e.g. conifers (living and fossil) such as *Araucaria,* pine, juniper, cypress, yew, etc.; *Lebachia, Walchia, Palissya,* etc., are found as fossils only.

The Palaeozoic. From the standpoint of evolution this age is immensely interesting inasmuch as it shows many special features of land vegetation. In the early period (Devonian or possibly the Silurian) of this age there was the first invasion of land by the aquatic flora of the sea. Such land plants were the Psilophytales, and they might have evolved from certain green algae. They were small herbaceous plants (hardly exceeding 0.3 m.), leafless and rootless but branched, bearing terminal sporangia. They, however, soon became extinct, but before they died out they had given rise to several groups of well-developed plants which permanently established themselves on land, and as a matter of fact dominated the surface of the earth during the Palaeozoic age, particularly the Carboniferous period. This period was warm and humid, and naturally favoured luxuriant vegetation. Thus big forests developed, covering wide areas. In such forests many tall trees, reaching a height of 20-40 metres, gained prominence, and there were also several herbaceous forms. Such groups were **Lycopodinae** (including giant lycopods), **Equisetinae** (including giant horsetails) and **Filicinae** (particularly Coenopteridales) among the pteridophytes, and **Cycadofilicales** and **Cordaitales** among the primitive gymnosperms. So the Palaeozoic may be called the 'age of ancient pteridophytes and ancient gymnosperms'. To summarize the above it may be stated that the five groups (leaving out Psilophytales) appeared first in the Devonian, flourished in the Carboniferous, and dwindled, rather rapidly, and almost disappeared in the Permian or a little later in some cases.

The Mesozoic. Heavy glaciation, upheaval of mountains, redistribution of water and land areas in the late Palaeozoic brought about marked changes in the vegetation of the earth. Palaeozoic ferns and giant lycopods and horsetails disappeared, and so did most of the primitive gymnosperms. In the Mesozoic the primitive pteridophytes became replaced by newer herbaceous forms in large numbers. Some of the common Mesozoic ferns were *Phlebopteris, Todites, Osmundites, Tempskya, Schizaeopsis,* etc. Similarly the primitive gymnosperms rapidly declined in the Permian—the Cordaitales became extinct, while the remnants of the Cycadofilicales somehow continued up to the early Mesozoic. Four new groups of gymnosperms, however, appeared in the late Palaeozoic (the Permian) or in the early Mesozoic in diversified forms from the dying Palaeozoic stock. Thus the Bennettitales (soon, however, altogether extinct) and Cycadales (both living and extinct) made their appearance from the Cycadofilicales stock, while the Coniferales (both living and extinct) and Ginkgoales (now with only one living species—*Ginkgo biloba*, a native of China but planted in some countries) evolved from the Cordaitales stock. These four groups of gymnosperms, particularly the Coniferales, flourished during the Mesozoic; they became very abundant, widespread and dominant in this age. So the Mesozoic may be called the 'age of advanced gymnosperms'. In the late Mesozoic, newer and more modern types of conifers, viz. pine (*Pinus*), cypress (*Cupressus*), yew (*Taxus*), spruce (*Picea*), fir (*Abies*), cedar (*Cedrus*), juniper (*Juniperus*), redwood (*Sequoia*), etc., made their appearance. Another important event of the late Mesozoic is the first appearance of angiosperms; their ancestry, however, could not be traced yet. Yet another event of the Mesozoic is the discovery of a new group of angiosperm-like plants from the mid-Jurassic rocks of Yorkshire—the Caytoniales—which at one time were regarded as the progenitors of angiosperms (the idea, however, was later discarded).

The Cenozoic. With the advent of the angiosperms and their rapid increase in number, diversification of forms, and widespread distribution in the Cretaceous, forming the dominant vegetation of the earth since then, the ferns and the gymnosperms rapidly dwindled and lost their dominant position both numerically and geographically. So the Cenozoic may be called the 'age of angiosperms'. Among the ancient angiosperms found in the Cretaceous and later, mention may be made of water lilies, predominantly, however, several arborescent types, such as *Populus, Quercus, Ficus, Juglans, Magnolia, Fagus, Salix*, etc., and among monocotyledons several grasses, sedges, aroids, *Typha, Smilax*, palms, etc. With the progress of evolution it is noted that the herbaceous forms have far outnumbered the woody forms; the former now dominate the surface of the earth, not only in their worldwide distribution but also in their specific and individual numbers.

Pteridophytes

A. Psilophytopsida or Psilophytinae. Members of this class were the most primitive vascular plants that were abundant and widespread during the Devonian period. There are 2 orders of Psilotopsida: Psilophytales—an altogether extinct order with *Psilophyton* as the typical genus, and Psilotales—a modern order with 2 genera, viz. *Psilotum* and *Tmesipteris*.

Order **Psilophytales.** The small plants of the Psilophytales (hardly exceeding one metre, often much smaller) had a dichotomously branched stem with no leaves except in *Asteroxylon*. Roots were altogether absent. The underground rhizome bore numerous hairy rhizoids in some genera. The stem was protostelic with the xylem consisting of tracheids only. The plants were sporophytes reproducing by one kind of spores (homosporous). The sporangia were terminal or in some cases lateral. The gametophytes are not known yet. Our knowledge of this order began with the discovery of *Psilophyton princeps* in Canada by Sir William Dawson in 1859. But his discovery did not attract much attention. Much later in 1917 Kidston and Lang first discovered numerous fossils of similar other plants (*Rhynia, Horneophyton, Asteroxylon,* etc.) in well-preserved forms from the middle Devonian rocks at Rhynie (Scotland), occurring there in thick deposits. With these discoveries the order Psilophytales came to be established. The members of this order were found to be abundant in the lower and middle Devonian when they had a worldwide distribution—Great Britain, Canada, North America and Australia. Possibly they appeared in the Silurian (early or late) but they disappeared in the upper Devonian. They were the earliest land plants and the oldest and most primitive of all vascular plants. Whatever might have been their origin (probably from some algal stock), they were generally believed to be the ancestors of lycopods, horsetails and ferns in divergent lines.

Psilophyton (20 sp.). *Psilophyton* (family *Psilophytaceae*) was widely distributed in Europe, North America and Canada. They grew in dense clumps in marshy places, and hardly exceeded 60 cm. in height and 1 cm. in diameter. *P. princeps* is the best known species. The plants were slender herbs with short slender rhizome bearing hairy rhizoids but no roots, and unequally dichotomous branches without leaves. There were, however, many small outgrowths (variously described as incipient leaves, spines and thorns) in the lower parts of the aerial shoots. The vegetative shoots were circinate, while the fertile shoots bore small oval sporangia (about 5 mm. long) singly at the tips of curved bifurcated branches. The sporangia were homosporous. The gametophyte is still unknown. The internal structure of the stem could not be properly assessed excepting that it was protostelic with the central xylem consisting of some annular tracheids, and the cortex possibly photosynthetic in nature.

Rhyina (2 sp.). *Rhynia* (family *Rhyniaceae*) was abundant in the middle Devonian rocks in Scotland.. *R. major* was about 50 cm. in height and more or less 6 mm. in diameter at the base. The other species was much smaller. The plants were extremely simple and naked (without leaves), having a slender branching rhizome and slender erect aerial shoots branching dichotomously. The rhizome bore hairy rhizoids but no roots. Some of the branches ended in solitary sporangia (about 12 mm. long). They were pear-shaped and thick-walled, and homosporous. The gametophyte is still unknown. The stem was protostelic, consisting of central xylem (made of annular tracheids only) surrounded by phloem (made of some elongated cells). The cortex was thick and carried on photosynthesis. The epidermis had a thick cuticle and some stomata.

B. Lycopsida or Lycopodinae. Members of this class had the sporophyte differentiated into the root, stem and leaves—the root and the stem often dichotomously branched, and the leaves mostly small and spirally arranged (regarded as evolutionary derivatives of the emergences of Psilotopsida). The vascular system was commonly a protostele. The sporangia were solitary on the upper surface of certain leaves called sporophylls. Often sporophylls bearing sporangia were grouped together at the end of branches forming cones or strobili. The order Lepidodendrales of this class is altogether extinct although other orders (see p. 621) have living representatives.

Order **Lepidodendrales.** The plants of this order numbering over 200 species are found as fossils only. Probably they originated from the Psilophytales stock in the late Devonian, or the early Carboniferous, and reached their climax of development in the upper Carboniferous. They soon rapidly declined and disappeared in the Permian. They were very tall trees with stout trunks known as giant lycopods. Some of them reached a height of 40 metres with a diameter of 2 metres, showing thus a considerable amount of secondary growth. They had linear ligulate leaves, spirally arranged, which left characteristic leaf-scars on the stem on falling off. The leaf-scars distinguish the different genera. The stelar structure ranged from protostele to ectophloic siphonostele. They produced spores of two kinds (heterosporous): the microsporangium produced numerous small microspores, while the megasporangium produced 4 to 16 large megaspores. In some cases only a single megaspore matured. A female gametophyte was produced in the megaspore while still retained within the megasporangium, and after fertilization the entire structure was shed like the seed of a 'flowering' plant. They were dominant plants of the coal age and hence important in the formation of coal. *Lepidodendron* and *Sigillaria* (both giant lycopods) are the best known genera.

Lepidodendron. This was a dominant Palaeozoic genus abundantly found in well-preserved fossilised forms in the Carboniferous coal formations. Most of the species had attained very large sizes (giant lycopods). The single tall trunk (40 m. or so) bore a crown of numerous dichotomous branches at the top, and had spirally arranged linear leaves of varying lengths, often up to 20 cm., sometimes even more. The whole trunk was covered with leaf-scars which showed a distinct *spiral* arrangement. Internally the stem showed a considerable amount of secondary growth, mostly consisting of very thick secondary wood with scalariform tracheids. The stem produced at its base usually four, sometimes many, very long root-like branches (rhizophores) which forked repeatedly. They bore roots at their tips.

Sigillaria. This was another important Palaeozoic genus which held a dominant position in the Carboniferous. Many species were tall (attaining a height of 20 m. or more) and arborescent like *Lepidodendron*. Some species were, however, much shorter. The stem was unbranched or sparingly branched, and bore long linear leaves (sometimes about a metre in length) at the top in a cone-like tuft. The leaf-scars showed a *vertical* arrangement. As in *Lepidodendron*, there was a considerable amount of secondary growth in *Sigillaria*. Like the former *Sigillaria* also bore rhizophores.

C. **Sphenopsida or Equisetinae.** Members of this class also had sporophytes with distinct roots, stems and leaves. The stem was jointed and ribbed, and the leaves were small and simple, occurring in whorls at the nodes and forming a sort of sheath—a feature characteristic of this group and distinguishing it from other pteridophytes. Another characteristic feature found among the Sphenopsida was the presence of sporangium-bearing axes (sporangiophores) in whorls. In many forms these sporangiophores were recurved so that the terminal sporangia were directed towards the axis. Strobili were formed in many genera. This group was contemporaneous with Lycopsida, being abundant in diversified forms during the Carboniferous, and then suffered a sharp decline. The order might have evolved from the Psilophytales stock.

Order **Hyeniales.** This order is the oldest of all Sphenopsida, which lived during the middle Devonian, and resembled the Psilophytales in some respects. *Hyenia* and *Calamophyton* were the typical genera.

Hyenia. The plants of this genus which developed during the middle Devonian were about 30 cm. in height and less than 1 cm. in diameter at the base. They had a horizontal stout rhizome with roots, from which arose erect aerial dichotomously branched shoots—sterile and fertile. The sterile axis bore several whorls of small slender leaves repeatedly forked. The fertile axis bore several whorls of sporangiophores similarly forked at the apex, and had the appearance of a long loose cone (strobilus) but no bracts (sporophylls) were present. Two of the several forks were, however, recurved and each of them ended in a pair of sporangia. They were homosporous.

Order **Sphenophyllales.** This order was represented by the fossil genus *Sphenophyllum* which first appeared in the upper Devonian and continued up to the lower Triassic, being very abundant in the Carboniferous. The plants were of small size, and lay prostrate on the ground. The slender stem hardly exceeding 1 cm. in diameter bore whorls of leaves at the nodes. They were wedge-shaped and generally lobed, and each whorl consisted of 3 leaves or any multiple of this number. The sporophylls also occurred in whorls but in long terminal cones (strobili). The sporangia were homosporous. The solid primary wood (protostele) was triangular in shape with the protoxylem at each corner (exarch). The stem showed some amount of secondary growth.

Order **Calamitales.** This order consisted of several arborescent (tree-like) species (giant horsetail) closely related to the Equisetales. It appeared in the Upper Devonian, became abundant and dominant in the swamp forests of the Upper Carboniferous, and finally dwindled and disappeared in the Upper Permian. The principal family was *Calamitacese* and it was represented by genera like *Protocalamites, Calamites, Calamostachys, Palaeostachya,* etc.

Calamites. Several species of *Calamites* were very abundant and dominant in the Upper Carboniferous (along with *Lepidodendron*) though their life span extended from the Upper Devonian to the Upper Permian. In habit the plants were mostly very tall, some attaining a height of 20-30 m. and a diameter of 60 cm., with a large hollow pith cavity. The tall erect stem developed from an underground rhizome, and was distinctly jointed and ridged. Branches and leaves appeared in whorls from the nodes. The leaves were simple and narrow but much larger than those of *Equisetum.* They appeared in widely varying numbers (4-60) at the nodes, and were mostly free or in some species united into a sheath. Anatomically the primary bundles were small and arranged in a ring, with the protoxylem having a carinal cavity, as in *Equisetum.* The bundles divided at each node and alternated with those of the next internode. But unlike *Equisetum* a considerable amount of secondary growth was formed outside the metaxylem, resulting in the large girth of the stem. Wood rays, often very wide, were present in both primary wood and secondary wood but annual rings were absent. The wood consisted of tracheids with scalariform thickening or with bordered pits. The strobili (cones) occurred singly at the node or in groups terminally or on special branches. Each cone was made of whorls of peltate sporangiophores bearing only four pendant sporangia. In several members of Calamitales (e.g. *Calamostachys*) whorls of sheathing bracts united at the base into a sort of disc occurred, alternating with the sporangiophores, while in others (e.g. *Calamites*) such bracts were absent. In them the overlapping sporangiophores afforded necessary protection to the young sporangia. Most of the species were homosporous, while a few showed heterospory. Excepting the size *Calamites* had a close resemblance with the Equisetales—*Equisitites* of the Mesozoic and *Equisetum* of today.

Order **Equisetales.** This order consists of herbaceous plants having close resemblance to modern *Equisetum*. The order, as represented by the typical fossil genus *Equisetites*, became prominent in the Mesozoic (Triassic) but it was also present in the Carboniferous. The slender simple leaves of *Equisetites* were in whorls but had lost photosynthetic activity. The strobili resembled those of *Equisetum*.

D. Pteropsida or Filicinae. The ferns are the largest group among the Pteridophyta. They have a stem, well-developed leaves and roots. The sporangia are borne mostly on the leaves in groups or sori. Ferns were no doubt abundant in the Palaeozoic, but many of them later proved to be 'seed ferns' or Cycadofilicales. Some of the Palaeozoic families of ferns such as *Gleicheniaceae, Marattiaceae, Schizaeaceae,* etc., have their living representatives today. With the close of the Palaeozoic the ferns declined and many of them disappeared altogether. In the Mesozoic new ferns appeared in diversified forms and soon became abundant and widespread. The descendants of many Mesozoic ferns have continued till today in increased numbers and forms. The order Coenopteridales of this class was extinct in the Palaeozoic.

[*Continued on page* 855

Era	Period (age in million years)	Dominant Plants (& Animals)
CENOZOIC	Recent (QUATERNARY)	Herbaceous plants dominate; forest trees decline. Civilization begins.
	Pleistocene [1-5] (QUATERNARY)	Many large trees die out; many hardy types of herbs survive and increase. Great mammals die out.
	Pliocene [12] (TERTIARY)	Restricted distribution of forests; rise of herbs. Appearance of man. Upheaval of the Himalayas—2nd phase.
	Miocene [20] (TERTIARY)	Forest areas decrease; polar flora retreat; grasses increase. Mammals reach zenith.
	Oligocene [35] (TERTIARY)	Tropical forests of angiosperms world-wide. Primitive mammals disappear; higher mammals and birds increase. Upheaval of the Himalayas—1st phase.
	Eocene [60] (TERTIARY)	Pteridophytes continue; mesozoic gymnosperms disappear except cycads and conifers which decline; angiosperms dominate; tropical forests in polar regions. Modern birds and higher mammals.
MESOZOIC	Cretaceous [135]	Pteridophytes continue; gymnosperms rapidly decline except conifers; angiosperms rapidly increase in a great variety of forms; modern forests begin; Giant reptiles extinct; primitive mammals increase.
	Jurassic [180]	Luxuriant forests of higher gymnosperms and modern pteridophytes; probable origin of angiosperms. Flying reptiles, dinosaurs abundant; primitive birds; higher insects.
	Triassic [230]	Change in flora due to heavy glaciation in Permian; most of the primitive forms disappear; new pteridophytes appear in herbaceous forms; higher gymnosperms (cycads and conifers) increase. Origin of mammals and giant reptiles (dinosaurs).
PALAEOZOIC	Permian [280]	Primitive pteridophytes and gymnosperms decline and almost disappear, while advanced groups (cycads, conifers, etc.) appear. Land vertebrates.
	Carboniferous [345]	Pteridophytes: giant lycopods and horsetails (*Calamites*) dominant and widespread; ferns abundant; primitive gymnosperms abundant; luxuriant growth of tall trees forming dense forests; extensive coal deposits. Amphibians increase; reptiles and insects.
	Devonian [395]	Early land plants; Psilophytales abound but soon disappear; possible origin of gymnosperms (Cycadofilicales and Cordaitales); giant lycopods and horsetails, and ferns. Origin of amphibians; fishes dominant.
	Silurian [425]	Marine algae dominant; possible origin of land plants: Psilophytales. Higher invertebrates increase. First vertebrates.
	Ordovician [500]	Marine algae abundant. Higher invertebrates.
	Cambrian [600]	Green algae, stoneworts and red algae. Primitive invertebrates: trilobites abundant.
	Proterozoic [2,000]	Bacteria, fungi and blue-green algae.
	Archeozoic [5,000]	No fossil records but unicellular life quite probable.

For description see next page

Continued from page 853]

Order **Coenopteridales**. The members of this ancient order extended from the Devonian to the Permian. They were, however, most abundant in the Carboniferous, and constituted the most primitive group of ferns (also called the Primofilices). Several species (about 65 sp.) have been found so far, and most of them were distinctly fern-like in appearance. They resembled the Psilophytales on the one hand by their large terminal sporangia (synangia), and in some cases by their small undivided leaves, and the ferns on the other by their often pinnately divided fern-like frond in many cases. The plants of this order were small or medium in size, often erect or sometimes prostrate, and bore single large pyriform sporangia or a group of sporangia united into a synangium, terminally at the apex of the pinnule of a frond or sometimes directly on the rachis. The frond was branched in many planes, less often in one plane. The stem, often much-branched, was protostelic, the stele being circular or lobed in outline. All the species were homosporous. *Botryopteris* and *Etapteris* were the two common genera of the order.

Gymnosperms

Order **Cycadofilicales** (or **Pteridospermales**). The members of this order were distinctly fern-like plants with pinnately compound leaves but they bore seeds. So they are called 'seed ferns'. They were the most primitive group of seed plants. They appeared in the upper Devonian, became abundant and widespread in the Carboniferous, rapidly declined in the Permian, and became extinct in the early or mid-Jurassic. They were of varying habits, usually slender and short unlike the Cordaitales. Some were like tree-ferns but much smaller in size. Detached parts of 'seed-ferns' were more or less perfectly and beautifully preserved showing roots, stems, leaves, pollen-bearing organs (even pollen grains) and seeds (without, however, any trace of the embryo). They had distinct fern-like fronds, sometimes very large, and for many years they were taken to be ferns. On the basis of their anatomical structure (intermediate between ferns and cycads) this group was named 'Cycadofilicales' by Potonie in the year 1899. About the year 1903 Oliver and Scott found seeds together with the stems and the leaves, and they proposed the name 'Pteridosperms' for such seed-bearing, fern-like plants.

The table on the previous page shows the possible origin and evolution of main groups of plants (and also animals) in different geological periods. Broadly speaking, lower Palaeozoic was the 'age of algae', and higher Palaeozoic the 'age of pteridophytes' with abundant primitive gymnosperms; Mesozoic was the 'age of advanced gymnosperms' with abundant advanced pteridophytes; and Cenozoic the 'age of angiosperms' (herbs dominating the recent period) with modern pteridophytes and modern gymnosperms.

The primary xylem in them was mostly mesarch (a fern character). They showed three kinds of stele—usually protostele, sometimes polystele, and also ectophloic siphonostele. Cortical bands of sclerenchyma were a constant feature in them. Some amount of secondary growth, thick or thin, was also present in them, and further they showed gymnospermic tracheids with bordered pits, mostly on their radial walls. Evidently they were gymnosperms, and their distinct fern-like appearance also suggests a relationship with the ferns. There was no strobilus formed in this group. The microsporangia (pollen-bearing organs) were borne on the margin or lower side of the fertile pinnule, as in many ferns. The ovule (seed) was borne at the end of the frond or modified frond (megasporophyll). A deep pollen-chamber was present in the nucellus, as in gymnosperms. The seeds were like those of other seed-plants but usually enclosed in a cupule.

The pteridosperms were no doubt the oldest seed plants. Their relationship with the ferns is evident from the nature of leaves, microsporophylls, primary wood, and microsporangia. Their relationship with the Bennettitales and the Cycadales, if not to other gymnosperms, is also close. But the origin of the Pteridosperms is only speculative, and cannot be traced with any amount of certainty because of missing links. Two alternatives are, however, probable: they might have been derived from some unknown Filicinian stock (Primofilices) of the early Palaeozoic age, or they might have originated from the Psilophytalean stock in the Devonian period. Arnold holds the view that the Pteridosperms with their diversified forms might not have arisen from a single (common) ancestral stock. Whatever might have been their origin the two groups (Pteridosperms and Filicinae) had developed later in parallel lines.

The order has been divided into seven families—the first three confined to the Palaeozoic, and the last four either confined to the Mesozoic or extended from the Palaeozoic to the Mesozoic.

1. *Lyginopteridaceae.* Straggling habit. Stems usually 1-5 cm. in diameter; protostelic. Large fronds. Limited amount of secondary growth. Integument fused with the nucellus except at the apex. Seed usually borne in a cupule. Common genera: *Lyginopteris* (stems—very common in coal-balls), *Heterangium* (stem), *Sphenopteris* (frond), *Crossotheca* (pollen-bearing organs), *Lagenostoma* and *Calymmatotheca* (seeds), etc.

2. *Medullosaceae.* Trunk-like stem but narrow (usually 2-6 cm. in diameter; much thicker in some species). Plants, pollen-bearing organs and seeds much larger than those of *Lyginopteridaceae*. Stem polystelic (sometimes a large number of steles). Secondary growth wide and distinct. The inner integument fused with the nucellus. Seed enclosed in a stony cupule. Common genera: *Medullosa* (stems), *Neuropteris* and *Alethopteris* (fronds), *Stephanospermum* and *Codonospermum* (seeds), *Codonotheca* (pollen-bearing organs), etc.

3. *Calamopityaceae.* This family is only imperfectly known. Leaves large and fern-like. Stems with solid protostele to circum-medullary groups of primary xylem with mixed pith (tracheids and parenchyma). Secondary wood either manoxylic (soft and sparse with broad rays), or pycnoxylic (dense and compact with very narrow and small rays). Common genera: *Calamopitys* and *Stenomyelon* (stems).

4. *Glossopteridaceae* Glossopteris flora is characteristic of the Gondwanaland, (see p. 860). Common genera: *Glossopteris* and *Gangamopteris* (leaves), *Vertebraria* (roots), *Ottokaria* and *Scutum* (reproductive organs) etc.

5. *Peltaspermaceae.* Fronds, pollen-bearing organs (but not stems) have been found in the Triassic rocks of Natal, Madagascar, Greenland, Argentina, Australia and China. Common genera: *Lepidopteris* (fronds), *Peltaspermum* (seed-bearing organs) and *Antevsia* (pollen-bearing organs), etc.

6. *Corystospermaceae.* Fronds, pollen-bearing and seed-bearing organs have been found in the Triassic rocks of Natal, Australia, Argentina and India. Common genera: *Xylopteris* and *Dicroidium* (fronds), *Umkomasia* (seed-bearing organs), *Pterucus* (pollen-bearing organs), etc.

7. *Caytoniaceae.* A group of angiosperm-like plants (once supposed to be ancestors of angiosperms; see p. 689) first discovered by Thomas in 1925 in the mid-Jurassic rocks of Yorkshire. Leaves (in *Sagenopteris*) with four leaflets in two pairs on a slender petiole; reticulate venation; shedding of leaves and leaflets by abciss-layer; prominent midrib. Pollen-bearing organs or microsporophylls (in *Caytonanthus*) on a pinnate type of rachis, each branch ending in a synangium; pollen-sacs four, dehiscing except at the apex. Seeds (in *Caytonia*) borne in fruits. Pollination mechanism, as in gymnosperms; pollen-chamber present.

Calymmatotheca hoeninghausi. Fossils of this plant found in the Coal Measures of Europe and America as detached parts and described under different names such as the stem as *Lyginopteris oldhamia*, frond as *Sphenopteris sp.*, seed as *Lagenostoma sp.*, root as *Kaloxylon sp.*, etc., were later assembled together under the name *Calymmatotheca hoeninghausi*. The plant grew to a diameter of 4 cm., and showed a ring of mesarch primary bundles, a fair amount of secondary growth, large tracheids with several rows of bordered pits, distinct medullary rays, masses of stone cells in the large pith, and conspicuous leaf-traces and leaf-gaps. Fronds were large, compound (much divided) and spirally arranged; the rachis was with a distinct gland. The fertile pinnule (of *Crossotheca*) was peltate in nature and bore on its undersurface a number of bilocular microsporangia or pollen-bearing organs. Ovules were also found attached to some of the fertile pinnules terminally. The seed remained enclosed in a seed cup or cupule which split into 4-6 segments..

Order **Bennettitales** (or **Cycadeoideales**). The Bennettitales formed an intermediate group between the Cycadofilicales and the Cycadales. They flourished during the Mesozoic, reaching their culmination in the Jurassic when they had a world-wide distribution, specially in India, Western Europe and North America, often forming forests in many areas. This order might have evolved from the Pteridospermales in the Triassic but they declined and became extinct in the upper Cretaceous. The Bennettitales seem to be closely related to the Cycadales on the one hand and to the Cycadofilicales on the other by their pinnate type of compound leaves, circinate vernation, same type of naked seeds, structure of ovules, loose pattern of secondary wood with large tracheids, numerous medullary rays, and large pith. So these three orders are together called 'Cycadophyta'. In the Bennettitales, however, the sporophylls were much more specialized. Cones (flowers) in this order were unisexual or bisexual, with numerous bracts. Microsporophylls, pinnate or entire, bore numerous microsporangia. Megasporophylls bore many stalked ovules on an elongated receptacle. Seeds had two cotyledons. The plants were of varying habits—most of them below 1 m. in height, while a few were 2 m. or a little higher. The two principal families of this order were as follows:

1. **Cycadeoideaceae.** This was represented by the only genus *Cycadeoidea* (=*Bennettites*) with over 30 species. It had usually short, stout and barrel-like or columnar stem, presence of scales and leaf-scars, short internodes, a crown of cycad-like pinnate leaves, and numerous bisporangiate (bisexual) strobili. Each strobilus was borne on a short axillary branch, and consisted of a number of sterile appendages or perianth-like bracts at the base, arranged spirally and overlapping each other, a whorl of 10-12 pinnately divided, leaf-like microsporophylls, each with numerous microsporangia or pollen-sacs, and several megasporophylls intermixed with scales on the top of the receptacle, forming a sort of compact cone; each megasporophyll, highly modified and unlike leaf, was represented by a long stalk with an ovule on its top. The seed with a large embryo (but no endosperm) of dicotyledonous type was also found. Thus *Cycadeoidea* had a flower-like appearance and construction, resembling the flower of *Magnolia*. It was, therefore, at one time suggested that the Bennettitales might have been the ancestors of the angiosperms. Considering, however, other grounds, particularly the specialized sporophylls (both micro- and mega-) it is more reasonable to suppose that the resemblance is only superficial and is the result of parallel development (evolution), and not an indication of direct relationship.

2. **Williamsoniaceae.** This was represented by the typical genus *Williamsonia*. It had a slender, branched or unbranched stem (tall or short), a dense covering of scales, presence of leaf-scars, short internodes, a crown of cycad-like pinnate leaves, and mostly monosporangiate (unisexual) strobili. A few other genera of this family were *Ptilophyllum, Pterophyllum, Otozamites, Dictyozamites*, etc.

Order **Cycadales.** The extinct Cycadales were a group of cycad-like gymnosperms which flourished during the mid-Mesozoic. They first appeared in the upper Triassic and reached their maximum development during the Jurassic-Cretaceous periods.

They had then a world-wide distribution. Among the 'Cycadophyta' it is only the Cycadales that have survived until the present day. They were contemporaneous with the Bennettitales and were closely related to them. Both the groups might have been derived from the Cycadofilicales, but independently of each other. Some of the extinct genera of the Cycadales were *Nilssonia, Beania, Palaeocycas,* etc., while the living Cycadales are represented by nine genera with about 100 species.

Order **Cordaitales.** The members of this extinct group of gymnosperms appeared in the late Devonian, reached their climax of development during the upper Carboniferous, and became practically extinct by the end of the Permian. Both the Cordaitales and the Cycadofilicales were contemporaneous but the former soon completely died out, while the latter lingered on for some time more (see FIG. 1). Both the orders were, however, abundant in diversified forms during the Carboniferous, but the Cordaitales dominated the gymnospermic forests. Detached parts of roots, shoots, leaves, strobili, and seeds (but no embryo) in well-preserved fossil forms were described before under different names but later Williamson (1851), Renault (1879), Scott (1900) and others brought them together under the order Cordaitales. The order is divided into two families, viz. (*a*) *Cordaitaceae,* e.g. *Cordaites* (a typical genus), and (*b*) *Poroxylaceae,* e.g. *Poroxylon.* The Cordaitales were mostly tall trees (unlike the pteridosperms) reaching a height of 30 m. but were not fern-like in their appearance and in the nature of leaves. On the other hand in their habit and in the structure of the secondary wood they approached the Coniferales, and are thus regarded as their ancestors. The Cordaitales bore a crown of branches at the top and also many large simple, parallel-veined leaves (often 1 m. long and 15 cm. broad); the leaves showed internal differentiation into palisade and spongy tissues and several ribs of sclerenchyma. The stem anatomy was very much like that of modern *Araucaria,* a coniferous genus. It showed a considerable amount of secondary growth, with well-developed tracheids having many rows of bordered pits on radial walls, medullary rays one cell thick, and a large pith. The plants were monoecious, evidently bearing monosporangiate (unisexual) strobili. Several strobili were borne on special branches developing from the axis, looking like inflorescences. Each strobilus (male or female) consisted of several spirally arranged sterile bracts and some sporophylls (micro- or mega-). The male strobilus consisted of a few long-stalked microsporophylls or stamens, each with 4-6 terminal microsporangia or pollen-sacs. The female (ovulate) strobilus consisted of a few (1-4) ovules on short (dwarf) branches.

The Cordaitales were an ancient group of 'seed' plants as old as the Cycadofilicales. Like the latter the Cordaitales were also abundant during the upper Carboniferous, forming extensive forests. Regarding their origin two alternatives are probable: both the groups might have a common origin from some extinct fern-like plants (may be the Psilophytales), or the Cordaitales might be an early offshoot of the Cycadofilicales. The former view is shared by many authorities. Whatever might have been their origin, it is assumed for valid reasons that the Cordaitales, before they died out in the Permian, gave rise to two well-known groups of gymnosperms—the Coniferales and the Ginkgoales— which flourished during the Mesozoic.

Order **Ginkgoales**. The members of this order were abundant and had world-wide distribution during the Mesozoic. Some of the Mesozoic genera were *Baiera, Ginkgoites, Windwardia*, etc. Now the only living representative of the order is the maidenhair tree (*Ginkgo biloba*) which is referred to as the 'living fossil'. The order might have been derived from the Cordaitales.

Order **Coniferales**. The members of this order can be traced back to the Permian but they were abundant and widely distributed during the Mesozoic when they appeared in diversified forms, and reached their maximum development during the Cretaceous. They might have been derived from the Cordaitales. The order has continued up to the present day with living members numbering 500 species. As in the Mesozoic so also in the present period this is the largest order of the gymnosperms. The earliest known Coniferales were *Lebachia, Walchia*, etc., of the Permian, and *Volziopsis, Pseudovoltzia*, etc., of the Triassic. Some of the modern conifers can be traced back to the Jurassic.

The Gondwana System. In the early Carboniferous the flora of the southern hemisphere and that of the northern were very much alike. But by the upper Carboniferous and lower Permian times the two hemispheres became separated by a vast sea called the Tethys (see p. 844). The land mass of the southern hemisphere comprising Africa, Madagascar, India, Australia, South America and Antarctica formed one huge southern continent, called the Gondwanaland by the geologists. Possibly from the Jurassic period the Gondwanaland- began to break up into separate components which gradually drifted to their present locations. This 'continental drift' theory holds that the northern part of the Gondwanaland, which has become the Indian subcontinent, collided with the Asian continent with such a tremendous force that it had pushed up huge land masses forming the present-day Himalayas. Further, according to the same theory, its southern part—Antarctica—drifted all the way down to the

South Pole. The name Gondwana was first introduced by Medlicott in 1872 after the kingdom of the Gond, an ancient tribe of Central India, while he was studying the geological formations of the area. The Gondwana had mainly developed in India in a triangular area: (a) from the Godavari Valleys to the Rajmahal Hills, (b) the Damodar, Sone and Narmuda Valleys extending east to west, and (c) along the Godavari extending north-west to south-east. Other formations have been found along the foothills of the Himalayas (Nepal, Bhutan and Assam), and also isolated formations in Salt Range, Kashmir, Garhwal, Kutch, Saurashtra and Madhya Pradesh. At that time the flora was similar in both northern and southern hemispheres. During the upper Carboniferous and the lower Permian, however, there was an extensive glaciation, known as the Permo-Carboniferous Ice Age, in the Gondwanaland. This heavy glaciation in the southern hemisphere resulted in the destruction of most of the older vegetation, and in the redistribution of land and sea, and the upheaval of mountains. Succeeding this, there was an appreciable change in the climate; it became humid and warmer. Under these conditions an almost entirely new type of vegetation sprang up from the meagre flora that survived the catastrophe, and this is known as the **Glossopteris Flora**. The Glossopteris flora was entirely different from that of North America and Europe. This flora was characterized by a small number of species and scarcity of woody plants. Numerous fossils of leaves have been found and these have been named *Glossopteris*, and the flora as a whole the Glossopteris flora. Leaves were simple and commonly tongue-shaped but were of varying shapes and sizes (some even 30 cm. or more in length); venation always reticulate, and the mid-rib distinct. There were two types of leaves, viz. (a) foliage leaves, as described above, and (b) small scale-leaves. *Glossopteris* probably belonged to the order Pteridospermales. *Gangamopteris*, another constituent of the Glossopteris flora, resembled *Glossopteris* excepting that it had no mid-rib. Other constituents were some lycopods and horsetails, and several ferns and gymnosperms. The Glossopteris flora was characteristic of the Lower Gondwana (Permo-Carboniferous) of the Gondwanaland continent. The Upper Gondwana (Mesozoic) on the contrary was predominated by the **Ptilophyllum flora**, the main constituents being *Ptilophyllum* (order Bennettitales) and *Thinnfeldia* (order Pteridospermales) together with several advanced ferns and gymnosperms. The Gondwanaland has been divided into two main divisions: Lower Gondwana and Upper Gondwana on the basis of fossil floras, and these further sub-divided into series and stages. During the Permo-Triassic transition the Glossopteris flora of the Lower Gondwana suffered a decline and a break, and was succeeded by the Ptilophyllum flora of the Upper Gondwana (Mesozoic). The Gondwana as a whole extended from the Upper Carboniferous to the Jurassic or Lower Cretaceous, but began to break up during the Cretaceous, and in the Tertiary period angiospermic vegetation began to dominate, as we find in the present times.

Fossils of the Main Gondwana Rocks of India[1]. The Lower Gondwana (Upper Carboniferous to Upper Permian or Lower Triassic) comprises the following series in the ascending order of formations: (1) Talchir, (2) Damuda (Raniganj and Barakar), and (3) Panchet; while the Upper Gondwana (Middle Triassic to Upper Jurassic or Lower Cretaceous) comprises the following series: (1) Mahadeva, (2) Rajmahal, and (3) Jabalpur. The following is an account of the fossil contents of these series.

Lower Gondwana.

(1) Talchir Series (Upper Carboniferous): **Pteridospermales**—*Glossopteris, Gangamopteris, Vertebraria*, etc.; **Cordaitales**—*Noeggerathiopsis*; **Incertae**—*Samaropsis*.

(2) Damuda Series (Permian): a good number of fossils have been found in Raniganj and Jharia coalfields: **Lycopodiales**—*Bothrodendron*; **Equisetales**—*Schizoneura* (a few species), *Phyllotheca*, etc.; **Sphenophyllales**—*Sphenophyllum*; **Filicales**—*Alethopteris, Actinopteris*, etc.; **Pteridospermales**—*Glossopteris* (several species), *Gangamopteris, Sphenopteris, Vertibraria*, etc.; **Cordaitales**—*Noeggerathiopsis* and *Dadoxylon*; **Cycadeoideales**—*Taeniopteris*; **Incertae**—*Dictyopteridium, Cordaicarpus* and *Samaropsis*.

(3) Panchet Series (Upper Permian to Lower Triassic): the lower beds of the Panchet Hill lying to the north-west of Asansol contain numerous animal fossils but few plant fossils such as *Glossopteris, Schizoneura*, and also *Pecopteris* and *Cyclopteris*.

Upper Gondwana. Fossils represent more advanced types of plants.

(1) Mahadeva Series (Middle to Upper Triassic): this series has been named after the Mahadeva Hills near Panchmari; plant fossils are not common in this series although some leaf impressions have been found in clayey soil.

(2) Rajmahal Series (Lower to Middle Jurassic): this series is rich in plant fossils and already a good number of them have been collected and identified; they are: **Lycopodiales**—*Lycopodites* and *Lycoxylon*; **Equisetales**—*Equisetites*; **Filicales**—*Marattiopsis, Gleichenites, Pecopteris*; etc.; **Pteridospermales**—*Thinnfeldia, Danaeopsis* and *Sphenopteris*; **Cycadeoideales**—*Ptilophyllum, Otozamites, Dictyozamites, Taeniopteris* (a few species), *Nilssonia* (several species), etc.; **Coniferales**—*Elatocladus, Brachyphyllum, Pagiophyllum*, etc.; also several gymnospermic stems and cones; **Caytoniales**—*Sagenopteris*; **Incertae**—*Rajmahalia* and *Podozamites*.

(3) Jabalpur Series (Upper Jurassic to Lower Cretaceous): **Filicales**—*Gleichenites, Cladophlebis*, etc.; **Pteridospermales**—*Thinnfeldia*; **Cycadeoideales**—*Ptilophyllum, Otozamites, Williamsonia, Dictyozamites, Taeniopteris, Nilssonia* (a few species); **Coniferales**—*Elatocladus, Brachyphyllum, Pagiophyllum, Araucarites* (a few species), etc.; **Ginkgoales**—*Ginkgoites*; **Incertae**—*Podozamites*.

---

[1] Mainly based on *Geology of India and Burma* by M. S. Krishnan, 1960 edition, and *Fifty Years of Science in India* (Indian Science Congress Association, 1963): *Progress of Botany* by P. Maheshwari and R. N. Kapil; and *Progress of Geology* by S. Ray.

Glossary of Names of Plants

Botanical name in *italics*; English name in Roman; Indian name in CAPITALS
A. for Assamese; B. for Bengali; G. for Gujarati; H. for Hindi; K. for Kannada; M. for Malayalam; M'. for Marathi; O. for Oriya; P. for Punjabi; T. for Tamil; and T'. for Telugu.

Abelmoschus esculentus—lady's finger; A., O. & M'. BHENDI; B., H. & P. BHINDI; G. BHINDA; K. BHENDE KAYI; M. & T. VENDAKKA; T'. BENDA

Abroma augusta—devil's cotton; A. BONKOPAHI; B. & H. ULATKAMBAL; G. GUMCHI; K. DEVVA HATHI; M'. OLAKTAMBOL; P. ULTKAMBAL; T. SIVAPPUT-TUTTI

Abrus precatorius—crab's eye or Indian liquorice; A. LATUMMONI; B. KUNCH; H. & P. RATTI; K. GULAGANJI; M. KUNNI; M'. GUNJ; O. KAINCHA, GUNJA; T. KUNDOOMONY; T'. GURUGINJA

Abutilon indicum—A. JAPAPETARI; B. PETARI; G. DABALI; H. KANGHI; K. THURUBI GIDA, SHREE-MUDRE GIDA; KISANGI; M. & T. PERINTHOTTY; M'. MUDRA; O. PEDIPEDIKA; P. PILIBUTI; T'. THUTIRIBENDA

Acacia nilotica (=*A. arabica*)—gum tree; A. TORUAKADAM; B. BABLA; G. KALOABAVAL; H. BABUL, KIKAR; K. KARI JALI; M. & T. KARUVELAM; M'. BABHUL; O. BABURI; P. KIKAR; T'. NALLATUMMA

Acacia catechu—catechu; A., B. & M'. KHAIR; G. KHER; H. & P. KATHA, KHAIR; K. KAGGALI, KACHU; M. KADARAM; O. KHAIRA; T. KADIRAM; T'. KHADIRAMU

Acalypha indica—B. MUKTOJHURI; G. VANCHI KANTO; H. KUPPI; K. KUPPI GIDA, TUPPAKEERE; M. & T. KUPPAMANI; M'. KHOKALI; O. INDRAMARISHA; P. KOKALI KUPPAMANI

Achras zapota—sapota or sapodilla plum; A. SAPHEDA; B. SHABEDA; H. & M. SAPOTA; K. CHIKKU, SAPOTA; M'. CHIKKU; O. SAPETA; P. CHIKU; T. SIMAIYILUPPAI; T'. SIMAIPPA

Achyranthes aspera—chaff-flower; A. UBTISATH; B. APANG; G. SAFED AGHEDO; H. LATJIRA; K. UTTARANI; M. KATALADY; M'. AGHADA; O. APAMARANGA; P. PUTH KANDA, KUTRI; T. NAHIROORVY; T'. ATTARENI

Acorus calamus—sweet flag; A., B. & H. BOCH; G. GODAVAJ; K. BAJE; M. VAYAMBU; M'. WEKHAND; O. BACHA; P. WARCH, BOJ, BARI; T. VASAMBOO; T'. VASA

Adhatoda vasica—A. BANHAKA; B. BASAK; G. ADALSO; H. ADALSA; K. ADUSOGE, KURCHI GIDA, ADDALASA; M. ADALODAKAM; M'. ADULSA; O. BASANGA; P. BANSA SUBJ, BASUTI; T. ADATODAY; T'. ATARUSHAMMU

Aegle marmelos—wood-apple; A. & B. BAEL; G. BILIVA-PHAL; H. SIRIPHAL; K. BILVA PATRE; M. KOOVALAM; M'. BEL; O. BELA; P. BIL; T. VILVAMARAM; T'. BILAMBU

Aeschynomene indica—pith plant; A. KUNHILA; B., H. & P. SHOLA; K. BENDU KASA; M. KADESU ATTUKEDESU; O. SOLA; T. ATTUNETTEE; T'. JILUGA

Agave americana—American aloe or century plant; B. & H. KANTALA; G. JANGLI-KANVAR; K. KATTALE; M. NATTUKAITA; M'. GHAYPAT; O. BARABARASIA; P. WILAYATI KANTALA; T. ANAKUTTILAI; T'. BONTHARAKASI

Albizzia lebbek—siris tree; A., B., H., H'., M'. & P. SIRISH; G. PITOSARSHIO; K. SHIIRSHA BAGE, HOMBAGE; M. VAGA; O. SIRISA; T. VAGAI; T'. DIRISANA

Allium cepa—onion; A. PONORU; B., H. & P. PIYAZ; G. DUNGARI; K. NEERULLI, ULLAGADDI; M. ULLI; M'. KANDA; O. PIAJA; T. VENGAYAM; T'. NEERULLI

Allium sativum—garlic; A. NAHARU; B. RASUN; G. LASAN; H. & P. LASHUN; K. BELLULLI; M. VELUTHULLI; M'. LASUN; O. RASUNA; T. VELLAIPOONDU; T'. VELLULLI

Alocasia indica—A. & B. MANKACHU; G. ALAVU; H. MANKANDA; K. MANAKA; M'. ALU; O. MANASARU; P. ARVI

Aloe vera—Indian aloe; A. CHALKUNWARI; B. GHRITAKUMARI; G. KUNVAR; H. GHIKAVAR; K. LOLESARA; M. KATTARVAZHA; M'. KORPHAD; O. GHEE-KUANRI; P. KAWARGANDAL, GHIKUAR; T. KUTTILAI; T'. KALABANDA

Alpinia allughas—A. TORA; B. TARA; K. DUMPARASME; M. CHITTARATHTHA; M'. TARAKA; O. GHODAGHASA; P. KALANJAN; T. PERIYARATHTHA

Alstonia scholaris—devil tree; A. CHATIAN; B. CHHATIM; H. SATIAN, SAPTA-PARNA; K. SAPTA PARNA, MADDALE, KODALE; M. EZHILAMPALA; M'. SATVIN; O. CHHATIANA, CHHANCHANIA; P. SATONA; T. ELILAIPILLAI; T'. EDAKULAPALA

Alternanthera amoena—B. SENCHI; G. JALAJAMBO; K. HONAGANE SOPPU; M. KOZHUPPA; M'. KANCHARI; O. MADARANGA; P. CHURA; T. PONNAN KANNI KEERAI; T'. PONA-GANTIKURA

Amaranthus spinosus—prickly amaranth; A. KATAKHUTURA; B. KANTANATE; G. TANJALJO; H. & P. CHULAI; K. MULLU KEERE (or HARIVE SOPPU); M. MULLANCHEERA; M'. KATE MATH; O. KANTANEUTIA, KANTAMARISHA; T. MULLUKKERAI; T'. MUNDLA THOTAKURA

Amorphophallus campanulatus—A. & B. OL; G. & M'. SURAN; H. ZAMIKAND, KANDA; K. SUVARNA (or CHURNA), GEDDE; M. CHAENA; O. OLUA; P. ZAMIN KANDA; T. KARUNAKILANGU; T'. THIYA KANDHA

Anacardium occidentale—cashewnut; A. KAJUBADAM; B. HIJLIBADAM; G., H., M'. & P. KAJU; K. GODAMBI, GERUPAPPU; M. KASHUMAVU; O. LANKA BADAM; T. MUNDIRI; T'. JIDIMAMIDI

Ananas comosus—pineapple; A. MATI-KOTHAL; B. ANARAS; G., H. & M' ANANAS; T. ANASSAPPALAM; T'. ANASAPANDU

Andrographis paniculata—A. KALPATITA; B. & H. KALMEGH, MAHATITA; G. KIRYATO; K. NELA BEVU, KALA MEGHA; M. KIRIYATHTHU; M'. PALEKIRAIET; O. BHUINIMBA; P. CHARAITA; T. NELAVEMBU

Anisomeles indica—B. GOBRA; K. MANGA MARI SOPPU, HENNU KARI THUMBE; M. POOTHACHETAYAN; M'. GOPALI; O. BHUTA-AIRI; T. PEYAMERATTI

Annona reticulata—bullock's heart; A. ATLAS; B. NONA; G., H., M'. & P. RAM-PHAL; K. RAMA PHALA; M. ATHA; O. NEUA, BADHIALA; T. & T'. RAMSITA

Annona squamosa—custard-apple; A. ATLAS; B. ATA; G. & M'. SITAPHAL; H. & P. SHARIFA, SITAPHAL; K. SEETHA PHALA; M. SEEMA-ATHA; T. & T' SEETHA

Anthocephalus indicus—A., B., H. & P. KADAM; G. & O. KADAMBA; K. KADAMBBA MARA, KADAVALA; M. KADAMBU; M'. KADAMB

Arachis hypogaea—groundnut or peanut; A., B. & O. CHINABADAM; G MAFFALI; H. & P. MUNGPHALI; K. NELAGADALE, SHENGA, KALLEKAI; M. & T. NILAKKADALAI; M'. BHUIMUG; T'. VERU SANAGA

Areca catechu—areca- or betel-nut; A. TAMBUL; B., G., M'. & P. SUPARI; H. KASAILI; K. ADIKE; M. ADAKKA; O. GUA; T. PAKKU; T'. POKA

Argemone mexicana—prickly or Mexican poppy; A. KUHUMKATA; B. SHEAL-KANTA; G. DARUDI; H. PILADHUTURA; K. DATTURADA GIDA, ARISINA UMMATTI; M. SWARNAKSHEERI; M'. PIWALA DHOTRA; O. AGARA; P. KAN-DIARI; T. BRAHMADANDU; T'. BRAHMADANDI

Aristolochia gigas—pelican flower; A., B. & O. HANSHALATA; K. KURI GIDA; M. GARUDAKKODI; M'. POPAT VEL; P. BATKH PHUL; T. ADATHINAPALAI

Aristolochia indica—Indian birthwort; A. ISWERMUL; B. ISHERMUL; G. & M'. SAPSAN; H. ISHARMUL; K. ESHWARI BERU TOPPALU; M. ISVARAMMULI; O. GOPOKORONI; P. ANANTMUL, ISHARMUL; T'. ESWARI

Artabotrys hexapetatus (= *A. adoratissimus*)—A. KOTHALICHAMPA; B., G. & H. KANTALICHAMPA; K. MANORANJINI, KANDALA SAMPIGE; M. & T.

GLOSSARY OF NAMES OF PLANTS

MANORANJINI; M'. HIRWA CHAPHA; O. CHINICHAMPA; P. CHAMPA; T'. MANORANJITHAM

Artocarpus heterophyllus—jack tree; A. KOTHAL; B. KANTHAL; G. MANPHA-NASA; H. KATAHAR; K. HALASU; M. & T. PILA; M'. PHANAS; O. PANASA; P. KATAR

Artocarpus lakoocha—monkey jack; A. CHAMA, DEWA; B. DEO, DEOPHAL; H. DEOPHAL, BARHAL; K. WATE GIDA; M'. LAKUCH; O. JEUTA; P. DEHEO

Asparagus racemosus—A. SHATMUL; B. SATAMULI; H. & P. SATAWAR; K. SHATAVARI; O. CHHATUARI; M., M'., & T. SATHAVARI; T'. SADAVARI

Asteracantha longifolia—B. KULEKHARA; G. EKHARO; H. GOKULA-KANTA; M'. KOLSHINDA; O. KANTAKALIA, KOILIKHIA; P. TALMAKHANA; T. NIRMULLI

Averrhoa carambola—carambola; A. KORDOITENGA; B. KAMRANGA; G. KAMA-RAKHA; H. & P. KAMRAKH; K. KAMARAXI; KAMARAK; M. IRIMPANPULI; M'. CAMARANGA; O. KARMANGA; T. KAMARANKAI; T'. TAMARTA

Azadirachta indica—margosa; A. MOHA-NIM; B., H. & P. NIM, NIMBA; G. LIMBA; K. OLLE BEVU; M. VEPPU; M'. KADU LIMB; O. NIMBA; T. VEMBU; T'. VEPA

Baccaurea sapida—A. LETEKU; B. LATKAN; H. LUTKO; K. KOLI KUKKE; P. KALA BOGATI

Balanites aegyptiaca—A. HINGOOL; B. HINGAN; G. HINGER; H. & P. HINGOL, HINGU; K. INGALADA MARA, INGLIKA; M. NANJUNTA; M'. HINGANBET; O. HINGU; T. NANJUNDAN

Bambusa tulda—bamboo; A. BANH; B., H. & P. BANS; G. KAPURA; K. HEBBI-DIRU, UNDE BIDIRU; M. MULAH; M'. BAMBOO; O. BAUNSA; T. MULAI

Barleria prionitis—A. NILBAGI; B. KANTAJHANTI; G. KANTAASHERIO; H. VAJRADANTI; K. MULLU GORANTI; M. KANAKABARAM; M'. KORANTI; O. DASKARANTA; P. PILA BANSA, GAT SARIYA; T. CHEMMULLI

Barringtonia acutangula—A. HIDOL; B. & H. HIJAL; G. SAMUDARPHAL; K. NEERU GANIGALU, DHATRI PHALA; M. &. T. SAMUNDRAKSHAM; M' DHATRIPHAL; O. HINJALA; P. SAMUNDURAPHAL

Basella rubra—Indian spinach; A. PURAI; B. PUIN; H., O. & P. POI; K. KEMPU BAYI BASALE; M. SAMPARCHEERA; M'. VELBONDI; T. SAMBARKEERAI

Bassia latifolia—see *Madhuca latifolia*

Batatas edulis—see *Ipomoea batatas*

Bauhinia variegata—camel's foot tree; A., B. & M'. KANCHAN; G. KOVIDARA; H. & P. KACHNAR; K. ULIPE, BILI MANDARA; M. MANDARUM; O. KANCHANA; T. TIRUVATTI; T'. ADAVIMANDARA

Benincasa hispida—ash gourd; A. KOMORA; B. CHALKUMRA; G. KOHWLA; H. & P. PETHA; K. BOODU GUMBALA; M. KUMPALAM; M'. KOHALA; O. PANIKAKHARU; T. KUMPALY; T'. PULLA GUMMUDI

Beta vulgaris—beet; A. BEET-PALENG; B. BEET-PALANG; G. & M'. BEET; H. & P. CHUKANDAR; K. BEET ROOT; O. PALANGA SAGA, BEET

Biophytum sensitivum—sensitive wood-sorrel; A. & B. BAN-NARANGA; G. JAHARERA; H. LAJALU; K. HORA MUNI; M. MUKKUTTI, THINDANAZHI; M' LAJARI

Blumea lacera—A. KUKURSHUTA; B. KUKURSONGA; G. KALARA; H. &. P. KOKRONDA; K. GANDHARI GIDA; M'. BURANDO; O. POKASUNGA; T. KATU-MULLANGI; T'. KARUPOGAKU

Boerhaavia diffusa—A. PONONUA; B. & M'. PUNARNAVA; G. GHETULI; H. THIKRI, GADHAPURVA; K. BALAVADIKE, GONAJALI, RAKTA PUNARNAVA; M. THAZHUTHAMA; O. GHODAPURUNI; P. BISKHAPPA, ITSIT; T. MUKKARATAI; T'. PUNARNABA

Bombax ceiba (=*Salmalia malabarica*)—silk cotton tree; A. SIMALU; B. SIMUL; G. RATOSHEMALO; H. & P. SIMAR, SIMBAL; K. BOORUGA, KEMPU-

BOORUGA; M. & T'. ELAVU, MULLILAVU; M'. KATE SAVAR; O. SIMULI; T' KONDABURAGA, SALMALI

Borassus flabellifer—palmyra-palm; A. & B. TAL; G. & M'. TAD; H. & P. TAR; K. TALE MARA, TATI NUNGU; M. KARIMPANA; O. TALA; T. PANAI; T'. THADI

Boswellia serrata—incense tree; A. DHUNA; B. DHUP, GUGGUL; G. DHUP-GUGALI; H. GUGUL; K. CHILAKA DHUPA, CHILAKADI, MADDI; M. MUKUNDAM; M'. DHUP; P. SALAI, SALER; T. ATTAM; T'. ANDUGA

Brassica campestris—mustard; A. SARIAH; B. SARISHA; G. SAFED-RAI; H. & P. SARSON; K. SASIVE; M. KATUKU; M'. MOHORI; O. SOROSHA; T. KARUP-PUKKADUGU; T'. AAVA

Bryophyllum pinnatum—sprout-leaf plant; A. PATEGAZA, DUPORTENGA; B. PATHURKUCHI; H. ZAKHM-I-HAYAT; K. KADU BASALE; M'. PANPHUTI; O. AMARPOI; P. PATHURCHAT; T. RANAKALLI; T'. SIMAJAMUDU

Butea monosperma—flame of the forest or parrot tree; A., B. & M'. PALAS; G. KHAKARA; H. & P. DHAK; K. MUTTUGA; M. CHAMATHA; O. PALASA; T. SAMITHU, PALASAM; T'. MODUGA

Caesalpinia crista (=*C. bonducella*)—fever nut; A. LETAGUTI; B. NATA; G. KAKACHIA; H. KATKARANJ; K. GAJJIGA; M. KAZHANCHIKKUROO; M' SAGARGOTA; O. GILA; P. BEL KARANJWA; T. KALAKKODI

Caesalpinia pulcherrima—dwarf gold mohur or peacock flower; A. KRISHNA-CHURA; B. RADHACHURA; G. SANDHESHARO; H. GULETURA; K. KENJIGE GIDA, RATNA GANDHI; M. RAJMALLI; M'. SHANKASUR; O. KRUSHNACHUDA, GODI-BANA; P. KRISHANACHURA; T. MAYIRKONRAI; T'. TURAYI

Caesalpinia sappan—sappan or Brazil wood; B., H. & P. BAKAM; G. PATANG; K. PATHANGA, SAPPANGA; M. PATRANGAM; M'. PATANG; T. PATANGAM; T'. PATANGA

Cajanus cajan—pigeon pea or red gram; A. RAHAR-MAH; B. ARAHAR; G. TUVARE; H. RAHAR; K. THOGARI KALU; M. THUVARA; M'. TUR; O. HARADA; T. THOVARAY; T'. KANDULU

Calamus viminalis—cane; A. BAT; B., H. & P. BET; K. NEERU HAMBU, BETTA; M. CHOORAL; M'. VET; O. BETA; T. SURAI; T'. BETTAMU

Calophyllum inophyllum—Alexandrian laurel; B. & H. SULTANA-CHAMPA, PUNNAG; K. SURA HONNE, PINNE KAI, PUNNAGA; M. & P. PUNNA; M'. UNDI; O. POLANGA; T. PUNNAGAM; T'. PUNNAGA

Calotropis gigantea—madar; A. AKON; B. AKANDA; G. AKADO; H. & P. AK; K. EKKADA GIDA; M. & T. ERUKKU; M'. RUI; O. ARKA; T'. JILLEDU

Canavalia ensiformis—sword bean; A. KANTAL-URAHI; B. MAKHAN-SHIM; H. BARA-SEM; K. THAMATE BALLI, SHAMBE; M. & T. VAALAVARAKKAI; M' ABAI; O. BADA SIMBA, MAHARADA; P. BARASEM, TALWAR PHALI; T'. TUMBAT-TAN KAYA

Canna orientalis (=*C. indica*)—Indian shot; A. PARIJAT-PHUL; B. & O. SARBAJAYA; G. KARDALI; H. SABBAJAYA; K. KYANA GIDA; M. KATIUVAZHA; M'. KARDAL; P. HAKIK; T. KALVAALAI

Cannabis sativa—hemp; A., B., H. & P. BHANG, GANJA; G., P. & T. GANJA; K. GANJA GIDA, BHANGI; M. KANCHAVU; M'. BHANG; O. BHANGA, GANJEI; T'. GANJA CHETTU

Capparis sepiaria—B. KANTA-GURKAMAI; G. KANTHARO; H. HIUN; K. OLLE UPPI GIDA, KADU KATTARI; M. THORATTI; M'. BASHINGI, KAKADANI; O. KANTIKAPALI; P. HIUS, HIUNGARNA; T. KARINDU; T'. NALLUPPI

Capsicum annuum—chilli; A. JOLOKIA; B. LANKA, MARICH; G. LALMIRICHI; H. & P. LAL-MIRCH; K. MENASINA KAI; M. MULAGU; M'. MIRCHI; O. LANKA-MARICHA; T. MILAGU; T'. MIRAPAKAYA

Cardiospermum halicacabum—balloon vine; A. KOPALPHOTA; B. KAPAL-PHUTKI, SHIBJHUL; G. KARODIO; K. BEKKINA BUDDE GIDA, ERUMBALLI; M.

GLOSSARY OF NAMES OF PLANTS

VALLIYUZHINJA; M'. KAPALPHODI; O. PHUTPHUTKIA; P. HAB-UL-KULKUL; T. MODAKATHAN; T'. BUDDAKAKKIRA, KASARITIGE

Carica papaya—papaw; A. AMITA; B. PAYPAY; G. PAPAYI; H. & P. PAPITA; K. PARANGI, PAPAYA, PAPALI; M. KARUTHA; M'. POPAI; O. AMRUTABHANDA; T. PAPALI; T'. BOPPAYI

Carissa carandas—A. KORJA-TENGA; B. KARANJA; H. KARAUNDA; K. KAVALI GIDA, KARANDA; M. ELIMULLU; M'. KARVANDA; O. KHIRAKOLI; P. GARANDA; T. KALAKKAI; T'. KALIVI VAKA

Carthamus tinctorius—safflower; A. & B. KUSUMPHUL; G. KUSUMBO; H. & P. KUSAM; K. KUSUBI, KUSUME; M. SINDOORIM; M'. KARDAI; O. KUSUMA; T. KUSUMBA; T'. AGNISIKHA

Carum copticum—see *Trachyspermum ammi*

Cassia fistula—Indian laburnum; A. SONARU; B. SHONDAL; G. GARMALA; H. & P. AMALTASH; K. KAKKE GIDA, HONNAVARIKE; M. & T. KONNAI; M'. BAHAWA; O. SUNARI

Cassia sophera—A. MEDELUA; B. KALKASUNDE; G. KASUNDARI; H. & P. KASUNDA; K. KASAMARDA; M. PONNARAN or PONNAMTHAKARA; M'. KALAKASBINDA; O. KUSUNDA; T. PONNAVEERAN

Cassytha filiformis—B. AKASHBEL; H. AMARBELI; K. AKASHA BALLI, MANGANA UDIDARA; M. AKASAVALLI; M'. AKASHVALLI; O. AKASHA BELA; P. AMIL, AMARBELI

Casuarina equisetifolia—beef-wood tree or she-oak; A., B., H. & P. JHAU; G. VILIYATI SARU; K. SARVE MARA, GALI MARA; M. CHOOLAMARUM; KATTADIMARUM; M'. KHADSHERANI; O. JHAUN; T. SAVUKKU; T'. SARAVU

Cayratia carnosa—see *Vitis trifolia*

Cedrela toona—toon; B., H. & P. TOON; K. NANDI URUKSHA, NANDURI, BELANDI; M. CHUVANNAGIL; M'. MAHANIM; T. MALAVEMBU; T'. GALIMANU

Celosia cristata—cock's comb; A. KUKURA-JOA-PHUL; B. MORAG-PHUL; G. LAPADI; H. JATADHARI; K. MAYURA SHIKHI; M. KOZHIPULLU; M'. KOMBADA; O. GANJA-CHULIA; P. KUKUR-PHUL

Centella asiatica—Indian pennywort; A. MANIMUNI; B. THULKURI; G. KARBRAHMI; H. & P. BRAHMI-BOOTI; K. ONDELAGA, BRAHMI SOPPU; M. KODANGAL, KOTAKAN; M'. BRAHMI; O. THALKUDI; T. VULLARAI

Cestrum nocturnum—queen of the night; A. & B. HAS-NA-HANA; H. RAT-KIRANI; K. RATRI RANI HOOVU

Chenopodium album—A. JILMIL-SAK; B. & H. BATHUA-SAK; G. CHEEL; K. HUNCHIK POLYA; M'. CHAKAVAT; O. BATHU SAGA; P. BATHU; T. PARUPUKKIRAI

Chrysanthemum coronarium—A. & B. CHANDRAMALLIKA; G. & H. GULDAUDI; K. SHAVANTIGE, SEVANTIGE; M. SHEVANTI; O. SEBATI; P. GULDADU; T. SHAMAN-TIPPU; T'. CHAMANTI

Chrysopogon aciculatus—love thorn; A. BONGUTI; B. CHORKANTA; K. GANJIGARIKE HULLU; O. GUGUCHIA; P. CHORKANDA

Cicer arietinum—gram; A. BOOTMAH; B. CHHOLA; G., H. & P. CHANA; K. KADALE, CHANA; M. & T. KADALAI; M'. HARABHARA; O. BUTA; T. SANIKALU

Cinnamomum camphora—camphor; A. & B. KARPUR; G., H. & M'. KAPUR; K. KARPURADA GIDA; M. KARPPURAVRIKSHAM; O. KARPURA; P. KAFUR; T. KARUPPURAM; T'. KAPPURAMU

Cinnamomum tamala—bay leaf; A. TEJPAT, MAHPAT; B. TEZPATA; G. & H. TEZPAT; K. KADU DALCHINNI; M'. TAMAL; O. & P. TEJPATRA; T. TALISHAPPATTIRI; T'. TALLISHAPATRI

Cinnamomum zeylanicum—cinnamon; A., B., G., M'., O. & P. DALCHINI; H. DARCHINI; K. DALCHINNI, LAVANGA CHAKKE; M. & T. ILLAVANGAM; T'. DALCHINA CHEKKA

Cissus quadrangularis—A., B. & H. HARHJORA; K. MANGARA VALLI, SANDU BALLI; M. PIRANTA; M'. KANDAWEL; O. HADAVANGA; P. GIDAR-DAK, DRUKRI; T. PIRANDAI; T'. NALLERU

Citrullus colocynthis—colocynth; A. KOABHATURI; B. MAKAL; G. & H. INDRAYAN; K. DODDA HALMEKKE, INDRAVARUNI; M. & T. KUMMATHIKKAI, PEYKUMMATTY; M'. KAVANDAL; O. INDRAYANA; P. TUMMA; T'. PATSAKAYA

Citrullus lanatus—water melon; A. KHORMUJA; B. TARMUZ; G. KARIGU; H. & P. TARBUZA; K. KALLANGADI BALLI; M. & T. KUMMATTIKKAI; M'. KALINGAD; O. TARABHUJA

Citrus aurantifola—sour lime; A. NEMU-TENGA; B. KAGJI-NEBU; G. LIMBU; H. NIMBOO; K. NIMBE; M. CHERUNARAKAM; M'. KAGADI LIMBU; O. LEMBU; P. GALGAL; T. ELIMICHCHAM; T'. NARINJA

Citrus grandis—pummelo or shaddock; A. REBAB-TENGA; B. BATABI-NEBU; G. OBAKOTRU; H. & P. CHAKOTRA; K. CHAKKOTHA; M. BAMBLEENARAKAM; M'. PAPANAS; O. BATAPI; T. BAMBALMAS

Citrus reticulata—orange; A. KAMALA-TENGA; B. KAMALA; G. SUNTRA; H. NARANGI; K. KITTALE; M. NARAKAM; M'. SANTRA; O. KAMALA; P. SANGTRA; T. NARANGAM; T'. NARANJI

Cleome—see *Polanisia*

Clerodendrum infortunatum—A. BHETTITA; B. & H. BHANT, GHENTU; K. MADARASA MALLIGE, IBBANE; M. PERU-VALLEM; G. & M'. KARI; O. KUNTI; P. KARU; T. KARUKANNI; T'. BASAVANAPADU

Clitoria ternatea—butterfly pea; A., B. & O. APARAJITA; G. GARANI; H. APARAJIT, GOKARNA; K. GIRI KARNIKE, SATUGARA GIDA; M. SANKHUPUSHPAM; M'. GOKARNA; P. APARAJIT, NILI LOEL; T. KAKKATAN; T'. SANGAPUSHPAM

Coccinia indica (=*C. cordifolia*)—A. BELIPOKA; B. TELAKUCHA; H. KUNDARU, BHIMBA; K. THONDE KAYI, KAGE DONDE; M. KOVEL; M'. TONDALE; O. KUNDURI, KAINCHI-KAKUDI; P. GHOL; T. KOVARAI; T'. KAKIDONDA

Cocos nucifera—coconut-palm; A. NARIKOL; B. NARIKEL; G., H. & P. NARIYAL; K. TENGU; M. THENGU, NALIKERAM; M'. NARAL; O. NADIA; T. THENGU; T'. TENKAYA

Coix lachryma-jobi—job's tears; A. KAURMONI; B. & P. KALA-KUNCH, GURGAR; G. KASAI; H. SANKRU; K. KALMATTU BEEJA, KOTHI BEEJA; M'. RAFI JONDHALA; O. GARAGADA; T. KATTU KUNDUMANI

Colocasia esculenta—taro; A. & B. KACHU; H. & P. KACHALU; K. KESAVINA GEDDE, SAVE GEDDE; M. CHEMPU; M'. KASALU; O. SARU; T. SAMAKILANGOO; T'. CHEMA

Commelina bengalensis—A. KONASIMOLU; B. KANSHIRA; G. MHOTUNSHUSH-MULIYUN; H. KANKIYA; K. GUBBACHI BALE, KANNE SOPPU; O. KANSIRI

Coriandrum sativum—coriander; A., B., H., O. & P. DHANIA; G. DHANE; K. KOTHAMBARI, HAVEEJA; M. & T. KOTTAMALLI; M'. KOTHIMBIR; T'. DHANIYALU

Crataeva nurvala (=*C. religiosa*)—A. & B. BARUN; G. VAYAVARNA; H. & P. BARNA; K. ADIRAJA, MAVALINGA, NERVALA; M. NIRMATHALAM; M'. WAYAWARNA; O. BARUNA; T. MAVALINGAM; T'. VOOLEMERI

Crinum asiaticum—B. & P. SUKHDARSHAN; M. POLATHALI; G. & M'. NAGDAUNA; H. PINDAR; K. VISHA MOONGILI, VISHA BIDURU; O. ARISA; T. VESHAMOONGHEE

Crocus sativus—saffron; A., B. & O. JAFRAN; G. & M'. KESHAR; H. & P. ZAFRAN; K. KUMKUMA KESARI; T. KUNGUMAPU; T'. KUNKUMAPUVU

Crotalaria juncea—Indian or sunn hemp; A. SHON; B. SHONE; G., H. & P. SAN; K. APSENA-BU, SANNA SENABU; M. THANTHALAKOTTI; M'. KHULKHULA; O. CHHANAPATA; T. SANAPPAI; T'. JANNAMU

Crotalaria sericea—rattlewort; A. GHANTAKORNA; B. ATASHI; H. JHUNJHU-

GLOSSARY OF NAMES OF PLANTS 869

NIA; K. GIJIGIJI GIDA; M. THANTHALAKOTTI; M'. GHAGRI; O. JUNKA; P. JHANJHANIAN

Croton tiglium—A. JOYPAL; B. JAIPAL; G. JAMAL GOET; H., M'. & P. JAMALGOTA; K. JAPALA; M. & T. NIRVALEM; O. BAKSA GACHHA

Cucumis melo—melon; A. BANGI; B. PHUTI; G. TARBUCH; H. & P. KHARBUZA, PHUTI & KAKRI; K. KARABUJA, KEKKARIKE; M. & T. THANNIMATHAI, M'. KHARBUJ; O. KHARBUJA

Cucumis sativus—cucumber; A. TIANH; B. SASHA; G. KAKRI; H., M'. & P. KHIRA; K. SOUTHE KAYI; M. MULLENVELLARI; O. KAKUDI; T. MULLUVELLARI

Cucurbita moschata—sweet gourd; A. RONGALAU; B. MITHAKUMRA; H. MITHAKADDU; K. SEEGUMBALA; M. MATHANGAI; M'. KALA BHOPALA; O. MITHA KOKHARU; P. HALWA-KADDU; T. POOSANIKAI

Curcuma amada—mango ginger; A. & B. AMADA; G. AMBA-HALDAR; H. AM-HALDI; K. MAVINA SHUNTI, KARPURA ARISINA; M'. AMBE HALAD; O. AMBA KASSIA ADA; P. AMBA HALDI; T'. MAMIDIALLAM

Curcuma lónga—turmeric; A. HOLODHI; B. HALOOD; G. & M'. HALAD; H. & P. HALDI; K. ARISINA; M. MANGAL; O. HALADI; T. MANJAL; T'. PASUPU

Cuscuta reflexa—dodder; A. AKASHILOTA, RAVANARNARI; B. SWARNALATA; G. AKASWEL; H. AKASHBEL, AMARBEL; K. BADANIKE, BANDALIKE, MUDITALE; M'. AMAR VEL; O. NIRMULI; P. AMARBEL

Cynodon dactylon—dog grass; A. DUBORIBON; B. DURBAGHAS; G. DURVA; H. & P. DOOB; K. GARIKE HULLU, KUDIGARIKE; M. & T. ARUGAMPULLU; M'. HARALI; O. DUBA GHASA; T'. GERICHA GADDI

Cyperus rotundus—sedge; A. MOTHA; B. & H. MUTHA; G. BARIK-MOTHA; K. TUNGE HULLU, KONNARI GEDDE; M. KORA; M'. & P. NAGAR-MOTHA; O. MUTHA GHASA; T. KORAI; T'. PURA GADDI

Dalbergia latifolia—Indian rosewood; B. SITSAL; G. SISAM; K. BEETE MARA, TODEGATTI; M. & T. ITTI; M'. SISSU; O. PAHADI SISU; T'. JITTEGI

Dalbergia sissoo—Indian redwood; A. SHISHOO; B. SISSOO; G. SHISHAM; H. & P. SHISHAM, TAHLI; K. BIRADI, BINDI, SHISSU; M. VEETI; M'. SHISAVI; O. SISSU

Datura stramonium—thorn-apple; A. DHOTURA; B. DHUTRA; G. DHATOORA; H. & P. DHUTURA; K. DATTURA, UMMATTI; M. UMMAM; M'. DHOTRA; O. DUDURA; T. OOMMATHAI; T'. UMMATHA

Delonix regia—gold mohur; A. RADHACHURA; B. KRISHNACHURA; G., H., M. & P. GULMOHR; K. SEEME SANKESWARA, KEMPU TURAI; M. MARAMANDARAM; O. RADHACHUDA; T. MAYILKONNAI

Dendropthoe falcata (=*Loranthus longiflorus*)—A. ROGHUMALA; B. MANDA; G. VANDO; H. BANDA; K. SIGARE BANDANIKE; M. ITHTHIL; M'. BANDGUL; O. MALANGA, MADANGA; P. PAND; T. PULLURUVI; T'. BAJINNIKI, BADANIKA

Desmodium gangeticum—B. SALPANI; G. SALVAN; H. SALPAN; K. SALAPARNI, KOLAKU NARU; M. PULLATI; M'. SALPARNI; O. KURSOPANI; P. SHALPURHI; T. PULLADI; T'. GITANARAM

Desmodium gyrans—Indian telegraph plant; A. & B. BANCHANDAL, GORACHAND; H. BAN-CHAL; K. NAGATAGARE, TELEGRAPH GIDA; O. GORA CHANDA, TELEGRAPH GACHHA; P. PAUDA TAR

Dillenia indica—A. OU-TENGA; B., H. & P. CHALTA; G. CARAMBAL; K. MUCHHILU, KALTEGA; M. VALLAPUNNA; M'. KARAMAL; O. OU; T. UVATTEKU; T'. UVVA

Dioscorea alata—white yam; A. KATH-ALOO, PATNI-ALOO; B. & H. CHUPRIALOO, KHAM-ALOO; K. MUDI GENASU, TOONA GENASU; M'. KONA; O. KHAMBOALOO; P. KNISS; T. KAYAVALLI; T'. GUNAPENDALAMU

Dioscorea bulbifera—wild yam; A. GOCH-ALOO; B. GACHH-ALOO; G. SAURIYA; H. & P. ZAMINKHAND; K. HEGGENASU, KUNTA GENASU; M. KATTUKACHIL;

M'. KADU KARANDA; O. DESHI-ALOO, PITA-ALOO; T. KATTUKKILANGU;
T'. PENDALAMU

Diospyros ebenum—Indian ebony; B. ABLOOSH; H. TENDU; K. BALE MARA;
M. KARU; M'. & P. ABNUS; T. KAKKAYITALI; T'. NALLAVALLUDU

Diospyros peregrina—wild mangosteen; A. & O. KENDU; B. & P. GAB; G.
TEMRU; H. TENDU; M. VANANJI; M'. TEMBURNI; T. TUVARAI; T'. TUMMIKA

Dolichos lablab—country bean; A. UROHI; B. SHIM; G. AVRI; H. & P. SEM;
K. AVARE BALLI; M. SIMA-PAYARU; M'. PAVATA; O. SIMA; T. AVARAI;
T'. CHIKKUDI

Duranta repens (=*D. plumieri*)—A. JEORA-GOCH; B. DURANTA-KANTA; H. &
P. NILKANTA; K. DURANTHA KANTI; M'. DURANTA; O. BILATI KANTA,
BENJUATI

Ecbolium linneanum—B. NILKANTHA; H. & P. UDAJATI; K. KAPPUKARNI,
KAPPUBOBLI; M. KURANTA; M'. RAN ABOLI; O. NILAKANTHA; T. NILAMBARI

Eclipta alba—A. KEHORAJI; B. KESARAJ; G. BHANGRA; H. & P. SAFED
BHANGRA; K. GARUGADA GIDA, GARUGALU; M. &. T. KAYYANYAM, KAITHONNI;
M'. MAKA; O. KESHDURA

Elephantopus scaber—elephant's foot; B. & H. HASTIPADA, GOBHI; G. BHOPA
THARI; K. HASTIPADA, HAKKARIKE; M. &. T. ANACHUVADI; M'. HASTI PAD;
O. GOBI; P. GAOZBAN

Eleusine coracana—B. & H. MARUA; G. NAVTO; K. & T'. RAGI; M. PANJAP-
PULLU; M'. NACHANI; O. MANDIA; P. KODRA, MANDWA; T. KOLVARAKU

Emblica officinalis—emblic myrobalan; A. AMLOKI; B. AMLA, AMLAKI;
G. AMBALA; H. AMLA, AMLIKA; K. NILLI-BETTADA NELLI, NELLI-ISNELLI;
M. & T. NELLIKKAI; M'. AWALA; O. ONLA; P. AMLA; T'. USIRI

Enhydra fluctuans—A. HELACHI-SAK, MONOA-SAK; B. & P. HALENCHA; H. HARUCH;
M'. HARKUCH; O. HIDIMICHI, PANI SAGA

Entada gigas (=*E. phaseoloides*)—nicker bean; A. GHILA; B., H., O. & P.
GILA; G. SUVALI-AMLI; K. GARDALA, HALLEKAYI BALLI; M. KAKKUVALLY;
M'. GARBI; T. CHILLU; T'. GILLATIGAI

Enterolobium saman—see *Pithecolobium saman*

Ervatamia coronaria—crepe-jasmine; A. KOTHONAPHUL; B. & M'. TAGAR;
H. & P. CHANDNI; K. NANDI BATLU, NANJA BATLU; M. & T. NANTHIAR
VATTAM; O. TAGARA

Erythrina variegata—coral tree; A. MODAR; B. MANDAR; G. PANARAWAS;
H. PANJIRA; K. HARIVANA, VARJIPE; M. & T. MURUKKU; M'. PANGARA;
O. PALDHUA; P. DARAKHT FARID, PANGRA

Euphorbia antiquorum—B. BAJBARAN or TESHIRA-MANSHA; G. TANDHARI;
K. BONTE GALLI, CHADARA GALLI; M. CHATHIRAKKALLI; M'. CHAUDHARI
NIWDUNG; O. DOKANA SIJU; P. DANDA THOR, TIDHARA SEHUD; T. SHADRAI-
KALLI; T'. BONTHAKALI

Euphorbia neriifolia—A. SIJU; B. MANSHASIJ; G. THOR; H. SIJ; K. ELE
GALLI; M. & T. ILAKKALLI; M'. CHAUDHARI NIWDUNG; O. PATARA SIJU;
P. GANGICHU; T'. AKUJEMUDU

Euphorbia nivulia—A. SIJU; B. SIJ; G. THOR KANTALO; H. SIJ, THOR; K.
GOOTA GALLI; M. & T. ILAKKALLI; M'. NIWDUNG; O. SIJU; T'. AKUJEMUDU

Euphorbia (=*Poinsettia*) *pulcherrima*—poinsettia; A. LALPAT; B., M'. & P.
LALPATA; K. POINSETTIA GIDA; O. PANCHUTIA; P. LAL-PATTI; T. MAYILKUNNI

Euryale ferox—A. NIKORI; B., H. & P. MAKHNA; M'. PADMA KANT, MAKHAN;
O. KANTA PADMA

Evolvulus alsinoides—G. JHINKIPHUDARDI; H. SANKHAPUSHPI, VISHNU-
KRANTA; K. VISHNUKRANTHI; M. VISHNUKTANTHI, KRISHNAKTANTHI;
M'. VISHNUKRANT; O. BICHHAMALIA; P. SHANKH-HOLI; T. VISHNUKIRANDI;
T'. VISHNUKRANTHI

Feronia limonia—see *Limonia acidissima*

GLOSSARY OF NAMES OF PLANTS

Ferula assa-foetida—asafoetida; A., B., G., H., M'. & P. HING; K. INGU, HINGU; M. KAYAM; O. HENGU; T'. INGUVA

Ficus bengalensis—banyan; A. BORGOCH; B. BOT; H. & P. BARH; G. & M' WAD; K. AALADA MARA; M. PEERALU; O. BARA; T. AALUMARAM; T'. MARRI

Ficus glomerata—A. DIMORU; B. JAJNYA-DUMUR; G. UMBARO; H. & P. GULAR; K. ATHI; M. & T. ATHTHIMARAM; M'. UMBAR; O. DIMURI; T'. BODDA

Ficus religiosa—peepul; A. ANHOT; B. ASWATHA; G. JARI; H. & P. PIPAL; K. ARALI, ASWATHA; M. ARAYALU; M'. PIMPAL; O. ASWATHA; T. ARASU; T'. ASWATHAM

Flacourtia jangomas (=*F. cataphracta*)—A. PONIAL; B. & H. PANIALA; G. TALISPATRA; K. GORAJI, CHANCHALLI, TALISAPATRE; M. & T. TALISAM; M'. JUGGUM; O. PANIONLA; P. PANIALA, PANIAUNLA; T'. TALISAPATRAMU

Flacourtia ramontchi & *F. indica*—B. BOINCHI; H. BOWCHI, BILANGRA; K. GAJABIRA, MULLUTARE, KUDAVALE; M'. BHEKAL; O. BAINCHA KOLI; P. KATAL, KUKAI; T. MALUKKARAI; T'. KANAREGU

Foeniculum vulgare—anise or fennel; A. GUAMOORI; B. PANMOURI; G. WARIARI; H. & P. SAUNF; K. DODDA JEERIGE, DODDA SOMPU; M'. BADISHEP; O. PAN MOHURI

Garcinia mangostana—mangosteen; B., H. & M'. MANGUSTAN; G. & K. MANGOSTEEN; M. SULAMPULI; O. MANGOSTEEN, SITAMBU; T. SULAMBULI

Gardenia jasminoides—cape jasmine; A. TOGOR; B., H. & P. GANDHARAJ; G. DIKAMALI; K. SUVASANE MALLE; M'. GANDHRAJ; O. SUGANDHARAJ

Girardinia zeylanica—A. SHORUCHORAT; B. BICHUTI; M. AANACHORIYANAM; O. BICHHUATI; P. BICHUTI, BHABHER

Gloriosa superba—glory lily; A. & B. ULATCHANDAL; G. & M'. KHADYANAG; H. KALIARI, KULHARI; K. SHIVASHAKTI, LANGULIKA; M. MANTHONNI, PARAYANPOOVA; O. PANCHAANGULIA; P. GURHPATNI, KULHARI; T. KALAPAI-KILANGU; T'. AGNISIKA

Glycosmis arborea—A. CHAULDHOA; B. ASHHOURA; H. BANNIMBU; M. PANAL; O. CHAULADHUA

Gossypium sp.—cotton; A. KOPAH; B., H. & P. KAPAS; G. RUI; K. HATHI; M. KURUPARATHY; M'. KAPUS; O. KOPA; T. PARATHY

Gynandropsis gynandra—A. BHUTMULA; B. HURHURE; G. ADIYA-KHARAM; H. HURHUR; K. NARAMBELE SOPPU; M. KATTUKATUKU; M'. TILVAN; O. ANASORISIA, SADA HURHURIA; P. HULHUL; T. NAIKADUGU; T'. VAMINTA

Helianthus annuus—sunflower; A. BELIPHUL; B. & O. SURJYAMUKHI; G. SURYAMUKHI; H. & P. SURAJMUKHI; K. SURYAKANTHI; M., T. & T' SURIYAKANTI; M'. SURYAPHUL

Heliotropium indicum—heliotrope; A. & B. HATISUR; G. HATHISUNDHANA; H. HATTASURA; K. CHELUKONDI GIDA, CHELUMANI GIDA; M. TEKKADA; M'. BHURUNDI; O. HATISUNDA; P. UNTH-CHARA

Hemidesmus indicus—Indian sarsaparilla; A. & B. ANANTAMUL; G. DURIVEL; H. ANANTAMUL, SALSA; K. SUGANDHI BERU, SOGADE BERU; M'. ANANTMUL; O. ANANTAMULA, KAPRI; P. DESI SARVA; T. NANNARI; T'. SUGANDIPALA

Hibiscus cannabinus—Madras or Deccan hemp; B. NALITA; G. BHINDI; H. AMBARI; K. PUNDI, GOGU; M. KANJARU; M'. AMBADI; O. KAUNRIA, NALITA; P. SAN-KUKRA; T. KACHURAI

Hibiscus esculentus—see *Abelmoschus esculentus*

Hibiscus mutabilis—A. & B. STHALPADMA; G. UPALASARI; H. GULIAJAIB; K. BETTA DAVARE, KEMPUSURYAKANTHI; M. CHINAPPARATTI; M'. GULABI BHENDI; O. THALAPADMA; P. GUL-I-AJAIB; T. SEMBARATTAI

Hibiscus rosa-sinensis—China rose or shoe-flower; A. JOBA; B. JABA; G. JASUNT; H. GURHAL, JASUM; K. KEMPU DASAVALA; M. CHEMPARATHY; M'. JASWAND; O. MANDARA; P. GURHAL, JIA PUSHPA; T. SAMBATHOOCHEDI; T'. DASANI

Hibiscus sabdariffa—rozelle; A. MESEKA-TENGA; B. MESTA; H. & P. PATWA;
K. KEMPU PUNDRIKE; M. PULICHI; M'. LAL-AMBADI; O. KHATA KAUNRIA
Hiptage bengalensis (=*H. madablota*)—A. MADHOILOTA; B. & O. MADHABI-
LATA; G. MADHAVI; H. MADHULATA; K. MADHABI LATHE; M. SITAPU;
M'. MADHUMALATI; P. MADHULATA, BANKAR; T. KURUKKATTI, MADAVI
Holarrhena antidysenterica—A. DUDKHORI; B. KURCHI; G. INDRAJAVANU;
H. KUTAJ, KARCHI; K. KODACHAGA, KODAMURUKA, KORJU; M. KODAKAPPALA;
M'. KUDA; O. PITA KORUA; P. INDER JAU, KAWAR
Hordeum vulgare—barley; A. & B. JOB; G. BAJRI; H. JAWA; K. BARLEY, JAVE
GODHI; M'. SATU; O. BARLEY, JABA; P. JAU; T. BARLIYARISI; T'. YAVAKA
Hydrocotyle—see *Centella*
Hygrophila spinosa—see *Asteracantha longifolia*
Impatiens balsamina—balsam; A. DAMDEUKA; B. DOPATI; H. GULMENDI;
K. GOURI HOOVU, BASAVANA PADA; M. & T. BALSAM; M'. TERADA; O. HARA-
GOURA; P. MAJITI, BANTIL, PALLU
Indigofera tinctoria—indigo; A., B., H. & P. NIL; G. GALI; K. OLLE NEELI,
HENNU NEELI; M. AMARY; M'. NEEL; O. NILA; T. AVARY; T'. AVIRI
Ipomoea batatas—sweet potato; A. & B. MITHA-ALOO; G. SHAKKARIA; H. &
P. SHAKARKAND; K. GENASU; M. MADHURAKI ZHANGU; M'. RATALA;
O. CHINI-ALOO, KANDAMULA; T'. KANDAMOOLA
Ipomoea aquatica (=*I. reptans*)—water bindweed; A. KALMAU; B. & H.
KALMI-SAK; G. NALINIBHAJI; K. BILI HAMBU; M'. KALAMBI, NAL; O. KALAMA
SAGA; P. NALI, KALMI SAG; T'. TUTICURA
Ipomoea pes-tigridis—B. LANGULI-LATA; K. ADAMBALLI; M. VELLATAMPU;
O. KANSARINATA; P. ISHOPECHAN
Ixora coccinea—A. & B. RANGAN; H. GOTAGANDHAL, RANJAN; K. MALE
HOOGIDA, KEPALE; M. & T. CHETHTHY, THETTY; M'. MAKADI; O. KHADIKA
PHULA, RANGANI; P. RUNGAN
Jasminum sambac—jasmine; A. JUTIPHUL; B. BELA; G. BATMOGRI; H. MUGRA;
K. GUNDU MALLIGE; M. MULLA; M'. MOGARA; O. MALLI
Jatropha curcas—physic or purging nut; A. BONGALI-ARA; B. BAGH-
BHARENDA; G. JEPAL; H. JANGLI-ARANDI; K. KADU HARALU; M. KATALA-
VANAKKU; M'. MOGALI ERAND; O. BAIGABA; P. JAMALGOTA, JABLOTA,
JAPHROTA
Jatropha gossypifolia—A. BHOTERA; B., H. & P. LAL-BHARENDA; K. CHIKKA
KADU HARALU, HATHI YELE HARALU; M'. VILAYATI ERAND; O. NALI BAIGABA,
VERENDA; T. ADALAI; T'. NEPALEMU
Jussiaea repens—A. TALJURIA; B. KESSRA; K. NEERU DANTU, KAVAKULA;
M. NIRGRAMPU; M'. PAN LAWANG; T. NIRKIRAMPU; T'. NIRUYAGNIVENDRAMU
Lagenaria siceraria—bottle gourd; A. JATI-LAU; B. & O. LAU; H. LAUKI;
K. EESUGAYI BALLI, HALU GUMBALA; M. & T. CHORAKKAI; M'. DUDHYA
BHOPALA; P. GHIYA; T'. ANAPA
Lagerstroemia speciosa (=*L. flos-reginae*)—A. AJAR; B., H. & P. JARUL;
K. HOLE DASAVALA, CHELLA, BENDEKA; M. NIRVENTEKKU; M'. TAMAN;
O. PATOLI; T. PUMARUTHU
Lantana aculeata & *L. indica*—lantana; G. GHANIDALIA; K. LANTAVANA GIDA;
M. PUCHEDI; M'. GHANERI; O. NAGA-AIRI; P. DESI LANTANA; T. ARIPPU;
T'. LANTANA
Laportea crenulata—devil or fever nettle; A. DOM-CHORAT; M. CHORIYANAM;
M'. & P. CHORPATTA; T'. OTTAPLAVU
Lathyrus aphaca—wild pea; A. & B. BAN-MATAR; G. JANGLI VATANA;
H. JANGLI MATAR; M'. VANMATAR; O. JANGALI MATAR; P. JANGLI MATAR,
RAWARI
Lathyrus sativus—A. KOLA-MAH; B., H. & O. KHESARI; G. MATER; K. CHIKKA
TOGARI, VISHA TOGARI, KESARI BELE; M'. LAKH; P. KISARI DAL

GLOSSARY OF NAMES OF PLANTS

Lemna paucicostata—duckweed; A. SORUPUNI; B. KHUDI-PANA; K. NEERU HASARU CHUKKE; M'. TIKLICHE SHEWALE; O. CHUNIDALA, BILATI DALA; P. BUR

Lens culinaris—lentil; A. MOSOORMAH; B., H., M'. & P. MASUR; G. MASURIDAL; K. MASURU BELE, LENTEL GIDA; O. MASURA

Leonurus sibiricus—A. RONGA-DORON; B. DRONA; H. HALKUSHA, GUMA; O. KOILEKHIA; T'. ENUGUTUMMI

Lepidium sativum—garden cress; A. & B. HALIM-SAK; G. ASALIYA; H. HALIM; K. KURTHIKE, KURATHIRUGI; M'. ALIV; O. HIDAMBA SAGA; P. HALON

Leucas lavandulaefolia (=*L. linifolia*)—A. DORON, DURUM-PHUL; B. SWET-DRONA; G. JHINA-PANNI KUBO; H. CHOTA-HALKUSA; K. GANTU THUMBE, KARJALI GIDA; M. THUMPA; M'. DRONAPUSHPI, GUMA; O. GAISA; P. GULDODA; T. THUMBAI; T'. TAMMA CHETTU

Limonia acidissima—elephant-apple; A. & B. KATH-BAEL; G. KOTHA; H. & P. KAITHA; K. KADU BILVA PATRE, NAYI BELA; M. BLANKA; M. KAWATH; O. KAINTHA; T. VELAMARUM; T'. VELAGA

Linum usitatissimum—linseed; A. TICHI; B. TISHI; G. JAVA; H. & P. ALSHI; K. SEEME AGASE BEEJA; M'. JAWAS; O. PESI; T. AALIVIRAI

Lochnera rosea (=*Vinca rosea*)—periwinkle; A. & B. NAYANTARA; H. SADA-BAHAR; K. KEMPUKASI-KANIGALU, TURUKU MALLIGE; M. KASITHUMPA; M'. SADAPHULI; O. SADABIHARI; P. RATTAN JOT

Loranthus longiflorus—see *Dendropthoe falcata*

Luffa acutangula—ribbed gourd; A. JIKA; B. JHINGA; G. SIROLA; H. & P. KALITORI; K. HEERE BALLI; M. PEECHIL, PEECHINGAI; M'. DODAKA; O. JAHNI; T. PEECHANKA

Luffa cylindrica—bath sponge or loofah; A. BHOL; B. DHUNDUL; H. & P. GHIYATORI; M'. GHOSALE; O. PITA TARADA

Lycopersicum sculentum—tomato; A. BELAHI-BENGENA; B. BILATI-BEGOON; H. & P. TAMATAR; G. TAMETA, TOMATO, RAKTAVURNTTANK; K. TOMATO; M. & T. THAKKALIKKAI; M'. TAMBETA; O. BILATI BAIGANA; T'. THAKKALI

Madhuca latifolia (=*Bassia latifolia*)—A., B. & H. MAHUA; G. MAHUDA; K. HIPPE, ALIPPE, M'. MOHA; O. MAHULA; P. MOHWA; T. ILLUPAI; T'. IPPA

Malva verticillata—mallow; A. & B. LAFFA; H. & P. SONCHAL; K. KADU KADDALE

Marsilea quardrifolia—A. PANI-TENGECHI; B. SUSHNI-SAK; K. NEERU PULLAM PARACHI-ELE GIDA; M. NALILAKKOTAKAN; O. SUNSUNIA; P. CHAUPATI; T. ARAKKODAI

Martynia annua (=*M. diandra*)—tiger's nail; A. & B. BAGHNAKHI; G. VICHCHIDA; H. SHERNUI; K. HULI NAKHA, GARUDA MOOGU; M. & T. KAKKA-CHUNDU, PULINAGAM; M'. WINCHAURI; O. BAGHA NAKHI; P. HATHAJORI; T'. GARUDA MUKKU

Mentha viridis—spearmint or garden mint; A. PODINA; B., G., H. & M'. PUDINA; M. PUTIYINA

Mesua ferrea—iron-wood; A. NAHOR; B. NAGESWAR; H. NAGKESAR; K. NAGA KESARI, NAGA SAMPIGE; M. & T. IRUMPARATHTHAN; M'. NAGCHAMPAKA; O. NAGESWARA; P. NAGAR KESAR; T'. NAGAKESARI

Michelia champaca—A. & P. CHAMPA-PHUL; B. CHAMPA or SWARNACHAMPA; G. RAE CHAMPAC; H. CHAMPAK; K. SAMPIGE; M. & T. CHEMPAKAM; M'. SONCHAPHA; O. CHAMPA; T'. SAMPAKA

Millingtonia hortensis—Indian cork tree; B., H., M. & P. AKASNIM; K. SEESE BIRATO MARA; O. RIALI

Mimosa pudica—sensitive plant; A LAJUKILOTA; B. LAJJABATILATA; G. LAJJAWANTI; H. LAJWANTI, CHHUIMUI; K. MUTTIDARE MUNI, MUDUGU DAVARE; M. THOTTALVADI; M'. LAJALU; O. LAJAKULI, LAJKURI; P. LAJ-WANTI; T. THOTTASINIGI; T'. PEDDA NIDRAKANTHA

Mirabilis jalapa—four o'clock plant or marvel of Peru; A. GODHULIGOPAL; B. KRISHNAKOLI; H. GULABBAS; K. SANJE MALLIGE, GULBAKSHI, BHADRAKSHI; M. NALUMANICHEDI; M'. GULBAKSH; O. RANGANI, BADHULI; P. GUL-E-ABBASI; T. ANDIMANDARAI; T'. CHANDRAKANTA

Momordica charantia—bitter gourd; A. TITA-KERALA; B. KARALA, UCHCHE; G., H. & P. KARELA; K. HAGALA KAYI; M. & T. PAVAL, PAVAKKAI; M'. KARLE; O. KALARA; T'. KAKARA

Moringa oleifera—drumstick or horse radish; A. & O. SAJANA; B. SAJINA; G. SARAGAVA; H. SAINJNA; K. NUGGE MARA, MOCHAKA MARA; M. MURINGA; M'. SHEVAGA; P. SAONJNA; T. MURUNGAI; T'. MUNAGA

Morus alba & *M. nigra*—mulberry; A. NOONI; B. TOONT; G. TUTRI; H. & P. SHAH-TOOT; K. KAMBALI GIDA, RESHME HIPPALI GIDA; M. MALBERRY; M'. TUTI; O. TUTAKOLI; T'. POOTIKAPALLU

Mucuna prurita—cowage; A. BANDARKEKOA; B. ALKUSHI; G. KIVANCH; H. & P. KAWANCH; K. NASAGUNNI, NAYI SONKU BALLI; M. NAIKORUNA; M'. KHAJ KUIRA; O. BAIDANKA

Murraya paniculata (=*M. exotica*)—chinese box; A. KAMINIPHUL; B. & O. KAMINI; H. MARCHULA; K. KADU KARI BEVU, ANGARAKANA GIDA; M. MARAMULLA; M'. PANDHARI KUNTI; P. MARUA; T. KATTUKARUVEPPILAI; T'. NAGAGOLUGI

Musa paradisiaca—banana; A. KOL; B. KALA; G. & H. KELA; K. BALE GIDA, BALE HANNU; M. VAZHA; M'. KADALI, KEL; O. KODOLI, ROMBHA; T. VAZHAI; T'. ARATI, KADALI

Myristica fragrans—nutmeg; B., H., M'. & P. JAIPHAL; G. JAYIPHAL; K. JATIKAYI, JAPATRE; M. & T. JATHIKKAI; O. JAIPHOLO; T'. JAJIKAYA

Nelumbo nucifera (=*Nelumbium speciosum*)—lotus; A. PODUM; B. & O. PADMA; G. & M'. KAMAL; H. & P. KANWAL; K. KAMALA, TAVARE; M. THAMARA; T. THAMARAI; T'. TAMARA

Nerium indicum—oleander; A. KORBIPHUL; B. KARAVI; G. & M'. KANHER; H. & P. KANER; K. KANIGALU; M. & T. ARALY; O. KARABI; T'. GANNERU

Nicotiana tabacum—tobacco; A. DHOPAT; B. TAMAK; G., H., M'. & P. TAMBAKU; K. HOGE SOPPU, TAMBAKU; M. & T. PUKAYILA; O. DHUANPATRA; T'. POGAKU

Nigella sativa—black cumin; B. & O. KALA-JIRA; G. KADU-JEEROO; H. KALOUNJI; K. KARI JEERIGE; M. & T. KARUN-JIRAGAM; M'. KALA JIRE; P. KALONGI, KALAJIRA

Nyctanthes arbor-tristis—night jasmine; A. SEWALI; B. SHEWLI, SHEPHALI; G. RATRANE; H. HARSHINGAR; K. PARIJATA; M. PAVIZHAMULLA; M'. PARIJATAK; O. SINGADAHARA; P. HARSANGHAR; T. PAVELAM; T'. PARIJATHAM

Nymphaea lotus—water lily; A. BHET; B. SHALOOK; G. NILOPAL; H. & P. NILOFAR; K. KENDAVARE, KANNAIDILE; M. & T. AMPAL; M'. LALKAMAL; O. KAIN, KUMUDA; T'. KALUVA

Ocimum sanctum—sacred basil; A. TULASHI; B., G., H. & P. TULSI; K. SREE TULSI, VISHNU TULSI; M. & T. THULASI; M'. TULAS; O. TULASI; T'. ODDHI.

Oldenlandia corymbosa—B. & P. KHETPAPRA; G. PARPAT; H. DAMANPAPPAR; K. HUCHHU NELA BEVU, KALLU SABBASIGE; M'. PITPAPADA; O. GHARPODIA

Opuntia dillenii—prickly pear; A. SAGORPHENA; B. PHANIMONSHA; G. NAGNEVAL; H. NAGPHANI; K. PAPAS KALLI, CHAPPATE KALLI; M. ELAKKALLI; M'. PHADYA NIWDUNG; O. NAGAPHENI; P. CHITARTHOR; T. SAPPATHTHIKKALLI; T'. NAGADALLY

Orobanche indica—broomrape; B. BANIABAU; H. & P. SARSON-BANDA; K. BENKI GIDA, BODU GIDA; T. POKAYILAI-KALAN

Oroxylum indicum—A. BHATGHILA; B. SONA; G. PODVAL; H. ARLU; K. PATAGANI, SONEPATTA, TIGUDU; M. PATHIRI; M'. TETU; O. PHANPHANIA, PHAPANI; P. SANNA; T. PAYYALANTHA; T'. PAMPINI

GLOSSARY OF NAMES OF PLANTS

Oryza sativa—paddy; A., B. & H. DHAN; G. CHOKHA; K. BHATHA, NELLU; M. ARI; M'. BHAT; O. DHANA; P. CHAWAL; T. ARISHI; T'. VARI

Oxalis repens (=*O. corniculata*)—wood-sorrel; A. SENGAITENGA, TENGECHI; B. AMRULSAK; H. CHUKATRIPATI, KHATTIPATTI; K. PUTTAM PURALE; M. PULIYARILA; M'. AMBOSHI; O. AMBILITI, AMLITI; P. KHATTIBUTI

Paederia foetida—A. BHEDAILOTA; B. GANDHAL; G. GANDHANA; H. GANDHALI; M. TALANILI; M'. PRASARUM; O. PASARUNI; P. GUNDALI; T'. SAVIRELA

Pandanus tectorius (=*P. odoratissimus*)—screwpine; A. KETEKI; B. & G. KETAKY; H. & P. KEORA; K. TALE HOOVU, KEDIGE; M. KAITHA; M'. KEWADA; O. KIA; T. THAZHAI; T'. MOGIL

Panicum miliaceum—Indian millet; B., H., O. & P. CHEENA; G. SAMLI; K. BARAGU; M. THENA; M'. WARAI; T. VARAGU; T'. VARAGI

Papaver somniferum—opium poppy; A. AFUGOCH; B. AFING; G. APHIM; H. & P. POST; K. GASA GASE, APPEEMU GIDA; M. & T. GASHAGASHA; M'. APHU; O. APHIMA

Passiflora foetida—passion flower; A. JUNUKA; B., H. & P. JHUMKALATA; K. KUKKI BALLI; M. KRISTHUPAZHAM; M'. KRISHNA KAMAL; O. JHUMUKA LATA; T. SIRUPPUNAIKKALI; T'. JUKAMALLE

Pedilanthus tithymaloides—jew's slipper; B. RANGCHITA; H. NAGDAMAN; M. VERAKKODI; M'. VILAYATI SHER; O. BILATI SIJU, CHITA SIJU; P. NAG DAUN

Pennisetum typhoideum—pearl millet; B., H., O. & P. BAJRA; K. SAJJE, KAMBU; M. & T. KAMPU, BAJRA; M'. BAJARI; T'. SAJJA, SAJJALU

Pentapetes phoenicea—noon flower; B. DUPOHRIA; G. DUPORIO; H. & P. GULDUPAHARIA; K. BANDURE; M'. DUPARI; O. DIPAHARIA

Phaseolus aureus—green gram; A. MOGU-MAH; B. & H. MOONG; G. MUGA; K. HESARU; M. CHERUPAYARU; M'. HIRAVE MUG; O. JHAIN-MUGA; P. MUNG; T. PACHAPAYARU; T'. PESALU

Phaseolus mungo—black gram; A. MATI-MAH; B. MASH, KALAI; G. UDAD; H. URID; K. UDDU; M. UZHUNNU; M'. UDID; O. MUGA; P. MASH; T. ULUNNU; T. UDDULU

Phoenix sylvestris—date-palm; A. & B. KHEJUR; G., H. & P. KHAJUR; K. EECHALU, KHARJURA; M. ITTA; M'. KHARIK; O. KHAJURI; T. ICHCHAM; T'. ITHA

Phragmites karka—A. KHAGRA; B. & P. NAL; H. NUDA-NAR; K. HULUGILA HULLU; O. JANKAI

Phyllanthus acidus—A. HOLPHOLI, PORAMLOKHI; B. NOAR; H. CHALMERI, HARFARAURI; K. NELLI-KIRUNELLI; M. NELLIPULI, ARINELLI; M'. RAY AWALI; O. NARAKOLI; T. ARUNELLI; T'. RATSAVUSIRIKI

Phyllanthus emblica—see *Emblica officinalis*

Piper betle—betel; A., B., G., H. & P. PAN; K. VEELE DELE, YELE BALLI; M. & T. VETHILA; M'. NAGWELI; O. PANA; T'. TAMALAPAKU

Piper cubeba—cubeb; B., H. & O. KABAB-CHINI; G. TADAMIRI; K. B. I.A MENASU; M. & T. THIPPLI; M'. KABAB CHINI, KANKOL; T'. TOKAMIRIYALU

Piper longum—long pepper; A. PIPOLI; B. PIPOOL; G. PIPARA; H. PIPLI; K. HIPPALI; M. THIPPALI; M'. PIMPALI; O. PIPALI; P. DARFILFIL, MAGHAN

Piper nigrum—black pepper; A. JALUK; B. GOLMARICH; G. KALOMIRICH; H. GOLMIRCH; K. KARI MENASU; M. KURUMULAGU; M'. KALI MIRI; O. GOLA MARICHA; P. KALI MARCH; T. MILAGOO; T'. SAVYAMU

Pistia stratiotes—water lettuce; A. BORPUNI; B. PANA; G. JALAKUMBHI; H. & P. JALKHUMBI; K. ANTARA GANGE; M. MUTTAPPAYAL; M'. GANGAVATI; O. BORA JHANJI; T. AGASATHAMARAI; T'. AKASATAMARA

Pisum sativum—pea; A. MOTOR; B., H., O. & P. MATAR; G. VATANA; K. BATANI, VATAGI; M. PAYARU; M'. WATANE; P. PATTANI; T'. GUNDUSANI-GHELU

Pithecolobium saman—rain tree; A. SIRISH GOCH; K. MALE MARA; M. URAKKAM-THOONGIMARAM; M'. SAMAN; O. BADA GACHHA CHAKUNDA, BANA SIRISHA

Plantago ovata—flea seed; A., B. & O., ISOBGUL; G. UTHAMUJEERUM; H., M' & P. ISOBGOL; K. ISPHA GOLU, ISAMGOLU; M. KARKATASRINGI; T'. ISHAPPUKOL

Plumbago zeylanica—A. AGYACHIT; B. CHITA; G. CHITRAMULA; H., M'. & P. CHITRAK; K. BILI CHITRA MOOLA; M. & T. KODUVELI; O. DHALACHITA

Plumeria rubra—temple or pagoda tree; A. GULANCHI; B. KATGOLAP; G. RHAD CHAMPO; H. & P. GOLAINCHI; K. HALU SAMPIGE; M. EEZHAVA-CHEMPAKAM; M'. KHUR CHAPHA; O. KATHA CHAMPA; P. GULCHIN

Polanisia icosandra—B. HALDE-HURHURE; G. TILVAN; H. HULHUL; M. &. T. NAIKADUGU; M'. PIWALI TILVAN; O. ANASORISIA; P. BUGRA, GANDHULI; T. KUKKA VAVINTA

Polianthes tuberosa—tuberose; A., B. & O. RAJANIGANDHA; H. & P. GULSHABO; K. SUGANDHA RAJA; M'. GULCHHADI; T. NILASAMPANGI; T'. SUKANDARAJI

Polyalthia longifolia—mast tree; A. & O. DABADARU; B. DEBDARU; G. ASHOPALO; H. & M'. ASHOK; K. PUTRAJEEVI, KAMBADA MARA; M. ARANAMARAM; P. DEVIDARI; T. NETTILINGAM; T'. DEVADARU

Polygonum sp.—A. BIHLONGONI; B. PANI MARICH; H. NARI; M. MOTHALA-MOOKA; O. MUTHI SAGA; P. NARRI; T. AATALARIE

Portulaca oleracea—purslane; A. HANHTHENGIA; B. NUNIA-SAK; G. LONI; H. & P. KULFA SAG; K. DODDA GONI SOPPU; M. KARICHEERA; M'. GHOL; O. BALBALUA; T. KARIKEERAI; T'. PEDDAPAVILIKURA

Pothos scandens—A. HATILOTA; G. MOTO PIPAR; K. ADKE BEELU-BALLI, AGACHOPPU; M. ANAPPARUVA; M'. ANJAN VEL; O. GAJA PIPALI; P. GAZPIPAL

Prosopis spicigera—A. SOMIDH; B., H., M'. & O. SHOMI; G. KANDO; K. VUNNE, PERUMBE; M. PARAMPU; P. JAND; T. PERUMBAI; T'. JAMBI

Psidium guayava—guava; A. MODHURI-AM; B. PAYARA; G. JAMFAL; H. & P. AMRUD; K. SEEBE, CHEPE, PERALA; M. PERAKKA; M'. PERU; O. PIJULI; T. KOYYA; T'. JAMA

Pterospermum acerifolium—A. KONOKCHAMPA; B. MOOCHKANDA; H. KANAKCHAMPA; K. MUCHUKUNDA GIDA; M'. MUCHKUND; O. MOOCHKUNDA; T. VENNANGU; T'. MUSHKANDA

Punica granatum—pomegranate; A. & B. DALIM; G. DADAM; H. & P. ANAR; K. DALIMBE; M. MATALAM; M'. DALIMB; O. DALIMBA; T. MADULAM

Quamoclit pinnata (=*Ipomoea quamoclit*)—A. KUNJALOTA; B. KUNJALATA, TORULATA; H. & P. KAMLATA; K. KAMALATHE; M'. GANESH PUSHPA; O. KUNJALATA; T'. KASIRATNAM

Quisqualis indica—Rangoon creeper; A. MADHABILOTA; B. SANDHYAMALATI; G. BARMA SINIVEL; H. & P. LAL MALTI; K. RANGOON KEMPUMALLE; M'. LAL CHAMELI; O. MODHUMALATI; T. RANGOON MALLI

Raphanus sativus—radish; A., B., M'. & O. MULA; H. & P. MULI; K. MOOLANGI; M. MULLANKI; T. & T'. MULLANGI

Rauwolfia serpentina—A. CHANDO; B., G. & T. SARPAGANDHA; H., M'. & P. SARPGANDH; K. SARPAGANDHI; SHIVANABHI BELLI; SUTRANABHI; M. AMALPORIYAN; O. PATALA GARUDA

Ricinus communis—castor; A. ERIGOCH; B. & P. ARANDA; G. ERANDI; H. RENDI; K. HARALU; M. & T. AVANAKKU; M'. ERAND; O. JADA; P. RENDI, ARANDA; T'. AMUDAMU

Rumex vesicarius—sorrel; A. CHUKA-SAK; B. CHUKA-PALANG; H. CHUKA, KHATTA-PALAK; K. CHUKKI SOPPU, SUKKE SOPPU; M'. CHUKA; O. PALANGA; P. KHATTA-MITHA; T'. CHUKKAKURA

Saccharum officinarum—sugarcane; A. KUNHIAR; B. AKH; G. SHERDE; H. GUNNA; K. KABBU; M. & T. KARIMPU; M'. USA; O. AKHU; P. GUNNA; T'. CHERUKU

GLOSSARY OF NAMES OF PLANTS

Saccharum spontaneum—A. KANHIBON; B. KASH; G. & H. KANS; K. DHARBE, KADU KABBU; M. NAINKANA; M'. BAGBERI; O. KASHATANDI; P. KAHI

Salmalia malabarica—see *Bombax ceiba*

Sansevieria roxburghiana—bowstring hemp; A. GUMUNI; B. MURGA, MURVA; H. MARUL, MURVA; K. MANJINA NARU, GODDUMANJI; M. PAMPINPOLA; O. MURUGA; T. MARUL

Santalum album—sandalwood; A., B., H., M'. & P. CHANDAN; G. SUKHADA; K. SREEGANDHA; M. & T. CHANNANAMARAM; O. CHANDANA; T'. CHANDANAMU

Sapindus mukorossi & *S. trifoliatus*—soap-nut; A. MONICHAL, HAITAGUTI; B., H., M'. & P. RITHA; G. ARITHA; K. ANTUVALA, NOREKAYI; M. URVANJI; O. RITHA, MUKTAMANJI; T. PONNANKOTTAI; T'. KUNKUDU

Saraca indica—asoka tree; A., B., & P. ASOKA; G. ASUPALA; H. SEETA ASOK; K. ASHOKADA MARA, KENKALI, ACHANGE; M. & T. ASOKAM; M'. SITECHA ASHOK; O. ASOKA; T'. ASAKAMU

Sesamum indicum—gingelly; A. TISI; B., H., M'. & P. TIL; G. MITHO TEL; K. YELLU; M. & T. ELLU; O. KHASA, RASHI; T'. NUVVULU

Sesbania grandiflora—A. & B. BAKPHUL; G. AGATHIO; H. & P. AGAST; K. AGASE, CHOGACHI; M. AGATHI; M'. AGASTA; O. AGASTI; T. AGATHYKKEERAI; T'. AVISI

Sesbania sesban—A. JOYANTI; B. JAINTI; G. RAYSANGANI; H. & P. JAINT; M. SHEMPA; M'. SEVARI; O. JAYANTI; T. SITHAGATHI

Setaria italica—Italian millet; A. KONIDHAN; B. KAUN; G. KANG; H. CHEENA, KAUNI; K. NAVANE, KONGU; M. NAVANA; M'. RALE; O. TANGUN; P. KANGNI; T. TENNAI; T'. KORRA, KORALU

Shorea robusta—sal tree; A., B., H. & P. SAL; G. RAL; K. BILE BHOGE, AASINA MARA, ASCHA KARNA; M. MARAMARAM; M'. SHALA, RALVRIKSHA; O. SALA; T. SHALAM; T'. GUGGILAMU

Sida cordifolia—A. BARIALA; B. BERELA; G. JANGLI METHI; H. BARIARA; K. HETHUTHI; M. KURUMTHOTTI; M'. CHIKANA; O. BISIRIPI; P. KHARENTI; T. KARUMTHOTTEE; T'. CHIRUBENDA

Smilax zeylanica—sarsaparilla; A. HASTIKARNA LOTA; B. KUMARIKA; H. CHOBCHINI; M'. GHOT VEL; O. KUMBHATUA, KUMARIKA; P. USHBA

Solanum ferox—A. BON BENGENA; B. RAM BEGOON; K. ANE SUNDE GIDA, HALADI GULLA; M. ANACHCHUNTA; M'. BHAJICHE WANGE; O. BHEJI BAIGANA; ANAICHUNDAI; T'. MULAKA

Solanum indicum—A. BHEKURI GOCH, TIT-BHEKURI; B. BRIHATI; G. UBHIRINGANI; H. BIRHATTA; K. KEMPU GULLA, HABBU GULLA; M. KATTUCHUNDA; M'. DORLI; O. KANTARA; P. BARI KANDIARI

Solanum melongena—brinjal; A. BENGENA; B. BEGOON; G. & O. BAIGANA; H. BAIGON; K. BADANE KAYI; M. VAZHUTHANA; M'. WANGE; P. BENGAN; T. KATHTHIRI; T'. VANGA

Solanum nigrum—black nightshade; A. POKMOU; B. GURKI; G. PILUDU; H. GURKAMAI, MAKOI; K. KARI KACHI GIDA, KEMPU KACHI, KAKA MUNCHI; M. MULAGUTHAKKALI; M. KANGANI; O. NUNNUNIA; P. MAKO; T. MANATHAKKALI; T'. KAMANCHICHETTU

Solanum surattense—A. KANTAKARI; B. KANTIKARI; G. BHOYARINGANI; H. KATELI, KATITA; K. RAMA GULLA; M. KANDAKARYCHUNDA; M'. KATERINGANI; O. ANKARANTI; P. KANDIALI; T. KANDANKATHTHIRI; T'. NELAVAKUDU

Solanum tuberosum—potato; A., B., H., O. & P. ALOO; G. PAPETA; K. ALUGEDDE; M. & T. URULAKKIZHANGU; M'. BATATA; T'. URULAGADDA

Sorghum vulgare—great millet; A. JOUDHAN; B. & G. JUAR; H. & P. JOWAR; K. BILI JOLA; M. & T. CHOLAM; M'. JAWAR; O. BAJARA; T'. JONNALU

Spinacia oleracea—spinach; A. MITHA-PALENG; B. PALANG, MITHA-PALANG; H. & M. & M'. PALAK

Sterculia foetida—A. BAN-BADAM; B., H., M'. & P. JANGLI BADAM; G. NAR-KAYA-UDA; K. PEE NARI, PATHALA MARA, BHETALA; M. ANATHTHONDI; O. JANGALI BADAM; T. PAEMARAM; T'. GUTTAPUBADAMU

Syzyzium aromaticum—clove; A., H. & P. LAUNG; B. LAVANGA; G. LAVANG; K. NAGE, LAVANGA; M. GRAAMPU; M'. LAWANG; O. LABANGA

Syzygium cuminii—A. JAMU; B. KALA-JAM; G. JAMDUDO; H. & P. JAMAN; K. NERILE; M. & T. NAAVAL; M'. JAMBUL; O. JAMUKOLI

Syzygium jambos—rose-apple; A. GOLAPI JAMU; B. GOLAP-JAM; H. & P. GULAB-JAMAN; K. JAMBU NERILE; M. PANINIRCHAMPA; M'. GULAB JAMB; O. GOLAP JAMU; T. NAAVAL; T'. NEEREDU

Syzygium malaccense—Malay apple; A. PANI-JAMU; B. JAMRUL; H. MALAY-JAMAN; K. PANNERILE; M'. SAFED JAMB; P. MALAY KA SEB

Tabernaemontana—see *Ervatamia*

Tagetes patula—marigold; A. NARJIPHUL; B. & H. GENDA; K. CHENDU HOOVA, SEEME SHAVANTIGE; M'. GULJAPHIRI; O. GENDU; P. GENDA, GUTTA

Tamarindus indica—tamarind; A. TETELI; B. TENTUL; G. AMIL; H. & P. IMLI; K. HUNISE MARA; M. & T. PULI; M'. CHINCH; O. KAINYA, TENTUIJ; T'. CHINTHA

Tamarix dioica—A. JHAU-BON; B. & H. BON-JHAU; K. SEERE GIDA; M'. JAO; O. DISHI-JHAUN, THARTHARI; P. PILCHI

Tectona grandis—teak; A. & B. SHEGOON; G. & H. SHAGWAN; K. TEGADA MARA, SAGUVANI; M. & T. THEKKU; M'. SAG; O. SAGUAN; P. SAGWAN; T'. TEKU

Tephrosia purpurea—wild indigo; A. BON-NIL; B. & H. JANGLI-NIL; G. JHILA; K. VAJRA NEELI, KOGGILI, KOLINJI; M. KOJHINGIL; M'. SHARAPUNKHA; O. BANA NILA; P. JHANA; T. KOLINGI; T'. VEMPALI

Terminalia arjuna—A. ARJUN GOCH; B. & H. ARJUN; G. SAJADAN; K. BILI MATHI, HOLE MATHI, TORA MATHI; M. VELLI-LAVU; M'. ARJUN-SADADA; O. ARJUNA; P. ARJAN; T. MARUTHU; T'. TELLA MADOI

Terminalia belerica—beleric myrobalan; A. BHOMRA-GUTI; B. & P. BAHERA; G. BERANG; H. BHAIRAH; K. TARE MARA, SHANTHI MARA; M. & T. THANNIKKAI; M'. BEHADA; O. BAHADA

Terminalia catappa—country almond—A. BADAM-GOCH; B., G., H., M'., O. & P. DESHI-BADAM; K. NADU BADAMI; M. ADAMARAM; T. NATTUVADUMAI; T'. BADAMI

Terminalia chebula—chebulic myrobalan; A. SHILIKHA; B. HARITAKI; G. PILO-HARDE; H. & P. HARARA; K. ALALE KAYI; M. & T. KADUKKAI; M'. HIRDA; O. HARIDA; T'. KARAKA

Thespesia populnea—portia tree; B. PARAS; G. PURUSA-PIPALO; H. & P. PARAS-PIPAL; K. BUGURI, HOOVARISI, JOGIYARALE; M. & T. POOVARASU; M'. BHENDICHA JHAR; O. HABALI; T'. GANGARAVI

Thevetia peruviana—yellow oleander; A. KARABI; B. KALKE-PHUL; G., H. & P. PILA-KANER; K. KADUKASI KANAGALU; M. & T. SIVANARALI; M' PIWALA KANHER; O. KANIARA, KONYAR PHULA; T'. PACHCHAGANNERU

Tinospora cordifolia—A. AMORLOTA, AMOILOTA; B. GULANCHA; G. GADO; H. GURCHA; K. AMRUTA BALLI, MADHU PARNI; M. AMRITHU; M'. GULVEL; O. GULUCHI; P. GALO; T. SINDHILKODI; T'. TIPPATIGE

Trachyspermum ammi—ajowan or ajwan; A. JONI-GUTI; B. JOWAN; G. AJAMO; H. & P. AJOWAN; K. OMU, AJAWANA; M. AYAMODAKAM; M'. OWA; O. JUANI; T. OMAM; T'. OMAMU

Tragia involucrata—nettle; A. CHORAT; B. BICHUTI; H. & P. BARHANTA; K. TURACHI BALLI, CHELURI GIDA; M. CHORIYANAM; M'. KHAJAKOLTI; O. BICHHUATI; T. KANJURI; T'. DULAGONDI

Trapa natans (=*T. bispinosa*)—water chestnut; A. SHINGORI; B. PANI-PHAL;

GLOSSARY OF NAMES OF PLANTS

G. SHENGODA; H. & P. SINGARHA; K. MULLU KOMBU BEEJA, SINGARA BEEJA; M. KARIMPOLA; M'. & O. SINGADA; T. SINGARAKOTTAI; T'. KUBYAKAM

Trewia nudiflora—A. BHELKORA; B. PITULI; H. BHILLAURA; K. KADU GUM BALA, KATAKAMBA, HEELAGA; M. THAVALA; M'. PITARI; O. JANDAKHAI, PANIGAMBHAR; P. TUMARI, KHAMARA; T. AATTARASU

Tribulus terrestris—B. GOKHRIKANTA; G. GOKHARU; H. GOKHRU; K. SANNA NEGGILU; M. NERUNJIL; M'. KATE GOKHRU; O. GOKHARA; P. BHAKHRA; T. NERINJI; T'. PALLERU

Trichosanthes anguina—snake gourd; A. DHUNDULI; B. CHICHINGA; G. PADAVALI; H. CHACHINDA; K. PADAVALA; M. PADAVALAM; M'. PADVAL; O. CHHACHINDRA; P. PAROL; T. PUDALAI; T'. POTLA

Trichosanthes dioica—A. & B. PATAL; H. & P. PARWAL; K. KADU PADAVALA; M. PATOLAM; G. & M'. PARWAR; O. PATALA; T. KOMBUPPUDALAI; T. KOMMUPOTLA

Trigonella foenum-graecum—fenugreek; A. MITHIGUTI; B., G., H., O., M' & P. METHI; K. MENTHYA SOPPU, MENTHYA PALYA; M. VENTHIAM; T. VENDAYAM

Triticum aestivum—wheat; A. GHENHU; B. GOM; G. GAHUN; H. & P. GEHUN; K. GODHI; M. KOTHAMPU; M'. GAHU; O. GAHAMA; T. GHODUMAI; T'. GOTHI, GODUMULU

Typha elephantina—elephant grass or bulrush; B. HOGLA; G. GHABAJARIN; H. PATER; K. ANEJONDA, APU, NARI BALA; M'. PAN KANIS; O. HAUDAGHASA, HOGOLA; P. PATIRA; T. CHAMBU

Typhonium trilobatum—A. SAMAKACHU; B. GHETKACHU; K. KANDA GEDDE; H. GHEKUL; M. CHENA; T. KARUNKARUNAI, ANAIKKORAI; T'. JAMMUGADDI

Urena lobata—A. BON-AGARA; B. BAN-OKRA; H. & P. BACHATA; K. DODDA BENDE, KADU THUTHI; M. OORPUM; M'. VAN-BHENDI; O. JATJATIA; T. OTTATTI

Utricularia sp.—B. JHANJI; K. NEERU GULLE GIDA, SEETHASRU BEEJA; M. MULLANPAYAL, KALAKKANNAN; M'. GELYACHI VANASPATI; O. BHATUDIA DALA

Vanda roxburghii—orchid; A. KOPOUPHUL; B., H. & P. RASNA; G. RASNA-NAI; K. VANDAKA GIDA; M. MARAVAZHA; M'. BANDE; O. RASHNA, MADANGA

Vangueria spinosa—A. KOTKORA, MOYENTENGA; B. & H. MOYNA; K. CHEGU GADDE, ACHHURA MULLU; M'. ALU; O. GURBELI; T. MANAKKARAI; T'. SEGAGADDA

Vernonia cineria—B. KUKSHIM; G. SADORI; H. & M'. SAHADEVI; K. KARE HINDI, GAYA DOPPALU, SAHADEVI; M. POOVAN-KURUNTHILA; O. POKASUNGA, JHURJHURI; P. SAHDEVI, KUKSHIM; T. PUVAMKURUNDAL; T'. GARITIKAMMA

Vigna sinensis—cow pea; A. NESERA MAH; B. BARBATI; H. BORA; K. ALASANDI, TADAGANI; M'. CHAVLI; O. BARGADA; P. RAUNG; T. THATTAPAYERU; T'. ALACHANDALU

Vinca rosea—see *Lochnera rosea*

Viscum monoicum—mistletoe; A. ROGHUMALA; B. BANDA; H. & P. BHANGRA, BANDA; K. HASARU BADANIKE; M. ITHTHIL; M'. JALUNDAR; O. MALANGA; T. OTTU

Vitis trifolia (=*Cayratia carnosa*)—B. AMAL-LATA; G. KHAT-KHATUMBO; H. & P. AMALBEL; K. NEERGUNDI, NOCHHI, NEERLAKKI; M. SORIVALLI; M'. AMBATVEL; O. AMARLATA

Vitis vinifera—grape vine; A., B., H. & P. ANGOOR; G. MUDRAKA; K. DRAKSHI BALLI; M. & T. MUNTHIRYVALLY; M'. DRAKSHA-VEL; O. ANGURA; T'. DRAKSHA

Wedelia calendulacea—A. BHIMRAJ; B. BHIMRAJ, BHRINGARAJ; G., H. & P. BHANGRA; K. KESHARAJA, GARGARI; M. PEE-KAYYANNYAM; M'. PIVALA-BHANGRA; O. BHRUNGA-RAJA

Withania somnifera—A. LAKHANA; B. ASWAGANDHA; G. ASUNDHA; H. & P. ASGANDH; K. ASWAGANDHI, PENNERU, HIRE MADDINA GIDA; M. & T. AMUKKIRAM; M'. ASKANDH; O. AJAGANDHA; T'. ASVAGANDHI

Wrightia tomentosa—A. DUDHKHOROI; B. INDRAJOB; G., H. & P. DUDHI; K. BILI GANAGALA, KADU JANAGALU; M. & T. NILAM-PALA; M'. KALA-INDERJAW; O. PHAOKURNI; T'. PALA

Xanthium strumarium—cockle-bur; A. AGARA; B. & H. OKRA; G. GADIYAN; K. MARALU UMMATHI; M'. SHANKESHVAR; O. CHOTA GOGHURU; P. GOKHRU KALAN; T. MARLUMUTTA; T'. MARULAMATHANGI

Zanthophyllum budrunga—A. BROJONALI; B. BAZINALI; H. BADRANG; K. MARALU MATHANGI; M. KATTUMURIKKU; M'. BUDRANJ; T. IRATCHAI; T'. RACHAMAM

Zea mays—Indian corn or maize; A. MAKOI-JOHA; B. BHUTTA; G. & P. MAKAI; H. MAKKA, BHUTTA; K. MUSUKINA JOLA, GOVINA JOLA; M. & T. MAKKA CHOLAM; M'. & O. MAKKA; T'. MOKKA JONNA

Zingiber officinale—ginger; A., B. & O. ADA; G. ADHU; H. ADRAK; K. SHUNTI, ALLA; M. INCHI; M'. ALE; P. ADARAK; T. INJI; T'. ALLAM

Zizyphus mauritiana (=*Z. jujuba*)—Indian plum; A. BAGARI; B. KUL; G. BORADI; H. & P. BER; K. ELACHI, BORE HANNU; M. & T. ELINTHAI; M'. BOR; O. BARKOLI; T'. REGU

Zizyphus oenoplia—A. BAN-BAGARI; B. SHIAKUL; K. SURI MOLLU, SONDLI GIDA; M. THODALI; M'. BURGI; O. BHUINKOLI; P. MAKOH; T. SOORAI; T'. BANKA

Glossary of Arabic (A) and Nepali (N) Names of Plants

Sincere and grateful thanks of the author are due to Dr P. K. Mukherjee and Mr Saeed Abdo Gabali, Head—Department of Biology, Higher College of Education, Khormaksar, Aden (PDRY) for kindly furnishing Arabic names of plants, and to Dr G. S. Yonzone and his colleague Professor K. K. Tamang—Darjeeling Government College, Darjeeling for Nepali names of plants.

Abelmoschus esculentus—lady's finger; A. BAMIA; N. BHENDI. *Abroma augusta*—devil's cotton; N. SANU KAPASI. *Abrus precatorius*—crab's eye or Indian liquorice; N. LAL GERI. *Abutilon indicum*—N. GHANTI PHUL. *Acacia catechu*—catechu; A. KAAT; N. KHAIR. *Acacia nilotica* (=*A. arabica*)—gum tree; A. SANAT ARABI; N. BABUL. *Achras zapota*—sapota or sapodilla plum; A. ABBASI; N. SAPHEDA. *Acorus calamus*—sweet flag; N. BOJHO. *Adhatoda vasica*—N. ASURU. *Aegle marmelos*—wood-apple; N. BEL. *Aeschynomene indica*—pith plant; N. GUDI RUKH. *Agave americana*—American aloe or century plant; A. SABER; N. HATI BAR, *Albizzia lebeek*—siris tree; A. LEBBAKH; N. HARRA SIRISH. *Allium cepa*—onion; A. BASAL; N. PYAZ. *A. sativum*—garlic; A. THOMA; N. LAHSUN. *Alocasia indica*—A. KAIKAS; N. MANE. *Aloe vera*—N. GHIU KUMARI. *Amaranthus spinosus*—prickly amaranth; N. KANRAY. *Ananas comosus*—pineapple; A. ANANAS; N. BHUI KATAHAR. *Annona reticulata*—bullock's heart; N. RAM PHAL. *A. squamosa*—custard-apple; A. ATT; N. SARIFA. *Arachis hypogaea*—groundnut or peanut; A. LOAZE or PHOOL SUDANEE; N. BADAM. *Areca catechu*—areca-nut or betel-nut; A. PHOPHAL; N. SUPARI. *Artocarpus heterophyllus*—jack tree; N. RUKH KATAHAR. *A. lakoocha*—monkey jack; N. BORHAN. *Asparagus racemosus*—A. KOSHK-ALMAZ; N. SAT MULI. *Averrhoa carambola*—carambola; N. KAMARAK. *Azadirachta indica*—margosa; A. MOREMIRA; N. NIM.

Baccauria sapida—N. KUSUM. *Bambusa tulda*—bamboo; A. GHAB or BAMBOO; N. RARANTI BANS. *Barringtonia acutangula*—N. HIJAL. *Basella rubra*—Indian spinach; N. POL. *Bauhinia variegata*—camel's foot tree; N. KOIRALO. *Benincasa hispida*—ash gourd; N. SETO PHARSI. *Beta vulgaris*—beet; A. BHANGAR; N. BEET. *Blumea lacera*—N. BABURI. *Bombax ceiba*—silk cotton tree; N. SIMAL. *Borassus flabellifer*—palmyra palm; N. TAL. *Bougainvillea spectabilis*—glory of the garden; A. GAHANNAMIYA. *Brassica campestris*—mustard; A. KHARDAL; N. SETO RAYO. *Butea monosperma*—flame of the forest or parrot tree; N. PALAS.

Caesalpinia crista (=*C. bonducella*)—fever nut; N. YANGKUP. *C. pulcherrima*—dwarf gold mohur or peacock flower; N. GOLMOHUR. *Cajanus cajan*—pigeon pea; N. RAHAR. *Calamus viminalis*—cane; A. KHAIZARAN; N. BET. *Calotropis gigantea*—madar; A. OSHAR; N. AUK. *Canna orientalis*—(=*C. indica*)—Indian shot; A. KANA. *Cannabis sativa*—hemp; A. KENNAB; N. BHANG. *Capsicum annuum*—chilli; A. BISBAS. *Carica papaya*—papaw; A. AMBA BABAL; N. MEWA. *Cassia fistula*—Indian laburnum; N. RAJBRIKSH. *C. sophera*—N. TAPRE. *Cassytha filiformis*—N. AMAR BEL. *Casuarina equisetifolia*—beef-wood tree or she-oak; A. KASARINA; N. KUR KURE RUKH. *Celosia cristata*—cock's comb; N. BHALE KO SIR. *Centella asiatica*—Indian pennywort; N. MADHESI DUNGRI JHAR; *Cestrum nocturnum*—queen of the night; N. RAAT KO RANI. *Chrysanthemum coronarium*—N. GADWARI. *Cicer arietenum*—gram; A. SOMBRA; N. CHANA; *Cinnamomum camphora*—camphor; N. KAPUR. *C. tamala*—bay leaf; N. TEZPATA. *C. zeylanicum*—cinnamon; A. KIRFA; N DALCHINI. *Cissus quadrangularis*—A. HALAS; N. PANI LAHARA. *Citrullus colo-*

cynthis—colocynth; A. HANDAL; N. INDRAYAN. *C. lanatus*—water melon; A. HABHAB; N. TARMUZA. *Citrus aurantifolia*—sour lime; A. LEIMONN; N. KAGATI NIMBU. *C. grandis*—pummelo or shaddock; N. SANGKATRA. *C. reticulata*—orange; A. PORTUKAL; N. SUNTALA. *Clitoria ternatea*—butterfly pea; N. APARAJITA. *Cocos nucifera*—coconut-palm; A. NARGEEL; N. NARIWAL. *Colocasia esculenta*—taro; N. MANE. *Cariandrum sativum*—coriander; A. KABZARA; N. DHANIA. *Crotalaria juncea*—Indian or sunn hemp; N. SAN. *C. sericea*—rattlewort; N. SUHUTUNG RUNG. *Croton tiglium*—A. KROTON; N. JAIPHAL. *Cucumis melo*—melon; A. SHAMMAM; N. KHARBUZA. *C. sativus*—cucumber; N. KAKRA. *Cucurbita pepo*—vegetable marrow; A. DUBBA; N. PHARSI. *Curcuma amada*—mango ginger; N. ANP HARDI. *C. longa*—turmeric; N. HARDI. *Cuscuta reflexa*—dodder; A. HAKOOL; N. SWARNALATA. *Cynodon dactylon*—dog grass; A. NAGEEL; N. DUBO.

Dalbergia latifolia—Indian rosewood; N. SATISAL. *D. sissoo*—Indian redwood; N. SISSAU. *Datura stramonium*—thorn-apple; N. DHUTURA. *Daucas carota*—carrot; A. GUZAR. *Delonix regia*—gold mohur; N. KRISHAN CHURA. *Desmodium gyrans*—Indian telegraph plant; N. SARKINU. *Dillenia indica*—N. PANCHAPHAL. *Dioscorea alata*—white yam; N. GHAR-TARUL. *D. bulbifera*—wild yam; N. BANTARUL. *Dolichos lablab*—country bean; A. PHOOL. *Duranta repens* (=*D. plumieri*)—A. & N. DURANTA.

Eleusine coracana—N. MARUA. *Emblica officinalis*—emblic myrobalan; N. AMALA. *Entada gigas*—nicker bean; N. PANGRA. *Ervatamia coronaria*—N. TAGAR. *Erythrina variegata*—coral tree; N. PHALEDO. *Euphorbia antiquorum*—N. SHIV RUKH. *E. neriifolia*—N. SIJ. *E.* (=*Poinsettia*) *pulcherrima*—poinsettia; N. LAL PATTA. *Euryale ferox*—N. MAKHNA.

Ferula assa-foetida—asafoetida; A. ALTIT; N. HING. *Ficus bengalensis*—banyan; A. TEEN BENGALI; N. BAR. *F. glomerata*—fig; N. DUMRI. *F. religiosa*—peepul; N. PIPUL. *Foeniculum vulgare*—anise or fennel; A. SHAMAR; N. MAURI.

Garcinia mangostana—mangosteen; N. CHUNYEL. *Gardenia jasminoides*—cape iasmine; N. GANDHA RAJ. *Gloriosa superba*—glory lily; N. KULHARI. *Gossypium sp.*—cotton; N. KAPAS. *Gynandropsis gynandra*—N. BAHU PATE.

Helianthus annuus—sunflower; A. ABBAD-AS-SHAMS; N. SURYA MUKHI. *Heliotropium indicum*—heliotrope; N. HATI SUR. *Hibiscus cannabinus*—Madras or Deccan hemp; N. AMBARI. *H. rosa-sinensis*—China rose or shoe flower; A. ABESCHUS; N. JAVA KUSUM. *H. sabdariffa*—rozelle; N. PATWA. *Holarrhena antidysenterica*—N. KHIRRA. *Hordeum vulgare*—barley; A. SHAIER; N. JAW.

Impatiens balsamina—balsam; N. PHAT PHATE. *Ipomoea batatas*—sweet potato; A. SHUKARIAKAND; N. SAKARKHANDA. *Ixora coccinea*—N. CHIWRIPAT.

Jasminum sambac—jasmine; A. YASMIN; N. HARE LAHARA. *Jatropha curcas*—N. BHARENDA. *J. gossypifolia*—N. CHAP CHAPE.

Lagenaria siceraria—bottle gourd; N. LAUKA. *Lagerstroemia speciosa* (=*L. flos-reginae*)—N. BORDERI. *Laportea crenulata*—devil or fever nettle; N. MORINGE. *Lawsonia alba*—A. HENNA. *Lens culinaris* (=*L. esculenta*)—lentil; A. AADAS; N. MASUR. *Lochnera rosea* (=*Vinca rosea*)—periwinkle; N. TARA PHUL. *Luffa acutangula*—ribbed gourd; A. TORI; N. JHINGENI. *L. cylindrica*—bath sponge or luffa; A. LUFFA; N. JHUTRA. *Lycopersicum esculentum*—tomato; A. TAMATA; N. RAMBHERA.

Marsilea quadrifolia—N. PANI SAK. *Martynia annua* (=*M. diandra*)—tiger's nail; N. BAGH NANGRE. *Mentha viridis*—spearmint or garden mint; N. PODINA.

BOTANY

Mesua ferrea—iron-wood; N. NAGESURI. *Michelia champaca*—N. AULE CHANP. *Mimosa pudica*—sensitive plant; A. ALMUSTAHIYA; N. BUHARI JHAR. *Mirabilis jalapa*—four o'clock plant; N. KRISHNA KALI. *Moringa oleifera*—drumstick or horse radish; N. SAJANI. *Morus alba* & *M. nigra*—mulberry; N. KIMBU. *Mucuna prurita*—cowage; N. KUACH. *Murraya paniculata*—Chinese box; N. KAMINI PHUL. *Musa paradisiaca*—banana; A. MOSZE; N. KERA.

Nelumbo nucifera (=*Nelumbium speciosum*)—lotus; N. KAMAL. *Nerium indicum* (=*N. oleander*)—oleander; A. DAFLA; N. KARAVI. *Nicotiana tabacum*—tobacco; A. TUMBAK; N. SURTI. *Nigella sativa*—black cumin; A. HABBAT ASSODA; N. KALO JIRA. *Nyctanthes arbor-tristis*—night jasmine; N. RAT GANDHA. *Nymphaea lotus*—water lily; N. SANO KAMAL.

Ocimum sanctum—sacred basil; N. TULSI. *Opuntia dillenii*—prickly pear; A. TEEN SHOKI; N. KARE SIWRI. *Oroxylum indicum*—N. TOTALA. *Oryza sativa*—paddy; A. ROZZ; N. DHAN. *Oxalis repens*—wood-sorrel; N. CHAKI AMILO.

Paederia foetida—N. GANDHALI. *Pandanus tectorius* (=*P. odoratissimus*)—screwpine; A. KADI; N. TARIKA. *Papaver somniferum*—opium poppy; A. KHASKHAS; N. APHIM. *Passiflora foetida*—passion-flower; N. GARANDEL PHUL. *Pennisetum typhoides*—pearl millet; N. BAJRA. *Pentapetes phoenicea*—noon flower; N. DEWSE. *Phaseolus aureus*—N. MOONG. *P. mungo*—N. KALO DAL. *Phoenix dactylifera*—date; A. TAMR; N. KHAJUR. *Phyllanthus acidus*—N. AMALE. *Piper betle*—betel; A. TUMBOL; N. PAN. *P. longum*—long pepper; N. PIPLA. *P. nigrum*—black pepper; A. PHIL-PHIL; N. GOLOMARICH. *Pistia stratiotes*—water lettuce; N. PANIKUMBI. *Pisum sativum*—pea; A. ATAR; N. MATAR. *Pithecolobium dulce*—A. DAIMEN; P. saman—rain tree; N. JHARE RUKH. *Plumeria rubra*—temple tree or pagoda tree; N. MANDIRE. *Polyalthia longifolia*—mast tree; N. DEBDARU. *Portulaca oleracea*—purslane; A. REGNA. *Psidium guayava*—guava; A. ZAITOON; N. AMBAK. *Pterospermum acerifolium*—N. HATIPAILA. *Punica granatum*—pomegranate; A. RUMMAN; N. DARIM.

Quamoclit pinnata (=*Ipomoea quamoclit*)—N. KUNJALATA. *Quisqualis indica*—Rangoon creeper; N. RANGOON LAHARA.

Raphanus sativus—radish; A. BAKL; N. MULA. *Ricinus communis*—castor; A. KHARWA; N. RERI. *Rumex vesicarius*—sorrel; N. MADISE HAL HALE.

Saccharum officinarum—sugarcane; A. SUKKAR; N. UKHU. *Saraca indica*—asoka tree; N. ASOK RUKH. *Sesamum indicum*—gingelly- A. GIL-GIL; N. TIL. *Sesbania grandiflora*—N. BAKULE. *S. sesban*—N. JAINTI. *Shorea robusta*—N. SAL. *Smilax zeylanica*—sarsaparilla; N. KUKUR DAINE. *Solanum indicum*—N. BIHI. *S. melongana*—brinjal; A. BEEDINGAN; N. BAIGOON. *S. nigrum*—black nightshade; N. KALE BIHI. *S. tuberosum*—potato; A. BATATA; N. ALU. *Sorghum vulgare*—great millet; N. JAWAR. *Spinacia oleracea*—spinach; N. PALANG SAK. *Syzygium aromaticum*—clove; A. KORON PHUL; N. LWANG. *S. cuminii*—N. JAMUNA. *S. jambos*—rose-apple; N. JAMUNA. *S. malaccense*—Malay apple; N. PANI JAMUNA.

Tagetes patula—marigold; A. MARIGOLD; N. SAYAPATRI. *Tamarindus indica*—tamarind; A. HOMAR; N. TITERI. *Tamarix dioica*—A. ETHL; N. JANGALI JHAU. *Tectona grandis*—teak; N. SHAHGOON. *Terminalia arjuna*—N. PANISAJ. *T. belerica*—N. BARRA. *T. catappa*—N. DESHIBADAM. *T. chebula*—N. HARRA. *Thespesia populnea*—portia tree; A. LAKEED; N. KAPHAIMUK. *Thevetia peruviana*—yellow oleander; A. TEVETIA; N. PAHELE BIKHU. *Tinospora cordifolia*—N. GURJO. *Trachyspermum ammi*—ajowan or ajwan; A. NAKHWA; N. JUWANO. *Tragia involucrata*—nettle; N. MADESHI SHISHNU. *Trapa natans*—N. PANI PHAL. *Trichosanthes anguina*

—snake gourd; N. CHICHINDA. *T. dioica*—N. POTAL. *Trigonella foenum-graecum*—fenugreek; A. HULBA; N. METHI. *Triticum aestivum*—wheat; A. KAMHH; N. GAHU. *Typha elephantina*—N. PATAR.

Utricularia sp.—N. SHIM PHAKUNDE.

Vanda roxburghii—N. VANDASUNAKHARI. *Vangueria spinosa*—N. TIKHE. *Vicia faba*—broad bean; A. PHOOL. *Vigna sinensis*—cow pea; A. DUGGRA; N. BORI. *Viscum monoicum*—mistletoe; N. HARCHUR. *Vitis vinifera*—grape vine; A. ENAP; N. ANGOOR.

Withania somnifera—N. ASGANDHA. *Wrightia tomentosa*—N. KHIRRA.

Xanthium strumarium—N. OKRE.

Zea mays—maize; A. DURRA HINDI; N. MAKAI. *Zingiber officinale*—ginger; N. ADUWA. *Zizyphus mauritiana* (=*Z. jujuba*)—A. DOME; N. BAER.

INDEX

(Numbers followed by f refer to illustration pages)

Abaca, 772
Abeitaceae, 670
Abelmoschus esculentus, 89, 717
Abroma (A. augusta), 128, 718
Abrus (A. precatorius), 44, 728
Absciss (-sion) layer, 266, 267f
Absorption, 298-303
Abutilon indicum, 717
Acacia, 37, 68, 174;
 Australian-, 34, 50, 50f; -spp., 729
Acalypha, 57, 66; -spp., 764
Acanthaceae, 752-4, 753f
Acanthus ilicifolius, 424, 754
Acer, 131, 137, 137f, 436
Acetyl coenzyme A, 367
Achania malvaviscus, 717
Achene, 130, 130f
Achlamydeous, 72, 699
Achyranthes (A. aspera), 66, 759
Acidity and alkalinity, 282-3
Aconite (Aconitum), 706, 829
Acorus (A. calamus), 224, 775
Acrogynous, 590
Actinomorphic, 78
Actinostele, 622
Adam's needle, see dagger plant
Adansonia digitata, 720
Adaptation, 787
Adhatoda (A. vasica), 66, 82, 753
Adhesion, 89, 90
Adiantum, 401, 401f
Adina cordifolia, 68, 738
Adnate, 89; -stipules, 36, 36f
ADP (adenosine diphosphate), 324, 325, 364, 365
Adventitious roots, 5, 6-10
Aeciospore (-cium), 572f, 573
Aegiceros majus, 432
Aegle marmelos, 721
Aerenchyma, 201, 201f
Aerides spp., 782
Aerua, 62, 214, 423; -spp., 759
Aeschonomene aspera, 728

Aestivation, 83-4, 84f
Aganosma dichotoma, 744
Agaricus, 573-6, 574f
Agave, 203, 217, 403, 423, 770
Ageratum conyzoides, 741
Ajowan (Ajwan), 737
Akinete, 446, 447f
Alae, 81f, 82
Albizzia spp., 729
Albugo, 538-40, 538f
Alburnum, 249
Alder (Alnus), 436
Aldrovanda, 346, 346f, 347f
Aleurites spp., 763-4
Aleurone grains, 170, 170f
Alfalfa, 728
Alhagi, 423
Alisma plantago, 58
Alkaloids, 62, 177
Allamanda (A. cathartica), 55, 744
Alleles (allelomorphs), 797
Allium cepa, see onion; A. sativum, see garlic; -spp., 768
Allmania nodiflora, 759
Allogamy, 99, 100-8
Allophylus cobbe, 723
Allopolyploids, 189
Allosomes, 805
Allspice, 731
Almond, 730; Country-, 14, 132
Alocasia (A. indica), 23, 774
Aloe, American-, see Agave; Indian- (Aloe vera), 50, 164, 422, 768
Alpine forests, 430, 437
Alpinia spp., 772
Alstonia (A. scholaris), 55, 744
Alternanthera amoena, 759
Alternation of generations, 439-41; Heteromorphic-, 503; Isomorphic-, 486, 499
Althaea rosea, 717
Aluminium, 298
Alyssum (A. maritimum), 129, 713

57

Amanita, 574
Amaranth (*Amaranthus*), 66; -spp., 759; -stem, 258-9, 258f
Amaranthaceae, 759
Amaryllidaceae, 769-70, 770f
Amaryllis, 770; -root-, 238-40, 239f
Amino-acids, 172, 338-9
Amitosis, 192, 192f
Amomum, 825; -spp., 772
Ammonification, 292
Amoora spp., 721
Amorphophallus, 24, 24f, 62, 63, 101, 101f, 775
Amphicribral, 224
Amphigastria, 603
Amphigynous, 537
Amphiphloic, 623
Amphitropous, 98, 98f
Amphivasal, 224, 264
Amplexicaul leaf, 35
Amylase, 359; -lose, 168
Amyloplast, 154, 168
Anabaena, 446-8
Anabolism, 377
Anacardiaceae, 723-4, 723f
Anacardium, 127, 127f, 130, 728
Anacrogynous, 590
Analogy, 59-60
Ananas comosus, see pineapple
Anaphalis, 437
Anaphase, 181, 186
Anatropous, 98, 98f
Androcytes, 604
Androecium, 72, 85-91
Andrographis paniculata, 63, 141, 753
Androgynophore, 73
Androphore 73, 73f
Androspore, 484, 484f
Andropogon squarrosus, 779
Aneilema spp., 770, 771
Anemone elongata, 707
Anemophily, 103-4, 104f
Anethum graveolens, 737
Aneupolyploidy, 189
Angelonia, 751
Angiosperms, 658, 688; Life-cycle of-, 690-1; Origin of-, 688-90
Angular divergence, 55-6

Anise, 46, 67, 74, 74f, 131, 737
Anisogamy, 439, 452
Anisomeles indica, 757
Annona, 134; -spp., 709
Annonaceae 709-10, 710f
Annual rings, 248-9, 248f
Annulated roots, 7, 7f
Annulus, 614, 619, 650
Anther, 72, 85
Antheridium, 483, 488, 532, 615, 650
Antheridiophore, 597, 598, 598f
Antherozoid, 408, 439
Anthocarp, 758
Anthocephalus indicus, 37, 68, 102, 738
Anthoceros, 605-10
Anthocyanins, 155
Anthophore, 73
Anthurium, 775
Antibiotics, 585
Antigonon, see *Corculum*
Antipodal cells, 96f, 97, 110
Antirrhinum majus, see snapdragon
Aphania danura, 723
Apium graveolens, 737
Aplanogamete, 451
Aplanospore, 450, 465
Apocarpous pistil, 91, 92f
Apocynaceae, 742-4, 743f
Apogamy, 409
Apomixis, 409
Apophysis, 620
Apospory, 410
Apostrophe, 391
Apothecium, 527, 563
Apple, 127, 127f, 132f, 133, 730
Apposition, Growth by-, 159
Apricot, 730
Aquilaria agallocha, 435
Aquilegia vulgaris, 707
Araceae, 774-5, 774f
Arachis hypogaea, see groundnut
Aralia, 827
Arceuthobium, 17
Archegoniophore, 597, 598, 599f
Archegonium, 616, 616f, 651, 651f
Archesporium, 87, 601
Archichlamydeae, 699
Arc indicator, 379, 379f

INDEX

Areca catechu, 776
Arenga saccharifera, 776
Areole, 734
Argemone, 49f, 50, 62, 95, 176, 712
Argyreia speciosa, 747
Aril, 115, 140
Arisaema, 63, 63f, 775
Aristida, 141, 142f
Aristolochia gigas, 138, 138f
Aroids, 41, 62, 66, 75, 176
Arrowhead, 40, 58, 59f
Arrowroot, 23, 773
Artabotrys, 14, 15f, 91, 133, 134f, 709, 710f
Artemisia spp., 741
Arthrocnemum, 760
Artichoke, Jerusalem-, 23, 741
Artocarpus chaplasha, 58f, 59; *A. heterophyllus*, 59; -spp., 765
Arum maculatum, 775
Arundina bambusifolia, 782
Arundo donax, 779
Asafoetida, 737
Ascent of sap, 314-18
Asclepiadaceae, 744-6, 745f
Asclepias, 69, 88, 138, 745
Ascobolus, 564-6, 564f
Ascocarp, 550
Ascogenous hyphae, 550, 553
Ascogonium, 550, 552
Ascomycetes, 525, 527, 554
Ascorbic acid, 387
Ascospores (-cus), 526, 550, 553
Ash, 286; -gourd, see *Benincasa*; -plant, see *Fraxinus*
Asparagus, 6, 29, 30f, 423; -spp., 768, 769; -stem, 234, 234f
Aspergillus, 551-3, 551f
Asphodel (*Asphodelus*), 769
Asplenium nidus, 421
Association, 415
Assimilation, 360
Aster, 741
Asteracantha, 82, 424, 753
Asteraceae, see *Compositae*
ATP (adenosine triphosphate), 324, 325, 364, 365
Atriplex hortensis, 760
Atropa belladonna, 177, 749, 830

Auriculate leaf, 35
Autoecious, 568
Autogamy, 99, 100
Autophytes (-trophic), 17, **342**
Autopolyploids, 189
Autosomes, 805
Auxanometers, 379-80
Auxins, 384
Auxospore, 495
Avena sativa, 778-9
Averrhoa, see carambola
Avicennia, 424, 755
Awn, 777
Azadirachta indica, 38, 63, 722
Azolla, 419
Azotobacter, 294
Azygospore, 451, 475, 544

Bacca, see berry
Baccaurea sapida, 115, 764
Bacillariophyceae, see diatoms
Bacillus, 515; Hay-, 520-21
Back cross, 798-9
Bacopa monnieria, 751
Bacteria, 515-21; Autotrophic-, **335**
Bacteriophage, 523, 523f
Bagasse, 823, 824, 841
BAJRA, see *Pennisetum*
Balanites (*B. aegyptiaca*), 45, 45f
Balanophora dioica, 17, 18f
Balausta, 133, 133f
Balloon vine, 16, 26, 27f, 115, **723**
Balsa, 720; -sam, 74, 82 83f, 140, **176**
Bamboo (*Bambusa*), 11, 104, **779**; Giant-, 779
Banana, 66, 75, 771f, 772, 821
Banyan, 7, 8f, 37, 62, 71, 103, **135**
Baobab tree, 720
Barberry (*Berberis*), 49f, 50
Barbula, 614
Barleria, 141; -spp., 753
Bark, 252; Functions of-, 256-7
Barley, 119, 778
Barringtonia acutangula, 434, **731**
Basella (*B. rubra*), 7, 164, 425, **760**
Basidiocarp, 527, 574, 576
Basidiomycetes, 525, 527, 566
Basidiospore (-dium), 526, 570, **574**
Basil, see *Ocimum*

Bast, see phloem; fibres, 208; Hard-, 219, 222
Batatas, 6 7f; see *Ipomoea batatas*
Bath sponge, 127, 138, 733
Batrachospermum, 512-14
Baccaurea sapida, 751
Bauhinia, 38; *B. vahlii,* 141, 141f; -spp., 729
Bay leaf, 42, 91, 825
Bean, 82, 128, 728; Broad-, see *Vicia;* Country-, see *Dolichos;* French-, see *Phaseolus;* Soya-, see *Glycine;* Sword-, see *Canavalia*
Beaumontia grandiflora, 744
Beech, see *Fagus*
Beet, 5, 166, 760, 824; -root, 262-3, 263f
Begonia, 12, 13f, 68, 402
Belamcanda, 55
Bell flower, 80
Benincasa hispida, 90, 90f, 733
Bennettitales, 856-7
Bennettites, 857
Bentham and Hooker's system, 696-8
Berberis, 49f, 50
Berry, 132, 132f
Beta vulgaris, see beet
Betel, see *Piper;* -nut, see *Areca*
Betula, 66, 109, 252, 436, 841
Beverages, 830-4
Bicollateral bundle, 223, 231
Bignoniaceae, 751-2
Bignonia unguis-cati, 48, 49f; *B. venusta* 48; -spp., 752 -stems,

Biophytum sensitivum, 398, 398f
Birch, see *Betula*
Bird of Paradise, 772
Bird's nest fern, 421; -orchid, 782
Bischofia javanica, 764
Bittersweet, 177, 749
Biuret reaction, 171
Blackberry, 731
Blackman reaction, 320
Bladder (-wort), 53, 53f, 347-8, 348f
Bleeding, 305, 315; -heart, 107, 108f
Blight, Fire-, 584; Late-, 535, 580
Blood flower, see *Asclepias*
Blumea lacera, 63, 741-2

Boehmeria nivea, 202, 766, 839
Boerhaavia, 11, 62, 130, 142, 142f, 211, 759; -stem, 259-60, 259f
Bombacaceae, 719-20
Bombax ceiba, 46, 61, 90, 720
Bonnaya, see *Lindernia*
Borage, Country-, see *Coleus*
Boraginaceae, 746
Borassus flabellifer, 125, 776
Bordeaux mixture, 583
Border parenchyma, 245
Boron, 297
Bottlebrush tree, 731
Bottle-palm, 776
Bougainvillea, 15, 15f, 75, 76f, 101, 759
Brachymeiosis, 565
Bract (-teole), 75-6; -scale, 675
Brassica, see mustard; -spp., 713
Breeding, 795; Economic importance of-, 807-9
Breynia rhamnoides, 764
Bridal creeper, 748
Bridelia retusa, 764
Brinjal, 79, 748
Bristles, 61
Broomrape, 17, 18f
Brosimum, 765
Broussonetia papyrifera, 765, 840
Brownian movement, 271
Bruguiera (*B. gymnorhiza*), 426, 432
Brunfelsia hopeana, 749
Bryonia, 733
Bryophyllum (*B. pinnatum*), 12, 13f, 402
Bryophyta, 590
Buchanania latifolia, 724
Bucklandia, 139, 436
Buckwheat, 37, 107, 130, 761
Bud, 12-13; -scale, 37
Budding, 193, 193f, 401, 545, 545f
Bulb, 23, 23f
Bulbil, 13, 30, 403, 403f
Bulbophyllum 782
Bulbostylis barbata, 780
Bullock's heart, 134, 709
Bulrush, see *Typha*
Burmannia, 20
Butcher's broom, see *Ruscus*

INDEX

Butea monosperma, 728
Buttercup, see *Ranunculus*
Butterfly lily, 772; -pea, see *Clitoria*
Butterwort, 345, 345f
Button flower, 759
Buttress roots, 8

Cabbage, 80, 297, 713
Cabomba, 711
Cactaceae, 733-5, 734f, 735f
Cactus (-ti), 29, 423, 734
Caducous, 47, 79
Caesalpinia, 65; -spp., 729
Caesalpinieae, 725-6; 727f, 728f, 729
Caffeine, 177, 832
Cajanus cajan, 728, 816
Cajeput, 731
Calabash, 752
Caladium, 63, 775
Calamites, 852
Calamus, 14, 15f, 776
Calathea, 773
Calciferol, 388
Calcium, 290; -carbonate, 175; -oxalate, 176
Calendula, 77, 741
Callicarpa spp., 756
Calligonum polygonoides, 761
Callistemon lanceolatus, 731
Callose (-lus), 207, 266
Calophyllum, (*C. inophyllum*), 41, 41f
Calotropis, 62, 88, 108, 108f, 138, 139f, 745, 745f
Calyculus, 761
Calymmatotheca, 856
Calyptra, 613, 617
Calyptrogen, 199, 199f
Calyx, 71, 79
Cambium, 221f, 222-23, 227; Interfascicular-, 247, -ring, 247, 254
Camellia, see *Thea*
Camel('s) foot climber, 141, 141f; -thorn, 423; -tree, 729
Campanula, 80
Camphor, 91, 825
Campylotropous, 98, 98f
Cananga odorata, 709

Canavalia ensiformis, 728
Candle tree, 752
Candytuft, 66, 80, 129, 713
Cane (*Calamus*), 14, 15f, 776
Cane-sugar, see sucrose
Canker, 583
Canna (*C. indica*), 23, 42, 773; -stem, 234-5, 235f
Cannabinaceae, 767
Cannabis (*C. sativa*), 43, 202, 767, 838
Cannaceae, 773, 773f
Canscora, 32, 35
Caperbush, 715
Capillarity, 281, 317; -ry water, 299
Capillitium, 530
Capitulum, see head
Capparidaceae, 713-5, 714f
Capparis, 37, 73; -spp., 714-5
Capsella (*C. bursa-pastoris*), 129, 713
Capsicum, 825; -spp., 749
Capsule, 129, 129f, 617, 649
Carambola, 398, 399
Caraway, 737
Carbohydrates, 165-70, 318, 353
Carbon, 291; -assimilation, 318; -cycle, 291
Carboxylases, 326, 360
Carcerule, 131f, 132
Cardamine debilis, 713
Cardamom, 772, 824-5
Cardanthera triflora, 58, 58f, 753
Cardiospermum, 16, 26, 27f, 115, 723
Carex indica, 780
Carica papaya, see papaw
Carina, 81f, 82
Carinal canal, 642
Carissa (*C. carandas*), 27, 28f, 744
Carnation, 715
Carnivorous plants, 343-8
Carotene, 154, 155; -noids, 155
Carpel, 71, 91
Carpellary scale, 675
Carpophore, 74, 74f, 131, 736
Carpogonium, 510, 514
Carpospore, 511, 513f, 514
Carrot, 46, 67, 131, 737
Carthamus, see safflower

Carum spp., 737
Caruncle, 115, 116
Caryophyllaceae, 715
Caryophyllaceous, 80, 80f
Caryopsis, 129, 130f, 777
Caryota urens, 776
Cashew-nut, 127, 127f, 130, 724
Cassava, 764, 818
Casparian strip, 219, 219f
Cassia, 37, 44, 65, 131; -spp., 729
Cassytha (C. filiformis), 17, 18, 343
Castanea, 130
Castilloa, 765
Castor, 90, 140, 763, 763f; -oil, 819; -seed, 116, 116f, 119, 120f
Casuarina, 14, 109, 767; -stem, 235, 235f
Casuarinaceae, 767
Catabolism, 377-8
Catalases, 359
Cataphoresis, 271
Cataphylls, 35
Catechu, 174, 729
Catkin, 65f, 66
Caytonia, 856
Cat's nail, 48, 49; -tail, 66
Cattleya, 782
Caudate, 37; -dex, 11; -dicle, 781
Caulerpa, 489
Cauliflower, 80, 297, 713
Cedar, 668
Cedrela (C. toona), 128, 722
Cedrus, 428, 668, 836
Ceiba (C. pentandra), 8, 720
Celery, 737
Cell, 144, 145f, 156, 156f
Cellulose, 162; -lase, 359
Celosia, 127; -spp., 759
Celtis australis, 764
Cenozoic, 848
Censer mechanism, 138
Centaurea, 102, 398; -spp., 741
Centella, 25, 67; -spp., 737
Centriole, 156
Centromere, 180, 183
Centrosome, 155-6, 156f
Century plant, see *Agave*
Cephaelis, see *Psychotria*
Cephalanthus, 738
Ceratophyllum, 104, 418

Ceratopteris, 419
Cerbera odollam, 744
Cereals, 119, 167, 778, 811
Cereus, 29, 734, 734f; - spp., 734
Ceriops (C. roxburghiana), 426, 432
Cestrum, 101; -spp., 749
Chaetophora, 467-8, 467f
Chaff-flower, see *Achyranthes*
Chalaza, 96, 96f
Chalazogamic, 109
Chantransia, 514
Chaplash, 58f, 59, 765
Chara, 149, 489-93, 490f
Chayote, 733
Cheiranthus, 66, 713
Chemosynthesis, 336-7
Chemotaxis, 391; -tropism, 396
Chenopodiaceae, 759-60
Chenopodium spp., 760
Cherry, 730
Chestnut, see *Castanea*; Water-, see *Trapa*
Chiasmata, 185
Chicory (*Cichorium*), 741
Chikrassia tabularis, 722
Chilli, 78, 749, 825
Chimaera (chimera), 406
China rose, 36, 76, 89, 164, 717, 717f
Chinese box, see *Murraya*; -hat, see *Holmskioldia;* -lantern, see *Achania*
Chitin, 164, 545
Chlamydomonas, 453-6, 454f, 455f
Chlamydospores, 526, 579
Chloromycetin, 585
Chlorenchyma, 200
Chlorophyceae, 450; Origin of-, 453; Sexuality in-, 451
Chlorophyll, 154-5, 321, 332-5
Chloroplasts, 154-5
Chlorotic, 334
Chloroxylon swietenia, 722
CHOLAM, see *Sorghum*
Chondriosome, see mitochondria
Chromatid, 179, 185; -tin, 151
Chromomere, 183, 185
Chromonemata, 179, 183
Chromoplasm, 442, 445; -plasts, 155
Chromosome; 151, 179, 802; Chemistry of-, 182; Sex-, 805; Struc. of-;

INDEX

182-3, 182f
Chrozophora plicata, 764
Chrysanthemum, 26, 26f, 741
Chrysopogon aciculatus, 141, 142f, 779
Cicca acidus, 764
Cicer arietinum, 727, 815-16
Cichorium spp., 741
Cinchona, 62, 136, 177, 738, 829
Cincinnus, 69
Cinnamon (Cinnamomum), 42, 91, 825; -spp., 825
Circinate, 53, 650
Circulation, 149, 150f
Circumnutation, 393
Cirrhose, 38, 38f
Cissus quadrangularis, 31, 32f
Citric acid cycle, see Krebs cycle
Citron, 404, 721
Citrullus spp., 733
Citrus, 34, 46, 133, 821-2; -spp., 721
Cladode, 29, 30f
Cladonia, 587, 587f
Cladophora, 484-6, 485f
Cladophyll, 29
Clamp connexion, 528-9, 529f
Clausena spp., 721
Claviceps, 560-2, 560f
Cleistogamy, 100, 100f
Cleistothecium, 527, 550, 551f, 553
Clematis, 16, 16f, 138, 139f; -spp., 707
Cleome, see Polanisia
Clerodendron thomsonae, 107, 108f; -spp., 755
Climax, 415
Climbers, 14-17
Clinogyne dichotoma, 773
Clinostat, 394, 395f, 396
Clitoria (C. ternatea), 82, 728
Clostridium, 294
Clove, 731, 827
Cocci, 131, 515
Coccinea indica, 733
Cockle-bur, see Xanthium
Cock's comb, 127, 759
Cocoa (-tree), 718, 833-4
Coco de mer, see Lodoicea
Cocoloba, 28f, 29, 761
Coconut (Cocos nucifera), 125f, 126,
139, 776, 839; -oil, 820; Double-(Lodoicea), 139, 140f, 776
Codiaeum (C. variegatum), 5, 764
Coenobium, 457, 458
Coenocyte, 193, 486
Coenogametes, 542
Coenopteridales, 852
Coenzyme, 357
Coffee (Coffea), 739, 834
Cohesion, 89, 93; -theory, 316-17
Coir, 839-40
Coix lachryma-jobi, 779
Cola acuminata, 718
Colchicum autumnale, 24, 768
Coleochaete, 468-70, 469f
Coleoptile (-orhiza), 118, 121, 125
Cileus (C. aromaticus), 4f, 5, 71, 757
Collenchyma, 201, 201f
Colloidal system, 148, 268-72
Colocasia escuienta, 24, 66, 90, 774
Colocynth, 733
Colony, 415
Columbine, 707
Columella, 541, 619
Commelina bengalensis, 100 100f; C. obliqua, 150; -spp., 770
Commelinaceae, 770-1
Commensalism, 418
Communities, 415
Complementary cells, 252
Composite, 739-42, 740f, 741f; Pollination in-, 102
Compression (or pan) balance, 310, 310f
Condiments, 824-6
Conduction, 303-18
Conceptacle, 504, 505f
Conidia (-diophore), 526, 550, 552, 558, 579
Coniferae (-rs), 669; Distribution of-, 669-10
Coniferous forests 428
Conjugation, 408, 472
Conjunctive tissue, 237, 238
Convolvulaceae, 746-8, 747f
Convolvulus spp., 748
Copper, 297
Coptis teeta, 435, 707

Coral root (*Corallorhiza*), 20, 782;
 -tree, see *Erythrina*
Corchorus, see jute; -*spp.*, 719
Corculum, 16, 26, 27f, 761
Cordaitales, 858-9
Cordaites, 858
Cordate, 39
Cordia spp., 746
Cordyline australis, 224, 769
Coriander (*Coriandrum*), 46, 67, 74, 131, 131f, 736f, 737; Wild-, 67, 737
Cork, 251-2; Functions of-, 256-7; Storied-, 265; -cambium, 250, 256
Cork oak, 252; -tree, Ind.-, 39, 752
Corm, 24, 24f
Cornflower, 741
Corn, Indian-, see maize
Corolla, 71, 79-84
Corona, 83, 83f, 91, 744
Corpusculum, 744
Cortex, 218, 226
Corymb, 66, 67f
Corypha, 776
Cosmarium, 478-80, 479f, 480f
Cosmos, 46, 68, 102, 741
Costus speciosus, 772
Cotton, 36, 76, 89, 716, 837;
 Silk-, see *Bombax*; White-, see *Ceiba*
Cotyledon, 116, 117
Cowage, 62, 728
Cow-tree, 765
Crab's eye, see *Abrus*
Crassula, 12, 403
Crataeva nurvala, 46, 715
Cremocarp, 131, 131f, 736
Crescentia cujete, 752
Crescograph, 380
Cress, Bitter-, Garden-, 713; Water-, 742
Crinum spp., 770
Cristae, 157
Crocus (*C. sativus*), 24
Crosier (crozier), 527-8
Crossandra undulaefolia, 753
Crossing over, 185, 804-5, 805f
Crotalaria, 82, 128, 203; -*spp.*, 728
Croton spp., 764

Croton, Garden-, 5, 14, 763
Crowfoot, Water-, 58, 707
Crown, 490f, 492
Cruciferae, 712-3, 713f
Cruciform, 80, 80f, 712
Cryptoblasts, 507
Cryptostegia grandiflora, 745
Crystalloid, 170, 170f
Cucumber (*Cucumis*), 733; -*spp.*, 733
Cucurbita, see gourd; -stem, 228-31, 229f, 230f; -*spp.*, 733
Cucurbitaceae, 731-3, 732f, 733f
Culm, 11
Cumin (*Cuminum*), 67, 74, 131, 737; Black-, 707
Cuneate, 40
Cupule, 597
Curculigo orchioides, 770
Curcuma amada, 7, 7f; -*spp.*, 772
Curry-leaf plant, 721
Curtis' theory, 351
Cuscuta, 17, 18f, 83, 83f, 748
Cuspidate, 37
Custard-apple, 134, 709
Cuticle, -tin, -inization, 163
Cyanophyceae, 442-8
Cyanotis spp., 771
Cyathium, 70, 70f, 762
Cyathula spp., 759
Cycad (*Cycas*), 664-9
Cycadeoideales, 856-7
Cycadeoidea, 857
Cycadofilicales, 854-6
Cyclosis, 149
Cymbidium aloifolium, 782
Cymbopogon, 174; -*spp.*, 779
Cyme (-mose), 31, 68-9
Cynodon dactylon, 779
Cynoglossum lanceolatum, 746
Cyperaceae, 779-80
Cyperus spp., 779; *C. papyrus*, 840
Cypripedium, 782
Cypsela, 130, 130f
Cyst, 450
Cystocarp, 511, 511f, 514
Cystolith, 175, 175f
Cystopus, see *Albugo*
Cytase, 359
Cytochrome, 360, 367

INDEX

Cytokinesis, 181-2
Cytoplasm, 145

Daedalacanthus, see *Eranthemum*
Daemia extensa, 134, 746
Daffodil, 83, 770
Dagger plant, 60-1, 61f, 101, 769
Dahlia, 6, 7f, 166, 741; Tree-, 741
Daisy, 741
Dalbergia spp., 728; *D. sissoo*, 834
Darwin's theory, 790-2
Date-palm, 50, 124, 125f, 776
Datura, 62, 177; -*spp.*, 749
Daucas carota, 46, 131, 737
Deciduous, 47, 79; -forests, 427, 434
Decompound leaf, 45f, 46
Decumbent, 11; -current, 35
Decussate, 54
Deeringia celosioides, 759
Defensive mechanisms, 60-3
Definitive nucleus, 96f, 97, 110
Dehiscence, 127, 128f
Dehydrogenases, 360
Delonix regia, 72, 72f, 729
Delphinium (*D. ajacis*), 77, 82, 707
Dendrobium, 781f, 782; -*spp.*, 782
Dendrocalamus, 779
Dendropthe falcata, 761
Denitrification, 292
Dentella repens, 739
Deodar, 428, 668, 836
Dermatogen, 198, 199
Derris spp., 728
Desmids, 478, 479f
Desmodium (*D. gyrans*), 36, 392, 392f; -*spp.*, 728
Devil nettle, see *Laportea*; -'s cotton, see *Abroma*; -'s spittoon, 63; -tree, see *Alstonia*
De Vries' theory, 793
Dextrin, 169
Diadelphous, 89, 89f, 725
Diakinesis, 185
Dianthus, see pink; -*spp.*, 715
Diastase, 166, 327, 359
Diatoms, 493-5, 494f
Dichlamydeous, 72
Dichogamy, 106; -otomy, 32
Dicliny (-nous), 106

Dicliptera roxburghiana, 753
Dicotyledons & monocotyledons, 703
Dictyostele, 624
Didynamous, 90, 90f
Dieffenbachia, 775
Diffusion, 272; -pressure deficit, 274; -theory, 351
Digera arvensis, 759
Digestion, 360
Digitalis purpurea, 436, 751
Digitate, 46, 46f
Dihybrid cross, 799-800
Dikaryon (-tic), 526
Dill, 737
Dillenia, 79, 127
Dimorphic, 90, 107, 107f
Dioecious, 72, 106
Dionaea, 345-6, 346f
Dioscorea, 17, 41, 136, 136f, 403, 403f, 818
Diospyros (*D. peregrina*), 72, 80, 115
Diplanetic (-tism), 532
Diplochlamydeous, 699
Diploid, 180, 440
Diplophase, 573
Diplotene, 185
Dipterocarpaceae, 715-6
Dipterocarpus, 130, 137, 137f; -*spp.*, 715-6
Dischidia (*D. rafflesiana*), 51, 52f, 745
Diseases, Plant-, 580-4
Distichous, 55, 55f
Dixon and Jolly's theory, 316
DNA (deoxyribonucleic acid), 152-5, 340, 802-4, 802f; -and protein synthesis, 340-1
Dodder, see *Cuscuta*
Dodonaea (*D. viscosa*), 137, 723
Dog grass, 779
Dolichos lablab, 116, 116f, 728
Dombeya spp., 718
Dominant, 796
Dorsiventral leaf, 47, 243-6, 244f
DPN (diphosphopyridine nucleotide), 324
Dragon plant (*Dracaena*), 769; -stem, 264-5, 265f
Dregea volubilis, 746

Drosera, 17, 344-5, 344f
Drumstick, see *Moringa*
Drupe, 132, 132f
Drymaria cordata, 715
Drynaria quercifolia, 421
Dryopteris, 646f, 647, 648
Duabanga, 434, 836
Duckweed, 1f, 3, 30
Dumb-cane, 775
Duramen, 249
Duranta repens (=*D. plumieri*), 27, 28f, 754f, 755
Dwarf, male, 484, 484f
Dysophylla verticillata, 757
Dysoxylum procerum, 722

Ecbolium linneanum, 754
Ecesis, 417
Echinocactus, 734
Echinops, 60, 70, 423, 742
Eclipta alba, 741
Ectocarpus, 497-500
Ectophloic, 623
Ectoplasm, 146
Egg-apparatus, 96f, 97; -cell, 97, 111, 408, 439; -plant, 749
Ehretia acuminata, 746
Eichhornia, see water hyacinth
Elaeis guineensis, 776
Elaioplasts, 154
Elaters, 602, 645
Electron microscope, 144
Electron transport system, 369
Elephant apple, see *Limonia*; -climber, see *Argyreia*; -ear plant, see *Begonia*; -'s foot, see *Elephantopus*; -grass, see *Typha*
Elephantopus scaber, 742
Elettaria cardamomum, 772, 824-5
Eleusine coracana, 779, 815
Elm (*Ulmus*), 436
Emarginate, 38
Emblica officinalis, 173, 764
Embryo, 111-3, 112f, 113f
Embryo-sac, 92, 96-8, 97f
Emetine, 827
Endarch, 228
Endemism, 438
Endive, 741

Endocarp, 127
Endodermis, 218, 218-9, 227, 237
Endogenous, 3, 243, 243f
Endoplasm, 146
Endoplasmic reticulum, 156f, 157
Endosperm, 110, 116, 117; Dev. of-, 113-15; -haustoria, 114-15
Endosporous, 636, 676
Engelhardtia, 434
Engler's system, 699-700
Enhydra fluctuans, 742
Entada (*E. gigas*), 131, 729
Enterolobium, see *Pithecolobium*
Entomophily, 100-3
Enzymes, 356-60
Ephedra (*E. foliata*), 431, 680
Epiblema, 214, 236, 237
Epicalyx, 76, 716, 731
Epicarp, 127; -cotyl, 120
Epidermis, 213-4, 226
Epigeal, 119-20
Epigyny, 75, 75f
Epinasty, 393
Epipetalous, 90; -phyllous, 90
Epiphyllum (*E. truncatum*), 28f, 29, 734
Epiphytes, 9, 9f, 19, 420-2
Epistrophe, 391
Epithelium, 118, 119
Epithem cells, 210
Equisetites, 852
Equisetum, 642-6,
Eranthemum nervosus, 753
Erepsin, 359
Ergastic substances, 164-5
Ergot, 560, 561
Eria, 782
Erigeron mucronatus, 741
Eriobotrya japonica, 53, 730
Eriodendron, see *Ceiba*
Eruca sativa, 713
Ervatamia coronaria, 32, 744
Eryngium (*E. foetidum*), 67, 737
Erysiphe, 553-4, 553f, 582
Erythrina (*E. variegata*), 728
Eschscholzia, 712
Essential oils, 174
Etaerio, 133-4
Etiolated, 381f, 382

INDEX

Eucalyptus, 14, 174, 731
Eucharis, 91, 770
Eudornia, 457-8, 458f
Euglena, 448-50, 449f
Eupatorium spp., 741, 742
Euphorbia, 29, 62, 70, 423; *E. pulcherrima*, see poinsettia; *E. splendens*, 423; *E. tirucalli*, 29, 29f; -*spp.*, 764
Euphorbiaceae, 762-4, 762f, 763f
Euphoria longana, 722
Eulploidy, 188
Eurotium, see *Aspergillus*
Euryale (*E. ferox*), 419, 711
Eusporangiate, 621
Evergreen forests, 428, 434
Everlasting flower, 741
Evolution, 782-93
Evolvulus, 747f, 748; -*spp.*, 748
Exarch, 237
Excoecaria (*E. agallocha*), 426, 432
Exine, 85, 88
Exodermis, 237, 241
Extrorse, 85
Exudation, 313, 314

Fagopyrum, see buckwheat
Fagus, 130, 436
Falcate, 40
Fats (and oils), 172-3, 341-2, 355
Fatty acids, 172, 342
Fehling's sol., 166
Fennel, see anise
Fenugreek, 728
Fermentation, 375-7; -vessel (Kuhne's), 376, 376f, 554
Fern, 647-54; Walking-, 401, 401f
Feronia, see *Limonia*
Fertilization, 108-11, 109f, 408; Double-, 110-11
Fertilizers, 285-6
Ferula assa-foetida, 737
Fever nettle, 62; -nut, 729
Fibres, 202, 836-40
Fig (*Ficus*), 62, 71, 71f, 103, 103f, 135, 135f, 175, 209; *F. pumila*, 14, 14f, 57; -*spp.*, 693, 765
Fimbristylis spp., 780

Fission, 407, 517, 546
Flacourtia, 27
Flame of the forest, see *Butea*
Flax, 202, 849-50
Flea seed, 164
Flemingia, 728
Fleurya, 62, 766
Flocculation, 271
Floral diagram and formula, 703-4
Floscopa scandens, 771
Flower, 71; -is a modified shoot, 76-8
Fluggea microcarpa, 764
Fluorescence, 154
Foeniculum vulgare, 46, 74, 74f, 131, 737
Follicle, 128, 129f, 742
Food, 165, 318, 353, 810-5
Forget-me-not, 746
Formation, 415
Fossils, 784, 845-61
Fountain plant, 751
Four o'clock plant, see *Mirabilis*
Foxglove, 751
Fragaria, 25f, 26, 76; -*spp.*, 730
Franciscea, see *Brunfelsia*
Fraxinus, 130, 136, 137f
Free cell formation, 192, 192f
Freesia, 69
Fritillaria, 110
Frond, 496
Fructification, 574
Fructose, 166, 324
Fruits, 126-35, 820-2; -dispersal, 136-42
Fucoxanthin, 496
Fucus, 503-6
Funaria, 614
Fungi, 524-33
Funicle, 96, 96f
Fusarium, 579
Fusiform foot, 5, 5f

Galium rotundifolium, 739
Gamete, 86, 108, 408, 439
Gametophore, 610
Gametophyte, 440, 630, 649
Gamopetalous, 79; -phyllous, 768
Gamosepalous, 79
Gangamopteris, 856, 860

Gardenia (*G. jasminoides*), 37, 738
Garlic, 24, 768, 827
Geitonogamy, 99
Gel, 148
Gemma, (cup), 598
Gemmation, 545f, 546
Gene, 801-4, 806; -mutation, 804
Generative nucleus, 85, 86f
Genetics, 794-809
Genetic spiral, 55
Genotype, 798
Genus, (-nera), 693
Geotropism, 394-6
Geranium, 74, 132
Gerbera, 399, 741
Germplasm, 792
Germ pore, 85, 109; -tube, 536, 562
Gibberellins, 386
Gibbous, 82, 83f
Gills, 574f, 575, 575f
Gingelly, see sesame
Ginger, 22f, 23, 772, 826; Mango-, 7, 7f, 772; Wild-, 772
Ginkgo (*G. biloba*), 662
Ginkgoites, 859
Girardinia zeylanica, 62, 766
Girdling, 349
Glabrous, 38
Gladiolus, 24, 24f
Glands (-dular tis.), 62, 209-11, 210f
Glaucous, 38
Globba bulbifera, 403, 403f, 772
Globoid, 170, 170f
Globule, 490f, 491
Gloeocapsa, 443-4, 443f
Glory-lily (*Gloriosa*), 16, 16f, 48, 48f, 769
Glory of the garden, see *Bougainvillea*
Glossopteris, 857, 861
Glucose, 165, 324, 364
Glucosides (glycosides), 178
Glumes, 66, 76, 776-7, 777f
Glycine max (=*G. soja*), 171, 728, 817
Glycogen, 169, 354, 525
Glycolysis, 364, 366
Glycosmis arborea, 721
Gmelina arborea, 755

Gnaphalium indicum, 62, 423, 742
Gnetum, 681-7
Golden rod, see *Solidago*
Gold mohur (tree), 65, 72f, 729
  Dwarf-, 65, 729
Golgi body, 156f, 157
Gomphrena globosa, 759
Gondwana system, 860-2
Gonidangium, see sporangium
Gonimoblasts, 513f, 514
Gonophore 73
Gooseberry, 79, 132, 749; Wild, 79, 749
Goosefoot, see *Chenopodium*
Gossypium, see cotton; -*spp.*, 693
Gouania, 26, 722
Gourd, 106, 115, 732-3f, 733; -seed, 119, 119f; -stem, 228-31, 229f, 230f; Ash-, Bitter-, Bottle-, Ribbed-, Snake-, Sweet-, Wax-, 733
Grafting, 404, 405f, 406f
Gram, 82, 727, 815-6; -root, 236-8, 237f; -seed, 115, 116f, 120, 120f; Black-, 728, 816; Green-, 728
Gramineae, 776-80, 778f
Grana, 153, 156f
Grape, 132, 166; -fruit, 721; -sugar, see glucose
Grass, 55, 66, 104, 779; Dog-, Lemon-, Saboi-, Spear-, and Thatch-, 779
Grassland, 428, 434
Grevillia, 115
Grewia spp., 719
Grit cells, 203
Groundnut, 37, 120, 394, 395f, 728; -oil, 819
Growth, 378-89; Grand period of-, 382-3; Phases of-, 382, 383f; -ring, 249
Guard cells, 215
Guava, 132, 731
Guazuma tomentosa, 718
Gum 174; -tree, see *Acacia*
Gunpowder plant, see *Pilea*
Guttation, 313, 314
Gymnosperms, 658-87; -and angiosperms, 659; -and cryptogams, 658; Dev. of seed in-, 661
Gymnostegium, 744, 781

INDEX

Gymnostemium, 781
Gynandrous, 90
Gynandropsis, 46, 73, 73f, 714, 714f
Gynobasic, 92, 92f, 756
Gynoecium, 72, 91-8
Gynophore, 73, 73f, 74
Gypsophila elegans, 715

Habenaria, 782
Hadrocentric, 224
Haemanthus multiflorus, 770
Haematoxylon, 729
Halophytes, 424-7, 424f, 425f
Hamelia patens, 68, 738
Haplochlamydeous, 699
Haploid, 183, 440
Haplophase, 573
Haplostele, 622
Haptera (-ron), 471, 477
Haptotropism, 393
Hastate, 39
Haustorium (-ia), 8, 17, 18f
Head, 67, 67f, 739
Healing of wounds, 266
Heart-wood, 249
Hedera helix, see ivy
Hedychium coronarium, 772, 772f
Helianthus, see sunflower; *-spp.*, 741
Helichrysum, 741
Helicoid, 31, 68, 70f
Helicteres isora, 718
Heliotrope (*Heliotropium*), 69, 92, 746
Heliotropism, 394
Helminthosporium, 578-9, 578f
Hemerocallis fulva, 68, 769
Hemicellulose, 169
Hemidesmus indicus, 745
Hemiphragma, 58f, 59, 751
Hemp, see *Cannabis*; Bowstring-, 63, 203, 769; Deccan-, 203, 716; Indian-, or sunn-, 203, 728, 838; Manila-, 772; Sisal-, 203
Henbane, 749
Hepaticae, 590
Herbaceae, 701
Heredity, 787, 806
Heritiera, 6, 122; *-spp.*, 718
Herkogamy, 107

Hermaphrodite, 72
Herpestis, see *Bacopa*
Hesperidium, 132f, 133
Heteroecious, 568
Heterocysts, 442, 446
Heterogametes, 408
Heterophylly, 58-9
Heterophytes (-rotrophic), 17, 342
Heterosis, 795
Heterosporous, 408, 635
Heterostyly, 107, 107f
Heterothallic, 451, 460, 543, 571
Heterozygous, 798
Hevea (*H. brasiliensis*), 176, 764, 840
Hibiscus, see China rose; *H. cannabinus*, 203; *H. mutabilis*, 78; *H. sabdariffa*, 79, 203; *-spp.*, 716-7
Hill reaction, 322-3
Hilum, 96, 96f, 115, 116f, 168
Hiptage, 17, 130, 131f, 137, 137f
Histogen theory, 196
Hogplum, 724
Hogweed, see *Boerhaavia*
Holarrhena, 138, 743
Hollyhock, 717
Holmskioldia (*H. sanguinea*), 37, 755
Homomorphic, 626
Homogamy, 100
Homology, 59-60
Homosporous, 408, 629, 643, 649
Homothallic, 451, 460, 488, 543
Homozygous, 798
Honeysuckle, Wild-, 35
Hopea, 130, 131f, 137, 137f, 716
Hordeum vulgare, 778
Hormogonia, 442, 444f, 445, 446f
Hormones, 383-5
Hornwort, 418
Horsetail, see *Equisetum*
Hound's tongue, 746
Hoya parasitica, 14, 88, 745
Humulus lupulus, 767
Humus, 284
Hutchinson's system, 700-2
Hyaloplasm, 146
Hybrid (-dization), 795; -vigour, 795
Hydathodes, 210, 210f

Hydrangea, 436
Hydrilla, 104, 149, 769
Hydrocharis, 419, 768
Hydrocharitaceae, 767-8
Hydrocotyle, see *Centella*
Hydrodictyon, 463-5, 464f
Hydrolysis, 165-6, 172
Hydrophily, 104-5, 105f
Hydrophytes, 417-19
Hydrosere, 415
Hydrotropism, 397, 397f
Hyenia, 851
Hygrophila spp., 753
Hygrophytes, 418-20
Hygroscopic, 299, 392
Hymenium, 575, 575f
Hyocyamus niger, 749
Hypanthodium, 71, 71f, 765
Hypertonic solution, 277
Hypha, 441, 526, 541
Hyphaene thebaica, 32, 776
Hypocotyl, 119
Hypocrateriform, 81
Hypodermis, 218, 226, 230
Hypoestis triflora, 753
Hypogeal, 120, 121
Hypogyny, 74, 75f
Hyponasty, 393
Hypophysis cell, 111
Hypsophylls, 35

Iberis (*I. amara*), 66, 129, 713
Ichnocarpus frutescens, 744
Imbibition, 272; -theory, 317
Imbricate, 84, 84f
Impatiens, see balsam
Imperata cylindrica, 779
Inarching, 405, 495f
Indian pipe, 20, 20f; -corn, see maize
India-rubber plant, 7, 175, 765
Indigo (*Indigofera*), 728; Wild-, 728
Indusium, 649
Inflorescence, 64-71
Infructescence, 134
Infundibuliform, 80, 81f
Inheritance, 788, 789; Laws of-, 797
Integuments, 96, 96
Interpetiolar stipules, 36f, 37

Interxylary pholem, 208, 260
Intine, 85, 88
Intraxylary pholem, 208
Introrse, 85
Intussusception, Growth by-, 158
Inulin, 166, 167f, 354; -lase, 359
Invertase, 359
Involucre, 75, 76f, 599
Ions, 301
Ipecac, 7, 7f, 404, 738, 827
Ipomoea, 17, 80; *I. botatas*, 818; *I. quamoclit*, 81; *-spp.*, 747, 748
Iris, 55, 107
Iron, 290
Iron-wood tree, see *Mesua*
Irritability, 389
Ischaemum angustifolium, 779
Isidia, 590
Isobilateral, 47, 246-7, 245f, 246f
Isoetes, 639-42
Isogametes (-my), 408, 439, 452
Isomorphic, 486, 499
Isotopes, 326
Ivory-palm, Vegetable-, 169, 776
Ivy, 14, 57; Indian-, 14, 14f, 57, 765
Ixora, 36f, 37, 55, 81; *-spp.*, 738

Jacaranda spp., 752
Jack, 12, 37, 120, 765; Monkey-, 765
Jacquemontia, 748
Jaculator, 140f, 141, 754
Jalap, 747; Indian-, 747
JARUL, 136, 835
Jasmine (*Jasminum*), 69, 81, 101; Cape-, 37; Night-, 69, 81, 101, 140
Jatropha, 62, 132; *-spp.*, 764
Jerusalem artichoke, 23; -thorn, 51, 51f, 729
Jew's slipper, 70, 764
JHUM cultivation, 813
Job's tears, 779
JUAR, see *Sorghum*
Juglans (*J. regia*), 109, 436
Jujube, see *Zizyphus*
Juncellus inundatus, 780
Juncus, 68, 88, 100
Juniper (*Juniperus*), 436, 437
Jussiaea, 9, 9f
Justicia, 753, 753f, 754

INDEX

Jute, 203, 719, 839

Kaempferia rotunda, 772
Kalanchoe, 12; -spp., 402, 402f
Kandelia (K. rheedei), 426, 432
Kapok, 8, 720
Karyogamy, 526
Karyokinesis, see mitosis
Karyolymph, 151, 180
Keel, 81f, 82
Kelps, 496
Kleinhovia hospita, 718
Knop's nor. cul. solution, 288
Kochia (K. tricophylla), 760
Kohl-rabi (or knol-kohl), 713
Krebs cycle, 367-9
Kuhne's ferment. vessel, 376, 376f
Kyllinga, 780

Labellum, 780
Labiatae, 756-8, 757f, 758f
Laburnum, Indian-, 65, 131, 729
Lactuca sativa, 741
Lady's finger, 76, 89, 164, 717; -slipper, 782; -umbrella, 37, 755
Lagarosiphon, 768
Lagenaria siceraria, 733
Lagerstroemia, 75, 136, 136f, 835
Laggera alata & L. pterodonta, 35
Lallemantia, 164
Lamarck's theory, 789-90
Lamellae, see gills
Laminaria, 500-3
Laminarin, 496, 504
Lantana spp., 755
Laportea spp., 62, 766
Larkspur, 77, 82, 83f, 707
Latex, 62, 176, 208; -cells & vessels, 208-9, 209f
Lathyrus, 16, 16f, 48, 48f; -spp., 728
Lavender (Lavandula), 757
Layering, 404, 405f
Leaf, 33-59, 243-7; -area cutter, 329, 329f, -clasp, 307, 307f; -fall, 266-1, 267f; -gap, 624; -mosaic, 57, 57f; -scar, 267; -trace, 624
Leek, 24, 768
Legume, 128, 129f
Leguminosae, 724-5

Lemma, 66, 777
Lemna, see duckweed
Lemon, 27, 46, 60, 90, 721; -grass, 174, 779
Lenticel, 252-3, 252f
Lentil (Lens culinaris), 48, 727, 816
Leonurus, 71, 82, 90, 757, 758f
Lepidium sativum, 713
Lepidodendrales, 850
Lepidodendron, 851
Leptadenia, 746; -stem, 266B-C
Leptocentric, 224
Leptosporangiate, 621
Leptotene, 185
Lettuce, 741; Water-, see Pistia
Leucas, 82, 90; -spp., 757
Leucoplasts, 153-4; -cophyll, 155
Lianes, 17
Lichens, 20, 587-91
Life tree, Child-, see Putranjiva
Light, 331, 333, 413
Light and dark reactions, 321
Light screen, 329, 329f, 335
Lignin (-fication), 163
Lignosae, 701
Ligulate, 82, 82f
Ligule, 35, 633, 776
Lilac, see water hyacinth; Persian-, 39, 722
Liliaceae, 768-9, 769f
Lilium bulbiferum, 403; -spp., 768, 769
Lily, 24, 768; Arum (or trumpet)-, 775; Butterfly-, 772; Day-, 769; Easter-, 770; Glory-, 769; Pincushion-, 770; Spider-, 89, 91, 770; Zephyr-, 770
Lime, see calcium; -tree, see Citrus
Limiting factors, 332-3
Limnanthemum, 419
Limnophila heterophylla, 59, 751
Limonia acidissima, 721
Linaria, 82, 102; -spp., 751
Lindenbergia (L. indica), 751
Linderhia spp., 751
Linkage, 801
Linnaean system, 695-6
Linseed (Linum), 94, 107, 164, 202, 839

Lipase, 172, 342, 359
Lippia spp., 755
Liquorice, Indian-, see *Abrus*
Liriodendron, 709
Litchi (*Litchi chinensis*), 115, 120, 723
Littoral forest, 433
Liverworts, 590; -and mosses, 620
Loam, 280
Lochnera rosea, 743f, 744
Lodicules, 777, 777f
Lodoicea maldivica, 139, 140f, 776
Logwood, 729
Lomentum, 130f, 131
Longan (*Euphoria longana*), 723
Lonicera flava, 35
Loofah, 138, 733
Loquat, 53, 730
Loranthaceae, 761
Loranthus, 17, 343; -spp., 761
Lotus, 74, 74f, 91, 92f, 419, 711
Love thorn, 141, 142f, 779
Lucerne, 728
Luffa, 127, 138; -spp., 733
Lupin (*Lupinus*), 37, 46, 728
Luvunga scandens, 721
Lycopersicum esculentum, 749
Lycopodium, 628-33
Lycopsida, 621, 849
Lyginopteris, 855
Lysigenous cavities, 194
Lysosome, 156f, 158

Mace, 826
Maceration, 225
Machillus (*M. bombycina*), 433
Madar, see *Calotropis*
Madder, 738
Madhuca latifolia, 858
Magnesium, 290
Magnolia, 12, 37, 73, 91; spp., 708
Magnoliaceae, 707-9, 708f
Mahogany, 722, 835
Maize, 104, 104f, 778, 778f, 814; -grain, 117, 118f, 121, 121f; -stem, 231-3, 232f, 233f
Malachra capitata, 717
Malay apple, 731
Mallotus philippinensis, 764

Mallow (*Malva*), 106, 717; Indian- (*Abutilon*), 717; Musk- (*Hibiscus*), 717
Maltose, 166, 359
Malus sylvestris, 730
Malvaceae, 716-7, 717f
Malvastrum, 717
Mammoth tree, 14
Manganese, 297
Manglietia, 709
Mango (*Mangifera*), 72, 132, 132f, 724, 820-1; -ginger, 7, 7f
Mangosteen (Wild-), 72, 80, 115
Mangrove, 425f, 427, 429, 433
Manihot, see tapioca; -spp., 764
Manometer, 304, 304f
Manubrium, 491, 491f, 492f
Manuring, 285
Maple, 131, 137, 137f, 436
Maranta, see arrowroot, -spp., 773
Marantaceae, 773
Marchantia, 597-603
Margosa, 38, 63, 722
Marica northiana, 403
Marigold, 68, 76, 102, 741
Marjoram, 757
Marking nut, see *Semecarpus*
Marsh vegetation, 426
Marsilea, 654-8

Martynia annua (=*M. diandra*), 141, 142f
MARUA, see *Eleusine*
Mason and Maskell's theory, 351
Massecuite, 824
Mast tree, see *Polyalthia*
Mazus japonicus, 750f, 751
Mecardonia dianthera, 751
Mechanical system (-tissues), 211-3
Meconopsis, 436, 712
Medicago sativa, 728
Medicinal plants, 826-30
Medulla, (-ry rays), 220, 227
Medullosa, 855
Meiosis, 183-7; Mitosis and-, 187
Melaleuca, 731
Melia azedarach, 39, 722
Meliaceae, 721-2
Melocactus, 734

INDEX

Melocanna, 779
Melochia corchorifolia, 718
Melon, 733; Musk-, Water, 733
Mendel's experiments, 796-800
Mentha, see mint; *-spp.*, 756
Mericarp, 131, 736
Meristematic tissues, 195-200
Mesocarp, 127
Mesophyll, 244, 246
Mesophytes, 420
Mesozoic, 847
Mesua (M. ferrea), 12, 37
Metabolism, 377-8
Metamorphoses, 26, 47
Metaphase, 180, 185
Metaphloem, 222
Metaxylem, 221f, 222, 228
Metroxylon rumphii, 776
Meyenia erecta, 753
Michelia, 73, 91, 92f; *-spp.*, 708-9
Micropyle, 96, 96f, 115
Microsomes, 146
Middle lamella, 159, 159f
Mikania scandens, 742
Mildews, 553, 582
Milkweed, see *Asclepias*; -tree, 765
Millets, 119, 167, 779, 811
Millingtonia hortensis, 39, 752
Millon's reaction, 171
Mimicry, 63
Mimosa (M. pudica), 37, 68, 393, 398f, 729, 729f; *-spp.*, 729
Mimoseae, 726, 727f, 729f, 729
Mimulus, 751
Mineral crystals, 175-6
Mint (*Mentha*), 25, 26; *-spp.*, 756
Mirabilils, 5, 6, 32, 130, 130f, 758; -stem, 260-1, 260f
Mistletoe, 17, 19f, 32, 142, 761
Mitochondria, 156-7, 156f
Mitosis, 178-82; -and meiosis, 187
Molar solution, 278
Moll's experiment, 330, 330f
Molybdenum, 298
Momordica, 63; *-spp.*, 733
Monadelphous, 89, 89f, 716
Moniliform roots, 7, 7f
Monkey flower, 751
Monk's hood, 706

Monochasial, 68
Monochlamydeous, 72
Monoecious, 72, 106
Monohybrid cross, 796-7
Monokaryotic, 577
Monophyletic, 688
Monopodial, 31
Monospore, 512, 513f
Monotropa uniflora, 20, 20f
Monsoon forest, 427-8
Monstera, 775
Montanoa, 741
Moon flower, 735
Moraceae, 765
Morinda tinctoria, 739
Moringa, 45, 73, 136, 136f, 404
Morning glory, 80, 748
Morphine, 62, 177
Morus, see mulberry; *-spp.*, 765
Mosaic disease, 522; Leaf-, 57
Moss, 615-22
Mould, 583; Black-, 544; Pink-, 540; Water-, 530
Movements, 389-400
Mucilage, 164
Mucor, 540-4
Mucronate, 38
Mucuna (M. prurita), 62, 728
Muehlenbeckia, 28f, 29, 761
Mulberry, 65f, 66, 72, 134, 135f, 765; Paper-, 765, 840
Munch's theory, 352
Murraya, 45; *-spp.*, 721
Musa, 66, 75, 771f, 772, 821; *-spp.*, 772
Musaceae, 771-2, 771f
Mushroom, 573
Mussaenda, 77, 77f, 101; *-spp.*, 738-9
Mustard, I/1, 65, 80, 90, 713; -oil, 819
Mutation, 793; Gene-, 804
Mycelium, 526, 541
Mycorrhiza, 20, 21
Myosotis pallustris, 746
Myristica fragrans, 115, 825-6
Myrobalans, 173, 764
Myrtaceae, 731, 731f
Myrtle (*Myrtus communis*), 731
Myxomycetes, 529-30

Myxophyceae, see **Cyanophyceae**

Naias, 104, 418
Nannorhops, 437
Napiform root, 5, 5f
Naravelia, 16, 48, 49f, 138, 139f, 707
Narcissus, 83, 770
Nardostachys jatamansi, 437
Nastic movements (Nasties), 397-9
Nasturtium, Garden-, 16, 47, 82, 83f
Nasturtium spp., 713
Natural selection, 790, 791
Necrosis, 523
Nectary, 83, 101, 781
Nelumbium, see *Nelumbo*
Nelumbo nucifera, 74, 74f, 91, 92f, 419, 711
Neottia, 782
Nepenthes, see pitcher plant
Nephelium litchi, see *Litchi chinensis*; *N. longana*, see *Euphoria longana*
Neptunia (*N. oleracea*), 398, 419, 729
Nerium (*N. indicum*), see oleander
Nettles, 61-2, 764; Devil-, 62, 766
Neurospora, 557-60, 557f
Nicker bean, see *Entada*
Nicotiana, 62, 177, 211; *-spp.*, 749
Nigella sativa, 707
Nightshade, Black-, 749, 749f; Deadly-, 177, 749, 830
Nipa-palm (*Nipa fruticans*), 432, 776
Nitella, 149, 489
Nitrification, 292
Nitrobacter & *Nitrosomonas*, 292
Nitrogen, 291-6; -cycle, 295-6, 296f
Nodes & Inter-, 3, 11
Nodules, 294, 294f, 295
Nodulose root, 6, 7f
Nomenclature, Binomial-, 693
Noon flower, 85, 718
Nopalia, 734
Normal culture sol., 287-8
Nostoc, 445-7, 445f; - and *Anabaena*, 448
Nucellus, 96, 96f
Nuclear spindle, 180

Nucleic acids, 151-2
Nucleolus, 151
Nucleoplasm, 151
Nucleoprotein, 147, 151
Nucleus, 144, 150-3, 151
Nucule, 499f, 491
Nut, 130; Marking-, 127, 127f
Nutmeg, 115, 825-6
Nutation (Circum-), 393
Nux-vomica, 62, 177, 827
Nyctaginaceae, 758-9
Nyctanthes, see Jasmine, Night-
Nyctinasty, 399
Nymphaea, see water-lily; *-spp.*, 711
Nymphaeaceae, 710-1, 711f

Oak, see *Quercus*
Oat, 119, 779
Obdiplostemonous, 720
Ochreate stipules, 36f, 37, 760
Ochroma pyramidale, 720
Ocimum, 71, 82, 90; *O. basilicum*, 757, 757f; *-spp.*, 756, 757
Odina wodier, 724
Oedogonium, 480-4, 481f
Oenanthe, 24; *-spp.*, 737
Oenothera lamarckiana, 793
Offset, 25f, 26
Oidia, 526
Oils, see fats; Essential-, 174; -seeds, 818-9
Oleander, 55, 62, 83, 83f, 217, 217f, 743f, 744; Yellow-, 62, 176, 744
Oldenlandia spp., 738, 739
Onion, 23, 23f, 768, 769f; -seed, 123-4, 124f
Onosma echioides, 746
Oogamy, 439, 452
Oogonium, 483, 488
Oomycetes, 524
Ooplasm, 539
Oosphere, 97, 408; -spore, 111, 408
Operculina turpethum, 747
Operculum, 619
Opium, 712, 830
Opuntia dillenii, 28f, 29, 423, 734
Orange, 46, 95, 133, 721, 821-2
Orchid, 9, 9f, 20, 90, 95, 138, 782; -root, 240-1, 240f

INDEX

Orchidaceae, 780-2, 781f; Pollination in-, 781-2
Orchis, 782
Oreodoxa regia, 776
Origanum vulgare, 757
Orobanche indica, 17, 18f
Oroxylum, 45, 136, 136f, 752
Orthostichy, 55
Orthotropus, 98, 98f
Oryza sativa, see rice
Oscillatoria, 443-5
Osmometer, 275
Osmosis, 273-6, 274f
Ostiole, 504, 557
Ottelia, 419, 768
Ovary, 72, 92-3
Ovule, 92, 96-8, 96f, 98f
Ovuliferous scale, 675, 675f
Ovum, 97, 408
Oxalis, 25, 25f, 107, 403, 403f
Oxidases, 359
Oxystelma esculentum, 745

Pachytene, 185
Padauk, 428
Paddy, see rice
Paederia foetida, 63, 738
Pagoda tree, see *Plumeria*
Palaeobotany, 844-62
Palaeozoic, 847
Palea, 66, 777
Palisade parenchyma, 244, 244f
Palm, 66, 75, 104, 775, 776
Palmae, 775-6
Palmatifid, (-tipartite, -tisect), 43
Palmella stage, 454-5
Palmyra-palm, 124, 125f, 776
Pancratium, 89, 91, 770, 770f
Pandanus, see screwpine
Pandorina, 456-7, 456f
Pangenesis, 791
Panicle, 69, 70f
Panicum spp., 779
Pansy, 100, 108
Papain, 176, 359, 822
Papaver, 62, 138, 177, 830; -spp., 712
Papaveraceae, 711-2
Papaw, 62, 72, 95, 135, 176, 822

Paper, 841-3
Papilionaceae, 725, 726f, 726-8
Papilionaceous, 81, 81f, 725
Pappus, 79, 137, 139f
Parachute mechanism, 137
Paragynous, 535
Paraphysis, (-ses), 504, 563, 615
Parasites, 17-19, 343; Facultative-, 526; Obligate-, 525
Paratonic movement, 390, 391
Parenchyma, 200-1, 201f
Parietal, 94f, 95
Paripinnate, 44
Parkia, 729
Parkinsonia aculeata, 51, 51f, 729
Parmelia, 586, 587f
Parmentiera cerifera, 752
Parsley, 737
Parsnip, 737
Parthenocarpy, 110, 409
Parthenogenesis, 409
Parthenospore, 451, 475, 544
Paspalum, 779
Passage cell, 219, 239, 241
Passion-flower (*Passiflora*), 15, 16f, 26, 27, 73, 73f, 83, 83f
Pastinaca sativa, 737
Patchouli, (-ly), 756, 757
Pathogens, 580
Pathology, 579-84
Pavement tissue, 684, 685
Pavetta indica, 37, 738
Pea, 16, 16f, 48, 48f, 81f, 82, 128, 728, 816; -seed, 115, 116f, 120, 120f; Butterfly-, Cow-, Pigeon-, Sweet-, 728; Wild-, 48, 48f, 728
Peach, 75, 132, 730
Peanut, see groundnut
Pear, 77, 133, 730; Prickly-, see *Opuntia*
Pectic compounds, 163
Pedate leaf, 40, 40f
Pediastrum, 462-3, 462f, 463f
Pedicel, 65, 72
Pedilanthus, 70, 764
Peduncle, 65
Peepul, 37, 62, 71, 103, 135, 765
Pelican flower, 138, 138f
Peltate leaf, 47

Peltophorum, 729
Penicillin, 549, 584
Penicillium, 548-51, 549f
Penicillus, 550
Pennisetum typhoides, 779, 815
Pennywort, Indian-, see *Centella*
Pentamerous, 79
Pentapetes (*P. phoenicea*), 85, 718
Pentastichous, 56
Peperomia, 39
Pepo, 132f, 133
Pepper, see *Piper*; Negro-, see *Xylopia*; Red-, see chilli
Peppermint, 756
Pepsin, 359
Perennation, 22
Pereskia bleo, 423, 735
Pergularia pallida, 745
Perianth, 72
Periblem, 198, 200
Pericarp, 127
Perichaetium, 599, 599f
Pericycle, 219-20, 227, 230, 237
Periderm, 252
Peridium, 550, 553, 565, 573
Perigynium, 600
Perigyny, 74, 75f
Perinium, 643
Periphyses, 559, 572
Periplasm, 534, 539
Perisperm, 115; -stome, 619
Perithecium, 527, 558, 562
Periwinkle, 62, 81, 743f, 744
Personate, 82, 82f, 750
Petal, 72
Petiole, 33-4
Petrea volubilis, 755
Petunia hybrida, 749
Peziza, 562-4, 563f
Phaeophyceae, 495
Phanerogams, 658, 695
Phaseolus mungo, 816; -spp., 728
Phaylopsis imbricata, 753
Phellem, 251, 256
Phelloderm, 251, 256
Phellogen, 250, 256
Phenotype, 798
Philodendron, 775
Phloem (or bast), 206-8, 221f, 222

Phlogocanthus spp., 754
Phoebe spp., 836
Phlox, 141
Phoenix, see date-palm
Pholidota imbricata, 782
Phormium tenax, 769
Phosphoglyceric acid, 326, 365
Phosphorus, 290
Phosphorylases, 360
Photolysis, 322
Photonasty, 399; -taxis, 391; -tonic, 381; -tropism, 394
Photoperiodism, 388B
Photosynthesis, 319-36
  Limiting factors in-, 332-3
Photosynthometer, Ganong's-, 327, 328f
Phragmites (*P. karka*), 779
Phrynium variegatum, 773
Phycocyanin, 445
Phycoerythrin, 508
Phycomyetes, 524, 526, 530
Phyla, see *Lippia*
Phyllanthus spp., 764
Phyllocactus, 29, 29f, 734
Phylloclade, 28-9, 28f, 29f
Phyllode, 50-1, 50f, 51f
Phyllotaxy, 54-7
Phylogenetic system, 698
Physalis, 80; -*spp.*, 749
Phytelephas macrocarpa, 169, 776
Phytogeographical regions of India, 429-38
Phytohormones, 384
Phytophthora, 535-8, 536f, 580
Pilea microphylla, 766
Pileus, 574, 574f
Piliferous layer, 214, 236, 238
Pimento (*Pimenta*), 731
Pimpinella anisum, 737
Pinaceae, see *Abeitaceae*
Pine (*Pinus*), 670-81, 836
Pineapple, 134, 135f, 403, 404f, 821
Pinguicula, 345, 345f
Pink, 80, 85, 94, 95, 715
Pinnatfid, (-tipartite, -tisect), 43
Piper spp., 8, 14, 98, 825; -stem, 261, 261f

INDEX

Pisonia aculeata, 14, 61, 759
Pistachio (*Pistacia*), 724
Pistia (*P. stratiotes*), 25f, 26, 210, 775
Pistil, see gynoecium
Pistillode, 92
Pisum sativum, see pea
Pitcher (Plant), 16, 16f, 51, 52f, 347
Pith (-ray), 220, 227
Pithecolobium, 115; -*spp.*, 729
Pits, 160-1, 161f
Placenta (-tation), 94-6, 94f
Plantain, 772; Water-, 58
Plantago, 164
Plant breeding, 795, 807-9
Plasma membrane, 146
Plasmodesma, 159, 159f
Plasmodium, 530
Plasmogamy, 526
Plasmolysis, 277-8, 278f
Plastids, 153-5
Plectostele, 622, 628
Plectranthus, 757
Plerome, 198, 200
Pleurococcus, see *Protococcus*
Plum, 75, 132, 730; Indian-, 37, 37f, 42, 722
Plumbago, 62, 94, 142, 211
Plumeria (*P. rubra*), 32, 744
Plumule, 112, 116, 116f
Pneumatophores, 6, 6f, 425f. 427
Pod, see legume
Podophyllum, 436
Pogostemon, 4, 4f; -*spp.*, 756, 757
Poinsettia (*Poinsettia pulcherrima*), 70, 70f, 75, 101, 762f, 764
Polanisia (*P. icosandra*), 46, 714
Polar nodule, 446
Polianthes tuberosa, see tuberose
Pollen, 85-8, 86f; Dev. of-, 86, 87f
Pollination, 99-108
Pollinia (-ium), 88, 88f, 744, 781
Polyadelphous, 89f, 90
Polyalthia (*P. longifolia*), 134, 709
Polycarpon, 95, 715
Polychasial, 69
Polyembryony, 410, 679
Polygamous, 72
Polygonaceae, 760-1

Polygonum, 37, 72, -*spp.*, 761
Polyhybrid cross, 800
Polymorphic, 569
Polyphyletic, 689
Polyphyllous, 768
Polyploidy, 188-92
Polyporus, 577
Polysiphonia, 509-12
Polytrichum, 614
Pome, 132f, 133
Pomegranate, 27, 60, 133, 133f
Pongamia pinnata, 728
Poplar (*Populus*), 436
Poppy, Garden-, see *Papaver*; Himalayan-, see *Meconopsis*; Opium, see *Papaver*; Prickly (or Mexican)-, see *Argemone*
Porana paniculata, 748
Porella, 603-5
Porogamic, 109
Porometer, Darwin's-, 308, 308f
Portia tree, 717
Portulaca, 127, 399
Potamogeton, 419
Potassium, 289-90
Potato, 22f, 23, 748, 817-18; Sweet-, 6, 7f, 747, 818
Potentilla (*P. fulgens*), 436, 730
Pothos (*P. scandens*), 8, 14, 775
Potometer, Darwin's-, 310, 311f; Ganong's-, 309, 309f; Garreau's-, 307, 308f
Pouzolzia indica, 766
Prefoliation, 53
Premna spp., 755
Press mud, 824
Prichardia grandis, 776
Prickles, 14, 15f, 27, 61
Primrose (*Primula*), 75, 95, 107, 436
Procambial strands, 221
Procambium, 197
Procarp, 510
Procumbent, 11
Proembryo, 680
Progametes, 542
Promeristem, 197
Promycelium, 544, 567
Prophase, 179, 185
Prosopis (*P. spicigera*), 61, 423, 729

Prosthetic group, 357
Protandry, 106
Protective tissues, 257
Proteins, 170-2, 337-41, 354-5
Prothallus, 650, 651f
Protococcus, 470-1, 470f
Protoderm, 197
Protogyny, 106
Protonema, 620
Protophloem, 222, 234
Protoplasm (-st), 144, 145-50
Protostele, 622
Protoxylem, 221f, 222, 228
Prune (*Prunus*), 27, 75, 132; spp., 730
Psalliota, see *Agaricus*
Pseudopodium, 149, 613
Psidium guayava, see guava
Psilophytales, 848
Psilophyton, 850
Psilotopsida, 621, 848-9
Psilotum, 625-7
Psoralea corylifolia, 728
Psychotria ipecacuanha, see ipecac
Pteridophyta, 622
Pteridosperm (-males), 855
Pteris, 646, 647
Pterocarpus, 428, 728
Pteropsida, 621, 852
Pterospermum, 73, 73f, 85, 718
Ptilophyllum, 861
Ptomaine, 519
Ptyxis, 53
Pubescent, 38
Puccinia, 568-73, 581
Pulsation, 392; -theory, 317-18
Pulses, 128, 727, 815-17
Pulvinus, 33, 34f
Pummelo, 46, 721
Pumpkin, 733
Punica granatum, see pomegranate
Pupalia, 141, 142f, 759
Putranjiva roxburghii, 764
Pycnia (-nospore), 571, 572
Pyrenoids, 454, 465, 472
Pyrethrum, 741
Pyrus, 77; -spp., 730
Pyruvic acid, 365, 366
Pythium, 533-5

Quamoclit pinnata (=*Ipomoea quamoclit*), 43, 748
Queen of the night, see *Cestrum*
Quercus, 66, 110, 130, 252, 436; -spp., 437, 438
Quillwort, see *Isoetes*
Quince, 730
Quinine, see *Cinchona*
Quisqualis, see Rangoon creeper

Raceme (-mose), 31, 65, 65f
Rachis, 44
Radicle, 116, 116f
Radish, 5, 5f, 65, 80, 90, 713
Rafflesia (*R. arnoldi*), 17, 19, 19f
Railway creeper, 17, 80, 748
RAGI, see *Eleusine*
Rain tree, 729; -forest, 427
Ramenta, 648
Ramie, see rhea
Randia fasciculata, 37, 738
Rangoon creeper, 17, 54, 101
Ranunculaceae, 706-7, 707f
Ranunculus, 58, 92; -root, 237-8, 238f; stem, 228, 229f; -spp., 707
Rape, 90, 713
Raphanobrassica, 189, 191
Raphanus sativus, see radish
Raphe, 96, 96f, 115, 116f
Raphides, 175f, 176
Raspberry, 134, 134f, 730
Rattlewort, see *Crotalaria*
Rauwolfia serpentina, 743, 827; *R. canescens*, 744
Ravenala, 55, 55f, 772
Receptacle, 65, 598
Receptive hyphae, 572
Recessive, 796
Reclinate, 53
Reduction division, see meiosis
Redwood, Indian-, 728, 834; -tree, 14
Reed, 779; Giant-, 779
Regma, 131, 131f
Rejuvenescence, 407, 407f, 482, 488
Reniform, 39, 716
Replum, 95, 129, 712
Reproduction, 400-10, 439
Resin, 174; -duct, 227
Respiration, 361-77; -and fermenta-

INDEX

-tion, 376; -and photosynthesis, 374
Respiratory cavity, 216, 244; -roots, 9, 9f
Respiratory quotient (R. Q.), 373
Respirometer, Ganong's-, 371, 372f
Respiroscope, 368
Retinaculum, 744
Rhamnaceae, 722
Rhamnus nepalensis, 722
Rhea, 202, 766, 839
Rheum spp., 761
Rhipsalis cassytha, 734
Rhizobium (R. radicicola), 295
Rhizoid, 593, 596, 614, 649
Rhizome, 22, 22f
Rhizophora, 6, 122, 427
Rhizophore, 634
Rhizopus, 544
Rhododendron spp., 436, 437-8
Rhodophyceae, 508
Rhoeo, 771
Rhubarb, 761
Rhus spp., 724
Rhyncostylis retusa, 782
Rhynia, 849
Riboflavin, 387
Ribosomes, 156f, 158
Ribulose diphosphate, 324
Riccia, 592-5, 593f
Rice, 778, 778f, 811-3; -grain, 117, 118f, 121, 121f
Richardia africana, 775
Ricinus communis, see castor
Ringing expt., 315, 349
Ring-porous wood, 249
Ringworm shrub, 729
RNA (ribonucleic acid), 152-5, 340, 802-3
Rivularia, 448, 448f
Root, 1-10, 235-41; -apex, 199-200, 199f; -cap, 2, 2f, 199; -hairs, 3, 3f, 236; -pocket, 1, 3; -pressure, 304, 316; -stock, 23
Rorippa indica, 713
Rosaceae, 730
Rose (*Rosa*), 14, 15f, 37, 74, 74f, 78; -*spp.*, 730; -apple, 731
Rosemary (*Rosmarinus*), 757
Rosewood, Indian-, 728

Rostellum, 781, 781f
Rotation, 149, 150f; -of crops, 296-7
Roupellia grata, 744
Roydsia suaveolens, 715
Rozelle, 79, 203, 716
Rubber, 841
Rubiaceae, 737-9
Rubia cordifolia, 739
Rubus, see raspberry; -*spp.*, 730
Rue, see *Ruta*
Ruellia, 6, 140f, 141; -*spp.*, 753
Rumex, 37, 98; -*spp.*, 761
Rungia spp., 753
Runner, 24, 25f
Ruscus aculeatus, 29, 30f, 769
Rush, 68, 88
Russelia equisetiformis, 751
Rust, see *Puccinia*
Rutaceae, 720-1, 721f
Ruta graveolens, 721

Sabai grass, 777, 779
Sabal palmetto, 776
Saccharomyces, see yeast
Saccharum, see sugarcane; -*spp.*, 779
Safflower, 68, 741
Saffron, 24; Meadow-, 24, 768
Sage, see *Salvia*
Sagittaria, 40, 58, 59f
Sagitate leaf, 40
Sago palm, Indian-, 776
Sal, see *Shorea*
Salicornia brachiata, 424, 760
Salmalia, see *Bombax*
Saltwort (*Salsola*), 424, 760
Salvadora stem, 262, 262f
Salvia, 71, 89, 103, 103f, 757
Salvinia, 419
Samara, 130, 131f; Double-, 131
Samaroid, 130, 131f
Sambucus, 436
Sandalwood (*Satalum*), 17, 428, 430; Red- (*Pterocarpus*), 428, 430, 728
Sandwich Is. climber, see *Corculum*
Sansevieria, 63, 203; *spp.*, 769
Santonin, 741
Sapindaceae, 722-3
Sapindus spp., 723
Saponaria vaccaria, 95, 715

Saponin, 178
Sapota (*Achras*), 863
Sappan, 729
Sapria himalayana, 17, 19
Saprolegnia, 530-3
Saprophytes, 20, 20f, 343
Sap-wood, 249
Saraca (*S. indica*), 31, 44, 729
Sarcodes, 20
Sarcolobus globulus, 746
Sarcostemma, 745
Sargassum, 506-8, 507f
Sarsaparilla, see *Smilax*; Indian-, see *Hemidesmus*
Satinwood, 722
Savannah, 428
Saxifraga, 436
Scale, 12, 50; Ovuliferous-, 675
Scape, 11, 65, 768
Schizanthus pinnatus, 749
Schizocarpic, 131
Schizogenous, 194
Schizomycetes, see bacteria
Schleichera trijuga, 723
Scilla indica, 12, 769
Schima, 433
Scindapsus officinalis, 421, 775
Scion, 405
Scirpus spp., 779, 780
Scitamineae, 771
Sclereids (-rotic cells), 203-4, 203f
Sclerenchyma, 202-3, 202f
Sclerotium, 526, 560f, 561
Scoparia dulcis, 751
Scorpioid, 31, 69, 69f
Screwpine, 7, 8f, 134
Scrophulariaceae, 749-51, 750f
Scrophularia elatior, 751
Scutellum, 118, 118f, 119
Sea-blite, 424, 760
Sechium edule, 733
Secondary growth, 247-57; Anomalous-, 257-65
Sedge, 56, 104, 779
Sedum, 91, 92f
Seed, 111-26; -dispersal, 136-42
Seismonasty, 397-9
Selaginella
Semecarpus, 127, 127f, 728

Senebiera, 129
Senecio, 740
Senna, Indian-, 729
Sensitive plant, see *Mimosa*; -wood-sorrel, see *Biophytum*
Sepals, 72, 79
Septicidal (-tifragal), 128
Sequoia spp., 14
Serjania, 27, 723
Sesame (*Sesamum indicum*), 819
Sesbania, 44; -spp., 728
Seta (tae), 601, 617
Setaria italica, 779
Sex chromosomes, 805
Shaddock, see pummelo
Shallot, 768
Shepherd's purse, 129, 713
Shield cells, 941, 941f
Shorea, 130, 131f, 137, 137f, 716, 835
Sida spp., 717
Sieve-plate & -tube, 206-7, 207f
Sigillaria, 850-1
Silene, 73, 715
Siliqua (-licula), 129, 129f, 712, 713
Silk cotton tree, see *Bombax*
Silverweed, 730
Sinuous, 732
Siphonostele, 623, 623f
Smilax, 34, 34f, 41, 768
Smut, see *Ustilago*
Snake plant, 63, 63f, 775
Snapdragon, 82, 82f, 102, 751
Soap-nut, 178, 723; -wort, 95, 715
Soils, 279-86
Solanaceae, 748-9, 749f
Solanum (*S. tuberosum*), see potato; *S. dulcamara*, 177, 749; *S. nigrum*, 749, 749f; *S. surattense* (=*S. xanthocarpum*), 423, 749; -spp., 748, 749
Solidago virgaurea, 741
Sonchus asper, 35, 62, 742
Somatic cell-division 178
Sonneratia (*S. apetala*), 122, 426, 432
Soredia (-ium), 587f, 588
Sorghum vulgare, 779, 814-5
Sorophore, 655
Sorosis, 134, 135f

INDEX

Sorrel, see *Rumex*; Wood-, see *Oxalis*
Sorus (-ri), 648, 648f
Sour sop, 709
Soya-bean, 171, 728, 817
Space marker disc, 383; -wheel, 383
Spadix, 66, 66f, 75, 76f, 771, 775
Spathe, 66, 66f, 75, 76f, 771, 775
Spathodea campanulata, 752
Spathulate, 39
Spear grass, see *Aristida*; -mint, see *Mentha*
Species, 692; Origin of-, 790
Spectrum analysis, 335
Spergula arvensis, 715
Spermatia, 510, 514, 571, 571f, 572
Spermatophytes, 658
Spermogonium, 571, 571f
Sphacelia stage, 561
Sphaero-crystals, 175f, 176
Sphagnum, 610-5
Sphenophyllum, 851
Sphenopsida, 621, 851
Sphenopteris, 855
Spices, 824-6
Spider lily, 91, 770, 770f; -wort, 91, 150
Spike, 65, 65f; -let, 65f, 66, 776-7, 777f; Sporangiferous- see strobilus
Spinach (*Spinacia*), 760; Indian-, 760
Spine, 50, 60
Spiraea cantoniensis, 730
Spirogyra, 471-6, 472f
Spondias spp., 724
Spongy parenchyma, 244, 244f
Sporangiophore, 542, 544
Spore (-rangium), 407, 526, 649; Mega-, 91, 635, 665; Micro-, 85, 635, 664
Sporidia, 568
Sporobolus, 431
Sporocarp, 654, 654f
Sporocyte, 608, 654
Sporodochium, 579
Sporogonium, 601, 602, 619
Sporophore, 574, 576
Sporophyll, 71, 629, 648; Mega-, 91, 634, 664; Micro-, 85, 634, 663
Sporophyte, 440; Dev. of-, 591-2

Sprout-leaf plant, see *Bryophyllum*
Spur, 82, 83f, 780
Spurges, see *Euphorbia*
Squash, 733
Stachytarpheta indica, 755
Stamen, 71, 85; -minode, 85
Stapelia grandiflora, 745
Starch, 167-9, 167f, 168f, 324, 354-4; -print, 328, 329f; -sheath, 218, 227; Uses of-, 169
Stele, 199; Types of-, 622-4, 623f
Stellaria media, 715
Stem, 10-33, 225-34; -apex, 197, 198f
Stephanotis floribunda, 745
Sterculiaceae, 717-8
Sterculia villosa, 59; -*spp.*, 718
Stereospermum, 136, 136f, 752
Stigma, 72
Sterigma (-mata), 526, 550, 575, 575f
Stilt roots, 7, 8f, 424, 425
Stinging hairs, 61, 62f, 766
Stipe, 574, 574f
Stipel, 36
Stipule, 33, 36-7; -lode, 490
Stolon, 25, 25f
Stoma (-ata), 213, 215-7, 216f
Stomium, 650
Stonecrop, see *Sedum*; -wort, see *Chara*
Storage, 353-6
Strawberry, 25, 37, 135, 730; Wild-, 25, 26, 730
Streblus asper, 765
Strelitzia reginae, 772
Streptomycin, 585
Striga (*S. lutea*), 17, 18
Strobilanthes spp., 753
Strobilus, 630, 636
Stroma (-ta), 153, 156f, 560f, 562
Struggle for existence, 790
Strychnos nux-vomica, see nux-vomica
Stylopodium, 736
Suaeda, 424; -*spp.*, 760
Suberin (-rization), 163
Succession, 415-17
Sucker, 26, 26f
Sucrose, 166, 822-3
Suction pressure, 274

Sugar, 166; see also glucose & sucrose
Sulphur, 290
Sundew, 344-5, 344f
Sunflower, 68, 76, 102, 740f, 741; Mexican-; 741; -stem, 225-8, 226f
Survival of the fittest, 790-1
Suspensor, 111, 542
Suture, 92, 95
Swamp forest, 434
Sweet flag, 224, 775; -sultan, 102-3, 398, 741; -William, 715
Swietenia mahagoni, 722, 835
Syconus, 134, 135f, 765
Symbionts (-biosis), 20-1, 343
Sympetalae, 699
Sympodial, 31, 32f, 69, 69f
Synandrium (-drous), 90, 90f, 732
Synangium, 625, 625f
Synapsis, 185
Syncarpous, 91, 91f; -carpy, 93
Synergids, 97, 110
Syngenesious, 89f, 90, 739
Syzygium aromaticum, 826; spp., 731

Tabernaemontana, see *Ervatamia*
Tagetes patula, 741
Talauma pumila, 708, 709
Talipot-palm, 776
Tamarind (*Tamarindus*), 44, 729; seed, 119, 119f
Tamarix, 50, 423, 432
Tannins, 173-4
Tapetum, 87
Tapioca, 43, 764, 818
Tapiria hirsuta, 724
Taro, see *Colocasia*
Taxillus vestitus 761
Taxisms, 390-1
Taxonomy, 691
Tea (*Thea*), 80, 106, 830-3
Teak (*Tectona grandis*), 69, 755, 834
Tecoma, 69, 136; -spp., 752
Tecomella undulata, 752
Tegmen, 115, 116
Telegraph plant, Ind.-, 36, 392, 392f, 728
Teleutospore (-liospore), 567, 570, 570f

Telium, 570, 570f
Telophase, 181, 186
Template hypothesis, 804
Temple tree, see *Plumeria*
Tendril, 15, 16f, 26, 27f, 47-8, 48f
Tentacles, 344
Tepals, 72
Tephrosia spp., 728
Terminalia, 39, 132, 136, 137f; -spp., 8, 173
Terramycin, 585
Testa, 115, 116
Test cross, 798
Tetradynamous, 90, 90f, 712
Tetrameles nudiflora, 433
Tetraploidy, 188, 191
Tetraspores (-ric plant), 510, 511f 511
Thalamus, 72, 73-5, 76
Thalictrum spp., 707
Thallophyta, 439
Thatch grass, 779
Theobroma cacao, 718, 833-4
Thermotonic, 396
Thespesia populnea, 717
Thevetia peruviana, 62, 176, 743
Thigmotropism, 393
Thistle, 60; Globe-, 60, 70, 423, 742
Thorn, 27, 28f, 60; -and prickle, 27-8; -apple, 749; -forests, 428, 431
Three bean experiment, 123, 123f
Thunbergia spp., 754
Thyme (*Thymus*), 756
Tiger's nail, see *Martynia*
Tiliaceae, 718-9
Tilia europaea, 719
Tillandsia, 421
Timber trees, 834-6; -of Assam, 836
Tinospora, 10
Tissue, 195-208; -system, 213-24
Tithonia tagetiflora, 741
Tobacco, 62, 177, 749
Toddalia aculeata, 721
Toddy-palm, 776
Tomato, 95, 132, 749
Tonoplasm, 146
Toon, see *Cedrela*
Torenia spp., 751
Torula, 544

INDEX

Torus, 73, 161, 161f
Tournefortia roxburghii, 746
TPN (triphosphopyridine nucleotide), 324
Trabecula (-lae), 634, 639
Trace elements, 289, 297-8
Trachea, see vessel
Tracheid, 204, 205f
Trachyspermum ammi, 737
Tractile fibres, 182
Tradescantia, 91, 150, 771
Tragia (*T. involucrata*), 62, 764
Trama, 575, 575f, 577
Transamination, 339
Transfusion tissue, 672
Transition from root to stem, 241-3, 242f
Translocation of food, 348-52
Transpiration, 307; -coefficient or ratio, 311; -and exudation, 314
Trapa natans, 10, 10f, 130, 419
Traveller's joy, see *Naravelia*; -tree, see *Ravenala*
Trema orientalis, 765
Trewia nudiflora, 764
Tribulus (*T. terrestris*), 142, 423
Trichodesma indicum, 746
Trichogyne, 510, 552, 558
Trichome, 214, 446
Trichosanthes spp., 733
Tridax (*T. procumbens*), 68, 741f, 742
Trigonella, 728
Trimerous, 79
Triple fusion, 110, 111
Tristichous, 56
Triticum aestivum, see wheat
Triumfetta rhomboidea, 719
Tropaeolum, see nasturtium, Garden-
Tropisms, 393-7
Truncate, 37
Trypsin, 359
Tube-nucleus, 85, 86f, 676, 684
Tuber, 22f, 23
Tubercles, 298
Tubercular roots, 5f, 6
Tuberose, 11, 24, 66, 770
Tulip tree, 709
Tunica-corpus theory, 197

Turgid (-dity), 276
Turgor pressure, 276; -move., 392
Turmeric, 23, 772
Turnip, 5, 5f, 90, 713
Turpeth, 747
Tylophora asthmatica, 745
Tyloses, 249, 266, 266f
Typha, 55, 88, 432
Typhonium, 66, 66f, 775

Ulmaceae, 764-5
Ulmus (*U. lancifolia*), 436, 764
Ulothrix, 465-7, 466f
Umbel, 66, 67f, 735
Umbelliferae, 735-7, 736f
Umbrella plant, 779
Uncaria, 14, 15, 739
Uncinula, 554-7, 555f, 582
Unona (*U. discolor*) 91, 709
Urea, 285-6
Uredium (-dospore), 569, 569f,
Urena (*U. lobata*), 141, 142f, 717
Urtica (*U. dioica*), 62, 766
Urticaceae, 766
Usnea, 588f, 589
Ustilago, 566-8, 566f, 581
Utricularia, 53, 53f, 347-8, 348f
Uvaria hamiltoni, 709

Vacuoles, 146; Contents of-, 177-8
Vallaris solanacea, 744
Vallecular canals, 642
Vallisneria, 104, 104f, 149, 768
Valvate, 84, 84f
Valvular, 91, 127
Vanda, 9f, 10, 20; -root, 240-1, 240f; -spp., 782
Vandellia, 751
Vangueria (*V. spinosa*), 27, 37, 55, 60, 739
Vanilla planyfolia, 782
Variation, 786
Vascular bundles, 221-4, 221f, 223f
Vateria indica, 716
Vatica lancaefolia, 716
Vaucheria, 486-8, 486f
Velamen, 9, 19f, 20, 240, 240f
Velum, 574, 574f

Venation, 40-43
Venter, 599, 616, 651
Ventilago maderaspatana, 722
Venus' flower, 48; -fly-trap, 345-6, 346f
Verbenaceae, 754-6, 754f
Verbena officinalis, 755
Vernalization, 388B
Vernation, 53
Vernonia spp., 741, 742
Veronica spp., 751
Versatile, 88
Verticillaster, 70f, 71, 756
Vessel, 193-4, 204-6, 206f
Vestigial, 786
Vetiveria zizanioides, 779
Vexillary, (-lum), 81, 81f, 84, 84f
Vicia (*V. faba*), 120, 728
Victoria amazonica (=*V. regia*), 419, 419f 711
Vigna sinensis, 728
Vinca (*V. rosea*), see periwinkle
Vine (*Vitis*), 15, 26, 31; Balloon-, 26, 27f; Wild-, 31
Viola, 100, 108
Virgin's bower, see *Clematis*
Viruses, 521-4
Viscum monoicum, see mistletoe
Vitamins, 387
Vitex spp., 755, 756
Vitis, see vine; *V. pedata*, 40, 40f; *V. trifolia* (=*Cayratia carnosa*), 31; -stem, 258, 259f
Vittae, 737
Vivipary, 121, 121f, 426f, 427
Volutin, 495
Volva, 574
Volvox, 458-61, 459f

Wallflower, 66, 713; -pressure, 276
Walnut, 107; see *Juglans*; Indian-, see *Aleurites*
Waltheria indica, 718
Walsura robusta, 722
Water culture expts., 287-9, 289f
Water hyacinth, 26, 34, 34f, 176, 418
Water lettuce, see *Pistia*
Water lily, 77, 77f, 78f, 96, 711, 711f; Giant-, 419, 419f, 711

Watson-Crick model, 802, 802, 802f
Water plantain, see *Alisma*
Wax gourd, see *Benincasa*; -plant, see *Hoya*
Wedelia calendulacea, 741
Weismann's theory, 792-3
Wheat, 119, 778, 813-4
Williamsonia, 857
Willow (*Salix*), 436
Willughbeia edulis, 744
Wind flower, 70
Withania somnifera, 749
Wolffia, 30, 419
Wood, see xylem; Autumn- & Spring-, 249; Heart- & Sap-, 249
Wood-apple, see *Aegle*
Woodfordia, 107
Wood-oil tree, see *Dipterocarpus*
Wood-rose, 17

Wood-sorrel, see *Oxalis*; Sensitive-, see *Biophytum*
Wormwood, Indian-, 741; -seed, 760
Wounds, Healing of-, 266
Wrightia tomentosa, 744

Xanthium, 141, 142f, 742
Xanthophyll, 154, 155
Xanthoproteic reaction, 171
Xenia (Meta-), 799
Xenogamy, 99
Xerophytes, 423-4
Xerosere, 416
Xylem, 204-6, 221, 221f, 228
Xylopia aromatica, 709

Yam, see *Dioscorea*
Yeast, 544-8, 545f
Yucca gloriosa, see dagger plant

Zalacca beccarii, 776
Zanthoxylum spp., 721
Zea mays, see maize
Zephyr lily (*Zephyranthes*), 770
Zeuxine, 782
Zinc, 297
Zingiberaceae, 772, 772f
Zingiber, see ginger, -spp., 772
Zinnia, 68, 741
Zizyphus, 37, 37f, 42; -spp., 722

Zoogloea, 521
Zoophily, 105
Zoospore, 407, 450, 526, 531
Zygnema, 477-8, 477f
Zygomorphic, 79

Zygomycetes, 525
Zygospore (-gote), 408, 474
Zygotene, 185
Zymase, 360, 548
Zymogen, 357